ENCYCLOPEDIA OF
science
technology
AND ethics

EDITORS AND CONSULTANTS

ENCYCLOPEDIA OF

science
technology
AND ethics

EDITED BY
CARL MITCHAM

volume

2

d–k

MACMILLAN REFERENCE USA
An imprint of Thomson Gale, a part of The Thomson Corporation

THOMSON

GALE

Detroit • New York • San Francisco • San Diego • New Haven, Conn. • Waterville, Maine • London • Munich

THOMSON

★

GALE

Encyclopedia of Science, Technology, and Ethics

Carl Mitcham, Editor in Chief

LIBRARY OF CONGRESS CATALOGING-IN-PUBLICATION DATA

Encyclopedia of science, technology, and ethics / edited by Carl Mitcham.
p. cm.
Includes bibliographical references and index.
ISBN 0-02-865831-0 (set, hardcover : alk. paper)—ISBN 0-02-865832-9 (v. 1) —
ISBN 0-02-865833-7 (v. 2)—ISBN 0-02-865834-5 (v. 3)—ISBN 0-02-865901-5 (v. 4)
1. Science—Moral and ethical aspects—Encyclopedias.
2. Technology—Moral and ethical aspects–Encyclopedias.
I. Mitcham, Carl. Q175.35.E53 2005
503—dc22   005006968

While every effort has been made to ensure the reliability of the information presented in this publication, Thomson Gale does not guarantee the accuracy of the data contained herein. Thomson Gale accepts no payment for listing; and inclusion in the publication of any organization, agency, institution, publication, service, or individual does not imply endorsement of the editors or publisher. Errors brought to the attention of the publisher and verified to the satisfaction of the publisher will be corrected in future editions.

This title is also available as an e-book.
ISBN 0-02-865991-0
Contact your Thomson Gale representative for ordering information.

Printed in the United States of America
10 9 8 7 6 5 4 3 2 1

D

DAMS

• • •

Dams, barriers to alter flowing bodies of water, are among the most ancient and powerful examples of the proclivity of humans to alter nature for their own benefit. (Dams are also a type of construction shared with other animals, that is, beavers.) Before the advent of written history, dams were already being built to provide water storage and irrigation. An earthen dam in the Orontes Valley in Syria was ancient when visited by the Greek geographer Strabo around the beginning of the Common Era. The oldest large dam of which traces survive today is at Sadd-el-Kafara, near Cairo. Ninety-eight meters long, there are indications that it was intended to stand 125 meters high. It is estimated that this structure was built around 2500 B.C.E.

Dam Engineering

Despite their ubiquity and importance, dams are a stepchild of traditional engineering. Premodern treatises on construction such as Vitruvius's *De architectura* (first century B.C.E.) do not mention dams, although Roman dam achievements were not to be matched for 1,500 years. The scientific engineering of dams begins in the 1800s and was one of the early achievements of civil engineering as it replaced trial-and-error intuition with empirical rules of thumb for dam design.

In terms of function, dams primarily supply water for irrigation or urban use, or serve as sources of power. In conjunction with closely related structures called dikes, dams may also protect from flooding and/or facilitate transportation by creating navigable bodies of water such as canals.

In terms of design, dams are of two basic types: earth- or rock-filled gravity embankment dams and masonry or concrete dams. The former take the general shape of a large-based equilateral triangle with sloping embankments facing both upstream and downstream; the latter have more the shape of a right-angle triangle with a perpendicular upstream face and a sloping downstream face.

It was not until the mid-1800s that French engineers designed the first dams using scientific procedures to determine such issues as the slope of repose for embankments. At the same time engineers began to consider the geological structures on which various types of dams might rest and to analyze the internal stresses of masonry and concrete dams. Such analyses promoted the design of arch dams, in which a vertical upstream face is given a convex horizontal curve to help transfer forces from the impounded water into the walls of a canyon. The engineering of auxiliary structures such as spillways, locks, and power conversion systems also became part of dam design.

Progressive demands for water and power together with advances in dam engineering led in the first half of the twentieth century to what may be called the golden age of dam construction. But the second half of the twentieth century witnessed a technical reassessment of dam engineering in terms of safety and ecology, social and natural.

Dam Debates

For most of human history, dams were conceived and built with an eye only to the task to be accomplished, such as water storage, irrigation, or more recently,

Hoover Dam. Constructed in 1935, the dam holds back twelve trillion gallons of water and generates enough hydroelectric power to serve 1.3 million people. (© *Corbis*.)

promotion of tourism, and without much concern for other implications, such as the impact on local populations or the environment. Of all major rivers in the United States, only the Salmon and Yellowstone are without dams. Half of the American wetlands that existed in 1790 have been flooded and destroyed by dam projects—up to 80 percent in river states such as Missouri, where one-third of all the water in the Missouri River is stored behind dams.

At the same time some experts argue that dams are often inefficient mechanisms for water storage, spreading water out over large areas in hot, dry desert climates where it evaporates. As much as 8 percent of Colorado River water may be lost to evaporation behind the Glen Canyon Dam in northern Arizona. Dams, by promoting water use, also contribute to the eventual depletion of aquifers.

In the modern world dams nevertheless continue to be seen as important symbols of human domination of the environment, sometimes outweighing all other issues. China's Three Gorges Dam, which will flood thousands of acres of agricultural land and displace more

than one million people, is nevertheless viewed by the Chinese government as a powerful symbol of mastery and progress.

DAM SAFETY AND FAILURES. Like other huge, complex human technology projects, dams can fail if ill-designed or negligently maintained. The most famous failure in the United States was that of the South Fork Dam in Johnstown, Pennsylvania, in May 1889. Over the years, successive owners of the dam made dangerous modifications, eliminating outlet pipes, reducing its height, and narrowing the spillway. During an unprecedented rainfall, the water rose 3 meters (10 feet) above the usual lake level, breaking the dam and inundating Johnstown, with the loss of almost 3,000 lives.

RELOCATING PEOPLE. Dam projects have often involved the removal of the populations least able to defend themselves politically. Most often the groups forced to relocate are poor members of minority groups, subsisting on small-scale agriculture.

In June 1957 Congress voted the creation of Kinzua Dam in western New York, flooding half of a Seneca Indian reservation. More than 500 Seneca were forcibly moved in the dead of winter to trailer camps. Without access to hunting grounds, and denied compensation for their homes, these already poor individuals were, according to the sociologist Joy A. Bilharz (1998), driven into greater poverty, which lasted for decades.

Organized political opposition to large dam projects was pioneered in India, where in the late 1940s important projects backed by the prime minister, Jawaharlal Nehru, made little provision for the relocation of affected villages. Large demonstrations and other opposition increased the costs unacceptably, causing the government to back away from some of these projects.

ENVIRONMENTAL CONCERNS. During the twentieth century, the environmental movement advanced the argument that natural beauty was a factor to be taken into account in dam construction. John Muir led an early campaign against the O'Shaughnessy Dam in Yosemite National Park's Hetch Hetchy Valley on the grounds that it would destroy a unique environment. Later came the related idea that wild species themselves had interests worthy of protection, interests that might be harmed by dam construction. Environmentalists went to court to end construction of the Tellico Dam on the Little Tennessee River, on the grounds that it would destroy the remaining population of snail darters, an endangered fish. In response, federal courts halted construction of a dam already 80 percent completed. In

1978 the U.S. Supreme Court affirmed the court order halting construction, stating that the Endangered Species Act unambiguously bars projects that threaten the continued existence of a listed species. Congress, however, later passed legislation exempting Tellico from the Endangered Species Act, and the dam was completed.

Egypt's Aswan High Dam has been argued to have caused an environmental disaster, starving the Mediterranean of nutrients, making croplands excessively salty, and creating a reservoir in one of the highest evaporation zones on Earth.

DAM REMOVALS. Because of changing views of the utility of dams and the relative importance of environmental considerations, more than 500 dam removal projects were undertaken in the United States during the last decades of the twentieth century. The first dam removed for purely environmental reasons was the Quaker Neck Dam on the Neuse River in North Carolina. Built in 1952 to provide cooling water for a steam-driven electrical generating plant owned by Carolina Power & Light Company, the dam prevented shad from migrating upstream. The shad catch, 318,000 kilograms (700,000 pounds) in 1951, was only 11,400 kilograms (25,000 pounds) by 1996.

Carolina Power & Light was glad to get rid of Quaker Neck. The dam was expensive to maintain and also created litter and liability problems. Instead of the dam, a canal between two channels of the river now provides cooling water. More than 1,600 kilometers (1,000 miles) of local rivers have since been reopened to fish.

As the political and psychological importance of dams has faded and other considerations have come to the fore, Americans have stopped building dams. Since the mid-1970s, there has not been a single major dam construction project commenced in the United States.

JONATHAN WALLACE

SEE ALSO Bridges; Environmental Ethics; Three Gorges Dam; Water.

BIBLIOGRAPHY

Bilharz, Joy A. (1998). The Allegany Senecas and Kinzua Dam: Forced Relocation through Two Generations. Lincoln: University of Nebraska Press. Examines impacts on the Seneca nation of forced relocation to accommodate the Kinzua Dam project.

Devine, Robert S. (1995). "The Trouble with Dams." Atlantic Monthly 276(2): 64–74. Influential article arguing that most dams serve no significant purpose.

Goldsmith, Edward, and Nicholas Hildyard. (1984). The Social and Environmental Effects of Large Dams. San Francisco: Sierra Club Books. Representative of arguments critically reassessing dams during the last half of the twentieth century.

Jackson, Donald C. (1995). Building the Ultimate Dam: John S. Eastwood and the Control of Water in the West. Lawrence: University Press of Kansas. A biography of Eastwood (1857–1924), who from 1906 until his death designed more than sixty dam projects in the western United States, that stresses the psychological attractiveness of the multiple arch dam.

Jackson, Donald C., ed. (1997). Dams. Brookfield, VT: Ashgate. Collects seventeen previously published articles highlighting technical and social developments.

Khagram, Sanjeev. (2004). Dams and Development: Transnational Struggles for Water and Power. Ithaca, NY: Cornell University Press.

Levy, Matthys, and Mario Salvadori. (2002). Why Buildings Fall Down, rev. edition. New York: Norton. Analyzes dam collapses from a structural engineering standpoint.

Petersen, Shannon. (2002). Acting for Endangered Species: The Statutory Ark. Lawrence: University Press of Kansas. Contains an account of the snail darter and Tellico Dam.

Schnitter, Nicholas J. (1994). A History of Dams: The Useful Pyramids. Rotterdam, Netherlands: A. A. Balkema. A technical history with some comments on social context that takes its subtitle from the praise of dams by the Roman Sextus Julius Frontinus (c. 35–c. 103 C.E.).

Smith, Norman. (1971). A History of Dams. London: Peter Davies. The first general history of dams.

World Commission on Dams. (2000). Dams and Development: A New Framework for Decision-Making. London: Earthscan. A report that grew out of a 1997 World Bank workshop, arguing for a critical assessment of large-scale dam projects in terms of technical, economic, environmental, and social impacts.

DAOIST PERSPECTIVES

• • •

The word Daoism (or Taoism) was coined in the early nineteenth century from the Chinese expression "dao jiao teachings" (tao), which encompasses both the intellectual activities and historical religious movements that shaped the various and changing meanings of the term Dao (or Tao), meaning, literally, "the Way." Modern scholars have claimed that the term specifically refers to Daoist schools or Daoist sects, though some European Daoism scholars contend that this distinction is unnecessary or even misleading. In contemporary academic circles the words religion and philosophy are inevitably applied to Chinese traditions; one must remember, however, that in the Chinese context these two words

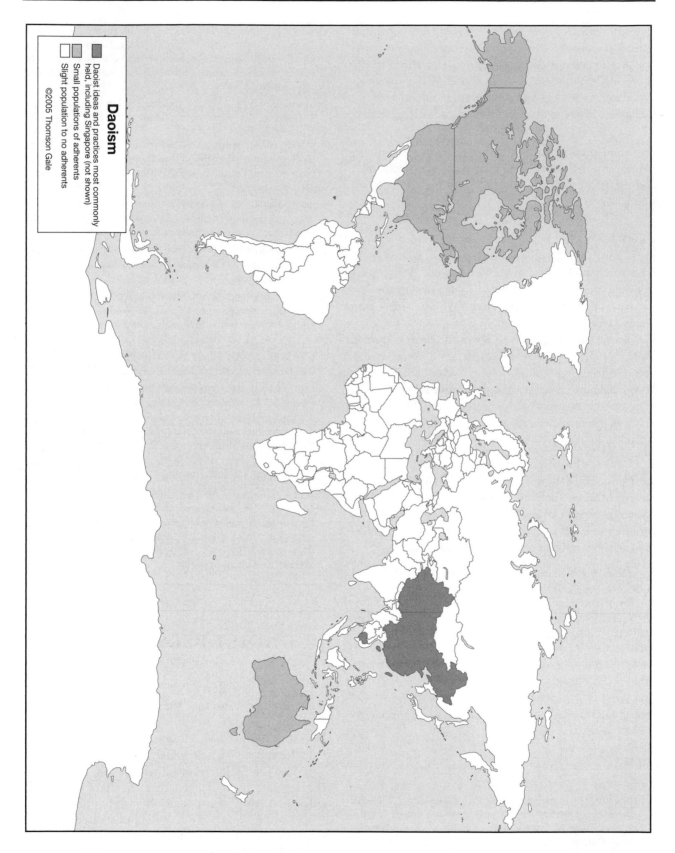

Daoism

Daoist ideas and practices most commonly held, including Singapore (not shown)

Small populations of adherents

Slight population to no adherents

©2005 Thomson Gale

diverge from their Western usages. Nevertheless, Daoism has suggestive importance as a perspective on science, technology, and ethics.

Daoist philosophy is attributed to Laozi, who, according to the ancient and authoritative *Records of History*, is believed to have been an elder contemporary of Confucius (551–479 B.C.E.) and the author of the *Laozi* (*Daode jing*, or *Tao-te-ching*), a work roughly 5,000 characters long. This traditional account has been challenged by skeptics, yet the three Guodian bamboo versions of the *Laozi* unearthed in 1997 prove that the text was extant and prevailing in the fourth century B.C.E. and may have been composed still earlier. Another founding thinker of Daoism was Zhuangzi. He and his followers created the *Zhuangzi*, a much longer work that is full of thought-provoking fables, stories, anecdotes, and inspiring ideas and arguments.

The religious worship of Laozi, together with the Buddha, is recounted in the official dynastic history in the first century C.E. Daoist religious movements, inspired by and combined with immortality beliefs, traditional medicine, yin–yang theories, *Yijing* (Classic of change) theories, and prognostication and apocrypha, developed in the following centuries. Regional Daoist religious activities, however, were not recognized by an independent royal court until the fifth century C.E. Because of its origination, Daoist religion had strong associations with folk and royal religious practices and beliefs, such as polytheistic worship, the pursuit of longevity, and the belief in immortality, physical or spiritual. Daoist priests and scholars may simultaneously be believers in Buddhism and practitioners of Confucianism.

A Philosophical Paradox

Daoism is commonly tagged as a sort of irrational mysticism. Actually, Daoist attitudes toward science and technology are mixed and varied. There are statements in the *Laozi* that seem directed against knowledge and artistry: "Eliminate knowledge, get rid of differentiation, and the people will benefit one hundredfold. Eliminate craftiness, get rid of profit, and there will be no robbers and thieves" (chap. 19, bamboo version). "The more cunning and skill a person possesses, the more vicious things will occur" (chap. 57).

In the *Zhuangzi*, one can find stories such as this one: Confucius's disciple Zigong while traveling saw an old man working in a garden. Having dug his channels, he made many trips to a well, returning with water in a large jar. This caused him a great expenditure of energy for very small returns. Zigong said to him, "There is a contrivance by means of which a hundred plots of

ground may be irrigated in one day. Little effort will thus accomplish much. Would you, Sir, not like to try it?" After hearing Zigong's description of the contrivance based on the lever principle, the farmer's face suddenly changed and he laughed, "I have heard from my master," he said, "that those who have cunning devices use cunning in their affairs, and that those who use cunning in their affairs have cunning hearts. ... I already knew all about it, but I would be ashamed to use it" (chap. 19). The farmer presents a typical Daoist criticism of technology and scientific invention. This is nevertheless a moral observation on the side effects of technological inventions, not an overall theory about technology and science.

Actually, the *Zhuangzi* contains many intriguing fables praising craftsmen who demonstrate fascinating artistry, such as boatmen, a butcher, sword makers, carvers of bell stands, arrow makers, and wheelwrights. A wheelwright once gave a lesson to the Duke Huan about the limitations of communication through the example of his artistry. He said:

> If my stroke is too slow, then the tool bites deep but is not steady; if my stroke is too fast, then it is steady but does not go deep. The right pace, neither too slow nor too fast, is the hand responding to the heart. But I cannot tell the skill by words to my son and he cannot learn it from me. Thus, it is that though in my seventieth year, I am still making wheels. The ancient author of the classic you are reading are dead and gone—so then what you are reading, is but the sages' dregs and refuse! (chap. 13)

This fable is not only a paean to the artisan and his artistry but also an ancient version of modern or postmodern theories of hermeneutics and linguistics.

In one chapter, the *Zhuangzi* raises questions about the natural world and its movements:

> How ceaselessly heaven revolves! How constantly earth abides at rest! Do the sun and the moon contend about their respective places? Is there someone presiding over and directing these things? Who binds and connects them together? Who causes and maintains them, without trouble or exertion? ... Then how does a cloud become rain, and the rain again form clouds?" (chap. 14)

These questions come from and in turn stimulate curiosity about the natural world, which inspires investigation into scientific and technological mysteries. Daoism considers human beings to be equally part of the natural world and has a strong interest in the ultimate origins of, reasons behind, mechanisms of, and mysteries of the universe, including human lives—especially in comparison with Confucianism and Buddhism.

One distinctively Daoist concept is *wuwei* (nonaction), which is often misunderstood as inactivity or literally doing nothing. But the *Huainanzi* (142 B.C.E.), a Daoist work of the early Han period, argues that this term does not mean inactivity. *Wuwei* actually suggests that no personal prejudice interferes with the universal Way and that no desires or obsessions lead the true courses of Daoist techniques astray. To undertake an enterprise one must follow reason, and to realize an achievement one must take account of surrounding conditions to be consistent with the principle of naturalness. For example, if one used fire to dry up a well or led the waters of the Huai River uphill to irrigate a mountain, these would be contrary to the principle of naturalness and be called taking action (*youwei*, the opposite of *wuwei*). Nevertheless, such activities as using boats on water or sledges on sand, making fields on high ground, and reserving low ground for a pond constitute Daoist *wuwei* or nonaction. This interpretation of *wuwei*, deriving from the *Laozi*'s idea of "assisting the naturalness of the ten thousand things without daring to act," promotes a rational and observant attitude in everyday life, which favors the scientific spirit.

Religious Pursuits

While Daoist thinkers presented reflective and inspiring ideas, religious scholars and priests, in their informal roles as inventors, practitioners, compilers, or distributors, made great practical and academic contributions to the development of science and technology in China. According to the first official 5,305-volume *Daoist Canon* (completed in 1445), Daoist scholarship and practice pursued knowledge and technology in various fields, such as chemistry, mineralogy, biology, botany, pharmacy, medicine, anatomy, sexology, physics, mathematics, astronomy, and cosmology. Ancient Daoists were not professional scientists or technicians, and their essential concern was attaining longevity and material immortality, rather than science and technology for their own sake. This pursuit makes Daoism distinct among religions and led Daoists to seriously observe and explore the natural world, including the human body and life, from generation to generation. Thus, religious enterprise provided fertile ground for the development of science and technology.

A good example of this confluence is the discovery of gunpowder. Joseph Needham (1981) contends that saltpeter (potassium nitrate) was recognized and isolated at least by the fifth century in China. This first compounding of an explosive mixture arose in the course of exploring the chemical and pharmaceutical properties of a great variety of inorganic and organic substances. It was the hope of realizing longevity and physical immortality that led to this discovery, one of the greatest technical achievements of the medieval Chinese world. One finds the first reference to it in the ninth century, toward the end of the Tang dynasty, in a description of the mixing of charcoal, saltpeter, and sulfur. This mention occurs in a Daoist book that strongly recommends not mixing these substances, especially with arsenic, because some alchemists who had done so had the mixture deflagrate, singe their beards, and even burn down the house in which they were working.

The fields of medicine and pharmacology were also directly shaped by the Daoist pursuit of longevity and immortality. Daoist scholars and priests advanced Chinese medical theory and compiled important herbal medicine classics. Tao Hongjing (451–536), a direct descendant of the founder of the Supreme Purity Sect, is the most prominent of these scholars. His eighty works involve astronomy, calendrics, geography, literature, arts, and the arts of war, in addition to medicine and pharmacology. He argued that humans control human destiny, not Heaven. The reason people die early is not because of fate, but because their way of living harms their spirits or bodies. A piece of semifinished pottery is made of earth, yet is different from earth. Still it will dissolve in water before it is fired, even though it has already dried. If it is not fired properly, it will not hold up. If it is fired well and becomes thoroughly strong, it will survive over vast stretches of time. Similarly, people who pursue immortality take drugs and elixirs to make the body strong, breathe in fresh air, and participate in gymnastic exercise.

All these practices complement each other without conflict. If the spirit and the body are refined together, as in a senior immortal, one can ride clouds and drive a dragon; if the spirit and the body become separated, as in a junior immortal, one can leave one's old body and take on a new one. To preserve spirit and body, Daoists emphasized the significance of moderation in desires and emotion. It is impossible for the average person to have no desires or do nothing, but they can keep their minds in a state of harmony and minimize concerns. The "seven kinds of emotion" (anger, anxiety, worry, sorrow, fear, aversion, and astonishment) and the "six desires" (for life and death, and of the eyes, ears, mouth, and nose) are all harmful to the spirit and should be controlled.

Tao Hongjing also argued that the harm caused by bad eating habits is more serious than that of lust, because people eat daily, and he urged restraint in taking food. To be healthy, he claimed, less food is better than being overly full; walking after meals is more helpful than lying down; and physical labor is preferable to an easy life. Most of this early Daoist's advice accords with suggestions from modern doctors and professional medical workers.

Furthermore, Tao Hongjing compiled the *Collected Commentaries on Medicinal Herbs*, without which the contents of the earliest Chinese medicine classics would have been lost forever. He was the person who created a typology of Chinese medicinal herbs and inorganic substances in the treatment of various diseases and symptoms; this became and remains the foundation of Chinese medical theory.

According to Daoist tradition, the technology of sexual life is related to prolonging youth and vigor, though it was rejected by some later Daoists. Ge Hong (283–364?) once argued that sexual intercourse was necessary to achieve longevity and immortality. Even if one were to take all the famous medicines, Ge claimed, without knowledge of how to store up the essence of life through sexual activity, attaining health, let alone longevity, would be impossible. While people should not give up sex entirely, lest they contract melancholia through inactivity and die prematurely from the many illnesses resulting from depression and celibacy, overindulgence can diminish one's life, and it is only by harmonizing the two extremes that damage can be avoided.

It was further held that foreplay and slow and complete arousal are important for healthy intercourse. Men should pay attention to women's reflexes step by step and delay climax to adjust for the differential in arousal time to ensure the woman's full satisfaction. Some of these theories seem to have been confirmed and adopted by modern sexologists. Kristofer Schipper (1993), a Dutch Daoist scholar, claims that Chinese sex manuals reflect an impressive knowledge of female anatomy and reflexes; they are the only ancient books on this subject that do not present sexuality solely from the male point of view. Indeed, compared to other traditions, Daoism includes much less discrimination against women, perhaps because of Daoists' strong belief in the harmony of yin and yang, which work in all things and processes in the universe.

Modern Resonance

Although Daoism is an indigenous Chinese cultural tradition of some antiquity, modern scientists have found that it resonates with certain aspects of the spirit of modern science and responds to modern social and environmental issues. Raymond J. Barnett (1986) found a surprising degree of similarity between Daoism and biological science in their views on death, reversion (cyclicity of phenomena), the place of humans in the universe, and the complementary interactions of dichotomous systems. The use of the terms yin and yang is similar to the way scientists describe the behavior of subatomic particles: One can say some things about these particles, but only if one realizes that what is said is a statement of statistical probability and that a certain modicum of uncertainty is unavoidable. And in the autonomic nervous system both the sympathetic and parasympathetic subsystems, like the yin and yang, affect most organs. The state of an organ is not a function of one system being totally "off" and the other totally "on." Rather, the health of an organ depends on the balance between the activities of both systems, with each able to change its input and alter the balance.

Similar parallels between Daoist ideas and science are too numerous to be discussed at length, but a few deserve brief mentions. James W. Stines (1985) demonstrated that the philosophy of science of British chemist Michael Polanyi (1891–1976), especially his theory of tacit knowledge, correlated with Daoist intuition. Hideki Yukawa (1907–1981), who in 1949 became the first Japanese physicist to receive a Noble Prize, claimed that his creativeness was greatly inspired by Laozi's and Zhuangzi's philosophical insights. The famous American humanistic psychologist Abraham H. Maslow (1993) found the advantage and complementary role of Daoist objectivity in scientific investigation. Fritjof Capra, in his best-seller *The Dao of Physics* (2000), revealed the parallel between Daoism (along with other Eastern traditions) and the notion of a basic "quantum interconnectedness" emphasized by the Danish physicist Niels Bohr (1885–1962) and the German physicist Werner Heisenberg (1901–1976). Norman J. Girardot and colleagues (2001) discuss broadly and significantly the relationship of Daoism and modern ecological issues. Finally, one should certainly not forget the pioneer researcher Needham, who contended that Daoist thought is basic to Chinese science and technology.

LIU XIAOGAN

SEE ALSO *Acupuncture; Buddhist Perspectives; Confucian Perspectives.*

BIBLIOGRAPHY

Barnett, Raymond J. (1986). "Taoism and Biological Science." *Zygon* 21(3): 297–317.

Capra, Fritjof. (2000). *The Tao of Physics: An Exploration of the Parallels between Modern Physics and Eastern Mysticism*, 4th edition. Boston: Shambhala.

Chan, Wing-tsit. (1963). "The Natural Way of Lao Tze." In *A Source Book in Chinese Philosophy*. Princeton, NJ: Princeton University Press. This is a philosophical translation of the Laozi with brief comments.

Girardot, Norman J.; James Miller; and Liu Xiaogan, eds. (2001). *Daoism and Ecology: Ways within a Cosmic Landscape*. Cambridge, MA: Harvard University Center for the Study of World Religions. A collection of papers authored by philosophical, religious, literary, and ecological scholars, as well as practitioners of Daoism.

Liu Xiaogan. (1993). "Taoism." In *Our Religions*, ed. Arvind Sharma. San Francisco: Harper San Francisco. A relatively comprehensive introduction to Daoism from its origination to modern development, including Daoist philosophy and religion, their relationship, and other perspectives.

Maslow, Abraham H. (1993). *The Farther Reaches of Human Nature*. New York: Arkana. Maslow presents the concept of "Daoist objectivity" as a development of his earlier concept of "Daoist science" discussed in his *The Psychology of Science* (Gateway edition, 1966).

Needham, Joseph. (1981). *Science in Traditional China: A Comparative Perspective*. Hong Kong: Chinese University Press.

Needham, Joseph, with Wang Ling. (1956). *Science and Civilization in China*, Vol. 2: *History of Scientific Thought*. Cambridge, UK: Cambridge University Press. This is an initiative comprehensive study with tremendous materials and references. Needham attributes many Chinese scientific and technological inventions and contributions to Daoist thought and practice.

Schipper, Kristofer. (1993). *The Taoist Body*, trans. Karen C. Duval. Berkeley: University of California Press. Schipper is not only a leading scholar, but also an ordained Daoist priest. This book is an introduction to aspects of Daoist religion, such as teachings, divinity, alchemy, immortals, ritual, and meditation.

Sivin, Nathan. (1968). *Chinese Alchemy: Preliminary Studies*. Cambridge, MA: Harvard University Press.

Sivin, Nathan. (1995). *Medicine, Philosophy, and Religion in Ancient China: Researches and Reflection*. Aldershot, Hampshire, UK: Variorum. Sivin strongly contends that many of Needom's Daoist scientists do not belong to Daoist communities.

Stines, James W. (1985). "I Am the Way: Michael Polanyi's Taoism." *Zygon* 20(1): 59–77.

Watson, Burton. (1968). *The Complete Works of Chuang Tzu*. New York: Columbia University Press. A popular and complete translation of the Zhuangzi.

Yukawa, Hideki. (1973). *Creativity and Intuition: A Physicist Looks at East and West*, trans. John Bester. Tokyo: Kodansha International.

DARWIN, CHARLES

• • •

Naturalist Charles Darwin originated the theory of evolution by means of natural selection. Darwin (1809–1882), who was born in Shrewsbury, England, on February 12, established the modern scientific understanding of humanity's place in nature. After his undergraduate education at Cambridge, Darwin served for nearly five years as a naturalist aboard a surveying ship, HMS *Beagle*, which traveled up and down the coasts of South America and then circled the globe. Darwin spent several years after his voyage publishing the results of his researches into fossils, botany, zoology, and geology. On the basis of this work, he formulated his initial ideas on evolution in the late 1830s and then spent two decades developing the theory of natural selection before publishing his chief work, *On the Origin of Species by Means of Natural Selection, or the Preservation of Favoured Races in the Struggle for Life* (1859). In *The Descent of Man, and Selection in Relation to Sex* (1871), Darwin explicitly included human beings within the theory of evolution and analyzed the biological basis of human social and moral behavior. Darwin died on April 19 in England and is buried at Westminster Abbey.

In his autobiography, Darwin says that the one book he most admired as an undergraduate was William Paley's *Natural Theology: or, Evidences of the Existence and Attributes of the Deity, Collected from the Appearances of Nature* (1802). Paley (1743–1805) was the best-known proponent of natural theology, a school of thought that combined providential theology with inquiry into adaptive structures in animals. From the perspective of natural theology, adaptive structure or *design* is evidence for the beneficent governance of the world by its creator. Darwin's theory of natural selection provided an alternative scientific explanation for adaptive structure. Within Darwin's theory, adaptive structure is the result of *natural selection*. Innate variations in physiology or anatomy regularly occur. Many such variations are neutral or harmful to an organism, but some variations offer advantages that enable an organism to survive or reproduce more effectively than its competitors. These favorable variations are inherited and transmitted, and over many generations inherited variations produce new species.

Darwin's theory of natural selection is not grounded in theology or ethics, but it has implications for metaphysical and ethical beliefs. In his later years, Darwin became a professed agnostic, but at the time of writing *On the Origin of Species*, he was still vaguely theistic and

Charles Darwin, 1809–1882. Darwin discovered that natural selection was the agent for the transmutation of organisms during evolution, a theory he presented in *Origin of Species*. (© *Bettmann/Corbis*.)

and to use that quality as a model or norm for human ethical behavior. The injunction to *follow nature* has been interpreted to mean either that one should imitate the supposedly benign character of the providential order or that one should ignore all conventional social constraints and seek only to satisfy one's own desire and ambition. Since the middle of the nineteenth century, many thinkers have rejected this approach and have argued that human morality is something separate from the natural order. In their view, humans should not follow nature but should instead cultivate their own specifically human moral sentiments independently of nature. Among Darwin's contemporaries, John Stuart Mill and Thomas Henry Huxley (1825–1895) advocated this moral philosophy, and in the later twentieth century it was advocated by prominent Darwinian thinkers such as George C. Williams (b. 1926), Richard Alexander (b. 1929), Richard Dawkins (b. 1941), and Donald Symons (b. 1942).

Darwin's own theory of human morality breaks away from the idea that one should take the larger order of nature as the model for human moral behavior, but Darwin does not argue that human morality is simply separate from the order of nature. He argues instead that human moral sentiments derive from the evolved and adapted structure of human psychology. The human capacity for moral behavior results from two aspects of our evolved psychology: our character as social animals, and our uniquely human ability to think abstractly. Our social nature enables us to feel sympathy for other humans, to feel pain at their suffering and pleasure at their happiness. Our ability to think abstractly makes it possible for us to rise above the present moment, to link the present with the past and future, and thus to take account of the long-term consequences of our behavior.

In typical Victorian fashion, Darwin hoped that humanity would progress steadily toward a higher state of moral consciousness, and he envisioned human moral progress as circles of sympathy expanding out from kin and tribe, to nations and cultures, to all human beings, and eventually to all life on earth. At the highest level of human development, Darwin hoped that humans would become ecological curators for the earth.

In *Descent of Man*, Darwin considered the issue of eugenics. He acknowledged that care of the weak has dysgenic effects, but he nonetheless rejected *social Darwinism* or ruthless social competition because, he felt, that sort of behavior would damage the more "noble" qualities of social sympathy on which all human moral behavior depends.

regarded the development of life on earth as the result of a divine creation. The evolutionary process that he explains nonetheless exhibits qualities of ruthlessness and cruelty. In order to describe this process, Darwin frequently uses metaphors such as the "Struggle for Life," the "battle of life," or the "war of nature." In all species, many more individuals are born than can ever survive or reproduce. This disproportion between birth rates and the rates of survival and reproduction provides the competitive situation within which natural selection operates. Individuals within and among species compete for food and other resources; individuals of one species prey on individuals of other species; and most species eventually become extinct and leave no successor species. In a letter of 1856 to his botanist friend Joseph Hooker (1817–1911), Darwin exclaims almost in despair over "the clumsy, wasteful, blundering, low, and horribly cruel works of nature!" In the last paragraph of the *Origin*, he declares that there is "grandeur in this view of life," but it is a grandeur that emerges out of "famine and death."

Both before and after Darwin, it has been common practice to invest the larger natural order with some moral quality, either of beneficence or of ruthlessness,

From the second through the sixth decade of the twentieth century, the adaptationist psychology that Darwin inaugurated in *Descent of Man* went into eclipse, supplanted by the belief that culture and society control behavior and are not themselves prompted and constrained by biology. The advent of human sociobiology in the 1970s brought Darwinian thinking back into psychology, anthropology, and the other human sciences. In sociobiology and related schools such as *human ethology, evolutionary psychology,* and *behavioral ecology,* the adaptationist view of human nature has had a deep and far-reaching influence on twenty-first century ethical thinking. For contemporary Darwinian theorists of human ethical behavior, the most significant issue under debate is a question about the level at which natural selection operates. Proponents of *selfish gene* theory argue that natural selection operates exclusively at the level of genes, and they extrapolate the idea of "selfishness" from the level of genes to the level of individual human motives. Proponents of *group selection,* in contrast, affirm the reality of altruistic or "unselfish" motives. Many theorists argue that selection operates at multiple levels and that these levels are interactive and interdependent. The idea of a genetically encoded "altruism" that ultimately subverts inclusive fitness would contradict the logic of natural selection, but a co-operative and interdependent structure is a fact of evolutionary history and manifests itself at the level of cells, organs, social groups, and ecosystems.

JOSEPH CARROLL

SEE ALSO *Aggression; Christian Perspectives; Evolutionary Ethics; Evolution-Creationism Debate; Gatton, Francis; Social Darwinism; Sociobiology; Treat, Mary.*

BIBLIOGRAPHY

Darwin, Charles. (1845). *Journal of Researches into the Geology and Natural History of the Various Countries Visited by H. M. S. Beagle, Under the Command of Captain Fitzroy from 1832 to 1836,* 2nd edition. London: John Murray.

Darwin, Charles. (1859). *On the Origin of Species by Means of Natural Selection, or the Preservation of Favoured Races in the Struggle for Life.* London: John Murray.

Darwin, Charles. (1871). *The Descent of Man, and Selection in Relation to Sex.* London: John Murray.

Darwin, Charles. (1958). *The Autobiography of Charles Darwin, 1809–1882. With Original Omissions Restored,* ed. Nora Barlow. London: Collins.

DATA MINING

SEE *Transaction Generated Information and Data Mining.*

DC-10 CASE

• • •

The troubled history of the DC-10 aircraft, especially in relation to questions raised as a result of its involvement in three major accidents between 1974 and 1989, provides a multidimensional case study in the ethics of engineering design and the uses of technology.

The DC-10 is a wide-bodied aircraft with two wing engines and a third engine distinctively placed in the tail fin. It was introduced into commercial service in 1972, during a time of unusually intense competition in the U.S. aviation industry. The market would support only two viable manufacturers, and because the Boeing 747 was well established, either Lockheed Corporation or McDonnell Douglas Corporation would have to withdraw and suffer a substantial financial loss. McDonnell Douglas won the competition, but evidence of its haste to beat Lockheed is reflected in these case studies.

Design Vulnerability

Because airliners fly at high altitudes, the passenger cabin must be pressurized, up to 38 pounds per square inch. Because a heavy floor able to withstand this force would not be economical, the cargo hold is also pressurized. Thus the floor has to be strong enough to support only the weight of passengers, crew, seats, and so on. If, however, either part of the aircraft experiences a sudden decompression, the loss of equalizing pressure would cause the floor to buckle or collapse, resulting in damage to the control system, which is located in the interior spaces of the floor beams.

The 1972 Windsor Incident

Less than a year after the DC-10 was in service, a rear cargo door was improperly closed on a flight from Detroit, Michigan, and it blew open over Windsor, Ontario, causing the floor above it to collapse downward. Only the skill of the American Airlines pilot and a very lightly loaded airplane enabled the plane to land safely.

Ordinarily a problem of this magnitude would result in the Federal Aviation Administration (FAA) issuing an Airworthiness Directive (AD), a public document that has the force of law, requiring owners of a particular aircraft to modify their airplanes within a certain time. But the FAA charter contains a dual mandate: The FAA must not only ensure aviation safety but also promote the aviation industry. An AD at this time would have given Lockheed a competitive advantage by drawing attention to the DC-10 problem. Instead, John Sheaffer, the head of the FAA, finessed these conflicting objectives by making a "gentleman's agreement" with McDonnell Douglas to develop a fix for the cargo door and implement it through service bulletins sent only to owners of DC-10s, thus avoiding harmful publicity.

Two weeks after Windsor, Dan Applegate, head of project engineering at Convair, a subcontractor for the DC-10 cargo doors, expressed grave doubts about the "Band-Aid" fixes being proposed for the cargo door lock and latch system. He took his concerns to higher management in an effort to have Convair contact McDonnell Douglas and develop a more secure fix. Although he wrote a strong memo, management felt its hands were tied by a "reliance clause" in the contract, which stated that if Convair disagreed with the design philosophy it must make its concerns known in the design stage or pay for any later required changes. Because DC-10s were already rolling off the production line, Convair was faced with the prospect of paying for expensive retrofits to the DC-10 if it raised questions now. No approach to McDonnell Douglas was made.

The 1974 Paris Crash

When the service bulletins were sent out, many DC-10s were sitting on the McDonnell Douglas lot awaiting delivery. Ship 29, later sold to Turkish Airlines, was recorded as having all service bulletins for the cargo door performed, but in fact a critical item was omitted. Critics believe that an AD would have been taken more seriously.

On a fully loaded flight from Paris to London, on March 3, 1974, Ship 29 lost its rear cargo door shortly after takeoff, and the floor collapsed. Deprived of its control system, the plane crashed: Six passengers from the rear of the aircraft were found, still strapped in their seats, nine miles away; the cargo door that failed was nearby. French investigators collected more than 20,000 human fragments of the 346 passengers and crew. At the time, it was the worst aircraft accident in history.

The 1979 Chicago Crash

On May 25, 1979, American Airlines DC-10 crashed shortly after takeoff from Chicago when a wing engine broke loose and damaged the leading edge of the wing. Loss of the engine and damage to the wing resulted in decreased lift: One wing was pushing up harder than the other. A photo shows the plane, wings vertical, plunging to the ground.

Had the pilots known that the wing was damaged, they would have been able to take corrective measures to control the plane. But they could not see the wing from the cockpit and had to rely on instruments. Ironically, the needed warning devices were powered by the engine that broke off, and there was no provision for a backup power supply. The crash killed all 271 persons onboard the DC-10 and two persons on the ground.

The separation of the engine was caused by a maintenance procedure designed to save more than 200 person-hours of work. The engine is held in place by a large pylon attached to the wing, and the McDonnell Douglas removal procedure required that the engine (weighing 5,000 kilograms) be removed first, followed by the pylon (900 kilograms). The new procedure used a forklift to bear the weight of the engine, allowing engine and pylon to be removed as a unit. The pylon is not designed for the stresses this procedure can introduce and developed cracks, which eventually led to it and the engine breaking away from the wing.

It is normal for airlines to develop innovative maintenance procedures without FAA approval. McDonnell Douglas knew that Continental Airlines and American were using the forklift procedure and that it required extreme precision in positioning. It also knew that Continental had reported two cases of cracks to the pylons that required repair. Neither the FAA nor American learned of these potential dangers because FAA regulations do not require such reporting. But an engineer's first professional obligation is to protect the public from harm, and engineers at McDonnell Douglas and Continental had clear evidence of the danger of this procedure and should have investigated further and warned others. For a professional, following the regulations is not good enough when there is clear evidence of danger.

The 1989 Sioux City, Iowa, Crash

On July 19, 1989, a United Airlines DC-10 tail engine disintegrated in flight, resulting in the loss of fluid in all three hydraulic systems. The 170-kilogram front fan disk, rotating at high speed, broke apart, and the fragments took out everything in their path. Without hydraulics, none of the control surfaces on the wings and tail could be operated. The plane could only be crudely maneuvered by varying the speed of the two wing

A McDonnell-Douglas DC-10. A string of highly publicized crashes doomed the aircraft to a short lifespan. (© *George Hall/Corbis.*)

engines. Remarkably, the pilots managed to crash-land at the Sioux City, Iowa, airport, with only 111 deaths among the 296 passengers.

The other wide-body jet with a large tail engine, the Lockheed L-1011, has four independent hydraulic systems, one of which has a shutoff valve forward of the engine. If there is a leak, the valve closes the line, preventing further fluid loss. After the accident, the FAA issued an airworthiness directive requiring a shutoff valve for the DC-10.

Assessment

All three DC-10 crashes were caused by failures that need not have resulted in the loss of the aircraft. The inadequately protected control system of the DC-10 allowed these otherwise predictable problems to cause the crashes that took 728 lives. It would be satisfying to find engineers and managers who clearly disregarded the safety of air travelers, but the reality is a complex and ambiguous interplay of engineering, design, financial, legal, historical, and organizational factors that allowed an underprotected aircraft to enter the stream of

commerce. Without the intense economic competition with Lockheed, there might have been more attention to the cargo door design, redundancy added to warning systems, and a shutoff valve placed in the hydraulic lines. Add to this Douglas Aircraft Company's complete dominance of the aviation industry from the 1930s to the 1950s, which may have fostered a climate of complacency about the problems with the DC-10. (McDonnell Douglas had been formed in 1967 from the merger of Douglas Aircraft and McDonnell Aircraft Corporation.) The regulatory safety net, as always, was catching up to the problems posed by the new generation of wide-body jets.

After each of these crashes the FAA required changes in design, procedures, or training. Critics call this "tombstone technology," meaning that safety changes are made only if there are enough deaths to prove the changes are needed. But safety is defined as "of acceptable risk," which changes over time, and often it takes a severe accident to determine what level of risk is socially acceptable. Safety entails higher costs, and regulators must try to balance the safety and cost factors in evaluating complex, sophisticated technology that has a substantial interface with large numbers of people. Inevitably, mistakes will sometimes be made and innocent people will die before adequate regulations are in place.

JOHN H. FIELDER

SEE ALSO *Airplanes; Aviation Regulatory Agencies; Engineering Ethics.*

BIBLIOGRAPHY

Fielder, John H., and Douglas Birsch, eds. (1992). *The DC-10 Case: A Study in Applied Ethics, Technology, and Society.* Albany: State University of New York Press.

DDT

• • •

DDT ranks among the most infamous acronyms in history. During the mid-twentieth century, its effectiveness at killing insects made it one of the miracle products of wartime investments in science and technology. Yet within thirty years, many industrialized countries banned the synthetic insecticide due to fears of its long-term effects on humans and wildlife. At the turn of the twenty-first century, the devastating resurgence of malaria across the developing world reignited debates over the ethics of using DDT.

The chemical compound that is DDT, dichloro-diphenyl-trichloroethane, was first synthesized in 1873, but not until 1939 did Swiss chemist Paul Müller discover its insecticidal properties. The U.S. military used DDT during World War II to protect soldiers and civilians from the destructive insect-borne diseases typhus and malaria. DDT's persistence and its broad spectrum of action made it extremely successful at killing insects over a long period, in small doses, and at low cost. In response to civilian demand, the U.S. government made the celebrated chemical available to the public in 1945, despite private concerns among federal scientists of potential long-term hazards. The agricultural and public health promise of DDT led to mass aerial spraying programs, and Müller won the 1948 Nobel Prize in physiology or medicine. Production by U.S. companies increased from 10 million pounds in 1944 to more than 100 million pounds in 1951.

Rachel Carson burst the bubble of confidence concerning the safety of DDT in 1962 with her best-selling exposé of the overuse of synthetic chemical pesticides, *Silent Spring.* The book publicized scientific evidence of the toxic effects of DDT on humans and animals, including nervous system dysfunction, reproductive abnormalities, and cancer. It explained how DDT's insolubility in water and fat-solubility enable it to persist in the soil and water, enter the food chain, and accumulate in the fatty tissues of non-target organisms such as the bald eagle, whose plummeting numbers were linked to DDT-induced eggshell thinning. *Silent Spring* also showed how mosquitoes and other target insect populations develop genetic resistance to DDT, thereby undermining its efficacy.

Carson criticized the arrogance of entomologists who presumed they could control pests by waging chemical warfare. She made a strong ethical argument for the need to respect the other creatures with which humans share the earth. Although some critics accused her of privileging wildlife over people, she testified to Congress on behalf of "the right of the citizen to be secure in his own home against the intrusion of poisons applied by other persons" (Lear 1997, p. 454). Spurred by increasing evidence of DDT's carcinogenicity, Congress banned the sale of DDT in the United States in 1972. Within three decades DDT was banned in thirty-four countries, and severely restricted in thirty-four others. It continued to be used in several developing nations, primarily in the *malaria belt.*

Since the 1970s, malaria has become one of the deadliest infectious diseases in the world, killing at least 1.1 million people each year. Children under age five comprise more than half the victims. Many environmentalists and health experts blame malaria's huge resurgence on the overuse of both chemical insecticides and anti-malarial drugs, which led their respective targets—anopheles mosquitoes and plasmodium parasites—to develop genetic resistance. Anti-DDT groups advocate preventive methods, including the use of mosquito nets dipped in the nontoxic insecticide permethrin and the cultivation of fish that consume mosquito larvae, as part of a systematic approach to the disease. In their opinion, DDT should be used only as a last resort due to its well-documented negative effects.

In contrast a strong opposition movement argues that DDT is still the cheapest, most effective anti-malarial measure, and that its declining use is responsible for the recent resurgence of malaria. Pro-DDT groups condemn environmentalists for scaring developing countries from using the chemical, and for caring more about bald eagles than suffering children. They point to scientific studies that fail to confirm evidence of long-term risks of exposure to DDT, and contend that it serves as a crucial insect repellent even in places

Two women are sprayed with DDT. Although widely used in pesticides in the 1940s and 50s, the compound has been banned in North America and most of Europe since the 1970s due to fears of detrimental long-term effects. (*The Library of Congress.*)

where mosquitoes have become resistant. From their perspective, it is unethical not to utilize DDT as a first resort against malaria, because its life-saving capacity for millions of people outweighs any potential negative environmental or human health effects.

Despite such conflicting outlooks, a compromise was struck in 2001, when delegates from 127 nations signed an international treaty to phase out twelve toxic, persistent, fat-soluble chemicals, including DDT. After intense debate, developing nations received exemptions permitting them to continue using DDT against the mosquito vectors of malaria until safer, affordable substitutes become available. The Stockholm Convention on Persistent Organic Pollutants entered into force on May 17, 2004.

CHRISTINE KEINER

SEE ALSO *Agricultural Ethics; Carson, Rachel; Food Science and Technology.*

BIBLIOGRAPHY

Carson, Rachel. (1962). *Silent Spring.* Boston: Houghton Mifflin. A must-read for anyone interested in the impact of synthetic pesticides on human and environmental health.

Lapkin, Ted. (2003). "DDT and the New Colonialists." *Quadrant* 47: 16–19. Argues for the use of DDT for malaria control.

Lear, Linda. (1997). *Rachel Carson: Witness for Nature.* New York: Holt. The definitive biography of Carson.

McGinn, Anne Platt. (2002). "Malaria, Mosquitoes, and DDT." *WorldWatch* 15: 10–17. Argues against the use of DDT for malaria control.

Russell, Edmund P. (1999). "The Strange Career of DDT: Experts, Federal Capacity, and *Environmentalism* in World War II." *Technology and Culture* 40: 770–796. Presents new historical evidence about early DDT research.

DEATH AND DYING

• • •

Death is defined as the irreversible loss of biological life functions, and occurs in all organisms. It is the inevitable conclusion of a finite existence, and is often applied by analogy even to geological features that contain life (the death of a river), social orders (death of a city), or machines (one's car or computer died). Science can study the phenomenon of death, technology may delay its approach, but medicine cannot cure humans of their mortality.

Dying, by contrast, is the process that leads to death, and is a distinctly human event, embedded in numerous moral traditions and well-circumscribed by prescriptions for appropriate conduct in its presence. No other animal attends as carefully to the dying process or accords it such significance. Modern medical technologies, including drugs and therapies, aim to delay the onset of this process (life-saving technologies), extend it (life-sustaining or life-support technologies), take control of it (technologies of euthanasia), or provide comfort (palliative technologies) as the time of death nears. Whereas premodern thought commonly interpreted dying in religious terms, viewing it as a process of transformation from one state to another and calling forth techniques of ritual engagement with larger orders of reality, contemporary technical achievements are the result of aspirations for control over the process that pose challenges in a moral framework.

Historical and Cultural Background

As described by the cultural historian Philippe Ariès (1991), the European experience of death has itself undergone significant transformations. Dying is not simply a basic feature of the human condition and the termination of an individual history; it has its own history. From the Graeco-Roman world up through the first millennium of the Christian era, death was so ever-present as to have been accepted as a normal aspect of human affairs. Theologically death was also often interpreted as a result of living in a fallen world marked by sin. When someone died, people paid their respects but did not dwell on the issue because of the greater importance of the community as a whole.

During the eleventh century in Europe, the rise of individualism brought with it a new perspective on death as a threat not to the community but to the self, which in turn gave rise to the development of *Ars moriendi* or treatises on the art of dying well. In the sixteenth century the emphasis shifted toward concern for the death of loved ones in a family. The romantic pathos at the death bed of family and friends ironically contributed to the transfer of the event of dying from home to hospital, where it acquired a higher public profile.

How is it, asks Ariès, that "the community feels less and less involved in the death of one of its members"? One answer is that the community "no longer thinks it necessary to defend itself against a nature which has

been domesticated once and for all by the advance of technology" (Ariès 1991, p. 612). Another is that individualism has fractured the sense of solidarity. As death ceases to be a public threat it is progressively transformed into an emotional issue and relegated to the realm of privacy. Medical science and technology are brought in to study the process and reduce the pain, but behind this rational management "the death of the patient in the hospital, covered with tubes, [becomes] a popular image, more terrifying than the ... skeleton of macabre rhetoric" (Ariès 1991, p. 614).

There are, however, other perspectives on death and dying that have likewise been impacted by scientific and technological change. Outside European traditions, for instance, the Hindu view embraces death as part of a great spiritual journey. The soul is in a state of continuous evolution and awareness from the limited to the limitless. Death separates the indestructible soul-body from the weakened physical-body. The soul-body reincarnates a physical-body, which matures and develops the soul-body in a life and death cycle. The act of reincarnation enables the soul to renew its work of resolving karma or moral effects arising from failures to conduct oneself in harmony with the dharma or moral order of the cosmos. The fulfillment of this process leads to liberation from the cycle of rebirth in a state called *moksha* and is characterized by *sat chit ananda* or limitless being, awareness, and bliss.

One influential Buddhist tradition *The Tibetan Book of the Dead* (from the fourteenth or fifteenth centuries), a guidebook for the deceased that echoes similar Egyptian, Daoist, and even Kabbalist literature, teaches how death is a process of moving toward pure truth. Once liberated from the confines of a mortal body, awareness enters a series of intermediate states called *bardos* in which it experiences various, sometimes frightening visions. To work through these *bardos* calls for assistance in order to attain either reincarnation or nirvana, a state in which confusion and suffering cease.

Across these various cultural perspectives one common theme is a distinction between natural and unnatural death and dying (Young). Natural deaths are associated with old age and disease, unnatural with accident and murder. In Hinduism, Jainism, Buddhism, Confucianism, and Daoism natural death has a normative value, thus implying criticism of the technological or artificial manipulations of the dying process, especially in the form of active euthanasia. Insofar as the natural is also seen as a manifestation of the spiritual, such a view reaffirms death as a gateway between the natural and the supernatural.

Encyclopedia of Science, Technology, and Ethics

477

It is in reaction against such metaphysical perspectives—perspectives that have much in common with the premodern Christian views—that the Enlightenment gave birth to the typically modern, empirical explanation of death and efforts to take instrumental control of dying. From the invention of the microscope and the discovery of a small pox vaccination emerged the germ theory of disease. According to germ theory, microbes enter a body and attack its metabolic functioning. Public health initiatives are thus required to develop external means of protection against such attacks. With the discovery of antibiotics physicians were able to go inside the body and kill germs there as well. Death is no longer considered a consequence of sin, but is simply the result of disease and old age.

The Psychology of Dying and Death Redefined

Further developments in scientific and technological medicine have made it possible not only to protect from and respond to disease, but to aggressively manage the dying process with lifesaving or life-support technologies and to provide high-tech palliative care—and even take control of it with techniques for euthanasia. Psychological models of the dying process (which faintly echo the traditional guidebooks for the dying) and efforts to redefine death itself highlight alternative responses to the challenges thus raised.

The most popular psychological analysis of the dying process is *On Death and Dying* (1969) by Elizabeth Kübler-Ross. In her thanatology, she distinguishes five stages of dying through which people progress when informed that they have a terminal illness: denial, anger, bargaining, depression, and acceptance. This analysis has become so influential that some advocate its application in other situations in which people suffer loss or experience traumatic change. Indeed in the face of issues arising from revolutionary new technologies such as human cloning, there is room to argue that a culture passes through stages in which there is public denial of the possibility, anger at its scientific creation, bargaining for how it is to be used, depression that it seems inevitable, and finally acceptance as just another aspect of the advanced technological condition. Others, however, advocate modification and revision of Kübler-Ross's model, although no one has rejected the general idea that in the course of dying people typically go through different stages. To what extent this description is a normative as well as a descriptive paradigm remains unclear.

Kübler-Ross and others (see Moody 2001) also claim that empirical studies of the dying process and especially near-death experiences confirm the existence of an afterlife. This is a highly contentious position that nevertheless to some extent makes common cause with traditional guidebooks for the art of dying. More widely accepted are manuals that provide psychological guidance that extend Kübler-Ross's original approach (see Byock 1998).

In contrast to psychological studies of the dying experience, which was simply an open research question inviting scientific scrutiny, the need to redefine death was made acute by techno-medical advances. Death was traditionally defined in metabolic terms and indicated by cardiac or pulmonary arrest. But with the invention of artifacts that are able to substitute for the functions of heart and lungs, the human metabolism can often be indefinitely sustained. The Karen Ann Quinlan case of 1975, in which parents were initially denied the right to remove their comatose daughter from life support, was only one of a series of related cases that brought this issue to public attention, and served as a powerful stimulus to the creation of the field of bioethics. In response the 1978 Presidential Commission on the Study of Ethical Problems in Medicine and Biomedical and Behavioral Research, in conjunction with the American Bar Association (ABA), the American Medical Association (AMA), and the National Conference of Commissioners of Uniform State Laws, proposed a Uniform Determination of Death Act (UDDA). According to this draft act, "An individual who has sustained either (1) irreversible cessation of circulatory and respiratory function, or (2) irreversible cessation of the entire brain, including the brain stem, is dead." The draft UDDA has in various forms been enacted in most of the United States and has become the most widely adopted standard.

Tensions in Autonomy

As dying and death are reconceptualized under the influence of ever-advancing science and technology, a host of related decisions become progressively problematic—especially with regard to autonomy, one of the fundamental principles of contemporary Western ethics. The individual right to choose or self-determine whether or not to be placed on life-sustaining treatment is typical. Traditionally people did not have the *right to choose* their deaths. Age, accidents, sicknesses, and disease determined it for them. Individuals were ultimately passive as death approached. The goal of the patient-physician relationship was no more than the easing of symptoms related to life-limiting illnesses or the aging process.

All this changed as personal autonomy became a moral ideal and medical procedures and therapies altered what constituted unpreventable death and sustainable life. For example, during the late-nineteenth and early-twentieth centuries many patients with life-threatening conditions such as polio were sent home to die. By the late 1920s, lifesaving treatments such as the iron lung enabled them to live for years in machines that breathed for them. Under such conditions, even mere acceptance of whatever contemporary medical procedure has the most professional momentum behind it constitutes a choice, and the traditional attitude of passive acceptance in the face of extreme illness increasingly means an acceptance not of nature but of science and technology.

The dying and those who care for them are thus faced with progressively difficult decisions regarding nutrition, hydration, antibiotics, ventilation, and a host of more aggressive medical technologies. Dilemmas become especially apparent when advanced techniques of life-support enable sustaining basic metabolic functions with little hope for full recovery. Attempts to reflect on the dimensions of autonomous decision making under such conditions involve at least four overlapping tensions: (a) informed consent and ignorance; (b) private decisions and public demands; (c) curative care and futile care; and (d) benefits of treatment and the burden of care.

The tension between informed consent and ignorance is particularly difficult to negotiate when considering high-tech medical treatments under the stress of illness and pain that may well be terminal. Although there is a widely shared consensus that patients must be informed about and freely consent to treatments, how much should patients be expected to understand? What are the communication responsibilities of medical professionals? Even when patients claim to understand, can they always be trusted? What if they want to avoid becoming knowledgeable about their condition or desire to abdicate decision making to others, whether family members or medical professionals?

When confronting tensions between private decisions and public demands in a pluralistic society, it is commonly argued that private decisions take precedence. Presumably personal values inform private decisions while public attitudes differ and reflect fragmented cultural, religious, and emotional biases. For example, although most Americans express public opposition to euthanasia or physician-assisted suicide, they are personally more accepting when faced with extended, terminal suffering and pain. Thus there is a tendency to promote

individual choice. Certainly most states and many countries give people the opportunity to make their wishes known with respect to possible life-limiting illnesses in the form of physician directives, medical powers-of-attorney, and do-not-resuscitate (DNR) orders.

A third tension arises between curative care and futile care. The desire to extend life is pervasive in Western cultures, but life support treatments sometimes prolong death more than they extend life. Curative care employs life-supporting technologies with the goal of recovery, but the consequences of living on continuous life-support, without hope of recovery, must be considered as well. The alternative of palliation or comfort care that allows the process of dying to take its course is another option. There are even instances when an otherwise curative procedure such as chemotherapy may be used as a palliative treatment.

Closely related to the curative versus futility tension is one between quality of life and quality of function. Quality of life issues generally revolve around physical or mental disabilities, which may impede but not destroy a person's ability to interact with and engage the environment. For example, persons with head injuries confined to hospital beds may be aware of the environment and engage the world around them. Although their quality of life is altered, such circumstance does not justify withholding or withdrawing life-sustaining treatment. Quality of function issues, however, may justify withdrawal of life-sustaining treatments when the ability to function is seriously impaired due to some significant pathology. Persons with head injuries who are in a persistent vegetative state and unaware of or unable to engage their environment have a low quality of functioning.

Finally tensions exists between the benefits of treatment and the burden of care. Life-support treatment always comes with both physical and financial costs. Therapies such as ventilation carry risks of infection, aspiration, and skin breakdown. When outcomes are successful, benefits are usually taken to outweigh burdens, although burn victim Dax Cowart has argued that this is not always so (Kliever 1989). Moreover while it is generally agreed that financial costs should not be the determinative factor when considering life-support, they must be considered. As the private burdens and public costs for funding life-support technologies rise, some argue that there are instances in which individuals may have a duty to die.

Burdens of care also often raise questions of equity and social justice. In principle all persons should have equal access to life-saving and life-sustaining technologies. In reality the poor, uneducated, and uninsured are far less likely to be treated or have the opportunity to make decisions about such treatment. At the same time, although it is commonly argued that death and dying decisions must relate to personal needs, some forms of this position have been challenged as excessively individualistic.

The Question of Modernity

Once viewed as spiritual experiences, in the early-twenty-first century death and dying are often considered no more than the cessation of metabolic or brain functioning. However, related issues continue to provoke strong responses because views on death and dying emerge in a variety of cultural contexts. It often appears that as technology and medicines advance, the fear of death increases, encouraging greater efforts to prolong life. Daniel Callahan (1973, 2000) argues that a fundamental rejection and fear of death is at the foundation of modern science and technology.

Certainly in Western scientific and technological culture, death has a negative connotation. Attempts to deny death are manifest both in the scientific efforts to map its physiological details and genetic basis as well as in technological efforts to hold it at bay as long as possible. Denial is further reinforced through the sequestering of the aged, ill, and dying, especially in the United States. Such individuals are institutionalized in hospitals, long-term nursing facilities, and retirement villages. Euphemisms such as *slumber*, *expired*, and *passed away* further reinforce the denial of death. Even contemporary religious customs enforce denial by limiting grieving time and trying to help families cope with personal losses.

Yet in an insightful reflection on death and dying in relation to the ideal of autonomy and technological mastery—what is termed the *instrumental activism* of the West—Talcott Parsons and Victor Lidz suggest an alternative interpretation. They identify a range of efforts to control and manage death and dying through "scientific medicine and public health services designed to protect life; insurance, retirement, and estate planning to manage the practical consequences of deaths; and mourning customs that emphasize recovery of survivors' abilities to perform ordinary social roles soon after the death of family members, friends, and associates" (Parsons and Lidz 2004a, p. 597). From such perspective, what others described as attempts to hide or ignore death are seen as techniques for its management under conditions in which there is no strong cultural consensus about the

meaning of either life or death, short of what can be concluded from the empirical evidence (that it is bound to occur) and postulated on the basis of scientific theory (its evolutionary benefits to both organisms and society).

To what extent might these same techniques be integrated into a different cultural context in which death and dying continue to be experienced as a spiritual transition between worlds? The question is not easy to answer, and may not be answerable at all apart from historical efforts at adaptation. But whether the standard bioethical efforts to promote patient autonomy, equity of access, and a quality hospital experience can transcend the contemporary cultural framework remains unclear.

ERIC T. HAMMER
CARL MITCHAM

SEE ALSO *Brain Death; Cancer; Euthanasia; Euthanasia in the Netherlands; Medical Ethics; Playing God; Right to Die.*

BIBLIOGRAPHY

Ariès, Phillipe. (1991). *The Hour of our Death*, reissue edition. Oxford: Oxford University Press. French original 1981. An influential study of the cultural experience of death.

Barley, Nigel. (1997). *Grave Matters: A Lively History of Death around the World*. New York: Henry Holt. A popular review of anthropological knowledge.

Bulger, Ruth Ellen; Elizabeth Meyer Bobby; and Harvey V. Fineberg, eds. (1995). *Society's Choices: Social and Ethical Decision Making in Biomedicine*. Washington, DC: National Academy Press.

Byock, Ira. (1998). *Dying Well: Peace and Possibilities at the End of Life*. New York: Penguin Riverhead Books.

Callahan, Daniel. (1973). *The Tyranny of Survival and Other Pathologies of Civilized Life*. New York: Macmillan.

Callahan, Daniel. (2000). *The Troubled Dream of Life: In Search of a Peaceful Death*. Washington, DC: Georgetown University Press.

Dunn, Hank. (2001). *Hard Choices for Loving People*. Herndon, VA: A & A Publishers.

Fox, Renée C., ed. (1980). "The Social Meaning of Death." *Annals of the American Academy of Political and Social Science* 447: 1–101.

Kliever, Lonnie D., ed. (1989). *Dax's Case: Essays in Medical Ethics and Human Meaning*. Dallas, TX: Southern Methodist University Press.

Kübler-Ross, Elizabeth. (1969). *On Death and Dying*. New York: Simon & Schuster.

Lo, Bernard; Delaney Ruston; Laura W. Kates; et al. (2002). "Discussing Religious and Spiritual Issues at the

End of Life: A Practical Guide for Physicians." *Journal of the American Medical Association* 287(6)(February): 749–754.

Moody, Raymond A. (2001). *Life after Life: The Investigation of a Phenomenon—Survival of Bodily Death.* San Francisco: Harper.

Parsons, Talcott, and Victor Lidz. (2004a). "Death, V: Death in the Western World." In *Encyclopedia of Bioethics*, 3rd edition, Vol. 4, ed. Stephen G. Post. New York: Macmillan.

Parsons, Talcott, and Victor Lidz. (2004b). "Postscript." In *Encyclopedia of Bioethics*, 3rd edition, Vol. 4, ed. Stephen G. Post. New York: Macmillan.

Pattison, E. M. (1990). "Biomedical Definition of Death." In *Dictionary of Pastoral Care and Counseling*, ed. Rodney J. Hunter. Nashville, TN: Abingdon Press.

President's Commission for the Study of Ethical Problems in Medicine and Biomedical and Behavioral Research. *Defining Death: Medical, Legal and Ethical Issues in the Determination of Death.* Washington, DC: Government Printing Office.

Tobin, Daniel R, M.D., with Karen Lindsey. (1999). *Peaceful Dying.* New York: Perseus Books.

Young, Katherine K. (2004). "Death, II: Eastern Thought." In *Encyclopedia of Bioethics*, 3rd edition, Vol. 4, ed. Stephen G. Post. New York: Macmillan.

INTERNET RESOURCE

"Withholding and Withdrawing Life-prolonging Treatments: Good Practice in Decision-making." General Medical Council. Available from http://www.gmc-uk.org/global_sections/search_frameset.htm.

DEATH PENALTY

• • •

There is an ongoing crucial debate within the criminal justice system as to the moral status of the death penalty. Retentionists hold that the death penalty is morally justified; abolitionists argue that it is not. Proponents of the death penalty justify it from either a retributive or a utilitarian framework, sometimes using both theories for a combined justification. Abolitionists reject these contentions arguing that the principle of the sanctity of human life gives each person an inalienable right to life and thus prohibits imposition of the death penalty. Scientific research and technological developments provide modest contributions to both arguments.

Retributive Arguments

The retributivist argues (1) that all the guilty deserve to be punished; (2) that only the guilty deserve to be punished; and (3) that the guilty deserve a punishment proportional to their crime. It follows that death is a suitable punishment for anyone who commits a capital offense (that is, those offenses such as murder and treason that are especially morally heinous). The concept is suggested in the Bible: "Thou shalt give life for life, an eye for an eye, a tooth for a tooth, a hand for a hand, burning for burning, wound for wound, stripe for stripe" (*Exod.* 21: 23–25).

A classic expression of the retributivist position on the death penalty is Immanuel Kant's statement that if an offender "has committed murder, he must *die.* In this case, no possible substitute can satisfy justice. For there is no *parallel* between death and even the most miserable life, so that there is no equality of crime and retribution unless the perpetrator is judicially put to death (at all events without any maltreatment which might make humanity an object of horror in the person of the sufferer)" (Kant 1887, p. 155).

For Kant, the death penalty was a conclusion of the argument for justice: just recompense to the victim and just punishment to the offender. As a person of dignity, the victim deserves (as a kind of compensatory justice) to have the offender harmed in proportion to the gravity of the crime, and as a person of high worth and responsibility, the offender is deserving of the death penalty. Accordingly the torturer should be tortured exactly to the severity that he tortured the victim, the rapist should be raped, and the cheater should be harmed to a degree equal to that suffered by the one cheated. Criminals deserve such punishment in accordance with the principle of proportionality.

The abolitionist disagrees. Putting the criminal to death only compounds evil. If killing is an evil, then the evil is doubled when the state executes the murderer, violating the latter's right to life. The state commits *legalized murder.* To quote the famous eighteenth-century abolitionist Cesare di Beccaria, "The death penalty cannot be useful because of the example of barbarity it gives to men ... it seems to me absurd that the laws ... which punish homicide should themselves commit it" (*On Crimes and Punishment*, 1764).

The retentionist responds that the abolitionist is mistaken. The state does not violate the criminal's right to life, for the right to life (more precisely, the right not to be killed) is not an absolute right that can never be overridden (or forfeited). If the right to life were absolute, one could not kill an aggressor even when such action is necessary to defend one's own life or the lives of loved ones. It is a *prima facie* or conditional right that can be superseded in light of a superior moral reason. The individual right to life, liberty, and property is connected to the societal duty to respect the rights of others to life, liberty, and property. A person can forfeit the right to liberty by violating the liberty rights of others. A person can forfeit property rights by violating the property rights of others. Similarly the right to life can be forfeited when a person violates the right to life of another. An individual's *prima facie* right to life no longer exists if that person has committed murder.

Utilitarian Arguments

The utilitarian argument for capital punishment is that it deters would-be offenders from committing first degree murder (that is, types of murder that seem especially vicious, brutal, or deleterious, such as assassination). However some studies that compared states that allow capital punishment to those that do not permit it concluded that imprisonment works as well as the death penalty in deterring homicide. Other studies purport to show that when complex sociological data (race, heredity, regional lines, standards of housing, education, and opportunities, among others) are taken into account, the death penalty does deter. Anecdotal evidence exists to support this. Abolitionists argue that isolated cases are poor indicia of the reality regarding deterrence.

Abolitionists point out that the United States is the only Western democracy to retain the death penalty. The retentionist responds that this is not an argument, but an appeal to popularity. Furthermore, in many Western countries that prohibit the death penalty (such as England, Italy, and France), there is evidence that the majority of citizens favor it.

Scientific and Technological Contributions

Science and technological issues are relevant to the debate in so far as some argue that neuroscience and psychology show that criminals, including murderers, commit crimes due to neurological dysfunction and are not responsible for what they do. However the same arguments could be used to deny human responsibility altogether.

An application of technology to the death penalty is illustrated by attempts to find more benign forms of execution, for example, replacing the electric chair (which some consider cruel and unusual punishment) with lethal injection. Such changes have caused debate about whether the condemned deserves humane treatment or should be subjected to some pain and anxiety as part of the punishment.

Other abolitionists argue, on the basis of social scientific studies, that the U.S. penal system is inherently unfair, is biased against the poor and minorities, and favors the rich who can afford to hire better legal counsel. Furthermore they contend that there is always the possibility that an innocent person will be executed. The retentionist, recognizing these dangers, responds, Amend it, don't end it.

LOUIS P. POJMAN

SEE ALSO *Crime; Justice; Police; Science, Technology, and Law.*

BIBLIOGRAPHY

Bedau, Hugo Adam, ed. (1982). *The Death Penalty in America*, 3rd edition. New York: Oxford University Press.

Kant, Immanuel. (1887). *The Metaphysics of Morals.* Edinburgh: T & T Clark.

Murphy, Jeffrie, ed. (1968). *Punishment and Rehabilitation.* Belmont, CA: Wadsworth Publishing.

Pojman, Louis P., and Jeffrey Reiman. (1998). *The Death Penalty: For and Against.* Lanham, MD: Rowman & Littlefield.

Sorrel, Tom. (1987). *Moral Theory and Capital Punishment.* Oxford: Blackwell.

DECISION SUPPORT SYSTEMS

• • •

Decision support systems (DSS) are tools meant to assist in human decision-making (Turban and Aronson 2001). In an increasingly complex and rapidly changing world where information from human, software, and sensor sources can be overwhelming, DSS tools can serve as a bridge between the social and technical spheres. DSS tools offer support based on formal, technical approaches, but do so within a context that is often largely socially mediated.

Most DSS tools are assembled out of hardware devices and software constructs. The hardware devices, in the early twenty-first century, are dominated by digital computers and peripherals such as sensors, network infrastructure, and display and alerting devices meant to interact with these. Historically, many DSS were hardwired to solve a specific task; control systems in nuclear power plants are an example. DSS hardware is increasingly dominated by physically distributed systems that make use of wired and wireless networks to gather and share information from and with remote sources (Shim et al. 2002). The convergence of remote sensing, sensor-networks, and distributed computational grids using the Internet as a foundation in the late 1990s–early 2000's reflects this trend.

The software, or algorithmic, component of DSS derives from historical research in statistics, operations research, cybernetics, artificial intelligence, knowledge management, and cognitive science. In early monitoring decision support systems the algorithms were typically hard-wired into the system, and these systems tended to be unchanging once built. Software-based decision support allows for multiple approaches to be applied in parallel, and for systems to evolve either through new software development or via software that "learns" through artificial intelligence techniques such as rule induction (Turban and Aronson 2001).

When used appropriately, DSS tools are not meant to replace human decision-making—they are meant to make it more effective (Sprague and Watson 1996). DSS tools do this by presenting justified answers with explanations, displaying key data relevant to the current problem, performing calculations in support of user decision tasks, showing related cases to suggest alternatives, and alerting the user to current states and patterns. In order to be a support rather than a hindrance, these tools must be constructed with careful attention to human cognitive constraints. As a result, DSS design is a prime area of human-computer interaction and usability research. In many cases, DSS tools make use of adaptive software interfaces; depending on the situation, different contents will be displayed on the interface, so as not to overwhelm the user with secondary or irrelevant information.

Decision Support Tools

Decision support tools fall into two broad classes: those that operate at the pace of the user (for example, to support planning decisions) and those that operate at or near the pace of real-time world events (such as air traffic control systems). The decision-making domain can be further divided into situations in which the system can be completely and accurately defined (in other words, closed and formal systems) and those where this is not feasible, desirable, or possible. The former is not normally considered a prime situation for decision *support* because a formal situation can be addressed without human intervention, while the latter requires the hybrid human-machine pairing found in DSS. In the case of open systems, heuristic approximations (rules of thumb) are needed in lieu of formal models; these may also be needed in cases in which a formal model exists but cannot be computed in a reasonable amount of time.

Systems that operate at the pace of the user provide support for such tasks as planning and allocation, medical and technical diagnosis, and design. Typical examples include systems used in urban planning to support the complex process of utility construction, zoning, tax valuation, and environmental monitoring, and those used in business to determine when new facilities are needed for manufacturing. Such tools include significant historical case-knowledge and can be transitional with training systems that support and educate the user. Formal knowledge, often stored as rules in a modifiable knowledge base, represent both the state of the world that the system operates on and the processes by which decisions transform that world. In the cases where formal knowledge of state and process are not available, heuristic rules in a DSS expert system or associations in a neural network model might provide an approximate model. DSS tools typically provide both a ranked list of possible courses of action and a measure of certainty for each, in some cases coupled with the details of the resolution process (Giarratano and Riley 2005).

Systems that operate at or near real time provide support for monitoring natural or human systems. Nuclear power plant, air traffic control, and flood monitoring systems are typical examples, and recent disasters with each of these illustrate that these systems are fallible and have dire consequences when they fail. These systems typically provide support in a very short time frame and must not distract the user from the proper performance of critical tasks. By integrating data from physical devices (such as radar, water level monitors, and traffic density sensors) over a network with local heuristics, a real-time DSS can activate alarms, control safety equipment semi-automatically or automatically, allow operators to interact with a large system efficiently, provide rapid feedback, and show alternative cause and effect cases. A central issue in the design of such systems is that they should degrade gracefully; a flood monitoring system that fails utterly if one cable is shorted-out, for example, is of little use in a real emergency.

History of Decision Support Research and Tools

As indicated above, DSS evolved out of a wide range of disciplines in response to the need for planning-support and monitoring-support tools. Management and executive information systems, where model and data-based systems dominated, reflect the planning need; control and alerting systems, where sensor and model-based alerting systems were central, reflect the monitoring need. The original research on the fusion of the source disciplines, and in particular the blending of cognitive with artificial intelligence approaches, took place at Carnegie-Mellon University in the 1950s (Simon 1960). This research both defined the start of DSS and also was seminal in the history of artificial intelligence; these fields have to a large degree co-evolved ever since. By the 1970's research groups in DSS were widespread in business schools and electrical engineering departments at universities, in government research labs, and in private companies. Interestingly, ubiquitous computer peripherals such as the mouse originated as part of decision support research efforts.

By the 1980s the research scope for DSS had expanded dramatically, to include research on group-based decision making, on the management of knowledge and documents, to include highly specialized tools such as expert-system shells (tools for building new expert systems by adding only knowledge-based rules), to incorporate hypertext documentation, and towards the construction of distributed multi-user environments for decision making. In the mid-1980s the journal *Decision Support Systems* began publishing, and was soon followed by other academic journals. The appearance of the World Wide Web in the early 1990's sparked a renewed interest in distributed DSS and in document- and case-libraries that continues in the early twenty-first century.

Outstanding Technical Issues with Decision Support Tools

DSS tools, as described above, integrate data with formal or heuristic models to generate information in support of human decision making. A significant issue facing the builders of these tools is exactly how to define formal models or heuristics; experts make extensive use of tacit knowledge and are notoriously unreliable at reporting how they actually do make decisions (Stefik 1995). If the rules provided by domain experts do not reflect how they actually address decisions, there is little hope that the resulting automated system will perform well in practice.

A second, related, issue is that some systems are by their very nature difficult to assess. Chaotic systems, such as weather patterns, show such extreme sensitivity to initial (or sensed) conditions that long-term prediction and hence decision support is difficult at best. Even worse, many systems cannot be considered in isolation from the decision support tool itself; DSS tools for stock market trading, for example, have fundamentally changed the nature of markets.

Finally, both the DSS tools and the infrastructure on which they operate (typically, computer hardware and software) require periodic maintenance and are subject to failure from outside causes. Over the life of a DSS tool intended to, for example, monitor the electrical power distribution grid, changes to both the tools themselves (the hardware, the operating system, and the code of the tool) and to their greater environment (for example, the dramatic increase in computer viruses in recent history) mean that maintaining a reliable and effective DSS can be a challenge. It cannot be certain that a DSS that performs well now will do so even in the immediate future.

Ethical Issues

Decision support rules and cases by their very nature include values about what is important in a decision-making task. As a result, there are significant ethical issues around their construction and use (see, for example, Meredith and Arnott 2003 for a review of medical ethics issues). By deciding what constitutes efficient use in a planning support system for business, or what constitutes the warning signs of cardiac arrest in an intensive care monitoring system, these tools reflect the values and beliefs of the experts whose knowledge was used to construct the system. Additionally, the social obligation of those who build DSS tools is an issue. On the one hand, these are tools for specific purposes; on the other hand, many social and natural systems are so interrelated that, in choosing to build an isolated and affordable system, many issues will be left unresolved.

The ruling assumption of efforts to build DSS tools is that decision-making is primarily a technical process rather than a political and dialogical one. The bias here is not so much intellectual as informational: It may overestimate the usefulness of information in the decision-making process. Rather than more information, or ever more elaborate displays, people might need more time to reflect upon a problem. Coming to understand another perspective on an issue is a matter of sympathy and open-mindedness, not necessarily information delivery. Delivering detailed information, cases, and suggested courses of action to a single user is opposed to the idea of community-base processes. While placing these issues outside of the scope of a system design might be a useful design decision from a technical position, it is a value-laden judgment.

In fairness, the decision support literature does occasionally recognize that the public needs a better understanding not only of technology but also of science. There is often little appreciation, however, that decision support is an ethical and political process as much as a technical one—or that the flow of information needs to involve the scientist, the engineer, and the public. Exactly how the political process can be engaged for systems that must by their very nature operate in real time is an open question. Certainly the process of knowledge and value capture for such systems could be much more open than is currently the norm.

A second pressing issue regarding DSS tools is the degree to which the data, knowledge, sensors, and results of their integration represent a limitation on individual freedoms and/or an invasion of privacy. DSS tools based on expert-systems approaches actively monitor every credit card transaction made. Semi-automatic face recognition systems are widespread. Radio-frequency identification tags built into price tags on consumer goods allow consumer behavior to be monitored in real-time. Cell-phone records provide not only who a person was speaking to, but where they were at the time. Decision support tools for national security, market research, and strategic planning integrate information, apply rules, and inform decisions that affect human freedom and privacy every day.

Conclusions

DSS tools will only become more common in the future. The widespread reach of Internet connections and the dramatic decrease in the cost of sensors is driving the creation of decision support tools within governments and industries worldwide. It remains to be seen how these systems may impact on human lifestyles, freedoms, and privacy, and whether these tools can continue to evolve to handle the difficult questions facing decision makers in a complex and changing world.

ROBIN M. HARRAP
ROBERT FRODEMAN

SEE ALSO *Choice Behavior; Decision Theory; Engineering Method; Information Society; Rational Choice Theory; Science Policy.*

BIBLIOGRAPHY

Giarratano, Joseph C., and Gary D. Riley. (2005). *Expert Systems: Principles and Programming,* 4th ed. Boston: Thomson Course Technology. A focussed introduction to expert systems technology from the authors of the open source expert system tool CLIPS, widely used in decision support systems.

Meredith, Rob, and David Arnott. (2003). "On Ethics and Decision Support Systems Development." In *Proceedings of the Seventh Pacific Asia Conference on Information Systems, 10–13 July 2003,* ed. Jo Hanisch, Don Falconer, Sam Horrocks, and Matthew Hillier. Adelaide: University of South Australia. An examination of the methods of decision support systems in medical practice.

Shim, J. P.; Merrill Warkentin; James F. Courtney; et al. (2002). "Past, Present, and Future of Decision Support Technology." *Decision Support Systems* 33(2): 111–126. A systematic overview of the development of decision support systems.

Simon, Herbert A. (1960). *The New Science of Management Decision.* New York: Harper and Row. An early and highly influential look at the automation of decision making in business from the leading researcher of the era.

Sprague, Ralph H., Jr., and Hugh J. Watson. (1996). *Decision Support for Management.* Upper Saddle River, NJ: Prentice Hall. A broad overview of decision support systems, emphasizing systems used in business practice for financial and organizational planning.

Stefik, Mark. (1995). *Introduction to Knowledge Systems.* San Francisco: Morgan Kaufmann. A detailed examination of the human and software problems in constructing software-based knowledge systems.

Turban, Efraim, and Jay E. Aronson. (2001). *Decision Support Systems and Intelligent Systems,* 6th edition. Englewood Cliffs, Upper Saddle River, NJ: Prentice Hall. A broad overview of decision support research and artificial intelligence approaches that support decision making. Includes extensive case studies from the history of DSS tools.

DECISION THEORY

• • •

Decision theory is the science of rational choice in situations in which there is uncertainty about the outcome. Rational choice theory asserts that individuals whose behavior satisfies a few plausible conditions (such as transitivity, which means that if A is preferred to B and B is preferred to C, then A is preferred to C) will behave as though they are maximizing a preference function defined over the choice outcomes. For instance, consider an agent with the preference function $u(a, n)$ defined over two goods, Apples and Nuts, and with an amount of money M to spend. Thus, $u(2, 5)$ is the "utility" the agent derives from consuming two apples and five nuts (for this reason, economists call a preference function a *utility function*). If the prices of Apples and Nuts are p_a and p_n, the individual will choose the amount of Apples a and the amount of Nuts n that will maximize $u(a, n)$, subject to the constraint that the total cost is not greater than M (i.e., $p_a a + p_n n \leq M$). Decision theory deals with such choices when there is uncertainty regarding the amount of Apples and Nuts that will be delivered.

Decision theory relies on probability theory, the development of which began in the seventeenth and eighteenth centuries, associated with scholars such as Blaise Pascal (1623–1662), Daniel Bernoulli (1700–1782), and Thomas Bayes (1702–1761). The analysis here will be confined to the case where there is only a finite set of outcomes A, written as $A = \{a_1, \ldots, a_n\}$. A *probability distribution* over A is called a *lottery* and consists of n numbers $p_1, \ldots, p_n$ such that each $p_i \geq 0$ and $p_1 + \cdots + p_n = 1$. In this case p_i is interpreted as the probability that outcome a_i occurs.

The Expected Value and Expected Utility of a Lottery

One of the first problems addressed by probability theorists was the determination of the *certainty equivalent* of a lottery: If someone offers to sell a person a lottery with monetary prizes for a certain amount of money, what is the maximum the buyer should be willing to pay? Early probability theorists suggested that that person should be willing to pay the "average" payoff of the lottery. However, it soon was shown that the average payoff of lottery x is equal to its expected value, which is defined as $\mathbf{E}x = p_1a_1 + \cdots + p_na_n$. For instance, the expected value of a lottery that pays $1,000,000 with a probability of 1/1,000,000 has an expected value of $1.

Daniel Bernoulli, however, developed a simple example, known as the *St. Petersburg Paradox*, that clearly showed that the idea that people will pay the expected value must be incorrect. Suppose a person is offered either a sum of money M or the following lottery: A coin is tossed a number of times until it turns up heads, after which the game is over. If the first toss is heads, the person is paid $1. If the first toss is tails and the second is heads, that person is paid $2, and so on, with each additional round paying twice as much ($4, $8, ...). Most people, if offered M = $20, will take this rather than play the coin-tossing game, yet the expected value of the game is infinite:

$$\mathbf{E} = \tfrac{1}{2}\times 1 + \tfrac{1}{4}\times 2 + \tfrac{1}{8}\times 4 + \ldots = \tfrac{1}{2} + \tfrac{1}{2} + \tfrac{1}{2} + \ldots = \infty.$$

Bernoulli suggested that the problem here is that the "utility" of each additional unit of money *decreases* as the amount increases, just as the additional utility of each additional scoop of ice cream decreases for a consumer. He suggested that the utility of money may be logarithmic and that people maximize the expected utility of a lottery, not the expected value. If the utility of an amount of money M is log(M), the expected utility of the St. Petersburg lottery is

$$\mathbf{E} = \tfrac{1}{2}\log(1) + \tfrac{1}{4}\log(2) + \tfrac{1}{8}\log(4) + \tfrac{1}{16}\log(8) + \ldots$$
$$= 0 + 0.17 + 0.17 + 0.13 + 0.09 + \ldots$$
$$\approx 1.66$$

The von Neumann–Morganstern Axioms

A general model of expected utility was not developed until centuries later. John von Neumann (1903–1957) and Oskar Morgenstern (1902–1976) developed decision theory as a model of rational choice in regard to lotteries. They supplied three conditions from which the expected utility principle could be derived. The first was, as in standard rational choice theory, that the agent has a weak preference relation $\succeq$ that is complete and transitive over the set of lotteries. By *complete* it is meant that for any two lotteries x and y, one is weakly preferred to the other (i.e., either $x \succeq y$ or $y \succeq x$). This is called a weak preference because any lottery x is weakly preferred to itself (i.e., for any lottery x, there is $x \succeq x$). Strong preference can be defined $\succ$ as $x \succ y$ means that "it is false that $y \succ x$." By *transitive* it is meant that if x is weakly preferred to y and y is weakly preferred to z, then x is weakly preferred to z ($x \succeq y$ and $y \succeq z$ implies $x \succeq z$).

Suppose x and y are lotteries and suppose p is a probability (i.e., a number between zero and one). One writes $px + (1 − p)y$ for the lottery that gives lottery x with probability p and lottery y with probability $1 − p$. For instance, suppose x is the lottery that pays off $20 with probability 0.25 and $10 with probability 0.75 and suppose y is the lottery that pays off $5 with probability 0.90 and $100 with probability 0.10. Then $0.33x + 0.67y$ is the lottery that pays off x with probability 0.33 and y with probability 0.67. This so-called compound lottery thus has payoffs $0.33x + 0.67y = 0.33[0.25($20) + 0.75($10)] + 0.67[0.90($5) + 0.10($100)]$, and so this is a lottery that pays $20 with probability (0.33)(0.25) = 0.0825, pays $10 with probability (0.33)(0.75) = 0.2475, pays $5 with probability (0.67)(0.90) = 0.6030, and pays $100 with probability (0.67)(0.10) = 0.067. Note that these probabilities add up to one, as they should. As an exercise, one may check to see that if $\mathbf{E}x$ and $\mathbf{E}y$ are the expected values of lotteries x and y, then $a\mathbf{E}x + (1 − a)\mathbf{E}y$ is the expected value of the lottery $ax + (1 − a)y$.

The second von Neumann–Morgenstern condition is that if $x \succ y$ and z is any lottery and $p < 1$ is a probability, then $px + (1 − p)z \succ py + (1 − p)z$. This is called the *independence* condition. It says that the value of a prize depends only on the prize and the probability of winning it, not on other payoffs or probabilities.

The third condition is that if x, y, z are lotteries and $x \succ y \succ z$, there are numbers p and q such that $px + (1 - p)z \succ y \succ qx + (1 - q)z$. This says that there is no lottery that is infinitely valuable or infinitely distasteful. This is called the *Archimedian condition*.

With these three conditions von Neumann and Morgenstern showed that the agent has a utility function $u(a)$ defined over the outcomes $a_1, \cdots, a_n$ such that for any two lotteries $x = p_1a_1 + \cdots + p_na_n$ and $y = q_1a_1 + \cdots + q_na_n$, $x \succ y$ if and only if

$$p_1u(a_1) + p_2u(a_2) + \cdots + p_nu(a_n) > q_1u(a_1) + q_2u(a_2) + \cdots + q_nu(a_n).$$

Note that the first sum is the expected value of the lottery x in which the payoffs are replaced by the *utility* of the payoffs, and this also applies to the second sum. This motivates the definition of the expected utility of a lottery x as

$$\mathbf{E}_pu = p_1u(a_1) + p_2u(a_2) + \cdots + p_nu(a_n).$$

The *expected utility theorem* thus states that an individual whose behavior satisfies the conditions listed above (complete transitive preferences that satisfy the independence and Archimedian conditions) chooses among lotteries to maximize expected utility.

Subjective Probability Theory

The purpose of decision theory is to explain and predict behavior, and an agent's behavior depends on that agent's *subjective assessments* of the likelihood of different outcomes. Modern *subjective probability theory* was developed in the twentieth century by Frank Ramsey (1903–1930) and Bruno de Finetti (1906–1985) and was applied to decision theory by Leonard Savage (1917–1971).

Savage begins with a set of all possible mutually exclusive "states of the world" that are relevant for an agent's decision. For instance, to decide whether to buy a new car, a couple may consider (a) possible changes in their employment and health status over the next few years, (b) whether they may increase their family's size, (c) whether next year's models will be better or more affordable than this year's, and (d) whether they can find a lower price elsewhere.

Savage then defines an action f such that $f(s)$ is an outcome or payoff for each state of the world $s \in S$. For instance, for the couple, $f(s) =$ "buy car" for some states of the world and $f(s) =$ "don't buy car" for the other states. Savage shows that if the decision maker has preferences for actions that satisfy certain plausible conditions, it is possible to infer a probability distribution p over states of the world S and a utility function u over

outcomes such that the decision maker maximizes expected utility $\mathbf{E}_pu(s) = \sum_{s \in S} p(s)u(s)$ (Kreps 1988).

Violations of Expected Utility Theory

The expected utility approach to decision theory is used widely in behavioral modeling, virtually to the exclusion of other approaches. However, laboratory studies of actual behavior have revealed consistent deviations from the application of the theory. For one thing, there are indications that the independence axiom may be violated, implying that the probability weights in the expression for $\mathbf{E}_pu(s)$ may be *nonlinear*. This fact was first discovered by Maurice Allais (b. 1911, winner of a Nobel Prize in economics in 1988), using the following schema.

Consider a choice between lotteries x_1, which offers $1,000,000 with probability one, and x_2, which offers a 10 percent chance for $5,000,000, an 89 percent chance for $1,000,000, and a 1 percent chance for $0. Consider a second choice between lotteries y_1, which offers a 10 percent chance for $5,000,000 and a 90 percent chance for $0, and y_2, which offers an 11 percent chance at $1,000,000 and an 89 percent chance for $0. An individual who prefers x_1 to x_2 prefers an 11 percent chance of $1,000,000 to a 10 percent chance of $5,000,000 plus a 1 percent chance of $0. If an 89 percent chance of $0 is added to both of these possibilities, this individual, if maximizing expected utility, must prefer an 11 percent chance of $1,000,000 to a 10 percent chance of $5,000,000 and therefore must prefer y_2 to y_1. However, in fact, most people prefer x_1 and y_2. An analysis of this and other violations of the independence axiom is provided in the work of Mark Machina (1989).

A second violation of the expected utility model is *loss aversion*, which first was proposed by Daniel Kahneman (b. 1934, winner of a Nobel Prize in economics in 2003) and Amos Tversky (1937–1996) in a 1991 paper. For example, if faced with the choice between a lottery that pays $5 with probability one and a lottery that pays $10 with probability ½ and $0 with probability ½, most people will choose the former (they are said to be *risk-averse* because they prefer the expected value of a risky lottery to the risky lottery). For instance, a risk-averse person will prefer a certain $5 to a risky lottery with expected value $5, such as either winning $10 or winning $0, each with probability ½. However, if they are given $10 (say, for showing up for an experimental session) and are faced with the choice between a lottery that loses $5 with probability one and another that loses $10 with probability ½ and loses $0 with probability ½, most people will choose the latter. In this case the subjects are *risk-loving*. Note that the subjects go home with the same amount of money in either lottery. This is certainly a violation of the expected utility theorem. Loss aversion explains many phenomena that defy explanation in traditional decision theory, including the so-called *endowment effect* (Kahneman, Knetch, and Thaler 1990) and the *status quo bias* (Samuelson and Zeckhauser 1988).

Assessment

The rational choice model and its subsidiary, rational decision theory, offer the most powerful analytic tools for modeling human behavior and the behavior of living organisms in general. The laboratory experiments of Allais, Kahneman and Tversky and others (for a summary, see Kahneman and Tversky, 2000) show that in some circumstances expressions more complex than expected utility are needed and that there are important parameters in an individual's preference function, such as the current time and the agent's possessions at the time when decisions are made. Decision theory has been criticized, but its critics have offered nothing that could replace it, and the criticisms generally have been mutually contradictory and often misinformed.

The most famous sustained critique was offered by Herbert Simon (1916–2001, winner of a Nobel Prize in economics in 1978), who suggested that agents do not maximize but instead satisfy "bounded rationality." Simon's observations are correct but are not incompatible with rational decision theory as long as one adds a cost of decision making and interprets probabilities as subjective, not objective. Several disciplines in the social sciences, including sociology, anthropology, and social psychology, implicitly critique the theory by ignoring it in formulating their underlying core theories. This may account for their relative lack of coherence compared with disciplines that embrace rational choice theory (Gintis 2004).

Perhaps the major implication of decision theory is that human beings have a declining marginal utility of money. This is evidenced by the ubiquity of risk aversion and the willingness of individuals to insure against loss. This has an important ethical implication: A dollar transferred from a rich person to a poor person will increase the well-being of the poor person much more than it will reduce the well-being of the rich person.

Some philosophers and philosophically minded economists have played word games in attempting to refute this obvious implication of declining marginal utility (e.g., by suggesting that welfare is not comparable across individuals, an implausible assertion), but its force remains. It implies that with everything else being equal, a more equal distribution of wealth would improve the general welfare. Of course, there are often considerations that act against this principle, such as the maintenance of effective economic incentives and the just treatment of and respect for the rights of the wealthy.

HERBERT GINTIS

SEE ALSO *Choice Behavior*; *Game Theory*; *Prisoner's Dilemma*; *Rational Choice Theory*; *Scientific Ethics*; *von Neumann, John*.

BIBLIOGRAPHY

Gintis, Herbert. (2004). "On the Unity of the Behavioral Sciences." In *Logic, Epistemology, and the Unity of Science*, ed. Dov Gabbay, Shahid Rahman, John Symons, and Jean Paul Van Bendegeme. New York: Kluwer.

Herbert Simon, Herbert. (1982). *Models of Bounded Rationality*. Cambridge, MA: MIT Press.

Kahneman, Daniel; Jack L. Knetch; and Richard H. Thaler. (1990). "Experimental Tests of the Endowment Effect and the Coase Theorem." *Journal of Political Economy* 98(6): 1325–1348.

Kahneman, Daniel, and Amos Tversky. (2000). *Choices, Values, and Frames*. Cambridge, UK: Cambridge University Press.

Kreps, David M. (1988). *Notes on the Theory of Choice*. London: Westview.

Machina, Mark J. (1989). "Dynamic Consistency and Non-Expected Utility Models of Choice under Uncertainty." *Journal of Economic Literature* 27: 1622–1668.

Samuelson, William, and Richard. Zeckhauser. (1988). "Status-Quo Bias in Decision Making." *Journal of Risk and Uncertainty* 1: 1–59.

Simon, Herbert. (1982). *Models of Bounded Rationality*. Cambridge, MA: MIT Press.

Tversky, Amos, and Daniel Kahneman, (1981). "Loss Aversion in Riskless Choice: A Reference-Dependent Model." *Quarterly Journal of Economics* 106(4) [Nov]: 1039–1061.

DEFORESTATION AND DESERTIFICATION

• • •

A common claim of defenders of tropical rain forests is that because of the shallowness of rain forest soils cutting down those forests for crops or cattle grazing will lead to massive soil erosion and eventually create deserts in areas where lush forests once grew and provided a high percentage of the earth's biodiversity (Sponsel, Headland, and Bailey 1996; Burch 1994; *The Burning Season* 1994).

Complexity of Causes

However, the causes of desertification are much more complex than this scenario would suggest. It is true, for instance, that in the Mediterranean Basin deforestation over centuries has been a significant factor in desertification from Spain and the western part of North Africa in the west to Lebanon and Palestine in the east. Nevertheless, cutting down forests was only one among several human factors that advanced desertification in that region, along with climatic factors:

> First and most fundamental are climate factors. Here is one summary: Conditions [for desertification] are common in the Northern and Southern hemispheres between 20° and 30° latitude.... The most common factor in determining climate is the intense equatorial solar radiation, which heats the air and generates high levels of humidity. Warm tropical air rises; as it does it cools, and the atmospheric moisture condenses. That results in high rainfall patterns in the equatorial region. The rotating earth causes these air masses to move away from the equator toward both poles, and the air begins to descend on either side of the Tropic of Capricorn and the Tropic of Cancer around the 30° latitudinal band. As the air descends it warms and relative humidity declines, resulting in a warm belt of aridity around the globe (Mares 1999, p. 169).

This is the explanation for the existence of deserts worldwide, but for many people concerned with science, technology, and ethics the term *desertification* has a different meaning:

> Desertification is the degradation of productive drylands, including the Savannas of Africa, the Great Plains and the Pampas of the Americas, the Steppes of Asia, the "outback" of Australia and the margins of the Mediterranean. Desertification is occurring to such a degree that some lands can no longer sustain life (Middleton and Thomas 1997, p. iv).

It is controversial whether humans can do anything about climate change, and so the basic formation of the world's deserts is of less interest here—specifically as an ethical or social problem to the mitigation of which science and technology might contribute—than is desertification in the latter sense. However, even with respect to desertification related to humans and their lifestyles over the millennia, the issue is enormously complex.

Attempts at Remediation

One area of increasing desertification is the Mediterranean Basin of southern and southeastern Europe, along with limited areas of western and eastern North Africa. The *World Atlas of Desertification* (Middleton and Thomas 1997) is a product of the United Nations Environment Program (UNEP) and is related to the United Nations Convention to Combat Desertification (CCD). The atlas contains a chapter, "Desertification and Land Use in Mediterranean Europe," that helps illustrate the complexities of the issue. For example, the atlas states: "The region has suffered from land degradation at least since the Bronze Age" (p. 129). There has been damaging "terrace construction over many centuries ... [and] in recent years major changes in the population distribution have occurred with ... the movement of people to the major cities and coastal areas [for tourism] and the development of irrigated agriculture and industry ... [with attendant] flooding and erosion, groundwater depletion, salinization and loss of ecosystem integrity" (p. 129). One of the hardest-hit areas is southeastern Spain, in a country that has seen all these impacts for centuries, including massive deforestation and extensive irrigated farming in the Valencia region.

One of the goals of the UNEP/CCD program is to utilize the latest science and technology, including remote sensing techniques to map desertification advances, and the Mediterranean Desertification and Land Use (MEDALUS) project includes the Guadalentin Target Area in southeastern Spain: "The most degraded and eroding areas are ... former common grazing lands that were taken into cultivation due to an expansion of mechanized agriculture in the 1960s and ... were abandoned, as systems failed" (p. 131).

All these factors have been at work to varying degrees throughout the Mediterranean Basin, where there is ongoing desertification. MEDALUS scientific studies and rehabilitative efforts are ongoing throughout the region, from Portugal, to Italy, to Greece and Asia Minor.

Desertification is increasing rapidly in the world's best-known desert, the Sahara, and particularly along its southern border, the Sahel region. Two major causes are overgrazing, especially after prolonged drought beginning in the 1960s, and the use of brushwood as fuel in homes (Middleton and Thomas 1997, pp. 46–48 and 68–69, 168ff).

The *World Atlas* includes reports on the Middle East, southern Asia, Australia, China, and Mexico. A United Nations CCD conference report, *Sustainable Land Use in Deserts* (Breckle, Veste, and Wucherer 2001) covers the Aral Sea reclamation effort, changing patterns of overgrazing in South Africa, the monitoring of desertification in Uzbekistan and Kazakhstan, and reclamation efforts in Israel, among many other topics.

Ethical Issues

The ethics of desertification reflects extremely diluted responsibilities. Since the Bronze Age in the Mediterranean Basin, for example, up to the present (such as in Spain), farmers have tried in numerous ways to eke out a hard living in arid lands. Some people would lay blame primarily on government planning agencies for overirrigation and groundwater depletion, salinization, and other impacts of population density and tourism in arid regions. However, in any particular case it is difficult to lay too much blame on individual agents, although some environmental ethicists would blame a culture that is and has been for centuries heedless of impacts on arid lands.

In regard to science, technology, and rehabilitation/ restoration projects such as those of UNEP/CCD, it may be too early to tell whether they will be effective in the long run against what is widely perceived to be rapidly advancing desertification.

PAUL T. DURBIN

SEE ALSO *Agricultural Ethics; Biodiversity; Ecology; Environmental Ethics; Environmentalism; Global Climate Change; National Parks; Rain Forest; Sierra Club; United Nations Environmental Program.*

BIBLIOGRAPHY

Breckle, Siegmar-W.; Maik Veste; and Walter Wucherer, eds. (2001). *Sustainable Land Use in Deserts*. Berlin and New York: Springer.

Burch, Joann Johnson. (1994). *Chico Mendes, Defender of the Rainforest*. Brookfield, CT: Millbrook. Intended for younger readers.

The Burning Season: The Chico Mendes Story (1994). Warner Home Video. Theatrically released in 1981. Story of a man dedicating his life to saving the Amazon rain forest. Based on a novel by Andrew Revkin.

Mares, Michael A. (1999). *Encyclopedia of Deserts*. Norman: University of Oklahoma Press.

Middleton, Nick, and David Thomas, eds. (1997). *World Atlas of Desertification*, 2nd edition. London: Arnold.

Sponsel, Leslie E.; Thomas N. Headland; and Robert C. Bailey, eds. (1996). *Tropical Deforestation: The Human Dimension*. New York: Columbia University Press. Scholarly coverage of the issue.

DEHUMANIZATION

SEE *Humanization and Dehumanization.*

DEMATERIALIZATION AND IMMATERIALIZATION

• • •

Dematerialization refers to technological production using less energy and fewer or lighter-weight materials. *Immaterialization* is a similar approach, militating against the consumption of material goods.

Dematerialization

The concept of dematerialization is strongly associated with the work of economist and planner Paul Hawken, who proposed that industry should recalibrate inputs and outputs to adapt to environmental constraints. "To accomplish this, industrial design would employ 'dematerialization,' using less material per unit of output; improving industrial processes and materials employed to minimize inputs; and a large scale shift away from carbon-based fuels to hydrogen fuel, an evolution already under way that is referred to as 'decarbonization'" (Hawken 1993, p. 63). Indeed, Hawken sees dematerialization as a long-term trend, because much contemporary technology—refrigerators, televisions, cars, even houses—already weigh less and use less material than they did in the 1970s. According to Hawken's calculations, during the ten year period from 1972 to 1982, the redesign of automobiles in the United States reduced annual resource use by 250 million tons of steel, rubber, plastic, aluminum, iron, zinc, copper, and glass. Hawken's approach thus implies a rejection of heavy industry as the foundation of a technological economy, and is allied with notions of industrial ecology, green design, and natural capitalism.

Hawken, however, credits Buckminster Fuller (1895–1983) with originating the concept of dematerialization, which Fuller called "ephemeralization." Fuller's own invention of the geodesic dome was an example of ephemeralization, because it weighed only three percent of what a traditional structure of equivalent size would weigh, while being even more earthquake- and fire-resistant. According to Fuller, ephemeralization had already triumphed in his day. "[B]etween 1900 and today," he said in 1968, "we have gone from less than one percent to more than forty percent of humanity living at a high standard [with] the amount of resources [consumed per person] continually decreasing . . ." This "came only as fall-out of the doing-more-with-less design philosophy" (Fuller 1970, p. 68).

Fuller also described a design curve under which technologies increase in size soon after their invention until they "reach a giant peak, after which miniaturization sets in" (p. 73). Subsequent developments in personal computer, cell-phone, and portable music technologies such as CD, MP3, and iPod players bear out Fuller's theory. The prospects of nanotechnology provided further confirmation. He concluded, playfully, that "Ephemeralization trends towards an ultimate doing of everything with nothing at all—which is a trend of the omniweighable physical to be mastered by the omniweightless metaphysics of human intellect" (p. 73).

Dematerialization is also operative in science. The replacement of field work and laboratory experimentation by computer modeling and simulation may be described as another type of dematerialization.

Immaterialization

The immaterialization of consumption, as a companion process to dematerialization in production, has weak and strong forms. (It should not be confused with immaterialism in metaphysics, regarding the reality of immaterial phenomena such as the mind or soul.)

In its weak form, immaterialization is simply the consumption of dematerialized consumer goods—the same ones purchased in the past, such as refrigerators or automobiles, but now manufactured using less energy and materials. These goods are designed to consume less energy when used, and to be more easily recyclable, so that there is reduced waste.

In its strong form, immaterialization of consumption refers to the replacement of material goods with immaterial ones such as services, information, and social relationships. The use of an electronic telephone directory is an immaterial alternative to the use of a large paperback telephone directory. The Finnish cell phone manufacturer Nokia, whose motto is "Connecting People," sees both dematerialization and immaterialization as ways to promote a sustainable consumer economy. Immaterialization thus reflects another aspect of the service economy and the information, or knowledge, society.

Immaterialization in the strong sense also points toward possible cultural transformations, including shifts in ideas about the good life. Material consumption is not a good in itself, but a means to the end of human well-being. When analyzed in terms of well-being rather than material goods, productivity may actually be decreasing; human beings may be consuming more, but enjoying it less. Certainly the marginal utility of another unit of material consumption has declined, suggesting cultural or spiritual goods such as music and meditation as more inherently fulfilling than the purchase of another television set, however dematerialized. Yet just as the paperless office has remained full of paper, so immaterialized goods seem always to be complemented with material, such as music posters, coffee table art books, designer wardrobes, and specialized furniture for those who practice meditation.

CARL MITCHAM
JONATHAN WALLACE

SEE ALSO *Ecological Economics; Environmental Economics; Fuller, R. Buckminster; Green Design; Information Society; Materialism.*

BIBLIOGRAPHY

Hawken, Paul. (1993). *The Ecology of Commerce: A Declaration of Sustainability.* New York: HarperBusiness.

Fuller, R. Buckminster; Eric A. Walker; and James R. Killian. (1970). *Approaching the Benign Environment.* University: University of Alabama Press for Auburn University. A collection of 1968 talks.

Ryden, Lars, ed. *Foundations of Sustainable Development: Ethics, Law, Culture, and the Physical Limits.* Uppsala, Sweden: Baltic University Programme, Uppsala University. A multi-authored work reflecting Scandinavian interests in dematerialization and immaterialization as central to sustainable development.

DEMOCRACY

• • •

Democracy poses problems for science and technology because it leads to potential conflicts between two strong sets of ethical values. Democracy prizes the ethics of inclusiveness and political equality. Within a democratic system all citizens have an equal say in collective decisions. The fields of science and technology embrace the ethics of autonomy and respect for scientifically established findings, regardless of how other citizens receive or are affected by those findings. Scholars and practitioners have proposed a variety of processes and institutions in an attempt to resolve these conflicts.

Historical Development

Over the centuries philosophers have developed various conceptions about the nature of democracy. These different versions of democracy pose distinct conflicts among ethical values linked to science and technology, as well as suggest different solutions to those problems.

The classic form of democracy or rule by the people is generally taken from Athens in the fifth century B.C.E., where a form of direct or participatory government was practiced by the free males of the city-state. In Rome from the fifth to the first centuries B.C.E. there developed a classic form of republican or representative democracy, in which individuals are elected by the people to handle governmental decision making. During the Middle Ages democratic forms of government were relegated to the margins of public life where they continued to play important roles in religious institutions such as monasteries; they reemerged into public affairs during the rise of modern nation–states. Indeed modern political philosophy is characterized by diverse and continuing arguments for the primacy and legitimacy of democratic institutions, and struggles with efforts to create appropriate functioning democratic organizations under historically unique conditions.

One common observation is that the development of modern forms of science, technology, and democracy have in fact gone hand in hand. Modern science itself asserted a radical democracy, although only among a scientifically educated elite. The industrial revolution was certainly associated with the extension of political rights—from white property owners to all men to women. Expansions of citizenship have in turn been associated with the expansion of consumer economies, which thereby influenced technological change. And in many instances expansions in democracy have been proposed as solutions to the problems caused by scientific or technological change. Reflecting such associations, many commentators on science-technology-democracy relations have tended to emphasize synergies rather than oppositions. Certainly this was true of Alex de Tocqueville's *Democracy in America* (1835 and 1840), a perspective repeated even more forcefully in Daniel Boorstin's *The Republic of Technology* (1978).

Especially since the early mid-twentieth century, however, questions and problems have become increasingly prominent. Taking the two basic forms of democracy in reverse order to their historical origins, one may describe these as related to representation and direct democracy.

REPRESENTATION. After World War II scientists gained a great deal of attention and prestige from the government. Due to the scientists' great success in developing technologies for the war, from radar to nuclear weapons, government officials hired them into agencies and national laboratories and put them on important advisory committees. These developments raised the issue of how best to bring scientists and engineers, and their expertise, into the decision-making processes of representative democracy. This political involvement of scientists threatened two important ethical values. First, how could scientists avoid compromising their scientific autonomy and integrity as they became more involved in politics? Would they be able to speak freely, unencumbered by motivations of the government officials for whom they worked? When they advised government about research budgets, which affected them directly, would they succumb to the conflicts of interest that such roles entailed?

Second, how would this new scientific elite affect democracy? Would scientific pronouncements simply trump other forms of advice and political input? If the subject at hand was purely technical, deference to technical advice might be appropriate. However most important scientific and technological policy issues are a complex mixture of technical and political or social considerations, and scholars have shown that, in practice, it is difficult to separate these two features, even if it is desirable in principle. This concern over scientists gaining excessive power was most famously stated in President Dwight D. Eisenhower's famous warnings about a military-industrial complex in his farewell address in 1961.

DIRECT DEMOCRACY. Most theories of democracy state that citizens need to do more than simply vote for officials every few years. A robust democracy requires that citizens be able to participate directly, either as groups or individuals, in political decision making. If the issue at hand involves extensive scientific or technological knowledge, how can nonscientific citizens participate in deciding such an issue, an important democratic value, while still respecting the technical competence of experts, an important scientific value?

Responding to Problems

Late-twentieth-century developments in democratic theory have included a broad spectrum of responses to perceived problems in the science-technology-democracy relationship. These responses have included analyses and criticisms of a number of phenomena related especially to representation and direct democracy centered around such issues as peer review, lobbying, advisory bodies, and deliberation.

REPRESENTATION. A number of government agencies, in contrast to the direct mission driven distribution of funds by a program director on the basis of personal assessment have adopted peer review as a means to distribute funds. After World War II the federal government dramatically increased its funding of scientific and technological research. Following the model that it had developed during the war, much of that research was performed outside of the government itself. Instead of becoming the dominant employer of scientists, the government decided to fund scientists who were employed by universities or businesses.

In peer-reviewed funding, scientists submit proposals to the government requesting funding for particular research projects. Peer review is a method for evaluating and ranking those proposals and deciding which ones to fund. The funding agency, such as the National Science Foundation (NSF) or the National Institutes of Health (NIH), identifies scientists outside of government who are experts in the field relevant to the proposed research, who are the peers of the scientist submitting the proposal. Those scientists then review and evaluate the proposal, providing an expert opinion of its technical merits. The government keeps the names of the reviewers confidential so that they feel free to be objective in evaluating the proposal without having to worry about reprisals from the people they are reviewing. These reviews powerfully influence who the government funds.

Peer review is not perfect and has engendered numerous controversies and studies. Scientists also try to influence the total size of the government research budget, often through individual or group lobbying of Congress or the executive branch. In addition, many scientists may adapt their research agendas to be responsive to growing parts of the budget, which means that they are not as autonomous as peer review may make them appear. However it is still a reasonable attempt to balance scientific and democratic values. Scientists independent of the government provide evaluations of the merit of the proposed scientific research, emphasizing the ethics of scientific independence and autonomy. However, in many cases government officials make the final decisions on funding and in all cases governmental institutions determine the total amount of money that the government gives out for research, which lets representative institutions influence the research as well, emphasizing the value of democratic accountability.

Second, the federal government has created a host of science and technology advisory bodies. These groups attempt to bring technical expertise into making and executing government policy in a manner that respects both scientific integrity and democratic accountability. Some of these bodies are part of the government itself, and its scientists are government employees, as in the congressional Office of Technology Assessment (OTA) (disbanded by Congress in 1995) or the Office of Science and Technology Policy, an advisory group to the president. In addition, the government employs numerous technical specialists in various agencies and national laboratories.

In addition, the federal government utilizes many advisory committees made up of scientists and engineers from outside the government. Numerous agencies have such advisory committees and the White House has the President's Committee of Advisors on Science and Technology (PCAST). The National Academy of Sciences (NAS) also has an elaborate system for providing technical advice to the government. NAS, a private, though congressionally chartered, organization, possesses a research arm, the National Research Council (NRC). NRC assembles experts in particular fields to prepare reports that summarize the state of the science related to some particular topic. These groups have no formal authority, but they give the agencies access to expertise that is outside of government agencies and so hopefully is independent of such agencies' agendas. Of course the effectiveness of these advisory groups depends on the quality of the people appointed to them. In addition, these advisory groups lack any democratic accountability.

SOLVING PROBLEMS OF DIRECT PARTICIPATION. Citizens participate in policy making in two ways, either as groups or as individuals. The process of participating in groups is often called interest group liberalism, or pluralism. The justification for pluralism assumes that citizens recognize their interests and how government policy affects those interests. To further their interests they organize themselves into private groups and those groups pool their resources so that they may influence government policy. Different groups have different resources, from large numbers of voters to large sums of money to social status to charismatic leaders.

Such groups often center around scientific and technological issues and are a major part of the policy process. They include environmental groups, organizations representing different scientific disciplines, groups that lobby for research on certain diseases, industry groups that seek support for particular technologies, and so on. Many scholars have written about such groups and the ways they try to influence policy. In terms of ethical values, interest group involvement in scientific issues reflects the values that underlie pluralist democracy more generally. Pluralism requires only that all groups have equal opportunities to participate in politics. Groups are only supposed to represent their interests, as they perceive them. While outright lying about relevant science violates a general ethic of honesty, this form of democracy has no process to resolve more subtle scientific and technological disagreements. In most public disputes over scientific and technological issues, experts will disagree about some of the scientific questions. Within interest group pluralism, the groups have no obligation to find ways to resolve those disagreements; the theory assumes that honest competition among the groups will lead to satisfactory resolution of the issues, scientific and otherwise.

Citizens may also participate in scientific and technological policy issues as individuals. Scholars have concluded that this sort of participation works best when it involves extensive deliberation. In other words, citizens do not simply give their off-the-cuff opinion on some issue, either through voting or responding to an opinion poll. Instead they become involved in a process that requires them to learn about the issue and discuss it with others.

Theories of deliberative democracy have stated that such a process not only informs citizens about the substance of an issue, but also gives them a broader outlook, making them think about the public interest as well as their narrow private interests. It is the process of learning about and debating an issue, in an environment that is conducive to friendly give-and-take, which not only lets citizens state their interests but also makes them better citizens in how they think about their interests. This development satisfies a democratic ethic important to this theory of democracy, that citizens learn to deliberate over the public interest instead of merely advocating private interests.

The Cambridge Experimental Review Board is a classic example of such deliberation. In 1976 two universities in Cambridge, Massachusetts wanted to build biotechnology laboratories in the city. People in and out of the biology discipline worried that genetically modified organisms might escape from the labs and harm people. Cambridge is a very densely populated city and the building of these labs, and the risks that might accompany them, became a highly charged political issue. The mayor decided to appoint a special review board, consisting of ordinary citizens, to decide whether and under what circumstances the universities should be allowed to build the labs. The board heard testimony from all concerned parties, including university scientists who wanted to build the labs and people who opposed them. In the end the board decided to let the labs be built, with certain safety procedures for their operation. Those procedures were very similar to the ones later adopted by NIH, the principal federal funding agency for such research. NIH could impose regulatory conditions on the universities that it funded. Almost all sides to the controversy praised the work that the board had done.

This process encountered all the ethical issues related to science and democracy. Citizens had to learn about the technical issues. They did not have to become scientists, but had to understand the issues well enough to make sensible policy decisions about them. In educating citizens, the process demonstrated respect for scientific integrity. The process also satisfied the norms of deliberative democracy, in that it involved citizens deeply in an issue that potentially affected their lives, gave them the means to learn about it, and gave them the power to actually decide about it. The downside to this process, and all deliberative processes, is that it involved directly only a few citizens out of the many that lived in the city and required that they spend a great deal of time on the issue. Deliberative processes always involve this tradeoff: In exchange for deep participation, one sacrifices broad participation.

Since the 1970s organizations have sponsored a host of experiments using different forms of deliberative participation. For example, deliberative polling combines traditional opinion polling with a deliberative process. The process begins with a representative sample of citizens taking an opinion poll on the issue at hand. After the poll, the same group then assembles for a weekend of deliberation on the issue, guided by facilitators and with experts available to answer questions. At the end of the deliberations they are polled again. In most cases, their opinions change, often significantly, as a result of the deliberation.

A deliberative form of participation closer to the Cambridge example is the consensus conference. Initially developed in Denmark, this process brings together a small number of citizens to deliberate and see if they can reach a consensus on some issue, usually related to science and technology. The group then reports their results. In Denmark consensus conferences provide important input to parliament. In the United States they have not yet attained any official status. The not-for-profit organization The Loka Institute and a few academic groups have sponsored consensus conferences.

Contemporary Issues

Previous discussion of the practical issues of representation and participation are complemented by general theoretical discussions of technology and democracy. Among these discussions, one of the more salient arguments has been that of Langdon Winner and Richard Sclove that technological design itself constitutes a kind of political constitution writing that can be more or less democratic. These scholars point out that particular configurations of technological systems can favor some groups and discourage others, politically and socially, as well as economically. These social effects may be designed into technologies or may be unintended consequences, but either way, the "artifacts have politics," as Langdon Winner put it. This scholarly work means that those who are concerned with the science-technology-democracy relationship have to focus on the actual designs of the technologies themselves, as well as the institutions that govern them.

In the early twenty-first century forms of direct participation like consensus conferences are limited in the United States and do not enjoy the formal authority that they do in places like Denmark. However they are growing in number and their advocates hope they will have effects on policy making by local or state governments by force of moral suasion if not by law. All forms of participation, via groups or individuals, are growing in the United States and elsewhere. Legislation mandates some form of participation in many policy areas and some private firms are taking public participation seriously. So large is this activity that the government and business officials who run such programs have started their own professional association, the International Association for Public Participation, an organization that now has more than 1,000 members from twenty-two countries. Many of the issues in which such participation occurs involves science and technology.

One of the most difficult issues to deal with at the intersection of democracy and science and technology is the problem of boundaries between science and politics. As indicated above, one important aspect of democratizing science and technology is respecting the scientific value of the autonomy and integrity of science. But what parts of issues belong to the realm of science and what parts to the realm of politics? At first glance, this seems like an obvious question. The scientific parts of

an issue are technical details about the issue, things that one would clearly ask of a technical expert, such as the existing reserves of oil, the toxicity of some pollutant, or the risks to patients of some new medical treatment. The political parts of an issue would seem to be questions like how much should oil be taxed and under what circumstances, or at what level is injury from pollution is politically acceptable. However these questions are not so neatly technical or political. Existing reserves of oil are uncertain, so for policy purposes should there be a high, low, or intermediate estimate? The toxicity of a pollutant may depend on whether the people exposed are healthy or more susceptible to it, as someone with asthma may be to air pollutants. Are we talking about toxicity for the average population or the most vulnerable members of the population? Answering these questions requires a complex mixture of technical and political decisions, which means that the boundaries between the technical and political parts of the issue are negotiated and often changing, not fixed and prompted by nature. An important part of participation is enabling participants to recognize and debate these boundaries. Only then can such participation satisfy both the scientific and democratic values involved in the process.

FRANK N. LAIRD

SEE ALSO *Conservatism; Direct Democracy; Discourse Ethics; Expertise; Liberalism; Libertarianism; Participation; Strauss, Leo; Tocqueville, Alexis de*.

BIBLIOGRAPHY

Boorstin, Daniel. (1978). *The Republic of Technology: Reflections on Our Future Community*. New York: Harper & Row.

Hill, Stuart. (1992). *Democratic Values and Technological Choices*. Stanford, CA: Stanford University Press. An in-depth case study that critiques the conventional notion of citizen participation examines citizens' values toward a controversial technology.

Laird, Frank N. (1993). "Participatory Analysis, Democracy, and Technological Decision Making." *Science, Technology, and Human Values* 18 (Summer): 341–361. A review and analysis of normative arguments for including citizens in technological policy making.

Renn, Ortwin; Thomas Webler; and Peter Wiedemann, eds. (1995). *Fairness and Competence in Citizen Participation*. Dordrecht, Boston, and London: Kluwer Academic Publishers. This edited volume provides both descriptions and analysis of numerous forms of citizen participation.

Sclove, Richard E. (1995). *Democracy and Technology*. New York: The Guilford Press. Develops an extensive normative theory of democracy and technology and spells out is implications for democratic practice, with numerous short cases.

Sclove, Richard E.; Madeleine L. Scammlee; and Brenna Holland. (1998). *Community-Based Research in the United States*. Amherst, MA: The Loka Institute. Available from the Loka Institute, www.loka.org. Contains twelve case studies of such participatory research, as well as comparisons with the Netherlands.

Von Schomberg, René, ed. (1999). *Democratising Technology: Theory and Practice of a Deliberative Technology Policy*. Hengelo, The Netherlands: International Centre for Human and Public Affairs. A series of essays by some of the leading American and European scholars on the subject.

Vig, Norman J., and Herbert Paschen, eds. (2000). *Parliaments and Technology: Tthe Development of Technology Assessment in Europe*. Albany: State University of New York Press. Provides cases of participatory technology assessment in different European countries.

Winner, Langdon, ed. (1992). *Democracy in a Technological Society*. Dordrecht, The Netherlands: Kluwer Academic Publishers. Contains essays by the major scholars of the time on democracy and technology.

Winner, Langdon. (1986). *The Whale and the Reactor: A Search for Limits in an Age of High Technology*. Chicago: University of Chicago Press. Develops some of the core ideas relevant to the justification for democracy in technology policy, including the notion that artifacts have politics and technologies as forms of life.

DEONTOLOGY

• • •

Deontology refers to a general category of ethical or moral theories that define right action in terms of duties and moral rules. Deontologists focus on the rightness of an act and not on what results from the act. Right action may end up being pleasant or unpleasant for the agent, may meet with approval or condemnation from others, and may produce pleasure, riches, pain, or even go unnoticed. What is crucial on this view is that right action is *obligatory*, and that the goal of moral behavior is simply that it be performed. The slogan of much of deontology is that the right is independent of the good. Deontology is opposed, therefore, to consequentialist or teleological theories in which the goal of moral behavior is the achievement of some good or beneficial state of affairs for oneself or others. For deontologists, the end of moral action is the very performance of it. For consequentialists, moral action is a means to some further end.

There are three interrelated questions that any deontological theory must answer. First, what is the content of duty? Which rules direct human beings to morally right action? Second, what is the logic of these duties or rules? Can their claims be delayed or defeated? Can they make conflicting claims? Third, why must human beings follow exactly those duties and rules, and not others? That is, what grounds or validates them as moral requirements?

The relevance of deontological ethics to issues in science and technology is not immediately obvious. Typical duties or rules in these theories are often quite abstract and sometimes address personal morality; hence they seem ill suited to broad and complicated questions in technical fields. As a matter of personal morality, deontologists might require one never to lie or steal, to give to charity, and to avoid unnecessary harm to people and animals. These rules are often internalized and are supported by religious, social, and civil institutions, and in some cases by enlightened self-interest. But is there a duty to support open source software, or to reject nanotechnology, or to avoid animal experimentation for human products? What list of rules is relevant to moral quandaries over cloning or information privacy?

Though the specific connection between ethical duties and scientific and technological practices may not be immediately obvious, it is clear that deontology can and should play an important role in evaluating these practices. Deontological theories give one a way to evaluate types of acts, so that one can judge a token of an act as obligatory, permissible, or forbidden even before the act is committed. Consequentialist evaluations, on the other hand, must await an accounting of the consequences of scientific and technological acts. Waiting on the consequentialist analysis may be perilous, because the long-term results of large-scale enterprises are often impossible to anticipate and very difficult to repair. As Edward Tenner (1997) has pointed out, modern technology often exacts a kind of revenge in the scope and severity of unintended consequences. Especially in fields such as bioethics, practitioners have often wanted bright lines between right and wrong acts in their ethical guidelines. That is, they want to have ethical rules or principals that are not wholly contingent on consequences. A form of deontological view in bioethics known as principalism focuses on the need for clear guidelines for action in order to avoid problems with unintended and far-reaching consequences of treatments and clinical practices. Even the basic and broadly applicable principle "Do no harm!" is deontological; it does not allow a tradeoff of benefit for some at the cost of harm to others.

Two deontological theories, from the works of Immanuel Kant (1724–1804) and W. D. Ross (1877–1971), serve as the foundations for much work in deontological ethics. Because they differ significantly in the content, logic, and ground of duties, it will be useful to examine them in modest detail before returning to questions of science and technology.

The Categorical Imperative

Kant developed the most important deontological ethical theory in Western philosophy. Scholars have come to agree that Kant provided not so much a list of duties as a procedure for determining duties. The procedure that specifies duty is the categorical imperative or *unconditional command* of morality. Kant articulated the categorical imperative in several distinct formulations. Even though these formulations provide different ways of generating duties, Kant maintained that his systematic ethic of duties was *rigorous*—in the technical sense that a "conflict of duties is inconceivable" (Kant 1997, p. 224). Indeed, a main feature of Kant's ethics is its reliance on consistency or harmony in action. This feature can be seen in the first formulation of Kant's categorical imperative, which goes as follows: "*Act only on that maxim through which you can at the same time will that it become a universal law*" (Kant 1997, p. 421).

Because a maxim in Kant's theory is a plan of action, the categorical imperative above provides an ethical test for intended actions, presumably to be used before one commits them. The point of the test is that one ought to be able to endorse the *universal* acceptability of the plans or intentions behind actions. People should not be partial to plans simply because they conceived such plans; the plans must be acceptable from any point of view. Maxims that cannot be universalized will produce logical contradiction or *disharmony* when they are run through the test of the categorical imperative. The grounding or validation of this principle lies in the universality of practical reason. For Kant, ethical duties arise from what is common to humans as rational beings. Humans have a kind of freedom that is gained in *creating* universal moral laws through intentional behavior. This moral and rational activity is, for Kant, what produces *self-legislation* or autonomy, and autonomy allows humans to transcend their animal nature.

The ability of humans to act from freely chosen moral rules explains the special moral status they enjoy; humans are, according to Kant, *ends-in-themselves*. Consequently this conception of a special status gives rise to another formulation of the categorical imperative: "*Act in such a way that you always treat humanity [yours or another person's] never merely as a means but always at the same time as an end-in-itself*" (Kant 1995a, p. 429).

This special moral status or intrinsic value implies that humans ought never to be valued as less significant than things that have merely instrumental value. Things of instrumental value are mere tools, and though they can be traded off with one another, they can never be more important than intrinsically valuable things. All technology is in some sense a mere tool; no matter how many resources society pours into technologies, the moral status of humans is supposed to trump the value of mere tools. Kantian duties are designed to protect that status.

The application of Kant's theory to issues in the ethics of technology produces intriguing questions. Do some technologies help persons treat others as mere means? The moral inquiry would have to consider aspects of the technologies and see whether technologies have "maxims" themselves—what Günther Anders called a "mode of treatment incarnated in those instruments" (Anders 1961, p. 134). These aspects might include the anonymity of online communities, the distributed effects of computer viruses, the externalizing of costs by polluting corporations, or the inherent destructiveness of a nuclear weapon. Further, one might ask whether some technologies *themselves* treat persons as mere means? Such a worry is related to Martin Heidegger's view that, under modern technology, humanity becomes a *standing reserve* to be exploited, and to Herbert Marcuse's claim that such a technological society debases humans by providing a *smooth comfortable unfreedom*. While these critics of technology do not always identify themselves as Kantians, the influence of Kant's humanistic account of duties has been so deep and broad that it is almost inescapable. Still there are deontologists who have parted ways with the Kantian tradition.

Prima Facie Duties

According to the British philosopher W. D. (Sir David) Ross, moral duties are not universal and unconditional constraints of universal practical reason. Rather they are conditional or prima facie obligations to act that arise out of the various relations in which humans stand to one another: neighbor, friend, parent, debtor, fellow citizen, and the like. This view gives content to duties based on a kind of role morality. It is through moral reflection that one apprehends these duties as being grounded in the nature of situated relations. Duty is something that, for Ross, arises between people, and not merely within the rational being as such. What exactly these prima facie duties are is not infallibly known until the problematic situations present themselves.

Nonetheless, Ross thinks, situated moral agents can grasp some obvious basic forms of duties. Fidelity, reparation, gratitude, justice, beneficence, self-improvement, and non-maleficence are what he identifies as nonreducible categories of duty—he admits that there may be others. Ultimately these duties are known by moral intuition and are objectively part of the world of moral relations and circumstances that humans inhabit. Much as one knows, in the right moment, what word *fits* in a poem, so too can one know what to do when duty makes demands. Sometimes an agent will intuit that more than a single duty applies, and in these cases must judge which duty carries more weight in order to resolve the conflict.

Ross's view is therefore both flexible and pluralistic, and is grounded in the actual roles of human lives. In these respects, it provides a foundation for a variety of professional codes of ethics, many of which are found in the scientific and technological community.

Hans Jonas and the Imperative of Responsibility

While Kant and Ross argued specifically against consequentialist theories in explaining their respective deontological views, other theorists are motivated by concerns over consequences in ways that influence the content of duties. Such is the case with the *imperative of responsibility* put forward by Hans Jonas (1984). Jonas calls for a new formula of duty because he thinks that traditional ethical theories are not up to the task of protecting the human species in light of the power of modern technology. His worry relates directly to the irreversible damage that modern technology could do to the biosphere, and hence to the human species. Because humans have acquired the ability to radically change nature through technology, they must adjust their ethics to constrain that power.

In language intentionally reminiscent of Kant's categorical imperative, Jonas gives his formula of duty as follows: "Act so that the effects of your action are compatible with the permanence of genuine human life" or so that they are "not destructive of the future possibility of such life" (Jonas 1984, p. 11). Referring to Kant's first version of the categorical imperative, Jonas criticizes its reliance on the test of logical consistency to establish duties. There is no *logical* contradiction, he notes, in preferring the future to the present, or in allowing the extinction of the human species by despoiling the biosphere. The imperative of responsibility, as a deontological obligation, differs from the ethics of Kant and Ross because it claims that humans owe something to others who are not now alive. For Jonas, neither the rational

nature nor the particular, situated relations of human beings exhaustively define their duties. Indeed one will never be in situated relationships with people in far-off generations, but remoteness in time does not absolve the living of responsibilities to them.

Are All Duties Deontological?

Most professional codes of ethics in science and engineering consist of duties and rules. Does it follow that their authors tacitly accept the deontological orientation in ethics? It does not, and there is an important lesson here about the choice between deontology and other ethical orientations. The primary difference between professional codes and deontological ethical theories is that, in the former, the duties or rules are put forth as instrumental for competent or even excellent conduct within the particular profession. Some duties are directed toward the interests of clients or firms, but ultimately the performance of these duties supports the particular profession. The grounding of duties in professional codes resembles the function of rules under rule utilitarianism.

These rules would not be morally required for the general public, as would the rules of a deontological ethics. Professional codes are tools to improve the profession; the end of right action, in this case, is dependent upon the good of the profession, and the content of duties will depend on the particular views of the authors concerning that good.

Further Applications and Challenges

Duty ethics have been applied with some success in technological fields where consequentialist or utilitarian reasoning seems inappropriate. In biomedical ethics there is general acceptance of the view that do-not-resuscitate orders and living wills are to be respected, even when doing so means death for the patient and possibly great unhappiness for loved ones. In computer ethics, the argument for privacy of personal data does not generally depend on the use to which *stolen* data would be put. It is the principle, and not the damage, that is at the heart of the issue. There also seem to be lines of a deontological sort that *cannot be crossed* when it comes to some forms of experimentation on animals and treatment of human research subjects. For some emerging technologies, there are well-grounded deontological reasons for opposing research and development, even though the technologies eventually could yield great benefits. No one denies the good of the end, but they do deny that the end justifies any and all means. Where the claims of duties are not well grounded, a deontological approach to ethics runs the risk of sounding reactionary and moralistic.

THOMAS M. POWERS

SEE ALSO *Consequentialism; Discourse Ethics; Engineering Ethics; Jonas, Hans; Kant, Immanuel; Scientific Ethics.*

BIBLIOGRAPHY

Anders, Günther. (1961). "Commandments in the Atomic Age." In *Burning Conscience: The Case of the Hiroshima Pilot, Claude Eatherly, Told in His Letters to Günther Anders.* New York: Monthly Review Press.

Beachamp, Tom L., and Childress, James F. (2001). *Principles of Biomedical Ethics,* 5th edition. New York: Oxford University Press.

Darwall, Stephen L., ed. (2002). *Deontology.* Oxford: Basil Blackwell Publishers.

Jonas, Hans. (1984). *The Imperative of Responsibility: In Search of an Ethics for the Technological Age* Chicago: University of Chicago Press. Originally published in 1979 as *Das prinzip verantwortung,* this book made Jonas famous.

Kant, Immanuel. (1995a). "Grounding for the Metaphysics of Morals." In *Ethical Philosophy* 2nd edition, trans. James W. Ellington. Indianapolis: Hackett Publishing. Originally published in 1785. Kant's most widely read and accessible work in ethics.

Kant, Immanuel. (1995b). "Metaphysical Principles of Virtue." In *Ethical Philosophy,* 2nd edition, trans. James W. Ellington. Indianapolis: Hackett Publishing. Originally published in 1797. Part of Kant's metaphysics of morals, the other half of which concerns the principles of political right

Marcuse, Herbert. (1992 [1964]). *One-Dimensional Man: Studies in the Ideology of Advanced Industrial Civilization,* 2nd edition. New York: Beacon Press. Popular treatise that inspired a progressive critique of technology.

Ross, W. D. 1965 (1930). *The Right and the Good.* London: Oxford University Press. Classic text by an important British philosopher.

Tenner, Edward. (1997). *Why Things Bite Back: Technology and the Revenge of Unintended Consequences.* Cambridge, MA: Harvard University Press.

DESCARTES, RENÉ

• • •

René Descartes (1596–1650) was born in La Haye (now Descartes), France, on March 31, and he died in Stockholm, Sweden, on February 11. Although of Roman Catholic heritage, he lived in a region controlled by Protestant Huguenots at a time when Protestants and Catholics were frequently at war. His inherited wealth

René Descartes, 1596–1650. Descartes ranks as one of the most important and influential thinkers in modern western history. His views on science and technology are similar to those of Francis Bacon. *(The Library of Congress.)*

allowed him freedom to study and travel around Europe. He made important contributions to metaphysics, mathematics, and physiology. In mathematics, he invented coordinate geometry, which combines algebra and geometry into a powerful tool for the mathematical study of the physical world. Although he offered proofs for the existence of God and the immortality of the soul, he was suspected of being an atheistic materialist, and lived in fear of persecution. When Galileo Galilei (1564–1642) was condemned in 1633 as a heretic for teaching that the earth revolved around the sun, Descartes suppressed any publication of his agreement with Galileo. After Descartes's death, his books were put on the Catholic Church's Index of Prohibited Books.

Because he broke away from scholastic Aristotelianism and thought through the philosophic implications of a new science of nature, Descartes is often called the founder of modern philosophy. Using six ideas—doubt, method, morality, certainty, mechanism, and mastery—he set the stage for modern science in a way that has had lasting impact while being subject to continuous debate.

Doubt and Method

Descartes's most famous book is the *Discourse on Method* (1637), which is divided into six parts, each developing one of the key ideas that run throughout his writing. In Part One, he presents the idea of doubt. He rejects all traditional thinking because it does not produce proven conclusions that can guide life. The traditional liberal arts education promotes philosophical disputes that are never resolved. Similarly, the moral customs of people around the world are contradictory, and there is no reliable way to resolve this confusion. So Descartes decides to turn inward, to seek within himself some source of conclusive knowledge.

Although modern science often seems to require doubting all traditional beliefs and customs, historians of science have noticed that modern science depends on intellectual traditions. Scientists tend to work within what Thomas S. Kuhn (1922–1996) called "paradigms," broad intellectual frameworks that organize research. To doubt everything received from one's society would deprive one of any starting point for inquiry. And insofar as science is a collective enterprise, it requires that scientists share social norms of thought and conduct. When scientists challenge a traditional belief, it is because they have found resources within their inherited traditions for doing so. Even Galileo's challenge to the traditional idea that the earth was the center of the universe arose from his appeal to an alternative, heliocentric theory that was thousands of years old.

In Part Two, Descartes presents the idea of method. He summarizes his method for scientific inquiry in four rules:

(1) accept only those ideas that are so clearly and distinctly present to the mind as to be self-evident,

(2) divide difficult problems into simple parts that are manageable,

(3) solve problems by moving in small steps from simple to complex,

(4) survey every part of the reasoning so that nothing is overlooked.

Descartes has formulated these rules of scientific method by generalizing from the procedures in geometrical demonstrations, in which one moves from self-evident principles (definitions and axioms) to solve complex problems by moving step by step from simple ideas to more complex propositions.

Many philosophers of science question the adequacy of the Cartesian method for explaining modern science. Michael Polanyi (1891–1976), for example,

argued that there is always a personal judgment in scientific discovery that cannot be reduced to the formalized procedures demanded by such a method. The insight for grasping fruitful ideas in scientific research does not arise from an impersonal method. Jacques Hadamard (1865–1963) surveyed the lives of some famous mathematicians to show that even mathematical reasoning depends on personal, intuitive judgments that go beyond formal logic.

Morality and Certainty

In Part Three, Descartes presents the idea of morality. He admits that his scientific method could not give him moral knowledge to guide his conduct. So he had to adopt a "provisional morality" by which he could live while working to complete his intellectual project. His provisional moral code consists of four rules:

(1) accept whatever customs, laws, and religious beliefs prevail in one's country;

(2) act decisively according to the most probable opinions as if they were absolutely certain;

(3) change desires rather than the world;

(4) realize that the pursuit of truth is the best life for an intellectual person such as himself.

If Cartesian scientists cannot derive morality from their science, then they have to accept whatever moral and religious customs happen to be traditional in their society. This suggests a fundamental problem with modern science—that progress in scientific knowledge does not bring progress in moral knowledge. Cartesian scientists cannot even provide a scientific argument for the moral worth of a life devoted to science. The life of Cartesian science is incoherent. On the one hand, Cartesian scientists doubt everything and refuse to accept anything that is not proven true. On the other, they must accept the moral and religious prejudices of their society because their science cannot produce moral and religious knowledge. Ultimately, this could lead to moral nihilism with the thought that moral value is beyond scientific knowledge and must be left to unexamined prejudice. One must wonder, therefore, whether a scientifically grounded morality is possible.

In Part Four, Descartes presents the idea of certainty. "I think, therefore I am." This most famous claim of Descartes captures his thought that while doubting everything, he cannot doubt his existence, because this is confirmed by his very act of doubting. To doubt is to think, and to think presupposes his existence as a thinker. Beyond this, another idea comes to him—the idea of a perfect being—and this leads him to infer that God's existence is a self-evident certainty. From having the idea of God as a perfect being, Descartes concludes that God must exist, because if he did not exist, he would not be perfect. Descartes derives this ontological argument for God's existence from Anselm of Canterbury (1033 or 1034–1109).

Few people have found the ontological argument a persuasive proof for God's existence, and Descartes's restatement of the argument is weak. This has led some readers to suspect that he is not serious about the argument, and that it is part of his provisional morality to profess belief in the religion of his country to protect himself from persecution. Some readers see this as an indication that modern science as Descartes conceives of it is inherently atheistic.

"I think, therefore I am." Is this an immediately self-certifying truth? Or does it rather, as Friedrich Nietzsche (1844–1900) argued in *Beyond Good and Evil* (1886), illustrate "the prejudices of philosophers"? How does Descartes know that if there is thinking, there must be an "I" to do the thinking? How does he even know what thinking is? Has he perhaps confused thinking with feeling or willing? One could easily continue asking such questions to point out the numerous assumptions buried in Descartes's seemingly simple intuition, assumptions that are not self-evident, assumptions in need of proof if the Cartesian method is to be upheld. One might conclude that even the most rigorous science cannot attain complete certainty, because every proof depends ultimately on some fundamental assumptions that cannot themselves be proven.

Mechanism and Mastery

In Part Five, Descartes presents the idea of mechanism. He expresses reluctance to fully state his mechanistic view of the world, because it would be unpopular. He sketches his physics, explaining how the universe could have emerged through purely mechanical laws. He explains how all life, including the human body, can be explained as governed by mechanical causes. He declares, however, that the "rational soul" of a human being cannot be derived from the mechanical laws of nature, and therefore it must have been specially created by God.

Historians of science have identified Cartesian mechanism as fundamental for modern science. Prior to the seventeenth century, people generally understood nature through the metaphor of the world as a living organism. The Earth was a nurturing Mother. But modern Cartesian science understood nature through the metaphor of the world as a dead machine. The earth was matter in motion.

This mechanical view of the world was criticized as atheistic materialism, because it seemed to deny the immaterial and immortal reality of God and the soul. Descartes defended himself against such criticisms by affirming his belief in God and the soul. He insisted that material body and immaterial soul were two utterly different substances. In his *Treatise of Man* (1664), Descartes explained the physiology of the human body and brain as matter in motion determined by mechanical forces. This was not published until after his death, because he feared it would be too unpopular. Later, Julien Offray de La Mettrie (1709–1751) argued in his book *Man a Machine* (1748) that Descartes had shown that all living beings—including human beings—were merely machines. La Mettrie suggested that Descartes's dualistic separation between body and soul was only a trick to protect himself against persecution from the theological authorities.

The view of the human mind as a computational mechanism has been a powerful influence in the modern science of the brain. This has led some computer scientists to the thought that sufficiently complex computers will eventually replicate or surpass human intelligence. In some stories by Isaac Asimov (1920–1992), robots become Cartesian thinkers, declaring "I think, therefore I am!" But some prominent scientists such as John C. Eccles (1903–1997) argue that human self-conscious thought manifests the uniquely human power of an immaterial soul. So the debate continues over whether science can fully explain the human soul as a material mechanism.

In Part Six, Descartes presents the idea of mastery. The general aim of scientific research should be conquering nature for human benefit. The specific aim should be making such advances in medical science that human health would be improved dramatically, perhaps even to the point of prolonging life and thus conquering death. In this way, human beings would become "the masters and possessors of nature."

Descartes thus joins the project of Francis Bacon (1561–1626) for directing modern science and technology to the mastery of nature for relieving human suffering and enhancing human life. In support of this project, Descartes offers a distinctly modern vision of human beings scientifically constructing and technologically manipulating nature so that they can become like God.

LARRY ARNHART

SEE ALSO *Nature; Newton, Isaac; Scientific Revolution.*

BIBLIOGRAPHY

Asimov, Isaac. (1950). *I, Robot*. New York: Doubleday.

Descartes, René. (1988). *Descartes: Selected Philosophical Writings*, trans. John Cottingham, Robert Stoothoff, and Dugald Murdoch. Cambridge, UK: Cambridge University Press.

Descartes, René. (2003). *Treatise of Man*, trans. Thomas Steele Hall. Amherst, NY: Prometheus Books. Originally published, 1664.

Hadamard, Jacques. (1954). *An Essay on the Psychology of Invention in the Mathematical Field*. New York: Dover.

Kuhn, Thomas S. (1996). *The Structure of Scientific Revolutions*, 3rd edition. Chicago: University of Chicago Press.

Lachterman, David Rapport. (1989). *The Ethics of Geometry*. New York: Routledge. Study of Descartes' modern idea that mathematics is a god-like activity of construction.

La Mettrie, Julien Offray de. (1994). *Man a Machine*, trans. Richard A. Watson and Maya Rybalka. Indianapolis, IN: Hackett. Originally published, 1748.

Lampert, Laurence. (1993). *Nietzsche and Modern Times: A Study of Bacon, Descartes, and Nietzsche*. New Haven, CT: Yale University Press. Shows how Descartes contributed to modern break from Socratic philosophy and biblical religion.

Lange, Frederick Albert. (1925). *The History of Materialism*, 3rd edition, trans. Ernest Chester Thomas. London: Routledge and Kegan Paul. Demonstrates the importance of Cartesian mechanism in the history of materialism.

Merchant, Carolyn. (1980). *The Death of Nature: Women, Ecology, and the Scientific Revolution*. San Francisco: Harper and Row.

Nietzsche, Friedrich. (1968). *Basic Writings of Nietzsche*, trans. and ed. Walter Kaufmann. New York: Random House.

Polanyi, Michael. (1958). *Personal Knowledge*. Chicago: University of Chicago Press.

Popper, Karl R., and John C. Eccles. (1977). *The Self and Its Brain: An Argument for Interactionism*. New York: Springer International. Scientific and philosophical arguments for immaterial soul interacting with material brain.

Watson, Richard A. (2002). *Cogito, Ergo Sum: The Life of René Descartes*. Boston: David R. Godine.

DES (DIETHYLSTILBESTROL) CHILDREN

• • •

The scientific world was shocked by the 1971 discovery of the devastating effects in young women of a drug, diethylstilbestrol (DES), taken by their mothers twenty years earlier. The story of DES, from its discovery and widespread marketing without adequate testing or proof of efficacy, to the banning of its use by pregnant women, provides a good example of the serious harm that can result from inadequately protective regulation of new drugs and technologies.

Historical Development

In 1938, Sir E. Charles Dodd formulated DES, the first orally active, synthetic estrogen. This (nonsteroidal) estrogen, estimated to be five times as potent as estradiol, was very inexpensive and simple to synthesize. Because it was not patented, the developing pharmaceutical industry quickly began worldwide production; it was ultimately marketed under more than two hundred brand names for a wide range of indications. DES underwent very limited toxicological testing, a fate common to pharmaceutical products at that time.

Experiments with high doses of DES in women threatening to abort were conducted a few years later. The use of DES for prevention of miscarriage was promoted by the work of Drs. Olive and George Smith, who conducted multiple (uncontrolled) trials of DES for use in pregnancy throughout the 1940s. Despite limited evidence of safety or efficacy, the drug was deemed effective for this purpose and safe for mother and fetus. In 1947, DES obtained market approval in the United States for use in pregnancy in cases of threatened abortion and hormonal inadequacy.

Following the first poorly supported claims of the effectiveness of DES for the prevention of miscarriage, several studies were carried out to assess its efficacy, with mixed results. As these studies became more rigorous, support for the use of DES declined. In 1953, W.J. Dieckmann and colleagues demonstrated the lack of efficacy when DES was compared to a placebo in a randomized trial of pregnant women. Although the authors concluded that DES was ineffective, the drug continued to be prescribed even to women without previous pregnancy problems or evidence of threatened pregnancy. A reanalysis of Dieckmann's data in 1978, which showed that DES actually increased the risk of miscarriage, noted that had the data been properly analyzed in 1953, nearly twenty years of unnecessary exposure to DES could have been avoided.

The dangers of DES were not discovered, however, until 1971. Dr. A.L. Herbst and colleagues identified seven cases of a rare vaginal cancer (vaginal clear cell adenocarcinoma) in a single hospital. Using a case-control study they linked this rare cancer to the young women's prenatal exposure to DES. The results were so overwhelming that the Food and Drug Administration (FDA), in its November 1971 bulletin, declared that DES was contraindicated for use in pregnancy. Subsequent data demonstrated DES to be teratogenic as well as carcinogenic, and showed extensive damage to the reproductive systems of both men and women who had been exposed prenatally.

Elsewhere in the world, DES continued to be sold to pregnant women, in some countries into the 1980s. The fact that DES was prescribed for so long after its lack of efficacy had been demonstrated and dangers recognized illustrates a massive drug system failure.

In fact, it was not the lack of efficacy that triggered the end of marketing of DES for use in pregnancy, but a fortuitous accident. The cancer that DES caused in young women is extremely rare. It is estimated to have occurred in less than one in a thousand exposed daughters. If the cancer cases originally detected by Herbst and his colleagues had been diagnosed in several different medical centers, rather than at a single hospital (Massachusetts General Hospital, where DES use had been high as the site of the Smiths early experiments, the dangers of DES might well have gone unrecognized. Thus, this cancer, its link to DES, and other consequences of DES exposure might well have gone undetected.

DES Case Lessons

The DES story demonstrates that long-term and hidden effects of hormonal exposure may result from prenatal exposure, and that such consequences may be devastating. Could the mishap have been prevented? Where did science, society, and technology fail?

First, no long-term toxicity tests were ever carried out. Ironically, Dodds, the discoverer of DES, wrote in 1965, "I suppose we have to be very thankful that [DES] did prove to be such a non-toxic substance," referring to the minimal testing it underwent before marketing. Six years later the dangers of DES were identified.

Second, DES was put on the market without adequate proof of efficacy. Adequate pre-market testing would have shown that DES was never effective for the prevention of miscarriage. Therefore, a properly conducted and analyzed clinical trial might have avoided the entire episode. This accident is less likely in the early twenty-first century for pharmaceuticals, where thorough toxicity testing and evidence of efficacy are required prior to marketing.

Third, the widespread use of DES was furthered by the faith, prevalent at the time, in the advances of science and human abilities to control nature. DES was believed to be safe and effective, and both "modern and scientific." Its use became fashionable and there was pressure on physicians from peers and patients to prescribe DES. In the Netherlands, for example, the use of DES was aided by endorsement of the Queen's gynecologist.

Pharmaceutical retailers and advertising promoted the effectiveness and safety of DES to doctors and consumers. In fact, some manufacturers promoted it as a panacea for use in all pregnancies. The eagerness of the pharmaceutical companies to sell this profitable, unpatented product was compounded by the failure of medical and regulatory agencies to react rapidly to the emerging evidence.

Even prior to marketing for use in pregnancy, DES was a known animal carcinogen, a suspect human carcinogen, and a drug that had been shown to produce observable changes in the offspring of women exposed in pregnancy. Moreover, after DES was proven to be ineffective for use in pregnancy in 1953, a review of its risks and benefits should have resulted in immediate contraindication of this use. Had DES been withdrawn for use in pregnancy at that time, the unnecessary and tragic exposure of millions of mothers, sons, and daughters could have been avoided.

Regulatory authorities are also more alert in the early 2000s to reporting of adverse drug reactions and more inclined to take action than they were in the 1960s and 1970s. However, it should be remembered that regulation of non-pharmaceuticals is far from rigorous, and prenatal exposure to non-pharmaceuticals may also convey serious risk. The DES lesson can serve to raise consciousness about the dangers of inadequately identifying those risks.

DOLORES IBARRETA
SHANNA H. SWAN

SEE ALSO *Abortion; Drugs; Medical Ethics.*

BIBLIOGRAPHY

Giusti, Ruthann M.; Kumiko Iwamoto; and Elizabeth E. Hatch. (1995). "Diethylstilbestrol Revisited: A Review of the Long Term Health Effects." *Annals of Internal Medicine* 122: 778–788.

Ibarreta, Dolores, and Shanna H. Swan. (2002). "The DES story: Long-term consequences of prenatal exposure." In *The Precautionary Principle in the 20th Century: Late Lessons From Early Warnings*, ed. Poul Harremoës, David Gee, Malcolm MacGarvin, et al. London: Earthscan.

Swan, Shanna H. (2001). "Long term human effects of prenatal exposure to diethylstilbestrol." In *Hormones and Endocrine Disruptors in Food and Water: Possible Impact on*

Human Health, ed. Anna-Maria M. Andersson, Kenneth N. Grigor, Eva Rajpert de Meyts, et al. Copenhagen: Munksgaard.

INTERNET RESOURCE

National Cancer Institute. (1999). "DES Research Update 1999: Current Knowledge, Future Directions." Available from http://planning.cancer.gov/whealth/DES/.

DESERTIFICATION

SEE *Deforestation and Desertification*.

DESIGN ETHICS

• • •

Design ethics concerns moral behavior and responsible choices in the practice of design. It guides how designers work with clients, colleagues, and the end users of products, how they conduct the design process, how they determine the features of products, and how they assess the ethical significance or moral worth of the products that result from the activity of designing. Ethical considerations have always played a role in design thinking, but the development of scientific knowledge and technology has deepened awareness of the ethical dimensions of design. As designers incorporate new knowledge of physical and human nature as well as new forms of technology into their products, people are increasingly aware of the consequences of design for individuals, societies, cultures, and the natural environment.

The design arts are important because they are the means by which scientific knowledge and technological possibilities are converted into concrete, practical form in products that serve the needs and desires of individuals and communities. Design is difficult to define because of its breadth of application. One can discuss the design of scientific experiments, of theories of nature and society, of political systems and individual actions, of works of fine art, and of the everyday products created by engineering and the other useful or practical arts. In all of these examples, design may be described generally as the art of forethought by which society seeks to anticipate and integrate all of the factors that bear on the final result of creative human effort.

Descriptive definitions have a useful place in explaining the nature of design for a general audience— for example, "design is the art of forethought," "design

is planning for action," "design is making things right." However a formal definition has the advantage of bringing together all of the causes or elements of design in a single idea so that their functional relationships are clear, and provides a framework for distinguishing and exploring the ethical dimensions of design. The following formal definition serves present purposes: *Design is the human power of conceiving, planning, and bringing to reality all of the products that serve human beings in the accomplishment of their individual and collective purposes.* There are four ethical dimensions represented in this definition, each identifying an area of ethical issues and potential moral conflict that often complicates the activity of designing but also enhances the value of the designer's work. These dimensions represent the web of means and ends that are the central concern of ethics and moral conduct in design.

Character and Personal Values

The first ethical dimension of design arises from the human power or ability to design. One may reasonably argue that design itself is morally neutral because the art is only an instrument of human action. However designers are not morally neutral. They possess values and preferences, beliefs about what is good and bad for human beings, and an array of intellectual and moral virtues or vices that constitute personal character. The power or ability to design is embedded in a human being, within the character of the designer. Personal accounts, written statements, manifestos, and biographies are the beginnings of the study of ethics in design. They provide direct and indirect evidence of individual character and personal values, and often include accounts of the moral dilemmas and decisions that individuals have made in the course of their careers. Thus the first ethical dimension of design is the character and personal morality of the designer.

Integrity of Performance

A second ethical dimension arises from the activity of conceiving, planning, and bringing products to reality. These activities are the immediate goal or purpose of design. The standard of performance demonstrates fidelity to the art of design itself and is a matter of personal and professional integrity. In the film *The Bridge on the River Kwai* (1957), a British colonel and his fellow prisoners of war are instructed by their Japanese captors to build a railway bridge for the transportation of troops and munitions. For the colonel, constructing the best bridge—*a proper bridge*— is a matter of personal and professional integrity, and he pushes his men harder than their captors to complete the work on schedule. The tragedy of his narrow commitment emerges at the end of the film when the colonel realizes that his obsession with achieving the immediate goal of professional performance in the prison camp conflicts profoundly with the ultimate goal of his service in the British army. Ultimate goals are another ethical dimension of design to be considered later, but this film, while a work of fiction, effectively illustrates the second ethical dimension of design.

Performing well raises other closely related ethical issues. Designers are responsible for relationships with others involved in performance of the art. In some cases the designer works alone and is responsible directly to a client. Ethical standards of fairness, honesty, and loyalty serve to guide the client relationship, as in any personal or business dealing. In most cases, however, the designer works with other individuals and has shared responsibility for maintaining those relationships according to ethical standards. For example because of the increasing complexity of products, technology, and other factors, designers work in teams with fellow designers or with technical specialists from a variety of disciplines and professions. There are also new practices of participatory design in which clients and even representatives of the end users of products participate directly in the design process. Finally there is an increasing emphasis in some forms of design on user research, requiring the ethical treatment of human subjects.

Guidance in these matters comes partly from personal morality, but also from professional codes of ethics formulated and established by professional societies. Because many of the branches of design are young— some were established as professions only in the early and middle decades of the twentieth century—designers turned to already established professional associations, such as those for medicine, law, business, engineering, and architecture, for guidance on many ethical issues, including how to formulate their codes. At the beginning of the twenty-first century, designers continue to look to those professions for sophisticated practical discussions of emerging ethical issues. The codes of ethics of national organizations such as the American Institute of Architects (AIA), the Industrial Designers Society of America (IDSA), and the American Institute of Graphic Arts (AIGA) and their international counterparts have evolved gradually. They began with issues of competence, integrity, and professionalism, emphasizing ethical standards in technical practice and education, in business matters, and in compliance with laws and regulatory codes associated with safety. They expanded to include intellectual property rights and the general area of service in the public interest, such as preservation of the cultural trust and sustainability of the human community. The evolution corresponds to the successive ethical dimensions of design.

Product Integrity

A third ethical dimension, product integrity, arises from the nature of the products created through the art of design. Product integrity should be distinguished from the end purpose or worth of products. It is the synthesis of form and materials by which one judges a product to be well or poorly designed. There are specific ethical issues of product integrity for each kind of design (engineering, communication, industrial, and architectural design), but in general the issues concern safety and reliability, compliance with laws and regulatory codes, sustainability in its various aspects, and service to the public good. Products are created to serve human beings in their various activities and pursuits. Anything that directly or indirectly harms a human being or harms someone or something for which a human being is responsible presents a serious problem of product integrity requiring both technical and ethical consideration.

Because of the complex nature of human-made products, it is important to distinguish three elements of form that identify design issues as well as their associated ethical considerations. These elements concern what is *useful, usable,* and *desirable* in all products. Their successful integration is one of the fundamental challenges of design thinking.

1. Structural Integrity of Form. This element involves technological reasoning that ensures the proper performance of a product so that it is useful in supporting an activity. In some products technological reasoning means employing mechanical and electrical principles in an efficient and safe relationship. In computer software the reasoning follows logical principles and best practices of program layout in order to create efficient and reliable computation and, increasingly, security of information. In graphic or communication design, the reasoning of form and content follows more general principles for the presentation of information and arguments about the subject that the designer seeks to communicate. Honesty and truth become serious ethical issues when communication design is employed in marketing, packaging, and instructional materials. Structural integrity of the physical form and of information is the frontline of safety and reliability.

2. Usability of Form. This element requires product features such as operating controls, control surfaces, information displays, seats, doors, and panels that allow human beings to access and operate a product—or deliberately prevent dangerous access or operation of a product—and maintain it in a safe and reliable condition. In design these are sometimes called *affordances*, because they afford a human being with access to the form in the way that doors provide access to a building. By analogy one can easily see the extension of the usability features of mechanical products into software and even products of visual and verbal communication. Software is accessed by means of a user interface, meaning all of the features presented on a computer screen that allow a human being to operate and control the software. In graphics and communication design, the size of fonts, the layout of information, and similar matters allow a person to understand what is being communicated. It is more than a technical matter when, for example, bus signs and timetables are printed in font sizes that are too small for elders to read. Unfortunately usability is often seen only in terms of the immediate use or functioning of a product. In reality usability issues affect the entire lifecycle of products. Can the product be produced efficiently and safely, can it be operated effectively, can it be maintained, and can it be disassembled and disposed of or recycled safely? These are technical issues with significant ethical implications for design thinking.

3. Aesthetics of Form. This element is sometimes a puzzling subject for scientists and engineers, but for the designer it is the final element in the creation of a complete product. The aesthetic element of form makes a product desirable to possess and use. Many products that are otherwise useful and even usable are incomplete and fail to be integrated into the everyday lives of human beings because the form is not aesthetically pleasing. This is a source of confusion and consternation to inventors and developers and sometimes to policy makers who seek to influence individual and social behavior through the adoption of certain products—for example, seat belts in automobiles or products that support recycling or sustainability.

Part of the misunderstanding of aesthetics rests with the term itself. In its original and broadest meaning, aesthetics refers to the pleasurable or painful sensations that human beings feel through their senses. In this meaning all products have an aesthetic element, by accident or by design. The sound of a door closing, the texture of a control surface, the visual appearance of information in a software interface, the smell of plastics and metals, the taste of medicine: All are examples of the aesthetic element of form. Over time aesthetics has taken on a second, more restricted meaning as the study and theory of beauty. The psychological, social, cultural, and philosophical significance of aesthetics is a complex and profound subject. One way to understand the place of aesthetics in design is how it leads a human being to identify with a product. Identification with a product—to imagine a product as a desirable part of one's lifestyle and a valuable extension of the user into the world—shows how important the aesthetic element of form may be in design thinking.

The complexity of aesthetics points toward several areas of ethical issues that the designer must consider. Aesthetics plays a subtle and important role in supporting the usability of products and, hence contributes to safety and accessibility. Aesthetics also concerns the social, cultural, and even political value placed on sensations of pleasure and pain. Economic necessity plays an important role in the degree of luxury that products provide, but local community values also influence what is acceptable in making products pleasurable. Adapting products to local values is an ethical consideration for the designer and the designer's client. It is closely related to the issue of *appropriate technology,* which concerns selecting the kind of technology for a product that is suited to the economic, environmental, and social or cultural conditions of people.

There are further ethical issues surrounding beauty: what it is, its value, its use as a political instrument to affect the development of society and culture, helping to achieve the goals of one or another cultural agenda. For some there is aesthetic delight in the intelligent working of a product such as a mechanical or electronic device. The beauty of an idea realized in concrete form may itself be captivating. However this and other forms of beauty often flow from individual delight into social and political movements, taking on further ethical and moral significance. For example the so-called *modernists* of twentieth-century design believed that creating a certain kind of formal beauty in their products would have a direct effect in improving the values and behavior of people. The *good design* movement of the 1950s is a specific example. In contrast the so-called *post-modernists* of the 1980s and early 1990s used other concepts of beauty and even anti-beauty to express cultural diversity and encourage alternative aesthetic values. In both cases the aesthetics of design was associated with moral values.

In addition to ethics of product form, there are ethical issues involved in the materials employed in bringing a form to reality. Traditional and new materials present hazards that the designer has a responsibility to understand and respect. The selection of proper materials literally supports structural integrity in engineering, industrial design, and architecture. There are also ethical implications when designers make excessive use of materials or of particularly precious materials, because this may be regarded as a waste of natural resources. Similarly there are ethical issues surrounding the long-term impact of materials on human beings and on the natural environment. Developments in science and technology are a source of the problem of sustainability, and play a role in society's efforts to create sustainable communities. Many people believe that the designer and the designer's client have a newly recognized responsibility for creating products that support the goal of sustainability.

The development of science and technology has had profound impact on products and product forms, an influence that will only grow through the development of designer materials by means of biotechnology, nanotechnology, and other methods. Perhaps most importantly it has broadened the understanding of what a product of design is. At the beginning of the twentieth century, a product was regarded simply as a tangible, physical artifact, whether a consumer good or industrial machinery or medical and scientific instruments or a building. At the beginning of the twenty-first century, these product categories remain but have been the object of much elaboration. The categories of the physical have also increased to include chemical and biological products as physical artifacts that result from design thinking. Furthermore people recognize that information products, visual communications, services and processes, and even organizations are products of design thinking, subject to forethought and requiring careful, responsible decision making in their creation.

The broadening of the general understanding of what a product is comes from several factors associated with the development of science and technology. One is the concept of a system, which depends on a rational ordering or relationship of parts to achieve some goal. Rationalization and standardization now play a fundamental role in design and product development, supporting mass production and mass communication. Another factor is the development of new materials and the machines to process and shape them. Closely related to both of these factors is the development of digital technology, with scientific and industrial applications as well as applications suited to the daily lives of human beings through personal devices as well as access to information and communication through the internet. Among the many factors that have changed the understanding of what a product is, perhaps the most important, from an ethical perspective, is assessment of the consequences of the product's creation on the lives of individuals, society, and the natural world. This has come through the application of the physical and biological sciences, tracing the impact of products far beyond the marketplace (Winner 1986). It has also come through the development and application of the psychological and social sciences. Base-line efforts in these sciences during the twentieth century have resulted in the gathering of information that allows informed discussion of social policy and the philosophical implications of science, technology, and design.

Ethical Standards and the Ultimate Purpose of Design

A fourth ethical dimension of design arises from the service nature of the design arts, and presents some of the most difficult ethical issues designers face. The design arts are fundamentally a practical service to human beings in the accomplishment of individual and collective purposes. That is, the end purpose of design is to help other people accomplish their own purposes. This is where the personal character and morality of the individual designer, as well as the other ethical dimensions of design, are inevitably placed in a larger social, political, religious, and philosophical context. What is the moral significance of the particular purposes that designers are asked to serve? What is the moral worth of particular products that seek to achieve these purposes? What consequences will products have for individuals, society, and the natural environment in the short and long terms? What ethical standards can designers employ in making decisions about the proper use of design?

Ethical guidance in these matters comes from several sources including personal morality, professional organizations, the institutions of government, religious teachings, and philosophy. The potential for moral conflicts and dilemmas is so great that in this fourth ethical dimension the ethical problems of design are essentially the same as the ethical problems of citizenship and practical living in general. It is difficult to distinguish design from politics, political science, and political philosophy. This reaffirms Aristotle's treatment of ethics and politics: They do not address different subject matters but the same subject matter from different perspectives.

Nonetheless there are grounds for continuing to treat design ethics as a distinct problem with a distinct perspective on individual and social life. For example the natural and social sciences study what already exists in the world, but design seeks to create what is possible and does not yet exist—design is concerned with invention and innovation and, generally, with matters that may be other than they are through human action. This is the basis for Herbert A. Simon's treatment of design as *the sciences of the artificial.* Whether one refers to design as an art or a science, most designers would agree with Simon that design is a systematic discipline involving choices that are "aimed at changing existing situations into preferred ones" (Simon 1981, p. 129). One implication has special significance for ethics. Following other philosophers, Caroline Whitbeck has observed that the traditional discourse of ethics tends to emphasize making moral judgments—the critique or evaluation of actions already taken. In contrast she argues that ethics may be considered from the perspective of the moral agent seeking to devise ethical courses of action (Whitbeck 1998). This argument—that ethics itself is a form of designing—is directed primarily toward the ethics of professional conduct, how designers relate to supervisors and clients, and how designers or any one else may respond creatively and responsibly to ethical and moral problems in their work.

The argument may be expanded in a direction that many designers would acknowledge: Not only is ethics a form of designing, but designing is a form of ethics. One aspect of the designer's creativity and responsibility is to devise ethical courses of action that navigate the moral dilemmas of practical life. This happens in the normal course of the design process when, for example, the designer studies the client's brief or charge and finds it inadequate or inappropriate for solving the problem that may be the real concern of the client. This leads to a rethinking and recasting of the initial purpose set by the client, often reached through negotiation over the nature of the product to be created.

In a broader sense, moral issues are addressed when the designer employs clear and well-articulated ethical standards in making decisions about the proper use of design in any particular situation. There is no single set of ethical standards in the field of design; the pluralism of the human community in general is mirrored in the design community in particular. However there are distinct ethical positions in the discussions of designers, and they bear a recognizable relationship to positions in the tradition of formal ethical theory. Two of these positions point toward a natural foundation of design ethics, and two others point toward conventional and arbitrary foundations established by human beings.

Designers whose ethical position is grounded on a natural foundation typically argue that the products of design should be *good,* in the sense that they affirm the proper place of human beings in the spiritual and natural order of the world. This position finds its strongest premises in spiritual teachings and some forms of philosophy (Nelson 1957). Alternatively they argue that products should be *appropriate and just,* in the sense that they are appropriate for human nature and the physical and cultural environment within which people live, and that they support fair and equitable relationships among all human beings. This position finds its strongest premises in human dignity and the development of human rights, encompassing civil and political rights, economic rights, and cultural rights (Buchanan 2001).

Designers whose ethical position is grounded on conventional and arbitrary foundations typically argue that products should satisfy the *needs and desires of human beings* within acceptable constraints. The constraints at issue are the simply conventional expectations of a community and what is considered normal in the physical, psychological, and social condition of human beings in a particular time and place. The strongest premises are drawn from the study of manners, taste, and prevailing laws, and by scientific study of what is normal and abnormal in the body and mind. Alternatively various designers argue that products are merely *instrumental,* in the sense that they are useful in enabling human beings to achieve any of their wants and desires, limited only by the power of individuals and the state to curb willfully destructive actions and turn creativity in acceptable directions. This position draws its strongest premises from the concept of the *social contract,* upon which it is argued that any state is created.

As observed earlier, the development of scientific knowledge and technology has had a profound effect on human understanding of the nature and consequences of the products created by the design arts, deepening consciousness of the ethical dimensions of design. Additionally the development of design thinking has made important contributions to discussions of science, technology, and ethics. Nowhere is this more evident than in the central concern of design to humanize technology and place the advancement of scientific knowledge in the context of practical impact on human life. The contributions are typically made through the concrete expression of design thinking in real products that influence daily life rather than through writing about design. As designers have ventured out from traditional products and product forms, their explorations and experiments in creating new products have provided the concrete cases that focus discussion of ethical issues and the limits of science and technology. In many instances, the design arts have been deliberately employed to provoke critical debate in the general public about the place of science and technology in community life.

Toward an Ethical History of Design

An ethical history of design would present the origins and development of design from the perspective of designers as moral agents, tracing the successive issues and ethical dimensions of design as they have arisen through individual and collective action. Such a history has not yet been written or even attempted because the formal study of ethics has received little attention among designers and scholars of design studies. Indeed there are grounds for arguing that the formal study of ethics in the philosophy of design began no earlier than the mid-1990s, with the publication of articles by authors such as Alain Findeli and Carl Mitcham. Mitcham's "Ethics into Design" draws from philosophical discussions of ethics, the philosophy of technology, and the development of ethics in engineering. He argues that the two traditions of design in the twentieth century—design as art and aesthetic sensitivity and design as science and logical process—"must be complemented by the introduction of ethics into design, in order to contribute to the development of a genuinely comprehensive philosophy of design" (Mitcham 1995, p. 174). Mitcham's essay is important because it gives disciplined philosophical focus to the many discussions of ethics, politics, and morality that have shaped design since the beginning of the twentieth century.

Several such discussions have made important contributions in opening up new lines of thinking. In the late-nineteenth century, the political writings of William Morris (1834–1896) introduced ideas about socialism that helped to shape the arts and crafts movement and questioned the value of industrialization. The documents of the Bauhaus in Germany—for example, the essays included in *Scope of Total Architecture* (1962) by Walter Gropius (1883–1969)—helped to set the moral agenda of modernism. Artist Laszlo Moholy-Nagy's (1895–1946) *Vision in Motion* (1947) developed these ideas further and contributed to a form of humanism in design. Work at the Ulm school of design, particularly under the influence of the *Frankfurt School* of social theory, showed a struggle between sociopolitical questioning and the introduction of scientific methods into the

design process. The writings of George Nelson (1908–1986) elevated discussions of *good design* to a higher moral concern for the responsibilities of the designer and true good in products. Kenji Ekuan's *Aesthetics of the Japanese Lunchbox* (1998) offered a Buddhist perspective on issues of ethics and morality in product design. Victor Papanek's *Design for the Real World: Human Ecology and Social Change* (1984) and *The Green Imperative* (1995) introduced the ideas of appropriate technology and sustainability to design thinking. In *Cradle to Cradle* (2002), William McDonough and Michael Braungart extend the theory of sustainability in a controversial discussion of industrial design and architecture. Beginning in 1982, the journal *Design Issues: History, Criticism, Theory* provided a venue for some of the most important discussions of design ethics. Authors such as Alain Findeli, Richard Buchanan, Ezio Manzini, Tony Fry, and Victor Margolin addressed practical as well as philosophical issues surrounding design ethics, and their work poses a challenge for a new generation of students of design. The continuing pace of scientific and technological development and the growing sophistication of reflections on design, supported by new doctoral programs and research in many universities, suggest that design ethics will become a progressively more important subject.

RICHARD BUCHANAN

SEE ALSO *Architectural Ethics; Building Codes; Building Destruction and Collapse; Engineering Design Ethics; Engineering Ethics; Participatory Design.*

BIBLIOGRAPHY

Attfield, Judy. (1999). *Utility Reassessed: The Role of Ethics in the Practice of Design.* Manchester, England: Manchester University Press.

Borgman, Albert (1987). *Technology and the Character of Contemporary Life: A Philosophical Inquiry.* Chicago: University of Chicago Press.

Buchanan, Richard. (2001). "Human Rights and Human Dignity: Thoughts on the Principles of Human-Centered Design." *Design Issues* 17(3)(Summer): 35–39.

Ekuan, Kenji. (1998). *Aesthetics of the Japanese Lunchbox,* ed. David B. Stewart, trans. Don Kenny. Cambridge, MA: MIT Press.

Findeli, Alain. (1994). "Ethics, Aesthetics, and Design." *Design Issues* 10(2)(Summer): 49–68.

Gropius, Walter. (1962). *Scope of Total Architecture.* New York: Collier Books.

Harries, Karsten. (1997). *The Ethical Function of Architecture.* Cambridge, MA: MIT Press.

Lindenger, Herbert, ed. (1991). *Ulm Design: The Morality of Objects*, trans. David Britt. Cambridge, MA: MIT Press.

Manzini, Ezio. (1995). "Prometheus of the Everyday: The Ecology of the Artificial and the Designer's Responsibility," trans. John Cullars. In *Discovering Design: Explorations in Design Studies*, eds. Richard Buchanan and Victor Margolin. Chicago: University of Chicago Press.

Margolin, Victor. (2002). *The Politics of the Artificial: Essays on Design and Design Studies*. Chicago: University of Chicago Press.

McDonough, William, and Michael Braungart. (2002). *Cradle to Cradle: Remaking the Way We Make Things*. New York: North Point Press.

Mitcham, Carl. (1995). "Ethics into Design." In *Discovering Design: Explorations in Design Studies*, eds. Richard Buchanan and Victor Margolin. Chicago: University of Chicago Press.

Moholy-Nagy, Laszlo. (1947). *Vision in Motion*. Chicago: Paul Theobald.

Molotch, Harvey (2003). *Where Stuff Comes From*. London: Routledge.

Nelson, George. (1957). *Problems of Design*. New York: Whitney.

Papanek, Victor. (1984). *Design for the Real World: Human Ecology and Social Change*, 2nd edition. New York: Van Norstrand Reinhold.

Papanek, Victor. (1995). *The Green Imperative: Natural Design for the Real World*. London: Thames and Hudson.

"Responsible Design: Managing the Ethical Choices in the Design Process." (1991). *Design Management Journal* 2(4)(Fall): 6–84.

Simon, Herbert A. (1981). *The Sciences of the Artificial*, 2nd edition. Cambridge, MA: MIT Press.

van Toorn, Jan. (1994). "Design and Reflexivity." *Visible Language* 28(4)(Autumn): 317–325.

Whitbeck, Caroline. (1998). *Ethics in Engineering Practice and Research*. New York: Cambridge University Press.

Winner, Langdon (1986). *The Whale and the Reactor: A Search for Limits in an Age of High Technology*. Chicago: University of Chicago Press.

OTHER RESOURCE

Bridge on the River Kwai. (1957). Directed by David Lean. Columbia TriStar Pictures.

DESSAUER, FRIEDRICH

•••

Friedrich Dessauer (1881–1963) was born in Aschaffenburg, Germany, on July 19, and died in Frankfurt am Main on February 16. He led an active life as an inventor, entrepreneur, politician, theologian, and philosopher who put forth a strong ethical justification of technology as being even more significant than science. On the basis of his experience with technological creativity Dessauer argued that the act of invention goes beyond appearance to provide contact with Kantian things-in-themselves and, in theological terminology, realizes the *imago dei* in which human beings have been created.

Early in his life Dessauer became fascinated with Wilhelm Röntgen's (1845–1923) discovery of X-rays (1895), which promised a penetration of appearances, and his design of high-energy X-ray power supplies earned him a doctorate in 1917. As an inventor and entrepreneur he developed techniques for deep-penetration X-ray therapy in which weak rays are aimed from different angles to intersect at a point inside the body where their combined energy can be lethal to a tumor while having less of an effect on the surrounding tissues. While continuing his work in biophysics, after 1924 Dessauer was a Christian Democratic member of the Reichstag until he was forced to leave Germany in 1933 because of his anti-Nazi stance. After World War II Dessauer returned to lead the Max Planck Institute for Biophysics until he died from cancer brought on by X-ray burns incurred during his experimental work.

Beginning in the 1920s, Dessauer also pursued a wide-ranging intellectual dialogue about the meaning of modern technology. Especially in *Philosophie der Technik* (1927) and *Streit um die Technik* (1956), Dessuaer defended a Kantian and Platonic theory of technology. In the *Critique of Pure Reason* Immanuel Kant (1724–1804) had argued that scientific knowledge is limited to appearances (the phenomenal world) and unable to grasp "things-in-themselves" (noumena). Subsequent critiques of moral reasoning and aesthetic judgment required the positing of a "transcendent" reality but precluded direct contact with it. In his "fourth critique" of technological making Dessauer argued for existential engineering contact with noumena:

> The Platonic idea descends into the imagination, recasting it. The airplane as thing-in-itself lies fixed in the absolute idea and comes into the empirical world as a new, autonomous essence when the inventor's subjective idea has sufficiently approached the being-such of the thing. …[And] it is possible to verify … [that] the thing-in-itself … has been captured [when] the thing works. (Dessauer 1927, p. 70)

Invention creates "real being from ideas," that is, engenders "existence out of essence" (Dessauer 1956, p. 234).

In conjunction with this metaphysics Dessuaer further articulated a moral assessment of technology that went beyond a simple consideration of practical benefits or risks. The autonomous, world-transforming consequences of modern technology bear witness to its transcendent moral value. Human beings create technologies, but the results, resembling those of "a mountain range, a river, an ice age, or a planet," extend creation.

> It is a colossal fate, to be actively participating in creation in such fashion that something made by us remains in the visible world, continuing to operate with inconceivable autonomous power. It is *the greatest earthly experience of mortals* (Dessauer 1927, p. 66).

For Dessauer invention is a mystical experience.

Although seldom stated as forthrightly as Dessauer put it, this view of technological activity as a supreme participation in the dynamics of reality arguably has influenced the ethos of cutting-edge engineering practice, as is discussed in David Noble's *The Religion of Technology* (1997). It is a view that merits more conscious examination in terms of both its strengths and its weaknesses than it has received.

CARL MITCHAM

SEE ALSO *Engineering Ethics; German Perspectives.*

BIBLIOGRAPHY

Dessauer, Friedrich. (1927). *Philosophie der Technik: Das Problem der Realisierung* [Philosophy of technology: the problem of realization]. Bonn: F. Cohen. Dessauer's most important book. Partial English version: "Philosophy in Its Proper Sphere," trans. William Carroll, in *Philosophy and Technology*, ed. by Carl Mitcham and Robert Mackey. New York: Free Press, 1972.

Dessauer, Friedrich. (1956). *Streit um die Technik* [The controversy of technology]. Frankfurt am Main: Klett. A revised and expanded edition of *Philosophie der Technik* (1927).

Noble, David. (1997). *The Religion of Technology: The Divinity of Man and the Spirit of Invention.* New York: Knopf. Criticizes religious faith in technology.

DETERMINISM

•••

Philosophical questions about determinism involve the nature of the causal structure of the world. Given the occurrence of some factor or factors C that cause an effect E, could E have turned out otherwise than it did? Determinists answer *no*: In a strictly deterministic world all things happen by necessity, as a direct function of their causal antecedents. Indeterminists hold that E might not have occurred, even with exactly the same initial conditions, because of the possibility of true randomness or free will.

General Forms of Determinism

Early religious versions of determinism were based on the belief that people's lives are supernaturally ordained. As exemplified in the tale of Oedipus, even actions taken to try to avoid what the gods have in store turn out to be the means of sealing that destiny. Predestinarianism, a view held by some Christian sects, states that God controls and foreordains the events of human lives so that it is determined in advance whether one will go to heaven or hell. A related view holds that determinism follows from God's omniscience; if the future is undetermined, God cannot be said to be all-knowing. Modern forms of determinism dispense with supernatural beings and hold that invariable laws of nature fix events.

Determinism sometimes is defined in terms of predictability. The philosopher Karl Popper (1902–1994) called this "scientific" determinism. In a commonly performed thought experiment one imagines a Cartesian demon who knows all the laws of nature and the complete, precise state of the world at some time T; if the world is strictly determined, the demon can use that information to predict any future or past event with any degree of accuracy. Real scientists lack perfect theories and perfect data, and so imperfect prediction in practice does not by itself speak against predictability in principle. (Prediction is still possible in an indeterministic world, but only probabilistically.) Classical Newtonian physics typically is thought to describe a deterministic world—though John Earman (b. 1942) identifies a possible exception) as does relativistic Einsteinian physics.

How is determinism relevant to ethics? Some philosophers argue that if universal determinism is the case, morality is impossible because personal ethical responsibility requires the possibility of free action: One cannot be blamed or praised for doing something if one could not have done otherwise. Such incompatibilists hold that morality requires undetermined free will. Compatibilists argue that morality is possible even in a deterministic world. Some go further and hold that the kind of free will that is essential to morality actually requires determinism. If the world is indeterministic and people's actions result from mere chance, people are no more moral than a flipped coin.

Specific Forms of Determinism

Even if one sets aside such global issues, questions about determinism remain ethically significant at other levels of explanation. Various specific forms of determinism posit one or another causal factor as the driving force of change in human life and can be considered separately.

Is biology destiny? Explaining the social roles and behavior of men and women by reference to their sex, for example, is a common form of biological determinism. To specify further that genes are the ultimate biological determinant is genetic determinism. Are all human behaviors, thoughts, and feelings determined by basic characteristics of human nature and individual past experiences? Psychological determinism was a basic assumption of the psychologist Sigmund Freud's (1856–1939) psychoanalytic theory, which held that nothing that human beings do is ever accidental but instead is the result of the forces of the unconscious. The nature versus nurture debate (e.g., regarding the cause of sexual orientation) often is couched in terms of a choice between biological determinism and social determinism.

Other forms of social determinism include economic determinism: the view that economic forces are the fundamental determinants of social and political change. This thesis commonly is attributed to the political philosophers Karl Marx (1818–1883) and Friedrich Engels (1820–1895), though their thesis was more focused, stating that the mode of production determines social consciousness. They argued that because the material forces of production are given at a certain stage in history and people have no choice about whether to enter into such relations of production, the broad structure of people's social, political, and intellectual life is set by forces beyond their control.

Technological determinism tries to explain human history in terms of tools and machines. In a classic example a simple advance in cavalry technology—the stirrup—changed military and political history. However, many people consider this to be too narrow a conception, arguing that technology properly includes the entirety of material culture or even nonmaterial technologies such as knowledge and processes. In reaction against this view advocates of cultural determinism or the related view of social constructivism emphasize that technology itself is human-made and carries the imprint of the social and historical circumstances that formed it.

One could extend this list of midlevel determinist theses, with each thesis being distinguished by a claim that some causal factor determines some general, social effect. All such determinist theses come in stronger or weaker versions, depending on the claimed autonomy of the cause. A hard technological determinist, for example, would argue that technology develops by its own internal laws with a one-way effect on social structures, whereas a soft technological determinist would allow that the development and influence of technology could be mediated by other factors.

This issue sometimes is conflated with questions about reduction. Strictly speaking, reduction is the explanation of one thing in terms of another (typically though not necessarily its components) with no implication of exclusivity. However, one sometimes speaks derogatorily of an explanation as being "reductionistic" when a factor is claimed to determine something without acknowledging other causes.

Qualifications

With the accumulation of scientific evidence and the advance of technology it is possible to modify assessments of particular determinist theses. For instance, it is not a foregone conclusion that the world is fully deterministic. Indeed, evidence from quantum mechanics indicates that chance processes are a part of the causal structure of the world. Some ethicists, such as Robert Kane (b. 1938), have argued that quantum indeterminacy is what allows the possibility of human free will. By contrast, evidence from biology, psychology, and cognitive science that reveals causes of behavior, thoughts, and feelings may be taken to weaken the plausibility of free will. Even these very general issues can play a role in discussions of practical ethical matters, such as penal policy.

Midlevel determinist theses may have other ethical implications. For instance, as science identifies some causal factor as a determinant of social change or another ethically salient effect, people acquire (or lose) moral responsibility for such effects to the degree that they can control (or not control) the cause. Thus, to the degree that in (re)making technology people (re)make the world, people bear a responsibility to make ethical choices about what forms of technology to pursue or reject. The philosopher Herbert Marcuse (1898–1979), for instance, argued that technology dominates all other forms of control and that although people designed machines to free themselves, those machines often determine people's lives for the worse. If this is the case, their value should be reexamined. Similarly, people may have a responsibility to pursue technologies that would improve their lives. The debate over genetic engineering and other biotechnologies involves all these issues. If it is possible to reengineer human nature, should that be done?

The global questions about the relationship between universal determinism, indeterminism, free will, and morality remain paradoxical. However, advances in science and philosophy can help resolve questions about midlevel determinist hypotheses. For instance, one can sort out many issues by moving from a simple two-place parsing of the causal relation C causes E to a four-place analysis: C causes E in situation S, relative to some alternative *a* (CaSE). This analysis recognizes that there are always multiple causal factors that produce a given effect and places that people choose not to focus on for a particular question—a pragmatic matter—in "situation S." (The specified alternative does not contribute to the effect but provides a baseline against which to measure whether C's effect is positive or negative and to what degree.)

This model makes it clear that no single factor determines an effect by itself and that an effect can have multiple explanations, all equally legitimate and objective, depending on the (pragmatically delimited) situation. For instance, it is reasonable to say that a trait is determined by a gene only if specific environmental factors are taken as given. Thus, the thesis of genetic determinism is seen to be incorrect if it is taken in an exclusive reductionistic sense, though it can be correct in particular cases (that is, if scientific evidence shows that a particular gene causes effect E in a given environmental situation) in the same way that the environment can be said to determine the effect (if science shows that it is an explanatory causal factor of E, given a set genetic situation). This more fine-grained causal analysis allows a more precise assessment of determinist theses and thus a better moral evaluation.

ROBERT T. PENNOCK

SEE ALSO *Autonomous Technology; Evolutionary Ethics; Freedom; Free Will.*

BIBLIOGRAPHY

Dennett, Daniel C. (2003). *Freedom Evolves.* New York: Viking.

Earman, John. (1986). *A Primer on Determinism.* Dordrecht, The Netherlands, and Boston: D. Reidel.

Libet, Benjamin; Anthony Freeman; and Keith Sutherland, eds. (1999). *The Volitional Brain: Towards a Neuroscience of Free Will.* Thorverton, UK: Imprint Academic.

Kane, Robert. (1996). *The Significance of Free Will.* Oxford and New York: Oxford University Press.

Marcuse, Herbert. (2000). "The Problem of Social Change in the Technological Society." In *Towards a Critical Theory of Society: Collected Papers of Herbert Marcuse,* Vol. 2, ed. Douglas Kellner. New York: Routledge.

Marx, Karl and Frederick Engels. (1948 [1848]). *The Communist Manifesto,* ed. and annotated by Frederick Engels. New York Labor News Co.

Obhi, Sukhvinder S., and Patrick Haggard. (2004). "Free Will and Free Won't." *American Scientist* 92(4): 334–341.

Popper, Karl. (1991). *The Open Universe: An Argument for Indeterminism.* London: Routledge.

DEVELOPMENT

SEE *Change and Development.*

DEVELOPMENT ETHICS

• • •

Since the mid-twentieth century development has been promoted as the process of overcoming the condition of deprivation that prevails in many regions of the world. Underdevelopment is, correspondingly, a situation from which people and governments want to remove themselves, using science and technology to increase efficiency and generate innovations in the production of goods and services. Social science plays a crucial role in explaining the causes of and finding solutions to underdevelopment.

Development discourse often acts like an ideology, either as an uncritical recipe for all kinds of social ills or as a way of justifying policies that benefit the powerful while speciously purporting to aid the poor. In their 1992 work *The Development Dictionary: A Guide to Knowledge as Power* Wolfgang Sachs, Ivan Illich, Vandana Shiva, Arturo Esteva, and others recommend dropping development discourse altogether as being part of a project based on the quantitative and global instead of the qualitative and local. They also consider development to be an imposition from outside and above. As an example, they explain that the countries dominated by the United States after World War II only became *underdeveloped* when Harry S. Truman in his 1949 inaugural speech announced a program aimed at improving what he called underdeveloped areas. Before that the label did not exist.

But the distinction between the two kinds of countries was already in place. Some were rich, powerful, and dominant; others were—and continue to be—poor, weak, and dependent. By using the categories of imperialism and neocolonialism instead of development and underdevelopment, Marxists point to the historical roots of the difference, although political strategies to fight neocolonial relations are obviously not the same as development plans, and success in the first aspect does not guarantee success in the second.

There is no controversy as to the description of underdevelopment in terms of lack of food, shelter, education, health care, job opportunities, rule of law, good governance, and political power. Developing nations—formerly known as the Third World and sometimes as the South—share similar problems although to different degrees. The countries consistently listed at the bottom of the United Nations Development Programme (UNDP) annual report suffer acutely from an overall condition of deprivation, often aggravated by civil strife and corruption. It is not coincidence that many of the countries at the top of the list were colonial powers and that *all* the nations at the bottom were colonies of those at the top until the late-twentieth century.

Defining Development

It is more difficult to define a developed country for at least three reasons. First there are several models of development. The United States and Canada, for instance, are both developed countries in the usual definition of the term. But they are not developed in exactly the same way. Their social security and health care systems operate differently and do not cover similar percentages of the population.

Second it is not contradictory to state that there are varying degrees of underdevelopment but no real development so far in the world. There is room for improvement even in countries such as Norway and Sweden with a human development index close to 1 according to the 2001 UNDP report.

Third development create new problems. Homelessness is more of a problem in countries at advanced stages of change than in societies devoted to subsistence agriculture where family ties are stronger. The connection between mass consumption and clinical depression has been documented by Yale psychologist Robert E. Lane.

Moreover the very idea of development has experienced an evolution as a consequence of both a deeper theoretical understanding of what *developed* means and because of the practical problems encountered by governments and international agencies. An asymmetry can thus be found between development and underdevelopment. Whereas underdevelopment has referred to similar facts and conditions since the term began to be used, development has taken on different meanings, so that the notion itself shows a history of development. From development as economic growth, the notion became more complex to include world peace (Pope Paul VI; growth with equity [Amin 1977]); satisfaction of basic needs (Streeten 1981); sustainable development (Brundtland Report of the World Commission on Environment and Development 1987); and development as freedom (Sen 1999), and human security measured in the index used by the Global Environmental Change and Human Security (GECHS) Project, based at the University of Victoria in Canada.

Ethics of Development

Among the most important tasks of the ethics of development is to work out an evolved notion of development and to propose alternative models to governments, international agencies, non-governmental organizations (NGOs), and communities. Louis Joseph Lebret (1959) and Denis Goulet (1965, 1971) are considered pioneers in this endeavor, and as a critical examination of the values underlying plans for social change, development ethics reached maturity when the International Development Ethics Association (IDEA) was founded in Costa Rica in 1987. IDEA has been active in this work since its inception through conferences held in the Americas, Europe, India, and Africa. Another important task is to assess technological innovation from an ethical perspective. New technologies and their implications for the well-being of humans and nature pose urgent ethical questions. Experience demonstrates that technology is a necessary condition for the improvement of human well-being, but that it can also do harm.

Harmful technologies are scientifically unsound, wasteful, unsustainable, or inappropriate for their declared purposes. Trofim Denisovitch Lysenko's agricultural methods imposed by Stalin in the Soviet Union had no scientific basis and led to widespread famine. Those opposed to Lysenko's ideas and methods were persecuted and many died in prison. China's backyard iron furnaces during the period known as the Great Leap Forward (1958–1963) were a great failure, a cause of starvation for many millions, and led to the destruction of the precious few forests remaining in China at the time. Bad technologies in principle can be corrected or abandoned as soon as their inadequacies are clearly known, but some political regimes seem reluctant to do that.

Evil technologies are designed to enslave or eliminate individuals and groups. They respond to irrational hate, lust for power, or blind ideological commitments.

Adolph Hitler's use of technology in the so-called Final Solution is an example. Torture instruments are widely used by repressive regimes, and terrorists employ different destructive technologies to wreak havoc among civilians.

Given the fact that technology can be used to do harm, it is important to discuss how to ethically assess it. According to some, technology is ethically neutral, and ethics becomes relevant only when dealing with applications of objects and processes. Responsibility would thus lie only with the users of technologies, not with the engineers who designed them. But because the possible uses and abuses of technology are already present at the design stage, questions of aims and purposes, of good and evil, arise even before artifacts come into being. This is especially true of highly specific technologies. Although a hammer can be used to drive a nail or to commit murder, electric chairs have very few possible uses. It seems contradictory to justify building a torture chamber on the grounds that some other possible benign uses may be found for it.

How technology modifies the environment is another question that can and should be answered at the design stage. Any answer implies values held either by individuals, corporations, governments, or societies. Who makes the decisions, and on what grounds, are likewise ethical issues of great importance. The ethical principles of inclusion and participation are relevant here: As a general rule, the opinions of those affected by decisions should be taken into consideration.

Underdevelopment and Asymmetrical Relations between Countries

A particular problem is posed by the asymmetry in power between developed and developing nations. Two examples of asymmetrical relations are often mentioned in this connection: the patent system and subsidies. First the patent system internationally enforced in the early-twenty-first century and as interpreted by many in developing nations and by the UNDP's *Human Development Report 2001* is so rigid that it stifles possibilities of implementing changes necessary for the improvement of conditions in developing nations. One of the consequences of the strict imposition of the patent system is to give legal status and political power to huge monopolies that render it difficult for weak countries to develop their own technologies and protect their citizens from disease and death. In this connection, the Human Development Report 2001 mentions an emerging consensus on the unfair redistribution of knowledge as a

consequence of intellectual property rights. It points out that since the late twentieth century the scope of patent claims has broadened considerably at the same time that the use of patents by corporations has become far more aggressive. Among those who may be interested in claiming patents, corporations are in the best position to do so because their focus on small improvements is geared to meet the required criteria for patenting. They also have the advantage of easy access to expensive legal advice in order to defend their patents under civil law. With such legal protection internationally enforced, companies use patent claims as a business asset to stake out their slice of the market. Although the report advocates fairness in international mechanisms for the protection of intellectual property, it also expresses concern because of the signals that the cards are stacked against latecomers. Another source of concern is the unequal relation between powerful corporations and the weak governments of developing countries. As pointed out by the UNDP report, advanced nations routinely issue compulsory licenses for pharmaceuticals and other products during national emergencies, and impose public, noncommercial use and antitrust measures. However by 2001 not a single compulsory license had been issued by a developing nation due to fear of the loss of foreign investments and the cost of possible litigation. Even the production of generic drugs is usually contested by advanced nations in trade negotiations with developing countries. Early-twenty-first century developed nations have profited enormously from the flow of information, discoveries, and inventions of previous eras and often have resorted to reverse engineering (procedures that are no longer available to developing nations because of the strict imposition of the patent system) to catch up with inventions. Yet they routinely oppose any such moves by developing nations.

Second the asymmetry among nations is also obvious in subsidies: In trade negotiations developed countries require developing nations to eliminate subsidies in the production of goods and services for export but refuse to abide by the same strictures. Marxists, dependency-theory scientists, and dependency-ethics theoreticians all denounce these unequal relations as an obstacle to the development of poor nations.

Ethics and Development Plans

Like technologies, development plans are designed for specific purposes and according to certain values, though implicit. Also as is the case in technologies, the selection of problems to be solved and the methods of solving them illustrate the values of decision makers. Those who formulate development plans often do so without consulting the people who may suffer the consequences of implementation of those plans.

Development ethics may follow two approaches. According to the first, which dates back to Plato and Aristotle and can also be found in the work of Hegel, justice is the main purpose of ethics and the state is the proper instrument by which to achieve a just society. Beyond commutative justice, in which personal differences are not taken into consideration in transactions between individuals, distributive justice aims at equality among people in unequal conditions. There must be an entity, which is greater than the individual, that is concerned with the interests of the many as opposed to the profits of the few; that entity is the state. Development ethics, in this perspective, is traditional ethics dressed in new clothes.

Some feel that a different approach is needed because there is little relation between public policies and distributive justice in modern states. They point out that politics is most often conceived of as the art of acquiring and keeping power. Rulers often stay in power by resorting to violence because they want to enrich themselves and their cronies or impose a particular ideology. Justice is the least of their concerns, and propaganda deflects the attention of the people from this fact. Consequently most people do not relate justice to the actions of the ruling classes and have many reasons to believe that governments are best described as instruments of injustice.

This view explains why ethics is often invoked against the rule of power and employed to overturn unjust laws. Because development plans in the hands of governments determined to impose a particular ideology or follow purely technocratic criteria often lead to suffering for the masses, development ethics, under this approach, is not simply traditional ethics in disguise. Rather development ethics is a critique of the unexamined ends and means that can form the basis of a new way of governing, a voice for the victims of development projects, and a call for accountability of those who consider themselves to be experts. Because development ethics risks placing too much importance on *development* and too little on *ethics*, it must have a strong theoretical foundation.

In light of the above discussion, an analysis of the connection between development and technology is useful. Development, in its social and economic sense and in the most general terms, is often conceived of as an increase in income or consumption per capita, plus social change. The first aspect is referred to as economic growth, which is easy to measure but can be used for purposes other than the improvement of conditions of the population, for instance, when a country fosters economic growth as a means to achieve military power. Social change is more difficult to define or measure, and has been described as the idea of development evolved.

The important role of technology in both aspects of development is obvious. W. W. Rostow (1987) argues that post-Newtonian science and technology are conditions for economic take-off, a means to break through the limits of per person output traditionally imposed on nontechnologically advanced societies. Technology applied to agriculture makes labor more productive, thereby preparing traditional societies for the transition to high consumption, and freeing large numbers of people for work in industry. Technology also makes possible large-scale industrial production. As an impetus for social change, advances in technology create new techniques, careers, jobs, opportunities, businesses, procedures, legislation, and even lifestyles. According to sociologist David Freeman in *Technology and Society* (1974), the social impact of technology follows four successive phases. First new technological products simplify daily tasks and chores. A pocket calculator is easier to handle than a slide rule; a word processor more versatile than a typewriter. Second job qualifications change. In the early-twenty-first century, secretaries are expected to use computers, instead of just type and file. Third allocation of authority and prestige also changes. Those who have expertise in cutting-edge technologies are in high demand and therefore make more money and enjoy greater social status than people working in older technologies. Finally values held in great esteem by society change. The values of traditional as opposed to industrial societies differ.

Thus changes in how human beings make things lead to cultural change. Even the valuation of change is subject to modification. As pointed out by Rostow, the value system of traditional societies ruled out major changes whereas modern societies incorporate the assumption that transformation and growth will occur. Commercial propaganda in high-consumption societies emphasize change as valuable in itself.

Because technology influences morality by changes in valuation, it is possible to perform an ethical analysis of social change brought about by technology. For example, a society that uses advanced technology to build weapons of mass destruction, and in which the military enjoys great prestige, is not morally the same as one that uses advanced technology to improve conditions for the poor. The fact that a technology is new, and even that it allows for greater productivity, does not mean that it is better. It may increase the gap between the rich and poor, or damage the environment. Increased productivity in agriculture due to new methods is usually associated with monoculture, whereas traditional agricultural practices, with their typical combination of different species, were safer both for human beings and for the environment.

One argument for preserving older technologies is that there is no way to tell when and how they may be needed as practical solutions in the future. In the event that certain technologies can no longer be used, knowing the old way of doing things may represent the difference between life and death. Each particular technology requires certain conditions for its functioning and more advanced technologies usually require more specific inputs. If such inputs (electricity, for instance) are not available, the ability to use alternative technologies is crucial. Because of the increasing dependency of technology on science, science is central to development. Government agencies dedicated to the promotion of scientific and technological research have existed in Latin America and other developing areas since the 1970s. The success of such agencies is not uniform, but the Latin American countries included in the 2001 UNDP report as countries with high human development (Argentina, Uruguay, Costa Rica, and Chile) also enjoy a long tradition of public support for science and technology.

Mastery of mathematics, physics, chemistry, and biology is the foundation of technological advancement. The social sciences also play an important role in developing nations. Because underdevelopment is a social condition, a scientific explanation could be found in social sciences. Development plans nevertheless tend to marginalize the importance of input from those disciplines.

Ethics of Science and Technology in Development

Before development economics existed, Francis Bacon (1561–1626) sought knowledge that could alleviate human misery. Gottfried Wilhelm Leibniz (1646–1716) struggled to develop a logical method to solve all kinds of theoretical and practical problems, which he employed in an attempt to alleviate the social ills he saw in Europe. David Hume (1711–1776) and Adam Smith (1723–1790) discussed the difference between rich and poor countries and whether it was morally desirable to bridge the gap. Their answer was affirmative.

After the Industrial Revolution in Great Britain in the late-eighteenth and early-nineteenth centuries, other countries experienced similar profound economic and social changes. In the nineteenth century, aspiration to better social conditions was summarized in the idea of *progress*. In the twentieth century, countries with different political regimes formulated and implemented far–reaching economic plans such as the Four-Year Plans in Germany and the Five-Year Plans in the Soviet Union in the 1930s, as well as Franklin D. Roosevelt's famous 100 Days, which included a number of measures designed to reverse the effects of the Great Depression. In the 1950s a clear distinction between the two kinds of countries entered the political arena, and plans were explicitly created to make change.

It became clear that development plans were not useful to large numbers of people forced to change their lives as part of the implementation of those plans. Several critics have examined the ends and means of development, the values implicit in plans, and the real beneficiaries of change. Denis Goulet (1965) devised a method to examine the choice of problems and solutions in light of values implicit and explicit. Because development ethics as conceived by authors such as Goulet and David A. Crocker aims at proposing alternative models to development, the question arises as to the feasibility of those models. Respect for cultural values is essential for these alternatives to succeed. Proponents hope that the social change brought about through such models will have a solid foundation and be, consequently, more sustainable for succeeding generations. The next generation should have at least the same natural, human-made and human capital than the previous one. If it has less than the previous generation, then development is not sustainable.

The connection of science and technology with development focuses on two questions: What, if any, is the relation between science and technology? and How do either or both relate to socioeconomic development? From the perspective of ethics, however, the basic question is not whether science and technology are subject to ethical analysis, but how science and technology can be used ethically for development.

A Development Ethics

At first glance, the answer to this question is simple. Development is morally justified when human beings are not mere *objects* in plans and projects but *subjects*, in the sense of being free agents who want to improve their condition. This position assumes that development plans are valid instruments by which to insure the universal right to an adequate standard of living, as expressed in Article of the Universal Declaration of Human Rights, and that each person is entitled to economic, social, and cultural entitlements as a member of society. Human development must be realized through national effort, but government plans and policies often ignore those who should benefit from them. Thus it is necessary to ground the legitimacy of plans and projects in the active role of development subjects. Without adequate living standards, human life cannot flourish, but an arbitrary imposition of change denies human beings the condition of being free agents.

Science and technology, whatever their relation to development, should be included in the process of making human beings actors instead of passive recipients. In addition to taking into consideration local knowledge, scientific theories must be relevant in the solution of the problems of the dispossessed. An economic approach that fails to appreciate the importance of unemployment and asymmetrical relations in trade is morally defective. An economics of development able to explain the difference between developed and developing regions of the world is needed. But, in addition, an economics *for* development must be created. The same is true for other social sciences, especially psychology and anthropology. Also, obviously, the resources of natural science should be harnessed in the effort to increase productivity and reduce poverty.

A second stage in the move from passive recipient to actor concerns the formulation of development plans and projects. Local knowledge, techniques, and technologies are usually more efficient and appropriate than imported ones, a point often made in Latin American fiction, for example in Jorge Amado's novel *Gabriela, Clove and Cinnamon* (1974). Local values embedded in cultural practices must respectfully be taken into account; mere lip service to those values, which is typical of political and social manipulation, should be condemned. When respect for a culture and its values is genuine, development plans are not arbitrary but are the outcome of consideration of the aspirations and desires of those who will be affected. Values that are deeply ingrained in cultures may be inimical to development and thus pose a challenge to development ethics. An ethics that includes not only values but also duties and obligations may counter antidevelopment sentiment.

However it is not enough for people to realize themselves as actors in development. Even when a project is rooted in local values and is the result of negotiation among individuals and groups, it may be morally indefensible or technically defective. Democracy guarantees public participation, but this in turn does not insure a morally correct result. Hence the interplay between insiders and outsiders in development is important, a point often made by Crocker.

Insiders are in a good position to incorporate local values into the process, whereas outsiders are not influenced by such values when assessing the rights and wrongs of plans and projects. Local experiences may be relevant but limited; outside expertise may be less relevant but wider in scope. For a fruitful collaboration to occur, insiders and outsiders must share some basic values and be committed to similar goals in connection with the improvement of human conditions. Thus development ethics can be conceived as a dialogue among cultures aimed at sharing valuable experiences in the struggle to overcome obstacles in the path of free social agents.

LUIS CAMACHO

SEE ALSO *Alternative Technology; Bhutan; Change and Development; Colonialism and Postcolonialism; Mining; Progress; Sustainability and Sustainable Development.*

BIBLIOGRAPHY

Adas, Michael. (1989). *Machines as the Measure of Men: Science, Technology and Ideologies of Western Dominance.* Ithaca, NY: Cornell University Press.

Amado, Jorge. (1974). *Gabriela, Clove and Cinnamon.* New York: Avon/Bard Books.

Aman, Kenneth, ed. (1991). *Ethical Principles for Development: Needs, Capacities or Rights. Proceedings of the IDEA / Montclair Conference* Montclair, NJ: Institute for Critical Thinking.

Amin, Samir. (1977). *Imperialism and Unequal Development.* Hassocks, England: Harvester Press.

Camacho, Luis. (1993). *Ciencia y Tecnología en el Subdesarrollo* [Science and technology in developing countries]. San Jose, Costa Rica: Editorial Tecnológica de Costa Rica.

Crocker, David A. (1991). "Insiders and Outsiders in International Development Ethics." *Ethics and International Affairs* 5: 149–174.

Crocker, David A. (1998). *Florecimiento Humano y Desarrollo Internacional* [Human flourishing and international development]. San Jose, Costa Rica: Editorial de la Universidad de Costa Rica.

Crocker, David A., and Toby Linden, eds. (1998). *Ethics of Consumption.* Lanham, MD: Rowman & Littlefield Publishers.

Escobar, Arturo. (1995). *Encountering Development: The Making and Unmaking of the Third World.* Princeton, NJ: Princeton University Press.

Freeman, David. (1974). *Technology and Society.* Chicago: Rand McNally.

Goulet, Denis. (1965). *Etica del Desarrollo* [Ethics of development]. Barcelona: Estela/IEPAL.

Goulet, Denis. (1971). *The Cruel Choice: A New Concept in the Theory of Development.* New York: Atheneum.

Goulet, Denis. (1989). "Tareas y métodos en la ética del desarrollo" [Tasks and methods in ethics of development]. *Revista de Filosofía de la Universidad de Costa Rica* 27(66): 293–306.

Goulet, Denis. (1995). *Development Ethics: A Guide to Theory and Practice.* New York: Apex Press.

Lane, Robert E. "The Road Not Taken: Friendship, Consumerism and Happiness." In *Ethics of Consumption,* eds. David A. Crocker and Toby Linden. Lanham, MD: Rowman and Littlefield.

Lebret, Louis Joseph. (1959). *Dynamique concrète du développement* [Concrete dynamics of development]. Paris: Les Éditions Ouvrières.

Rapley, John. (2002). *Understanding Development: Theory and Practice in the Third World.* Boulder, CO: Lynne Rienner Publishers.

Robinett, Jane. (1994). *This Rough Magic: Technology in Latin American Fiction.* New York: Peter Lang.

Rostow, W. W. (1960). *Stages of Economic Growth.* Cambridge, England: Cambridge University Press.

Rostow, W. W. (1987). *Rich Countries and Poor Countries: Reflections on the Past, Lessons for the Future.* Boulder, CO: Westview Press.

Sachs, Wolfgang, et al., eds. (1992). *The Development Dictionary: A Guide to Knowledge as Power.* London; Atlantic Highlands, NJ: Zed Books.

Sen, Amartya. (1999). *Development as Freedom.* New York: Anchor Books.

Streeten, Paul. (1981). *First Things First.* Oxford: Oxford University Press.

United Nations Development Programme. (2001). *Human Development Report 2001: Making New Technologies Work for Human Development.* New York: Oxford University Press.

United Nations Development Programme. (2003). *Human Development Report 2003: Millenium Development Goals: A Compact among Nations to End Human Poverty.* New York: Oxford University Press.

DEWEY, JOHN

• • •

Born in Burlington, Vermont, on October 20, John Dewey (1859–1952) lived a long and productive life as a psychologist, social activist, public intellectual, educator, and philosopher. Educated at the University of Vermont and Johns Hopkins, Dewey taught philosophy at the universities of Michigan, Minnesota, and Chicago, and Columbia University. He initiated the progressive

John Dewey, 1859–1952. During the first half of the 20th century, Dewey was America's most famous exponent of a pragmatic philosophy that celebrated the traditional values of democracy and the efficacy of reason and universal education. *(The Library of Congress.)*

laboratory school at the University of Chicago, where his reforms in methods of education could be put into practice. He was instrumental in founding the American Association of University Professors (AAUP), helped found the National Association for the Advancement of Colored People (NAACP), and was active in the American Civil Liberties Union (ACLU). Dewey remained active until shortly before his death in New York City on June 1.

Dewey's philosophical pragmatism, which he called "instrumentalism," is both an extended argument for and an application of intelligence-in-action. Intelligence-in-action, human reasoning understood as fallible and revisable, aims to ameliorate existing problems (ethical, scientific, technical, social, aesthetic, and so on). It is rooted in the insights and methodologies of modern science and technology.

Intellectual Influences

At Vermont, Dewey studied the work of Charles Darwin (1809–1882) and evolutionary theory, from which he learned the inadequacy of static models of nature, and the importance of focusing on the interaction between an organism and its environment. For Darwin, living organisms are products of a natural, temporal process in which lineages of organisms adapt to their environments. These environments are significantly determined by the organisms that occupy them. At Johns Hopkins, Dewey studied the organic model of nature in German idealism, the power of scientific methodology, and, with Charles Sanders Peirce (1839–1914), the notion that the methods and values of the natural sciences and technology (technosciences) should serve as a model for all human inquiry. Strongly influenced also by William James (1842–1910), Dewey became a proponent of philosophical naturalism. For Dewey, knowledge and inquiry develop as adaptive human responses to environing conditions which aim at reshaping those conditions.

Inquiry as Scientific and Technological

Along with Peirce and James, Dewey took the open, experimental, and practical nature of technoscientific inquiry to be the paradigmatic example of all inquiry. For Dewey, all inquiry is similar in form to technoscientific inquiry in that it is fallibilistic, resolves in practice some initial question through an experimental method, but provides no final absolute answer. In *Studies in Logical Theory* (1903), Dewey identifies four phases in the process of inquiry. It begins with the *problematic situation*, a situation in which one's instinctive or habitual responses to the environment are inadequate to fulfill needs and desires. Dewey stresses throughout his work that the uncertainty of the problematic situation is not inherently cognitive, but also practical and existential. The second phase of the process requires the *formulation of a question* that captures the problem and thus defines the boundaries within which the resolution of the initial problematic situation must be addressed. In the third, *reflective phase* of the process, the cognitive elements of inquiry, such as ideas and theories, are evaluated as possible solutions. Fourth, these solutions are *tested in action*. If the new resulting situation resolves the initial problem in a manner conducive to productive activity, then the solution will become part of the habits of living and thus a part of the existential circumstances of human life.

This method of inquiry works because, as Dewey points out in *Experience and Nature* (1925), human experience of the world includes both the stable, patterned regularity that allows for prediction and intervention and the transitory and contingent aspects of things. Hence, although for Dewey people know the

world in terms of causal laws and mathematical relationships, such instrumental value of understanding and controlling their situations should not blind them to the sensuous characteristics of everyday life. Thus, not surprisingly, the value of technoscientific understanding and practice is most significantly realized when humans have sufficient and consistent control over their circumstances that they can live well.

Science, Technology, and the Good Life

Dewey rejects the distinction between moral and nonmoral knowledge because all knowledge has possibilities for transforming life, and arises through inquiry into a problematic situation. Thus, all knowledge has moral dimensions. Throughout his more explicitly aesthetic, ethical, and social writings, Dewey stresses the need for open-ended, flexible, and experimental approaches to problems, approaches that strive to identify means for pursuing identifiable human goods ("ends-in-view") and that include a critical examination of the consequences of these means.

For Dewey, people live well when they cultivate the habits of thinking and living most conducive to a full flourishing life. In *Ethics* (1932) he describes the flourishing life as one in which individuals cultivate interests in goods that recommend themselves in the light of calm reflection. In works such as *Human Nature and Conduct* (1922) and *Art as Experience* (1934), he argues that a good life is one characterized by (a) the resolution of conflicts of habit and interest within the individual and within society; (b) the release from rote activity in favor of enjoying variety and creative action; and (c) the enriched appreciation of human culture and the world at large. Pursuing these ends constitutes the central issue of individual ethical concern. The paramount goal of public policy is nurturing the collective means for their realization. Achieving these goals requires intelligence in action, best cultivated through democratic habits in everyday life, and education and practice in technoscientific modes of inquiry.

In the late twentieth and early twenty-first centuries, Dewey's ideas have had increasing influence in areas of applied philosophy such as philosophy of technology, bioethics, and environmental ethics. Nonetheless, Dewey has often been criticized as a mere apologist for the status quo and for a narrow straight-line instrumentalism that leaves no room for reflection on, or critical evaluation of, ends. Others criticize his work by noting that technoscience has unleashed great horrors on the world (such as nuclear weapons and environmental degradation), and increased the possibilities of social control and manipulation (Taylorism, mass media, surveillance, and so on). Dewey does not deny that technoscience has sometimes failed, but this has not been due to something intrinsic to science and technology. Failures in the use of science and technology are rather failures to consistently employ intelligence-in-action; failures of inquiry, failures to be sufficiently experimental, reflective, and open.

Among the influential interpreters of Dewey's work, especially as it applies to science, technology, and ethics, are Paul Durbin (b. 1933) and Larry A. Hickman. For some years Durbin has argued what has come to be known as the "social worker thesis," that philosophers dealing with science, engineering, and medicine have obligations similar to social workers not simply to analyze problems but to become socially and politically engaged in their solution. Hickman, director of the Center for Dewey Studies (Southern Illinois University at Carbondale), argues that Dewey's pragmatism offers the best account of how to develop moral intelligence and then bring it to bear in the context of an advancing technoscientific culture.

J. CRAIG HANKS

SEE ALSO *Expertise; Pragmatism; Social Engineering.*

BIBLIOGRAPHY

Campbell, James. (1995). *Understanding John Dewey: Nature and Cooperative Intelligence.* Chicago: Open Court.

Dewey, John. (1967–1991). *The Collected Works of John Dewey,* 37 volumes, ed. Jo Ann Boydston. Carbondale: Southern Illinois University Press.

Dewey, John. (1981). *The Philosophy of John Dewey,* ed. John J. McDermott. Chicago: University of Chicago Press.

Dewey, John. (1998). *The Essential Dewey,* 2 vols., ed. Thomas M. Alexander and Larry A. Hickman. Bloomington: Indiana University Press.

Durbin, Paul T. (1992). *Social Responsibility in Science, Technology, and Medicine.* Bethlehem, PA: Lehigh University Press.

Eldridge, Michael. (1998). *Transforming Experience: John Dewey's Cultural Instrumentalism.* Nashville, TN: Vanderbilt University Press.

Hickman, Larry A. (1990). *John Dewey's Pragmatic Technology.* Bloomington: Indiana University Press.

Hickman, Larry A. (2001). *Philosophical Tools for Technological Culture: Putting Pragmatism to Work.* Bloomington: Indiana University Press.

Hickman, Larry A., ed. (1998). *Reading Dewey: Interpretations for a Postmodern Generation.* Bloomington: Indiana University Press.

DIAGNOSTIC AND STATISTICAL MANUAL OF MENTAL DISORDERS

• • •

The Diagnostic and Statistical Manual of Mental Disorders (DSM) represents the most influential effort in the field of mental health to identify psychological and psychiatric abnormalities for the purposes of treatment. The extent to which this effort has been pursued in a rigorously scientific manner, and the ethical issues surrounding the distinction between normal and abnormal mental functioning, are important questions for clarification and debate.

The DSM, which has been compiled and published by the American Psychiatric Association (APA) since its first publication in 1952, is intended to serve as a standard tool for mental health professionals in the diagnosis of mental illness. In addition to providing the field with a definition of the term *mental disorder,* the fourth edition of the manual (DSM-IV-TR; APA 2000) contains a catalog of the clinical symptoms of 365 different mental disorders (for example, obsessive–compulsive disorder, borderline personality disorder), which are organized into sixteen major diagnostic classes (such as anxiety disorders, personality disorders, and so on).

With each subsequent edition, the classifications provided by the DSM have become more widely referenced in the field of psychopathology. In addition, the DSM system of diagnosis has become increasingly central to the communication between mental health professionals and those outside the field, such as lawyers, insurance companies, and the media. Nevertheless, the system remains highly controversial, even among those who have contributed to its development. Some of this controversy surrounds the general issue of whether or not the diagnosis of mental illness is a scientific endeavor at all. More specific criticism has also been leveled, however, at the specific approaches the DSM has taken over its history to describe or explain mental disorders. In both cases, the debate over the DSM has often raised fundamental questions about the nature and diagnosis of mental illness.

The Origins and History of the DSM

As Gerald N. Grob (1991) has detailed, psychiatrists of the early 1900s were largely uncomfortable with the idea that the symptoms of mental illness could be broken down into any meaningful classification scheme. The professionals of this period tended to view the individual case as highly unique and subject to a wide variety of interrelated personal and environmental variables. Various classification schemes were proposed between 1900 and 1920, including a collaborative effort of the U.S. Bureau of the Census and the existing version of the American Psychiatric Association that produced a taxonomy of twenty-two categories of mental disorder, most of which were predicated on a particular form of biological abnormality. Such systems, however, were largely irrelevant in clinical practice or research. Instead, they served primarily to provide gross survey categories for hospitals and local governments to use in the compilation of statistics on rates of mental illness among different demographic and ethnic groups and on standards of care across different communities.

During World War II, however, mental health professionals began caring for large numbers of patients (i.e., soldiers) who did not require long-term confinement in a hospital. These patients showed psychophysical, personality, and acute stress disorders that were not well documented and that added significant variety to the existing classifications of mental illness. Inspired by these circumstances, the APA formed a committee of experts to establish a diagnostic system that expanded upon systems developed for the U.S. armed forces and adapted the international statistical classification of diseases, injuries, and causes of death, developed by the World Health Organization, for use in the United States. This process involved a significant expansion and reorganization of the existing systems and culminated in the publication of the first DSM (DSM-I) in 1952. The DSM-I (and the subsequent DSM-II, published in 1968) represented a major turning point in the nature and purpose of a taxonomy of mental disorders. For one thing, it was the first attempt to standardize psychiatric diagnoses according to a particular theory of mental illness (that is, psychoanalytic theory). Moreover, the DSM was proposed to advance the science, and not just the administration, of mental health services. By providing mental health professionals with a common diagnostic language and by grounding the descriptions of the disorders in the prevailing psychoanalytic theory, the DSM was intended to further stimulate and synthesize research into the nature of mental illness.

These first two editions of the DSM, however, were not received with unequivocal support. The two primary complaints mental health professionals voiced against the DSM concerned the lack of evidence for the distinctions it made among various disorders and the small number of experts involved in determining the classification scheme. The reliance on psychoanalytic concepts was also increasingly questioned given the rise of more empirical and behavioral approaches in clinical settings. In response the APA took a distinctively different approach to developing the third (DSM-III, 1980) and fourth (DSM-IV, 1994) editions. For each of these editions, expert researchers and clinicians were organized into work groups for each category of disorders (e.g., anxiety disorders, substance-abuse disorders). These groups conducted reviews of the available literature to determine whether or not the criteria for each disorder and the distinctions among disorders were supported by empirical evidence. Although the findings from the work groups continued to be compiled and reviewed by committee, the emphasis on research increased both the objectivity of the decision-making process and the number of professionals who could influence the final product. The manual also became accessible to a wider range of professionals by abandoning a central theoretical perspective and adopting a focus on clinically observable symptoms such as thoughts of suicide or repetitive behaviors.

Current Issues in the Development and Application of the DSM

The primary purpose for the development of the DSM has always been its use as a clinical tool for guiding the assessment and treatment of mental disorders. Perhaps the greatest strength of the DSM is its usefulness in differential diagnosis. For example, a patient's complaint of feeling down or depressed can be evaluated in light of other clinical symptoms that are present and compared with the criteria for disorders such as Major Depressive Episode and Adjustment Disorder with Depressed Mood. Although disorders such as these share some common features, distinctions among them with regard to etiology (cause) and prognosis may provide important guidance for treatment planning. In fact, some clinicians argue that the future of mental health as a science depends heavily on the ability of professionals to distinguish among treatments that are or are not effective for specific diagnoses. Such an approach ultimately leads to the matching of treatment with diagnosis based on support from available research.

An important problem with such an approach, however, is that patients with distinctly different symptoms and clinical presentations can receive the same diagnosis. In the DSM-IV, for example, each disorder is characterized by a set of equally weighted criteria. Patients need not meet all criteria for a given disorder in order to fit the diagnosis. This flexibility allows for more reliable diagnosis across clinicians, but it can also lead to minimal overlap in symptoms between any two patients with the same diagnosis. Often, symptoms that these patients do not share play a major role in treatment planning and clinical management.

A similar problem concerns the frequency with which patients meet criteria for more than one disorder at a given time (known as comorbidity). As Lee Anna Clark, David Watson, and Sarah Reynolds (1995) have noted, more than half of all individuals with a DSM diagnosis also meet criteria for another disorder. In many cases, the presence of a second disorder is a significant issue that has a dramatic effect on a patient's response to a given treatment.

A third problem with the use of the DSM in treatment planning is the lack of a coherent theoretical framework for understanding the causes and progressions of the various disorders. This limitation is ironic, given that a descriptive, symptom-focused approach was deliberately adopted in the DSM-III and DSM-IV to make the manual accessible to a range of professionals with different theoretical orientations. Clinicians inevitably rely on a particular theoretical framework in assessment and treatment planning, however, and so a purely descriptive manual cannot help but appear removed from reality in clinical settings.

Beyond its clinical utility, the DSM has also been developed to facilitate research and communication among professionals regarding the nature of mental illness. Prior to the development of the DSM, clinicians developed colloquial classification schemes that did not generalize far beyond their immediate setting. Although many professionals of the time considered such an approach to be unavoidable, they also recognized the difficulties this posed for efforts to increase or disseminate their base of knowledge. The DSM has certainly increased systematic research into mental illness and placed that research in a framework that is accessible to a broader scientific community. A prominent example is the dramatic increase of research in personality disorders that has occurred because these disorders were given special emphasis in the DSM-III (Widiger and Shea 1991).

The question remains, however, whether this proliferation in research has resulted in any real increases in scientific knowledge concerning mental illness. For instance, critics have noted that as the DSM classifications have become more widely adopted, they have begun to take on the nature of assumptions rather than scientific problems to be investigated. Thus, researchers may rely on DSM criteria instead of independent, theoretically driven criteria in selecting research participants.

In this way, the DSM has become a somewhat self-perpetuating framework.

In addition, it is important to keep in mind that the ultimate decisions about making changes to the manual are not purely empirical exercises. Such decisions must appeal to fundamental assumptions about principles concerning the nature of mental illness and the goals of the system itself. Along these lines, Arthur C. Houts (2002) has argued that it is unlikely that the continual expansion of the DSM from 106 disorders in 1952 to 365 disorders in 1994 represents real scientific advance in the ability to detect and diagnose mental illness. In particular, this expansion of labels has not occurred alongside the necessary solidification of a limited number of "covering" or "synthesizing" laws that would explain how all these new disorders relate to one another. A more specific and highly publicized example of this problem concerns the removal of homosexuality as a mental disorder in the third edition of the DSM. Regardless of whether or not homosexuality should be included in the DSM as a mental disorder, even the leadership of the revision process has admitted that the decision was ultimately based more on social pressures than on the weight of scientific evidence (Spitzer, Williams, and Skodol 1980).

A third central purpose for the development of the DSM concerns the justification of professional services and judgments. Particularly in the arenas of insurance reimbursement and legal proceedings, mental health professionals are expected to demonstrate that their evaluations and treatment plans meet some standard of common practice in the profession. With respect to insurance, however, it continues to be difficult to justify treatment decisions based on a particular diagnostic picture. Because of the heterogeneity of patients who can share a given diagnosis and because the DSM continues to explicitly require clinical judgment in assigning a diagnosis and planning treatment, the assignment of a particular diagnosis to a patient can have very little impact on the clinical services provided to that patient. Furthermore, clinicians often use DSM diagnoses for purely instrumental reasons (e.g., to promote or protect the relationship with the patient, to obtain services from a resistant insurance provider). In a common example, clinicians, in order to avoid stigmatizing or scandalizing a patient, will often diagnose the person with adjustment disorder (which connotes a more transient and normative reaction to stress) instead of a more serious disorder even though the patient meets the criteria for the latter. With regard to the courts, mental health professionals cannot assume that a particular DSM diagnosis of mental illness bears any correspondence to the legal definition of "mental disease" or "mental defect." Thus, questions of diagnosis are often abandoned altogether in the courtroom in favor of more straightforward comparisons of symptoms and states of mind with legal definitions of sanity.

Future Directions for the DSM

Shortly after publication of the DSM-IV, clinicians began expressing hopes that future editions would address several fundamental flaws in the current classification scheme, in addition to those mentioned above. Perhaps the most common desire is to move away from categorical (i.e., yes or no) diagnosis of mental disorders and toward a system of rating patients on a small number of basic, personological dimensions (e.g., personality traits). Proponents argue that such a system would have greater value in guiding differential diagnosis, would help consolidate the growing number of disorders, and would more validly reflect the dynamic nature of the individual. A less radical revision that, presumably, would also reduce the degree of disparate diagnoses is John F. Kihlstrom's (2002) proposal that the current phenomenological groupings of symptoms be replaced with diagnoses based on laboratory findings such as characteristic cognitive or affective deficits. Finally, Thomas A. Widiger and Lee Anna Clark (2000) recommend that greater attention be paid to the most basic element of the diagnostic system: the establishment of meaningful boundaries between normal and abnormal psychological functioning. Common to all these proposals is a need for the DSM to develop a more unified and coherent framework of mental illness that is more validly rooted in the fundamental nature of the human person.

Central to the ongoing debates surrounding the DSM, then, is the role of values and metaphysical assumptions in defining psychological normality and, thus, providing a foundation for the identification and treatment of abnormality. Whereas empirical science may be invaluable in describing the mental, emotional and physical processes underlying psychological disorders, interpretation of these descriptions inevitably proceeds from a framework containing statements about the nature of human abilities, the kind of life worth living, and the ideal form of relationships among persons or between persons and their environment. As Daniel Robinson (1997) has argued, any theory of psychological disorder and therapy that is divorced from these questions fails to answer the question, "therapy for what?" because it will fail to account for the kind

of healing or remediation that is necessary. These kinds of metaphysical issues, which are also assuming an increasingly central role in the ethics of the biological and genetic sciences, bring into clearer relief the nature and limits of a scientific attempt to identify problems or deficiencies in the psychological life of persons.

CARLETON A. PALMER

SEE ALSO *Homosexuality Debate; Psychology.*

BIBLIOGRAPHY

American Psychiatric Association (APA). (2000). *Diagnostic and Statistical Manual of Mental Disorders: DSM-IV-TR*, 4th edition, text revision. Washington, DC: Author. This most recent revision updates the text of the DSM-IV to include the results of research conducted after 1992. The revision is intended to maintain the usefulness of the DSM as an educational resource between major editions.

Clark, Lee Anna, David Watson, and Sarah Reynolds. (1995). "Diagnosis and Classification of Psychopathology: Challenges to the Current System and Future Directions." *Annual Review of Psychology* 46: 121–153. The authors review the purposes of a diagnostic taxonomy and consider alternatives to the current, descriptive approach. Research failing to support categorical diagnosis of psychopathology is highlighted.

Grob, Gerald N. (1991). "Origins of DSM-I: A Study in Appearance and Reality." *American Journal of Psychiatry* 148(4): 421–431. Examines the social and political context in which the DSM-I was developed.

Houts, Arthur C. (2002). "Discovery, Invention, and the Expansion of the Modern Diagnostic and Statistical Manuals of Mental Disorders." In *Rethinking the DSM: A Psychological Perspective*, ed. Larry E. Beutler and Mary L. Malik. Washington, DC: American Psychological Association. Analyzes the assumption that the proliferation of DSM diagnostic labels across editions represents actual scientific progress in the assessment and differentiation of psychological disorders.

Kihlstrom, John F. (2002). "To Honor Kraepelin ... : From Symptoms to Pathology in the Diagnosis of Mental Illness." In *Rethinking the DSM: A Psychological Perspective*, ed. Larry E. Beutler and Mary L. Malik. Washington, DC: American Psychological Association. Argues that a taxonomy based on laboratory observations would advance the science of psychopathology and return the practice of psychological diagnosis to its historical roots.

Robinson, Daniel N. (1997). "Therapy as Theory and as Civics." *Theory and Psychology* 7(5): 675–681. The author suggests that fitness for the moral and civic aspects of human life provides the grounding for any theories of mental disorder and any attempts at psychotherapy.

Spitzer, Robert L., Janet B. W. Williams, and Andrew E. Skodol. (1980). "DSM-III: The Major Achievements and an Overview." *American Journal of Psychiatry* 137(2): 151–164. Leaders from the task force on DSM-III describe the origins of its development and the reasons for the major transition toward an atheoretical approach to defining and classifying mental disorders.

Widiger, Thomas A., and Lee Anna Clark. (2000). "Toward DSM-V and the Classification of Psychopathology." *Psychological Bulletin* 126(6): 946–963. Discusses the more significant issues of disagreement and debate surrounding the DSM-IV and recommends continued exploration of alternatives to the current perspective.

Widiger, Thomas A., and Tracie Shea. (1991). "Differentiation off Axis I and Axis II Disorders. *Journal of Abnormal Psychology* 100(3): 399–406. Suggests that the distinction between personality and symptomatic disorders first introduced in DSM-III ignores important relationships between them and instances of significant overlap in symptomatology.

DIGITAL DIVIDE

• • •

The *digital divide* refers to the gap between those who can effectively benefit from information and computing technologies (ICTs) and those who cannot. The term is a social construction that emerged in the latter half of the 1990s, after the Internet came into the public domain and the World Wide Web (Web) exploded into the largest repository of human knowledge that has ever existed. For those who can both contribute and retrieve information from the Web, ICTs hold the promise of broad collaborations in science and technology, transparency in government, rationality of markets, and shared understandings between peoples. Sadly this utopian promise applies only to an elite few. As of 2003, less than ten percent of the world's 6.4 billion people have had access to the Web (NielsenNetRatings, February 2003). While information poverty is rarely blamed as a direct cause of human suffering, the digital divide raises ethical questions of universal access. Like access to food or clean water, access to essential information has moral and ethical implications that merit consideration in the formation of public policy.

Differing Divides

The digital divide is a problem of multiple dimensions. In 1999 Rob Kling summarized the problem from (a) a technical aspect referring to availability of the infrastructure, the hardware, and the software of ICTs, and (b) a social aspect referring to the skills required to manipulate technical resources. Pippa Norris (2001) described (a) a global divide revealing different capabilities between the industrialized and developing nations; (b) a social divide referring to inequalities within a given population; and (c) a democratic divide allowing for different levels of civic participation by means of ITCs. And Kenneth Keniston (2004) distinguished four social divisions:

1. those who are rich and powerful and those who are not;

2. those who speak English and those who do not;

3. those who live in technically well-established regions and those who do not;

4. those who are technically savvy and those who are not.

From a global perspective, a high concentration of access to ICTs is observed in North America, Europe and the Northern Asia Pacific while access is noticeably sparse in the southern regions of the globe, particularly in Africa, rural India, and the southern regions of Asia. The poorer nations, plagued by multiple burdens of debt, disease, and ignorance are those least likely to benefit from Internet access.

The entry costs to secure equipment and to set up services are far beyond the means of most poor communities. Startup costs and expenses of technical maintenance compete with resources needed for essential human survival. Policy makers are challenged to find justification for investment in ICTs when local and national resources are limited and where the urgent needs of people for basic nutrition, health care, and education remain unsatisfied. If ICT development is justified in these countries, it is on the belief that ICTs are instruments to be wielded in order to meet essential human needs.

Overcoming Divides

One formidable obstacle to ICT diffusion is language. There is a self-perpetuating cultural hegemony associated with ICTs (Keniston 2004). By the year 2000, only 20 percent of all Web sites in the world were in languages other than English, and most of these were in Japanese, German, French, Spanish, Portuguese, and Chinese. But in the larger regions of Africa, India, and south Asia, less than 10 percent of people are English-literate while the rest, more than 2 billion, speak languages that are sparsely represented on the Web. Because of the language barrier the majority of people in these regions have little use for computers. Those who do not use computers have little means to drive market demands for computer applications in their language. Left simply to the market, this Anglo-Saxon hegemonic cycle will continue unhindered.

If the digital divide is viewed purely as a technical problem, the solution is within reach. European and North American capitalism has the means to intervene where market forces lack the power to bridge the divide. It is not an unrealistic task to tie every nation, every tribe, and every community, no matter how isolated, into a common interconnected information infrastructure. It is within technical means to manufacture low-cost, durable computers for wide distribution. It is within fiscal means to distribute these devices to places where computers are most lacking. Gifted programmers and translators can be recruited to convert existing online resources into many different languages.

Beyond Technical Issues

While such technical solutions can be conceived, the problem of the digital divide is not primarily a technical problem. Expenditure of monies for ICTs comes with no guarantee that problems that plague the poor of the world will be addressed. Policy makers cannot simply thrust technology into people's hands with any expectation that it will be used. Experimentation has shown that new initiatives tend to fail unless they are built from existing social and economic structures (Warshauer 2003). ICT projects must be conceived from an assessment of actual needs defined locally by target populations. Planners must pay attention to existing human networks and social systems, taking into account local language and cultural factors, literacy and educational levels of users, and institutional and social structures of the community.

M. S. Swaminathan, one of India's best-known scientists, suggests that if technological and information empowerment is to reach the unreached, then policy makers must focus their "attention to the poorest person" (Swaminathan 2001, p. 1). This concept, coined by Gandhi as *antyodaya*, provides a model for technical development using a bottom-up approach. Digital initiatives of the Swaminathan Research Foundation have demonstrated how ICTs can change the lives of the poor in remote villages by strategies that enlist local involvement from their inception. Projects begin from assessments of specific local needs and by instituting practices that rely entirely on local villagers rather than distant agencies and technical experts. Including the excluded in the empowerment brought by knowledge and skills is the most effective approach to harnessing technologies in the interests of the poor. The divide

may never be fully closed, but where a bridge is to be spanned, it will be constructed by active participants from both sides.

MARTIN RYDER

SEE ALSO *Information Society; Internet; Networks; Poverty.*

BIBLIOGRAPHY

Keniston, Kenneth (2004). "Introduction: The Four Digital Divides." In *IT Experience in India: Bridging the Digital Divide*, ed. Kenneth Keniston and Deepak Kumar. Delhi: SagePublishers, pp. 11–36. This book examines the history of the "digital divide" as seen from observers on both sides of the abyss. A collection of essays from social scientists, policy analysts, and technologists, the book lends special focus to the social and technical problems that emerge with introduction of new technologies in remote and impoverished regions. It reviews the fundamental issues facing policy makers whose decisions impact the balance between expert direction and local control, between modern goals and traditional values, and between future hopes and immediate needs.

Kling, Rob. (1999). "Can the 'Next Generation Internet' Effectively Support 'Ordinary Citizens'?" *The Information Society* 15: 57–63. Rob Kling (1944–2003) was professor of information systems at the School of Library and Information Science at Indiana University. This article was originally delivered as a white paper in 1988 to the Clinton Presidential Advisory Committee on High Performance Computing and Communications, Information Technology, and the Next Generation Internet (NGI). Kling advocated a focus on social value in planning and resource allocation for the NGI. Policies should privilege methods of Internet access for ordinary people over exotic media advances that benefit only technological elites. NGI initiatives should seek to reduce cost and improve value for ordinary citizens and strengthen IT infrastructures for the public commons.

Norris, Pippa. (2001). *Digital Divide: Civic Engagement, Information Poverty, and the Internet Worldwide.* New York: Cambridge University Press. Pippa Norris is Associate Director of Research at the Shorenstein Center on the Press, Politics and Public Policy at Harvard University. This book examines the digital divide from case studies involving Internet access of users from 179 countries. Norris describes the ways in which people in different cultures make use of the Internet for information and civic engagement and she attempts to provide a descriptive model to explain patterns of participation on the Internet. Norris reviews the manner that institutional entities have created or enabled opportunities for public engagement and she analyzes how the various publics have responded to these opportunities.

Warschauer, Mark. (2003). "Demystifying The Digital Divide." *Scientific American* 289(2): 42–48. Mark Warschauer teaches Informatics in the Department of Education at the University of California at Irvine. He has conducted qualitative research on technology adoption in Egypt, China, India, Brazil, Singapore. In a three-year longitudinal case study in Egypt, Warschauer examined the steps that governmental agencies took to introduce ICTs in the schools and he observed the various impacts that were manifested from those initiatives. From his observations, Warschauer concluded that policies which promote the mere presence of new technologies in schools without a corresponding emphasis on social mobilization and transformation can squander limited resources while leaving problems of sociotechnical inequities in tact.

INTERNET RESOURCES

NielsenNetRatings. (2003). "Global Internet Population Grows an Average of Four Percent Year-Over-Year"" Available from http://www.nielsen-netratings.com/pr/pr_030220.pdf. Nielsen NetRatings provide ongoing measurement of Internet usage. Detailed information collected by Nielsen is intended primarily for commercial use, however summaries are published monthly and may be accessed at http://www.nielsen-netratings.com/news.jsp?section=dat_gi.

Swaminathan, Monkombu Sambasivan. (2001). "Antyodaya Pathway of Bridging the Digital Divide." Available from http://www.worldbank.org/wbi/B-SPAN/docs/antyoday.pdf. (Lecture delivered before the Tilak Smarak Trust, Pune, India, August 1, 2001.) Through his involvement in hybrid grain research in the 1960s, geneticist Monkombu Sambasivan Swaminathan played a principle role in the introduction of high-yielding wheat varieties that significantly contributed to India's "Green Revolution" (1967–1978). In 1987 the celebrated scientist created the Swaminathan Foundation, with the aim of supporting fledgling technological development projects that directly address hunger, poverty and environmental concerns. The foundation is known for its emphasis on bottom-up participatory approaches to economic development where local control is the key element to the introduction of new technologies.

DIGITAL LIBRARIES

• • •

Digital libraries are organized collections of information resources and associated tools for creating, archiving, sharing, searching, and using information that can be accessed electronically. Digital libraries differ from traditional libraries in that they exist in the "cyber world" of computers and the Internet rather than in the "brick and mortar world" of physical buildings. Digital libraries can store any type of information resource (often referred to as documents or objects) as long as the resource can be represented electronically. Examples include hypertexts, archival images, computer simulations, digital video, and, most uniquely, real-time scientific data such as temperature readings from remote meteorological instruments connected to the Internet.

The digitization of resources enables easy and rapid access to, as well as manipulation of, digital library content. The content of a digital library object (such as a hypertext of George Orwell's novel, *1984*) includes both the data inherent in the nature of the object (for example, the text of *1984*) and metadata that describe various aspects of the data (such as creator, owner, reproduction rights, and version). Both data and metadata may also include links or relationships to other data or metadata that may be internal or external to any specific digital library (for instance, the text of *1984* might include links to comments by readers derived from a literary listserv or study notes provided by teachers using the novel in their classes).

The concepts of organization and selection separate digital libraries from the Internet as a whole. Whereas information on the Internet is chaotic and expanding faster than either humans or existing technologies can trace accurately, the information in a digital library has been organized in some manner to provide the resource collection, cataloging, and service functions of a traditional library. In addition, the resources in digital libraries have gone through some sort of formal selection process based on clear criteria, such as including only resources that come from original materials or authoritative sources. Digital libraries are thus an effort to address the problem of information overload often associated with the Internet.

Origins

Although the concept of digital libraries has been traced back to nineteenth-century scientific fiction writers such as H. G. Wells, most library historians credit Vannevar Bush's description of the memex in the July 1945 edition of *Atlantic Monthly* as the original source. Despite being limited to analog technologies such as microfilm that seem crude in the early twenty-first century, Bush anticipated several key features of digital libraries, including rapid and accurate access to scientific and cultural information.

Contemporary conceptions of digital libraries developed in tandem with the rapid growth of the Internet and especially the widespread, flexible access to digital information afforded by the development of World Wide Web browsers in the early 1990s. For example, in the United States, Phase One of the Digital Libraries Initiative was launched in 1993 when the National Science Foundation (NSF), the Defense Advanced Research Projects Agency (DARPA), and the National Aeronautics and Space Administration (NASA) provided six universities with nearly $25 million to develop digital library test-beds. Another pioneer digital library effort was the U.S. Library of Congress's American Memory project. This groundbreaking digital collection of historical artifacts was first made available on interactive videodiscs, later on CD-ROMs, and most recently via the Internet. Related digital library projects have been underway in Europe, Canada, and elsewhere since the mid-1990s.

In 1998 Phase Two of the Digital Libraries Initiative (DLI2) was launched with funding from NSF, DARPA, NASA, the Library of Congress, the National Library of Medicine, the National Endowment for the Humanities (NEH), and the Federal Bureau of Investigation (FBI). The seemingly strange bedfellows supporting DLI2 suggests some of the ethical issues surrounding digital libraries. These include privacy (who can find out about the resources someone has accessed via digital libraries?), security (who decides what information should or should not be freely accessible?), intellectual property (who owns what information?), hegemony (who controls the access to information?), and globalization (who assures that cultural identity is not weakened or even destroyed by digital libraries?).

Challenges: Technical and Ethical

The technical challenges confronting librarians, computer scientists, cognitive psychologists, and others working on the frontiers of digital libraries are formidable. These include interoperability (what protocols and standards are needed to ensure that distributed digital libraries will provide widespread interconnected access?), access (what types of user interfaces are most effective in providing easy access to diverse communities of users seeking information for different reasons?), preservation (what technologies are needed to assure the long-term survival of digital information resources?), and sustainability (what financial resources are needed to support the maintenance of digital libraries, and how can they be procured?).

In a manner similar to the science of genetics and the Human Genome Project, ethical debates about the ultimate status and value of information science and digital libraries may be even more complex than the technological challenges. It is inevitable that much information will be primarily available through digital technologies in the foreseeable future, a result that leads to complex social and ethical questions that must be addressed. How can traditional library values such as providing all people with free access to high-quality information be upheld when large corporations increasingly seek to profit by selling the information they control? Will the "digital divide" (that is, the unequal access to information technologies currently inherent in the growth of the Internet, which is largely controlled by Western powers such as the United States and the European Union) be decreased or increased by the development of digital libraries? How can the validity of information resources be established when increasingly sophisticated technologies threaten fundamental concepts such as authorship and copyright? How can digital libraries be designed to improve education at all levels?

In his 2000 book *Digital Libraries*, William Y. Arms concludes that "a dream of future libraries combines everything that we most prize about traditional methods with the best that online information can offer. In some nightmares, the worst aspects of each are combined" (p. 272). Although the future of digital libraries is unclear, digital libraries will nevertheless influence the future.

THOMAS C. REEVES

SEE ALSO *Education; Hypertext; Information.*

BIBLIOGRAPHY

Arms, William Y. (2000). *Digital Libraries.* Cambridge, MA: MIT Press. This volume remains the definitive description of digital library concepts and structures.

Bush, Vannevar. (1945). "As We May Think." *Atlantic Monthly* 176(1): 101–108. This article has influenced several generations of the designers of hypertext, hypermedia, and digital libraries.

INTERNET RESOURCES

Arms, William Y. *Digital Libraries.* Available from http://www.cs.cornell.edu/wya/.

Bush, Vannevar. (1945). "As We May Think." *Atlantic Monthly* 176(1): 101–108. Available from http://www.theatlantic.com/unbound/flashbks/computer/bushf.htm.

"Digital Libraries Initiative Phase Two." National Science Foundation. Available from http://www.dli2.nsf.gov/.

U.S. Library of Congress. "American Memory: Historical Collections for the National Digital Library." Available from http://memory.loc.gov/.

DIGNITY

• • •

Dignity in modern Europe and North America is that quality of an individual human person that warrants treating him or her as an end, never merely as a means to some further end. Many things have a price; they are exchangeable for something of equal or greater value.

A human person has no price and is not exchangeable; nothing has more value. Philosopher Immanuel Kant (1724–1804) gave voice to the Enlightenment view by saying that dignity is "an intrinsic, unconditioned, incomparable worth or worthiness" (Kant, p. 36). In a context of expanding technological ability to treat many topics, including persons, as means, the concept of dignity has been associated with the setting of boundaries on such treatments.

In common parlance society distinguishes between expressing dignity and having dignity. To *express dignity* is to behave in a dignified manner, to retain composure and a sense of self-worth in a difficult situation. To *have dignity* is a status independent of any behavior. It is to be treated by others as a person of worth or with respect. It is the second of these, having dignity, that carries moral weight. Every person has dignity regardless of his or her wealth, class, education, age, gender, or demonstrated abilities. Dignity is said to be inherent, inborn.

What is the warrant for the assumption that each person has dignity? The capacity to reason or to make moral judgments are Enlightenment criteria by which human beings are distinguished from other sentient creatures. The theological tradition shared by Jews and Christians locates the ground of dignity in the *imago dei*, the image of God within the human race; and Christians add the incarnation, according to which God enters into the humiliation of becoming human in order to exalt the human race. These provide justification for belief in dignity plus modern commitments to human rights and social equality.

Metaphysically dignity is innate or inborn—that is, dignity applies universally to all human beings regardless of distinctive personal characteristics. Phenomenologically, however, dignity is relational—that is, dignity is first conferred and then claimed. When a family treats infants and young children as persons of worth, these children grow up to see themselves as worthy, as valuable in themselves. Then they are able to express dignity by claiming their rights in society. One way to view the ethical task of persons in free societies is to affirm our responsibility to confer dignity upon persons who are marginalized politically or economically or socially, so that they will be able to rise up and claim equal rights. To be treated by others as having dignity enables one to rise up and express that dignity.

Societal Threats to Dignity

Human dignity today faces four threats. First, quite obviously, totalitarian governments and repressive religious regimes deny a sense of final value to their citizens. Problems of how to deal with such governments or religious traditions, especially in a world increasingly linked by technological means of communication and scientific research, remains a serious political issue.

Second, animal rights groups accuse European and North American society of *speciesism* and seek to confer dignity on nonhuman creatures and, in some cases, on the environment. The extent to which dignity applies to animals, plants, or even certain artifacts such as works of art, remains a debated issue.

Third, modern industrial economics appears to treat individuals impersonally, as part of a mass. Karl Marx (1818–1883) reflected this threat when describing factory workers as flesh and blood appendages to machines of steel. Science and technology also are frequently seen as the instruments whereby bureaucratized industry is given the power to destroy traditional family values and undermine the personal relationships necessary for dignity to enjoy a conferring context.

In recent years the Roman Catholic Church has become one of the world's champions of human dignity against this third threat. Social forces enhanced by biomedical technologies appear to compromise social commitments to protect human life at all costs. Abortion—both therapeutic and elective—seems to threaten life at the beginning; and certain forms of euthanasia seem to threaten life at the end. Ethical debates over pregnancy termination and end-of-life medical practices appear to Vatican eyes as a hardening of hearts against those who cannot protect themselves from the economics of an increasingly technology-dependent civilization. Pope John Paul II referred to this as the *culture of death*. The Pope believes that at conception God places a newly created immortal soul in the conceptus; and the presence of this soul establishes morally protectable dignity. This translates into an ethics that will not allow society to put to death a person with a soul, whether prior to birth or when suffering from a terminal illness. In our culture of death "the criterion of personal dignity—which demands respect, generosity and service—is replaced by the criterion of efficiency, functionality and usefulness: others are considered not for what they are, but for what they have, do and produce. This is the supremacy of the strong over the weak" (Pope John Paul II, p. 42).

The fourth threat, at least in the eyes of the public, comes from genetic research and biotechnology. This is because DNA has become associated with the essence of a human person. DNA is said to be the so-called *blueprint*. Manipulation of one's genes, then, appears to subordinate one's essence to some further end. Proposals for *designer children* or *perfect children* through genetic selection and genetic engineering appear to subordinate the welfare of the children to the images and ends of the parents. Proposals for human reproductive cloning, resulting in two persons with identical genomes, appear to violate the individuality of both for purposes exacted by those making the cloning decision. Such proposals elicit public anxiety over the possible loss of dignity.

This fourth threat to human dignity is more apparent than real. It is a mistake to identify DNA with human essence. No matter how significant one's genome may be, genes alone do not constitute a person. Even identical twins, who share the same genome, develop their own private self-awareness and express their own individual claims to worth. Dignity is not lodged in DNA. Any person coming into the world having been influenced by genetic technologies will enter into the same sets of relationships that confer or deny dignity. Metaphysically no amount of genetic manipulation will reduce a person's dignity

As a belief held by a culture, dignity is a conviction that must be rearticulated in the face of threats. Even though built into this conviction is the idea that human worth is innate or inborn, social ethics requires that it be conferred, cultivated, enhanced, and fought for. The doctrine that each person already has dignity is actually a hope that some day all people will realize—and express—dignity.

TED PETERS

SEE ALSO *Abortion; Embryonic Stem Cells; Euthanasia; Freedom; Genetic Research and Technology; Holocaust; Human Cloning; Humanization and Dehumanization; Human Nature; Human Rights; Posthumanism.*

BIBLIOGRAPHY

Kant, Immanuel. (1953). *Groundwork of the Metaphysic of Morals*, trans. H. J. Paton. New York: Harper.

Peters, Ted. (1996). *For the Love of Children: Genetic Technology and the Future of the Family*. Louisville, KY: Westminster/John Knox Press.

Pope John Paul II. (1995). Encyclical letter *Evangelium Vitae* (March 25, 1995). Vatican City: Vatican.

DIRECT DEMOCRACY

• • •

The modern, mainstream democratic ideal has been republican or representative democracy, but the original Greek ideal was direct democratic participation in all major decisions by all citizens. To some extent even administrative actions were directly democratic, insofar as various executive and judicial functions were determined by lot. Along with direct democracy, two general terms around which efforts to theoretically and practically promote such broad contemporary involvement of citizens in their own governance are those of participatory and anticipatory democracy. In as much as both are argued to be especially facilitated by advanced telecommunications technologies such as television and the Internet, terms of choice range from digital and e-democracy to teledemocracy.

Background

The modern roots of contemporary direct democracy ideals are nineteenth-century anarchist experiments in Europe and populous and progressive movements in the United States. Populism, which reflected agrarian interests, and progressivism, more urban based, sought to institutionalize the citizen legislative initiative, the referendum, and the recall. The participatory democracy movement itself has been closely associated with theories of strong, radical, grass roots, deliberative, and consensus democracy. Anticipatory democracy gives direct democracy a futurist spin. Bioregional democracy is a related notion stressing environmental or ecological issues. Cyberdemocracy stresses virtual reality both as means and as end.

The unifying thread in such diverse direct democracy movements is that all citizens, not just their periodically elected representatives, have rights and responsibilities to contribute to collective decision making. Independent of arguments for such rights and responsibilities, and analyses of the strengths and weaknesses of participatory democracy, one of the most well-developed efforts to promote citizen participation through advanced telecommunications is the Direct Democracy Campaign (DDC) in Great Britain. Using the motto *Let the people decide*, the DDC has advanced a number of specific proposals. The popular initiative would require the government to hold a binding referendum on an issue if 2 percent of the electorate submitted a petition to this effect. The popular veto would allow 1 percent of the electorate to challenge any existing legislation and call for a binding referendum. Moreover according to the DDC web site, "the era of pencil crosses on paper must give way to an age of secure electronic communication."

In the United States, although the theory of participatory democracy emerged in left-wing political circles during the 1960s, proposals for the utilization of advanced telecommunications technologies were promoted more in right-wing political circles during the 1970s. Post 1960s left-wing work moved in the direction of trying to get citizens directly involved in processes of scientific and engineering design decision making, using such means as consensus conferences and by often questioning the adequacy of electronic or virtual participation (Sclove 1995).

Liberal theorists have on occasion utilized measurement theory, especially as applied in psychology by S. S. Stevens (1946), to make some critical assessment of representative democracy and propose reforms that might serve to attract more citizen involvement or enhance the justice of decision making. Among various mathematical or scientific models for enhancing the influence of minority viewpoints or interests are, for example, possibilities for proportional representation, which again might be facilitated by technological means.

Right-wing work, by contrast, has been more populist and positive about electronic democracy. Indeed conservative futurist Alvin Toffler has argued that technological change at once demands intensified, anticipatory democracy as a "continuing plebiscite on the future" (Toffler 1970, p. 422) and provides the "imaginative new technologies" (Toffler 1970, p. 424) to make this possible. Clem Bezold (1978) further advanced the idea of anticipatory democracy. Brian Martin (demarchy) and Robin Hanson (futarchy) have proposed other related ideas appealing to or utilizing market theory.

Outside Great Britain and the United States, efforts to promote and practice participatory democracy utilizing advanced telecommunications technologies exist in, among other countries, Switzerland, Canada, Australia, and New Zealand. Often these efforts exist most vigorously at the local or regional levels. As expected, they have also sponsored numerous web sites.

Questions

Historically there have been three main objections to direct democracy. One is that it provides for no check on emotional responses to complex situations. Another is that most people do not have time or interest enough to become sufficiently educated in the issues to participate intelligently. A third is that there is simply no technical means by which it could work in a modern, large-scale nation-state.

According to Toffler (1980) all three objections can be met. There could be structural or constitutional requirements for a cooling-off period or a second vote

on certain issues. Increased affluence and leisure give people more time for politics, and in fact when offered the chance many citizens take advantage of opportunities to become informed about an issue. Social learning generally takes place through trial and error. Finally contemporary communications technologies, especially the Internet, make direct electronic democracy realistically feasible.

More specific questions about the utilization of advanced technological means of communication have also been raised. Has C-SPAN improved democratic intelligence? To what extent can the Internet promote critical reflection and engagement with the nonvirtual world in which political action ultimately takes place?

CARL MITCHAM

SEE ALSO *Democracy; Participation.*

BIBLIOGRAPHY

Bezold, Clem, ed. (1978). *Anticipatory Democracy.* New York: Random House.

Sclove, Richard E. (1995). *Democracy and Technology.* New York: Guilford Press. Widely influential work in liberal circles.

Stevens, S. S. [Stanley Smith]. (1946). "On the Theory of Scales of Measurement." *Science* 103: 677–680.

Toffler, Alvin. (1970). *Future Shock.* New York: Random House.

Toffler, Alvin. (1980). *The Third Wave.* New York: William Morrow. Influential in conservative political circles.

DISABILITY

• • •

Science and technology are pursued for human benefit. But the particular benefits of scientific research and technological development are the result of human activities embodying various cultural, economic, and ethical frameworks as well as the perspectives, purposes, and prejudices of any given society and of powerful groups within it. One group that should benefit includes disabled people. But such benefit will in large measure depend on the governance of science and technology, the involvement of disabled people themselves in their governance, and on the very concept of disability, an issue that is more contentious than commonly recognized. With regard to science, technology, and disability, there are at least two ethical issues that deserve more consideration than they are usually given: What perception of disabled people guides scientific research and technological development? What role do disabled people play in this process?

Solutions Follow Perceptions

Science and technology for disabled people depend on how so-called *disabilities* and disabled people are perceived. Definitions of disability range from the biomedical and economic to the liberal, social-political, minority rights, and universalist models (Penney 2002). These may nevertheless be reduced to three main perspectives:

The *medical individualistic perspective* (MP) sees disabled people as patients in need of being treated so that their level of functioning and appearance approaches that of the so-called non-disabled people (the norm). It assumes that disabled people perceive themselves as patients, and their own biological reality as defective or subnormal. It promotes the use of science and technology for the development of normative therapies for disabled people.

The *transhumanist perspective* (TP) is similar to the medical perspective with the modification that it sees both disabled and non-disabled people as patients. The human body in general is judged to be defective and in need of indefinite enhancement. Even normally existing abilities are subject to being raised above the norm or complemented with new abilities. The transhumanist (or *posthumanist*) perspective does not accept a given norm, and thus does not accept the subnormal/normal distinction between the disabled and non-disabled. The human body in general is seen as subnormal and in need of scientific-technological enforcement. It is no accident that Ray Kurzweil, inventor of the computer voice synthesizer the Kurzweil Reader for the blind, is also a strong proponent of transhumanist technological transformations for everyone.

The *social justice perspective* (SP) does not see disabled people as patients in need of treatment or enhancement so much as society in need of transformation. It assumes that most problems faced by disabled people are not generated by their non-normative bodies or capabilities but by the inability of society to fully integrate, support, and accept individuals with different biological realities and abilities. The social justice perspective encourages the use of science and technology to alter the physical environment so that disabled people may more easily interact with non-disabled people. The focus is on social, not medical, cures.

The social perspective also allows *able-ism* to be seen as analogous to racism or sexism, with able-ism constituting a network of beliefs and practices that yield a particular kind of self and body (the corporeal standard) that is then projected as the perfect, species-typical, and therefore essential and fully human. From the viewpoint of able-ism, disability becomes a diminished state of being human.

The social perspective on disability does not deny that disabled people possess certain biological realities (such as having no legs), which make them different in their abilities and cause them to deviate from a norm. However, it views the "need to fit a norm" as the main problem, and questions whether all deviations from the norm require a medical solution (adherence to the norm). Maybe in some cases a social solution (change or elimination of norm) would be just as appropriate. Neither does the social perspective deny the existence of symptomatic acute medical problems that should be treated. It simply questions the increasing medicalization of non-normative characteristics and sees many so-called diseases, defects and impairments not as acute medical problems but as societal constructions (Wolbring 2003b). It questions whether medical solutions are always the best response.

Scientific research and technological development may emphasize fixing the disabled (on an MP basis); science and technology may also seek to enhance the disabled (on a TP basis). Or technology especially can be used to reconstruct the world in ways that allow the disabled to interact freely with others without altering their biological identity/reality (on an SP basis). "Barrier-free access" is, for example, an SP program.

Society has a long history of adopting the MP approach to seeing disabilities. Many legal instruments describe a disabled person as someone with subnormal or diminished functioning in need of *special* care. They do not see disabled people as having a biological reality leading to *different* sets of abilities, *different* ways of functioning and *different* needs. The medical understanding of disabilities is essential for the acceptance and marketability of many scientific and technological applications such as genetic and non-genetic prebirth testing and genetic and non-genetic therapies and enhancements.

However, this traditional focus is being replaced by a transhumanist focus on science and technology as a means to not just meet norms but to enhance existing abilities and add new ones. It is becoming increasingly difficult to draw a line between therapy and enhancement. If one believes someone is defective without legs, and finds it acceptable to develop artificial legs that function like normal biological legs restoring the normative characteristics of walking (medical cure), one has a hard time justifying the denial of artificial that improve on the natural capabilities of biological legs (running faster, jumping higher) and that might add capabilities beyond the scope of normal biological legs such as climbing walls (transhumanist enhancements).

Scientific and technological research with an SP justification to develop software and hardware that allows the usage by clients with the widest range of abilities is rare. This reflects the fact that most product development is geared toward the largest common denominator in the market so as to ensure the highest profit. Products are seldom developed for disabled people because their numbers are not big enough to make a profit. Without a change in social policies and dynamics, this will not change significantly. In the United States, the Americans with Disabilities Act (ADA), which resulted from the lobbying of the disabled as a public interest group, created a new market for barrier free access and other technological developments.

Roles For Disabled People

Can disabled people influence the indicated dynamics? Do they have to accept the medical or transhumanist perspectives on disabilities? Is it possible for them to promote the social justice perspective?

The disabilities rights movement and emergence of disability activism in the early 1980s provides one kind of answer to such questions (see Shapiro 1994). Disabled persons, no matter how they are defined, worked together in ways that led, in the United States for instance, to passage of the Americans with Disabilities Act (ADA) of 1990, the focus of which was on changing the environmental parameters of the lives of disabled people. However the ADA has been siege ever since its passing, and although it was somewhat successful with access issues, the ADA does not sufficiently cover emerging technologies; nor has it decreased the development of technologies that focus on the medical perception of disability. An increase of the presence of disabled persons among the ranks of scientists, engineers, and ethicists would be another means of more specifically influencing scientific and technology policy.

Certainly some negative consequences of science and technology can be avoided by integrating ethics into the governance of science and technology. But what kind of ethics? Ethical guidelines are not always free from biases that reflect the perspectives, purposes, and prejudices of the most powerful social groups. Much debate about science policy and the legal regulation of

technology appears to accept the medical perspective of disability, with some qualified influence of the transhumanist perspective, but little appreciation of the social justice perspective (see, e.g., Harris 2000; Singer 2001; UNESCO 2003; and Wolbring 2003a and 2003c). One of the most effective ways for disabled persons who might object to this situation to counter it is to themselves become involved in policy national and international formation.

At the same time, disabled persons need not shy from inviting non-disabled people of power to ask themselves the following questions:

- Does scientific and technological decision making lead to further marginalization of disabled people?

- Does scientific and technological practice allow disabled people to freely choose their self-identity?

- Do science and technology by themselves have similar impacts on disabled and non-disabled people?

- Does the transhumanist perspective force non-disabled people to enhance their abilities and does it encourage society to make these new abilities eventually the new norm ("normative creep") that makes society less accepting of differences?

- What policy guidelines are needed to promote science and technology for the common good in an inclusive society?

- How the governance of science and technology be made more inclusive and diverse?

GREGOR WOLBRING

SEE ALSO Bioethics; Informed Consent; Medical Ethics.

BIBLIOGRAPHY

Albrecht, Gary L.; Katherine D. Seelman, and Michael Bury, eds. (2001). Handbook of Disability Studies. Thousand Oaks, CA: Sage. A collection of thirty-five articles on the shaping of disabilities studies, experiencing disability, and disability in context.

Barton, Len, ed. (2001). Disability Politics and the Struggle for Change. London: David Fulton. Twelve essays on disability, mostly in the British context.

Harris, John. (2000). "Is There A Coherent Social Conception of Disability?" Journal of Medical Ethics 26(2): 95–100. Harris denounces the social justice model of disability and the claim that disability is a medical issue.

Johnston, David. (2001). An Introduction to Disability Studies, 2nd edition. London: David Fulton. A synoptic overview of disability studies in both the United Kingdom and United States.

Shapiro, Joseph P. (1994). *No Pity: People with Disabilities Forging a New Civil Rights Movement*. New York: Three Rivers Press. A journalistic account of the disabilities civil rights movement in the United States that led to the passage of the Americans with Disabilities Act (1990).

Singer, Peter. (2001). "Response to Mark Kuczewski." *American Journal of Bioethics* 1(3): 55–57. Singer denounces the social jusice model of disability and a disability rights aproach to bioethics issues.

Wolbring, Gregor. (2003a). "Disability Rights Approach to Genetic Discrimination." In *Society and Genetic Information: Codes and Laws in the Genetic Era*, ed. Judit Sándor. Budapest: Central European University Press. Uses the three models (MP, SP, TP) together for the first time and shows that anti-genetic discrimination laws are discriminating against disabled people. Provides examples of disability discrimination within the usage of advances of human genetic interventions .

Wolbring, Gregor. (2003b). "The Social Construct of Health and Medicine." *Health Ethics Today* 13(1): 5–6. Looks at the social construction of health and medicine.

Wolbring, Gregor. (2003c). "Disability Rights Approach towards Bioethics." *Journal of Disability Policy Studies* 14(3): 154–180. Shows that the ethical divide that approves the prohibition of sex selection while allowing disability (MP) deselection is untenable. Questions the two-tiered ethical approach of medical versus social reasons

INTERNET RESOURCES

Harris, John. (2000). "Is There A Coherent Social Conception of Disability?" *Journal of Medical Ethics* 26(2): 95–100. Available from http://www.ncbi.nlm.nih.gov.

Penney, Jonathan. (2002). "A Constitution for the Disabled or a Disabled Constitution? Toward a New Approach to Disability for the Purposes of Section 15(1)." *Journal of Law and Equality* 1: 83–94. Available from http://www.jle.ca/files/v1n1/JLEv1n1art3.htm.

Penney, Jonathan. (2002). "A Constitution for the Disabled or a Disabled Constitution? Toward a New Approach to Disability for the Purposes of Section 15(1)." *Journal of Law and Equality* 1: 83–94. Describes different possible models and perception of disability and disabled people. Available from http://www.jle.ca/files/v1n1/JLEv1n1art3.htm

UNESCO. (2003). "Report of the IBC on Pre-implantation Genetic Diagnosis and Germ-line Intervention." *UNESCO Publication SHS-EST/02/CIB-9/2 (Rev. 3) Paris*, 24 April 2003. Available from http://portal.unesco.org./shs/en/file_download.php/1f3df0049c329b1f8f8e46b6f381cbd1ReportfinalPGD_en.pdf. This report sees disability totally within a medical framework. It advances the notion that preimplantation genetic diagnostic and germ-line interventions are ethically permissible if done for medical reasons. It denounces sex selection on the grounds that intervention is done for social reasons.

DISCOURSE ETHICS

• • •

Discourse ethics (DE) has two aims: to specify the ideal conditions for discourse, and to ground ethics in the agreements reached through the exercise of such discourse. DE thus instantiates the intuition that if people discuss issues in fair and open ways, the resulting conclusions will be morally binding for those appropriately involved in the conversation. Such a view of ethics has special relevance in a scientific and technological world characterized by expanding means of communication. DE may also arguably provide the best framework for understanding the ethics of scientists and engineers operating within their professional communities.

Theoretical Framework

Discourse ethics is primarily associated with the work of Karl-Otto Apel (1980) and Jürgen Habermas, who conjoins his own theory of communicative rationality and action (1981) with Apel's insights (Habermas 1983, 1989). Apel and Habermas root DE in Immanuel Kant's emphasis on the primacy of moral autonomy for both the individual and the moral community (Apel 2001) and in Aristotle's understanding of the importance of human community praxis as the crucible in which all theory must be tested. More broadly DE includes the work of John Rawls (1971), Stephen Toulmin (Jonsen and Toulmin 1990), and Richard Rorty (1989). As Robert Cavalier notes, each of these thinkers argues for "widening reflective equilibrium by embedding empathy and detailed reciprocity into moral reflection and by placing the deliberative process within the intelligent conduct of communal inquiry" (Cavalier Internet site). DE has deeply influenced not only philosophy and sociology—but also, in keeping with its praxis orientation, such applied fields as business ethics (Blickle et al. 1997) and nursing (Marck 2000).

Apel-Habermasian DE seeks to circumscribe—and justify—the *ideal speech situation* in which members of a *democratic* community, free of domination (*herrschaftsfreie*), engage in a rational dialogue or debate in order to achieve consensus about the fundamental rules of the community. Drawing on the Kantian understanding that rules are morally legitimate only as free human beings consent to follow them, Habermas argues that such community rules may emerge from discourse that meets certain necessary (but not sufficient) conditions—the first of which is freedom and equality for participants. In his essay "Justice and Solidarity" (1989), Habermas summarizes the basic intuition of discourse ethics with the statement that "under the moral point of view, one must be able to test whether a norm or a mode of action could be generally accepted by those affected by it, such that their acceptance would be rationally motivated and hence uncoerced" (Habermas 1989, p. 6).

In *"Diskursethik: Notizen zu einem Begründungspro-gram"* [Discourse Ethics: Notes on Philosophical Justifi-cation] (1983), Habermas further emphasizes the impor-tance of *perspective-taking* on the part of all discourse participants: Possible norms for a community can be legitimate only if they emerge from a discourse setting that "constrains all affected to adopt the perspectives of all others in the balancing of interests" (Habermas 1990, p. 65). These conditions of free but rational debate, shaped by such perspective-taking, issue in legit-imate *universal norms*—meaning that (a) all who are affected by a proposed norm are willing to accept the consequences and side effects likely to follow from observing that norm, and (b) these consequences are preferred over those of other possible norms under consideration.

Seyla Benhabib notes that such norms are better characterized as *quasi*-universal. They are morally legiti-mate for the specific discourse community whose debate and dialogue generates them. But diverse communities, shaped by different histories, traditions, and contexts, may come to agree upon a range of possible norms rather than a single monolithic set (Benhabib 1986). In this way, consistent with its Aristotelian and Kantian roots, DE establishes an *ethical pluralism*—in contrast with both monolithic ethical *dogmatism* (asserting that only a single set of norms can be right) and *relativism* (asserting that any set of values and norms is as acceptable as any other).

To circumscribe such discourse more carefully, Habermas refines a set of rules first proposed by Robert Alexy (1978). According to Habermas (1990, p. 86), these are:

1. Every subject with the competence to speak and act is allowed to take part in a discourse.

2a. Everyone is allowed to question any assertion whatever.

2b. Everyone is allowed to introduce any assertion whatever into the discourse.

2c. Everyone is allowed to express his (or her) atti-tudes, desires, and needs.

3. No speaker may be prevented, by internal or external coercion, from exercising his (or her) rights as laid down in (1) and (2).

Finally, partly in response to feminist and postmodernist critiques that his discourse ethics exhibits a masculine form of rationalism, especially because of the exclusion of emotion, Habermas argues that a sense of *solidarity* is also required between participants.

In short the conditions for the practical discourse out of which (quasi-) universally valid norms may emerge include the free participation and acceptance of all who are affected by such norms, as such norms meet their interests—where such participation is shaped by rational debate, perspective-taking, and solidarity.

Discourse Ethics in Technology and Science

Discourse ethics thus intends to define the conditions of a free and democratic discourse concerning important norms that affect all members of a community. It aims to do so in ways that are directly *practical* for the real and pressing problems facing both local and more com-prehensive communities. In this light, DE would seem well-suited for circumscribing discourse concerning pressing issues provoked by science and technology.

Indeed DE can be seen to be implicitly at work in a first instance in the Technology Assessment (TA) movement. Beginning in the 1970s in the United States, and then developing further in Europe, TA seeks to develop ways for programmatically assessing the risks and benefits of proposed or emerging technologies, in order to determine whether the technology *ought* to be developed and deployed in light of central social values, such as protecting both human life and the larger envir-onment. Rather than having decisions regarding new technologies made solely by a relatively narrow circle of scientists and market-dependent corporations, one ver-sion of TA has sought to *democratize* technology devel-opment by enlarging the circle of decision-makers to include non-technical citizens' representatives. One dramatic instantiation of such democratic technology assessment emerged in the consensus conferences devel-oped by the National Institutes of Health (NIH) in the United States in the 1970s (Jacoby 1985) and then expanded upon, initially in Denmark in 1985 (Klüver 1995). Such consensus conferences were occasioned by the issues raised by the Human Genome Project and genetically modified (GMO) foods, and were composed of carefully structured dialogues involving scientific and technological experts, policy experts, political represen-tatives, and lay or non-technical citizens. Subsequently held throughout Scandinavia and Europe, they have also been applied to issues raised by emerging informa-tion technologies (for example, see Anderson and Jæger 1997).

Although not explicitly developed as such, consensus conferences are clearly consistent with the goals and sensibilities of DE, beginning with the intuition that democratic control of science and technology depends on citizens' discourse intended to generate consensus on those values and norms affecting the members of a community—in this case, with regard to the possible development and implementation of technologies with both obvious and not-so-obvious benefits and risks for human beings and their environment. Indeed Barbara Skorupinski and Konrad Ott (2002) have argued that the European consensus conferences, as efforts to develop what they call *participatory Technology Assessment* (pTA), are rooted not only in basic notions of democratic governance, but also precisely in the work of Habermas. They review six examples of such consensus conferences from the 1990s—including a Danish conference on GMO food, as well as Swiss and German conferences on genetic technology—to argue that these represent a sometimes imperfect implementation of DE. Similarly Richard Brown (1998) has argued that the environmental justice movement, including the specific history of Love Canal, can be evaluated in DE terms. To make his case, however, Brown develops a notion of science as narration in order to fit science more directly into the rhetorical and communicative DE frameworks.

Along with its ability to provide a framework for promoting the external democratic discussion of technology, DE may in a second instance also illuminate the internal structure of the scientific and technical communities—especially in terms of professional ethics. Robert Merton, the mid-twentieth-century founder of the sociology of science, analyzed the ethos of the scientific community as producing knowledge that is universal, commonly owned, not tied to special interests, and fallible (Merton 1942). Since Merton there has been considerable debate about the status of these norms, especially insofar as detailed case studies in the history and sociology of science have revealed the often parochial, egotistic, self-interested, and dogmatic behavior of scientists. Using DE, however, it might be possible to reconstruct the norms of professional science as precisely those principles that promote technical communication, and thus properly articulated and taught by means of professional ethics codes.

Pragmatic Discourse Ethics

Although discourse ethics has not been applied explicitly to analyzing or interpreting professional ethics in science or engineering, the explicit work in relation to

TA has been carried forward in special areas. For example, Matthias Kettner (1999) has elaborated additional conditions for moral discourse, such as *bracketing of power differentials* and *nonstrategic transparency* (that is, avoiding lies of omission), especially as applied to issues in bioethics. Similarly Jozef Keulartz and his colleagues, in *Pragmatist Ethics for a Technological Culture* (2002), sought to bridge Habermasian DE and pragmatism to deal with issues in agricultural ethics. In particular Paul Thompson (2002) draws on the American pragmatist tradition to avoid what he argues is a crucial failure of DE in Habermas—namely, that the emphasis on ideal speech situations tends to focus on debate *about* ethics (meta-ethics), rather than, as needed, move forward consensus-building about pressing issues.

DE has further played both a theoretical and practical role in connection with the Internet and the World Wide Web. For example, DE has been used to structure online dialogues regarding important but highly controversial social issues such as abortion. These dialogues in fact realize the potential of DE to achieve consensus on important community norms, insofar as they bring to the foreground important normative agreements on the part of those holding otherwise opposed positions, agreements that made a *pluralistic* resolution of the abortion debate possible (Ess and Cavalier 1997). In 2002 DE served as the framework for the ethics working committee of the Association of Internet Researchers (AoIR), as they sought to develop the first set of ethical guidelines designed specifically for online research—and with a view toward recognizing and sustaining the genuinely *global* ethical and cultural diversity entailed in such research. The guidelines stand as an example of important consensus on ethical norms achieved by participants from throughout the world.

The Future of Discourse Ethics?

Despite its promotion of a pluralistic universalism—namely, one that recognizes a wide range of possible discourse resolutions as shaped, for example, by diverse cultural traditions—discourse ethics is more prominent on both theoretical and practical levels in the Germanic cultures of Northern Europe than elsewhere. This regionalized predominance reflects a still larger cultural divide between the United States and Europe in terms of how to take up important ethical issues in science and technology. Thus Jeffrey Burkhardt, Paul Thompson, and Tarla Peterson (2000) note that European analysis and discussion of agricultural and food ethics is marked by a strong preference for *deontological* approaches to ethics, in contrast with the U.S. preference for *utilitarian* approaches. This same contrast can be seen in European approaches to data privacy protection and research ethics (as more deontological) versus American approaches (as more utilitarian).

That is, deontological approaches are associated with Kant and his emphasis on duties to individuals, which is required by their status as rational, autonomous beings. Kantian deontology is a central influence in DE. By contrast, utilitarian approaches—long associated with the Anglo-American philosophical tradition shaped by Jeremy Bentham and John Stuart Mill—seek instead to determine *the greatest good for the greatest number* through a kind of moral calculus that prefers those acts which maximize benefits and minimize costs. Markets, in particular, are justified on utilitarian grounds: While individuals and groups will inevitably lose out in market competition, such competition is justified as leading to greater economic efficiency and thus greater good for at least the greater number. Deontologists are wary of such strictly utilitarian approaches, precisely because they can result in the rights and interests of a minority being sacrificed for the ostensible benefit of the majority.

The Germanic reliance on DE in consensus conferences is thus consistent with the larger preference for deontological approaches. Indeed the European Commission continues to fund important initiatives concerning the ethical dimensions of emerging technologies such as GMO foods, human cloning, stem cell research, and therapeutic cloning research. By contrast the United States abolished the Office for Technology Assessment in 1995. Paul Riedenberg's observation about data privacy protection appears more generally true: The United States pursues a market-oriented (and thus more utilitarian) approach, in contrast with the European reliance on "socially-protective, rights-based governance"—an emphasis on the role of government to protect deontological rights and values (Reidenberg 2000, p. 1315).

In particular the success of consensus conferences in Europe—especially Scandinavia—appears tied to a well-defined set of conditions, beginning with the commonly held value that "democracy is only possible in a society where all citizens are enlightened enough to make an informed and conscious choice" in electing their representatives and voting—where such enlightenment further requires high levels of general education (Anderson and Jæger 1997, p. 150). Moreover the frameworks for Danish consensus conferences explicitly note that "market forces should not be the only forces involved" in deciding the design and deployment of information technology, which should further serve such fundamental deontological values as "free access to information and exchange of information" and "democracy and individual access to influence" (Anderson and Jæger 1997, p. 151). Consensus conferences thus exemplify what Reidenberg describes as the European emphasis on socially-protective, rights-based governance, in contrast with the U.S. utilitarian preference for market-oriented approaches.

Insofar as consensus conferences approximate DE ideals, societies must be committed to citizen enlightenment, as fostered by a strong educational system, and to citizen involvement in democratic processes, including those such as consensus conferences, as fostered by free access to information. In the twenty-first century, however, budgets for education systems continue to shrink and countries around the world are increasingly influenced by the U.S. emphasis on market forces alone to resolve important social issues. This is clearly a move away from socially-protective, rights-based governance in general, and from a belief that government should foster citizen assessment and possible regulation of technological development and deployment in particular. Spending taxpayers' funds on consensus conferences for the assessment of emerging technologies is explicitly criticized. Such circumstances are hardly promising for the application of DE to pressing issues in science and technology.

Nevertheless more promising conditions for DE as applied to democratic procedures for assessing science and technology may emerge in the future. Indeed such conditions are necessary for the sake of democratic procedures in TA. In addition the human, social, ethical, and financial resources required for DE and consensus conferences are the resources needed to realize and further more broadly the Enlightenment project of liberation and democracy.

CHARLES ESS

SEE ALSO *Consensus Conferences; Constructive Technology Assessment; Democracy; Deontology; Habermas, Jürgen; Internet; Kant, Immanuel; Rhetoric of Science and Technology; Technology Assessment: Germany and Other European Countries; Office of Technology Assessment.*

BIBLIOGRAPHY

Alexy, Robert. (1978). "Eine Theorie des praktischen Diskurses" [A Theory of Practical Discourse]. In *Normenbegründung–Normendurchsetzung*, ed. Willi Oelmüller. Paderborn, Germany: Schöningh. An English translation by

David Frisby is included in *The Communicative Ethics Controversy*, eds. Seyla Benhabib, and Fred Dallmayr (Cambridge, MA: MIT Press [1990]).

Anderson, Ida-Elisabeth, and Birgit Jæger. (1997). "Involving Citizens in Assessment and the Public Debate on Information Technology." In *Technology and Democracy: Technology in the Public Sphere—Proceedings from Workshop 1*, eds. Andrew Feenberg, Torben Hviid Nielsen, and Langdon Winner. Oslo: Center for Technology and Culture. A seminal discussion of consensus conferences as applied to information technology: Co-author Anderson remains a leader in organizing consensus conferences both in Denmark and abroad, including the United States.

Apel, Karl-Otto. (1980). "The A Priori of the Communication Community and the Foundations of Ethics." In *Towards a Transformation of Philosophy*. London: Routledge.

Apel, Karl-Otto. (2001). *The Response of Discourse Ethics to the Moral Challenge of the Human Situation as Such and Especially Today*. Leuven, Belgium: Peeters. Mercier Lectures, Louvain-la-Neuve, March 1999. A recent and accessible overview of Apel's understanding of discourse ethics and its contemporary relevance.

Benhabib, Seyla. (1986). *Critique, Norm, and Utopia: A Study of the Foundations of Critical Theory*. New York: Columbia University Press. A magisterial study of and central feminist contribution to critical theory from Hegel and Kant through Habermas.

Blickle, Gerhard; Sabine Hauck; and Wolfgang Senft. (1997). "Assertion and Consensus Motives in Argumentations." *International Journal of Value-Based Management* 10: 193–203.

Brown, Richard Harvey. (1998). *Toward a Democratic Science: Scientific Narration and Civic Communication*. New Haven, CT, and London: Yale University Press.

Burkhardt, Jeffrey; Paul B. Thompson; and Tarla Rae Peterson. (2002). "The First European Congress on Agricultural and Food Ethics and Follow-up Workshop on Ethics and Food Biotechnology: A U.S. Perspective." *Agriculture and Human Values* 17(4): 327–332.

Ess, Charles, and Robert Cavalier. (1997). "Is There Hope for Democracy in Cyberspace?" In *Technology and Democracy: User Involvement in Information Technology*, ed. David Hakken and Knut Haukelid. Oslo, Norway: Center for Technology and Culture.

Fixdal, Jon. (1997). "Consensus Conferences as Extended Peer Groups." In *Technology and Democracy: Technology in the Public Sphere—Proceedings from Workshop 1*, eds. Andrew Feenberg, Torben Hviid Nielsen, and Langdon Winner. Oslo: Center for Technology and Culture.

Habermas, Jürgen. (1981). *Theorie des Kommunikativen Handelns* [The Theory or Communicative Action], 2 vols. Frankfurt: Suhrkamp. English translation: *The Theory of Communicative Action*, Vol. 1: *Reason and the Rationalization of Society* and *The Theory of Communicative Action*; Vol. Two: *Lifeworld and System: A Critique of Functionalist Reason*, both translated by Thomas McCarthy (Boston: Beacon University Press [1984/1987]). Habermas's central effort to develop a systematic theory of communicative rationality that might resolve the "legitimation crisis" of enlightenment reason in the twentieth century—an effort further elaborated in the following two works.

Habermas, Jürgen. (1989). "Justice and Solidarity: On the Discussion Concerning Stage 6." In *The Moral Domain: Essays in the Ongoing Discussion Between Philosophy and the Social Sciences*, ed. Thomas Wren. Cambridge, MA: MIT Press.

Habermas, Jürgen. (1990). "Discourse Ethics: Notes on Philosophical Justification." In *Moral Consciousness and Communicative Action*, trans. Christian Lenhardt and Shierry Weber Nicholson. Cambridge, MA: MIT Press. Published in Germany as "Diskursethik: Notizen zu einem Begründungsprogram," in *Moralbewusstsein und kommunikatives Handeln* (Frankfurt: Suhrkamp [1983]).

Jacoby, Itzhak. (1985). "The Consensus Development Program of the National Institutes of Health." *International Journal of Technology Assessment in Health Care* 1: 420–432.

Jonsen, Albert R., and Stephen E. Toulmin. (1990). *The Abuse of Casuistry: A History of Moral Reasoning*. Berkeley: University of California Press.

Keulartz, Jozef; Michiel Korthals; Maartje Schermer; and Tsjalling Swierstra, eds. (2002). *Pragmatist Ethics for a Technological Culture*. Dordrecht, The Netherlands: Kluwer. A central collection of important contributions towards ethical frameworks for technology that include discourse ethics.

Kettner, Matthias. (1999). "Discourse Ethics: A Novel Approach to Moral Decision Making." *International Journal of Bioethics* 19(3): 29–36. One of several significant extensions and revisions of discourse ethics by a contemporary German philosopher.

Klüver, Lars. (1995). "Consensus Conferences at the Danish Board of Technology." In *Public Participation in Science: The Role of Consensus Conferences in Europe*, eds. Simon Joss, and John Durand. London: Science Museum.

Lagay, Faith L. (1999). "Science, Rhetoric, and Public Discourse in Genetic Research." *Cambridge Quarterly of Healthcare Ethics* 8(2): 226–237.

Marck, Patricia B. (2000). "Recovering Ethics After *Technics*: Developing Critical Text on Technology." *Nursing Ethics* 7(1): 5–14.

Merton, Robert. (1942). "Science and Technology in a Democratic Order." *Journal of Legal and Political Sociology* 1: 115–126. Reprinted as "The Normative Structure of Science" in *The Sociology of Science: Theoretical and Empirical Investigations* by Robert K. Merton (Chicago: University of Chicago Press [1973]).

Rawls, John. (1971). *A Theory of Justice*. Cambridge, MA: Belknap Press. One of the most significant contributions to political philosophy in the twentieth century—one that shares with Habermasian discourse ethics an emphasis on consensus procedures as the means of developing legitimate moral norms.

Reidenberg, Joel R. (2000). "Resolving Conflicting International Data Privacy Rules in Cyberspace." *Stanford Law Review* 52(May): 1315–1376.

Rorty, Richard. (1989). *Contingency, Irony, and Solidarity*. Cambridge, UK: Cambridge University Press.

Skorupinski, Barbara, and Konrad Ott. (2002). "Technology Assessment and Ethics." *Poiesis & Praxis: International Journal of Technology Assessment and Ethics of Science* 1(2)(August): 95–122.

Thompson, Paul B. (2002). "Pragmatism, Discourse Ethics and Occasional Philosophy." In *pragmatist Ethics for a Technological Culture*, eds. Josef Keulartz, Michiel Korthals, Maartje Schermer, and Tsjalling Swierstra. Dordrecht, The Netherlands: Kluwer.

INTERNET RESOURCES

Association of Internet Researchers. "Ethical Decision-making and Internet Research: Recommendations from the Aoir Ethics Working Committee." Available from www.aoir.org/reports/ethics.pdf. The first interdisciplinary, international ethical guidelines for Internet research, developed through an explicit application of discourse ethics.

Cavalier, Robert. Academic Dialogue on Applied Ethics. Carnegie-Mellon. Available from http://caae.phil.cmu.edu/Cavalier/Forum/ethics.html. See in particular "Abortion: Religious Perspectives," available from http://caae.phil.cmu.edu/Cavalier/Forum/abortion/abortion.html.

DISEASE

SEE *Health and Disease*.

DISTANCE

• • •

One of the well-recognized benefits of science and technology is that they reduce distance across both space and time. Science looks back in time toward the origins of the cosmos and provides information about microscopic phenomena and distant planets. Technologies of transportation and communication reduce the significance of distance limitations on human travel and personal interaction, making globalization a commonplace experience. But while celebrating the ways in which science and technology bring the far near, some thought must also be given to the ways science and technology can make the near far.

The social critic Ivan Illich (1973) was among the first to note some cultural and political implications of distance reversal. The automobile, for instance, brings the suburbs within a daily commuting distance of the central city, while simultaneously placing a living interaction with the city itself outside the bounds of a simple stroll. Illich argued that automobiles "can shape a city into its image," practically ruling out other forms of locomotion. He coined the term *radical monopoly* to des-ignate this type of exclusivity in rendering a service. Something analogous occurs when the telephone, the Internet, and cell phone enhance interactions with distant relatives and friends, while tending to situate immediate neighbors in other worlds. Such technologies invite people to virtually traverse distances at the same time that they might be contributing decisively to the impoverishment of local collectives, communities, and urban spaces. The advent of online education likewise tends to obscure the importance of nearness in knowledge acquisition (Huyke 2001).

As science attaches to the knowledge of distant places and times a kind of exotic glamour, one has to work hard to pay attention to what is immediately at hand. As people get used to online education, for instance, the illustrations brought forth by distant experts may outshine local experience and events. With the advent of biotechnology, high-yielding herbicide-resistant plants of major commodity crops become available throughout the world, shackling farmers to the patented plants and herbicides of a few multinational conglomerates, while also diverting them from local forms of agriculture and a more diverse produce.

Other commentators highlight the positive potential of such transformations in the character of distance. From the perspective of critical social theory, Andrew Feenberg (2002) calls for the democratic design and control of systems that facilitate self-organizing, nonterritorial communities throughout the globe. He likewise defends online education (which used to be called "distance education"), as long as it is "shaped by educational dialogue rather than the production-oriented logic of automation" (p. 130). The phenomenologist Don Ihde (1990) acknowledges an inevitable overwhelming of near "monocultural lifeworlds"—that is, ingrown German or Italian cultures, and especially indigenous cultures—but argues that independent of political efforts to limit the damage, such lifeworlds will become "pluricultural" through selective adoptions and incorporations. With the use of image-technologies, future traditions will inevitably be characterized by multiplicity and abundance, or what Ihde calls plurality. The local adaptation of global trends, a bringing of the far near sometimes known as "glocalization," can free individuals from the limitations of too specifically conceived traditions.

A third response seeks to identify those conditions that allow for personal, political, and cultural flourishing in the context of sciences and technologies that will continue to bring the far close and make the near distant. One insightful representative of this approach is the philosopher Albert Borgmann. In his 1984 book, *Technology and the Character of Contemporary Life*, Borgmann argued that the key to the good life is engagement with what he calls "focal things and practices" that order and intensify human experience, such as playing music or cross-country running. Contemporary technology, however, exhibits a guiding pattern, which he terms the "device paradigm," that is at odds with such experiences. Rather than needing to be played, music is able to be consumed by CDs and other devices, and running easily becomes an activity that takes place on a running machine rather than in nature.

The abstract problem of distance reversal is made concrete in the technological device itself, which increasingly hides its own near inner workings in favor of unhindered delivery of some commodity. The traditional hearth called forth ordered engagement in cutting wood and tending fire, and how it produced heat was transparent for all to see. The central heating system reduces engagement to a maintenance contract and is more or less mysterious to the consumer. Other examples permeate contemporary life: Few people know how digital clocks work, but such devices unambiguously state the time. Without the burdens of cooking, processed food is everywhere and available at any time. Humans progressively construct a world monopolized by the prominent availability of goods and a parallel disappearance of things and practices that might engage and challenge. Genuine nearness that could lead to "the unity of achievement and enjoyment, of competence and consummation" is replaced by the easy consumption of commodities that in the past would have required the expenditure of time or the traversing of space (Borgmann 1984, pp. 202–203). In the case of virtual reality, the line between the real and the virtual gets blurred in the context of "a deceptive sense of ease and expertise" that comes with digitalized cultural information about things (Borgmann 1999, p. 176).

Borgmann argues for a distinctive reform of technology. He has repeatedly called for the design of technologies that engage people bodily, socially, and politically. In opposition to Illich before him, Borgmann believes that more appropriate or enabling technologies will not constitute the deciding difference for a reformed future, because technological devices exhibit their own perfections and attractiveness. Instead he calls for a two-sector economy that would limit production with devices and of devices, leaving room for and encouragement of focal things and practices. To what extent such a project is politically feasible remains at issue. How it might help meet the challenges of time and space displacements found in scientific knowledge and technological tendencies is yet to be explored.

HÉCTOR JOSÉ HUYKE

SEE ALSO *Material Culture; Place; Space.*

BIBLIOGRAPHY

Borgmann, Albert. (1984). *Technology and the Character of Contemporary Life: A Philosophical Inquiry*. Chicago: University of Chicago Press. A comprehensive inquiry and critique of the technological pattern that prevails in contemporary life.

Borgmann, Albert. (1992). *Crossing the Postmodern Divide*. Chicago: University of Chicago Press. A reflection on the importance of the local and the communal for articulating a vision of postmodern culture.

Borgmann, Albert. (1999). *Holding On to Reality: The Nature of Information at the Turn of the Millennium*. Chicago: University of Chicago Press. A reflection on the history and limits of technological information or information as reality

Borgmann, Albert. (2001). "Opaque and Articulate Design." *International Journal of Technology and Design Education* 11(1): 5–11. A philosophical discussion of technological design from the perspective of the things and practices people value the most.

Feenberg, Andrew. (2002). *Transforming Technology: A Critical Theory Revisited*. Oxford: Oxford University Press. An inquiry on the social values that frame contemporary technology and the possibility of a change in direction based on genuine democratic values.

Huyke, Héctor José. (2001). *Anti-profesor: Reflexiones contra el profesor y su estudiante, con particular atención en la sociedad, el conocimiento y las tecnologías que se promueven en el salón de clases* [Anti-professor: Reflections against the professor and the student with particular attention to the society, the knowledge, and the technologies that are promoted in the classroom]. Río Piedras: Editorial de la Universidad de Puerto Rico. A critique of the university professor's role in reproducing some of the most troubling predicaments of contemporary conditions of life.

Ihde, Don. (1990). *Technology and the Lifeworld: From Garden to Earth*. Bloomington: Indiana University Press. An exploration of some of the most crucial cultural issues presented by the tools and instruments of the contemporary age.

Illich, Ivan. (1973). *Tools for Conviviality*. Berkeley, CA: Heyday. A critical examination of the institutions that dominate modern life and obstruct creativity and joy.

DOMINANCE

• • •

For students of animal behavior, *dominance* refers to the phenomenon by which individuals of a social species organize themselves with regard to access to resources. Although some social species appear to be egalitarian in many respects, close observations reveal differential access among individuals in nearly all cases, especially when resources are in short supply. These resources may include food, nest sites, mates, or any other considerations that have consequences for evolutionary success, or fitness; a dominance hierarchy is one of the most common patterns whereby access to these resources is established.

Dominance Hierarchies and Relationships

Although dominance relationships have in the past been seen as a species characteristic in themselves, they most importantly reflect differences in size, aggressiveness, and/or motivation among individuals, with these differences generating, in turn, a hierarchy of access to fitness-enhancing opportunities. It also appears to be beneficial to individuals to recognize their competitive relationship with respect to others, because without such recognition considerable time and energy might be wasted re-establishing priority, not to mention risking injury if a confrontation results in actual fighting. A signal characteristic of dominance hierarchies is that despite their aggressive underpinnings, animal societies characterized by rigid dominance relationships tend to experience relatively little actual fighting.

Most specialists maintain that—as with other biological phenomenon—there are no ethical implications of animal dominance relationships per se. While human observers may be inclined to deplore the *unfairness* whereby some individuals achieve disproportionate access to resources while others are comparatively excluded, dominance relationships, by definition, are not egalitarian. Indeed, during the late-nineteenth and early-twentieth centuries, when social Darwinism was especially influential, dominance relationships among human beings were considered admirable, as a working out of natural law. In the early twenty-first century, biologists acknowledge that dominance relationships among animals do indeed reflect the working out of natural tendencies and inclinations, as do predator-prey relationships, or the patterns of energy flow among different levels of natural communities. Just as neither eagles nor decomposing bacteria are good or bad, the same is true of dominance hierarchies. They are part of natural life, and as such, ethically neutral.

From an evolutionary perspective, dominance relationships among individuals develop because individuals are selected to maximize their fitness, their success in projecting copies of their genes into the future. Natural selection rewards those who succeed in doing so, and, in certain cases, this success is achieved by establishing one's self in a clearly defined situation of social superiority over others.

This is not to say that dominance relationships develop by some sort of intentional decision process on the part of the animals themselves, in which the latter get together and agree to establish a hierarchy. Rather, individuals who are somewhat larger, more aggressive, smarter, or who may have enjoyed such advantages in the past, simply assert themselves and, by virtue of that circumstance, succeed in gaining priority. Natural selection, in turn, supports those who achieve this success insofar as priority to food, mates, and nesting sites, among other things, correlates positively with ultimate reproductive success. Gene combinations that lead to success in such competition are favored in succeeding generations.

In some cases—barnyard chickens are the classic example—individuals end up establishing a *pecking order* whereby individual 1 dominates individual 2 and all those below, individual 2 dominates individual 3 and all those below, with that pattern continuing. However, dominance relationships are not always linear, nor are they always transitive: In many territorial species, for example, individual 1 may dominate individual 2, and individual 2 dominates individual 3, but individual 3 may dominate individual 1! In others—harem-keeping or polygynous species, such as elk, for example—there may be a single dominant individual (the dominant bull), who is clearly *number one*, with a less clear hierarchy among the remaining subordinate males.

Dominance relationships among animals depend upon an often tacit acknowledgment of the existing situation, on the part of dominants and subordinates alike. Thus once a dominance relationship is established, it is typically unnecessary for the various participants to fight—or even, in most cases, to engage in elaborate threat and subordination behavior—in order to maintain the pattern. When a dominance pattern is well established, individuals promptly respond to their mutual relationships by recognizing each other as individuals. (Indeed, this rapid, tacit response can be taken as powerful evidence of the participants' capacity to recognize individuals in the first place.)

Traditionally, dominance hierarchies have been seen as relatively immutable. More recent studies, however, have shown that they are not. Even though hierarchical relationships among animals tend to be resistant to change, they are subject to modification, as when a dominant male harem-keeper among langur monkeys is overthrown by one of the previously subordinate bachelors. Similarly, dominance hierarchies among female animals commonly vary as a function of hormonal and reproductive state: Breeding females and those in estrous often experience a temporary increase in their dominance status.

Correlation to Human Dominance Patterns

There is considerable variation in the nature of dominance relationships among different animal species, even some that are closely related. Chimpanzee social behavior, for instance, is generally oriented along lines of male dominance whereas the dominance system of bonobos (formerly known as pygmy chimpanzees) is primarily structured by the interactions of females. This, in turn, leads to question as to which animal system—if any—is most appropriate for understanding social dominance among human beings. Nonetheless biologists as well as increasing numbers of social scientists believe that in some complex way the biological nature of human beings underlies the nature of human politics just as that of other species underlies their pattern of social interactions.

Status signaling has also received considerable research attention. Although it seems legitimate to distinguish between physical characteristics (such as elaborate crests, ruffles, and antlers) used to achieve success in sexual selection by generating greater attractiveness to members of the opposite sex, such traits often also contribute to success in same-sex competition, and thus, with regard to dominance relationships. Would-be competitors are themselves more fit if they respond appropriately to indicators of probable physical or even mental superiority rather then subject themselves to possible injury or time wastage finding out *who is successful relative to whom*. Additionally it is probably adaptive for potential mates to employ the same traits that are used to establish and maintain same-sex dominance relationships as signals that generate success in between-sex courtship. This is because such traits—if genuinely connected to health and vitality—would lead to more successful offspring and hence be appropriate signals for an individual of one sex to employ in choosing a potential mate, and also because any offspring of such a union, insofar as they possessed these characteristics, would likely to be attractive to the next generation of *choosers*.

Among human beings dominance is a function of many things, including physical characteristics, intellectual qualities, and the control of material resources. Social dominance typically goes beyond the merely physical ability to intimidate a would-be rival, and carries with it signifiers of social rank such as clothing, make of automobile, speech patterns, and self-confidence. As in the case of animals, it is difficult —and perhaps impossible—to separate intrasexual from intersexual aspects of dominance. There is evidence that mastery of technology contributes to social dominance, and moreover, that the pursuit of technological and scientific success is generated, albeit unconsciously, by an underlying pursuit of social dominance (which itself is pursued because of its ultimate connection with reproductive success). The fact that such connections and motivations—if they exist—are not consciously pursued, does not make them any less genuine. At the same time, even as biologists are agreed that dominance and the pursuit of dominance is *natural*, there is no evidence that it is either ethically privileged or, by contrast, to be disparaged.

DAVID P. BARASH

SEE ALSO *Ethology; Selfish Genes; Social Darwinism.*

BIBLIOGRAPHY

Ardrey, Robert. (1971). *The Social Contract.* New York: Doubleday. An early and rather speculative attempt to examine the role of dominance hierarchies in human beings.

De Waal, Frans B. M., and Peter Tyack, eds. (2003). *Animal Social Complexity: Intelligence, Culture, and Individualized Societies.* Cambridge, MA: Harvard University Press. A technical work that examines the inter-relationships of animal social behavior—including but not limited to dominance relationships—and aspects of culture and intelligence.

Masters, Roger D. (1989). *The Nature of Politics.* New Haven, CT: Yale University Press. A political scientist considers the likely role of biological considerations in influencing complex patterns of human behavior.

Schein, Martin W. (1975). *Social Hierarchies and Dominance.* New York: Dowden, Hutchinson & Ross. A technical consideration of how dominance hierarchies form among animals and how they can be measured.

Smuts, Barbara. (1999). *Sex and Friendship in Baboons.* Cambridge, MA: Harvard University Press. A primatologist examines prosocial behavior among free-living baboons, showing that dominance relationships allow for benevolent interactions.

DOUBLE EFFECT
AND DUAL USE

• • •

In moral philosophy the principle of double effect traditionally refers to conflict situations in which an action or series of actions will result in multiple effects, some good and some bad. It parallels the contemporary policy concept of dual use: the idea that scientific knowledge or technological artifacts can serve multiple purposes, depending on the context. Dual use targeting and dual use research are areas that sometimes raise ethical dilemmas about the production and use of scientific knowledge and technologies but on other occasions provide multiple justifications for a single policy. Double effect seldom is referred to explicitly in those situations, but its general conditions may provide conceptual clarity with regard to moral permissibility. However, at the level of practical political decision making activities such as risk assessment, technology assessment, and scenario building provide better guidance for handling the ethical problems posed by dual use situations than does double effect reasoning.

Double Effect

Still widely discussed in the bioethics literature, the principle of double effect originated in Catholic scholastic moral philosophy, specifically in the discussion by the theologian Thomas Aquinas (1224–1274) of killing in self-defense:

> A single act may have two effects, of which only one is intended, while the other is incidental to that intention. But the way in which a moral act is to be classified depends on what is intended, not what goes beyond such an intention. Therefore, from the act of a person defending himself a twofold effect can follow: one, the saving of one's own life; the other the killing of the aggressor. (*Summa theologiae*, IIaIIae, q.64, a.7)

This raises the central distinction in double effect reasoning between intention and foresight (Aulisio 1995). In a morally acceptable case of killing in self-defense, the death of the aggressor is a foreseeable effect but the intention is to preserve one's own life. If, however, the killing was intended and not merely foreseen, it is considered homicide.

Originally formulated in slightly more complex terms, the principle of double effect commonly is stated as follows: An action with multiple effects, good and bad, is permissible if and only if (1) one is not committed to intending the bad effects either as the means or the end and (2) there is a proportionate reason for

bringing about the bad effects (Bole 1991). The proportionality clause arises from Thomas's insistence that one should not use more violence than necessary in defending oneself: "An act that is properly motivated may, nevertheless, become vitiated, if it is not proportionate to the end intended" (*Summa theologiae*, IIaIIae, q. 64, a. 7). Subsequent interpreters saw this condition as referring more broadly to the overall balance of good and bad effects.

Paradigm applications of double effect in Catholic bioethics pertain to cases of maternal-fetal conflict and distinctions between palliative care and euthanasia. Double effect also has been used in debates about the use of embryos in medical research. Many theorists question the relevance of double effect reasoning outside the Catholic moral framework (Boyle 1991). Some have argued that although the distinction between intention and foresight is difficult to apply practically, double effect nonetheless applies in any of the multiple moral frameworks that incorporate deontological constraints (in the form of intention) on consequentialist considerations (Kagan 1989). (Deontology asserts that certain acts are intrinsically right or wrong, whereas consequentalism asserts that the rightness or wrongness of an act depends on its consequences.) Traces of double effect reasoning can be seen even in Anglo-American law, for example, in the distinction between first-degree murder and manslaughter.

Double Effect and Dual Use

The concept of dual use is not well formulated for general use but can be understood in light of the principle of double effect as referring to any activity, artifact, or body of knowledge that is intended to bring about good effects but also has foreseeable negative consequences. This definition, however, excludes one of its most common applications: cost-sharing research programs involving industry and the military. For example, the U.S. Department of Defense operates a Dual Use Science and Technology Program to fund jointly with industry partners technologies that can be of use both on the battlefield and in the market. Defined in this sense, dual use is somewhat difficult to consider under the principle of double effect because there is no admitted or foreseen bad result, only multiple good ones. It merely refers to basic research with the potential for positive benefits in more than one sector of the economy and thus offers multiple justifications for governmental support. It is often the case that if political support for a research program cannot be marshaled with one argument (knowledge production alone), scientists have few qualms about appealing to others, such as military or health benefits and economic competitiveness. However, in this case ethical questions arise about whether both uses are equally sound or valid and whether rhetorical appeals to one may contaminate the other.

Insofar as dual use implies both good and bad outcomes, the concept presents even more fundamental challenges for social policies in regard to public support of science and technology. Stanley Rosen introduces the problem by noting that "all fundamental aspects of the natural sciences ... may lead to the destruction of the human race. ... Whereas no one would argue the wisdom of attempting to prevent a nuclear holocaust or the biochemical pollution of the environment, not many are prepared to admit that the only secure way in which to protect ourselves against science is to abolish it entirely" (Rosen 1989, p. 8). Security requires not only the abolition of science but also the destruction of all children because it is impossible to be certain who eventually may produce knowledge that threatens human existence. Rosen calls this the "ultimate absurdity of the attack against the enlightenment" (Rosen 1989, p. 9).

This absurdity follows from the notion of dual use because nearly all knowledge and artifacts, despite good intentions, could produce foreseeable bad effects. Examples can be as exotic as the "grey goo" (uncontrolled replication of nanotechnology) envisioned by Bill Joy (2000), as mundane as using a pen as a stabbing instrument, or as horrifying as the deadly use of commercial airplanes by terrorists on September 11, 2001. Rosen's point is that the only way to guarantee safety is to ban science and its technological products entirely.

Of course, society does not follow this absurd logic because most people feel that the benefits provided by science and technology (the intended good effects) make it worthwhile to risk some of the foreseeable bad effects. People seek a judicious regulation of scientific inquiry and technological progress, and it is in this middle ground that the major ethical questions are raised by dual use phenomena: Do the foreseeable bad effects outweigh the intended positive ones? Are there ways to minimize the negative effects without compromising the positive ones? Are there some foreseeable consequences that are so appalling that people should ban the production or dissemination of knowledge in a certain area altogether?

These questions show the importance of the proportionality condition of the principle of double effect. In fact, proportionality is disclosed through activities such as risk assessment, technology assessment, and scenario

building. Those activities involve processes of weighing the good and bad effects of research and technology in light of uncertainty about their relative probabilities. The distinction between intention and foresight is less difficult to apply, at least in theory, because if someone is attempting intentionally to bring about bad effects, say, by engineering a supervirulent pathogen, it seems obvious that there should be intervention to end that work. Indeed, in the realm of biotechnology dual use situations are difficult to deal with precisely because bad effects are not intended (cures, vaccines, and other good effects are intended) but nonetheless are foreseeable. Dual use situations present practical challenges to regulate research and ensure the proper use of technology in cases in which double effect analysis provides some insight and conceptual clarity. Dual use can be conceived of more broadly than can the conditions of double effect, however, because some bad effects of science and technology may be unforeseeable, let alone unintended.

Conduct of War and Biological Research

Precision-guided munitions and satellite-aided navigation have enhanced the accuracy of aerial bombardment. Although this has improved the ability of military planners to minimize collateral damage, it has raised an ethical dilemma: Military leaders are faced with questions of the legitimacy of dual use targeting, or the destruction of targets that affect both military operations and civilian lives. An example of such dual use targeting was the destruction of Iraqi electrical power facilities by the U.S. military in Operation Desert Storm in 1991.

Under the principle of double effect such activity would be deemed morally acceptable if the intention was not to harm or kill civilians (a bad effect that is foreseen but unintended) and the good effects outweighed the bad. This application of the principle of double effect relates to the idea of the just war that can be traced back to the theologian Augustine 354–430). Thomas expanded Augustine's idea that one cannot be held accountable for unintended effects caused by chance by applying that principle to include even foreseeable unintended effects that are not due entirely to chance. Like all versions of morality in terms of principles or formulas, however, the principle of double effect only establishes basic guidelines, and the majority of the work lies in deciding how and by whom such judgments about good and bad effects should be made.

Nuclear science provides the paradigmatic case of dual use summarized in the tension between physicists' initial hopes of "atoms for peace" and the grim reality of international proliferation of nuclear weapons. The dual nature of civilian and military use of nuclear science and technology poses grave problems in international relations, as witnessed by suspicions that Iran and other nations were developing nuclear weapons while claiming that such research was intended for civilian use only. The added possibility that terrorists could acquire weapons-grade nuclear material raises the stakes even higher.

The same concerns have surfaced around nanotechnology but have taken on a more mature form in regard to biological research. In 2004 the U.S. National Research Council (NRC) issued a report titled *Biotechnology Research in an Age of Terrorism*. Presenting recommendations to minimize the misuse of biotechnology, the authors warned: "In the life sciences ... the same techniques used to gain insight and understanding regarding the fundamental life processes for the benefit of human health and welfare may also be used to create a new generation of [biological warfare] agents by hostile governments and individuals" (U.S. National Research Council 2004, p. 19). Attention was paid to the risk that dangerous research agents could be stolen or diverted for malevolent purposes and the risk that research may result in knowledge or techniques that could facilitate the creation of novel pathogens. The report characterizes the central tension as one of reducing the risks of the foreseeable unintended bad effects while allowing for the continuation of the good effects yielded by biomedical research. One major dilemma is the trade-off between national security and scientific freedom of inquiry.

The distinction between intention and foresight and the proportionality condition are reasonable concepts for understanding the nature of this dual use situation. Clearly, mechanisms must be in place to ensure that researchers are not working intentionally toward bad effects either directly in the laboratory or covertly by sharing information with terrorists or other enemies. The more difficult questions, however, are left even when the assumption is made that no malevolent intentions exist.

The U.S. government established the National Science Advisory Board for Biosecurity (NSABB) to provide advice to federal departments and agencies on ways to improve biosecurity, which refers to practices and procedures designed to minimize the bad effects of biological research while maximizing the good effects. The U.S. Patriot Act of 2001 and the Bioterrorism Preparedness and Response Act of 2002 established the statutory and regulatory basis for protecting biological

materials from misuse. The NSABB develops criteria for identifying dual use research and formulates guidelines for its oversight and the public presentation, communication, and publication of potentially sensitive research. It works with scientists to develop a code of conduct and training programs in biosecurity issues. NSABB rules apply only to federally funded research. A possible avenue for the oversight of dual use research is Institutional Biosafety Committees (IBCs) for case-by-case review and approval.

The mechanisms fashioned by the NSABB for the regulation of dual use research are a good example of how the general spirit of double effect analysis is manifested in specific actions, raising political issues such as the proper balance of self-regulation by the scientific community and outside intervention. Members of IBCs and those involved in implementing other NSABB rules face the challenge of interpreting and applying the general guidelines provided by the principle of double effect in the sense that they must wrestle with difficult ethical dilemmas posed by good intentions and their foreseeable bad effects.

ADAM BRIGGLE

SEE ALSO *Consequentialism; Cultural Lag; Normal Accidents; Unintended Consequences.*

BIBLIOGRAPHY

Aulisio, Mark P. (1995). "In Defense of the Intention/Foresight Distinction." *American Philosophical Quarterly* 32(4): 341–354. Defends the distinction against challengers who claim that foresight of a probable consequence of one's actions is sufficient to consider that consequence part of one's intentions.

Bole, Thomas J. (1991). "The Theoretical Tenability of the Doctrine of Double Effect." *Journal of Medicine and Philosophy* 16: 467–473. Contends that the principle of double effect is relevant to different moral frameworks.

Boyle, James. (1991). "Further Thoughts on Double Effect: Some Preliminary Responses." *Journal of Medicine and Philosophy* 16: 467–473.

Joy, Bill. (2000). "Why the Future Doesn't Need Us." *Wired* 8(4): 238–262. Also available from http://www.wired.com/wired/archive/8.04/joy.html. A pessimistic outlook on the impending loss of human control as genetics, nanotechnology, and robotics become integrated research programs.

Kagan, Shelly. (1989). *The Limits of Morality.* New York: Oxford University Press. Argues that the ordinary understanding of limits imposed by morality and limits on what morality can demand of people cannot be defended adequately. Contains a section on intending harm.

Rosen, Stanley. (1989). *The Ancients and the Moderns: Rethinking Modernity.* London: Yale University Press.

U.S. National Research Council: Committee on Research Standards and Practices to Prevent the Destructive Application of Biotechnology. (2004). *Biotechnology Research in an Age of Terrorism.* Washington, DC: National Academies Press.

DRUGS

• • •

Drugs are notoriously difficult to define and yet present some of the most difficult ethical issues for the science and technology on which they are based. At the simplest level, drugs are molecules whose biochemical effects have been classified as socially desirable or undesirable in different times and places. *Dorland's Medical Dictionary* defines a drug as a "chemical compound that may be used on or administered to humans or animals as an aid in the diagnosis, treatment, or prevention of disease or other abnormal condition, for relief of pain or suffering, or to control or improve any physiologic or pathologic condition" (p. 510). But this ignores so-called *recreational drugs,* which may be described as substances used mainly for their psychoactive properties and pleasurable effects.

Historically drugs have been derived from plants and other natural materials and thus their production relied on indigeneous forms of knowledge and premodern techniques, often appropriated for modern applications. Over half of drugs in clinical use today continue to be derived from natural sources—including the excretions of insects, animal organisms, or microbes—from which they are extracted through direct or indirect processes (Aldridge 1998). The other half is synthesized through chemical processes that are now industrialized.

International Regulation

Ethical issues relating to human exploitation of indigenous knowledge and resources—sometimes called *bio- prospecting*—became central with the rise of a multinational pharmaceutical industry in the mid- to late-twentieth century. International treaties now provide safeguards that guarantee countries sovereign right over their genetic resources and a share of pharmaceutical profits derived from them. Yet such treaties also make national drug policy inflexible, inhibit innovation, and do not necessarily guarantee that indigenous groups that provide genetic materials are fairly compensated.

Given the high profit margins and relatively recession-proof nature of the pharmaceutical industry, drug production, marketing, distribution, and consumption are tightly regulated through a complex series of legal protocols and social controls that start from a set of international treaties that are coordinated through the United Nations Single Convention (1961), still the foundational document of international drug control (McAllister 2000). Prior to 1961 nine separate international treaties governed illicit or addictive drugs, primarily narcotics (opium and its derivatives), coca and its derivatives, and marijuana/hashish. The Single Convention defined the boundary between licit and illicit drugs, as well as legitimate medical and illegitimate, nonmedical, or recreational use, granting expert committees of the World Health Organization (WHO) authority for adding or altering *drug schedules,* which define how strictly drugs are regulated according to the level of their abuse liability. The Single Convention also mandated that national governments create and maintain drug-control agencies, and otherwise required signers to conform their domestic drug policies to its international mandate.

Regulation in the United States

Regulatory regimes are divided in the United States between illicit drugs, regulated by the Federal Bureau of Narcotics (FBN), the Treasury Department unit responsible for enforcing the Volstead Act (alcohol prohibition) and the Harrison Act, which transmuted into the Drug Enforcement Administration (DEA) in 1973; and licit drugs, regulated by the Food and Drug Administration (FDA) following the Food, Drug, and Cosmetic Act (1938). The 1938 Act granted the FDA authority to designate which drugs would be available only with a physician's prescription (Swann 1988). The liberalization of prescription laws in the 1960s and 1970s culminated in direct-to-consumer (DTC) advertising briefly becoming legal in the United States in the early 1980s prior to an FDA-imposed moratorium finally lifted in 1997. Supranational organizations such as the European Union have limited the spread of direct-to-consumer advertising (DTC) of prescription drugs. New Zealand is currently the only other country besides the United States that allows DTC, although it is under consideration elsewhere.

Well into the twentieth century, *proprietary medicines* were unregulated in terms of production, advertising, marketing, or distribution. Heavy advertising of commercial compounds emerged in the mid-nineteenth century United States, as patent medicine manufacturers were among the first to market their products nationally. Total advertising expenditures for proprietary medicines soon exceeded those of all other products combined; it was not unusual for nineteenth-century advertising budgets to exceed $100,000 per year, and some reached the million-dollar mark.

Narcotic drugs were restricted to prescription by the Harrison Act (1914), an outcome of growing international concern about widespread use and abuse of opiates, which were one of the few effective drugs then considered part of the medical armamentarium. An ongoing search for a non-addicting analgesic mounted by the National Research Committee propelled early-twentieth century innovation in the U.S. pharmaceutical industry in the context of concerns about addiction liability (Acker 2002). Addiction remains a classically public problem to which a coordinated federal response is understood as necessary, despite disagreement over the form that it should take. The United States remains the largest consumer of illicit drugs and has continued to struggle against what has proven a largely intractable problem. Basic research efforts into the neurobiochemistry of addiction led to the visualization of multiple opiate receptors, long hypothesized to exist, in the early 1970s. Federally funded studies of drug addiction have shifted away from the social and health consequences of abuse, and toward the use of molecular and animal models in establishing the basic neurobiological and now biogenetic mechanisms of drug action.

Drug Evaluation

Classified according to what is known about their mechanism of action, as well as their predominant effects on human and animal populations, both prescription and over-the-counter drugs must now be evaluated for safety, efficacy, abuse liability, and therapeutic effects. First their metabolic effects must be determined in animal models. Clinical trials in healthy human volunteers take place after pharmacokinetic studies in animals. Trials are divided into four phases, three of which take place before licensing and one of which occurs once patients are prescribed the drug by participating physicians. The large-scale clinical trials system in place in the United States since the 1960s is complex, lengthy, and expensive. Both the FDA and the National Institutes of Health (NIH) reluctantly became involved in regulating and coordinating the testing and licensure system for new drugs (Hertzman and Feltner 1997). Ethical concerns relevant to clinical trials include determinations of the capacity for informed consent of experimental human subjects; balancing rights to privacy and confidentiality with public access to information; the design and execution of double-

blind, placebo-controlled studies; and how to go about occasionally halting a trial as adverse effects become clear.

Ethical questions are raised both in terms of the type of drug development, production, marketing, and distribution being promoted; and the conditions of use. Drugs play a different role depending upon whether they are administered within allopathic or homeopathic therapeutic regimes. Homeopathy involves the administration of minute dosages of drugs designed to produce symptoms in healthy persons that mimic the symptoms of the disease for which the person is being treated. Developed by Samuel Hahnemann (1755–1843), homeopathy has been the target of many conflicting claims concerning its safety and efficacy in the face of the dominant practice in western medicine, allopathic treatment, which seeks to produce conditions that are incompatible or antagonistic to the disease. Many aspects of complementary and alternative medicine (CAM) are now being explored through large-scale clinical trials, since tremendously high percentages of U.S. patients now seek alternative practitioners in conjunction with allopathic practitioners, leading to a vast and less regulated market for so-called nutraceuticals, off-label use of pharmaceutical drugs, and herbal remedies untested by scientific regimes.

One of the major events in twentieth century history of drugs was the coincidence between the trend toward deinstitutionalization of mental hospitals that began in the 1940s with the psychopharmacological revolution that occurred upon introduction of a major tranquilizer, chlorpromazine (CPZ, marketed as Thorazine in the United States and Largactil in Europe), in the 1950s. This was followed by the first popular use of pep pills (amphetamines) and the mass marketing of minor tranquilizers such as Miltown in the late 1950s, which brought advances in psychopharmacology to popular attention (Smith 1991). Since that time periodic concerns have surfaced as to the social value of drugs used for performance enhancement or marketed widely as lifestyle drugs in ways that have changed the meaning of *medical use*. Pfizer Pharmaceutical's introduction of Viagra, a drug used to temporarily correct erectile dysfunction and targeted toward relatively affluent male consumers, brought to light disparities in insurance coverage of lifestyle drugs such as the lack of insurance coverage for female contraceptives, whose coverage has been restricted due to the abortion controversy. This *Viagra gap* illustrates one of the persistently troubling ethical issues in the domain of drugs, namely that of research and development targeted toward developing or widening markets among the affluent through life-style or look-alike drugs that are simply a means for drug companies to gain market share, compared to the relative lack of attention to drugs for treating orphan diseases that seriously affect small numbers of individuals, or those diseases—such as malaria or schistosomiasis—that mainly affect individuals in the developing world.

While the FDA is often regarded as an agency that largely serves the needs of the pharmaceutical industry, three major reproductive health controversies of the 1960s and 1970s propelled the FDA into taking a somewhat proactive regulatory role. These were the development of hormonal methods of contraception; widespread prescription of Thalidomide to pregnant women in Europe, while the drug was still experimental in the United States when it was demonstrated to cause severe birth defects; and prescription of diethylstilbestrol (DES), which caused *in utero* defects and increased rates of cancer in the children of women who took it. These controversies arose simultaneously with interrelated social movements that targeted health and physician-patient relationships, including the patients' rights movement, the consumer rights movement, the women's health movement, and, later, the HIV/AIDS movement. These social movements sought to limit the use of certain classes of therapies such as electroshock (ECT) and drugs such as the major tranquilizers or benzodiazepines (Valium) among certain populations. They also agitated for increased inclusion in clinical trials, earlier and more democratic access to experimental drugs, and expanded patients' rights including privacy, confidentiality, and informed consent.

Pharmacogenomics

Pharmacogenomics is the attempt to identify individual, genetic variation in drug response—metabolism, transport, and receptors—and to extend those findings to population genetics through a variety of information and visualization technologies. Pharmacogenomics promises individually tailored medications that would likely decrease adverse drug reactions, currently the fourth leading cause of death in the United States.

Projects in this research arena raise novel ethical and legal issues related to the creation of sample repositories or banks of genetic materials that would enable hypothesis-driven research on statistically significant differences in the phenotypes of human subjects. Such research could help establish the safety, efficacy, and compatibility of certain classes of drugs for particular individuals or populations; and could be used to create a complex set of biomarkers that describe the particular complement of different neuroreceptors that an individual has that may make him or her more or less responsive to a range of addictive substances (tobacco, opiates, and others) or prescribed medications. Pharmacogenetic databanking is potentially useful for avoiding adverse consequences but could also create a rationale for genetic and health-related discrimination. As with previous advances in the field of pharmacology, pharmacogenetics presents a double-edged sword, and its meaning and ethical value will be determined by the social contexts in which it is deployed.

NANCY D. CAMPBELL

SEE ALSO *Bioethics; Clinical Trials; Complementary and Alternative Medicine; DES (Diethylstilbestrol) Children; Food and Drug Regulatory Agencies; HIV/AIDS; Medical Ethics; Sports.*

BIBLIOGRAPHY

Acker, Caroline. (2002). *Creating the American Junkie: Addiction Research in the Classic Era of Narcotic Control.* Baltimore, MD: Johns Hopkins University Press. Thorough historical study of addiction research up to the 1940s, including the ongoing search for a non-addicting analgesic.

Aldridge, Susan. (1998). *Magic Molecules: How Drugs Work.* Cambridge, England: Cambridge University Press. Highly accessible general introduction to drugs.

Balance, Robert, Janos Pogany, and Helmut Forstner. (1992). *The World's Pharmaceutical Industries.* Brookfield, VT: Edward Elgar.

Campbell, Nancy D. (2000). *Using Women: Gender, Drug Policy, and Social Justice.* New York: Routledge. Social and cultural history of the representation of different populations of women who have used illicit drugs in the twentieth-century United States.

Courtwright, David T. (2001). *Dark Paradise: A History of Opiate Addiction in America,* enlarged edition. Cambridge, MA: Harvard University Press. Classic historical demography on opiate addiction in the United States.

Cozzens, Susan E. (1989). *Social Control and Multiple Discovery in Science: The Opiate Receptor Case.* Albany: State University of New York Press.

Dorland's Medical Dictionary. 1980. Philadelphia: Saunders Press.

Healy, David. (2002). *The Creation of Psychopharmacology.* Cambridge, MA: Harvard University Press.

Hertzman, Marc, M.D., and Douglas E. R. Feltner, M.D. (1997). *The Handbook of Psychopharmacology Trials: An Overview of Scientific, Political, and Ethical Concerns.* New York: New York University Press.

Liebenau, Jonathan, Gregory J. Higby, and Elaine C. Stroud, eds. (1990). *Pill Peddlers: Essays on the History of the*

Pharmaceutical Industry. Madison, WI: American Institute of the History of Pharmacy.

McAllister, William B. (2000). *Drug Diplomacy in the Twentieth Century*. New York: Routledge. Detailed diplomatic history of the global regulation of drugs.

Musto, David F. (1999). *The American Disease: Origins of Narcotic Control Policy*, 3rd edition. New York: Oxford University Press. Foundational text in the field of drug policy history.

Smith, Mickey C. (1991). *A Social History of the Minor Tranquilizers*. New York: Haworth Press.

Swann, John P. (1988). *Academic Scientists and the Pharmaceutical Industry: Cooperative Research in Twentieth-century America*. Baltimore, MD: Johns Hopkins University Press.

Terry, Charles, and Mildred Pellens. (1970 [1928]). *The Opium Problem*. New York: Bureau of Social Hygiene. Reprinted by special arrangement with The American Social Health Association, Montclair, NJ: Patterson Smith. Fascinating compendium of over 4,000 nineteenth- and early twentieth-century studies on the effects of opiates.

White, William L. (1998). *Slaying the Dragon: The History of Addiction Treatment and Recovery in America*. Bloomington, IL: Lighthouse Institute. The best of very few historical works on drug treatment regimens.

René Dubos, 1901–1982. Dubos pioneered in the development of antibiotics and was an important writer on humanitarian and ecological subjects. (© *Bettmann/Corbis*.)

DUBOS, RENÉ

• • •

René Jules Dubos (1901–1982), the French-American microbiologist and Pulitzer Prize-winning author, was born in Saint-Brice-sous-Forêt, France, on February 20. At the age of twenty-three, after completing his undergraduate training in agronomy, he used the money he made from translating scientific writings to travel to the United States. There he spent the rest of his prolific career, making groundbreaking contributions to antibiotic development, tuberculosis research, and environmental philosophy. René Dubos died in New York City on his eighty-first birthday.

Dubos's early work as a translator exposed him to the research of the Russian microbiologist Sergei Winogradsky, who stressed the importance of studying soil microbes in their natural setting, not just the sterile conditions of the laboratory. As Dubos reminisced late in life, "I have been restating that idea in all forms ever since. The main intellectual attitude that has governed all aspects of my professional life has been to study things, from microbes to man, not per se but in their complex relationships" (quoted in Kostelanetz 1980, p. 195). He earned his doctorate in agricultural microbiology from Rutgers University in 1927, and soon after won a fellowship from the Rockefeller Institute for Medical Research to find a way to disarm the microbe that causes pneumonia by destroying its protective polysaccharide coating. His unconventional approach entailed collecting dozens of soil samples in search of a bacterium that could decompose the material in question. Dubos's success led to his 1939 discovery of gramicidin, the first commercially produced antibiotic, which in turn stimulated efforts by other researchers to develop the antibacterial drugs that revolutionized medicine during the mid-twentieth century.

Dubos's ecological approach enabled him to predict the development of bacterial resistance to antibiotic drugs in the early 1940s, decades before antibiotic drug failure became a global health crisis. His subsequent research on the bacterium that causes tuberculosis, which killed his first wife, sharpened his appreciation of the social determinants of the disease, and his growing conviction that controlling microbial diseases required much more than eradicating the responsible microbes. In *The Mirage of Health* (1959) and *Man Adapting* (1965), Dubos challenged the dominant paradigm of scientific medicine by emphasizing the environmental determinants of disease, and the impossibility of vanquishing infectious diseases due to the constant flux of environmental conditions. A colleague at the Rockefeller University, Walsh McDermott, later hailed Dubos as "the conscience of modern medicine."

During the late 1960s and 1970s, Dubos's long career studying the links between environment, health, and disease facilitated his transformation into "the philosopher of the earth," as the *New York Times* called him near the end of his life. Dubos won the Pulitzer Prize for his book *So Human an Animal* (1968), in which he presents a holistic critique of modern civilization:

> Most of man's problems in the modern world arise from the constant and unavoidable exposure to the stimuli of urban and industrial civilization, the varied aspects of environmental pollution, the physiological disturbances associated with sudden changes in ways of life, the estrangement from the conditions and natural cycles under which human evolution took place, the emotional trauma and the paradoxical solitude in congested cities, the monotony, boredom and compulsory leisure—in brief, all the environmental conditions that undisciplined technology creates. (Dubos 1968, p. 216–217)

In later publications, Dubos elaborated his philosophy that humans can overcome such problems by creating what he called humanized environments that meet modern physiological, emotional, and esthetic human needs. His argument that humans can improve on nature by applying ecological insights to the built environment set him apart from the prominent pessimists of the burgeoning environmental movement, attracting widespread attention. The United Nations commissioned Dubos to chair a group of experts for the landmark 1972 United Nations Conference on the Human Environment, and to coauthor its influential background report, *Only One Earth* (1972).

Dubos's experience with the environmental megaconferences of the 1970s convinced him that solving global environmental problems requires dealing with them at the regional level, with respect to their unique physical, technological, economic, and cultural contexts. His practical approach spawned the famous phrase *Think globally, act locally*, which continues to inspire environmental activists around the world. He linked the maxim with his ecological insights and ethical concerns in *The Wooing of Earth* (1980): "Global thinking and local action both require understanding of ecological systems, but ecological management can be effective only if it takes into consideration the visceral and spiritual values that link us to the earth." Therefore "ecological thinking must be supplemented by humanistic value judgments concerning the effect of our choices and actions on the quality of the relationship between humankind and earth, in the future as well as in the present" (Dubos 1980, p. 157).

To promote such ideas in the policymaking arena, in 1975 he cofounded what later became the internationally recognized René Dubos Center for Human Environments. For reasons that include his prescient warnings against the overuse of antibiotics to his humanistic perception of environmental problems, Dubos deserves a central place among the foremost twentieth-century scholars of science, technology, and ethics.

CHRISTINE KEINER

SEE ALSO *Environmental Ethics*.

BIBLIOGRAPHY

Dubos, René. (1968). *So Human an Animal*. New York: Scribner. A reissue of Dubos's Pulitzer Prize-winning 1968 analysis of technologically-induced dehumanization.

Dubos, René. (1976). "Symbiosis of the Earth and Humankind." *Science* 193: 459–462.

Dubos, René. (1980). *The Wooing of Earth*. New York: Scribner.

Eblen, Ruth A. (1990). "Preface." In *The World of René Dubos: A Collection from His Writings*, eds. Gerard Piel and Osborn Segerberg Jr. New York: Holt.

Eblen, Ruth A. (1994). "René Jules Dubos." In *The Encyclopedia of the Environment*, eds. Ruth A. Eblen and William R. Eblen. Boston: Houghton Mifflin. This reference work embodies Dubos's humanistic approach to environmental concepts.

Kostelanetz, Richard. (1980). "The Five Careers of René Dubos." *Michigan Quarterly Review* 19: 194–203. A brief overview of Dubos's multifaceted career.

Moberg, Carol L. (1999). "René Dubos, A Harbinger of Microbial Resistance to Antibiotics." *Perspectives in Biology and Medicine* 42: 559–581. An analysis of Dubos's prophetic warnings of antibiotic overuse.

Piel, Gerard. (1980). "Foreword." In *The World of René Dubos: A Collection from His Writings*, eds. Gerard Piel, and Osborn Segerberg Jr. New York: Holt. A compilation of excerpts from Dubos's numerous publications regarding science, medicine, civilization, microbiology, ecology, and environmental management.

DURKHEIM, ÉMILE

• • •

Émile Durkheim (1858–1917), the son and grandson of rabbis, was born in the Alsatian town of Épinal, Vosges, France, on April 15. In 1887 he married Louise Julie Dreyfus, and the death of their son in World War I hastened Durkheim's own premature end in Paris on November 15.

Émile Durkheim, 1858–1917. Durkheim was one of the founders of 20th-century sociology. (*The Library of Congress.*)

In 1870, when Durkheim was twelve, German troops occupied his home during the Franco-Prussian War, forcing him to confront a normless, anomic (unstable) social environment and loss of collective well-being that was later to figure as a theme in his sociological research. He attended the École Normale Superieure (1879–1882), France's best teachers' college, and formed an early friendship with Jean Jaurès (1859–1914), later a leading socialist, which broadened Durkheim's academic and political interests to include philosophy and political action. In 1887 he was named professor at the University of Bordeaux, where he became the first person to teach social sciences in France, and from which he moved to the University of Paris in 1902. As a youth he had been schooled in the traditional education of male Jews, but when still young found himself attracted to Catholic mysticism, eventually dispensing with formal religion altogether. Nevertheless, a deeply religious and ethically alert sensibility shaped virtually all his mature scholarship, though skillfully recast in secular, scientific terms.

Durkheim's central sociological argument, which extends from his earliest to his final works, holds that a scientifically crafted theory of societal morality could prevent the sort of "anomie" that he thought afflicted citizens within France's Third Republic (1870–1940), and that extended as well to all rapidly industrializing nations. He treated this topic in his dissertation, *The Division of Labor in Society* (1893), a book with now almost biblical significance in sociology. Durkheim posed this question: How might morally binding norms be promulgated within a secularized and diversified society? His answer was that such norms would have to be shaped through professional groups, each of which would be responsible for guiding and monitoring the behavior of its members.

Other important works include *Suicide* (1897), which demonstrates that killing oneself is as much a sociological as a psychological event, and *The Rules of Sociological Method* (1895), which points to the "social fact" as the foundation of social research, thus separating sociology from the work of the other social sciences. The book he regarded his masterpiece, *The Elementary Forms of the Religious Life* (1912), is an exhaustive study of aboriginal religious practices compared with their modern progenies. With his nephew Marcel Mauss (1872–1950), Durkheim also cowrote *Primitive Classification* (1903), an innovative study in what came to be called "the sociology of knowledge." Highlighting as examples Australian aboriginals, the Zuni, Sioux, and Chinese, the two authors showed that the contrasting ways different societies arrange knowledge is a direct reflection of their particular forms of social organization; that is, they concluded, mental categories repeat social configurations. This was a direct attack on conventional epistemology, which held that all humans comprehend and analyze their environment in roughly the same way.

What gives Durkheim a unique status in the living tradition of classical social theory is his ability to blend science with ethics, as part of his lifelong effort to create what he called a "science of morality." To twenty-first-century ears this seems a quixotic venture, because science and ethical maxims have been severed one from the other (at least since Max Weber wrote "Science as a Vocation" in 1917, if not before), particularly among researchers whose principal allegiance is to scientific procedure. Yet even in his *Rules of Sociological Method* (still a key text for apprentice sociologists), he showed that identifying "social facts" is never an end in itself, but rather a realist propaedeutic (preparatory study) to understanding how norms operate in various societies, and how deviant behavior is curtailed or controlled.

In a famous essay, "The Dualism of Human Nature and Its Social Conditions," Durkheim invoked "the old formula *homo duplex*," explaining that "Far from being simple, our inner life has something that is like a double center of gravity. On the one hand is our individuality—and more particularly, our body in which it is based; on the other is everything that is in us that expresses *something other than ourselves*" (1973 [1914], p. 152; emphases added). Durkheim's deeply ambivalent relation to "pure" science originates in his divided loyalties as expressed in this essay: On one side stands the scientist looking for "laws" of social life; on the other is the ethicist and philosopher of culture, whose main goal is to identify, albeit via strictly scientific methods, the "something other" that encourages people to lay aside their natural egocentricity and embrace values that often conflict with their own best, individualized interests.

From his earliest work in *Division of Labor* and *Suicide* up through his masterly *Elementary Forms*, Durkheim always sang the praises of modern science and insisted that sociology be imbued with rigorous positivism. Yet never far away from his gaze were the "larger questions" that had troubled ethicists since Plato and Confucius, culminating in Leo Tolstoy's famous question: "What constitutes a life worth living?" To this pressing query, science has no answer, as Durkheim well knew.

In addition to his virtuosic sociological research, Durkheim also established the first scholarly journal of sociology in France, trained an entire generation of anthropologists and sociologists (many of them, along with his son, slaughtered in World War I), and wrote a posthumously published history of education in France that remains a standard work. Given all these scholarly achievements, many argue that Durkheim is indeed the father of modern sociology and the first to lay out in exact terms how the sociological viewpoint differs from that of its allied disciplines.

ALAN SICA

SEE ALSO *Communitarianism; French Perspectives; Parsons, Talcott; Professions and Professionalism; Sociological Ethics.*

BIBLIOGRAPHY

Durkheim, Émile. (1915 [1912]). *The Elementary Forms of the Religious Life*, trans. Joseph Ward Swain. London: Allen and Unwin.

Durkheim, Émile. (1933 [1893]). *The Division of Labor in Society*, trans. George Simpson. New York: Free Press.

Durkheim, Émile. (1938 [1895]). *The Rules of Sociological Method*, 8th edition, trans. Sarah A. Solovay and John H. Mueller, and ed. George E. G. Catlin. Chicago: University of Chicago Press.

Durkheim, Émile. (1951 [1897]). *Suicide*, trans. John A. Spaulding and George Simpson. Glencoe, IL: Free Press.

Durkheim, Émile. (1973 [1914]). "The Dualism of Human Nature and Its Social Conditions." In *Émile Durkheim on Morality and Society*, ed. Robert Bellah. Chicago: University of Chicago Press.

Durkheim, Émile, and Marcel Mauss. (1963 [1903]). *Primitive Classification*, trans. and ed. Rodney Needham. Chicago: University of Chicago Press.

LaCapra, Dominick. (1972). *Emile Durkheim: Sociologist and Philosopher*. Chicago: University of Chicago Press.

Lukes, Steven. (1972). *Emile Durkheim: His Life and Work*. San Francisco: Harper and Row.

Poggi, Gianfranco. (2000). *Durkheim*. New York: Oxford University Press.

INTERNET RESOURCE

Weber, Max. (1917). "Science as a Vocation." In *From Max Weber: Essays in Sociology*, ed. H. H. Gerth and C. Wright Mills. New York: Oxford University Press (1946). Available from http://www2.pfeiffer.edu/∽lridener/DSS/Weber/scivoc.html.

DUTCH PERSPECTIVES

• • •

In the Netherlands, various styles of applied ethical research can be distinguished. They have resulted in "best practices" that formerly regarded each other as competitive, but tend to see themselves as complementary in the early twenty-first century.

Two Preliminary Observations

A first general observation is historical. Twenty centuries ago, the border of the Roman Empire followed the Rhine, thus dissecting the area that later was to become the Netherlands into a southern part (inside the empire) and a northern and western part (outside the empire). This division has written itself into the Dutch cultural landscape in an astonishingly obstinate manner. It is still noticeable today, in terms of dialect, culture, manners, ethics, and religion. Whereas before the onset of secularization the south was predominantly Catholic (that is, oriented toward "Rome"), the north and west were predominantly Protestant.

This difference in cultural geography continues to be visible in the domain of ethics. In the south, ethical research tends to be oriented toward and influenced by

Continental (notably German and French) intellectual developments and trends. Thus, ethicists from this area are influenced mainly by hermeneutical or phenomenological approaches. Ethicists from the northern and western part, however, are more likely to be influenced by analytical approaches and debates. They often subscribe to theories and views that dominate the Anglo-American spheres of influence. Although the difference has become less obvious than it was in the 1980s, the two ethical profiles remain distinguishable.

A second observation has to do with the international status of Dutch ethics. It has been said that Dutch philosophy is the philosophy of the country that possesses the largest harbor in the world, namely Rotterdam (Nauta 1990). And because ethics is a special discipline within the broader field of philosophy, this goes for ethics as well. What does it mean? One might say that Dutch ethicists are better at importing and exporting than at producing philosophy. In terms of style, the Dutch are neither as "profound" as the Germans nor as sensitive to new trends as the French. They do have a special talent, however, for intellectual transfer. Their mastery of international scholarly languages such as English, German, and French also plays a role here. Dutch philosophers often serve as intellectual intermediaries. This is, of course, a generalization, but a systematic review of academic performance will show that as a rule the Dutch tend to focus on assessing, processing and connecting ideas rather than on originating them.

Three Styles of Ethical Research

Three styles of ethical research exist in the Netherlands. They start from different understandings of what ethics is.

(1) ethics = analyzing and solving moral problems

(2) ethics = intellectual reflection

(3) ethics = moral conflict management

According to the first option, which is based on a more or less Anglo-American approach, an ethicist is someone who analyzes moral problems and formulates possible solutions, usually by applying a set of moral principles (ethical input) to problem cases (solutions as output).

The second option reflects a more hermeneutical or Continental way of thinking. An ethicist is seen as someone who tries to interpret certain forms of moral discourse by situating them in a broader cultural and historical perspective. The focus is on understanding, rather than on solving, problems. The philosophical ethicist works toward a "diagnosis" rather than a "solution."

The third option entails a more pragmatic approach. The ethicist identifies stakeholders and value perspectives, and works toward consensus formation, based on stakeholder participation, by means of interviews, workshops, and similar techniques.

These three ways of doing ethical research entail different views on the relationship between expert knowledge and public knowledge. According to the first option, ethicists are experts, perhaps even "ethical engineers" (Van Willigenburg 1991). They have learned to analyze moral problems in a professional manner. Consistency is important, even if this means that ethicists distance themselves from common intuitions and conventional morality.

According to the second option, however, the ethicist's expert knowledge is knowledge of moral traditions, of types of discourse, or of fundamental cultural attitudes that are noticeable in the ways in which moral debates evolve and problem cases are being framed and presented (Van Tongeren 1994). The ethicist relies on erudition rather than analytical tools. The attention is directed toward fundamental issues rather than concrete problems. In other words, the problem cases at hand are regarded as exemplifications of broader, cultural issues.

According to the third option, it is not the ethicist's job to add new insights, but rather to build on the knowledge, values, and intuitions of the stakeholders involved. Rather than performing desk research, the ethicist enters into dialogue with others, inviting them to articulate and clarify their (tacit) views. The ethicist's expertise is of a pragmatic and intermediary nature (Keulartz et al. 2002). Ethicists have at their disposal a toolbox for moral deliberation and moral conflict management. Their input in the decision-making process does not come from ethics as such, but from the views and experiences of stakeholders themselves.

Through the late 1990s, the first style of doing ethical research dominated (the public image of) institutionalized ethics in the Netherlands, whereas the second style was more prevalent in academic circles. Since the early 2000s, the pragmatist approach is gaining ground. In fact, Dutch ethicists tend to be flexible when it comes to method in the early twenty-first century. To some extent, they are willing and able to use all three models, depending on context. Congenial with the pragmatist turn, but not exactly identical with it, is the *empirical turn* in ethics. More and more often, research in applied ethics involves the collection of empirical data and the use of tools borrowed from the social sciences such as interviews, questionnaires, and participant observation.

Ethics of Science and Technology: Examples

In the Netherlands, as elsewhere, moral disputes tend to arise in response to technological changes. Initially, the growing interest in ethical research was associated with medical or clinical ethics. An interesting case is the famous Dutch euthanasia debate that started around 1970 in response to the dramatic increase of medical technology and therefore of treatment options with which many lives, that previously would have had no chance of survival, could now be saved, or at least prolonged. The debate was triggered by Jan Hendrik van den Berg (1978), a physician who was also trained as a phenomenologist, and therefore a representative of Continental philosophy. Moral problems involved in end-of-life decision were interpreted as indications that something was fundamentally wrong with current views and attitudes toward life and death *as such*. Soon, however, the debate was taken over by applied ethicists who subscribed to an analytical approach. On the basis of the principle of autonomy, they argued in favor of the patients' right to refuse treatment or even to request that physicians end their lives. Eventually, the ethical debate over euthanasia shifted toward a more pragmatic and empirical approach: How are end-of-life decisions actually taken, and by whom, how often, and on what grounds? Last but not least, what kind of technical contrivances co-influence decisions of this type?

During the 1990s, the attention of professional ethicists in the Netherlands drifted away from euthanasia. Reproductive technologies, biotechnology, genetic modification of organisms, and animal research became important items of concern. Even more so than in the case of medical ethics, moral disputes arose in response to technological change. These debates thus exemplified the ways in which technological developments influence ethical controversies. After the introduction of recombinant DNA techniques in the 1970s and 1980s, the genetically modified research animal became an important object of research, and knockout experiments (deleting genes) became an important research tool.

This new technology had a major impact on ethical debates concerning laboratory animals. It caused the focus of the debate to shift away from traditional concerns (animal suffering and animal welfare) to issues involved in the recently acquired power of biologists to modify—to *change*—their laboratory animals, and to adapt them to research requirements. Concepts such as integrity and intrinsic value, borrowed from medical and environmental ethics, respectively, were used to articulate new moral concerns over genetic engineering.

Furthermore, the three styles of ethical research distinguished above are recognizable here as well, although demarcations are somewhat less rigid than before. The majority of contributions to animal ethics and biotechnology ethics since 2000 adhere to a more or less analytical approach. Their usual aim is to enrich a traditional, consequentionalist view (focusing on animal welfare and animal suffering) with deontological elements, using concepts such as integrity and intrinsic value (Heeger and Brom 2001). A more Continental and phenomenological approach, however, is represented here as well. Its aim is to elucidate the different ways in which animals are perceived. Thus, the scientific understanding of animalhood is confronted with life-world perspectives and artistic perspectives. In other words, this line of research studies the various conditions under which relationships with animals (notably in the context of research practices) evolve (Zwart 2000). Finally, promising examples of empirical and pragmatic approaches have begun to enter the animal ethics scene as well.

Early Twenty-First-Century Developments

Genomics, the most recent chapter in the history of the life sciences and their technological applications, is what occupies the majority of ethicists in the Netherlands in the early twenty-first century. The basic trend is toward establishing large, multidisciplinary programs in the domain of ethical, legal, and social issues (ELSI) research. In the context of such programs, ethicists (of various styles and backgrounds) collaborate, not only together, but also with experts coming from various other disciplines, such as the social sciences, psychology, cultural studies, communications, economics, and law. This trend is sometimes referred to as the "elsification" of science and technology.

During the 1990s, the focus of applied ethicists tended to be on the individual or institutional level (the micro- and meso-level) rather than on the societal (or macro-) level. The empirical turn in ethics likewise tended to restrict itself to research on a relatively small scale. But in the early 2000s it became clear that the most challenging issues involved in so-called "enabling technologies," such as genomics, will present themselves on a much broader, cultural, and societal scale. Rather that providing information on discrete monogenetic defects (relevant for specific target groups), for example, genomics is expected to inundate the public realm with genetic information on multifactorial health risks that will be relevant for virtually everybody.

Although the ethics of science and technology in the Netherlands tends to focus on the life sciences and biotechnology, and on genomics in particular, this is but one example of "enabling technologies" that are emerging in research laboratories in the early twenty-first century. Other technologies, notably Information and Communication Technologies (ICT) and nanotechnology, are items of concern as well (Van den Hoven 1999; De Mul 1999). They are regarded as enabling technologies in the sense that they will give birth to a wide variety of applications. As ethical debates tend to reflect technological developments, the agenda of ethics will no doubt continue to orient itself toward these three major scientific and technological breakthroughs of the past and coming decades.

Genomics, ICT and nanotechnologies will give birth to a wide variety of new and yet unanswered questions. How will new technologies in these fields change existing roles and responsibilities of professionals and citizens? How can the knowledge and information that is generated in these fields be evaluated and used; how can abuse be prevented? In answering these questions, ethicists will find themselves no longer alone, but in the company of (in particular) scholars from Science and Technology Studies (STS) and from the Philosophy of Technology (who often are members of the STS community in a broader sense).

STS scholars study the ways in which science and technology are intertwined (in terms of content and organization, but also socially) with the development of modern societies and cultures. Science and technology are regarded not as the producers or influencers of society and culture, but both science and technology on the one hand and society and culture on the other are seen as interacting with one another and as *co-producing* one another. While STS formerly focused on the deconstruction of epistemological claims, thereby underpinning the idea that there are different ways to perceive nature or reality, the field in the early 2000s tends to move towards a more normative and hence ethically oriented approach. Constructive Technology Assessment (CTA) for example, geared towards the "management of technology in society," aims at early feedback and learning cycles in the development of new technologies, particularly with respect to the societal use and entrenchment of new technologies (Rip et al. 1995, Schot et al. 1997).

The ambition of STS scholars to put on the agenda the political question "how to help shape the technological culture we live in" has influenced the landscape of STS into a more normative direction (Bijker 1995

among others). Large technological "projects," and the transformations they are expected to induce, such as nanotechnology, genomics, and ICT, thus have increased the interest for ethical and normative questions from different fields and disciplines. Ethical questions have become the domain of an interdisciplinary research field. Put differently, "elsification" (entrenchment of ethical, legal and social projects in large technological programs) has enhanced new forms of ethical research, characterized by interdisciplinary collaboration, proximity to scientific consortia, and sensitivity to social change. The development of new interdisciplinary modes of doing ethical research also gives rise to new networks and institutions. Interesting examples are Nanonet and the establishment of the Centre for Society and Genomics (CSG) at the University of Nijmegen.

Institutionalization

It is to be expected that in the near future collaboration between philosophers and ethicists on the one hand and social science researchers on the other will continue to increase. At the moment, they still can be seen as separate domains. Research in the Netherlands is organized on the basis of research schools that assemble experts from various universities into common programs. With regard to research into the societal aspects of science and technology, two research schools are particularly relevant: the Onderzoekschool Ethiek (the Netherlands School for Research in Practical Philosophy) and the Onderzoeksschool Wetenschap, Technologie en Moderne Cultuur (the Netherlands Graduate Research School Science, Technology, and Modern Culture, WTMC). Both research schools were established in 1994. In the Netherlands School for Research in Practical Philosophy the analytical style is dominant, but pragmatic and Continental approaches are represented as well. Methodology and epistemology of ethics have been important issues from the very outset, and the "empirical turn in ethics" is a major item of concern. The Netherlands Graduate Research School Science, Technology, and Modern Culture brings together researchers from the interdisciplinary field of science and technology studies (STS). In the Netherlands, STS emerged in the late 1960s as a result of new interactions between history, philosophy and sociology of science. The focus of WTMC is on the interrelatedness and interpenetration of science, technology, and society. The membership list of WTMC indicates that the school recruits scholars from the sociology of science, history of technology, philosophy of technology, philosophy of science, arts and culture, psychology, political sciences, science dynamics and policy and innovation studies.

Although demarcations in terms of style have become less obvious than in the past, the Netherlands School for Research in Practical Philosophy is dominated by ethicists who come from an analytical background, although Continental and phenomenological approaches to technology are present as well. The Netherlands Graduate Research School Science, Technology, and Modern Culture is oriented more toward pragmatism and constructionism. Yet, as was already noticed, within the Dutch STS community, interest in normative (ethical) issues has increased in the past five years. See for example Verbeek (2003), who analyzes the ways in which artifacts influence human experience, while new technologies are interpreted as material answers to ethical questions.

The Future

Until recently, bioethics and the philosophy of technology were seen as separate fields. As has been indicated, this will no longer hold in the near future. Bioethics increasingly will have to regard itself as an ethics of science and technology. A broader understanding of the coevolution of science and technology thus will have to become an integral part of bioethics. The emphasis (within applied ethics and bioethics) on the micro-level will shift towards the development of science and technology *at large* and towards ethical and philosophical questions concerning the role of science and technology in modern societies. The focus on (and the interest for) the moral aspects of (for example) the interaction between physicians and patients, or between laboratory researchers and laboratory animals, will be increasingly overshadowed by the need to address the social dynamics of technological change. These broader issue will dominate the future agenda of bioethics, applied ethics and—as it often does already—the philosophy of technology.

Ethics can be expected to broaden its perspective and become an increasingly interdisciplinary endeavor. And while ethicists will "discover" the importance of the broader social and cultural impact of technological innovations, social scientists already working on these questions will increasingly acknowledge the importance of the normative issues they tended to avoid in the past.

HUB ZWART
ANNEMIEK NELIS

SEE ALSO *Applied Ethics; Engineering Ethics; European Perspectives.*

BIBLIOGRAPHY

Bijker W. E. (1995). *Democratisering van de Technologische Cultuur Inaugurale rede* [Democratisation of technological culture: inaugural lecture]. Maastricht: University of Maastricht.

De Mul, Jos. (1999). "The Informatization of the Worldview." *Information, Communication, and Society* 2(1): 69–94.

Heeger, Robert, and Frans Brom. (2001). "Intrinsic Value and Direct Duties: From Animal Ethics towards Environmental Ethics." *Journal of Agricultural and Environmental Ethics* 13(1): 241–252.

Keulartz, Jozef; Michiel Korthals; Maartje Schermer; and Tsjalling Swierstra, eds. (2002). *Pragmatist Ethics for a Technological Culture.* Dordrecht, Netherlands: Kluwer Academic.

Nauta, Lolle. (1990). "De subcultuur van de wijsbegeerte: Een privé-geschiedenis van de filosofie" [The subculture of philosophy: a private history of philosophy]. *Krisis* 38: 5–19.

Rip, Arie; Thomas Misa; and Johan Schot, eds. (1995). *Managing Technologies in Society: The Approach of Constructive Technology Assessment.* London: Printer Publishers.

Schot, Johan and Arie Rip. (1997). "The past and Future of Constructive Technology Assessment. *Technological Forecasting and Social Change* 54(1997): 251–268.

Van den Berg, Jan Hendrik. (1978). *Medical Power and Medical Ethics.* New York: Norton. Originally published in Dutch, 1969.

Van den Hoven, M. J. (1999). "Ethics, Social Epistemics, Electronic Communication, and Scientific Research." *European Review* 7(3): 341–349.

Van Tongeren, Paul. (1994). "Moral Philosophy as a Hermeneutics of Moral Experience." *International Philosophical Quarterly* 34(2): 199–214.

Van Willigenburg, Theo. (1991). *Inside the Ethical Expert: Problem Solving in Applied Ethics.* Kampen, Netherlands: Kok Pharos.

Verbeek, Peter-Paul. (2003). *De daadkracht der dingen* [The activity of things]. Amsterdam: Boom.

Zwart, Hub. (2000). "The Birth of a Research Animal: Ibsen's *The Wild Duck* and the Origin of a New Animal Science." *Environmental Values* 9(1): 91–108.

DYING

SEE *Death and Dying.*

DYSTOPIA

SEE *Utopia and Dystopia.*

E

EARTH

• • •

In science and philosophy earth (German *Erde*, Greek *ge*) can refer to one in a set of primordial material elements (earth, air, fire, and water, for the Greeks; wood, fire, earth, metal, water, for the Chinese) and to the physical body on which humankind lives. As physical home, the Earth serves as the reflective horizon or framework for human self-awareness and as a contingent unity among the array of individual entities they encounter. The Earth, defined by an elemental earthiness of rock and soil, is that which grounds the identity of humans in both physical and psychological senses, independent of wherever they may venture in information networks or outer space, while serving as a fund of resources available for exploitation. The tensions between these various approaches are imaged in the diagrammatic icon of the atom and the photo of the blue planet taken from space: matter that is mostly space and a life-giving sphere that appears more water than rock and calls perhaps for technological management.

Earth Science and Engineering

As soil and matter, earth has become a distinctive object for science and technology. The material out of which all things are made has itself become subject to chemical processing, synthesis, and nuclear engineering. The scientific study of matter existing independent of humans has expanded to examining those new forms of matter intentionally and unintentionally designed by humans and the interactions between the two, especially insofar as they may impact on humans themselves. As a planet the Earth is a body in space with a stable orbit at a distance from the sun suitable for the origin

and evolution of life. During its 4.56 billion years the Earth has given rise to an abundance of organisms, first in the sea, and then diversifying and evolving to occupy land and air. As recipient of heat energy from both the sun and its own core, the Earth is a site of dynamic terrestrial behavior. The seven major tectonic plates comprising its rocky outer crust diverge and then compensate through convergence; its land masses, ocean basins, islands, and other prominent features such as volcanoes, mountains, plains, and valleys have gone through continual development—producing new materials essential of life and humans. Hominids appeared on the Earth 7 million years ago and Homo sapiens about 200,000 years ago. Humans began to till the Earth about 10,000 years ago.

This early-twenty-first-century perspective on earth and the Earth sees them as dynamic complexities inviting examination and provoking manipulation. Especially with regard to the Earth, it is now perceived as a nexus of interactions between the solar system and its own atomic and subatomic foundations, as well as of exchanges between its own landmasses, oceans, atmosphere, and living organisms. Through earth system science these have in turn become, because of human technological powers and their commercial development, also subject to speculative engineering management. Earth system science is complemented by the possibilities of earth systems engineering and thus challenged to reflect ethically on both ends and means.

Philosophies of the Earth

The Earth has throughout human history been a focus of philosophizing, central to ethics, and a framework for self-understanding. For the Greeks, the Earth was implicated

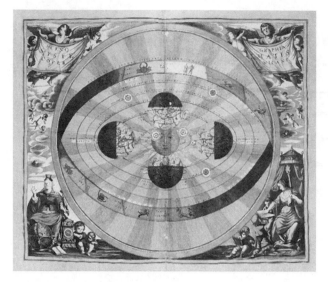

A print by Andea Cellario entitled "Harmonia Macrocosmica," showing the Ptolemaic system. Proposed by Claudius Ptolemy in the 2nd century A.D., the system postulated that the earth was at the center of the universe, and was accepted for more than 1000 years. (© Enzo & Paolo Ragazzini/Corbis.)

in their cosmology not only as planet and home of humanity, as focus of the gods who lived above its plane, and in relation to the heavens; its core constituent, earth, was also one of the four elements, earth, water, air, and fire. The Earth itself was a compound of the four elements. For the Chinese, the earth and heaven are the two forces responsible for engineering and completing nature and all its aspects. Earth and heaven also work together to create the five Chinese elements: wood, fire, earth, metal, and water. As an element, the Earth is located at the center and is the cauldron, with the other four elements located in the four outer directions, east, west, north and south. Earth is also the element of the "naked" animal, the human, of the actions is representative of "thought." The element earth is also associated with the sense of touch, the sound of singing, the organ of the spleen, and the virtue of good faith. There have been two related controversies about the place of the Earth within cosmological, metaphysical, and ethical visions: whether the Earth is the center of a given scheme of existence or is only an element in a vaster cosmos, and whether the earth is a site of corruption at a distance from a purer realm or is a unique locus of corporeal and spiritual development.

Plato approaches both issues in his atypical dialogue, the *Timaeus*. He describes an original Demiurge who takes the elements of earth, fire, air, and water and "out of such elements which are four in number, the body of the world was created, and it was harmonized by proportion, and therefore has the spirit of friendship; and having been reconciled to itself, it was indissoluble

by the hand of any other than the framer" (33, c). This picture of the Earth as a model of balance, harmony, and fairness is complemented by a *world soul* infusing the world with a vitality and rationality of fair proportion: "The world became a living creature truly endowed with soul and intelligence by the providence of god" (30, c). The Earth as an entity in the cosmos is described as located at the center and surrounded by the moon, the sun, and five planets in circular orbits. This picture from the *Timaeus* is opposed by another from the *Phaedo*. There Plato writes of the "earth and the stones and all the places [as] corrupted and corroded, as things in the sea are by brine so that nothing worth mention grows in the sea, and there is nothing perfect there, one might say, but caves and sand and infinite mud and slime wherever there is any earth, things worth nothing" (110, b). He condemns the passions and senses for "nailing" people to the Earth that, by its attractive power, can "drunkenly" estrange human souls from their true home in the aether beyond (83, d). His emphasis on the immaterial nature of the soul and its kinship with the intelligible structure underlying reality leads to a condemnation of the earthly as tempting snare.

Aristotle, by contrast, observes the Earth and catalogues its differences in beings—animate and inanimate—embracing "the delight we take in our senses," especially the sense of sight as indicating that "this, most of all the senses, makes us know and brings to light many differences between things" (980, a). His vision of the Earth as a nexus of beings defined by their *for sake of which*—their purpose as fully actualized—working in concert with other beings' drive to actualization, renders a grandeur to the dynamism and wholeness of the Earth and the totality of its excellences fully realized. His cosmological vision in *On the Heavens* further emphasizes this foundation status because "the earth does not move and does not lie elsewhere than at the center." Aristotle's placing of the Earth at the center of the cosmos around which the sphere of the fixed stars daily rotates, carrying with it the spheres of the sun, moon, and planets, is the authority cited by Ptolemy (85–165 C.E.) in working out his plan of the Earth in relation to the heavens.

The shift in perspective known as the Copernican Revolution began when the Polish astronomer Nicolaus Copernicus (1473–1543) wrote his *Little Commentary* (1514). He argued that there was no one center to the universe, the Earth's center is not the center of the universe, the rotation of the Earth accounts for the apparent daily rotation of the stars, and the Earth revolves in a vast space. These ideas helped inaugurate the thinking that Galileo Galilei (1564–1642) would confirm a

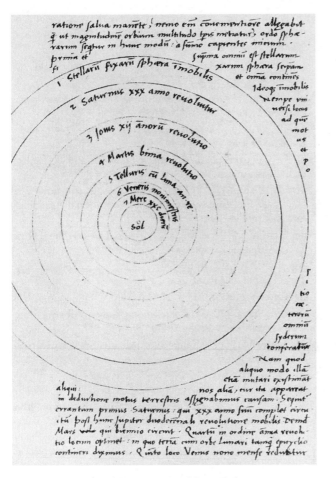

Page from Copernicus's *De Revolutionbus Orbium Coelestium*, showing a sun-centered solar system. This conception of the universe represented a historical shift in thinking from an earlier view in which the Earth was seen as the center of the cosmos. (© *Hulton/Archive. Reproduced by permission.*)

century later. This philosophy not only displaced the Earth from its central position in religious cosmologies, but made the planet itself into a composite of more basic materials to be analyzed and manipulated.

The Cartesian method of analysis led several European scientists in the 1860s to articulate how the basic constituents of all chemical compounds could be broken down into their simplest components. These elements, as measured and compared by their atomic weight, were arrayed in a chart, the periodic table, that both presented them sequentially (giving them an atomic number based on their atomic mass) and grouped them according to their electron configuration, which gives them similar chemical behavior such as the group of inert gases or that of alkali metals, for example. The table as presented in 1869 by the Russian chemist Dmitrii I. Mendeleev is still used with little revision other than filling in spaces left blank for predicted new elements.

Earths and Ethics

The approach to earth as a composite or collection of discrete units has informed one dominant modern philosophical perspective. Seen in terms of external relations among material constituents, this perspective tends toward a utilitarian approach to ethical problems. If the greatest number of people benefit from some alteration or use of an environment, or if some part of the environment which occurs naturally can be functionally replaced through technological advance, then utilitarianism allows for these alterations of the earth, even if they might involve a diminution in its diversity or degradations in its ecological viability. This approach has nevertheless promoted the rights of excluded social groups in arguments for environmental justice, as well as suggested that animals have rights as part of the earth (Singer 1990).

A contrasting philosophical perspective contends that the Earth has a distinct holistic identity, perhaps inseparably intertwined with the collective identity of humanity. Explorations of this option often rely on James Lovelock's *Gaia: A New Look at Life on Earth* (1979), in which he posited the Earth is an evolving, self-regulating organism. In this view, the planet through its temperature, gaseous constituents, minerals, acidity, and many other factors maintains a homeostasis by active feedback processes operating in the biota. Other philosophical views, while not seeing the planet itself as a living being, do envision human identity as internally related to aspects of the earth in such a way that these relationships themselves constitute the identity of both, such as in the work of Arne Ness, Dave Abram, Glen Mazis, or Freya Matthews. From such a perspective a utilitarian ethics fails to adequately safeguard either the Earth or humanity and all those parts of the biosphere due respect for their intrinsic work.

Returning to the question of the Earth as the horizon for humanity's sense of meaning and purpose, one path in philosophy is that first proposed by Friedrich Nietzsche in *Thus Spoke Zarathustra* (1883) and its claim that the nihilism of modern culture can only be undercut by a reevaluation of values and a reidentification of humanity as no disembodied spirit but as an animal of passion, body, sensuality, *and* reason—whose greatest challenge is to create value and meaning while obeying the injunction to "remain faithful to the earth." Edmund Husserl called the Earth the *foundation* [*Boden*] of the sense of being human and likened the planet to an ark that would always be with humanity as its abiding sense of identity no matter where humanity ventured.

For Martin Heidegger, humans open up a horizon for meaning and purpose through the way that art and other institutions open "the strife between earth and world," as he articulated in "the Origin of the Work of Art." The artist, like other creators, must initiate a dynamic struggle between the context of meaning and value, which makes up the "world" of various epochs and cultures with the opacity and resistance of the sheer materiality of the earth. The earth both anchors and occludes this birth of meaning, so it is literally "grounded" and yet never fully fathomable, but suggestive. In his analysis of the elements fire, water, air, and earth, Gaston Bachelard saw the Earth as the dimension which gives humanity a *rootedness*, and a sense of infinite depth, as well as a resistance against which meaning is forged in action. The resistance of the earth is the "partner of the will." Humans are motivated to create and shape in response to the earth. Differing visions and temperaments respond to the continuum of earth in its span from hardness to softness. Humanity is motivated to forge the earth into creations as well as struggle against earth's gravity towards flight.

Increasingly, too, there is a growing interchange of Western philosophy with global philosophical perspectives that suggests the inseparability of humanity and earth. These ideas include the Buddhist emphasis on the ontological interdependence of all living and non-living beings expressed through the concept of "emptiness," which might be better evoked as the relativity among all beings, as well as the depiction of the Buddha's original inspiration to seek enlightenment after shedding tears at seeing the worms and insects cut up by the plows making furrows in the fields with the same grief he would have had for the death of his family. There is the Daoist sense of *nonacting* [Wu wei] in which the beings of the Earth act through humans or as the Way [Dao] itself is the dynamic interplay of the entities of the Earth—"the ten thousand things"— as well as the Earth itself as a larger field of energy. Within North America, there is the Native American sense that all beings are part of Mother Earth or the Great Spirit, living on *turtle island*, the community of two-legged, four-legged, and all other beings of the four directions.

A challenge ahead is whether these philosophical perspectives can integrated with earth system science and engineering at the micro and macro scales in which they are now being practiced in order to give some basis for ethical decision making and a coherent perspective.

GLEN A. MAZIS

SEE ALSO *Air; Earth Systems Engineering and Management; Environmental Ethics; Fire; Gaia; Green Ideology; Nature; Water; Wilderness.*

BIBLIOGRAPHY

Abram, Dave. (1996). *The Spell of the Sensuous.* New York: Random House.

Aristotle. (1941). *The Basic Works of Aristotle.* New York: Random House.

Bachelard, Gaston (2002). *Earth and Reveries of Will.* Dallas: Dallas Institute.

Connelly, Dianne M. (1994). *Traditional Acupuncture: The Law of the Five Elements.* Columbia: Traditional Acupuncture Institute.

Feng, Gia-Fu, and Jane English. (1972). *Tao Te Ching.* New York: Alfred A. Knopf.

Heidegger, Martin. (1971). *Poetry, Language, Thought.* New York: Harper and Row.

Luhr, James F. (2003). *Earth.* New York: DK Publishing.

Matthews, Freya. (2005). *Rehabilitating Reality.* Albany: State University of New York Press.

Mazis, Glen A. (2002). *Earthbodies.* Albany: State University of New York Press.

Naess, Arne. (1989). *Ecology, Community and Lifestyle.* Cambridge: Cambridge University Press.

Plato. (1961). *The Collected Dialogues of Plato,* ed. Edith Hamilton and Huntington Cairns. Princeton: Princeton University Press.

Singer, Peter. (1990 [1975]). *Animal Liberation: A New Ethics for Our Treatment of Animals,* 2nd edition. New York: New York Review.

Ward, Peter, and Donald Brownlee. (2003). *Rare Earth: Why Complex Life is Uncommon in the Universe.* New York: Copernicus Books.

EARTHQUAKE ENGINEERING

• • •

Earthquake engineering is the collective effort of earth scientists, geotechnical engineers, structural engineers, and public policymakers to provide a built environment that is safe in the event of an earthquake. A significant part of this effort and the focus here is related to structural engineering, which involves the design and construction of structures and the anchorage of nonstructural building contents. Additionally structural evaluations and targeted retrofit of existing structures can be utilized to mitigate the risk of human and economic loss from an expected maximum probabilistic earthquake at a given site due to building collapse, loss of building contents, or economic downtime. Earthquake engineering thus constitutes a case study in specific relations between science, technology, and ethics.

FIGURE 1

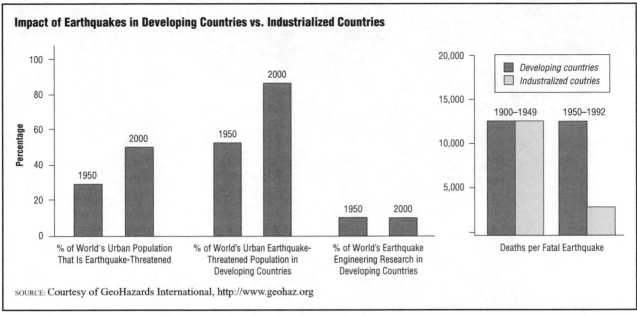

Impact of Earthquakes in Developing Countries vs. Industrialized Countries

SOURCE: Courtesy of GeoHazards International, http://www.geohaz.org

Historical and Technical Background

Interest in constructing buildings to provide greater resistance to earthquakes arose in association with the scientific and professional development of engineering, especially from the late 1800s and early 1900s, in response to large earthquake damages that occurred in Japan, Italy, and California. For instance, the earthquake near San Francisco, in April 1906 (magnitude M = 7.8 on the Richter scale, 3,000 fatalities) destroyed structures in an area 350 miles long by 70 miles wide, and was the most expensive natural disaster in U.S. history until hurricane Andrew in 1992, with $500 million in damages (equivalent to $10 billion in 2004 dollars).

In order to defend investments and continue growth, initial press reports from San Francisco minimized the quake itself and focused instead on the fires started by downed electrical wires, cracked gas lines, and broken stoves (Geschwind 2001). Yet shaken by this and related events, California has become one of the most progressive states in the public reduction of earthquake risk through engineering design. More recent major losses in August 1999 in Izmit, Turkey (M = 7.6, 17,000 fatalities); January 2001 in Gujarat, India (M = 7.7, 20,000 fatalities); and December 2003 in Bam, Iran (M = 6.6, 43,000 fatalities) have promoted recognition of the need to deal systematically with earthquakes in the regions affected.

Despite the length of time since public attention was first drawn to earthquake risks, earthquake engineering remains a young science because of the relative infrequency of large quakes and the tremendous number of variables involved. Since the 1960s, earthquake-engineering development has made important progress by moving to incorporate knowledge from the pure geosciences with structural engineering, moving even toward multidisciplinary efforts to include sociology, economics, lifeline systems, and public policy (Bozorgnia et al. 2004). The scientific study of earthquakes or *seismology* is also relevant (see Bolt 1993).

Complete or partial structural collapse is the major cause of fatalities from earthquakes worldwide; earthquakes themselves seldom kill people, collapsing buildings do. Earthquake energy causes structures not sufficiently designed to resist earthquakes to move laterally. At this point, a building may lose its load carrying capacity and collapse under its own weight. Portions of buildings (such as roof parapet walls) or the interior nonstructural contents (refrigerators, bookshelves, and so on) can topple onto inhabitants inside or outside the building. Directly adjacent buildings can pound into each other, serving sometimes to stabilize each other when neighboring structures are on both sides (termed bookends) or to cause additional damage if a neighboring structure is on one side only or the floors do not align. Buildings on corners of city blocks are known to perform poorly, being pushed into the street due to one-sided pounding. Tsunamis, or tidal waves triggered by seafloor seismic movements, are another source of damage. Fires can be initiated from broken gas or electrical lines. Water saturated soil can lose its strength during dynamic shaking, and landslides or soil liquefaction may cause buildings to slide, be buried, or sink as if into quicksand.

A security officer stands next to a seismic brace inside the Diablo Canyon Nuclear Power Plant. Braces like these are representative of the technological advances in earthquake engineering. (© Roger Ressmeyer/Corbis.)

To affect a structure, earthquake energy must first transmit through the bedrock from the *epicenter*, or the fault rupture location, through the soil above the bedrock (if any), through the foundation system, and then up through the building itself. All of these elements between the epicenter and building structure affect the level of lateral force (termed *base shear*) used for structural design. Frequency of ground motion can vary with distance from the epicenter, directivity, type of fault rupture, and magnitude. In the United States, the U.S. Geological Survey (USGS) maintains probabilistic earthquake demand topography maps based on statistical analyses of seismicity, referenced by building codes and design standards and used by structural engineers for design.

Engineering judgment, based on experience and observation of damage during past earthquakes, is relied on heavily in approximating earthquake demand, structural analysis, and overall structural design. Geotechnical engineers determine site soil conditions and site-specific seismic hazard. Structural engineers model the structural mass and stiffness, or how much a building moves when pushed laterally, based on the earthquake-resisting structural system used in design. Dynamic force and displacement limits are assumed based on structural detailing of connections and experimental testing results. Though material standards are used to set minimum criteria for material properties, there still exists some variability in material strength and ductility, requiring designing for a range of properties. Due to these many variables, two identical structures at different locations may require quite different earthquake-resisting systems.

After an earthquake, it is often difficult to know immediately if a building is severely damaged. The structure is typically covered by finishes, suspended ceilings, and fireproofing that need to be removed for visual investigation of connections, cracking, and other damage. In the United States, structural engineers may travel thousands of miles to aid in the initial building tagging and reconnaissance efforts, to quickly assign a red (no entry, evacuate), yellow (limited entry), or green (functional) placard at the entrance points. Developments in instrumentation have allowed for real-time building motions to be streamed over the Internet, which facilitates accuracy in initial tagging, but visual observation remains the primary basis for evaluation. In the case of a large office building, red tagging means the loss of weeks or months of revenue. In the case of hospitals and emergency response centers, a decision to evacuate means disruption of critical care during an emergency situation, increasing the death rate. For such reasons, engineers have an ethical responsibility to be extremely careful about a recommendation to evacuate a damaged building.

As architecture, construction materials, technology, and economics of construction evolve, seismic engineering evolves as well. Assumptions made during design are put to the test in future earthquakes, both validating previous thinking and exposing flaws. After the January 1994 earthquake in Northridge, California (M = 6.7, 60 fatalities, $40 billion in damage), it was found that many steel beam-to-column connections in relatively new structures had fractured at yield stress in buildings across the city, much different than the ductile behavior assumed in design. The structural engineering community initiated a six-year research project funded primarily by the U.S. Federal Emergency Management Agency (FEMA) to determine the cause of the poor performance, devise repair schemes, determine new design procedures that would produce desired ductile behavior, and modify building codes to avoid similar failures in future earthquakes (SAC 2000).

Building Codes, Economics of Construction, and Seismic Loss

In general the purpose of building codes is to protect public safety. But building codes and design standards, like the structures and societies in which they exist, are not permanent static entities, but dynamic and evolving to meet the demands and knowledge of changing times.

To minimize construction costs, building codes function as minimum requirements to permit damage from an earthquake but prevent collapse of the main structure, structural attachments, or contents. New buildings are expected to be repairable after a major earthquake, but some may be too costly to repair. Engineers have a responsibility to inform clients that building codes are not intended to preserve a structure, but do provide opportunities to increase the structural capacity or add special elements such as supplemental energy dissipation devices (viscous dampers and friction dampers, among others) or base isolation to reduce damage permissible by design codes.

Building owners are thus able to increase a building's earthquake *performance level* if they are willing to pay the additional construction and design costs. Generally, however, it is difficult to sell higher performance engineering and construction costs to owners in the United States. In Japan and New Zealand, by contrast, higher performance

structural elements are more frequently used. Building codes increase earthquake demand for critical structures, such as hospitals, schools, and communications hubs, with the intent that less damage occur during a major earthquake allowing the structure to remain operational afterward. In capitalist societies, history has shown that either economic incentives (tax breaks) or the threat of a facility being closed are often required to make building owners decide to retrofit. Both tactics are used in California (Geschwind 2001).

It is cheaper by far to allow for seismic forces during initial design than to incur damage or to retrofit later. Considering seismic forces initially may increase construction costs by 2 to 5 percent. Retrofit costs are typically on the order of 20 to 50 percent of original construction costs, excluding design fees and business interruption costs (Conrad 2004). Though seemingly inexpensive in comparison with the potential loss of the entire structure, there is major resistance to a 5 percent increase in construction cost from building owners, developers, and engineers not familiar with seismic design, especially in areas where the earthquake return period is longer than 100 years, when building codes (as in the United States) assume the typical building life to be fifty years. The area along the Mississippi River between St. Louis, Missouri, and Memphis, Tennessee, experienced three magnitude 7.8 to 8.1 earthquakes in 1811 and 1812, which reportedly moved furniture in James Madison's white house and rang church bells in Boston, yet many in the local communities maintain that designing for earthquakes is too costly. Money not spent on seismic retrofit for public facilities could theoretically be spent on the salaries of police and teachers, better hospital care, highway upgrades, and social programs. However probabilistic risk analysis demonstrates that ignoring earthquakes in design is often much more costly in the long run than short-term benefits of construction savings or budget reallocations.

In addition to loss of life, earthquake damages can significantly affect the local and world economies. The January 1995 earthquake in Kobe, Japan (M = 6.9, 5,502 fatalities) caused more than $120 billion in economic loss. It is estimated that a similar earthquake in a major metropolitan area in the United States could result in a comparable loss (House Committee on Science, Subcommittee on Research, 2003). In the United States, earthquakes pose significant risk to 75 million Americans in 39 states. Averaging single event losses over the time between events, total annualized damages in the United States have been estimated at approximately $4.4 billion (House Committee on Science, Subcommittee on Research, 2003). When industrial transportation and utility losses are considered,

TABLE 1

Magnitude and Intensity of Significant Earthquakes

Date	Time (GMT)	Place	Latitude	Longitude	Fatalities	Intensity	Magnitude
January 23, 1556		Shensi, China	34.5	109.7	830,000	–	~8
November 1, 1755	10:16	Lisbon, Portugal	36.0	−11.0	70,000	–	~8.7
December 16, 1811	08:00	New Madrid, MO, USA	36.6	−89.6		–	~8.1
January 23, 1812	15:00	New Madrid, MO, USA	36.6	−89.6		12	~7.8
February 7, 1812	09:45	New Madrid, MO, USA	36.6	−89.6		12	~8
August 31, 1886	02:51	Charleston, SC, USA	32.9	−80.0	60	–	~7.3
June 15, 1896	19:32	Sanriku, Japan	39.5	144.0		–	~8.5
June 12, 1897	11:06	Assam, India	26.0	91.0	1,500	–	~8.3
April 18, 1906	13:12	San Francisco, CA, USA (San Andreas fault from Cape Mendocino to San Juan Bautista)			3,000	11	7.8
August 17, 1906	00:40	Valparaiso, Chile	−33.0	−72.0	20,000	11	8.2
December 16, 1920	12:05	Ningxia-Kansu, China	36.60	105.32	200,000	–	8.6
September 1, 1923	02:58	Kanto, Japan	35.40	139.08	143,000	–	7.9
May 22, 1927	22:32	Tsinghai, China	37.39	102.31	200,000	–	7.9
March 2, 1933	17:31	Sanriku, Japan	39.22	144.62	2,990	–	8.4
March 11, 1933	01:54	Long Beach, CA, USA	33.6	−118.0	115	–	6.4
December 26, 1939	23:57	Erzincan, Turkey	39.77	39.53	32,700	11	7.8
May 22, 1960	19:11	Chile	−38.24	−73.05	5,700	11	9.5
March 28, 1964	03:36	Prince William Sound, AK, USA	61.02	−147.65	125	–	9.2
February 9, 1971	14:00	San Fernando, CA, USA	34.40	−118.39	65	11	6.7
July 27, 1976	19:42	Tangshan, China	39.61	117.89	255,000*	10	7.5
September 19, 1985	13:17	Michoacan, Mexico	18.44	−102.36	9,500	9	8.0
October 18, 1989	00:04	Loma Prieta, CA, USA	37.14	−121.76	63	9	6.9
January 17, 1994	12:30	Northridge, CA, USA	34.18	−118.56	60	9	6.7
January 16, 1995	20:46	Kobe, Japan	34.57	135.03	5,502	11	6.9
August 17, 1999	00:01	Izmit, Turkey	40.77	30.00	17,118	–	7.6
January 26, 2001	03:16	Gujarat, India	23.39	70.23	20,085	–	7.7
December 26, 2003	01:56	Bam, Iran	29.00	58.34	26,200	9	6.6
December 26, 2004	00:58	offshore Sumatra, Indonesia	3.31	95.85	225,000 (est.)	–	9.0

*Fatalities estimated as high as 655,000.

SOURCE: U.S. Geological Survey, Earthquakes Hazards Program. Available from http://earthquake.usgs.gov; National Geophysical Data Center. Available from http://www.ngdc.noaa.gov/seg/.

Earthquake Intensity is a measure of earthquake size based on observed damage of buildings and other structures on the earth's surface. Intensity is measured on a scale of 1 to 10+, with 10+ representing the most damage. Intensity is a different measurement than earthquake Magnitude, a measure of the strain energy released over the area of fault rupture. Magnitude is not a linear scale; each 1.0 increase in magnitude number represents greater than a factor of 30 times total energy released. Values of Intensity and Magnitude do not numerically correlate between different earthquake events due to local geology, depth of fault rupture, existing construction, and many other factors.

the estimated annual damage approaches $10 billion (Bonneville 2004). The September 11, 2001, terrorist attacks in the United States caused approximately 3,000 deaths and $100 billion in losses, roughly the same proportions as a major earthquake. Just as insurance, travel, and security measures have been increased throughout the world in response to the attacks of September 11, 2001, preparing for the next major earthquake would lessen the worldwide economic effects of future events.

Seismic Risk Analysis and Societal Response

Since 1990 financial risk management analysis has been increasingly utilized by various levels of government, large corporations, and universities to understand and work toward reducing the financial impacts of major earthquakes. For example, a small one-story structure storing landscaping equipment may not be as important to a client as a one-story structure that houses emergency generators and an essential communications antenna. If the one-story structure is a collapse hazard, the owner may decide to strengthen the structure or move essential components to reduce risk.

Risk analyses use various loss estimating measures. The most common is the probable maximum loss (PML) due to a major earthquake, presented as a percentage of the building value. A 50-percent PML anticipates that half of the building will be damaged beyond repair in a major earthquake. Risk assessments need to be periodically updated to show progress and to reevaluate a client's portfolio with the ever-improving tools available to structural engineers produced through new research, analysis software, code developments, and observed damage. Values of PML studies need to be defined and investigated carefully as each methodology or computer program assumes slightly different parameters (Dong 2000).

Three requirements must be satisfied for a successful earthquake resistant design protocol. First, there must be practical structural design standards that reflect current observations and research, standards that are used by engineers and legally enforced as minimum requirements. Second, there must be thorough structural engineering performed by qualified and licensed engineers that leads to clear and explicit construction drawings. Third, construction must be monitored by qualified inspectors or by the designing engineers to ensure that the intended materials are used and construction proceeds as shown in the drawings and specifications. In case of unforeseen construction difficulties, the structural engineer must be involved in a solution that meets the intent of the design without compromising the structure, but also is as economic as possible.

If one or two of these three requirements are satisfied, the protocol is not successful. For example, after the 1999 earthquake in Izmit, Turkey, reports focused on shoddy construction and unenforceable building codes. Building codes are quite good in Turkey, closely following the standards published in the United States. However, for cultural reasons the building codes are frequently not enforced when a design is reviewed, and the contractor is held neither to building to the design standard nor to having an engineered design (EERI 1999).

Due to the Izmit earthquake, efforts to mitigate current and future earthquake risk in Europe are underway in Turkey, as well as Greece, Portugal, Italy, and the rest of the European Union (Spence 2003). All countries with moderate or high earthquake risk have their own cultural, financial, and political barriers toward earthquake risk mitigation. However, as has been demonstrated in the United States, Japan, and elsewhere when the three requirements of practical codes, sound structural design, and construction monitoring work together, earthquake risk is decreased as new buildings replace older ones.

It is extremely difficult for developing countries to mitigate seismic risk. Priorities are on more immediate needs such as food, clean water, and disease prevention and on the effects of poverty and war. Construction uses available materials and follows traditional methods without structural calculations. While economic losses in developing countries may not be as high as in the United States, loss of life is much more severe, potentially approaching the proportions of the July 1976 earthquake in Tangshan, China (M = 7.5), where between 250,000 and 655,000 people were killed and more injured when nearly the entire city was razed.

Population expansion, and hence the rate of construction using traditional (seismically unsafe) methods, is at a much higher rate in countries such as India or Nepal than in the United States, exponentially increasing the earthquake risk in these countries. It is estimated that the risk of fatalities in developing countries compared to industrialized countries is 10 to 100 times greater—and increasing. This trend is the largest ethical and functional difficulty worldwide with regard to earthquake risk. In addition to moral obligations to reduce earthquake risk in developing countries, there are financial reasons as well. Due to economic globalization, a major disaster in a developing country has direct immediate and long-term financial impact on the world economy.

JAMES P. HORNE
DAVID R. BONNEVILLE

SEE ALSO Earth; Earth Systems Engineering and Management; Safety Engineering.

BIBLIOGRAPHY

Bolt, Bruce A. (1993). Earthquakes, newly revised and expanded. W. H. New York: Freeman. Contains information about seismology, structural engineering, and public response to historical earthquakes.

Bonneville, David. (2004). "Securing Society Against Earthquake Losses." Structural Engineers Association of Northern California Newsletter 59(2004): 1–3.

Bozorgnia, Yousef, and Vitelmo V. Bertero, eds. (2004). Earthquake Engineering: From Engineering Seismology to Performance-Based Engineering. Boca Raton, FL: CRC Press. Describes the current knowledge of practicing engineers' and researchers' techniques in structural engineering for earthquake resistance.

Earthquake Engineering Research Institute (EERI). (2000). Financial Management of Earthquake Risk/EERI Endowment Fund White Paper. Oakland, CA: EERI. An excellent description of the key aspects of seismic risk analysis techniques after a decade of development and private practice.

Geschwind, Carl-Henry. (2001). California Earthquakes: Science, Risk, and the Politics of Hazard Mitigation. Baltimore, MD: Johns Hopkins University Press. An extremely well developed text from start to finish of seismic legislation and public policy related to curbing seismic loss in California.

SAC Joint Venture. (2000). Federal Emergency Management Agency (FEMA) Publication 354. The SAC joint venture is a very thorough research project that employed a great deal of coordination and cooperation between academia and practicing engineers, in an effort to better understand the poor behavior that was observed after the 1994 Northridge earthquake, which was exactly the opposite of the intention of the building codes which produced the results. Engineers and researchers took it upon themselves to publicize the problem, obtain funding, and publish these documents to avoid the poor behavior in future buildings, as well as publish ways to retrofit existing structures with similar characteristics.

Spence, Robin. (2003). "Earthquake Risk Mitigation in Europe: Progress towards Upgrading the Existing Building Stock." In *Proceedings of the Fifth National Conference on Earthquake Engineering*. Istanbul, Turkey: Fifth National Conference on Earthquake Engineering. A discussion on the European view of earthquake engineering, the cultural resistance, and potential for future progress.

INTERNET RESOURCES

Applied Technology Council. Available from www.atcouncil.org. Organization web site.

Conrad, Katherine. (2004). "Seismic Retrofit May Limit Bomb Damage." *East Bay Business Times*. Available from http://eastbay.bizjournals.com/eastbay/stories/2004/04/05/focus1.html

Consortium of Universities for Research on Earthquake Engineering. Available from www.curee.org. Organization web site.

Earthquake Engineering Research Institute (EERI). "Learning From Earthquakes: The Izmit (Kocaeli), Turkey Earthquake of August 17, 1999." Available from: http://www.eeri.org/lfe/pdf/turkey_kocaeli_eeri_preliminary_report.pdf.

House Committee on Science, Subcommittee on Research. *The National Earthquake Hazards Reduction Program: Past, Present and Future*. 108th Cong, 1st sess. May 8, 2003, 108–114. Available from http://commdocs.house.gov/committees/science/hsy86870.000/hsy86870_0.HTM. An interesting interpretation of where and how funds in the United States are appropriated, and the information this is based on.

Structural Engineers Association of California. Available fromwww.seaoc.org. Organization web site.

EARTH SYSTEMS ENGINEERING AND MANAGEMENT

• • •

The biosphere, at levels from the landscape to the genome, is increasingly a product of human activity. At a landscape level, islands and mainland regions are affected by agriculture, resource extraction, human settlement, pollution, and invasive species transported by humans. Few biological communities can be found that do not reflect human predation, management, or consumption. At the organism level, species are being genetically engineered by humans to increase agricultural yields; reduce pesticide consumption; reduce demand for land for agriculture; enable plant growth under saline conditions and thereby conserve fresh water resources; produce new drugs; reduce disease; and support a healthier human diet. At the genomic level, the human genome has been mapped, as has that of selected bacteria, yeast, plants, and other mammals.

Moreover too little of the discussion about the potential effects of advancements in cutting-edge fields, such as nanotechnology, biotechnology, and information and communication technology (ICT), is focused on their global impacts on integrated human-natural systems. Major human systems, from urban to economic to philosophic systems, increasingly are reflected in the physical behavior and structure of natural systems, yet there is little study and understanding of these subtle but powerful interactions.

A planet thus dominated by the activities, intentional and unintentional, of one species is a new historical phenomenon. This species is affecting a complex, dynamic system of which it is a part. Changes in such systems cannot be predicted by linear causal models; witness the continuous debate over the extent global warming is occurring, and its likely consequences. Probabilistic models and continuous data collection can help human beings enter into a dialogue with these coupled human-technological-environmental systems.

Appropriate data-gathering, modeling and dialogue is impeded by the absence of an intellectual framework within which such broad technological trends, and their cumulative impact on global human-natural systems, can be conceptualized. The current base of scientific and technical knowledge, governance institutions, and ethical approaches are inadequate to this challenge (Allenby 2001). Managing these highly complex systems requires an integration of the physical and social sciences that is difficult for both cultural and disciplinary reasons, and the institutional structures that would foster this understanding, and enable its implementation, do not yet exist.

Emergence of Earth Systems Engineering and Management

The challenge of the anthropogenic Earth drove Brad Allenby to propose Earth Systems Engineering and Management (ESEM), an interdisciplinary framework for perceiving, understanding, and managing complex, coupled human-natural-technological systems. It reflects not just the need to respond to, and manage, systems at scales of complexity and interconnection that current practices cannot cope with, but also to minimize the risk and scale of unplanned or undesirable perturbations in coupled human or natural systems. It does not replace traditional scientific, engineering, and social science disciplines or study; rather it draws on and integrates them to enable responsible, rational, and ethical response to the relatively new phenomenon of the anthropogenic Earth. Therefore, ESEM draws heavily on related work in multiple fields (Clark 1989, Turner et al. 1990).

ESEM is a response to a broad set of multidisciplinary questions that are relatively intractable to twenty-first-century disciplinary and policy approaches: How, for example, will people cope with the potential ramifications for environmental systems of nanotechnology, biotechnology, and ICT? How can they begin to redesign human relationships with complex ecosystems such as the Everglades; engineer and manage urban centers to be more sustainable; or design Internet products and services to reduce environmental impact while increasing quality of life?

The Ethics of ESEM

Dealing responsibly with the complex web of interconnections between human and natural systems will thus require experts skilled in new approaches and frameworks, capable of creating policy and design options that protect environmental and social values while providing the desired human functionality. Such an ESEM approach requires both a rigorous understanding of the human, natural, and technological dimensions of complex systems, and an ability to design inclusive strategies to address them, all the while recognizing that no single approach or framework is likely to be able to capture the true complexity of such systems.

Even at this nascent stage, it is possible to begin to establish a set of principles applicable to ESEM (Allenby 2002):

(a) Try to articulate the current state of a system and desired future states, consulting with multiple stakeholders. Establish a process for continuous sharing of knowledge and revision of system goals, based on continuous monitoring of multiple system variables and their interactions. Anticipate potential problematic system responses to the extent possible, and identify markers or metrics by which shifts in probability of their occurrence may be tracked.

(b) The complex, information dense and unpredictable systems that are the subject of ESEM cannot be centrally or explicitly controlled. ESEM practitioners will have to be reflective, seeing themselves as an integral component of the system, closely coupled with its evolution and subject to many of its dynamics.

(c) Whenever possible, engineered changes should be incremental and reversible. In all cases, scale-up should allow for the fact that, especially in complex systems, discontinuities and emergent characteristics are the rule, not the exception, as scales change. Lock-in of inappropriate or untested design choices, as systems evolve over time, should be avoided.

(d) ESEM projects should support the evolution of system resiliency, not just redundancy. In a tightly coupled complex system, a failure of one component can be fatal, and it is virtually impossible to build in sufficient redundancy for every component (Perrow 1984). The space shuttle is an example. Resilient systems are loosely coupled; the system as a whole can adapt to failures in one component. The Internet is an example, as are many natural systems. However, even in resilient systems, there are tipping points where the amount of disruption exceeds the ability of the system to adapt, and a major transformation occurs. Therefore, even resilient systems require monitoring and management.

To succeed, ESEM depends on the development of an Earth Systems Engineer (ESE) who would have a core area of expertise, perhaps environmental science or systems engineering or social psychology, and be able to take a global systems view of environmental problems. The ESE would have to be what Collins and Evans call an interactional expert, capable of facilitating deep, thoughtful conversations across disciplinary boundaries (Collins and Evans 2002) that enable productive trading zones (Galison 1997) for managing complex environmental systems (Gorman and Mehalik 2002). The ESE would also be involved in the creation of new data monitoring and modeling tools that would add rigor to ESEM. To assess its value, the ESEM approach needs to be piloted on several complex systems, and the results described in detailed case-studies from which others can learn. The ESEM framework has the potential to facilitate intelligent management of trading zones centered on converging technologies: nanotechnology, biotechnology, information technology and cognition (Gorman 2003).

BRADEN R. ALLENBY
MICHAEL E. GORMAN

SEE ALSO *Engineering Ethics; Environmental Ethics; Environmentalism; Management: Models.*

BIBLIOGRAPHY

Allenby, Braden R. (2001). "Earth Systems Engineering and Management." *IEEE Technology and Society Magazine* 19(4): 10–21.

Clark, William C. (1989). "Managing Planet Earth." *Scientific American* 261(3): 46–54. This article introduces and gives an overview of an issue of *Scientific American* that points the way toward ESEM.

Collins, H. M., and Robert Evans. (2002). "The Third Wave of Science Studies." *Social Studies of Science* 32(2): 235–296.

Galison, Peter Louis. (1997). *Image and Logic: A Material Culture of Microphysics.* Chicago: University of Chicago Press.

Gorman, Michael E. (2003). "Expanding the Trading Zones for Convergent Technologies." In *Converging Technologies for Improving Human Performance: Nanotechnology, Biotechnology, Information Technology and Cognitive Science*, ed. Mihail C. Roco and William S. Bainbridge. Dordrecht, Netherlands: Kluwer.

Gorman, Michael E., and Matthew M. Mehalik. (2002). "Turning Good into Gold: A Comparative Study of Two Environmental Invention Networks." *Science, Technology and Human Values* 27(4): 499–529.

National Academy of Engineering. (2000). *Engineering and Environmental Challenges: A Technical Symposium on Earth Systems Engineering.* Washington, DC: National Academy Press.

Perrow, Charles. (1984). *Normal Accidents: Living With High-Risk Technologies.* New York: Basic Books.

Turner, B. L., William Clark; Robert W. Kates; et al., eds. (1990). *The Earth as Transformed by Human Action.* Cambridge: Cambridge University Press.

INTERNET RESOURCE

Allenby, Braden R. (2002). "Observations on The Philosophic Implications of Earth Systems Engineering and Management." Available from http://www.darden.virginia.edu/batten/pdf/WP0010.pdf. This is the most comprehensive account of ESEM, including its relationship to multiple disciplines, perspectives and precursors.

ECOLOGICAL ECONOMICS

• • •

Economics is frequently defined as the science of the allocation of scarce resources among alternative desirable ends. The first question this implies—What are the desired ends?—is ultimately a question of values and ethics. Most economists would agree that while the ultimate desired end is too difficult to define, increasing social welfare serves as a reasonable placeholder. Seeking to establish itself as an objective, value-free science, mainstream (neoclassical) economics strives to maximize welfare as measured by the dollar value of market goods plus the imputed dollar value of nonmarket goods and services produced. Therefore neoclassical economists, including natural resource and environmental economists, devote most of their attention to markets, which under certain strict conditions efficiently allocate resources toward uses that maximize dollar values. Taking an explicitly ethical position, ecological economics asserts that ecological sustainability and just distribution take priority over efficient allocation as prerequisites to increasing social welfare. Markets cannot be relied upon unless these first two priorities have been met.

Once the desired ends have been determined, ecological economists rely on insights from physics and ecology to assess the nature of the scarce resources. Only then do they seek appropriate allocative mechanisms, drawing from mainstream economics as well as other social sciences. Ecological economics embraces the full complexity of the economic question, and the full range of inquiry necessary to answer it. It lays no claim to being a value-free science, but rather works to be a transdisciplinary field, integrating knowledge and skills from both the humanities and sciences. (Costanza, Daly, and Bartholomew 1991; Norgaard 1989).

As an emerging transdiscipline, ecological economics has an exceptionally broad scope of inquiry, and has not yet achieved the level of consensus that characterizes an established science. This overview leaves out much brilliant work, and not all ecological economists will agree with all it says.

The Resources of Nature and the Nature of Resources

An understanding of scarce resources begins with hard science and the laws of thermodynamics. The first law states that the quantity of matter-energy cannot be created or destroyed and remains constant in a closed system. Everything produced by humans (human-made capital) must come from raw materials supplied by nature (natural capital). Any waste produced by the economy must return to the ecosystem. In contrast, most standard microeconomics textbooks argue that through specialization and trade, society can "increase production with no change in resources" (Parkin 2003, p. 42).

The second law of thermodynamics states that entropy never decreases in an isolated system. From the perspective of economics, entropy can be thought of as a measure of used-up-ness, or the extent to which the capacity of matter-energy to perform work or be useful has been exhausted. When oil is burned to run an engine or heat a house, the energy it contains is not destroyed in performing this work, but it cannot be used again for the same purpose. When the steel in cars rusts and flakes off, it does not disappear but is scattered about the ecosystem so randomly one cannot gather it back up. The quantity of matter-energy is constant in a system, but the quality is constantly deteriorating. These laws suggest that human-made capital will inevitably be used up or worn out and return to the ecosystem as high entropy waste. A constant flow of low entropy natural capital is required simply to maintain the economy.

Fortunately the Earth is not an isolated system, because the sun provides a daily source of low entropy energy. But it is this solar inflow that limits the physical size of the economy in the long run, not the nonrenewable stock of fossil fuels. While fossil fuels can be used up as quickly as one chooses, solar energy comes at a fixed rate. People can therefore use fossil fuels to achieve rapid physical growth of the economic system, but not to create a sustainable system (Georgescu-Roegen 1971).

Humans depend not only on raw materials provided by nature, but like all other species on the planet, are sustained by the solar-powered life support functions of healthy ecosystems. All of human technology simply cannot provide the climate stability, waste absorption capacity, water regulation, and other essentials that more than 6 billion people require to survive. In other words, natural capital has two components. Ecosystem goods are the raw materials provided by nature, as well as the structural components of the ecosystem. Ecosystem services are the valuable functions that emerge when those structural components interact in a complex ecosystem to create a whole greater than the sum of the parts. When humans remove low entropy raw materials from nature to build the economy and return high entropy waste, they must pay an opportunity cost measured in both ecosystem goods and services lost.

These laws of thermodynamics are responsible for the core vision of ecological economists: The human system is sustained and contained by the global ecosystem. When the physical size of the economic system increases, it does not expand into a void, but must instead consume and displace the natural capital on which humans depend for survival (Daly and Farley 2003).

Scale, Distribution, and Allocation

As a consequence of the ecological economists' core vision, their primary concern is with *scale*—the physical size of the human economy relative to the ecosystem that contains and sustains it. The scale of the economy cannot exceed the capacity of the ecosystem to sustain it. This priority emerges from an understanding of the laws of physics combined with an ethical responsibility to future generations.

Sustainable scale is necessary, but inadequate. Virtually all economists accept the law of diminishing marginal utility—the more one has of something, the less an additional unit is worth. As human-made capital increases, its marginal utility diminishes. A corollary is the law of increasing opportunity costs—as natural capital dwindles, the opportunity costs of continued losses increase. Increasing opportunity costs must eventually surpass diminishing marginal utility. At this point, an economic system has reached its optimal scale, and the physical growth of the economy should stop—though economic development, as measured by improvements in social welfare, can still continue.

Two hundred years ago when market economies were emerging, human-made capital was relatively scarce and natural capital abundant. Economists logically focused on allocating the former. In the early twenty-first century, however, it is natural capital that constrains economic development. If people need more fish or timber, the problem is depleted fish stocks and forests, not a shortage of boats or chainsaws. It is likely that humans have exhausted nearly half the planet's supply of conventional petroleum in less than 150 years (Campbell and Laherrère 1998), threatening to destabilize the global climate in the process. Yet natural capital does not increase in fecundity or quantity in response to an increase in price—the driving force behind markets.

However while natural capital does not respond to price signals, technology does: As a resource becomes scarce, its price goes up, and people can either use it more efficiently or create a substitute, leading many conventional economists to conclude that resource scarcity imposes no limits on economic growth. At one extreme, economists such as Julian Simon deny that natural resources are finite and argue that a growing human population brings more brainpower to solve society's problems (Simon 1996). Similar claims from statistician Bjørn Lomborg (2001), supposedly based on evaluation of empirical data, have received considerable publicity, but the quality of his scholarship raises serious concerns (Rennie 2002). For example, he accepts without question a doubling and even tripling of estimated oil reserves in several member states of the Organization of the Oil Exporting Countries (OPEC) that took place shortly before their quota negotiations in 1988, while rejecting as implausible four out of five scenarios for climate change from an intensively peer-reviewed report by leading scientists working with the International Protocol on Climate Change (Schneider 2002). Nonetheless more credible technological optimists such as Amory Lovins are actively creating pollution reducing, resource and energy efficient technologies such as the hydrogen powered *hyper-car*.

While not denying its importance, ecological economists are leery of undue faith in technological advance for both practical and ethical reasons. In practical terms few ecosystem services even have a price to signal market scarcity and thus induce technological innovation, and even imputed prices cannot capture the fact that most ecosystem services do not have clear substitutes (Gowdy 1997). While there is a greater capacity to develop substitutes for ecosystem goods than for services, efficiency improvements have physical limits, and continued economic growth must eventually lead to more resource use, more waste output, and diminishing marginal utility—a growing fleet of hyper-cars will still require more roads and parking lots and induce more traffic jams. The fact is that efficiency in resource use rarely stimulates frugality, but frugality quite often stimulates efficiency (Daly and Farley 2003). From the viewpoint of ethics, no one can say for certain what technologies will emerge and when, and the gamble is whether or not future technologies will create substitutes for critical resources before they are exhausted. Ecological economists weigh the gains from winning against the costs of losing. If the technological optimists are wrong, continued increases in the rate of resource use could lead to the irreversible loss of vital ecosystem life support functions. If the optimists are right, then limiting resource extraction and waste emissions will impose only short term costs to standards of living while technological innovation develops substitutes.

Thus ecological economists operate on the assumption that natural capital has become the scarcest resource required to achieve the desired ends, and recognize that markets fail to respond to this scarcity and cannot be relied on as a mechanism for determining desirable scale. Environmental economists in contrast believe markets can determine desirable scale if they calculate the dollar value of ecosystem services then feed this information back into the market system. However all economic production degrades ecosystem services through resource extraction and again through waste emissions. Two prices must be calculated for every price the market detects. This defeats the whole purpose of a market whose virtue is its reliance on decentralized information. Ecological economists believe scale should

be determined by a participatory democratic process informed by appropriate experts and the ethical values of citizens. Stakes are high, decisions are urgent, and facts are uncertain. Society must act quickly, but should err on the side of caution and leave room to adapt as it learns more (Funtowicz and Ravetz 1992; Prugh, Costanza, and Daly 2000). The Endangered Species, Clean Air, and Clean Water acts in the United States and the Montreal Protocol on ozone depleting substances are only a few examples of this approach. In sum, while environmental economists in contrast strive to calculate prices first, and then allow scale to adjust, ecological economists strive to determine the desirable scale first, and then allow prices to adjust.

The second priority for ecological economists is just distribution, which emerges in part from their concern with scale. What ethical system would allow a concern for the welfare of people not yet born, and ignore the welfare of those alive and suffering today? If a finite planet imposes finite limits on the size of the economy, then society cannot grow its way out of poverty, and alleviating poverty requires redistribution. On practical grounds, no one living in poverty can really afford to think about the future—hungry people around the world will sacrifice essential natural capital for immediate needs. Unjust distribution is therefore incompatible with ecological sustainability.

How markets allocate resources depends on the initial distribution. For example, a society with highly unequal distribution will allocate resources toward both slums and yachts, while one with more equal distribution will allocate resources toward neither. A given market allocation is therefore no more desirable than the initial distribution that produced it. Nonetheless the tradition in neoclassical economics is to leave the distribution question to other disciplines or policymakers, while ecological economists consider just distribution a prerequisite to desirable allocation.

Distribution should also be decided by a participatory democratic process. Three principles can guide the decision. Wealth created by nature and society as a whole should be equally distributed. Those who degrade that wealth, through pollution or resource depletion, for example, should compensate society for its loss. Those who benefit from society should provide compensation in proportion to their gains.

The third priority for ecological economists is efficiency. Once society has ensured the preservation of enough natural capital to sustain the system, and that remaining resources are justly distributed, those resources should be allocated toward uses that generate as much welfare as possible. Markets can be an efficient allocative mechanism when resources are privately owned, use by one person precludes used by another, and production and consumption have minimal impacts on others. When these conditions do not hold, markets alone will fail to generate efficient outcomes, and society must again rely on participatory democratic decision making to allocate resources, complemented when appropriate by market mechanisms.

JOSHUA C. FARLEY

SEE ALSO *Environmental Economics; Environmental Ethics; Sustainability and Sustainable Development.*

BIBLIOGRAPHY

Campbell, Colin J., and Jean H. Laherrère. (1998). "The End of Cheap Oil." *Scientific American* March: 78–83. This article offers a solid, multi-tiered analysis predicting that oil production will peak within the next few years, leading to a dramatic increase in oil prices.

Costanza, Robert; Herman Daly; and Joy Bartholomew. (1991). "Goals, Agenda and Policy Recommendations for Ecological Economics." In *Ecological Economics: The Science and Management of Sustainability*, ed. Robert Costanza. New York: Columbia University Press. This chapter is an excellent brief introduction to the field of ecological economics.

Daly, Herman, and John J. Cobb, Jr. (1989). *For the Common Good: Redirecting the Economy Toward Community, the Environment, and a Sustainable Future*. Boston: Beacon Press. This award winning book remains one of the best and most complete introductions to ecological economics, and the first description of the index of sustainable economic welfare.

Daly, Herman H., and Joshua Farley. (2003). *Ecological Economics: Principles and Applications*. Washington, DC: Island Press. Textbook that provides a comprehensive and readable introduction to ecological economics, suitable for upper level undergraduates and graduate students.

Funtowicz, Silvio O., and Jerome R. Ravetz. (1992). "Three Types of Risk Assessment and the Emergence of Post-Normal Science." In *Social Theories of Risk*, ed. Sheldon Krimsky and Dominic Golding. Westport, CT: Praeger. This chapter describes a general approach for solving ecological economic problems when facts are uncertain, decisions are urgent, stakes are high and values matter.

Georgescu-Roegen, Nicolas N. (1971). *The Entropy Law and the Economic Process*. Cambridge, MA: Harvard University Press. A classic work in ecological economics; the first comprehensive analysis of the importance of the entropy law to economics.

Gowdy, John. 1997. "The Value of Biodiversity: Markets, Society, and Ecosystems." *Land Economics* 73: 25–41. Pre-

sents a fine analysis of different types of value and clarifies the limits to monetary valuation.

Heilbroner, Robert, and Lester Thurow. (1981). *The Economic Problem*. Englewood Cliffs, NJ: Prentice-Hall.

Lomborg, Bjørn. (2001). *The Skeptical Environmentalist: Measuring the Real State of the World*. New York: Cambridge University Press. An influential but poorly researched example of the cornucopian world view that natural resources and waste absorption capacity are limitless, and the laws of thermodynamics are essentially irrelevant to economics.

Norgaard, Robert. (1989). "The Case for Methodological Pluralism." *Ecological Economics* 1(1): 37–58. Explains why complex ecological economic problems demand a transdisciplinary approach.

Parkin, Michael. (2003). *Microeconomics*, 6th edition. Boston: Addison Wesley. A standard microeconomics textbook.

Prugh, Thomas T.; Robert R. Costanza; and Herman H. E. Daly. (2000). *The Local Politics of Global Sustainability*. Washington, DC: Island Press. This short, readable book describes participatory democratic processes, and presents them as an effective mechanism for the sustainable and just allocation of resources.

Rennie, John. (2002). "Misleading Math about the Earth: Science Defends Itself against *The Skeptical Environmentalist*." *Scientific American* 286(1): 61. A brief editorial that introduces a series of articles describing the factual and analytical errors in Lomborg's *The Skeptical Environmentalist*.

Schneider, Stephen. (2002). "Global Warming: Neglecting the Complexities." *Scientific American* 286(1): 60–63. One in the series of articles introduced by John Rennie (see above).

Simon, Julian Lincoln. (1996). *The Ultimate Resource 2*, revised edition. Princeton, NJ: Princeton University Press. One of the first, best written and most influential descriptions of the cornucopian world view (see annotation following *The Skeptical Environmentalist*).

ECOLOGICAL FOOTPRINT

• • •

In the early 1990s, Dr. William Rees and a graduate student, Mathis Wackernagel, developed and quantified the first "ecological footprint" for the city of Vancouver, Canada. Fundamental to this research was answering the question, "how large an area of productive land is needed to sustain a defined population indefinitely, wherever on earth that land is located?" Ecological footprints build on earlier studies, all designed to quantify the natural resources used by humans and compare that to those that are available. However, footprints are distinguished, according to leading practitioners, by the many categories of human activity included in the ana-

TABLE 1

Ecological Footprint Results 1999

	Total Footprint [global hectares/pers] (1999)	Biocapacity [global hectares/pers] (1999)	Ecological Deficit [global hectares/pers] (if negative)	Total Footprint [global acres/pers] (1999)	Biocapacity [global acres/pers] (1999)	Ecological Deficit [global acres/pers] (if negative)
World	**2.3**	**1.9**	**−0.4**	**5.6**	**4.7**	**−0.9**
Argentina	3.0	6.7	3.6	7	16	9
Australia	7.6	14.6	7.0	19	36	17
Austria	4.7	2.8	−2.0	12	7	−5
Bangladesh	0.5	0.3	−0.2	1.3	0.7	−0.6
Belgium & Luxembourg	6.7	1.1	−5.6	17	3	−14
Brazil	2.4	6.0	3.6	6	15	9
Canada	8.8	14.2	5.4	22	35	13
Chile	3.1	4.2	1.1	8	10	3
China	1.5	1.0	−0.5	4	3	−1
Colombia	1.3	2.5	1.2	3	6	3
Costa Rica	2.0	2.3	0.4	5	6	1
Czech Republic	4.8	2.3	−2.5	12	6	−6
Denmark	6.6	3.2	−3.3	16	8	−8
Egypt	1.5	0.8	−0.7	4	2	−2
Ethiopia	0.8	0.5	−0.3	1.9	1.1	−0.8
Finland	8.4	8.6	0.2	21	21	0
France	5.3	2.9	−2.4	13	7	−6
Germany	4.7	1.7	−3.0	12	4	−7
Greece	5.1	2.3	−2.8	13	6	−7
Hungary	3.1	1.7	−1.3	8	4	−3
India	0.8	0.7	−0.1	1.9	1.7	−0.2
Indonesia	1.1	1.8	0.7	3	5	2
Ireland	5.3	6.1	0.8	13	15	2
Israel	4.4	0.6	−3.9	11	1	−10
Italy	3.8	1.2	−2.7	9	3	−7
Japan	4.8	0.7	−4.1	12	2	−10
Jordan	1.5	0.2	−1.4	4	0	−3
Korea (Republic of)	3.3	0.7	−2.6	8	2	−6
Malaysia	3.2	3.4	0	8	8	1
Mexico	2.5	1.7	−0.8	6	4	−2
Netherlands	4.8	0.8	−4.0	12	2	−10
New Zealand	8.7	23.0	14	21	57	35
Nigeria	1.3	0.9	−0.4	3.3	2.2	−1.1
Norway	7.9	5.9	−2.0	20	15	−5
Pakistan	0.6	0.4	−0.2	2	1	−1
Peru	1.2	5.3	4.2	3	13	10
Philippines	1.2	0.6	−0.6	2.9	1.4	−1.5
Poland	3.7	1.6	−2.1	9	4	−5
Portugal	4.5	1.6	−2.9	11	4	−7
Russia	4.5	4.8	0.4	11	12	1
South Africa	4.0	2.4	−1.6	10	6	−4
Spain	4.7	1.8	−2.9	12	4	−7
Sweden	6.7	7.3	0.6	17	18	2
Switzerland	4.1	1.8	−2.3	10	4	−6
Thailand	1.5	1.4	−0.2	4	3	0
Turkey	2.0	1.2	−0.7	5	3	−2
United Kingdom	5.3	1.6	−3.7	13	4	−9
United States	9.7	5.3	−4.4	24	13	−11

SOURCE: World Wildlife Fund (2002).

Ecological footprint and biocapacity figures for representative countries around the world. Ecological deficit refers to the extent that a country's footprint exceeds its biocapacity.

lysis, and by the measure's ability to compare current demand with current ecological limits (biocapacity).

The ecological footprint is an environmental accounting tool that measures human impact on nature, based on the ability of nature to renewably produce the resources that humans use and absorb the ensuing waste. Footprinting provides a way to aggregate into a single composite measure many of the ecological impacts associated with built-up land (i.e., roads and buildings), food, energy, solid waste, and other forms of waste or consumption. The result represents the impact or footprint. Using an area-based measure, such as hectares or acres, the size of a footprint can be compared to the renewable services the Earth's biocapacity can produce in a given year. The footprint methodology can be used to evaluate a population's progress toward ecological sustainability.

The footprint has been criticized on a variety of fronts, primarily related to the complex methodology that underlies the measure, as well as the applications for which it is appropriate. Along with other aggregate indicators, the footprint has been criticized for obscuring the components and assumptions that comprise the measure. While the methodology behind the measure is readily available, it is complicated and therefore not approachable without some technical background. Other critics argue that the premise of living within resource limitations can be overcome with technological innovation. It is true that in many ways the footprint is a worst-case scenario because it describes the situation if there are no technological improvements; but the converse, counting on improvements, could be risky in the long run as well.

When a country or community uses more renewable resources than are available, it has exceeded ecological limits. It will not be sustainable over an indefinite period of time. Such a situation can occur over a relatively short time-span because natural capital can be depleted to fill the renewable resource gap. Imports can also meet society's needs, but may simply shift depletion of natural capital around the globe. Over time, global stocks may be depleted to the point where they cannot regenerate or require significant human intervention to do so.

The *Living Planet Report 2002* contains footprints of countries with populations greater than one million. Estimates for the year 1999 show that the average American required approximately 9.6 hectares (24 acres) of ecologically productive land to sustain his or her lifestyle. In comparison, the average Canadian lived on a footprint that was nearly one-third smaller (6.9 global hectares or 17 acres), while the average Italian lived on an ecological footprint that was less than half the size (3.8 global hectares or 9 acres) of the American's. Each of these footprints can be compared to the amount of ecologically productive land area available locally or to the amount available globally on per person basis (1.9 hectares or 4.7 acres). See Table 1.

Footprint Methodology

The basic procedure for the footprint methodology is to determine annual global productivity and assimilation capacity (biocapacity) of major land areas. Then, this biocapacity is compared to the demands placed on it by human consumption and waste production. Productive lands are aggregated as cropland, pasture, for-est, fisheries, and built-up land. Built-up land is generally assumed to occupy former cropland, as this is the predominant settlement pattern in human history. The present footprint methodology holds that less than one quarter of the Earth's surface provides sufficiently concentrated biomass to be considered biologically productive—leaving out deep ocean areas, deserts, frozen tundra, and other less productive parts of nature. Biocapacity can change: both negatively, due to land alterations such as desertification; and positively, due to improvements in technology that result in higher yields.

Ecological footprints can be calculated using two basic approaches: component and compound. Component footprinting is a bottom-up approach consisting of calculating the ecological footprints of individual parts of a system and then adding them up. Compound footprinting, on the other hand, is a top-down approach using aggregate figures such as production, imports, and exports of agriculture, energy, and other commodities, usually for nations.

Using either methodology, human consumption and waste components of a footprint are attributed to the final point of utilization (where a product is used up and enters the waste stream), regardless of where the output is actually assimilated. For example, some waste products, such as carbon dioxide, may be assimilated well outside the boundaries of the place where they are actually emitted, either because the wastes are carried away from the point of use or because the wastes are generated at a remote production site.

The final footprint results from the comparison of global biocapacity to consumption and waste. High available biocapacity allows for more or larger footprints, and higher levels of consumption require more biologically productive land. Consumption beyond renewable levels of biocapacity requires the depletion of natural capital and is considered unsustainable if it draws resources down to the point at which they cannot regenerate.

Measuring the ecological footprint of energy is a particularly significant and complex challenge that can be addressed in a variety of ways. A primary question that arises concerns the type of energy that is being used. Highly renewable forms of energy production, such as wind and solar power, typically have footprints equivalent to the land area they occupy plus the materials embodied in the collection mechanism. At the other extreme, nuclear energy is inherently unsustainable both because the resources it utilizes are non-renewable and extremely toxic, and because the potential destruction from nuclear accidents produces a dramatic increase in footprint area. The current approach is to convert nuclear energy to the equivalent fossil fuel impact. The footprint of fossil fuels can be calculated as either the amount of land area that would be required to grow and harvest an equivalent amount of fuelwood, or as the amount of land area required to assimilate associated carbon dioxide emissions. The latter approach is the most typically used in footprint accounts.

Footprint calculations through the beginning of the twenty-first century have assumed optimistic yield factors for foods and forests (making them conservative) and have left unmeasured many of the impacts associated with pollution, water use, and habitat and species decline. Though improvements are being made in the methodology, the ecological footprint cannot be considered a definitive measurement of humanity's ecological impact without significant additions.

Applications

Footprinting provides a methodology to evaluate potential tradeoffs among alternative actions, designs, energy sources, policies and products. It can be used as a yardstick for measuring humanity's impact on the earth in terms of ecological sustainability. Research in the field has provided the stimulus and foundation for academics at universities throughout the world. The ecological footprint has informed discussions and debates from the global to local level in national governments, meetings of the United Nations, research institutes, and municipal sustainability initiatives.

Footprints change over time, as populations change, consumption patterns shift, and biocapacity increases or decreases. The changes allow humanity to see its progress toward sustainability, at a global, national, state, and local level.

DAHLIA CHAZAN
JASON VENETOULIS

SEE ALSO *Ecology; Ecological Economics; Ecological Restoration; Sustainability and Sustainable Development.*

BIBLIOGRAPHY

Chambers, Nicky; Craig Simmons; and Mathis Wackernagel. (2000). *Sharing Nature's Interest: Ecological Footprints as an Indicator for Sustainability.* London: Earthscan.

Costanza, Robert. (2000). "Commentary Forum: The Ecological Footprint. The Dynamics of the Ecological Footprint Concept." *Ecological Economics,* 32: 341–345.

Rees, William. (1992). "Ecological Footprints and Appropriated Carrying Capacity: What Urban Economics Leaves Out." *Environment and Urbanization* 4(2): 121–130.

Rees, William, and Mathis Wackernagel. (1994). "Ecological Footprints and Appropriated Carrying Capacity: Measuring the Natural Capital Requirements of the Human Economy." In *Investing in Natural Capital: The Ecological Economics Approach to Sustainability,* ed. AnnMari Jansson et al. Washington, DC: Island Press.

Venetoulis, Jason. (2001). "Assessing the Ecological Impact of a University: The Ecological Footprint for the University of Redlands." *International Journal of Sustainability in Higher Education* 2(2): 180–196.

Wackernagel, Mathis, and William E. Rees. (1996). *Our Ecological Footprint: Reducing Human Impact on the Earth.* Gabriola Island, BC: New Society Publishers.

Wackernagel, Mathis et al. (2002). "Tracking the Ecological Overshoot of the Human Economy." *Proceedings of the National Academy of Sciences of the United States of America* 99(14): 9266–9271.

INTERNET RESOURCES

"The Ecological Footprint Project." Sustainable Sonoma County. Available from http://www.sustainablesonoma.org/projects/scefootprint.html.

"Living Planet Report 2002." World Wildlife Fund. Available from http://www.panda.org/news_facts/publications/general/livingplanet/index.cfm.

ECOLOGICAL INTEGRITY

• • •

Ecological or biological integrity originated as an ethical concept in the wake of Aldo Leopold (1949) and has been present in the law, both domestic and international, and part of public policy since its appearance in the 1972 U.S. Clean Water Act (CWA). Ecological integrity has also filtered into the language of a great number of mission and vision statements internationally, as well as being clearly present in the Great Lakes Water Quality Agreement between the United States and Canada, which was ratified in 1988.

The generic concept of integrity connotes a valuable whole, the state of being whole or undiminished, unimpaired, or in perfect condition. Integrity in common usage is thus an umbrella concept that encompasses a variety of other notions. Although integrity may be developed in other contexts, wild nature provides paradigmatic examples for applied reflection and research.

Because of the extent of human exploitation of the planet, examples are most often found in those places that, until recently, have been least hospitable to dense human occupancy and industrial development, such as deserts, the high Arctic, high-altitude mountain ranges, the ocean depths, and the less accessible reaches of forests. Wild nature is also found in locations such as national parks that have been deemed worthy of official protection.

Among the most important aspects of integrity are the autopoietic (self-creative) capacities of life to organize, regenerate, reproduce, sustain, adapt, develop, and evolve over time at a specific location. Thus integrity defines the *evolutionary and biogeographical processes* of a system as well as its parts or elements at a specific location (Angermeier and Karr 1994). Another aspect, discussed by James Karr in relation to water and Reed Noss (1992) regarding terrestrial systems, is the question of what spatial requirements are needed to maintain native ecosystems. Climatic conditions and other biophysical phenomena constitute further systems of interacting and interdependent components that can be analyzed as an open hierarchy of systems. Every organism comprises a system of organic subsystems and interacts with other organisms and abiotic elements to constitute larger ecological systems of progressively wider scope up to the biosphere.

Ecological Integrity and Science

Finally ecological integrity is both "valued and valuable as it bridges the concerns of science and public policy" (Westra et al. 2000, pp. 20–22). For example, in response to the deteriorating condition of our freshwaters, the CWA has its objective: "to restore and maintain the chemical, physical, and biological integrity of the Nation's waters" (sec. 101[a]). Against this backdrop, Karr developed the multimetric Index of Biological Integrity (IBI) to give empirical meaning to the goal of the CWA (Karr and Chu 1999). Karr defines ecological integrity as "the sum of physical, chemical, and biological integrity." Biological integrity, in turn, is "the capacity to support and maintain a balanced, integrated, adaptive biological system having full range of elements (genes, species, and assemblages) and processes (mutation, demography, biotic interactions, nutrient and energy dynamics, and metapopulation processes) expected in the natural habitat of a region" (Karr and Chu 1999, pp. 40–41). Scientists can measure the extent to which a biota deviates from integrity by employing an IBI that is calibrated from a baseline condition found "at site with a biota that is the product of

evolutionary and biogeographic processes in the relative absence of the effects of modern human activity" (Karr 1996, p. 97)—in other words, wild nature. Degradation or loss of integrity is thus any human-induced positive or negative divergence from this baseline for a variety of biological attributes (Westra et al. 2000). Noss's Wildlands Project, which aims to reconnect the wild in North America, from Mexico to Alaska (Noss 1992, Noss and Cooperrider 1994) utilizes the ecosystem approach to argue the importance of conserving areas of integrity.

But the most salient aspect of ecosystem processes (including all their components) is their life-sustaining function, not only within wild nature or the *corridor surrounding wild areas* although these are the main concerns of conservation biologists. The significance of life-sustaining functions is that ultimately they support life everywhere. Gretchen Daily (1997), for instance, specifies in some detail the functions provided by *nature's services*, and her work is crucial in the effort to connect respect for natural systems integrity and human rights.

Arguments against the value of ecological integrity for public policy have identified the concept as *stipulative* rather than fully scientific (Shrader-Frechette 1995). In a similar vein even the concept of ecology as such has been criticized as *not robust* enough to guide public policy (Shrader-Frechette and McCoy 1993). But ecological integrity is already a part of public policy, thus requiring consideration of its meaning and the role its inclusion should play in policy, rather than arguing for its rejection. Further to maintain that "we need a middle path—dictated in part by human not merely biocentric theory" (Shrader-Frechette 1995, p. 141) ignores how humans do not exist apart from other organisms: Biocentrism is life-oriented, and this principle is increasingly accepted not only by science, but in the law.

The routine use of Karr's IBI to reach general conclusions illustrates the ethical effectiveness of the scientific concept of ecological integrity in public policy. The law analyzes a crime or victim under a particular set of circumstances. But public policy must abstract from specifics. Disintegrity (or lack of integrity) and *environmental crime* (Birnie and Boyle 2002) are global in scope and need international fora and broad concepts to ensure that they will be proscribed and possibly eliminated.

In addition, there is mounting evidence to connect disintegrity or *biotic impoverishment* (Karr 1993) in all its forms, from pollutions, climate change, toxic wastes, and encroachment into the wild (Westra 2000) to human morbidity, mortality, and abnormal functioning. International law has enacted a number of instruments to protect human rights (Fidler 2001) and the World Health Organization (WHO) invited the Global Ecological Integrity Project (1992–1999) to consult with it. This collaboration eventually produced a document titled "Ecological Integrity and Sustainable Development: Cornerstones of Public Health" (1999) (Soskolne and Bertollini).

The Ethics of Integrity

Because of this global connection between health and integrity, and the right to life and to *living* (Cançado Trindade 1992), a true understanding of ecological integrity reconnects human life with the wild, and the rights of the latter with those of the former. The ethics of integrity primarily involves respect for *ecological rights* (Taylor 1998) without limiting these to the human rights that are the primary focus of the law. The main point of an ethic of integrity is that it is a *new* ethic (Karr 1993), one founded on recent science demonstrating the interdependence between humankind and its habitats. Environmental ethicists may prefer to focus on one or the other aspect of this interconnected *whole*— biocentrism or anthropocentrism. While biocentrists accept the presence of humankind as such within the rest of nature, anthropocentrists attempt to separate the two, in direct conflict with ecological science.

If, as argued, human health and function are both directly and indirectly affected by disintegrity (Soskolne and Bertollini Internet article), then no theory can properly separate one from the other. The strength of the proverbial canary-in-the-mine example is based on the fact that the demise of the canary anticipates that of the miner. Hence it is necessary to accept a general imperative of respect for ecological integrity. Onora O'Neill makes this point well:

> The injustice of destroying natural and man-made environments can also be thought of in two ways. In the first place, their destruction is unjust because it is a further way by which others can be injured: systematic or gratuitous destruction of the means of life creates vulnerabilities, which facilitate direct injuries to individuals. ... Secondly, the principle of destroying natural and man-made environments, in the sense of destroying their reproductive and regenerative powers, is not universalizable. (O'Neill 1996, p. 176)

In addition, the *vulnerability* that follows the destruction of integrity links this concept to environmental justice. The *principle of integrity* together with appropriate sec-

ond order principles would ensure (a) the defense of the *basic rights* of humankind (Shue 1996) as well as (b) the support of environmental justice globally, because it would ensure the presence of the *preconditions* of agency and thus the ability of all humans to exercise their rights as agents (Gewirth 1982, Beyerveld and Brownsword 2001).

Ecological integrity is thus not an empty metaphor or a grand theory of little utility. It is a concept robust enough to support a solid ethical stance, one that reinstates humans in nature while respecting the latter, thus permitting clear answers in cases of conflicts between (present) economic human interests and (long-term) ecological concerns.

Ecological Integrity and the Law

It is reasonable to conceive of humanity as being morally responsible to protect the integrity of the whole ecosystem, and for that responsibility to be translated into such mechanisms as standard setting in a manner that is cognizant of ecological thresholds (Taylor 1998). Insofar as such responsibility is justified as a protection of human life and health, breaches of environmental regulations deserve not just economic penalties but criminal ones. Nevertheless there is a growing parallel movement to recognize the intrinsic value of both the components and the processes of natural systems, not only in philosophy (Westra 1998, Callicott 1987, Stone 1974, Leopold 1949), but also in the law (Brooks et al. 2002).

A number of international legal instruments also reflect the emerging global ecological concerns, and thus include language about respect for the intrinsic value of both natural entities and processes. This point is illustrated by a project involving the justices of the world's highest courts, which is funded by the United Nations Environment Programme (UNEP). The project's biocentric goal, as outlined by Judge Arthur Chakalson of South Africa, is one of the most important results of the Johannesburg meeting (also known as "Rio+10"). The 2000 Draft International Covenant on Environment and Development incorporates the mandates of the Earth Charter, which was adopted by a United Nations Economic, Scientific, and Cultural Organization (UNESCO) resolution on October 16, 2003, in its language, and includes articles on ecological integrity and the intrinsic value of nature.

Although the positions advanced in these international initiatives are present in law, economic interests often obscure the opposition between the basic rights of persons and peoples and the property rights of legal entities and institutions. In the process courts tend to weigh these incommensurable values as though they were equal. But the right to life and the survival of peoples is not comparable to economic benefits or even the survival of corporate and industrial enterprises.

An additional connection arises from a consideration of *ecological integrity* a complex concept that, after several years of funded work, the Global Ecological Integrity Project eventually defined in 2000 (Westra et al. 2000). The protection of basic human rights through recognition of the need for ecological integrity, as Holmes Rolston (1993) acknowledges, is a step in the emerging awareness of humanity as an integral part of the biosphere (Westra 1998, Taylor 1998).

On the basis of the biocentric foundation for ecological integrity, it is necessary to move toward the twin goals of deterrence and restraint, as is done in the case of assaults, rapes, and other violent crimes. Laws that restrain unbridled property rights represent a first target; but efforts should not be limited to action within the realm of tort law. The reason is obvious: Economic harms are transferable, thus acceptable to the perpetrators of such harms, although the real harms produced are often incompensable. As Brooks and his colleagues indicate in reference to U.S. law, science is now available to support appeals to interdependence. "Not only has conservation biology as a discipline and biodiversity as a concept become an important part of national forest and endangered species management, but major court cases reviewing biodiversity determinations have been decided" (Brooks et al. 2002, p. 373). In addition, Earth System Science increasingly provides "multidisciplinary and interdisciplinary science framework for understanding global scale problems," including the relations and the functioning of "global systems that include the land, oceans and the atmosphere" (Brooks et al. 2002, p. 345). In essence, the ecosystem approach and systematic science of ecological integrity have contributed support to what Antonio A. Cançado Trindade terms "the globalization of human rights protection and of environmental protection" (Cançado Trindade 1992, p. 247).

As noted these ideals are contained in the language and the principles of the Earth Charter. The global reach of these ethics and charters, to be effective, must be supported by a supranational juridical entity such as the European Court of Human Rights. As the case for environmental or, better yet, ecological rights, becomes stronger and more accepted in the international law, the best solution as suggested by Patricia Birnie and Adam Boyle could be to empower the United Nations (UN). It might be desirable "to invest the UN Security Council, or some other UN organ with the power to act in the interests of 'ecological security,' taking universally binding decisions

in the interests of all mankind and the environment (Birnie and Boyle 2002, p. 754). Empowering the United Nations in this way would foster support for programs based on the abundant evidence linking ecology and human rights and could become the basis for a new global environmental/human order (Westra 2004).

LAURA WESTRA

SEE ALSO *Ecology; Research Integrity.*

BIBLIOGRAPHY

Angermeier, Paul L., and James R. Karr. (1994). "Protecting Biotic Resources: Biological Integrity versus Biological Diversity as Policy Directives." *BioScience* 44(10): 690–697.

Beyleveld, Derek, and Roger Brownsword. (2001). *Human Dignity in Bioethics and Biolaw.* Oxford and New York: Oxford University Press.

Birnie, Patricia, W., and Adam E. Boyle. (2002). *International Law and the Environment,* 2nd edition. Oxford: Oxford University Press.

Brooks, Richard; Ross Jones; and Ross A. Virginia. (2002). *Law and Ecology.* Aldershot, Hants, UK: Ashgate Publishing.

Brunnée, Jutta. (1993). "The Responsibility of States for Environmental Harm in a Multinational Context—Problems and Trends." *Les Cahiers de Droit* 34: 827–845.

Brunnée, Jutta, and Stephen Toope. (1997). "Environmental Security and Freshwater Resources: Ecosystem Regime Building." *American Journal of International Law* 91: 26–59.

Cançado Trindade, Antonio A. (1992). "The Contribution of International Human Rights Law to Environmental Protection, with Special Reference to Global Environmental Change." In *Environmental Change and International Law,* ed. Edith Brown-Weiss. Tokyo: United Nations University Press.

Daily, Gretchen C., ed. (1997). *Nature's Services: Societal Dependence on Natural Ecosystems.* Washington, DC: Island Press.

Fidler, David, P. (2001). *International Law and Public Health.* Ardsley, NY: Transnational Publishers.

Gewirth, Alan. (1982). *Human Rights Essays on Justification and Applications.* Chicago: University of Chicago Press.

Hurrell, Andrew, and Benedict Kingsbury, eds. (1992). *The International Politics of the Environment.* New York: Oxford University Press.

Karr, James R. (1993). "Protecting Ecological Integrity: An Urgent Societal Goal." *Yale Journal of International Law* 18(1): 297–306.

Karr, James R. (1996). "Ecological Integrity and Ecological Health Are Not the Same." In *Engineering Within Ecological Constraints,* ed. Peter C. Schulze. Washington, DC: National Academy Press

Karr, James R. (2000). "Health, Integrity and Biological Assessment: The Importance of Whole Things." In *Ecological Integrity: Integrating Environment, Conservation and*

Health, ed. David Pimentel, Laura Westra, and Reed F. Noss. Washington, DC: Island Press.

Karr, James R. (2003). "Biological Integrity and Ecological Health." In *Fundamentals of Ecotoxicology*, 2nd edition, ed. Michael C. Newman and Michael A. Unger. Boca Raton, FL: CRC Press.

Karr, James, R., and Ellen W. Chu. (1999). *Restoring Life in Running Waters*. Washington, DC: Island Press.

Kiss, Alexandre Charles. (1992). "The Implications of Global Change for the International Legal System." In *Environmental Change and International Law*, ed. Edith Brown-Weiss. Tokyo: United Nations University Press.

Leopold, Aldo. (1949). *A Sand County Almanac, and Sketches Here and There*. New York: Oxford University Press.

Noss, Reed F. (1992). "The Wildlands Project: Land Conservation Strategy." *Wild Earth*, Special Issue: *The Wildlife Project*: 10–25.

Noss, Reed F., and Allen Y. Cooperrider. (1994). *Saving Nature's Legacy: Protecting and Restoring Biodiversity*. Washington, DC: Island Press.

O'Neill, Onora. (1996). *Towards Justice and Virtue*. Cambridge, MA: Cambridge University Press.

Pogge, Thomas. (2001). "Priority of Global Justice." In *Global Justice*, ed. Thomas Pogge. Oxford: Blackwell Publishers.

Rolston, Holmes, III. (1993). "Rights and Responsibilities on the Home Planet." *Yale Journal of International Law* 18(1): 251–275.

Shrader-Frechtte, Kristin. (1995). "Hard Ecology, Soft Ecology, and Ecosystem Integrity." In *Perspectives on Ecological Integrity*, ed. Laura Westra and John Lemons. Dordrecht, The Netherlands: Kluwer Academic Publishers.

Shrader-Frechette, Kristin, and Earl D. McCoy. (1993). *Method in Ecology: Strategies for Conservation*. New York: Cambridge University Press.

Shue, Henry. (1996). *Basic Rights: Subsistence, Affluence and American Public Policy*. Princeton, NJ: Princeton University Press.

Sterba, James. (1998). *Justice Here and Now*. Cambridge, MA: Cambridge University Press.

Stone, Christopher. (1974). *Should Trees Have Standing?: Towards Legal Rights for Natural Objects*. Los Altos, CA: W. Kaufmann.

Taylor, Prudence. (1998). "From Environmental to Ecological Human Rights: A New Dynamic in International Law?" *Georgetown International Environmental Law Review* 10: 309–397.

Westra, Laura. (1998). *Living in Integrity Toward a Global Ethic to Restore a Fragmental Earth*. Lanham, MD: Rowman and Littlefield.

Westra, Laura. (2000). "Institutionalized Violence and Human Rights." In *Ecological Integrity: Integrating Environment, Conservation and Health*, ed. David Pimentel, Laura Westra, and Reed F. Noss. Washington, DC: Island Press.

Westra, Laura. (2004). *Ecoviolence and the Law (Supranational, Normative Foundations of Ecocrime)*. Ardsley, NY: Transnational Publishers, Inc.

Westra, Laura; Peter Miller; James R. Karr, et al. (2000). "Ecological Integrity and the Aims of the Global Ecological Integrity Project." In *Ecological Integrity: Integrating Environment, Conservation and Health*, ed. David Pimentel, Laura Westra, and Reed F. Noss. Washington, DC: Island Press.

INTERNET RESOURCE

Soskolne, Colin, and Roberto Bertollini. "Global Ecological Integrity and 'Sustainable Development': Cornerstones of Public Health." World Health Organization, Regional Office for Europe. Available from http://www.euro.who.int/document/gch/ecorep5.pdf.

ECOLOGICAL RESTORATION

• • •

Ecological restoration (hereafter restoration) is "the process of assisting the recovery of an ecosystem that has been damaged, degraded or destroyed" (Society for Ecological Restoration Science & Policy Working Group). *Restoration ecology* and ecological restoration are terms often interchanged: The former is the scientific practice that is contained within the broader embrace of the latter, which incorporates both science *and* many varieties of technological and political practice.

Restoration refers to an array of salutary human interventions in ecological processes, including the elimination of weedy species that choke out diverse native assemblies, prevention of harmful activities (such as excess nutrient loads), rejuvenation of soil conditions that foster vigorous plant communities, reestablishment of extirpated species, and rebuilt webs of social participation that foster ecologically rich and productive ecosystems. The metaphor of healing is often used to describe what restorationists do.

However not everyone regards restoration as a fully positive practice. Some view it as a technological response to ecological damage, while others worry that restoration deflects attention from avoiding harm in the first place. There is also concern that restored ecosystems may be simply pale imitations of nature, and that ecosystems are always more complicated than those seeking to restore them can truly understand. Restoration practice is driven by the tension between a technological approach to restoration—technological restoration—and a participatory, humble, culturally aware approach, or what this author terms "focal restoration." The furious debates among practicing restorationists regarding these issues and others provide particular perspectives on relations between science, technology, and ethics. Moreover, conceptual clarity offers practitioners a guide to pitfalls and opportunities for good restoration.

Concept and Origins

Restoration is practiced in all regions of the world, although what counts as restoration varies according to cultural perspective and socioeconomic condition. This has complicated the creation of a precise definition of this relatively new field, especially because international conversation and cooperative projects have become more common in the early-twenty-first century. In North America, the aim is typically to restore an ecosystem to its predisturbance condition under the presumption that reversion to a pristine, original state is the ideal end point. In Europe and other regions, long and continuous human occupation has resulted in landscapes that present a distinctively cultural benchmark. In many regions of the southern hemisphere, and especially in areas where poverty and civil disruption prevail, the focus is on restoration of productive landscapes that support both ecological and cultural ideals.

No comprehensive history of restoration is available, especially one that treats diverse international perspectives. North Americans often claim to be the founders of restoration, in part because of a tradition in the twentieth century of supporting scientific and practical restoration capacity including the formation of the premier organization devoted to restoration, the Society for Ecological Restoration International (founded 1987). Prairie restoration projects at the University of Wisconsin Arboretum under the direction of Aldo Leopold, Theodore Sperry, and Henry Greene in the 1930s are often cited as inaugural moments in modern restoration. Important as these efforts are, there were prior influential developments in applied ecology, rehabilitation (the recovery of a landscape to productive capability), revegetation, and naturalistic gardening that made the Wisconsin projects possible (Perrow and Davy 2002; Mills 1995; Jordan, Gilpin, and Aber 1987). Restoration was being practiced under different guises in North America, Europe, and other regions of the world prior to the twentieth century, and, as historical accounts of these efforts are written, a tangled and interconnected lineage will undoubtedly be revealed.

Points at Issue

A spate of articles written since the 1980s has positioned restoration as one of the most hotly contested issues in environmental philosophy. Why is this? Philosophers, many environmentalists, and some restorationists are uneasy about claims that ecosystems can in fact be restored. Much turns on the standards set for restoration, most prominently the demands for historical accuracy. If the aim is to reset ecosystems to some prior time or sequence, then restoration is by definition an austere and limited practice, depending on a limited ranges of options and choices.

If the demands for historical fidelity are relaxed, the practice opens up, although enlargement of scope creates other problems. What are appropriate boundaries on restoration? How much history is necessary? How precise ought be the demands for ecological integrity? (Ecological integrity is an umbrella term that describes the capacity of an ecosystem to adjust to change—resiliency, elasticity, stress response, and so on [Kay 1991]). How much should human agency matter? How much should human participation in ecological processes matter? Without much digging, restoration turns into a conceptual quagmire, which is occasionally vexing for practitioners and always intriguing for philosophers (Throop 2002).

Arguably what has proved most contentious is the instrumental character of restoration. At worst, some would argue, restoration is a mere technological fix, that is a forgery of nature, and deflects attention from pressing and underlying environmental problems (Eliot 1997). While few hold such a dim view and most acknowledge that restoration creates value, there is a fundamental concern that restoration is a practice that grew up and thrives in a technological culture. Indeed restoration is always a series of deliberate interventions in ecological processes. As restorative capacity rises, so does the risk that such capacity will be used as a justification for destruction or careless modification of ecosystems. The challenge is to keep restoration from becoming an apologia for environmental destruction while manifesting a powerful will to repair the damage that continues to be done. Hence most restorationists operate under the belief that their actions benefit nonhuman species and enrich the social engagement between people and ecosystems. Limiting human will and ensuring that restoration does not become an end in itself is a central challenge.

The Future of Restoration

The tendency to think of restoration in technological terms is abetted by increasingly large projects—restoration megaprojects such as Florida's Everglades restoration—that are driven by typically top-down imperatives and serve primarily as emblems of environmental responsibility. The dominant tradition in restoration encompasses relatively small-scale projects that depend on bottom-up participation; these projects are deeply embedded in locality and enliven human communities. While the appearance to some observers that restoration is a set of prescriptions imposed on nature, in fact most restoration projects to date are modest in intention, self-reflexive, and tentative; exactly the opposite of what one might think of as large-scale technologically constituted practices.

Restoration practitioners are approaching a crossroads at which they will have to choose between *technological* and *focal* restoration, which focuses on community and participation (Higgs 2003). Focal restoration is one term for describing the alternative or antidote to technological restoration, and derives from Borgmann's (1984) formulation of "focal practice," in which the relations between "things" and practices are brought to the center and given priority. When focusing on something that truly matters to a community—an ecosystem to be restored for instance—the values of that community and the integrity of the thing are given heightened respect. Other terms such as "ecocultural" restoration are found in the restoration literature with roughly the same intention, but this author prefers the identification of focal restoration with its robust commentary about, and philosophy of, technology. The choice between technological and focal restoration may not be exclusive or stark, but reflective practitioners must decide which vision of restoration is appropriate. Scholarly and popular criticism has raised awareness of the risks that restoration will become thoroughly enmeshed in technological culture. The challenge is to steer along the road of participation, with respect for ecological process, modesty, and humility.

Ecological restoration has stirred profound debates about the constitution of nature in a technological society and human relations with ecosystems. Perhaps as much as any other practice, restoration has brought a conceptual spotlight to issues that arise in environmental management, conservation biology and other related endeavors. In particular, restoration demands attention to the social, economic and political relationships people have with places, which inspires a broader perspective on the scientific and technical dimensions. It is, therefore, insufficient to discuss "restoration ecology" without "ecological restoration; both matter to achieving the socially constituted goal of good restoration. The dynamic character of ecosystems also poses some fascinating challenges to other uses of the term restoration, such as those found in art, architecture, and literature.

ERIC HIGGS

SEE ALSO *Acid Mine Drainage; Ecological Footprint; Ecology; Environmental Economics; Rain Forest; Sierra Club; Wilderness.*

BIBLIOGRAPHY

Borgmann, Albert. (1984). *Technology and the Character of Contemporary Life*. Chicago: University of Chicago Press. Borgmann's original book on technology—he has since published two others—has become a foundation for contemporary philosophy of technology.

Eliot, Robert. (1997). *Faking Nature: the Ethics of Environmental Restoration*. London: Routledge. The title derives from Eliot's oft-cited paper from 1982 that assailed restoration practice and sparked a significant portion of the philosophical debate about restoration.

Higgs, Eric. (2003). *Nature By Design: People, Natural Process, and Ecological Restoration*. Cambridge, MA: MIT Press. Sets ecological restoration in a technological culture, and identifies the perils and responses to those perils that restorationists should follow.

Higgs, Eric; Andrew Light; and David Strong. (2000). *Technology and the Good Life?* Chicago: The University of Chicago Press.

Jordan, William R., III; Michael E. Gilpin; and John D. Aber, eds. (1987). *Restoration Ecology: A Synthetic Approach to Ecological Research*. New York: Cambridge University Press. Among the first synthetic accounts of restoration theory and practice.

Kay, James J. (1991). "A Non-Equilibrium Thermodynamics Framework for Discussing Ecosystem Integrity." *Environmental Management*. 15(4): 483–495.

Mills, Stephanie. (1995). *In Service of the Wild: Restoring and Reinhabiting Damaged Landscapes*. Boston: Beacon Press. A heartfelt and lyrical discussion of restoration.

Perrow, Martin R., and Anthony J. Davy, eds. (2002). *Handbook of Ecological Restoration*, 2 Vols. Cambridge, UK: Cambridge University Press. A good initial if limited attempt at summarizing restoration practice internationally.

Throop, William, ed. (2002). *Environmental Restoration: Ethics, Theory, and Practice*. Amherst, NY: Humanity Books.

INTERNET RESOURCE

Society for Ecological Restoration Science & Policy Working Group. "The SER Primer on Ecological Restoration." Available from www.ser.org/content/ecological_restoration_primer.asp.

ECOLOGY

• • •

The word *ecology* is derived from the Greek *oikos*, "household," and *logos*, "reason," thus indicating the logic of living creatures in their homes. Although *oikos* originally indicated only human households, as a term coined in 1866 by Ernst Haeckel, ecology names a biological science such as molecular biology or evolutionary biology, though often thought to be less mature, that studies organism–environment relations. Closely related to ecology in this sense are conservation biology and environmental science. Ecology, the science, studies ecosystems at multiple levels and scales in space and time. Ecosystems have proved to be often quite complicated and resist analysis. Experiments in the field are difficult, and the systems may be partly chaotic.

In part because of such complications ecology has become the focus of a particular set of discussions related to science, technology, and ethics. The term *ecological ethics* may, for instance, call for doing ethics in the light of what ecologists have found in their studies of the world. Perhaps it is appropriate, at times, for humans to imitate the way ecologies themselves function, or look toward ecosystems as fundamental goods to be appreciated and preserved. Given these associations, ecology can also feed into a worldview or philosophy.

What has been called the environmental or ecological crisis seems to rest on assumptions about or commitment to the goodness of ecosystems in the face of threats to their continuing vitality from pollution or other phenomena. Ecology thus becomes mixed with ethics in urging that humans ought to find a lifestyle more respectful or harmonious with nature. As the founder of wildlife management, Aldo Leopold, argued: "A thing is right when it tends to preserve the integrity, stability, and beauty of the biotic community. It is wrong when it tends otherwise" (Leopold 1968 [1949], pp. 224–225). More recently, since the United Nations Conference on Environment and Development (1992), the focus has been a sustainable economy based on a sustainable biosphere.

Leading Concepts

Leading concepts in ecology involve ecosystems (a term coined by Arthur G. Tansley in 1935), a succession of communities rejuvenated by disturbances, energy flow, niches and habitats, food chains and webs, carrying capacity, populations and survival rates, diversity, and stability. A main claim is that every organism is what it is where it is, its place essential to its being, the "skin-out" environment as vital as "skin-in" metabolisms. Early ecologists described organism–environment relations in terms of homeostasis, equilibrium, and balance. Contemporary ecologists give a greater role to contingency, flux, dynamic change, or even chaos. Others emphasize self-organizing systems (autopoiesis).

As subsequent studies have shown, any ecological stability is not simply homeostatic but quite dynamic, and may differ with local systems, the level of analysis, and over time. There are perennial processes—wind, rain, soil, photosynthesis, competition, predation, symbiosis, trophic pyramids, and networks. Ecosystems may wander or be stable within bounds. When unusual disturbances come, ecosystems can be displaced beyond recovery of their former patterns. Then they settle into new equilibria. Ecosystems are always on a historical trajectory, a dynamism of chaos and order entwined.

Ecology, Technology, Management

How far can human environmental policy be drawn from ecology? The question raises classical is/ought concerns about moving from facts to values, and worries about the naturalistic fallacy. Perhaps ecology, a "piecemeal" science, can offer no more than generalizations of regional or local scope, and supply various concepts (such as eutrophication of lakes, keystone species, nutrient recycling, niches, and succession) for analyzing particular circumstances. Humans could then step in with their management objectives and reshape ecosystems consonant with cultural goals.

Certainly humans have always had to rest their cultures upon a natural life-support system. The human technosphere is constructed inside the biosphere. In the future this could change; the technosphere could supersede the biosphere. The natural sciences would be increasingly replaced by the sciences of the artificial, as in computer science, or materials science (as with Teflon), or engineered biotas. Edward Yoxen (1983) has celebrated the prospect: "The living world can now be viewed as a vast organic Lego kit inviting combination, hybridisation, and continual rebuilding. . . . Thus our image of nature is coming more and more to emphasise human intervention through a process of design" (pp. 2, 15).

Ecosystem management (if not more global, planetary management) appeals alike to scientists, who see the need for understanding ecosystems objectively and for applied technologies, as well as to landscape architects and environmental engineers, who see nature as redesigned home, and finally to humanists, who desire benefits for people. A good thing in nature may not be a good in culture, and vice versa. Viruses kill people; people's cities kill wild animals. The combined ecosystem/management policy promises to operate at systemwide levels, presumably to manage for indefinite sustainability, alike of ecosystems and their outputs. Such management sees nature as "natural resources" at the same time that it has a "respect nature" dimension. Christian ethicists note that the secular word *manager* is a stand-in for the earlier theological word *steward*, and also that the biblical "dominion" involves more cultivating a garden Earth than conquering and controlling it.

At the same time, ecosystem management has been criticized as an umbrella idea under which different managers can include almost anything they wish, because what one is to manage ecosystems for is left unspecified. They might be managed for maximum sustainable yield, for equal opportunity in the next generation, for maximum biodiversity, or for quick profit. Nevertheless there usually is the idea of fitting human uses into ongoing ecosystem health or integrity. There is less overconfidence than with those who view nature as a vast Lego kit and seek to redesign the planet. This is often a matter of managing human uses of their ecosystems with as much care as one is managing, or revising, wild nature.

Editing a 1989 *Scientific American* issue on "Managing Planet Earth," William C. Clark identified two central questions: "What kind of planet do we want? What kind of planet can we get?" (Clark 1989, p. 47). Over great stretches of Earth, evolutionary and ecosystemic nature has been diminished in favor of engineered design. Nature is at an end. The principal novelty of the millenium is that Earth will be a managed planet. Humans will make it a better home for themselves.

Ecological Limits?

Such claims raise concerns about how far nature can and ought to be transformed into humanized nature. Ecologists are likely to fear the arrogance rather than to celebrate the expertise of such planetary engineers. Much transformation is the positive result of human managerial successes: widespread irrigation, agricultural production, electric power. But just as often there are unintended, undesired results: The seeds of exotic weeds are carried afar on ships and trains; the landscape is increasingly weedy. Toxic, nondegradable agricultural chemicals seep into the nooks and crannies of all nature. Industrial production and mass consumption produces global climate change. The "dominion" mentality is what led to the ecological crisis; more clever dominion, the ultimate technological fix, is a dangerous myth. Rather people should think of humans as fitting themselves into a sustainable biosphere, as members of a larger community of life on Earth, as a better logic of our being at home on Earth.

But, critics rejoin, the community of life on Earth is already human-centered; this is the fact of the matter. The end of nature may be, in its own way, a sad thing; but it is inevitable, and the culture that replaces nature has many compensating values. Humans too belong on the planet. With the arrival of humans, and their technologies, pristine nature vanishes. Nature does not vanish equally and everywhere, but there has been loosed on the planet such a power that wild nature will never again be the dominant determinant of what takes place on the inhabited landscapes.

Should this rebuilding of humanity's Earth home be thought of as a sort of dialectic: nature the thesis, culture the antithesis, and the synthesis a humanized nature? Possibly, but there is a still better ecological model: that of an ellipse with two foci. Some events are generated under the control of a culture focus: society, its economics, its politics, its technologies. Under the other focus, nature, some events take place in the absence of humans—wild, spontaneous, ecological, evolutionary nature (in parks, reserves, and wilderness areas).

From a larger ecological perspective, a domain of hybrid or synthetic events is generated under the simultaneous control of both foci, the result of integrated influences from nature and culture. Human labor and craft put natural properties to use in culture, mixing the two to good effect in agricultural, industrial, scientific, medical, and technological applications. *Symbiosis* is a parallel biological word.

Lest technologists become too arrogant, there is a sense in which nature has not ended and never will. Humans stave off natural forces, but the natural forces can and will return, if one takes away the humans. Nature is forever lingering around. Nature bats last. In, with, and under even the most technologically sophisticated culture, there is always this once and future nature.

Ecological Is and Ought

Scientists and ethicists alike have traditionally divided their disciplines into realms of the is and the ought, facts and values. No study of nature, it has been argued, will tell humans how they ought to behave. But this neat division is challenged by ecologists and their philosophical and ethical interpreters. There may be goods (values) in nature that humans ought to consider and care for. Animals, plants, and species, integrated into ecosystems, may embody values that, though nonmoral, count morally when moral agents encounter them. Ecology invites human beings to open their eyes and to appreciate realities that are valuable in ways humans ought to respect.

Ecological or environmental science may thus inform environmental technology and environmental ethics in subtle ways. Scientists describe the order, dynamic stability, and diversity in biotic communities. They analyze interdependence, or speak of health or integrity, perhaps of resilience or efficiency. Scientists describe the adapted fit that organisms have in their niches. They analyze an ecosystem as flourishing, as self-organizing. Strictly interpreted, these are only descriptions; and yet they embody already quasi-evaluative terms, perhaps not always but often enough that by the time the descriptions of ecosystems are in, some values are already there, putting constraints on what we think might be appropriate human technological development of such areas.

Ethicists can with considerable plausibility also claim that neither conservation, nor a sustainable biosphere, nor sustainable development, nor a well-managed planet, nor any other harmony between humans and nature can be gained until persons learn to use Earth both justly and charitably. These twin concepts are found neither in wild nature nor in any science that studies nature, nor in any technology as such. One needs human ecology, humane ecology, and this requires insight more into human nature than into wild nature. True, humans cannot know the right way to act if they are ignorant of the causal outcomes in the ecosystems they modify. And they cannot act successfully without technology. But there must be more, and here ethics is required to keep science, technology, and life human and humane on this, humanity's home planet.

HOLMES ROLSTON III

SEE ALSO *Biodiversity; Deforestation and Desertification; Ecological Economics; Ecological Footprint; Ecological Integrity; Ecological Restoration; Environmental Ethics; Rain Forest; Sustainability and Sustainable Development; United Nations Environmental Program.*

BIBLIOGRAPHY

Barbour, Ian G. (1980). *Technology, Environment, and Human Values.* New York: Praeger Publishers. Older study, but remains a useful analysis.

Clark, William C. (1989). "Managing Planet Earth." *Scientific American* 261(3): 46–54. Introduces a theme issue on managing the planet.

Cooper, Gregory J. (2003). *The Science of the Struggle for Existence: On the Foundations of Ecology.* Cambridge, UK, and New York: Cambridge University Press. Examines the philosophical foundations of ecosystem science.

Golley, Frank B. (1998). *A Primer for Environmental Literacy.* New Haven, CT: Yale University Press. Presents leading themes in ecology and how understanding these might change individual lifestyles.

Gotelli, Nicholas J. (2001). *A Primer of Ecology,* 3rd edition. Sunderland, MA: Sinauer Associates.

Grumbine, R. Edward. (1994). "What Is Ecosystem Management?" *Conservation Biology* 8(1): 27–38. Discusses ecosystem management and whether it is feasible.

Leopold, Aldo. 1968 (1949). *A Sand County Almanac.* New York: Oxford University Press.

Odum, Eugene P. (1997). *Ecology: A Bridge between Science and Society,* 3rd edition. Sunderland, MA: Sinauer Associates. Ecology and its social implications.

Real, Leslie A., and James H. Brown, eds. (1991). *Foundations of Ecology: Classic Papers with Commentaries.* Chicago: University of Chicago Press. The classic papers in the founding of ecology as a science.

Yoxen, Edward. (1983). *The Gene Business: Who Should Control Biotechnology?* New York: Harper and Row. Examines genetic modification of foods for human benefit and who makes what kind of choices.

Zweers, Wim, and Jan J. Boersema, eds. (1994). *Ecology, Technology, and Culture.* Cambridge, UK: The White Horse Press. Essays from a European perspective.

ECONOMICS AND ETHICS

• • •

Economics often is regarded as the most successful of the social sciences in developing a scientific theory of social behavior. Therefore, economics is a science with manifest ethical implications.

General Equilibrium Theory

Contemporary economic theory is based on the *general equilibrium model* first outlined by the nineteenth-century Swiss economist Léon Walras (1834–1910) and perfected in the post–World War II era by Kenneth Arrow (b. 1921; winner of a Nobel Prize in economics in 1972) and others. The Walrasian general equilibrium model includes firms, which transform production inputs (land, labor, natural resources, capital goods such as buildings and machines, and intermediate goods produced by other firms) into outputs (including consumer goods and services) by using a technologically determined production function that summarizes the most technically efficient way to transform a specific array of inputs into a particular output or array of outputs. The only other actors in the general equilibrium model are individuals and government. Individuals supply labor to firms and own the land, natural resources, and capital, which they supply to firms, and also are consumers who use the income they derive from supplying inputs to production to purchase goods and services that they then consume. The government enforces property rights and contracts and intervenes to alter economic outcomes that are considered inefficient or inequitable.

The general equilibrium model assumes that there are many firms competing to supply each good desired

FIGURE 1

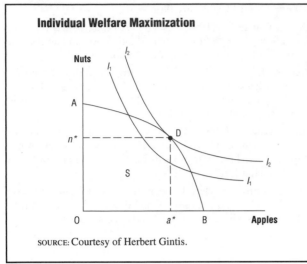

Individual Welfare Maximization

SOURCE: Courtesy of Herbert Gintis.

by consumers. Equilibrium takes the form of a set of prices for the production inputs and outputs so that supply equals demand for each good as well as for labor, land, capital, and natural resource inputs. General equilibrium theory shows that once the equilibrium prices are known, if individuals and firms are allowed to trade in competitive markets, the equilibrium allocation of production inputs and outputs will emerge. This process often is called *market clearing*.

General equilibrium theory assumes that each individual has a *preference function* that reflects that individual's labor supply and consumption rankings, as described by *rational choice theory* and *decision theory*. The central property of preferences in the theory is that they are self-regarding; this means that individuals care only about their personal labor supply and commodity consumption. It also means that individuals are completely indifferent to the welfare of others and never willingly sacrifice on behalf of other market participants. To make this assumption more palatable, the individuals in general equilibrium theory often are described as families, thus allowing for nonmarket altruistic interactions among nuclear family members.

Consumer Sovereignty

The most important ethical judgment in general equilibrium theory is that involving *consumer sovereignty*: A state of affairs A is normatively better than a state of affairs B for individuals if, with everything else being equal, these individuals prefer the labor and consumption bundles they have in state A over those they have in state B. For a graphic illustration, assume that there are only two goods, Apples (*a*) and Nuts (*n*). Suppose the consu-

mer is restricted to choosing from the Apples-Nuts bundles depicted by region S in Figure 1, bounded by OADB.

In this figure I_1I_1 and I_2I_2 *represent indifference curves*, which are sets of points along which the consumer is equally well off. These curves exhibit *diminishing marginal rates of substitution*; this means that the greater the ratio of nuts to apples is, the more the individual values apples over nuts, and vice versa. Note that the indifference curve I_1 intersects the interior of region S, and so an agent may increase his or her consumption of both Apples and Nuts. Thus, that individual can shift out his or her indifference curve, and hence increase his or her utility, as long as that indifference curve continues to intersect region S. The consumer is thus best off with indifference curve I_2, which intersects S at the single point D, at which point the indifference curve is tangent to the constraint set S. Consumer sovereignty judges consumption point D, at which the individual consumes a^* units of Apples and n^* units of Nuts, to be a welfare optimum for the individual.

Consumer sovereignty is a problematic ethical judgment in at least three ways. First, it ignores the distribution of economic benefits across individuals. If individual I_1 is very rich and all the other individuals in the economy are very poor, it can be said that society as a whole is normatively better if I_1 is made even richer as long as this is not done at the expense of the other individuals. Assuming that individuals are self-regarding, this is a plausible ethical statement, but if the poor care about equity and are hurt when their relative deprivation is exacerbated, the consumer sovereignty judgment will be flawed. In fact, it appears that individuals do care not only about their own consumption but about how it compares with that of others as well, and so improving the consumption opportunities of one group can hurt another group (Lane 1993).

A second problem with the consumer sovereignty principle lies in its failure to recognize that individuals may prefer things that are not in their own interest. For instance, it is in the nature of the addiction that a cigarette smoker prefers smoking to abstaining, but even smokers recognize that they would be better off if they abstained. Consumer sovereignty at one time was an inviolable article of faith for economists, who considered evaluating people's preferences an insulting and socially undesirable form of paternalism. Widespread phenomena such as obesity, recreational drug use, and substance addiction have convinced many economists that there is a role for government intervention to curb consumer sovereignty in such spheres. However, these sentiments are restricted to a few well-defined areas.

The values promoted by economic theory are generally hostile to the notion that scientists and the educated elite (e.g., teachers, preachers, and social workers) know best what is good for everyone else.

The third problem is that consumer sovereignty implies that individuals care only about their own well-being, whereas people often care about each other. In fact, people often positively value contributing to the welfare of the less well off and to the punishment of social transgressors.

Pareto Efficiency

Consumer sovereignty leads to a very simple but powerful means of comparing the normative worth of two economic situations. One says that state A is *Pareto superior* to state B if at least one person is better off in state A than in state B and no one is worse off in state A than in state B, where *better off* and *worse off* are synonyms for *higher up* and *lower down* on one's preference ordering according to consumer sovereignty. It then can be said that state A is *Pareto efficient* if there is no other state that is Pareto superior to it. These conditions are named after the nineteenth-century Italian engineer and sociologist Vilfredo Pareto (1848–1923).

The important point here is that the Pareto efficiency condition expresses the very weak ethical judgment that society is better off when one member is better off and none worse off, and an individual is better off when he or she has more of what he or she prefers. Any maximally ethically desirable state of the economy will be Pareto efficient because otherwise, by definition, there would be a normatively superior state. Thus, one can separate the normative question "Who deserves to get what?" completely from the positive, nonethical question "What are the conditions for Pareto efficiency?"

The relationship between Pareto efficiency and the normative question of the distribution of welfare among individuals was diagramed by the English economist Francis Ysidro Edgeworth (1845–1926) in what has come to be known as the Edgeworth Box diagram (see Figure 2). One can consider a simple economy with two individuals (I_1 and I_2), two goods (Apples and Nuts), no labor, and no firms—the two individuals simply trade with each other. The width of the rectangle represents the total amount of Apples, and the height represents the amount of Nuts.

Suppose point C represents the initial wealth of the two individuals so that I_1 has FG Apples and 1E Nuts and I_2 has G2 Apples and EF Nuts. The curve

$$I_1^c$$

represents an indifference curve for I_1, a locus of points (combinations of Apples and Nuts) among which I_1 is indifferent, preferring all the points to the northeast to points on the curve and preferring all points on the curve to points to the southwest of the curve. Similarly,

$$I_2^c$$

is an indifference curve for I_2. Note that point D lies on both curves, and it is easy to see that D is Pareto efficient because any move away from it will make either I_1 or I_2 worse off. Clearly, the initial point C makes both agents worse off than they are at D, and so it would benefit them to trade, with I_1 increasing the amount of Apples in his bundle by getting them from I_2 and I_2 increasing the amount of Nuts in her bundle by getting them from I_1.

The locus of points 1ADB2 is called the *contract curve* and is the set of Pareto efficient points for this economy. Note that at point 1 individual 2 gets everything, whereas at point 2 individual 1 gets everything. The points between represent different distributions of the benefits of the total supply of Apples and Nuts in the economy. Of course, I_1 prefers C to most of the points on the contract curve below C and I_2 prefers C to most of the points on the contract curve above C. To find out exactly which point or points each individual prefers, one can draw the indifference curves for the two agents that go through C and see where they hit the contract curve. Suppose they hit at C_1 and C_2 (not shown in the figure). Then the two agents will be willing to trade at any point on the contract curve between C_1 and C_2.

Implications for Ethics

The general equilibrium model has several important implications for ethical theory. The First Fundamental Theorem of Welfare Economics states that any equilibrium of the market economy is Pareto efficient. Note that this conclusion depends on the assumption of self-regarding preferences. If, for instance, above a certain income level people care only about their relative position in the distribution of material benefits, a market-interfering law that prohibited people from working more than a certain number of hours per week could increase the welfare of all people.

Suppose that the various production sectors have production functions that do not depend on one another and that efficient firm size is sufficiently small that there

FIGURE 2

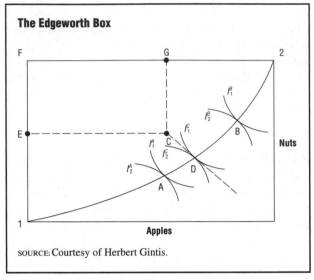

The Edgeworth Box

SOURCE: Courtesy of Herbert Gintis.

can be many firms producing each good in equilibrium. Then a Second Fundamental Theorem of Welfare Economics holds. This theorem states that if the economy satisfies the conditions stated above and a few technical conditions, any Pareto efficient allocation can be supported by a suitable initial distribution of ownership rights in land, natural resources, capital goods, and labor. This theorem successfully separates the positive (technical, scientific) issues of Pareto efficiency from the normative issue of who deserves to get what.

Perhaps the most distinctive normative characteristic of the Walrasian general equilibrium model is its strong commitment to separating considerations of technical efficiency from considerations of normative distribution. This separation is completely justified only if there is a mechanism to distribute initial ownership rights in a way that achieves an ethically desired distribution of welfare. The separation nevertheless often is defended by saying that if the economy attends to the efficiency side of the dichotomy rather than sacrificing efficiency in the name of equity, in the long run most individuals will be better off. This is doubtless a defensible position, although there are often government interventions that promote efficiency and satisfy egalitarian goals as well (Bowles and Gintis 1996).

Several aspects of the general equilibrium model render it an imperfect basis for making judgments about social policy and ethics. First, people are not entirely self-regarding. Rather, they are what may be called strong reciprocators who prefer to reward those who help them and contribute to social goals and to punish those who hurt them or act in an antisocial manner (Gintis, Bowles,

Boyd, and Fehr 2003). Strong reciprocators prefer to redistribute resources to the needy if the recipients are considered worthy but not otherwise. This leads to social policies that would not be envisioned under the assumption of the general equilibrium model that people are self-regarding (Fong, Bowles, and Gintis 2004).

In addition, the idea of achieving social equity by means of an initial distribution of wealth among individuals in society followed by market exchange ignores the problem that with incomplete knowledge of the future the process of egalitarian redistribution away from the wealthy and toward the needy will have to be repeated time and time again as the economy moves away from a condition of basic equality to one of severe inequality. That type of redistribution may be infeasible because of the ensuing individual disincentives to accumulate wealth and income-earning capacity.

To see this one must remember that the general equilibrium model assumes that all goods and services are marketable and can be the subject of contracts that are enforced costlessly by a third party such as the judicial system. For instance, several behaviors that are critical to high levels of productivity—hard work, maintenance of productive equipment, entrepreneurial risk taking, and the like—are difficult to monitor and thus cannot be specified fully in any contract that is enforceable at a low cost. As a result key economic actors, namely, employees and managers, must be motivated by incomplete contracts in which monetary rewards are contingent on their performance. However, when incentive rewards are necessary to motivate behavior, egalitarian redistribution works against those who supply a high level of effort, leading to a dampening of the incentive system. Hence, it may be impossible in practice to separate efficiency from equity issues.

Another problem with periodic egalitarian redistribution is that it may violate the principles of justice that many people hold. According to the English philosopher John Locke,

> every man has a property in his own person.... The labour of his body, and the work of his hands, we may say, are properly his. Whatsoever then he removes out of the state that nature hath provided, and left it in, he hath mixed his labour with, and joined to it something that is his own, and thereby makes it his property (*Second Treatise on Government* (*Of Property* Chapter 5, Section 27).).

Such values would preclude the involuntary redistribution of wealth even if it furthered widely approved egalitarian ends.

In short, technical efficiency and normative issues concerning justice and equality cannot be separated in the manner intended in the Walrasian general equilibrium model. Moreover, because individuals are not completely self-regarding, social policies based on this model will fail to tap the genuine egalitarian motives of voters and citizens. This said, it would be folly to use these shortcomings to override completely the assumption in the Walrasian model that in the long run economic efficiency and efficiency-oriented technical change are more likely to help the less well off. Insofar as this is the case, issues of egalitarian reform should be biased as much as possible toward efficiency-enhancing redistributions such as education, training, and the financing of small business and small-scale farming.

Contracts

Another important set of issues arises when it is recognized that the neoclassical assumption that contracts can be written and enforced costlessly generally does not hold for either labor or capital. In the case of labor an employer can offer workers a legally binding wage, but a worker cannot offer the employer a legally binding amount of effort and care. This is the case because effort and care are not sufficiently measurable that a violation would hold up in a court of law. Therefore, employers generally enter into long-term agreements with their employees, using the threat of termination and the promise of promotion to elicit a high level of performance. However, this practice will motivate employees only if dismissal is costly to an employee, and this will be the case generally only if it is difficult to obtain comparable employment with another firm. That will be the case only if there is *equilibrium unemployment* in the economy. It can be shown that if employers follow this strategy of worker motivation, there indeed will be unemployment in equilibrium (Gintis 1976, Shapiro and Stiglitz 1984, Bowles 1985, Gintis and Ishikawa 1987, Bowles and Gintis 2000).

This situation accounts for the fact that employers generally have power over their employees in the sense that employers can use the threat of dismissal to induce employees to bend to their will, whereas the converse is not true. Although this power may be used benignly, it also may be used in an unethical manner, as occurs when employers force employees to accept unhealthy working conditions or subject them to sexual harassment and other forms of personal humiliation and discrimination.

In the case of capital the difficulty in contract enforcement arises because the borrower cannot make an easily enforced promise to repay a loan. Of course, a wealthy borrower can offer collateral in the form of valuable assets that the lender has the right to seize if the borrower defaults. Nonwealthy borrowers who lack collateral thus are frozen out of many capital markets. Special credit institutions have arisen to give nonwealthy individuals access to credit for home and automobile ownership as well as credit cards for consumer purchases. In the case of home and automobile purchases the asset itself provide collateral, and requiring the buyer to provide a sizable down payment assures the lender against sustaining a loss. In the case of credit cards the threat of a loss of one's credit rating and hence future access to consumer credit serves to protect lenders against loss (Bowles and Gintis 2000).

The absence of costlessly enforced contracts in capital markets has several important social implications. First, demand generally exceeds supply, leading to credit rationing (Stiglitz and Weiss 1981) in which wealthy agents have access to loans whereas nonwealthy agents do not. Second, banks and other lending agencies have the same sort of power over borrowers that employers have over employees by virtue of their superior "short-side" market position. This power is subject to abuse by lenders, although large borrowers have a counterbalancing power to injure lenders so that in effect it is only the small borrower who must be protected against the arbitrary actions of lenders (Bowles and Gintis 2000).

HERBERT GINTIS

SEE ALSO *Consequentialism; Efficiency; Political Economy; Rational Choice Theory.*

BIBLIOGRAPHY

Bowles, Samuel. (1985). "The Production Process in a Competitive Economy: Walrasian, Neo-Hobbesian, and Marxian Models." *American Economic Review* 75(1):16–36.

Bowles, Samuel, and Herbert Gintis. (1996). "Asset Based Redistribution: Improving the Tradeoff between Allocative Gains and Dynamic Inefficiency Losses." MacArthur Economics Initiative Working Paper, University of California at Berkeley.

Bowles, Samuel, and Herbert Gintis. (2000). "Walrasian Economics in Retrospect." *Quarterly Journal of Economics*, November, pp. 1411–1439.

Fong, Christina M.; Samuel Bowles; and Herbert Gintis. (2004). "Reciprocity and the Welfare State." In *Moral Sentiments and Material Interests: On the Foundations of Cooperation in Economic Life*, eds. Herbert Gintis, Samuel Bowles, Robert Boyd, and Ernst Fehr. Cambridge, MA: MIT Press.

Gintis, Herbert. (1976). "The Nature of the Labor Exchange and the Theory of Capitalist Production." *Review of Radical Political Economics* 8(2): 36–54.

Gintis, Herbert, and Tsuneo Ishikawa. (1987). "Wages, Work Discipline, and Unemployment." *Journal of Japanese and International Economies* 1: 195–228.

Gintis, Herbert; Samuel Bowles; Robert Boyd; and Ernst Fehr. (2003). "Explaining Altruistic Behavior in Humans." *Evolution & Human Behavior* 24: 153–172.

Lane, Robert E. (1993). "Does Money Buy Happiness?" *The Public Interest* 113: 56–65.

Locke, John. (1965 [1690]). *Two Treatises of Government*, ed. Peter Laslett. New York: New American Library.

Shapiro, Carl, and Joseph Stiglitz. (1984). "Unemployment as a Worker Discipline Device." *American Economic Review* 74(3): 433–444.

Stiglitz, Joseph, and Andrew Weiss. (1981). "Credit Rationing in Markets with Imperfect Information." *American Economic Review* 71: 393–411.

ECONOMICS: OVERVIEW

• • •

In economics, issues of science, technology, and ethics are more diverse than in any other scientific or technological discipline. In the first instance, like all the sciences, economics is both dependent on and independent of ethics. Its methods involve internal commitments of an ethical character (e.g., truth telling) but are subject to external ethical oversight (e.g., with regard to the proper treatment of human participants in empirical research). At the same time, as the entry on "Economics and Ethics" points out, the content of the science may have ethical implications in ways that physics, for instance, does not.

In the second instance, insofar as economics constitutes a technique or technology, it may provide guidance for how to achieve externally determined ends. As such it exhibits multiple interactions with various ethical, legal, and policy perspectives. Such interactions are referenced in entries such as those on "Capitalism," "Market Theory," "Political Economy," and "Science Policy."

Modern Economics

Economics in the modern sense (also called "neoclassical economics") is the science of the allocation and utilization of resources under conditions of scarcity, that is, when there are not enough resources or goods to satisfy all human needs or wants. In a widely adopted definition, for example, the British economist Lionel Robbins

(1932) describes economics as "the science which studies human behavior as a relationship between ends and scarce means which have alternatives uses" (p. 16). Insofar as economics assumes that most goods and services are scarce or insufficient to satisfy human wants, and that by and large all human wants are legitimate, economics places the free satisfaction of individual human desires at the top of its own internal moral hierarchy. This may be described as the ethics of economics, one that further provides a basic justification for modern technology as a means to increase efficiency in exploitation, production, and distribution, and has been subject to extended historicophilosophical assessment and some criticism (see, e.g., Polanyi 1944, Dumont 1977, Rhoads 1985, Achterhuis 1988, Nelson 2001).

The science of economics is divided into two main overlapping branches dealing with smaller scale and larger scale economic phenomena. The economic analysis of scarcity and the pursuit of productive efficiency in the sense of maximizing satisfaction (or utilities) among individuals at the level of consumers, firms, and markets is called microeconomics. The economic analysis of scarcity at the national level, usually in terms of policies that promote or hinder gross economic productivity, employment, investment, or inflation, is called macroeconomics. There is a greater consensus about the principles operative in and recommendations for behaviors in microeconomics than in macroeconomics.

A strong consensus at the level of microeconomics is exhibited around what are known as the first and second theorems of welfare economics. It is universally agreed that both theorems follow logically from the general equilibrium model—even while there are disagreements about the plausibility of the assumptions necessary for the theory to hold that make problematic any policy recommendations based on it. Again, see "Economics and Ethics."

The first theorem states that the general equilibrium in a competitive economy is Pareto efficient. A special kind of efficiency, Pareto efficiency (as formulated by the Italian engineer economist Vilfredo Pareto) is that situation in which it is not possible to make anyone better off without making someone else worse off. A competitive general equilibrium refers to the outcome in an ideal setting in which consumers are able, in a free and well-informed manner, to exchange goods and services in an open market with multiple independent producers. Of course, this ideal is not always the reality.

The second theorem states that any feasible allocation of welfare to economic actors can be achieved by the appropriate assignment of property rights to agents, followed by competitive production and exchange. What this means is that any desired Pareto-efficient outcome can be achieved simply by an appropriate initial distribution of property rights followed by free-market activity.

It is important to note that because of debates about the assumptions behind the model on which these theorems rely, they do not in themselves fully justify the market economy. The market economy based on private property upheld by the state simply seems to work better in achieving popularly approved welfare goals than alternative systems.

Insofar as economics involves both scientific theory about decision-making and techniques (or technologies) for decision-making, it has further implications for ethics. Indeed, those special economic analyses found in "Game Theory" and its generalization known as "Rational Choice Theory" have on occasion been presented as scientific assessments of some aspects of human behavior that also have normative force.

The less than strong consensus at the level of macroeconomics is reflected in extended debates about how science and technology contribute to national economic productivity, employment, investment, or inflation. These debates are reviewed in the entries on "Innovation," "Invention," "Political Economy," and "Science Policy." They are also related to a host of studies in the history and sociology of science, technology, and economic change that are relevant but not considered at length (see, e.g., Rosenberg 1976, 1982, 1994; Mokyr 1990, 2002; Rosenberg, Landau, and Mowery 1992; Mirowski and Sent 2002).

Still a third main branch of economic analysis concerns development. This field of economics and its special relations to science, technology, and ethics is considered in the entry on "Development Ethics."

Postmodern Economic Issues

Along with these three main branches of economics, there are a number of closely related specialized forms that qualify or extend the modern economic framework. Two of these have been given special entries: "Ecological Economics" and "Environmental Economics."

Environmental economics, which began to be recognized as a special field in the 1970s, seeks to adapt the principles of micro- or welfare economics to satisfying individual environmental desires for clean water and clean air by seeking to identify the best market mechanisms to promote pollution or emission reductions and waste management. To some extent it is often

argued that this requires the social scientific management of markets.

Ecological economics, which emerged in the 1980s, especially contends that market mechanisms are insufficient to evaluate ecological phenomena. As a result, it seeks new ways to conceptualize, for instance, the carrying capacity of the environment and the economic value of natural goods and services.

Both environmental and ecological economics, because they require experts to adjust or correct markets to make them reflect social values, must deal with the problem formulated by social choice theory. Social choice theory concerns the question of whether societies—rather than individuals—can be said to have preferences, and if so, how these preferences relate to the preferences of the individual members of a society. The core result of social choice theory is an impossibility theorem, formulated by the economist Kenneth J. Arrow (1970), that challenges the notion that a society can rank its options in a coherent way. Arrow's theorem states that if everyone in a society has individual preferences that satisfy some basic principles of consistency, and applies these preferences to rank-order a set of options, unless everyone has the same preferences (or agrees to appoint a dictator) there will be no way to add up the individual preferences to achieve a social preference ranking that retains the consistency observed in individual preferences.

CARL MITCHAM
ROSS MCKITRICK

SEE ALSO *Capitalism; Development Ethics; Ecological Economics; Economics and Ethics; Environmental Economics; Game Theory; Innovation; Invention; Market Theory; Political Economy; Rational Choice Theory; Science Policy.*

BIBLIOGRAPHY

Achterhuis, Hans. (1988). *Het rijk van de schaarste: Van Thomas Hobbes tot Michel Foucault* [The reign of scarcity: From Thomas Hobbes to Michel Foucault]. Baarn, Netherlands: Ambo.

Arrow, Kenneth J. (1970). *Social Choice and Individual Values,* 2nd edition. New Haven, CT: Yale University Press.

Dumont, Louis. (1977). *From Mandeville to Marx: The Genesis and Triumph of Economic Ideology.* Chicago: University of Chicago Press.

Mirowski, Philip, and Esther-Mirjam Sent, eds. (2002). *Science Bought and Sold: Essays in the Economics of Science.* Chicago: University of Chicago Press. A collection of nineteen texts: one (by Charles Sanders Peirce) from the 1800s, three from the 1950s and 1960s, ten from the 1990s and 2000s, with five previously unpublished. Good introduction.

Mokyr, Joel. (1990). *The Lever of Riches: Technological Creativity and Economic Progress.* New York: Oxford University Press.

Mokyr, Joel. (2002). *The Gifts of Athena: Historical Origins of the Knowledge Economy.* Princeton, NJ: Princeton University Press.

Nelson, Robert H. (2001). *Economics as Religion: From Samuelson to Chicago and Beyond.* University Park: Pennsylvania State University Press.

Polanyi, Karl. (1944). *The Great Transformation.* New York: Rinehart.

Rhoads, Steven E. (1985). *The Economist's View of the World: Government, Markets, and Public Policy.* Cambridge, UK: Cambridge University Press. A widely regarded and often reprinted introduction by a political scientist.

Robbins, Lionel. (1932). *An Essay on the Nature and Significance of Economic Science.* London: Macmillan.

Rosenberg, Nathan. (1976). *Perspectives on Technology.* Cambridge, UK: Cambridge University Press.

Rosenberg, Nathan. (1982). *Inside the Black Box: Technology and Economics.* Cambridge, UK: Cambridge University Press.

Rosenberg, Nathan. (1994). *Exploring the Black Box: Technology, Economics, and History.* Cambridge, UK: Cambridge University Press.

Rosenberg, Nathan; Ralph Landau; and David C. Mowery, eds. (1992). *Technology and the Wealth of Nations.* Stanford, CA: Stanford University Press.

EDISON, THOMAS ALVA

• • •

Inventor and entrepreneur Thomas A. Edison (1847–1931) was born in Milan, Ohio, on February 11, and became the most prolific inventor in U.S. history, with a record 1,093 patents. Through his technological innovations and companies, "The Wizard of Menlo Park" (in New Jersey, where his laboratory was located) helped found the electric light and power, sound recording, and motion picture industries, and contributed substantially to the telecommunications, battery, and cement industries. He was also close friends with Henry Ford, the pioneer of mass production. Edison established the first industrial laboratories devoted to inventing new technologies and recast invention as part of a larger process of innovation that encompassed manufacturing and marketing. The philosopher Alfred North Whitehead famously credited him with the invention of a method of invention. Edison died in West Orange, New Jersey, on October 18.

Thomas Alva Edison, 1847-1931. The American inventor held hundreds of patents, most for electrical devices and electric light and power. Although the phonograph and incandescent lamp are best known, perhaps his greatest invention was organized research. (© UPI/Corbis-Bettmann.)

The Invention Process and Intellectual Property

After working as a telegraph operator in the mid-1860s, Edison began his inventive career by becoming a contract inventor in the telegraph industry. At a time when general incorporation laws were just beginning to reshape American business, these companies were learning how to deal with technological innovation. Concerns over conflict of interest were also just beginning to emerge, and Edison saw no conflict in working for companies in direct competition with each other.

Perhaps the best-known conflict of interest in Edison's early career arose over his most important telegraph invention—the quadruplex telegraph, which enabled four messages to be sent simultaneously over one wire. Edison worked on this invention under an informal arrangement with Western Union Telegraph. At the same time he was working under more formal contracts with officials of the Automatic Telegraph Company to develop a competing system that used machinery rather than human operators to send messages at high speeds. After successfully demonstrating his quadruplex in the fall of 1874 on Western Union lines, Edison sought payment from the company, but

Western Union did not act promptly on what he believed were relatively modest demands for payment. Facing the loss of his house and shop in Newark due to the general economic depression caused by the Panic of 1873, Edison felt free to sell his rights in the invention to railroad financier Jay Gould, who was in the process of creating a competing telegraph network by combining several small competing firms including Automatic Telegraph. Although Western Union had to sue to assert its rights to the invention, the company nonetheless agreed to retain Edison's services to continue work on multiple telegraph systems, but this time under a formal contract. Later Edison signed another agreement with Western Union that secured all his work related to landline telegraphy, including the new telephone technology.

Edison entered into this latter contract in early 1877 in an effort to secure support for his new Menlo park laboratory, the first devoted to the creation and commercialization of new technologies. Edison's *invention factory* played a key role in the creation not just of specific devices but of methodological research and development leading to market innovation. Indeed, in order to make the incandescent light bulb commercially viable, Edison created a system for the distribution of electricity and designed the manufacturing technology for producing lamps.

As the laboratory and its workforce grew, Edison depended more and more on the assistance of a large staff of experimenters and machinists who made important contributions to his inventive efforts. As a consequence, he was faced with finding ways to give appropriate credit and financial awards for their work. At the time employees entered the laboratory they were made to understand that they were working on Edison's ideas, and that their work on his inventions would be credited to him.

Nonetheless the issue of credit remained a tricky one. While Edison and his assistants perceived their role as working on his ideas, he gave general directions and relied on their abilities to work out important details. Edison thus generally made it a policy to take out the key patents, while permitting assistants to take out ancillary patents he considered to be primarily their contribution. At the time, U.S. patent law gave priority to an employer in disputes with employees and discouraged joint inventions unless a true partnership in the invention could be demonstrated. In lieu of joint patents or other credit for their inventive assistance, Edison gave his chief experimenters an interest in royalties and other profits. He also placed many of them in management positions in his companies, and some became partners. Edison continued

these policies at the larger laboratory he opened in West Orange, New Jersey, in 1887.

The issue of credit was also a significant one for Edison's competitors, particularly because of the popular image of him as the primary inventor of several new technologies. Edison's reputation was partially a consequence of the fact that he had a much more sophisticated understanding of invention than his contemporaries. Edison saw invention as just the first stage of a larger process of innovation. Thus he took a leading role in marketing the inventions he developed through companies he established and that bore the Edison name. Because his name was associated with the technology he continued to make improvements to insure its reputation as well as his own. This kept him in the public eye as reporters wrote stories about his latest improvements.

Edison's public image was also a result of his skill at public relations. He had developed an understanding of the newspaper business while working as a press-wire telegraph operator and, after becoming famous for inventing the phonograph, he had established close relationships with several reporters in New York City who found Edison a ready source of news, opinion, and human interest. Thus even when other inventors made important technical contributions, the public credited Edison first.

While Edison's willingness to make announcements through the press aided his marketing efforts, it created problems for his scientific reputation. When Edison claimed that he had observed a new natural phenomenon and termed it *etheric force* in 1875, he made his first announcements through the newspapers and continued to press his claims through press interviews rather than through the scientific journals as did his opponents. Similarly after British inventor David Hughes's claim to the invention of the microphone appeared in the scientific journal *Nature*, Edison launched a public attack through the New York City newspapers rather than responding in the scientific press. In both cases, Edison's claims in the scientific community were weakened by his failure to adhere to the norms of scientific publication and debate.

While Edison saw himself as a member of the larger scientific community and presented papers before the American Association for the Advancement of Science (AAAS) and the National Academy of Sciences (NAS) in the 1870s and 1880s, he was foremost an inventor and more interested in attracting public interest in his work than advancing scientific knowledge. Nonetheless when his inventive work produced devices that were primarily useful for scientific research he was willing to forego royalties in their manufacture and make them available to the scientific community. This occurred with a heat measurer he called the tasimeter in 1878, when he gave some early light bulbs to scientific researchers in 1880, and with his work on X-ray technology in 1886.

Public Policy Issues

Because the public saw Edison as a leading figure of science and technology, his comments on important public issues could carry significant weight. In two instances his reputation proved crucial to the enactment of public policy.

In the first and more controversial instance, Edison was asked in 1888 for his expert opinion on the establishment of electrocution as a more humane form of execution than hanging. Although opposed to the death penalty, Edison agreed to support this position and also allowed Harold P. Brown, a self-taught electrician, to conduct experiments on animal electrocutions at his laboratory. These experiments in support of electrocution were undertaken in part due to Edison's firm belief in the dangers of high-voltage electricity, and thus his ethical opposition to its public use.

But it also stemmed from the increasing competition his low-voltage direct-current (DC) electrical system was receiving from the high-voltage alternating current (AC) system being marketed by George Westinghouse. The debate on electrocution thus became wrapped up in this commercial struggle. Edison's strong opposition to high-voltage and the demonstrations at his laboratory that showed high-voltage AC to be more dangerous than high-voltage DC led him to champion the electric chair and testify on behalf of the state in the appeals of the first death penalty case involving electrocution. Edison would later regret his role in the development of the electric chair but never gave up his opposition to high-voltage electricity.

Edison's other significant involvement in public policy came as the result of a 1915 *New York Times* interview in which he urged greater military preparedness and the need for a national research laboratory to develop new military technologies for defense. This led Secretary of the Navy Josephus Daniels to ask Edison to establish and head a new Naval Consulting Board. The Board was made up of leading inventors, engineers, and industrial research scientists. Edison would eventually lose the larger debate within the Board over the nature of the research laboratory. Based on the newer style of industrial research laboratories, the new Naval Laboratory, which was not

established until after World War I and was headed by naval officers rather than civilians, focused on science-based research leading to the development of small-scale prototypes. It was not a works laboratory like Edison's, equipped with extensive machine shop facilities for turning prototypes into commercial technology.

The differences over the Naval Laboratory were also reflected in Edison's own contribution to research during World War I. Although Edison developed forty-two inventions that he believed could contribute to the war effort, the Navy adopted none. Instead the Navy officers responsible for introducing new technology turned to the efforts of those researchers whose approach included the mathematical rigor and theoretical basis that their university educations had taught them were the foundations of modern research.

The growing differences between Edison and more youthful researchers marked a shift in the nature of scientific and technical training. This shift became more evident by the end of Edison's life, when news accounts treated him as the last of the lone cut-and-try inventors rather than the creator of the first industrial research laboratory. A closer study of his life, however, reveals that in the course of reshaping the ways in which invention took place, including at his laboratories, Edison was faced with many of the same ethical issues encountered by twenty-first-century inventors and industrial researchers.

PAUL ISRAEL

SEE ALSO *Entrepreneurism; Invention; Technological Innovation.*

BIBLIOGRAPHY

Bazerman, Charles. (1999). *The Languages of Edison's Light.* Cambridge, MA: MIT.

Carlson, W. Bernard, and A. J. Millard. (1987). "Defining Risk within a Business Context: Thomas A. Edison, Elihu Thomson, and the AC-DC Controversy, 1885–1900." In *The Social and Cultural Construction of Risk,* ed. Branden B. Johnson and Vincent T. Covello. Boston: Reidel Publishing Co.

Carlson, W. Bernard, and Michael E. Gorman. (1990). "Understanding Invention as a Cognitive Process: The Case of Thomas Edison and Early Motion Pictures, 1888–1891." *Social Studies of Science* 20: 387–430.

Essig, Mark. (2003). *Edison and the Electric Chair: A Story of Light and Death.* New York: Walker and Company.

Hughes, Thomas P. (1983). *Networks of Power: Electrification in Western Society, 1880–1930.* Baltimore, MD: Johns Hopkins University Press.

Hughes, Thomas P. (1958). "Harold P. Brown and the Executioner's Current: An Incident in the AC-DC Controversy." *Business History Review* 32: 143–165.

Israel, Paul. (1998). *Edison: A Life of Invention.* New York: John Wiley and Sons.

Jonnes, Jill. (2003). *Empires of Light: Edison, Tesla, Westinghouse, and the Race to Electrify the World.* New York : Random House.

Millard, Andre. (1990). *Edison and the Business of Innovation.* Baltimore, MD: Johns Hopkins University Press.

Musser, Charles. (1991). *Before the Nickelodeon: Edwin S. Porter and the Edison Manufacturing Company.* Berkeley: University of California Press.

Pretzer, William S., ed. (1989). *Working at Inventing: Thomas A. Edison and the Menlo Park Experience.* Dearborn, MI: Henry Ford Museum and Greenfield Village.

EDUCATION

• • •

Any regular practice, for example, agriculture, craft production, navigation, or scholarship, requires learning opportunities for novice practitioners, which have often been provided in workplaces, or through informal instruction and self-directed study. This survey, however, will be limited to *formal* education, that is, to teaching and learning in institutions such as colleges and universities established exclusively for these purposes.

A broad historical account (to be elaborated below) of scientific and technical education in relation to ethics runs as follows.

Science and ethics initially were intimately related in ancient education, while technology was explicitly excluded. Medieval Christians were ambivalent about ancient pagan science, because they held an opposing notion of moral perfection. Greek science nonetheless retained a minor place in medieval education, though its intimate association with ethics was weakened insofar as morality was religiously based. When classical learning was recovered in Western Europe by the mid-thirteenth century, and the scholastics sought to render it consistent with church teachings, natural philosophy (science) and moral philosophy were added to the curriculum as standard, but distinct, university subjects.

Renaissance scholarship facilitated the scientific revolution of the sixteenth and seventeenth centuries. By the start of the eighteenth century, however, modern science had become divorced from teaching in the English universities, and had forged new institutional links with technology and commerce. Ethics, however, was revitalized as a university subject by the contributions of Renaissance humanists. Natural science was reestablished as a teaching field in the early-nineteenth century. Technology remained excluded from North American colleges, and moral philosophers attempted to harmonize the new science with the prevailing Protestant worldview and morality. New German universities, however, rejected all association with religious creeds and devoted themselves to the free study and teaching of science.

Engineering, agriculture, and other technical and professional fields became university degree subjects late in the nineteenth century in both England and the United States. Politics, economics, and sociology became separated from moral philosophy as positive sciences, and ethics attained autonomy from moral theology, becoming merely one academic discipline among others in the secular multiversity. The resulting civilization of science, technology, business enterprise, and the nation-state gained unprecedented control of nature and social life in the twentieth century.

The technoscientific civilization has nevertheless experienced a profound ethical crisis as a result of the atomic bomb, environmental pollution, resource depletion, nuclear accidents, and other *techno-shocks*. Many scholars and social leaders came to fear that technoscience had outstripped social capacities for its ethical control, and that it even threatened human survival. Some science, technology, and ethics professors, therefore, collaborated in forging closer relationships between their fields, in an attempt to bring ethical judgment to bear on further developments in science and technology.

From Greek *Paedeia* to Medieval Scholasticism

Systems of education in science and technology that have taken on worldwide influence have their root influences in classical school experiences. It is thus appropriate that the present survey should highlight this historical background.

SCIENCE AND ETHICS CONJOINED. Formal elementary education was established for free males in several Greek city-states by the fifth century B.C.E., taking place in the *didaskaleion*, the area set aside for teaching. After learning to read and write, older children learned classic literature and music. Youths attended public gymnasiums for physical and character training and military preparation. Beyond this level, education in the fifth century was not standardized. The gymnasiums, however, became sites for discussion groups and lectures given by sophists.

Science and ethics were esoteric subjects taught informally by masters to a few chosen disciples.

Science and ethics, however, were not distinct subjects. Pythagoras of Samos (ca. 580–500 B.C.E.), and Democritus of Abdera (460–370 B.C.E.), are important examples. Pythagoras of Samos studied arithmetic, astronomy, geometry and music at Miletus, and possibly also at Babylon and Egypt. Pythagorean ethics held that virtue was a harmony of the soul that mirrored the harmony of the spheres, and that mathematics is the pathway to moral perfection. In contrast Democritus conceived the natural world, including the soul, as a machine behaving in accordance with laws of matter, so that freedom of action is an illusion to be overcome by reflection on the determinism in nature. A state of tranquil acceptance of mechanistic reality is thus the ethical good.

By the fourth century B.C.E., more formal philosophical schools evolved from the informal learning at the gymnasiums, and Athens became the recognized center of learning. The school of Isocrates, based on the rhetorical teachings of the sophists, opened in 390 B.C.E.. In 387 B.C.E., Plato (428–347 B.C.E.) established a school with a program of study similar to that of Pythagoras. It came to be called the Academy because of its location near the Groves of Academus. Aristotle (384–322 B.C.E.) gave lectures after 335 B.C.E. at the gymnasium dedicated to Apollo Lyceios; his school became known as the Lyceum. Aristotle distinguished between theoretical and practical studies; science and ethics were taught as distinct subjects. Aristotle nonetheless agreed with Pythagoras and Plato that the highest good is contemplative knowledge, and thus he taught that the highest ethical life is not practical action in the polis, but theoretical contemplation.

Both Plato and Aristotle distinguished technical arts from those suitable for liberal education. In the Philebus (55e–56a) Plato argued that when the mathematics was abstracted from technical arts such as navigation and architecture, what remained was intellectually trivial. The educational program Plato laid out for the Guardians in Book 7 of the *Republic* (380 B.C.E.) was based on the four Pythagorean mathematical arts: arithmetic, geometry, astronomy, and harmonics. All of these he conceived entirely in abstract terms, with sensory observations and utilitarian applications removed. (*Real* astronomy, for example, had no concern for the sun, moon, or stars, but only with solids in revolution.) Aristotle considered technical arts degrading and slavish. (*Politics* Bk 3, 1277a5-a12, 1277b34-1278a14). As handcrafts workers engage in repetitive acts, they "are like certain lifeless things that act … without knowing what they do, as fire burns"(*Metaphysics* Bk 1, 981a13–b9).

The Athenian schools continued under new leaders (or *scholarchs*) after the death of the masters. While Aristotle's school devoted itself almost exclusively to natural science, the other schools continued to offer a program in which science and ethics were intertwined, but ethics soon became predominant. In 306 Epicurus (341–270 B.C.E.) established a school that followed the teachings of Democritus. Zeno of Citium in Cyprus (c. 335–263 B.C.E.), teaching at the painted column or stoa, taught a *stoic* ethic of rational preferences ordered according to nature, mastery of passions, and indifference to fate.

The theory of the liberal arts attained a definite form by the first century B.C.E. By that time the circle of learning, the *enkyklos paedeia,* had come to include logic, rhetoric, grammar (literature), and the four Pythagorean mathematical sciences. In the Roman Latin schools, however, the literary arts dominated. Mathematical subjects were recognized, but taught cursorily, if at all.

ROMAN AND MEDIEVAL SCHOOLING. The educated classes during both the Roman and medieval periods admired but exhibited a certain ambivalence regarding classical learning. Neoplatonic philosophers of the Roman period, for instance, preserved the Pythagorean and Platonic program—mathematical study for ethical perfection—but were more oriented toward education that would serve overt political ends, as with oratory. More than their pagan peers, perhaps, early Roman Christians admired Neoplatonism because of its unworldly and ascetic emphases. It made a deep impression on Augustine of Hippo (354–430 C.E.), and inspired the grand educational project of Boethius (480–525 C.E.).

Boethius, noting that prevailing Latin textbooks in grammar and rhetoric were adequate, sought to revitalize the *enkyklopaideia* by preparing Latin handbooks on logic and the four mathematical disciplines, for which he invented the name *the quadrivium*. Following the Neoplatonists, Boethius conceived these studies as pathways from the sensible world to supersensible reality as a means of ethical perfection. Boethius's manuals (on arithmetic, logic, and music) became standard school and university textbooks for almost 1,000 years.

The death of Boethius in 525 C.E. and the closing of the Platonic Academy of Athens, by the Eastern Emperor Justinian in 529 C.E., mark the end of Greek learning in the West. Barbarians gradually also destroyed the Latin schools, and eventually even Latin classics were unavailable in Western Europe. Greek classics were preserved at Byzantium and then entered the stream of Islamic learning. Latin classics were preserved in Ireland.

LATIN SCHOOLS REVIVED AND UNIVERSITIES BORN. Charlemagne, crowned Holy Roman Emperor in 800, sought to revive learning in order to provide educated clergy and administrators for his realm, and ordered his cathedrals to establish schools. An organized program of teaching, however, requires textbooks, and in logic and the sciences only those of Boethius, preserved in Ireland, were available. The Church retained a deep ambivalence about pagan learning, which contained a view of moral perfection at odds with its own as best expressed in Tertullian's famous question, "What has Athens to do with Jerusalem?"

Nonetheless the cathedral schools were, in theory, organized along classical lines: a grammar school for logic, rhetoric and grammar (for the first time called the *trivium*), followed by a higher school for the *quadrivium*. In practice, while the *trivium* provided useful training for clergy and administrators, the *quadrivium* was often neglected. Most schools could manage only practical arithmetic for calculation, geometry for architecture and surveying, and astronomy to calculate Easter. Science education improved in some cathedral schools in the eleventh century. At Reims Gerbert of Auillac (955–1003), acquainted with Arabic scholarship in Spain, refreshed the *quadrivium* by using Arabic numerals, the abacus for calculation, and the astrolobe for astronomical observation.

In 1079 Pope Gregory VII issued a papal decree ordering all cathedrals and monasteries to open schools for the training of clergy. As schooling expanded it became necessary to regulate teacher preparation and licensure. The church claimed a monopoly over teaching licenses (*licencia docendi*). Municipal chancellors offered these licenses only to those intending to teach in their districts.

Some municipalities, however, attracted students from many regions, and gained recognition as *studia generale*, whose degrees (licenses) were recognized throughout Europe. These *universities* were divided into lower schools for the seven liberal arts plus schools for law, theology, and medicine. Two models for the university emerged: one at Paris, Oxford, and Cambridge, where the arts course predominated; the other at Bologna and Salerno, where, contrary to the dictates of Plato and Aristotle, the arts course became merely a minor preliminary to technical education in the professions.

Science education in the arts course remained grossly inadequate. By papal decree, lectures on the *quadrivium* could be offered only on public holidays. By the last third of the twelfth century, however, the importation of classical texts from Muslim Spain reached its peak. Adelade of Bath had translated Euclid's *Elements*, and the Aristotelian corpus was made available in Latin translation.

SCHOLASTICISM: SCIENCE, ETHICS, AND RELIGION. In the thirteenth century the challenging task of assimilating the classical inheritance began. In geometry, for example, the study of Euclid prompted new discoveries in optics by Robert Grosseteste (c. 1170–1253) and his student Roger Bacon (c. 1220–1292). Grosseteste, chancellor of Oxford (1215–1221), made optical studies, wrote a commentary on Aristotle's *Posterior Analytics*, and championed empirical inquiry. Bacon, who said he had learned more from simple craftsmen than from famous professors, carried on Grosseteste's empirical studies of lenses as aides to natural vision.

Assimilation of Aristotle's writings in natural and moral philosophy was among the greatest challenges faced by the thirteenth century universities. Pope Innocent III banned the study of Aristotelian natural philosophy in 1210. A committee was formed in 1231 to expunge all heretical ideas from his texts so they might be suitable for teaching, and by 1255, Aristotle's works returned to the syllabus. Scholasticism, the project of rendering the classical inheritance compatible with church teachings, came to dominate university studies. Thomas Aquinas (1224–1274), the greatest of the scholastics, saw that with the recovery of ancient learning, the seven liberal arts had become inadequate as a pattern of study, and the arts course was expanded to include the *three philosophies*: metaphysics, natural philosophy (empirical science), and moral philosophy.

The scholastic method of education stressed formal definition and logical argument. The scarcity of books dictated its primary tools: lectures (where books were publicly read and interpreted), recitations (where students demonstrated their familiarity with the books), and public disputations (where students presented public arguments in syllogistic form). Scholastic natural philosophy thus remained confined to theory, logical argument, and thought experiment. Controlled observations and technical applications were rare. The old textbooks dominated the syllabus for centuries. Scholasticism, while useful as a method of organizing official knowledge and conveying it in a standard form as preparation for professional studies, failed to encourage systematic and creative scientific studies, and eventually became bogged down in fruitless verbal controversies.

Early Modernity: Science, Technology, Humanism, and the Reformation

By the fifteenth century medieval institutions no longer provided Europe with either social order or a rational world picture. Renaissance humanists, working outside the universities and in opposition to scholasticism, sought inspiration in the pagan classics for reshaping learning and civic life. They praised Aristotle's ethics, and placed moral philosophy at the center of their curriculum. Claiming that moral virtue grew from emulation of classical authors and orators, they tied ethics closely with rhetoric, history, literature, and classical languages in a complex that became the humanities.

The humanists, however, rejected Aristotelian logic as artificially formal. They promoted a practical, natural logic based on study of the arguments of the great orators, thereby incorporating logic within rhetoric. They also rejected Aristotle's qualitative natural philosophy in favor of Plato's quantitative approach, thus easing the path for Nicolaus Copernicus, Johannes Kepler, and Galileo Galilei. The latter's aphorism that *the book of nature is written in mathematical characters* might have been taken directly from Plato. Humanists made few direct contributions to scientific scholarship, as the recovery of pagan scientific classics had been completed, but their intellectual independence and daring established a new spirit of learning congenial to later modern scientific inquiry.

Martin Luther and John Calvin, the leaders of the Protestant Reformation, were themselves humanist scholars. Their encouragement of the close reading of scriptural texts stimulated close reading of the book of nature. Protestantism directly undermined scholasticism, as it eliminated the need to square classical authorities, including Aristotle, with Catholic Church teachings. The new Protestant universities of Northern Europe could start afresh, and thereby became leaders in incorporating modern science into their curricula.

THE SCIENTIFIC REVOLUTION AND ITS SOCIAL INSTITUTIONS. The fifteenth and sixteenth centuries were periods of rapid developments in commerce, navigation and ship construction, instrument making, mining, and mechanics. These new conditions, when conjoined with the mathematical knowledge brought to the Christian West from Byzantium and Muslim Spain, illuminated new pathways for the growth of scientific knowledge.

Until the seventeenth century Europe possessed no scientific societies or journals to stimulate or publish reports of new investigations. To develop an infrastructure for science required a vision, a site for meetings of scientists and technical experts, a critical mass of expert scientific workers, and an organization to stimulate and assess significant scientific achievements and make them widely known through its publications. The coordination of these factors in England led to the establishment of the Royal Society in 1660.

Francis Bacon (1561–1626) framed the vision. He maintained, against both classical authorities and the scholastics, that the only useful knowledge was based on empirical study of nature, and that a clear method for scientific work would provide human mastery over the natural world. Under such conditions, *knowledge is power*. His inductive method, though a technical failure, shaped an agenda for practically useful science that included close study of mechanical crafts.

Thomas Gresham (1519–1579), a wealthy London merchant, provided the site, by endowing a college for merchants and craftsmen that opened in 1598. Gresham College offered no degrees, but provided free public lecture courses in rhetoric, astronomy, geometry, music, divinity, medicine (physic), and law. The Gresham professors were selected from among the most eminent scholars of their time. The first Gresham Professor of Geometry, Henry Briggs, developed logarithmic tables and popularized their use. The college's central location in London provided the ideal meeting place for scientists and technicians. Briggs also made it the central clearing house for scientific and technical information.

Oxford provided the critical mass of scientific experts. When Briggs was appointed the first Savilian Professor of Geometry at Oxford in 1619, he strengthened ties between Gresham College and the university. In the 1640s a group of distinguished natural philosophers including John Wilkins (of Wadham College), Seth Ward (later the Savilian Professor of Astronomy), Robert Boyle, William Petty (later professor of anatomy), and Jonathan Goddard (of Merton College) frequently attended Christopher Wren's lectures at Gresham College and then met with coworkers, including navigators and instrument makers. Boyle called this group the *invisible college*.

In 1651 the Oxford Philosophical Society was formed and began publishing transactions. A similar organization failed at Cambridge because no scholars were willing to perform experiments. In 1660 members of the invisible college formed a national society, which was incorporated by royal charter as the Royal Society in 1662. It soon established its offices and meetings at Gresham College, and published its own transactions. Other nations established parallel societies. In France, Jean-Baptiste Colbert founded the Academie des Sciences in 1666, which was reorganized with royal approval in 1699. In Germany Frederick the Great founded the Academie der Wissenschafften in 1700, with Gottfried Wilhelm Leibniz as the first president. By the beginning of the eighteenth century the infrastructure for European science was in place.

SCIENCE EDUCATION. While English university scholars were central in the scientific revolution, science teaching retained its medieval character. The colleges at Oxford and Cambridge were at first mere residence halls, whose tutors were simply older men taking responsibility for the conduct and finances of younger students. By tradition recent graduates (regents masters) of the colleges were required to lecture. By the sixteenth century, however, the regents lecture system had broken down, and the universities recognized the need for a new organization of teaching, including appointment of permanent lecturers.

Lady Margaret, mother of Henry VII, endowed professorships of theology at both Oxford and Cambridge (1497–1502). Sir Robert Rede provided in his will for lectureships at Cambridge in philosophy, logic, and rhetoric. Henry VIII added royal patronage to this trend after conducting visits to the universities in 1535, following his break from the Roman church. Henry's reforms, reflecting the humanist spirit of the sixteenth century; replaced scholastic textbooks with humanist commentaries on Aristotle's natural and moral philosophy. Henry also endowed Regius professorships in classical Greek and Hebrew, as well as divinity, medicine, and civil law. A series of similar endowments and appointments include, for example, the Henry Lucas Professorship of Mathematics at Cambridge, which Isaac Newton held from 1669 to 1701.

These distinguished professorships had almost no impact on teaching, however, because the colleges, which were wealthier and more powerful than the universities, completely dominated teaching. Students were required to live in colleges, where tutors were assigned to lecture and conduct recitations on authorized texts. The tutors were generalists offering instruction on the *ordinary* subjects required for disputations and exams. Professors lectured only on *extraordinary* subjects outside the mandated curriculum. Since colleges prevented universities from examining students on extraordinary subjects for several centuries, few students attended the professorial lectures, and eventually few professors even bothered to deliver them. Not one of the three Regius professors of physic (medicine) at Cambridge from 1700 to 1817 gave a single lecture.

The situation was different in Germany. The first modern university opened at Halle in 1694. Gottingen rivaled Halle as a center of learning after its opening in 1736. The University of Berlin was established in 1800, under the direction of William von Humboldt. Berlin adopted the Platonic ideal, training leaders as philosophers. Professors combined original research with teaching, and students worked closely with professors on research projects. Students thus acquired the cultural and scientific heritage in the very process of working alongside those who knew it best. Berlin rejected attachments to religious creeds and schools of thought, accepting subservience only to science and learning. It thus added to the arts and professional universities of the middle ages a third model, the research university, which soon dominated Protestant Europe.

Science and Ethics in Eighteenth-and Nineteenth-Century American Colleges

The first colleges in the New World, Harvard (1636) and William and Mary (1693) based their statutes on those of the colleges of Cambridge and Oxford. As in England, while mathematics and science were given lip service, the seventeenth-century teachers lacked knowledge of current developments. In 1700 in North America the modern scientific subjects were still associated with navigation and mechanical arts, not with college education. The lack of constraint by an entrenched teaching elite, however, eased the way for their introduction into colleges.

ACADEMIC SCIENCE AND ETHICS IN THE EIGHTEENTH CENTURY. In the mid-eighteenth century Yale acquired some scientific apparatus, and introduced Newton's fluxions to the math curriculum. John Winthrop, the Hollis Professor of Mathematics and Natural Philosophy at Harvard (1738–1779), removed the last traces of Aristotle from the course in natural philosophy and introduced the new science of Galileo and Newton. When the American Philosophical Society for Promotion of Useful Knowledge was established in Philadelphia in 1769, with Benjamin Franklin serving as president, however, the founding members were amateur investigators rather than teachers. The society had

650 members by 1800, but only fifteen of 124 noted college teachers of the period ever became members.

Moral philosophy underwent a more profound revolution. John Locke's "Essay Concerning Human Understanding" (1690), with its consideration of the foundations of moral knowledge, was in the curriculum by 1720. By mid-century moral philosophy was central to both the college curriculum and public discourse, attracting the attention of such Enlightenment leaders as Franklin, Thomas Jefferson, and Benjamin Rush. The American Revolution further invigorated the Enlightenment spirit. William Paley's widely adopted textbook *Moral and Political Philosophy* (1785) presented Christian utilitarianism as a natural science based on empirical observations and first principles, which included the natural rights of man as expressed by Locke. Paley, in his *Natural Theology* (1802), based the existence of God, as the divine intelligence governing the universe, on the argument from design. A course on natural theology was added to the curriculum.

ACADEMIC SCIENCE AND ETHICS IN THE FIRST HALF OF THE NINETEENTH CENTURY. By 1820, as rapid advances took place in U.S. commerce and industry, leaders demanded that mathematics and science in the colleges be improved and made more practical. Mark Hopkins of Amherst called Francis Bacon's notion of knowledge as power the single most influential idea in the popular mind in the early-nineteenth-century. Science teaching, as a result, got a large boost between 1820 and 1850, though reformers had to contend with inadequate textbooks, untrained teachers, and lack of apparatus.

By 1836 adequate textbooks were available in algebra, geometry, trigonometry, analytical geometry, and calculus, and by 1840 calculus was a standard part of the liberal arts course. By 1850 natural philosophy had been reorganized as physics, chemistry, and natural history (biology and geology, formally subjects for amateur naturalists). Physics had been further divided into mechanics, hydrostatics, electricity and magnetism, and chromatics, and textbooks treating these topics in sufficient mathematical detail were widely available. By 1860 five clearly defined courses in natural science—astronomy, physics, chemistry, biology, and geology (including mineralogy)—were part of the liberal arts curriculum.

Prospective teachers began to study in the Protestant universities of Northern Europe. In 1802 President Timothy Dwight of Yale urged Benjamin Silliman (1779–1864), a distinguished graduate, to prepare himself for a professorship in chemistry by studying at the University of Edinburgh. He emerged in 1805 with current knowledge in theoretical and experimental chemistry, as well as practical knowledge of geology, mineralogy, and zoology. George Tinknor (1791–1871) was among the first Americans to prepare for a professorship at a German university; on his return from the University of Gottingen in 1817, he became a professor at Harvard and introduced German methods of study and research. Foreign preparation of college teachers remained the norm until U.S. research universities were established late in the nineteenth century.

The value of laboratory apparatus expanded twenty fold between 1820 and 1850. The introduction of the blackboard transformed mathematics teaching; professors were for the first time able to exhibit spontaneous thoughts and invite groups of students to the board, where the former could watch assigned problems worked out and probe methods of reasoning. Laboratories introduced similar changes in science teaching.

By 1850 the revolution in the liberal arts college was complete. The question of technology—mechanical, agricultural, and mercantile arts—now had to be faced due to popular demand for practical training. In 1828 a famous *Yale Report* maintained that mathematics and science belonged in the college, but not technology. Amherst issued an almost identical report. Despite public pressures, these sentiments prevailed among college leaders until the Civil War, although their marketing efforts emphasized the practical value of college education as early as 1850.

The college curriculum, however, had by then become seriously overcrowded due to the expansion of science and mathematics. Indeed demands for new scientific courses continued to proliferate. National surveys and requirements of the mineral industries created demands for separate mineralogy courses; developments in medical education led to demands for courses in organic chemistry and physiology. The colleges attempted to resolve these conflicts by introducing practical *partial* scientific courses outside the required curriculum, and diploma programs in *parallel* scientific schools such as the Lawrence School of Science at Harvard (1847) and the Sheffield School at Yale (1854). These efforts, though marketed as practical alternatives to the college course, failed because the programs lacked the prestige of a college education and provided no specific qualification for any position in the U.S. economy of that era.

Moral philosophers writing under the influence of New School Calvinism, a doctrine emphasizing that theology was completely compatible with human standards of reason and morality, did not resist the expansion of science. Instead they discarded the enlightenment empiricism of Locke and Paley, which they saw as undermining religion by making human reason self-sufficient. In its place they adopted the Scottish common sense philosophy of Thomas Reid and Dugald Stewart, which held that sense experience must always be actively assimilated by active powers instilled by God in the human mind, and that moral duties were presented immediately as intuitions.

The typical class in moral philosophy was taught as a capstone course by the college president, who deployed common sense principles as bases for conclusions on the ethical issues of the day. The *Elements of Moral Science* (1835) by Francis Wayland, president of Brown, was based on common sense realism, and became the most widely adopted textbook of its era. While the presidents did not even wish to dictate the results of science, they could strive to contain them within the consensus Protestant worldview and framework for moral action. This period in moral philosophy has thus been perceptively labeled as the era of *Protestant scholasticism*.

The Birth of the Multiversity

In the 1860s U.S. colleges faced three main criticisms: The curriculum was overcrowded with science and mathematics, which most students found daunting and unrelated to career plans, and threatened to marginalize the traditional humanities; the college was seen as elitist in neglecting technical and professional subjects; and due to the absence of research, colleges were failing to advance knowledge in the sciences and useful arts. The Massachusetts Institute of Technology's (MIT) founder and first president, William Barton Rogers, was among those who argued that practical studies and research would require a new kind of institution beyond the traditional network of liberal arts colleges.

Union College introduced a bachelor of science degree, parallel to its bachelor of arts degree, by 1828. Wayland, at Brown, proposed an elective system to alleviate the crowding of the prescribed curriculum as early as 1850. Charles Elliot made free election of courses central to his reform of Harvard after assuming the presidency in 1869, and other colleges soon followed Harvard's lead.

Technical and professional studies gained a foothold when Rensselaer Polytechnic Institute (RPI) opened in 1824 to provide engineering education. RPI was reorganized in 1849 and established the template for the engineering curriculum: humanities, physical science, mathematics, and hands-on shop training. The Morrill Land Grant Act of 1861 provided federal funding for colleges in the agricultural and mechanical arts. MIT, founded in 1861, was reorganized in 1865 under the provisions of the act, while Cornell, chartered in 1862, used funds from both the act and private sources for programs primarily in the agricultural and mechanical arts. Cornell soon became a model institution for maintaining harmony among its humanities, sciences, and technical-professional curricula.

The reorganization of industry into a national mass-production system created a vast demand for engineers, and universities reorganized engineering education to meet the need. In the early 1870s Victor Della Vos of the Moscow Imperial Technical School developed a new approach to practical training based on a careful sequencing of skills in shop-like classrooms. His method became widely adopted for practical elements of engineering education after John Runkle of MIT saw it exhibited at the Philadelphia Centennial Exposition in 1876. Engineering science, however, became increasingly dominant in the engineering curriculum. Universities soon added similar science-based degree programs in architecture, forestry, and veterinary medicine.

Because of the lack of opportunities for research training in the United States, many scholars before the 1870s traveled to Scotland or Germany for advanced degrees. Yale awarded the first Ph.D. in the United States in 1861, but graduate education was only institutionalized with the establishment of new research universities: Johns Hopkins (1876), Clark (1889), Stanford (1891), and the University of Chicago (1892).

At Oxford and Cambridge, from 1820 to 1850, the colleges felt the same pressures as their U.S. counterparts, and made modest accommodations. At Cambridge earning *distinction* in *both* classics and mathematics was dropped as a requirement for an honors degree in 1850. The moral science tripos and natural science tripos were introduced in 1851, providing students with two distinct areas of concentration. In the 1850s, and again in the 1880s, however, royal commissions were appointed to suggest education reforms to parliament. By 1890 the established colleges had lost their control over teaching. Instead the colleges provided lecturers for courses open to all university students, and cooperated in funding university-wide professorships. Early in the twentieth century the teaching staffs were reorganized, on an all-university basis, into branches corresponding to the main divisions of study. Civic university colleges opening in Leeds Manchester and Liverpool in the 1880s, like their Morrill Land

Grant Act counterparts, introduced technical and professional studies.

ETHICS IN THE LATE-NINETEENTH-CENTURY COLLEGE. Prior to the 1850s zoology, botany, and geology were offered primarily as bases for natural theology, by providing evidence of intelligent design in the universe. Charles Darwin (1809–1882) found natural theology, based on Paley's intricate demonstrations of the perfect adjustment of organisms and their environments, the most valuable course he attended at Cambridge. But like many others, he found it implausible that predesigned organisms would be dropped, ready-made, into preexisting niches. His theory of natural selection, as a mechanical explanation of organism-environment compatibility, helped to undermine the harmony between biological science and belief in divine intelligence.

Moral philosophy, or *moral science* as it was frequently called, was by 1900 becoming subdivided, as natural philosophy had been a half century earlier, into its component sciences: ethics, politics, sociology, and economics. Some college presidents continued to teach ethics as a senior course, though theological foundations were now downplayed in favor of exhortations to moral leadership in society. Gradually, however, ethics became merely one of the many specialist subjects students could elect to study.

By the turn of the twentieth century scientific rationality was the one core value of higher education. The Protestant worldview, with its religious constitution for moral action, had lost its intellectual and social authority. Free from all authoritative constraints, university-trained professionals began the march of progress through an allegedly 'value-free' technoscience that by the end of the century had transformed the way humans lived.

TECHNOSCIENTIFIC CIVILIZATION IN CRISIS As early as 1923, however, Henry Churchill King of Oberlin, one of the few clergyman-presidents still teaching moral philosophy, noted a "strange contradiction" between the arrogance of the new scientists with their acclaimed alliance with the forces of nature, and the pervasive sense that modern civilization is unleashing forces that are irresistible and inevitable. By the mid-twentieth century this contradiction was sharpened, as seductive new technologies pushed many areas of human activity beyond effective control. Medical technology gave us wonder drugs that cured age-old diseases, but drug resistant strains appeared and some diseases returned. Nuclear scientists created power plants that used *atoms for peace*, but nuclear power programs

contributed to the proliferation of nuclear weapons. Humans conquered space, but created a race for space-based weapons. Computers dramatically increased human productivity, and also led to increased surveillance of all human activities.

Dramatic events—the atom bomb, DDT and asbestos, thalidomide babies, environmental pollution, napalm in Vietnam, the 1970s oil shocks, urban smog, electricity blackouts, Bhopal, the *Challenger* disaster, Three Mile Island and Chernobyl, Exxon-Valdez, global warming, the ozone hole, cloning, job outsourcing through network technologies, and child addictions to violent computer games—keep forcing this contradiction upon public consciousness. Science and technology can contribute to human life, but also create problems that challenge ethical guidelines and problem-solving capabilities.

The idea of continuous progress through science and technology is no more plausible in the early-twenty-first century than the idea of a fixed universe designed by divine intelligence was after Darwin. Scholars and public leaders are thus called upon to shape a new postmodern ethical vision for technological civilization.

Since the 1970s some science, technology, and ethics scholars have initiated collaborative efforts to forge a closer relationship among their fields. Applied ethical studies of agriculture, engineering, biomedical science, and computer technology have provided a knowledge base for mandated courses in professional ethics. Science and technology courses explicitly tied to social issues have been introduced in general education at universities and secondary schools to enable graduates to participate as democratic citizens in the ethical modulation of science and technology.

A significant problem is that contemporary moral philosophers, unlike those of Catholic or Protestant scholasticism, have neither a specific moral authority of their own, nor the backing of institutions with broad-based social authority. Their authority rests upon their positions in the multiversity, whose core value of unconstrained technoscientific rationality is precisely what is now in question.

At the start of the twenty-first century, therefore, significant questions remain about both the capacity of the human community to constrain scientific and technological developments within ethical bounds, and the role of institutions of formal education in fostering and maintaining that capacity.

LEONARD J. WAKS

SEE ALSO *Digital Libraries; Museums of Science and Technology; Robot Toys; Rousseau, Jean-Jacques.*

BIBLIOGRAPHY

Fiering, Norman. (1981). *Moral Philosophy at Seventeenth Century Harvard: A Discipline in Transition.* Chapel Hill: University of North Carolina Press. A detailed description of moral philosophy in the early colonial college, valuable in presenting the background for the development of moral philosophy as a modern discipline in the eighteenth century.

Frank, Thomas Edward. (1993). *Theology, Ethics and the Nineteenth-Century American College Ideal: Conserving a Rational World.* San Francisco: Mellen Research University Press. The essential study of moral philosophy in the nineteenth century liberal arts college curriculum, valuable for tracing the rise and fall of 'protestant scholasticism', the attempt to preserve the harmony of protestant Christian ethics and modern science.

Geiger, Roger. (1986). *To Advance Knowledge: The Growth of American Research Universities, 1900–1940.* New York: Oxford University Press.

Guralnick, Stanley M. (1975). *Science and the Ante-Bellum American College.* Philadelphia: The American Philosophical Society. An essential, highly detailed study of the evolution of science education in the American college in the period from 1820 to 1850, during which modern science gained a central place in the liberal arts curriculum.

Kimball, Bruce A. (1995). *Orators and Philosophers: A History of the Idea of Liberal Education,* 2nd edition. New York: College Entrance Examination Board. A useful history of the concept of liberal arts education, particularly useful in clarifying the shifting relationships between logic and rhetoric.

Kintzinger, Martin. (2000). "A Profession But Not a Career? Schoolmasters and the *Artes* in Late Medieval Europe." In *Universities and Schooling in Medieval Society,* ed. William Courtenay and Jürgen Miethke. Leiden, The Netherlands: Brill.

Leff, Gordon. (1992). "The Trivium and the Three Philosophies." In *A History of the University in Europe,* Vol. I: *Universities in the Middle Ages,* ed. Hilde De Ridder-Symoens. Cambridge, England: Cambridge University Press.

Lyons, Henry. (1968). *The Royal Society 1660–1940: A History of its Administration Under its Charters.* New York: Greenwood Press. Originally published in 1944.

Mallet, Charles Edward. (1968a). *A History of the University of Oxford* Vol II: The Sixteenth and Seventeenth Centuries. New York: Barnes and Noble. Originally published in 1924.

Mallet, Charles Edward. (1968b). *A History of the University of Oxford* Vol III: *Modern Oxford (1688–1887).* New York.: Barnes and Noble. Originally published in 1924.

Marenbon, John. (2003). *Boethius.* Oxford: Oxford University Press.

Mourelatos, Alexander. (1991). "Plato's Science: His View and Ours of His." In *Science and Philosophy in Ancient Greece,* ed. Alan Bowen. New York: Garland.

North, John. (1992). "The Quadrivium." In *A History of the University in Europe,* Vol. I: *Universities in the Middle Ages,* ed. Hilde De Ridder-Symoens. Cambridge, England: Cambridge University Press.

Porter, Roy. (1996). "The Scientific Revolution and the Universities." In *A History of the University in Europe,* Vol. II: *Universities in Early Modern Europe (1500–1800),* ed. Hilde De Ridder-Symoens. Cambridge, England: Cambridge University Press

Smith, Wilson. (1956). *Professors and Public Ethics: Studies of Northern Moral Philosophers before the Civil War.* Ithaca, NY: Cornell University Press.

Tuplin, Christopher, and Tracey Elizabeth Rihill, eds. (2002). *Science and Mathematics in Ancient Greek Culture.* Oxford: Oxford University Press.

Winstanley, Denys. (1935). *Unreformed Cambridge.* Cambridge, England: Cambridge University Press.

Winstanley, Denys. (1947). *Later Victorian Cambridge.* Cambridge, England: Cambridge University Press.

EFFICIENCY

• • •

In the fields of technological innovation, economic development, business management, and public policy planning, as well as in everyday life, efficiency is a pivotal criterion that guides the behavior of both individuals and institutions. The widespread utilization of this criterion, however, raises serious epistemological, methodological, and practical questions, along with ethical challenges. Although efficiency may seem to be a clear, morally neutral concept, difficulties arise in conjunction with its extremely abstract character, the vast array of interpretations involved in concrete applications, and the fact that its pursuit may crowd out or obscure other important values.

Origins and Abstractions

The term *efficiency* is derived from the Latin *efficere* ("to produce, effect, or make"). In his *Physics,* Aristotle sees *causa efficiens* as one of the four factors (along with formal, material, and final causation) that explain change. Traditionally, efficiency has been understood as the agency or power of something or someone to bring about results, to produce a desired effect. In this sense there was no clear distinction between *efficiency, effectiveness,* and *efficacy* until the second half of the nineteenth century, when the term was given a technical meaning in the field of engineering.

The contemporary technical concept of efficiency arose from analyses of engine performance, or what is known as thermodynamic efficiency. The performance of an engine was defined as a ratio of the useful work obtained to the energy (heat) used. At best, the maximal amount of energy obtained would be the same as the energy consumed in the process. The concept then was used in economic theory, disseminated through the work of an engineer turned social scientist, Vilfredo Pareto (1848–1923), and other influential economists and engineers (Mitcham 1994). Economists saw themselves as engineers who were managing the scarce resources devoted to promoting social welfare, just as engineers attempted to find economic solutions to technological problems. The concept of efficiency moved into the political and public domains during the twentieth century, becoming a universally applied value. In the twenty-first century it is widely accepted that to be effective—that is, to obtain the intended goals—is not enough. It is also necessary to be efficient, that is, to obtain the intended results without wasting resources.

There are several definitions of the concept in its widest scope. The most common usages define efficiency as a ratio of results to resources or, alternatively, of ends to means or outputs to inputs. An activity, process, design, or system is said to reach maximum efficiency if (1) a desired result (output) is obtained through the use of the minimal possible amount of resources (input), (2) the maximal amount of results from a given resource is obtained, or in general (3) a combination of results and resources is obtained in such a way that it is not possible to increase any of the results or reduce any of the resources without reducing some other result (or amount of a result) or increasing some other resource (or amount of a resource).

Multiple Meanings

In its various usages the concept of efficiency gives rise to multiple meanings when this abstract idea is given specific applications: technical efficiency, energy efficiency, economic efficiency, resource efficiency, productive efficiency, market efficiency, and ecological efficiency, among others. Imprecise use, lack of agreement among experts, different backgrounds of expertise and technical traditions, hidden assumptions, the mathematical and practical complexities involved in making measurements, and other factors often make it extremely difficult to know the extent to which these terms should be taken as mere delimitations of a more general concept or, rather, suggest different but related concepts. The situation becomes more complicated when one considers the wide array of uses of the concept in heterogeneous fields such as energy technology, agricul-

ture, health care, business management, public administration, and academic or personal performance. As a consequence, there is a huge technical literature dealing with these problems. Philosophical analyses that do not take the definition of efficiency for granted are uncommon, although there are exceptions, such as the work of Mario Bunge (1989), Stanley Carpenter (1983), Miguel Angel Quintanilla (1989), and Henryk Skolimowski (1966).

Initially it seems easy to distinguish between purely technological or engineering and economic conceptualizations. The engineering solution to a problem is efficient when it uses the smallest amount of technological means independently of economic constraints. In real-life situations, however, technological means often must be measured in economic terms. For instance, although it is technically feasible to obtain gold from other elements, the cost of the procedure is so high compared with the value of the results that any attempt would be inefficient because of the excessive resources that must be used to achieve the objective.

Even in economics assessments generally are not equivalent. Narrowly productive points of view and the quest for personal profit repeatedly conflict with efficiency requirements in terms of social welfare. To harmonize legitimate aspirations with both personal gain and social benefit, economists, acting in effect as political assessors, resort to cost-benefit analyses that are supposed to identify a so-called Pareto optimum, which is defined as that situation in which it is not possible through any reallocation of resources to make any person better off without making someone else worse off without compensation.

Critics such as Amartya Sen (2002) have exposed the weaknesses in this conception and attempted to find more rigorous and fair alternatives. Because it does not take into account the problems associated with the fair distribution of public spending, the application of the Pareto optimum maintains unjust situations. Suppose there is a fixed public spending budget of $100 to be distributed between education and airport infrastructure. If the education budget is increased by $10 to make it easier for the poorest members of society to attend university on scholarship, that amount must be subtracted from the budget to improve airport infrastructure. Therefore, some benefit at the expense of others. The Pareto optimum no longer is reached because it would advise against any change in the assigned budgets. However, it would be difficult to defend denying scholarships to qualified students so that people who can afford a good education can reach their favorite vacation spots more easily.

Because efficiency is essentially context-dependent, many of the problems that arise in discussing it are caused by attempts to decide what counts as a resource or result and what is considered valuable, desirable, feasible, or even possible (Carpenter 1983, de Cózar 2000). Efficiency is not determined by preexisting conditions but is constructed by deciding which factors to consider in defining the problem and frequently by actively modifying the physical, economic, and legal environment in which an intervention is made to change the state of things. Geographical limits, the temporal vector, side effects, and other elements further frame the context of an intervention.

Aware of the practical problems raised by seeking the most efficient solution, or optimization, Herbert Simon (1982) proposed the concept of *satisficing*: the attempt to achieve a good, if not perfect, solution. A large telecommunications company may decide not to develop a radically new system of communication even though it is faster, more powerful, and easier to use if it has no clear estimation of the cost of gathering the information required to predict the success of the new technology in the market. This cost, together with the cost of research and development, can surpass the profits the innovation is projected to generate for the company. In other words, the company would be content with satisficing its behavior by making less ambitious improvements. Alternatively, a company might gamble on this major innovation if it were confident in its ability to influence, among other aspects of its social environment, public regulations and the perceived needs of the consumers.

Obscuring Other Values

Contextuality is a key issue in understanding the conflict between the modern technological project and the criticisms leveled at it by many philosophers of technology. The technological impulse is tied intimately to the design of increasingly more efficient machines, devices, tools, systems, and processes. In the course of this activity, which has contributed much to humanity in terms of safety, health, and welfare, the technological mind typically delimits problems in the narrowest possible way and then searches for basically quantitative solutions. Its success depends on this strategy, and much of the attraction of efficiency for experts and nonexperts alike lies in the perception that it always provides (as it really does in some cases) a mathematical, *automatic* comparison between alternatives that can be used to determine the best path to follow.

In this manner a descriptive concept becomes a prescriptive one. Arguments for efficiency appear to derive prescriptions from a dispassionate description of objects and situations, thus hiding the often conflicting values that lie behind decision making in real-world situations. For instance, Amory Lovins (1977) has argued in effect that proposals to build more efficient power plants ignore the possible desirability of reducing energy consumption by increasing the insulation of buildings.

It is important to remember that despite its familiarity, the concern for efficiency is relatively recent. The novelty of the current situation is that efficiency is being converted into an absolute criterion for decisions in many facets of life. As Jacques Ellul (1954) observed, if modern technological activity becomes indistinguishable from the pursuit of absolute efficiency, an ethics of nonpower is also conceivable. Such an ethics can, and indeed must, pose a limit to the *cult of efficiency* and its abuses. As Carl Mitcham suggests, there is a parallel between the well-known *naturalistic fallacy* and an *efficiency fallacy*. The philosopher David Hume (1711–1776) argued that *ought-statements* cannot be inferred from *is-statements*, and G. E. Moore (1873–1958) warned against equating goodness with some natural property. In regard to something that is said to be natural, with the implication that it is good, one may still reasonably ask, But is it good? Similarly, after one says that something is efficient, it makes sense to ask, But is it good? Twentieth-century history exhibits a long list of cases in which unethical goals were pursued with bloodcurdling efficiency. Therefore, one should define the goals one judges as good and only then, if appropriate, look for the means to achieve them efficiently.

JOSÉ MANUEL DE CÓZAR-ESCALANTE

SEE ALSO *Critical Social Theory; Economics and Ethics; Neutrality in Science and Technology; Pareto, Vilfredo; Taylor, Frederick W.; Work.*

BIBLIOGRAPHY

Bunge, Mario. (1989). *Treatise on Basic Philosophy*. Vol. 8. Dordrecht, Netherlands: Reidel. Bunge's extensive work contains one of the first attempts at dealing with efficiency as a key concept to consider in philosophy of technology but his approach remains too close to the neopositivistic paradigm.

Carpenter, Stanley R. (1983). "Alternative Technology and the Norm of Efficiency." *Research in Philosophy and Technology* 6: 65–76. A pioneer paper that contributes to the clarity and precision of discussions on technological practice that invoke efficiency, while at the same time adopting a critical stance on the abuse of this criterion.

De Cózar-Escalante, José Manuel. (2000). "Toward a Philosophical Analysis of Efficiency." *Research in Philosophy and Technology* 19: 87–100. This essay identifies two problems of the concept of efficiency (unicity and applicability) and argues for a critical reconsideration of efficiency in social life.

Ellul, Jacques. (1954). *La Technique, ou l'enjea du siècle*. Paris: A. Colin. Trans. by John Wilkinson as *The Technological Society* (1964). New York: Knopf. In this book Ellul defines technique as the totality of methods rationally arrived at and having absolute efficiency (for a given stage of development) in every field of human activity. One of the main aims of Ellul's work is to denounce the fixed end of technological design and activity, by which the multiplicity of means and ends are reduced to only one: the most efficient.

Mitcham, Carl. (1994). *Thinking through Technology: The Path between Engineering and Philosophy*. Chicago: University of Chicago Press. In his comprehensive revision of the most important analyses of the technological phenomenon, Mitcham includes useful information as well as insightful remarks about the meaning of efficiency in technological design and everyday life.

Quintanilla, Miguel Angel. (1989). *Tecnología: Un enfoque filosófico* [Technology: A philosophical approach]. Madrid: Fundesco. In a brief section of this book the author works out a formal definition of efficiency as a basis upon which to further distinguish other technological criteria.

Sen, Amartya. (2002). *Rationality and Freedom*. Cambridge, MA: Belknap Press. The reputed economist Amartya Sen is known, among other contributions to economic, social and ethical theory, for his sharp criticism of standard models of rational behaviour. Being reductionist and ill-conceived, those models tend to obscure crucial aspects of an individual's choice in real life situations.

Simon, Herbert A. (1982). *Models of Bounded Rationality*. Cambridge, MA: MIT Press. During his long career, Simon did impressive work to establish the strength and limits of rational design and management. One of his more interesting findings was the concept of bounded rationality: even if often is not realistic to search for the most efficient solution in absolute terms, a solution that is good enough can be achieved within the bounds set by the specific context in which the problem is stated.

Skolimowski, Henryk. (1966). "The Structure of Thinking in Technology." *Technology and Culture* 7(3): 371–383. Reprinted in *Philosophy and Technology: Readings in the Philosophical Problems of Technology* (1972), ed. Carl Mitcham and Robert Mackey. New York: Free Press. An early attempt at philosophically clarifying the concept of efficiency and its role as an ideal in engineering. However, Skolimowski's analysis has some weaknesses, especially because it does not clearly distinguish efficiency from effectiveness.

EINSTEIN, ALBERT

• • •

Albert Einstein (1879–1955) was born in Ulm, Germany, on March 14 into a middle-class assimilated German-Jewish family; by the time of his death on April 18 in Princeton, New Jersey, he was recognized as being equal in accomplishment to Isaac Newton, but one significantly more publicly involved in human affairs.

Life

Einstein showed precociousness in science and mathematics, with mixed accomplishment in other areas. He spent his early professional years in Switzerland working in the patent office. At age twenty-six—in the *miracle year* of 1905—he published several papers on special relativity, on the particle (photon) nature of light, resulting in the *complementary* idea that light was both a wave and a particle, and seminal papers in statistical physics. His general theory of relativity, first conceived in about 1907, achieved its final form in 1915. This theory was dramatically confirmed by its successful explanation of the hitherto mysterious precession of the perihelion of the planet Mercury, and with the observation during a solar eclipse of predicted bending of starlight by the Sun's gravitational field in 1919. It was especially this latter event that led to world fame.

Einstein's private life was not very dramatic; he was married twice and had two children. He emigrated to the United States at the time of Adolf Hitler's ascent in 1932, settling at the Institute for Advanced Studies at Princeton. He remained there for the rest of his life, continuing his physics research unabated, particularly his search for a *unified field theory*.

Achievements in Science

Einstein is best known for his theories of special and general relativity, although he also made enormous contributions to quantum mechanics, statistical mechanics, condensed matter physics, and cosmology. Through his contributions to the understanding of the nature of light and atomic structure, his revolutionary concepts of space and time, and his famous equation of $E = Mc^2$ (Energy equals mass times the speed of light squared) that shows the equivalence of mass and energy and led the way to the creation of controlled nuclear reactors and nuclear weapons, his impact on contemporary society and culture touches everyone.

Albert Einstein, 1879–1955. The German-born American physicist revolutionized the science of physics. He is best known for his theory of relativity. *(The Library of Congress.)*

Other less commonly appreciated impacts derive from early work on radiation theory, which led to the concept of stimulated light emission, the basis for the laser. In the 1990s his prediction, inspired by an earlier paper of Satyendra Nath Bose, of what is now known as Bose-Einstein Condensation, led to an entirely new field of physics that studies the macroscopic effects of quantum mechanics on extremely cold gaseous systems. The theory had been used earlier to help explain superconductivity and superfluidity.

Even this list of achievements does not adequately describe Einstein's involvement with the world of physics. Throughout his life he was in continual touch with numerous colleagues; he read voraciously and was fully involved with the developing conceptual framework of the new views of nature required by quantum physics and relativity. He was generous with his contemporaries, freely offering and taking suggestions from correspondents throughout the world, while eagerly conducting ongoing dialogues with the other great physicists then active, including Niels Bohr, Max Planck, Werner Heisenberg, Wolfgang Pauli, and Erwin Schrödinger. Although he was one of the original formulators of quantum mechanics he was never satisfied that it represented a complete theory, because it assumed the statistical nature of microscopic events, while he firmly believed in the Newtonian idea of causality in nature. Accordingly he always felt that quantum mechanics was incomplete, awaiting a deeper explanation for the statistical nature of the wave function in terms of a more causal theory. Einstein's often quoted statement, "God does not play dice," reflects this view. His *minority opinion* has resulted in an enormous literature on the interpretation of quantum theory, continuing with non-diminishing intensity into the twenty-first century.

Einstein's vision of a unified field theory that would unite all the known forces of nature into a single theoretical structure drove his research efforts during the last thirty years of his life. Although this incomparable challenge led to only limited success in his own hands, this holy grail of modern physics continues to inspire future generations of theoretical physicists.

Politics and Ethics

Einstein's overarching goal in physics was to formulate unifying principles for all phenomena in nature. This philosophy extended itself to other aspects of his life, including personal habits, and his deep involvement with issues such as world peace, human rights, and social justice. He was an implacable foe of militarism, even during his residence in Germany in World War I. He became the victim of intense anti-Semitism in Germany during the inter-war period, when his physics, especially relativity, was attacked as being *Jewish physics*. Although he espoused many liberal causes, he was never attracted to Communism and opposed Stalinist Soviet Russia as strongly as Nazi Germany. He was an unswerving advocate of international government and international control of armaments, including nuclear weapons. His advocacy of such positions often resulted in conflicts with authority, including the U.S. government. The Federal Bureau of Investigation (FBI) dossier on him consists of 1,427 pages. Although a non-practicing Jew, he was a strong supporter of the state of Israel, and was even proposed, at the time of Chaim Weizmann's death in 1952, to be its next President (although he swiftly turned down the invitation).

Additionally Einstein's name is indelibly connected with the atomic bomb, not only because of his famous formula for energy-mass equivalence but also because of the letter he signed in 1939, written by his friend Leo Szillard, alerting President Franklin Delano Roosevelt to the possibility that Germany might be working on the development of such a weapon. In his later years he regretted this action. Indeed, after Hiroshima and Nagasaki he argued that "everything has changed, save our modes of thinking" and that "the bomb [presents] a problem not to physics but of ethics." In 1955, in response to development of the hydrogen bomb, he co-signed with Bertrand Russell a public manifesto calling on all scientists to become involved in helping to reverse the nuclear arms race.

It is important to remember, however, that Einstein's concern for the social implications of science and technology always remained central to his core of beliefs. In 1931, for instance, in a talk at the California Institute of Technology, he told students that "concern for man himself and his fate must always form the field interest of all technical endeavors."

The literature on Einstein—his life, science, and beliefs—is overwhelming. There are more than 4 million Internet sites containing his name. As one noteworthy example, see the American Institute of Physics History site. At the end of the twentieth century *Time* magazine called him the *person of the century*. He remains the personification of the scientist. Einstein's combination of pure brilliance, high ideals, personal integrity, as well as human weaknesses yield the picture of a human being at the highest level of achievement.

BENJAMIN BEDERSON

SEE ALSO *Atomic Bomb; Energy; Pugwash Conferences.*

BIBLIOGRAPHY

Brian, Denis. (1996). *Einstein, A Life.* New York: John Wiley & Sons, Inc. Readable and anecdotal.

Einstein, Albert. (1931). *Address before the Student Body, the California Institute of Technology, February 16, 1931.* Original manuscript held at Pasadena, CA: Caltech Institute Archives. Quoted in many speeches and sources.

Einstein, Albert. (1954). *Ideas and Opinions,* ed. Carl Seelig. New York: Bonanza Books.

Einstein, Albert. (1987). *The Collected Papers of Albert Einstein,* ed. P Havas et al. 8 vols. Princeton, NJ: Princeton University Press. English translation.

Pais, Abraham. (1982). *Subtle Is the Lord: The Science and Life of Albert Einstein.* Oxford: Oxford University Press. The definitive scientific biography, highly recommended.

ELLUL, JACQUES
• • •

Jacques Ellul (1912–1994) was born in Bordeaux on January 6 and spent his academic career as Professor of the History and Sociology of Institutions at the University of Bordeaux Law Faculty and Professor in its Institute of Political Studies. His more than fifty books and hundreds of articles range across Christian theology, ethics, and biblical studies as well as sociological analysis and critique of mass media and communication, bureaucracy, and modern law and politics. He died in Bordeaux on May 19.

Technique: Ellul's Central Thesis

At the heart of his sociological works is his study of technology or, the term he preferred, Technique (*la technique*). Indeed Ellul initially became widely known in the English-speaking world for *The Technological Society* (1964). Its intellectual significance and originality derives in part from its argument being conceived twenty years before the original French edition (*La Technique* [1954]) when, after reading Karl Marx's *Capital*, Ellul (a law student in his early twenties) concluded that Technique, not capital, was central to modern civilization. This seminal idea was subsequently developed with Bernard Charbonneau in the French personalist movement of the 1930s.

Ellul was adamant that *la technique* "does not mean machines, technology, or this or that procedure for attaining an end" (Ellul 1964, p. xxv). He defined it as "the totality of methods rationally arrived at and having absolute efficiency (for a given stage of development) in every field of human activity" (Ellul 1964, p. xxv). Technique is, in other words, a universal category (Ellul compares it to *dog* rather than *spaniel*) embracing all the various self-consciously developed means found in art, politics, law, economics, and other spheres of human life. Central to these means is a quest for efficiency that is the defining characteristic of Ellul's account of Technique.

Two theses drive Ellul's analysis. First that "no social, human or spiritual fact is so important as the fact of technique in the modern world" (Ellul 1964, p. 3). Second that the contemporary "technical phenomenon . . . has almost nothing in common with the technical phenomenon of the past" (Ellul 1964, p. 78). Whereas previously Technique was limited and diverse, social changes in the eighteenth and nineteenth centuries led to its dominance and totally changed the relationship between Technique and society.

Jacques Ellul, 1912-1994. Ellul was a French thinker, sociologist, theologian, and Christian anarchist. He wrote several books against the "technological society," and some about Christianity and politics. (© Patrick Chastenet.)

In addition to its *rationality* and *artificiality*, Ellul proposes five more controversial characteristics of modern Technique. First there is *automatism* of technical choice because inefficient methods are eliminated and *the one best way* predominates. Second *self-augmentation* exists as technical developments automatically engender further innovations. Third Technique is characterized by *monism* as different techniques form an interconnected whole. This means that individual technologies must not be isolated and analyzed apart from an understanding of the wider technical phenomenon. Fourth there is a technical *universalism* that is both geographic (here Ellul offers an analysis that anticipates globalization) and qualitative (as every area of life is subordinated to technical efficiency). Fifth, and decisive for the novelty and hegemony of Technique, is its *autonomy*. This means Technique is no longer controlled by economics, politics, religion, or ethics; the common belief in Technique as a neutral means is false.

These five features are returned to in *The Technological System* (1980) where Ellul argues they characterize an elaborate technical system within society. The characteristic of *uncertainty*—seen in such factors as the ambivalence of technical progress and the unpredictability of its development—is then added in his *The Technological Bluff* (1990) that critiques contemporary discourse about Technique.

Technique in Society and Criticisms of the Analysis

Ellul's analysis leads him to conclude that whereas previous societies developed through the dialectical play of different social forces, it is now dominated by Technique. Most of *The Technological Society* is an account of the society that Technique is creating in relation to economics, politics, law, the state, and human affairs such as education, entertainment, sports, and more. Both there and in such works as *Propaganda* (1965) and *The Political Illusion* (1967) the prescience and power of Ellul's analysis remain striking at the beginning of the twenty-first century and explain why some describe him as a prophet. Having initially claimed Technique no longer belongs within human civilization but has established a technical civilization, Ellul later extended this, arguing that the social environment that had earlier replaced humanity's natural environment has in turn now been replaced by a technical *milieu*: Technique provides humans with what they need to live, is that which now threatens and endangers them, and is most immediate to them.

Most seriously, Ellul believed that Technique was incompatible with a truly human civilization. Technique focuses on quantitative improvements and facts rather than qualitative change based on values. It is a means of power—a central Ellul theme—and not subject to human values. Although it originally enhanced human freedom, building civilization by enabling people to overcome natural and social constraints and necessities, Technique has become human fate and a form of necessity. What used to be a means to freedom for humans has become a condition of slavery. In the terms of Ellul's theological writings, Technique is a contemporary idol that attracts human faith, hope, love, and devotion; a locus of the sacred in a supposedly secular society (Ellul 1975).

The criticism constantly made against Ellul is that he is a technophobe and a fatalist. Although the all-embracing nature of Technique in his work and his often caustic style of writing creates problems, Ellul's desire was "to arouse the reader to an awareness of technological necessity and what it means" and present "a call to the sleeper to awake" in order to challenge the destruction of human civilization. In later writings, Ellul sketched a new ethics of response to the dominance of Technique. This comprises the need for proper recognition of the other person (reflecting Ellul's personalism) and of nature (Ellul was an early environmentalist) and an ethics of voluntary limitation, rejecting the technical mindset that whatever can be done therefore should be done. The practice of such an *ethics of non-power* is very difficult in a world dominated by technological power and is a central part of Ellul's ethic for Christians that he suggests.

From the early 1930s, Ellul's aim was to help people understand and preserve a sense of criticism vis-à-vis technical civilization. More than half a century after it was written, his *Technological Society* remains an insightful, even if at times infuriating, analysis of modern Technique and its effects on contemporary society.

ANDREW J. GODDARD

SEE ALSO *Autonomous Technology; Efficiency; Freedom; French Perspectives*.

BIBLIOGRAPHY

Ellul, Jacques. (1964). *The Technological Society*. New York: Knopf. This is Ellul's classic text, written in the late 1940s, which was first published as *La technique ou l'enjeu du siecle* (Paris: Armand Colin, 1954).

Ellul, Jacques. (1975). *The New Demons*. New York: Seabury. Ellul's study of the new forms of the sacred in contemporary society.

Ellul, Jacques. (1980a). *The Technological System*. New York: Continuum. This, Ellul's second major study of technique, first appeared in 1977 as *Le systeme technicien* (Paris: Calmaan-Levy)

Ellul, Jacques. (1980b). "Nature, Technique and Artificiality." In *Research in Philosophy and Technology* 3: 263–283.

Ellul, Jacques. (1998). *Jacques Ellul on Religion, Technology and Politics: Conversations with Patrick Troude-Chastenet*. Atlanta, GA: Scholars Press. This important and wide-ranging collection of interviews with Ellul originally appeared as *Entretiens avec Jacques Ellul* (Paris: La Table Ronde, 1994).

Goddard, Andrew. (2002). *Living the Word, Resisting the World: The Life and Thought of Jacques Ellul*. Carlisle, UK: Paternoster Press.

EMBRYONIC STEM CELLS

• • •

In 1998 a team of researchers reported that they had isolated and removed stem cells from the inner cell mass of human embryos that had been donated by couples undergoing fertility treatment (Thomson et al. 1998).

The embryos had divided for several days to reach the blastocyst stage of approximately 100 cells. At this stage embryos have a hollow sphere in the middle, an outer layer of cells committed to forming the placenta and other cell lines, and a mass of undifferentiated cells pushed to one side (inner cell mass). The cells in the inner mass have, for a short time, the capacity to develop into all cells in the human body, and are known as embryonic stem (ES) cells. The researchers' announcement that they had isolated ES cells in human embryos generated considerable interest because it suggested that the cells could be removed, cultured, and coaxed to differentiate for use in medical therapy. Among other things, it was thought that large supplies of specialized cells could be widely available and used to replace cells destroyed by Parkinson's disease, Alzheimer's disease, neural cord injuries, and other diseases and conditions. The announcement also generated controversy because the act of removing ES cells destroys the embryo. In the years since, research has been limited to pre-clinical studies. Numerous safety issues must be addressed before clinical trials ethically can be conducted.

Ethical Issues

Of the many ethical issue raised by ES cell research, four are described here. First what is the moral status of human embryos? Some individuals argue that embryos have the same moral status as persons, which means it would be purposefully unethical to destroy embryos for any reason. Others argue that embryos are potential human beings that do not share the same rights as children and adults. For them, the destruction of embryos may be warranted under certain circumstances.

Second, independent of the particular moral status of human embryos, will ES cell research contribute to a mindset that treats embryos as commodities? Some express concern that using embryos for medical purposes will turn embryos into merchandise and diminish the dignity of humans in the process. Others counter that strict rules overseeing ES cell research protect human dignity while respecting the interests of patients who need therapies

Third, are ES cells necessary for medical therapies? Proponents of ES cell research claim that ES cells are versatile and easy to work with, and that they raise significant hope for effective medical therapies. Opponents claim that adult stem cells, found in human tissues and not requiring the destruction of human embryos, also hold the potential for medical therapies and provide a viable alternative form of research.

Fourth, what impact does the source of the embryos have on the ethics of ES cell research? The embryos used by James Thomson and his colleagues were donated for research by couples who were patients at in vitro fertilization (IVF) programs and who no longer needed stored embryos for their conception efforts. Arguably these embryos were created for an ethical purpose (reproduction) but were not needed; therefore it would be appropriate to secure some good from them before their inevitable destruction. It has also been advocated that embryos may need to be created solely for ES cell removal in order to secure a sufficient number of healthy and genetically diverse embryos to meet research and therapeutic needs. Critics, however, contend that this would be less ethical than using donated embryos, because the embryos would be created with the intention of destroying them. Still another possible source of ES cells is from the creation of embryos through a cloning technique (somatic cell nuclear transfer), in which the intended patient's nucleus would be used to create an embryo for deriving genetically compatible ES cells. Creating cells specifically for a patient would presumably eliminate the need for anti-rejection drugs. Critics, however, argue that therapeutic cloning would tempt individuals to use the embryos for reproduction rather than therapy.

Policy issues

Policy issues for human ES cell research have revolved around whether governments should fund studies involving ES cells. The issue became volatile immediately after the announced isolation of human ES cells in 1998.[1] In the United States, the U.S. Congress held hearings on the question, numerous interest groups lobbied both for and against funding, and policy advisory bodies convened to make recommendations (Bonnicksen 2002). For example, the National Bioethics Advisory Commission concluded it would be ethical to fund the removal and use of ES cells from donated embryos (National Bioethics Advisory Commission 1999). A working group convened by the National Institutes of Health (NIH), however, concluded it was appropriate to fund only the use of ES cells (National Institutes of Health 1999). The removal of cells (and hence destruction of the embryo) would have to be funded privately. Both groups agreed the government should not fund research creating embryos solely for ES cell removal.

Following intense lobbying by, among others, right-to-life groups opposed to federal funding and scientific associations and patient advocacy groups supporting funding, President George W. Bush announced a lim-

ited compromise position on August 9, 2001 (Vogel 2001). Under the new policy, the federal government would consider funding a narrow range of proposals in which (a) ES cells had been removed with private funds prior to the date and time of Bush's speech, and (b) the embryos were donated with informed consent by couples in IVF programs. At the time it was thought approximately sixty ES cell lines worldwide met these conditions. Within a couple of years, however, it became clear that fewer than fifteen cell lines were available for research.

Opponents argue that the government should not fund research that many people regard as immoral. Advocates argue that governmental funding is necessary for the potential of this research to succeed and for new therapies to become available to help persons with presently untreatable illnesses. Funding also has the benefit of opening the door to federal oversight of the research. Assuming ES cell research lives up to its potential, more studies will be conducted in the future and new cell lines will be needed to meet the standards required for clinical tests of medical therapies. If funding remains strictly limited, research will be conducted with private sector funding outside the public eye. Inasmuch as ES cell issues generate intense discourse, it is ironic that ES cell research will proceed without the public scrutiny that comes with significant federal funding.

Debates over the ethics of cell research are ongoing in nations worldwide. For example, in Europe differences among nations have precluded funding for ES cell research by the European Union (Vogel 2003). Research is proceeding in individual nations with accommodating governmental policy, such as the United Kingdom where, among other things, a UK stem cell bank has been set up with government backing (nibsc.ac.uk/divisions/dbi/stemcell.html).

ANDREA L. BONNICKSEN

SEE ALSO *Dignity; Fetal Research; Human Cloning; In Vitro Fertilization and Genetic Screening; Medical Ethics; Research Ethics.*

BIBLIOGRAPHY

Bonnicksen, Andrea L. (2002). *Crafting a Cloning Policy: From Dolly to Stem Cells.* Washington, DC: Georgetown University Press.

Jones, Howard W., Jr., and Jean Cohen. (2004). "IFFS Surveillance 04." *Fertility and Sterility* 81(Supp. 4): S1–S54.

National Bioethics Advisory Commission. (1999). *Ethical Issues in Human Stem Cell Research: Report and Recommen-*

dations of the National Bioethics Advisory Commission. Rockville, MD: National Bioethics Advisory Commission.

National Institutes of Health. (1999). "Draft National Institutes of Health Guidelines for Research Involving Human Pluripotent Stem Cells." *Federal Register* 64(231): 67576–67579.

Ruse, Michael, and Christopher A. Pynes, eds. (2003). *The Stem Cell Controversy: Debating the Issues.* Amherst, NY: Prometheus Books.

Thomson, James A.; Joseph Itskovitz-Eldor; Sander S Shapiro; et al. (1998). "Embryonic Stem Cell Lines Derived from Human Blastocysts." *Science* 282: 1145–1147.

Vogel, Gretchen. (2001). "Bush Squeezes Between the Lines on Stem Cells." *Science* 293: 1242–1245.

Vogel, Gretchen. (2003). "E.U. Stem Cell Debate Ends in a Draw." *Science* 302: 1872–1873.

EMERGENT INFECTIOUS DISEASES

• • •

Emergent infectious diseases (EIDs) are conditions caused by pathogenic microorganisms or parasites that have recently appeared or reappeared in human or animal populations. Typically, EID agents have begun to change the range of their infection, spread through new vectors or the movement of preexisting vectors, rely on shifts in patterns of host susceptibility, or have only recently been identified as the causes of existing diseases. This includes reemerging disease agents once thought to have been eradicated, but that have returned in resistant strains, or as a result of disintegrating public health infrastructure. Emergent diseases have tremendous impact on human health, and the health of pets and livestock. Furthermore, they pose a threat to biodiversity because many wild animal species are also at risk.

Science and Origins

An emerging infection can be caused by such viral agents as Ebola virus, HIV, or the SARS-associated corona-virus (SARS-CoV) identified as the cause of severe acute respiratory syndrome (SARS); bacteria such as methicillin-resistant *Staphylococcus aureus* (MRSA); or prions responsible for bovine spongiform encephalopathy (BSE, or "mad cow disease"), scrapie in sheep, chronic wasting disease in wild and domesticated deer and elk, and variant Creutzfeldt-Jakob disease (vCJD) in humans.

Emergence of an infectious agent is a two-step procedure: introduction into a new host species, followed by dissemination into a population. Varied origins include the evolution of a new virus or variant, bacteria, or prion; the introduction from another species (zoonoses); or dissemination from a smaller into a larger population. The latter two are usually the result of some environmental, social, or political disturbance bringing the naive host population into contact with the infectious agent.

Emergence can be illustrated through the case of Ebola virus, a virus of zoonotic origin. In 1995, a Swiss scientist died from Ebola while studying a chimpanzee population in the Côte d'Ivoire. In January of 1996, twenty-nine of thirty-seven confirmed cases of Ebola in a Gabon village were traced to contact with a dead chimpanzee.

Viruses and bacteria often mutate and adapt through the exchange of genetic material that can select for traits such as virulence, adaptability to different host organisms, and resistance to antiviral drugs or antibiotics. Viruses are ephemeral entities that undergo antigenic mutation, and adapt to new ranges of host, or vector.

A viral example that captures many of the characteristics of an EID is influenza caused by the influenza virus. Many influenza epidemics threatened public health throughout the twentieth century. A mutated influenza virus that originated in swine or avian hosts caused the 1918 Spanish flu pandemic, which killed more than twenty million people worldwide. It is thought that mixed variants of human and avian strains of influenza virus caused the 1957 Asian flu and the 1968 Hong Kong flu pandemics. In these cases, a preexisting swine or avian influenza virus either infected human beings directly and became adapted to the new hosts, or else previously-existing human variants of the influenza virus obtained genetic information from animal viruses. In some of these cases, the virulence of the newly adapted influenza virus was great enough to explode in the newly-acquired human host population, expanding throughout the global population.

Between 1998 and 1999, the previously unknown Nipah virus claimed 105 lives and resulted in the slaughter of 1.1 million hogs in Malaysia. Nipah virus exemplifies many of the characteristics of a newly emergent virus. It is carried by flying foxes (*Pteropus vampyrus*). The emergence of Nipah virus from the flying fox reservoir resulted from environmental changes in the infected hosts' environment. A drought allowed fires, set to clear land, to run out of control, destroying the flying foxes' habitat and food source. Many flying foxes set up new residence in orchards, often run in conjunction with swine husbandry operations. It has been

A crowd of people in Zaire watch health workers who have come to deal with an epidemic of the Ebola virus. Since its discovery in 1976, different strands of Ebola have caused epidemics with 50 to 90 percent mortality in several countries in Africa. (© *Patrick Robert/Corbis.*)

suggested the bats contaminated fruit that was then fed to the pigs, who then infected their caretakers. Changing environmental conditions and agricultural practices create conditions where the hosts of infectious diseases come into contact with new, potential hosts, with sometimes tragic consequences.

Of particular concern to public health is the emergence of infections by bacteria that have developed resistance traits to a variety of antibiotics. Antimicrobials are perceived as essential for combating both human and animal bacterial infections. In 1945, penicillin discoverer Alexander Fleming (1881–1955) warned of the danger of antibiotic resistance when bacteria in his lab developed resistance traits to penicillin through mutations and a process of natural selection. Resistance also develops through the transference of genes from resistant to non-resistant bacteria. An early case of the danger recumbent in the transference of genes for resistance is ampicillin-resistant *Haemophilus influenzae* and *Neisseria gonorrhoeae,* which appeared in the early to mid 1970s. Both diseases shared genetic material thought to have been transferred from one species of bacteria to another (Levy 2002).

It is now considered an item of scientific faith that the use of antimicrobials will favor the selection of resistant strains for most bacterial species. Some authors stipulate increasing resistance is the inevitable outcome of antimicrobial use in both human health and agricultural contexts (Levy 2002). Stuart Levy has coined the expression "the antibiotic paradox" to characterize the intertwined promises and threats of antimicrobial use.

Impacts

Antimicrobial resistant bacteria are increasingly ubiquitous, and their costs are immense and growing. One overview of the human health literature on resistance notes MRSA has been reported in community-based infections at rates from twenty to sixty-two percent of cases in the 1990s. This study reports widespread rising resistance to second and third generation cephalosporins in *Enterobacter* species, suggesting antimicrobial resistance has "become a fact of hospital life and is so common that it often goes unnoted until it is either extreme or epidemic" (Weinstein 1998, p. 215).

A 2001 Center for Disease Control and Prevention (CDC) publication notes "as we enter the 21st century, many important drug options for the treatment of common infections are becoming increasingly limited, expensive, and in some cases, nonexistent."

In a 1969 speech, the U.S. Surgeon General proclaimed that the frontiers of infectious diseases had been reached, remaining problems in the United States were marginal, and that it was the responsibility of the medical establishment to focus on chronic illness. Antibiotics were proclaimed miracle drugs (Levy 2002). They were understood to be an increasingly effective weapon in the armamentarium against bacterial infections and, with the promise of many new viral vaccines, it was believed the technology existed to eradicate disease worldwide.

With a growing awareness of the vastness of epizootic reservoirs of infectious agents, endemic changes in environment and agriculture, increasingly rapid global movement of animal and human populations, and growing problems with antibiotic resistance, this era of optimism is at an end.

Ethics and Policy Issues

One consequence of the emergent character of these diseases is the burden of uncertainty under which policy makers must function—far greater than the uncertainties faced by policy makers dealing with well understood risks such as cigarette smoking or automobile driving. The next EID may be innocuous, or it may be a deadly pandemic. This uncertainty makes it difficult to compare the risks of EIDs to other, more certain public health hazards. Indeed, one reason the media often covers EIDs more closely is because of this uncertainty.

Contemporary efforts to defend against EIDs follow the stages of prevention, detection, and response. Optimal allocation of resources among these stages is a question that has generated much controversy.

Conventional approaches to public health are incapable of preventing many of the factors that are presently increasing the rate of disease emergence. Public health institutions rarely have the resources or the mandates to put a halt to rapid environmental change or to changing patterns of agriculture, let alone to control the increased global movement of human and animal populations.

In the case of resistant bacteria, restrictions and judicious use guidelines on antibiotic use in human and animal health have been suggested as well as a curtailment of their use as growth promotants in the animal industries (Rollin 2001). Rising levels of resistance have fueled a debate over responsibility between representa-

tives of the human and animal medical fields. For example, in 2004 the U.S. Food and Drug Administration was revising its drug approval and labeling procedure for antimicrobials to be used in animal agriculture, and a number of European countries have banned their use as growth promotants under the precautionary principle. The CDC and a variety of private initiatives are instituting educational programs encouraging patients and medical professionals to curtail their use of antibiotics. In the United States as of 2004, there was no legislation to further restrict doctors' prescriptions of antibiotics.

In the United States, responsibility for managing emerging infectious diseases is distributed widely. The CDC often takes the lead, but in instances of food-borne disease, the Food and Drug Administration (FDA) and the Department of Agriculture (USDA) are also involved. In the case of antimicrobial resistance, a U.S. federal government interagency task force initiated in 1999 involved more than eleven agencies and departments.

The implementation of vaccines as a means of prevention is hindered by the contemporary market in pharmaceuticals. Pharmaceutical companies tend to focus research and development monies on profitable repeatable treatments for chronic ailments. Vaccines and antibiotics—which will only be used once or a few times per individual over their lifetime—do not provide the same return on investment. Incentives, regulatory assistance, or an alternative drug research and development system is needed to address these gaps in the preventive armamentarium.

Around the turn of the twenty-first century, greater emphasis has been placed on understanding the role of industrial agri-food practices in the spread of infectious diseases. In the United States, rapid progress in the development of new techniques for managing industrial animal agriculture for food safety concerns have been hindered by the sometimes conflicting mandates of the principal governmental agencies involved in dealing with emerging diseases among food animals, including the CDC, the FDA, and the USDA.

Some policy suggestions have focused on early detection of aberrant syndromes through disease surveillance, on the anticipation of new host and virus interactions brought about by changing ecological and agricultural conditions, and on the control of new diseases through planned response. Similar surveillance tactics have been suggested to deal with antimicrobial-resistant bacteria.

There is limited but growing international coordination of emerging infectious disease surveillance and response. Most surveillance and response systems are national in scope. This includes the CDC, particularly

A health officer checks the temperature of an arriving passenger at Kuala Lumpur International Airport as part of a screening for Severe Acute Respiratory Syndrome (SARS). After the first outbreak occurred in China, the disease spread rapidly, reaching other countries via international travellers. (© Reuters/Corbis.)

the National Center for Infectious Diseases, which often responds to emerging disease emergencies outside the United States. Another CDC program, jointly run with the USDA, is the National Antimicrobial Resistance Monitoring System (NARMS).

At the international level, there are two institutions of note. The United Nations World Health Organization (UN-WHO) Communicable Disease Surveillance and Response is the principal international organization that identifies, verifies, and responds internationally to epidemics of infectious disease. This organization is overworked and underfunded. Animal diseases are monitored and managed by the Office Internationale des Epizooties (OIE), organized under the World Trade Organization (WTO) to maintain animal health and welfare worldwide. The OIE publishes trade standards on the presence of epizootic diseases, animal health, and food safety that govern the importation of animals and animal products between WTO member countries.

Responses to EIDs are administered by the above agencies and organizations and relevant agencies within a particular nationality's borders. Responses range from quarantine of humans and animals to radical eradication programs such as the slaughter of infected animals, vaccination programs, and mass-treatment with a variety of antiviral and antibiotic drugs.

The development of new antiviral and antibiotic medication suffers from the same market-induced problems as the development of new vaccines. Incentives or the creation of new, perhaps not-for-profit, institutions of drug research and development could alleviate the current dearth in treatment options. The National Institute of Health (NIH) supports research in drug development, but the costs of bringing these new drugs to market are still deemed prohibitive by many pharmaceutical companies.

Quarantines and mass animal slaughter wreak emotional, moral, social, and economic havoc. The

2003 SARS epidemic shut down international trade and travel and damaged the economic lives of cities as far apart as Toronto and Hong Kong. Foot and mouth disease and bovine spongiform encephalopathy (BSE) eradication programs in the United Kingdom resulted in massive animal slaughter, and in movement bans that eventually necessitated the welfare slaughter of even more animals as feed stocks were depleted. This devastated the British rural economy, and seriously affected British agricultural trade with Europe. Quarantines, animal slaughter, and animal movement bans are currently the most effective means of coping with an epidemic, but there is much research that needs to be done on how to lessen the impact of these techniques on the affected populations of humans and animals.

WESLEY DEAN
H. MORGAN SCOTT

SEE ALSO *Antibiotics; Health and Disease; HIV/AIDS; Medical Ethics.*

BIBLIOGRAPHY

Centers for Disease Control. (1995–). *Emerging Infectious Diseases.* Unless otherwise cited, all facts in this piece were taken from this flagship journal in the field.

Levy, Stuart. (2002). *The Antibiotic Paradox: How the Misuse of Antibiotics Destroys their Curative Powers,* 2nd edition. Cambridge, MA: Perseus Press.

Morse, Stephen, ed. (1993). *Emerging Viruses.* New York: Oxford University Press. Includes three articles of special note: Donald Henderson's "Surveillance Systems and Intergovernmental Cooperation," Morse's "Examining the Origins of Emergent Viruses," and Edwin Kilborne's "After word: A Personal Summary Presented as a Guide for Discussion."

Rollin, Bernie. (2001). "Ethics, Science and Antimicrobial Resistance." *Journal of Agricultural and Environmental Ethics* 14(1): 29–37.

Weinstein, Robert, and Mary Hayden. (1998). "Multiply Drug-Resistant Pathogens: Epidemiology and Control." In *Hospital Infections,* 4th edition, ed. John V. Bennett and Philip S. Brachman. Philadelphia, PA: Lippincott-Raven Publishers.

INTERNET RESOURCE

Center for Disease Control. (2001). *Preventing Emerging Infectious Diseases: Addressing the Problem of Antimicrobial Resistance, A Strategy for the 21st Century.* Available from http://www.cdc.gov.

EMOTION
• • •

The role of emotions in moral behavior has been debated by ethicists since ancient Greece. The scientific study of emotions is much more recent, yet the advances in twenty-first century understanding of the neural mechanisms that subserve emotions take on added meaning in the context of these ancient debates. New developments in *emotional technologies* add further nuances to these old questions. This entry provides a brief account of what emotions are, outlines the way emotions have been viewed in some major philosophical traditions, and discusses the ethical questions raised by some forms of technology.

What Are Emotions?

Emotions may be defined in a number of ways (Evans 2001). From a neurobiological perspective, emotions are defined in terms of the neural mechanisms that implement emotional processes in the brains of humans and other animals. In all mammals most emotional processes are mediated by a set of neural structures known collectively as the *limbic system.* The limbic system is an ill-defined term, but usually refers to a variety of subcortical structures, including the hippocampus, the cingulate gyrus, the anterior thalamus, and the amygdala (see Figure 1).

Neurobiological definitions of emotion can be regarded as parochial, because they exclude organisms that lack brains like those of humans from having emotions. A less chauvinistic alternative would be to define emotions in functional terms—that is, as dispositions to behave in certain ways. Fear, for example, disposes the organism to mobilize defensive and flight behaviors, and to focus attention on possible threats.

Functional definitions of emotion have been criticized on the grounds that they leave out feelings. Feelings—the conscious awareness of emotional states—have often been regarded as the central component of emotions, but some neuroscientists such as Antonio Damasio prefer to distinguish between emotions, which they regard as objectively measurable processes, and subjective feelings.

Emotions may be distinguished from other affective phenomena such as moods and personality traits by temporal duration. Many psychologists regard emotions as relatively rapid and brief processes, lasting no more than a minute or two, and class longer-lasting affective states

FIGURE 1

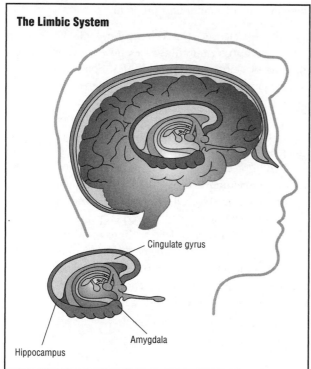

The Limbic System

Cingulate gyrus

Amygdala

Hippocampus

The human brain, with some of the limbic structures highlighted.

as moods, although this distinction is not universally accepted.

Emotions in Philosophy and Ethics

At the risk of over-simplification, it is useful to distinguish three main traditions in Western thought regarding the role of emotions in moral behavior. First many thinkers, such as Plato (c. 428–348 B.C.E. and Immanuel Kant (1724–1804), have regarded emotions principally as obstacles to good conduct. Plato compared the rational mind to a charioteer whose task was to keep his horses (his emotions) under a tight rein. Kant argued that good actions were only truly moral when performed purely out of concern for the moral law, and not motivated by any emotion.

A second tradition, exemplified by thinkers such as Aristotle (384–322) B.C.E. and economist Adam Smith (1723–1790), has regarded emotions as vital ingredients in generating moral behavior. Aristotelian ethics is based on the idea of virtue, which is a *golden mean* halfway between opposing vices. Because many vices are defined as deficits or excesses of particular emotions, Aristotelian virtues may be regarded as optimal midpoints between emotional extremes. This ancient approach to ethics finds many echoes in the modern concept of *emotional*

intelligence, which also stresses the need for cultivating emotional self-regulation. Smith argued that certain social emotions such as sympathy, which he called *moral sentiments*, lay at the heart of all ethical conduct.

Finally, a third tradition takes issue with both of the preceding positions, arguing that all moral judgments are merely an expression of the speaker's emotions. According to this view, championed by philosopher David Hume (1711–1776), when someone says that a certain action is right or wrong, what is meant is that the speaker has a feeling of approval or disapproval toward the action. This is sometimes referred to as the *emotive theory of ethics*.

More recently the philosophy of emotions has begun to address other questions besides the role of emotions in moral behavior. Contemporary philosophers such as Paul Griffiths, for example, have argued that emotions are such a heterogenous bunch of phenomena that they cannot constitute a single natural kind. Others have attempted to clarify the complex relationship between emotions and rationality (de Sousa 1991).

Emotions and Technology

Since its very beginning, much human technology has been driven by the desire to exert greater control over the emotional states. Many human inventions, from cooking to music, may be viewed as *technologies of mood,* in the sense that they are designed primarily to induce certain emotions in the user. Modern developments such as psychotherapy and antidepressants, therefore, may increase the effectiveness of the ability to manipulate emotions by artificial means, but the ethical questions they raise are not new. The objections raised by critics such as Francis Fukuyama (2002) to the possibility of *cosmetic psychopharmacology*, in which people manipulate their emotional states at will by means of sophisticated new drugs, have many echoes in ongoing debates about authenticity. Such objections seem to some to smack of *psychopharmacological Calvinism*, the niggardly belief that happiness must be earned the hard way—that is, without the help of drugs (Kramer 1994). The inflammatory rhetoric that has so far characterized such debates needs to be eliminated if people are to have a mature and reasoned discussion about the benefits and dangers of developing more powerful technologies for influencing moods and emotions.

Other modern technologies raise ethical questions that had not previously been considered. The advent of neuroimaging techniques and other means of monitoring emotional processes that were previously thought to be irreducibly private and subjective raises new issues of neural privacy. How much of one's emotional life should others be able to assess by means of these technologies, and how reliable are they? These questions will become more urgent as new technologies such as *sensitive clothing* (garments with embedded sensors that monitor physiological changes) and brain-machine interfaces permit further intrusion into the emotional lives of others.

Another technology that raises new ethical questions concerning emotions is affective computing, a branch of artificial intelligence that attempts to build emotional machines (Picard 1997). One line of thinking in robotics argues that robots will need emotions if they are to be truly autonomous, and some commercially available entertainment robots already come programmed with a repertoire of basic emotions. It is arguable, of course, whether such mechanisms constitute genuine emotions or merely simulated emotions, but this distinction may be irrelevant because people tend to react to such robots as if they possessed genuine emotions. The ethical problems raised by such developments have been explored in great detail in science fiction, from Isaac Asimov's (1920–1992) short story "The Bicentennial Man" (1976) to Arthur C. Clarke's famous novel *2001: A Space Odyssey* (1968). While such fictional scenarios often postulate devices that are far in advance of available technology, the questions they raise are deep and sometimes disturbing. In Asimov's short story, for example, a robot redesigns his own circuitry so that he may experience the whole range of human emotions. Because some rights are often held to be contingent on having certain emotional capacities, a robot with human-like feelings might have to be accorded a moral status equivalent to that of a human. There are parallels here with the animal rights movement, which has placed great emphasis on the capacity of certain animals for pain and suffering in its attempts to provide them with greater legal protection.

More sinister scenarios envision emotional machines turning against their human creators, raising the question of whether efforts in affective computing should be curtailed. In Clarke's novel, for example, an onboard computer called HAL turns against the crew of the spaceship Discovery I, killing all but one of the astronauts. Asimov has suggested that such dangers might be avoided by programming machines to obey certain principles such as his *three laws of robotics*, stated in several of his stories and novels from 1950 onwards (including "The Bicentennial Man"), of which the first is that "a robot may not injure a human being or, through inaction, allow a human being to come to harm." Yet it is hard to see how such principles could be implemented in a computer program.

This list of issues raised by technologies of emotion is not exhaustive, but illustrates how a greater understanding of emotions impacts ethics. Science and technology have powerful emotional dimensions that are often ignored by those involved in developing them. Yet it is vital to think about these dimensions, because adverse emotional reactions to new technologies among the general public can have serious consequences. From the Luddites, eighteenth-century English artisans who destroyed machinery during the industrial revolution, to twenty-first century environmentalists who oppose the planting of genetically engineered crops, new technologies have often inspired deep feelings of mistrust. Those developing future technologies risk provoking similar reactions unless they engage the public at large in open and informed debate in which emotional dimensions are addressed as well as the scientific facts.

DYLAN EVANS

SEE ALSO *Artificial Intelligence; Emotional Intelligence; Evolutionary Ethics; Psychopharmacology; Neuroethics; Risk and Emotion; Robots and Robotics; Scientifc Ethics.*

BIBLIOGRAPHY

Asimov, Isaac. (1976). "The Bicentennial Man. In *The Bicentennial Man and Other Stories*. New York: Doubleday, 1976. In this story Asimov manages to explore many of the moral dilemmas of giving computers emotions more effectively than any non-fiction account.

Clarke, Arthur C. (1968). *2001: A Space Odyssey.* New York: ROC. A science-fiction classic, featuring HAL, the onboard computer.

Damasio, Antonio. (2003). *Looking for Spinoza: Joy, Sorrow and the Feeling Brain.* London: Heinemann. A prominent neuroscientist explores the distinction between emotions and feelings.

De Sousa, Ronald. (1991). *The Rationality of Emotion.* Cambridge, MA: MIT Press. A philosophical discussion of the role of emotions in rational behavior.

Evans, Dylan. (2001). *Emotion: The Science of Sentiment.* Oxford and New York: Oxford University Press. A good overall introduction to the science of emotion.

Fukuyama, Francis. (2002). *Our Posthuman Future: Consequences of the Biotechnology Revolution.* London: Profile Books. An emotional polemic against modern technology.

Griffiths, Paul E. (1997). *What Emotions Really Are: The Problem of Psychological Categories.* Chicago and London: University of Chicago Press. A philosophical discussion of the concept of emotion.

Kramer, Peter D. (1994). *Listening to Prozac*. London: Fourth Estate. A forceful advocate for the view that drugs can improve our emotional lives.

Picard, Rosalind W. (1997). *Affective Computing*. Cambridge, MA, and London: MIT Press. Discusses recent attempts to build emotional computers and robots.

EMOTIONAL INTELLIGENCE

• • •

Emotional Intelligence (EI) or Emotional Quotient (EQ) is a concept that challenges the assumption that the Intelligence Quotient (IQ) is the best predictor of professional success. Unlike IQ, which proposes to be a measurement of innate potential that is relatively stable, the proponents of EI maintain that it is a continuously developing ability, competency, or skill in which "the sky is the limit" (Segal 1997, p. 19). The same proponents claim that developing one's EQ is the key to succeeding in activities from academics, sales, customer service, and management to improving marriages, mental and physical health, lowering crime, and even an individual's spiritual relationship with God. Research on EI and attempts to apply it constitute extensions of science and technology into the ethical realm. In contrast the critics of EI argue that the concept is too all-encompassing, with EI measurements contributing little beyond existent constructs and its predictive claims largely unverified (Matthews et al. 2003).

The Scientific and Ethical Concept of EI

EI is conceptually related to Howard Gardner's (1985) theory of multiple intelligence, which criticizes the overemphasis on IQ and argues for the possibility of affective and social modes of intelligences. Peter Salovey and John Mayer (1990) first proposed the term *emotional intelligence* to describe a kind of ability to monitor, discriminate, and use the information of one's own and other's emotions to guide thinking and action. However it was Daniel Goleman's 1995 book *Emotional Intelligence* that popularized EI as a general capacity to motivate and persist at goals, to delay gratification, to regulate one's own emotions and those of others, to empathize, and to hope. In general the concept of EI is vague and there is no precision in attempts to clarify, define, or measure it. Some literature refers to EI as a type of sensitivity to emotions in self or others (Lam and Kirby 2002). Other literature understands it as an overarching term for any non-rational skill or ability, such as optimism, manners, empathy, or self-efficacy,

that contributes to social and professional success beyond rational skills (Brown 2003).

The underlying scientific theory of EI relies on research, such as that by Antonio Damasio (1994), on the neuropsychology of emotions. This research has challenged the idea that emotions are irrelevant or an impediment to rational decision-making. Instead it suggests that the emotional circuitry of the brain (i.e., the amygdala, cingulate gyrus, hippocampus, and ventromedial prefrontal region) is interconnected with the higher cognitive areas (i.e., the neocortex) and indispensable for rational and social decision-making. Damasio's book *Descartes' Error* examined patients with damage to areas associated with emotional processing and found that they could successfully engage in rational abstract tests, such as those that measure IQ, but were unable to make even trivial social decisions. This research has also shown that, although innate emotional responses can function independently, the neocortical area of the brain works with emotions and can modulate emotional responses to environmental circumstances. This degree of plasticity of emotions supports the claim of EI as a life-long developing capacity.

Conceptually EI also has implications for ethical theory and related educational policies. EI can trace its ethical roots to Aristotle's analysis of emotions in the *Nicomachean Ethics*. Aristotle's ethical theory relies on the development of ethical dispositions or character traits in which both reason and emotion are habituated to deliberately choose the ethical action. Certain ethical theorists, for example Martha Nussbaum (2001), also reject ethical theories that understand ethics as purely a rational activity; instead, similar to Aristotle, Nussbaum stresses the importance of emotions as an integral aspect of ethical judgment and normative appraisals. In this view, ethical development does not depend on rational evaluation, but relies on learning how to check impulses and using USE emotional information to guide behavior. The practical implication of this ethical theory has been to implement educational curriculum and staff training that emphasizes the development of EI skills (Goleman 1995, Brown 2003).

Review of Research

EI literature spans many disciplines from the popular psychology self-help genre that has virtually no scientific evidence for its claims, to more scientific analysis in neuropsychology, clinical psychology, education, management, business, and behavioral economics. In addition many collaborators in the field of psychometrics have devoted attention to developing reliable and consistent standards in the attempt to measure and explain individual differences in EI.

Similar to ethical theory, certain avenues of research in the social sciences, such as behavioral economics, reject standard models of human decision-making, such as utility theory, that minimize or ignore the role of emotions in decision-making (Sanfey et al. 2003). This research focuses on the analysis of EI as a relationship between rational and emotional processes in decision-making. Adopting methods, such as functional magnetic resonance imaging (fMRI) from neuroscience into game theory, behavioral economics seeks to explain how and why individuals will often reject a purely rational decision when this decision is seen as unfair. In management research, Brown (2003) has also examined the role of emotions in enhanced service provision and profitability. In other disciplines, such as political science, George Marcus and colleagues (2000) have attempted to understand the role of emotion in political learning and decision-making.

The typical research analyzing EI as a type of aptitude focuses mainly on developing psychometric tests to measure EI for both scientific understanding and potential commercial applications (Matthews et al. 2002). One of the most popular measurement is the performance test, such as the Multifactor Emotional Intelligence Scale (MEIS) or the modified Mayer-Salovey-Caruso Emotional Intelligence Test (MSCEIT) developed and tested by David Caruso, John Mayer, and Peter Salovey, which measures the management and regulation of emotions by predetermined consensual, expert, or target scoring. The main difficulty with predetermined criteria is that, unlike IQ tests that have definite right or wrong evaluations, EI criteria are open to criticism of personal and cultural norms. Another type of measurement is a simple self-reporting questionnaire of competency, such as the Emotional Quotient Inventory (EQ-i) developed and tested by Bar-On and collaborators. Other tests used to measure EI include Goleman's Emotional Competency Inventory (ECI) and Nicola S. Schutte and collaborators' Schutte Self-Report Inventory (SSRI). Although self-reporting tests are less costly than performance tests, they are highly susceptible to response bias due to respondent's lack of awareness or even deliberate attempts to reflect expected social norms.

Assessment

Despite the popular and commercial appeal of EI as holding an indefinite possibility to improve an individual's personal and private life, there is little scientific evidence for such claims. Many of the popular claims of EI proponents offer little more than commonsense advice, such as proposing that children who are taught manners are more liked by their teachers (Shapiro 1997) or standard yoga meditation techniques for calming emotions (Segal 1997). Beyond such problems of popular accounts, the main difficulty of a scientific understanding of EI is the lack of a clear, concise concept. EI is often a catch-all term of any list of qualities or character traits that could explain why individuals with high IQs do not necessarily succeed professionally or why those with lower IQs often are more successful. But a simple negative categorizing of EI as any trait that is not measured by IQ does not provide for any clear scientific evaluation of EI or its popular claims. In addition, as Gerald Matthews NAME] and collaborators (2003) point out, many of the valuable aspects of EI, such as those reliably measured by the psychometric tests, have much in common with already established personality tests. The concept of EI is vague, imprecise, and in many cases redundant.

The most valuable aspect of EI is when it is conceptually understood not as a character trait such as optimism or self-efficacy, but as a concept reflecting the importance of emotion as a type of cognition that functions together with reason in social and ethical decision-making. This understanding of EI connects it with research in the neuroscience of emotion, which has focused on understanding how the brain receives and processes information. Unlike the popular version of EI that conceives it as an ability to use, manage or, control emotions, this version of EI rejects the notion of any simple mastery over emotions. Instead EI represents emotions as making a cognitive contribution essential to practical, non-abstract decision-making.

This conceptualization of EI has implications for possible research in various disciplines, from decision-making in the social sciences to ethical theory. The concept of EI suggests that because emotions are involved in social decision-making, understanding emotions is an essential aspect to understanding political, economic, and other social behavior. In addition EI understood as a necessary aspect of cognitive decision-making has practical ramifications for developing education and training policy that include more than simply teaching abstract, rational knowledge. However, before any useful practical application of EI-based programs, more clarification of the concept and measurement tests need to be developed to avoid the problems of unevaluated claims or measurement redundancy.

MARLENE K. SOKOLON

SEE ALSO *Aristotle and Aristotelianism; Emotion; IQ Debate; Risk and Emotion.*

BIBLIOGRAPHY

Aristotle. (1934). *The Nicomachean Ethics*, trans. Harris Rackham. Cambridge, MA: Harvard University Press.

Brown, Randall B. (2003). "Emotions and Behavior: Exercises in Informational Intelligence." *Journal of Management Education* 27(1): 122–134.

Damasio, Antonio R. (1994). *Descartes' Error*. New York: Harper Collins Publishers. A highly readable challenge to the assumption that emotions necessarily hinder social and ethical decision-making. Drawing on the experiences of patients who have had damage to emotional processing areas of the brain, this neuroscientific work provides scientific evidence that without emotional responses human beings are unable to assess ethical and social dilemmas.

Gardner, Howard. (1985). *Frames of Mind: The Theory of Multiple Intelligences*. New York: Basic Books.

Goleman, Daniel. (1995). *Emotional Intelligence*. New York: Bantam Books. The classic account of emotional intelligence as self-awareness and impulse control. A clear explanation of the brain and behavioral research underlying EI and limitations of IQ as a predictor of success. Less compelling is the commonsense guidance for encouraging the development of emotional awareness.

Lam, Laura Thi, and Susan L. Kirby. (2002). "Is Emotional Intelligence an Advantage? An Exploration of the Impact of Emotional and General Intelligence on Individual Performance." *Journal of Social Psychology* 142(1): 133–143.

Marcus, George, et al. (2000). *Affective Intelligence and Political Judgment*. Chicago: University of Chicago Press.

Matthews, Gerald, et al. (2002). *Emotional Intelligence: Science and Myth*. Cambridge, MA: MIT Press.

Nussbaum, Martha C. (2001). *The Upheavals of Thought: The Intelligence of Emotions*. New York: Cambridge University Press. An exhaustive account of the role of emotions in ethical decision-making, touching on such wide-ranging subjects as classical and modern moral philosophy, anthropology, child development, music, and religion. The conclusion of the lengthy volume is that emotions facilitate rather than impede human morality.

Salovey, Peter, and John Mayer. (1990). "Emotional Intelligence." *Imagination, Cognition, and Personality* 9: 185–211.

Sanfey, Alan G., et al. (2003). "The Neural Basis of Economic Decision-Making in the Ultimatum Game." *Science* 300: 1755–1758.

Segal, Jeanne. (1997). *Raising Your Emotional Intelligence*. New York: Henry Holt and Company.

Shapiro, Lawrence E. (1997). *How to Raise a Child with High EQ*. New York: Harper Collins.

ENERGY

• • •

Energy, from the Greek *energeia* or activity, denotes the capacity of acting or being active. Aristotle used the term to denote the activity of tending toward or enacting a goal, which differs from the modern understanding of energy as the capacity to do work. To a certain degree energy functions as the abstract equivalent of fire, one of the Aristotelian four elements. The modern concept of energy can engender either physical or psychological activity and be analyzed in one or more of three senses: scientific, technological, and ethical.

Science of Energy

In modern science, the term *energy* has become a precise technical concept with such distinctions as kinetic (energy related to the motion of a body) and potential (stored energy of position). Other important distinctions pertain to the different forms of energy, including thermal, mechanical, electrical, chemical, radiant, and nuclear.

The history of the modern science of energy reveals that developing a precise technical concept of energy is a convoluted process, one that raises controversial tensions between constructivist and realist interpretations of scientific knowledge (Crease 2004). To what extent did the phenomenon of energy precede the development of the concept itself? And to what extent do the cultural and technological contexts in which energy came to be represented actually shape that natural phenomenon in terms of intersubjective agreement? The modern concept of energy arose through both purely ahistorical theories and a changing social context, marked especially by the development of different energy technologies. This means that the contexts of discovery and justification cannot be isolated from one another, because energy cannot be justified without the use of historically given concepts (e.g., work and heat) and technologies (e.g., steam engines). Energy is at once real (i.e., not an artifact of language and culture) and constructed (i.e., inextricably embedded in human history).

The modern science of energy originated with the development of thermodynamics in the nineteenth century and efforts to understand the dynamics of steam engines and other mechanical devices. In 1842 Julius Robert von Mayer (1814–1878) calculated the caloric equivalent of mechanical work. This *Kraft* (force or power) was the precursor of energy as a scientific concept that denoted the quantitative equivalence between physiological heat and mechanical work. By the mid-nineteenth century, it was experimentally well established that such physical phenomena as electricity, heat, electromagnetism, and even light were interconvertible at determinate rates of exchange (Kuhn 1959). To German scientists in particular, the fixed rates of exchange governing the conversion of diverse phenomena suggested the existence of a single underlying substance. They postulated a metaphysical *Arbeitskraft* (workforce) behind physical manifestations.

In 1847 Hermann von Helmholtz (1821–1894) formulated the first law of thermodynamics by stating that *Arbeitskraft* can be neither created nor destroyed. So enshrined in the "law of energy conservation," energy denotes an unknowable substance manifest in the transformations of matter and measurable in units of work. Rudolf Clausius (1822–1888) formulated the second law of thermodynamics on the notion of entropy (a measure of disorder or the quality of energy) in 1850. The scientific concept of heat was reduced to the kinetic energy of theoretically postulated particles and divorced from the commonly experienced primal element of fire.

Work by Bernhard Riemann (1826–1866) and others further removed the concept of energy from common experiences, but Ernst Mach (1838–1916) argued that energy and other concepts in physics ought to be grounded in practical and experimental experience rather than theoretical abstractions. Albert Einstein (1879–1955) utilized Riemann's mathematically constructed curved "space-time" to formulate an "energy-momentum tensor" according to which mass and energy are interconvertible in the equation $E = mc^2$;, where c is the speed of light. This means that a small amount of matter (mass) is the equivalent of a large amount of energy, so that matter can be thought of in scientific terms as frozen energy.

Thus, E began as a principle of equivalence between the phenomena of physiological heat and mechanical work. First forged as a bridge between incommensurable domains, E slowly shed any reference to everyday experience. The scientific elaboration of an insensible E occurred through the interplay of mathematically formulated theories and controlled experiments set within evolving social and technological contexts.

Technologies of Energy

As an engineering concept, energy may be related to the primal element of fire, and insofar as fire has played a key role in civilizing human beings (as described in the myth of Prometheus), so energy development is described as central to human progress.

Although water mills and windmills have been in use for well over a thousand years, ancient and medieval technologies of energy were primarily animate (human and animal) in nature. Indeed many in the ancient Greek world viewed slavery as an indispensable means of providing the necessities of a civilized life. The domestication of draft animals roughly 10,000 years ago spurred the agricultural revolution. The transition from wood to coal, made first in England beginning in the sixteenth century, heralded vast social and technological changes. Coal powered the Industrial Revolution and its attendant energy technologies, especially the steam engine. Oil and natural gas were developed extensively in the nineteenth century, and nuclear energy for civilian and military purposes developed after World War II. These changes have led to the widespread use of modern energy technologies, including the heat engine, fossil fuel and nuclear-powered electricity generating plants, and dams, wind turbines, photovoltaic cells, and other forms of renewable energy generation.

The use of these technologies raises important distinctions among the terms *energy*, *power*, and *work* in their mechanical or technical senses. Energy (E) is the capacity for doing work. Work (W) is defined as the energy transferred to an object by a force as that object moves; it is the result of converting energy from one form to another. Power (P) is the rate at which work is done, that is, the rate at which energy is converted. So, $E = Pt$, and $P = dE/dt$, where t is time. In terms of electricity generation and consumption, the most common units for power (demand or capacity) are the watt (equal to one joule per second) and kilowatt, and the most common unit for energy (consumption) is the kilowatt-hour. For example, a 100 watt lightbulb left on for ten hours will use 1 kilowatt-hour of energy.

Power and energy are central to the classical definition of engineering, which the English architect and engineer Thomas Tredgold (1788–1829) formulated as "the art of directing the great sources of power in nature for the use and convenience of man." This highlights the fundamental human condition that in order to accomplish one's ends, energy must be exerted. The hardships endured have long fueled the utopian dream of infinite energy availability. Modern engineering has undoubtedly unlocked vast stores of energy for human use and convenience. But the quest for limitless energy has yielded dangers in the form of pollution and threats of nuclear war. This quest is apparent in the past hoax and future hope of cold fusion and the development of renewable energy technologies to replace nonrenewable forms.

Ethics and Politics of Energy

Engineer and physicist William Rankine (1820–1872) popularized *energy* as a technoscientific term in the mistaken belief that the Greek *energeia* meant work. In fact, in contradistinction to slave labor and craftwork, *energeia* originally indicated political and moral activity (Arendt 1958). But once the term was defined scientifically in the early 1800s as the power to do work, the lived meaning was relegated, against its own etymology, to secondary or metaphorical status. References to personal energy, psychic energy (e.g., Sigmund Freud's libido), spiritual energy (e.g., Hindu *prana*, Hebrew *rauch*, and Daoist *qi*), aesthetic energy, social or political energy, and more are all thought of as less rather than more concrete, and often interpreted in technological terms. Thus the meaning of the term was somewhat purged of its original ethical and political connotations.

But contemporary issues surrounding energy extraction and use have refocused attention on the fundamental connection between energy and ethics. Energy cannot be considered a neutral instrument, but rather an integral component of political and ethical ends. As the Industrial Revolution and countless other events in history demonstrate, the availability and use of different energy sources reciprocally interacts with social and technological developments. One major practical consequence of this derives from the heterogeneous global distribution of energy reserves (e.g., oil fields) and the unequal demands for energy consumption. Stores of energy and the resulting wealth generated by their extraction and sale can contribute to unequal wealth distribution, violence, war, corruption, and coercion both within and between nation-states.

Within this context, national energy policies inevitably manifest ethical values about distributive justice, health, and equity and raise geopolitical concerns about national security. The disproportionate energy consumption by developed countries causes transboundary environmental problems. Most controversially, the carbon dioxide produced from the combustion of fossil fuels contributes to rising sea levels, which negatively affect many developing countries that have not benefited from the goods and services provided by those fuels. Many of these countries cannot afford the adaptation measures necessary to mitigate their vulnerability, and the question becomes to what extent developed nations are responsible for helping the rest of the world cope with the consequences of their large energy appetites. Another political and ethical dilemma posed by proposals to shift away from fossil fuels is the status of nuclear energy. Do its attendant risks and benefits present an acceptable tradeoff as a transitional source of energy in the move from fossil fuels to renewables?

Questioning the dominant assumption that social progress depends on increases in per capita energy consumption raises deeper ethical issues about the good life. It is commonly believed that high civilization depends on high energy use, which explains the modern quest for new and greater reserves of energy. There is a correlation between quality of life, as measured by the Human Development Index, and per capita energy consumption, but this is not a linear relationship. Indeed the improvement in quality of life levels off when per capita electricity consumption equals 4,000 kilowatt-hours. Yet some countries have per capita consumptions over 20,000 kilowatt-hours. This relates to issues in development ethics (e.g., neocolonialism and cultural homogenization), because metrics of progress are often tied to energy consumption.

Although the rise of "energy slaves" (the use of mechanical or inanimate energy sources to replace animate forms) has brought enormous benefits (including the replacement of human slaves), it has also created risks and concerns about environmental sustainability. Furthermore it contributes to the questionable assumption that living well requires increasing dependence on these energy slaves. A. R. Ubbelohde (1955) characterized the modern ideal society as based on a large proportion of inanimate energy slaves as the "Tektopia." The Tektopia brings both new possibilities and new moral dilemmas resulting from such factors as increased luxury, changes in the administrative state, displacement of workers by machines, and difficulty in controlling, regulating, and distributing energy.

Ivan Illich (1974) also critiqued this image of the good life by noting that as the number of energy slaves increases, so rises not only inequity but also social control and personal stress, alienation, and meaninglessness. He challenged the energy crisis focus "on the scarcity of fodder for these [energy] slaves," preferring instead "to ask whether free men need them" (p. 4). He argued that energy policies (whether capitalist or socialist) focused on high energy consumption will lead to technocracies that degrade cultural variety and diminish human choice. For Illich, "only a ceiling on energy use can lead to social relations that are characterized by high levels of equity.... Participatory democracy postulates low-energy technology" (p. 5). Beyond a certain threshold, increased energy affluence can come only through greater concentration of control, and thus greater inequality.

Failure to differentiate the technoscientific concept of energy from its older political meaning can lead to dangerous ideologies that reduce the plural, lived energies of human interaction to manipulable technical constructs. People are reflected as mere human motors in the mirror of energy slaves (Rabinbach 1990). The technical notion of energy begins to blur distinctions between nature and machines, living organisms and persons, mechanical work and human action. Efficiency subverts more human goals. The resulting blindness to the distinction between the technoscientific and political versions of energy partially maimed moral judgments about the use of the atomic bomb. Consideration of ethical and political issues associated with energy thus becomes an opportunity to redistinguish what may have been improperly united: energy as a basic concept in science, as a resource, and as an ethical issue.

JEAN ROBERT
SAJAY SAMUEL

SEE ALSO *Automobiles; Einstein, Albert; Fire; Oil.*

BIBLIOGRAPHY

Arendt, Hannah. (1958). *The Human Condition.* Chicago: University of Chicago Press.

Crease, Robert P. (2004). "Energy in the History and Philosophy of Science." In *Encyclopedia of Energy,* ed. Cutler J. Cleveland. Amsterdam: Elsevier.

Illich, Ivan. (1974). *Energy and Equity.* New York: Harper and Row.

Kuhn, Thomas S. (1959). "Energy Conservation as an Example of Simultaneous Discovery." In *Critical Problems in the History of Science,* ed. Marshall Clagett. Madison: University of Wisconsin Press.

Rabinbach, Anson. (1990). *The Human Motor: Energy, Fatigue, and the Origins of Modernity.* New York: Basic.

Ubbelohde, A. R. (1955). *Man and Energy.* New York: George Braziller.

ENGINEERING DESIGN ETHICS

• • •

Engineering design ethics concerns issues that arise during the design of technological products, processes, systems, and services. This includes issues such as safety, sustainability, user autonomy, and privacy. Ethical concern with respect to technology has often focused on the user phase. Technologies, however, take their shape during the design phase. The engineering design process thus underlies many ethical issues in technology, even when the ethical challenge occurs in operation and use.

Engineering Design

Engineering design is the process by which certain goals or functions are translated into a blueprint for an artifact, process, system, or service that can fulfill these functions. The function of cutting bread, for example, can be translated into a knife. A car fulfills the function of transportation. Engineering design is different from other forms of design—such as fashion design or the design of policy—in that it results in artifacts and systems grounded in technical knowledge.

The character of the engineering design process has been much debated, but for present purposes it may be described as an iterative process divided into different phases. The following phrases are the simplest and most accepted (Pahl and Beitz 1996):

- Problem analysis and definition, including the formulation of design requirements and the planning for the design and development of the product, process, system, or service.

- Conceptual design, including the creation of alternative conceptual solutions to the design problem, and possible reformulation of the problem.

- Embodiment design, in which a choice is made between different conceptual solutions, and this solution is then worked out in structural terms.

- Detail design, leading to description that can function as a guide to the production process.

In each phase, engineering design is a systematic process in which use is made of technical and scientific knowledge. This process aims at developing a solution that best meets the design requirements. Nevertheless, the final design solution does not simply follow from the initially formulated function because design problems are usually ill-structured. Nigel Cross (1989) has argued that proposing solutions often helps clarify the design problem, so that any problem formulation turns out to be partly solution-dependent. It is impossible to make a complete or definite list of all possible alternative solutions to a problem. It is also extremely difficult to formulate any criterion or set of criteria with which alternatives can be ordered on a scale from "good" or "satisfactory" to "bad" or "unsatisfactory," even though any given feature of the design may be assessed in terms of some given criterion such as speed or efficiency.

Ethical Issues

Design choices influence how ethical issues are addressed in technology. Because such choices are differentially manifested in the different phases of the design process, ethical issues themselves take on distinctive forms in each case.

PROBLEM FORMULATION. Problem definition is of special importance because it establishes the framework and boundaries within which the design problem is solved. It can make quite a difference—including an ethical difference—from whose point of view a problem is formulated. The problem of designing an Internet search engine looks different from the perspective of a potential user concerned about privacy than from the perspective of a provider concerned about selling banner advertisements. The elderly or physically disabled will have different design requirements than the young or healthy.

An important ethical question in this phase concerns what design requirements to include in the problem definition. Usually design requirements will be based on the intended use of the artifact and on the desires of a client or user. In addition, legal requirements and technical codes and standards play a part. The latter may address, if only implicitly, ethical issues in relation to safety or environmental concerns. Nevertheless, some ethical concerns may not have been adequately translated into design requirements. Engineering codes of ethics, for example, require that engineers hold "paramount the safety, health and welfare of the public," an obligation that should be translated into design requirements.

The idea that morally relevant values should find their way into the design process has led to a number of new design approaches. An example is eco-design or sustainable design, aimed at developing sustainable products (Stitt 1999). Another example is value-sensitive design, an approach in information technology that accounts for values such as human well-being, human dignity, justice, welfare, and human rights throughout the design process (Friedman 1996).

Ethical issues may arise as well during the operationalization of design requirements. Take for example a design criterion such as minimizing global warming potential, which may arise from a moral concern about the greenhouse effect. The global warming potential of substances can be measured on different time scales potentially resulting in different rankings of these substances (Van de Poel 2001). The choice of different time scales is ethically relevant because it relates to the question of how far into the future the current generation's responsibility extends.

CONCEPTUAL DESIGN. Design is a creative process, especially during the conceptual phase. In this phase the designer or design team thinks out potential solutions to a design problem. Although creativity is not a moral virtue in itself, it is nevertheless important for good design, even ethically. Ethical concerns about a technology may on occasion be overcome or diminished by clever design.

One interesting example is the design of a storm surge barrier in the Eastern Scheldt estuary in the Netherlands (Van de Poel and Disco 1996). In the 1950s, the government decided to dam up the Eastern Scheldt for safety reasons after a huge storm had flooded the Netherlands in 1953, killing more than 1,800 people. In the 1970s, the construction plan led to protests because of the ecological value of the Eastern Scheldt estuary, which would be destroyed. Many felt that the ecological value of the estuary should be taken into account. Eventually, a group of engineering students devised a creative solution that would meet both safety and ecological concerns: a storm surge barrier that would be closed only in cases of storm floods. Eventually this solution was accepted as a creative, although more expensive, solution to the original design problem.

EMBODIMENT DESIGN. During embodiment design, one solution concept is selected and worked out. In this phase, important ethical questions pertain to the choice between different alternatives.

One issue is tradeoffs between various ethically relevant design requirements. While some design requirements may be formulated in such terms that they can be clearly met or not —for example, that an electric apparatus should be compatible with 220V—others may be formulated in terms of goals or values that can never be fully met. Safety is a good example. An absolutely safe car does not exist; cars can only be more or less safe. Such criteria as safety almost always conflict with other criteria such as cost, sustainability, and comfort. This raises a question about morally acceptable tradeoffs between these different design criteria. Is there a minimum level of safety each automobile should meet, or is it acceptable to design less safe cars if they are also cheaper?

Formal engineering methods—such as cost-benefit analysis and multiple criteria design analysis—exist to deal with design criteria tradeoffs. The question, however, is whether these methods result in morally acceptable tradeoffs. These methods often treat different design criteria and the moral values on which they are based as if they are commensurable, which may be problematic.

Alternative designs cannot only be compared in terms of the original design criteria, but also in terms of the risks they imply. In engineering, a host of methods exist to assess the risks of new technologies, and increasingly such methods also inform design choices. In general, one may prefer a design with minimal risks, but the acceptability of risks also depends on such issues as their distribution and the degree to which they are accepted voluntarily (Shrader-Frechette 1991). Free and informed consent can be an issue in engineering design, just as in the design of medical research experiments with human subjects.

Whereas an evaluation in terms of risks usually focuses on minimizing potential harm or justly distributing potential harm, other evaluations may focus on the possibility of doing good. An approach that may prove interesting in this respect focuses on the so-called "scripts" of technological artifacts. Authors such as Bruno Latour have used the notion of a script to describe the built-in use and moral presuppositions of an artifact (Latour 1992). The automatic or passive seat belt is a case in point. This artifact contains a script that forces the driver to use the seat belt before the car engine can be started, which raises an interesting ethical question. To what degree is it acceptable to limit user autonomy in order to achieve other moral goods such as safety? It is usually argued that a failure to use a seat belt will impose hardships and costs to others in the event of an accident.

DETAIL DESIGN. During detail design, a design solution is further developed, including the design of a production process. Examples of ethical issues addressed at this phase are related to the choice of materials: Different materials may have different environmental impacts or impose different health risks on workers and users. Choices with respect to maintainability, ability to be recycled, and the disposal of artifacts may have important impacts on the environment, health, or safety. The design of the production process may invoke ethical issues with respect to working conditions or whether or not to produce the design, or parts of it, in low-wage countries.

Design as a Social Process

Engineering design is usually not carried out by a single individual, but by design teams embedded in larger organizations. The design of an airplane includes hundreds of people working for several years. Organizing such design processes raises a number of ethical issues.

One is the allocation of responsibilities. What is the best way to allocate responsibility for safety in the design process? One option would be to make someone in particular responsible. A potential disadvantage of this solution is that others—whose design choices may be highly relevant—do not take safety into account. Another approach might be to make safety a common responsibility, with the danger that no one in particular feels responsible for safety and that safety does not get the concern it deserves.

A second issue is decision-making. During design, many morally relevant tradeoffs have to be made. Sometimes such decisions are made explicitly, but many times they occur implicitly and gradually, evolving from earlier decisions and commitments. Such patterned decision making may lead to negative results that never would have been chosen if the actors were not immersed in the problematic decision-making pattern (Vaughan 1996). This raises ethical issues about how to organize decision making in design because different arrangements for making decisions predispose different outcomes in ethical terms (Devon and van de Poel 2004).

A third issue is what actors to include. Engineering design usually affects many people with interests and moral values other than those of the designers. One way to do right to these interests and values is to give different groups, including users and other stakeholders, a role in the design and development process itself. Different approaches have been proposed to this issue, such as participatory design in information technology development (Schuler and Namioka 1993). Constructive technology assessment likewise aims to include stakeholders in the design and development process in order to improve social learning processes at both the technical and normative levels with respect to new technologies (Schot and Rip 1997).

As the heart of the process of technological development and future use, engineering design must likewise be at the core of ethical reflection on technology. Major ethical issues in engineering design include what requirements, values, and actors to include in the design process and how to trade off different requirements and values. Major issues also arise with respect to organizing the design process in such a way that moral responsibilities are adequately and fairly allocated.

IBO VAN DE POEL

SEE ALSO *Design Ethics; Engineering Ethics.*

BIBLIOGRAPHY

Cross, Nigel. (1989). *Engineering Design Methods.* Chichester, UK: Wiley. Discusses engineering design in general and the ill-structured nature of design problems.

Devon, Richard, and Ibo van de Poel. (2004). "Design Ethics: The Social Ethics Paradigm." *International Journal of Engineering Education.* 20(3): 461–469. Discusses ethical issues related to design as a social process.

Friedman, Batya. (1996). "Value-Sensitive Design." *Interactions* 3: 17–23. An introduction to the value-sensitive design approach in information technology.

Latour, Bruno. (1992). "Where Are the Missing Masses?" In *Shaping Technology/Building Society; Studies in Sociotechnical Change,* ed. Wiebe Bijker and John Law. Cambridge, MA: MIT Press. Introduces the notion of script or built-in morality in artifacts.

Pahl, G., and W. Beitz. (1996). *Engineering Design: A Systematic Approach,* 2nd edition, trans. K. Wallace, L. Blessing, and F. Bauert. London: Springer-Verlag. A general work on engineering design and the phases of the design process.

Schot, Johan, and Arie Rip. (1997). "The Past and Future of Constructive Technology Assessment." *Technological Forecasting and Social Change* 54: 251–268. An introduction to the approach of constructive technology assessment.

Schuler, Douglas, and Aki Namioka, eds. (1993). *Participatory Design: Principles and Practices.* Hillsdale, NJ: Lawrence Erlbaum.

Shrader-Frechette, Kristin S. (1991). *Risk and Rationality: Philosophical Foundations for Populist Reform.* Berkeley: University of California Press. Deals with ethical issues with respect to risk assessment and risk acceptance.

Stitt, Fred A., ed. (1999). *Ecological Design Handbook: Sustainable Strategies for Architecture, Landscape Architecture, Interior Design, and Planning.* New York: McGraw-Hill.

Van de Poel, Ibo. (2001). "Investigating Ethical Issues in Engineering Design." *Science and Engineering Ethics* 7: 429–446. Overview and illustration of main ethical issues in engineering design.

Van de Poel, Ibo, and Cornelis Disco. (1996). "Influencing Technology: Design Worlds and Their Legitimacy." In *The Role of Design in the Shaping of Technology,* ed. Jacques Perrin and Dominique Vinck. Luxembourg: Office for Official Publications of the European Communities. On social controversies about engineering designs. Contains the Eastern Scheldt example.

Vaughan, Diane. (1996). *The Challenger Launch Decision.* Chicago: The University of Chicago Press. Detailed description of design and research decisions eventually leading to the Challenger disaster.

ENGINEERING ETHICS

• • •

Overview
Europe

OVERVIEW

Engineering ethics is concerned with the ethical responsibilities of engineers, both as individual practitioners and organizational employees, and as members of a profession with obligations to the public. The issues in engineering ethics range from micro-level questions about the everyday practice of individual engineers to macro-level questions about the effects of technology on society (Herkert 2001). Because engineers are the primary creators of science-based technology, engineering ethics is one of the most important intersections between science, technology, and ethics.

Development of Engineering and Engineering Ethics

Compared to the clergy, law, and medicine, engineering is a relatively young profession, having acquired something like its present form in France in the eighteenth century. In the United States, the United States Military Academy at West Point graduated its first engineers in 1817. The first private engineering college in the United States was Rensselaer Polytechnic Institute, founded in 1823. By the mid-nineteenth century, the land grant colleges in the United States had programs in civil engineering. In 1850, the first year the United States census counted engineers, only one in 10,000 persons identified themselves as engineers (for 2,000 total). By 1900, however, the numbers were increasing dramatically and the fields of engineering multiplying because of new discoveries and inventions in electricity, power generation, chemical processing, automobile development, and flight. The emerging large corporations also required increasing numbers of engineers. At the end of the twentieth century, about one in one hundred Americans was an engineer (Davis 1998).

Codes of ethics appeared in England in the middle of the nineteenth century and in the United States early in the twentieth century. In 1912 the American Society of Mechanical Engineers (ASME) proposed to the American Society of Civil Engineers (ASCE) and the Institute of Electrical Engineers (IEE) that a code for all three societies be constructed. The attempt was unsuccessful due to differences in the disciplines and their different relationships to business. The societies agreed that a code of ethics was desirable, and each society wrote its own. Not surprisingly, the codes had many similarities (Layton 1986).

Early codes focused on such issues as limiting professional advertising, protecting small businesses and consulting firms from underbidding, and the primacy of the obligation of engineers to their clients and employers. After several decades of relative neglect of the codes, a major change occurred in 1974, when the Engineers' Council for Professional Development (ECPD) adopted a new code of ethics that held that the paramount obligation of engineers was to the health, welfare, and safety of the public. Virtually all engineering codes of the early twenty-first century identify this as the primary obligation of engineers, not the obligation to clients and employers.

The emergence of engineering ethics as an academic subject also began in the 1970s. From this period to the present, there has been a growing emphasis on including engineering ethics in some form in the engineering curriculum. The emergence and continuing growth of this new discipline is due to a number of factors. One is a series of high-profile disasters, such as the problems of the Ford Pinto and the crash of the DC-10 outside Orly Field in Paris in 1974. In the intervening years, such events as the *Challenger* and *Columbia* space shuttle disasters have reinforced the need for engineers to be both technically competent and ethically responsible.

In 1985, the Accreditation Board for Engineering and Technology (ABET, Inc.), which accredits engineering colleges, reached a decision to require engineering programs to provide students with "an understanding of the ethical characteristics of the engineering profession and practice," supplying still more impetus to the development of engineering ethics. The ABET 2000 requirements were even more specific with regard to the ethics dimension of engineering education, requiring engineering graduates to have not only an understanding of ethical and professional issues related to the practice of engineering, but also an understanding of the impact of engineering on larger social issues.

Finally, the increased emphasis on ethics in large business organizations, where most engineers work, has also reinforced the importance of engineering ethics. Ethics codes have proliferated in business organizations, as has the creation of "ethics officers" to interpret and implement the codes. In 1992 the Ethics Officers Association (EOA) was founded. The organization had almost 900 organizations as members at the beginning of its second decade. Business organizations may increasingly expect engineers to have some knowledge and sophistication in the area of ethics and professionalism.

In order to promote the development of the emerging field of engineering ethics and to develop material for classroom use, in the late 1970s both the National Endowment for the Humanities (NEH) and the National Science Foundation (NSF) sponsored a series of workshops to develop teaching materials and provide pedagogical advice for faculty who wanted to introduce engineering students to ethics. Led by Robert Baum and Vivian Weil, these workshops brought together engi-

neering faculty and ethics teachers. One early fruit of these collaborations was the first edition of the textbook *Ethics in Engineering* (1996) by philosopher Mike Martin and engineer Roland Schinzinger, who came as a team to Baum's NEH workshop.

Because much of the impetus for the development of engineering ethics as an academic area came from the need for educational materials, some early publications focused on teaching. For example, Robert Baum's monograph, *Ethics and Engineering* (1983) included a statement of the goals of ethics education endorsed by a large group of educators across the curriculum who, sponsored by the Hastings Center, met over a three-year period to discuss the goals of ethics instruction in higher education. Adapted to each academic area, the five goals were:

1. to stimulate the moral imagination of students;
2. to help students recognize ethical issues;
3. to help students analyze key moral concepts and principles;
4. to stimulate a sense of moral responsibility; and
5. to help students deal constructively with moral ambiguity and disagreement.

Case studies have proven one of the most popular and effective ways of pursuing these goals. Since its early support of Vivian Weil's workshop, the NSF has consistently funded engineering ethics projects, particularly those designed to develop case studies for classroom use. In addition to Martin and Schinzinger, the first editions of a number of engineering ethics textbooks followed Baum's monograph. There was Unger (1994), Harris, Pritchard, and Rabins (2000), Whitbeck (1998), and Fleddermann (1999). Baum and Flores (1983), Schaub and Pavlovic (1983), Johnson (1991), and Vesilind and Gunn (1998) have published anthologies in engineering ethics, and Davis (1998) and Cook (2003) have published single-authored texts on aspects of engineering ethics.

Articles on engineering ethics began appearing frequently in engineering periodicals and philosophical journals such as *Business and Professional Ethics* and *Professional Ethics*. In 1995 *Science and Engineering Ethics*, a periodical that regularly publishes articles across a wide spectrum of issues in engineering ethics, began publication. With the support of NSF, Caroline Whitbeck initiated the Online Center for Ethics in Science and Engineering, which includes diverse resources for engineering ethics educators.

Although the emergence of engineering ethics as an academic area is especially evident in the United

States, serious interest is by no means confined to it. The editorial board of *Science and Engineering Ethics* is represented by Canada, the United Kingdom, Russia, Germany, Poland, Romania, Italy, Norway, France, Belgium, Sweden, and Japan, and it has had guest editors from the Netherlands. European educators have collaborated to produce a volume edited by Philippe Goujan and Bertrand Heriard Dubreuil (2001). The Martin and Schinzinger and Harris, Pritchard, and Rabins texts have been translated into Japanese. Shuzo Nakamura has also published an original textbook in Japanese, *Practical Engineering Ethics* (2003).

The rise of engineering ethics is not without its critics. Engineer Samuel Florman agrees that engineers should avoid being inaccurate, careless, or inattentive. For him, engineering ethics is about reliability; people count on engineers to do their work well and not make mistakes. However, cautions Florman, "We do not leave it to our soldiers to determine when we should have war or peace. Nor do we leave it to our judges to write our laws. Why, then, should we want our engineers to decide the uses to which we put our technology?" (Florman 1983, p. 332). Responses to Florman typically claim that engineers are in the best position to inform the public about the possible uses and likely consequences of technology, to alert employers and (if necessary) the public of defects and possible disasters associated with technology, to participate in the setting of engineering standards, and to help investigate problems, such as the *Challenger* and *Columbia* space shuttle disasters, or the collapse of the World Trade Towers in New York City on September 11, 2001. This does not necessarily mean that it should be left to engineers to decide all the uses for a technology. It only means that responsible decisions require information that engineers are in the best position to provide.

Topics in Engineering Ethics

Engineering experience as well as public responses to technological developments to which engineers contribute raise topics in engineering ethics. A review of key issues easily begins with the codes of ethics of professional engineering societies, which attempt to identify the major areas of ethical concern for engineers. Reflection on the nature and function of the codes themselves has itself produced considerable discussion. Some writers argue that the codes are coercive and should therefore be thought of as codes of conduct rather than codes of ethics (Ladd 1991, Luegenbiehl 1991). Others think of codes of ethics as guides and expressions of commitment that enable engineers, their clients, and the public to

know what to expect rather than instruments of coercion (Davis 1998, Unger 1994). Even so, there are issues about the range of applicability of codes. Professional societies adopt engineering codes of ethics, but most engineers do not belong to professional societies. Do the standards, rules, principles, and ideals contained in the codes still apply to them?

A related issue is professional registration. Most U.S. engineers do not have the Professional Engineer (P.E.) license. This means that most engineers cannot cite the possibility of losing their P.E. registration as a way to resist pressures to engage in unethical conduct. There is considerable resistance in the engineering profession to making the P.E. license mandatory. Should the requirements for engineering registration be changed to make licensure more acceptable to most engineers? Short of P.E. registration, are there other ways of ensuring quality in engineering work and protecting engineers from undue pressure to be unethical?

As has already been noted, prior to the 1970s most engineering codes of ethics held that the first obligation of an engineer is loyalty to a client or employer. The codes said little about obligations to the public. By the turn of the twenty-first century, most codes gave pride of place to the so-called paramountcy clause, which requires engineers to hold paramount the safety, health, and welfare of the public. However, there has been surprisingly little discussion of what, specifically, this requires engineers to do. Most attention has focused on whether whistle-blowing is either morally required, or at least permissible, when violations of the paramountcy clause are observed (DeGeorge 1981, James 1995, Davis 1998).

The issue of whistle-blowing has been central to some classic cases in engineering ethics, such as the Bay Area Rapid Transit case (Anderson, Perucci, Schendel et al. 1980), the DC-10 case (Fielder and Birsch 1992) and, above all, the *Challenger case* (Boisjoly 1991, Vaughn 1996). Important as such cases are, however, they touch on only one aspect of engineers' responsibility for public safety, health, and welfare. Whistle-blowing typically occurs only when something bad is imminent or has already occurred. The codes have little, if anything, to say about engineers' attempting to anticipate and resolve problems before they get out of hand. This deficiency is also reflected in the engineering ethics literature, which tends to focus on wrongdoing and its prevention, rather than on steps that should be taken to promote public safety, health, and welfare.

Questions involving conflicts of interest produce dilemmas for engineers, especially those in private practice (Davis 1998). A conflict of interest in the professions is a situation in which some professional or personal interest threatens to interfere with professional judgment, rendering that judgment less trustworthy than it might otherwise be. One of the topics that often arises in discussions of conflicts of interest is accepting gifts and bribes. An offer of a bribe creates a conflict of interest, because it may corrupt professional judgment, even when rejected. While it may be easy to say that accepting bribes is unethical, offers of gifts and favors from vendors can produce more subtle dilemmas. Such offers are likely to pose the first ethical issues that engineers face in their professional careers. These issues lend themselves especially well to treatment by the method of casuistry. For example, a case where accepting a gift from a vendor would usually be considered permissible (such as accepting a cheap plastic pen) and a case where accepting a gift from a vendor would usually be considered impermissible (such as accepting a gift worth several thousand dollars) can be compared with a more difficult case. By determining whether the case in question is more analogous to the permissible or impermissible case, the engineer can decide on the moral status of the questionable case (Harris, Pritchard, and Rabins 2000). Of course, identifying legitimate and illegitimate cases will in part be guided by the particular culture in which one is working.

The issue of confidentiality arises most commonly for engineers in private practice (Armstrong 1994). Although engineers ordinarily owe strong obligations of confidentiality to clients, the primacy of the obligation to the safety, health, and welfare of the public can be overriding in some situations. Suppose an engineer is hired by a client to assess the structural soundness of a building and finds fundamental flaws that threaten the safety of the present occupants. The engineer may be obligated to violate engineer/client confidentiality in order to inform authorities or tenants of the danger. Again, the method of casuistry can be used effectively to deal with troublesome cases of confidentiality.

Computer ethics is a rapidly developing area of interest, raising a host of questions, such as the control of pornography and spam, privacy, intellectual property rights, the legitimacy of sending unsolicited and unwanted cookies, the proper uses of encryption, selling monitoring software to totalitarian states, the proper uses of Social Security numbers, national ID cards, identity theft, whether Internet sites for making bombs or holocaust denial should be allowed, the legitimacy of downloading music, and software piracy. Interesting conceptual issues can be raised about the status of such entities as computer programs. Are they more like books, where copyright would be the appropriate form of protection, or like inventions, where patents would be the more appropriate form of protection? (Johnson 2000, Johnson and Nissenbaum 1995).

Engineers have more effect on the environment than any other professional group; yet engineers are only gradually assuming environmental responsibilities. Provisions relating to engineers' responsibility for the environment appeared only in the codes of the Institute of Electrical and Electronics Engineers (IEEE), the American Society of Civil Engineers (ASCE), and the American Society of Mechanical Engineers (now ASME International). Vesilund and Gunn (1998) explore a number of religious and philosophical bases for engineers' directly embracing environmental concerns, and Gorman, Mehalik, and Werhane (2000) have published a wide range of case studies that pose environmental challenges for engineers. For those who support the notion that engineers have direct responsibility for the environmental effects of their work, the basis and extent of that responsibility is still under debate. A key question is whether accepting responsibility only in areas where there is a clear threat to the health or well-being of human beings is sufficient, or whether a concern for the environment for its own sake is needed.

Another area where engineering work directly affects the public is in the imposition of risk as a result of technology. Martin and Schinzinger have suggested that engineering work is a kind of social experimentation and, as such, imposes risks on those on whom the "experiment" is performed, namely the public. What is acceptable risk? Who should determine it? Answers to the first question strongly affect answers to the second. Scientists and engineers tend to take a somewhat consequentialist or utilitarian approach. Defining risk as the product of the probability and magnitude of harm, they find a risk acceptable if the potential benefits outweigh the potential harms. Because they believe the public is often irrational and ill-informed about risk, scientists and engineers may be inclined to say that the determination of acceptable risk should be left to them. Representatives of the public, however, tend to link acceptable risk to free and informed consent and the equitable distribution of risks and benefits. This position is more congruent with an approach that emphasizes respect for individual rights (Shrader-Frechette 1985, 1991).

Engineers increasingly have work assignments in host countries with different practices, traditions, and values from an engineer's home country, raising still other issues. What criteria are appropriate in determining when engineers should adopt the values and practices of the host country? For example, when, if ever, is it appropriate to make "grease payments" and to exchange rather substantial gifts with customers and potential customers, where this is commonly practiced? (Harris 2000).

Future Directions

As an academic discipline in an early phase of its evolution, engineering ethics can be expected to show further maturation in every area, but the following areas seem particularly in need of further cultivation and growth.

METHODOLOGY. As in many areas of practical ethics, methodology needs further development. In practical ethics there are at least three different methodologies, each with characteristic strengths and weaknesses. One is to turn to traditional philosophical theories, especially consequentialist or utilitarian and deontologist or person-respecting theories, a "top-down" approach. Traditional ethical theories serve several useful functions in engineering ethics. First, they help identify relevant moral considerations in a dilemma. For example, knowledge of moral theory is useful in identifying the different moral perspectives of scientists and engineers (who often take a consequentialist approach) and the lay public (who often take a deontological approach) with respect to risk, and in confirming that both perspectives have deep and legitimate moral roots.

Second, moral theories often allow one to construct and even predict the arguments that will be made for or against certain policies or courses of action. Suppose one is considering whether there should be strong or weak protections of intellectual property. Utilitarian arguments for strong protections point out that such protections give incentive for technical advancement by insuring that those who are responsible will reap the economic rewards. Utilitarian arguments against strong protections point out that severe restrictions can impede the advance of technology by restricting the flow of information. Arguments based on a respect for persons typically point out that respect for individual rights of the creators of new technology requires that their creations be protected from unauthorized use. These lines of thinking do in fact reflect the discussion in the courts and scholarly literature.

Third, moral theories are often useful in assessing whether an argument has been resolved satisfactorily. If arguments from the two perspectives agree, there is good reason to accept the conclusion. If they disagree, there is a clearer basis for identifying morally relevant differences and determining which arguments are the most persuasive. Despite these advantages, however, many writers and teachers find that theories are often not useful for analysis and resolution of many of the concrete dilemmas that engineers face. Furthermore, engineering students are often not sympathetic to grand theories.

In contrast, a "bottom-up" approach emphasizes the need for careful analysis of the particulars of a given situation and makes much less use of broad moral principles. One version of this approach is the ancient method of casuistry, which has also been revived in medical ethics (Jonsen and Toulmin 1988). As has already been pointed out, paradigms of acceptable (or unacceptable) action are first identified. Then the salient ethical features of the paradigms are compared with those of the case under consideration. The casuist must then determine whether the case in question more closely resembles the paradigm of acceptable behavior or the paradigm of unacceptable behavior. For this method to work effectively, appropriate paradigms of acceptable and unacceptable behavior must be identified and generally accepted by the profession. The critical question is whether this can be done without relying on just the sorts of principles those sympathetic to the top-down approach take as their starting point.

An approach that falls somewhere between the top-down and bottom-up approaches proceeds from what might be called "mid-level" moral rules and principles, such as: "keep your promises and agreements"; "don't cheat"; "don't harm others"; "be truthful"; and "minimize the influence of conflicts of interest." Engineering codes of ethics tend to operate at this level. Questions about the appropriate grounding of such mid-level rules and principles remain, however, as do questions about their application to particular circumstances. If, for example, one asserts that exceptions to the rules or principles are justified as long as a rational person would be willing to have others make the same exception, one must give reasons for taking this position. Do the reasons make reference to principles of a still broader nature, perhaps even general moral theories? Whether all three approaches are useful, or only one, or some other approach such as "virtue ethics," is still a matter of debate.

GOOD WORKS AND CHARACTER. Many of the cases that have driven research and teaching in engineering ethics have dealt with engineering disasters and the responsibilities of engineers to prevent them or respond to them adequately after they have occurred. Some writers, however, have begun to stress the importance of going beyond basic duties to protect the public from the disastrous effects of technology to the duty to promote the public good (Pritchard 1992, 1998). The General Electric (G.E.) engineers who in the 1930s worked together against odds and with relatively little managerial support to develop the sealed-beam headlight exemplified good works. Some writers have stressed the importance of character and personal ideals in motivating such good works (Pritchard 2001, Martin 2002). Physicians who are members of "Physicians Without Borders" and engineers in "Engineers Without Borders" exemplify this kind of activity, but there are many less dramatic examples, such as the G.E. engineers. Many believe that the place of good works and the motivations for them deserve more emphasis in teaching, in research, and in the engineering profession itself.

Taking on responsibilities that go beyond standard job requirements in order to improve public safety is not unusual for engineers. Beyond the efforts of individual engineers or small groups of engineers, professional societies can make important contributions. The rapid emergence of the boiler industry in the late nineteenth and early twentieth centuries provides an illustration of the constructive role engineers can play in the face of serious risks arising from technological development. Initially ill-understood and without a set of regulations to guide their safe construction and use, boilers frequently exploded, injuring and killing untold numbers of people. Through the efforts of the leadership and dedicated work of a large number of mechanical engineers in the ASME, guidelines and regulations for the construction and safe use of boilers were eventually put in place (Cross 1990).

SOCIAL POLICY ISSUES. Most cases in engineering ethics have focused on the decisions of individual engineers in the context of a particular situation, but the effect of technology on society is often more a function of larger social policy issues—what some have called "macro-issues" as opposed to "micro-issues" (Herkert 2001). The legal and medical professions often make policy statements in areas of their expertise. Engineers, perhaps because of the absence of a unified professional society that can represent the profession to the public, have been much less conspicuous in public debates related to technology. In the light of engineers' responsibility to hold paramount the safety, health, and welfare of the public, what are their responsibilities (if any) in this area?

Some believe that engineers should step forward to help the public reflect on what future technological development might hold in store—both positive and negative (Fouke 2000). These developments will have an impact on the quality of our environment, the availability and distribution of needed resources, the quality of life that is possible, and the ability to live in peace or conflict. Many of these questions have to do with the appropriate laws and governmental regulations.

Several such issues have already been suggested. Others include the relationship of bio- and related engineering to cloning and genetic engineering. Still others have to do with nanotechnology, national defense, and the use of cell phones. The proper decisions in these areas, as well as the extent to which engineers should have responsibility for making policy statements or informing the public, is a matter that deserves more consideration in the engineering profession and in engineering ethics.

Engineers must also be concerned with codes and laws that are important in protecting the public. The ASME has long been associated with the code governing boilers and pressure vessels. Some engineers have incurred considerable personal risk and liability by promoting requirements for trench boxes to protect workers in deep trenches. Engineers have been involved in promoting improvements in building codes that protect buildings from earthquake damage, damage due to subsoil shifting, and hurricane and wind damage. Yet the extent and nature of engineers' obligations in these areas has received scant attention in the literature of engineering ethics.

CONTEXT OF ENGINEERING DECISIONS. The context in which engineering decisions are made and the implications of this context for ethical analysis are insufficiently explored. Engineers commonly make recommendations and decisions about design and other issues in the context of incomplete knowledge and considerable uncertainty. Often their work is limited to only a part of the total project or product design, and managers, not engineers, sometimes make crucial decisions. Assessments of individual responsibility in such contexts and the proper criteria for making decisions under conditions of uncertainty have yet to be fully analyzed.

ENGINEERS, MANAGERS, AND RIGHTS IN THE WORKPLACE. The relationship of engineers to managers is an especially sensitive area. On the one hand, managers can overrule the decisions of engineers, even when professional issues are at stake. On the other, managers control the jobs of engineers, and many engineers aspire to management positions. Engineers do not want to jeopardize careers by unnecessarily offending managers. Some attention has been devoted to the question of when decisions are properly made by engineers and by managers, and to the professional rights of engineers in the workplace (Harris 2000, Martin 2000). The issues that arise between engineers and managers and how they should be dealt with have been insufficiently studied, however, and no engineering code of ethics has raised the question of the rights of engineers as professionals in the workplace. This is a topic that merits further study in academic engineering ethics and by professional engineering societies.

INTEGRATION WITH OTHER AREAS. Engineering ethics may need further integration with several other areas, such as the philosophy of technology, law, management theory, and the philosophy of engineering. Engineers, as well as teachers and writers in engineering ethics, need to be more aware of the nature of technology and its influence on society, the impact of law on ethical decisions, the relationship of engineering decisions to management decisions, and the important differences between the way engineers and scientists use scientific knowledge. This can help bring ethical analysis more closely in line with engineering practice. How this integration will affect the evaluation of professional decisions is not yet clear, but the need for this integration seems obvious (Mitcham 2003).

CHARLES E. HARRIS, JR.
MICHAEL S. PRITCHARD
MICHAEL J. RABINS

SEE ALSO *Architectural Ethics; Bay Area Rapid Transit Case; Bioengineering Ethics; Building Codes; Building Destruction and Collapse; Chinese Perspectives: Engineering Ethics; Codes of Ethics; Computer Ethics; Conflict of Interest; Consequentialism; DC-10 Case; Deontology; Design Ethics; Earth Systems Engineering and Management; Engineering Design Ethics; Engineering Method; Environmental Ethics; Ford Pinto Case; Institute of Electrical and Electronics Engineers; Preventive Engineering; Professional Engineering Organizations; Safety Engineering: Practices; Space Shuttle Challenger and Columbia Accidents; Whistleblowing.*

BIBLIOGRAPHY

Anderson, Robert M., et al. (1980). *Divided Loyalties: Whistle-Blowing at BART*. West Lafayette, IN: Purdue Research Foundation.

Armstrong, M. B. (1994). "Confidentiality: A Comparison Across the Professions of Medicine, Engineering, and Accounting." *Professional Ethics* 3(1): 71–88.

Baum, Robert J. (1983). *Ethics and Engineering Curricula*. Hastings-on-Hudson, NY: The Hastings Center.

Baum, Robert R., and Albert Flores, eds. (1978). *Ethical Problems in Engineering*. Troy, NY: Center for the Study of the Human Dimensions of Science and Technology.

Boisjoly, Roger. (1991). "The Challenger Disaster: Moral Responsibility and the Working Engineer." In *Ethical Issues in Engineering*, ed. Deborah Johnson. Englewood Cliffs, NJ: Prentice-Hall.

Cook, Robert Lynn. (2003). *Code of Silence: Ethics of Disasters*. Jefferson City, MO: Trojan Publishing.

Cross, Wilbur. (1990). *The Code: An Authorized History of the ASME Boiler and Pressure Vessel Code*. New York: American Society of Mechanical Engineers.

Davis, Michael. (1998). *Thinking Like an Engineer*. New York: Oxford University Press.

De George, Richard T. (1981). "Ethical Responsibilities of Engineers in Large Organizations: The Pinto Case." *Business and Professional Ethics Journal* 1(1): 1–14.

Fielder, John H., and Douglas Birsch, eds. (1992). *The DC-10*. New York: State University of New York Press.

Fleddermann, Charles B. (1999). *Engineering Ethics*. Upper Saddle River, NJ: Prentice Hall.

Florman, Samuel. (1983). "Commentary." In *The DC-10*, ed. John H. Fielder and Douglas Birsch. New York: State University of New York Press.

Fouke, Janie, ed. (2000). *Engineering Tomorrow: Today's Technology Experts Envision the Next Century*. New York: IEEE Press.

Gorman, Michael E.; Matthew M. Mehallik; and Patricia Werhane. (2000). *Ethical and Environmental Challenges to Engineering*. Englewood Cliffs, NJ: Prentice-Hall.

Harris, Charles E.; Michael S. Pritchard; and Michael J. Rabins. (2000). *Engineering Ethics: Concepts and Cases*. Belmont, CA: Wadsworth/Thompson Learning.

Herkert, Joseph R. (2001). "Future Directions in Engineering Ethics Research: Microethics, Macroethics and the Role of Professional Societies." *Science and Engineering Ethics*. VII(3): 403–414.

James, Gene G. (1995). "Whistle-blowing: Its Moral Justification." In *Business Ethics*, 3rd edition, ed. W. Michael Hoffman and Robert E. Frederick. New York: McGraw-Hill.

Johnson, Deborah G. (2000). *Computer Ethics*, 3rd edition. Englewood Cliffs, NJ: Prentice-Hall.

Johnson, Deborah G., ed. (1991). *Ethical Issues in Engineering*. Englewood Cliffs, NJ: Prentice-Hall.

Johnson, Deborah G., and Helen F. Nissenbaum, eds. (1995). *Computers, Ethics, and Social Values*. Englewood Cliffs, NJ: Prentice-Hall.

Jonsen, Albert R., and Stephen E. Toulmin. (1988). *The Abuse of Casuistry*. Berkeley: University of California Press.

Ladd, John. (1991). "The Quest for a Code of Professional Ethics: An Intellectual and Moral Confusion." In *Ethical Issues in Engineering*, ed. Deborah G. Johnson. Englewood Cliffs, NJ: Prentice-Hall.

Layton, Edwin T. (1986). *The Revolt of the Engineers*. Baltimore, MD: Johns Hopkins University Press.

Luegenbiehl, Heinz C. (1991). "Codes of Ethics and the Moral Education of Engineers." In *Ethical Issues in Engineering*, ed. Deborah G. Johnson. Englewood Cliffs, NJ: Prentice-Hall.

Martin, Mike W. (2000). *Meaningful Work*. New York: Oxford University Press.

Martin, Mike W. (2002). "Personal Meaning and Ethics in Engineering." *Science and Engineering Ethics* 8(4): 545–560.

Martin, Mike W., and Roland Schinzinger. (1996). *Ethics in Engineering*, 3rd edition. New York: McGraw-Hill.

Mitcham, Carl. (2003). "Co-Responsibility for Research Integrity." *Science and Engineering Ethics* 9(2): 273–290.

Mitcham, Carl, and R. Shannon Duball. (1998). *Engineer's Toolkit*. Upper Saddle River, NJ: Prentice-Hall.

Nakamura, Shuzo. (2003). *Practical Engineering Ethics*. Kyoto, Japan: Kagaku-Dojin Publishing.

Pritchard, Michael S. (1992). "Good Works." *Professional Ethics*. 1(1 and 2): 155–177.

Pritchard, Michael S. (1998). "Professional Responsibility: Focusing on the Exemplary." *Science and Engineering Ethics*. 4(2): 215–233.

Pritchard, Michael S. (2001). "Responsible Engineering: The Importance of Character and Imagination." *Science and Engineering Ethics* 7(3): 391–402.

Schaub, James H., and Karl Pavlovic, eds. (1983). *Engineering Professionalism and Ethics*. New York: John Wiley & Sons.

Schrader-Frechette, Kristen. (1985). *Science Policy, Ethics, and Economic Methodology*. Boston: Kluwer.

Schrader-Frechette, Kristen. (1991). *Risk and Rationality: Philosophical Foundations for Populist Reforms*. Berkeley: University of California Press.

Unger, Stephen H. (1994). *Controlling Technology*, 2nd edition. New York: Holt, Rinehart & Winston.

Vaughn, Diane. (1996). *The Challenger Launch Decision: Risky Technology, Culture, and Deviance at NASA*. Chicago: The University Of Chicago Press.

Vesilind, P. Aarne, and Alastair S. Gunn. (1998). *Engineering, Ethics, and the Environment*. New York: Cambridge University Press.

Whitbeck, Caroline. (1998). *Ethics in Engineering Practice and Research*. New York: Cambridge University Press.

INTERNET RESOURCE

Case Western University. Online Ethics Center for Engineering and Science. Available from http://www.onlineethics.org.

EUROPE

In most European countries engineering ethics is increasingly conceived as an interdisciplinary reflection at the crossroads of professional ethics, the human and social sciences, and the philosophy of technology (especially the ethics of technology). This is in marked contrast with the situation in the United States, where engineering ethics is a form of professional ethics.

Europe nevertheless includes countries with diverse cultural, juridical, professional, and educational traditions of engineering, something that has promoted efforts within the European Union to harmonize technical education, including its nontechnical requirements in the humanities, social sciences, and professional ethics. European integration has further required the development of professional guidelines for the mutual recognition of diplomas and titles. Thus any comparison between engineering ethics in Europe and in the United States cannot ignore a diversity of professional traditions. Engineering ethics in Europe requires a contextualist approach referencing the perceptions of the various engineers who formulate them.

Engineering Education: British versus Continental Models

Histories of engineering education frequently begin with France, ignoring that the first engineering schools in the world were the Moscow School for Military Engineers (established 1698) and the Apprenticeship School for Civil and Military Engineers (founded in Prague in 1707). As for Western Europe, from its commonly accepted origination with the Bureau des dessinateurs du roi (Bureau of the King's Draftsmen), established in France in 1744 (and the forerunner of l'École royale des ponts et chaussées, or Royal School of Bridges and Roads, founded in 1747), it is still a long way to engineering education as known in the twenty-first century, with its strong theoretical and practical content. The role of the bureau was primarily to provide a tutorial to guide new recruits in their first projects.

The creation of the Bureau des dessinateurs du roi was followed by l'École du génie de Mézières (School of Military Engineering, Mézières) in 1748 and l'École royale des mines (Royal School of Mines) in 1783. If these are among the oldest engineering schools in Western Europe, the one that has most influenced the engineering educational system is l'École polytechnique. The polytechnique was founded in 1794, one year after the dissolution of the French universities, and soon after the failure of the school at Mezières, from which it borrowed the idea of a formal curriculum, rather than imitating the ancien régime tutorship in place at l'École royale des ponts et chaussées. The polytechnique's formalized theoretical curriculum with its emphasis on mathematics became an influential model for engineering education throughout France and beyond. It also contributed to the establishment of a high scientific and technical education outside university.

Engineering in the United Kingdom adopted a different approach and only later established a structured education for engineers. Engineering degrees were not offered in the United Kingdom until 1838, when King's College, London, began to teach civil engineering. Indeed, Oxford and Cambridge Universities did not offer engineering degrees until the first decade of the twentieth century. Instead, British engineers were for a long time given occupational training exclusively in workshops; apprenticeship promotion is what truly integrated them into their peer group. For this same reason, Britain is the uncontested birthplace of industrial technology. These engineers were at the heart of the Industrial Revolution and played a major role in the development of both the steam engine and its uses.

It is also noteworthy that because of their habit of meeting in clubs in order to exchange ideas and proposals—and above all to capitalize on their experiences and projects—these British engineers prepared the ground for professional engineering organizations well before their Continental colleagues. It may also be significant that when engineering degrees did begin to be offered in the United Kingdom this was done not in independent institutions but in universities that already offered degrees in the liberal arts and sciences.

Professional Engineering Associations in Europe

With regard to France, historians of the engineering profession often cite the long existence of a particular organized group of engineers. Indeed, since 1676 there existed in France a Corps du génie (Engineering Corps) that was in fact a military organization. This particularly early institutionalization thus had little to do with those professional organizations that arose later in the majority of countries. The primary difference is that engineers of the Corps du génie were exclusively engineers of the state, that is, royal functionaries. Because of this state service the Corps du génie did not constitute a truly free organization of professionals, such as was established by "civil" engineers in Great Britain as an outgrowth of the previously mentioned informal clubs, notably the Society of Civil Engineers (founded 1771), later renamed "Smeatonians" after John Smeaton (1724–1792), one of its original members. Another of these societies, the Institution of Civil Engineers (ICE) was founded in 1818 by a small group of young engineers. In 1828, it obtained a royal charter and became a leader in the profession, with 80,000 members in the early twenty-first century.

From the middle of the nineteenth century, several European countries followed the British model, beginning with France (the Société des ingénieurs civils de France, founded in 1848), Germany (Verein Deutscher Ingenieure [Association of German Engineers], or VDI, 1856), and Spain (Asociación de ingenieros industriales [Association of Industrial Engineers], 1861). But while the prestigious British Institution of Civil Engineers was a club for practitioners, the French, German, and Spanish organizations were all created by a group of certified engineers coming from a single school in each country: the l'École centrale des arts et manufactures de Paris (Central School of Arts and Manufacturing in Paris), Berlin Gewerbeinstitut (Berlin Technical Institute), and the Escuela de ingenieros industriales de Madrid (School of Industrial Engineers in Madrid), respectively. Each association was only later open to qualified persons from other institutions or even to autodidacts (the self-taught).

By contrast, in the United Kingdom there still exist no institutions of higher education devoted exclusively to engineering such as those found on the Continent. The closest approximations are the British "polytechnics," founded in the mid-twentieth century, which include the education of technicians as well as engineers. Great Britain is also different from its neighbors in regard to another important point: It is the only European country in which the engineering associations were for a long time given a monopoly over designating who was an engineer and who was not. Since the 1920s this power has been limited to the power of the ICE to determine the legitimacy of the title "chartered engineer."

From Professional Organizations to Professional Ethics

This historical review shows that the early institutionalization of engineering education did not directly lead to the early establishment of professional engineering organizations. Instead, it was the autonomous organization of practitioners that promoted the initial affirmation of a collective identity and the formalization of a collective moral framework for professional conduct. It is not therefore by chance that the first code of professional ethics written by and for engineers was formulated in Great Britain.

Indeed, historians of the professions commonly consider the "professional code of conduct" adopted by the ICE in 1910 as the model for engineering ethics codes first in the United States and subsequently throughout the world. In 1911 the American Institute of Consulting

Engineers became the first U.S. association of engineers to adopt a code of ethics, a code composed of five articles strongly inspired by that of the British ICE, with seven supplementary articles.

On the Continent again, in 1604 in France, even prior to the creation of the Corps du génie in 1676, the prime minister of King Henry IV (1553–1610), who was also superintendent of fortifications, proclaimed a "Great Regulation" for all royal engineers. This set of directives and general rules was applied until the end of the seventeenth century, but had more the character of administrative law than of a code of professional ethics.

With regard to contemporary codes of professional ethics in Europe, their development is not the same in every country and they are much less important than in North America. Generally speaking, the presence of professional codes of ethics for engineers is stronger in those countries more influenced by Anglo-American cultural models, as is equally true throughout the rest of the world. Indeed, it is striking to note that when the Fédération européenne d'associations nationales d'ingénieurs (European Federation of National Engineering Associations, created in 1951) decided during the 1990s to formulate a code of ethics, it began by studying documents coming exclusively from anglophone countries (the United States, Canada, Australia, and New Zealand) rather than the few existing European codes, which were little known.

Among the little-known European codelike documents were three from Scandinavia and one from Germany. The Scandinavian documents were a "Code of Honor" from the Samlar Sveriges Ingenjörer (Swedish Association of Graduate Engineers), first adopted in 1929 and revised in 1988; a similar "Code of Honor" of the Tekniska Föreningen i Finland (Association of Swedish-Speaking Engineers in Finland) from 1966; and an "Ethical Code for Members of the Norwegian Civil Engineers Association" from 1970. In 1950 the VDI had adopted the "Engineer's Confession," which was more a quasi-religious statement than a professional code.

The European situation thus remains different from that of the United States, where the profusion of codes and of successive revisions within the different branches of the profession constituted a first fundamental phase of engineering ethics. This internalist phase ended during the 1970s, when ethical reflection began to take into account considerations external to the profession and thus challenged a hierarchy of values in which the public interest sometimes gave way to professional prestige. In Europe, however, the public interest has from the beginning been more pronounced, although in a different way than in countries that have had to deal with an ethos of individualism and competition influenced more strongly by American culture.

Engineering Ethics in Twenty-First-Century Europe

Contrary to the situation in the United States, contemporary European reflection on engineering ethics did not arise from a will to renew an existing and explicit reflection at the heart of the profession, and to open it to other actors such as scholars and academics. In the United States engineering ethics found new inspiration in the collaboration among engineering professionals, on one side, and philosophers, historians, and more recently social scientists, on the other. But in Europe engineering ethics was not heir to a prior internalist approach. Instead, its heritage was more that of a professional conscience intuitively sensitive to social responsibilities and to legal expectations for professional conduct associated with the Code Napoléon (the first modern legal code of France, promulgated by Napoléon Bonaparte in 1804).

Certainly, there existed at the end of the twentieth century, in some European countries, some more or less obsolete ethical codes. But there was no formalized ethical reflection, with one exception. In Germany, World War II led engineers to a painful crisis of conscience over the use of science and technology in the service of a monstrous program, and the postwar period saw a strong engagement of the VDI in reflection on the proper ends of technology and the moral responsibility of engineers. But even in Germany no formal code of ethics existed until 2001.

In France, a country with a long engineering tradition, the first ethics code dates back only to 1997, with a 2001 revision. But the two versions of this code, especially the first, are more French adaptations of the North American manner of formulating an ethical framework. A different dynamic, independent from that of the formulation of these initial codes, began in the 1990s to introduce ethical reflection into engineering education in courses (often under different names) dealing with the questions relevant to engineering ethics—courses on philosophy, on epistemology, and on the sociology of sciences and technologies, aroused by contemporary intellectual and social debates.

It is thus not surprising that the first European handbook on engineering ethics, *Technology and Ethics* (2001), which was the product of a team of thirty-seven researchers from ten different European countries, adopted an approach different from U.S. textbooks on the same topic. This volume, which provides one perspective on the state of ethical reflection in European engineering practice, distinguishes three levels of analysis. The first deals with the microsocial level and concerns ethical problems encountered by individual engineers (dilemmas and cases of conscience). The second focuses on the mesosocial level, where the technical systems and institutions are in competition. A third emphasizes the macrosocial level, and therefore technical development in general as a societal question.

Whereas textbooks from the United States are often centered on a code of professional ethics for the profession—that is, on the roles, responsibilities, decisions, and attitudes of engineers individually confronted by ethical dilemmas—*Technology and Ethics* situates this dimension within a more comprehensive framework. To some extent it makes engineering ethics more complex by situating it within the institutional and social context in which engineers participate with other actors (scientists, entrepreneurs, end users, and others) in the development of technologies. At the same time it strives to be more realistic and place less emphasis on individual moral heroism as the best response to ethical problems.

The contextualist approach taken here suggests two sets of questions. First, engineering ethics in Europe may be handicapped by the absence of strong and dynamic professional organizations. This weakness is partially compensated by the growing internationalization of technological universities and professional organizations such as the Institute of Electrical and Electronics Engineers (IEEE), the International Federation for Information Processing (IFIP), and others. But what is their influence with respect to the large, multinational corporations that employ the great majority of engineers? Is a collaboration possible with business ethics?

Second, the freshness of European ethical reflection has permitted it to adapt more rapidly to questions posed by those engineers who develop and maintain the new technological systems (within computer, nuclear, and biotechnological engineering). Engineers are indeed only one of several groups of agents who must articulate and address within their fields the new social and societal questions posed by the development of these techniques. On this point a collaboration with the Science, Technology, and Society (STS) studies movement is greatly desirable.

CHRISTELLE DIDIER
BERTRAND HÉRIARD DUBREUIL
TRANSLATED BY JAMES A. LYNCH

SEE ALSO *Tradeoffs.*

BIBLIOGRAPHY

Didier, Christelle. (1999). "Engineering Ethics in France: A Historical Perspective." *Technology in Society* 21(4): 471–486.

Goujon, Philippe, and Bertrand Hériard Dubreuil, eds. (2001). *Technology and Ethics: A European Quest for Responsible Engineering.* Leuven, Belgium: Peeters. The first textbook incorporating an interdisciplinary European perspective.

Grelon, André, ed. (1986). *Les ingénieurs de la crise: Titre et profession entre les deux guerres* [The engineers of crisis: Title and profession between the two wars]. Paris: Editions de l'École des hautes études en sciences sociales. A basic study of the engineering profession in France.

Ropohl, Günter. (1996). *Ethik und Technikbewertung* [Ethics and technical value]. Frankfurt am Main: Suhrkamp. The first engineering ethics textbook in Germany.

ENGINEERING METHOD

• • •

Since the early modern period natural science has been defined in terms of method. The two major approaches to scientific method are those of rationalist deduction and empirical experimentation, analyses of which are often traced back to René Descartes (1596–1650) and Francis Bacon (1561–1626), respectively. Both methods have been argued to have ethical components or to be applicable to ethics. Engineering has been much less described in terms of some distinctive method. In fact, it was only in the mid-twentieth century that discussions of engineering method came to the fore. Interpretations of engineering method are, however, more varied than with science, with less effort to draw connections to ethics, although on both counts the negligence is unwarranted. What follows is a modestly polemical assessment of engineering method that seeks to redress previous oversights by defining engineering method, comparing it with alternative definitions, and establishing the nexus between engineering method and engineering ethics.

The *engineering method* is "the use of heuristics to cause the best change in a poorly understood situation within the available resources" (Koen 2003, p. 28). Two words in this definition, *heuristic* and *best*, are used in an engineering sense. A *heuristic* is anything that provides a plausible aid or direction in the solution of a problem but is in the final analysis unjustified, incapable of justification, and fallible. Engineering heuristics include mathematical equations, graphs, and correlations as well as the appropriate attitudes for solving problems or minimizing risk in an engineering design. Such attitudes

FIGURE 1

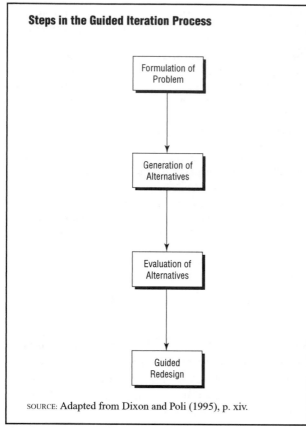

Steps in the Guided Iteration Process

Formulation of Problem

↓

Generation of Alternatives

↓

Evaluation of Alternatives

↓

Guided Redesign

SOURCE: Adapted from Dixon and Poli (1995), p. xiv.

cover or devise some means of succeeding in a difficult task." Contemporary dictionaries concur by authorizing the verb *to engineer* as "to contrive or plan usually with more or less subtle skill or craft" and by giving examples such as "to engineer a daring jailbreak." The word *engineer* is used daily in a similar fashion on radio, television, and in the newspaper.

The engineering term *state-of-the-art* (and its acronym, *sota*) refers to the collection of heuristics that were appropriate for a specific engineering project at a designated time. Thus, a state-of-the-art CD player will be one that is consistent with the set of heuristics that represented "best engineering practice" at the time it was made.

Derivative of the research in general problem solving (Polya 1945), the most frequent alternate definition of engineering used by engineers involves trying to establish a morphology or structure through which the design process is believed to pass (Dixon and Poli 1995, Pahl and Beitz 1995, Shigley and Mitchell 1983). This morphology is often presented in a flow diagram as in Figure 1. In addition to their multiplicity, engineering morphologies must fail as definitions of engineering because no one argues that the engineer can simply pass through the proposed steps; rather, engineers always back-track, iterate, and expand each step guided by heuristics.

Applied science is the most popular non-engineering definition of the engineering method. For the engineer, however, scientific knowledge has not always been available, and is not always available now, and even if available, it is not always appropriate for use. Some historians credit the Ionian natural philosophers of the sixth century B.C.E. as the founders of science, but undeniably homes, bridges, and pyramids existed before then. Precise scientific knowledge is still unavailable for many of the decisions made by the modern engineer. Although it cannot be said that engineering is applied science, engineers do use science extensively as a heuristic when appropriate.

The related claim that engineering is a branch of science called *design science* (Hubka and Eder 1996), similar to the social sciences, does not really advance a definition of engineering method. Although the much stronger view that engineering is a branch of science on a par with physics or chemistry is sometimes encountered (Suh 2001), this view implies that there are facts and axioms of design immutable and normed against an eternal truth, just as the facts of physics are said to be undeniably true. By contrast, most practicing engineers agree with the words of the noted engineer Theodore Von Kármán (1881–1963): "scientists explore what is

obviously have ethical dimensions. Suggestions to "allocate resources to the weak link," "complete a design by successive approximations," and "make small changes in the state-of-the-art," are also engineering heuristics. Engineers frequently use the synonyms *rule of craft, engineering judgment,* or *rule of thumb* to express these experience-based aids that, although helpful, are nonetheless fallible. In France, engineers use the near synonym *le pif* (the nose); in Germany, *faustregel* (the fist); in Japan, *menoko kanjo* (measuring with the eye), and in Russia, *na paltsakh* (by the fingers).

The engineer's word *best,* usually called the optimum, refers to the most desirable tradeoff of the design variables in a multi-variant space in which each criterion has been given its relative importance. This procedure differs from the ideal or best of Plato that is almost universally used in the Western tradition outside of engineering.

This definition of engineering method is consistent with the etymology of the word *engineer,* its formal definition in the dictionary, and common usage. According to one of England's most noted nineteenth-century engineers, Sir William Fairbairn, the term *engineer* comes from an old French word *s'ingenieur* meaning "anyone who sets his mental powers in action to dis-

and engineers create what has never been" (Krick 1969, p. 36).

Some, identifying the engineering method with trial and error (Petroski 1994), imply that engineers try random problem solutions and discard those that do not work. Contrary to this view, thousands of design decisions are made worldwide by engineers every day resulting in very few failures because the engineer usually modifies a previously assured sota in creating a new design.

While these alternate definitions are useful in expanding an understanding of engineering, they fail to be convincing as a comprehensive description of engineering method for the reasons specified, and because each can be subsumed into the definition given initially as simply additional engineering heuristics.

Because the engineering method applies to situations that contain uncertainty, some risk of failure is always present. The success or failure of an engineering design is, therefore, not a sufficient basis for judging whether an engineer has acted ethically. The *Rule of Judgment* in engineering is to evaluate an engineer against the sota that defines best engineering practice at the time the design was made (Koen 2003). This sota must contain all of the appropriate ethical, as well as technical, considerations.

When engineering is recognized as a pluralistic utilization of heuristics to bring about the best change in a limited resource situation that remains to be fully understood, then not only are ethical principles available as useful heuristics but the engineering method can itself become a reasonable description of ethical problem solving in general.

BILLY V. KOEN

SEE ALSO *Choice Behavior; Decision Theory; Engineering Design Ethics; Engineering Ethics; Rational Choice Theory.*

BIBLIOGRAPHY

Dixon, John, and Corrado Poli. (1995). *Engineering Design and Design for Manufacturing: A Structured Approach.* Conway, MA: Field Stone. Textbook for upper division engineering students.

Hubka, Vladimir, and W. Ernst Eder. (1996). *Design Science: Introduction to the Needs, Scope and Organization of Engineering Design Knowledge.* Berlin: Springer-Verlag. Extensive survey of the literature in engineering design.

Koen, Billy Vaughn. (2003). *Discussion of The Method: Conducting the Engineer's Approach to Problem Solving.* New York: Oxford University Press. This book is the basis of the analysis given here.

Krick, Edward V. (1969). *An Introduction to Engineering and Engineering Design.* New York: Wiley. One of the original descriptions of engineering design for beginning engineering students.

Petroski, Henry. (1994). *Design Paradigms: Case Histories of Error and Judgment in Engineering.* Cambridge, UK: Press Syndicate of The University of Cambridge. One of several popularizations of engineering method by this author.

Pahl, Gerhard, and Wolfgang Beitz. (1995). *Engineering Design: A Systematic Approach.* Berlin: Springer-Verlag. Definitive German morpholophy of engineering design.

Polya, George. (1945). *How to Solve It: A New Aspect of Mathematical Method.* Princeton, NJ: Princeton University Press. Classic book on general problem solving.

Shigley, Joseph, and Larry Mitchell. (1983). *Mechanical Engineering Design,* 4th edition. New York: McGraw-Hill. Classic textbook on engineering design for mechanical engineers.

Suh, Nam. (2001). *Axiomatic Design: Advances and Applications.* New York: Oxford University Press. Defines engineering in terms of axioms as is often done for science.

Taguchi, Genichi; Subir Chowdhury; and Shin Taguchi. (1999). *Robust Engineering.* New York: McGraw-Hill. The Taguchi method of engineering design is widely used in industry.

ENGINEERS FOR SOCIAL RESPONSIBILITY

• • •

The New Zealand engineering profession has a strong tradition of social responsibility, and many engineers have worked voluntarily on engineering projects in the Pacific Islands and in Southeast Asia. In keeping with this tradition, Engineers for Social Responsibility (ESR) was founded in 1983 and was the first such organization in the world. The driving force in its foundation was Gerald Coates, a Wellington-based electrical engineer. Its objectives are "to encourage and support social responsibility and a humane professional ethic in the uses of technology, to inform the engineering profession, general public and public policy makers about the impact of technology" (ESR). It is based in Auckland and has branches in Wellington and Christchurch, with a combined membership of around 200. Membership is open to all engineers and related professionals. Branches sponsor seminars and presentations that are open to the public.

ESR's focus has always been international: Initially it was concerned with nuclear and peace issues, and most of the papers at its first conference in Hastings in 1984 were on this topic. Its focus broadened after the end of the cold war to include a wide range of national and international issues, including an association with Water for Survival, an engineers' organization that provides technical advice and assistance for water supply and wastewater projects in poor countries.

ESR was initially criticized as a fringe organization, especially by the Institute of Professional Engineers New Zealand (IPENZ). But after a relatively short period, the temperate profile of ESR led to its general acceptance. Indeed, ESR has maintained a close association with IPENZ and become a model for similar organizations in other countries such as American Engineers for Social Responsibility (founded 1988) and Architects and Engineers for Social Responsibility in the United Kingdom (founded 1989, as a transformation of Engineers for Nuclear Disarmament, which began seven years earlier). ESR is also linked with the International Network of Engineers and Scientists for Global Responsibility (INES). Other related but not directly linked organizations include Computer Professionals for Social Responsibility and Physicians for Social Responsibility.

ALASTAIR GUNN

SEE ALSO *Engineering Ethics.*

BIBLIOGRAPHY

Coates, Gerald. (1983). "The Responsibility of Engineers." *New Zealand Engineering* 38(8): 23–24.

INTERNET RESOURCE

Engineers for Social Responsibility (ESR). Available from http://www.esr.org.nz/.

ENHANCEMENT

SEE *Therapy and Enhancement.*

ENLIGHTENMENT SOCIAL THEORY

• • •

Enlightenment social theory is important to science, technology, and ethics because it represents one of the first venues in which human activities were widely studied from a scientific perspective, and in which utilitarian and naturalistic ethical systems were offered to replace the religiously-based deontological, or duty-oriented, ethical systems which had dominated premodern society.

One of the most frequently stated goals of the Enlightenment of the eighteenth century was the creation of a science of human nature and society incorporating deterministic laws of behavior to match the spectacular successes of the physical sciences. David Hume (1711–1766), for example, announced his intention to become "the Newton of the Moral sciences." But eighteenth-century social theorists did not agree on which model from the physical sciences social theories should emulate.

Generally speaking, one can identify three classes of natural scientific models for the social sciences. The first stressed the approach of natural history and Hippocratic medicine, emphasizing the observation of phenomena in their situated complexity (empiricism). The second emulated the characteristics of rational mechanics, emphasizing the derivation of effects from a small number of well-defined a priori principles. The third attempted to apply the methods of the newly emerging experimental sciences, which insisted upon the isolation of salient variables whose relationships were established empirically, through their controlled manipulation. Within the social sciences, those who viewed themselves as introducing *experimental* approaches did emphasize the isolation of relevant variables; but their notion of experiment was generally different from that used in the natural sciences. Hume explained that difference very clearly:

> We must glean up our experiments in this science from a cautious observation of human life, and take them as they appear in the common course of the world, by men's behavior in company, in affairs, and in their pleasures. Where experiments of this kind are judiciously collected and compared [for example, from histories and travel accounts], we may hope to establish on them a science, which will not be inferior in certainty, and will be much superior in utility to any other of Human comprehension (1969, p. 46).

With few exceptions, those eighteenth-century scientists and philosophers who derived their approaches largely from natural history—such as Charles Louis de Secondat, Baron de la Brede et de Montesquieu (usually known simply as Montesquieu), Adam Ferguson, and Edmund Burke—usually focused on humans as habitual and emotional beings and ended up toward the conservative end of the political spectrum. Those who derived their approaches principally from the rational mechanics tradition—such as the physiocrat Jean Claude Helvétius,

Mercier de la Rivière, Anne-Marie Condorcet, and the feminist Mary Wollstonecraft—focused on humans as rational beings and ended up at the radical end of the political and social spectrum. Those who saw themselves as synthesizing empirical and rational approaches—such as David Hartley, Adam Smith, and Etienne Condillac—tended to see humans as expressing both emotional and rational characteristics and ended up in the liberal portion of the political and social spectrum. Regardless of what model they adapted from the natural sciences, Enlightenment social theorists tended to reject deontological approaches to ethics in favor of consequentialist ones, though the utilitarian ethical theories of the radical and liberal thinkers were vastly different from those of the more thoroughly empirical conservatives.

In 1749 Montesquieu published his *Spirit of the Laws* in an attempt to explore how different legal systems developed. Though he was inclined to think that humans were pretty much identical everywhere, as the president of a local judicial body that often found itself in conflict with the central authority of the French crown, he was painfully aware of the immense variations in local customs and laws, and he took as his task the explanation of those variations. To classical republican arguments that laws had to be suited to the principles attached to the form of government of a people, Montesquieu added three kinds of arguments that were to have immense long-term significance.

First, he argued that the laws and customs of a country will depend upon the dominant mode of subsistence of that country, classifying modes of subsistence as hunting, herding, agricultural, and commercial. Hunting societies, for example, will have much less complex laws that herding societies because the complication of private ownership of animals is added in herding societies. Laws will be even more complex in agrarian societies in which heritable real property becomes important; and they will be even more complex in commercial societies in which it is critical to have legal means for enforcing a wide variety of contracts. Montesquieu felt that trade promoted mutual dependence and therefore increased tolerance for cultural differences among trading partners; so it promotes peace among nations. Within a given nation, however, Montesquieu argued that trade promoted competition and egotism rather than cooperation and altruism.

Second, Montesquieu argued for a kind of environmental determinism that made customs and laws suitable to one region quite unsuitable to others. For example, he argued that the high temperatures in the tropics made men lazy, justifying the practice of slavery so that

work would get done. Similarly he thought that women aged more rapidly in tropical regions, justifying the practice of male plural marriage with women of different ages. Neither slavery nor plural marriage was, however, justifiable in temperate regions. This *situational* ethics that derived from Montesquieu's environmental determinism illustrates how attempts a social science could undermine deontological ethics.

Finally, Montesquieu was one of the first serious social theorists to articulate a principle that would become the hallmark of conservative political theory through the twentieth century. This principle is often called the principle of unintended consequences, and Montesquieu openly appropriated it from Bernard Mandeville's "Fable of the Bees" of 1705, though he gave it much greater currency. The particular example used by both Mandeville and Montesquieu was that of how the vanity of the wealthy produced the rise of fashion in clothing, which in turn provided jobs for textile workers. The vice of pride thus produced the unintended consequence of promoting commerce and industry. There was even a business in providing the baubles on which hierarchy could be seen to be based—beads, cosmetics, physical distinctions such as tattoos, and so forth.

In the long run, the principle of unintended consequences became the foundation for virtually all conservative claims that society cannot be successfully reformed by design: For every positive intended consequence there is likely to be a negative unintended one. It is better from this perspective to simply let society develop naturally. In the words of Adam Ferguson, one of Montequieu's most able admirers, "nations stumble upon establishments which are indeed the result of human action, but not the execution of any human design. ... The establishments of men ... were made without any sense of their general effect; and they bring human affairs to a state of complication which the greatest reach of capacity with which human nature was ever adorned, could not have projected" (Ferguson 1966, pp. 122, 182).

Taking his cue from Montesquieu, Ferguson attempted to write a "natural history of man" in *An Essay on the History of Civil Society* in 1767, but Ferguson made a number of new arguments that were widely adopted by subsequent social theorists. First, he temporalized Montesquieu's four modes of existence, creating a dynamic theory in which hunting, herding, agriculture, and commerce represented progressive stages in a temporal development that was repeated at different times in different places. Next, he emphasized the fact that people band together into societies not out of some rational expectation of meeting selfish needs, as Thomas Hobbes had proposed in the seventeenth century, but rather out of "a propensity to mix with the herd and, without reflection, to follow the crowd of his species" (Ferguson 1966, pp. 16–17). Finally, Ferguson argued that conflict, even to the extent of war, is often the vehicle for social advances: "Their wars ... their mutual jealousies, and the establishments which they devise with a view to each other, constitute more than half the occupations of mankind, and furnish materials for their greatest and most improving exertions" (Ferguson 1966, p. 119).

Against the tradition of *philosophical history* initiated by Montesquieu and Ferguson, a second group of Enlightenment social theorists claimed that to argue for particular social arrangements from the simple fact of their historical existence was to grant the past far too much power over the future. Rivière, spokesman for a group of theorists known as *économistes* or physiocrats (persons who favored government according to the nature [phusis] of things, rather than aristocrats who advocated government by an elite, or democrats who favored government by all) made their point particularly clearly in 1767:

> I do not cast my eye on any particular nation or sect. I seek to describe things as they must *essentially* be, without considering what they have been, or in what country they may have been. .... By examining and reasoning we arrive at knowing the truth self-evidently, and with all the practical consequences which result from it. Examples which appear to contrast with these consequences prove nothing (Hutchinson 1988, p. 293).

Among the most important social theorists to adopt this rational mechanist model were Claude-Adrien Helvetius and his utilitarian followers, including Jeremy Bentham in Britain and Cesare Beccaria in Italy. According to this group, all social theory must begin from the fundamental insight that humans are motivated solely by a desire to be happy; so the goal of political and moral philosophy should be to create the greatest net pleasure for the greatest number in society. Because members of the utilitarian school generally assumed that the private happiness of one person was likely to diminish the happiness of others, they proposed to establish sanctions that would offer pleasurable rewards to those who acted for the general good and punish those who acted in opposition to it.

Among those who advocated a more experimental approach to social theory, the tradition initiated by

Francis Hutcheson, David Hartley, and Adam Smith was undoubtedly most important in terms establishing a new foundation for ethics and morality. This group generally found strong evidence that humans acted not only out of self-interest, but also out of a social instinct or sense of sympathy. For most of these social theorists, there seemed to be a natural accommodation between the well-being of the individual and that of the group that was nicely articulated in Smith's image of the "invisible hand" that ordered economic activity for the general benefit if each actor worked to forward his own interests. This approach led to a *laissez faire* or naturalistic approach to moral and ethical behavior.

The heritage of Enlightenment social theory remains current in virtually all disagreements among different groups concerned with policies relating to science and technology. The principle of unintended consequences, as directly derived from Ferguson, for example, was still being appealed to by conservative social theorists such as Friederich A. von Hayek in the late nineteen-sixties (Hayek 1967). It later became the foundation for arguments by the often politically liberal or radical critics of rapid technological development. The consequentialist ethical tradition established among eighteenth-century utilitarians continues to inform policy makers at the beginning of the twenty-first century in the form of cost-benefit analyses so favored by advocates of development. And the *laissez faire* admonitions of the Smithian school continue to resonate in the market-driven analyses of public choice economic theorists.

RICHARD OLSON

SEE ALSO *Human Nature; Hume, David; Modernization; Shelley, Mary Wollstonecraft; Smith, Adam; Unintended Consequences.*

BIBLIOGRAPHY

Brandon, William. (1986). *New Worlds for Old: Reports from the New World and Their Effect on the Development of Social Thought in Europe, 1500–1800.* Athens: Ohio University Press. The title characterizes this excellent work beautifully. The discussions of how descriptions of indigenous American societies produced a change in traditional roman notions of "liberty," divorcing them from particular corporate status, are particularly interesting

Ferguson, Adam. (1966 [1767]). An *Essay on the History of Civil Society*, ed. Duncan Forbes. Edinburgh, Scotland: Edinburgh University Press. Important example of the empirical and "conservative" approach to society in the enlightenment.

Fox, Christopher; Roy Porter; and Robert Wokler, eds. (1995). *Inventing Human Science: Eighteenth Century Domains.* Berkeley and Los Angeles: University of California Press. Very good set of essays considering enlightenment discussions of topics that in the early 2000s would be identified with the social sciences.

Gay, Peter. (1966–1969). *The Enlightenment: An Interpretation,* 2 volumes. New York: Vintage Books. Still the most comprehensive introduction to the enlightenment. Contains a very extensive and valuable bibliographical essay.

Hayek, Friedrich A. (1967). "The Results of Human Action but Not of Human Design." In his *Studies in Philosophy, Politics, and Economics.* Chicago: Chicago University Press. Illustrates the explicit appropriation of Ferguson's ideas by a key twentieth century conservative theorist.

Hont, Istvan, and Michael Ignatieff, eds. (1983). *Wealth and Virtue: The Shaping of Political Economy in Eighteenth Century Scotland.* Cambridge, UK: Cambridge University Press. An outstanding collection of essays on Scottish social thought in the enlightenment.

Hume, David. (1969 [1739]). *A Treatise of Human Nature,* edited by Ernest C. Mossner. New York: Penguin. Hume's initial attempt to establish a science of humanity.

Hutchinson, Terence. (1988). *Before Adam Smith: The Emergence of Political Economy.* Oxford: Basil Blackwell. Excellent account of economic thought in the seventeenth and eighteenth centuries.

Kelley, Donald R. (1990). *Social Thought in the Western Legal Tradition. Explores the significance of eighteenth century legal education and legal reform for social theory.* Cambridge, MA: Harvard University Press.

Moravia, Sergio. (1980). "The Enlightenment and the Sciences of Man." *History of Science* 17: 247–248. A good short account of the importance of Locke's epistemology for Enlightenment social theorists.

Olson, Richard. (1990). "Historical Reflections on Feminist Critiques of Science: The Scientific Background to Modern Feminism." *History of Science* 28: 125– 147. Explores the use of arguments from eighteenth century associationist psychology to justify feminist arguments in the works of Condorcet and Mary Wollstonecraft

Olson, Richard. (1993). *The Emergence of the Social Sciences, 1642–1792.* New York: Twayne. Short argument that the key roots of most modern approaches to social scientific disciplines are to be found in the seventeenth and eighteenth centuries.

Reil, Peter. (1975). *The German Enlightenment and the Rise of Historicism.* Berkeley and Los Angeles: University of California Press. A good introduction to the centrality of organic and developmental models of social theory in the eighteenth and early nineteenth centuries.

Smith, Roger. (1997). *The Norton History of the Human Sciences.* New York: Norton. Outstanding and comprehensive account of the development of the social sciences with a strong section on the enlightenment.

ENQUETE COMMISSIONS

• • •

Enquete commissions are temporary groups established periodically by European parliaments in order to guide public discourse and decision making in complex areas. Commissions have focused on questions such as economic globalization, environmental sustainability, and the formation of new religious and ideological groups. Roughly half of the enquete commissions to date have addressed the use and regulation of emerging science and technology. In these cases, the commissions serve as forums for joint scientific and political consultation designed to inform decision makers, involve the public, and articulate recommendations and strategies for future action. Each commission is unique in terms of membership, topic, and mandate, so general evaluations of the enquete commission as an overarching system for improving democratic discourse and decision making are difficult to formulate. Although they have had mixed results and need improvement, enquete commissions are important innovations in the relationship between politics and science in democratic societies.

Background

Parliaments, as elected representative bodies, should play a key role in guiding public discourse about the proper development of society. There are doubts, however, about how well parliamentary bodies can fulfill this leadership position given the complex problems presented by the modern world. Decision makers are inundated with competing demands for investment in science, technology, and the military. They also deal with conflicting reports about economic, educational, environmental, and health care policies. In these areas, parliaments must rely upon the superior knowledge of experts and the bureaucratic structure of specialized departments and agencies. Yet mechanisms for delegating authority to specialists tend to alienate government officials from the very discourse they should guide and shape. Thus the legislative function of parliament becomes disengaged from the debate on essential issues of societal development.

Enquete commissions are designed to reengage the governmental body regarding these complex issues. They serve as independent agencies that support the parliament, thereby counterbalancing the institutional inertia toward bureaucratization and the delegation of decision making to experts who have no fiduciary or other responsibility to the public.

One of the most important roles of an enquete commission is to serve as a common institutional forum where scientific knowledge and political judgment meet. Several enquete commissions have been charged with the task of evaluating issues regarding the proper use and regulation of technologies and the proper conduct of scientific research. In these cases, especially, enquete commissions provide common ground for decision makers, the public, and experts. Cooperation between scientists and politicians is of particular importance when the knowledge of experts is contested or uncertain and when political party lines are ill-defined with regard to an issue. In many countries scientific advice issues from special institutions such as the Parliamentary Office of Science and Technology (POST) in the United Kingdom, the Parliamentary Office for Evaluation of Scientific and Technological Options (OPECST) in France, and the disbanded Office of Technology Assessment (OTA) in the United States, but these do not serve as institutions of joint scientific-political consultation.

Enquete commissions are partially modelled on various review commissions that are periodically appointed to investigate alleged failures by public officials or public institutions (for instance Royal Commissions in the United Kingdom and Congressional Committees in the United States). Enquete commissions, however, are usually established by parliamentary mandate in order to develop scenarios, strategies, and recommendations with respect to potential problems areas. Yet only a few parliaments—most notably France, Germany, Sweden, and Italy—have established rules for the membership and operations of such committees, and only these countries have significant experiences with the process of forming and evaluating enquete commissions.

German Experience

Because Germany has the most elaborate model with the broadest variety of applications, it is appropriate to include an in-depth discussion of German enquete commissions. Since 1969 the German parliament has, by standing order, permitted enquete commissions to be established by the approval of at least one quarter of its members for the purpose of providing information relevant to *extensive and important* issues. In practice a broader quorum distributed over the parties in power and opposition is necessary for any chance of successful work. The enabling legislation leaves open what qualifies problem areas as extensive and important.

Since the order was implemented, two to five commissions have been created in each electoral term.

Roughly half have focused on topics in the fields of science, technology, and the environment. Some commissions that have been authorized by the German parliament include The Future of Atomic Energy Policy (1979–1982), New Information and Communication Technology (1981–1983), Prospects and Risks of Genetic Technology (1984–1986), Assessment and Evaluation of the Social Consequences of New Technology: Shaping the Conditions of Technological Development (1985–1990), Precautionary Protection of the Earth's Atmosphere (1987–1994), Protection of Human Beings and the Environment: Evaluation Criteria and Perspectives for Environmentally Acceptable Circular Flow Substances in Industrial Society (1992–1998), The Future of the Media in Economy and Society (1996–1998), Sustainable Energy Supply in the Modern Economy (2000–2002), and Law and Ethics of Modern Medicine (2000–2002; reinstated 2003).

These and other commissions have received a correspondingly wide set of mandates, but there are a few general purposes that underlie the task of all enquete commissions. These include:

- Establishing a political discourse with the intent of assuring, if not the preeminence, at least the influence of political and social concerns in shaping technological change.

- Searching for a consensus or well-founded dissent comprising knowledge, interests, values, and norms, and thereby preparing for compromise in the negotiation process.

- Elaborating long-term foundations for decisions and making concrete recommendations to parliamentary legislators.

- Enhancing public awareness of an issue by involving the media and by reporting to the public either as individual members or through official reports.

COMPOSITION AND STRUCTURE. Enquete commissions are unique institutions for the treatment of specific societal issues because of their consciously crafted representative mix of political parties and external experts. Each party nominates representative parliamentary members according to their relative political power (they are able to elect between four and fifteen members). Because all parties with parliamentary status participate, normative and ideological perspectives are represented in a manner that mirrors the larger legislative body. Each commission reflects the proportionality of power and perspective found in parliament. External experts are chosen either by an iterative process of nomination, rejection, and acceptance or they are simply appointed in a manner proportional to the power of each party. Representation on the science side is usually fairly well balanced because the selection of experts by the parties covers the spectrum of competing paradigms and can even include extreme opinions. Parliamentary and external members have the same voting rights.

The goal of every commission is to present a report, which serves as the basis for a general parliamentary discussion, before the end of the electoral term. As a rule, recommendations for legislative decisions are also expected. Usually additional experts without voting rights are also included and their opinions are commissioned. The commissions often organize public hearings and other public dialog. Initially governmental and department officials did not participate in the process at all. The main advantage of including parliamentary members with experts is to make commissions better equipped to structure and convey recommendations pertinent to the needs of decision makers. The corresponding disadvantage is a tendency to politicize scientific findings. Another impact of incorporating scientific experts and politicians in such tightly structured dialogue is the addition of more focused, problem-orientated discourse to traditional negotiations between majority and minority parties.

Each commission serves as a working group for intense research and reflection in a particular subject area. A research staff assists each commission by procuring and processing information. One member of the commission serves as the chair and is vitally important for ensuring the integrity and overall success of the commission. The privileged position of the chair is sometimes misused to serve individual or political ends.

ANALYSIS AND EVALUATION. The fairly long history of enquete commissions in Germany points to the importance of comprehensive and exhaustive dialogue at the intersection of politics and science. Though political maneuvering unavoidably comes into play when choosing members, setting an agenda, and negotiating reports, the underlying purpose is to hold open dialogues on problems and alternative solutions before party lines are settled and decisions reached. The goal is to transform solid party positions into negotiable interests. Scientific and political members agree to seek consensus and compromise, which can be further shaped by party leaders and wider government involvement. Nevertheless the commissions face constant pressure from both political and scientific interest groups, which often seek to use enquete commissions to achieve their own special interests rather than the common interest. Politicians

often press particular agendas, whereas scientists repeatedly cloak their agendas in the guise of disinterested objectivity. Commissions require strong leadership if they are to bridge the differences among various stakeholders. When such leadership is present, the enquete commission is a successful model for crafting an improved and more democratic relationship between science, government, and the general public.

Whether and to what extent enquete commissions enrich and aid political culture is an open question. Case studies and empirical analyses of their mode of operation have generated serious criticisms. Party tactics repeatedly threaten the efforts of members to achieve mutual understanding and common perspectives. Often commissions are used as instruments of symbolic politics, giving the impression of governmental action that conceals an unwillingness to make real progress. For example, the commission on the future of media in economy and society was allegedly misused in this way. Its charge was to "pave Germany's way toward the information society," but one of its most distinguished members, Wolfgang Hoffmann-Riem, professor of law and judge at the constitutional court, commented in an essay that it was only "creeping along secret paths to non-decision." The chance was wasted to develop guidelines for new technologies (especially telecommunication and the associated changes of the occupational field), higher education, infrastructure, and the media.

By contrast the commission on genetic technology thoroughly influenced legislation on safety regulation at the work place, rules of liability, and restrictions on research with human embryos. The commission on technology assessment did not have direct impacts, but its work indirectly supported the foundation of the German Office of Technology Assessment in 1990, which offers recommendations about science and technology to parliament. The commission's report on the protection of the atmosphere was influential even at the international level. It played a decisive part in regulating and outlawing various ozone depleting chemicals within the European Union (EU) and assisted in the Montreal Protocol process.

Due to the variability in enquete commissions, it is not possible to make a general evaluation of their success in crafting consensus, aiding legislation, and guiding public discourse. Several criticisms, however, have suggested that the priorities of the enquete commission system need to be rethought in order to maximize its strengths. Critics argue that the indirect inputs into public discourse are more important functions of the commissions than direct impact on legislation. If the commissions are able to address and include important associations, not-for-profit organizations, nongovernmental organizations, the media, and influential individuals, then their procedures and reports can demonstrate parliament's ability to guide public discourse on important questions about the future development of society.

Scientists involved in enquete commissions often resent abandoning their position as (supposedly) neutral, outside analysts by engaging in the political system. Nonetheless most are able to maintain their reputations within the scientific community by crafting and supporting high quality, balanced reports. The politicians involved also have misgivings, especially concerning mandates for cooperation and consensus building with actionable recommendations. In addition engaging in long-term, complex issues usually does not offer the political payoff of involvement in more pressing, short-term issues. Usually, however, there is sufficient individual initiative among politicians to overcome these concerns.

No critics advocate abandoning the enquete commission model altogether. Some place deficiencies in the system on the early twenty-first century style of party politics and its focus on personalities and media resonance. More theoretically minded observers note a permanent overburdening of the commissions due to their hybrid structure. These critics argue that increasing the management skills and capacities of commission leaders is the only way in which improvements can be made. In the end, the continued existence of the commissions and the fact that both majority and opposition factions have initiated roughly the same number of them over ten election periods speaks for their value. Enquete commissions can be an important ingredient in the public culture of politics. They can increase understanding, elevate public discussion, and evaluate and respond to societal problems. They are especially useful in evaluating the risks and benefits presented by complex, emerging technologies.

Enquete commissions need to be used in conjunction with other procedures, such as lobbying, hearings, and stakeholder conferences, which often represent and consider interests in different ways. In spite of determined attempts to strike consensus or compromise and frame political programs, the complex and contested nature of many problems sometimes prohibits workable solutions. Yet even in these cases, enquete commissions can help improve and clarify public discourse, venture models for risk assessment, develop scenarios and options for the future, map the landscape of social values, and make tentative preparations for legislative action.

WOLFGANG KROHN

SEE ALSO *Bioethics Commitees and Commissions; Royal Commissions.*

BIBLIOGRAPHY

Altenhof, Ralf. (2002). *Die Enquete-Kommissionen des Deutschen Bundestages* [The enquete-commissions of the German parliament]. Wiesbaden, Germany: Westdeutscher Verlag.

Weaver, Kent, and Paul Stares, eds. (2001). *Guidance for Government: Comparing Alternative Sources of Public Policy Advice.* Tokyo and New York: Japan Center for International Exchange.

Vig, Norman, and Herbert Paschen, eds. (2000). *Parliaments and Technology: The Development of Technology Assessment in Europe.* Albany: State University of New York Press.

ENTERTAINMENT

• • •

Entertainment is a ubiquitous phenomenon that has been transformed extensively by science and technology. To some extent that transformation has ethical dimensions that merit more consideration than they usually receive.

The Historical Spectrum

There is evidence that human beings have found ways to amuse themselves since the beginning of history. Ancient Mesopotamians reserved six days a month for designated holidays, half of which were tied to religious lunar festivities. Hunting was a favorite pastime of Assyrian kings, as wall reliefs attest; that pastime was shared by Egyptian pharaohs, as is affirmed by the decorations on their tombs. Sports such as boxing and wrestling were practiced widely in the ancient world, sometimes between divine beings and men as in the struggle between Gilgamesh and Enkidu in the *Epic of Gilgamesh* and that between Jacob and an angel in the Hebrew Bible. Black-figure vases and amphorae indicate the Greeks' love for those two sports as well as the others featured in the ancient Olympic games and their imitators throughout the ancient Aegean world. A variety of board games from ancient times (e.g., serpent, dog-and-jackal, and senet from Egypt) challenge contemporary people to discern what the rules might have been, whereas games such as chess, go, and various others involving stone, bone, clay, or glass dice can be recognized by modern players in their earliest written, engraved, and stone forms from China, India, Mesoamerica, Africa, and the Near East. Children's model houses with miniature furniture and figures and model ships, wagons, chariots, and carts from sites across the ancient world indicate that toys are also of ancient origin.

People entertained one another on musical instruments, as many ancient literary and sacred texts attest, including ancient songs that survive in the form of the Psalms and the "Song of Miriam" in the Hebrew Bible, the *Iliad* and *Odyssey*, and hymns to Osiris and other ancient gods in addition to love songs and songs that express the challenges and triumphs of daily life. Singers, snake charmers, bear trainers, jesters, and acrobats—all the roles that later would be revived in vaudeville, traveling carnivals, and circuses—can be located among ancient peoples. String, wind, and percussion instruments, many trimmed with rare metals or precious stones, have been described in print and discovered in situ by archaeologists, allowing a better appreciation of the tonal systems and musical compositions that the ancients created as a source of creativity and for amusement. Ancient plays from the Greeks give voice to many modern concerns about life, meaning, and human affairs.

On an even wider scale one thinks of the grand public spectacles of ancient Babylon and ancient Rome, cities whose rulers spared no expense in putting on public entertainment for the masses, drawing on vast human, animal, and fiscal resources for events that could last for months and involve extensive human and animal carnage. Assurnasirpal II of Assyria, when inaugurating his palace at Calah, claims on a palace relief to have hosted a banquet for 47,074 people, who consumed, among other items, more than 1,000 cattle, 10,000 sheep, 15,000 lambs, 10,000 fish, 10,000 loaves of bread, and 100 containers of beer. Roman emperors staged banquets, games, and entertainments for the masses that sometimes bankrupted the state treasury.

Technological Presence

Pervasive in all these ancient forms of entertainment is the presence and necessity of technology. Natural materials have been reshaped to create implements and means by which human beings can amuse themselves, opening up vast areas for enjoyment beyond those afforded by nature. Technological innovations in chariot wheels and steering mechanisms, the raising and lowering of massive platforms through the use of advanced hydraulics, springs and hinges that could be opened and closed at a distance with precision and split-second timing, and many other inventions and improvements contributed to the crowd-pleasing spectacles of Greek and Roman theaters and the battles in the Colosseum in Rome between beast and beast, person and beast, and person and person.

Continuous innovation in designing and defining amusements of various kinds for particular classes of individuals and entire societies was accelerated with the advent of the printing press and then the Industrial Revolution as mass production of what had been luxury goods for the wealthy began to spread to other levels of society. Greater leisure time for a widening segment of the population created new opportunities for amusements to pass the time. Entertainment itself, however, always has manifested an ability to penetrate social barriers. Shakespeare's plays, for example, appealed not only to the masses but also to extremely wealthy and influential persons.

The modern era brought with it an array of new means of entertainment, including radio, television, video, computer games, virtual reality, film, e-mail, and chat rooms. However, even the older forms of entertainment underwent major changes as sports, for example, moved from the realm of mainly part-time amateur pursuits to a specialized, professional status (there were limited numbers of professional athletes in ancient times).

Within a generation, American football became a multi-billion-dollar television- and media-saturated semiglobal industry and football players became cultural heroes. Technological innovations transformed football from a game played by college students on dirt fields or cow pastures with no equipment to multi-million-dollar weekly gridiron contests in which each side employs advanced scouting technologies, sophisticated weight training and conditioning regimens, carefully managed nutrition programs, lightweight materials for protection, advanced telecommunications equipment to relay commands and insights, rapid-response medical treatments designed to keep players on the field as long as possible, complex ticketing systems, coordinated crowd control, prescheduled advertising breaks, and many other techniques and processes to induce fans to spend thousands of dollars to support their favorite teams.

Miniaturization and Combination

The latest miniaturization and communications technologies allow entertainment to be fully mobile. As they get increasingly smaller and more powerful in terms of resolution quality, camera phones have become the bane of many schools, fitness facilities, and other public venues where some people use them to take and transmit photographs of people in various stages of undress. Students have attempted to use them to film examination questions and send them to others, and similar problems arise with text messaging devices. At the same time users have employed them to film robberies, hit and run incidents, and other criminal acts that have led to court convictions that probably would not have been possible without the visual evidence they provide. Families and individuals have derived enjoyment from camera phone photographs they have taken of special moments and then downloaded into more permanent forms of storage for retrieval when desired. Cam-phone sites have joined the range of types of websites on the Internet, and it is estimated that 260 million camera phones were sold in 2004.

The pervasiveness of computers that are increasingly more powerful yet smaller with each new generation has spawned an enormous industry in designing sophisticated online games. A number of universities have established programs, and others are increasing the number of courses they offer in this area. The most advanced current form of these games are Massive Multiplayer Online Role-Playing Games (MMORPG) that involve thousands of players in a constantly evolving scenario that is affected directly by the self-selected roles and self-assigned personas of the players.

Blogs (web logs) and vlogs (video blogs) are a recent technological innovation in which individuals create self-published websites that feature video clips, running texts of observations or other materials, photographs, and sound to communicate their thoughts or express themselves. Originally pioneered in the late 1990s by sites such as Pop.com and Digital Entertainment Network, they initially failed to catch on but are having a resurgence though sites such as Undergroundfilm and Ourmedia. The more pervasive blogs, which often feature only text, are exerting a growing influence on mainstream media as bloggers democratize and decentralize journalism, news reporting, and information dissemination in entertaining forms.

Various forms of technology are being combined in new ways with the new media to create full-body experiences for people. In a way similar to the manner in which "surround sound" immersed a listener in a piece of music, people can experience a video in three dimensions while simultaneously feeling sensations on their skin and hearing things as if they were fully immersed in the environment they are seeing.

This ability to "experience the world" without really experiencing it raises important issues. Certainly there are training applications in which being able to experience an environment safely and learn how to react successfully within it could save lives in the future as pilots and others in high-risk situations can practice in a simulated world that looks, feels, smells, and tastes like the real thing. At the same time it is easy to imagine situations in which ethical issues should preclude exchanging the real thing for a simulated experience that mimics it exactly, for example, engaging in sexual experiences that one never could or would do in one's normal life.

Preliminary Assessment

The many forms of entertainment available today and the various means by which one can obtain and experience them can lead to a retreat from the world and oneself so pervasive that a person can focus only on the next thrill. Countries with a broad array of entertainment options suffer from what Gregg Easterbrook (2003) terms "the paradox of progress" because despite overwhelming numbers of possessions and experiences, real as well as vicarious, a sense of personal satisfaction and happiness elude people.

Some people have learned that certain forms of media can produce addictions as powerful as those caused by illicit drugs. This is the case in part because one never just uses technology; one also experiences it. This sensory, intellectual, and emotional interplay affects the user in both predictable and unpredictable ways. Reality shows on television have extended this impact more fully to the "actors" themselves as they create live, unscripted drama that others get to enjoy voyeuristically and register their pleasure or displeasure with a particular person on the show just as the emperor and the crowd determined the ultimate fate of ancient gladiators; the difference is that now the phone or mouse click rather than the thumb is the determining signal. Online chat rooms have led some people to alter the course of their lives; although some of the end results appear to be positive, they seem to be outweighed by media and professional counselors' stories of poor decisions and damaging consequences. Many people struggling with personal issues seek escape and relief in a fantasy world that makes them incapable of facing their problems.

Modern people's ancient ancestors would recognize most of the dilemmas that modern entertainment presents. They undoubtedly also would recognize that these ethical and moral challenges have multiplied over time and space.

DENNIS W. CHEEK

SEE ALSO *Movies; Music; Museums of Science and Technology; Popular Culture; Radio; Robot Toys; Science, Technology, and Society Studies; Special Effects; Sports; Television; Video Games; Violence.*

BIBLIOGRAPHY

Bender, Gretchen, and Timothy Druckrey, eds. (1994). *Culture on the Brink: Ideologies of Technology.* Seattle, WA: Bay Press. A set of essays on the meaning of technology that challenges notions of progress and determinism.

Bray, John. (2002). *Innovation and the Communications Revolution: From the Victorian Pioneers to Broadband Internet.* London: Institution of Electrical Engineers. A historical look at how global communications have developed and have affected the way people live.

Brende, Eric. (2004). *Better Off: Flipping the Switch on Technology.* New York: HarperCollins. A technology guru reveals his thoughts about the pervasive and unhealthy influence of technology on life in the twenty-first century.

Easterbrook, Gregg. (2003). *The Progress Paradox: How Life Gets Better While People Feel Worse.* New York: Random House. A hard look at one of the main dilemmas of modernity.

Hakken, David. (2003). *The Knowledge Landscapes of Cyberspace.* New York: Routledge. An anthropologist looks at the many aesthetic, ethical, and political questions posed by the uses and abuses of information technology.

McCarthy, John, and Peter Wright. (2004). *Technology as Experience.* Cambridge, MA: MIT Press. A thoughtful monograph that explores the emotional, intellectual, and sensual aspects of people's interactions with technology.

Pesce, Mark. (2000). *The Playful World: How Technology Is Transforming Our Imagination.* New York: Ballantine. A clever analysis of the way the use of distributed intelligence, engineered structures that interact with natural structures, and instantaneous access to information will transform culture and create a new Babel.

Peters, John Durham. (1999). *Speaking into the Air: A History of the Idea of Communication.* Chicago: University of Chicago Press. An evocative history of communication in Western thought, showing how it shapes and has been shaped by people.

Postman, Neil. (1986). *Amusing Ourselves to Death: Public Discourse in the Age of Show Business.* New York: Penguin. A well-known critique of contemporary life and culture.

Schultze, Quentin J. (2002). *Habits of the High-Tech Heart: Living Virtuously in the Information Age.* Grand Rapids, MI: Baker Book House. A Christian critique of contemporary high-tech culture in the United States.

Squier, Susan Merrill, ed. (2003). *Communities of the Air: Radio Century, Radio Culture.* Durham, NC: Duke University Press. A pioneering analysis of radio as both a material and a cultural production.

Sterne, Jonathan. (2003). *The Audible Past: Cultural Origins of Sound Reproduction.* Durham, NC: Duke University Press. A look at the technological and cultural precursors of telephony, phonography, and radio.

Stewart, David. (1999). *The PBS Companion: A History of Public Television.* New York: TV Books. A look at an American cultural icon that is frequently mentioned but infrequently watched.

ENTREPRENEURISM

• • •

"[An] entrepreneur is a person who habitually creates and innovates to build something of recognized value around perceived opportunities" (Kotelnikov Internet article). This "recognized value" should incorporate social and ethical concerns, as well as economic ones. There are moral dimensions to all forms of entrepreneurship.

Conceptual Distinctions

Entrepreneurs include both scientists seeking to advance research and engineers seeking new design opportunities. Entrepreneurship is not the same as invention. Alexander Graham Bell obtained a broad patent that included the transmission of speech, but he was not an entrepreneur—others took his patent and used it to create a corporate giant (Carlson 1994). Thomas Edison, in contrast, supervised invention, manufacturing, and marketing of a new electric lighting system (Hughes 1983); therefore, he is both inventor and entrepreneur. Classic theorists and economists also have developed and expressed their own opinions concerning entrepreneurship and its influence on economic development. In 1928, economist Joseph Schumpeter stated that the "essence of entrepreneurship lies in the perception and exploitation of new opportunities in the realm of business ... it always has to do with bringing about a different use of national resources in that they are withdrawn from their traditional employ and subjected to new combinations" (Filion 1997, p. 3).

Entrepreneurs must promote their ideas relentlessly. They have, however, an obligation to be honest with themselves and others about their prospects.

Unethical Entrepreneurship

In the early twenty-first century, America watched companies such as Enron and WorldCom collapse. Enron was formed by the merger of Houston Natural Gas, a regional pipeline company, and InterNorth, a Nebraska-based pipeline owner, which was organized by Kenneth Lay in 1985. The beginnings of Enron's downfall can be traced to the late 1980s: When federal regulations allowed gas prices to fluctuate naturally, Enron saw this as an opportunity to add gas trading to its list of business endeavors. Then beginning in the mid-1990s, "Enron tried to duplicate its initial success at energy trading in new fields—coal, paper, plastics, metals and even Internet bandwidth. Many of these ventures went badly wrong, so executives turned to the tried-and-true method of big business—hide the problem and hope that everything gets better" (Maass 2002, pp. 6–7) Maass then notes that "Enron hid its mounting losses and skyrocketing debt, both in little-examined nooks and crannies of official statements and in off-the-record *partnerships* run by Enron executives. By hiding debt in the partnerships, Enron's official bottom line continued to look healthy—while executives raked in millions in fees for administering them" (Maass 2002, pp. 6–7). Enron lied to its own employees and shareholders, many of whom were left with virtually worthless stock. Joe Lieberman, Senator from Connecticut, commented that "Enron has become a grand metaphor for the real human problems that profit pressure can produce when it goes to gross extremes because it is unchecked by personal principles or business ethics" (Lieberman 2002). WorldCom, an entrepreneurial telecommunications company, masked losses by clever, but dishonest accounting schemes.

The environment in which new companies enter may be responsible for the ethics dilemmas companies encounter. Arthur Levitt, former SEC chairman, states "fierce competition in the marketplace is healthy, but we've seen that the corporate race to beat analyst projections can breed disdain for investors' interests and the law" (Lieberman Internet site). Jennifer Lawston also writes that "the entrepreneurial world—particularly the high-tech entrepreneurial world—is living through a time of high temptation. The devil on one shoulder tells you to make the numbers and set projections to make investors feel good, the angel on the other says to tell the story like it really is" (Lawston 2003 Internet article).

Entrepreneurship requires truth-telling—to investors and the public. The Enron and WordCom cases illustrate the consequences of lying. Entrepreneurs also need to be honest with themselves.

The history of dot-com company failures reveals the dangers of self-delusion. Peter Coy suggests thinking of "dot-com startups not as companies but as hypoth-

eses—economic hypotheses about commercial methods that needed to be tested with real money in the real world. Nobody was forced to fund the experiments, but plenty of people who hoped to get rich quickly were happy to thrust money into the hands of entrepreneurs such as Walker, Jeff Bezos of Amazon, Tim Koogle of Yahoo!, and Candice Carpenter of iVillage, a Website for women" (Coy Internet article). The dot-coms suffered from confirmation bias (Gorman 1992)—they believed that because their stock was rising, their hypothesis was right, and the *old* economic laws did not apply to their situation.

Doing Well by Doing Good

Entrepreneurs are pioneers who open new territory. C.K. Prahalad and Allen Hammond (2002) have used a pyramid metaphor to describe the global market. Tier 1 consists of roughly 100 million people whose earnings are greater than $20,000 per year. Tier 2 consists of the poor in developed countries and Tier 3 consists of the rising middle class in the developing world, amounting to approximately 1.75 billion people whose earnings fall between $2,000 and $20,000 per year. Tier 4 includes the majority of the Earth's population, about 4 billion people earning less than $2,000 per year. As one goes down the pyramid, the proportions of people in each tier shift from the developed to the developing world.

In *Development As Freedom* Amartya Sen (1999) argues that "economic unfreedom, in the form of extreme poverty, can make a person helpless prey in the violation of other kinds of freedom" (p. 8). Sen believes the development of a competitive market system in poverty-stricken countries will, in time, improve the economic condition, which will in turn create numerous freedoms for their inhabitants.

The Tier 4 market therefore represents a new frontier that most established businesses shun—where an entrepreneur could make a profit while improving the quality of life. In 1969 Karsanbhai Patel, a factory chemist dissatisfied with his job and low income, decided to create and manufacture an affordable detergent for the Tier 4 market in India. Patel mixed a powder and began selling it to neighboring towns on his bicycle. Distributors eventually showed an interest in the product, and Patel's product spread nationwide.

Patel created a cottage industry that allowed individuals from Tier 4 markets to make money manufacturing and selling his product, but this cottage industry structure meant he did not have to pay his employees benefits. His efforts inspired Hindustan Lever Limited, the former leaders in market share, to enter this Tier 4 territory, thereby providing Tier 4 consumers with a choice between products.

Another example of an entrepreneur who wanted to benefit women around the world and also make a profit is Mary Ann Leeper. She bought the rights to a prototype female condom, but modified it, figured out how to manufacture it, and made it available on a global basis. Leeper created the Female Health Company, which "has focused its marketing efforts on establishing a presence in major world markets and building relationships with key world health agencies and programs. The female condom has been introduced in Japan, Africa, Latin America, the United Kingdom, the United States and Europe. ("The Female Health Company Biography: Mary Ann Leeper" Internet article). The female condom has been "hailed as a way of giving women increased power to protect themselves from sexually transmitted diseases" (Baille 2001).

Entrepreneurs have the ability to choose whether ethics will be a priority in their fledgling companies. Ben Cohen, a founder of Ben & Jerry's Ice Cream wrote in 1976 that "Business has a responsibility to give back to the community from which it draws its support" (Mead 2001). Cohen and Jerry Greenfield developed what they called a *values-led* company, which for them "meant a commitment to employees, the Vermont community, and social causes in general" (Mead 2001). In 1985 Cohen and Greenfield established the Ben & Jerry's Foundation to help disadvantaged groups, social change organizations, and environmentalists, donating 7.5 percent of the company's annual pre-tax profits. Ben & Jerry's became a subsidiary of Unilever, a multinational corporation that is also the parent company of HLL and is dedicated to measuring success via a triple bottom-line, in which environmental and social progress is just as important as financial gain (Gorman, Mehalik, and Werhane 2000).

Conclusions

For scientists and engineers, entrepreneurship represents an opportunity to discover and even create markets (Gorman and Mehalik 2002). Attention to social and ethical impacts will actually increase the likelihood that an innovation will be accepted.

The entrepreneur needs to:

- Be truthful with potential customers and investors.

- Consider whether a new technology is more likely to benefit or harm the global environment.

- Consider the impact of a new technology on the Tier 4 market. Will it increase the gap between rich and poor, or give the poor the opportunity to improve their situation?

- Measure progress using social and environmental metrics, as well as economic.

MICHAEL E. GORMAN
EMILY LYNN BRUMM

SEE ALSO *Business Ethics; Management: Models; Technological Innovation; Work.*

BIBLIOGRAPHY

Ahmad, Pia Sabharwal, et al. (2004). *Hindustan Lever Limited (HLL) and Project Sting: Case A.* Charlottesville: University of Virginia Darden School. This case-study, and others below, are available as Darden School publications; see www.darden.edu.

Carlson, W. Bernard. (1994). Entrepreneurship in the Early Development of the Telephone: How did William Orton and Gardiner Hubbard Conceptualize This New Technology? *Business and Economic History* 23(2): 161–192.

Gorman, Michael E. (1992). *Simulating Science: Heuristics, Mental Models and Technoscientific Thinking.* Bloomington: Indiana University Press. Describes when confirmation is and isn't a bias.

Gorman, Michael E., Matthew M. Mehalik. (2002). "Turning Good into Gold: A Comparative Study of Two Environmental Invention Networks." Science Technology and Human Values, 27(4): 499–529. This paper provides an STS framework for entrepreneurs in science and technology.

Gorman, Michael E., Matthew M. Mehalik, and Patricia Werhane. (2000). *Ethical and Environmental Challenges to Engineering.* Englewood Cliffs, NJ: Prentice-Hall. This book contains case-studies of engineers and businesses that encountered ethical dilemmas while they were conducting entrepreneurial activities.

Hughes, Thomas P. (1983). *Networks of Power: Electrification in Western Society: 1880–1930.* Baltimore: Johns Hopkins University Press.

Mead, Jenny, Robert J. Sack, and Patricia H. Werhane. (2001). "Ben & Jerry's Homemade, Inc. (A): Acquisition Suitors At The Door." Publication E-0225. Charlottesville: University of Virginia Darden School Foundation.

Prahalad, C.K., and Allen Hammond. (2002). "Serving the World's Poor, Profitably," *Harvard Business Review* 80(9): 48–57.

Sen, Amartya K. (1999). *Development as Freedom.* New York: Oxford University Press.

INTERNET RESOURCES

Baille, Andrea. "Women Offer to Participate in Female Condom Study." Canoe. Available from http://www.canoe.ca/Health0110/26_condom-cp.html.

Coy, Peter. "Rise & Fall." Cornell Magazine. Cornell University. Available from http://cornell-magazine.cornell.edu/Archive/2001mayjun/RiseFall.html.

"The Female Health Company Biography: Mary Ann Leeper." Availiable from http://www.femalehealth.com/PressReleasesAug99toCurrent/PressReleaseBiography7.19.2000.htm

Filion, Louis Jacques. (1997). "From Entrepreneurship to Entreprenology." Availiable from http://www.usasbe.org/knowledge/proceedings/1997/P207Filion.PDF.

Kotelnikov, Vadim. (2004). "The Entrepreneur: The Key Personality, Environmental and Action Factors. Available from http://www.1000ventures.com/business_guide/cross-cuttings/entrepreneur_main.html.

Lawston, Jennifer. (2003). "Ethics and the Entrepreneur." USA Today. Available from http://www.usatoday.com/money/jobcenter/2003-01-03-entre-ethics_x.htm.

Lieberman, Joe (2002). "Business Ethics in the Post-Enron Era: What Government Can Do And What Business Must Do." 1 April 2002. Available from http://www.southerninstitute.org/Resources-GoodBusiness-Content(1).htm.

Maass, Alan, and Todd Chretien. (2002). "A Tale of Greed, Deceit, and Power Politics." Socialist Worker Online. Available from http://www.socialistworker.org/2002-1/390/390_06_EnronTaleOfGreed.shtml.

ENVIRONMENTAL ECONOMICS

• • •

As the entry on "Economics: Orientation" points out, welfare economics puts the "satisfaction of individual human desires at or near the top of its own internal moral hierarchy." Two economists observe, "The basic premises of welfare economics are that the purpose of economic activity is to increase the well-being of the individuals that make up the society, and that each individual is the best judge of how well off he or she is in a given situation" (Stokey and Zeckhauser 1978, p. 277).

Environmental economics builds on the theory of welfare economics (or microeconomics) and in particular the view—prepresented as an ethical theory—that the satisfaction of preferences taken as they come ranked by the individual's willingness to pay (WTP) to satisfy them is a good thing because (by definition) this constitutes welfare or utility. According to economist David Pearce (1998, p. 221), "Economic values are about what people want. Something has economic value—is a benefit—if it satisfies individual preferences." This approach uses maximum WTP to measure how well off the individual believes a given situation makes her or him. A representative text states, "Benefits are the sums of the maximum amounts that people would be willing to pay to gain outcomes that they view as desirable" (Boardman, Greenberg, Vining, et al. 1996).

Preference Satisfaction

The attempt to link preference satisfaction (and therefore WTP) with well-being or benefit, however, encounters four problems. First, one may link preference with welfare by assuming that individuals prefer what they believe will make them better off. Research has shown, on the contrary, that with respect to environmental and other policy judgments, people base their values and choices on moral principles, social norms, aesthetic judgments, altruistic feelings, and beliefs about the public good—not simply or even usually on their view of what benefits them. The basis of environmental values in moral principle, belief, or commitment rather than self-interest severs the link between preference and perceived benefit or welfare.

In recent decades, environmental economists have put a great deal of effort into developing methodologies for measuring the benefit associated with goods—sometimes called "non-use" or "existence" values—that people care about because of moral beliefs, aesthetic judgments, or religious commitments, rather than because of any benefit or welfare change they believe those goods offer them. According to economist Paul Milgrom (1993, p. 431), for existence value to be considered a kind of economic value, "it would be necessary for people's individual existence values to reflect only their own personal economic motives and not altruistic motives, or sense of duty, or moral obligation." The difference between what people believe benefits them (economic motives) and what they believe is right (moral obligation) divides economic value from existence or non-use value. The attempt to translate moral beliefs and political judgments into economic benefits—principled commitments into units of welfare and thus into data for economic analysis—may continue to occupy economists for decades to come, because many logical, conceptual, and theoretical conundrums remain.

Second, the statement that the satisfaction of preference promotes welfare states a tautology if economists define "welfare" or "well-being" in terms of the satisfaction of preference, as generally they do. Concepts such as "welfare," "utility," and well-being are mere stand-ins or proxies for "preference-satisfaction" and so cannot justify it as a goal of public policy.

Additionally, if "well-being" or "welfare" refers to a substantive conception of the good, such as happiness, then it is simply false that the more one is able to satisfy one's preferences, the happier one becomes. That money (or income—a good surrogate for preference satisfaction) does not buy happiness may be the best-confirmed hypothesis of social science research. Thus, the thesis that preference satisfaction promotes welfare appears either to be trivially true (if "welfare" is defined as preference satisfaction) or empirically false (if "welfare" is defined as perceived happiness).

Third, if preferences are mental states, they cannot be observed. If they are inferred or "constructed" from behavior, they are also indeterminate, because there are many ways to interpret a person's actions as enacting a choice, depending on the opportunities or alternatives the observer assumes define the context. For example, the act of purchasing Girl Scout cookies could "reveal" a preference for eating cookies, supporting scouting, not turning away the neighbor's daughter, feeling good about doing the right thing, avoiding shame, or any of a thousand other possibilities. Choice appears to be no more observable than preference because its description presupposes one of many possible ways of framing the situation and determining the available options.

Fourth, few if any data indicate maximum WTP for any ordinary good. When one runs out of toothpaste, gets a flat tire, or has to buy the next gallon of milk or carton of eggs, one is unlikely to know or even have an idea about the *maximum* one is willing to pay for it. Instead, one checks the advertisements to find the *minimum* one has to pay for it. It is not clear how economists can estimate maximum WTP when all they can observe are competitive market prices. Competition drives price down to producer cost, not up to consumer benefit. For example, one might be willing to pay a fortune for a life-saving antibiotic, but competition by generics may make the price one actually pays negligible.

The difference between price and benefit is clear. People usually pay about the same prices for a given good no matter how much they differ in the amount they need or benefit from it. People who benefit more and thus come first to the market may even pay less, for example, for seats on an airplane than those who are less decided and make later purchases. Thus, maximum WTP, which may correlate with benefit, cannot be observed, while market prices, which can be observed, do not correlate with benefit.

Market Prices

Environmental economists also propose that the outcome of a perfectly competitive market—one in which property rights are well defined and people do not encounter extraordinary costs in arranging trades and enforcing contracts—defines the way environmental assets are most efficiently allocated. Market prices constantly adjust supply and demand—the availability of goods to the wants and needs of individuals. As the "Orientation" entry observes, a perfectly competitive market may be used to define the idea of economic efficiency—the condition in which individuals exhaust all the advantages of trade because any further exchange would harm and thus not gain the consent of some individual.

Economists often explain the regulation of pollution not in moral terms (trespass, assault, violation of rights or person and property) but in terms of the failure of markets to "price" goods correctly. Suppose for example a factory emits smoke that causes its neighbors to bear costs (such as damage to property and health) for which they are not compensated. The factory, while it may pay for the labor and materials it uses, "externalizes" the cost of its pollution. When only a few neighbors are affected, they could negotiate with the factory, either paying the owner to install pollution-control equipment (if the zoning gave the factory the right to emit smoke) or by accepting compensation. The factory owner and the neighbors would bargain to the same result; the initial distribution of property rights determines not the outcome but the direction in which compensation is paid. This is an example of the second theorem described in the "orientation" entry, according to which the initial distribution of goods and services does not really matter in determining the outcome of a perfectly functioning market.

Where many people are affected, as is usually the case, however, the costs of bargaining ("transaction costs") are large. Economists recommend that the government tax pollution in an amount that equals the cost it "externalizes," that is, imposes on society. The industry would then have an incentive to reduce its emissions until the next or incremental reduction costs more than paying the tax—the point where in theory the cost (to the industry) of reducing pollution becomes greater than the benefits (to the neighbors). Such a pollution tax would "internalize" into the prices the factory charges for its products the cost of the damage its pollution causes, so that society will have the optimal mix of those products and clean air and water.

Many economists point out, however, that the government, in order to set the appropriate taxes or limits, would have to pay the same or greater costs as market players to gather information about WTP for clean air or water and willingness to accept (WTA) compensation for pollution. Pollution taxes, to be efficient, "should vary with the geographical location, season of the year, direction of the wind, and even the day of the week ..." (Ruff 1993, p. 30). The government would be "obliged to carry out factual investigations of mind-boggling complexity, followed by a series of regulatory measures that would be both hard to enforce and valid only for a particular, brief constellation of economic forces" (Kennedy 1981, p. 397). Thus, regulation is unnecessary when transaction costs are small (because people can make their own bargains) and unfeasible when they are great (because the government would have to pay them).

By arguing that emissions be optimized on economic grounds—rather than minimized on ethical grounds—economists reach an impasse. According to Ronald Coase, "the costs involved in governmental action make it desirable that the 'externality' should continue to exist and that no government intervention should be undertaken to eliminate it"(1960, p. 25–26).

Maximum WTP

Economists regard the ubiquitous and pervasive failure of markets to function perfectly as a reason that society, in order to achieve efficiency, should transfer the power to allocate resources to experts, presumably themselves, who can determine which allocations maximize benefits over costs. By replacing market exchange with expert opinion to achieve efficiency, however, society would sacrifice many non-allocatory advantages of the market system. For example, by making individuals responsible for decisions that affect them—rather than transferring authority to the government to act on their behalf—markets improve social stability. People have themselves, each other, or impersonal market forces to blame—not the bureaucracy—when purchasing decisions do not turn out well for them.

Economists have encountered logical and conceptual hurdles, moreover, in their efforts to develop scientific methods for valuing environmental assets and thus for second-guessing market outcomes. First, there is little evidence that economic experts are able to assemble information about WTP and WTA any better than market players when the costs of gathering that information are high. Second, economic estimates of benefits and costs when made by government agencies become objects of lobbying, litigation, and criticism. Experts can be hired on both sides of any dispute and then produce dueling cost-benefit analyses (Deck 1997). Third, when society transfers power to scientific managers, even if they are

trained welfare economists, it courts all the problems of legitimacy that beset socialist societies, which likewise may rely on scientific managers to allocate resources.

Institutional Approaches

Pollution control law, from a moral point of view, regulates pollution as a kind of trespass or assault, on analogy with the common law of nuisance. Statutes such as the Clean Air Act and Clean Water Act, moreover, explicitly rule out a cost-benefit or efficiency test and pursue goals such as public safety and health instead (Cropper and Oates 1992). For this reason, the government often limits to "safe" levels the maximum amount of various pollutants industries and municipalities may emit into the water and air. To determine what levels are "safe enough" legislators and regulators have to consider the state of technology and make ethical and political judgments. To help society attain the mandated levels in the most cost-effective ways, economists have made an important contribution to environmental policy by urging government to create market-like arrangements and thus to generate price signals for allocating environmental goods for which markets do not exist.

For example, the Environmental Protection Agency, by creating pollution permits or allowances that firms can buy and sell under an aggregate total ("CAP"), gave industries incentives to lower emissions of lead, smog, and other pollutants to below permitted levels, because they could sell at least part of the difference to other companies that find emissions more expensive to reduce. Tradable rights in environmental assets (from emission allowances to rights to graze the public range) show that incentives matter; marketable permits can reduce pollution more effectively and at lower cost than "command and control" policies. In addition, market arrangements decentralize decisions by encouraging industries to make their own bargains to attain the overall "CAP" rather than to conform to one-size-fits-all regulation.

Environmental economics has enjoyed success in helping society construct market-like arrangements for achieving in the most cost-effective ways environmental goals, such as pollution-reduction, justified on moral, political, and legal grounds. Environmental economics as a discipline has been less successful in finding scientific methods to second-guess or replace markets in order to achieve goals it itself recommends, such as preference-satisfaction or efficiency, that are not plainly consistent with moral intuitions, legislation, or common law traditions.

MARK SAGOFF

SEE ALSO *Ecological Economics; Economics and Ethics; Environmental Ethics; Market Theory.*

BIBLIOGRAPHY

Anderson, Terry L., and Donald L. Leal. (1991). *Free Market Environmentalism*. San Francisco: Pacific Research Institute for Public Policy.

Boardman, Anthony; David H. Greenberg; Aidan R. Vining; and David L. Weimer. (1996). *Cost-Benefit Analysis: Concepts and Practice*. Upper Saddle River, NJ: Prentice Hall.

Coase, Robert H. (1960). "The Problem of Social Cost." *Journal of Law and Economics* 3(1): 1–44.

Cropper, Maureen L. and Wallace E. Oates (1992), "Environmental Economics: A Survey." *Journal of Economic Literature* 30: 675–740.

Daily, Gretchen C., ed. (1997). *Nature's Services: Societal Dependence on Natural Ecosystems*. Washington, DC: Island Press.

Deck, Leland. (1997). "Visibility at the Grand Canyon and the Navajo Generating Station." In *Economic Analysis at EPA: Assessing Regulatory Impact*, ed. Richard D. Morgenstern. Washington, DC: Resources for the Future.

Kennedy, Duncan. (1981). "Cost Benefit Analysis." *Stanford Law Review* 33: 387–421.

Milgrom, Paul. (1993). "Is Sympathy an Economic Value?" In *Philosophy, Economics, and the Contingent Valuation Method*, in Contingent Valuation: A Critical Assessment," ed. Jerry A. Hausman. Amsterdam: Elsevier North-Holland.

Pearce, David. 1998. *Economics and the Environment*. Cheltenham, UK: Edward Elgar.

Ruff, Larry. (1993). "The Economic Common Sense of Pollution." In *Economics of the Environment: Selected Readings*, 3rd edition, ed. Robert Dorfman and Nancy Dorfman. New York: Norton.

Stokey, Edith, and Richard Zeckhauser. (1978). *A Primer for Policy Analysis*. New York: W.W. Norton.

ENVIRONMENTAL ETHICS

• • •

Modern science and technology have brought about a unique, human-caused transformation of the Earth. Although humans have for thousands of years had measurable terrestrial impacts with fire, agriculture, and urbanization, since the Industrial Revolution the scope, scale, and speed of such impacts have exceeded all those in the past (Kates, Turner, and Clark 1990) and promise to become even more dramatic in the future. Humans have become what the Russian scientist V. I. Vernadsky in the 1920s called a *geological force*, in a sense even more strongly than he imagined it. Environmental ethics and, more generally, environmental philosophy comprise a variety of philosophical responses to the concerns raised by the magnitude of this transformation.

Basic Issues

Since the noxious clouds and pollution-clogged rivers of the Industrial Revolution, society has generally agreed that many modern technological activities, due to their potentially devastating impact on nature and people, are in need of regulation. Recognition of the fact that humans can foul their own nests is now widely accepted and often politically effective. Indeed concerns about the counterproductivity of scientifically and technologically enhanced human conduct in the early-twenty-first century extends to discussions of population increase, environmental costs borne by the poor and minorities, and responsibilities to future generations. But it is not clear that addressing such anthropocentric worries adequately encompasses all properly human interests. Beyond working to live within the environmental limits for the production of resources and absorption of wastes—which can be pursued both by moderating activities and transforming technologies—questions arise about whether nonhuman or extrahuman considerations have a role to play. What are the ethical responsibilities, the moral duties, of humans to nonhuman animals, plants, populations, species, biotic communities, ecosystems, and landforms? Should environmental outcomes for these, or at least some of these, not mean something in their own right quite apart from their mere resource value for human exploitation? Ought humans not to *respect* nature to some degree for what it is intrinsically? Much of the professional field of environmental ethics has been exercised with articulation and debate regarding the relative weight of human- and extrahuman-centered concerns.

Moreover early-twenty-first-century humans are often ambivalent about the place of nature in human life. Human beings of all times and places have needed and wanted freedom from many of the harsh conditions of the natural world, but earlier humans also celebrated the grace of nature in art, song, story, and ceremony. Modern technological efforts perfect the former and neglect the latter. Could it be that human flourishing is connected to nature flourishing? If so, over and above respect for nature, a *celebration* of things natural and the natural world in human lives and communities throughout the world is necessary. Although their numbers remain comparatively small, many people are stirred by passionate feelings about nature in communities that have long *turned their backs to the river*. While some critics argue the pathetic fallacy of such positions, the question of how much say nature and natural things will have over human life and the planet remains.

Any philosophical criticism of the anthropogenic transformation of the natural world must ultimately lead to an assessment of human culture. Perhaps humans have been at some level mistaken about the fundamental payoffs of environmental exploitation. In many instances, the technological control of nature that displaces its celebration leaves people numb, mindless, or out of shape. It seems necessary to coordinate a critique of technological damage with discovery of new ways for living with nature in order simultaneously to save the planet from environmental degradation and society from cultural impoverishment. The quality of the environment and of human life —questions of environmental and interhuman ethics (the good life)—may be inseparable.

The extensive transformation of the Earth deserves to be seen from the perspective of both the natural world and culture. This transformation could not have taken place without widespread agreement underlying the fundamental orientation of the modern technological project. There may be several ways of understanding this agreement, and there is debate among scholars on this issue (Borgmann 1984; Higgs, Light, and Strong 2000; Zimmerman et al. 2000). Despite differences, there is nevertheless a consensus that unless people unite concerns for nature and culture, environmental ethics will prove to be inconsequential. In other words, an effective environmental ethics and philosophy must include as well a philosophy of technological culture.

Historical Development

Historically environmental ethics is associated with a certain unease about the unbridled exploitation of nature that is typical of post-Industrial Revolution society. As Roderick Nash (2001), among others, points out, such uneasiness was first evidenced in post-Civil War United States concerns over the loss of both wilderness and natural resources—concerns that led to the creation of the first U.S. national park (1872) and then forest service (1905). After World War II, the creation of a second wave of environmental concern centered around the wilderness movement of the 1950s that led to the Wilderness Act (1964) and Rachel Carson's *Silent Spring* (1962), which argued that aggressive technology in the form of the extensive use of chemical pesticides, especially DDT, was killing millions of songbirds and could eventually have a much broader impact on plant, animal, and even human life. Nuclear weapons and energy

production, technological disasters (such as the Santa Barbara oil spill of 1969), wasteful extraction and use of resources, the rise of consumerism, the population explosion, oil shortages (in 1973 and 1977), pollution, and a host of related environmental problems combined to establish in popular consciousness what can be called an environmental or ecological crisis about the health of the Earth as a whole. Existing conservation measures, with their many successes, were nevertheless judged too weak to respond to the new problems, leading to the enactment of the National Environmental Policy Act (NEPA 1969) and to the establishment of the first national Environmental Protection Agency (EPA) by President Richard Nixon in 1972. At the same time, some began to question whether enlightened self-interest was a sufficient basis for assessing the contemporary state of environmental affairs and argue that nature mattered in ways beyond its strictly human utilities and should be protected with an eye for more than human safety and health.

Among figures such as Ralph Waldo Emerson, Henry David Thoreau, John Muir, and Albert Schweitzer, wildlife biologist and ecologist Aldo Leopold advocated this position before environmental ethics became a popular movement. Early in his career as a professor of wildlife management, Leopold thought that nature could be reorganized for human ends (enhancing wildlife populations by eliminating wolves, for instance) if one took a long-range view and was scientifically informed. However in his mature work *A Sand County Almanac and Sketches Here and There* (1949), Leopold criticized reliance on the conservationist position, and argued that *the land*, what is now called an ecosystem, must be approached holistically and with love and respect, that is, with what he calls *the land ethic*. He said, "A thing is right when it tends to preserve the integrity, stability, and beauty of the biotic community. It is wrong when it tends otherwise" (Leopold 2000, pp. 224–225). Leopold articulated the emerging new ethic concerning both living things as individuals and the natural system itself, including the *community concept* generated by the relatively new sciences of evolution and ecology. This was in contrast to earlier theories that invited human domination of nature, which Leopold believed were encouraged by the older hard sciences.

The groundwork for environmental ethics as such was laid in the 1970s. During the first half of that decade, four independent philosophical works launched the academic field. Arne Naess's "The Shallow and Deep, Long-range Ecology Movement: A Summary" (1973) called for a radical change in the human-nature relationship, and became a seminal work of the deep ecology movement. Richard Sylvan's (then Routley) "Is There a Need for a New, an Environmental, Ethic?" (1973) argued that modern ethical theories were inadequate for the full range of moral intuitions regarding nature. Peter Singer's "Animal Liberation" (1973) reanimated Jeremy Bentham's proposal for including sentient members of nonhuman species in the utilitarian calculus. And Holmes Rolston III's "Is There an Ecological Ethic?" (1975) distinguished between a *secondary sense* environmental ethic in which moral rules are derived from concerns for human health or related issues, and a *primary sense* environmental ethic, in which nonhuman sentient animals as well as all living things, ecosystems, and even landforms are respected because they are intrinsically valuable *apart* from any value to humans. For Rolston, the secondary ethic is anthropocentric and not truly environmental, whereas the primary is truly environmental and ecocentric. Subsequently Kenneth Goodpaster (1978) developed the fertile concept of *moral considerability* to discuss more generally who and what, if anything nonhuman, counts ethically. In 1979 Eugene C. Hargrove founded *Environmental Ethics*, the first journal in the field.

Mainstream environmental ethics matured over the next decade. Animal liberation and rights discussions flourished and became a separate field from, and often in conflict with, environmental ethics, because ecosystems sacrifice the welfare of individual animals. Within mainstream environmental ethics, ethical theories regarding individual *lives* of animals and plants, usually called biocentric or life-centered, began to be distinguished from holistic approaches that dealt with preserving entire ecosystems, called ecocentric or ecosystem-centered ethics. Leopold's earlier vision developed in different ways in the systematic works of J. Baird Callicot, deep ecologists, and others.

During the 1980s and 1990s, the field witnessed remarkable growth. Currently there are numerous journals, two professional organizations (the International Society for Environmental Ethics and the International Association for Environmental Philosophy), and an array of Internet sites devoted to it. Colleges and universities routinely teach courses in environmental ethics.

Environmental ethics theorists, in the early twenty-first century, believe they are taking a radically new direction because they are informed by scientific insight and philosophical prowess. Many aspire to produce universal claims about humans and the environment. They argue the urgent need for a new environmental ethic governing the duties of people toward nature, and reject the view of nature, which started with the rise of modern science, as value-neutral stuff that humans can manipulate as they please.

However recent developments in science complicate and challenge environmental ethics. Ecology has always accepted change, but modern ecology has moved away from early ecology's notion of stable, climax communities (usually pre-Columbian) reached by moving at a steady pace through successive stages. The notion that nature tends toward equilibrium conditions, a *balance of nature*, has become largely rejected in favor of the view that ecological processes are much more unruly and undirected. Catastrophic, episodic, and random events may be more responsible for the ecological condition than ordinary cycles. Ecological settings, once disturbed, do not automatically return to their predisturbance state. What were thought to be symbiotic relationships between members of an ecological community are often better understood as assemblages of individuals acting opportunistically. The assumed relationship between biodiversity and stability does not always hold up to scientific scrutiny. Added to these are complicating human and cultural influences such as the role of Native Americans in shaping ecosystems, the European introduction of horses, and global climate change. In light of these factors, environmental ethics theorists must again consider the acceptability of the control and maintenance of nature for human benefit. Those who have been inspired by ecology generally, and Leopold in particular, struggle to revise their theories. Most of these revisions turn on protecting dynamic processes rather than fixed-states and on considering the relative magnitude of anthropogenic transformation. Modern human-caused ecological changes differ dramatically from natural-caused changes in terms of rates (for instance, of extinction or of climate change), scope, and scale.

Beginning in the late 1980s and following a direction initiated by deep ecology, environmental ethics, with its focus on elaborating moral duties to nature, was felt by many to be too constrictive to address the questions of humankind's place in nature or nature's place in the technological setting. Nature seems to count in ways that were neither exclusively exploitive nor independent of humans. Others find that environmental ethics too often stops short of cultural critique. For instance, criticism of the modern transformation of the Earth from a predominantly technological and cultural standpoint is considered to be an inappropriate subject for the journal *Environmental Ethics*. Third, concern with a new environmental ethic in a primary sense was denounced for diverting attention from developing sophisticated and effective anthropocentric positions. Whereas environmental ethics is popularly understood as being synonymous with environmental philosophy, philosophers often conceive of environmental philosophy an alternative field, distinguished by its philosophical broader concern regarding the human and cultural relationship with nature (Zimmerman et al. 2000).

Scope and Central Issues of Environmental Ethics

Human beings are expected to act morally, but no such expectation exists for animals and plants; ethics is limited traditionally to the sphere of moral agents, those capable of reciprocity of rights and duties. No one in environmental ethics argues that anything in nature is a moral agent and morally responsible. If human beings can overcome the problem of extending moral duties beyond moral agents, other issues become central for environmental ethics including: (a) What duties should constrain human actions on the part of other beings who can suffer or are subjects of a life, that is, sentient animals? and (b) Should sentient animals be ranked, for example, primates first, followed by squirrels, trout, and shrimp, depending on the degree to which an animal can be pained or the complexity of their psychological makeup? Human duties toward different kinds of animals may be clarified with advances in neuroscience and animal psychology.

However ranking animals according to these hierarchies may simply be an imposition of anthropocentric norms, that is selecting paradigmatically human characteristics as a basis for rank. What duties do humans have toward those who are *alive*, usually defined as nonsentient animals and plants? Do all living things possess biological needs, even when there are no psychological interests? If some duties obtain, how should these take into consideration the natural order where life feeds on life and might makes right, for instance?

Other issues also merit consideration. Can *moral extensionism* by analogy apply beyond *individualistic accounts*, beyond selves, to other parts of the natural world? Do humans have duties to microbes and to mere things such as rocks, rivers, landforms, and places? A final crucial issue to be considered by environmental ethics is whether humans have duties to species and ecosystems that are not only not alive and do not suffer, but are not individual beings at all?

Many environmental ethicists believe that humankind can answer these questions only when the question of whether nature possesses intrinsic value is answered (Light and Rolston 2002). Normally intrinsic value is distinguished from *instrumental value*. Instrumental value is use-value, that is, something is valued merely for its utility as a means to some other end beyond itself. Exploiting nature for its instrumental value is seen as the root cause of the ecological crisis. If nature is value-neutral, then humans can dispose of it anthropocentrically as they please. The only alternative to instrumental value, it may seem, is a kind of hands-off, nonrelational respect for nature. If nature matters intrinsically, independently of humans, then its value may prescribe moral consideration.

For instance, all life forms seek to avoid death and injury and to grow, repair, and reproduce themselves by using elements (including other life forms) of the environment instrumentally. These elements have instrumental value for the life-form. On the other hand, in order to generate instrumental value these life-forms must be centers of purpose—growth, maintenance, and reproduction. They must have *sakes* or *intrinsic value* which they pursue. Because the organism's intrinsic value and the instrumental values derived from the organism's pursuit of its own well-being exist whether or not there are humans, such values are independent of humans. For example, grizzlies have a stake in the use of pesticides to control army cutworms in the Midwest because scientific studies show that migrated cutworm moths constitute a significant portion of their diet.

Yet, it is argued, this cannot be a complete account of intrinsic value because the individual may not be a *good kind,* for example, a nonnative species such as spotted-knapweed in North America. Should the life of such a species be respected and allowed to be a good of its kind, or should humans seek to eliminate it? Good kinds need to be weighed in relation to a natural ecosystem. Can a species that uses the forces of evolution to improve itself have intrinsic value, even though it has no *self?* Having wolves cull elk herds will help the species by assuring that elk maintain a *good* gene pool that is better adapted; however, no individual elk welcomes these wolves. Are any species more valuable than others? By genetic standards alone, there is more biodiversity among the microbes in some Yellowstone Park hotpots than the rest of the larger life-forms of the Greater Yellowstone ecosystem. Finally should the concept of intrinsic value be reserved for the *products* of the ecosystem and evolution and not the *processes* themselves, the source of these products?

Significant philosophical (and practical) problems as suggested occur here. How do humans adjust intrinsic values found in nature with people (Schmidtz and Willott 2002)? Second, can humans speak of intrinsic value apart from human minds? Environmental ethicists provide conflicting answers, ranging from conventional anthropocentric to nonanthropocentric *value subjectivist* versus *value objectivist* debates between Callicot and Rolston (Rolston 1993). Third, by focusing so much energy on the nonrelational, intrinsic value of the autonomy of natural things and of nature, environmental ethics tends to concentrate on nature disengaged from humans: particularly on wild nature and the independent natural order. Initial unease with the environment, however, is likely caused in part by the disruption of humanity's bonds of engagement with nature. Between the instrumental resource value of nature and the nonrelational intrinsic value of nature, between the misused and the unused, lies a third alternative: the well used and the well loved. Nature has *correlational value* for humans in the sense that nature's flourishing is bound up with human flourishing in a kind of *correlational coexistence* (Strong 1995, p. 70). Allowing consideration for nature to more strongly influence the design and maintenance of cities may make them more livable, enjoyable, and attractive.

Environmental Philosophy

Turning from environmental ethics to environmental philosophy, agreement about the limitations of conservation measures and analysis of nature solely in terms of its exploitive-value exists, but with an argument for broader reflection. Deep ecologists, for instance, call for metaphysical, epistemological, ethical, political, and cultural changes. What philosophers find most troubling—anthropocentrism, patriarchy, class struggle, placelessness, the technological project itself, and so on—colors the nature of the environmental philosophy.

Deep ecology has focused on anthropocentrism as the source of ecological problems. To overcome this, deep ecologists, such as Naess, advocate a new sense of self-realization (Fox 1995). Anthropocentric self-realization is atomistic, selfish behavior. From an ecological understanding of humankind as part of a larger whole, deep ecologists argue that human beings can reconceive of themselves as extending to that larger whole; reframed, human realization is tantamount to the realization of that larger whole, usually written as *Self-realization* in contrast to anthropocentric self-realization. From this perspective or reframing of human life, nature no longer seems like a resource to be used for a separate human good, but rather as its own good. Other ways of overcoming this anthropocentrism emphasize Naess's eight-point platform that includes a call for decreases in consumption. Yet to rail against consumerism and its destructiveness is not to understand its motivation and attraction. Without understanding those aspects (a topic for technology studies and ethics), can humans become genuinely liberated from it?

Stepping beyond strictly scientific accounts of ecosystems, humans historically and across cultures—for example, Greek, Chinese, and Incan temples—have understood profoundly, cared for, respected, revered, and celebrated the natural world. There is a good deal to learn from how some cultures prescribe the human relationship with nature, and for recent developments in environmental philosophy, such as bioregionalism, an understanding of cultures of place plays a central role their theories and practices (Abram 1996, Jamieson 2003, Snyder 2000). The intuitive and eclectic nature of deep ecology and bioregionalism, as well as their activist emphasis, has made these environmental philosophies especially popular.

Ecofeminism is another promising version of environmental philosophy. Common to different kinds of ecofeminism is the idea of, as Karen Warren puts it, the "twin domination of women and nature," and that both forms of domination ought to be overcome (Zimmerman et al. 2000, p. 325). Some ecofeminists distance themselves from other forms of environmental ethics by arguing that the latter are dominated by male voices and male-centrism, or androcentrism. Some ecofeminists, inspired by Carol Gilligan's work and postmodernism generally, criticize notions of abstraction and detachment, reason, and universality as pretentious and arrogant. They attempt to replace such concepts with an ethics of care, which is highly contextual, particular, and more focused on relationships than on formal rules and individuals (such as earlier philosophies that focused on animals and plants). For instance, ecofeminists have characterized the notion of Self-realization as a means of eradicating the differences between humans and the natural world rather than as an instrument that fosters recognition and acceptance of the differences of these *others*.

Val Plumwood in particular has shown that concerns about anthropocentrism can be addressed in the same ways concerns with androcentrism are dealt with by her, without giving up personal points of view, which she argues is impossible (Plumwood 1999). More specific analyses of anthropocentrism allow people to devise more alternatives to it. However arguments for an end to the domination of nature entirely are too general. One can criticize a limitless technological domination of nature without claiming that *all* human domination is unwarranted. Even though Warren rappels down a cliff as opposed to climbing and dominating it, she uses technological devices that lessen the risk involved and insure that the activity is performed safely. Whether humans use bicycles, public transportation, or SUVs to reach such cliffs, they use technology that dominates nature to some extent, albeit almost imperceptibly. What human beings must learn is to *carefully limit* technology and technological domination.

As with ecofeminist views of patriarchy, many social and political ecologists, inspired by socialist economic perspectives (and some inspired by the work Lewis Mumford), locate the source of human unease as social hierarchy, placing dominance of economic power at the root of social injustice and the ecological crisis. If social hierarchy does not end, humans cannot expect a substantive change in their relationship with nature.

Consideration of social issues opens environmental philosophy to social and philosophical theories of technology (and vice versa) in ways that remain largely undeveloped. The question concerns the earlier issue of how deeply human culture's fundamental orientation toward nature lies. Would a change of social hierarchy alone result in a sufficiently radical change of orientation or does society need to outgrow its current technological orientation, as is posited in the social theories of technology based on the work of Martin Heidegger and Jacques Ellul? Locating the center of gravity with technology, these theories of technology call for prescriptions that differ with other social theories. To use technology in a different way may call for a sea change for nature *and* material culture.

A related political issue is environmental justice. Some argue that it would take the resources of at least two more Earths to bring all people on this planet up to the standard of the developed nations. The environmental cost of the transformation of the Earth is borne disproportionately, both within and outside of the United States, by the poor, minorities, and women. Moreover the cost of environmental legislation often falls disproportionately on these groups, giving rise to a charge of elitism against environmentalists. Finally, those in developed countries who take modern conveniences for granted often callously disregard the genuine hardships suffered by those in developing countries where such technological relief is unavailable or inadequate. The challenge for environmental philosophy is to meet moral concerns for social justice and nature. What conditions are required to put a life of excellence within everyone's reach?

A Consequential Environmental Ethic?

What are the practical achievements of environmental ethics? While environmental ethics is not simply applied ethics (a body of traditional normative ethical theories is not being applied to specific ethical issues as is often the case in medical or business ethics), it is important to apply traditional theories of interhuman ethics to environmental problems in a secondary sense. Arguably the most valuable contributions to policy decisions have been made in terms of risk assessment and related issues (Shrader-Frechette 2002). Animal rights and liberation theories indirectly influence legislation such as the laboratory care and use of animals. Forest Service Employees for Environmental Ethics explicitly advances Leopold's land ethic in a quest for a new resource ethic; departments of natural resources, as they move toward ecosystem approaches of management, seem to be attentive to these discussions of Leopold and those philosophers influenced by him. More progressive hunting and fishing regulations sometimes mimic ecosystem processes by reducing the number of trophy animals harvested. Environmental philosophers have proven most effective publicly when they, like Rolston, listen and speak in intelligible ways to a broad spectrum of people including those in the fields of technical philosophy and science, activists, and ordinary people (Mitcham et al. 1999, Rolston 1993).

All major pieces of environmental legislation preceded the development of environmental ethics, and as yet there is no effective green party in the United States. In particular there is none inspired by philosophers. Environmental pragmatists criticize environmental ethics and environmental philosophy for missing opportunities to make significant contributions to policy because they are too impractical and dismissive of activism. In their view, environmental ethicists divert attention from actual environmental problems by being overly concerned with theoretical issues such as intrinsic value, whereas environmental philosophers do the same by concentrating on the impossible task of radical reform. Pragmatists urge philosophers to apply their unique abilities and resources to solving concrete environmental problems. Distinct problems, in their view, call for different approaches (ranging from the economic to the aesthetic); no single approach is the correct one. In fact, the same problem may require a solution that includes approaches from incompatible theoretical positions. Thus, as pluralists, they call for cooperation between environmental philosophers.

Although they would be wary of any absolutist tendencies, pragmatists also call for more cooperation between these philosophers and other kinds of, normally anthropocentric, reformist positions that have been developing simultaneously with the field. Some approaches are based on deeply held values, such as the difference between consumers and citizens, in conservative and liberal traditions in order to get people to change attitudes, behaviors, and polices toward nature. Others, such as environmental libertarianism, have developed free-market approaches to resolving environmental problems. This kind of thinking has been used to show that government subsidies to the forest service have been the impetus for much logging and road building that would not have otherwise occurred. More liberal economic approaches demonstrate that while the market is effective for resolving some environmental problems, it is limited with regard to ensuring environmental protection. Many of the market's shortcomings have to do with the limits of economic value, cost benefit analysis itself, or how ethically and scientifically sound solutions to environmental problems are ignored based on economic considerations (Sagoff 1984, Schmitz and Willott 2002). Alternatively green capitalism, in order to avoid ecological catastrophe, advocates government regulations and policies, such as "green taxes," that develop the economy in ecologically sustainable ways (Hawken, Lovins, and Lovins 1999, Thompson 1995).

Is an environmental ethic needed or is ecological prudence sufficient? Apart from meeting people's moral concerns with nature, many advocates argue that an environmental ethic is imperative in order to save the planet from catastrophe. Such pessimism invites detractors who contend that scientists and engineers are making progress with environmental problems and that some fears regarding the environment are unwarranted; moreover unfounded fear is cited as part of the problem (Baarschers 1996, Lomborg 2001, Simon 1995). Certainly informed debate, critical thinking, scientific literacy, and pragmatism are called for.

However even *if* some consensus were achieved and catastrophic outcomes could be ruled out safely, this debate is a diversion from a submerged but central environmental and ethical issue: Will a *saved planet* be worth living on? Those who would continue the technological project unimpeded except for refinements and adjustments are quite sanguine in their answer—often assuming that the indisputable early achievements of technology are analogous with later postmodern ones—whereas those concerned about survival are often covertly more concerned about the quality of human life. What level of environmental quality is correlatively important to the quality and excellence of human life? Where is nature's place in a technological setting? How *tamed* should nature be? Contemplating these questions requires the use of science, technology, ethics, and environmental ethics. These reflections will involve not only specialists, but also each and every person, in a public conversation that considers facts and fallacies, but ultimately ponders alternative visions of life. As Langdon Winner writes, "we can still ask, how are we living now as compared to how we want to live?" (Winner 1988, p. 163). Human beings need to reflect on whether to continue to seek prosperity and happiness entirely through affluence and goods provided by the technological project, or, alternatively, through a new engagement with, among other things that matter, the nonhuman world. In the former vision, the technological project is prudently modified to be environmentally sustainable and shared equally with all people, and nature is controlled as a mere resource and commodity. In the latter vision, nature plays a much greater role in a reformed technological setting.

Unreflective consensus threatens to subvert any substantive environmental ethic because most of the ethical claims of the natural world are overridden when they conflict with consumption as a way of life (Strong 1995). Quite often environmentalists and environmental academics want environmental protection and are attracted to affluence and full-scale technological development. Can both exist? Most people uneasily muddle ahead simply assuming they can. Humans need a vision that values the natural world and includes an understanding of why the planet is being transformed in the way that it is. Only then can people hope to attain some clarity with regard to the real environmental and social consequences of personal, collective, and material choices.

Environmental philosophers must remember the original environmental and cultural problems that caused them to reflect and measure their overall successes in terms of how far human culture has come in dispelling those concerns. In the early-twenty-first century, it is clear that the full autonomous, independent, and nonrelational character of nature has changed (McKibben 1989). Often in restoration work and matters concerning nature in urban settings, the questions are more clearly focused. Will nature be respected and celebrated as having dignity and a commanding presence, expressive of the larger natural and cultural world of particular places, and be correspondingly cared for in that way (will it have correlational rather than intrinsic value alone)? Or will nature be entirely demeaned as a mere resource for humans to control and modify for the convenience of consumption (Borgmann 1995; Higgs 2003)?

DAVID STRONG

SEE ALSO Acid Mine Drainage; Biodiversity; Carson, Rachel; Dams; Deforestation and Desertification; Ecology; Engineering Ethics; Environmental Ethics; Environmentalism; Environmental Justice; Environmental Regulatory Agencies; Environmental Rights; Global Climate Change; Mining; National Parks; Nongovernmental Organizations; Oil; Pollution; Rain Forest; Scandinavian and Nordic Perspectives; Sierra Club; Thoreau, Henry David; United Nations Environmental Program; Waste; Water.

BIBLIOGRAPHY

Abram, David. (1996). The Spell of the Sensuous: Perception and Language in a More-Than-Human World. New York: Pantheon Books.

Baarschers, William H. (1996). Eco-Facts & Eco-Fiction: Understanding the Environmental Debate. New York: Routledge.

Berry, Wendell. (1995). "The Obligation of Care." Sierra 80(5): 62–67, 101.

Borgmann, Albert. (1984). Technology and the Character of Contemporary Life. Chicago: University of Chicago Press.

Borgmann, Albert. (1995). "The Nature of Reality and the Reality of Nature." In Reinventing Nature?: Responses to Postmodern Deconstructionism, ed. Michael Soulé and Gary Lease. Washington, DC: Island Press.

Callicot, J. Baird. (1989). In Defense of the Land Ethic. Albany: State University of New York Press.

Carson, Rachel. (2002). Silent Spring. New York: Houghton Mifflin Mariner Books.

Fox, Warwick. (1995). Toward a Transpersonal Ecology: The Context, Influence, Meaning, and Distinctiveness of the Deep Ecology Approach to Ecophilosophy. Albany: State University of New York Press.

Goodpaster, Kenneth E. (1978). "On Being Morally Considerable." Journal of Philosophy 75(6): 308–325.

Hawken, Paul; Amory Lovins; and L. Hunter Lovins. (1999). Natural Capitalism: Creating the Next Industrial Revolution. Boston: Little, Brown.

Higgs, Eric. (2003). Nature by Design: People, Natural Process, and Ecological Restoration. Cambridge, MA: MIT Press.

Higgs, Eric; Andrew Light; and David Strong, eds. (2000). Technology and the Good Life? Chicago: University of Chicago Press.

Jamieson, Dale. (2003). A Companion to Environmental Philosophy. Oxford: Blackwell Publishers.

Kates, Robert W.; Billie Lee Turner; and William C. Clark. (1990). "The Great Transformation." In The Earth As Transformed by Human Action, eds. Billie Lee Turner, William C. Clark, Robert W. Kates, et al. Cambridge, UK: Cambridge University Press.

Leopold, Aldo. (2000 [1949]). A Sand County Almanac and Sketches here and There. Oxford: Oxford University Press.

Light, Andrew, and Eric Katz, eds. (1998). Environmental Pragmatism. London: Routledge Press.

Light, Andrew, and Holmes Rolston III, eds. (2002). Environmental Ethics: An Anthology. Oxford: Blackwell Publishers.

Lomborg, Bjørn. (2001). The Skeptical Environmentalist: Measuring the Real State of the World. Cambridge, UK: Cambridge University Press.

McKibben, Bill. (1989). The End of Nature. New York: Random House.

Mitcham, Carl; Marina Paola Banchetti-Robino; Don E. Marietta Jr.; and Lester Embree. (1999). Research in Philosophy and Technology: Philosophies of the Environment and Technology, Vol. 18. Stamford, CT: JAI Press.

Naess, Arne. (1973). "The Shallow and the Deep, Long-Range Ecology Movements." Inquiry 16: 95–100.

Nash, Roderick Frazier. (2001). Wilderness and the American Mind, 4th edition. New Haven, CT: Yale University Press.

Plumwood, Val. (1999). "Paths Beyond Human-Centeredness: Lessons from Liberation Struggles." In An Invitation to Environmental Philosophy, ed. Anthony Weston. New York: Oxford University Press.

Rolston III, Holmes. (1975). "Is There an Ecological Ethic?" Ethics 85: 93–109.

Rolston III, Holmes. (1993). Conserving Natural Value. New York: Columbia University Press.

Routley, Richard. (1973). "Is There a Need for a New, an Environmental Ethic?" Proceedings of the XV World Congress of Philosophy 1: 205–210. This author now publishes under the name Richard Sylvan.

Sagoff, Mark. (1984). The Economy of the Earth. Cambridge, UK: Cambridge University Press.

Schmitz, David, and Elizabeth Willott. (2002). Environmental Ethics: What Really Matters, What Really Works. Oxford: Oxford University Press.

Shrader-Frechette, Kristin. (2002). Environmental Justice: Creating Equality, Reclaiming Democracy. Oxford: Oxford University Press.

Simon, Julian L., ed. (1995). The State of Humanity. Cambridge, MA: Blackwell Publishers.

Singer, Peter. (1974). "All Animals Are Equal." Philosophic Exchange 1(5): 243–257.

Snyder, Gary. (2000). The Gary Snyder Reader. New York: Counterpoint Press.

Strong, David. (1995). Crazy Mountains: Learning from Wilderness to Weigh Technology. Albany: State University of New York Press.

Thompson, Paul S. (1995). Spirit of the Soil: Agriculture and Environmental Ethics. New York: Routledge.

Winner, Langdon. (1988). The Whale and the Reactor: A Search for Limits in an Age of High Technology. Chicago: The University of Chicago.

Zimmerman, Michael E.; J. Baird Callicot; George Sessions; et al., eds. (2000). *Environmental Philosophy: From Animal Rights to Radical Ecology*, 3rd edition. Upper Saddle River, NJ: Prentice-Hall.

ENVIRONMENTAL IMPACT ASSESSMENT

• • •

An environmental impact assessment (EIA) is a means for understanding the potential effects that a human action, especially a technological one, may have on the natural environment. It allows for the inclusion of environmental factors in making decisions by mandating a process for determining the range of environmental issues related to a particular action. The underlying assumption of an EIA is that all human activity has the potential to affect the environment to some degree, so that all major decisions should include environmental, as well as economic and political, factors. Understanding the potential environmental effects of an action helps policymakers choose which actions should proceed and which should not.

Many governments perform EIAs at the national, state, and local levels. Probably the best-known form of the EIA is the Environmental Impact Statement (EIS) of the United States government. The National Environmental Policy Act of 1969 (NEPA) mandates an EIS to accompany every major federal action or nonfederal action with significant federal involvement. NEPA tries to ensure that U.S. federal agencies give environmental factors the same consideration as other factors in decision making.

Most EIAs follow a process similar to the one mandated for the EIS. The first step is the preparation of an environmental assessment (EA) to determine whether the environmental impact of the action requires a complete EIS. The actual EIS begins by identifying issues and soliciting comments on the scope of the action, alternatives, and various impacts that the EIS should address. Then the lead agency collects and assimilates all the environmental information required for the EIS. In the United States, the Council on Environmental Quality (CEQ) regulations outline the recommended format for the EIS. The EIS must include public involvement throughout the process. All mitigation measures to address identified harms must be included in the EIS.

The primary problem of environmental impact assessments is that once the environmental factors have been analyzed there is little to force the actors to actually use the information in decision making. When the EIA is complete, the action can go forward regardless of any negative environmental consequences. In the case of the EIS, NEPA provides no enforcement provisions, though various court decisions have developed some such mechanisms. Decision makers are informed of potential environmental problems and can include environmental issues in making their decisions, but nothing requires them to nor is there any penalty for ignoring the environmental impact.

This is not to say that identifying environmental issues has no effect on the process. The fact that the information exists means it plays a role. Decision makers must elect to include or exclude it from their project. If they choose to ignore the information, others have a right to bring pressure on them. The identification of potential problems has sometimes motivated public criticism of planned actions and led to their rethinking. The existence of the information creates a better situation than not having the information at all.

FRANZ ALLEN FOLTZ

SEE ALSO *Environmental Regulation; Pollution; Waste; Water.*

BIBLIOGRAPHY

Sullivan, Thomas F. P. (2003). *Environmental Law Handbook*, 17th edition. Rockville, MD: Government Institutes.

INTERNET RESOURCE

Council on Environmental Quality. Available from http://www.whitehouse.gov/ceq/.

ENVIRONMENTALISM

• • •

Environmentalism is a broad term used to describe the ideology of social and political movements that emerged in the 1960s around concerns about pollution, population growth, the preservation of wilderness, endangered species, and other threatened non-renewable resources such as energy and mineral deposits. As such it is a vivid nexus for science, technology, and ethics interactions. Since the 1970s, environmentalism has proved to be one of the most powerful and successful of contemporary ideologies, although this very success has generated so many strains of environmentalist ideas as to threaten the meaningfulness of the term itself.

Intellectual Roots

Although modern environmentalism can be traced to multiple intellectual roots, in the United States there are three primary influences. The first are the U.S. romantic and transcendentalist movements, which found moral and artistic inspiration in the natural world. The greatest representative of these ideas is the nineteenth-century writer Henry David Thoreau (1817–1862), whose *Walden* (1854) uses the natural world as a philosophical vantage point from which to evaluate and criticize U.S. society and politics. From this tradition, which was developed by John Muir (1838–1914) and others, environmentalism gains a focus on the value of preserving wilderness and non-human species.

A second major intellectual source for environmentalism is the U.S. conservation tradition. The most important founders of this tradition are Theodore Roosevelt and his close adviser and the first head of the U.S. Forest Service, Gifford Pinchot (1865–1946). These and like minded progressive reformers from the early-twentieth century led a movement to regulate and conserve natural resources and preserve some spectacular wilderness areas as national parks. The overall concern of the conservation movement was to maintain a sustainable supply of natural resources for a growing economy, which was believed to be essential for the health of a democratic society. From this tradition, environmentalism has inherited concerns about sustainability, the impact of the economy on the natural world, and human equity and justice issues concerning the distribution of environmental benefits and risks.

A third intellectual source for environmentalism is found within the scientific community of the 1950s and 1960s, when scientists became alarmed by the worldwide impact of nuclear weapons use and testing, chemical pollution of the environment by modern economic activity, and the stress on the environment caused by the sharp growth in human population during the twentieth century. The three greatest representatives of this tradition are biologists who wrote highly popular and influential books that caused broad-based alarm about environmental problems, Rachel Carson (*Silent Spring* [1962]), Paul Ehrlich (*The Population Bomb* [1969]), and Barry Commoner (*The Closing Circle* [1971]). Inspired by such works, environmentalism has gained a focus on public health problems that grow from modern productive processes and military technology.

Although these three traditions are responses to different types of problems and have generated different sets of concerns, environmentalism weaves them loosely together. Environmentalist thinkers and organizations stress different strains of environmentalism, but concerns as disparate as wilderness preservation, reducing environmental pollution and addressing the health problems it causes, and evaluating and protesting the injustice of unequal environmental impacts of various public policies and economic activities on disadvantaged subgroups in U.S. society (such as the poor, or people of color), are all recognized as part of the environmentalist agenda.

Two key facts about environmentalism must be stressed. First it is simply one of the most remarkably successful of all contemporary social and political ideologies. What was a marginal set of concerns and views during the 1960s has become part of the social and political mainstream. Public opinion polls consistently demonstrate wide-ranging public support for environmentalist values and policies, even if the saliency of environmentalist concern is somewhat less than that found for other issues such as the economy. Not surprisingly, candidates from across the political spectrum have found it necessary to profess environmental values, even if there is reason at times to doubt their sincerity. The corporate world has discovered that it is increasingly good business to market products and services as *green, natural, organic,* or environmentally responsible. Academic disciplines, from law to ethics to the natural sciences to engineering to economics and beyond, have been influenced by environmentalist concerns and have developed sub-disciplines focusing on environmental issues. Vast rivers of private financial donations flow into the coffers of a variety of environmental organizations found on the local, national, and international levels. In short, in the course of a single generation, environmentalism has grown to be one of the most visible and important ideologies in contemporary life. Rarely has an ideology enjoyed this level of achievement in such a short period of time.

The second key fact to note is that this very success, coupled with the diverse intellectual roots that nourish it, has made the intellectual content of environmentalism ambiguous, perhaps even incoherent if one is looking to find a unified ideology.

Three Types of Environmentalism

In light of this ambiguity, it is helpful to divide the universe of environmentalist views into three broad categories. First *liberal* environmentalists think of environmental problems in the political and social context of conventional liberal ideals and social policy. Drawing primarily, but not only, on the conservation tradition, liberal environmentalists have been successful in promoting extensive environmental regulation of industry and other polluting activities. The environmental justice movement, as well as increased interest in applying the philosophical tools of pragmatic philosophy to the study of environmental ethics, are also fundamentally liberal developments in environmentalism; the first demands respect for liberal equity in the distribution of environmental risk, and the second draws on the liberal tradition of U.S. philosophical pragmatism in order to evaluate the ethical implications of particular human behaviors in relation to the natural world. Much of the growing field of environmental economics may also be included in the category of liberal environmentalism, because it applies conventional liberal economic principles and tools to the study of environmental policy. What liberal environmentalists share is a perspective that views environmental problems within the context of recognized liberal philosophical, political, and social values.

Radical environmentalism can be thought of as an array of environmentalist ideas that challenge the philosophical and political underpinnings of liberal democratic society. The greatest unifying theme among radical environmentalists is the insistence that the anthropocentrism of liberalism, the assumption that human beings are the source and measure of all value, be rejected in favor of a moral perspective more inclusive of values intrinsic to the non-human world, a view that is sometimes called biocentrism or ecocentrism. The claim is that conventional moral perspectives are incapable of appropriately appreciating non-human things, and therefore there is a need to discover fundamentally new ways of thinking about the natural world and its relationship to people. Beyond these claims, radical environmentalists quickly part company, pursuing a multitude of philosophical paths. Eco-feminists, for example, suggest that women have natural connections with and insights into nature that men are less likely to experience, and that are lost or suppressed within a patriarchal society; fighting patriarchy is therefore related to not only freeing women from men, but to the reconnection of human beings with nature more generally. Rather than emphasizing gender, deep ecologists promote what they understand to be more primal, unified understandings of the proper relationship of humans to the natural world than they find in modern social and political theory and practice. Social ecology, a form of eco-anarchism, claims that humans could naturally live in just, non-hierarchical social organizations, and that environmental problems grow out of and reflect the oppression of humans by humans in unjust,

hierarchical societies. Some would include eco-socialists among radical environmentalists, because they promote a political vision contrary to contemporary liberal democracy. Not all radical environmentalists, however, believe the socialist political program is sufficiently biocentric to be truly radical or environmentalist.

As an illustration of the huge growth in the ideological power of environmentalism, the late-twentieth and early-twenty-first centuries began to see the emergence of new forms of *conservative* environmentalism. While it is true that there have always been conservation groups that have historically appealed primarily to hunters and other groups not conventionally thought to be liberal or radical, these have been on the margins of environmentalism. Historically, conservatives have more often than not been hostile to environmentalism, on the grounds that it threatened to over expand the government's regulatory powers (in the case of liberal environmentalism) or, even worse, that it attacked the moral foundations of conventional society (in the case of radical environmentalism). There is a new and growing *free market environmentalism*, however, that is attacking the liberal environmental regulatory programs, and defending private property rights and conventional capitalist economic organization as the best way to promote environmental health and resource conservation. There is also some growth of a less militantly free market conservative sympathy for environmentalism that emphasizes the continuity of community traditions and religious piety toward what is understood to be a created universe.

Conclusion

Beyond the U.S. context, environmentalism has become a powerful force throughout the world, both within other countries and in the international order. The diversity of environmentalist views explodes within this broader context, from the demands of indigenous peoples to control local ecosystems in the face of pressure by international markets and corporations, to the growth of Green political parties (most importantly and successfully in Germany), to the attempt to design international policies for contending with world-wide environmental issues such as global climate change, to attempts to address wildly inequitable resource allocation between the rich and poor, the developed and developing, nations. In different contexts, and with different aims and intentions, environmental politics has become a factor in local, national, and international politics, and as such contexts have proliferated, so too has the breadth of environmentalist ideology expanded almost beyond measure and clear focus.

Given this array of environmentalist views and projects, it is clear that the very notion of environmentalism is being stretched to include incompatible ideas. The single unifying theme, to the degree that it can be found, is simply the attention paid to the human relation to the natural world and the promotion of ideas and policies intended to protect the health and fecundity of nature.

In light of the diversity of environmentalist views, it is difficult to clearly assess the implications of this ideology for modern science, technology, and ethics. It is clear, for example, that there have been elements of misanthropy and hostility toward science and technology in some strains of radical environmentalism, a kind of primitivism that views modern society in all its facets as a plague on the natural world to be resisted, even turned back, as much as possible. It is also true, however, that this is a marginal set of attitudes even within the radical environmentalist camp. Radical environmentalism does indeed insist on an ethical reorientation toward non-human things, but this by no means always reflects misanthropic views. On the contrary, the claim more often includes a presumption that humans will find their lives more meaningful if they learn to live harmoniously with nature, that radical environmentalism is a positive good for both people and nature. Likewise, even while much radical environmentalism distrusts science and technology, it often draws heavily on the science of ecology to inform its own analysis of problems, and often promotes what it considers to be environmentally friendly technologies.

Liberal and conservative environmentalisms usually appeal to conventional ethical categories (for example, the weighing of public goods against individual rights), and tend to work within the conventions of mainstream science and technology to promote their ends. The debates they engage are more often about the proper balancing of environmental goods against other important values, than about the need for such a balance in the first place. Liberal environmentalism also tends to be committed to using modern science to closely evaluate the overall environmental impact of existing technologies, and to producing the most environmentally benign technologies currently feasible.

Although it is difficult to generalize about environmentalism, given the great diversity of ideas and concerns found within the movement, the very power and popularity of environmentalist ideas reflects a growing sensitivity to and concern about the natural world. While environmentalists often worry about different issues, from wilderness preservation to public health to social justice, and often see the world in different ways, from radical biocentrists to conservative free market advocates to almost an infinity of variations in between, environmentalism reflects a rich diversity of attempts to think seriously about the appropriate relationship between people and the rest of nature. It is clear, from the popularity of environmentalist ideas, that there is a broad and growing sense of the importance of this overall project.

BOB PEPPERMAN TAYLOR

SEE ALSO *Air; Carson, Rachel; Conservation and Preservation; Deforestation and Desertification; Earth; Environmental Ethics; Fuller, R. Buckminster; Rain Forest; Sierra Club; Sustainability and Sustainable Development; Thoreau, Henry David; United Nations Environmental Program; Water.*

BIBLIOGRAPHY

Anderson, Terry Lee, and Donald R. Leal. (2001). *Free Market Environmentalism Today.* New York: Palgrave. A good representative of the free market environmentalism literature.

Bliese, John. (2001). *The Greening of Conservative America.* Boulder, CO: Westview Press. An example of a conservative environmentalism that is distrustful of a simple faith in the free market.

Bookchin, Murray. (1982). *The Ecology of Freedom: The Emergence and Dissolution of Hierarchy.* Palo Alto, CA: Chesire Books. A classic social ecology text.

Dobson, Andrew. (1998). *Justice and the Environment.* Oxford and New York: Oxford University Press.

Guha, Ramachandra. (2000). *Environmentalism: A Global History.* New York: Oxford University Press.

Luke, Timothy. (1999). Urbana: University of Illinois Press. *Capitalism, Democracy and Ecology.* Luke weds environmentalism to a leftist populism.

Merchant, Carolyn. (1996). *Earthcare: Women and the Environment.* New York: Routledge. A historian's view of the relationship between women and the environment.

Norton, Bryan. (1991). *Toward Unity Among Environmentalists.* New York: Oxford University Press. Among the most sophisticated works in the pragmatic school of environmental ethics.

Sagoff, Mark. (1988). *The Economy of the Earth: Philosophy, Law and the Environment.* New York: Cambridge University Press. The best liberal critique of the use of cost benefit analysis as the primary tool of environmental policymaking.

Sessions, George, ed. (1995). *Deep Ecology for the Twenty-first Century.* Boston: Shambhala. A standard deep ecology text.

Taylor, Bob Pepperman. (1992). *Our Limits Transgressed: Environmental Political Thought in America.* Lawrence:

University Press of Kansas. A survey and discussion of some of the primary strains of environmental thought in the United States.

ENVIRONMENTAL JUSTICE

• • •

Environmental justice encompasses distributive and political justice to address the interlocking relationship between environmental issues and social justice. Environmental justice can include a myriad of struggles experienced by local communities whose concerns include protecting the environments where people live, work, play, and pray. A central focus is on the environmental burdens of modern industrial society including, but not limited to, issues of toxic waste, pollution, workplace hazards, and unequal environmental protection. Another focal point involves the equal political representation of diverse groups in environmental values and decision-making processes. Environmental justice has served to effectively criticize the inequitable distribution of environmental benefits and harms that can be associated with many technological developments, often employing science to identify and assess these benefits and harms.

Historical Emergence

Because many of these issues are tied to specific grassroots organizations and networks, environmental justice fundamentally pertains to a larger social phenomenon referred to as the environmental justice movement (EJM). The EJM emerged in the 1980s when people of color formed grassroots responses to the location of environmental burdens, particularly toxic waste facilities and point production pollution sources. Luke W. Cole and Sheila R. Foster (2001) identify six intersecting social movements as the undercurrents of the EJM: the civil rights movements, labor movements, Native American movements, the anti-toxic movement, movements in academic scholarship, and the mainstream environmental movement. Although not included in their six undercurrents, the women's movement must be considered a seventh tributary, because it serves as a historical linchpin to the sciences currently used in environmental justice cases and because 70 to 80 percent of grassroots leaders in the EJM are working-class women, many of them women of color.

As early as the work of Jane Addams (1860–1935) and Alice Hamilton (1869–1970), when Hull-House pushed bacteriology and the new sciences of toxicology and epidemiology into connections between health, environment, and politics, women have been critical to the scientific knowledge of the neighborhood. As a result, new methods of data collection and analysis were created by these early environmental reformists to improve the industrial living conditions of the modern city. The attention given to women's health issues and environmental dangers from industrialization carries a direct thread between Hull-House and the contemporary EJM. Contemporary science and policy agendas, like those found in the U.S. Superfund Act (the Comprehensive Environmental Response, Compensation, and Liability Act of 1980), were spawned by the activism of women such as Lois Gibbs in Love Canal, New York. Thus the early advances in toxicology and epidemiology were partly due to environmental justice struggles led by women, and from that time policies to address environmental justice have had their origins in the activism of these community leaders.

Other important precursors that relate to Cole and Foster's six movements are identifiable as early as the 1960s when Martin Luther King Jr. and other civil rights leaders observed that people of color suffer higher pollution and more denigrated environments. By the end of the 1970s a series of studies had again drawn the historical inference that different human environments are directly related to social stratification. In a chapter of their seminal book addressing environmental justice, *Race and the Incidence of Environmental Hazards*, Paul Mohai and Bunyan Bryant compare studies dating from 1971 to 1992 that assess the correlation of toxics, including air pollution, hazardous waste, solid waste, and pesticide poisoning, with the impact on people of low income and racial minorities (Mohai and Bryant 1992). Two critical findings from this comparative study are worth highlighting. First, the study clearly proves that government agencies observed the relationship between social stratification and environmental burdens as early as 1971. Second, the comparisons provide empirical evidence that in the United States the distribution of environmental burdens has a strong correlation to race and socioeconomic class.

In addition to the Mohai and Bryant comparative study, the federal government in 1978 released a brochure called *Our Common Concern* that described the disproportionate impact of pollution on people of color. The struggle of César Chávez and the United Farm Workers to protect the health, environment, and rights of farmworkers was a vital precursor to the environmental justice movement. Studies of rural Appalachian living conditions were revealing the connection between poverty and environmental burdens, providing further evidence of trends of environmental injustice. Environmental justice also pervaded the struggles of Native Americans dealing with issues stretching from land rights to the hazardous industries of uranium mining, coal mining, and nuclear waste depository.

Addressing shared interests in environmental justice, the City Care Conference, held in Detroit in 1979, was jointly sponsored by the National Urban League and the Sierra Club. The intended purpose of this conference was to bring the civil rights movement and the environmental movement together for a dialogue to reconceptualize the very meanings of the terms *environment* and *environmental issues*. By the late 1980s and early 1990s, *environmental justice* became a newly established term used by scholars and policymakers. "Environmental justice" was first used in book and article titles by 1990, and the first environmental justice college course was offered in 1995. The latter came a year after President Bill Clinton signed Executive Order 12898, titled "Federal Actions to Address Environmental Justice in Minority Populations and Low-Income Populations," which introduced environmental justice as a federal mandate by White House fiat.

Founding Events

Although the EJM in the United States is not bound by a single event, many scholars and activists regard the 1982 protests in Warren County, North Carolina, as a historical launching point. These protests marked the first major civil rights–style response to an environmental issue. It involved nonviolent civil disobedience blocking trucks hauling PCB-laced soil from entering a newly placed toxic landfill, leading to over 500 arrests and drawing national media attention. The Afton site in Warren County prompted many questions about the direct correlation between African-American communities and hazardous waste sites. It incited District of Columbia Delegate Walter E. Fauntroy, who was himself arrested in the protest, to initiate the 1983 U.S. General Accounting Office study of hazardous waste landfill siting, which found a strong correlation between sitings of hazardous-waste landfills and race and socioeconomic status.

Fauntroy's study spawned later comprehensive studies, including the United Church of Christ's Commission for Racial Justice's frequently cited *Toxic Wastes and Race in the United States* (1987), a national study not only confirming the disparate environmental burdens suffered by minorities and lower socioeconomic groups nationwide, but also centrally locating race in the

disparity: "Race proved to be the most significant among variables tested in association with the location of commercial hazardous waste facilities" (p. xiii). At the presentation of *Toxic Wastes and Race* to the National Press Club in 1987, Benjamin Chavis, then director of the United Church of Christ, described the phenomenon as: "racial discrimination in environmental policy making and the enforcement of regulations and laws, the deliberate targeting of people of color communities for toxic waste facilities, the official sanctioning of the life-threatening presence of poisons and pollutants in our communities, and the history of excluding people of color from leadership in the environmental movement" (U.S. House 1993, p. 4).

Environmental Racism

Numerous studies concerning what came to be called environmental racism followed. In 1992 Marianne Lavelle and Marcia Coyle published their seven-year study of the U.S. Environmental Protection Agency in the *National Law Journal*, which revealed that polluters were fined more in white communities, responses were slower in people of color communities, and scientific solutions differed between the communities. The same science that would be used to determine the toxicity of a facility to a community was used differently between white communities and minority communities. Likewise, the same science that would determine the technological and economic responses, such as the technology of soil washing or soil removal or the shutting down of the polluting facility itself, would be compared to the economic assessment of community relocation because the implications of dangerous conditions involve costly relocations that make the project too expensive. Lavelle and Coyle revealed that different technological solutions would be used when the same scientific data described the health threats to the community. The different responses follow the trend that white communities receive more expensive and updated technological solutions and also receive higher compensation for health and property damage, and that polluters pay greater fines for damages to white communities than to minority communities even though scientifically, with regard to the pollution, the circumstances do not warrant these dramatic differences.

Further sociological and legal studies responded to the environmental racism charges by addressing fundamental methodological questions: Did the community or the environmental burden arrive first? Are there other categories to consider, such as age? How should a community be defined? Vicki Been (1994) argues that market forces drive the location of toxic facilities and the choice of many workers to come to a highly industrial sector. Admitting of racism in many social institutions, Been's study challenges the main measuring units used by earlier studies and raises important temporal questions about the relationship between minorities and environmental burdens. Other studies that altered the measuring unit of what constitutes a community found less disparity in the distribution of environmental burdens with regard to race than was initially claimed by the earlier studies defending the environmental racism charge. Numerous studies responded to this debate, thus generating a community of scientists, scholars, and activists to help deepen the ethical questions and broaden the scope of environmental sciences. What are the proper characteristics for determining the community that will host the environmental burden? What procedures will be used? Which scientific perspectives would best measure the risk of danger? How will race and socioeconomic background be considered in these risk assessments?

Discriminatory Environmentalism

The ethical considerations of these questions pertain to the discrimination undermining distributive justice and fair compensation for health or property loss. While environmental racism is indicative of actions considered illegal under federal laws such as the Civil Rights Act of 1964 and the Fourteenth Amendment to the Constitution, environmental discrimination on the basis of socioeconomic factors is not specifically illegal. Ethically, however, fundamental principles of distributive justice are violated. Peter S. Wenz has studied the environmental racism debate as a form of double effect in which race may be incidental to the socioeconomic target (Wenz 2001). Even if the market forces argument is true, he argues, distributing the environmental burdens onto the poor violates the principle of commensurable benefits and burdens, which stipulates that unless there are morally justifiable reasons, persons receiving the benefits of modern industrial technology should also receive the commensurate burdens. Those who receive an abundance of consumer goods should therefore be the targets of hazardous waste facilities and polluting industries, whereas those who receive noticeably fewer benefits, the poor residents, should be relieved of this incommensurable burden. Compensatory justice would follow the same moral foundation for redistributing benefits for incommensurable burdens.

Environmental justice also pertains to the principles of equality that require respect for the basic rights of all individuals. The most pronounced right in environmental justice is the right to a safe environment, which has assignable duty holders in the public (government) and private (corporate) sectors. In addition, the principle of self-determination, which honors the autonomy of individuals and their moral capacities to direct the activities that impact them the most, is of vital importance in the participatory justice dimension of environmental justice. The principle of self-determination entails that citizens ought to participate in the process of siting hazardous waste, as well as the procedures for determining fair compensation. Direct political participation, however, is not available for many residents in the burden-affected neighborhoods. The environmental decision-making is typically made prior to the time when community members are able to voice their opinions in the public review-and-comment meetings that are standard political mechanisms in the siting process.

The lack of representation in the mainstream environmental movement or the vital decision-making sectors can be referred to as discriminatory environmentalism. In discriminatory environmentalism, representation and participation in mainstream environmental groups, participation in environmental policymaking, representation in federal, state, and local environmental agencies, and decision-making power over the location of environmental burdens and benefits are either intentionally or unintentionally exclusionary. Underrepresentation in the mainstream environmental movement is also a fundamental contention of injustice against political recognition and participatory justice. In an effort to establish a genuine voice that would better represent the environmental concerns of people of color in the United States, alternative environmental caucuses were created. Often highlighted is the First National People of Color Environmental Leadership Summit, held in Washington, DC, in 1991, which symbolizes two important foundations of the environmental justice movement. The summit represents the lack of political representation of people of color in the greater environmental movement, and it generated seventeen "Principles of Environmental Justice."

Discriminatory environmentalism also identifies the ways in which mainstream environmental ethics has considerably overlooked the poorest and most disenfranchised peoples of the world in its efforts to securely ground moral obligations to nonhuman nature. In particular, the biocentric and ecocentric approaches of land ethic philosophy and "deep ecology" received criticisms for discriminatory environmentalism. The Indian ecologist Ramachandra Guha (1989) argues that the broad-sweeping universalist claims of deep ecologists would cause further distribution of resources for biological protection and environmental improvement away from poor nations to the wealthy nations. Various expressions of misanthropy emerged from deep ecology, which served to undermine environmental struggles of the poor and failed to distinguish between those who hold institutional control over our resource use and those who are subjected to the worse side effects of resource depletion and consumption. By making all human responsible for ecological impacts, deep ecologists overlooked not only the dramatic distinctions between the rich and the poor, but also who has consumed and controlled the use of the natural resources.

Originators of the deep ecology philosophy fundamentally distinguished this non-anthropocentric ethic from "shallow forms" of environmentalism that reflected anthropocentric ethics directed at pollution, work place hazards, and public health. This distinction between anthropocentric (shallow) and non-anthropocentric (deep) environmental ethics overlooked the populations of people struggling with the intersection between shallow and deep ecology. An irony of the split between non-anthropocentric environmental ethics and anthropocentric environmental ethics—a split that is often used to characterize the EJM as a shallow environmentalism—is that while the Principles of Environmental Justice reflect a challenge to the discriminatory environmentalism of mainstream environmentalism in the 1990s, it also shares fundamental values that clearly echo deep ecological sentiments. The first principles states, "Environmental justice affirms the sacredness of Mother Earth, ecological unity and the interdependence of all species, and the right to be free from ecological destruction" (Lee 1992). Although the mainstream environmental movement maintains an affluent, white membership, many of the mainstream environmental groups, such as Greenpeace, Ancient Forest Rescue, and the Sierra Club, have addressed discriminatory environmentalism by fusing environmental justice dimensions to their respective environmental agendas.

Greater pollution, cumulative climatic impacts, and mass consumption of resources have tremendous environmental consequences for the poorest and marginalized populations in the world. Many technological advances have been introduced around the world as strategies for economic development; the introduction of technologies, however, does not necessarily entail the introduction of environmental safety. In 1969 Union Carbide Corporation expanded its global production of pesticides, specifically methyl isocyanate, to Bhopal in central India. A technological disaster occurred in 1984 when a chain reaction of pressure, leaking hydrogen cyanide, and other lethal chemicals exploded and enveloped 40 square kilometers with a poisonous cloud. Failure to maintain safety systems and poor community communication led to the deaths of more than 2,000 residents and over 200,000 further injuries in the region (Applegate, Laitos, and Campbell-Mohn 2000). This tragedy, the worst chemical disaster in world history, is linked directly with global environmental justice in terms of transnational corporate responsibility, distribution of the most dangerous products and conditions to the least well-off, and the violation of public participation in the environmental issues that most affect the local residents. According to S. Ravi Rajan (2001), the Bhopal disaster should be considered "technological violence" because design engineers and executives at Union Carbide decided against a common corporate practice of keeping methyl isocyanate storage tanks underground. The high storage capacity and above-ground tanks at Bhopal aggravated the potential dangers to the environment and local residents, and the failure to install common safety features, when greater safety was warranted under the design conditions, made the corporation accountable for the massive technological disaster. Sophisticated modern technology involved in chemical manufacturing and petrochemical production, and even systems such as those found in military and space programs, involve numerous technological and scientific uncertainties. Basic safety precautions do not address this range of possibilities, and the level of disaster that can follow accidents makes risk assessment a statistical gamble for the local residents.

The magnitude of technological disasters such as that in Bhopal, the global reach of transnational corporations, and the existence of a select group of powerful global scientists and policymakers has given global environmental justice a dramatic scope. Issues pertaining to indigenous land rights and compensation for damages from technological expansion fall directly under the study of environmental justice. New technologies such as genetically modified foods and the ability to acquire and patent the traditional environmental knowledge of indigenous people have emerged as environmental justice concerns. Compensation and donor policies between the global North and South, as well as the environmental and economic consequences of global trade agreements, spark the distributive and participatory justice dimensions of the EJM. Transcontinental pollution and environmental impacts to the global commons find their ethical implications in the EJM. And across the globe there are localized environmental justice movements, such as Japan's Soshisha movement to address victims of Minamata disease, a debilitating neurological disorder caused by the dumping of mercury oxide into the public water supply, or Nigeria's Movement for the Survival of the Ogoni People (MOSOP), struggling against military aggression in a region of petrochemical corporate neglect. All this provides evidence of the expansive scope of global environmental justice, which Lois Gibbs has declared to be the fastest-growing, largest social movement in the world.

Basic Issues

The environmental justice movement has generated a host of ethical questions regarding environmental benefits and technological advances: To what extent is industrial technology implicated in the underlying struggle for the fair distribution of environmental burdens? What is the appropriate relationship between scientific analysis and environmental policies? What technological solutions are available and to whom? How can environmental burdens and benefits be fairly distributed to Earth's populations? What kinds of risks and social conditions constitute an unfair distribution of environmental burdens?

The movement has also produced ethical questions concerning the fair representation and inclusion in the decision-making and social dynamics surrounding environmental hazards: Do marginalized groups receive their proper voice in the process that is likely to affect them the most? How are racial dynamics related to environmental decision-making and environmental harms? What role does gender play? Is it morally acceptable to environmentally discriminate against communities, such as working-class and poor neighborhoods, if it is legal? To what extent are all interests represented in the process? Is the process appropriate for understanding the social and scientific relationships, and the community perception of risk compared to the scientifically acceptable range of risk?

Environmental justice has given scholars and activists the tools to address the environmental conditions of social justice. A vocabulary and conceptual framework now exists to discuss the relationship between environmental values and institutional racism. The political underpinnings of dominant environmental movements are now more easily exposed by the lens of environmental justice. False distinctions between social problems and environmental problems, which caused the splintering of movements such as the civil rights movement and the environmental movement, are now

confronted by environmental groups, civil rights groups, and the numerous grassroots groups that have formed to address environmental injustices in their communities. The movement has broadened the possible interpretations of justice itself by combining distributive justice with political justice and economic justice with cultural justice, under a new rubric of environmental empowerment for the least-well-off populations around the world. Indeed, the dimensions of nature and environment are being revised and transformed by the closer scrutiny that the environmental justice perspective entails. The contention that environmental justice brings new rigor to anthropocentric environmental ethics is an underestimation of the potential critique forged by environmental justice.

ROBERT MELCHIOR FIGUEROA

SEE ALSO Environmental Ethics; Justice; Pollution; Race; Sierra Club; United Nations Environmental Program.

BIBLIOGRAPHY

Applegate, John S.; Jan G. Laitos; and Celia Campbell-Mohn. (2000). The Regulation of Toxic Substances and Hazardous Wastes. New York: Foundation Press.

Been, Vicki. (1994). "Locally Undesirable Land Uses in Minority Neighborhoods: Disproportionate Siting or Market Dynamics?" Yale Law Journal 103(6): 1383–1422.

Cole, Luke W., and Sheila R. Foster. (2001). From the Ground Up: Environmental Racism and the Rise of the Environmental Justice Movement. New York: New York University Press.

Ferris, Deeohn, and David Hahn-Baker. (1995). "Environmentalist and Environmental Justice Policy." In Environmental Justice: Issues, Policies, and Solutions, ed. B. Bryant. Washington, DC: Island Press.

Gottlieb, Robert. (1993). Forcing the Spring: The Transformation of the American Environmental Movement. Washington, DC: Island Press. A revolutionary environmental history revealing various origins of the environmental justice movement and its early parallel path with the mainstream environmental movement.

Guha, Ramachandra. (1989). "Radical American Environmentalism and Wilderness Preservation: A Third World Critique." Environmental Ethics 11(1): 71–83.

Lavelle, Marianne, and Marcia Coyle. (1992). "Unequal Protections: The Racial Divide in Environmental Law." National Law Journal 15(3): S2.

Lee, Charles, ed. (1992). Proceedings: The First National People of Color Environmental Leadership Summit. New York: United Church of Christ, Commission for Racial Justice. A crucial moment in the environmental justice movement, and a defining document of the breadth and depth of the movement.

Mohai, Paul, and Bunyan Bryant. (1992). "Environmental Racism: Reviewing the Evidence." In *Race and the Incidence of Environmental Hazards*, ed. Bunyan Bryant and Paul Mohai. Boulder, CO: Westview Press. A seminal study of race and the disparate distribution of hazardous waste.

Rajan, S. Ravi. (2001). "Toward a Metaphysic of Environmental Violence: The Case of the Bhopal Gas Disaster." In *Violent Environments*, ed. Nancy Lee Peluso and Michael Watts. Ithaca, NY: Cornell University Press.

Schlosberg, David. (1999). *Environmental Justice and the New Pluralism: The Challenge of Difference for Environmentalism.* Oxford: Oxford University Press.

Shrader-Frechette, Kristin. (2002). *Environmental Justice: Creating Equality, Reclaiming Democracy.* Oxford: Oxford University Press.

United Church of Christ. Commission for Racial Justice. (1987). *Toxic Wastes and Race in the United States: A National Report on the Racial and Socio-economic Characteristics of Communities with Hazardous Waste Sites.* New York: Author. First national study of the correlation between race and class and the distribution of hazardous waste sites. Concluding that race plays the stronger role, this study sets off the environmental racism debate.

U.S. General Accounting Office. (1983). *Siting of Hazardous Waste Landfills and Their Correlation with Racial and Economic Status of Surrounding Communities.* Washington, DC: Author.

U.S. House. (1993). Committee on the Judiciary. Subcommittee on Civil and Constitutional Rights. *Environmental Justice: Hearings before the Subcommittee on Civil and Constitutional Rights of the Committee on the Judiciary.* 103rd Cong., 1st sess. A testimony of the environmental racism debate which influenced a number of federal and state agencies.

Wenz, Peter S. (1988). *Environmental Justice.* Albany: State University of New York Press. The first book titled "environmental justice."

Wenz, Peter S. (2001). "Just Garbage." In *Faces of Environmental Racism: Confronting Issues of Global Justice*, 2nd edition, ed. Laura Westra and Bill E. Lawson. Lanham, MD: Rowman and Littlefield.

ENVIRONMENTAL PROTECTION AGENCY

SEE *Environmental Regulation*.

ENVIRONMENTAL REGULATION

• • •

The regulation of human interactions with the environment has taken shape in various political institutions, policies, and market mechanisms that have evolved over time according to changes in social, cultural, and technological conditions. Forms of environmental regulation differ among nations and continue to emerge on the international level as industrialization and globalization create transboundary issues.

From the liberal or socialist perspective, in which the state is understood as a legitimate extension of the community, environmental regulation is regarded as a state activity representing effective public administration. But the conservative or libertarian perspective, in which the state should intervene as little as possible in the lives of its citizens, holds that market mechanisms or private agencies can provide environmental benefits more effectively. The complexity of environmental regulatory efforts also arises from questions about the proper role of scientific knowledge and various mechanisms for handling scientific uncertainty. Environmental regulation is a complex interdisciplinary effort involving ethical principles, political interests, scientific knowledge, and technological capacities. This broad scope of considerations ensures that several worldviews, with their attendant values and recommendations, will interact in regulatory efforts.

Environmental Regulation in the United States

The history of U.S. environmental and natural resource regulation can be categorized into three phases. The first phase, lasting roughly from 1780 to 1880, saw the evolution of legislation that promoted the settlement of the West and the extraction and use of its natural resources (Nelson 1995). Defining laws of this period are the General Land Ordinances of 1785 and 1787, the Homestead Act of 1862, the Mineral Lands Act of 1866, and the Timber Culture Act of 1873.

The success of western expansion spurred a second phase of environmental regulations. Generally termed the conservation movement, this period lasted from roughly the 1880s to the early 1960s. Policies of this period shifted the government's role from simply disposing of public lands to managing them. This management was informed by a philosophy of *wise use*, which held that resources should be managed for the greatest good, for the greatest number, for the longest time. This philosophy was enacted by a rising scientific elite, including Gifford Pinchot (1865–1946) and John Wesley Powell (1834–1902), who argued that the scientific management of natural resources must guide economic development in order to accomplish sustained yield and maximum efficiency. This placed the conservationists in conflict with John Muir (1838–1914) and other preservationists, who sought to maintain environments in their natural state (Caulfield 1989). The second phase witnessed the creation of the national park and national forest systems (for example, Yellowstone National Park in 1872; and the Organic Act [Forest Management Act] in 1897). The 1964 Wilderness Act, which sought to preserve pristine wilderness "untrammeled by man, where man himself is a visitor who does not remain," represents the culmination of this era.

The third phase marks the beginning of modern environmentalism, and received its greatest impetus from consciousness-raising works such as Rachel Carson's *Silent Spring* (1962) and Stewart Udall's (b. 1920) *Quiet Crisis* (1963). These books along with social changes wrought by modernizing technologies, industrialization, and urbanization triggered increased awareness of environmental problems and focused environmental policies on the regulation of air and water pollution, toxic chemicals, solid waste, and other impacts of the growing industries fueled by advances in science and technology. A later concern developed over global issues such as biodiversity and climate change. The modern environmental movement initiated an expanded role for the federal government in environmental regulation, which is especially evident in the major pieces of legislation passed in the 1970s: the National Environmental Policy Act in 1969; the creation of the Environmental Protection Agency (EPA) in 1970; Clean Air Act amendments in 1970 and 1977; the Clean Water Act in 1972 and amended in 1977; the Endangered Species Act in 1973; and the Toxic Substances Control Act in 1976.

By the end of the 1970s, federal and state governments had greatly expanded their environmental roles from public lands management to public health, industrial health and safety, agricultural development, and urban planning. The EPA took charge of a number of federal environmental responsibilities. Although independent of other federal agencies, the EPA is still a part of the executive branch and reports to the president. It operates within a context of other major federal agencies, including those housed under the Department of the Interior (DOI) (such as Fish and Wildlife Service, National Park Service, Bureau of Land Management, and Bureau of Reclamation) as well as the Department of Agriculture and the National Marine Fisheries Service. An enormous amount of regulatory activity continued to occur at the regional, state, and local levels. Governmental entities at every level have their own environmental regulations, constrained by the fact that they cannot defeat the purpose of federal regulations.

The 1980s, during the Ronald Reagan and George H. W. Bush presidencies, witnessed some weakening of environmental regulations, as an extension of more general deregulation policies that argued the inefficiencies of bureaucratic or command-and-control mechanisms as well as the need to perform cost-benefit analyses on regulatory activities. These changes were matched by the creation and strengthening of many nongovernmental organizations (NGOs) and other environmental activist and lobbying groups.

The Bill Clinton era (1992–2000) witnessed a modest revival of federal regulatory efforts. The George W. Bush presidency once again sought the de-federalization of environmental regulation as well as the more active extraction of energy resources on federal lands.

Other Nations and International Efforts

Other countries institutionalized environmental regulation by creating ministries of the environment (for example, Great Britain), or placed environmental responsibilities in existing ministries (such as West Germany). Eventually most European countries established environmental ministries, even though other ministries (such as agriculture, energy, or urban planning) continued to manage some environmental regulatory activities. Austria, France, Germany, Ireland, Italy, Sweden, and the United Kingdom eventually created more or less independent environmental regulatory agencies. At the European Union (EU) level, the European Environment Agency (EEA) is charged with generating and disseminating environmental information.

In Latin America, the process of introducing environmental regulation followed the European model. Until the 1990s, in many Central and South American countries there existed various national environmental commissions charged with coordinating different environmental protection activities. The 1992 Rio Conference (United Nations Conference on Environment and Development Earth Summit) provided an important impulse for administrative reforms in Latin America related to environmental protection and led to the creation of ministries of the environment throughout the Spanish- and Portuguese-speaking countries of the Americas.

As globalization continues, an increasing number of environmental problems present transboundary issues. Global climate change, invasive species and biodiversity, water use, and air and water pollution are just some of the problems that raise environmental regulation into the realm of international law and policy. The United

Nations has played a leading role in two of the more prominent instances of international collaboration around environmental issues. First, the UN Environment Programme established the international legal framework known as the Vienna Convention on the Protection of the Ozone Layer in 1985. This led to the Montreal Protocol on Substances that Deplete the Ozone Layer in 1987, which required industrialized countries to reduce their consumption of chemicals that harm the ozone layer. Second the United Nations Framework Convention on Climate Change (Framework) established in 1992 provides a forum for governments to gather and exchange information and adapt to the effects of climate change. An international meeting in Kyoto, Japan, held under the Framework, produced a document (the 1997 Kyoto Protocol) that established binding limitations on greenhouse gas emissions by developed nations. Russia's ratification of the protocol in 2004 fulfilled the participation requirements for developed nations, thus allowing the treaty to become effective.

However such international agreements generally just set basic guidelines that require domestic legislation. This is usually difficult to achieve, and in the case of the Kyoto protocol, monitoring compliance is complex and there is no international enforcement authority. Furthermore international negotiations usually involve several governmental bodies, such as agencies, ministries or departments. For example, the State Department (not the EPA) controls U.S. involvement in international climate negotiations. The proliferation of bureaucratic agencies can create political gridlock.

Types of Environmental Regulation

Environmental regulation is plagued by two intrinsic challenges. First, because many environmental regulations involve the protection of public (common) goods, they often conflict with individual rights (especially property rights). Second, environmental problems often occur over long time periods and wide physical areas, whereas most individuals involved in regulatory processes have short-range, narrow interests, especially concerning economic growth. For both reasons, traditional environmental regulations usually entailed the implementation of strict controls on the otherwise unrestrained expression of personal and economic interests in the free market. As John Baden and Richard Stroup point out:

> The dawn of the environmental movement coincided with an increased skepticism of private property rights and the market. Many citizen activists blamed self-interest and the institutions that permit its expression for our environmental and natural resource crises. From there it was a short step to the conclusion that management by professional public "servants," or bureaucrats, would significantly ameliorate the problems identified in the celebrations accompanying Earth Day 1970. (Baden and Stroup 1981, p. v)

What followed during the 1970s was a command-and-control approach to environmental regulation, wherein the government set strict legal limits and enforced sanctions against violators.

Although this top-down and sometimes heavy-handed approach resulted in important successes, it also revealed a crucial element of regulatory practices: There are governmental failures just as there are market failures. Several reasons for governmental failures exist. Bureaucrats, like all people, are self-interested, and when governmental structures are not designed to link authority with responsibility for program outcomes, "decision makers have few incentives to consider the full social costs of their actions" (Baden and Stroup 1981, p. v). Furthermore decision makers have only a limited capacity to comprehend complex social and environmental interactions, which can limit their ability to make wise regulatory decisions.

One response has been to improve the structure of government, but another reaction has been to improve the structure of markets by implementing what Terry Anderson and Donald Leal term *Free Market Environmentalism* (1991). The underlying philosophy of this regulatory approach is that markets and environmental concerns can be made compatible by internalizing costs and establishing the proper incentives. This perspective also challenges the common assumption that environmental degradation is inherently linked to economic growth. It should also be noted that the relationship between environmental regulations and job loss or economic downturns is controversial, and no such correlation may exist (Goodstein 1999).

Anderson and Leal claim that the approach of free market environmentalism is founded on a core assumption of human nature: Humans are self-interested. They write, "Instead of intentions, good resource stewardship depends on how well social institutions harness self-interest through individual incentives" (Anderson and Leal 1991, p. 4). Examples of utilizing market mechanisms for environmental regulations include green taxes, marketable emissions permits (for example, cap-and-trade systems), and the elimination of harmful government subsidies.

Command-and-control and free market regulatory strategies are not incompatible and can often be used in conjunction to achieve desired environmental outcomes. Free market mechanisms obviously also have social dimensions insofar as they influence levels of public service, consumer rights, minority interests, and more. Social regulations likewise have economic implications in that they provide a framework within which economic activities can take place. Public or private institutions may advocate for both types of regulation. At the public level, environmental agencies such as the EPA are often subject to enormous political pressures that can complicate their mission and even compromise their integrity (Landy, Roberts, and Thomas 1994).

Many environmental regulations involve statutes, which often include a citizen suit provision or other appeals procedures that allow citizens to challenge an agency's action (or inaction) when it appears to be out of compliance with the law. In the United States, suit can also be filed under the Administrative Procedures Act, which is another mechanism for holding federal employees and agencies accountable for properly exercising their authority. Many environmental statutes specify the basis on which decisions must be made. In the United States, public input at the scoping stage is usually mandatory, and notice and comment periods through the Federal Register are always required. Some statutes require protection of the environment, while others focus primarily on human health. Some mandate cost-benefit analysis, while others call for decisions based on the best available science alone, with no consideration given to economic cost.

Science and Environmental Regulation

For all environmental problems, a certain amount of scientific understanding of natural systems and their interaction with human social systems is a necessary component of any regulatory action. This partially explains the preeminent importance of scientific advice in the crafting of environmental regulation or *science for policy*. The role of scientific expert knowledge is independent of the type of administrative process. Establishing an independent agency raises further questions of democratic legitimacy and accountability. This is true especially in relation to the problems of scientific advisers turning into policy makers and policy makers delaying action while continuing to fund more scientific research (Jasanoff 1990).

In theory, the process of environmental regulation depends on two factors: the definition (by democratically legitimized institutions) of the public goods to be protected, including the degree and costs of protection; and the scientific knowledge necessary to determine how an action may impact those public goods. But it is erroneous to assume that these two factors alone define the regulatory framework. Also, in this view, moral and political considerations play a role only during the definition of regulatory aims; and the justification for adopting certain regulations is based solely on expert knowledge. However, as regulatory practice demonstrates, this position has to be complemented by other considerations, because the facts and values components of environmental regulations are engaged in an iterative dialectic.

The different regulatory approaches created to safeguard public health and the environment from the effects of a large number of technological applications have stimulated new kinds of scientific activity, among them environmental impact and risk assessment. The scientific evaluation of risks and impacts has spawned various types of cost-benefit and risk-cost-benefit analyses (National Research Council 1996). These management tools permit a limited comparison of the environmental and economical effects of various alternative technologies and production processes, as well as different regulatory approaches. They can also be used to analyze risk-tradeoffs, where the regulation itself may lead to the emergence of other risks and negative impacts.

The Role of Science

Such predictive models are often limited by lack of data and the impossibility of modeling complex, higher-order interactions. For example, identifying the environmental impacts and risks presented by a chemical substance is made difficult by long term, cumulative interactions (sometimes called the *cocktail effect*) that cannot be mimicked in a laboratory setting. In some cases, the environmental degradation may be patent but establishing the pertinent causal relations may nevertheless be extremely difficult. In the case of global climate change, this type of persistent uncertainty has tended to sidetrack political discussion and hamper the process of producing alternatives for decision makers and stakeholders. So, even though scientific understanding is indispensable, it is not the only ingredient in formulating and implementing sound environmental regulations. There are very few instances where science provides enough clarification to clear away politically charged, open-ended environmental problems. This has led some policy analysts such as Daniel Sarewitz (2004) to suggest that the values bases of disputes must be fully articulated and adjudicated before science can play an effective role in resolving environmental problems.

Scientific investigation is certainly crucial to crafting wise regulations, but also presents several challenges (Cranor 1993). First is the issue of burden of proof. Generating all the necessary scientific information can be a time and resource intensive task. This can delay any decision, which in turn means that a harmful activity continues unregulated. In such case, putting the burden of proof on those who try to demonstrate that an environmental impact indeed exists tends to favor the environmentally harmful activity instead of the protection of the environment. This situation has led those social groups most concerned about environmental protection to demand, at least for certain technologies, the inversion of the burden of proof (that is, the need for demonstrating the absence of important environmental impacts).

A related problem concerns the standards of proof, which determine if a technological activity is harmful for the environment or human health. A number of factors can make environmental risk and impact analysis a very complex activity. If standards are rigorous, regulatory action may be excessively delayed. The debate on global warming and its relation to the emission of greenhouse gases provides a good example. In many cases it may be more effective for the protection of the environment to synthesize all available information from different sources and make decisions based on cumulative weight instead of trying to identify and quantify with precision any single environmental impact or risk. This highlights the fact that the choice of a standard of proof is as much a political and ethical dilemma as a scientific question (Shrader-Frechette 1994).

A third problem is the indeterminacy that is inherent in any environmental impact or risk assessment (Wynne 1992). Indeterminacy can only be reduced through methodological choices (for instance, about different available mathematical models that establish the relationship between the presence of a substance and environmental effects). Any choice that affects the scientific methodology leads either to an increase of false positives (reaching the conclusion that the activity is harmful for the environment even though it is not) or of false negatives (reaching the conclusion that the activity is not harmful even though it is). In other words, any methodological choice has important regulatory consequences. This leads inevitably to the conclusion that scientists must take into account the consequences of the methodologies they choose, while society and decision makers must be aware of the uncertainties inherent in scientific knowledge about impacts and risks (Funtowicz and Ravetz 1992).

Since the 1990s, an important field in the debates on environmental regulation has focused on the so-called precautionary principle, proposed by some environmentalists as a means to face those problems posed by scientific uncertainties regarding environmental impacts (Raffensperger and Tickner 1999). A number of agreements and international treaties have adopted this principle. However, so far no commonly accepted definition exists. One of the more popular definitions is the one to be found in the 1992 Rio Declaration on Environment and Development: "Where there are threats of serious irreversible damage, lacks of full scientific certainty shall not be used as a reason for postponing cost-effective measures to prevent environmental degradation." Besides the discussion about its definition, there also exists a debate about when to invoke the precautionary principle, about its general meaning as well as its scope.

A Typology of Worldviews

John Dryzek and James Lester (1989) have created a typology of environmental worldviews that serves as one way of organizing the variety of problem definitions and prescriptions for regulatory policies and institutions. Six worldviews are distinguished according to their particular blend of two different dimensions: the locus of value (individuals, anthropocentric communities, or biocentric communities) and the locus of solutions (centralized or decentralized). Each worldview thus supports different policy recommendations.

The first three worldviews all agree that solutions must be centralized. First are the *Hobbesians and structural reformers*, who believe in modern liberal individualism, but argue that it must be checked by a certain degree of political centralization. This is still the dominant worldview, and most of its adherents are moderates, convinced that "more laws to regulate polluters, more funds for enforcement, and minor structural reforms" will suffice (Dryzek and Lester 1989, p. 318). Second are the *guardians*, who still value centralization, but argue that an elite group of scientific and technical experts should monopolize power. Examples include Alvin Weinberg's proposal to create a *permanent priesthood* of nuclear technologists to oversee energy systems and William Ophul's *class of ecological mandarins*. The third group of centralizers is the *reform ecologists*, who argue that ecological values must be represented in the highest echelons of government. Reform ecologists (for example, Eugene Odum, Paul Ehrlich, and Lester

Brown) are usually less concerned with the structure of political and economic institutions than with their scientifically defended ecocentric values.

The other three worldviews find the locus of solutions in decentralization. First are the *free market conservatives*, who, like Anderson and Leal, believe that government intervention in environmental problems has gone too far and self-regulating market systems can work much better. Second are the *social ecologists*, who base their decentralized vision not on the market but rather on the ideal of a cooperative community. Murray Bookchin represents the main stem of this worldview, but it also applies to ecofeminists and other groups that call for classless, stateless, and decentralized societies far removed from capitalism. Finally the *deep ecologists* take little interest in human communities (like the reform ecologists) and stress the importance of the realization of the self within the greater *Self* of the biotic community. Although it can verge on misanthropic antipolitics, deep ecology is also represented by such luminaries as Henry David Thoreau and Aldo Leopold and other insightful theorists such as Arne Naess, Bill Devall, and George Sessions.

Although not without its gaps and ambiguities, Dryzek and Lester's typology can be used as a heuristic to organize the complex and contested nature of environmental regulations. It captures the various roles that science can play (for instance, informing modest reforms or monopolizing entire discourses) according to the dominant worldview in the particular topic. It distinguishes between various forms of centralized and decentralized regulations. The typology also hints at the alternative futures that can occur as worldviews rise and fall from social and political dominance, thus leading to different regulatory mechanisms and philosophies. Finally it highlights the constructed nature of reality as participants bring different worldviews to the political agenda, which in turn opens up the dialogue over which values ought to be represented and which regulatory mechanisms can best deliver the valued outcomes.

JOSÉ LUIS LUJÁN
ADAM BRIGGLE

SEE ALSO *Environmental Ethics; National Parks; Pollution; Regulation; Science, Technology, and Law; United Nations Environmental Program; Waste.*

BIBLIOGRAPHY

Anderson, Terry, and Donald Leal. (1991). *Free Market Environmentalism*. San Francisco: Westview Press. The first comprehensive treatment of the idea that free markets can achieve environmental goals.

Anderson, Terry, and Donald Leal. (1997). *Enviro-Capitalists: Doing Good While Doing Well*. Lanham, MD: Rowman & Littlefield. An updated version of the central idea that free markets and environmentalism are not fundamentally antagonistic.

Baden, John, and Richard Stroup. (1981). *Bureaucracy vs. Environment: The Environmental Costs of Bureaucratic Governance*. Ann Arbor: University of Michigan Press. Outlines the failures of bureaucratic solutions to environmental problems and offers suggestions for reforms based on aligning incentives with valued outcomes.

Brunner, Ron; Toddi A. Steelman; Lindy Coe-Juell; et al. (2005). *Adaptive Governance: Integrating Natural Resource Science, Decision Making and Policy*. New York: Columbia University Press. Focuses on community-based initiatives to overcome the gridlock that can ensue from scientific management.

Caulfield, Henry. (1989). "The Conservation and Environmental Movements: An Historical Analysis." In *Environmental Politics and Policy: Theories and Evidence*, ed. James Lester. London: Duke University Press. Utilizes elite theory to explain the evolution of political movements and provides a historical analysis of these two movements.

Carson, Rachel. (1962). *Silent Spring*. Boston: Houghton Mifflin.

Cranor, Carl. (1993). *Regulating Toxic Substances: A Philosophy of Science and the Law*. New York: Oxford University Press. A detailed analysis of the political dimension of the epistemological characteristics of scientific research on impacts and risks. Its main focus is U.S. regulatory agencies.

Davis, Charles, and James Lester. (1989). "Federalism and Environmental Policy." In *Environmental Politics and Policy: Theories and Evidence*, ed. James Lester. London: Duke University Press. Contends that a state's ability to implement environmental programs depends on its institutional capacity and its degree of dependence on federal grants.

Dryzek, John, and James Lester. (1989). "Alternative Views of the Environmental Problematic." In *Environmental Politics and Policy: Theories and Evidence*, ed. James Lester. London: Duke University Press. Creates a typology of world views based on locus of value and locus of solutions.

Funtowicz, Silvio O., and Jerome R. Ravetz. (1992). "Three Types of Risk Assessment and the Emergence of Post-Normal Science." In *Social Theories of Risk*, ed. Sheldon Krimsky and Dominic Golding. West Port, CT: Praeger. In order to manage the uncertainty inherent in environmental risk assessment, especially in the case of global environmental risks, the authors propose the introduction of a methodologically diverse science, the so-called post-normal science, based on an extended peer community.

Goodstein, Eban. (1999). *The Trade-Off Myth: Fact and Fiction about Jobs and the Environment*. Washington, DC: Island Press. Argues that there is no jobs-environment trade-off.

Jasanoff, Sheila. (1990). *The Fifth Branch: Science Advisers as Policymakers*. Cambridge, MA: Harvard University Press.

An analysis of the policy role of science advisers in the United States, especially in the case of agencies such as the EPA or the Food and Drug Administration (FDA).

Landy, Marc; Marc Roberts; and Stephen Thomas. (1994). *The Environmental Protection Agency: Asking the Wrong Questions: From Nixon to Clinton*. Oxford: Oxford University Press.

Lash, Scott; Bronislaw Szerszynski; and Brian Wynne. (1996). *Risk, Environment & Modernity: Towards a New Ecology*. London: Sage Publications. Includes several articles of European scholars on the concept of risk society.

National Research Council. (1996). *Understanding Risk: Informing Decisions in a Democratic Society*. Washington, DC: National Academy Press. An analysis of the characterization of risk that defends a very broad definition of this concept.

Nelson, Robert. (1995). *Public Lands and Private Rights: The Failures of Scientific Management*. Lanham, MD: Rowman & Littlefield. Provides historical context for U.S. public land management debates. The first chapter examines the unexpected consequences of early public land laws.

Raffensperger, Carolyn, and Joel Tickner, eds. (1999). *Protecting Public Health & the Environment: Implementing the Precautionary Principle*. Washington, DC: Island Press. Includes a number of analyses of different aspects of the precautionary principle.

Sarewitz, Daniel. (2004). "How Science Makes Environmental Controversies Worse." *Environmental Science and Policy* 7(5): 385–403. Explains why environmental problems tend to become *scientized*, and offers some recommendations for improved decision making.

Shrader-Frechette, Kristin. (1994). *Ethics of Scientific Research*. Lanham, MD: Rowman & Littlefield. A study of ethical science-related problems, with special attention to the analysis of scientific knowledge in the area of environmental regulation.

Tesh, Sylvia Noble. (2000). *Uncertain Hazards: Environmental Activists and Scientific Proof*. Ithaca, NY: Cornell University Press. Shows the ways in which pressure exerted by activist groups leads to changes in scientific research on impacts and risks.

Udall, Stewart. (1963). *Quiet Crisis*. New York: Holt, Rinehart, and Winston.

Wynne, Brian. (1992). "Uncertainty and Environmental Learning: Reconceiving Science and Policy in the Preventive Paradigm." In *Global Environmental Change* 2(2): 111–127. An analysis of the indeterminacy in environmental risk assessment, as well as its consequences for environmental policy.

ENVIRONMENTAL RIGHTS

• • •

Often referred to as part of the *third generation* of human rights, the concept of environmental rights is unclear in meaning and content. Environmental rights are elusive because there is no universal definition, and they are controversial because they hybridize the ecocentric perspectives of environmentalists and the anthropocentric perspectives dominant among human rights activists (Apple 2004). No binding international agreement has had environmental rights as its primary focus because such rights fail to fit neatly into either of these two groups. This fact combined with the scarcity of binding international legal instruments has prevented environmental rights from becoming international law. Nonetheless progress on defining and enforcing environmental rights continues on the international, regional, and national levels.

Background

Throughout the late-1950s and early-1960s serious environmental disasters occurred in various regions of the world: oil spills at sea (for example, the tanker *Torrey Canyon* in the English Channel in 1967), the release of toxic substances from chemical industries (such as mercury in Minamata Bay, Japan, in 1968), and nuclear disasters (for instance, the nuclear center Kytchym, in the former Soviet Union, in 1957). Such accidents, repeated over the years, demonstrated the dangers of incorporating technology into human activity without including some regulation. People also became increasingly aware of risks to human health and the environment due to high-tech industrial and agricultural activities. Emblematic of this concern was Rachel Carson's *Silent Spring* (1962), which argued the presence and persistence of toxic substances in living organisms as a consequence of the massive use of pesticides.

Legal measures to control unhealthy and dangerous activities and to protect the environment from the abuses of human intervention followed. In 1970, on the date of the first Earth Day celebration, the U.S. government enacted the National Environmental Policy Act, which submitted major development projects to environmental review. Since then laws concerning the environment have multiplied around the world.

Many in the ecological and human rights movements argued that these legal measures were insufficient to guarantee a healthy environment for present and future generations. Some proposed the proclaiming a new human right: the right to a healthy environment. This right does not fit within the category of civil and political or *first generation* rights, nor of economic, social, and cultural or *second generation* rights. For this reason environmental rights (along with others, such as rights to development) are sometimes described as *third generation* rights. Just as the first generation aspired to guarantee individual liberties, and the second equality, the third aims to guarantee solidarity across national boundaries and between present and future generations. Third generation human rights are conceived as collective rather than individual, and they tend to challenge the sovereignty of the modern nation-state.

Formulations at Different Political Levels

The appearance and development of the right to a healthy environment is traceable on three levels: global, through three world conferences on the environment organized by the United Nations; regional, through some agreements on the subject of human rights; and national, through the inclusion of environmental rights in the constitutions of some countries. (Rachel Carson had in fact proposed consideration of an amendment to the U.S. Constitution guaranteeing the right to a clean environment.)

GLOBAL. During the First World Conference on Human Development (Stockholm 1972), the Declaration of the Human Environment was approved, proclaiming the right to a clean environment for the first time at the international level: "Man has the fundamental right to freedom, equality, and adequate conditions of life, in an environment of a quality that permits a life of dignity and well-being, and he bears a solemn responsibility to protect and improve the environment for the present and future generations" (principle 1). This was followed ten years later by the U.N. World Charter for Nature (1982), which proclaimed that, in recognition of the fact that humankind is part of nature, "Nature shall be respected and its essential processes shall not be impaired" (principle 1).

Twenty years after the Stockholm meeting the World Conference on the Environment and Development, known as the Earth Summit, took place in Rio de Janeiro. One of the documents approved at this conference was the Declaration of Rio, which affirmed: "The right to development must be fulfilled so as to equitably meet developmental and environmental needs of present and future generations" (principle 3). The declaration accepted the idea of sustainable development, a concept that had been defined by the World Commission on the Environment and Development in *Our Common Future* (1987) as "development that meets the needs of the present without compromising the ability of future generations to meet their own needs." The 2002 Johannesburg Summit unfortunately had neither the level of state participation nor world impact of the two prior conferences.

REGIONAL. Environmental rights are mentioned more explicitly at the regional level. In 1981 the African Charter on Human and People's Rights was approved in Banjul, Gambia, West Africa. The charter states: "All peoples shall have the right to a general satisfactory environment favorable to their development" (article 24). Similarly the additional protocol to the American Convention on Human Rights in the area of economic, social, and cultural rights, the Protocol of San Salvador (1988), affirms in article 11 that (1) Everyone shall have the right to live in a healthy environment and to have access to basic public services; and (2) The States' Parties shall promote the protection, preservation, and improvement of the environment.

In Europe the 1950 European Convention on Human Rights did not include environmental rights. Nevertheless the European Tribunal on Human Rights has included demands for the protection of the environment in some of the articles from the European Convention on Human Rights, such as the right to private and family life (article 8) and the right to information (article 11).

Also in Europe another important advance came in the form of the Aarhus Convention on Access to Information, Public Participation in Decision-making, and Access to Justice in Environmental Matters. Negotiated by the UN Economic Commission for Europe (UNECE), it was adopted in 1998 and implemented on October 30, 2001. Its first article expresses the object of the convention: "In order to contribute to the protection of the right of every person of present and future generations to live in an environment adequate to his or her health and well-being, each Party shall guarantee the rights of access to information, public participation in decision-making, and access to justice in environmental matters in accordance with the provisions of the Convention."

NATIONAL. At the national level are many constitutions passed in the seventies and eighties that include a mention of human rights to a sound environment. But those references do not specify jurisdictional guarantees, so some authors deny that they are real rights and consider them only as guidelines for the public powers.

Characteristics of Environmental Rights

There is no consensus on how to define environmental rights. First, it is difficult to define the environment: Is it physical, social, cultural, or all of these? Does it pertain only to nature or also to urban spaces, workplaces, and homes? Second, there is debate as to whether the holders of these rights are individuals, contemporary human communities, future generations, or even ecosystems. Third, there is no agreement about whether environmental rights can be exercised before a juridical organ or simply constitute a mandate to public powers that they develop policies to protect the environment. Finally, doubts arise as to whether environmental rights can also involve duties, as has been proclaimed in some constitutions.

Environmental rights present a challenge to the concept of human rights as they are formulated in the early-twenty-first century. Seriously considering the grant of these rights questions the modern world model that promotes unlimited growth for the rich and permits unjust environmental burdens on the poor, both within countries as well as among different nations. The concept may even be interpreted as challenging the assumed hierarchy of humans over nature that underlies so much economic and social activity.

Environmental rights have a double dimension: juridical and political. The strictly juridical can be narrowed down to a set of powers that individuals or communities can exercise: the right to participate in the making of development policies, the right to information on environmental matters, the right to access tribunals in order to make demands in matters related to the environment, and the right to environmental education. In the United States a number of parties have sued multinational corporations for environmental rights abuses under a federal statute, the Alien Tort Claims Act (ATCA). While the ATCA has been used successfully to prosecute first generation human rights abuses (torture, for example), it has not provided a legitimate basis for environmental rights claims. Environmental wrongs resulting in human harm are not interpreted as violations of international law in the early-twenty-first century (Apple 2004). The applicability of ATCA to non-state actors such as corporations also remains unclear.

The political dimension of environmental rights has both a national and international manifestation. At the national level it involves assuring that political leaders take action to protect and promote the environment. At the international level it extends to the set of endeavors that states undertake in order to achieve sustainable and shared development for the entire world. Environmental rights not only aspire to preserve nature, but also to achieve the conditions necessary for a more just and healthy life for all persons and all peoples on Earth.

Such broad ambitions, however, contribute to the ambiguity of the concept and hinder attempts to realize these goals in particular contexts. Jorge Daniel Taillant (2004) argues that it is unclear whether the term environmental rights refers to human rights with respect to the environment, the human obligation to respect nature for its own sake, or something else. He contends that a conceptual framework based on development and more traditional forms of human rights, rather than environmental rights, can bring better practical results.

Assessment

Any assessment of environmental rights in relation to science, technology, and ethics must recognize the tenuous status of even first and second generation human rights. The Universal Declaration of Human Rights itself is simply a declaration that establishes a common standard, urging individuals and organizations to strive to promote respect for human rights and freedoms. However there do exist many environmental treaties that have well-defined, binding clauses, such as the Law of the Sea Treaty. The extent to which such environmental treaties influence the governance of science and technology is a subject deserving of further examination and development.

VICENTE BELLVER CAPELLA
TRANSLATED BY JAMES A. LYNCH

SEE ALSO *Environmental Ethics; Environmental Justice; Environmental Regulation; Human Rights.*

BIBLIOGRAPHY

Bellver Capella, Vicente. (1994). *Ecología: de las razones a los derechos* [Ecology: From reasons to rights]. Granada: Comares. Analyzes the relevance of ecophilosophies in the configuration of the human right to the environment.

Jordano Fraga, Jesús. (1994). *La protección del derecho al medio ambiente* [The protection of the right to the environment]. Barcelona: Bosch.

Kiss, Alexandre, and Shelton, Dinah. (2003). *International Environmental Law.* Ardsley, NY: Transnational Publisher.

Loperena Rota, Demetrio. (1994). *El derecho al medio ambiente adecuado* [The right to a sound environment]. Madrid: Civitas. The author maintains that the human right to a sound environment is not only a principle to inform the political activity but a right in the strict sense of the term.

INTERNET RESOURCES

Apple, Betsy. (2004). "Commentary on the 'Enforceability of Environmental Rights.'" Carnegie Council on Ethics and International Affairs. Available from http://www.carnegiecouncil.org/viewMedia.php/prmTemplateID/8/prmID/4464.

Taillant, Jorge Daniel. (2004). "A Nascent Agenda for the Americas." Carnegie Council on Ethics and International Affairs. Available from http://www.carnegiecouncil.org/viewMedia.php/prmTemplateID/8/prmID/4461.

EPIDEMIOLOGY

• • •

Epidemiology is the study of the frequency, distribution, and determinants of disease in humans. Its aim is the prevention or effective control of disease. The term originated in the study of epidemics, rapidly spreading diseases that affect large numbers of a population (from the Greek *epi* meaning upon and *demos* meaning people). Epidemiology touches on ethics in two key areas: The need for competent and honest use of its information, and questions of responsibility raised by the global picture it presents of the health of humanity.

Speculation about the nature and causes of disease dates back to antiquity. The formal history of epidemiology, like that of statistics, begins with the systematic official recording of births and deaths in the seventeenth century, proceeding to the quantitative investigation of diseases with the emergence of scientific medicine in the nineteenth. Based on the theory of probability, statistical inference reached maturity in the early-twentieth century and gradually spread into a wide range of disciplines. Its application to medical research gave rise to biostatistics and contemporary epidemiology.

There is no clear division between the two fields. Epidemiology focuses more on public health issues and the need for valid population-based information, but it uses the theory and methods of biostatistics. Its practitioners tend to be individuals with primary interest and training in medicine or a related science, whereas biostatisticians come from mathematics. They work together as members of the medical research team, in the dynamic context of scientific advances and the latest information technology.

Modern Epidemiology

The mathematical approach to medicine, with the methodical tabulation of patient information on diseases and treatment outcomes, was introduced in the 1830s by the French physician Pierre C. A. Louis (1787–1872). As a notable result of his researches in Paris hospitals, his Numerical Method revealed the uselessness of bloodletting. Inspired by Louis, his British student William Farr (1807–1883) became the central figure in the development of vital statistics in England and the use of statistics to address public health concerns. Farr worked with John Snow (1813–1858), the physician who investigated the cholera epidemic sweeping through London in 1854. Snow's finding that the *cholera poison* was transmitted in contaminated water from the Broad Street pump was a milestone event in epidemiology and public health. Farr also provided guidance in statistics for Florence Nightingale (1820–1910) to support her work in hospital reform.

The existence of microbes was discovered in the late-seventeenth century by the Dutch lens grinder Antonie van Leeuwenhoek (1632–1723), who saw "animalcules, more than a million for each drop of water" through his microscope (Porter 1998, p. 225). The role of germs as causes of disease was established by Louis Pasteur (1822–1895), French chemist and founder of microbiology. Pasteur invented methods to isolate and culture bacteria, and to destroy them in perishable products by a heat treatment now called *pasteurization*. He found that inoculation by a weakened culture provided *immunity*, protection against the disease. This explained the earlier discovery of the English physician Edward Jenner (1749–1823) that vaccination with the milder cowpox protected against smallpox. (Vaccination comes from the Latin *vacca* meaning cow.) The German physician Robert Koch (1843–1910), founder of bacteriology, further developed techniques of isolating and culturing bacteria. He identified the germ causing anthrax in 1876, tuberculosis in 1882, and cholera in 1883. He contributed to the study of other major diseases, including plague, dysentery, typhoid fever, leprosy, and malaria.

Extensive public health measures of hygiene and immunization, along with the introduction of the sulfonamide drugs in the late 1930s and antibiotics in the 1940s, brought most infectious diseases under control. Attention turned to chronic diseases, by then the leading causes of morbidity and mortality—multicausal diseases with a long latency period and natural course. Two historic discoveries of the mid-twentieth century were tobacco use as a cause of lung cancer, and risk factors for heart disease. From the study of infectious and chronic diseases epidemiology has evolved into a multidimensional approach, defined by disease, exposure, and methods, with focus on new developments in medical science. Its many specialties include cancer, cardiovascular, and aging epidemiology, environmental, nutritional, and occupational epidemiology, clinical and pharmaco-epidemiology, and molecular and genetic epidemiology. With the sequencing of the human genome,

TABLES 1–3

Table 1: Some Basic Terms of Epidemiology

Measures of Morbidity and Mortality

- PREVALENCE (Burden of disease): Number of existing cases of a disease at a given point in time divided by the total population.
- INCIDENCE (Cumulative incidence, risk): Number of new cases of a disease during a given time period divided by the total population at risk.
- INCIDENCE RATE (Incidence density): Number of new cases of a disease during a given time period divided by the total person-time of observation.
- PERSON-TIME (usually person-years): Total disease-free time of all persons in the study, allowing for different starting dates and lengths of time observed.
- CRUDE DEATH RATE: Number of deaths during a given time period divided by the total population.
- STANDARDIZED DEATH RATE: Crude death rate adjusted to control for age or other characteristic to allow valid comparisons using a standard population.

Example of Age-Adjusted Death Rates (2000 US Standard Population)

	Alaska	Florida	United States
Crude death rate/1,000 population (in 2000):	4.6	10.3	8.9
Percent of population over age 65 (in 2000):	5.7	17.6	12.4
Age-adjusted death rate/100,000 population (avg. for 1996–2000):			
Breast cancer	25.2	25.6	27.7
Prostate cancer	24.2	28.4	32.9

Prevalence, incidence, and death rates are expressed in units of a base (proportion mulitplied by base), usually per 1,000 or 100,000 population.

SOURCE: Courtesy of Valerie Miké. Data in example from U.S. Census Bureau website and American Cancer Society (2004).

Table 2: Case-Control Study of Lung Cancer and Smoking

Smokers		Lung Cancer		Controls		Odds Ratio (ad/bc)
Males:	Yes	647	(a)	622	(b)	14.0
	No	2	(c)	27	(d)	P<.00001
	Total	649		649		
Females:	Yes	41	(a)	28	(b)	2.5
	No	19	(c)	32	(d)	P<.05
	Total	60		60		

Historic study showing the association between cigarette smoking and lung cancer. No association would correspond to an odds ration of 1. P-values obtained by chi-square test for 2x2 tables.

SOURCE: Data from Doll and Hill (1950).

Table 3: Cohort Study of Risk Factors for Coronary Heart Disease: Systolic Blood Pressure

Systolic BP (mmHg)	Age 35–64		Age 65–94	
	Men	Women	Men	Women
<120	7	3	11	10
120–139	11	4	19	13
140–159	16	7	27	16
160–179	23	9	34	15
>180	22	15	49	31
Total Events	516	305	244	269

Average annual incidence per 1,000 persons of coronary heart disease, by systolic blood pressure. Example of relative risk (RR): For men (35-64), systolic BP>180 relative to <120 mmHg, RR=22/7=3.1. No association would correspond to RR=1. Results of 30-year follow-up in historic Framingham Heart Study of risk factors for cardiovascular disease.

SOURCE: Adapted from Stokes et al. (1989).

---

genetics is assuming increasing importance across all lines of inquiry. In its principles of studying human populations, epidemiology is related to psychology, sociology, and anthropology, all of which employ statistical inference.

Basic Concepts and Methods

Epidemiology may be *descriptive* or *analytic*. Descriptive epidemiology reports the general characteristics of a disease in a population. Its methods include *case reports*, *correlational studies* (to describe any association between potential risk factors and disease in a given database) and *cross-sectional surveys* (to determine prevalence of a disease and potential risk factors at a given point in time). Analytic epidemiology uses *observational* and *experimental* studies. The latter are *clinical trials* to test the effectiveness of interventions to treat or prevent a disease. But experimentation on humans is not ethically feasible for studying causes of disease. Observational research designs are thus the primary tools of epidemiology, the main types being *case-control* and *cohort* studies. After definition of some basic terms, these are discussed further below.

TABLE 4

Interpreting a Statistical Association

Possible Reasons for an Observed Statistical Association

1. CHANCE: This is precisely the meaning of P-value, the probability that the observed outcome is due to chance.

2. BIAS: Systematic errors that distort the results, such as selection bias, recall bias, and observation bias.

3. CONFOUNDING: There is an extraneous, confounding variable (perhaps as yet unknown) that is related to the risk factor being studied and is an independent risk factor for the disease.

4. CAUSE-AND-EFFECT: The risk factor in the observed association is a cause of the disease.

SOURCE: Courtesy of Valerie Miké.

Careful study is required to assess potential biases and confounding variables. General guidelines for establishing causality are provided by Hill's Criteria (Table 6).

MEASURES OF MORBIDITY AND MORTALITY. Some basic concepts of epidemiology are listed in Table 1. It is important to distinguish between the *prevalence* of a disease and its *incidence*. Prevalence signifies the amount of disease present at a point in time, such as the proportion of people with adult-onset diabetes in the United States on January 1, 2005. Incidence refers to new cases diagnosed during a given period of time, such as the proportion of U.S. adults diagnosed with diabetes in 2005. The denominator of *incidence rate* is *person-time*, a useful concept that allows for inclusion of subjects with different starting dates and lengths of time observed in a study. Causes of a disease can be investigated by observing incidence in a well-defined group of subjects without the disease, and patterns of disease incidence can be compared over time or populations.

Mortality is measured in terms of *crude death rate*, the actual proportion observed, or the *standardized death rate*, which involves adjustment for some characteristic. The example shows *age-adjusted* cancer death rates for the states of Alaska and Florida. Alaska has a much lower crude death rate than Florida, but its population is much younger. Both breast and prostate cancer are associated with older age, but after age-adjustment the two states are seen to have similar death rates for these two sites, both lower than the national average. The adjusted figures are meaningless in themselves, but provide for valid comparison of rates across groups and time. U.S. cancer death rates have been adjusted using the 2000 U.S. age distribution to make them comparable back to 1930 and ahead to the future.

OBSERVATIONAL RESEARCH DESIGNS: CASE-CONTROL AND COHORT STUDIES. A case-control study is *retrospective*: It identifies a group of people with the disease (cases) and selects a group as similar as possible to the cases but without the disease (controls). The aim is to determine the proportion of each group who were exposed to the risk factor of interest and compare them. Table 2 shows results of the case-control study of lung cancer and smoking reported in 1950 by Sir Richard Doll (b. 1912) and Sir Austin Bradford Hill (1897–1991), British pioneers of epidemiology and biostatistics. They identified 649 men and sixty women with lung cancer in twenty London hospitals and matched them with controls of the same age and sex but without lung cancer. The information they collected on all participants included their smoking history. The observed association, measured by the so-called *odds ratio* (the odds of smoking in cases over the odds of smoking in controls), was clearly statistically significant.

A cohort study is usually *prospective*. (It may be *historical*, if based on recorded past information.) It identifies a large group (cohort) of individuals who do not have the disease but for whom complete information is available concerning the risk factor(s) of interest; the cohort is then observed for the occurrence of the disease. A noted cohort design was the Framingham Heart Study, initiated by the U.S. Public Health Service in 1948 to identify risk factors for heart disease. Over 5,000 adult residents of Framingham, Massachusetts, men and women with negative test results for cardiovascular disease, agreed to join the study and undergo repeat testing at two-year intervals. The age and test measures at the start of each two-year period were used to classify subjects. Results of a thirty-year follow-up evaluation (part of a multivariate analysis including other risk factors and cardiovascular outcomes) are shown in Table 3, demonstrating a strong association between systolic blood pressure and incidence of coronary heart disease. Other suitable groups for cohort studies are members of professional groups, like doctors and nurses.

There are advantages and disadvantages pertaining to each research design, and the choice depends on the circumstances of the scientific question of interest. Any observed association then requires careful interpretation.

Association or Causation?

Possible reasons for an observed statistical association are listed in Table 4. *Chance* is simply the meaning of

TABLE 5

Koch's Postulates for Establishing the Causes of Infectious Diseases, with Molecular Update

Koch's Postulates	Molecular Koch's Postulates*
1. The microorganism should be found in all cases of the disease in question, and its distribution in the body should be in accordance with the lesions observed.	1. The phenotype or property under investigation should be significantly associated with pathogenic strains of a species and not with nonpathogenic strains.
2. The microorganism should be grown in pure culture in vitro (or outside the body of the host) for several generations.	2. Specific inactivation of the gene or genes associated with the suspected virulence trait should lead to a measurable decrease in pathogenicity or virulence.
3. When such a pure culture is inoculated into susceptible animal species, the typical disease must result.	3. Reversion or replacement of the mutated gene with the wild type gene should lead to restoration of pathogenicity or virulence.
4. The microorganism must again be isolated from the lesions of such experimentally produced disease.	

*In addition, guidelines for establishing microbial disease causation in terms of the prevalence of the nucleic acid sequence of a putative pathogen in relation to disease status are given in the third column of the table from which this is taken.

SOURCE: Brooks et al. (2001), p. 134.

Proposed in 1884 by Robert Koch for bacteria, the original wording has been modified to include other microbes. Further versions use molecular biology as a tool to associate microbial agents with disease. Table is adapted from a leading textbook of medical microbiology.

the P-value, the probability that the association is due to chance. *Bias* refers to systematic errors that do not cancel out with larger sample size, but distort the results in one direction. For example, in a case-control study patients with the disease may be more likely to recall exposure to the risk factor than the controls, leading to *recall bias*. Bias is a serious problem in observational studies and needs to be assessed in the particular context of each research design. *Confounding* is the effect of an extraneous variable that is associated with the risk factor, but is also an independent risk factor for the disease. For example, an association between birth rank and Down's syndrome, the genetic disorder Trisomy 21 (an extra copy of chromosome 21) does not imply causality; the confounding variable is maternal age, which is associated with birth rank and is a known risk factor for the disease. There may also be confounding variables as yet unknown, but their potential effects must always be considered.

The establishment of causation is a long-debated problem in the philosophy of science. In the practical field of medicine, where life-and-death decisions must be made every day, there are guidelines to help assess the role of agents in the etiology of disease. When microbes were being identified as causes of devastating diseases in the late-nineteenth century, Robert Koch formulated postulates to prove that a particular microbe causes a given disease. Anticipated by his teacher Jacob Henle (1809–1885), these are also called Henle-Koch Postulates. They are shown in Table 5, along with current updates using molecular biology. The original ver-

sion claims only necessary causation, not sufficient; the microorganism needs a *susceptible* host. Even more general, the molecular guidelines are expressed in terms of statistical association. But they are the organizing principle in contemporary studies of microbial etiology, crucial for the identification of newly emerging pathogens that may pose serious threats to public health.

Guidelines for establishing causality in observational studies are listed in Table 6. Formulated by Sir Austin Bradford Hill, they are based on criteria employed in the 1964 U.S. Surgeon General's Report to show that smoking causes lung cancer. Applied in a wider context, they are to be used primarily as an aid to exploration. In general there is no necessary or sufficient condition to establish causality from an observed association. Such conclusions result from a consensus of the scientific community.

Epidemiology and Ethics

The complex, probing methods of epidemiology yield tentative, partial, often conflicting results, replete with qualifications. Taken out of context by interest groups or the media, they can mislead and have harmful consequences. Their correct use requires professional competence and integrity. But beyond these issues of immediate concern, epidemiology plays a larger role. With its adjusted measures allowing comparison of health patterns over space and time, it provides a quantitative aerial video of the globe. Some of the images it presents are troubling.

TABLE 6

Hill's Criteria for Establishing Causality in Observational Studies

Aspects of Association to Consider

1. STRENGTH: Stronger associations more likely to be causal.

2. CONSISTENCY: Association is observed repeatedly in different populations under different circumstances.

3. SPECIFICITY: Disease outcome is specific to or characteristic of exposure.

4. TEMPORALITY: Exposure precedes disease.

5. BIOLOGIC GRADIENT: Monotone dose-response relationship (increase in exposure corresponds to increase in disease).

6. PLAUSIBILITY: Causal hypothesis is biologically plausible.

7. COHERENCE: Causal interpretation does not conflict with what is known about the natural history and biology of the disease.

8. EXPERIMENTAL EVIDENCE: Removal of putative cause in an intervention or prevention program results in reduction of disease incidence and mortality.

9. ANALOGY: Drug or chemical structurally similar to a known harmful agent may induce similar harmful effects.

SOURCE: Hill (1965).

Formulated in 1965 by Sir Austin Bradford Hill, these are very general, tentative guidelines, with numerous exceptions and reservations. Aside from temporality, which may be considered part of the definition of causation, there is no necessary or sufficient criterion for establishing the causality of an observed association.

There are now more obese than undernourished people living on earth, and their number is increasing rapidly in developing nations. According to a 2000 estimate of the World Health Organization (WHO), there are 220 million adults with Body Mass Index (BMI) <17, classified as undernourished, and over 300 million with BMI > 30, defined as obese. (BMI is weight in kilograms divided by height squared in meters.) This global epidemic of obesity, called *globesity*, brings with it the related conditions of diabetes, hypertension, and heart disease, and the problem is equally serious for children.

The harmful effects of tobacco have been known for half a century, and while the prevalence of smoking has been slowly declining in most industrialized nations, it has been rising steadily in the developing world. It is estimated that the number of smoking-related premature deaths worldwide, 5 million in 2000, will rise to 10 million per year by 2030, with 70 percent occurring in developing countries. Tobacco use will kill more people than the combined mortality due to malaria, pneumonia, tuberculosis, and diarrhea.

In the area of infectious diseases, after decades of exuberant optimism reality set in with the appearance of acquired immune deficiency syndrome (AIDS) in

the 1980s. Homo sapiens lives in a sea of microbes and will never have total control. Vigilance for the emergence of disease-causing strains must be the aim, to detect outbreaks, identify pathogens and their mode of transmission, and seek control and prevention. Knowing the cause may not eliminate the disease, even when possible in principle, if (as with smoking) it hinges on human behavior. AIDS, for example, is preventable. Ongoing threats include new diseases from mutation or isolated animal reservoirs (Ebola, West Nile, severe acute respiratory syndrome [SARS]), resurgence of older strains, drug-resistance, targeted release through bioterrorism, and rapid spread through global travel.

At a WHO conference held in Geneva in November 2004, experts issued an urgent appeal for greater international cooperation, and called on governments to make pandemic preparedness part of their national security planning. Of particular concern was the new bird influenza strain A(H5N1), which could mutate and cause a pandemic on the scale of the influenza epidemic of 1918 that killed more than 20 million people. It is estimated that a new pandemic virus could spread around the world in less than six months, infecting 30 percent of the population and killing about 1 percent of those infected. The drug industry would have to prepare billions of doses of the influenza vaccine within weeks of an outbreak to halt its course. There are questions of what could possibly be feasible technologically, the huge investment needed, and the driving force to motivate the effort when it cannot be a matter of fiscal gain.

In March 2005 the British medical journal *Lancet* published four articles reporting on the appalling state of global infant health care. Four million babies die each year in the first month of life, nearly all in low- and middle-income nations. The highest numbers occur in south-central Asian countries, while the highest rates are generally in Sub-Saharan Africa. It is estimated that three-quarters of these deaths could be prevented with low-lost interventions. A similar number of babies are stillborn and 500,000 mothers die from pregnancy-related causes each year. The moral implications of this public health tragedy are overwhelming.

The problems humanity faces at the start of the twenty-first century are inseparable from dominant worldviews and the interplay of powerful economic and political forces. Epidemiology provides health-related information as a guide to action. Its proper use is an essential component of the *Ethics of Evidence*, proposed for dealing with the uncertainties of medicine in the framework of contemporary culture (Miké 1999, 2003).

The *Ethics of Evidence* calls for integrating the best evidence of all relevant fields to promote human well-being, anchored in an inescapable moral dimension. Looking to the future, it urges all to be aware, to be informed, and to be responsible.

VALERIE MIKÉ

SEE ALSO *Biostatistics; Health and Disease.*

BIBLIOGRAPHY

Altman, Lawrence K. (2004). "Experts Urge Greater Effort on Vaccine for Bird Flu." *New York Times*, November 13, p. A3.

American Cancer Society. (2004). *Cancer Facts and Figures 2004.* Atlanta, GA: Author. Includes report on worldwide effects of tobacco use.

Brooks, George F.; Janet S. Butel; and Stephen A. Morse. (2001). *Jawetz, Melnick, & Adelberg's Medical Microbiology,* 22nd edition. New York: McGraw-Hill.

Doll, Richard., and Austin Bradford Hill. (1950). "Smoking and Carcinoma of the Lung: Preliminary Report." *British Medical Journal* 2: 739–748.

Gail, Mitchell H., and Jacques Benichou, eds. (2000). *Encyclopedia of Epidemiologic Methods.* New York: John Wiley & Sons.

Hill, Austin Bradford. (1965). "The Environment and Disease: Association or Causation?" *Proceedings of the Royal Society of Medicine* 58: 295–300.

Hennekens, Charles H., and Julie F. Buring. (1987). *Epidemiology in Medicine.* Boston: Little, Brown.

Lawn, Joy E.; Simon Cousens; and Jelka Zupan. (2005). "Neonatal Survival 1: 4 Million Neonatal Deaths: When? Where? Why?" *Lancet* 365: 891–900. First report in the four-part series.

Lilienfeld, David E.; Paul D. Stolley; and Abraham M. Lilienfeld. (1994). *Foundations of Epidemiology,* 3rd edition. New York: Oxford University Press.

Miké, Valerie. (1999). "Outcomes Research and the Quality of Health Care: The Beacon of an Ethics of Evidence." *Evaluation & the Health Professions* 22: 3–32. Commentary by Edmund D. Pellegrino, "The Ethical Use of Evidence in Biomedicine," is included in this issue.

Miké, Valerie. (2003). "Evidence and the Future of Medicine." *Evaluation & the Health Professions* 26: 127–152.

Porter, Roy. (1998). *The Greatest Benefit to Mankind: A Medical History of Humanity.* New York: W. W. Norton.

Rothman, Kenneth J., and Sander Greenland. (2000). "Hill's Criteria for Causality." In *Encyclopedia of Epidemiologic Methods,* eds. Mitchell H. Gail, and Jacques Benichou. New York: John Wiley & Sons.

Stokes, Joseph III; William B. Kannel; Philip A. Wolf; et al. (1989). "Blood Pressure as a Risk Factor for Cardiovascular Disease: The Framingham Study—30 Years of Follow-up." *Hypertension* 13(Supplement I): 113–118.

INTERNET RESOURCE

Controlling the Global Obesity Epidemic. World Health Organization (WHO). Updated September 2003. Available from http://www.who.int/nut/obs.htm. From the organization's Nutrition pages.

EQUALITY

•••

Equality is a key concept in both ethics and politics, one that influences personal and public self-understandings, and provides guidelines for relations between individuals and for state action. Insofar as scientific knowledge and technological change can either diminish or increase inequalities, and scientific research influences the understanding of what it means to be human, issues of equality exercise important ethical influences on the uses of science and technology. The ideal of equality also presents a special challenge within science and engineering, insofar as peers are supposed to be treated as equals at the same time that expertise makes claims to special influence.

Background

It is an empirical given that human beings are in many respects unequal. They are of different shapes, sizes, and sex; different genetic endowments; and different abilities. From the earliest age, some children manifest gregariousness, others pugnacity, some pleasant dispositions, others dullness and apathy. Take almost any characteristic—health, longevity, strength, athletic prowess, sense of humor, ear for music, intelligence, social sensitivity, ability to deliberate or do abstract thinking, sense of responsibility, self-discipline, or hormonal endowment (for example, levels of testosterone and endorphins)—and there are major differences among humans. Yet it is one of the basic tenets of almost all contemporary moral and political theories that humans are in some fundamental respect equal, and that this truth should be reflected in economic, social, and political structures.

Historically this was not always the case. In Plato's *Republic* Socrates argues for equal opportunity for women and men among the guardians, but some of his interlocutors contest the possibility of this ideal. Aristotle rejects it outright, holding to strong differences between males and females, free men and slaves. "It is manifest that there are classes of people of whom some are freeman and others slaves by nature, for these slavery is an institution both expedient and just" (*Politics* 1.5.1255). Indeed, for many Greeks, Romans, and premodern cultures, the primary challenge was not to treat equals as equal, but to avoid treating unequals as equals.

In the Jewish, Christian, and Islamic traditions all humans are seen as possessing equal worth because they are created in a common relation to God. In Hinduism and Buddhism people have unequal worth based on their karmic status, that is, depending on how well they have carried out their *dharma* (duty), but they have equal opportunity to progress to higher modes of existence and eventually to attain nirvana.

With the Enlightenment equality became a political ideal. In the words of the U.S. Declaration of Independence (1776): "We hold these truths to be self-evident, that all men are created equal, that they are endowed by their Creator with certain unalienable Rights, that among these are Life, Liberty and the pursuit of Happiness.—That to secure these rights, Governments are instituted among Men, deriving their just powers from the consent of the governed." The first article of the French Declaration of the Rights of Man and of the Citizen (1789) likewise stipulates: "Men are born and remain free and equal in rights. Social distinctions may be founded only upon the general good." As has been often noted, however, there is a tension between the ideals of liberty and equality. Inequality is not only produced by inheritance and traditional social orders; it is also produced anew by liberty, as people freely distinguish themselves from each other. Thus one is forced to inquire more precisely what kind of equality ought to be protected.

In relation to what should be "equalized" and the arguments that ground various egalitarian claims, one discovers both limited consensus and a plethora of competing ideas with regard to citizenship, law, opportunity, welfare, resources, opportunity, and capabilities. For instance, there is a measured consensus in support for equality in the areas of civil liberties, political participation, and opportunity. In the twentieth century, however, levels of social and welfare equality as a base for the exercise of individual liberty became contentious in the extreme. Moreover, together with debates between egalitarians about which version of egalitarianism is correct, there exists an even more fundamental argument between egalitarians and nonegalitarians, who question the moral significance of equality.

Conceptual Analysis

The first step in addressing such debates is to analyze more carefully the concept of equality. To begin, it is important to note that equality is sometimes interpreted as equity or fairness, but the two concepts are distinct.

Whether or not and in what ways treating people as equals is equitable or fair is subject to argument.

Equality involves a triadic relationship. A is equal to B with respect to some property P. Except with abstract ideas, such as numbers, there is no such thing as equality per se. Two objects are always different in some respect—even two Ping-Pong balls are made up of different pieces of plastic and exist in different places. Two things A and B, if they are equal, are also equal with respect to something. Two trees are of equal height, two baseball players have equal batting averages, two workers have produced the same amount of widgets in the same time frame, and so forth. So descriptive equality always must answer the question, "Equal in what respect?"

When equality has a normative dimension, the relationship is quadratic: If A and B are equal with respect to the normative (or merit-ascribing) property P, then A and B deserve equal amounts of dessert D. Two persons A and B who are equal with respect to the law deserve equal treatment by the law. Two scientists or engineers who are equally competent professionals and performing equal services deserve equal compensation. Determining equality with respect to P in such cases is, of course, difficult.

Normative egalitarian theories fall into two types: *formal* and *substantive*. A formal theory states a formula or policy but includes no specific content. A substantive theory identifies a criterion or metric by which egalitarian policies are to be assessed.

Aristotle's notion that "injustice arises when equals are treated unequally and also when unequals are treated equally" (*Nicomachean Ethics* 5.3.23–24) is the most common statement of a formal normative theory. If two things are equal in some respect, then if one of them is treated one way based on that respect, it is wrong to treat the other differently based on that same respect. When applied to distributive justice, the formula of formal equality stipulates giving equals equal shares and unequals unequal shares based on some criterion left unspecified. Formal equality is simply the principle of consistency, and Aristotle, who articulated it, was substantively what in the early twenty-first century would be called an inegalitarian, because he defended class, racial, and sexual inequalities.

Substantive normative theories of equality either identify a criterion in the formula for equality in relation to which people should be treated equally or simply assume that all people should receive equal shares of some good(s). But because people are unequal in many respects, the first question concerns which respects are morally indefensible. One of the major controversies of the modern period has been the degree to which class, wealth, race, and sexual differences are legitimately recognized as bases for inequalities in various treatments.

A second question concerns whether the state should do anything to delimit inequality or promote equality. Socialists and liberals, for instance, tend to be interventionists, calling for government action to redistribute goods when a moral case can be made for mitigating the effects of inequality. Conservatives and libertarians tend to limit the governmental role, leaving such matters to individual or voluntary action.

Debating Substantive Equality

Returning to the first question, a few idealists, such as the radicals of the French Revolution, have called for the abolition of virtually all distinctions between persons. Graccus Babeuf's "Manifesto of the Equals" (1796) suggested even the elimination of the arts, because they reveal the difference between a Rembrandt or Michelangelo and everyone else. Sports and academic grades would have to be abolished for the same reason.

Most egalitarians nevertheless agree that not all inequalities are morally repugnant. Candidates for those sorts of inequalities that are morally wrong and thus subject to correction include primary goods, resources, economic benefits, power, prestige, class, welfare, satisfaction of desire, satisfaction of interest, need, and opportunity. Some egalitarians emphasize great differences in wealth as the most morally repugnant item and propose various redistribution policies such as the regressive income tax. Other egalitarians emphasize political power as the item to be equalized.

Certainly there is no doubt that the ideal of equality has inspired millions to protest undemocratic forms of government, monarchies, oligarchies, despotisms, and even republicanism. The sense that each individual is of equal worth has been the basis for rights claims from the English Civil War (1642–1648) to women's suffrage (granted in the United Kingdom, 1918; United States, 1920) and the civil rights movements in the United States (1960s) and South Africa (1980s). Who is not moved by the appeal of Colonel Thomas Rainsborough of Oliver Cromwell's Parliamentary Army, petitioning in 1647 for political equality?

> I think that the poorest he that is in England hath a life to live, as the greatest he; and therefore truly, sir, I think it's clear, that every man that is to live under a government ought first by his own consent to put himself under that government; and I do think that the poorest man in England is not at all bound in a strict sense to that government that he hath not had a voice to put himself under. (Putney debates, October 29, 1647)

But the ideal of equality has dangers too. The French aristocrat Alexis de Tocqueville, in his visit to the United States in the 1830s, was amazed at Americans' passion for and preoccupation with equality. He saw in it both the promise of the future and a great danger. Its promise lay in the prospect of full citizenship, political participation, and economic equality. Its danger lay in the tendency to mediocrity and the envy of those who stood out from the crowd.

Contemporary egalitarians most commonly divide on whether *resources* or *welfare* is the primary good to be equally distributed. Resource egalitarians, such as John Rawls, Ronald Dworkin, and Eric Rakowski, hold that in societies of abundance human beings are entitled to minimally equal shares of the resources or opportunities. Welfare egalitarians, such as Kai Nielsen, R. M. Hare, and Richard Norman, go further and maintain that in such societies people should receive equal welfare, interpreted in terms of fulfillment, outcomes, or preference satisfaction.

The strongest pro-equality consensus concerns equality of opportunity, of which there are two versions. The first is weak equal opportunity (sometimes called "formal equal opportunity"), which holds that offices should be open to talent. This was classically set forth by Plato and in postrevolutionary France by Napoleon Bonaparte, who chose officers not by class but by ability ("*la carriere ouverte aux talents*" [the tools to him that can handle them]) It is meritocratic equal opportunity, but does not address the advantages people have because of natural or family resources, thus leaving the matter of initial starting points untouched.

The second is strong equal opportunity (sometimes called "substantive equal opportunity"), which holds that individuals ought to have equal life chances to fulfill themselves or reach the same heights. It calls for compensation for those who had less fortune early in life to bring them to the level of those who had advantages. This kind of equal opportunity would support affirmative action programs and other compensatory policies. At the extreme, such equal opportunity would have to result in groups succeeding in obtaining coveted positions in proportion to their makeup in the population. Insofar as equal opportunity would be equivalent to equal outcomes, it might be called "superstrong equal opportunity."

Justifying Equality

One important theoretical issue in the debate over equality concerns whether or not equality of whatever substance is an intrinsic or an instrumental good. Thomas Nagel (1979), for instance, after making the distinction, affirms its intrinsic value for providing an independent reason to favor economic equality as a good in its own right.

Even more strongly, Christopher Jencks, in *Inequality: A Reassessment of the Effect of Family and Schooling in America* (1972), a report on U.S. education, maintains that "for ... a thoroughgoing egalitarian, inequality that derives from biology ought to be as *repulsive* as inequality that derives from early socialization" (p. 73). And Richard Watson (1977) argues that equality of resources is such a transcendent value, at least for many purposes, that if equal distribution of food were to result in no one getting enough to eat, this annihilation of the human race should nevertheless be chosen rather than an unequal distribution.

By contrast, it can be argued that equality is not a value in itself but only in relation to its potential effects. Utilitarians commonly argue that total happiness in a society is best maximized by means of equality. And although economists often argue that certain kinds of equality are in the interest of market efficiency, they also criticize efforts to achieve strong equality as themselves being too costly for the marginal utility they may introduce.

Science, Technology, and Equality

As science and technology have become increasingly important goods, inequalities in distribution within and between nations have become public issues. Indeed, scientific exchanges and communications technology, by making people more aware of disparities, intensify the discussion. Under appropriate circumstances, the same scientific and technological activities can also serve as means for the more effective promotion of equality. Ethical and political issues arise in relation to considerations of the extent to which this may be appropriate or feasible.

During the latter third of the twentieth century, as extensions of the civil rights and women's movements, equality within science and engineering became topics of intense debate. What was the cause of the underrepresentation of minorities and women in such sciences as physics or in engineering as a whole? To what extent was this the result of natural differences in interest or ability, or of inequality in access and opportunity?

During this same period scientific research, while not rejecting numerous well-recognized differences between individuals, tended to challenge if not minimize their importance. For instance, genetics points to minimal differences not only between races but also between the sexes, and even between human beings and some higher animals. What significance, if any, does this have for the egalitarian versus libertarian debate? On the one hand, it might well be argued that egalitarianism is so well established at the genetic level that nothing more need be done. On the other, it could also be argued that basic genetic equality is grounds for a more vigorous promotion of social equality.

Finally, increasing possibilities for the technological manipulation of human physiology open doors to radically new forms of the promotion of equality. Should science be used to alter individual genetic endowments through genetic modification? Even before such powers become generally available, it is already known that when parents have the power to choose the sex of their children, there exist strong tendencies in some cultures to choose males over females, thus creating a new kind of radical sexual equality. Moreover, the use of plastic surgery, performance-enhancing drugs, and eventually genetic engineering may be able to undermine inequalities among the gifted and the nongifted in many areas of physical appearance, athletic ability, and perhaps mental achievement. In such cases there may be dangers not only in the top-down or government-sponsored promotion of equality but even in the bottom-up initiatives of individuals practicing personal liberties. The decentralization of scientific and technological powers may alter the theory and practice of equality in unexpected ways.

Finally, ideals of equality pose challenges for relations between democratic practice and scientific or technical expertise. To what extent are scientists and engineers properly to be given special influence in decisions regarding such issues as the control of nuclear weapons, environmental pollution, or global climate change? Is technocracy an antiegalitarian danger in an economy that is dependent on scientific and engineering expertise? Such questions constitute important dimensions of any general reflection on science, technology, and ethics.

LOUIS P. POJMAN

SEE ALSO *Justice.*

BIBLIOGRAPHY

Abernethy, George L., ed. (1959). *The Idea of Equality*. Richmond, VA: John Knox Press.

Arneson, Richard. (1989). "Equality and Equal Opportunity for Welfare." *Philosophical Studies* 56(1): 77–93.

Clayton, Matthew, and Andrew Williams, eds. (2002). *The Ideal of Equality*. New York: Palgrave Macmillan.

Cohen, Gerald A. (1989). "On the Currency of Egalitarian Justice." *Ethics* 99(4): 906–944.

Dworkin, Ronald. (2000). *Sovereign Virtue: The Theory and Practice of Equality*. Cambridge, MA: Harvard University Press.

Jencks, Christopher. (1972). *Inequality: A Reassessment of the Effect of Family and Schooling in America*. New York: Basic.

Nagel, Thomas. (1979). "Equality." In *Mortal Questions*. New York: Cambridge University Press.

O'Neill, Onora. (1977). "How Do We Know When Opportunities Are Equal?" In *Feminism and Philosophy*, ed. Mary Vetterling-Braggin, Frederick A. Elliston, and Jane English. Totowa, NJ: Rowman and Littlefield.

Pojman, Louis P., and Robert Westmoreland, eds. (1996). *Equality: Selected Readings*. New York: Oxford University Press.

Rae, Douglas; Douglas Yates; Jennifer Hochschild; et al. (1981). *Equalities*. Cambridge, MA: Harvard University Press.

Rakowski, Eric. (1991). *Equal Justice*. Oxford: Clarendon Press.

Roemer, John E. (1998). *Equality of Opportunity*. Cambridge, MA: Harvard University Press.

Sen, Amartya. (1992). *Inequality Reexamined*. New York: Russell Sage Foundation; Cambridge, MA: Harvard University Press.

Temkin, Larry S. (1993). *Inequality*. New York: Oxford University Press.

Walzer, Michael. (1983). *Spheres of Justice: A Defense of Pluralism and Equality*. New York: Basic.

Watson, Richard. (1977). "World Hunger and Equality." In *World Hunger and Moral Obligation*, ed. William Aiken and Hugh LaFollette. Englewood Cliffs, NJ: Prentice-Hall.

Westen, Peter. (1990). *Speaking of Equality*. Princeton, NJ: Princeton University Press.

ERGONOMICS

• • •

Ergonomics (used by many interchangeably with such terms as human factors, human engineering, engineering psychology, and the like) can be thought of as the field in which the social and biological sciences are applied to various problems related to the use of products, equipment, or facilities by humans in the performance of specific tasks or procedures in a variety of natural and artificial environments. Ergonomics attempts to evaluate and design the things people use, in order to better match their capabilities, limitations, needs, or physical dimensions (Sanders and McCormick 1993). General elements of the ergonomics field may include the study of humans as (technology-based) system components, design of human-machine interfaces, and consideration of the health, safety, and well-being of humans within a system. Specific areas of study may examine human sensory processes and information processing or anthropometric data to allow professionals in this field to design more effective displays or controls for an engineered system.

Examples

There are many examples of the kinds of successes that the ergonomics field has achieved over the years. As the military is one of the primary users of ergonomic advances, the evolution of military equipment serves as an excellent example of how ergonomics has changed the way things are. The development of the infantry helmet from a shallow "steel pot" to a protective device fabricated from advanced materials formed into a highly functional shape demonstrates the efficacy of ergonomic design.

Ergonomic advances are, by no means, limited to the military. The changes over the years in consumer products such as snow shovels, electric razors; or even more recently, cellular telephones establish the role of human factors in people's everyday lives.

Background

The term *ergonomics* is a combination of the Greek *ergon*, work, and *nomos*, law. The term was created in 1857 by the Polish scientist Wojciech Jastrzebowski (1799–1882) as a name for the scientific study of work. More than a century earlier, however, the Italian physician Bernardino Ramazinni (1633–1714) had initiated the study of work-related illness in the second edition of his *De Morbis Artificum* (1713). And it was not until a century later, in 1952, that the name was given official status in the formation of the British Ergonomic Society.

In the United States, the development of the principles of scientific management by Frederick W. Taylor (1865–1915) and his followers Frank Gilbreth (1868–1924) and Lillian Gilbreth (1878–1972) initiated similar research. It was out of this tradition that the Human Factors Society was founded in 1957. What began as research on work in the civilian sector became during the 1950s and thereafter heavily associated with the military, especially the Air Force Research Laboratory Human Effectiveness Directorate.

Ethical Issues

Given that one of the objectives of this field is to adapt technological systems to the needs, capabilities, and limitations of human beings, there is an inherent ethical dimension in ergonomics. Certainly the members of this profession must consider their ethical responsibilities. For example, practitioners should not function outside their areas of competence. They should have the proper education, professional training, and work experience. They should avoid and must disclose any actual or perceived conflicts of interest (Human Factors and Ergonomics Society 1989). While these principles seem obvious, they may prove to be problematic for those in the ergonomics field.

Because there is limited formal training in ergonomics and many practitioners come from other disciplines (for example, experimental psychology, industrial engineering), care must be taken so that individuals engaged in ergonomics truly understand their own professional "capabilities and limitations." This is especially true because ergonomics is such a broad and diverse field. For example, someone who works primarily in the area of visual perception may be qualified to work in the allied area of visual cognition, but not be qualified to perform work in the area of bioacoustic protection (that is, mitigating the effects of harmful noise).

Experts in many professions provide forensic testimony that goes beyond the mere recounting of facts. These experts are retained primarily to offer opinions regarding certain elements of a case. This is no different in the ergonomics field. The conduct of ergonomic experts in these types of proceedings should be governed by their professional ethics. The principles they should follow in these matters cover subjects such as the objectivity of their testimony; respect of the integrity of other witnesses; discretion regarding the disclosure of details about the case with outside parties; or discernment if making any public statements regarding the matter, as imprudence here may influence the judicial proceedings or be harmful to the litigant's interests (Human Factors and Ergonomics Society 1989).

As with many fields where the recruitment and use of experimental subjects is a key component in the performance of much of the work (such as in sociology and medicine), the treatment of subjects is of paramount importance and lapses in this area could lead to serious ethical criticisms. Approval of the work and the qualifications of the professionals involved by an institutional review board (IRB) is an important concern. Further, complete disclosure regarding the general nature of the work that the subjects will be involved in and specific risks they may be exposed to are requisite elements of any methodology involving humans.

Examining ethical issues entirely within the realm of ergonomics, Yili Liu (2003) considers several questions. Can ergonomically-based approaches be used to address ethical issues in general? This could also be thought of as whether a better understanding of humans from a psycho-physical standpoint can contribute to a greater understanding of ethical issues. An example of this might be whether providing avionics to fighter pilots that extend their ability to identify a friendly or enemy aircraft is helpful when considering the morality of war. Can ergonomics make human-machine systems more ethical? This might seem obvious given the objectives of the field; however, is an improvement in an individual assembly line process that reduces a worker's exposure to hazardous conditions (for example, the mechanization of a manual chemical dipping process to treat a material), but also speeds up the assembly line, which may cause increased levels of stress for all of the workers, really "ethical"?

Such questions point toward moral responsibilities for those working in product planning, design, or evaluation—with "product" including systems, processes, and more. Most professionals engaged in ergonomics work for paid compensation. Most of the products they plan, design, or evaluate are used by others. There would seem to be a compelling moral responsibility on the part of those employed in these practices to inform employers or clients if they know of an inherent danger or serious hazard associated with the use of a certain product. However, if the ergonomicist knows that use of the product would be inconvenient, inefficient, or difficult, and the cost to correct or change the product so that any problems could be ameliorated might be sizeable, what then is the proper course of action? Does the designer give allegiance to the client or the consumer? If one thinks of the ultimate user as the controlling factor here, how would one's opinion change if the inconvenience were characterized as slight and the cost as monumental? Specifics of a case often make it difficult to reach a final decision.

The advent of ergonomics in the twentieth century brought about great improvements in the design of technological systems from the standpoint of the user or the person in the system. Ergonomics has contributed to the improved safety and usability of technology. Given that this specialized field of knowledge holds the keys to understanding the soft boundary between humans and technology, it must be applied within a moral and ethical framework that, in many respects, is still evolving.

MARTIN T. PIETRUCHA

SEE ALSO *Taylor, Frederick W.*

BIBLIOGRAPHY

Liu, Yili. (2003). "The Aesthetic and the Ethic Dimensions of Human Factors and Design." *Ergonomics* 46(13/14): 1293–1305.

Sanders, Mark S., and Ernest J. McCormick. (1993). *Human Factors in Engineering and Design*. New York: McGraw-Hill.

INTERNET RESOURCE

Human Factors and Ergonomics Society. (1989, amended 1998). "Human Factors and Ergonomics Society Code of Ethics." Available from http://www.hfes.org/About/Code.html.

ETHICAL PLURALISM

• • •

Pluralism is a term used to describe a number of positions from different fields. This entry will confine itself to a discussion of ethical—as opposed to political, social, or metaphysical—pluralism.

Basic Definition and History

Ethical pluralism (also referred to as value pluralism) is a theory about the nature of the values or goods that human beings pursue, and the pursuit of which make up the substance of their moral lives. Most simply ethical pluralism holds that the values or goods legitimately pursued by human beings are plural, incompatible, and incommensurable. That is, there are many genuine human values, which cannot all be reduced to, or described in terms of, a single overriding value or system of values. This is because certain human values, by their very nature, come into conflict with other, equally valid, human values. Individual liberty, for instance, can conflict with equality, public order, or technological efficiency; impartial justice with compassion and mercy; scientific truth with public utility; and so on.

Sometimes compromises between values can be achieved, or solutions to value conflicts found; at other times, one is forced to choose between values. Such a choice may entail the sacrifice of a genuinely important, attractive, binding value or good, and so a moral loss. Finally, pluralism holds that values are incommensurable in that they cannot be ranked: There is no single most important or ultimate value, nor can values be ranked in a stable or universal hierarchy, nor is there a single principle or source of truth—such as utility, or a rational principle of moral duty, or natural law or the will of God—that can serve as a sure guide in making choices or compromises between values. Whether there can be any comparison between values of a less general and more practical sort is an issue that divides exponents of pluralism.

The first self-avowed pluralist was the U.S. philosopher and psychologist William James (1842–1910), who applied pluralism to the theory of knowledge and metaphysics. An early, forceful application of pluralism to ethics was made in 1918 by the German sociologist, historian, and philosopher Max Weber (1864–1920). The first full exposition of ethical pluralism under that name and in the form in which it is now known was given by the U.S. philosopher Sterling P. Lamprecht (1890–1973) in 1920. The thinker who did the most to develop and popularize ethical pluralism was the British historian, philosopher, and political theorist Isaiah Berlin (1909–1997), and it is from his work that most contemporary discussions of pluralism take their bearings.

Contemporary Problems and Debates

The theory of pluralism expounded by Berlin contained a number of ambiguities and possible weaknesses, and these have been the basis for recent debates among the proponents and opponents of pluralism. One of the most persistent debates concerns the meaning of the claim that values are *incommensurable*. Berlin used the term to suggest that there is no single standard by which all values can be ranked, or that can be used to determine which value should be chosen in a particular case; and that no eternal scale or hierarchy of values exists—liberty is not inherently more valuable than equality, or spontaneity than dependability, or beauty than practicality. But Berlin also suggested that human beings can, at least sometimes, compare the relative importance and desirability of different courses of action or different values in particular circumstances; and that sometimes, at least, this comparison will lead to the conclusion that one value or course of action is more valid or desirable than another.

Other theorists have given a more radical account of incommensurability, holding that different values are wholly *incomparable*—they cannot be compared, or rationally chosen among, in any circumstance or way. This could lead to the conclusion that choices among values must be arbitrary, because, values being incomparable, there is no way to give a reason for regarding one value as inherently more important or better than another in any circumstance.

Many critics of pluralism maintain that it is no different from *relativism*—a claim that is difficult to evaluate in part because such critics rarely define exactly what they mean by relativism. Berlin insisted that pluralism is different from relativism by defining relativism in terms of a denial of common human understanding and common rules and values. Relativism, in Berlin's definition, holds that a Homeric Greek's admiration of ferocity, pride and physical prowess as moral attributes, for example, is as difficult for a person living in the early-twenty-first century to understand or share, as it is for one person who strongly dislikes peaches to understand another person's enjoyment of peaches. A person in the early 2000s may not admire Homeric heroism. Tastes simply differ; and values are ultimately a matter of taste. Berlin's pluralism holds, on the contrary, that one can understand the attractiveness and value of the Homeric ethic, even if one ultimately rejects that ethic in favor of other values, which are of greater importance to that particular individual.

One problem with this argument is that it rests on a distinctive and tailor-made definition of relativism that not everyone would accept. Another more common definition of relativism is the view that there simply is no inherently right or good course of action or true answer. On this definition, too, pluralism is opposed to relativism, because it holds that there are such things as inherently right or good courses of action and true values and answers; but right and goodness and truth are not singular. This is why pluralism insists that there are genuinely tragic moral dilemmas and conflicts, while relativism cannot allow that such dilemmas are genuinely tragic—they may be frustrating for individuals who feel pulled in different directions, but those individuals need not feel so conflicted.

Relativism can also be defined as holding that certain things are valuable or good solely in relation to their context. This, too, is opposed to pluralism, in at least two ways. First it can be taken to mean that, relative to a particular context, there is a correct value or way of being, which is not appropriate to a different context; pluralism holds that there are a variety of values that remain valid regardless of context, and that in many cases there will not be a single value which is obviously best or most important in a given context. Thus the relativist might say that social cohesion is of greater importance than individual freedom in, say, a traditionalist, pre-industrial society, while the opposite is the case in a modern, *advanced* society; while the pluralist would hold that both values are important to both sorts of society, and that people in both societies will be drawn to, and torn between, and have to choose or find a balance between both values.

Finally relativism can be interpreted as denying the existence of a universally valid, binding, and morally limiting core or horizon of human values; yet, Berlin—and other writers after him—have insisted on the existence of such a core or horizon as part of pluralism.

The most lively debate among political theorists about pluralism is the connection between pluralism and political liberalism. While Berlin attempted to link pluralism with liberalism, arguing for liberalism on pluralist grounds, the British political theorist John Gray (b. 1948) has argued that pluralism actually undermines the authority of liberalism. Liberalism is a theory of government that privileges, and seeks to promote, certain values—primarily individual liberty—against and above other, non-liberal values. Gray asserts that if one take ethical pluralism seriously, one cannot assert the superiority, or impose on others, a single form of life, political system, or culture, because these embody and promote certain values to the detriment or exclusion of others.

While liberalism is certainly a valid choice for certain societies that, given their historical development and present situation, are more oriented toward the values that are central to liberalism, other forms of social and political organization have their own validity, and people in liberal societies must respect the claims and rights of societies that pursue other, non-liberal values. Other political theorists have tried to show that, while pluralism may not entail allegiance to liberalism, liberalism is a preferable political system to others because it better recognizes a genuine plurality of values, and allows for and protects greater freedom and variety of individual choice in pursuing these values.

The Relevance to Science, Technology, and Ethics
Pluralism presents a radical and important challenge to most traditional ethical doctrines, such as Kantianism and Utilitarianism, as well as an alternative to relativism, while also offering a distinctive and versatile perspective on moral experience.

One of the few major exponents of pluralism to address the ethics of the use of science and technology is Gray. Earlier pluralists have generally shared an anthropocentric perspective, treating pluralism as a theory concerned with human values. Gray, however, has expanded pluralism beyond the human sphere, arguing that anthropocentrism—and thus humanism—are misguided. The world should be viewed as a whole—a *biosphere*—with human beings counting as but one species among many. Human-centric conceptions of humanity's place and stature are akin to monism in their denial of the incommensurability and conflict between human and non-human goods. Moral philosophers and ethicists should cease to always put human beings first, and should denounce humanity's arrogant subordination and abuse of nature—Gray has remarked that *homo sapiens* should be re-christened *homo rapiens*—in favor of a moral outlook that takes into account the whole of the earth. Few other pluralists have followed Gray's lead.

Much work remains to be done in applying pluralism to the ethical consideration of science and technology. A pluralistic ethics would suggest that people be aware of the varied and sometimes conflicting values that science and technology seek to serve. A pluralist perspective would recognize the inherent value of scientific research as conducive to the acquisition of knowledge—a genuine value in itself—as well as the value of applied science and technology in increasing human happiness, physical well being, and power. But it would also recognize the costs of the scientific quest, and of the employment of technology. It is a further reminder that, in using science and technology in the pursuit of other values, human beings are faced with choices between the competing values that science and technology may serve.

In doing so pluralism does not provide answers, but rather affirms the validity, difficulty, and intractability of the problems. A pluralist will, for example, recognize that both sides in the debate over the use of animals for medical experiments appeal to genuine values, and that a victory for either side would mean a serious moral sacrifice. A pluralist might also see the conflict between economic growth and environmental safety as embodying a genuine conflict of values—between the well being provided by jobs, economic expansion, and greater human control over nature, and thus comfort, versus the health of the environment, the existence of other species, and, ultimately, human health as well. A pluralist will advocate deciding between contending parties advocating conflicting values on a case-by-case basis, and will be wary of the use of monistic ethical the-

ories (such as utilitarianism) to derive authoritative answers to such conflicts.

Pluralism thus provides ethicists, scientists, political activists, and policy makers with no certain answers to their moral problems, or sanction for their agendas. But it may inspire an increased awareness of, and respect for, the importance of such problems, promoting a greater moral seriousness and honesty in confronting the conflicts of values and possible moral sacrifices and losses that are involved in the pursuit and use of scientific knowledge and technology, and fostering a spirit of greater deliberation, humility, and respect for the priorities and perspectives of others.

JOSHUA L. CHERNISS

SEE ALSO *Berlin, Isaiah; Values and Valuing; Weber, Max.*

BIBLIOGRAPHY

Berlin, Isaiah. (1997). *The Proper Study of Mankind: An Anthology of Essays*, eds. Henry Hardy and Roger Hausheer. London: Chatto and Windus. A collection of Berlin's major essays, including his lengthiest expositions of pluralism.

Crowder, George. (2002). *Liberalism and Value Pluralism.* London and New York: Continuum. An argument for the compatibility of pluralism with a liberal political theory.

Galston, William. (2002). *Liberal Pluralism: The Implications of Value Pluralism for Political Theory and Practice.* Cambridge, UK: Cambridge University Press. An attempt to bring the insights of pluralism to bear on questions of political theory and public policy; argues that pluralism can support and improve liberalism, rather than undermining it.

Gray, John. (1996). *Isaiah Berlin.* Princeton, NJ: Princeton University Press. An exposition of Berlin's thought that offers a fuller, more systematic, and possibly more radical account of pluralism than Berlin himself did.

Gray, John. (2002). *Straw Dogs: Thoughts on Humans and Other Animals.* London: Granta Books.

Hampshire, Stuart. (1983). *Morality and Conflict.* Oxford: Blackwell. A set of terse and withering reflections by a major exponent of pluralism, condemning humanity's use of science and technology to master nature.

Kekes, John. (1993). *The Morality of Pluralism.* Princeton, NJ: Princeton University Press.

Lamprecht, Sterling P. (1920). "The Need for a Pluralistic Emphasis in Ethics." *Journal of Philosophy, Psychology and Scientific Methods* 17: 561–572. The first explicit articulation of what has become familiar as the theory of ethical pluralism.

Lukes, Steven. (1998). "Berlin's Dilemma." *Times Literary Supplement* March 27, 8–10. Discusses the problem of whether pluralism is distinguishable from relativism.

Walzer, Michael. (1983). *Spheres of Justice: A Defence of Pluralism and Equality.* New York: Basic Books. Develops a pluralistic theory of distributive justice, thus applying pluralism to questions of government policy.

Weber, Max. (1946 [1918]). "Politics as a Vocation." In *From Max Weber: Essays in Sociology,* trans. and eds. Hans H. Gerth and C. Wright Mills. New York: Oxford University Press.

Weber, Max. (1946 [1918]). "Science as a Vocation." In *From Max Weber: Essays in Sociology,* trans. and eds. H. H. Gerth and C. Wright Mills. New York: Oxford University Press.

INTERNET RESOURCE

Hardy, Henry. "Writings about pluralism before/independently of Isaiah Berlin." The Isaiah Berlin Virtual Library. Available from http://berlin.wolf.ox.ac.uk//lists/pluralism/onpluralism.htm.

ETHICS AND PUBLIC POLICY CENTER

SEE *Public Policy Centers.*

ETHICS ASSESSMENT RUBRICS

• • •

The introduction of new engineering accreditation criteria that includes "an understanding of professional and ethical responsibility" has firmly established the teaching of ethics as an important component of undergraduate education (Engineering Accreditation commission 2003, Herket 2002) yet, in establishing this outcome criterion, the commission also required its assessment. This is a particularly challenging proposition because ethics education is concerned not only with learning content but equally important, with developing problem solving skills. Further, such problems, or dilemmas, are rarely clear-cut and consequently do not have a definitive resolution, making traditional forms of assessment of limited value. One promising approach to this challenge is the development and use of scoring rubrics, a process that has been used for a broad range of subjects when a judgment of quality is required (Brookhart 1999). As opposed to checklists, a rubric is a descriptive scoring scheme that guides the analysis of a student's work on performance assessments. These formally defined guidelines consist of pre-established criteria in narrative format, typically arranged in ordered categories specifying the qualities or

processes that must be exhibited for a particular evaluative rating (Mertler 2001, Moskal 2000). A valid rubric would allow educators to assess their students learning to date, and identify areas of weakness for further instruction.

There are two types of scoring rubrics: holistic and analytic. A holistic rubric scores the process or product as a whole, without separately judging each component (Mertler 2001). In contrast, an analytic rubric allows for the separate evaluations of multiple factors with each criterion scored on a different descriptive scale (Brookhart 1999). When it is not possible to separate the evaluation into independent factors—that is, when overlap between criteria exists—then a holistic rubric with the criteria considered on a single descriptive scale may be preferable (Moskal 2000).

Further, rubrics are intended to provide a general assessment rather than a fine-grained appraisal (such as in a 1–100 grading scale). For example, a rubric might include levels from one ("shows little or no understanding of key concept") to five ("shows full understanding of key concept; completes task with no errors"). Among the advantages of using rubrics are: (1) assessment can be more objective and consistent; (2) the amount of time faculty spend evaluating student work is reduced; (3) valuable feedback is provided to both students and faculty; and (4) they are relatively easy to use and explain (Georgia Educational Technology Training Center 2004).

Generally, rubrics are best developed starting from a desired exemplar learning outcome and working backward to less ideal outcomes, preferably using actual student work to define the rubric's various levels. The scoring system should be objective, consistent, and relatively simple, with a few criteria sets and performance levels; three to five evaluative criteria seem to be appropriate (Popham 1997).

Extensively used in K–12 education assessment, higher education areas such as composition and art, and, increasingly, engineering education (Moskal, Knecht, and Pavelich 2001), rubrics have yet to be widely adopted for assessing ethics tasks. An example is Holt et al. (1998) who developed an analytical rubric for assessing ethics in a business school setting, identifying five categories:

(1) Relevance: Analysis establishes and maintains focus on ethical considerations without digressing or confusing with external constraints;

(2) Complexity: Takes into account different possible approaches in arriving at a decision or judgment;

TABLE 1

Analysis Component of Scoring Rubric for Assessing Students' Abilities to Resolve Ethical Dilemmas

Level 1	Level 2	Level 3	Level 4	Level 5
• No analysis provided. • Defaults to a superior or authority without further elaboration. • Takes a definitive and unambiguous position without justification. • Any analysis appears to have been done without reference (explicit or implicit) to guidelines, rules or authority.	• Authoritative rule driven without justification. Position may be less definitive (e.g., "should do" vs. "must do"). • Minimal effort at analysis and justification. • Relevant rules ignored. • May miss or misinterpret key point or position. • If ethical theory is cited, it is used incorrectly.	• Applies rules or standards with justification, notes possible consequences or conflicts. • Correctly recognizes applicability of ethical concept(s). • Recognizes that contexts of concepts must be specified. • Coherent approach.	• Applies rule or standard considering potential consequences or conflicts. • Uses an established ethical construct appropriately. Considers aspects of competence and responsibility of key actors. • May cite analogous cases. • Incomplete specification of contexts of concepts.	• Correctly applies ethical constructs. • May offer more than one alternative resolution. • Cites analogous cases with appropriate rationale. • Thorough evaluation of competence and responsibility of key actors. • Considers elements of risk for each alternative. • Explores context of concepts.

SOURCE: Courtesy of Larry J. Shuman, Barbara M. Olds, and Mary Besterfield-Sacre.

Shown is the Analysis Component (one of five components) of the rubric. Note that the rubric gives the rater criteria to classify the student's response into one of five levels with five being the highest. The rater should choose the criteria set that most closely matches the student's response.

(3) Fairness: Considers most plausible arguments for different approaches;

(4) Argumentation: Presents a well-reasoned argument for a clearly-identified conclusion, including constructive arguments in support of decision and critical evaluation of alternatives;

(5) Depth: Shows an appreciation of the grounds or key moral principles that bear on the case.

These categories were rated from 1 for "non-proficient" to 6 for "excellent" according to each level's criteria.

Although not developed specifically for assessing ethical problem solving, the widely used Holistic Critical Thinking Scoring Rubric (HCTSR) with its four criteria could be adapted for a holistic assessment of students' ethical problem solving ability (Facione and Facione 1994). One recent effort along these lines has resulted in the development and validation of a rubric designed to measure engineering students' ability to respond to ethical dilemmas using case scenarios, for example, a case based on the first use of an artificial heart (Sindelar, Shuman, Besterfield-Sacre, et al. 2003). To a certain extent, the rubric follows the case analysis process of Charles E. Harris, Michael S. Pritchard, and Michael J. Rabins (1999). It consists of five components each with five levels (See Table 1):

(1) Recognition of Dilemma (relevance): Levels range from not seeing a problem to clearly identifying and framing the key dilemmas.

(2) Information (argumentation): At the lowest level, pertinent facts are ignored and/or misinformation used. At the high end, assumptions are made and justified; information from student's own experiences may be used.

(3) Analysis (complexity and depth): At the lowest level no analysis is performed. Ideally, thorough analysis includes citations of analogous cases with consideration of risk elements with respect to each alternative.

(4) Perspective (fairness): The lowest level is a lack thereof; that is, a wandering focus. The ideal is a global view of the situation, considering multiple perspectives.

(5) Resolution (argumentation): At the base level only rules are cited, possibly out of context. The ideal considers potential risk and/or public safety, and proposes a creative middle ground among competing alternatives.

Using such a rubric holds out the promise of being able to assess the learning of ethics reasoning skills in a more objective manner than has previously been the case. Indeed, there is the possibility that, given new developments in technology and learning, such rubrics could be programmed into computer-based learning modules that would be comparable to some of those developed for the self-guided teaching and learning of technical subjects.

LARRY J. SHUMAN
BARBARA M. OLDS
MARY BESTERFIELD-SACRE

BIBLIOGRAPHY

Brookhart, Susan M. (1999). *The Art and Science of Classroom Assessment: The Missing Part of Pedagogy.* Washington, DC: George Washington University, Graduate School of Education and Human Development.

Harris, Charles E., Jr.; Michael S. Pritchard; and Michael J. Rabins. (2000). *Engineering Ethics: Concepts and Cases,* 2nd ed. Belmont, CA: Wadsworth.

Herkert, Joseph R. (2000). "Continuing And Emerging Issues In Engineering Ethics Education," *The Bridge* 32(2): 8–13.

Holt, Dennis; Kenneth Heischmidt; H. Hammer Hill, et al. (1998). "When Philosophy And Business Professors Talk: Assessment Of Ethical Reasoning In A Cross-Disciplinary Business Ethics Course," *Teaching Business Ethics* 1(3): 253–268.

Moskal, Barbara M.; Robert D. Knecht; and Michael J. Pavelich. (2001). "The Design Report Rubric: Assessing the Impact of Program Design on the Learning Process." *Journal for the Art of Teaching: Assessment of Learning* 8(1): 18–33.

Popham, W. James. (1997). "What's Wrong—and What's Right—with Rubrics." *Educational Leadership* 55(2): 72–75.

INTERNET RESOURCES

Engineering Accreditation Commission. (2003). *Criteria for Accrediting Engineering.* Baltimore, MD: Accreditation Board for Engineering and Technology (ABET). Available from http://www.abet.org/criteria.html.

Facione, P.A., and N.C. Facione. (1994). "Holistic Critical Thinking Scoring Rubric." Available from http://www.insightassessment.com/HCTSR.html.

Georgia Educational Technology Training Center. (2004). "Assessment Rubrics." Available from http://edtech.kennesaw.edu.

Mertler, Craig A. (2001). "Designing Scoring Rubrics for Your Classroom." *Practical Assessment, Research & Evaluation* 7(25). Available from http://PAREonline.net.

Moskal, Barbara M. (2000). "Scoring Rubrics: What, When, and How?" *Practical Assessment, Research & Evaluation* 7(3). Available from http://PAREonline.net.

Sindelar, Mark; Larry J. Shuman; Mary Besterfield-Sacre; et al. (2003). "Assessing Engineering Students' Abilities to Resolve Ethical Dilemmas." *Proceedings, Frontiers in Education Conference.* Available from http://fie.engrng.pitt.edu/fie2003/index.htm.

ETHICS OF CARE

• • •

The ethics of care is a distinctive approach to moral theory that emphasizes the importance of responsibility, concern, and relationship over consequences (utilitarianism) or rules (deontologism). The concept of care is inherent to professions that care for individuals and this approach to ethics has therefore been a central part of professional ethical issues in both nursing and medical ethics, but in fact has much broader applications in relation to science and technology. "Due care" has for example, been a part of statements in engineering and has been used to include such typically technical activities as the maintenance and repair of an engineered system.

Origins and Development

As a moral theory the ethics of care originated during the 1970s and 1980s in association with challenges to the standard moral theories of utilitarianism and deontologism, primarily by women philosophers. The original work was Carol Gilligan's, conducted in the early 1970s and articulated in *In a Different Voice* (1982). Gilligan argued in response to the psychology of moral development formulated by Lawrence Kohlberg (1927–1987). Kohlberg himself built on the ideas of Jean Piaget (1896–1980), who did preliminary work on moral development as one facet of cognitive growth.

In his research Kohlberg posed moral dilemmas to males of various ages and compared the kinds of reasoning with which they responded. The dilemmas tended to be shorn of details about the people involved. The responses moved from self-centered thinking, emphasizing the importance of physical pleasure through thinking under the influence of peer pressure, to a moral orientation toward justice and abstract appeals to universal rights (Kohlberg 1984). Gilligan, on the basis of alternative research with both men and women, discovered a contrasting tendency, predominantly but not exclusively among women, to interpret "the moral problem as a problem of care and responsibility in relationships rather than as one of rights and rules" (p. 73). "While an ethics of justice proceeds from the premise of equality—that everyone should be treated the same—an ethic of care rests on the premise of nonviolence—that no one should be hurt" (p. 174).

Like Kohlberg, however, Gilligan sees an ethics of care emerging in three phases. In the early phase individuals care more for themselves than for others. In a middle phase care comes to emphasize concern for others over care for oneself. Finally, in its mature form the ethics of care seeks a balance between care for oneself and care for others. What nevertheless remains primary in each case is personal relationships: of others to oneself, of oneself to others, or mutually between oneself and others.

This new ethics of care was developed further by Nel Noddings (1984) in relation to education, and given a more philosophical formulation by Annette C. Baier (1985). According to Baier, Gilligan exemplifies a strong school of women philosophers that includes Iris Murdoch (1919–1999) and G. E. M. Anscombe (1919–2001), out of which have developed moral theories that stress living relationships over abstract notions of justice illustrated, for example, by the work of Immanuel Kant (1724–1804). Indeed, three decades prior to Gilligan, Anscombe had already suggested the need for a philosophical psychology as the gateway to any moral philosophy that might be adequate to issues arising in relation to science and technology.

Baier herself criticizes the rationalist individualism that rests content with establishing a minimalist set of traffic rules for social interaction as inadequate on a multitude of counts. Historically, it has failed to oppose injustices to women, the poor, and racial and religious minorities. While most human relations are between unequals, it has focused almost exclusively on relations between alleged equals. Despite the fact that many morally significant relations are not freely chosen, it has emphasized freedom of choice and rational autonomy. And although emotions are often as important as reasons, it has persistently stressed the rational control of behavior. At the same time Baier is careful to emphasize how an ethics of care complements rather than discards an ethics of justice. A good moral theory "must accommodate both the insights men have more easily than women, and those women have more easily than men" (Baier 1985, p. 56).

Applications in Biomedicine

From her empirical studies of people faced with difficult moral decisions, Gilligan identified a distinct approach—one of care, responsibility, concern, and connection, based on personal relations. This care orientation forms the basis of the ethics of care, "grounded in responsiveness to others, that dictates providing care, preventing harm and maintaining relationships" (Larrabee 1993, p. 5). It was natural that such an approach to ethics would be applied in the field of medicine, especially in nursing, where caregiving is already a defining characteristic. It is often argued that care is dis-

torted by the dominance of scientific and technological practices in the practice of medicine.

In this regard one can note, for instance, how care has come to play an increasingly prominent role in such an influential text as Tom L. Beauchamp and James F. Childress's *Principles of Biomedical Ethics*. From its first edition (1979), this representative of the "Georgetown School" of bioethics emphasized a deontological "system of moral principles and rules" that highlighted four principles: autonomy (of the patient), nonmaleficence, beneficence, and justice. In neither the first nor the second edition (1983) did the ethics of care play a role. In the third edition (1989) and subsequent editions care has nevertheless been acknowledged especially in conjunction with an account of criticisms of principlism.

> Although [principled] impartiality is a moral virtue in some contexts, it is a moral vice in others. [Principlism] ... overlooks this two-sidedness when it simply aligns good and mature moral judgment with moral distance. The care perspective is especially meaningful for roles such as parent, friend, physician, and nurse, in which contextual response, attentiveness to subtle clues, and the deepening of special relationships are likely to be more momentous morally than impartial treatment. (Beauchamp and Childress 2001, p. 372)

The authors go on to note the centrality of two themes in the ethics of care—mutual interdependence and emotional responsiveness. For the ethics of care, "many human relationships involve persons who are vulnerable, dependent, ill, and frail [and] the desirable moral response is attached attentiveness to needs, not detached respect for rights" (p. 373). The ethics of care further corrects a "cognitivist bias [in principlism] by giving the emotions a moral role" (p. 373) and encouraging attention to aspects of moral behavior that might otherwise be ignored.

In the field of nursing, in which care exercises an even more defining role than in other medical professions, the ethics of care has been accorded even more significance. Helga Kuhse's *Caring: Nurses, Women, and Ethics* (1997) provides a good overview in this area.

Criticisms

Beauchamp and Childress also summarize key criticisms of the ethics of care in the biomedical context. First, the ethics of care is incompletely developed as a theory. Second, one can easily imagine situations in which relatives or medical professionals are called on to override emotional responses and to abide by principles. Third, the ethics of care can be distorted by cultural expectations. Indeed, some feminist critics have argued that care is easily distorted by contemporary interests, as in cases in which the terminally ill request to be allowed to die because they do not want to continue to be a burden to those around them. Finally, still others have challenged the empirical basis for some of conclusions advanced by Gilligan and others, and questioned popular associations between the ethics of care and female experience.

More constructively, it is unnecessary to maintain an essentialist connection between the ethics of care and female experience. In fact, Gilligan herself argues that the connection may be only historical. It may just be that those who are marginalized in a rule-governed scientific and technological culture have a natural tendency to emphasize alternatives. But this possibility reinforces rather than diminishes the need to attend to the claims in ethics of care. In a culture that values competition and efficiency the ethics of care also promotes such activities as conflict resolution and dispute mediation when dealing with ethical and other conflicts.

Application to Technology and Engineering

The most salient definition and framework of care to apply to the contexts of science and technology is that of Joan C. Tronto and her colleague Berenice Fisher. Tronto and Fisher suggest that caring be viewed as "a species activity that includes everything that we do to maintain, continue and repair our world so that we can live in it as well as possible. That world includes our bodies, our selves, and our environment, all of which we seek to interweave in a complex, life-sustaining web" (Tronto 1993, p. 103).

The ethics of technology and of science needs to be a system ethics to be followed by a system of actors, doers, and stakeholders. It needs to work in the context of the science and technology enterprises, which are distinct. The justice and rights perspective gives an abstract, universalizable goal as Kohlberg, and indeed Kant before him, intended, but the praxis of science and technology calls for a guide for action in terms denoting action. This is what the ethics of care provides. Care in this sense is larger than care implied by familial and close community relationships. Care, too, is universalizable, but not abstract.

The Fisher-Tronto definition provides the actions—maintain, continue, and repair—that care demands, words closely associated with engineering, the action element of technology. This definition of care also recognizes that human existence is intricately woven into the web of the natural environment and that the ethics of care must apply to nature as well as to humans and their communities. In this perspective, care is well positioned as an ethics for a sustainable world, a prime challenge to today's technology. In her analysis of care, Tronto recalls David Hume's understanding of justice, an artificial passion, as a necessary complement to the natural passion of benevolence, which alone may not be sufficient as a moral basis in a human society. These ideas also hark back to Aristotle who sees practical deliberation as the means of achieving the ethical good and praxis as the end of ethics.

Marina Pantazidou and Indira Nair (1999), who have examined care particularly in the context of engineering, identify care as a value-guided practice, not a system of values. Care emerges in response to a need. Meeting human needs is indeed the ideal for technology. Tronto has provided a framework for practicing care that is particularly suited for application to technology and indeed to science. Tronto identifies four phases of care that parallel closely stages identified with the process of engineering design.

(1) Attentiveness, or "caring about," is the phase of recognizing the correct need and realizing care is necessary. This is parallel to the need identification stage in design.

(2) Responsibility, or "taking care of," is the phase that involves "assuming responsibility for the identified need and determining how to respond to it" (Tronto 1993, p. 106). This is parallel to the conceptualization phase of design.

(3) Competence, or "caregiving," is the phase in which the need is met with the expertise needed. This is parallel to the actual design and production.

(4) Responsiveness, or "care receiving," is the phase in which "the object of care will respond to the care it receives" (Tronto 1993, p. 107). This is parallel to the acceptance (or rejection) of the designed product.

Total care requires an attuned caregiver, who through commitment, learning, and experience has an understanding of the process as well as the competence and skills and watches the response of the one cared for. Tronto introduces a fifth component to complete the process. She calls this the Integrity of Care, requiring

FIGURE 1

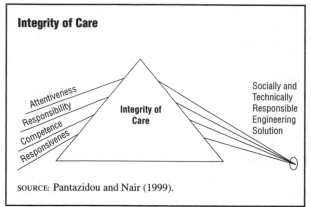

Integrity of Care

Attentiveness
Responsibility
Competence
Responsivenes

Integrity of
Care

Socially and
Technically
Responsible
Engineering
Solution

SOURCE: Pantazidou and Nair (1999).

"Engineering with care" depicted as a prism that integrates the four elements of care into the resulting engineering product. For a sharp focus, the proportion of attentiveness, responsibility, competence, and assessment (responsiveness), represented by the rays and angles of incidence, has to be accurate. Engineering with less than accurate angles will result in a smeared focus in this prism analogy (Pantazidou and Nair 1999).

"that the four moral elements of care be integrated into an appropriate whole."

Figure 1 is the representation of this process by Pantazidou and Nair with the Integrity of Care as a prism that focuses the four care components to a socially and technically responsible technological product. Carrying the prism analogy forward, a technology that has no room for any error will require extremely fine tuning of the four angles of the phases of care to yield a sharp focus. One may argue that in general an ethics of care applied to a technology will say that such a technology poses high risk and may be best avoided. Where such precision is not required, there may be more tolerance of how the phases come together. In some cases, a single focused solution may not be possible or it might not be critical. Then, a range of perhaps suboptimal solutions—a smeared focus—may be sufficient or even necessary for pragmatic reasons.

Figure 2 shows how the ethics of care and the description of the engineering design process compare.

Care in Science

Science in general is not as easily mapped into such a scheme unless it is science done expressly for the purpose of answering a technology-derived question or problem. In this case, Figure 1 applies directly, because the science is done in response to a need.

In the case of science in general, the ethics of care can provide some ethical tests attuned to each phase.

(1) Attentiveness: Is the science being done in response to a perceived need? Or, are needs being scientifically assessed so that a given technology is likely to be the best response? As human needs are perceived, are scientific resources being directed toward those?

(2) Responsibility: What is the science that determines if a technological process or product is the answer to the need? Does new scientific knowledge direct action toward the appropriate human need?

(3) Competence: This is perhaps the one phase toward the accomplishment of which the current scientific ethic is almost solely directed.

(4) Responsiveness: The science of the consequence of a technology is a requisite. This would include predictive science. Hans Jonas (1984) has suggested that one imperative of human technological power is that "knowledge (science) must be commensurate with the causal scale of our action ... that predictive knowledge falls behind the technical knowledge that nourishes our power to act, itself assumes ethical importance" (p. 8).

This last corollary is perhaps the most important result that the ethics of care can yield in the case of science—that science to reduce the uncertainty of human technological actions take on importance in the scientific enterprise.

Care in Engineering

"Reasonable standard of care" has been common parlance in product specifications separate from the consideration of any ethical standard. Product liability issues assess whether "due care" was taken. Thus care has become an inherent notion in technological products spurred over time by legal demands. A working definition of the standard of care for engineering, set by legal precedent, has been proposed by Joshua B. Kardon (2002) as "That level or quality of service ordinarily provided by other normally competent practitioners of good standing in that field ... under the same circumstances." While proposed as an ethic for the engineer to follow, this standard does not fully address all the elements of the ethics of care.

Moreover, challenged by the requirements of sustainability, technological planning has begun to consider system characteristics such as environmental impacts of a product life cycle in the design of a product or a process. With technology intertwining with everyday lives in

FIGURE 2

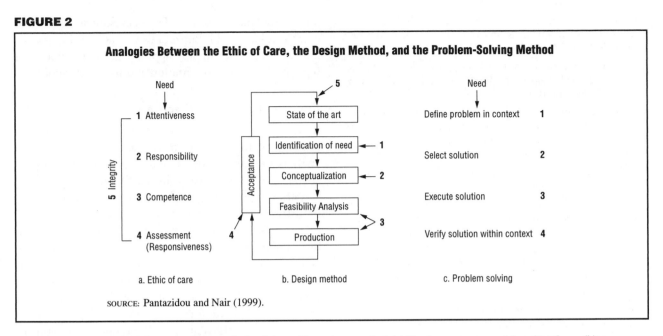

Analogies between the ethic of care, the design method and the problem solving method. (a) The five components of the ethic of care; (b) Mapping between steps of the design process (after Dieter [24]) and components of the ethic of care.

intricate ways, interface design of all sorts of technology has become important. Industrial ecology, green design, green chemistry, and humane design are some of the trends that illustrate the ethics of care at work (Graedel and Allenby 2003; Collins Internet article).

A systematic application of the ethics of care to science and technology is yet to be done and may indeed benefit practice. Such an analysis and a synthesis of standards of the practice of science and technology with the ethics of care may yield a framework that is realistic enough for the handling of the complexity of technological and scientific progress. The ethics of care may aid in this by responding to Jonas's condition of the sustainability of humanity as a technological imperative, Manfred Stanley's call for placing human dignity on par with species survival (1978), and Anthony Weston's observation that tough ethical problems be treated as problematic situations and not as puzzles (1992).

INDIRA NAIR

SEE ALSO *Anscombe, G. E. M.; Bioethics.*

BIBLIOGRAPHY

Anscombe, G. E. M. (1958). "Modern Moral Philosophy." *Philosophy* 33: 1–19. Calls for an adequate philosophy of psychology before attempting moral philosophy.

Baier, Annette C. (1985). "What Do Women Want in a Moral Theory?" *Nous* 19(1): 53–63. Reprinted in the author's *Moral Prejudices: Essays on Ethics* (Cambridge, MA: Harvard University Press, 1994).

Beauchamp, Tom L., and James F. Childress. (2001). *Principles of Biomedical Ethics*, 5th edition. New York: Oxford University Press. First edition, 1979; 2nd edition, 1983; 3rd edition, 1989; 4th edition, 1994. Standard text in the field of biomedical ethics.

Dieter, George E. (1991). *Engineering Design: A Materials and Processing Approach*, 2nd edition. Boston: McGraw-Hill. Standard text in engineering design.

Flanagan, Owen. (1991). *Varieties of Moral Personality*. Cambridge, MA: Harvard University Press. An exploration of the relationship between psychology and ethics, arguing for constructing ethical theory and theory of moral development from an understanding of cognitive psychology.

Gilligan, Carol. (1982). *In a Different Voice: Psychological Theory and Women's Development*. Cambridge, MA: Harvard University Press. A classic work that set forth the care perspective in moral reasoning.

Gilligan, Carol; Janie Victoria Ward; and Jill McLean Taylor, eds. (1988). *Mapping the Moral Domain*. Cambridge, MA: Harvard University Graduate School of Education. Collection of essays discussing research and research strategies for exploring care and justice perspectives in diverse populations.

Graedel, Thomas E., and Braden R. Allenby. (2003). *Industrial Ecology*, 2nd edition. Upper Saddle River, NJ: Prentice Hall. First textbook covering the emerging field of industrial ecology.

Jonas, Hans. (1984). *The Imperative of Responsibility: In Search of an Ethics for the Technological Age*. Chicago: University of Chicago Press. Jonas, a pioneer of bioethics, explores constraints that technology should follow in order to be able to preserve life on the planet.

Kohlberg, Lawrence. (1971). "From Is to Ought." In *Cognitive Development and Epistemology*, ed. Theodore Mischel. New York: Academic Press.

Kohlberg, Lawrence. (1981). *Essays on Moral Development*, Vol. 1: *The Philosophy of Moral Development*. San Francisco: Harper and Row.

Kohlberg, Lawrence. (1984). *Essays on Moral Development*, Vol. 2: *The Psychology of Moral Development*. San Francisco: Harper and Row. Standard works on moral psychology by pioneer Lawrence Kohlberg.

Kuhse, Helga. (1997). *Caring: Nurses, Women, and Ethics*. Malden, MA: Blackwell.

Larrabee, Mary Jeanne, ed. (1993). *An Ethic of Care*. New York: Routledge. Collection of essays concerned with the ethics of care over a decade following Gilligan's first work.

Noddings, Nel. (1984). *Caring: A Feminine Approach to Ethics and Moral Education*. Berkeley and Los Angeles: University of California Press. 2nd edition, 2003. Applies the care ethic to education.

Pantazidou, Marina, and Indira Nair. (1999). "Ethic of Care: Guiding Principles for Engineering Teaching and Practice." *Journal of Engineering Education* 88(2): 205–212. First paper on applying care ethics to engineering.

Stanley, Manfred. (1978). *The Technological Conscience*. New York: Free Press. A sociologist's call for an ethic of technology that values human dignity above species survival.

Tronto, Joan C. (1993). *Moral Boundaries: A Political Argument for an Ethic of Care*. New York: Routledge. Presents care as a central and sustaining activity of human life; examines the moral boundaries set by the politics of power in devaluing care professions.

Weston, Anthony. (1992). *Toward Better Problems: New Perspectives on Abortion, Animal Rights, the Environment, and Justice*. Philadelphia: Temple University Press. An application of the philosophy of pragmatism acknowledging the context and the multiplicity of values in ethical problems.

INTERNET RESOURCES

Collins, Terrence. "Ethics." Institute for Green Oxidation Chemistry, Carnegie Mellon University. Available from http://www.chem.cmu.edu/groups/collins/ethics/. Discusses the responsibility of chemists to promote sustainability.

Kardon, Joshua B. (2002). "The Structural Engineer's Standard of Care." Online Ethics Center for Engineering and Science, Case Western Reserve University. Available from http://onlineethics.org/cases/kardon.html. Points out the role care plays in the practice of engineering.

ETHICS: OVERVIEW

• • •

From the perspective of science, technology, and ethics, ethics itself—that is, critical reflection on human conduct—may be viewed as a science, as a technology, and as providing multidimensional independent perspectives on science and technology. The encyclopedia as a whole constitutes manifold illustrations for each of these possibilities. It is nevertheless appropriate to provide in a separate entry some orientation within the manifold.

Ethics as Theory and Practice

In the works of Plato (c. 428–347 B.C.E.), dialogues rather than treatises, ethics is interwoven with logical analysis and theories of knowledge, reality, and political affairs so as to resist clearly distinguishing these different branches of philosophy. What came to be called ethics nevertheless clearly serves as first or primary philosophy. In Socrates's autobiography (*Phaedo* 96a ff.) it is not the foundations of nature but the ideas of beauty, goodness, and greatness that act as the basis of philosophical inquiry. The search for a full account of ethical experience calls forth an appreciation of different levels of being and different forms of knowing appropriate to each—although the highest reality is once again ethical, the good, which is beyond being (*Republic* 509a–b).

According to Aristotle (384–322 B.C.E.), however, philosophy originates when discourse about the gods is replaced with discourse about nature (compare, e.g., *Metaphysics* 1.3.983b29 and 1.8.988b27). It is the study of natures, as distinguishing functional features of the world, that both constitutes natural science and provides insight into the *telos* or end of an entity. For Aristotle the various branches of philosophy themselves become more clearly distinguished, and ethics functions as the systematic examination of *ethos*, as constituted by the customs or behaviors of human beings. More than any other type of entity, humans have a nature that is open to and even requires further determinations. At the individual level these supplemental determinations are called character; at the social level, political regimes. Their very multiplicity calls for systematic (that is, in the classical sense, scientific) analysis and assessment.

Such analysis and assessment takes place on three levels. In the first instance it is descriptive of how human beings in fact behave. As Aristotle again notes, human actions by nature aim at some end, and the end pursued can be of three basic types, defining in turn the lives of physical pleasure, of public honor, and of intellectual investigation.

In the second instance, ethics compares and contrasts these ends and seeks to identify which is superior and for what reasons. For Aristotle the life of reason is superior because it is that which humans by nature do only or best, and is itself the most autonomous way of life. Humans share with other animals the pursuit of pleasure; the pursuit of honor is dependent on recognition by others and the historical contingency of having been born into a good regime.

Finally, in the third instance the ethical life itself becomes a striving simply for knowledge of human behavior. It seeks conceptual clarification regarding different forms of perfection (virtue) and imperfection (vice), synthetic appreciation of the relations between human nature and other forms of nature, and ultimately a transcendence of the subordinate dimensions of human experience. Ethics in this final form becomes science in the most general sense, concerned not with the part (humans) but the whole (cosmos).

Yet as Aristotle also notes, humans undertake ethical inquiry not simply to know about the good but also to become good (*Nicomachean Ethics* 2.2). Ethics is not just a science but a practice, a technique for self- and social improvement. Insofar as this is the case, ethics provides guidelines for development of character and counsel for political organization and rule. Ethics leads to politics, meaning not just political action but political philosophy (*Nicomachean Ethics* 10.9).

Roman philosophers, continuing the Greek tradition, likewise examined the *mores* (Latin for *ethoi*, the plural of *ethos*) of peoples, in what came to be called moral theory. Thus ethics is to ethos as moral theory is to morals. *Ethics* and *moral theory* are but two terms for the same thing: systematic reflections on human conduct that seeks to understand more clearly and deeply the good for humans.

During the Middle Ages these articulations of ethics or moral theory (science) and ethical or moral practice (technique) were enclosed within the framework of revelation. For instance, according to the argument of Augustine (354–430 C.E.) in *On True Religion*, revelation takes the truths of philosophy, known only by the few, and makes them publicly available to the many. By so doing religion makes the practical realization of the good more effective than was previously possible, at both personal and political levels.

According to Thomas Aquinas (c. 1225–1274), the supernatural perspective allows Christians to provide more accurate descriptions, more sure assessments, and more perfect insight into the ultimate nature of reality and the human good than was possible for pagans. What for Aristotle could be no more than the counsels of practical wisdom became for Thomas natural laws of human conduct, laws that gear down the cosmic order and are manifest in human reason as a "natural inclination to [their] proper act and end" (*Summa theologiae*

I–II, ques. 91, art. 2). The self-evident first principle of ethics that "good is to be done and promoted and evil is to be avoided" is given content by the natural inclinations to preserve life, to raise a family, and to live in an intelligence-based community (*Summa theologiae* I–II, ques. 94, art. 3).

The traditional forms of ethics as science and as technique acted to restrain the independent pursuit of science and technics. As entries on "Plato," "Aristotle," "Augustine," and "Thomas Aquinas" further suggest, these traditions provide continuing resources for the critical assessment of modern science and technology. Indeed, the contentious character of these often alternative assessments may be one of their most beneficial aspects, in that they call for reconsidering the assumptions that now animate scientific and technological activity.

Ethics as Science and Technology

In the modern period a basic transformation occurs in the understanding of ethics, one related to a transformation in science and technology. The scientific understanding of nature came to focus no longer on the natures of different kinds of entities, but on laws that transcend all particulars and kinds. The knowledge thus promoted the merger of technics into technology, the systematic power to control or reorder matter and energy. Technological knowledge became the basis for a technological activity that produced artifacts in greater regularities and quantities than ever before possible.

In like manner, the science of ethics sought to elucidate rules for human action. Divides emerged in the details of different ethical systems, but the major approaches nevertheless all pursued ethical decision-making processes that could be practiced with competence and regularity on a scale to cope with the new powers first of industrialization and then of globalization. The modern period thus witnessed the development of ethics as a science with a unique intensity and scope.

From their origins science and technology were supported with fundamentally ethical arguments—by Francis Bacon (1561–1626), René Descartes (1596–1650), and others—for a new vision of human beings as deserving to control the natural world and dominate it. The Enlightenment and the Industrial Revolution flourished in conjunction with the progressive articulation of ideas about how humans might, through science and technology, remake both the physical and social worlds. Romanticism served as a critical response to the difficulties, threats, and complications inherent in such a reshaping of human experience, but in ways that were ultimately incorporated into the emerging cultural transformation. (Encyclopedia entries on "Bacon, Francis," "Descartes, René," and the "Industrial Revolution," among others, explore such issues in more detail.)

The systematic development of the modern science of ethics itself emerged in two major traditions. One was the consequentialist utilitarian tradition as elaborated by Jeremy Bentham (1748–1832), John Stuart Mill (1806–1873), and their followers. The other was a deontological or duty-focused tradition with roots in the thought of Jean-Jacques Rousseau (1712–1778) but most closely associated with the work of Immanuel Kant (1724–1804). For consequentialists, rules for ethical decision-making are best determined by end uses or the effects of actions; for deontologists rules are grounded in the intentional properties of the actions themselves. Leading twentieth-century representatives of these two traditions include Peter Singer (b. 1946) and John Rawls (1921–2002), respectively. Both traditions are efforts to deal with the moral challenge created by the loss of nature as a normative reality within and without human beings. (Encyclopedia entries on the traditions of "Consequentialism" and "Deontology" are complemented by separate entries on such thinkers as "Rousseau, Jean-Jacques," "Kant, Immanuel," and "Rawls, John.")

Prior to the modern period, natural entities were understood as possessed of functional tendencies toward internal and external harmonies. When they function well and thereby achieve their *teloi,* plural of *telos* or ends, acorns grow up into oak trees, human beings speak and converse with one another in communities. Furthermore, both oak trees and humans fit in with larger natural orders. Because these harmonies are what constitutes being itself, they are also good, which is simply the way that reality manifests itself to, draws forth, and perfects the appetite. Although the first name of the good may be that which is one's own, a second and superior name is form or being, the different instances of which themselves come in an ascending order. For the Platonic and Aristotelian traditions, ethics as practice was thus constituted by the teleological perfection of human nature, realizing ever-higher states of functional potential. Such a view has obvious affinities with religious traditions as diverse as Hinduism, Buddhism, and Christianity. But insofar as nature comes to be seen as composed not of entities with natures to be realized, but as constructions able to be used one way or another and modified at will, fundamental questions arise about the foundations of the good as an end to be pursued as well as the rightness of any means to be employed in such pursuit.

The fundamental problem for modern ethics is not just what the good is, but its basis. In simplified terms, for the consequentialist tradition the good is what human beings need or want, and there are no limits on actions as means other than what might be at odds with perceived wants; for the deontological tradition right means are those whose intentions may be consistently pursued or universalized, with no limits on the goods that might flow from them.

Efforts to make consequentialist and deontological systems truly scientific have been pursued both formally and substantively. In the first half of the twentieth century a pursuit of formal rigor led to the development of metaethics. Eschewing any normative goals, metaethics simply aspires to clarify the structure of ethical language and reasoning. In its radical form metaethics has tended to reduce the meaning of ethical statements to forms of emotional approval; in more moderate forms it has simply disclosed the complexities of ethical judgments, sometimes pointing up and seeking to rectify inconsistencies. In the second half of the twentieth century the inadequacies of metaethical analysis for the substantive issues faced in the creation and use of science and technology brought about development of applied ethics. The term is somewhat anomalous, because all traditional ethics applied to real life. Applied ethics is applied only in contrast to metaethical formalist aspirations.

Across the twentieth century efforts to make ethics scientific in more substantive ways developed in two tracks. One was to try to base ethics on evolutionary theory. This approach commonly takes those behaviors that are descriptively given moral value (such as altruism) and shows how and why such approval could have been the outgrowth of the processes of evolutionary selection.

Another effort to make ethics substantively scientific has been to elucidate the rationality of ethical behavior through the mathematics of game and decision theory. In the same spirit as game and decision theory, parallel efforts to supply practical wisdom with the strengths of quantitative methods have given rise to operations research and risk–cost–benefit analysis. Much more than evolutionary theory, such efforts have produced ethical techniques for dealing with the complexities of the advanced scientific and technological world, especially in relation to public policy analysis.

Continuing efforts to model ethics on the authority of modern science and the powers of technology, including the computer modeling of artificial ethics, have proved selectively suggestive and insightful. Despite significant achievements, however, neither scientific nor technological ethics has proved able to capture the richness of ethical reflection that is spread across the diversity of ethical traditions, ancient and modern.

Ethical Perspectives on Science and Technology

A different approach to the ethics of science and technology eschews making ethics into a science or a technology but to consider science and especially technology as new fields requiring ethical analysis and reflection. Here there has been a divide between those who seek to bring ethics to bear on science and technology as a whole, and those who choose to limit their ethical reflection to specific sciences or technologies.

With regard to the holistic approach, the work of Hans Jonas (1903–1993) may be taken as representative. For Jonas the powers of modern technology, which are more extensive across space and time, on the macro- and the microscale, than all previous human abilities, require a new ethics of responsibility. In his words, "Modern technology has introduced actions of such novel scale, objects, and consequences that the framework of former ethics can no longer contain them" (Jonas 1984, p. 6). In response Jonas formulated the new imperative of responsibility as: "Act so that the effects of your action are compatible with the permanence of genuine human life" (p. 11). (For more detail, see the entry on "Jonas, Hans.")

With regard to approach that focuses on specific technologies, this is well represented by the various fields of applied ethics such as agricultural ethics, bioethics, business ethics, computer ethics, engineering ethics, environmental ethics, and more (each of which is given its own entry). Further specificity can be found in many of the case studies included in the encyclopedia, from "Abortion" to "Zoos."

In both holistic and particularist approaches, however, there are at least two common themes. One is whether on balance science and technology—or some particular science or particular technology—should be encouraged or in any way restrained. Another is whether existing ethical traditions are adequate to deal with the ethical challenges of science and technology, or whether instead wholly new ethical concepts and frameworks need to be developed.

Finally, even when the adequacy of existing traditions is assumed or defended, there are a number of dis-tinctive concepts and principles that tend to recur in the ethical examinations of science and technology—or particular fields therein. Examples include the principles of respect for human autonomy and the exercise of responsibility and public participation, along with the concepts of safety and risk, the environment, and expertise. Each of these, along with a number of closely related terms, are thus also accorded encyclopedic entries.

The Limitations of Ethics

Any overview of ethics, especially one that highlights the way ethics attempts to deal with the dangers and challenges of science and technology, should not fail to mention the danger of ethics itself. These dangers come in three forms: economic, personal, and philosophical.

First, the economic danger in bringing ethics to bear on science and technology will limit scientific and technological progress, which in turn will limit economic development.

Second, there is what may be called the personal temptation to false righteousness. Turning a technical problem into an ethical one can make it more difficult to discuss, because the discussants now address it in terms of emotionally loaded senses of right and wrong or good and bad rather than the less loaded senses of more or less efficient or effective. Because of such emotional investments, social and political discussions can become intractable when ethical principles are invoked and people become unwilling to compromise. When the NIMBY ("not in my backyard") syndrome is justified not simply on the basis of practical concerns but by appeal to fundamental rights or other principles, it can become almost impossible to find common ground solutions. The opposition between fundamentalist religious beliefs about abortion and zealous commitments to women's rights provide another example of the problems that can be created by assessing science or technology in ethical terms.

Third, philosophers from Karl Marx to Michel Foucault have argued that morality is often simply a disguised form of self-interest. A modern tradition of the philosophical criticism of ethics has highlighted numerous ways that morality has been used to justify human oppression and exploitation, from racism to gender discrimination. Ethics can be simply another name for lack of self-knowledge, a kind of false consciousness.

Finally, another philosophical issue with ethics is that to define a problem as one of ethics can obscure not only its scientific and technical aspects but also its epistemological, metaphysical, aesthetic, and even theological dimensions. As philosopher Robert Frodeman

(2003) has argued with regard to an extended examination of problems in the geosciences, environmental ethics is not enough. The issues of environmental ethics are often as much aesthetic and ontological as they are ethical. The category of the ethics must not be allowed to obscure other equally significant categories of reflection that are called forth by efforts to understand and assess science and technology.

CARL MITCHAM

SEE ALSO *Consequentialism; Cosmology; Deontology; Humanization and Dehumanization; Nature; Virtue Ethics.*

BIBLIOGRAPHY

Chadwick, Ruth, ed. (2001). *The Concise Encyclopedia of the Ethics of New Technologies.* San Diego, CA: Academic Press. Thirty-seven articles on various technologies (biotechnology, genetic engineering, nuclear power) and related issues (brain death, intrinsic and instrumental value, precautionary principle) selected from the *Encyclopedia of Applied Ethics* (1998).

Frodeman, Robert. (2003). *Geo-Logic: Breaking Ground between Philosophy and the Earth Sciences.* Albany: State University of New York Press.

Jonas, Hans. (1984). *The Imperative of Responsibility: In Search of an Ethics for the Technological Age,* trans. Hans Jonas and David Herr. Chicago: University of Chicago Press. Combines two German books published first in 1979 and 1981.

Kaplan, David M., ed. (2004). *Readings in the Philosophy of Technology.* Lanham, MD: Rowman and Littlefield. Includes substantial sections on ethics and politics.

Keulartz, Jozef, Michiel Korthals, Maartje Schermer, and Tsjalling Swierstra, eds. (2002). *Pragmatist Ethics for a Technological Culture.* Dordrecht: Kluwer Academic. A collaborative argument for pragmatism as the best ethical tradition for dealing with science and technology.

Mitcham, Carl, and Robert Mackey, eds. (1983). *Philosophy and Technology: Readings in the Philosophical Problems of Technology.* New York: Free Press. The "Ethical-Political Critiques" section includes a number of classic texts; other articles in this early collection are also relevant.

Scharff, Robert C., and Val Dusek, eds. (2003). *Philosophy of Technology: The Technological Condition.* Malden, MA: Blackwell. Part V, "Technology and Human Ends," includes thirteen relevant contributions.

Scott, Charles E. (1990). *The Question of Ethics: Nietzsche, Foucault, Heidegger.* Bloomington: Indiana University Press. A critical and sympathetic examination of the modern tradition of the ethical questioning of ethical thought.

Tavani, Herman T. (2003). *Ethics and Technology: Ethical Issues in an Age of Information and Communication Technology.* Hoboken, NJ: Wiley.

Winner, Langdon. (1986). *The Whale and the Reactor: A Search for Limits in an Age of High Technology.* Chicago: University of Chicago Press.

INTERNET RESOURCE

"The Online Ethics Center for Engineering and Science." Case Western Reserve University. Available from http://onlineethics.org.

ETHOLOGY

• • •

Ethology is the biological study of animal behavior. It derives from the Greek root *ethos,* which, in normal English usage, refers to the manner of living, or customary behavior, of a social entity. One may therefore speak of the ethos of a particular sports club, small town, or professional organization, for example. By the same token, ethologists are concerned with the ethos of animals: their way of behaving.

Ethology traces its history to the early decades of the twentieth century, especially the work of the Austrian physician Konrad Lorenz (1903–1989), Dutch biologist Niko Tinbergen (1907–1988), and German entomologist Karl von Frisch (1886–1982); in recognition of their achievements, these three shared the Nobel Prize in physiology or medicine in 1973. The characteristics of ethology as a scientific discipline can be appreciated by comparing it to one of its well-known counterparts, comparative psychology.

Whereas comparative psychology is primarily concerned with understanding human behavior, such that animal research is conducted with an eye to better understanding *Homo sapiens,* ethology focuses specifically on the behavior of animals for its own sake. Similarly, comparative psychologists study a small range of animal species—particularly laboratory rats, macaque monkeys, and pigeons—as easily manipulated substitutes for human beings. By contrast, ethologists study the diversity of animal species, especially invertebrates, fish, and birds. Because of their underlying concern with understanding human behavior, researchers in comparative psychology are especially interested in examining the various concomitants of learning (which have a notable impact on human beings). Ethologists pay considerable attention to behavior that is loosely described as "instinctual," which tends to be more prevalent in the species with simpler nervous systems that are typically the subject of ethological research. Ethologists also emphasize the study of animal behavior in its natural context; that is to say, under field conditions where the organisms normally live and to which they are adapted by natural selection. By contrast, comparative psychologists typically conduct their research in a laboratory setting within which they can carefully control for extraneous factors while focusing on the role of various aspects of experience.

Some Aspects of Classical Ethology

Ethology, as the study of how organisms conduct their lives, long has been especially concerned with compiling careful, detailed descriptions of actual behavior patterns, known as *ethograms*. These detailed records (including verbal descriptions, photographs, and sonograms of vocal communications, for example) are not generally considered ends in themselves, but are fundamental to a rounded, ethological understanding of any species: Ethologists emphasize that they must first know what the animals in question do before they can pose meaningful questions.

According to Niko Tinbergen, those questions are especially concerned with the following:

(a) How does the behavior in question influence the survival and success of the animal? In modern evolutionary terms, what is its adaptive significance; or, how does it contribute to the inclusive fitness of the individual and the genes responsible, recognizing that inclusive fitness involves not only personal, Darwinian reproductive success but also the effect of each behavior on the fitness of other genetic relatives.

(b) What actually makes the behavior occur at any given moment? This might include the role of hormones, brain mechanisms, prior learning, and so forth.

(c) How does the behavior in question develop as the individual grows and matures? What is its developmental trajectory, or *ontogeny*?

(d) How has the behavior evolved during the course of the species' evolutionary history? In short, what is its *phylogeny*?

It is worth noting that of these, question *a* has become the special province of *sociobiology*, a research discipline closely allied to ethology and that emphasizes matters of adaptive significance and evolutionary—often called *ultimate*—causation. By contrast, question *b* is associated in the public mind with research into animal behavior more generally; it is often called *proximate* causation. Ideally, a complete understanding of animal behavior will involve both ultimate and proximate considera-

tions, as well as attention to matters of ontogeny and phylogeny.

Through their research, early ethologists developed a number of concepts now considered part of "classical ethology." These include, but are not limited to, the following. *Fixed action patterns* are the fundamental building blocks of behavior, consisting of simple, relatively unvarying movements that are more or less independent of prior experience. Once initiated, fixed action patterns generally continue to completion even if the initiating stimulus is no longer present; this emphasizes the unthinking nature of these acts, which are the products of natural selection rather than complex cognition or daily experience. Fixed action patterns, in turn, are evoked by *releasers*, features of the environment or other animals to which the receiving animal is delicately attuned. The situation is analogous to a lock-and-key mechanism: a lock is carefully adjusted (in the case of animal behavior, by natural selection rather than by a locksmith) to the specific characteristics of a key. In ethological terminology, the lock is an *innate releasing mechanism*, a characteristic of the receiving animal—usually but not necessarily located in the animal's central nervous system—that responds to the traits of the releaser. Continuing the analogy, when the key fits the lock, a door opens; this is equivalent to the fixed action pattern. And, just as a door moves along a fixed, predetermined pathway, so do the behavior patterns with which ethologists have traditionally been most concerned.

Although it may appear that this schema is only capable of generating simple behaviors (a simple releaser evokes a comparably simple fixed action pattern), ethologists demonstrated that these connections can be "chained," such that fixed action patterns by one individual, for example, can serve as a releaser for another, whose fixed action pattern, in turn, serves as a releaser for another fixed action pattern in the first; and so on. In the process, complex sequences of courtship, parental care, or communication can be constructed.

In the courtship behavior of the three-spined stickleback fish—a species that has been intensively studied by ethologists—males develop a bright red abdomen in response to the warming water and increased day length of spring; females react to this releaser by their own fixed action pattern, a "head-up" display which in turn reveals their abdomens, swollen with eggs; the male, in turn, responds by his own fixed action pattern, a "zigzag dance," which involves swimming rapidly toward a nest made of algae that he would have previously constructed and then swimming quickly toward the female; the female responds by following the male; the male lays on his side in a characteristic posture "showing" the nest entrance to the female; she enters; he rhythmically prods the base of her tail with his snout, whereupon she deposits her eggs; she swims away; he enters the nest, fertilizes the eggs, and continues to care for them until they hatch. Throughout this complex sequence, each situation or behavior by one animal serves as a releaser for a fixed action pattern by the other, and so on in turn.

Ethologists also developed descriptive models for the control of behavior. Two notable models are the hydraulic model of Lorenz and Tinbergen's hierarchical schema. Lorenz proposed that a kind of motivational pressure—which he labeled *action specific energy*—builds up within the central nervous system of an individual. This energy is dissipated when the appropriate fixed action pattern is performed. In some cases, if the behavior in question is blocked, the motivational energy spills over into another channel, generating a seemingly irrelevant behavior, known as a *displacement activity*. For example, shorebirds known as avocets, when engaged in a dispute at a territorial boundary, may tuck their heads into their wing feathers, in a posture indistinguishable from that normally assumed during sleep.

Lorenz's scheme is also consistent with vacuum activities, whereby an animal may suddenly perform a fixed action pattern in the absence of any suitable releasing stimulus; in this case, presumably the energy associated with a given fixed action pattern has built up to such a level that it essentially overflows its neuronal banks and the relevant brain centers discharge in an apparent vacuum. Although the hydraulic model does not have many current devotees, it still serves as a useful heuristic model.

Tinbergen proposed a similar perspective, one somewhat more consistent with known neurobiological mechanisms. He suggested that various major instinctive tendencies (e.g., reproduction, migration, food-getting) were organized hierarchically, such that reproduction, for instance, was subdivided into fighting, nest-building, mating, and care of offspring, each of which, in turn, was further subdivided. Thus, depending on the species, fighting might involve chasing, biting, and threatening, whereas care of offspring might involve provisioning the young, feeding them, defending them from predators, and providing various kinds of learning opportunities.

Ethology in the Twenty-first Century

Despite the ethological focus on animal behavior that can loosely be labeled "instinctive," an important reali-

zation characterizes all studies of behavior, whether conducted by ethologists, sociobiologists, or comparative psychologists: Behavior always derives from the interaction of genetic and experiential factors. Variously labeled instinct/learning, genes/experience, or nature/nurture, contemporary researchers widely acknowledge that these dichotomies are misleading. Just as it is impossible for an organism to exist or behave without some influence from its environment (the extreme case of "pure instinct"), it is impossible for environmental factors acting alone to produce behavior (the extreme case of "pure learning"). Every situation must involve both factors: there must always be an organism to do the behaving, and, moreover, organisms with different experiences exposed to the same situations always respond somewhat differently.

Ethologists have branched out substantially from their earlier focus on careful naturalistic descriptions of animal behavior, increasingly blurring the distinction between ethology and various related disciplines. Thus, neuroethologists concern themselves with the brain regions and precise neuronal mechanisms that govern, for example, animal communication as well as the reception of auditory, olfactory, visual, and even tactile signals. Behavior genetics incorporates an amalgam of ethology and precise genetic techniques to unravel the genetic influence on various behavior patterns; such research may range from the creation of cross-species hybrids to the detailed analysis of DNA sequences in identified genes responsible for specific behavioral tendencies. Behavioral endocrinology investigates the role of hormones in predisposing animals toward courtship, aggressive, migratory, and other behaviors, as well as the environmental and social situations responsible for releasing the relevant hormones. Mathematically inclined ethologists have been increasingly interested in applying concepts derived from game theory in seeking to understand how behavior has evolved, especially in situations such that the benefit to each individual depends not only on what he or she does, but also on the behavior of another individual.

Ethology and Ethics

Researchers increasingly have been applying the basic ethological techniques of detailed, objective, nonjudgmental observation to human behavior as well. Human ethology is essentially an organized form of "people watching," whereas human sociobiology (sometimes called *evolutionary psychology*) seeks to apply the principles of evolution by natural selection to *Homo sapiens*. Critics assert that the former approach consists of a kind of empty empiricism, lacking powerful theoretical roots; others attack the latter for being overly driven by theory, occasionally lacking in adequate empirical findings.

The conduct of ethology occasionally raises ethical issues concerning the treatment of animal subjects, but generally such matters are more controversial in the laboratory-oriented disciplines such as comparative psychology and neuroethology. Because ethologists study undisturbed, natural populations, or—when conducting laboratory research—strive to maintain their subjects under naturalistic conditions, the major ethical dilemma facing ethologists tends to center around whether or not to intervene in the events of the normal lives of their study animals. For example, is it appropriate to prevent a forthcoming act of predation? (Ethologists nearly always answer in the negative, because they are typically committed to nonintervention on the lives of their subjects.) Moreover, because the goal of ethological research—unlike that of comparative psychology—is to understand behavior rather than to control it, and because—unlike evolutionary psychology—ethologists generally do not directly employ controversial assumptions about evolutionary factors currently operating on human behavior, ethology is generally free of the moral conundrums often attending its sister disciplines.

DAVID P. BARASH

SEE ALSO Aggression; Animal Tools; Dominance; Nature versus Nurture; Selfish Genes; Sociobiology.

BIBLIOGRAPHY

Alcock, John. (2001). Animal Behavior: An Evolutionary Approach. Sunderland, MA: Sinauer. The standard college textbook of animal behavior, which includes a good, if brief, treatment of ethology.

Barash, David P. (1982). Sociobiology and Behavior. New York: Elsevier. A wide-ranging introduction to sociobiology, emphasizing its distinction from classical ethology.

Goodenough, Judith; Betty McGuire; and Robert A. Wallace. (1993). Perspectives on Animal Behavior. New York: Wiley. A solid introduction to animal behavior that seeks to introduce both ethological and sociobiological approaches.

Gould, James L. (1982). Ethology: The Mechanisms and Evolution of Behavior. New York: Norton. A technical but rewarding early account of animal behavior, especially strong on physiological mechanisms.

Lorenz, Konrad Z. (1982). The Foundations of Ethology. New York: Simon and Schuster. Technical articles by one of the founders of ethology.

Tinbergen, Niko. (1989). The Study of Instinct. Oxford, UK: Clarendon Press. An early and now-classic account of ethology as a scientific discipline distinct from comparative psychology.

EUGENICS

• • •

Eugenics was an ideology that arose in the late nineteenth century to promote improving human heredity. It posed as a scientific enterprise, but combined ethical presuppositions and political action with research on human heredity. For example, there was no scientific way to determine what constituted "improvement." Progress and improvement, however, were the watchwords of many nineteenth-century intellectuals who often failed to recognize how such concepts can be culturally loaded. Indeed, many eugenicists supposed that health, strength, intellectual acuity, and even beauty were undeniably favorable traits and should be promoted in human reproduction. Another closely related ideology was that of Social Darwinism, which nevertheless has its own distinctive if interactive history. While Social Darwinism stressed natural selection and thus human competition, eugenics focussed on artificial selection. Though some eugenicists saw eugenics as a way to evade Social Darwinism, others were avid Social Darwinists.

Classic Eugenics

The basic idea of eugenics came to Francis Galton (1822–1911), the father of the eugenics movement, in the 1860s while reading Charles Darwin's The Origin of Species. Galton claimed that Darwin's theory "made a marked epoch in my own mental development, as it did in human thought generally" (Gillham 2001, p. 155). In 1869 Galton published his most famous book, Hereditary Genius, in which he traced the lineages of prominent men in British society in order to demonstrate that not only physical characteristics but also mental and moral traits were hereditary. Galton coined the phrase "nature and nurture" to describe the conflict between biological determinism and environmental determinism, and came down decidedly on the side of nature.

Galton's views on heredity not only drove him to engage in scientific research, but also motivated him to propose conscious planning to help speed up human evolution. He stated, "What nature does blindly, slowly, and ruthlessly, man may do providently, quickly and kindly" (Gillham 2001, p. 328). He favored measures to encourage the "most fit" people to reproduce. This is called positive eugenics. However, he also advocated negative eugenics: restricting the reproduction of those deemed "inferior." He thought inferior people should be branded enemies of the state and "forfeited all claims to kindness" if they procreated. Further, he believed that "inferior races always disappear before superior ones" (Gillham 2001, p. 197). Galton, like subsequent eugenicists, stressed human inequality and devalued the life of those considered inferior. When Galton died, he left a bequest to endow a chair in eugenics at the University of London, which was filled by Karl Pearson (1857–1936), his hand-picked successor as leader of the eugenics movement in Britain.

The eugenics movement blossomed in the 1890s and early twentieth century, partly fueled by fears of biological degeneration. By the 1890s many Darwinists were concerned that some of the improvements of modern civilization were a mixed blessing. Ernst Haeckel (1834–1919), the leading Darwinist in Germany, already warned in the 1870s that modern medical advances allowed those with weaker physical conditions to survive and reproduce, while in earlier ages they would have perished without leaving progeny. Other Darwinists also warned that the weakening of natural selection by modern institutions would bring biological decline. However, while embracing Darwinian principles, eugenicists did not want to abandon scientific, technological, and medical progress. Rather they sought to escape the negative consequences by consciously controlling human reproduction.

Simultaneous with this fear of biological decline, many psychiatrists by the 1890s were abandoning earlier optimistic beliefs that they could provide cures for many mental illnesses. Instead, they began viewing mental illnesses as often hereditary and beyond influence. Many psychiatrists began to push for control of human reproduction as the most effective means to prevent mental illness. August Forel (1848–1931), a famous psychiatrist at Burghölzi Clinic in Zurich, began promoting eugenics in the late nineteenth century, and he decisively influenced many other psychiatrists and physicians. One medical student in Zurich who imbibed eugenics from Forel was Alfred Ploetz (1860–1940), who in 1904 began editing the first eugenics journal in the world. The following year he founded the Gesellschaft für Rassenhygiene (Society for Race Hygiene), an organization dedicated to improving human heredity. He quickly recruited many leading scientists, psychiatrists, and physicians to the cause.

Eugenics in the Early Twentieth Century

In the ensuing two decades, eugenics organizations also formed in many other countries, not only in the United States and Europe, but also in Latin America and Asia. The prominent geneticist Charles Davenport (1866–1944) founded the Eugenics Record Office in Cold Springs Harbor, New York, which became one of the leading institutions in the United States promoting eugenics by compiling family medical histories. Many wealthy patrons, including Andrew Carnegie and John D. Rockefeller, funded eugenics organizations.

Eugenics also stimulated the rise of birth control organizations. Indeed, one of the primary goals of the pioneers in the birth control movement—including Margaret Sanger (1879–1966) in the United States and Marie Stopes (1880–1958) in Britain—was to diminish the reproductive rates of those members of society they considered inferior. In 1919 Sanger stated, "More children from the fit; less from the unfit—that is the chief issue of birth control" (Paul 1995, p. 20). Nonetheless, most eugenicists opposed the easy availability of birth control, because they feared it would lead to a decline in natality rates among the upper and middle classes, which they wanted to increase. They wanted birth control, of course, but under the control of physicians making decisions in the interests of society, not freely available to individuals.

The eugenics movement had clout far greater than reflected by the small number of people in eugenics organizations, because its influence in the medical profession, especially among psychiatrists, was strong. In some countries the eugenics movement exerted enough influence to pass legislation aimed at restricting reproduction of individuals considered "defective." The first eugenics legislation in the world was a compulsory sterilization law passed by the state of Indiana in 1907. Other states followed suit, allowing doctors to sterilize patients who had various hereditary illnesses, especially mental illnesses. On the basis of these laws, from the 1920s to the 1950s, about 60,000 people were compulsorily sterilized in the United States. The Supreme Court upheld the right of states to sterilize those with hereditary illness in the *Buck* v. *Bell* case in 1927. Denmark was the first European country to enact a sterilization law in 1929, but it was voluntary until new legislation in 1934 made it compulsory in some cases.

Nazi Eugenics and Afterward

The Nazi regime passed the most sweeping eugenics measures in the world, because Adolf Hitler and other leading National Socialists were fanatical about trying to produce a healthy master race in Germany. In 1933 the Nazis passed a compulsory sterilization law that resulted in more than 350,000 sterilizations during their twelve years in power. In 1939 Hitler secretly ordered the beginning of a "euthanasia" campaign, killing 70,000 mentally handicapped Germans within two years. The Nazis also considered the mass killing of those of races they deemed inferior—especially Jews, but also Gypsies and others—part of their eugenics program, because they believed that this would improve the human race. Many German physicians, imbued with eugenics ideals, participated in the Nazi euthanasia program and the Holocaust.

Since the Nazi era many people have mistakenly associated eugenics with right-wing, reactionary politics. However, in its early phases, most eugenicists were progressive politically, and eugenics was popular in leftist circles. Most of the early German eugenicists were non-Marxian socialists or at least sympathetic with socialism. Many anarchists, such as Emma Goldman (1869–1940), promoted eugenics, as did most Fabian socialists in Britain. The Danish government that enacted the 1929 and 1934 sterilization laws was socialist. Many liberals and conservatives supported eugenics as well, so it cut across political lines.

Despite the movement's successes, many countries rejected attempts to enact eugenics legislation, and critics of eugenics arose, challenging its premises. The Catholic Church was the staunchest adversary of eugenics, and the pope issued an encyclical in 1930 opposing eugenics, especially measures such as compulsory sterilization. Catholics and some conservative Protestants recognized that eugenics contradicted the traditional Christian attitudes toward sexual morality, compassion for the handicapped, and human equality. However, most liberal Protestants jumped on the eugenics bandwagon, seeing it as a progressive, scientific movement. By the late 1920s many German Protestant leaders supported eugenic sterilization, and Protestants in the United States sponsored prizes for the best sermons on eugenics.

By the 1960s the eugenics movement seemed dead, and the term itself had negative connotations. Eugenics suffered from its association with Nazism, but this was only one factor. The decline of biological determinism in most scholarly fields, especially psychology and the social sciences, made people suspicious of the claims of eugenics. Also, the individualism of the 1960s, along with calls for reproductive autonomy, undermined the collectivist mentality of eugenics and its desire to control reproduction. As the abortion debate heated up in the 1970s, pro-life forces and pro-choice advocates both opposed eugenics, the former because they saw it as devaluing human life, and the latter because it violated reproductive freedom.

The New Eugenics

However, advances of medical genetics in the late twentieth century led to a "new eugenics." New reproductive technologies, including amniocentesis, ultrasound, in vitro fertilization, sperm banks, genetic engineering, and cloning, opened up new possibilities to control human fertility and heredity, especially because the human genome project has now mapped human DNA. Some proponents want to use these new technologies not only to rid the world of congenital disabilities, but also to produce "designer babies." Intense debates are raging in the early 2000s over "designer babies" and reproductive cloning, because most people consider these unethical interventions in reproduction. The legalization of abortion in most countries and the widespread practice of infanticide, even though illegal, are other factors fostering the new eugenics, because this allows parents the opportunity to decide whether they want a child with particular characteristics. The big difference between this new eugenics and the old eugenics is that in most countries the decision making about human heredity is in the hands of the individual (though physicians and society often apply pressure). However, in 1995 China passed a eugenics sterilization law, which was ostensibly voluntary; especially in light of that government's one-child policy, the pressure to abort fetuses deemed defective is great.

RICHARD WEIKART

SEE ALSO Birth Control; Chinese Perspectives: Population; Galton, Francis; Health and Disease; Holocaust; Human Cloning; In Vitro Fertilization and Genetic Screening; Nazi Medicine; Race; Rights and Reproduction; Sanger; Margaret; Social Darwinism; Wells, H. G.

BIBLIOGRAPHY

Broberg, Gunnar, and Nils Roll-Hansen, eds. (1996). Eugenics and the Welfare State: Sterilization Policy in Denmark, Sweden, Norway, and Finland. East Lansing: Michigan State University Press. Five scholars examine eugenics sterilization policies in twentieth-century Scandinavia.

Buck v. Bell, 274 US 200 (1927). This was the Supreme Court decision allowing compulsory sterilization for eugenic purposes.

Dikötter, Frank. (1998). *Imperfect Conceptions: Medical Knowledge, Birth Defects and Eugenics in China*. New York: Columbia University Press. Discusses attitudes toward the disabled and eugenics in medical circles and its impact on governmental policy from the nineteenth century to the present.

Gillham, Nicholas Wright. (2001). *A Life of Sir Francis Galton: From African Exploration to the Birth of Eugenics*. Oxford and New York: Oxford University Press.

Kevles, Daniel J. (1985). *In the Name of Eugenics: Genetics and the Uses of Human Heredity*. Berkeley: University of California Press. This is the best comprehensive survey of eugenics ideology and policy in the Anglo-American world.

Paul, Diane B. (1995). *Controlling Human Heredity: 1865 to the Present*. Atlantic Highlands, NJ: Humanities Press. The best brief survey of the eugenics movement, focussing primarily on the United States, but also touching on other countries.

Proctor, Robert. (1988). *Racial Hygiene: Medicine under the Nazis*. Cambridge, MA: Harvard University Press. Proctor shows how the German medical community embraced eugenics and collaborated with Nazi eugenics policies, including sterilization and genocide.

Schneider, William H. (1990). *Quality and Quantity: The Quest for Biological Regeneration in Twentieth-Century France*. Cambridge, UK: Cambridge University Press. Provides excellent coverage of the eugenics movement in France in the first half of the twentieth century.

Stepan, Nancy L. (1991). *The Hour of Eugenics: Race, Gender and Nation in Latin America*. Ithaca, NY: Cornell University Press. Stepan shows that medical elites in Latin America embraced eugenics, just as their counterparts in Europe and the United States.

Weikart, Richard. (2004). *From Darwin to Hitler: Evolutionary Ethics, Eugenics, and Racism in Germany*. New York: Palgrave Macmillan. Demonstrates how the German eugenics movement (and Hitler) used Darwinian principles to devalue the life of the disabled and those of non-European races.

Weindling, Paul. (1989). *Health, Race and German Politics between National Unification and Nazism, 1870–1945*. Cambridge, UK: Cambridge University Press. Weindling shows that the collectivist and technocratic thrust of the German eugenics movement helped prepare the way for the authoritarian policies of the Nazi regime. This is the most detailed study of the German eugenics movement available in English.

EUTHANASIA

• • •

Strictly speaking, *euthanasia* is Greek for "good death," but it has come to be applied to cases of an ill or disabled person being helped to die or deliberately killed by another for the ill or disabled person's benefit. It is thus distinguished from murder. Euthanasia is also to be distinguished from mercy killing. Whereas *mercy killing* normally refers to an act on the part of a friend or relative, euthanasia is typically discussed in relation to health care professionals. A number of further distinctions are drawn between different types of euthanasia: between active and passive euthanasia, and between voluntary, non-voluntary, and involuntary euthanasia. Whereas active euthanasia implies a deliberate act of killing, passive euthanasia means causing death by *not* doing something: allowing to die by withdrawing or withholding treatment. Not all forms of withdrawing treatment count as euthanasia, as when the treatment is futile or constitutes an "extraordinary" means of maintaining life.

Indeed, advances in medical science and technology have intensified concern for euthanasia because of the increased power to keep persons alive who nevertheless become dependent on various treatments. Examples range from cases of feeding tubes and artificial respiration to kidney dialysis and organ transplants. In all such instances, science and technology sometimes lead to deteriorations in the quality of life or costs that lead patients, those closest to them, and health care givers and policy makers to raise questions about continuation of treatment. Such questions often focus on whether and under what circumstances euthanasia might be a proper alternative.

Voluntary euthanasia is a response to a request on the part of a competent individual who regards death as preferable to continuing to live: The individual in question must be in a position to understand the nature of what he or she is asking and to consent to it. Non-voluntary euthanasia occurs in cases where the individual is not in a position to make a euthanasia request, such as because of a lack of competence. Competence is context-specific, so it is not necessarily the case that the individual in question is unable to make any decisions at all. As a matter of fact, several of the most-discussed issues in euthanasia do concern cases of such total incompetence, as illustrated by the following.

Non-voluntary cases fall into different types. There are adults who have lost the capacity to make an informed choice—for example, because they are in a coma or persistent vegetative state. For such persons, the living will or advance directive is one way of stating a preference should such an eventuality arise. These are subject to criticism, though, on the grounds that people may be unable to anticipate how their preferences might change over time.

Alternatively, there are those who have not yet developed to the stage at which they have acquired the capacity to state a preference, such as infants. In some cases, newborns who are born with severe medical problems are rejected by their parents, and then decisions have to be made by health care professionals about how to deal with this situation. High profile cases such as the Arthur case in the United Kingdom (see Kuhse and Singer 1985) have dealt with the question of whether it is appropriate to allow such infants to die.

In contrast to non-voluntary euthanasia, involuntary euthanasia refers to ending people's lives against or in spite of their wishes. The distinction between involuntary euthanasia and murder is more difficult to draw than in the case of other types of euthanasia, but there are some possible instances; for example, where someone is critically injured on a battlefield, cannot be saved, and a military doctor who is present, having no morphine, shoots the injured person dead.

Arguments for Euthanasia

There is controversy both about the ethics of euthanasia *per se* and about the moral distinction between active and passive versions. The moral argument for euthanasia is normally put in terms of voluntary active euthanasia—that is, when persons are terminally ill and no longer have any hope of recovery, are in pain or distress, are considered competent, and ask for someone to end their lives, the argument is that these individuals have a right to die based on respect for their autonomy. Surely, the argument goes, if individuals should have control over anything, they should have it over their own bodies, although it may be argued that there is an inconsistency in using an autonomy argument to bring to an end the conditions of exercise of autonomy, namely bodily life.

Euthanasia can also be argued for on the grounds of beneficence, or on consequentialist grounds (that is, as a means to reduce suffering by removing conditions of the possibility of that suffering persisting). This type of argument can be used to justify non-voluntary euthanasia as well as voluntary euthanasia. Where the individual is unable to express a choice, then the incompetent individual could be denied access to help if autonomy were the only type of argument appealed to; they may be granted help if an argument of beneficence or consequentialism is relied upon. Whether the consequentialist argument could be used to justify involuntary euthanasia is much more contentious.

Arguments Against Euthanasia

The autonomy and beneficence arguments are strong, but may be deployed in a different way on the other side of the debate. If what is regarded as important is autonomy, then the autonomy of persons asked to carry out euthanasia must also be considered. They have to agree that this is a course of action they are prepared to undertake. So the fact that someone of sound mind requests euthanasia does not settle the question if the person who is being asked to perform the action does not agree.

From a consequentialist perspective, it has to be admitted that while in individual cases the best result may appear to be achieved by euthanizing someone whose life has become not worth living, this judgment is fraught with difficulty. In the case of non-voluntary euthanasia, a judgment is made in the absence of a person's own request, when it might well be argued that the benefit of the doubt should count for life. To make the judgment that another person's life is not worth living invites the charge of "playing God."

Partly for this reason, involuntary euthanasia has few supporters: The autonomy argument speaks against it. In the battlefield case it is necessary to assume, if it is to count as involuntary, that the doctor is acting against the express wishes of a soldier, who may be begging for help to save his life. In such a case the doctor may be presumed to make a decision based on the realization that this is not possible.

Apart from the consequences of euthanasia of whatever type for the individual killed, there may be side effects on others. These include worries about hardening the attitudes of those involved in the killing and a gradual lessening in society of respect for life. If euthanasia were widely practiced (as has been the case in some societies), there might be pressure on some people (for example, the elderly and infirm) to agree to request "voluntary" euthanasia. This may be regarded as evidence of a "slippery slope" from voluntary to what could be construed as involuntary euthanasia, because if people feel pressured to consent then their voluntariness is undermined.

In addition, there is the argument based on professional roles. Should health care professionals, who have been trained to cure and to care, use their skills for killing?

The strongest objection to euthanasia, however, derives from the view that killing is wrong *in itself*, on the grounds of the sanctity of life. If life is sacred, then it is wrong deliberately to take a human life. Of course this principle is very difficult to uphold in all circumstances, although people differ about the nature of potential exceptions such as self-defense, war, and capital punishment. Because of the difficulty of upholding an absolute prohibition on taking the life of another, various distinctions have been proposed, including the active-passive distinction already mentioned, and the doctrine of double effect.

A Moral Distinction Between Active and Passive Euthanasia?

According to the active-passive distinction, there is a moral difference between a deliberate act to end someone's life and allowing them to die. To a certain extent the issues here have increased in complexity with medical knowledge and scientific advance, for example, in the light of greater sophistication of use of drugs to control pain, which may at the same time hasten death. From a consequentialist perspective it has been argued that there is no moral difference between killing and allowing to die, because the ultimate outcome is the same—the person is dead. In fact, when side effects are taken into account, the consequences of allowing to die rather than killing might be worse in terms of distress to all concerned. If what is aimed at is a kind and peaceful death, a quick deliberate act may be more merciful than a long-drawn out "allowing to die."

From the perspective of a deontological tradition, however, the quality of the act is what is important. One case is a deliberate killing. The other allows nature to take its course. In some cases, however, it is clear that more than "allowing nature to take its course" is involved, even where it is claimed that deliberate killing is avoided. It is here that the doctrine of the double effect becomes relevant.

Doctrine of the Double Effect

The doctrine of the double effect presupposes that an action can have two kinds of effect: intended and foreseen. Whereas it is claimed that it is always wrong intentionally to do a bad act, it is sometimes permissible to do an act *foreseeing* that bad consequences will ensue. As applied to euthanasia, the point would be that while it is always wrong intentionally to kill, it may be permissible to give drugs to relieve pain, even foreseeing that death will be hastened as a result.

While in some cases this doctrine may appear to give intuitively the right result, and while it has indeed influenced medical practice to a considerable extent, it poses several problems. First, how does one know what constitutes the class of bad acts to be absolutely prohib-

ited? There is no agreement that deliberate killing is ruled out in all circumstances. Second, how does one distinguish between an intended and a foreseen consequence, and indeed between an act and its consequences? There need to be some limits to the freedom to describe the action in certain ways, otherwise it could be open to an agent to deny that any undesirable consequences were intended. The British case of Dr. Cox (Rv Cox [1992] 12 BMLR 38) concerned a physician who administered potassium chloride to a patient who was suffering from intractable pain. That drug does not have pain-relieving properties, so it was not open to the doctor to claim that it was given for that purpose. Had he administered morphine, he might have been able to rely on the doctrine of double effect. While philosophers have heavily criticized the doctrine, it has been influential in law.

Conclusion

From a moral point of view the strongest arguments in favor of euthanasia are to benefit an individual, whether to respect their autonomy or to prevent suffering. At the societal level, however, there are serious concerns about abuse of euthanasia for the benefit not of those killed, but of third parties or society especially in the light of issues about scarce health resources and health inequalities and concerns about these as influencing factors. So controversy continues about the practice of voluntary euthanasia according to guidelines such as in the Netherlands, for example. Despite widespread acknowledgement of the benefit to those with intractable suffering, there are also historical precedents of abuse, which lead opponents to argue that one should err on the side of preserving life. These concerns have to be taken into account in considering proposals for legalization.

RUTH CHADWICK

SEE ALSO *Death and Dying; Dignity; Euthanasia in the Netherlands; Nazi Medicine; Right to Die.*

BIBLIOGRAPHY

Dworkin, Ronald. (1993). *Life's Dominion: An Argument about Abortion, Euthanasia, and Individual Freedom.* New York: Knopf.

Foot, Philippa. (1978). *Virtues and Vices: And Other Essays in Moral Philosophy.* Oxford: Blackwell.

Kuhse, Helga. (1991). "Euthanasia." In *A Companion to Ethics*, ed. Peter Singer. Oxford: Blackwell.

Kuhse, Helga, and Peter Singer. (1985). *Should the Baby Live?: The Problem of Handicapped Infants*. Oxford: Oxford University Press.

Rachels, J. (1986). "Active and Passive Euthanasia." In *Applied Ethics*, ed. Peter Singer. Oxford: Oxford University Press.

Singer, Peter. (1995). *Rethinking Life and Death: The Collapse of Our Traditional Ethics*. New York: St. Martin's Press.

EUTHANASIA IN THE NETHERLANDS

• • •

In the Netherlands, euthanasia is understood to mean termination of life by a physician at the request of a patient. It is to be clearly distinguished from withdrawing from treatment when further medical intervention is pointless, allowing nature to take its course. The latter is normal and accepted medical practice, as is the administration of drugs necessary to relieve pain even in the knowledge that they may have the side effect of hastening death. It should be emphasized that both termination of life upon request and assisting at a suicide are prohibited in the Netherlands. But in the Dutch penal code a special ground for exemption from criminal liability has been developed for physicians who terminate a patient's life on request or assist in a patient's suicide, provided they satisfy the due-care criteria formulated in an act that went into effect in April 2002. This regulation on euthanasia—called the Termination of Life on Request and Assisted Suicide (Review Procedures) Act—is clearly a political compromise between Dutch liberals and Social Democrats, on the one hand, and the Christian Democrats, on the other. If this act had wholly decriminalized euthanasia it would not have received Christian Democrat support. In Belgium, the second country with legislation on euthanasia, the practice is not very different from that in the Netherlands with one exception: Premature termination of life is not considered a criminal act.

Theory and Practice

Pain, degradation of life, and the longing to die with dignity are the main reasons why patients request euthanasia. The initiative is on the part of the patient. To put it bluntly, without such a request it is a matter of murder. People in the Netherlands, as in other advanced countries, are living longer lives, so that, for example, cancer and its pains claim a rising proportion of victims. It should be emphasized that people in the Netherlands do not request euthanasia out of concern at the cost of treatment, because everyone is fully insured under the social security system.

When dealing with a patient's request for euthanasia, physicians must observe the following due-care criteria. They must (1) be satisfied that the patient's request is voluntary and well-considered; (2) be satisfied that the patient's suffering is unbearable and that there is no prospect for improvement; (3) inform the patient of his or her situation and further prognosis; (4) discuss the situation with the patient and come to the joint conclusion that there is no other reasonable solution; (5) consult at least one other physician with no connection to the case, who must then see the patient and state in writing that the attending physician has satisfied the due-care criteria listed in the four points above; and (6) exercise due medical care and attention in terminating the patient's life or assisting in his or her suicide.

Regional review committees (appointed by the Minister of Justice and the Minister of Health, Welfare and Sport) assess whether physicians' actions satisfy these criteria. If the assessment is positive, the Public Prosecution Service will not be informed and no further action will be taken. But if a review committee finds that a physician has failed to satisfy the statutory due-care criteria, the case will be referred to the Public Prosecution Service and the Health Inspectorate. These two bodies will then consider whether the physician should be prosecuted. The existence of a close physician–patient relationship is taken as premise. Physicians may perform euthanasia only on patients in their care. They must know their patients well enough to be able to determine whether the request for euthanasia is both voluntary and well-considered, and whether the suffering is unbearable and without prospect for improvement.

Even in cases in which patients are receiving care of the highest quality, they may still regard their suffering as unbearable and plead with their physicians to terminate their lives. In such cases, euthanasia could represent a dignified conclusion to good palliative care. There is, however, no requirement that physicians comply with the requests for euthanasia. Physicians can refuse to terminate life; after all it is not a normal medical procedure. The ability to refuse a request for euthanasia or assisted suicide guarantees physician's freedom of conscience. If a physician does not want to be involved, he or she is obligated to refer the patient to a colleague.

It is the task of the physician to try to imagine what the patient is feeling and based on his or her medical experience attempt to assess the patient's suffering objectively. Unbearable suffering also includes psychological suffering. If a patient has a psychological illness and his or her suffering is not primarily caused by a physical complaint, it is difficult to assess objectively whether a request for euthanasia is voluntary and well-considered. In such cases, the attending physician should consult two independent specialists, at least one of whom must be a psychiatrist, and they must personally examine and interview the patient. The presence of dementia or some other such condition is not in itself a reason to comply with a request for termination of life or assisted suicide. For some people, however, the very prospect of one day suffering from dementia and the eventual associated loss of personality and dignity is sufficient reason to make an advance directive covering this possibility. Each case needs to be individually assessed to decide whether, in the light of prevailing medical opinion, it can be viewed as entailing unbearable suffering for the patient with no prospect for improvement. In response to questions on this subject in the Dutch Parliament, the Minister of Health, Welfare and Sport stated that dementia can make the patient's quality of life unacceptable if the patient him- or herself regards his or her condition in this way, but that even then the physician must decide whether the patient's suffering is unbearable and without prospect for improvement in the light of prevailing medical opinion.

The aim of the Dutch policy is to bring matters into the open, to apply uniform criteria in assessing cases of euthanasia, and hence to ensure that maximum care is exercised in such cases. The price for this openness is a lot of formalistic procedures with no respect to content or guarantee of care. In this area the regional review committees function quite adequately. But not all end-of-life issues are covered by the issue of euthanasia. In the concentration on euthanasia and assisted suicide all forms of sedation with and without consent of the patient fall outside the scope and competence of the assessing committees. Palliative care is not concentrated on recovery but on alleviation of pain and other symptoms. Palliative and in particular terminal sedation may come close to euthanasia. For physicians who do not want to get involved with euthanasia for religious, bureaucratic, or whatever reasons these forms of sedation are a refuge.

Reflective Implications

Traditionally, as specified in the Hippocratic oath, physicians ended their care at the deathbed. Once death was inevitable, the office of the physician—which was to help people avoid the evils of sickness, physical deficiencies, the ailments of old age, and a premature death—had come to an end. In modern society the physician's task has been enormously extended so that the entire life of a human has been brought under a medical regime. Medical examinations are the order of the day. It is impossible to avoid the physician when going to school, participating in sports, holding down a job, taking out life insurance, and so on. Whomever is unwell hurries to make at least a short visit to the doctor or hospital in order to make use of the paraphernalia of modern medicine. Human health is controlled as a matter of routine. Not only has life undergone medicalization, but dying has also been brought under the medical regime. The result is that human death has become artificial. Many bitter deaths might be a product of modern medical science because postponement of death as a result of medical monitoring in a sense requires its toll. In the early twenty-first century, a natural death is likely an exception to the norm. Thus in normal cases the physician swings the scepter at a person's last bed by prompting the possibilities and impossibilities left to him or her. In some municipalities in the Netherlands in the early twentieth century, more than half of all deceased had no medical intervention while dying. That is now inconceivable.

Physicians play the role of experts in the end-of-life decisions. In some sense they act as examiners, while their patients attempt to pass an exam. A physician scrutinizes whether the wish to die is voluntary, whether it is well considered, whether the wish has been long-standing, whether it is not liable to emotions, whether the suffering is unacceptable to the patient, and so on. What at first sight seems to be a matter of self-determination turns out to be a matter of complete dependency. It is not surprising that well-educated people stand a better chance of having their request granted than those who are less educated. Physicians—in former times absent at the deathbed but now prominently present—find themselves in the position of the expert only because they have access to lethal drugs. This technologically privileged position maneuvers them at the same time into the role of moral examiner. For patients, the inaccessibility of lethal drugs makes the whole procedure into a technological adventure in which they are incompetent. Being alienated from nature, patients have no knowledge about the herbs and fruits in their own garden. Confronted with these final questions they have to throw themselves into the arms of the experts. Tried and tested methods out of ancient times have been blotted out.

The issue of unbearable pain on the deathbed is often technologically transformed into a mild death. Physician and patient talk about pain and how to get rid of it along technological lines. In contemporary technological society humans cannot deal with pain in another way. The opinion that pain should be tolerated, alleviated, and interpreted is no longer widely held. The medicalization of pain robs a culture of an integrative program of pain treatment. In traditional societies opium, acupuncture, or hypnosis were means of alleviating pain, but they were always put into practice in combination with language, rites, and myth. Most people who are morally against euthanasia support sedative treatment. Their position shows how difficult it is to leave the technological society behind, because from a technological point of view euthanasia is not very different from sedative treatment. In practice the outcome is often the same, but only in the mind of the physician does one find the difference between euthanasia and sedative treatment.

A society that denies a patient's request for euthanasia would best abstain from modern technological medical care. Living in a technological society may be compared with climbing mountains. People who have ascended too high must descend very carefully. Under some circumstances a descent may be more difficult than the ascent. When patients cannot tolerate pain any longer, who dares to ask them to interpret the meaning of their pain? Has the modern technological society not abandoned such questions or left them to personal decisions?

PIETER TIJMES

SEE ALSO *Bioethics; Death and Dying; Euthanasia; Medical Ethics; Right to Die.*

BIBLIOGRAPHY

Griffiths, John; Alex Bood; and Heleen Weyers. (1998). *Euthanasia and Law in the Netherlands.* Amsterdam: Amsterdam University Press.

van der Heide, Agnes; Luc Deliens; Karin Faisst; et al. (2003). "End-of-Life Decision-Making in Six European Countries: Descriptive Study." *Lancet* 362(9381): 345–350. Includes discussion of euthanasia in the Netherlands.

van Hees, Martin, and Bernard Steunenberg. (2000). "The Choices Judges Make: Court Rulings, Personal Values, and Legal Constraints." *Journal of Theoretical Politics* 12(3): 305–324.

Weyers, Heleen. (2003). *Euthanasie: Het proces van rechtsverandering* [Euthanasia: The process of legal change]. Amsterdam: Amsterdam University Press. The best survey of the Dutch discussion on euthanasia.

EVOLUTIONARY ETHICS

• • •

Evolutionary ethics rests on the idea that ethics expresses a natural moral sense that has been shaped by evolutionary history. It is a scientific understanding of ethics as founded in human biological nature.

The first full development of evolutionary ethics came from Charles Darwin (1809–1882) and Herbert Spencer (1820–1903) in the nineteenth century. At the beginning of the twentieth century, the Darwinian theory of ethics was renewed and deepened by Edward Westermarck (1862–1939). At the end of the twentieth century, this Darwinian tradition of ethical philosophy was reformulated by Edward O. Wilson, Robert McShea, Frans de Waal, and others.

Philosophers arguing over the ultimate grounds of ethics have been divided into Aristotelian naturalists and Platonic transcendentalists. The transcendentalists find the ground of ethics in some reality beyond human nature, while the naturalists explain ethics as grounded in human nature itself. In this enduring debate, proponents of evolutionary ethics belong to the Aristotelian tradition of ethical naturalism, while their strongest opponents belong to the Platonic tradition of ethical transcendentalism. (Of course, Aristotelians who reject evolutionary reasoning would also reject evolutionary ethics.)

The history of evolutionary ethics can be divided into three periods, with Darwin initiating the first period, Westermarck the second, and Wilson the third.

Darwin's View

As part of his theory of the evolution of life by natural selection, Darwin wanted to explain the evolution of human morality. From his reading of Adam Smith (1723–1790), David Hume (1711–1776), and other philosophers who saw morality as rooted in moral emotions or a moral sense, Darwin concluded that this moral sense could be understood as a product of natural selection. As social animals, human beings evolved to have social instincts. As rational animals, human beings evolved the rational capacity to reflect on their social instincts and formulate those moral rules that would satisfy their social instincts. Human survival and reproduction required that parents care for their offspring, and the social nature of human beings could be explained as an extension of parental feelings of sympathy to embrace ever larger groups of individuals. In his *Descent of Man* (1871), Darwin concluded: "Ultimately our moral sense or conscience becomes a highly complex sentiment—originating in the social instincts, largely guided by the approbation of our fellow-men, ruled by reason, self-interest, and in later times by deep religious feelings, and confirmed by instruction and habit" (Darwin 1871, Vol. 1, pp. 165–166).

Herbert Spencer (1820–1903) generally agreed with Darwin's evolutionary ethics, yet Spencer put more emphasis than did Darwin on evolution through the inheritance of acquired traits. And unlike Darwin, Spencer saw all of evolutionary history as moving toward a pre-determined end of perfection in which human societies would become so cooperative that they would achieve perpetual peace.

When *The Descent of Man* was published, Darwin's naturalistic theory of morality was attacked by biologist George Jackson Mivart (1827–1900), who claimed that there was an absolute separation between nature and morality. Although Darwin's theory of evolution could explain the natural origins of the human body, Mivart insisted, it could not explain the human soul as a supernatural product of divine creation, and therefore it could not explain human morality, which depended on the soul's freedom from natural causality. Mivart followed the lead of Immanuel Kant (1724–1804) in arguing that the realm of moral duty must be separated from the realm of natural causality, thus adopting a version of the distinction between values and facts.

This dispute between Darwin and Mivart shows the conflict between the naturalistic tradition of moral thought and the transcendentalist tradition that runs throughout moral philosophy and throughout the debate over evolutionary ethics. According to Plato (in *The Republic*), one cannot know what is truly good until one sees that all of the diverse goods of life are only imperfect imitations of the Idea of the Good, which is universal, absolute, and eternal. In Plato's theological version of this teaching, God as the Creator of the cosmos is said to be a providential caretaker of human affairs who judges human beings after death, rewarding the good and punishing the bad. Aristotle (in the *Nicomachean Ethics*) rejected this Platonic Idea of the Good, because he could not see any sense in saying there is a transcendent good separated from all the diverse natural goods that human beings seek. Looking to the common-sense experience of human beings, Aristotle thought that the ultimate end for which human beings act is happiness, and happiness would be the human flourishing that comes from the harmonious satisfaction of human desires over a whole life. Like Smith and Hume, Darwin followed the Aristotelian tradition in rooting morality in natural desires and emotions. Like Kant, Mivart followed the Platonic tradition in positing a moral *ought* belonging to a transcendent world of moral freedom beyond the empirical world of natural causes.

Thomas Huxley (1825–1895), one of Darwin's most fervent supporters, initially defended Darwin's evolutionary ethics against Mivart's criticisms. But eventually, in his 1893 lecture on "Evolution and Ethics," Huxley adopted Mivart's transcendentalist position. Because of the "moral indifference of nature," Huxley declared, one could never derive moral values from natural facts. He argued that "the ethical process of society depends, not on imitating the cosmic process, still less in running away from it, but in combating it," and thus building "an artificial world within the cosmos (Paradis and Williams 1989, pp. 117, 141)."

Westermarck's Views

After Huxley's attack, Darwin's naturalistic ethics was kept alive in the early-twentieth century by philosophers such as Westermarck. In his *History of Human Marriage* (1889), Westermarck explained the desires for marriage and family life as founded in moral emotions that had been shaped by natural selection as part of the biological nature of human beings. His most famous idea was his Darwinian explanation of the incest taboo, which can be summarized in three propositions. First inbreeding tends to produce physical and mental deficiencies in the resultant offspring, which lowers their fitness in the Darwinian struggle for existence. Second, as a result of the deleterious effects of inbreeding, natural selection has favored the mental disposition to feel an aversion toward sexual mating with those with whom one has been an intimate associate from early childhood. Third this natural aversion to incest has been expressed culturally as an incest taboo. Consequently, in all human societies, there is a strong tendency to prohibit fathers marrying daughters, mothers marrying sons, and brothers marrying sisters, although there is more variation across societies in the rules governing the marriage of cousins and others outside the nuclear family. (In 1995 Anthropologist Arthur Wolf surveyed the growing evidence confirming Westermarck's Darwinian theory of incest avoidance.)

Westermarck believed all of the moral emotions could be ultimately explained in the same way he had explained the abhorrence of incest. As animals formed by natural selection for social life, humans are inclined to feel negative about conduct perceived as painful, and positive toward conduct perceived as pleasurable. The mental dispositions to feel such emotions evolved in animals by natural selection because these emotions promote survival and reproductive fitness: Resentment helps to remove dangers, and kindly emotion helps to secure benefits. For the more intelligent animals, these dispositions have become conscious desires to punish enemies and reward friends.

Moral disapproval, Westermarck argued, is a form of resentment, and moral approval is a form of kindly emotion. In contrast to the non-moral emotions, however, the moral emotions show apparent impartiality. (Here he shows the influence of Smith's idea that the moral sentiments arise when we take the perspective of the *impartial spectator*.) If a person feels anger toward an enemy or gratitude toward a friend, these are private emotions that express personal interests. In contrast, if a person declares some conduct of a friend or enemy to be good or bad, he or she implicitly assumes that the conduct is good or bad regardless of the fact that the person in question is a friend or enemy. This is because it is assumed that when conduct is determined to be good or bad, a person would apply the same judgment to other people acting the same way in similar circumstances, independently of the effect on that individual. This apparent impartiality characterizes the moral emotions, Westermarck explained, because "society is the birth-place of the moral consciousness" (1932, p. 109). Moral rules originated as tribal customs that expressed the emotions of an entire society rather than the personal emotions of particular individuals. Thus moral rules arise as customary generalizations of emotional tendencies to feel approval for conduct that causes pleasure and disapproval for conduct that causes pain.

Although Westermarck stressed the moral emotions as the ultimate motivation for ethics, he also recognized the importance of reason in ethical judgment. "The influence of intellectual considerations upon moral judgments is certainly immense" (1932, p. 147). Emotions, including the moral emotions, depend upon beliefs, and those beliefs can be either true or false. For example, a person might feel the moral emotion of disapproval toward another that he or she believes has injured a friend, but if that same person discovers by reflection that an injury was accidental and not intentional, or that an action did not actually cause any injury at all, the disapproval vanishes. Moreover, because moral judgments are generalizations of emotional tendencies, these judgments depend upon the inductive use of human reason in reflecting on emotional experience.

Wilson's View

By the 1970s, however, there was little interest in the ethical naturalism of people such as Westermarck, and the transcendentalist tradition had largely conquered the intellectual world of philosophers and social scientists. Ethics and politics were assumed to belong to an autonomous human realm of reason and culture that transcended biological nature. This could be explained as a reasonable reaction against the morally repulsive conduct associated with "Social Darwinism" in the first half of the twentieth century.

This also explains why the publication of Wilson's book *Sociobiology* in 1975 provoked great controversy. Wilson defined *sociobiology* as the scientific study of the biological bases of the social behavior of all animals, including human beings. On the first page of the book, he claimed that ethics was rooted in human biology. He asserted that the deepest human intuitions of right and wrong are guided by the emotional control centers of the brain, which evolved through natural selection to help the human animal exploit opportunities and avoid threats in the natural environment.

One of the first serious responses to Wilson's proposal for sociobiological ethics was a conference in Berlin in 1977 titled "Biology and Morals." The material from this conference was later published as a book edited by Gunther Stent. In his introduction, Stent began by contrasting the "idealistic ethics advocated by Plato" and the "naturalistic ethics advocated by Aristotle." He suggested that those people who belonged to the *idealistic* tradition would reject Wilson's sociobiological ethics, while those belonging to the *naturalistic* tradition would be more inclined to accept it.

In this book Thomas Nagel, a philosopher, showed the reaction of the Platonic transcendentalist. He rejected sociobiological ethics because it failed to see that ethics is "an autonomous theoretical subject" (Nagel 1978) such as mathematics that belongs to a transcendent realm of pure logic. On the other side of this debate, Robert McShea, a political scientist, independently welcomed Wilson's sociobiological ethics as providing scientific confirmation for the insight of Aristotle and Hume that ethics is rooted in the emotions and desires of human biological nature (Mcshea 1978). All writing on this subject that followed, as of 2004, fell into one of these two intellectual camps.

The transcendentalist critics of evolutionary ethics include most of the leading proponents of evolutionary psychology, which applies Darwin's theory of evolution in explaining the human mind as an adaptation of human nature as shaped in evolutionary history. Evolutionary psychologists such as George Williams (1989) claim that ethics cannot be rooted in human nature because of the unbridgeable gulf between the selfishness of our natural inclinations and the selflessness of our moral duties. As the only rational and cultural animals, human beings are able to suppress their natural desires and enter a transcendent realm of pure moral duty. Like Huxley, Williams and other theorists of evolutionary psychology reject Wilson's sociobiological ethics because they think that ethics requires a transcendence of human biology through culture and reason. Unlike Wilson and Darwin, therefore, the proponents of evolutionary psychology do not believe that biological science can account for the moral conduct of human beings.

Objections and Replies

There are at least three major objections to this Darwinist view of morality. One common criticism of evolutionary ethics is that it promotes genetic determinism. If all choices are ultimately determined by genetic causes, that would seem to deny that human actions can be freely chosen, which would deny the fundamental presupposition of moral judgment that people can be held responsible for their moral choices.

But if genetic determinism means that behavior is rigidly predetermined by genetic mechanisms, so that neither individual learning nor social culture has any influence, then defenders of evolutionary ethics are not genetic determinists. What the genes prescribe, Wilson would say, is certain propensities to learn some behaviors more easily than others. Human nature, Wilson explains in his 1998 book *Consilience,* is not a product of genes alone or of culture alone. Rather, human nature is constituted by "the epigenetic rules, the hereditary regularities of mental development that bias cultural evolution in one direction as opposed to another, and thus connect the genes to culture" (p. 164). Consequently human behavior is highly variable across individuals and across societies, but the genetic nature of the human species is manifested in the general pattern of behavior.

So, for example, the natural human propensity to incest avoidance is actually a propensity to learn a sexual aversion to those with whom one has been raised. The precise character of the incest taboo will vary greatly across societies depending on the diversity in family life and kinship systems. For instance some societies will forbid marrying first cousins, while others will not. Yet the tendency to forbid the marriage of brother

and sister or of parent and child will be universal or almost universal. Moreover one can deliberate about the rules of incest avoidance by reflecting on the relevant facts and emotions. When the incest taboo is formally enacted in marriage law, legislators must decide what counts as incest and what does not.

Proponents of evolutionary ethics would say that people are not absolutely free of the causal regularities of nature. Exercising such absolute freedom from nature—acting as an uncaused cause—is possible only for God. But human beings are still morally responsible for their actions because of the uniquely human capacity for reflecting on motives and circumstances and acting in the light of those reflections.

A second criticism of evolutionary ethics is that it promotes a crudely *emotivist* view of ethics as merely an expression of arbitrary emotions. After all, from the first paragraph of *Sociobiology,* Wilson speaks of ethics as controlled by "the emotional control centers in the hypothalamus and limbic system of the brain" (1975, p. 31). He repeatedly identifies the ultimate foundation for ethical codes as "our strongest feelings of right and wrong" (Ruse and Wilson 1994, p. 422). "Murder is wrong" might be just another way of saying "I don't like murder." Does that deny the sense of moral obligation as something more than just an expression of personal feelings?

People might also wonder how an emotivist ethics would handle the response of those with deviant emotions, such as that of psychopaths who do not show the normal emotions of guilt, shame, or sympathy. How can society condemn them if there are no objective moral norms beyond emotion? Moreover, how does society resolve the emotional conflicts that normally arise within and between individuals? How does society rank some emotional desires as higher than others? Such problems lead many philosophers to dismiss emotivist ethics as incoherent.

In reply to this criticism, the defender of evolutionary ethics might again consider the case of the incest taboo. If Westermarck is right, moral condemnation of incest arises from an emotion of sexual aversion toward those with whom one has been raised in early childhood. This personal emotion of disgust becomes a moral emotion of disapprobation when generalizing emotional experience into an impartial social rule: People judge that incest is bad not just for themselves but for all members of society in similar circumstances. Reason plays a part in generalizing these emotions. By reason people must formulate what counts as incest. Generally society condemns the sexual union of siblings or of parents and children. But whether one condemns the marriage of cousins will depend on the circumstances of kinship and judgments about whether the consequences are good or bad for society.

Normally most human beings will feel no sexual attraction to their closest kin. Those who do will usually feel a conflict between their sexual desire and their fear of violating a social norm that expresses deep emotions, and this fear of social blame will usually override their sexual interest. Those who do violate the incest taboo will be punished by a disapproving society. A few human beings might feel no emotional resistance to incest at all. They might be psychopathic in lacking the moral emotions of guilt and shame that are normal for most people. If so then society will treat them as moral strangers, as people who are not restrained by social persuasion, and who therefore must be treated as social predators.

The main point for those favoring evolutionary ethics is that although the moral emotions are relative to the human species, they are not arbitrary. One can easily imagine that if other animal species were to develop enough intellectual ability to formulate moral rules, some of them might proclaim incest to be a moral duty, because the advantages of inbreeding for bonding between kin might be greater than the disadvantages. But human beings are naturally inclined to acquire an incest taboo, and therefore to condemn those individuals who deviate from this central tendency of the species.

Emphasizing emotion in moral experience denies the transcendentalist claim that morality depends on pure reason alone. The 1994 work of Antonio Damasio and that of other neuroscientists suggests that the emotional control centers of the brain are essential for normal moral judgment. Psychopathic serial killers can torture and murder their victims without feeling any remorse. Yet they are often highly intelligent people who suffer no deficits in their cognitive capacities. Their moral depravity comes not from any mistakes in logical reasoning but from their emotional poverty in not feeling moral emotions such as guilt, shame, love, and sympathy.

A third objection to evolutionary ethics is that it fails to recognize the logical gap between *is* and *ought*, between natural facts and moral values. Determining that something is the case does not say that it ought to be so. A scientific description of a behavior is not the same as a moral prescription for that behavior.

In reply to this objection, proponents of evolutionary ethics might agree with Hume's interpretation of the is/ought dichotomy, which claims that pure reasoning about factual information cannot by itself move people to moral judgments. Moral motivation requires moral emotions. Those moral emotions, however, manifest propensities of human nature that are open to scientific study.

The incest taboo illustrates this. The factual information about inbreeding does not by itself dictate any moral judgment. If society did not feel moral emotions of disgust toward inbreeding among human beings, it would not be condemned as immoral. Even the factual information about the deleterious effects of inbreeding would not incur moral condemnation if people did not feel sympathy for human suffering.

The move from facts to values is not logical but psychological. Because people have the human nature that they do, which includes propensities to moral emotions, they predictably react to certain facts with strong feelings of approval or disapproval, and the generalization of those feelings across a society constitutes moral experience.

If society decided that evolutionary ethics was correct about ethics being grounded in emotions, this would influence assessment of the technologies of emotion. People might decide, as many science fiction authors have suggested, that robots could become moral beings only if they could feel human emotions. Society might also wonder about the moral consequences of new biomedical technologies for manipulating emotions through drugs and other means. People might question whether the technology of birth control could obviate the need for the incest taboo.

LARRY ARNHART

SEE ALSO *Aristotle and Aristotelianism; Darwin, Charles; Determinism; Emotion; Evolution-Creationism Debate; Plato; Sociobiology; Spencer, Herbert.*

BIBLIOGRAPHY

Arnhart, Larry. (1998). *Darwinian Natural Right: The Biological Ethics of Human Nature*. Albany: State University of New York Press.

Arnhart, Larry. (2001). "Thomistic Natural Law as Darwinian Natural Right." In *Natural Law and Modern Moral Philosophy*, eds. Ellen Frankel Paul, Fred D. Miller, and Jeffrey Paul. Cambridge, UK: Cambridge University Press.

Casebeer, William D. (2003). *Natural Ethical Facts: Evolution, Connectionism, and Moral Cognition*. Cambridge, MA: MIT Press.

Damasio, Antonio R. (1994). *Descartes' Error: Emotion, Reason, and the Human Brain.* New York: G. P. Putnam's Sons.

Darwin, Charles. (1871). *The Descent of Man,* 2 vols. London: John Murrray.

Farber, Paul Lawrence. (1994). *The Temptations of Evolutionary Ethics.* Berkeley: University of California Press.

McShea, Robert J. (1978). "Human Nature Theory and Political Philosophy." *American Journal of Political Science* 22: 656–679.

Mivart, George Jackson. (1973). "Darwin's *Descent of Man.*" In *Darwin and His Critics,* ed. David Hull. Chicago: University of Chicago Press.

Nagel, Thomas. (1978). "Morality as an Autonomous Theoretical Subject." In *Morality as a Biological Phenomenon,* ed. Gunther Stent. Berkeley: University of California Press.

Paradis, James, and George C. Williams, eds. (1989). *Evolution and Ethics.* Princeton, NJ: Princeton University Press.

Ruse, Michael, and Edward O. Wilson. (1994). "Moral Philosophy As Applied Science." In *Conceptual Issues In Evolutionary Biology,* ed. Elliott Sober. Cambridge, MA: MIT Press.

Stent, Gunther, ed. (1978). *Morality as a Biological Phenomenon.* Berkeley: University of California Press.

De Waal, Frans. (1996). *Good Natured: The Origins of Right and Wrong in Humans and Other Animals.* Cambridge, MA: Harvard University Press.

Westermarck, Edward. (1922). *The History of Human Marriage,* 5th edition, 3 vols. New York: Allerton.

Westermarck, Edward. (1932). *Ethical Relativity.* London: Kegan Paul, Trench, Trubner and Company.

Williams, George. (1989). "A Sociobiological Expansion of Evolution and Ethics." In *Evolution And Ethics,* eds. James Paradis and George C. Williams. Princeton, NJ: Princeton University Press.

Wilson, Edward O. (1975). *Sociobiology: The New Synthesis.* Cambridge, MA: Harvard University Press.

Wilson, Edward O. (1998). *Consilience: The Unity of Knowledge.* New York: Alfred A. Knopf.

Wolf, Arthur P. (1995). *Sexual Attraction and Childhood Association: A Chinese Brief for Edward Westermarck.* Stanford, CA: Stanford University Press.

EVOLUTION–CREATIONISM DEBATE

• • •

The evolution-creationism debate deals with attempts to explain the ultimate causes of order in the living world. Some people think that order arose from natural evolutionary causes. Others think it arose from divine creative intelligence. A third group thinks it arose from divine intelligence working through natural causes.

Nature of the Debate

This debate can be traced back as far as ancient Greece, where it appears in Plato's philosophical dialogues. More recently the debate has been between followers of the Bible and followers of the scientist Charles Darwin (1809–1882). The opening chapters of the Bible relate how God created the world in six days and created human beings in his image. In *The Origin of Species* (1859) and *The Descent of Man* (1871) Darwin discusses how all the forms of life could have evolved by natural law, in which the heritable traits that enhanced reproductive success were naturally selected over long periods. The evolution-creationism debate entails comparing these two scenarios of the origins of life. Some people believe that both histories are true and therefore can be compatible. Some believe that if one of the two is true, the other must be false.

This becomes a debate over the ethical implications of modern science because much of the disagreement turns on judgments about the ethical consequences of accepting one or both views as true. On one side many of those who defend creationism fear that Darwinian evolution promotes a materialistic view of the world that is ethically corrupting, because it denies the moral dignity of human beings as created in God's image. On the other side, some see creationism as promoting fundamentalist religion and attacks on science.

This has also become a legal and political debate, particularly in the United States, where people have argued about whether creationism should be taught to students in public schools as an alternative to Darwinian evolution. Some public opinion surveys have reported that about half the people in the United States believe that human beings were created by God approximately 10,000 years ago; that would deny the Darwinian belief that the human species evolved from an apelike ancestral species millions of years ago.

History of the Debate

In Plato's dialogue *The Laws* (Book 10) the Athenian character warns against natural philosophers who teach that the ultimate elements in the universe and the heavenly bodies were brought into being not by divine intelligence or art but by natural necessity and chance. These natural philosophers teach that the gods and the moral laws attributed to the gods are human inventions. That form of scientific naturalism appeared to subvert the religious order by teaching atheism, subvert the moral order by teaching moral relativism, and subvert the political order by depriving the laws of religious and moral sanction. Plato's Athenian character responds to that threat by arguing for divine intelligent design as the ultimate source of order.

In a later period those influenced by biblical religion adopted Plato's arguments to defend the claim that the divinely intelligent designer of the world was the God of the Bible. However, in the nineteenth century Darwin's theory of evolution by natural selection seemed to explain the apparent design in the living world as arising from purely natural causes without the need for divine creation. This led to the modern debate between evolution and creationism.

In the United States that debate falls into three periods. The first period began in the 1920s when William Jennings Bryan (1860–1925) launched a Christian fundamentalist attack on Darwinism. Bryan was a leading politician, having run three times for the presidency as the Democratic Party's candidate. In 1925 the state legislature in Tennessee made it illegal for any teacher in a public school "to teach any theory that denies the story of the Divine creation of man as taught in the Bible, and to teach instead that man has descended from a lower order of animals" (Larson 1997, p. 50). When John Scopes, a public high school teacher in Dayton, was charged with violating this law, Clarence Darrow (1857–1938), a prominent lawyer who promoted scientific atheism, led the legal team defending Scopes, and Bryan joined the lawyers prosecuting Scopes.

The trial in July 1925 drew public attention around the world. Although Scopes was convicted, his conviction was overturned by a higher court on a technical issue. Bryan died shortly after the trial. Creationist opponents of Darwinian evolution continued to argue their case, although many of them, like Bryan, argued that the six days of Creation in the Bible were not literally six days but rather "ages," so that long periods of time could have elapsed. Some creationists followed Bryan in accepting Darwin's account of evolution by natural law as generally true but still insisted that the emergence of human beings required a miraculous intervention by God to endow them with a spiritual soul that made them superior to all animals.

The second period of the debate was initiated by the publication in 1961 of John Whitcomb and Henry Morris's *The Genesis Flood*. Those authors interpreted the biblical story of Creation as occurring during a literal six-day period that occurred no more than 10,000 years ago. They also argued that the geological record of fossils had been laid down during the worldwide flood reported in the Bible in the story of Noah's ark. Morris and others identified themselves as "scientific creationists," claiming that the Bible as literally interpreted was scientifically superior to Darwin's theory. They supported legislation in some states to require the teaching of "creation science" in public high schools. However, when this was done in Arkansas and Louisiana, federal courts struck down those laws as violating the constitutional separation of church and state because the biblical story of Creation seemed to be a religious doctrine rather than a scientific theory.

The third period of the debate began in 1991 with the publication of Phillip Johnson's *Darwin on Trial*. Johnson, a lawyer and law professor, argued that the scientific evidence is against Darwin's theory and that Darwinians believe the theory only because it supports their atheistic belief that the order in life can be explained by natural laws without the need for divine creation. Johnson also claimed that the complexity of the living world can only be explained as the work of an "intelligent designer" such as the God of the Bible.

Other writers joined this intellectual movement for "intelligent design" as an alternative to Darwinian evolution. In 1996 the biologist Michael Behe published *Darwin's Black Box*, in which he surveyed the evidence for "irreducibly complex" mechanisms in the living world that could not have evolved gradually by Darwinian evolution but could show the work of an "intelligent designer." Later the mathematician and philosopher William Dembski elaborated the formal criteria by which "design" could be detected in nature (Dembski and Kushiner 2001). Since the late 1990s proponents of "intelligent design" have tried to convince public school boards that "intelligent design theory" should be taught in high school biology classes as an alternative to Darwinian science or at least that the weaknesses in the Darwinian arguments should be discussed in schools.

Four Arguments

Beginning with Bryan, the creationist critics of Darwinian science have made four types of arguments: a scientific argument, a religious argument, an ethical argument, and a political argument. Similar kinds of arguments can be found in Plato's *Laws*.

The scientific argument of the creationists is that Darwin's theory is not truly scientific because it is based not on empirical evidence but on a dogmatic commitment to materialistic naturalism. They also claim that creationism is a more scientific view because the complex functional order of the living world provides evidence for an intentional design by a divinely intelligent agent. The irreducible complexity of life cannot be explained through the unintelligent causes of random contingency and natural necessity.

The common mousetrap is Behe's primary example of an irreducibly complex mechanism. It requires at least five parts—a platform, a spring, a hammer, a catch, and a holding bar—and those parts must be arranged in a specific way. If one part is missing or if the arrangement is wrong, the mechanism will not achieve its functional purpose of catching mice. It is known that such a device did not arise by chance or natural necessity; human intelligent agents designed it to catch rodents. Behe claims that many biological mechanisms show the same purposeful arrangement of parts found in human devices such as the mousetrap. This, he thinks, points to an intelligent designer outside nature.

Darwinians would agree with Behe that from an apparently well-designed mousetrap one plausibly can infer the existence of a human intelligent designer as its cause because people have common experience of how mousetraps and other artifacts are designed. However, Darwinians would insist that from an apparently well-designed organic process or entity one cannot infer the existence of a divinely intelligent designer as its cause, because people have no common experience of how a divine intelligence designs things for divine purposes. Religious belief depends on faith in a supernatural reality beyond the world, whereas scientific knowledge depends on reasoning about humankind's sense experience of the natural world. Furthermore, Darwinians would note that creationists or intelligent design theorists never explain the observable causal pathways by which the divine intelligence creates irreducibly complex mechanisms.

The religious argument of the creationists is that Darwinism promotes dogmatic atheism and therefore must be rejected by religious believers. This argument seems to be confirmed by the bold declarations of Darwinian scientists such as Richard Dawkins (1986) that Darwinian science proves the truth of atheism. But it is hard to see how explaining the world through natural causes denies the possibility that God is the ultimate ground of those natural causes. Some Darwinians present evolution as a substitute for religion. Even such a strong defender of evolution as Michael Ruse (2003) has admitted that museums of science that promote evolutionary theory often function as secular temples.

Creationists assume that God was unable or unwilling to execute his design through the laws of nature as studied by Darwinian biologists. However, Christian evolutionists such as Howard Van Till (1999) and others have argued that the Bible presents the divine designer as having given his Creation from the beginning all the formational powers necessary for evolving into the world as it is today. Catholic theologian John Haught (2001) has defended a "theology of evolution" based on ideas from the French Jesuit priest Pierre Teilhard de Chardin (1881–1955) and the British philosopher Alfred North Whitehead (1861–1947). In Haught's theology, evolution suggests that the universe is always in the process of being created as God allows a self-creating world to evolve towards him through time. If this is so, Darwinian science and religious belief are compatible.

The ethical argument of the creationists is that the reductionistic materialism of Darwinian science is ethically degrading. If Darwinians persuade people that they are nothing but animals and therefore are not elevated above other animals by having been created in God's image, people will not respect God's moral law or see the unique moral dignity of human beings. Instead they will become selfish hedonists in the pursuit of their animal desires.

Darwinians respond to this argument by noting that Darwin thought his account of human evolution supported a biological theory of morality rooted in a natural moral sense. As naturally social and rational animals human beings have social instincts that incline them to care for others and have a rational capacity to deliberate about the moral rules that would satisfy their social needs. For example, the human species could not survive if children were not cared for by their parents or by people assuming parental roles. Therefore, one can understand how natural selection has endowed human beings with a natural desire for parental care that supports the moral bond between parent and child. Consequently, Darwinian science sustains morality by showing that it is rooted in human nature.

The political argument of the creationists is that teaching Darwinism in public schools without teaching the creationist criticisms of Darwinism denies the freedom of thought required in a democratic society. Surely, creationists claim, promoting an open discussion in the public schools of the scientific, religious, and ethical debates surrounding Darwinian evolution would help students think for themselves about those important issues.

Some Darwinians reject this argument by claiming that creationism is not science but religion and that the teaching of science should be kept separate from the teaching of religion. However, other Darwinians welcome an open debate. If high school students were free to read writers who defend Darwin's theory along with writers who criticize it, the students could make up their own minds. In the process students might learn how to think through scientific debates and weigh the evidence and arguments for themselves rather than memorizing the conclusions given to them by textbooks and teachers.

LARRY ARNHART

SEE ALSO *Christian Perspectives; Darwin, Charles; Evolutionary Ethics.*

BIBLIOGRAPHY

Beckwith, Francis J. (2003). *Law, Darwinism, and Public Education: The Establishment Clause and the Challenge of Intelligent Design.* Lanham, MD: Rowman and Littlefield. An argument for the constituionality of teaching intelligent design theory in public schools.

Behe, Michael. (1996). *Darwin's Black Box: The Biochemical Challenge to Evolution.* New York: Free Press.

Dawkins, Richard. (1986). *The Blind Watchmaker.* New York: Norton.

Dembski, William A., and James M. Kushiner, eds. (2001). *Signs of Intelligence: Understanding Intelligent Design.* Grand Rapids, MI: Brazos Press. A collection of essays by proponents of intelligent design theory.

Forrest, Barbara, and Paul R. Gross. (2004). *Creationism's Trojan Horse: The Wedge Of Intelligent Design.* New York: Oxford University Press. A history of the intelligent design movement by two opponents.

Haught, John F. (2001). *Responses to 101 Questions on God and Evolution.* New York: Paulist Press.

Johnson, Phillip. (1991). *Darwin on Trial.* Downers Grove, IL: Intervarsity Press.

Larson, Edward J. (1997). *Summer for the Gods: The Scopes Trial and America's Continuing Debate over Science and Religion.* New York: Basic Books.

Miller, Kenneth R. (1999). *Finding Darwin's God: A Scientist's Search for Common Ground between God and Evolution.* New York: HarperCollins. A Christian and a biologist, Miller defends Darwinism as compatible with religious belief.

Numbers, Ronald L. (1992). *The Creationists: The Evolution of Scientific Creationism.* New York: Knopf. A comprehensive history of scientific creationism in the United States up to 1990.

Pennock, Robert T., ed. (2001). *Intelligent Design Creationism and Its Critics: Philosophical, Theological, and Scientific Perspectives.* Cambridge, MA: MIT Press. A collection of statements by both proponents and opponents of intelligent design theory.

Ruse, Michael. (2003). "Is Evolution A Secular Religion?" *Science* 299: 1523–1524.

Van Till, Howard. (1999). "The Fully Gifted Creation." In *Three Views on Creation and Evolution,* ed. J. P. Moreland and John Mark Reynolds. Grand Rapids, MI: Zondervan.

Whitcomb, John C., and Henry Morris. (1961). *The Genesis Flood.* Grand Rapids, MI: Baker Books.

EXISTENTIALISM

• • •

Existentialism came to prominence shortly after World War II as a philosophical and literary movement stressing individual human experience in a hostile or indifferent world and highlighting freedom of choice and personal responsibility. As a word, *existentialism* has roots in the Latin *existere,* meaning to stand forth. Indeed existentialists argue that human beings stand out from other things because of the way humans stand consciously and freely in relation with things and with one another. Existentialists developed criticisms of science and technology especially insofar as they deny or obscure this uniqueness.

Historical Development

In the nineteenth century, Søren Kierkegaard (1813–1855) first used the word *existence* to designate a deep individuality that escaped the grip of bourgeois society and religion, and rationalistic philosophy. Though Friedrich Nietzsche (1844–1900) did not use the word, his radical analyses and demands for self-creation influenced later existentialist thinkers. Nineteenth-century Romanticism can be seen as proto-existentialist, and writers such as Ralph Waldo Emerson (1803–1882) and Fyodor Dostoevsky (1821–1881) (who both influenced Nietzsche) sought to redefine the self and called for new levels of choice and new social relations.

In part this was a response to industrial and social revolutions that shook traditional values. Writers were aghast at poverty and social dislocation amid the optimistic complacency of a society that seemed to offer no place to be fully human. The dislocations and wars of the twentieth century increased this tension, and the triumphs of technological rationality and the growth of the psychological and social sciences threatened those dimensions of human existence that cannot be reduced to relations among law-governed objects. The twentieth-century tone is more despairing in authors such as Franz Kafka (1883–1924) who chronicle human imprisonment and lack of possibilities.

Along with Nietzsche and Kierkegaard, the German philosopher Edmund Husserl (1859–1938) also influenced the French generation that created existentialism as an explicit philosophical school. Husserl tried to reveal the acts and necessities that lay beneath and make possible our ordinary perceptions and actions. Seeking to go behind science to reveal it as a construction within a more fluid lived experience, Husserl showed how science's power could nonetheless transform human life and be readily accepted. Max Scheler (1874–1928), Martin Heidegger (1889–1976), and others extended Husserl's analyses in more practical and dramatic directions. The most influential work before World War II was Heidegger's *Being and Time* (1927), which proclaimed a new mode of analysis of the self and a new conception of our relation to time and history. After the war, existentialism as such manifested itself in the work of Jean-Paul Sartre (1905–1980). Soon the label "existentialist" was also given to the work of Gabriel Marcel (1889–1973), Albert Camus (1913–1960), Maurice Merleau-Ponty (1908–1961), and others, though Marcel and Merleau-Ponty later rejected the term. Although these thinkers had been forming and writing their ideas before the war, the experience of the Nazi occupation and the problems of postwar reconstruction intensified the urgency of their thought.

Common Threads

The common thread of the existentialist critiques of science and technology is that human existence has dimensions that cannot be scientifically or technologically grasped. In a technoscientific world, humans are in danger of being imprisoned in an impoverished mode of living that denies their deepest possibilities. This situation calls for a deeper analysis of the structures of human experience, and for the assertion of human freedom through new ethical values and new projects, or avant-garde art, or political action, or religion; these all escape an everydayness that hides who human beings really are or can be.

Existentialists refuse technological determinism even while they admit that for the most part humans may be determined by received values and orientations that deny them the chance to revise basic choices. Rational calculation is an inadequate approach to policy issues because it avoids questioning the framework within which calculations will be performed.

The existentialists demand self-creation that goes beyond everyday and rationally analyzed frameworks. To bring their message of a more than rational criticism and creativity, existentialists produced novels, plays, autobiographies, journals, and literary criticism as well as philosophical tracts. Some were politically radical, some conservative, some religious, and some atheistic, but they shared a sense that self and society faced a crisis that was all the more serious for its general invisibility. Crucial dimensions of selfhood and social life were being ignored, and the need for self-creative decision was being denied even while such decisions were made but covered over in what Sartre called *bad faith*.

Contra Science and Technology

For issues relevant to science and technology the two most important existentialist writers are Sartre and Heidegger. Sartre demands that human individuals realize that their freedom is the sole source of meaning, and act resolutely in an inherently meaningless world. Though Sartre himself did not write extensively about science or technology until his later more Marxist period, his early existentialist ideas fit well with technological ambitions to control the world and decide its significance. Sartre refuses any appeal to social roles or to a given human nature. Things acquire meaning when humans project possible courses of action and language involving them. Human selves and personalities acquire meaning in the same way, within a projected net of values and activities, that projection is totally free and need not be consistent with the past; people are bound only by how they choose to bind themselves. Individuals fear this totally open freedom, and cling to rigid self-definitions as if they were natural things with a fixed nature. Sartre's ideas resemble those technological optimists and some posthumanists who find no limits to what people might make of themselves.

In his later writings Sartre saw the expansion of science and technology as part of a larger thinning of life and denial of freedom due to the capitalist mode of production, which attempts to reduce humans to docile subjects of serial processes. The image of technological progress seduces people away from collective free responsibility for the future. Social processes seem fixed and unavoidable; changing them requires cooperative revolutionary action, not just Sartre's earlier individualistic choice.

Denying that objects dictate their own meaning and human possibilities, the existentialists denied the adequacy of reductions of human activity to physiology, and the reduction social connections to economic and technological relations. They saw science as science reducing experience to static abstractions and collected data. They saw capitalist industrial systems as increasing the dominance of impersonal routine in human life, and condemned the technologization of war, as in the atom bombing of Hiroshima, associating it with the mechanization of death in the holocaust.

Heidegger feared the technological impulse to control and wrote in opposition to it. He wrote not about the choice of values but about finding creative and resolute new paths within the network of projects and significations that make up the lived world. No free Sartrean choice will allow individuals to escape their time's overall basic meanings, but they can invent creative responses that find unexpected possibilities within those basic meanings.

Heidegger argues that people are mistaken when viewing technology as a neutral tool or as an application of disinterested science. Scientific research and technology are expressions of a more basic way of interpreting-revealing things as raw material to be manipulated efficiently. He claimed that this differs from older ways of understanding the being and meaning of things. It also differs from any simple anthropocentric view, because in the completed technological world human persons too join the *standing reserve* ready for manipulation and service. No one profits from this and no one escapes it.

Heidegger protests the spoliation of the environment and the technologization of life. Yet for Heidegger there is no return to an earlier world. Any active human choice will replay the technological game. Individuals can only wait for some new way of valuing and interpreting to come about. In that waiting, though, they are redefining themselves as resolutely receptive and creatively open to the coming of a new basic meaning of reality, which brings a deeper sense of human existence than the image of themselves as manipulated manipulators that technology offers.

Between them Heidegger and Sartre raise the question of how projects for the future link to past frameworks and values. Both deny that the past merely continues due to inertia; they argue that open temporal existence means that the influence of the past is carried on in human freedom, so the future is open to more authentic choices. They deny that rational analysis of the past can legislate future values. For Sartre human choices are always separated from the past by a moment of indeterminate freedom. For Heidegger human choices are always within a net of meanings and projects that individuals did not originate and cannot eliminate, but which they can creatively reread and reform by discovering new depths and new possibilities.

Both these alternatives stand opposed to the idea that a completed social and psychological science could provide a whole explanation of human life and a guide to its values. The project for such a complete explanation threatens to create a society where other dimensions of self or society can neither be expressed nor thought of, a society that has lost the ability to question its own values and directions.

Other existentialists who rejected Sartre's pure freedom followed Kierkegaard in seeing authentic choices arising in free receptivity to a call from beyond the ordinary, from God, one's deepest self, or the unrevealed possibilities of a particular time and tradition. Camus struggled to develop a position that was more socially engaged than the early Sartre while still affirming individual freedom in a world devoid of both traditional religious and scientifically rational meanings. Gabriel Marcel stressed interpersonal encounter and dialogue, arguing that freedom and true personhood happen amid the active receptivity of mutual commitment, fidelity, and hope. This space of mutual encounter is fundamentally open to include God. Scientifically objectivist and technologically manipulative approaches to humanity deny the deepest human possibilities when they reduce persons to calculable units and human excellence to "having" rather than "being."

Maurice Merleau-Ponty developed existentialist issues through dialogue with scientific developments in biology and experimental psychology. He used ideas from Gestalt psychology and added his own analysis of the relation of animal to environment and perceiving body to objects. He claimed that scientific materialism paradoxically reinforces a split between subject and object when it mistakenly presumes that perception is the presentation of discrete data that is then subjectively interpreted. He argued that the perceived world and the perceiving bodily person are intertwined, revealing each other in perception and practical activity, without the need for a middle layer of data or representations. His ideas have become part of attacks that question the adequacy of computer models for the mind and fault cognitive theory for clinging to a theory of mental representations.

Merleau-Ponty's ideas about embodiment have been taken up by those trying to develop an environmental ethics that questions any purely manipulative approach to nature and seeks to foster more connectedness with non-human creatures.

Human Nature and Authenticity

Existentialists encourage choosing more *authentic lives*. The English *authentic* comes from the French *autentique*, meaning authored. An authentic life is not one attained through social conditioning or everyday expectations but is authored by the individual's own deep choice and self-creation. An authentic choice need not be restricted to the social roles commonly available. While Kierkegaard thought that individuals might choose to lead authentic lives that were to all outside appearances totally humdrum and ordinary, Sartre and especially Heidegger thought that authenticity could require dramatic new commitments and modes of action.

Existentialism denies traditional pictures of a fixed human nature, and also denies programs for a rational foundation of values derived from Kantian, Hegelian, or Marxist philosophy, or in a different way from economics and game theory. Existentialists agree with Max Weber (1864–1920) that ultimate values cannot have a rational foundation, but they make these choices subject to the criterion of authenticity, rather than arbitrarily. The crucial question becomes just what ethical import the criterion of authenticity can have. Can it provide limits on self-invention? Can one say that some authentic choices would be *wrong*? Could individuals make *authentic* choices to be fully conscious Nazis? Could people sacrifice others to their own projects? Must every situation be approached with the possibility that it may call for extreme measures that will seem unethical?

Nietzsche thought so, and he took seriously the idea that individuals would have to move beyond standard notions of good and evil. Facing this issue and wanting to find some limits through a sense of justice, Sartre and Camus both wrote dramas where characters confronted violence and the choice of becoming assassins and terrorists. These plays derived from the demand for self-sacrifice in the French Resistance against the Nazis, and from the terror on both sides of the 1950s Algerian liberation struggle. Twenty-first-century society faces this issue not only in its struggles with violent movements, but also in making decisions about the use of powerful weapons, and about the biotechnology revolution that will allow humankind to redefine itself, perhaps reshaping human potentials with no consideration for freedom and authenticity. Existentialists would argue that such issues demand active choice, lest humanity be carried along an unthinking path of automatic supposed "progress" that avoids the central choices of humans as self-making.

Existentialists ask about the limits of rationality in fundamental decisions. How do individuals determine the values that should guide their ethical choices about the limits of technology, or its application in situations of scarcity? They also urge reevaluating the success of social scientific explanations of self and society. Could a total scientific explanation really guide human choices, or would its application depend upon values that are not the outcome of scientific investigations? This leads to more general questions that get overlooked in the technological rush for efficiency and comfort: What is science for? Can one have choices about its meaning? Are there directions built into technology that ought to be questioned? Heidegger argues that individuals are caught within the technological dynamic and must learn to resist its onward rush while understanding themselves more deeply and waiting for a new basic meaning. Sartre argues that individuals should shake off the past, take the future in their hands, and choose anew. Merleau-Ponty urges a reexamination of the basic experience of bodily inhabiting the world and a consequent redefinition of individuals and their possibilities.

For all existentialists, the real question is: What will people choose to become? Do they have more freedom than they imagine in relation to the past, traditions, and social conditions? Modernist writers extol freedom, but think of it mainly in terms of linear progress in already obvious directions. Existentialists argue that issues of authentic choice open more possibilities than such standard options.

DAVID KOLB

SEE ALSO *Alienation; Heidegger, Martin; Husserl, Edmund; Kierkegaard, Søren; Nietzsche, Friedrich W.; Neutrality in Science and Technology; Values and Valuing.*

BIBLIOGRAPHY

Abram, David. (1966). *The Spell of the Sensuous: Perception and Language in a More-Than-Human World.* New York: Pantheon Books. Ecological discussion influenced by Merleau-Ponty's ideas.

Camus, Albert. (1972). "The Just Assassins" [*Les justes*]. In *Caligula and Three Other Plays,* trans. Stuart Gilbert. New York: Vintage Books. Camus's play examines the experience of self-righteous terrorists.

Heidegger, Martin. (1996). *Being and Time,* trans. Joan Stambaugh. Albany: State University of New York Press. Originally published in German in 1929. Heidegger's most influential work, rethinking humans' relation to time and history, and the need for authenticity.

Heidegger, Martin. (1977). *The Question Concerning Technology, and Other Essays,* trans. William Lovitt. New York: Harper and Row. Includes: "The Question Concerning Technology," "The Turning," "The Word of Nietzsche: God is Dead," "The Age of the World Picture," "Science and Reflection." Heidegger's major critical writings on science and technological civilization.

Husserl, Edmund. (1970). *The Crisis of European Sciences and Transcendental Phenomenology*, trans. David Carr. Evanston, IL: Northwestern University Press. Posthumously published in German in 1954. Husserl's examination of science and the origin of the modern experience of nature out of the more basic flow of bodily and temporal perception and meaning.

Kierkegaard, Søren. (1974). *Fear and Trembling, and the Sickness unto Death*, trans. Walter Lowrie. Princeton, NJ: Princeton University Press. Anxiety, despair, and their healing in an authentic selfhood in relation to God.

Marcel, Gabriel. (1965). *Being and Having: An Existentialist Diary*, trans. Katharine Farrer. New York: Harper and Row. Originally published in French in 1935.

Merleau-Ponty, Maurice. (1962). *Phenomenology of Perception*, trans. Colin Smith. New York: Humanities Press. Originally published in French in 1945. Merleau-Ponty's study of the bodily intertwining of subject and object.

Nietzsche, Friedrich. (2001). *The Gay Science*, trans Adrian Del Caro. New York: Cambridge University Press. Nietzsche's most accessible presentation of his views; in the fifth part he discusses and critiques science.

Sartre, Jean-Paul. (1956). *Being and Nothingness*, trans. Hazel Barnes. New York: Philosophical Library. Originally published in French in 1942. The most influential existentialist treatment of individual freedom.

Sartre, Jean-Paul. (1973). "Dirty Hands" (*Les mains sales*). In *No Exit, and Three Other Plays*, trans. Stuart Gilbert and Lionel Abel. New York: Vintage Books. In the play Sartre asks whether there can be an authentic act of violence.

Sartre, Jean-Paul. (1983). *The Critique of Dialectical Reason*, trans. Alan Sheridan-Smith. New York: Schocken Books. Originally published in French in 1960, with a second volume published posthumously in 1982. In this work Sartre reformulated existentialism, taking more account of social and economic relations in modern routinized society.

Zimmerman, Michael E. (1994). *Contesting Earth's Future: Radical Ecology and Postmodernity*. Berkeley: University of California Press. Deep ecology that borrows from Heidegger's ideas.

EXPERIMENTATION

• • •

Experimentation is a foundational activity in modern science. Although several Renaissance thinkers prepared the way toward modern concepts of experimentation, Francis Bacon's *Novum Organum* (1624) was the first systematic attempt to articulate and justify and articulate the proper method of experimental scientific inquiry. Bacon envisioned scientific experimentation as a form of recursive knowledge production that both interprets nature and intervenes in it. Yet efforts to fully define experimentation in a consistent, comprehensive, and prescriptive way have been unsuccessful because of the diverse subject matter and disciplines, as well as instrumental developments, that continually create new variants. An alternative conception of experimentation construes it as an integral part of the actual formation and development of modern society, rather than as just a series of operations conducted in laboratories. Experimentation in the real world requires public participation; risk and uncertainty replace the ideal of an experimental world isolated from society.

Renaissance Roots of Experimentation

Two intellectual sources of Renaissance culture nurtured the idea of experimentation: humanistic values and the practices of superior artisans. In her historical-philosophical study *The Human Condition* (1958), Hannah Arendt demonstrated a deep break between Renaissance thinking and the received preeminence of the contemplative life in classical and medieval traditions. Claims for the superiority of *theoria* over utility were rooted in the Platonic and Christian visions of an eternal, unchanging world that could be known in the futile human life-world only by intuitive reason or spiritual contemplation. In the prosperous and independent city republics of the Renaissance, however, humanist writers questioned this hierarchical order and proposed a more balanced appraisal of the *vita active* in relation to the *vita contemplativa*. Beginning with the Florentine chancellors Coluccio Salutati (1331–1406) and Leonardo Bruni (1369–1444), humanists became advocates of worldly learning and dispensers of fame and glory in the services of cities, merchant families, princes, and popes.

This humanistic resurgence in *vita activa* was modest and not concerned with understanding or conquering nature but simply with rediscovering the great deeds of antiquity. But its ideals of austere republican virtue, participatory citizenship, and Machiavellian power communicated to the *vita activa* a new value of its own, paving the way for the Baconian *scientia activa*. Pico della Mirandola's famous oration "On the Dignity of Human Beings" (1486) is the literary highlight of the attempt to define humans not by some fixed location in the great chain of being, but by their ability and duty to determine their position outside the natural order *as a free and extraordinary shaper* of themselves. This is echoed in Arendt's interpretation of the *vita activa* as part of the "rebellion against human existence as it has been given" (Arendt 1958, p. 2). At the heart of the urge toward modern experimentation is a restless overturning of the primacy of the *vita contemplativa*, which holds that "no work of human hands can equal in beauty and truth the physical kosmos" (Arendt 1958, p. 15).

The unpolished vernacular writings of craftsmen, artist-engineers, instrument makers, and other practitioners who tried to escape the constraints of the guilds provide a different and clearer origin for experimentation and—again in the services of cities and princes—offered new devices, procedures, and designs apt to increase the power, fame, and delight of the patrons. Leonardo da Vinci (1452–1519) was the outstanding genius of this new social stratum of technological intellectuals. In a letter to the Duke of Milan, documented in the *codex atlanticus*, he offered new military, civil, and artistic technologies, concluding that "if any of the aforesaid things should seem impossible or impractical to anyone, I offer myself as ready to make a trial of them in your park or in whatever place shall please your Excellency" (Da Vinci 1956, p. 1153).

Renaissance texts show that the design of new technologies was viewed as an achievement with its own merits and reputation. William Norman, a mariner and instrument maker, wrote a treatise, "The New Attractive" (1581), on magnetic experiments that greatly influenced William Gilbert's "De Magnete" (1600). For the historian of science Edgar Zilsel (1881–1944), this episode served as a solid illustration of his general thesis that modern science developed from breaking down the barriers between three distinct strata of intellectuals (Zilsel 2000). While the university scholars contributed conceptual strength and logical argument, the humanists promoted a reappraisal of worldly affairs and secular thinking, and artisans supplied the experimental spirit in their intent to discover new and useful things. However the first outstanding and most fruitful field shaped by these components was not science proper, but Renaissance art, which brought together the Pythagorean-Platonic understanding of the world, the technical skills of the artists, and the humanist values of glory and fame (Panofsky 1960).

Francis Bacon on Experimentation and Modernization

Philosophers have since struggled with the question of whether experimental action is a subservient function of discovering the laws of nature, or a powerful strategy for giving unforeseen features to nature. For Francis Bacon (1561–1624), this interplay of conceptual understanding and experimental intervention signifies a recursive learning process termed *scientia active* (or *operative*). This kind of knowledge production would profoundly alter technology, nature, and society. The most provocative

pronouncement Bacon offered was that approval of the experimental method in philosophy and science implied turning society itself into an experiment, a proposition developed in his fragmentary *Great Instauration* (1620).

When Bacon was unable to use his position in the highest administrative ranks of the British Empire to advance the new science, he resorted, in the Preface to *Novum Organum*, to publicity: "I turn to men; to whom I have certain salutary admonitions to offer and certain fair requests to make." After having pondered the pros and cons of the new experimental method, he declared: "Lastly, even if the breath of hope ... were fainter than it is and harder to perceive; yet the trial (if we would not bear a spirit altogether abject) must by all means be made" (*Novum Organum* book I, aph. 114). The Latin original is *experiendum esse*. Society should give the experimental method an experimental chance. The promises of gains cannot be justified by anticipatory argument, but only by the outcomes of a test. Skeptics are invited to consider the deal in terms of risk assessment: "For there is no comparison between that which we may lose by not trying and by not succeeding; since by not trying we throw away the chance of an immense good; by not succeeding we only incur the loss of a little human labor. ... It appears to me ... that there is hope enough ... not only to make a bold man try [*ad experiendum*], but also to make a sober-minded and wise man believe." (*Novum Organum*, xxbook I, aph. 114).

Bacon's assessment of the societal risks of politically authorizing the experimental method was founded on an important assumption about the relationship between science and society: Experimental failure as well as errors of hypothetical reasoning are acceptable because they affect only the internal discourse of science, not its social environment. Mistakes in the laboratory can be easily corrected and society is only affected by its choice of options offered by approved scientific knowledge. In this sense, Bacon's notion of experimentation foreshadowed latter distinctions between basic and applied research.

Such conditioning of experimental science became institutionalized in the founding charters of scientific academies and learned societies, and has served as the backbone of the dominant ideology for supporting scientific progress. It makes scientific research and technological invention central aspects of organizing and modernizing society and its institutions. In other words, Bacon's conception of experimental science was the foundational element in the contract between science and society (Gibbons et al. 1994) and between society and nature (Serres 1995).

It is pointless to deny the epistemic and institutional advantages of laboratory science. But they have their price. Epistemologically laboratory science tends to develop ideals of constraint, abstraction, simplicity, and purity that are at odds with the course of nature and society, and give rise to a worldview that interprets space, time, things, and people as faint approximations of the abstractions that make up the laboratory world (Cartwright 1999). It fosters a view of scientific knowledge as objective, neutral, disposable, and instrumental, and research as socially independent and pure. However from the early beginnings of industrial society through the most recent development of the knowledge society, there is evidence of a recursive rather than a linear relation between the trials and errors experienced in the social dynamics of change and the failures and successes of experimental strategies. Both the intended and unintended consequences of scientific experimentation impact the development of society, which in turn influences scientific research. This has sparked several reinterpretations of the contract between science and society.

The Experimental Mode of Industrial Society

John Dewey (1859–1952) was prominent in this quest to reenvision the recursive relationship between the experimental production of knowledge and the activities of society: "The ultimate objects of science," he wrote, "are guided processes of change," and truths are "processes of change so directed that they achieve an intended consummation" (Dewey 1925, p. 133–134). In this way, Dewey married the search for certainty in knowledge to the struggle for reliability in action. Influenced by the epistemology of William James (1842–1910), Dewey asserted that truth is something that happens to an idea as it is tried out successfully in practical situations.

This vindicated Bacon's supposition that the experimental method (as one of the key features of science) would be writ large and institutionalized as societal experimentation. However Bacon's neatly drawn boundary separating pure knowledge experiments from an experimental society mobilized by and mobilizing new technologies has become increasingly blurred. Controversies about the legitimate basis of scientific experimentation arose. Among the most fiery and permanently debated vivisection, in support of which Claude Bernard (1813–1878) wrote his famous "Introduction to the Study of Experimental Medicine" (1865). While he declared vivisection indispensable for progress in medical research and proclaimed that mutilating living beings is justified by the noble goals of science, his opponents considered such research to be driven by perverse instincts intolerable to a humane society. Shortly thereafter the public discussion extended to questioning the scientific practice of victimizing ethnic minorities, criminals, patients, pregnant women, prostitutes, and soldiers. (Foucault 2003).

In the industrialization process of the nineteenth century, scientific experimentation became closely linked with experimental practices of innovation in various economic sectors. The distinguished chemist Justus von Liebig (1803–1873) promoted agricultural chemistry. His experiments clarified the chemical cycles involved in biological reproduction. Liebig applied this knowledge in agriculture to improve productivity. He realized that laboratory chemistry needed to be complemented by experiments located in complex natural systems. His seminal *Chemistry in its Application to Agriculture and Physiology* (1862) states: "Our present research in natural history rests on the conviction that laws of interaction not only exist between two or three, but between all the phenomena of the animal, vegetable, and mineral spheres which determine life on the surface of the earth" (Liebig 1862, p. 167–168). Louis Pasteur (1822–1895) attempted to convince farmers and ranchers of the efficiency and usefulness of animal vaccination. Under both Liebig and Pasteur, scientific experiments became closely allied with practical applications. The recursive learning process depends on opening the laboratory to the complexity of the world and, in turn, targeting scientific knowledge to relatively narrow applications.

Agriculture became standardized through the application of chemistry and microbiology. Similar processes of intertwined experimental learning can be observed in the fields of electrical and mechanical engineering, communication technology, and industrial chemistry. In all these areas, laboratories continue to be important sources of inventions, but are no longer the exclusive domain of the academic sciences. Science has permeated industry, commerce, and the military and is inextricably linked with market forces, production processes, and governmental decisions. Thomas Edison's (1874–1931) *invention factories* at Menlo Park and other places have served as models for modern industrial research laboratories.

Experimental Society

The social sciences have brought another aspect of societal experimentation into focus. Sociologists in the United States interpreted the dramatic growth of cities as collective self-experiments, guided both by planning and design and by unforeseen outcomes and surprises. Albion W. Small (1854–1926) described his *Introduction to the Study of Society* (1894) as a *laboratory guide*, whereby settlements and cities are ready-made experiments that are available to the sociological observer:

> All the laboratories in the world could not carry on enough experiments to measure a thimbleful compared with the world of experimentation open to the observation of social science. The radical difference is that the laboratory scientists can arrange their own experiments while we social scientists for the most part have our experiments arranged for us. (Small 1921, p. 187–88)

Small located the idea of experimentation in social life, not the scientific method. This notion of experimentation became influential in American sociology, especially within the Chicago School developed by Robert Park (1864–1944), but it lacked a precise specification of the societal and cultural conditions that give social life its experimental characteristics.

Donald Campbell (1969) presented an elaborated methodology of sociological real-world experiments. Reliable prediction of the success of social reform projects in areas such as education, youth delinquency, taxes, and housing is not possible, but a careful design of reforms as experiments would allow planners to learn about the acceptance and efficiency of strategies so that outcomes could be used to adjust future reforms. Although objections have been raised against the technocratic attitude of this approach (as reforms are more or less superimposed on the people concerned), it has also had great influence in the field of adaptive management.

Later discussions of real-world experimentation centered on the notion of acceptable risk, that is, the paradox of not knowing before the experiment whether the social and ecological risks are acceptable. One good example is the large-scale release experiments involving genetically modified organisms. The increased power of modern science and technology qualifies Bacon's original optimism about societal experimentation because the losses involved in failed experiments are potentially much greater than *a little human labor*. Experimentation in the real world unavoidably leads to surprises, which causes problems and provides opportunities for learning. Science involved in such endeavors renders the ideal of detached and austere knowledge production obsolete and makes public involvement necessary in order to enhance acceptance and legitimation of projects. Ecological experimentation in particular has gained support by incorporating local knowledge and by making the risks and uncertainties of theoretical models more transparent. Hearings, volunteer and stakeholder groups, and other methods of making experiments participatory entail costs in time and money yet fail to guarantee support or consensus. But the risks of experimentation can no longer be hidden from view. The production of knowledge in a democratic society requires public discourse and participatory involvement, and these are the features with which real-world experimentation must experiment.

WOLFGANG KROHN

SEE ALSO *Bacon, Francis.*

BIBLIOGRAPHY

Arendt, Hannah. (1958). *The Human Condition.* Chicago: University of Chicago Press.

Bacon, Francis. (1857–1874). *Works of Francis Bacon,* ed. James Spedding, Robert Leslie Ellis, and Douglas Denon Heath. London: Longman.

Bernard, Claude. (1957). *An Introduction to the Study of Experimental Medicine,* trans. Henry Copley Green. New York: Dover. First analysis of the meaning of experimentation with living substances; originally published in1865.

Campbell, Donald. (1969). "Reforms as Experiments." *American Psychologist* 24 (4): 409–429.

Cartwright, Nancy. (1999). *The Dappled World: A Study of the Boundaries of Science.* Cambridge, UK: Cambridge University Press. Offers a sceptical view of the scientific prospect to understand the reality completely by the means and models of natural laws.

Da Vinci, Leonardo. (1956). *The Notebooks of Leonardo da Vinci, Arranged, Rendered into English and Introduced by Edward MacCurdy.* New York: Braziller.

Dewey, John. (1925). *Experience and Nature.* Chicago: Open Court. Dewey's comprehensive attempt to understand knowledge, values, action, and facts in their natural connectivity.

Foucault, Michel. (2003). *Abnormal: Lectures at the College de France, 1974–1975.* New York: Picador. The lectures refer to Foucault's central theme of power, knowledge, and social exclusion.

Gibbons, Michael; Limoges, Camille; Nowotny, Helga; et al. (1994). *The New Production of Knowledge: The Dynamics of Science and Research in Contemporary Societies.* London and Thousand Oaks, CA: Sage.

Gilbert, William. (1958). *De magnete,* trans. P. Fleury Motteley. New York: Dover. Considered to be the first textbook on experimental natural science, published 1600; Gilbert was physician to Elizabeth I and James I.

Lederer, Susan E. (1995). *Subjected to Science: Human Experimentation in America before the Second World War.* Baltimore: Johns Hopkins University Press.

Liebig, Justus, Freiherr von. (1855). *Principles of Agricultural Chemistry: With Special Reference to the Late Researches made in England.* New York: Wiley.

Liebig, Justus, Freiherr von. (1862). *Die organische Chemie in ihrer Anwendung auf Agricultur und Physiologie* [Chemistry in its application to agriculture and physiology]. Braunschweig: Vieweg.

Panofsky, Erwin. (1960). *Renaissance and Renascenses in Western Art.* Stockholm: Almquist & Wiksell. Classical studies of the symbolic meaning and relevance of the arts at the beginning of the modern era.

Pico della Mirandola, Giovanni. (1986). *De hominis digitate*, trans. Douglas Carmichael. Lanham, MD: University Press of America. Pico lived from 1463–1494. The oration was first published in 1496 and reckons as the "manifesto" of the humanist movement.

Radder, Hans, ed. (2003). *The Philosophy of Scientific Experimentation.* Pittsburgh, PA: University of Pittsburgh Press.

Serres, Michel. (1995). *The Natural Contract*, trans. Elizabeth MacArthur and William Paulson. Ann Arbor: University of Michigan Press.

Small, Albion Woodbury, and George E. Vincent. (1971 [1894]). *An Introduction to the Study of Society.* New York: American Book Co.

Small, Albion Woodbury (1921). "The Future of Sociology." *Publications of the American Sociological Society* 15: 174–193.

Zilsel, Edgar. (2000). *Social Origins of Modern Science*, ed. Diederick Raven, Wolfgang Krohn, and Robert S. Cohen. Dordrecht, Boston and London: Kluwer Academic Publishers. Classical work proposing the relevance of superior artisans for the origins of experimental science.

EXPERTISE

• • •

The question of expertise—its nature, scope, and application—is one of the most urgent issues in the modern world. The recognition of expertise as an important issue and the analyses of its problems are firmly embedded in the Western tradition. Plato's discussions of *techné* and of the difference between philosophy and sophistry, for instance, are best characterized as discussions of expertise. "When Socrates seeks moral knowledge," Julia Annas writes, "it is only to be expected that this will be seen on the model of practical expertise, since this is the model for knowledge in general" (Annas 2001, p. 245).

In its modern usage, the word *expert* derives from the Latin *expertus*, the past participle of *experiri*, "to try"; an expert is one who has been tested and become skilled or knowledgeable through experience. Although this definition seems straightforward, in the real world experts are not always easy to identify or deal with. Although they are a familiar and indispensable element of the contemporary world, experts are also the object of widespread controversy and hostility; experts are capable of generating both trust and skepticism.

Reliance on Experts

In practical matters modern life is permeated by experts and expertise, a situation that is also central to scientific disciplines. Contemporary scientific research depends on evidence being generated, integrated, disseminated, evaluated, and reviewed by overlapping networks of investigators (Hardwig 1985). Nonscientific professions also are constituted by the need to reproduce, maintain, and supervise expertise. The defining character of both the public and private spheres thus is determined largely by the kinds of experts who are deferred to (including self-professed experts, "hired gun" experts, and faux experts), the circumstances in which such deference occurs, and the reasons that can be provided to justify that deference.

Experts shape not only professional disciplines but also everyday life. Citizens routinely defer to experts not only in issues involving a scientific-technological dimension but in "all sorts of common decisions" about anything and everything (Walton 1997, p. 24). The extent of routine deference to experts is staggering. Politicians, judges, businesspersons, and ordinary citizens rely on experts. Many activities once left as a matter of nature or common sense to clan, community, or culture, such as childbearing and child rearing, have become the province of experts (Hulbert 2004). As the cultural critic Neil Postman notes: "[E]xperts claim dominion not only over technical matters, but also over social, psychological, and moral affairs. There is no aspect of human relations that has not been technicalized and therefore relegated to the control of experts" (Postman 1993, p. 87).

Contemporary reliance on experts has a historical dimension. Around the beginning of the twentieth century demographic changes such as the massive influx of immigrants, the concomitant weakening of the authority of traditional cultural practices, and the accompanying fascination with being "modern" helped foster the view that scientific approaches could make many human activities previously governed by culture, community, and religion more effective and efficient. Meanwhile, new technologies arose whose principles could not be mastered by nonexperts and thus had to be delegated to specialists. Inevitably, with that new reliance on experts controversies arose over who was a genuine expert, how an expert was trained and legitimated, and the objectivity of certain fields of expertise. Thus, whereas the problem of expertise is as old as the ancient quarrel between philosophy and sophistry, the permeation by and dependence of modern life on expertise has made this question increasingly important.

Domains of Expertise

A brief look at the ways in which controversies have arisen in different domains can help illuminate different aspects of the issue of expertise.

GOVERNMENT. Democracy depends not only on an educated citizenry but also on educated decision making. Most countries attempt to establish this by incorporating experts into government operations through agencies, regulatory and review panels, committees, and advisory capacities. From the governmental perspective the use of expertise generally implies a distinction between the social and technical aspects of policy and its instruments: Although decisions about the social aspects are the province of elected representatives of the public, decisions about the technical aspects are relegated to experts. However, this separation is never clean because technical aspects are seldom neutral with respect to social ones. The sometimes murky boundary between the social and technical aspects of policy periodically leads to controversies over the governmental selection of experts and the advice they provide, along with attempts to reduce the influence of experts on policy.

A dramatic and instructive episode was the 1954 hearing on the scientist J. Robert Oppenheimer's (1904–1967) security clearance (Thorpe 2002). In his role as chairman of the General Advisory Committee (GAC), which was charged with advising the Atomic Energy Commission (AEC) "on scientific and technical matters," Oppenheimer had opposed the development of the "Super," an early impractical attempt to build a hydrogen bomb. Oppenheimer was not the only GAC member to oppose it, but his influence galvanized adversaries to seek his removal from a position in which he could influence the government, and his clearance was suspended.

At the end of a four-week hearing AEC counsel Roger Robb said bluntly to Oppenheimer, "You of course don't conceive yourself to be an expert in war, do you, or military matters?" No, was the reply. Then, continued Robb, did you not perhaps go "beyond the scope of your proper function as a scientist in undertaking to counsel in matters of military strategy and tactics?"

Would this not, Robb added, be as absurd as deeming John Ericsson to be qualified in naval strategy merely because he had designed the *Monitor?* Robb was challenging not only Oppenheimer's authority to address social issues such as military policy but in effect that of any scientist.

That challenge went unanswered and highlights the contentious nature of the border between technical and social issues as well as the discretionary power and potential ideological biases involved both in the selection of experts and in the advice they offer. Although controversies over such issues arise in almost every administration, the handling of experts and expertise by the government became a salient campaign issue in the U.S. presidential election of 2004. Organizations such as the Union of Concerned Scientists hosted websites that documented instances in which the Bush administration was declared guilty of abusing, distorting, and suppressing the advice of experts on issues ranging from abortion to stem cell research.

MEDIA. The use of experts in the media entails a different set of issues. The media not only rely heavily on experts for information but also frequently quote or interview them in the process of conveying content to the public. The experts who gain "standing" thus acquire an influential role in shaping public perception about what information is authoritative and in generating, perpetuating, and even resolving controversies. Media-designated experts, however, often are chosen to a large extent because of factors such as accessibility, skill at communicating, charisma, and even the particular positions they have adopted. The result is that these experts are not necessarily the ones who would be recognized by most or even many professionals of the field in question; the positions they advocate also may not be shared generally.

Moreover, the qualities required to gain standing vary from medium to medium: The kind of person cited as an expert in the print media differs from the kind who appears on television. A major difference between media-appointed and other kinds of experts is a sharply diminished incentive to define and rigorously police the difference between real experts and charlatans. The media often are encouraged to promote "balanced" voices, particularly colorful and charismatic ones, that advocate positions outside the mainstream. As a result individuals who have questionable credentials, who are being promoted by those with certain agendas, or whose conduct or methodology is not generally representative of those in their professions can be anointed experts.

A classic illustrative episode is Bailey Willis's role in the controversy over the construction of the Golden Gate Bridge. Willis had worked for the U.S. Geological Survey from 1898 to 1916 and then became a professor of geology at Stanford University, retiring as an emeritus professor in 1922. In the early 1930s he joined a fierce controversy about the Golden Gate Bridge, whose construction was in progress, when he claimed that the collapse of the bridge was inevitable because the bedrock of the south tier was too soft. The bridge had been opposed strongly by the local ferry company, the shipping industry, and landowners afraid of declining property values and increasing tourism. To those groups Willis was a godsend, and they used him as a point person. He was flamboyant and quotable, preached a doomsday scenario, and was credentialed as a professor and ex-employee of the U.S. Geological Survey. It thus is not surprising that he was cited regularly as an authority on the front pages of newspapers. To his scientific colleagues, however, Willis's methodology and behavior were abominable: His arguments were easy to refute, he was shown to have misread maps, and he refused to inspect the rocks firsthand. To those colleagues his credentials did not matter. Expertise requires possession of the appropriate skill, Willis lacked that skill, and any claim made for his having it was fraudulent.

Similar complaints about media-designated experts frequently surface in more recent controversies involving a scientific-technological dimension, such as those over breast implants, the dangers of chemical toxins, and the health effects of low levels of radiation. These episodes highlight the question of whether it is possible to describe and recognize what is involved in the "intuitive," first-person possession of expertise.

LAW. In the modern world controversial social issues often wind up in the law courts, which are forced to impose a cease-fire on terms that are frequently tentative, vague, imperfect, and open to revision. Nevertheless, these flawed practical resolutions often contain signposts indicating why it is so difficult to integrate conceptual and practical issues, a situation in which the use of experts is no exception (Golan 2004, Feigman 2004, Foster and Huber 1997).

Experts play a pivotal role in the courtroom, where their use turns on the distinction between evidence and opinion; nevertheless, what constitutes an expert in science and in law "is as far apart as day and night" (Angell 1996, p. 116). The introduction of expert testimony by the prosecution tends to increase conviction rates (Brekke and Borgida 1988, Kovera, Levy, Borgida, and Penrod 1994), whereas testimony from a defense expert tends to lessen the likelihood of a conviction (Hosch 1980, Schuller and Hastings 1996) even though jurors have proved themselves incapable of understanding the implications of much scientific testimony (Selinger 2003).

Tal Golan (2004), for instance, traces controversies involving the use of experts in court back to a late eighteenth-century case concerning the causes of the decline of the harbor in Wells, England, in which each party hired expert witnesses. The result was a much wider use of scientists as expert witnesses in the courtroom, and this paved the way to abuses. By the mid-nineteenth century Attorney General Sir Alexander Cockburn was expressing a widely held view, at a poisoning trial in 1856, when he remarked, "I abhor the traffic in testimony to which I regret to say men of science sometimes permit themselves to condescend."

Whereas poisoning trials were a common forum for clashes between scientific experts, twentieth-century medical technologies have expanded the opportunities for scientific expert testimony vastly not only in the areas mentioned here but also in lie detector evidence, insanity defenses, and DNA analysis. The use of expert witnesses in the courtroom has burgeoned, along with the burden placed on the legal system. One early decision, *Frye v. U.S.* (1923), stated that expert testimony must assist the jury in its decision making, that the testimony must be based on scientific principles that are accepted generally in the field, and that an expert witness must be suitably qualified. However, the rules established by *Frye* were found to be too broad, and the continuing legal controversy culminated in a landmark and still controversial 1993 decision by the U.S. Supreme Court, *Daubert v. Merrell Dow Pharmaceuticals*.

The *Daubert* decision sought a practical solution to these difficulties, attempting to take the role of assessing scientific testimony out of the hands of juries and putting judges in the role of gatekeepers for the admissibility of scientific evidence in federal courts. It also attempted to lay out guidelines of reliability and relevance for judges in evaluating technical data possibly beyond their expertise for courtroom use. Questions about its effectiveness, however, remain.

Although informed by both the philosophy of science (particularly the work of the philosophers Karl Popper [1902–1994] and Carl Hempel [1905–1997]) and the sociology of science, the *Daubert* decision has had a tendency to produce expensive and time-consuming pretrial hearings that have been viewed as discouraging the kind of sound gatekeeping that the decision was intended to establish. It has been elaborated by two further cases, *General Electric v. Joiner* (1997) and *Kumho Tire v. Carmichael* (1999), and the issue continues to generate much discussion and writing.

The continuing controversy over expertise in the courtroom has served to highlight in particular the question of how to integrate the possessors of scientific expertise with the needs of a particular arena, such as the courtroom, in which it is required.

Interdisciplinary Structure

Each of these controversies involving the use of experts in government, media, and the law poses a different set of questions involving expertise that call for conceptual clarification. Those questions include the following: How does one become an expert? Can experts be recognized by nonexperts? Is it possible for a consumer of expertise to detect the presence of hidden agendas, biased or tainted testimony, and incompetence in expert testimony? Is it possible to train experts in such a way that these contested problems do not arise? The inability to answer such questions definitively, especially in high-profile controversies, has contributed to a general skepticism regarding experts and to doubts about whether it is possible to achieve a pragmatic, effective, and permanent solution to the problem of expertise.

An essential first step would be greater conceptual clarification of the problem. Recent technological issues highlight the need for such clarity: Debates about the value of shifting expertise away from individual and credentialed content experts to a community of self-policing but not necessarily credentialed contributors have plagued *Wikipedia*, an online encyclopedia whose entries can be altered by essentially anyone who desires to change them; debates about reports occurring on blogs have brought traditional reporting to the threshold of a crisis; and debates about the collaborative categorization of information through the use of simple tags in social software have raised powerful questions concerning who has the right to manage data. (Social software designates software that is designed to support one or more of the following goals: (1) support conversational interactions between individuals or groups, (2) supports social feedback (i.e. rating of goods and services to create digital reputation), and (3) support and manage social networks (i.e., programs such as Friendster which allow you to network with people who you do not know, but who are acquainted with people you do know.)

No single key can unlock the problem of expertise all at once. Its analysis requires crossing several disciplinary boundaries: philosophical, sociological, political,

and even rhetorical. Philosophically, the question of expertise broaches the philosophy of mind—of what it means to know something and to be someone capable of acquiring knowledge—and is inextricably interwoven with issues of embodiment, apprenticeship, and artificial intelligence, among others. However, expertise has a social character as well inasmuch as the question of who is an expert is not a matter of training or skill alone but of definition and recognition. Politically, the authority conferred on experts collides with participatory democracy, with the democratic and antielitist urge to accord equality to all citizens. As media experts reveal, who "counts as" an expert often depends on rhetorical ability. Thus, "expertise" rarely is addressed comprehensively from more than one perspective at a time. It lurks, implicitly and usually uncritically, beneath discussions of concepts such as authority, colonization, power, skill, and even science. Nevertheless, anything short of a full interdisciplinary analysis runs the risk of producing a naive and overly simplistic account.

The practical and the conceptual problems of expertise are clearly related, and it is hard to imagine that a better, more synoptic understanding of expertise would not shed light on pragmatic decision making about expertise. This requires the recognition that expertise is not a simple property or relation but arises from a dynamic set of interactions whose two poles are the production and the consumption of knowledge: At one pole expertise is produced or possessed, at the other it is consumed or used, and a dynamic interaction takes place between the two. Literature on expertise has adopted different approaches to integrating these elements. Some research studies have emphasized the discretionary power and ideology of expertise, others its intuitional and interactive nature, and still others its distributive character.

Discretionary Power and Ideology

In a society strongly shaped by and dependent on advanced technology the most commonsense approach to expertise is via the idea that experts possess a special kind of knowledge and skill that nonexperts do not have but need for ordinary and extraordinary activities. Not only do nonexperts routinely find themselves needing expert advice, the thought continues, but nonexperts would be acting irrationally if they failed to recognize the value of interacting with experts to acquire such epistemic counseling and defer to such advice. Thus, the philosopher John Hardwig argues, "The rational layman will recognize that in matters about which there is good reason to believe that there is expert opinion, he ought (methodologically) not to make up his own mind. His stance on these matters will—if he is rational—usually be rational deference to the epistemic authority of the expert" (Hardwig, 1985, p. 343).

A host of issues arise concerning how and in what conditions a nonexpert can decide which expert to trust. After all, the epistemic inequality that seems to distinguish experts from nonexperts in principle prevents nonexperts from making a justified epistemic decision. A nonexpert could choose who among available experts has the best credentials. However, that decision would be of limited value; it would not address adequately the potential differences between the quality of an institution and the quality of an individual. Hence, in "Experts: Which Ones Should You Trust?" the philosopher Alvin Goldman contends that looking for a track record of success is the best way for a nonexpert to make a sound decision when selecting an expert to turn to for advice (Goldman 2001).

Steve Fuller, Paul Feyerabend, and Herbert Marcuse, among others, have countered that this commonsense position fails to address the way expert knowledge and skill is tainted by special interests, conceptual biases, and ideology and link the production of expertise to discretionary power and even to the aims of technocracy. In "The Constitutively Social Character of Expertise," for instance, Fuller (1994) contends that the significant dimensions of expertise can be specified when a social field is circumscribed. Fuller's work suggests that normative and epistemological implications would follow if people focused their attention on the ways experts create, maintain, and reinforce an interface in which their claims to cognitive authority are bolstered through networking and rhetorical persuasion. A consumer's apparent need for an expert's knowledge or skills could turn out to be a manufactured desire, created and maintained by a class of experts who want their services to be perceived as necessary or useful. Expert authority would be seen to emerge from nontransparent and sometimes deceptive interactions with consumers. If Fuller's account of discretionary power is accurate, the prestige and deference accorded to experts from every field must be tempered.

The philosopher of science Paul Feyerabend (1999) characterized modern scientific experts as "ideologues." From Feyerabend's perspective the more time and energy experts devote to advancing a position that accords with the tenets of Western science, the more difficult it becomes for them to be open-minded to points of view that call their core beliefs into question.

Still more radically Herbert Marcuse (1998) combines a Marxist approach to expertise with the Frankfurt School's use of Freudian psychological insights to critique the role of expertise in the aspirations and methods of technocracy. Modern occupations are characterized by an absence of socialization, in contrast with traditional skilled work, which involved socialization into a craft culture, and technical professions are geared toward producing instruments to serve the state. This process is made acceptable and desirable by the introjection of social demands in personality structures through processes of sublimation, reinforcement, and rationalization. Technical skills are not added to a preformed personality; instead, the personality is altered at its seemingly private core, the subject's very basis of self-understanding. This alteration of subjectivity, Marcuse finds, is integral to the perpetuation of the technological state. Experts and other trained professionals not only contribute to specific tasks and particular jobs but serve the "interest of autocratic power"; they assume the role of "social leaders" and "captains of industry" by virtue of being "technological leaders" (Marcuse 1998, pp. 54–55). Expert training is only one of the many factors in the environment of advanced capitalism that reduce the capacity for individuality in this positive sense, reducing the subject's capacity to exercise free judgment and proffer original or subversive criticism.

Intuitive and Interactive Experts

In contrast to an understanding of expertise in terms of discretionary power, ideology, and capitalist production, other approaches seek to access expertise through the process by which individuals acquire and maintain it. Hubert Dreyfus (1990), for instance, has analyzed expertise from a first-person perspective and, along with his mathematician brother Stuart (Dreyfus and Dreyfus 1986), has produced a general model of skill acquisition that details the cognitive and affective changes that typical learners experience as they make the transition from having little skill (being novices) to making domain-specific decisions intuitively (being experts). Dreyfus's work makes it clear how extensively the question of skill acquisition is connected with human embodiment and the interaction between human beings and the world.

According to Dreyfus, human beings are not passive objects in or omnipotent manipulators of the world but are caught up in it, even and all the more so in regard to skilled behavior. This perspective reflects the basic phenomenological tenet that all practical and theoretical activities, no matter how abstract their outcomes, need to be understood on a continuum with basic lifeworld practice. Experts, Dreyfus insists, act the way all people do when performing mundane tasks: "We are all experts at many tasks and our everyday coping skills function smoothly and transparently so as to free us to be aware of other aspects of our lives where we are not so skillful" (Dreyfus 1990, p. 243). In other words, just as everyday drivers act intuitively when driving (i.e., their actions are not guided by explicit or implicit rule following, but they develop a contextually sensitive capacity for recognizing and responding to patterns that allows them to respond immediately and effortlessly to changes in traffic and road conditions), all professional experts act intuitively when making decisions in their fields: Fighter pilots act intuitively when engaged in combat situations; nurses act intuitively when caring for their patients; environmental scientists act intuitively when assessing whether building a dam will affect the local wildlife in a particular way; and judges act intuitively when deciding which precedent it is appropriate to appeal to in a case.

This phenomenological position potentially has profound implications for the ways in which experts should be trained, communicated with, and utilized. First, if experts solve problems intuitively, educational programs that fail to train students to make intuitive decisions will fail to produce expert graduates. The bias against treating intuition as a serious epistemic resource—one that permeates much of Western intellectual and scientific history and underwrites much of modern management theory—thus is called into question precisely because it impedes the cultivation of the highest form of problem solving and fosters a misleading sense that the human mind can be modeled on computational machines. Similar suspicion is cast on technologically mediated forms of pedagogy that inhibit instructors in relating to their classes intuitively. From Dreyfus's perspective instructors who are trained to view teaching primarily as an opportunity to convey content on the Internet will not be able to develop the expertise that emerges from face-to-face educational interaction, such as learning to read a class's body language to discern whether the presented material has been found to be comprehensible, interesting, or useful. Instructors also will be discouraged from developing the wisdom that comes from dealing reflexively with finitude (e.g., looking a student in the eye and admitting that one does not know the answer to a well-posed question). Students subjected to such an educational process will be trained inadequately.

Second, if experts solve problems intuitively, the social policies and expectations that require experts to translate intuitive decisions into general procedural rules, such as the protocols followed routinely by expert witnesses, should be reevaluated. According to Dreyfus's model, those protocols force experts to provide misleading narratives that distort the ways in which their judgments were formed. Not only does such distortion threaten to transform experts into an "endangered species," it also places the United States at an economic disadvantage: "Demanding that its experts be able to explain how they do their job can seriously penalize a rational culture like ours, in competition with an intuitive culture like Japan's" (Dreyfus and Dreyfus 1986, p. 196).

Third, if experts act intuitively, attempts to export human expertise into nonintuitive technologies such as expert computer systems will fail. This issue will become more important as an increasing amount and variety of medical decisions are delegated to expert computer systems.

Others, however, have noted that despite its benefits Dreyfus's account seems to downplay or even ignore the possibility that ideology and hidden agendas can creep into expert opinion. It therefore is critical to correct this account by exploring how such things are possible (Selinger and Crease 2002). Dreyfus's account also overlooks the different varieties of expertise in performers, critics, and sociologists. For instance, although an expert musician such as a first violin would have to play well, an expert in music might be a musicologist who did not play music at all.

The sociologists Harry Collins and Robert Evans (2002) distinguish between two types of expertise: "interactional expertise" and "contributory expertise." A contributory expert is a practitioner who learns to make contributions to the field by being physically immersed in its corresponding "form-of-life." Medical doctors, for example, develop medical expertise by attending medical school; they then contribute to the development of medicine by publishing medical papers that are based on their clinical experiences. By contrast, an interactional expert is someone who can talk competently about aspects of a field (e.g., pass on information, assume a devil's advocate position, understand and tell insider jokes, and make judgments on a peer review committee) but learns about the field only by talking with people who have acquired contributory expertise. In other words, whereas interactional experts have quite a bit of tacit (nonpropositional) knowledge, they are not direct practitioners in the fields they study. This means that someone who lacks full physical immersion in a field can become so conversant about that field through linguistic socialization that under the conditions of a Turing test (two people who have not met face to face communicating to one another by typing electronic text messages back and forth) it would be hard for authorities to decide whether that person was an interactional expert or a contributory one. A sociologist of medicine who never performed surgery could become so conversant about surgical procedures as to have the kinds of conversations that could convince practicing surgeons that the sociologist was actually a physician.

This position on interactional expertise has implications for the ways in which experts should be identified and treated. First, if interactional expertise fits the criteria Collins and Evans provide, many of the social scientific and humanities disciplines that typically are looked down on by practitioners of the natural sciences (as well as critics and coaches who are looked down on by primary practitioners) should be viewed in a new light; these are indeed real experts, albeit experts who possess interactional expertise.

Second, there may have to be additional legal discussions about who qualifies as an expert witness. Collins and Evans discuss the case of their sociologist colleague Simon Cole (Collins and Evans 2002). Although Cole does not analyze fingerprints, he has studied the methods and conventions of fingerprint analysis rigorously and, as a result of his sociological work, has come to serve as an expert witness. However, Cole's credibility could be contested by the opposing lawyers; after all, he is not a contributory expert. The key consideration, Collins and Evans insist, is that Cole's interactional expertise should be understood as entitling him to make authoritative pronouncements on fingerprinting.

Third, if interactional expertise fits the criteria Collins and Evans provide, political activists who are linguistically socialized into an expert discourse, such as AIDS activists who are socialized in that manner into medical discourse, have a new vocabulary from which they can justify their demands for social change (Selinger and Mix 2004).

Distributed Expertise

Yet another approach to expertise is to focus neither on the forces that shape it nor on its acquisition and various forms but on how it is distributed. Although other accounts address different types of agents as experts and different types of contexts that influence such displays of agency, they fail to reckon with the ways in which expertise is "distributed"—externalized into a network of tools and practices in particular settings such as the laboratory and social networks, standardized in technologies, and more (Mialet 1999).

Bruno Latour's discussion of the Association Française contre les Myopathies (AMF, or French Muscular Dystrophy Association) is a case in point (Latour 1998). The AFM acquired enough funds through charitable donations to contribute more than did the French government to basic research on the human genome. The supported scientists became world players in molecular biology and published some of the first genomic maps in the journal *Nature*. Once their basic research was completed, the AFM-sponsored scientists disbanded the mapping laboratories and turned their attention to the risky field of gene therapy. Latour describes AFM's headquarters as follows:

> The very building at Ivry, south of Paris, where the AFM has its headquarters, illustrates the limit of a metaphor that would separate science from a society left outside: on the first floor, patients in wheelchairs; on the next floor, laboratories; on the third, administration. Everywhere the posters mark the next telethon while contributors visit the premises. Where is the science? Where is the society? They are now entangled to the point where they cannot be separated any longer" (Latour 1998, p. 208).

Latour's point is that what is happening at AFM is neither pure knowledge spilling over into application nor social pressure generating scientific research but represents a far more complex process in which expertise is inextricably bound up in network activity and a wide variety of events and interests can create and destroy its social stability.

The AFM research network may or may not prove to be a suitable model for science elsewhere or even a suitable model for molecular biology in the long run. Nevertheless, Latour claims that this case illustrates the fact that theorists of expertise need to be attentive to cases in which experts prove themselves capable of being (1) flexible because they are willing and able to adapt to, compromise with, and even change their intentions on the basis of the way unexpected network occurrences influence their initial goals; (2) selective because they are able to discern which elements of adaptation and compromise are important and which are inconsequential; (3) perseverant because they are able to endure the setbacks that occur when they attempt to enlist allies; (4) tactful because they are able to maximize other people's interests while remaining unobtrusive; (5) communicative because they are able to transcend the technical jargon concerning their specialization in order to bridge the gap between different people's interests; (6) creative because they are able to recruit likely as well as unlikely allies; and

(7) cooperative because they are willing to accept compromise as an essential feature of network interactions.

Solutions

Far more work needs to be done in exploring the role of discretionary power and potential ideological biases, the role of intuitive and interactive elements, and the distribution of expertise not only in analyzing the issues that arise within each of these perspectives but in seeing how these approaches overlap and differ. More work also needs to be done in approaching expertise in the light of other issues, such as participation.

However, the greatest obstacle to elucidating the nature, scope, and application of expertise is not the complexity of the process; complex phenomena are still amenable to description and analysis. The difficulty stems from how tightly expertise is woven into contemporary life and in how many different ways, making it difficult to place the subject at a manageable distance.

On the one hand, it is tempting but impossible to approach experts as one social class nested among others, such as engineers or doctors or lawyers, or as a subgroup within each social class whose activities may be defined and classified neatly and whose members are governed and disciplined by legislative bodies or citizen groups. However, the use of expertise ripples through modern life in so many forms that this is impossible. On the other hand, it might seem that the scope of expertise is too wide and too protean to lend itself to meaningful analysis. These problems have led some scholars to back away from the subject or to claim that it is ultimately without conceptual substance. Michel Callon, for instance, has described his research as challenging the distinction between expert and layperson, whereas Latour suggests that the concept of the expert is outmoded and is being replaced by that of the spokesperson. However, such cooptations, subversions, and replacements of the concept are unlikely to succeed. The problem of expertise will remain one of pressing issues of the twenty-first century.

ROBERT P. CREASE
EVAN M. SELINGER

SEE ALSO *Authoritarianism; Citizenship; Constructive Technology Assessment; Democracy; Dewey, John; Participation; Peer Review; Pragmatism; Profession and Professionalism; Science, Technology, and Law; Technocracy.*

BIBLIOGRAPHY

Angell, Marcia. (1996). *Science on Trial: The Clash of Medical Evidence and the Law in the Breast Implant Case.* New York: W.W. Norton.

Annas, Julia. (2001). "Moral Knowledge as Practical Knowledge." *Social Philosophy & Policy* 18: 236–256. This article examines Socrates' conception of expertise, and in so doing, shows why contemporary moral theorists have found it difficult to interpret ancient conceptions of ethics properly.

Brekke, Nancy, and Eugene Borgida. (1988). "Expert Psychological Testimony in Rape Trials: A Social Cognitive Analysis." *Journal of Personality and Social Psychology* 55: 372–386.

Collins, Harry. (2004). "Interactional Expertise as a Third Kind of Knowledge." *Phenomenology and the Cognitive Sciences* 3: 125–143.

Collins, Harry, and Robert Evans. (2002). "The Third Wave of Science Studies: Studies of Expertise and Experience." *Social Studies of Science* 32(2): 235–296. The authors analyze some of the problems that have arisen from the pervasive historical and sociological deconstructions of expertise.

Collins, Harry, and Robert Evans. (2003). "King Canute Meets the Beach Boys: Responses to the Third Wave." *Social Studies of Science* 33(3): 435–452.

Dreyfus, Hubert. (1990). "What Is Morality? A Phenomenological Account of the Development of Ethical Expertise." In *Universalism vs. Communitarianism: Contemporary Debates in Ethics*, ed. David Rasmussen. Cambridge, MA: MIT Press. In this article Dreyfus argues that ethics is a skill, one that can best be explained according to his developmental model of expertise. On this basis, he takes issue with leading philosophical and psychological accounts of moral decision-making.

Dreyfus, Hubert, and Stuart Dreyfus. (1986). *Mind over Machine: The Power of Human Intuition and Expertise in the Era of the Computer.* New York: Free Press. In this book the Dreyfus brothers provide a developmental model of human expertise, and in so doing, make the case that expert intuition is endangered by both contemporary practices of simulating expertise in computer programs and by contemporary expectations of what kinds of explanations experts can offer.

Feigman, David. (2004). *Laboratory of Justice: The Supreme Court's 200-Year Struggle to Integrate Science and the Law.* New York: Times Books.

Feyerabend, Paul. (1999). "Experts in a Free Society." In *Knowledge, Science and Relativism, Philosophical Papers*, vol.3, ed. John Preston. Cambridge, MA: Cambridge University Press.

Foster, Kenneth, and Peter Huber. (1997). *Judging Science: Scientific Knowledge and the Federal Courts.* Cambridge, MA: MIT Press. The main goal of this book is to reconcile the law's need for workable rules of evidence with the views of scientific validity and reliability that emerge from the sciences as well as from social scientific and humanities disciplines that have contribution to our understanding of scientific validity and reliability.

Fuller, Steve. (1994). "The Constitutively Social Character of Expertise." *International Journal of Expert Systems* 7: 51–64.

Golan, Tal. (2004). *Laws of Men and Laws of Nature: The History of Scientific Expert Testimony in England and America.* Cambridge, MA: Harvard University Press.

Goldman, Alvin. (2001). "Experts: Which Ones Should You Trust?" *Philosophy and Phenomenological Research.* 63(1): 85–110.

Hardwig, John. (1985). "Epistemic Dependence." *Journal of Philosophy* 82: 335–349.

Hosch, Harmon. (1980). "A Comparison of Three Studies of the Influence of Expert Testimony on Jurors." *Law and Human Behavior* 4: 297–302.

Hulbert, Ann. (2004). *Raising America: Experts, Parents, and a Century of Advice about Children.* New York: Knopf.

Kovera, Margaret; Robert Levy; Eugene Borgida; and Steven Penrod. (1994). "Expert Testimony in Child Abuse Cases: Effects of Expert Evidence Type and Cross Examination." *Law and Human Behavior* 18: 653–674.

Latour, Bruno. (1998). "From the World of Science to the World of Research?" *Science* 280(5361): 208–209.

Marcuse, Herbert. (1964). *One Dimensional Man.* Boston: Beacon Press Books.

Marcuse, Herbert. (1998). "Some Social Implications of Modern Technology." In *Technology, War, Fascism: Collected Papers of Herbert Marcuse,* vol. 1, ed. Douglas Kellner. New York: Routledge.

Mialet, Hélène. (1999). "Do Angels Have Bodies: The Cases of William X and Mr. Hawking." *Social Studies of Science* 29: 551–582.

Postman, Neil. (1993). *Technopoly: The Surrender of Culture to Technology.* New York: Vintage. This book is a polemic against many of the dangers of technology, including the dangers of allowing experts to have more authority than they sometimes claim to deserve.

Selinger, Evan. (2003). "Feyerabend's Democratic Argument against Experts." *Critical Review* 15(3–4): 359–373.

Selinger, Evan, and Robert P. Crease. (2002). "Dreyfus on Expertise: The Limits of Phenomenological Analysis." *Continental Philosophy Review* 35: 245–279.

Selinger, Evan, and Robert P. Crease, eds. (2006). *The Philosophy of Expertise.* New York: Columbia University Press. This is the first edited volume in which the topic of the philosophy of expertise is treated as a central concern.

Selinger, Evan, and John Mix. (2004). "On Interactional Expertise: Pragmatic and Ontological Considerations." *Phenomenology and the Cognitive Sciences* 3: 145–163.

Schuller, Regina, and Patricia Hastings. (1996). "Trials of Battered Women Who Kill: The Impact of Alternative Forms of Expert Evidence." *Law and Human Behavior* 16: 597–620.

Thorpe, Charles. (2002). "Disciplining Authority and Liberal Democracy in the Oppenheimer Case." *Social Studies of Science* 32: 525–562.

Walton, Douglas. (1997). *Appeal to Expert Opinion: Arguments from Authority.* University Park: Pennsylvania State University Press. This book examines the appeal to expert authority from a historical as well as logical perspective.

Weinstein, Bruce. (1993). "What Is an Expert?" *Theoretical Medicine* 14: 57–73.

EXPOSURE LIMITS

• • •

Exposure limits specify the maximal allowed exposures for individuals to chemical substances or other obnoxious influences such as noise or radiation. Such limits are usually expressed as environmental concentrations (e.g., 0.1 mg/m3 [milligrams per cubic meter] of atmosphere). Biological limits, expressed as blood concentrations, are used for some substances.

Exposure limits apply to all persons in regard to the environment, food, water, and consumer products. Public health, agricultural, and environmental protection agencies in most countries determine public exposure limits covering a wide variety of natural and nonnatural circumstances. The exposure limits that most affect us in our daily lives are probably those that limit the intake of toxic substances through food and drinking water. The area in which exposure limits have been most fully developed, however, are in relation to occupational health regulations.

Occupational Exposure Limits

Occupational exposure limits depend on specific theories about relations between exposures and harms, and on empirical data that can be brought to bear on particular cases. In some cases harms or responses do not begin until a certain threshold of exposure or dose is reached. The other argues that response is continuous from the most minimal exposure (see Figure 1).

There are also different ethical viewpoints about the degree to which workers should be protected in the workplace. One view, for instance, argues that workers are compensated for their exposure to certain possible harms by their wages and salaries, and that the only issue at most is educating them about their exposures. Another view is that workers should be no more exposed to environmental harms in the workplace than out of it. Disagreements between these two ethical views, combined with disagreements about dose–response relations, can lead to quite different interpretations of empirical data relevant to the establishment of occupational exposure limits.

The first occupational exposure limits were proposed by individual researchers in the 1880s. In the

FIGURE 1

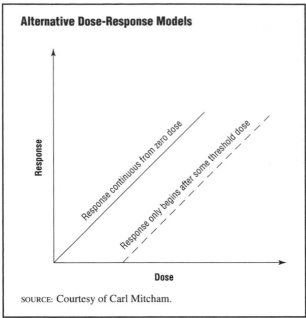

Alternative Dose-Response Models

SOURCE: Courtesy of Carl Mitcham.

1920s and 1930s several lists of exposure limits were published in both Europe and the United States, not always with clear identification of the dose–response relations or ethical views on which they were based. The term *occupational exposure limit* (OEL) was introduced in 1977 by the International Labour Organization (ILO). Other names for occupational exposure standards include threshold limit value (TLV), maximum allowed concentration (MAC), and permissible exposure limit (PEL).

Threshold Limit Values

The American Conference of Governmental Industrial Hygienists (ACGIH) was founded in 1938. In 1941 it set up the Threshold Limit Values for Chemical Substances Committee, which in 1946 issued the first list of TLVs covering around 140 chemical substances. This annually revised list has a dominant role as a standard reference for official lists all over the world.

The first TLV committee was dominated by industrial hygienists and included no physicians. Gradually, medical and scientific expertise was incorporated in the committee. In 1962 the first "Documentation of the TLVs" was published. It contained, for each substance on the list, a brief summary of its effects, with references and with grounds for the TLV that had been chosen.

In the 1940s and 1950s the ACGIH and the American Standards Association (ASA; now the American National Standards Institute [ANSI]) competed for the

position of leading setter of occupational exposure limits. The exposure limits of the ASA and those of the ACGIH did not differ much in numerical terms, but the ASA values were ceiling values below which all workplace concentrations should fluctuate, whereas the ACGIH values were (and still are, with few exceptions) upper limits for the average during a whole working day. Therefore, the ASA standards were more expensive for industry but provided greater protection for exposed workers.

The ACGIH won the struggle and emerged in the early 1960s as virtually the only source of exposure limits that practitioners looked to for guidance. In 1969 the U.S. federal Occupational Safety and Health Administration (OSHA) adopted the ACGIH's exposure limits as an official standard. Because of the sluggishness of legal processes, however, OSHA has not always adopted the updated values subsequently issued by the ACGIH.

In the 1980s, the ACGIH was again challenged. The National Institute for Occupational Safety and Health (NIOSH) criticized its TLVs for being too high, and therefore not protecting workers against potential disease. The alternative values proposed by NIOSH were often many times lower than the TLVs. At the same time, OSHA was criticized for being too harsh on industry. Once again, attacks on the ACGIH were unsuccessful, and the organization retained its position as the leading setter of occupational exposure limits.

Several explanations have been given about why the ACGIH and its TLVs have been so successful. The ACGIH was first with a comprehensive listing of all-important chemicals for which measurement methods were available. As a voluntary body it has been able to update its list annually without the time- and resource-consuming legal procedures that precede revisions of OSHA standards. Furthermore, the comparative ease with which the TLVs can be implemented has probably contributed to their success. Most competing exposure limits, such as those of NIOSH, are more costly and therefore give rise to more opposition from industry.

At the same time, the TLVs have been criticized for being insufficiently protective. Examples of harmful effects at levels below the TLVs are easily found. Grace E. Ziem and Barry I. Castleman (1989) reviewed the contents of four major peer-reviewed journals in occupational medicine for thirty-three months, from January 1987 to September 1989, and found thirty-one papers that described harmful effects at or below the TLVs.

Another common criticism is that the ACGIH has relied too much on unpublished corporate information. Many values have been based on information from a company to the effect that a certain level has been found to be safe, or that no evidence of damage to health has been found at a certain level. This type of information does not satisfy modern criteria for science-based health assessment. Nevertheless, ACGIH's list of TLVs would have covered many fewer substances if such corporate information had not been used. The present policy of the ACGIH is that TLVs "represent a scientific opinion based on a review of existing peer-reviewed scientific literature" (ACGIH website).

Exposure Limits in Other Countries

Since the 1970s most industrialized countries have official lists of occupational exposure limits. In many cases, these lists developed out of the ACGIH list. Because of the less litigious legal culture in Europe, many European countries have national lists of occupational exposure limits that are updated regularly. In some countries such as Sweden and Denmark, the national list has significantly lower values than the ACGIH list.

Developing countries often use the ACGIH list with few or no modifications. As the ACGIH has itself pointed out, however, some TLVs may be unsuitable for use in countries with different conditions from those in the United States, for instance in terms of the nutritional status of workers. The ACGIH also points out that the TLVs "are not developed for use as legal standards and ACGIH does not advocate their use as such."

Some countries, such as the Netherlands and Sweden, have developed an elaborate bipartition of the regulatory task into scientific and policy components. The scientific component is performed by experts in the relevant scientific fields. It derives its legitimacy from the expertise of those who perform it. The policy component is performed by decision makers in government agencies. This component of the process can, in a democratic society, derive its legitimacy only from the same source as other political or administrative processes, meaning that those who perform it must represent the people.

Difficulties in Setting Exposure Limits

The two major sources of knowledge for setting exposure limits are epidemiological studies and animal experiments. In an epidemiological study, groups of humans are statistically compared in search of associations between disease incidence and environmental or other causal factors. The effects of major workplace hazards, such as asbestos, lead, and vinyl chloride, have been convincingly identified and quantified in epidemiologi-cal studies. At the same time many epidemiological studies are inconclusive because of the multiplicity of factors that can influence the prevalence of disease in human populations. Epidemiology also has the crucial disadvantage that the toxic effects of a substance can be discovered only when workers have already been subjected to these effects.

In animal experiments, the health status of exposed animals is compared to that of an unexposed control group. Because of the high degree of biochemical and physiological similarity between humans and the common experimental animals, animal experimentation has predictive power, but unfortunately the predictions are far from perfect. There are substances to which humans are much more, or much less, sensitive than the common laboratory animals.

Because of the uncertainty inherent in both epidemiology and animal experiments, it is in practice virtually impossible to determine with certainty absolutely safe nonzero levels of toxic exposure. Furthermore, for genotoxic carcinogens, it is generally believed that although the risk diminishes with the exposure, it is not completely eliminated until the exposure has been reduced to zero. Accordingly, the ACGIH has stated that the TLVs "represent conditions under which ACGIH believes that nearly all workers may be repeatedly exposed without adverse health effects. They are not fine lines between safe and dangerous exposures, nor are they a relative index of toxicology" (ACGIH website). Other setters of exposure limits have made similar statements.

To set occupational exposure limits is no easy task. Workers exposed to potentially dangerous substances expect exposure limits to fully protect their health. Employers expect the exposure limits to impose only such costs as are necessary to protect employee health. It is in practice impossible to set OELs that fully satisfy both demands. The task of standard setters is to find a reasonable compromise. To achieve this is a science-based enterprise in the sense of making use of scientific information, but not in the sense of being based exclusively on science. It is in fact both science-based and value-based.

As already indicated, the determination of exposure limits involves not just empirical data but also scientific theories about how this data should be interpreted and ethical views about how it should be applied. In some cases the application of very safe exposure limits can put an industrial operation out of business, so that workers are fully protected but only at the cost of losing their jobs. In other cases, not to apply strong exposure limits can have deadly consequences. The adjudication of exposure limits in the workplace, as outside the workplace, is an issues that involves scientific and ethical education on the part of workers, employers, politicians, and citizens.

SVEN OVE HANSSON

SEE ALSO *Limits; Radiation; Risk-Cost-Benefit Analysis.*

BIBLIOGRAPHY

Cook, Warren A. (1985). "History of ACGIH TLVs." *Annals of the American Conference of Governmental Industrial Hygienists* 12: 3–9.

Hansson, Sven Ove. (1998). *Setting the Limit: Occupational Health Standards and the Limits of Science*. New York: Oxford University Press. Comparative study of practices in Germany, Sweden, and the UK.

Paull, Jeffrey M. (1984). "The Origin and Basis of Threshold Limit Values." *American Journal of Industrial Medicine* 5(3): 227–238. Early history of TLVs.

Ziem, Grace E., and Barry I. Castleman. (1989). "Threshold Limit Values: Historical Perspectives and Current Practice." *Journal of Occupational Medicine* 31(11): 910–918. Critical appraisal of TLVs.

INTERNET RESOURCE

American Conference of Governmental Industrial Hygienists (ACGIH). 2005. "Statement of Position Regarding the TLVs and BEIs" Available from http://acgih.org/TLV/PosStmt.htm.

F

FACT/VALUE DICHOTOMY

• • •

Representatives of modern science and its social institutions have repeatedly claimed that science is *value free*, and this claim has contributed to marginalizing serious discussion of the relations among science, technology, and values. Lying behind this claim is the philosophical view that there is not just a distinction but a sharp separation, an unbridgeable gap or *dichotomy*, between *fact* and *value*. The supposed fact/value dichotomy arose at the beginning of the seventeenth century, accompanying the early works of modern science, underpinning an interpretation of their character and epistemic status and became part of the mainstream tradition of modern science (Proctor 1991). Prior to that, it was not a major issue in philosophical thinking about science.

Science and Technology as Value Free

The claim that science is value free is that science deals exclusively with facts and—*at its core*—admits of no proper place for ethical (and social) values. This is not to deny that important relations between science and values exist—for example, that scientific knowledge is a value (even a universal one), that the conduct of scientific research requires the commitment of scientists to certain virtues—such as honesty and courage to follow the evidence where it leads (Merton 1973), and that experimental activities are subject to ethical restraint. Rather, elaborating what it is to keep values out of the core of science, it is to affirm four theses: (1) Scientific knowledge is *impartial*: Ethical values should not be among the *criteria* for accepting or rejecting scientific theories and appraising scientific knowledge. (2) Ethical values have no fundamental role in the *practices* of gaining and appraising scientific knowledge, because the broad characteristics of scientific methodology should be responsive only to the interest of gaining understanding of phenomena. (3) Similarly, research priorities should not be shaped systematically by particular values. The point of both (2) and (3) is that scientific practices are *autonomous*. (4) Scientific theories are *neutral*: Value judgments are not among the logical implications of scientific theories (*cognitive neutrality*); and, on application (e.g., in technology), in principle these theories can evenhandedly inform interests fostered by a wide range of value outlooks (*applied neutrality*) (Lacey 1999). The theses of impartiality and applied neutrality have counterparts regarding the claim that technology is value free. This claim involves the theses: (1) The characteristic criterion of appraisal for technological objects is *efficacy*, the factual issue of whether they work or not. (2) Technology progressively makes it possible to effectively achieve more ends, but it does not privilege any particular ends; its products are available to be used to serve the interests of a wide range of value outlooks (Tiles and Oberdiek 1995).

Sources of the Fact/Value Dichotomy

Materialist metaphysics constitutes one source of the fact/value dichotomy. In the words of Alexandre Koyré (1957), one of the most authoritative historians of early modern science, it—by rationalizing the mathematical and experimental character of science—led to the "discarding by scientific thought of all considerations based upon value-concepts, such as perfection, harmony, meaning and aim, and finally the utter devalorization of being, the divorce of the world of value from the world of facts" (p. 4).

According to materialist metaphysics, the "world of facts" is identical to the "world as it really is in itself." This world consists of the totality of the underlying (normally unobservable) structure and its components, processes, interactions, and mathematically expressed laws, whose generative powers explain phenomena, in a way that dissociates them from any relation to human experience, social and ecological organization, or values—the totality of *bare facts*, purely material facts. On this view, because its aim is to gain understanding of the world, science will attend to grasping the bare facts. Thus, scientific theories should deploy only categories that are devoid of evaluative connotations or implications, such as the quantitative ones (force, mass, velocity, etc.) characteristically used in physical theories. No value judgments follow, for example, from Isaac Newton's law of gravitation, and it makes no sense to ask whether it is good or bad, or whether one ought to act in accordance with it or not. Newton's law expresses a bare fact; faithful to the way the world is, it makes an *objective* statement.

Representatives of modern science often argue that value judgments, by contrast, do not make true or false statements about objects of the world. Rather they serve as expressions of *subjective* preferences, desires, or utilities (perhaps grounded in emotions). In this way, the fact/value dichotomy is reinforced by the objective/subjective dichotomy. Science deals with facts; it is objective. Ethics deals with preferences; it is subjective. The efficacy of technological objects, attested to by confirmed scientific theories, stands on the side of facts. Legitimating their uses, however, involves ethical judgments, which cannot be derived from the bare facts that account for the technology's efficacy and the material possibilities that it makes available.

Epistemology is a further source of the fact/value and objective/subjective dichotomies. Scientific epistemologies identify facts—*confirmed facts*—with what is well supported by empirical data, and the results of established scientific theories. Those that inform technological practices are exemplary instances. Confirmed facts derive from *intersubjectivity*, that is, replicability and agreement, which cuts across value outlooks and cultural norms. Value judgments are not considered intersubjective. Whereas from the metaphysical source, objectivity derives from faithfully representing objects of the world in statements that express bare facts; from the [other] from the epistemological, it derives from the intersubjectivity of confirmed facts. In practice the two notions of fact tend to fuse together and, from both sources, value judgments appear to be subjective,

unlike scientific results that are objective. Hilary Putnam (2002) reviews and criticizes much of the vast philosophical literature on the subjectivity of value judgments.

Finally, *logic* constitutes a third source of the fact/value dichotomy, and for many philosophers it is the principal one. David Hume, in *A Treatise of Human Nature* (1739–1740), is argued to have demonstrated an unbridgeable logical gap between fact and value, because factual statements cannot logically entail value judgments; *ought* is not logically entailed by *is*. The mark of a fact in Hume's argument is a linguistic one: the role of *is* and grammatically related verbs, and the absence of such terms as *good* and *ought*. Less discussed is the complementary thesis, defended by Francis Bacon in *The New Organon* (1620), with his famous injunction to avoid "sciences as one would," to avoid inferring *is* from *ought*, or *good*, or from what serves one's interests; for example, it may serve the interest of legitimating the use of a particular technology that it not occasion serious risks to human health, but that interest is irrelevant to determining what the facts are about the risks.

The Entanglement of Fact and Value

Many criticisms have been made of the fact/value dichotomy, including those of pragmatists and critical theorists. But they all come down to one basic argument, that rather than dichotomy there is some kind of *entanglement* (Putnam 2002) between facts and values. Some (not all) aspects of the entanglement, most of which were discussed by Dewey (1939), are identified below.

NO UNBRIDGEABLE GAP. Many significant factual statements are articulated in scientific theories (such as Newton's law of gravitation). Whether or not a theory is rationally accepted, and thus whether or not statements articulated in it represent confirmed facts, depends on the satisfaction of criteria that require that certain relations obtain between the theory and relevant *observed facts*. Exactly what these relations should be (inductive, abductive) remains disputed; nevertheless, it is clear that the theories are not logically entailed by the observed facts. the criteria that must be satisfied are those for *evaluating* the scientific knowledge and the understanding of phenomena represented in theories.

These criteria have been called *cognitive values* (McMullin 2000); they are a species of values in general, and include empirical adequacy, explanatory power, and consilience. Cognitive values are held to be distinct from ethical, social, and other kinds of values (Lacey 2004), although this is disputed (Longino 1990). Cognitive value judgments concern how adequately cognitive values are manifested in a theory in the light of available observed facts. Soundly accepting that a statement represents a confirmed fact amounts to making the cognitive value judgment that the cognitive values are manifested in the theory to a high enough degree. Far from there being an unbridgeable gap between fact and value, confirmed facts are partly constituted by cognitive value judgments.

FACTS AS PRESUPPOSITIONS AND SUPPORT FOR VALUES. Hume's argument by itself does not rule out that factual statements may provide support for value judgments; otherwise, it would also rule out that observed facts can provide evidential support for facts confirmed within scientific theories, for the fundamental hypotheses of scientific theories are not logically entailed by facts. Logical entailment need not be a particularly important relation in analyzing how facts may support other facts or other kinds of judgments. Consider, for example, the statement: "Recently enacted legislation is the principal cause of the current increase in hunger and child mortality rates." This is a factual statement, because it has the relevant linguistic marks, and empirical inquiry may confirm it to be true or false. At the same time, accepting that it is well confirmed would support holding the value judgment that the legislation should be changed, because, unless there are other factors to consider, it would make no sense to deny that the legislation should be changed, if it is accepted that the factual statement about the causes of hunger has been confirmed. Linked to this, the ethical value of the legislation *presupposes* that it does not have ethically undesirable causal consequences such as increased hunger (Lacey 2004; for a variant of this argument, see Bhaskar 1986).

SOME SENTENCES MAKE BOTH FACTUAL STATEMENTS AND VALUE JUDGMENTS. Declaring that legislation is the cause of hunger may be intended as the statement of a confirmed fact. Alternatively it may serve to express a value judgment, that is, ethical disapproval of the legislation. The logical and linguistic form of the declaration permits it to be used in either role, showing that there is an overlap of the predicates used in factual and ethical discourse. What have been called *thick ethical terms*, terms such as *honest* and *unjust* (also *hunger* and *high child mortality*)—in contrast to *thin ethical terms*, such as *good* and *ought*—may be used simultaneously to serve factual and evaluative ends (Putnam 2002).

Declaring that legislation causes hunger is simultaneously to describe it and normally to criticize it ethically. Using thick ethical terms in factual discourse is no barrier to arriving at results that are well confirmed in the light of the cognitive values and available empirical data; and when such results are obtained, the ethical appraisal is strengthened. Theories that contain such results are not cognitively neutral; they lend support to particular ethical appraisals. Of course, the ethical values of the investigators may explain why they engaged in the relevant research and used the thick ethical terms as their key descriptive categories. Ethical values may influence what facts a person comes to confirm; but they have nothing to do with their appraisal as facts.

SCIENTIFIC APPRAISAL MAY INEXTRICABLY INVOLVE EMPIRICAL CONSIDERATIONS AND VALUE JUDGMENTS. Empirical appraisal never provides certainty; in principle, even the best confirmed statements might be disconfirmed by further investigation. Thus, when a hypothesis is applied, the appraisal made is that it is sufficiently well confirmed by available empirical evidence so that, in considerations about the legitimacy of its application, it is not necessary to take into account that it might be disconfirmed by further investigation, and that, if it were, it might occasion negatively valued outcomes. In the light of this, the standards of confirmation that need to be satisfied depend upon how valuatively significant are these outcomes (Rudner 1953).

MODERN SCIENCE HAS FOSTERED THE VALUE OF EXPANDING HUMAN CAPACITIES TO EXERCISE CONTROL OVER NATURE. Because there are confirmed facts—that is, facts that reliably inform human action—that deploy thick ethical terms, not all confirmed facts are bare facts. This challenges the metaphysical view that the world "as it really is" is identical to the totality of bare facts; and, indeed, it is neither a bare fact, nor a confirmed fact, that the world "really" is that way. Scientists may make the choice to attend only to bare facts. Although this is not the only way to gain factual knowledge, it has generated an enormous amount of knowledge of inestimable social and technological importance. Moreover, because its categories are (by design) chosen to describe facts without the use of thick ethical terms, this knowledge has no ethical judgments at all among its implications. Approaching scientific research, attending only to bare facts, produces results that are cognitively neutral.

At the same time, the contribution of scientific knowledge to enhancing human capacities to exercise control over nature has been highly valued throughout the modern scientific tradition. It has been argued (Lacey 1999) that the approach to scientific research that attends principally to bare facts gained virtual hegemony because of its dialectical links with according high ethical value to enhancing human capacities for control, as well as the exercise of these capacities in ever more domains of life. Bare facts are especially pertinent for informing projects of technological control. Furthermore, sometimes the results of modern science (for example, the developments that have produced transgenic crops) have little application where competing values are held (such as the values of simultaneously gaining high productivity, ecological sustainability, protection of biodiversity, and empowerment of local producers [Altieri 2001]). Thus, while the results gained in this approach of modern science are cognitively neutral, they do not, on the whole, display applied neutrality. (For a variant of this point, see Kitcher 2001.) Humans have considerable knowledge of bare facts (in part) because the values about control are widely held in society and shape scientific institutions. It is not the nature of the world that leads humans to search out such facts but, contrary to the claim of autonomy, a choice highly conditioned by social and ethical values, one that Robert N. Proctor (1991) refers to as "political."

Assessment

Not all the components of the claim that science is value free can be sustained. While there are important results that are cognitively neutral, that is, results that do not logically entail value judgments, in general results do not fit applied neutrality; that is, they are not evenhandedly applicable for a wide variety of value outlooks. Moreover, because applied neutrality does not hold in general, the claim that technology is value free cannot be sustained. There is no objection, then, to engaging in research for the sake of obtaining results that could inform one's ethically favored projects. What confirmed facts are actually obtained reflect these values. That they are confirmed facts does not. The ideal of impartiality remains intact. Ethical values are not among the cognitive values, so that ethically laden commitments (ideological, religious, political, entrepreneurial) are irrelevant to appraising knowledge claims. Science does not need the strong separation of facts and values in order to protect the ideal of impartiality. It needs only a nuanced account of their entanglement.

HUGH LACEY

SEE ALSO *Scientific Ethics.*

BIBLIOGRAPHY

Altieri, Miguel. (2001). *Genetic Engineering in Agriculture: The Myths, Environmental Risks, and Alternatives.* Oakland, CA: Food First. Provides an example showing that developments of transgenic crops do not fit the thesis of applied neutrality.

Bhaskar, Roy. (1986). *Scientific Realism and Human Emancipation.* London: Verso. Qualified argument that factual statements may logically imply value judgments.

Dewey, John. (1939). *Theory of Valuation,* in *Foundations of the Unity of Science: Toward an Encyclopedia of Unified Science,* ed. Otto Neurath, Rudolf Carnap and Charles Morris. Chicago: University of Chicago Press. Primary source of arguments for the entanglement of fact and value.

Kitcher, Philip. (2001). *Science, Truth, and Democracy.* New York: Oxford University Press. Analysis of the interaction between the cognitive and value aspects of science.

Koyré, Alexandre. (1957). *From the Closed World to the Infinite Universe.* New York: Harper Torchbooks. Influential history of the changes that inaugurated modern science.

Lacey, Hugh. (1999). *Is Science Value Free? Values and Scientific Understanding.* London: Routledge. Comprehensive critique of the claim that science is value free.

Lacey, Hugh. (2004). "Is There a Significant Distinction between Cognitive and Social Values?" In *Science, Values, and Objectivity,* ed. Peter Machamer and Gereon Wolters. Pittsburgh, PA: Pittsburgh University Press.

Longino, Helen E. (1990). *Science as Social Knowledge.* Princeton, NJ: Princeton University Press. Important analysis of the social and value dimensions of science.

McMullin, Ernan. (2000). "Values in Science." In *A Companion to the Philosophy of Science,* ed. W. H. Newton-Smith. Oxford: Blackwell. Important source on cognitive values.

Merton, Robert, ed. (1973). *The Sociology of Science.* Chicago; University of Chicago Press. Contains the classic statement of the "scientific ethos," the virtues required for scientific research.

Proctor, Robert N. (1991). *Value-Free Science? Purity and Power in Modern Knowledge.* Cambridge, MA: Harvard University Press. A comprehensive critique of the thesis of neutrality of science.

Putnam, Hilary. (2002). *The Collapse of the Fact/Value Dichotomy and Other Essays.* Cambridge, MA: Harvard University Press. Provides many arguments, in the spirit of Dewey, against the fact/value dichotomy.

Rudner, Richard. (1953). "The Scientist *qua* Scientist Makes Value Judgments." *Philosophy of Science* 20: 1–6. Famous brief article that argues that value judgments cannot be kept out of science.

Tiles, Mary, and Oberdiek, Hans. (1995). *Living in a Technological Culture.* London: Routledge. Presents a detailed critique of the claim that technology is value free.

FAMILY

• • •

The family is one of a number of basic social institutions that have been subjected to scientific study and affected by changes in science and technology. Because of the fundamental role the family plays in socialization, including the inculcation of moral behavior and ethical attitudes, it merits consideration in relation to science, technology, and ethics.

Throughout human history there has been a strong relationship between the family and technology. That relationship can be understood by tracing the successive technological revolutions that began with hunting and gathering societies and the discovery and use of tools and progressed through a series of technological societies, such as horticultural, agricultural, industrial, and postindustrial societies (Ribeiro 1968). Each successive step has altered ways of thinking and doing things by human beings, and this progression has been made possible primarily by new means of environmental adaptation. In the past families provided the organizational structure needed to develop tools and techniques to meet basic human needs, and this has continued in many ways into the present.

Defining the Family

A family may be defined as a group of people linked by descent. However, because descent can be understood in biological or nonbiological terms and is subject to narrow or broad interpretations, the scientific study of the family has led to the recognition of a number of basic distinctions. Indeed, the family has taken different forms throughout history and across cultures, related to diverse functions. In hunting and gathering and horticultural societies the kin group performed all religious, economic, and political functions. Kinship groups were broad enough to include relationships with almost everyone with whom a person interacted (Radcliff-Brown 1930). The kin group remained the major socializing agent, and the production and consumption of material goods continued to be centered in the family.

With the advent of agrarian families inheritance of property, primarily along male lines, became a central concern. The evidence from several studies (Gough 1971) indicates that because of land ownership and a more settled way of life the power of males increased (compared with the situation in hunting societies) at the expense of females. In agrarian families parents had considerable control over their children. However, agrarian families still were concerned about alliances with immediate and distant relatives. Those alliances are known as the extended family.

Industrialization brought the rise of the conjugal family unit. The nuclear family was becoming less embedded in the extended family, bringing a host of changes (Goode 1963). The major changes, according to William J. Goode, included social mobility, specialization, and geographic mobility. The family was no longer economically a producing unit, but its function as a consuming unit was heightened. In addition, many functions were outsourced from the family unit, resulting in greater dependence on the larger society.

The relationship between families and society thus has undergone major changes throughout history. In hunting societies institutions such as the economy and religion were embedded in kin groups. In agricultural society various institutions still were embedded in the extended family, although some institutional differentiation started to appear. In industrial society, disembedding reached a peak and the nuclear family became one of the many institutions that served individuals.

These distinctions are especially important in understanding interactions among families, science, and technology in relation to three social functions: producing and consuming material goods, information technology, and human reproduction.

Producing and Consuming Material Goods

The earliest families used hunting and gathering as their modes of production. Family members were producers of food for sustenance, and most tools were associated with the basic activities of survival: spear and bow and arrow for hunting, stone ax for skinning animals to make clothing, and basketry for food gathering. Hunting tools were made from stone, bone, and wood. Hunting and gathering societies were small and migrated frequently. In the family gender roles were defined clearly: Men had a monopoly on hunting, and women gathered food and raised children. In addition, men, being physically stronger, were expected to defend the tribe, thus accumulating more decision-making power. Once those gender roles became traditional, they were considered not only practical but "natural." Some scholars believe that this is where sexism or gender superiority began, although not in as pronounced a form as was to occur later.

Families continued to be producers of food in horticultural societies. Horticultural societies, a precursor of agricultural societies, were based on hoe agriculture: small-scale farming using a hoe and a digging stick.

Some copper tools and sickles made from clay fired at a high temperature also were used. Horticulture allowed populations to settle and provided some permanence in people's lives, something that had not been possible in foraging societies.

Simple horticultural societies gave way to agrarian societies around 3000 B.C.E. The family remained the primary producer, but agriculture provided the means to move from an existence that was dependent on what was given by nature to one of active participation, utilizing the environment to enhance the potential for a better life. One of the most important innovations in that period was the introduction of plow cultivation, which Gudmund Hatt (1961, p. 218) has called "the prerequisite of civilization." Animals were used to pull a plow. With the introduction of iron, superior plows, weapons, and tools were produced. Male dominance increased because agrarian tasks required greater strength and more intensive labor. Women's status declined further because of economic dependence, which was a result of a lack of direct contribution to the economic activities required in large-scale agriculture.

Families in industrial societies lost many of their production functions and became little more than a source of labor. Gerhard Lenski and Jean Lenski (1987) divided the Industrial Revolution into four phases on the basis of technological innovation. The first phase (1760–1850) began in England with major developments in the textile, iron, and coal industries. The second phase (1850–1900) saw expansion throughout most of Europe and North America. The steam engine was adapted for transportation by railroads and steamships. Agricultural production increased with the use of new kinds of machines and chemical fertilizers. Family ownership of companies began to give way to corporations, and the number of industrial workers increased substantially.

The third phase (1900–1940) was characterized by major advances in energy technology. The use of automobiles increased in most industrialized countries, and with it the demand for petroleum. Most homes were electrified and were connected to others by telephones. The fourth phase (1940–1970) saw major changes in the aviation industry spurred by World War II. The war economy also saw the expansion or development of nuclear power, plastics, and aluminum. Entertainment industries such as television, radio, and films experienced tremendous growth. The industrial sector became automated, and the nature of labor changed considerably. The most important innovation in this phase was the development of electronic computers.

All this technological innovation had a substantial impact on the structure and functions of the family. Home and work were separated. Family members—mostly men—had to work outside the home to purchase goods and services. In the early phase of industrialization the status of women reached its lowest point because women had no role in the economy outside the family. However, that changed after World War II as women entered the workplace in increasing numbers. By contrast, children had to wait longer to enter into the labor market because industrial economies required specialized skills. Hence, their economic dependence on parents increased compared with that of their counterparts in agricultural economies.

Information Technology and the Family

The concept of a "postindustrial society," as developed by Daniel Bell (1973), refers to a new mode of technological and economic production that is based increasingly on information and services. Information technology (IT) has revolutionized almost all aspects of human life.

Although its effects on families vary, family members tend to use IT as often in managing home life as in regulating work-home relationships. Pagers, faxes, cell phones, telephone answering systems, and computers are used to keep track of children, spouses, and other family members. Paging children to find out if they have arrived home safely from school demonstrates parental responsibility. E-mail and cell phones keep family members in constant contact. In addition, family members use answering machines, cell phones, and palm pilots to coordinate complex household schedules. However, not every effect of the use of IT has been positive. The colonization of home time by work is one obvious negative. The technical ability to work from home has blurred the distinction between workplace and family life.

Human Reproduction

Technology also has altered human reproduction, initially by means of birth control. Artificial contraception has disembedded conception from sexual intercourse and made fertility dependent on personal decisions. At the same time, with the growth of genetic research infertile couples now have many options for having a child through assisted reproductive technologies (ARTs). Among the many types of ARTs, artificial insemination is the most common.

The other commonly used procedure is known as in vitro fertilization (IVF), in which a woman's eggs are removed surgically and placed in a petri dish with sperm from her husband or a donor. One or more fertilized eggs then can be implanted directly into the woman's uterus. Sometimes extra fertilized eggs are frozen for possible later use. Surrogate pregnancy is available for women who cannot bear children.

Ethical Considerations

Science and technology not only have altered family life, they have generated fundamental ethical questions. The most controversial are associated with reproductive technology.

Are new technologies redefining previously held notions of family, parent, mother, and father? In embryo transplantation a fertilized egg from a female donor is implanted into an infertile woman. The developing embryos may be tested for genetic abnormalities before implantation. Critics (Benokraitis 2002) have raised concerns about parental rights to reject imperfect fetuses and create "designer babies." In 1999 a wealthy couple placed an ad in the newspapers of universities such as Harvard, Princeton, and Stanford, offering $50,000 for the eggs of a woman who was intelligent (SAT score 1,400 and higher), athletic, and tall. More than 200 college women responded to the advertisement (Weiss 2000). The implications are that rich people can "manufacture" babies and women who are in debt may offer their bodies.

A surrogate mother may decide to keep the baby. In 1987, in the celebrated "Baby M" case, Mary Beth Whitehead was artificially inseminated with the sperm of William Stern. Although Whitehead signed a surrogate contract with Stern, she changed her mind and turned down the $10,000 she had contracted to receive. She lost the case, but the court granted her visitation rights. Many critics who object to surrogacy argue that it exploits poor women because rich couples can afford to "rent a womb" (Benokraitis 2002).

Barbara Rothman raises a larger question in regard to motherhood: Should it always be defined in biological terms? Her answer is no. For Rothman, "Every woman is the mother of the child she bears, regardless of the source of the sperm and regardless of the source of the egg. The law must come to such an explicit recognition of the maternity relationship" (Rothman 1994, p. 201). In another situation surrogacy raised a complicated question of kinship. In 1991 Annette Schwartz served as a surrogate for her daughter, Christa, and gave birth to twins who became Christa's legal children. What kinship term should be applied to the relationship between the twins and Schwartz (Benokraitis 2002)?

Questions also arise about the impact of IT on the family. Information technology has had many unintended consequences. The issues of privacy and security breaches have become a major problem in most advanced countries. For instance, users of cellular phones can have their location tracked and hackers can get into family computers and remove private information. Parents can use IT to keep track of their children.

Does IT create stronger social relationships or distract people because it does not promote face-to-face relationships? Harold Rheingold (1993) and Sherry Turkle (1995) argue that computers and telephones provide emotional support and a sense of belonging. However, skeptics such as Mark Slouka (1995) and Clifford Stoll (1995) think that online relationships are narrow and lacking in quality. Those relationships are also manipulative because making affiliations in an electronic medium teeming with strangers is dangerous for young people. Moreover, "In the office, in their cars, and in their houses, the demands of work come pouring in. Work is so pervasive that conventional boundaries between work and home have all but collapsed" (Rheingold 2002, p. 191).

The fact that the use of technology is never neutral is demonstrated clearly by its impact on the family. Technology has been considered the hallmark of civilization; it has enabled humans to overcome the inertia and entropy of a harsh physical environment. However, dependence on technology has created cultural disorder as well, calling forth ethical reflection and responses.

MURLI M. SINHA

SEE ALSO *Education; Eugenics; Feminist Perspectives; Genetic Counseling; In Vitro Fertilization and Genetic Screening; Natural Law; Psychology.*

BIBLIOGRAPHY

Bell, Daniel. (1973). *The Coming of Post-Industrial Society.* New York: Basic Books. Deals with changes in social structure and the influence of science and technology and society and polity.

Benokraitis, Nijole V. (2002). *Marriages and Families: Changes, Choices, and Constraints.* Upper Saddle River, NJ: Prentice-Hall. A comprehensive analysis of major issues facing families in the twenty-first century.

Goode, William J. (1963). *World Revolution and Family Pattern.* New York: Free Press. Deals with the Industrial Revolution and how it was associated with predictable changes in family patterns.

Gough, Kathleen E. (1971). "The Origin of the Family." *Journal of Marriage and the Family* 33: 760–770. Traces the evolution of the family in various cultures.

Hatt, Gudmund. (1961). "Farming of Non-European People." In *Plough and Pasture: The Early History of Farming*, ed. E. Cecil Curwen and Gudmund Hart. New York: Collier. Studies the history of farming in various cultures.

Lenski, Gerhard, and Jean Lenski. (1987). *Human Societies: An Introduction to Macrosociology*. New York: McGraw-Hill. Attempts to classify human societies from hunting and gathering to industrial societies.

Radcliff-Brown, A.R. (1930). "The Social Organization of Australian Tribes." *Oceania* 1: 44–46. The organization of earlier societies in terms of economy, religion, and polity.

Rheingold, Harold. (1993). *The Virtual Community: Homesteading on the Electronic Frontier*. Reading, MA: Addison-Wesley. Examines the social phenomena emerging from the early days of the Internet.

Rheingold, Harold. (2002). *Smart Mobs*. New York: Basic Books. Studies the impact of mobile communication technology and how people use it.

Ribeiro, Darcy. (1968). *The Civilization Process*. Washington, DC: Smithsonian Institution Press. Presents stages of change from earlier society to the present.

Rothman, Barbara. (1994). "Artificial Means of Production and Our Understanding of the Family." In *Contemporary Issues in Bioethics*, ed. Tom Beauchamp and LeRoy Walters. Belmont, CA: Wadsworth. How technology has changed the reproductive system.

Slouka, Mark. (1995). *War of the Worlds: Cyberspace and the High Tech Assault on Reality*. New York: Basic Books.

Stoll, Clifford. (1995). *Silicon Snake Oil: Second Thoughts on the Information Highway*. New York: Doubleday

Turkle, Sherry. (1995). *Life on the Screen: Identity in the Age of the Internet*. New York: Simon & Schuster.

Weiss, R. (2000). "Limited Pay for Egg Donors Advised." *Washington Post*, August 4.

FASCISM

• • •

Fascism played a major role in twentieth-century European and world history, especially in its attempt to develop a particular nonliberal and nonhumanistic modern perspective on science and technology. Fascism, in power, was a form of rule where key societal resources were monopolized by the state in an effort to penetrate and control many aspects of public and private life, through the state's use of propaganda, terror, and technology.

Fascism also remains a highly complex and illusive political phenomenon. Classical fascism (the small *f* for comparative purposes) can be described in terms of a number of loosely-related early-twentieth-century political parties, movements, and regimes, especially in Germany (Adolph Hitler's National Socialism), Italy (Benito Mussolini's *Fascism* proper, from which the generic term fascism is derived), and Spain (Francisco Franco's more radical wing of Falangism).

All fascisms oppose communism, the values of liberal democracy, rationalism, and scientific positivism, with assertions of bellicose nationalism, and each variety of fascism has sought in its own manner and cultural context to adopt advanced military, penal (including in the Nazi variant genocidal), and communication (propaganda) technologies, while criticizing the universalism and humanism of liberal science and technology. Some suggest that a *palingenetic* and inherently revolutionary mythology of rebirth ultimately binds all authentic forms of generic fascism together and separates them from authoritarian and reactionary military dictatorships, and totalitarian regimes such as that of Stalin (Griffin 1993).

In spite of fascisms' ritualistic invocation of an idealized past—the Nazi Aryan myth; the Italian myth of Rome, the preoccupation with the glorious age of Elizabeth I among interwar British fascists—fascism actually emerged from a background steeped in pseudoscience and social Darwinism, and the high-tech myths of futurism, and as such can be seen as an authentically *modern* movement, especially in terms of its attitudes toward, and application of, science and technology.

Fascism has also persisted since the collapse and defeat of the Mussolini and Hitler regimes, in various manifestations of *neo-* or *post-* fascism, operating as a sometimes influential, but often marginalized, opposition movement within liberal democracy. Latter day fascists often deny their fascist roots, or operate clandestinely, because of the negative and reviled nature of fascism because of its well known and understood connection with the systematic process of Nazi war and genocide. Others have been partially absorbed into liberal democracy and deradiclalized.

Fascism in Italy (1922–1943)

In 1932 the fascist dictator Mussolini, with the considerable help of the neo-Hegelian philosopher Giovanni Gentile (1875–1944), contributed an entry to the *Encyclopedia Italiana* on the definition of fascism. Italian conceptions of the work of Georg Hegel derived largely from Benedetto Croce (1866–1952), a philosopher of international repute. Mussolini asked Croce to write this doctrine of fascism for him, but Croce refused. But in Gentile's writings, Mussolini discovered a serviceable philosophical peg of neo-Hegelian idealism on which to hang his brutal, vitalistic doctrines.

The famous entry contains elements of Gentile's personal criticisms of liberal and post-enlightenment science and technology, depicting the state as the source of all ethics, individual as well as collective. A key passage reads:

> The State, as conceived and realized by Fascism, is a spiritual and ethical entity. . . . which in its origin and growth is a manifestation of the spirit. The State . . . safeguards and transmits the spirit of the people, elaborated down the ages in its language, its customs, its faith. The State is not only the present; it is also the past and above all the future. . . . the State stands for the immanent conscience of the nation. The forms in which it finds expression change, but the need for it remains. . . . it transmits to future generations the conquests of the mind in the fields of science, art, law, human solidarity; it leads men up from primitive tribal life to that highest manifestation of human power, imperial rule. . . . Whenever respect for the State declines and the disintegrating and centrifugal tendencies of individuals and groups prevail, nations are headed for decay. (Mussolini 1932, p.26)

Aside from this perverted Hegelian notion that the state equals life itself, the entry usefully emphasizes other core themes of Italian Fascism: a firm belief in the concrete reality of life, anti-individualism, antiliberalism and liberal democratic sentiments, antisocialism, the call for *action* and revolution, a denial that happiness is achieved through comfort and *well-being*, the belief that fascism is ultimately a *spiritual force*, and the idea that fascist ideology was a far stronger ethical basis for existence than any mere rule of law (on this last point see the writings of Carl Schmitt).

A major strand of Italian fascist technologism emerged from the prewar futurist movement in art, founded in 1909 by the poet Filippo Marinetti. Futurism arose as part of the general modernist artistic ferment that characterized the intellectual life of Europe, and particularly France and Italy, in the period before 1914. The futurists' goal was to celebrate modern technology and to free Italian art from the psychology of the past. In 1910 Umberto Boccioni published the *Manifesto of the Futurist Painters*. The cult of the machine age was central to futurism and from the beginning futurist ideology was saturated with violence and aggression. The infatuation with speed, change, and modernity soon became intertwined with ultranationalism and in 1915 Marinetti published *War—the Sole Hygiene of the*

World, placing science and technology at the service of war and brutal imperialism. What had started as the rejection of stagnation in art became an all-encompassing, authoritarian political message, in which all *decadent* (code for liberal and leftist) manifestations of the old Italy were to be overthrown. Under Mussolini's regime, futurism lost its radical edge and was largely confined to producing extravagant plans for buildings in the futurist style, very few of which were actually built.

But the tenor and relentless propaganda of the regime remained focused on placing the latest science, technology, and management techniques at the disposal of the Italian people—hence grand public buildings such as the Milan and Florence railway stations, the Autostrada, and electrification of the main railways network. There was also a ceaseless drive to embrace the dynamism of the second industrial revolution embodied in the Fascists' Third Rome, the exploitation of hydroelectric power, the propaganda surrounding the launch of any new Fiat vehicle, and Italo Balbo's daring flying antics in the United States. Fascist propagandists also strove tirelessly to emphasize the link between technology, science, modernization, and the regime. In addition, attempts were made to create new institutions for managing the modernization process, institutions of an authoritarian, technocratic character such as the Confederazione Generale dell'Industria Italiana and the Gruppi di Competenza. In addition, genuinely innovative institutions were created to manage the modernization process: Confederazione Generale dell'Industria Italiana, Gruppi di Competenza, Consigli Tecnici, and Istituto per la Ricostruzione Industriale. The highly technocratic Guiseppi Bottai, Minister of Corporations and also editor of *Critica Fascista*, used every opportunity to emphasize the technocratic and scientific core of the "New Italy" and its third way "Corporate State"—a process that rapidly ran out of steam when he ceased to be Minister in 1932. (For the best account of fascist modernism, see Griffin 1994.)

Germany (1933–1945)

Fascism in Germany was, in almost every aspect, the most radical and extreme manifestation of fascist ideology, putting science and technology to the most unethical of uses, including mass genocide achieved through Ford-style, efficient factory methods The German regime eventually waged a brutal, and for the period, high-tech war through the development of weapons of mass destruction and rocket-propelled delivery systems. From the road construction of the Todt Organization to the development of the V3 rocket bomb, there is no

Benito Mussolini addressing troops. The Fascist dictator was head of the Italian government from 1922 to 1943 and led Italy into three successive wars, the last of which overturned his regime. (© Bettmann/Corbis.)

question that Nazi Germany promoted a culture of advanced technology (Griffin 1994). As Roger Griffin cogently puts it, "the Third Reich was saturated with technocratic values.... The V3 rocket bomb could hardly have been developed by an anti-technological culture" (Griffin 1994, p.10).

Part of the reason for this was that Nazism emerged from a cultural climate imbued with the idea that the West was degenerating, a fear dating back to Edward Gibbon's *Decline and Fall of the Roman Empire* (1776–1787), and an associated sense of the urgency of the task of regeneration and rebirth. The rise of Nazism also coincided with the period of the most influential writings of Martin Heidegger (1889–1976) and Schmitt (1888–1985). Heidegger favored a form of antimodern rule that would restore *Being* to its proper role in Western affairs. Mistaking the Nazis as the political basis for

a rebirth of technology and humanity, he threw in his lot with Hitler's regime, and never repudiated Nazism, continuing to speak of its *inner truth and greatness.*

It was Friedrich Nietzsche who famously fathered a distinctly Germanic critique of decadent European society. He affirmed that a regenerative instinct for the *Will to Power* realized through the *blond beast* could destroy weak institutions and beliefs, blazing a trail for *the vital, the powerful, and the creative.* German interwar thinkers and public intellectuals adopted Nietzsche's Zarathustra in a bastardized and popularized form, as a symbol for a rejuvenating *Kultur* capable of overcoming the effects of decadent commercial and wasteful technological civilization. Nietzsche's writings influenced many of the leading German cultural pessimists, especially Ernst Junger (1895–1998) and Oswald Spengler (1880–1936) who, in turn, influenced Heidegger.

As Michael Zimmerman has argued, "Jünger claimed that the soft, decadent, and unmanly European bourgeoisie was being displaced by *der Arbeiter* (the Worker), a new type of humanity combining the steely hardness of modern technology with the iron will of a proto-Nietzschean blond beast. Jünger foresaw a powerful new upsurge of Will in the face of Western decrepitude" (p. 14). Adopting Junger's rhetoric of struggle and hardness, Heidegger appeared to the less sophisticated minds to be exhorting all Germans to submit to a technological Will to Power in order to overcome decline and despair (*uberwinden*).

But in fact he criticized the technological Will to Power and argued for its transcendence through Volk politics and *Gelassenheit* (detachment)—a highly misplaced hope in the case of the Nazis. Heidegger's ontological language was pitched at such a high level of linguistic and philosophical abstraction that it was impenetrable to most intellectuals, and his solution was an equally obscure and backward-looking spiritual renewal far too abstract for his Nazi masters to grasp. He was naturally predisposed toward Nazi ultranationalism through the special destiny he assigned to the German people because of their language, which he saw as the natural heir to classical Greek—a pure philosophical language, a quality that had disappeared from all other Western European languages. In addition, Heidegger's key concept of *Dasein* (a combination of the words *being* [*sein*] and *here* [*da*]) was based on the belief that the real is also rational, and, after 1933, the here was nazism and the obvious concrete power of National Socialism was, for him, an *uncovering* of authentic Being.

Heidegger's initial enthusiasm for nazism was soon reduced by the complete lack of interest the Nazis showed in his philosophy. As rector of Freiburg University in 1933, Heidegger delivered a famous address in which he announced that he had the correct philosophical understanding of National Socialism, but the Nazis did not understand him. His exclusive form of philosophical National Socialism was not based on any concept of race or imperial conquest and was, therefore, completely irrelevant to his political bosses.

Schmitt, a pupil of Max Weber, was a leading German thinker on constitutional law who wrote several seminal studies during the Weimar period and became known as *the enemy of liberalism*. Like Heidegger he entered the Nazi university establishment after 1933. Schmitt rejected *cosmopolitan* ideals and the intrinsic goodness of humankind and argued that the law was ultimately subservient to politics. Liberalism, meanwhile, offered a false universalism, which obscured the existentially paramount nature of politics and replaced it with the struggle for abstract notions of *rights*. Political *reality* ultimately transcended all legal norms for Schmitt, who supported the existential over the theoretical. Thus war lacks any normative justification, its reason lying not in ideals of justice, democracy, or economic prosperity, but in preserving the very existence of the sovereign and sacred polity when it is threatened—in this case a Germany threatened by decadent liberals, Jews, and communists.

Despite his openly Nazi ideals, Schmitt's work proved influential on later authoritarian conservatism outside Germany; Raymond Aron referred to him as a great social philosopher in the tradition of Weber. His writings continue to influence the left, as demonstrated by the content of the journal *Telos*, and to fascinate poststructuralists, including Chantal Mouffe (1999) and Jacques Derrida.

Spanish Falangists (1936–1975)

The Falange was a quasi-fascist political organization, which constituted the single official party in Spain between 1939 and 1975, making it the longest-lasting fascist-style regime. This minor party was founded in 1933 by Jose Antonio Primo de Rivera, and, with other parties, became the Spanish Phalanx of the Assemblies of National-Unionist Offensive.

During the Spanish Civil War, the Falangists fought on the *nationalist* side against the left-led Republicans. When Franco seized personal power, he united the Falange with the Carlist monarchists, forming the Movimiento Nacional—thus purging the Falange of its more radical and modernizing fascistic elements. After the war, moderate Falangist ministers had an important role in Francoism, but Franco turned increasingly to younger politicians thus allowing Spain be dominated by the technocratic wing of Falangism, whose policies arguably promoted a return to democracy.

Non-European Fascisms

Minor potential examples of generic fascism and neofacism have existed elsewhere in Europe and around the world both before and since 1945. But the only other continent that has witnessed significant concentrations of quasi-fascist parties, movement, and regimes is Latin America, principally Paraguay, Argentina, and Chile.

Between 1954 and 1989, Alfredo Stroessner's authoritarian Colorado Party made Paraguay a safe haven for Nazi war criminals such as Josef Mengele. However the most significant neofascist regime existed in Argentina under Juan Perón (president from

1946–1955 and 1973–1974) who fostered a powerful populist, authoritarian personality cult initially with the assistance of his beautiful but ill-fated wife Eva.

Islamic Fascism

Some claim that Islamic Fascism exists and is also a phenomenon of modernism (see Wistrich 2001). In this thesis it is basically a twentieth-century totalitarian movement—like fascism and communism. Islam existed before Islamic Fascism, and will exist after it.

Islamic Fascism is designed—like fascism and communism—to appeal to idealistic young people with a utopian future where the world will be *cleansed*. It really started with the Iranian revolution in 1979, and was formerly called *Islamic fundamentalism*. Other names for it include *Islamofascism* or *Islamism*. And—like fascism and communism—its only solution, according to its adherents, is the total and utter destruction of western liberal and Christian culture and philosophy. This may, of course, require a long cold war, lasting for perhaps the next two or three decades, punctuated by perhaps one or two more hot wars, but Islam will prevail.

Broader Issues for Science, Technology, and Ethics

Among other things, the rapid rise of fascism illustrates the severe problem of cultural disorder created by radical and rapid scientific and technological change in the early-twentieth century and the associated difficulty of moving from essentially premodern traditional societies to modern rationalistic, scientific, and technological societies with mass democratic systems.

By nature fascism is clearly opposed to those aspects of modernity linked with decadence, particularly cultural-pluralism, liberalism, and materialism. There are obvious examples of premodern thought within fascism—for instance the Blood and Soil movement, ideals in both Germany and Italy of regeneration of the peasantry and the restoration of the ancient bond between Germans and Italians and the land. Yet fascism is by no means entirely antimodern, as Gentile suggested:

> ... as a descendant of early twentieth-century modernist nationalism, fascism does not identify with anti-modernism, but in its own way ... it had a certain passion for modernity not inconsistent with its harking back to the traditions of the past ... The fascists saw themselves as the modern "Romans" ... compatible with the myth of the future and with fascism's ambition of revising modernity in order to leave its mark on the new

civilization in the age of the masses. (Gentile 1993, p. 24–25)

At one level fascism clearly represented a rejection of liberal scientific positivism. But equally, as Roger Griffin (1994) argues, it contained a readiness to employ the latest scientific and technological techniques to destroy liberalism and communism and achieve its irrationalist and dystopian ends.

Many varieties of fascism also tried to replace orthodox religion with a perverted secularized and spiritualized modernism, based in part on developing and deploying the dazzling potential gains of modern science and technology and offering the chimera of an economics of plenty—a technological heaven on earth. Indeed Gentile has depicted Italian Fascism as the first and most highly developed form of modern mass political religion—offering a new ideology to fill the void left by the decline of traditional religion in Italy. Earlier cults and myths of Italian ultranationalism forged the basis of a civic religion that was then colonized and adapted by the Fascist party. As such, Italian Fascism was a vital catalyst for contemporary Italian mass politics (Gentile 1996).

Fascism clearly demonstrates the considerable negative as well as liberating functions of modern science and technology, with the state entirely taking over its promotion, direction, and end use for the deeply unethical purposes of brutal imperialist wars and, in the case of Nazi Germany, systematic mass genocide. It is, perhaps, useful to speculate on what latter-day Nazis would do with current cloning techniques and biotechnology, or with the latest weapons of mass destruction—chemical, biological and nuclear based. And with regard to the miracles of modern mass communications: the frightening image of tall men in stylish black Nazi uniforms waiting at Heathrow, or JFK, talking animatedly into their exclusive SS-issue mobile phones and opening their sleek black SS laptops in a wireless-zone, to contact the web and read their encrypted emails, comes all too readily to mind.

DAVID BAKER

SEE ALSO *Arendt, Hannah; Authoritarianism; Critical Social Theory; Socialism; Totalitarianism.*

BIBLIOGRAPHY

De Grand, Alexander. (1995). *Fascist Italy and Nazi Germany: The Fascist Style of Rule.* Cambridge, UK: Cambridge University Press.

Eatwell, Roger. (1992). "Towards a New Model of Generic Fascism." *Journal of Theoretical Politics* 4(2): 161–194.

Eatwell, Roger. (1996). "On Defining the Fascist Minimum: The Centrality of Ideology." *Journal of Political Ideologies* 1(3): 303–319.

Gentile, Emilio. (1993). "Impending Modernity: Fascism and the Ambivalent Image of the United States. *Journal of Contemporary History* 28(1): 7–29.

Gentile, Emilio. (1996). *The Sacralization of Politics in Fascist Italy*. Cambridge, MA: Harvard University Press.

Griffin, Roger. (1993). *The Nature of Fascism*. London: Routledge.

Griffin, Roger. (1994). *Modernity under the New Order: The Fascist Project for Managing the Future*. Oxford: Thamesman Publications. Part of the series *Modernity and Postmodernity: Mapping the Terrain*.

Griffin, Roger. (2002). "The Primacy of Culture: The Current Growth (or Manufacture) of Consensus within Fascist Studies." *Journal of Contemporary History* vol. 37(1): 21–43.

Herf, Jeffrey. (1984). *Reactionary Modernism: Technology, Culture, and Politics in Weimar and the Third Reich*. Cambridge, UK: Cambridge University Press.

Heidegger, Martin. (1977). "Only a God Can Save Us Now," trans. D. Schendler. *Graduate Faculty Philosophy Journal* 6(1): 5–27. Interview originally published in *Der Spiegel*.

Heidegger, Martin. (1985). "The Self-Assertion of the German University," trans. Karsten Harries. *Review of Metaphysics* 38(March): 470–480.

Kershaw, Ian. (2004). "Hitler and the Uniqueness of Nazism." *Journal of Contemporary History* 39(2): 239–254.

Laqueur, Walter. (1996). *Fascism: Past, Present, and Future*. New York: Oxford University Press.

Linz, Juan J. (2000).*Totalitarian and Authoritarian Regimes*. Boulder, CO: Lynne Rienner.

Mosse, George L. (1979). "Towards a General Theory of Fascism." In *Interpretations of Fascism*, ed. George L. Mosse. London: Sage.

Mosse, George L. (1999). *The Fascist Revolution: Toward a General Theory of Fascism*. New York: H. Fertig.

Mouffe, Chantal, ed. (1999). *The Challenge of Carl Schmitt*. London: Verso.

Payne, Stanley G. (1980). *Fascism: Comparison and Definition*. Madison: University of Wisconsin Press.

Payne, Stanley G. (1995). *A History of Fascism, 1914–1945*. Madison: University of Wisconsin Press.

Paxton, Robert. (1998). "Five Faces of Fascism." *Journal of Modern History* 70(1): 1–23.

Prowe, Diethelm. (1994). "Classic Fascism and the New Radical Right in Western Europe: Comparisons and Contrasts." *Contemporary European History* 3(3): 289–313.

Renneberg, Monika, and Mark Walker, eds. (1994). *Science, Technology, and National Socialism*. Cambridge, UK: Cambridge University Press.

Zimmerman, Michael E. (2001). "The Ontological Decline of the West" In *A Companion to Heidegger's "Introduction to Metaphysics*," ed. R. Polt and G. Fried. New Haven: Yale University Press.

INTERNET RESOURCES

Mussolini, Benito. (1932). "The Doctrine of Fascism." Complete text available from http://www.worldfuturefund.org/wffmaster/Reading/Germany/mussolini.htm.

Wistrich, Robert S. (2001). "The New Islamic Fascism." Israel Science and Technology Homepage. Available from http://www.science.co.il/Arab-Israeli-conflict/Articles/Wistrich-2001-11-16.asp.

FAUST

• • •

The story of Faust has been widely used in literature and popular discussions to reflect on the ethics of science and technology. The Faust myth first appeared in 1587 when it was published by an unknown German Protestant in a popular chapbook. In 1592, the book was translated into English under the title *The Historie of the Damnable Life and Deserved Death of Doctor John Faustus*. There have been several famous interpretations of the myth since the original publication, including works by Christopher Marlowe (1564–1593), Gotthold Ephraim Lessing (1729–1781), Johann Wolfgang von Goethe (1749–1832), and Thomas Mann (1875–1955). All of the interpretations are united by the central theme of one man's insatiable quest for knowledge and its implications for his world and his own soul.

Historical Roots

Dr. Johann Georg Faust (c. 1480–1540) is the historic figure on which the myth has been built. An astrologer and alchemist, Dr. Faust was born in Knittlingen, Württemberg (southwest Germany); studied at Wittenberg, Erfurt, and Ingolstadt universities; and later became a lecturer. Often accused of practicing black magic, Dr. Faust was repeatedly banished from villages. An elusive and mysterious figure, he reportedly admitted of pledging himself to the devil with his own blood. Dr. Faust was put to death in Staufen, Breisgau.

The original German publication was titled *Historia von D. Johann Fausten dem weitbeschreyten Zauberer und Schwartzkünstler* (History of Dr. Johan Faust, the notorious black-magician and necromancer). In this book, details of Faust's life are connected with speculative ideas about black magic and pacts with the devil. The first part of the book describes Faust's childhood and his studies in Wittenberg, which ends in a pact with the devil, because he wanted *"alle Gründ am Himmel und Erden erforschen"* (to probe all causes in heaven and on earth) and *"die Elementa speculieren"* (to speculate on the elements). This cannot be achieved through mere scholarship, but only with the aid of demonic powers. The second part describes Faust's travels—thanks to the power of the devil—through Earth, Heaven, and Hell. It also relates how he finally beholds paradise. The third part is composed of various tales, magic, and conjuring tricks. In the last part, an old man tries in vain to convert Faust's soul, but Faust renews his pact with the devil. In front of his students, Faust conjures Helena, the beautiful daughter of Zeus. He marries her, and they have a son, Faustus Justus. The book concludes with Faust's agonizing death and his descent into Hell in accordance with the rules of his satanic pact. Helena and their son disappear after his death.

This original tale is a moral and theological warning to live a God-fearing, modest life. Importantly, Faust's pact with the devil was not made out of a desire for material wealth, as was the case in most of the similar myths from that time, but rather from a desire for knowledge. Faust thus personifies the scientific, inquisitive intellect that is opposed to both the Catholic tradition founded upon papal authority and the humility and consciousness of sin found in the followers of Martin Luther (1483–1546).

From Marlow to Goethe

Marlowe was captivated by the English translation of Faust's story and used it as the basis for his play, *The Tragical History of Doctor Faustus*. Two versions of his play exist, one dated to 1604 and the other to 1616. It is believed to be the first dramatic interpretation of the Faust tale, and it follows the original story closely in terms of the proportions of comedy and tragedy. Marlowe's Faust is a complex character and a renaissance person who is driven by an overwhelming intellectual curiosity. Always striving for power and seeking beauty, Faust signs a pact with Mephistopheles (the devil) because the sciences of his time could bring him neither godlike knowledge nor superhuman talents and power. The punishment for this hubristic bargain is eternal damnation.

Marlowe's play became one of the most successful dramas of the Elizabethan epoch. An adaptation for the puppet theater was brought to Germany by traveling artists and became an indirect inspiration for Goethe's drama *Faust*, because he watched the puppet play as a boy. A German translation of Marlowe's drama was published in Germany at the beginning of the nineteenth century. Upon reading it, Goethe reportedly remarked, "How greatly it is all planned!"

Goethe's *Faust* is possibly the most important drama in the German language, and many quotes have been adapted into colloquial usage and proverbial sayings. Goethe's tragedy has two parts, the first was published in 1808 and the second in 1832. Goethe's Faust character is distinguished from earlier variants by his rich inner complexity. The drama raises questions across the spectrum of human knowledge from philosophy and theology to anthropology and history to ethics and aesthetics.

The play opens with a wager between God and Mephistopheles. God gives permission to the devil to lure the soul of Faust, a scholar and alchemist, and maintains that Faust would be saved despite his reliance on reason and sorcery rather than faith. Later, Faust complains that *"Wir nichts wissen können!"* (we cannot know anything!). All science stays in the dark, because it lacks a secure and certain foundation. This is why Faust devotes himself to magic: *"Daß ich erkenne was die Welt / Im Innersten zusammenhält"* (That I may know what the world / holds at its very core.)

Faust is not interested merely in power, pleasure, and knowledge, but longs to take part in the divine secrets of life. He conjures up an Earth-Spirit, but it refuses to help him slake his insatiable thirst for knowledge. Faust becomes depressed and wants to kill himself. But it is Easter and the church bells tell of the resurrection. He is overcome by childhood memories: *"Die Botschaft hör' ich wohl, / allein mir fehlt der Glaube"* (I hear the message clearly, / but I alone lack the faith). He does not commit suicide, but his inner tensions heighten. He is both sick of life and unbearably hungry to know and experience its deepest offerings. He hunts ravenously for knowledge but he also yearns to satisfy his bodily desires for action. In this situation, Mephistopheles makes an appearance and offers to fulfill Faust's every desire—for the price of his soul.

In both parts of the drama, innocent people become victims of Faust's pact with the devil. In the first part, the victims are the girl Margarete (nicknamed Gretchen), her mother, and her brother. With the help of Mephistopheles, Faust seduces Margarete, but the narcotic he gives to her mother has a lethal effect. Margarete's brother attempts to take revenge for his mother and the lost honor of his sister in a duel with Faust, but he falls by Mephistopheles's intervention. Gretchen gives birth to Faust's child, kills it, and ends up in jail.

In the second part, Faust's megalomaniac enterprise demands human sacrifices. He wishes to wrest land from the sea in Greece, so he begins the engineering construction on a system of dykes—thus becoming an archetype not just of one pursuing scientific knowledge, but also of someone intent on technological power. The henchmen of Mephistopheles burn down the home of an old couple who had cared for him as a young man, which was the only thing that the enormously wealthy yet discontented Faust did not own. The fire kills the old couple. Faust as an engineer does not foresee the unintentional consequences of his work but finally accepts them approvingly.

Goethe's *Faust* is a tale of reckless striving for boundless love, knowledge, and power. In the end, this culminates in the blind and maniacal pursuit of an engineering project that breeds outrage, destruction, and doom. Nonetheless, Faust's soul ascends to heaven with the angels singing: "Whoever strives in ceaseless toil / Him we may grant redemption." And it seems that the moral is that as long as we struggle toward greatness, God will grant salvation, even if we stray into excesses and sin.

ADAM BRIGGLE
VOLKER FRIEDRICH

SEE ALSO *Frankenstein; Playng God; Prometheus; Science, Technology, and Literature*.

BIBLIOGRAPHY

Anonymous. (1587). *Historia von D. Johann Fausten: dem weitbeschreyten Zauberer unnd Schwartzkünstler*. Frankfurt am Main.

FEDERAL AVIATION ADMINISTRATION

SEE *Aviation Regulatory Agencies*.

FEDERATION OF AMERICAN SCIENTISTS

• • •

Founded in 1945 by scientists involved in the Manhattan Project to create the atom bomb, the Federation of American Scientists (FAS) is a nonprofit organization of more than 2,000 scientists, engineers, and other citizens dedicated to the responsible use of science and technology. Originally known as the Federation of Atomic Scientists, FAS continues to focus much of its efforts on nuclear arms control and security, but it also addresses issues involving information technologies, science policy, and the environment. To achieve its goals of informed debate and the application of science and engineering to national problems, FAS utilizes several strategies including research, advocacy, outreach, and grassroots organizing.

Membership and Finances

The composition of FAS, originally dominated by physicists, has slowly diversified. A 2002 in-house survey found that nearly thirty percent of the respondents identified themselves as physicists. The next largest fields represented were medicine (18%), biology (15%), engineering (15%), and chemistry (13%). Members receive *Secrecy News*, an informal electronic publication on government secrecy, security, and intelligence policies.

The FAS budget for fiscal year 2004 was $3 million, 70 percent of which directly funded projects, while the remaining 30 percent covered overhead expenses. Approximately two-thirds of the budget was derived from private foundation contributions and one-third from government grants. Membership dues in 2004 amounted to $125,000.

Origins and History

After World War II a minority of U.S. scientists (roughly 3,000) formed the loose "scientists' movement" that sought not just to create new technologies that had an impact on social and political change, but "tried to direct that change toward a particular end" (Smith 1965, p. 528). FAS was the most important element of this movement in the early post-war years. Roughly ninety percent of the Manhattan Project scientists supported the FAS mission. Ernest O. Lawrence, however, discouraged participation by scientists in organizations devoted to non-scientific ends. FAS, originally dubbed the "scientists lobby," emerged in the same spirit as the *Bulletin of the Atomic Scientists* (also founded in 1945 by members of the Manhattan Project) and the 1955 Russell-Einstein Manifesto, which led to the first Pugwash Conference on Science in World Affairs. In all these cases, scientists gathered to appraise the perils of science and technology, prevent their misuse, and advance solutions in the name of peace and prosperity.

Three topics dominated the early FAS agenda: the need for domestic and international control of atomic energy, the need to educate the public on the promises and perils of atomic energy, and the harmful effects of secrecy on international trust and scientific growth. One of the biggest battles waged by the early FAS members and other concerned scientists was over civilian versus military control of nuclear energy. Many scientists distrusted the military, and envisioned limitless, clean energy if only the proper civilian controls could be established.

FAS did play at least a minor role in the international monitoring and control of atomic energy and weapons. Although it is difficult to assess FAS impact on the process, the formation of the U.S. Atomic Energy Commission in 1946, a civilian entity that regulated nuclear energy and controlled national research, was a major success in the battle for civilian control of atomic energy (Hewlett and Anderson 1972). In general, however, FAS members always faced limits on what their technical data and scientific knowledge could contribute to international and domestic nuclear politics.

A period of disenchantment and diminished influence ensued after the early post-war years. Members defended the integrity of science and civil liberties of scientists vigorously, while their demands for a positive role in policy making waned. Although not a member of FAS, the judgment of J. Robert Oppenheimer as a security risk in 1953 further weakened the political clout of scientists by attacking their image of trustworthiness and independence. After the McCarthy era, members adopted more modest expectations about the contributions of scientists to public life. The increased incorporation of scientists into government also forced FAS to adjust its role.

By 1969, FAS had reached its lowest ebb, with an annual budget of roughly $7,000 and a mostly volunteer staff. The greatly defunct organization was rejuvenated with the appointment of Jeremy J. Stone as president in 1970. For the next five years, FAS was heavily influenced by Stone, because he was the only staff member. He began revitalizing and promoting FAS with his monthly newsletters. Membership grew rapidly over the next two decades (including a 450 percent increase between 1970 and 1974) and by the 1990s FAS was able to support a staff of roughly a dozen (Stone 1999). From its inception, FAS had been composed of local associations or chapters, which occasionally met but primarily worked independently of one another. In the 1950s, there were approximately thirty chapters, but by 1970 only two remained—one of which, the Boston chapter, called itself the Union of Concerned Scientists (Stone 1999). Stone disbanded the chapter system in 1970 and the two remaining chapters became independent organizations. In 1974, FAS established a permanent headquarters in Washington, DC, something it had not had since the late 1940s.

By the mid 1980s, FAS relied more heavily on journalists, professional staff, and policy analysts than famous scientists. FAS has maintained a sizable influence despite the increasingly crowded security-oriented public interest community and science lobby movement. Its mission has also steadily expanded to include other areas of science and technology. In 2000 Henry Kelly became the new president and further bolstered FAS under the overarching goals of strengthening science in policy and using science to benefit society.

Assessment

In its early years (1945–1948), FAS played an important role in efforts to maintain civilian control of atomic energy. Alice Kimball Smith (1965) argues that "By guarding the rights of a particular profession in a dangerous period in the 1950s the FAS contributed to the general cause of civil liberties" (p. 531). It has also served as an effective watchdog over the relations of science and public policy. This has afforded some protection for scientists and science against attacks and, by providing a forum for self-criticism, has prevented scientists from being dangerously seduced by their own successes (Smith 1965). Its website is a comprehensive source of information pertaining to global military technologies, intelligence, terrorism, and other areas of science and society. It is a valuable educational and research tool that enhances military and government transparency.

FAS publishes an "Occasional Paper Series" to inform and stimulate debate on current science and security policy issues. The second paper in the series (Kelly et al. 2004) takes up the state of science policy advice in the United States, and argues that the infrastructure for providing science and technology advice to Congress and the President is in a state of crisis. It asserts that sound policy needs sound science advice. However, this claim raises the question of where scientists stop acting as advisers (that is, providing balanced, "objective" information) and start acting as advocates (promoting a course of action that serve the scientific community but may not align with common interests).

Since its inception, this has been the central question about FAS and other professional science and engineering organizations concerned to play an active role in shaping how science is used in politics and how policies affect the practice of science. Should scientists have a privileged voice in public decision making? What is their proper role in the value-laden, political questions raised by science and technology? These are the more subtle questions about the dual role of scientist and citizen underlying the FAS mission to focus the energies of scientists and engineers on issues of critical national importance.

ADAM BRIGGLE

SEE ALSO *American Association for the Advancement of Science; Nongovernmental Organizations; Pugwash Conferences; Union of Concerned Scientists.*

BIBLIOGRAPHY

Hewlett, Richard G., and Oscar E. Anderson. (1972). *A History of the United States Atomic Energy Commission.* Vol. 1: *The New World, 1939/1946.* University Park: Pennsylvania State University Press.

Smith, Alice Kimball. (1965). *A Peril and a Hope: The Scientists' Movement in America: 1945–47.* Chicago: The University of Chicago Press. Chronicles in detail the early years of scientists' efforts to affect public policies regarding atomic energy and weapons. Provides an extensive early history of FAS as part of the overall tension between science and government bureaucracy.

Stone, Jeremy, J. (1999). *Every Man Should Try: Adventures of a Public Interest Activist.* New York: Public Affairs. An autobiography that contains information on the rejuvenation of FAS beginning in 1970.

INTERNET RESOURCE

Kelly, Henry; Ivan Oelrich; Steven Aftergood; and Benn H. Tannenbaum. (2004). *Flying Blind: The Rise, Fall, and Possible Resurrection of Science Policy Advice in the United States.* Federation of American Scientists Occasional Paper Two, December. Available from http://www.fas.org/main/content.jsp?formAction=297&contentId=346.

FEMINIST ETHICS

• • •

Feminist ethics has developed in response to feminist attention to androcentric and sexist limitations of traditional Western ethics.

Broadly speaking, the perspectives advanced by feminist ethics may have implications for the understanding of professional ethics in science and engineering, and in such science and technology related areas of applied ethics as biomedical, environmental, or computer ethics. In the sciences and technologies themselves, feminist ethics may have more direct impact on the theoretical structures of some disciplines (such as anthropology and psychology) than others (such as physics and chemistry).

Feminist critiques have focused on three interrelated concerns. First, that women's moral capacities frequently have been seen as less developed than those of men. Second, that accounts of moral capacities have privileged traits historically identified as masculine, such as reason, autonomy, and independence, over traits identified as feminine, such as caring, community, and interconnection, and that moral theories have similarly emphasized reason, principles, and impartiality over emotion, situatedness, and relationships—again reinforcing the primacy of the culturally masculine. Third, that the majority of moral theorizing has focused on the public realm, with primary attention to men's interests, and has neglected moral concerns arising in the private realm, as well as often overlooking women's rights and interests in the public realm.

The Virtues of Women

Attending to the gendering of ethics has a long history. The tenet that women's virtues differed from those of men was canonized in Aristotelian philosophy. Based on the commonly held belief in women's relative physical weakness and rational inadequacies in comparison to men, Aristotle (384–322 B.C.E.) argued in *The Nichomachean Ethics* that the virtues of women would be different than those of free men. Whereas free men were charged with developing such virtues as courage, temperance, honesty, and justice, women's virtues were viewed as those that would best suit their domestic role and rational capacities, namely, industry and self-command.

The question of women's specific virtues or moral abilities was an often-debated tenet in the history of philosophy. Whereas many non-feminist philosophers continued to insist upon women's inferior moral abilities, many philosophers whose works can be seen as concerned with women's rights questioned both the alleged difference in moral virtues and the imputed inferiority. Mary Wollstonecraft (1759–1797) and John Stuart Mill (1806–1873), for example, argued that there were no fundamental differences between women's and men's morality. But many others argued that although there was a difference in the virtues of women and men, this difference did not imply that women's moral capacities were inferior to those of men. A few such theorists, however, such as Elizabeth Cady Stanton (1815–1902) reversed the traditional position by arguing that women's morality was both different than and superior to that of men.

Few contemporary philosophers hold the type of gender essentialism that was historically the basis for viewing morality as gendered; however, an attention to gendered differences in moral reasoning and habits once again became popular with the work of psychologist Carol Gilligan. In her widely read *In a Different Voice: Psychological Theory and Women's Development* (1982), Gilligan critiqued the developmental psychologist Lawrence Kohlberg's widely-used model of moral development as privileging masculine approaches to moral reasoning, and, in particular, the privileging of universal principles such as justice, and ignoring more typically feminine concerns with relationships, caring, and responsibilities that are richly situated and not amenable to formulation through universal rules. Gilligan argued that the empirical finding that women typically did not develop to the higher stages of Kohlberg's scale of moral development (often only achieving the third stage of his six-stage process, whereas men often reached stages four and five) was a reflection not of women's moral inadequacy but of a biased methodology.

Some have interpreted Gilligan's work as supporting the view that women's morality is different in kind than that of men, but the most significant impact of her work has been to turn scholarly attention to an analysis of moral frameworks that include attention to care, community, and relationships: analyses previously inhibited by accounts of moral capacities that privileged traits historically identified as masculine.

An Ethics of Care

Gilligan's emphasis on an ethics rooted in relationships was the catalyst for a profound rethinking of ethics within feminist philosophy. The ethics of care is seen by many as either an alternative to or a complement to principle-based universalistic ethical theories. Unlike universal rules, which require that behavior be impartial and the same for everyone in like circumstances, an ethics of care is a richly situated ethics that sees each caring relationship as unique. Rather than the rational and emotional detachment predicated by deontological or utilitarian moral theories, care ethicists argue that ethical practice includes the emotions as well as reason.

A feminist ethics of care has been developed by a number of theorists, including Virginia Held, Eva Feder Kittay, Nel Noddings, Sara Ruddick, and Joan Tronto. Care ethicists do not all share the same definition of care, but there are areas of agreement among them. A caring relationship is seen as one in which an individual is both attentive to the specific needs and interests of another, as well as acts to advance them. Hence care involves both knowledge and motivation. It is also seen as an interactive relationship in which the one caring must be attentive to the responses of the one cared for, and modifies his or her efforts to care based on how the

other responds to the caring actions. Care ethicists have noted that many human relationships are not between equals but rather between individuals in very different positions—parent/child, doctor/patient, teacher/student. Based on this, care ethicists have argued that moral theories would be best constructed not from contract models that often assume relatively equally positioned individuals, but rather from models that recognize the range of relationships possible between what Noddings calls the "one-caring" and the "cared-for" (1984).

One relationship feminist care ethicists often turn to is that of parenting and, more specifically, mothering. Ruddick focused on "maternal practice" as a form of thinking that although traditionally overlooked by moral theorists provides an excellent model of relationships that strive to foster the goals of preservation, growth, and social acceptability (1989). Ruddick argues that the virtue of "attentive love" involves both reason and emotion and is key to good maternal practice. Ruddick does not limit what she calls maternal practice to women or even to parents, but rather argues that maternal practice should extend to the public world and argues that the skills of maternal thinking will enable society to move towards a politics of peaceful cooperation.

Kittay extends such insights to consider the situation of long-term care, including care of those who are too young, or too ill or impaired to take care of themselves (1999). She argues for the need for what she calls a globally pertinent ethics of long-term care. Kittay, like other care theorists, argues that much of traditional ethical theorizing, as well as liberal political theory with its distinction between public and private, gains legitimacy only through a deep denial of the inevitability of human dependency. Society's conception of justice and much of social policy is structured around the myth of the able-bodied, independent individual. Acknowledging the centrality and ever-changing nature of dependency to human life then demands a reassessment of issues of equity and justice that takes both dependency and the complex natures of care relationships seriously.

Critics of feminist ethics of care worry that care relationships have been "gendered" feminine in such a way that the traits privileged in caring relationships are and will continue to be seen as less valued than the traits of independent agents, and will result in women continuing to be seen as "naturally" more fit for such labors and thereby trapped in low-paid service occupations. Clearly a transformation of ethics that takes seriously the centrality of relationships must go hand-in-hand with social reorganization that recognizes and supports the value of caring labor.

Attention to Women's Concerns

Held has argued that dominant moral theories and the specific issues that have been at the heart of contemporary western ethical analyses have privileged men's experiences and have focused far more on the public realm than the private. Through their attention to the concerns of women, feminist ethicists have introduced new issues to ethics and social theory such as affirmative action, sexual harassment, and comparable worth, and have brought new insights to more traditional issues—for example, discussions of reproductive rights and technologies, and the institutions of marriage, sexuality, and love, as well as caring labor.

Feminist philosophers have also argued that attention to gender cannot be done in isolation from other axes of oppression such as sexuality, race, ability, or class. The work of feminist philosophers such as Linda Martín Alcoff, Claudia Card, Nancy Fraser, Marilyn Frye, Sarah Lucia Hoagland, Eva Feder Kittay, María Lugones, Anita Silvers, Elizabeth Spelman, Iris Marion Young, and Naomi Zack reveal the importance of analyses that identify the structure and consequences of the interaction between different forms of discrimination or subordination.

Sensitivity to women's concerns has had special influence in science and engineering education, where women have historically been underrepresented. It has led to reforms in the educational practices and in the structures of professional practice. In the area of professional ethics, for instance, feminist ethicists have argued for more emphasis on trust behavior development over moral rule following and have emphasized care and communities over rights and individuals. In some special areas of applied ethics related to science and technology, feminist ethics has even produced distinctive perspectives—as in the contribution of ecofeminism to environmental ethics and cyberfeminism to computer ethics.

NANCY TUANA

SEE ALSO *Consequentialism; Deontology; Ethics of Care; Sex and Gender; Virtue Ethics.*

BIBLIOGRAPHY
Baier, Annette C. (1991). *A Progress of Sentiments: Reflections on Hume's Treatise.* Cambridge, MA: Harvard University Press.

Card, Claudia. (1995). *Lesbian Choices.* New York: Columbia University Press.

Fraser, Nancy. (1989). *Unruly Practices: Power, Discourse, and Gender in Contemporary Social Theory.* Minneapolis: University of Minnesota Press.

Frye, Marilyn. (1983). *The Politics of Reality: Essays in Feminist Theory.* Trumansburg, NY: Crossing Press.

Gilligan, Carol. (1982). *In a Different Voice: Psychological Theory and Women's Development.* Cambridge, MA: Harvard University Press.

Held, Virginia. (1993). *Feminist Morality: Transforming Culture, Society, and Politics.* Chicago: University of Chicago Press.

Hoagland, Sarah Lucia. (1988). *Lesbian Ethics: Toward New Value.* Palo Alto, CA: Institute of Lesbian Studies.

Kittay, Eva Feder. (1999). *Love's Labor: Essays on Women, Equality, and Dependency.* New York: Routledge.

Lugones, María. (2003). *Pilgrimages = Peregrinajes: Theorizing Coalition Against Multiple Oppressions.* Lanham, MD: Rowman and Littlefield.

Noddings, Nel. (1984). *Caring: A Feminine Approach to Ethics and Moral Education.* Berkeley: University of California Press.

Nussbaum, Martha C. (1999). *Sex and Social Justice.* New York: Oxford University Press.

Ruddick, Sara. (1989). *Maternal Thinking: Towards a Politics of Peace.* Boston: Beacon Press.

Tong, Rosemarie. (1993). *Feminine and Feminist Ethics.* Belmont, CA: Wadsworth.

Tronto, Joan C. (1993). *Moral Boundaries: A Political Argument for an Ethic of Care.* Routledge.

Young, Iris Marion. (1990). *Justice and the Politics of Difference.* Princeton, NJ: Princeton University Press.

FEMINIST PERSPECTIVES

• • •

The term *feminism* encompasses various social movements, from the late-nineteenth-century women's rights movement to the mid-twentieth-century women's movement in Europe and the United States, as well as referring to theories that identify and critique injustices against women such as Mary Wollstonecraft's *A Vindication of the Rights of Woman* (1792) or Harriet Taylor Mill's *Enfranchisement of Women* (1868). A core connotation of "feminism" is thus a commitment to revealing and eliminating sexist oppression.

In the early twenty-first century, the label "feminist ethics" is used to signify a method or focus of attention for ethical theory and practice. Many scholars have marked the genesis of contemporary feminist philosophy and ethics with Simone de Beauvoir's *The Second Sex*

(1993 [1953]), which provides one of the first sustained analyses of the lived experience of "becoming woman." Beauvoir opened her classic text with a critique of theories contending that there are basic biological differences between women and men that explain women's secondary status in society. She concluded that "one is not born a woman: one becomes one" (p. 249), that is, that women and femininity are "produced" through complex disciplinary practices such as marriage, motherhood, and sexuality. In this way, Beauvoir's work foreshadowed contemporary work in the area of feminist science and technology studies. ·

Women in Science

Feminist investigations of science and technology emerged in the 1970s, but their origins can be traced to concerns over the low numbers of women in science. Feminists argued that it is a moral imperative to determine the causes of women's underrepresentation in the sciences and to remove those that unjustly block their participation. Because feminists soon realized that sexism also intersects with other axes of oppression, this move to understand the causes of women's underrepresentation in the sciences was followed by efforts to include similar studies of the impact of racism, and more recently of abilism (discrimination against persons with disabilities).

While the numbers of women have been improving in the biological and life sciences since the 1970s, the numbers of women receiving degrees in engineering, physics, and computer science continue to raise concerns. A study conducted by the U.S. National Science Foundation (NSF) found that while women received 57 percent of the doctoral degrees awarded in non-science and engineering fields in the United States in 2001, only 19 percent of the doctoral degrees in computer sciences and 17 percent of the doctoral degrees in engineering were earned by women (NSF 2004). The American Institute of Physics also reported that only 12 percent of the doctoral degrees in physics in 1997 were awarded to women. In addition both studies found that women scientists who worked in the academy were more likely to hold positions at the lower ranks in less prestigious institutions.

Given that overt barriers to women training in science had virtually disappeared by the 1950s, yet the number of women in science remained low, feminists began to explore features of science itself that might account for this disparity. Some of the more liberal approaches argued that the sole cause of the problem was that girls and women were not being encouraged to enter science. This approach led to proposals for science education reform designed to improve the education of girls and young women in science and mathematics. The American Association for the Advancement of Science's *Science for all Americans* (1989) and the National Research Council's *National Science Education Standards* (1996) are two examples.

Many feminist scholars nevertheless argued that solving the problem of science for women would take steps more far-reaching than simply reforming the education system. They began to examine the ways in which sexist and androcentric biases had marked the very topics that were of interest to scientists and had permeated research design as well as the interpretation of research findings. From this perspective, feminists began to propose a transformation of the themes and practices of science itself.

Gender Bias in Science

As feminists began to attend to the role of gender in science they identified a number of examples, particularly in the biological and medical sciences, of scientific practice that was either androcentric, that is, focused on male interests or male lives, or sexist, that is, manifested a bias that women and/or their roles are inferior to those of males.

One classic example of gender bias in science emerged out of feminist investigations of theories of human evolution. Feminists argued that theories of evolution, in providing accounts of the origin of the family and of the sexes and their roles, turned on widely accepted biases about sexual difference. "Man, the hunter" theories of human evolution were analyzed and critiqued not only for focusing primarily on the activities of males but also for the assumption that only male activities were significant to evolution. Hunting behavior alone was posited as the rudimentary beginnings of social and political organization, and only males were presumed to be hunters. Language, intellect, interests, emotions, tool use, and basic social life were portrayed as evolutionary products of the success of the hunting adaptation of males. In this evolutionary account, females were portrayed as following natural dictates in caring for hearth and home, and only male activities were depicted as skilled or socially oriented.

Feminist primatologists, among them Linda Marie Fedigan (1982), Sarah Blaffer Hrdy (1981), Nancy Tanner (1976), and Adrienne Zihlman (1978), not only exposed the gender bias of "man, the hunter" theories, their research led to an alternative account of evolution now accepted as more accurate. By questioning the assumption that women's actions were instinctual and thus of little evolutionary importance, these scientists began to examine the impact of women's activities, in particular the evolutionary significance of food gathering. From this focus, an alternative account of evolution emerged that posited food-gathering activities, now of both women and men, as responsible for increased cooperation among individuals, which resulted in enhanced social skills as well as the development of both language and tools (Haraway 1989).

Examples of androcentrism or sexism in science are numerous and frequently shown to result in poor science and, in many cases, ethically problematic beliefs or practices. The following list provides just a few examples identified by feminists: the exclusion of women in clinical drug trials, attributions of gendered cognitive differences in which female differences are posited to be deviations from the norm, the imposition on women of a male model of the sexual response cycle on women, and the lack of attention to male contraceptive technologies.

Objectivity and Situated Knowledges

Feminist perspectives on gender bias in science and technology led to an appreciation of the link between ethics and epistemology. Feminists such as Donna Haraway, Sandra Harding, and Helen E. Longino argued that nonfeminist accounts of scientific objectivity were inadequate because they provided no method for identifying values and interests that are unquestioningly embraced by the scientific community and that impact theoretical assumptions or the design of research projects. Careful analysis of the history of science documented systematic assumptions about women's biological, intellectual, and moral inferiority that were not the idiosyncratically held opinions of individual scientists but widely held beliefs imbedded in social, political, and economic institutions, as well as scientific theories and practices (Schiebinger 1989, Tuana 1993). Given this, no account or practice of scientific objectivity that does not control for community-wide biases and values could be sufficient.

Feminist science and technology theorists thus argue for a "strengthened objectivity" by developing methods for uncovering the values and interests that constitute scientific projects, particularly those common to communities of scientists, and developing a method for accessing the impact of those values and interests (Harding 1991). In developing such an account, feminists gave up the dream of a "view from nowhere"

account of objectivity with its axiom that all knowledge, and in particular scientific knowledge, can be obtained only using methods that completely strip away all subjective components such as values and interests. Feminists, rather, argue that all knowledge is situated, that is, emerges from particular social, economic, or political locations. Strengthened objectivity requires attention to particularity and to partiality, with the goal not to strip all bias from knowledge, but to assess the impacts of "beginning knowledge from different locations." On this account human knowledge is inherently social and engaged. The goal, then, of any quest for objectivity is to examine how values and interests can either limit or enlarge one's knowledge practices.

As just one of many examples analyzed by feminists, consider the emphasis on recombinant DNA technologies that has been proposed since the late twentieth century as a unifying principle for molecular biology (Lodish et al. 2003). Feminists have argued that rather than the lauded neutrality and objectivity, this position reflects numerous values and interests. Recombinant DNA technologies emphasize the centrality of DNA as a "master molecule" that controls life, and ignore or view as less important the organism's environment or the organism's history. In this way, such an allegedly "neutral" technology actively frames a sharp division between genetic and nongenetic factors, trivializes the role of environments, and reinforces biological determinism. Feminists have argued that efforts to cement molecular genetics as the foundation of the science of biology leads to a perception of life, including behavior and social structures, as "gene products."

This situated knowledge practice of contemporary molecular biology is clearly linked to the emergence of "big science" and its support by venture capital. Funding for the Human Genome Project has emphasized a hierarchical, centralized organization of scientific research. And venture capital, following the promise of marketable discoveries in biomedical research, has similarly fueled the growth of such science.

Insofar as molecular genetics becomes the focus of biology, it embeds ideologies concerning the functions and significances of genes and environments that carry with them a renewed emphasis on genetic factors in disease. For example, although the vast majority of all cancers, including breast cancer, are attributable to environmental factors, there is an increasing emphasis in scientific research and medical practice on genetic factors, a move that has been sharply criticized by feminists (Eisenstein 2001). Another concern of feminists and race theorists is that this "geneticization" of human health has also led to a renewed interest in biological difference between groups, which is reinscribing a biological basis to racial classifications (Haraway 1997).

These shifts in research focus can have dramatic effects on resource allocations. Occupational hazards and environmental carcinogens have been clearly implicated in cancer rates, and the effects of environmental racism on the health of minorities have been well documented. Yet funding for research into or cleanup of modifiable environmental factors is shifting to research on genetic inheritance.

Given feminist perspectives on the interaction between biology and environment in the constitution of sex (as well as gender) and sexual identity, this reemergence of biological determinism is in conflict with feminist values and interests. Strengthened objectivity calls attention to the different values and interests guiding research and asks for examination of their roles in contributing to more effective and liberatory practices of science and technology as well as an investigation of how practices of science and technology affect values and interests.

Feminist Technology Studies

Such attention to the values and interests guiding scientific practice also influenced feminists working in the field of technology studies. Feminists came to understand that historians of technology had been accepting gender stereotypes such as "man, the producer" and "woman, the consumer," which had biased the field. In the words of Judith A. McGaw (1989), theorists working in technology studies had "looked *through* masculine ideology at the past rather than looking *at* masculine ideology in the past" (p. 177). Following Harding's call for a strengthened objectivity, feminist investigations of the history of technology recovered the histories of women who both produced and employed a technology, that is, women architects, engineers, and inventors, as well as women workers and their experiences of technological change.

But an attention to sexist or androcentric ideology revealed other types of biases in the field. Technology studies often focused on only certain types of inventions and specific kinds of work as worthy of study. The work of women in textiles and food production, for example, was either ignored or labeled "consumption." Ruth Schwartz Cowan (1983) argued that technology studies had overlooked the fact that female experiences of technology and technological change were often markedly different than male experiences. Studies such as those of McGaw, for example, demonstrated that the

mechanization of industrialization often differentially affected men and women, keeping women in the lowest paying jobs where their skills were denied and they had no opportunity for advancement. Feminists also argued that attention to women's most common relationships to technology, namely through use, maintenance, and redesign, revealed an overemphasis in technology studies on the design of technology rather than its use. In critiquing the dichotomy commonly embraced in technology studies between production and consumption, feminists revealed how gender formation and technological development are co-constitutive, meaning that gendered norms are encoded into technological design and use, and that gender roles themselves emerge out of interactions with technologies (See, for example, Wajcman 1991 and 2004, and Rothschild 1983).

Medical Technologies

There is no more obvious arena for mapping the interactive emergence of gender and technology than in the science of medicine. Indeed this interaction can be found at its most literal instantiation, along with all the attendant ethical dilemmas, in the case of the intersexual child (that is, a child born with genitalia and/or secondary sexual characteristics of indeterminate sex, or which combine features of both sexes). In *Sexing the Body* (2000), Anne Fausto-Sterling argues that the U.S. and European medical practice of "fixing" intersexual individuals by assigning a specific sex and offering surgical and other medical

Such practices rest, of course, upon a series of technological advances including advances in plastic surgery originally developed to return to "normal" those bodies that had been deformed by war, accident, birth defects, or illness. But because they also rest upon a series of values, these practices provide a window into the ways in which beliefs about sex and gender affect medicine and also raise a complex series of ethical concerns. Whereas many in the medical community view infant genital surgery as being designed to fix or "cure" an abnormality, which they believe would then allow the individual to lead a "normal" and healthy life, many feminists and Lesbian, Gay, Bisexual, Transexual scholars have argued that such surgery is performed to achieve a social result, namely to make sure all bodies conform to a two-sex system. They also contest the belief that such surgery is necessary for either physiological or psychological health, citing the many cases of intersexuals whose lives were not negatively impacted by this physiological difference. While the medical community views early genital surgery as a medical imperative, critics note that such surgery is frequently a "failure," often requiring numerous additional surgeries, extensive scarring, and a decrease or elimination of sexual pleasure (Fausto-Sterling 2000). Ethical issues abound in this area of medical practice from questions of autonomy (Who decides what is best for an intersexual child?), to issues concerning sexual identity and current societal regulations concerning same-sex relations (Does an intersexual individual who has both a vagina and a penis "count" as a woman or a man in the prevailing two-sex legal economy?).

Ethical issues also permeate the new reproductive technologies, another focus of feminist analysis. Feminists have addressed the risks of various types of reproductive technologies as well as the fact that such technologies are available only to certain women, identifying the way that class issues as well as sexuality and marital status have been limiting factors in the availability of such technologies. Issues of "normalcy" are also central to feminist analyses of reproductive technologies. Many feminists have, for example, critiqued the ways in which prenatal testing intersects with societal biases concerning disability, noting that whereas prenatal testing and selective abortion for the purposes of sex selection are decried in many countries, this practice is widely accepted for fetuses with disabilities such as Down syndrome. Feminists have also investigated how new reproductive technologies are reshaping what is seen as "natural" and affecting the ways women and men experience their bodies. As women and men "bank" their eggs and sperm, as postmenopausal women become pregnant through technological interventions, as lesbian couples give birth to their own biological children, the nature/culture divide shifts and alters.

Global Issues

Feminist investigations of the impact of Western science on women in non-Western societies reveal the Eurocentric and undemocratic nature of Western science. Western scientific "voyages of discovery" were often part of colonialist efforts to mine other cultures for resources, both human and material, and maintain the forms of social control necessary to do so. Feminist and postcolonial science studies have documented how European expansion has contributed to the destruction or devaluation of the scientific practices of the colonized cultures, leading to the false belief in the superiority of Western science, indeed to the false but pervasive belief that Western science is "generic" and not itself "local," that is, not situated in particular economic and social practices (See, for example, Adas 1989).

Feminist scholars have also mapped the continuing de-development of other cultures and their scientific and technological practices through so-called development policies such as the "green revolution" and the more recent impact of biotechnology in agriculture. Feminists have examined who benefits and who is made worse through such practices, paying close attention to the profit margins of those chemicals companies, such as Novartis, AgrEvo, and Dupont, that sell the fertilizers, pesticides, and genetically engineered seeds of this revolution. Although economic impact is a key factor in such analyses, feminists pay close attention to the impact on diversity—both human diversity as well as biodiversity. Vandana Shiva (1997) has argued that the marginalization of women and the destruction of biodiversity through monocultures go hand in hand because women provide the majority of the agricultural labor in many Third World countries. Shiva examines how the biodiversity-based technologies of Third World societies have been viewed as backward and have been systematically displaced by monocultures biased toward commercial interests.

Feminists and postcolonialist science and technology theorists have argued for a democratized science/technology practice that acknowledges the importance of biological as well as cultural diversity as a way to undo the harms of colonialist science practices, including many of the current capitalist-generated practices. While this vision of science and technology emerged from feminist-inspired investigations, it is a moral vision of the intricate interactions between humans and the more than human world, between natures and cultures, and between organisms and environments that should inspire everyone.

NANCY TUANA

SEE ALSO *Assisted Reproduction Technology; Abortion; Homosexuality Debate; Juana Iñez de la Cruz; Race; Sex and Gender.*

BIBLIOGRAPHY

Adas, Michael. (1989). *Machines as the Measure of Men: Science, Technology, and Ideologies of Western Dominance.* Ithaca: Cornell University Press.

American Association for the Advancement of Science. (1989). *Science for All Americans: A Project 2061 Report on Literacy Goals in Science, Mathematics, and Technology.* Washington, DC: Author.

Beauvoir, Simone de. (1952). *The Second Sex.* New York: Bantham.

Birke, Lynda, and Ruth Hubbard. (1995). *Reinventing Biology: Respect for Life and the Creation of Knowledge.* Bloomington: Indiana University Press.

Cowan, Ruth Schwartz. (1983). *More Work for Mother: The Ironies of Household Technology from the Open Hearth to the Microwave.* New York: Basic.

Creager, Angela N. H.; Elizabeth A. Lunbeck; and Londa Schiebinger, eds. (2001). *Feminism in Twentieth-Century Science, Technology, and Medicine.* Chicago: University of Chicago Press.

Eisenstein, Zillah. (2001). *Manmade Breast Cancers.* Ithaca, NY: Cornell University Press.

Fausto-Sterling, Anne. (2000). *Sexing the Body: Gender Politics and the Construction of Sexuality.* New York: Basic.

Fedigan, Linda Marie. (1982). *Primate Paradigms: Sex Roles and Social Bonds.* Montreal: Eden Press.

Haraway, Donna. (1989). *Primate Visions: Gender, Race, and Nature in the World of Modern Science.* New York: Routledge.

Haraway, Donna. (1997). *Modest_Witness@Second_Millennium.FemaleMan©_Meets_ OncoMouse™: Feminism and Technoscience.* New York: Routledge.

Harding, Sandra. (1991). *Whose Science? Whose Knowledge? Thinking from Women's Lives.* Ithaca, NY: Cornell University Press.

Harding, Sandra. (1998). *Is Science Multicultural? Postcolonialisms, Feminisms, and Epistemologies.* Bloomington: Indiana University Press.

Hrdy, Sarah Blaffer. (1981). *The Woman that Never Evolved.* Cambridge: Harvard University Press.

Ivie, Rachel, and Katie Stowe. (2000). *Women in Physics, 2000.* Melville, NY: American Institute of Physics.

Lodish, Harvey, et al. (2003). *Molecular Cell Biology,* 5th edition. New York: W.H. Freeman.

Longino, Helen E. (1990). *Science as Social Knowledge: Values and Objectivity in Scientific Inquiry.* Princeton, NJ: Princeton University Press.

McGaw, Judith A. (1987). *Most Wonderful Machine: Mechanization and Social Change in Berkshire Paper Making, 1801–1885.* Princeton, NJ: Princeton University Press.

McGaw, Judith A. (1989). "No Passive Victims, No Separate Spheres: A Feminist Perspective on Technology's History." In *In Context: History and the History of Technology—Essays in Honor of Melvin Kranzberg,* ed. Stephen H. Cutcliffe and Robert C. Post. Bethlehem, PA: Lehigh University Press.

National Research Council. (1996). *National Science Education Standards.* Washington, DC: National Academy Press.

National Science Foundation. Division of Science Resources Statistics. (2004). *Women, Minorities, and Persons with Disabilities in Science and Engineering, 2004.* Arlington, VA: Author.

Rothschild, Joan, ed.. (1983). *Machina Ex Dea: Feminist Perspectives on Technology.* New York: Pergamon Press.

Schiebinger, Londa L. (1989). *The Mind Has No Sex? Women in the Origins of Modern Science.* Cambridge, MA: Harvard University Press.

Shiva, Vandana. (1997). *Biopiracy: The Plunder of Nature and Knowledge.* Boston: South End Press.

Tanner, Nancy. (1976). *On Becoming Human.* Cambridge, UK: Cambridge University Press.

Taylor Mill, Harriet. (1868). *Enfranchisement of Women.* St. Louis: Missouri Woman's Suffrage Association.

Tuana, Nancy. (1993). *The Less Noble Sex: Scientific, Religious, and Philosophical Conceptions of Woman's Nature.* Bloomington: Indiana University Press.

Wajcman, Judy. (1991). *Feminism Confronts Technology.* University Park, PA: Penn State University Press.

Wajcman, Judy. (2004). *TechnoFeminism.* Cambridge, UK, Malden, MA: Polity Press.

Westra, Laura, and Bill E. Lawson. (2001). *Faces of Environmental Racism: Confronting Issues of Global Justice,* 2nd edition. Lanham, MD: Rowman and Littlefield.

Wollstonecraft, Mary. (1792). *A Vindication of the Rights of Woman.* London: Joseph Johnson.

Zihlman, Adrienne. (1978). "Women and Evolution." *Signs* 4(1): 4–20.

FETAL RESEARCH

• • •

Fetal research encompasses a broad array of research activities and potential clinical applications. It is ethically controversial because, although it may yield beneficial results, it involves the human organism at a stage of development where its moral status is contested and informed consent is not possible.

Distinctions and Benefits

One key distinction centers on the stage of development of the human organism when the research is conducted, from pre-implantation to late fetal stages. The general sources of fetal material include tissue from dead fetuses; pre-viable or nonviable fetuses in utero prior to an elective abortion; nonviable living fetuses ex utero; or embryos, either in vitro or pre-implantation. Another distinction is that between *investigational* research that cannot benefit the subject fetus and *therapeutic* research that might benefit the fetus subject or is likely to benefit future fetuses. *Clinical* applications including transplantation using fetal material such as tissues, cells, or organs represent another form of fetal research. Moreover, among the many non-clinical uses of human embryos are the development of contraceptives and abortifacients and the study of abnormal cell growth and chromosomal abnormalities.

Moral Issues

One of the most controversial types of fetal research involves embryonic stem cells. Stem cells are primitive cells that have the capacity to divide for indefinite periods in culture and give rise to specialized cells. Most attention is focused on pluripotent cells, those capable of giving rise to most tissues of an organism but not all types necessary for fetal development. The source of these cells can be adult humans, modified stem cells from other species, or most promising (and controversial), cells from aborted fetuses or human embryos (either "spare" embryos from in vitro fertilization or embryos created specifically for research purposes). Some of the potential benefits of stem cell research being discussed are growing new organs, reversing paralysis and other neural/spinal damage, reversing the effects of neuro-degenerating diseases such as Alzheimer's, repairing heart damage, and treating diabetes, cancer, and other diseases.

Potential benefits of other types of fetal research include both knowledge and therapeutic applications. Among the many non-clinical uses of fetal and embryo research are: (1) investigation of abnormal cell growth including various cancers; (2) studying the development of chromasomal abnormalities and other birth defects; (3) understanding implantation problems and miscarriages; and (4) increased knowledge of cancer and aids.

Historical Background

Fetal research first appeared on the national policy agenda in the early 1970s after widely publicized exposés on several gruesome experiments conducted on still-living fetuses. In response, the U.S. Congress passed the 1974 National Research Act (Public Law 93–345) establishing the National Commission for the Protection of Human Subjects of Biomedical and Behavioral Research, whose first charge was to investigate the scientific, legal, and ethical aspects of fetal research. In 1976, regulations for the federal funding of such research were promulgated (45 CFR 46.201–211). Under the regulations, certain types of fetal research are fundable, with constraints based on parental consent and the principle of minimizing risk to the pregnant woman and the fetus.

In 1985, Congress passed a law (42 U.S.C. 289) forbidding federal conduct or funding of research on viable ex utero fetuses with an exception for therapeutic research or research that poses no added risk of suffering, injury, or death to the fetus and leads to important knowledge unobtainable by other means. Federal regulations on fetal research, then, appear to be quite clear in allowing funding within boundaries. However, in two key areas—embryo research and fetal tissue transplantation research—there in effect was a moratorium on federal funding between 1980 and 1995, which, according to the Institute of Medicine, severely hampered knowledge in medically assisted reproduction because the reliance on private funding shifted emphasis from essential basic research to rapid and often risky clinical applications. In addition, many states have laws regulating or even prohibiting certain types of fetal research, often either part of or attached to abortion statutes (for details see NCSL 2003).

At the center of the controversy over fetal research is disagreement over the moral and legal status of the fetus. Questions also center on who may consent for the use of fetal materials, under what circumstances the abortion procedure can be modified to meet the needs of the research, and what type of compensation, if any, for fetal tissues should be allowed. As a result of these issues, fetal research has been elevated to the public agenda and has become a highly volatile moral and political issue (Vawter, Kearney, and Gervais 1990).

Ongoing Debate

There is an ongoing debate over whether the use of fetal tissue from elective abortion encourages or legitimizes abortion. On the one hand, opponents argue that research using aborted fetuses gives abortion greater legitimacy and contend that use of embryos and fetuses for research exploits them and reduces them to biological commodities (Brown 2003). Moreover, because the fetus is unable to consent, there is concern over what type of consent and by whom is sufficient, and how to balance research needs with interests of the pregnant woman. On the other hand, supporters argue that research using human fetal and embryonic materials is critical for progress in many areas of medicine (Fletcher 1993).

As already suggested, perhaps the most controversial area of fetal research involves the use of fetal cells for transplantation to adult patients to treat a wide range of disorders (Stein and Glasier 1995). Such tissue can come from spontaneous abortions, induced abortions on unintended pregnancies, induced abortions on fetuses conceived specifically this purpose, or from embryos produced in vitro. A dependence on spontaneously aborted fetuses for research is impractical because of the limited number available and the inability to control the timing. The major supply of fetal tissue, therefore, is likely to come from induced abortions, but this raises vehement objections on moral grounds by groups opposed to abortion (Brown 2003).

Privatization and Related Issues

In light of the extent to which stem cell research might revolutionize health care, there is a huge commercial stake in fetal/embryo research. Already, marketing and advertising has started in the area of umbilical cord stem cell preservation and this is certain to be followed by broader efforts to market the fruits of this research. Moreover, once the benefits start to materialize, demand for cell lines will intensify, thus putting pressures on potential suppliers. Both of the two options, production of designated research embryos or increasing the supply of spare embryos, bring risks to women and raise questions concerning the consent process and proprietary rights over what promise to be very lucrative human materials.

There is also concern that increased pressures for these scarce resources could lead to exploitation of poor women paid to conceive solely to provide fetal material or an international market for multi-national drug companies.

Moreover, the continually expanding field of potential uses of fetal tissues, complicated by the difficulty of ensuring cooperation from abortion clinics and obstetricians in making fetal tissue available, has raised concerns for maintaining an adequate supply of fetal tissue. The availability of RU-486 and other abortifacients might actually diminish the supply of usable fetal tissue at a time when demand is increasing.

Assessment

Clearly some important areas of fetal research have been explicitly constrained on moral rather than scientific grounds. The presence of abortion politics continues to exert strong influence on research funded by the government across a wide range of substantive areas. In the process, some argue that long-term scientific goals are being compromised by immediate, pragmatic political objectives. The spirited debate over stem cell research in the 2004 election and the decision of California voters to invest $3 billion demonstrates that these issues will not dissipate. Given the sensitivity of human embryo and fetal research and its interdependence with abortion, this should not be surprising. Fetal research raises moral red flags for many persons and thus will remain a political as well as moral issue.

ROBERT H. BLANK

SEE ALSO *Abortion; Embryonic Stem Cells; Genethics; Genetic Counseling; Human Subjects Research; In Vitro Fertilization and Genetic Screening; Playing God; Research Ethics.*

BIBLIOGRAPHY

Fletcher, John C. (1993). "Human Fetal and Embryo Research: Lysenkoism in Reverse—How and Why?" In *Emerging Issues in Biomedical Policy*, Vol. 2, ed. Robert H. Blank and Andrea L. Bonnicksen. New York: Columbia University Press.

Stein, Donald G., and Marylou M. Glasier. (1995). "Some Practical and Theoretical Issues Concerning Fetal Brain Tissue Grafts as Therapy for Brain Dysfunction." *Behavioral and Brain Sciences* 18: 36–45.

Vawter, Dorothy E.; Warren Kearny; Karen G. Gervais, et al. (1990). *The Use of Human Fetal Tissue: Scientific, Ethical, and Policy Concerns.* Minneapolis: Center for Biomedical Ethics.

INTERNET RESOURCES

Brown, Judie. (2003). "Recycling Babies: The Practice of Fetal Tissue Research." American Life League, Inc. Available from http://www.all.org/issues/eg99y.htm.

National Conference of State Legislatures (NCSL). (2003). "State Embryonic and Fetal Research Laws." Available from http://www.ncsl.org/programs/health/genetics/embfet.htm.

FILMS

SEE *Movies*.

FIRE

• • •

From the prehistorical era, fire has generated the energy that allowed human beings to warm themselves and their surroundings, illuminate the darkness, prepare food, and create artifacts with both utilitarian and aesthetic value. More recently, fire has powered transportation and manufacturing and served as an object of and means for research.

These dual aspects of fire are represented in its deeply symbolic character. From the earliest periods fire has served as a symbol for moral and intellectual achievement. Many religions use fire in ceremonies, as in candles and funeral pyres. Fire is also common in celebrations, such as birthday candles and fireworks.

Although fire is indispensable to human beings and civilization, it also can kill and destroy or be used as a means for intentional destruction and warfare. The great library at Alexandria was destroyed by fire in 47 B.C.E. and the Chicago fire of October 1871 forced the city to rebuild itself. The World War II firebombings of Dresden and Tokyo were more destructive than the atomic bombings of Hiroshima and Nagasaki.

The Science of Fire

Chemically, fire is an exothermic reaction involving the rapid oxidation of fuel. For example, the burning of red oak could be approximated as follows:

$$CH_{1.7}O_{0.72}N_{0.001} + 1.065O_2 \rightarrow$$
$$CO_2 + 0.85H_2O + 0.0005N_2$$

This reaction liberates approximately 12.7 megajoules of energy for each kilogram of red oak burned. (One joule of energy is equal to one watt of power generated for one second; a megajoule represents one million joules.) This liberated energy raises the temperature of the reaction products and emits thermal radiation. Most fires occur in a normal air atmosphere, which is approximately 21 percent oxygen and 78 percent nitrogen. Thus, the burning of red oak in air is written more correctly as follows:

$$CH_{1.7}O_{0.72}N_{0.001} + 1.065(O_2 + 3.76N_2) \rightarrow$$
$$CO_2 + 0.85H_2O + 4.0049N_2$$

Because the nitrogen in the air plays no role in the reaction, the temperatures of flames in air are lower than the temperatures in pure oxygen; some of the energy that is liberated by the fire heats the ambient nitrogen.

Fires can be characterized as two different types of flaming: premixed and diffusion. Premixed flaming occurs when fuel and air are mixed before combustion; a diffusion flame exists when burning occurs as fuel and air are being mixed. Both types can be illustrated with a standard laboratory Bunsen burner. Typically, when a Bunsen burner is used, a blue flame is desired. That flame is premixed because air is entrained into the fuel stream through openings in the burner before the flame zone is established. However, if the openings where air is entrained into the burner are closed, the flame loses its regular shape and changes color from blue to yellow. This is a diffusion flame, where the gas feeding the burner must mix with the surrounding air in the area of flaming.

In general, all fires involve the combustion of gaseous fuel. If the fuel is a gas, such as the fuel that feeds a Bunsen burner, it only needs to mix with air for burning to occur. If the fuel is a liquid, it must be heated sufficiently to release vapors. The temperature at which a liquid fuel releases sufficient vapors for combustion to occur is called its *flashpoint,* and it occurs at a temperature lower than the boiling point of the liquid.

Solid fuels similarly must be heated to a temperature at which sufficient vapors are released to support combustion; however, solid fuels do not necessarily vaporize as liquids do. Some solid fuels melt and then subsequently vaporize before combustion. Others, such as wood, do not melt before releasing combustible vapors. Upon heating, these solids decompose into simpler compounds that are distinct from the original material in a process called *pyrolization.* The temperature at which a solid fuel releases sufficient vapors for combustion is called its *ignition temperature.*

Ignition of a fire requires the introduction of energy. For a gaseous fuel or a liquid fuel that is at a temperature above its flashpoint, a spark or small flame may be sufficient to provide that energy. For solid fuels or liquid fuels that are at a temperature below their flashpoint, the fuel first must be heated to a temperature at or above its flashpoint or ignition temperature.

Once a fuel is ignited, the heat liberated form the fire can transfer back to the fuel, causing the fire to sustain itself or grow. However, if the energy feedback to the fuel is not sufficient, the fire will decay and eventually go out. This can be illustrated by looking at logs in a fireplace. If there is only one wood log in a fireplace, so much of the energy liberated by burning the log is lost to the environment that the fire will go out. However, if more logs are added, some of the energy that would have been lost is transferred to the other logs, and the fire will be sustained.

Fire and Technology

Humans are the only creatures that have the ability to control and harness fire. Fire has been indispensable to technological progress. It is no accident that Prometheus is said to have stolen fire from the gods to make it possible for human beings to live. Early humanity learned how to start fires that could serve very simple uses. As humanity evolved, the heat generated by fire was used for more complex tasks, such as hardening clay and molding metals. Later, the energy liberated by fires created steam to power moving equipment.

As the understanding of fire increased, so did the efficiency with which it was used to generate energy.

Instead of using fires to heat water to create steam, fires could be ignited under controlled conditions in cylinders, allowing the potential energy in the fuel to be converted more directly to kinetic energy. It is through these types of processes that fire can be used to power generators that create electricity or to create mechanical energy to power airplanes and boats directly.

However, just as fire can serve benign purposes, it can be used in more destructive tasks. Fire applied to people or their environments either intentionally or unintentionally can cause death, injury, and the loss of property. Entire cities have been lost to fire; although the rate of death, injury, and destruction caused by fire has decreased steadily, fire continues to take a serious toll. It is likely this dichotomy of purpose that prompted fire as a symbol of attractive self-destruction, as when the moth flies into the flame.

The Ethics of Fire

The technological advances that have been made possible by the ability to harness fire also have created risks to society. As the Industrial Revolution created an environment in which manufacturing facilities were placed closely together, the flammability of the items inside those facilities, coupled with the closeness of buildings, allowed for fires that could destroy entire cities. This caused society to look for ways to protect people and the community from the hazards of fire.

As a result of fires that caused the loss of whole cities, people began to look for means of limiting the effects of a fire to a single building. This was accomplished by controlling the materials from which buildings were constructed, the spacing of buildings, and the types of openings, such as windows, installed in buildings.

As people learned ways to limit the impact of a fire to a single building, the goal of fire safety changed to limiting that impact to a single portion of a building. As methods of protection against fire improved, the maximum tolerable effects of fire became smaller.

Society has devised a number of ways to prevent fire that can be divided into two broad areas: prevention of fire ignition and management of fire impact. By controlling the methods of creating and distributing energy, it is possible to make it less likely that a fire will be ignited. Examples of means to prevent fire ignition include the electrical protections that typically are required in a home or business: the use of minimum wire sizes, electrical insulation on wires, and the use of fuses or circuit breakers.

Management of fire impact can involve means of managing a fire once it begins or ways to manage the things that are intended to be protected from fire. Fires can be managed by controlling the fire combustion process, suppressing fires, and controlling fires by means of the types of construction used. Examples of these means include controlling the type of fuel present, the use of fire suppression systems such as automatic sprinkler systems, and the use of building materials that resist the spread of fires, such as fire walls and doors.

In light of the fact that almost all human pursuits create fire risk, society has an obligation to ensure that the risks that are created are controlled. Developing better means of fire protection requires the development of a better understanding of fire and of the way fire affects buildings, people, and property. To this end there is a branch of science that is dedicated to the study of fire. Fire science involves scientific study of how fires start, how they grow, how they can be extinguished or suppressed, and the amount of heat and chemical compounds that are created when fires occur. Fire scientists also create models, or methods of simulating, fires. Fire models range in complexity from sophisticated computer programs to relatively simple equations that can be solved with a calculator.

Similarly, there is a branch of engineering that is dedicated to the application of scientific principles to protect people, property, and the environment. Fire protection engineers apply the scientific understanding of fire to reduce the risks of fire to reduce the likelihood of unwanted fires and manage the impact to society when unwanted fires occur.

MORGAN J. HURLEY

SEE ALSO *Air; Earth; Environmental Ethics; Water.*

BIBLIOGRAPHY

Bachelard, Gaston. (1964 [1938]). *The Psychoanalysis of Fire*, trans. Alan C. M. Ross. Boston: Beacon.

Drysdale, Dougal. (2002). "Chemistry of Physics and Fire." In *Fire Protection Handbook*, ed. by Arthur E. Cote. Quincy, MA: National Fire Protection Association.

Lyons, John. (1995). *Fire*. New York: Scientific American Books. Provides an overview of the hazards that fire pose to society, the science of fire, and the methods that are used to combat unwanted fire.

National Fire Protection Association Publication 550. (2002). *Guide to the Fire Safety Concepts Tree*. Quincy, MA: National Fire Protection Association. The fire safety concepts tree provides a method of communicating fire safety concepts to people who do not have specialized knowledge of fire protection.

Pyne, Stephen J. (2001). *Fire: A Brief History*. Seattle: University of Washington Press.

Tewarson, Archibald. (2002). "Generation of Heat and Chemical Compounds in Fires." In *SFPE Handbook of Fire Protection Engineering*. Quincy, MA: National Fire Protection Association.

FOOD AND DRUG ADMINISTRATION

SEE *Food and Drug Agencies*.

FOOD AND DRUG AGENCIES

• • •

Because foods and drugs are intimately involved with the quality of life, their purity and safety have been of deep concern to many citizens and the governmental agencies dedicated to human welfare. Throughout the world the purpose of food and drug agencies is to certify that foods are safe and drugs effective. Consequently these agencies have as one of their chief goals the prevention of adulteration—debasing foods or drugs by diluting them with less valuable ingredients or adding substances to make the food or drug appear to be what it is not. Adulteration has ethical consequences; for example, the dilution of a cancer drug may hasten rather than hinder death. Corrupt companies can use scientific knowledge and chemical techniques to thwart detection of their adulterated products, forcing food and drug agencies to develop advanced techniques to ferret out fraudulent drugs and thereby protect the public from harm. Science and technology are thus inextricably involved in the ethics of food and drug agencies and industries.

Early History

During the Latin Middle Ages writers of herbals and medical treatises expressed ethical qualms about adulteration and proposed remedies. These writers found that scarcity of supply played a role in fraudulent practices. In 1202 King John instituted the first English food law, which prohibited the admixture of inferior ingredients in publicly sold bread. In Germany and France rulers passed statutes that fined brewers for doctoring beer and wine. Arabs of medieval Islam appointed police officers to test the genuineness of foods and drugs

in markets. Medicinal compounds had to be prepared before a supervisor, who was the guarantor of the drug's purity.

With the European voyages of discovery in the fifteenth and sixteenth centuries, new foods and herbal drugs became part of an expanding global marketplace. To preserve foods on long journeys, producers and transporters used chemicals to retard spoilage and color foods. These practices led to abuses, and some European governments passed laws to prevent and punish harmful or deceptive practices. In the seventeenth century Robert Boyle, a British physicist and chemist, invented a device for determining specific gravities, which gave pharmacists a new way to detect drug adulteration. With the increasing sophistication of scientific knowledge in the eighteenth century, technical books by Adolph Gottlob Richter, Jean-Baptiste-Augustin Vanden Sande, and others appeared on adulteration and its detection and eradication. Although these authors used the new knowledge of chemistry and highly developed apparatus of the Scientific Revolution, they also analyzed the ethics underlying nefarious practices by merchants, pharmacists, and physicians. They suggested such remedies as better education and more effective laws to correct the injuries being done to customers and patients.

Food and Drug Agencies

As the first country to combat food and drug fraud through a comprehensive set of laws, Great Britain became the model for many other nations. Beginning in the eighteenth and early nineteenth centuries with laws on the adulteration of tea, wine, and beer, the British were able to protect the integrity of these and other important commodities and, through revenue officers, enhance state income. New technologies such as the microscope helped scientists detect the adulteration of coffee with chicory, but scandals associated with injurious foods and drugs forced legislators, in a series of new laws, to shift from noninjurious adulteration to the illegal addition of substances to foods and drugs that caused physical harm.

In colonial America the earliest food adulteration laws closely followed British examples, but in the late eighteenth century the first U.S. food law clearly targeted those persons, corrupted by greed, who sold unwholesome food in Massachusetts. Once convicted, such persons could be fined, imprisoned, or pilloried. The first U.S. federal drug law was passed in 1848, and it prohibited the importation of adulterated drugs. In 1862 President Abraham Lincoln signed legislation creating the U.S. Department of Agriculture (USDA), which included a Division of Chemistry (renamed the Bureau of Chemistry in 1901). This agency, which was a precursor of the Food and Drug Administration (FDA), employed chemists to identify adulterants in foods. During the rapid growth in population and industry after the Civil War (1861–1865), interstate traffic in foods and drugs also increased, as did tragedies associated with the addition of harmful dyes and preservatives to food and drink. An outraged public clamored for remedies, and between 1880 and 1906 more than a hundred bills were introduced in the U.S. Congress, but not one passed both houses.

The person largely responsible for breaking this deadlock was Harvey Wiley (1844–1930), chief chemist at the USDA from 1883 to 1912 and the "Father of the Pure Food and Drug Law." Convinced that many food and drug businesses were placing profits ahead of public health, Wiley hired idealistic young chemists, who were nicknamed the "Poison Squad," to study how chemical additives in foods affected health. Reports of their results aroused public concern, but "Wiley's Law" would never have been realized were it not for Upton Sinclair (1878–1968), whose novel *The Jungle* (1906) dramatized the repulsive practices in the Chicago meatpacking industry. The Pure Food and Drug Act of 1906 prohibited the "manufacture, sale, or transportation of adulterated or misbranded or poisonous or deleterious foods, drugs, medicines, and liquors." The Bureau of Chemistry administered the law, and Wiley and his successors developed an organization that won many victories for pure foods and drugs in the courts.

During the three decades after passage of the new law, weaknesses in its provisions appeared, because unscrupulous manufacturers were able to use advances in scientific knowledge and techniques to circumvent the statute. Muckraking journalists charged the food, drug, and cosmetics manufacturers with using 100 million Americans as "guinea pigs," and they provided examples of cosmetics that blinded women and drugs that caused children to suffer agonizing deaths (Kallet and Schlink 1933). Although administrative modifications were made (the Agricultural Appropriation Act of 1930 changed the agency name to the Food and Drug Administration), it was not until 1938 that Congress passed the Federal Food, Drug, and Cosmetic Act, which required manufacturers to provide scientific proof, through tests on animals and humans, that all their products were safe *before* they were put onto the market.

To isolate the FDA from advocacy groups, it was transferred from the USDA to the Federal Security Agency in 1940. World War II expanded the FDA workload, and during and after the war the number, variety, and power of new drugs increased dramatically. Food and drug companies grew in size and influence, which precipitated both abuses and legislative remedies. During the 1950s and 1960s, the 1938 act was periodically amended, and after the FDA became part of the Department of Health, Education, and Welfare in 1953, it used these new laws to give control of new drugs to doctors and FDA officials. The Delaney Clause (1958), which prohibited the use of substances in food if they caused cancer in laboratory animals, led to the controversial ban on saccharin, an artificial sweetener and weak carcinogen (this clause was replaced, in 1996, with the less stringent standard that "no harm will result from pesticide residues on raw and processed foods").

In the late 1950s, because of the widespread use of the sedative thalidomide by pregnant women in Europe, thousands of deformed infants were born, which eventually led to stronger drug laws in many countries. This drug was not widely available in America because of the valiant efforts of Frances Kelsey, an FDA examiner, whose suspicions about thalidomide led to her repeated rejections of applications to market it in the United States. Congress responded to the thalidomide tragedy by passing the Kefauver-Harris Amendment in 1962. This law changed the ways in which drugs were created, rested, developed, prescribed, and sold. The burden was now on the companies sponsoring a new drug to show that it was safe and effective. The FDA also issued new regulations that made the drug review process extremely stringiest, leading to criticisms that drug approval became glacially slow, because FDA officials, fearful of another thalidomide-like calamity, required study after study.

Criticizing the FDA—and Ethics

During the last four decades of the twentieth century the FDA came under attack by industry executives, congressional subcommittees, and public action groups. These FDA critics proposed that 200 million Americans were now being used as guinea pigs, because they were ingesting drugs and food additives that were even more deadly than those of the 1930s (Fuller 1972). Congress responded with a series of laws that, for example, strengthened FDA authority to regulate medical devices and commercial baby foods. These changes did not prevent the generic drug scandals of the 1980s, and in the 1990s the FDA continued to be an agency struggling to regain its credibility as the guardian of national health.

As with similar agencies in other countries, the FDA is charged with protecting public health, and in doing so it is often entangled in controversial ethical issues made more complex by advances in science and technology. For example, the FDA is responsible for regulating investigational new drugs (INDs). These drugs, not yet approved for sale, must be scientifically tested on animals, because an adverse effect on an animal often correlates with a similar effect on humans. Animal rights advocates have objected to this phase of IND development as unethical, whereas other groups have objected to the next three phases of IND testing, because humans are involved. In Phase I, small groups of healthy volunteers are given the IND to help researchers study its effectiveness, dosage, and metabolism. In Phase II, one to 200 patients with the drug-targeted disease are monitored for drug safety, efficacy, and side effects. In Phase III, even larger numbers of patients take the drug to refine optimum dosages, and placebos are given to some patients to make sure that IND effects are not due to chance or a developer's optimism.

Critics have raised doubts about FDA procedures on INDs. For example, in the 1970s, the General Accounting Office (GAO) studied ten of the more than 6,000 drugs then classified as INDs, concluding that in eight cases the FDA failed to halt human tests after learning that the new drugs were unsafe. Furthermore, the GAO found that drug companies delayed reporting adverse drug effects to the FDA. Other critics have attacked the FDA for approving too many drugs too quickly, thereby increasing risks, whereas still others blamed the FDA for approving drugs too slowly, thus depriving people of beneficial treatments.

Because of physicians' professional involvement with nutrition, prescription and nonprescription drugs, and medical technologies, they have a vested interest in food and drug companies as well as the FDA. This interest can raise ethical conflicts. According to some studies, physicians are protective of their independence and the integrity of the doctor–patient relationship, and many doctors are wary of governmental intrusions into how they practice medicine. Some critics have nevertheless pointed out the dangers of the close relationship that has developed between many doctors and drug companies.

Congressional subcommittees have questioned the ethics and legality of certain drug company activities. For example, in 1988 generic drugs became the focus of interest when investigators discovered that three generic drug companies were receiving accelerated approval of their drug applications in exchange for payoffs to FDA employees. By the time this scandal was over, federal courts had convicted ten companies and forty-two

people of corruption. This scandal also revealed a potentially corrupting collusion between FDA workers and pharmaceutical companies, because FDA employees often leave their government jobs for highly paid positions at the companies they have formerly regulated.

Other controversies associated with science, technology, and ethics have involved artificial hearts, genetically engineered foods, and pediatric drugs. In these and other cases some ethicists claim that the law of the marketplace has contaminated the ethic that has generally guided scientists. For instance, they believe that profits rather than a genuine concern for humans and the environment have guided research on genetically modified plants and animals. Because many newborns have died after receiving drugs unsuited to their undeveloped organs, ethicists have pleaded with pharmaceutical companies and the FDA to do more research on proper drug doses for infants and children. Other ethicists have been troubled by the predominant use of white males in many drug studies, to the exclusion of women and minorities, whose genetic make-up and susceptibility to certain diseases are different from white males.

Assessment of FDA Influence and Future Prospects

Food and drug industries are among the largest and most lucrative in the world, and the governmental agencies that have evolved to regulate them have also become massive and complex. Because of the accelerating growth of science and technology, some predict that these industries and agencies will continue to expand, but others have warned that pharmaceutical companies are not creating new drugs at a rate necessary to maintain their viability in the marketplace. Although global funding for drug research doubled in the last decade of the twentieth century, the number of new drugs declined by 50 percent. The reasons for this decline are controversial. Some blame an industrial emphasis on "blockbuster drugs" that generate huge profits. But others blame the gargantuan costs required to develop new drugs. At the end of the twentieth century it typically took fifteen years and $900 million to develop a new drug, but only a very small percentage of these drugs actually become commercial successes. Still other critics have proposed replacing governmental food and drug agencies with free-market certification agencies, arguing that economic incentives are more conductive to effective results than bureaucratic incentives. Furthermore, simple drug solutions to such complex diseases as cancer and Alzheimer's have proven to be illusory.

Optimists believe that new technologies will be able to lessen these skyrocketing costs. For example, some have predicted that the sequencing of the human genome will revolutionize drug creation, but so far the mass of new data has confused rather than clarified future prospects. Scientists have used rational drug design, combinatorial chemistry, and high-throughput screening to accelerate the development of new drugs, but pessimists point out that, although the quantity of potential new drugs has increased, their quality has not. These critics also emphasize a fundamental ethical conflict between commercial interests and human needs for life-enhancing foods and lifesaving drugs.

As the need for safe, high-quality foods and drugs has grown, possible solutions to the problems posed by pessimists have been offered. Some believe that the reason so many useless drugs are generated is poor understanding of basic life processes. These analysts hope that, with more research in molecular and cell biology, the information needed to create precisely targeted drugs will become available. Others believe that computers will be able to predict how certain "new molecular entities" will bond to target compounds in human cells, thus fulfilling Paul Ehrlich's dream of "magic bullets." Still others believe that the solutions will be found in the plants populating the rain forests of the world. For a growing number of scientists, the future of healthful foods and safe and effective drugs lies in combining all of these solutions together, but with a realization that new foods and drugs must be created, developed, and marketed in ways that are compatible with the most profound ethical ideals of the human family.

ROBERT J. PARADOWSKI

SEE ALSO DDT; Food Science and Technology; Organic Foods; Regulation; Science, Technology, and Law.

BIBLIOGRAPHY

Abraham, John. (1995). *Science, Politics, and the Pharmaceutical Industry.* New York: St. Martin's Press. The author, an English sociologist, argues that corporate bias creates political problems, causing patients to take ineffective or unsafe drugs.

Burkholz, Herbert. (1994). *The FDA Follies.* New York: Basic. Analyzes in lurid detail how the FDA failed in its mission to protect the health of Americans during the Ronald Reagan and George Bush administrations.

Fuller, John G. (1972). *200,000,000 Guinea Pigs: New Dangers in Everyday Foods, Drugs, and Cosmetics.* New York: Putnam. A sequel to *100,000,000 Guinea Pigs* in which the author claims that, forty years later, the American public is being exposed to new and more dangerous poisons in their food, drugs, and cosmetics.

Grabowski, Henry G., and John M. Vernon. (1983). *The Regulation of Pharmaceuticals.* Washington, DC: American Enterprise Institute for Public Policy Research. The authors argue that the FDA does not deal with new drugs neutrally, because there exists an imbalance of benefits and risks among manufacturers, patients, and FDA officials.

Hilts, Philip J. (2003). *Protecting America's Health: The FDA, Business, and One Hundred Years of Regulation.* New York: Knopf. A general history of the FDA, its founding, development, and involvement in many controversies.

Hinich, Melvin J., and Richard Staelin. (1980). *Consumer Protection Legislation and the U.S. Food Industry.* New York: Pergamon Press. After an introductory history of federal regulation of the food industry, the authors analyze the regulatory mechanisms of such agencies as the FDA and USDA.

Jackson, Charles O. (1970). *Food and Drug Legislation in the New Deal.* Princeton, NJ: Princeton University Press. A political and historical analysis of the complex struggle leading to the Food, Drug, and Cosmetic Act of 1938.

Kallet, Arthur, and F. J. Schlink. (1933). *100,000,000 Guinea Pigs: Dangers in Everyday Foods, Drugs, and Cosmetics.* New York: Grosset & Dunlap. The authors inform readers of the dangers of many foods, drugs, and cosmetics, and also make suggestions about how ordinary people can defend themselves against these dangers.

Lamb, Ruth deForest. (1936). *American Chamber of Horrors: The Truth about Food and Drugs.* New York: Grosset & Dunlap. Despite the provisions of the Copeland Food and Drugs Bill, Americans continued to suffer from dangerous foods, drugs, and cosmetics, according to this author, who pointed to the need for further legislation.

Marsa, Linda. (1997). *Prescription for Profits: How the Pharmaceutical Industry Bankrolled the Unholy Marriage between Science and Business.* New York: Scribner. A history of the rise and fall of the American pharmaceutical industry from the Second World War through the 1980s, showing how the lure of large profits corrupted what was once a "golden age" of medical research.

Nestle, Marion. (2003). *Safe Food: Bacteria, Biotechnology, and Bioterrorism.* Berkeley and Los Angeles: University of California Press. The author argues that safe foods intimately involves politics, since billions of dollars and the safety of hundreds of millions of people are at stake.

Patrick, William. (1988). *The Food and Drug Administration.* New York: Chelsea House. The author analyzes how science, technology, and politics interacted in helping the British cope with serious problems of drug adulteration.

Stieb, Ernst W. (1966). *Drug Adulteration: Detection and Control in Nineteenth-Century Britain.* Madison: University of Wisconsin Press.

Temin, Peter. (1980). *Taking Your Medicine: Drug Regulation in the United States.* Cambridge, MA: Harvard University Press. Both a narrative of drug regulation and an analysis of how federal regulators came to distrust the ability of doctors and consumers to choose drugs for themselves.

Turner, James S. (1970). *The Chemical Feast: The Ralph Nader Study Group Report on Food Protection and the Food and Drug Administration*. New York: Grossman. A critique of the FDA for acting to benefit the gigantic American food industry at the expense of consumers.

Young, James Harvey. (1989). *Pure Food: Securing the Federal Food and Drugs Act of 1906*. Princeton, NJ: Princeton University Press. A historical study of how pure food advocates, chemists, and politicians cooperated to make the Pure Food and Drug Law into a reality.

FOOD SCIENCE AND TECHNOLOGY

• • •

Among the concerns of food science and technology are postharvest changes in substances that nourish human beings. Food science examines everything that can happen to food between harvest and consumption. Food technology is used to develop and manage the processes by which food is transformed from raw harvest to edible goods purchased by individual consumers. Almost all foods are modified before consumption. Only some fruits, nuts, vegetables, meats, milk, and eggs may be eaten raw. About three-quarters of all the calories consumed by humans worldwide are derived from rice, wheat, and corn (maize)—truly the staff of life in almost all societies—all of which must be processed to make their delivery of nutrients feasible.

Food science and technology draw on chemistry, microbiology, engineering, physiology, toxicology, nutrition, dietetics, economics, marketing, and law; therefore, food science and food technology are inherently interdisciplinary subjects rather than narrow disciplines. Because of the importance of food, this topic also raises a host of ethical issues, including professional responsibility, equity of availability, determination of levels of safety in regard to public health, risk to workers' rights, and informed consent among consumers.

Background

Along with the making of shelter and clothing, the securing and preparing of food constitute one of the oldest technical activities, being coeval with the emergence of *Homo sapiens*. Because of its importance, from the beginning of human society food appears to have been associated with a number of ethical judgments in the form of rituals and taboos. Gender differences in regard to food procurement evolved for natural reasons: Males were the hunters, and females were the gatherers and subsequently the crop cultivators. Anthropologists also focus on cultural aspects such as food as a means of asserting identity or group membership; the reciprocal effects of class or caste systems on foodways; communal eating and food as a means of bonding and hospitality; ritual aspects of food, for example, at funerals and weddings; and food taboos and food eaten for religious reasons—these so-called ceremonial foods include bread, wine, and oil, the first manufactured foodstuffs.

Two major changes allowed human populations to shift from nomadic hunting and gathering, which they had engaged in for hundreds of thousands of years, to living in settled communities. The first was the domestication of animals, probably beginning with that of the Asiatic wolf as an aid in hunting, around 13,000 years ago after the end of the last ice age. More significant was the keeping of lactating animals such as goats and sheep to guarantee a regular supply of milk, meat, and nonfood products. By approximately 10,000 years ago sheep had been domesticated in the area that is now Iraq, as were goats. Pigs were domesticated a thousand years later, and it took another thousand years before the wild aurochs had been transformed into cattle in the Balkan area.

The second achievement was the recognition of the relationship between plants and their seeds. This allowed a previously nomadic clan to settle in an appropriate landscape. With the receding ice, fields of wild grain or grasses with edible seeds appeared, and eventually women began to plant seeds in cleared areas.

Those two achievements were the key elements in what has come to be known as the Agricultural or Neolithic Revolution, which occurred during the New Stone Age, a period that began 11,000 years ago in southern Asia and 9,000 years ago in the Tigris and Euphrates river valleys, from where the new techniques began to spread. The agricultural revolution provided more and better food, promoting improved human fertility and longevity, and therefore increased human population numbers.

Differentiating between life-sustaining and harmful foods is probably an instinctive human behavior. People are drawn to carbohydrate-rich foods, which are generally sweet, and usually are repelled by alkaloidal products, which contain bitter toxic chemicals. An important discovery was that heat, such as that provided in cooking by fire or hot water, can alter the characteristics of food. The transformation of food materials by heat to make them consistently and predictably edible, flavorful, and spoilage-resistant developed into a practice that preceded techniques for deliberately changing inorganic materials, as in the making of pottery from clay some 30,000 years ago and then the use of metallurgy about 6,000 years ago, both of which contributed to cookery.

According to Harold McGee (1990), chemistry began with the "food chemistry" of ancestral cooks. The molecules those cooks transformed and manipulated were food molecules. Each time contemporary people prepare food for eating, whether in a large food-processing plant or in a kitchen, they replicate the origins of an art practiced since the harnessing of fire 125,000 years ago.

It was not until the Enlightenment and well into the Industrial Revolution that food became a focus of scientific study. It was the modern period as well that witnessed the related developments in public health, medical nutrition, and mechanization in food processing, especially for mass production. The adaptation of mass production technologies to agricultural production and food processing radically transformed human-food relations. Those processes made it possible for smaller numbers of food workers to support larger numbers of food consumers, thus promoting urbanization on an unprecedented scale. That urbanization led to new technologies of preservation, transportation, and marketing; inspired scientific studies of nutrition (because in many instances the new technologies altered the balances in traditional diets); and raised ethical issues about the treatment of food processing workers as well as equity in access and distribution (which previously had been subject to the negotiations characteristic of traditional cultures).

Nevertheless, the basic objectives of assuring a satisfactory supply of food did not change. Those objectives only become more visible, controllable, and subject to management. Indeed, only new insights and improved techniques can assure a continuing stream of food products for the growing human population.

The Perennial Vital Objectives

All functioning modern societies attempt to provide people with foods that are readily available, abundant, affordable, appealing, appetizing, nutritious, and safe. Agriculture (including fisheries), along with food science and food technology, is essential in meeting those goals. Since prehistoric times the objectives related to feeding a clan or a larger community have been optimization of harvest yields, prevention of losses, achievement of edibility, and protection of food integrity factors such as flavor, texture, color, and nutrition.

The food system—the path from soil to mouth or from farm to fork—is a precarious one. Numerous technologies are involved in the modern effort to bring food to consumers. Much can go wrong, and much depends on climate and other natural forces. However, human ingenuity, a multitude of tools, and technological interventions are the critical factors in seizing life-sustaining products from nature. Then all foods must be protected during the transfer from their production habitats to their final destination. The notion of a carefree dependence on the abundance of nature is far removed from reality.

Each food product on the shelves of grocery stores can be traced through its passage from harvest (including slaughtering and fishing) to channels of transportation and then to storage, packaging, and distribution until it is purchased for preparation in a consumer's kitchen or an efficient mass-feeding facility. About half of all dollars spent on food consumption in the United States at the beginning of the twenty-first century was expended in eating away from home.

Other animals compete with humans for the products of nature. The biblical scourges of locusts are a familiar example, but it is mainly invisible competitors that take the most. Bacteria, molds, yeasts, and even viruses consistently make foods unavailable, inedible, or the cause of disease. Only a few microorganisms have been put to positive use, mainly in fermentation. Because eradication is impossible, pest control is a major activity and expenditure for farmers and food processors and even for the food service industry and some householders. This war against microscopic competitors is waged most effectively with chemical weapons and must be affordable and properly done.

Current agricultural pesticides are largely products of the 1950s. As with all technological interventions, it soon was realized that there was a side effect in that pesticide residues on and in foods could be harmful to human health and to the environment. A typical quandary is the war against food pests. This battle involves powerful weaponry to assure an abundance of crops and may do damage to people as a side effect.

In addition to rodent, insect, and microbiological losses numerous chemical changes occur in foods that have unpleasant results. Soured milk, bitter rice, rancid fatty food, and other unpalatable edibles are thrown away. Not even animals are fed with them because their owners suspect the presence of toxic substances. The losses to the "food system" and to society are obvious. Equally obvious is the fact that such losses, along with food deterioration overall, can be avoided to a large extent through the judicious application of food technologies. That constitutes the major preoccupation of modern food processors and handlers, the custodians who take possession of food after harvesting and deliver it to end users in the expected qualities and quantities.

Food losses and food waste are enormous, although no accurate data exist. Ironically, in places where food crops are usually scarce, often because of a lack of technological intervention but also as a result of natural disasters, personal wastage is rare. In the developed parts of the world, where technology assures an abundance of food, there is usually gross disregard for optimal personal food utilization. Examples include tray waste in institutional facilities and careless housekeeping practices.

Food protection spans the spectrum from seeds to the moment of consumption. The initial responsibility lies with food producers. Agricultural research began in the nineteenth century. It has always been devoted mainly to production studies that have culminated in the use of chemical, mechanical, computer, and more recently bioengineering technologies. Each technology has had opponents, has sparked heavy discussion, and has been improved as a result. One insight has become clear: Without science and appropriately applied technologies improvement of the human condition would be slow, difficult, and painful.

Food Processing

From cutting to gamma-irradiation, the subject of food processing involves dozens of operations. Only a few can be mentioned in this brief overview. At the heart of food technology are several processing operations that are used to modify foods primarily to preserve them for later consumption. Water removal is one way to preserve a food: Raisins last longer than grapes, cheese and sausages can be stored for long periods, fruits can be converted to fermented beverages, and grains can be made into beer. In all these cases, the original food disappears but the nutritive value is preserved.

Another method of preservation is the use of chemicals, such as acids, that are antagonistic to spoilage microorganisms. During the 1990s about 5,000 people died every year in the United States from bacterial food poisoning. The human toll from poisoned food was almost unbelievably high until the advent of food technology, along with hygienic measures and medical advances. Vinegar, yogurt, and pickled foods are examples of acid-preserved foods.

The pickling of vegetables has a long history, especially in China, and has relied primarily on the use of salt (sodium chloride). The history of salt, which is considered the first "food" of commerce, is interwoven with that of food preservation (Kurlansky 2002). A high sugar content also preserves food, as in the case of candied fruit and confectionery products. The inspiration must have come from honey, the original natural preserved food.

Modern food markets provide evidence that almost everything people eat is modified before consumption. The rationale of most processing is to protect a food until it is consumed, and an understanding of food chemistry and microbiology is essential in that endeavor.

The simplest way to defeat microorganisms is to remove the water that is vital to them. Most foods that are not dried properly spoil very quickly, but substances antagonistic to microorganisms can be added directly or indirectly, as in lactic and alcoholic fermentations. The result is not only protection but also better nutrient availability and palatability. Lactic acid fermentation utilizes the destructive and digestive ability of certain microorganisms for human advantage, as in the cases of fermented cabbage and yogurt. The production of vinegar, beer, and wine provides examples of acetic and alcoholic fermentation. Other preservatives are microbial inhibitors such as spices, herbs, and salts.

Inhibition of oxidation is achieved mainly by means of the addition of antioxidants. Foods that are rancid or have lost flavor or color are considered spoiled. The mechanism is driven largely by enzymes native to foods but also by oxygen in the air. Consequently, air exclusion is a preservative technique. The first efforts at producing and sealing sterilized food were not made until the late 1700s, and plastic wraps and packaging under nitrogen were not used until the mid-1900s.

Canning is the most noteworthy achievement in food technology. It was invented by Nicolas Appert, who in 1790 in Paris preserved heated foods in bottles. Twenty years later the food-canning industry was born when the first "tin" cans were produced in England. Only with the 1864 work of Louis Pasteur on bacteria and asepsis did it become possible to understand the principles behind this food preservation technology. It was not until 1928 that Charles Olin Ball worked out the mathematical formula that made the thermal processing of foods possible. All heat sterilizing procedures in food and pharmaceutical industries in the early twenty-first century rely on Ball's work.

Legal and Ethical Issues

In 1939 in the United States the Institute of Food Technologists (IFT) was created. Similar professional associations now exist in most major countries. This represented the beginning of the coordination of all research activities and industrial development work involving foods. By 1960 several university food science departments had emerged. In the early 2000s there are nearly fifty in North America, and the IFT, headquartered in Chicago, has almost 30,000 members. This professional association publishes a number of journals and organizes well-attended annual meetings and expositions. Its mission is to establish and promote standards of professional excellence at local as well as international levels. The IFT fosters communication, contributes to public policy, and helps individuals achieve career goals. Along with its counterparts in other countries, it embraces objectives such as combating hunger, enhancing the quality of foods, and stimulating progress in the food technology industries.

IFT lists six core values in its current strategic plan:

- Act with integrity
- Foster inventive and adaptive leadership
- Demonstrate responsible stewardship
- Focus on members
- Value diversity
- Chanmpion sound science in the interest of public well-being

IFT's counterpart in the UK, the Institute of Food Science and Technology, has a somewhat more explicit code relating to ethics. Its 10 professional conduct guidelines are entitled:

- Wholesomeness of food
- Relations with the media
- Confidentiality of information
- Conflicts involving professional ethics
- Duties towards subordinates
- Scientific issues and food promotion
- Responsibilities towards students
- Responsibilities towards the environment
- Members' business interests
- Responsibility to the profession

A number of activist groups have emerged with an interest in food technology. Greenpeace International is probably the best-funded and declares to "exist because this fragile earth deserves a voice, it needs solutions, it needs change. It needs action."

The Food Ethics Council was established in 1998 in England as a charitable trust. It has reported on such ethical issues ranging from drug use in farm animals to intellectual property in agriculture research.

It was inevitable that governments would take an interest in the food supply. Modern American food law began with the Food and Drug Act of 1906, also called the Pure Food Law. In 1938 it was redone as the Food, Drug, and Cosmetic Act, with amendments. The U.S. Food and Drug Administration (FDA) enforces this law through an elaborate set of regulations. Other agencies share this responsibility, including the U.S. Department of Agriculture, the Federal Trade Commission, and the Bureau of Alcohol, Tobacco, Firearms, and Explosives.

Food regulatory work often is subject to criticism. The public can get involved in the rulemaking process, but it is mainly consumer advocates along with trade associations and only occasionally individuals that participate.

At one time mainly unprocessed and raw foods were consumed, but then cookery, pasteurization, and sterilization created the category of mildly processed foods. Milling, brewing, refining, dairy processing, and many other food operations that frequently relied on the use of so-called food additives and blending with other ingredients provided what often is termed highly processed or reformulated foods.

The newest category in this area is synthetic food, which can be thought of as engineered edible systems. An imitation orange drink powder that could be reconstituted with water at home or during space flight was the first example, appearing in the 1970s. Except for the sugar in it there is no agricultural ingredient, and the sugar could be replaced with a synthetic sweetener to make it a diet beverage or a food for diabetic persons.

It can be said that a gradual merging of the food and pharmaceutical industries is under way. The word *nutraceutical* was coined in the 1990s, and with it came many foods and food components, including beneficial bacteria, that are claimed to have health-providing properties beyond those of traditional essential nutrients such as vitamins, amino acids, and certain minerals. Opportunities to defraud the public with scientifically unproven benefits are tempting; the subject of nutritional claims is debated hotly and is only in the early stages of governmental supervision.

Since biblical times human societies and their leaders have been interested in regulating trade and safeguarding foods. Food protection has economic and public health implications: People must be protected from cheaters and poisons. Because misrepresentation and adulteration can be inadvertent as well as deliberate, a legal and regulatory framework was needed to address these concepts and allow modern societies to function smoothly and safely.

The English Assize of Bread and Beer (*assisa panis et cerevisiae*) of 1266 attempted to control the quantity (weights and volumes) of food sold by merchants, not its quality. That law established strict penalties whose basic principles would be adopted by settlers in North America hundreds of years later. Adulteration was rampant, and the tools to detect it were lacking. In 1820 Frederick Accum, a German chemist and pharmacist living in London, published his *Treatise on the Adulteration of Food and Culinary Poisons*. Microscopy was an emerging technology that became the first analytical tool to verify food adulteration, mainly in the detection of rodent hairs and feces, insect fragments, and foreign objects such as dirt and unwanted plant matter. Chemical analysis has become a more powerful tool since that time, and the food laws of many nations stipulate the employment of food analysts and analytical methods. It is now possible to detect the presence of objectionable environmental chemical contaminants in trace amounts that are not significant in physiological terms, that is, amounts considered inconsequential.

Just as the law does not concern itself with trifles, the law of Paracelsus states that a small amount of a toxin is not worth considering because it has no effect. Paracelsus taught that "the dose makes the poison," and it can be demonstrated that a grain of salt has no effect on a living organism but that a cupful is deadly. Similarly, too much of a good thing may be harmful, as evidenced by the contemporary overconsumption of calories, especially in affluent societies. Sixty-five percent of Americans were considered obese at the start of the twenty-first century, and obesity is becoming the number one human health hazard. Discussion has begun about where to lay the blame for this phenomenon. Some have pointed to the "fast-food" industry as the primary culprit, ignoring free will, discipline, and responsibility.

The concept of American fast food also touches on ethical issues and may have spawned the "slow food" movement that arose in Italy in the late 1990s, presumably to resist the replacement of culinary traditions and the disappearance of local food varieties; however, it also might have been the product of anti-Americanism, anticapitalism, and antiglobalization. All over the world, especially in developing areas, the introduction of "Western food" constitutes a threat to indigenous food crops and processing operations that have been practiced by women for centuries. The enrichment of a

local diet is welcome from a nutritional standpoint, but it also is believed to undermine the potential for self-sufficiency and the value of indigenous knowledge. Entomophagy is widely accepted and always has been: Some five hundred insect species serve as food sources worldwide. The subject of underutilized species has been dealt with by the many organizations, and as a result new foods have been "unearthed." The fungal protein quorn, manufactured in the United Kingdom and sold as a meat substitute, is an example. Other potentials are seen in leaf protein concentrate, processed plankton or cellulose, and recycled waste products.

The Future

The newest trend in the food field is genetic engineering. Apart from drug manufacturing it is applied mainly in production agriculture and involves recombinant DNA and cell fusion techniques. The driving force behind this food biotechnology is the creation of higher yields from plants and animals. Critics argue that the driving force here is not a humanitarian spirit but corporate greed. Related objectives of the new biotechnology are foods with improved nutritional properties, such as the Swiss-originated vitamin A-enriched rice that is claimed to combat childhood blindness in Asian areas and the production of crops with better utilization/processing potential, including better flavor. Many of these products are already on the market. However, there has been vigorous political and even religious debate over these genetically modified (GM) crops and foods, even over GM drugs such as insulin.

New enzymes derived from GM microorganisms are being used in food processing. Indeed, knowledge about genetic maps and the amino acid sequences of proteins makes it possible to tailor-make food ingredients with specific desirable functions and properties. Among the 150 microbial enzymes in use for food production more than 40 are produced from GM microorganisms. It is surprising to many people that practically every item on an American restaurant menu has been subject to genetic modification. Since the introduction of GM foods in the 1980s a quiet revolution in the food supply has been under way. Worldwide, 46 percent of soybean acreage and 7 percent of corn fields were sowed with transgenic crops in 2001. No transgenic animals are used in food, mainly because of ethical barriers.

Disagreement about the safety of GM foods is rooted in the differences between American and European regulatory principles: regulation of the nature of the product in the United States versus regulation of the manner in which a product is produced in Europe. One consequence of the debate was the refusal in 2002 by the Zambian government to receive food aid from the United States because it involved GM food.

All new technologies seem to be accompanied by early resistance. GM crops have been embraced in the developing parts of the world, as was discussed during the Twelfth World Congress of Food Science and Technology in 2003. Food scientists are bracing themselves as the era of GM foods is unfolding. One challenge is to develop analytical methods that will differentiate between a GM species and a conventional one. The current debate seems to indicate that consumers wish to have a choice in selecting one or the other, and regulators may be charged by policymakers to monitor the trade in and consumption of these foods.

Food technology has improved the lot of humankind, but the work is far from over. Better tools will be designed, and it will be necessary to engage in transfers of food technology and institute governance, education, and transportation infrastructures so that no needy individual is left behind.

MANFRED KROGER

SEE ALSO *Agricultural Ethics; Biotech Ethics; DDT; Food and Drug Agencies; Green Revolution; Genetically Modified Foods; Nutrition and Science; Organic Foods.*

BIBLIOGRAPHY

Davidson, Alan. (1999). *The Oxford Companion to Food.* Oxford: Oxford University Press. Considered to be the opus magnum from one of the world's great authorities on the history and use of food. It has 2,560 A-Z entries on 892 pages including 40 feature articles on staple foods.

Encyclopaedia of Food Science, Technology and Nutrition. (1993). London: Academic Press. About 1,000 entries written by international experts. The eight volumes contain 5,364 pages; agricultural aspects are not covered.

Kass, Leon R. (1994). *The Hungry Soul—Eating and the Perfecting of Our Nature.* New York: Free Press. An exploration of the natural and cultural act of eating; how homo sapiens has humanized eating, even though it is an urgent and basic animal necessity.

Kurlansky, Mark. (2002). *Salt—A World History.* Toronto: Alfred A. Knopf Canada. The long and intriguing history of the "only rock we eat" is also a very important part of humankind.

McGee, Harold. (1990). *The Curious Cook: More Kitchen Science and Lore.* San Francisco: North Point Press. An investigation into culinary problems and dogma, telling in plain English what science has discovered about the food we eat.

Pyke, Magnus. (1970). *Food Science and Technology,* 3rd edition. London: John Murray. Throughout his career the author has drawn attention to the importance of food technology and nutrition via a dozen books and numerous radio broadcasts and public lectures.

Toussaint-Samat, Maguelonne. (1992). *A History of Food,* trans. Anthea Bell. Cambridge, MA: Blackwell. A comprehensive 801-page reference history of foodstuffs, the story of cuisine, and the social history of eating, from the origins of mankind to the modern-day technological era.

Trager, James. (1996). *The Food Chronology.* London: Aurum Press. A sweeping and entertaining 783-page overview of the cultural development of food and food availability throughout human history.

Wiley Encyclopedia of Food Science and Technology, 2nd edition. (2000). New York: Wiley. The four volumes contain articles by 368 contributors around the world with information useful to food engineers, chemists, biologists, ingredient suppliers, and other professionals.

FORD, HENRY

• • •

The American automobile manufacturer Henry Ford (1863–1947) is, along with Thomas Edison and the Wright Brothers, one of those who best symbolized the use of technology to transform human life in the early twentieth century. Ford himself recognized the social orientation of his efforts. As he explained in his 1922 autobiography, he believed that successful manufacturing was rooted in public service rather than in money making. He was equally clear about his own public service goal: "To lift farm drudgery off flesh and blood and lay it on steel and motors has been my most constant ambition." Somewhat unexpectedly, however, his focus shifted when he discovered "that people were more interested in something that would travel on the road than in something that would do the work on the farms".

Ford was born on a farm in Wayne County, Michigan, on July 30 and died in Dearborn, Michigan, on April 7. As a boy he experienced the agrarian way of life that once had dominated the American economy but that during his lifetime, in part as a result of his efforts, would be replaced by manufacturing. Among the relevant features of his youth were his education in rural schools (1871–1879), the early death of his mother (1876), and his fascination with machinery. That interest led to an apprenticeship in nearby Detroit (1879–1882) and a traveling job servicing steam traction engines. After his marriage in 1888 Ford's father gave him a forty-acre farm, but rather than take up farming,

Henry Ford and his wife moved to Detroit, where he became an engineer for the Edison Illuminating Company.

Automobile Manufacturing

By the early 1890s, when Ford turned his attention to using internal combustion engines to power road vehicles, the effort to develop automobiles had been under way for several decades. By that time American manufacturers had incorporated the general principles of machine production, interchangeable parts, and cost-based management, along with other practices of the factory system and large-scale business. Thus, Ford began neither the specific process of creating automobiles nor the overall process of industrialization. However, he would achieve lasting fame as well as notoriety by helping bring both processes to full maturity.

Ford's historic achievement was twofold. First, he rethought the basic idea of the automobile (making him more an innovator than an inventor), by aiming not for a large luxury vehicle but for one that was light and sturdy enough for unimproved rural roads and inexpensive enough for the average family. Second, he, along with the mechanics and engineers he employed, redesigned the manufacturing process to allow for the mass production of a product of unprecedented complexity.

The main features of this frequently told story include the completion of Ford's first experimental car (1896), his early interest in building race cars (driven by Barney Oldfield), the formation of the Ford Motor Company (1903), the introduction of the Model N (1906), and the successful challenge of the Seldon patent (1911), which ruled that George B. Seldon, a Rochester lawyer who was issued a patent in 1896 for the horseless carriage, was not entitled to a royalty for each car manufactured. However, looming over everything else was the Model T. First sold in 1908 for $825, the Model T remained in production until 1927, by which time 15 million had been made and the price had dropped to $290.

To lower costs and increase output, the company adopted the practices of progressive assembly at its Highland Park plant. The capstone of that effort was the continuously moving assembly line for attaching the various components to the chassis, which was put in place during the winter of 1913–1914. Although not a direct application of scientific management, Ford's system bore similarities to it, including the dramatically higher pay rate of "the Five Dollar Day" (1914). During and after World War I the company went on to construct the River Rouge plant, where production of the Model T achieved a high degree of vertical integration.

Henry Ford, 1863–1947. After founding the Ford Motor Company, the American industrialist developed a system of mass production based on the assembly line and the conveyor belt which produced a low-priced car within reach of middle-class Americans. (*AP/Wide World Photos.*)

This system was widely admired, copied, detested, and critiqued. Its place in the modern psyche can be seen in widely different cultural products, such as Charlie Chaplin's performance in the film *Modern Times* (1936) and the convention for numbering years that Aldous Huxley devised in *Brave New World* (1932): "A.F." for "After Ford."

Achievements and Criticism

Those achievements must be attributed to many people in addition to Henry Ford. Nevertheless, Ford personally led the enterprise. Before World War I the result was a highly favorable public image. However, "the Five Dollar Day" was accompanied by the systematic investigation by the Ford Motor Company of individual workers outside the plant, and after World War I that arrangement was replaced for the most part by a more traditional approach involving company spies and threats of violence. Meanwhile, with wealth and power also came the expression of personal idiosyncrasies. A newspaper Ford owned, for example, propounded anti-Semitic views that later struck a resonant chord in Nazi Germany.

From the vantage point of the present, however, probably the most significant of Ford's shortcomings was his failure to give up personal control of the company he had founded. He consolidated that control after World War I and held on to it until almost the time of his death. One result was continued production of the Model T until the company had saturated its market, making more difficult the conversion to other models (the Model A in 1928 and the V-8 engine in 1932). Limitations also can be seen in other products the company attempted to produce: submarine chasers during World War I and farm tractors and trimotor commercial aircraft in the interwar years. Even when the products were well conceived, problems arose with production or marketing; those problems could be traced back in part to Ford's personal control of the company.

Although the Ford Motor Company was his primary achievement, Henry Ford created other organizations of lasting importance, including the Ford Foundation and The Henry Ford (formerly the Henry Ford Museum and Greenfield Village) in Dearborn, Michigan.

THOMAS D. CORNELL

SEE ALSO *Automobiles; Edison, Thomas; Taylor, Frederick W.*

BIBLIOGRAPHY

Ford, Henry, with Samuel Crowther. (1922). *My Life and Work*. Garden City, NY: Doubleday. The first of several co-authored books that presented Ford's views and life story to the public.

Hounshell, David A. (1984). *From the American System to Mass Production, 1800–1932: The Development of Manufacturing Technology in the United States*. Baltimore and London: Johns Hopkins University Press. Places Ford's technological achievement in the context of the larger trend toward mass production.

Meyer, Stephen III. (1981). *The Five Dollar Day: Labor Management and Social Control in the Ford Motor Company, 1908–1921*. Albany: State University of New York Press. Explores the practices of Ford's company as a social system of production.

Nevins, Allan, and Frank Ernest Hill. (1954). *Ford: The Times, the Man, the Company*. New York: Charles Scribner's Sons.

Nevins, Allan, and Frank Ernest Hill. (1957). *Ford: Expansion and Challenge, 1915–1933*. New York: Charles Scribner's Sons.

Nevins, Allan, and Frank Ernest Hill. (1963). *Ford: Decline and Rebirth, 1933–1962*. New York: Charles Scribner's Sons. These three books by Nevins and Hill remain the most detailed treatment of Ford and the company he founded.

Wik, Reynold M. (1972). *Henry Ford and Grass-Roots America*. Ann Arbor: University of Michigan Press. Explores Ford's impact on popular culture.

FORD PINTO CASE

• • •

Events in the 1970s related to the Ford Pinto automobile illustrate some of the ethical issues related to technology and safety. In an effort to produce a stylish but affordable subcompact automobile with a low operating cost, Ford Motor Company management made a questionable decision regarding the positioning of and protection for the fuel tank. A safer gas tank and tank location were technologically feasible, but consumer affordability and style took precedence over safety. Ford engineers were constrained by design and cost limitations, and the case therefore illustrates how engineering decisions are often made in the context of marketing strategies. For example, the car was designed to have a short rear-end, perhaps in imitation of the extremely popular Ford Mustang. This limited the engineers' alternatives for fuel tank safety and placement. The tank was placed behind the rear axle instead of over-the-axle, a safer location that had been used in the Ford Capri. Critics charged that this decision was a result of the reduction in trunk space caused by the over-the-axle placement. Another example of a limitation on the engineers was that management apparently mandated that the car cost no more than $2000 and weigh no more than 2000 pounds. If these limitations were really stipulated, then the engineers would have been constrained in many areas related to safety. Given these design and cost limitations, is it fair to hold the engineers morally responsible for the preventable pinto fire injuries and deaths? Other issues illustrated by the Pinto events relate to the definition of *safety*, the appropriate responsibilities and professional obligations of engineers, the interactions between different parts of organizations, ethical management decision-making, and effective government safety policies.

For example, "safety" can be understood to mean "acceptable risk of harm," but how much risk is acceptable in a subcompact automobile? Additionally, did the engineers have a professional obligation to reject the Pinto design elements and management directives that seriously compromised safety? Should Ford management

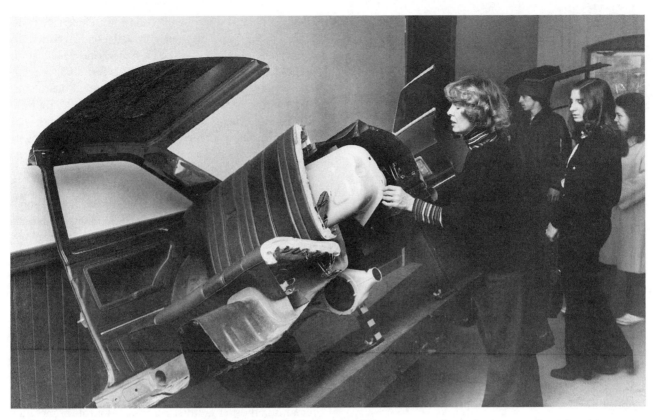

People examining close-ups of a Ford Pinto wagon in the basement of a courthouse. The wagon was used as evidence in a murder trial resulting from a fatal accident in which the gas tank exploded upon collision. (*Art Shay/Getty Images.*)

have had the final word on the Pinto design or should the engineers have had a "veto" related to safety? If management really placed marketing considerations above safety, was that objective ethical and are members of management morally responsible for the preventable Pinto fire deaths? Finally, was the National Highway Traffic Safety Administration ineffective or unethical because the Ford Pintos always complied with all the government standards?

Ford produced the Pinto automobile from 1971 to 1980. Initially the car sold well, but a defect in early models made Pintos prone to leaking fuel and catching on fire after relatively low-speed, rear-end collisions. The Pinto's gasoline tank was located behind the rear axle. A rear-end collision of about twenty-eight miles per hour or more would crush the car's rear end, driving the fuel tank against the differential housing and causing it to split and the filler pipe to break loose. Sometimes the spilled fuel and sparks from the crash caused fires that produced fatalities or serious burns. Many such victims or their relatives filed civil suits against Ford Motor. This litigation generated damaging publicity for Ford and for the Pinto, and it increased public concern over fuel system integrity in general. In 1976 the

National Highway Traffic Safety Administration (NHTSA) implemented a rear-impact safety regulation. The 1977 Pintos were in compliance with this standard, but earlier Pintos were not required to be in compliance and did not meet the standard. Responding to publicity about the Pinto's poor safety record, the NHTSA crash-tested some early Pintos and in 1978 announced that a safety defect existed in the fuel systems of 1971–1976 Pintos. With an NHTSA public hearing scheduled, Ford recalled the 1971–1976 Pintos to upgrade fuel system integrity.

The improvements to the 1977 and subsequent model-year Pintos and the recall of the earlier ones should have solved Ford's Pinto fuel system problems. In September 1978, however, an Indiana grand jury indicted Ford on three felony counts of reckless homicide. This indictment was related to an accident in which, after a van rear-ended a Pinto in an allegedly low-speed collision, three young women burned to death. In contrast to the previous Pinto cases, this one was a criminal trial, not a civil suit. Ford was found not guilty on all the charges because the corporation's lawyers persuaded the jury that the crash was not, in fact, a low-speed one, and hence the deaths did not result from

Ford's having kept a lethal vehicle in production in spite of an obvious fatal flaw. Ford stopped producing the Pinto after 1980, having sold about 3 million of the vehicles.

DOUGLAS BIRSCH

SEE ALSO *Automobiles; Engineering Ethics*.

BIBLIOGRAPHY

Birsch, Douglas, and John Fielder. (1994). *The Ford Pinto Case*. Albany: State University of New York Press. Includes the basic documents needed to understand the case and articles that discuss the ethical issues related to it.

Cullen, Francis; William Maakesstad; and Gray Cavender. (1987). *Corporate Crime under Attack: The Ford Pinto Trial and Beyond*. Cincinnati, OH: Anderson Publishing. Contains a section on corporate crime and the movement to control it and another part on the Indiana criminal trial of ford motor company related to three pinto fire fatalities.

Strobel, Lee Patrick. (1980). *Reckless Homicide? Ford's Pinto Trial*. South Bend, IN: And Books. Discusses the Indiana criminal trial in which Ford Motor Company was found not guilty on three counts of reckless homicide.

FORENSIC SCIENCE

• • •

The word forensic is derived from the Latin word *forensic*—a reference to Roman court forums in which evidence of wrongdoing was presented. Modern use of the term forensics refers to scientific principles and processes that are applied in the analysis of evidence for legal purposes. Alternatively known as *criminalistics*, forensics involves using sophisticated techniques and tools to identify, collect, analyze, preserve, and present evidence of crimes or civil wrongdoing in legal proceedings, as well as to verify identification of deceased individuals. The essential goal of forensics analysis is to verify connections between two or more physical items, for example, the blood of a homicide victim to that found on clothes worn by a suspect. Forensics involves analysis of many other types of evidentiary items such as prescription and illicit/illegal drugs, metals, glass, plastics, fuels, paints, tire/shoe prints, tool/tool marks, and latent substances such as synthetic fibers, human hair, and animal fur, among others.

Modern forensics began with nineteenth-century efforts of Alphonse Bertillon (1853–1914), director of the Bureau of Criminal Identification of the Paris (France) Police Department, to classify and identify criminals on the basis of their physical characteristics. In 1888 Francis Galton proposed a fingerprint classification method after which fingerprinting was first used for criminal identification by Scotland Yard investigators in 1901, and by New York City detectives in 1902. By 1930 the Federal Bureau of Investigation (FBI) of the U.S. Department of Justice had established a national fingerprint classification system, and in 1946 the FBI created its Identification Division that relied extensively on burgeoning fingerprint records for suspect identification in criminal cases. Since then the FBI lab has helped solve thousands of criminal cases using many forensics analysis methods, and is among the largest and most technologically capable forensic laboratories in the world.

Types of Forensics Evidence and Analysis

There are many types of forensic methods, each of which corresponds to the kind of evidence analyzed. For example, ballistics is the study of firearms, ammunition, bombs/explosives, bullets, and other projectiles. Forensic anthropology attempts to reconstruct the likeness of decomposed or dismembered bodies based on skeletal remains and other factors. Forensic odontology matches bite marks with teeth or dental records; and forensic entomologists study corpses infested with insects to determine the approximate time of death and other information. Forensic psychology and psychiatry seek to profile criminals, and also apply social work and mental health counseling practices to investigative situations. Forensic toxicology involves analysis of intoxicants, drugs, and poisons. Forensic taphonomy pertains to the examination of dead and decaying human, animal, and plant remains.

The most modern, prominent, and scientifically promising form of forensics is DNA analysis profiling which involves comparison of deoxyribonucleic acid found in human body tissue or fluids such as blood, perspiration, urine, semen, or vaginal secretions. In addition, biometrics analysis is used in forensics to verify identification of people by comparing biological traits such as finger/palm prints and iris or retina cell patterns. Other forms of forensics involve toxicology (the study of poisons and their harmful effects), computer forensics, voiceprint identification, and polygraph examinations (lie detector testing). In addition to determining the sources of criminal evidence and matching these to known sources, forensics also involves crime scene reconstruction—examining evidence to determine the nature of activities and physical dynamics of interactions among perpetrators and crime victims, series of events, directions of travel, angles and relative forces of impact, pre/post impact trajectories, and primary versus secondary causes of harm, and more.

Fundamental and Ethical Challenges in Forensics

Primary challenges in forensics pertaining to the overall validity, reliability, and credibility of evidence presented in court cases involves:

1. protecting evidence from harm before, during, and after its collection at crime scenes and in laboratories and evidence storage facilities;

2. accurately analyzing evidence and truthfully presenting findings in legal proceedings to help explain how crimes occurred and the possible guilt or innocence of individuals accused of crimes;

3. developing and maintaining expertise of forensics professionals through training;

4. acquiring, certifying, and maintaining laboratory equipment;

5. providing managerial oversight to ensure accurate analyses and truthful reporting of findings in legal proceedings;

6. truthfully testifying about analytical methods, findings, and credentials of examiners;

7. achieving laboratory accreditation by one or more nationally recognized professional membership associations.

Criticism of and concern about forensics analysis has involved all the challenges listed above. In addition, so-called *voodoo science* or *junk science* refers to the reality that all forms of forensics analysis require professional judgment in determining whether evidence collected at crime scenes matches known-source samples to the exclusion of all other possibilities. In many types of forensics analysis there is no scientific basis for employing statistical probability modeling to accurately estimate the chances that one or more evidentiary items are *not* a perfect match. Fingerprint analysis, for example, although long accepted by courts as a type of scientific evidence is actually a technical art predicated on the belief that no two people have exactly the same print patterns and that professionals conducting tests sought exculpatory evidence in addition to match points. This fundamental problem extends to other types of forensics analysis, and when combined with numerous legal cases in which forensics experts lied about their analytical findings and/or professional credentials, has resulted in considerable controversy about the reliability of evidence collection and forensics analysis procedures, and the trustworthiness of testimony in legal proceedings about forensic analysis/laboratory findings.

In *Daubert v. Merrell Dow Pharmaceuticals, Inc.* (1993), the U.S. Supreme Court scrutinized the field of forensics and established new legal standards regarding the admissibility of scientific evidence and expert witness testimony provided by forensics professionals. Standardized DNA evidence gathering and analysis championed by the National Institute of Justice of the U.S. Department of Justice, and acceptance of this form of truly scientific evidence by federal, state, and local level criminal justice systems, is important to the future of forensics, as are quality control standards such as those established by the American Society of Crime Lab Directors/Laboratory Accreditation Board. Ultimately the usefulness and reliability of forensics evidence in legal proceedings will depend on ethical (and potentially government regulated) use of forensics technologies in the public sector and in privately owned or operated laboratories.

SAMUEL C. MCQUADE, III

BIBLIOGRAPHY

Bevel, Tom. (1997). *Bloodstain Pattern Analysis: With an Introduction to Crime Scene Reconstruction.* Boca Raton, FL: CRC Press.

Cole, Simon A. (2001). *Suspect Identities: A History of Criminal Identification and Fingerprinting.* Cambridge, MA: Harvard University Press.

Daubert v. Merrell Dow Pharmaceuticals, Inc., 509 US 579 (1993).

Fisher, David. (1995). *Hard Evidence: How Detectives Inside the FBI's Sci-Crime Lab Have Helped Solve America's Toughest Cases.* New York: Simon & Schuster.

Galton, Francis. (1888). "Personal Identification and Description." *Nature* 201–202.

James, Stuart H., and Jon J. Nordby, eds. (2003). *Forensic Science: An Introduction to Scientific and Investigative Techniques.* Boca Raton: CRC Press.

National Research Council. (1992). *DNA Technology on Forensic Science.* Washington, DC: National Academy Press.

Schmalleger, Frank. (2005). "The Future of Criminal Justice." In *Criminal Justice Today: An Introductory Text For The Twenty-First Century*, 8th edition. Upper Saddle River, NJ: Pearson/Prentice Hall.

FOUCAULT, MICHEL

• • •

Michel Foucault (1926–1984), who was born in Poitiers, France on October 15 and died in Paris of AIDS on June 25, was a controversial philosopher whose interdisciplinary work has important if indirect implications for

Michel Foucault, 1926–1984. The French philosopher, critic, and historian was an original and creative thinker who made contributions to historiography and to understanding the forces that make history. (© Corbis-Bettmann. Reproduced by permission.)

science, technology, and ethics. His research often changed directions—archaeology and genealogy as ideas, history of the present, problematization, and modes of subjectification were prominent. In his final years he viewed these directions as theoretical tools to examine three perennially related but distinct relations: to truth, to power, and to self. Foucault was sufficiently intrigued by various sciences and technologies to devote much of his work (and personal involvement) to analyzing and questioning how they increasingly engage formative and dangerous aspects of human life.

Four themes with ethical implications highlight this intrigue. They are space, vision, biopolitics, and art of the self. Among humans space is seldom only a natural given. People instead design, build and defend, or violate a variety of spaces. Some illuminate ideals (utopias), many are ordinary (common domains), while others are designed for extraordinary times or unfamiliar figures (heterotopias). Asylums, hospitals, schools, and military camps are built to distinguish rituals and events (treating the mentally ill or sick, transforming adolescents or enlistees) that specifically aim to change our body, conduct, and self-understanding. Foucault studied how these spaces emerged, but also questioned their effect on human freedom, individuality, and justice.

Related to the technology of space are innovative kinds of vision. Obviously instruments such as the microscope introduce surprising ways to diagnose the body. Institutions repeatedly introduced strategies for observing the human body. Employing these different visions has two effects. First it renders individual subjects silent, because they are observed at a distance while their own words are discounted. Second this distance ushers in an allegedly more scientific understanding of human beings.

These effects are strikingly presented in the 1975 landmark book, *Discipline and Punish*. The book opens by juxtaposing an elaborate torture spectacle in 1757 Britain with a prison scene in 1838 France. A sign of moral progress in modern Europe? Not entirely. While English philosopher Jeremy Bentham's (1748–1832) design of an ideal prison, the Panopticon (literally, *all seeing*) was itself a practical failure, it paved the way for a radical shift from punishing the criminal to focusing on potentially deviant or abnormal persons—anyone, in principle. The result is a disciplinary society, one bent on surveillance and control. With typical rhetorical flair, Foucault asked, "Is it surprising that prisons resemble factories, schools, barracks, hospitals, which all resemble prisons?" (Foucault 1977, p. 228) Foucault acknowledged, however, that his portraits of modern society were occasionally hyperbolic.

The formation of new kinds of knowledge and their cultural effects has extensive political repercussions and culminates in what Foucault called *biopolitics*. This term refers to a political rationality in which specific knowledges and administrative technologies are used by a government to understand and regulate not only individuals but also groups or populations. Hence the ongoing links between, say, longevity and social security, health and insurance, risky behavior and family assistance, or poverty and education programs. Ian Hacking, an insightful extender of Foucault's approach, describes these relations as having looping effects, loosely but evidently intertwined in terms of a development of an expertise and its gradual influence on how human beings subsequently understand (accepting or resisting) new ideas about themselves.

During work on *The History of Sexuality* (1978–1984), Foucault began focusing on technologies of the self. Here technology is not so much about instruments or tools, but it is more a craft or care for oneself insofar as one uses available knowledge and experiences (such as diet, love, physiology, dream analysis, and structure of home life) to practice a moral life. While his scholarly attention surprisingly turned to texts of the early Greeks and Christians, Foucault cautioned against emulating them. Address the possibilities, he argued, rather than succumb to one's own blind spots.

Foucault was reluctant to spell out a theory of normative ethics. Not only was such an endeavor impossible for modern thought (see *Order of Things*, p. 328), he believed intellectuals should be wary of imposing solutions for those involved in specific struggles. In this light Paul Rabinow nevertheless identifies a four-fold of Foucault's ethics as comprising a will to truth, stylization of one's self, critical thought, and a telos or purpose that involves a dissembling of the self. Be prepared, in other words, that leading an ethical life amid scientific and technological changes will not confirm your identity, but transform you.

The work of Michel Foucault is daring in its range and depth. Although he builds on the approaches of phenomenology, Marxism, and existentialism, he takes the twentieth century European intellectual tradition into a new historical critical phase. As different strains of scientific discovery and technological innovations continue to emerge, his conceptual tools demand that one ask: How is it true? Where is its power? How might it change individuals and their relations to others?

ALEXANDER E. HOOKE

SEE ALSO *Regulation; Monitoring and Surveillance; Science, Technology, and Literature; Space.*

BIBLIOGRAPHY

Burchell, Graham, Colin Gordon, and Peter Miller, eds. (1991). *The Foucault Effect: Studies in Governmentality.* Chicago: University of Chicago Press.

Connolly, William. (1999). *Why I Am Not A Secularist* Minneapolis: University of Minnesota Press. Uses Foucault's style of analyzing discourses to examine controversies of contemporary politics. Connolly undercuts familiar oppositions—such as religion versus secularism, liberalism versus conservatism—to present new options in political pluralism.

Donzelot, Jacques. (1979). *The Policing of Families*, trans. Robert Hurley. New York: Pantheon.

Dumm, Thomas (2002). *Michel Foucault and the Politics of Freedom.* Lanham, MD: Rowman and Littlefield.

Eribon, Didier. (1991). *Michel Foucault*, trans. Betsy Wing. Cambridge, MA: Harvard University Press. The most thorough and balanced treatment of Foucault's personal life and intellectual development. Recounts philosophical and political climate of his life, interviews numerous associates and friends, and carefully describes his books and projects.

Foucault, Michel. (1965). *Madness and Civilization: A History of Madness in the Age of Reason*, trans. Richard Howard. New York: Vintage Books.

Foucault, Michel. (1970). *The Order of Things: An Archaeology of the Human Sciences.* New York: Vintage Books.

Foucault, Michel. (1977). *Discipline and Punish: The Birth of the Prison*, trans. Alan Sheridan. New York: Vintage Books.

Foucault, Michel. (1978–1984). *The History of Sexuality*, trans. Robert Hurley. 3 vols. New York: Pantheon.

Foucault, Michel. (1997). *Essential Works: Volume I—Ethics: Subjectivity and Truth*, ed. Paul Rabinow. New York: The New Press.

Fraser, Nancy, and Mary Gordon. (1997). "Genealogy of *Dependency* in Tracing a Keyword of the U.S. Welfare State." In *Justice Interruptus*, by Nancy Fraser. New York: Routledge. Insightful and concise application of Foucault's genealogy. Fraser and Gordon discuss how changes in the definitions of dependency correlate with cultural shifts (from benign to negative) on the value and status of being dependent.

Hacking, Ian. (1995). *Rewriting the Soul: Multiple Personality and the Sciences of Memory.* Princeton, NJ: Princeton University Press. Investigates the emergence or discovery of contemporary pathologies in light of popular perceptions, scientific analysis, and the growing prominence of psychology.

Hacking, Ian. (1999). *The Social Construction of What?* Cambridge, MA: Harvard University Press. Sometimes playful but always enlightening discussion of how specific ideas and discourses arise and the disputes they foster among various political and scientific representatives, particularly those who claim to ground their views either in natural kinds or social construction.

Martin, Luther; Huck Gutman; and Patrick Hutton, eds. (1988). *Technologies of the Self: A Seminar With Michel Foucault.* Amherst: University of Massachusetts Press. Collaborative project among scholars during Foucault's visit to University of Vermont. Historians, religious scholars, and philosophers develop their own themes sparked by Foucault's work. He has an essay and an interview with the hosts.

FRANKENSTEIN

• • •

Frankenstein, or The Modern Prometheus (1818) by Mary Shelley provides the most potent, characteristic, and uniquely modern myth of science gone fatally awry. The common association of the name Frankenstein, thanks to many popular movies, is with the ugly, lumbering, murderous monster whom the book never names. In his many film versions, this lurching omen reflects the eras of his creation, from the dazed, scorned and feared working-class creature played by Boris Karloff in James Whale's depression-era *Frankenstein* (1931) to the slyly silent and sexually potent creature played by Peter Boyle in the me decade's *Young Frankenstein* (1974). But while

Boris Karloff as Frankenstein's monster in the 1931 film verison of *Frankenstein*. Karloff's portrayal of the creature is perhaps the most well-known. (© *Bettmann/Corbis.*)

movies have spread the image of *Doctor* Frankenstein and associated his name with the manlike monster he created, the novel carefully never names his creation which is, in fact, a doppelganger, a dramatic double of the obsessive undergraduate who made him.

The Modern Prometheus

The ancient myth of Prometheus took two forms: Prometheus *pyrphoros* (fire-bringer) and Prometheus *plasticator* (shaper). In the first the god steals divine fire, emblematic of the combined good and bad potentials of all technologies, for humans; in the second he shapes humans from clay and breathes life into them. In both Zeus makes Prometheus suffer endlessly for his disobedience. In the modern myth, Frankenstein shapes his creation from *charnel matter* and reanimates it (rather than *creating life*) with electricity, an occurrence, as Shelley writes in her preface, "supposed by Dr. [Erasmus] Darwin, and some of the physiological writers of Germany, as not of impossible occurrence." The bounds that Frankenstein transgresses are those of obedience to community. He makes himself a monster in two senses. The price is death not only for himself but for his family and potentially all humanity.

As Gothic novels of the supernatural became stale, authors added a twist, revealing at the end some realistic explanation for the fantastic occurrences. By moving that explanation to the beginning of *Frankenstein,* Shelley created the genre that has explored human fears of science ever since: science fiction.

Structure and Narrative of the Novel

This early science fiction is composed of letters from an explorer, Robert Walton, to his sister back in England. He cannot send the letters because his ship is mired in the arctic where he seeks to confirm the ancient Hyperborea myth of a land of warmth beyond the far north, but he writes nonetheless. On a passing floe he discovers the debilitated Victor Frankenstein whom he rescues. During Victor's recuperation, Robert remarks that "I begin to love him as a brother" (1969, p. 27). In some sense, Robert and Victor, too, are doppelgangers.

The book is a series of nested narratives. The outermost, Robert's own, contains Victor's story that tells of his pursuit of greatness and withdrawal into feverish, isolated work. He finally succeeds, but one look at his stirring creation shows him instantly that the creature is evil. He would kill it, but it flees. The reader comes to learn that the creature is the strongest, smartest, most articulate character in the book, a fit embodiment of science. He confronts Victor on a glacier (the ice imagery mirroring the situation of Robert and Victor, all three males surrounded by frozen fertility) and pleads for paternal help, requesting a bride so that he, universally shunned for his ugly exterior, can find community. Victor reports the creature's narrative which includes his plea and his reported story of Felix (happiness) and Safie (wisdom), Christian-Muslim lovers who are promised help against prejudice and the opportunity to marry by Safie's father, but are betrayed by him. The creature learns the lovers' tale overhearing them in a cottage through a knothole in the wall of the outer shed he has been occupying while altruistically providing firewood for the blind old man who lives there. With the couple on the scene, the creature learns to read just by watching their sharing aloud three books: Milton's *Paradise Lost,* which concerns disobedience and provides *Frankenstein's* epigraph, fallen Adam's plea to God ("Did I request thee, Maker, from my clay / To mould Me man? Did I solicit thee / From darkness to promote me?) (Book X, lines 743–745); *Plutarch's Lives,* a classic collection of exemplary biographies; and *The Sorrows of Young Werther,* Goethe's famous tale of unrequited love ending in death. The creature, initially the most virtuous character in the book, is driven away when the blind cottager's guests see him. Readers believe him when he says to Victor that "My vices are the children of a forced solitude that I abhor; and my virtues will necessarily arise when I live in communion with an equal" (line 1470). At the heart of *Frankenstein's* nested narratives is the betrayal by Safie's father. The rupture of community echoes throughout the book.

When Victor first absents himself to work, his father sends a letter that says, quite rightly, "I regard any interruption in your correspondence as a proof that your other duties are equally neglected" (p. 55). Victor destroys his creature's unfinished bride in sight of the monster, who then begins murdering Victor's family to force him to start again. Instead they chase each other north. While Victor never writes, Robert always writes. Robert heeds his frightened crew and turns back from his quest, saving all their lives. Victor dies, and the monster (from the Latin for *warning*) carries him further north for a funeral pyre, knowing that with his *father* dead, his hopes for any family have died, too.

Science Unbound

At the heart of *Frankenstein* is the tension between the power science confers on individuals and the just restraints of community. Frankenstein, both creator and creature, stands not for science in general but for the acquisition of scientific power foolishly pursued without the wisdom of the world. As such, Frankenstein has represented, in the films of the Great Depression, the isolation of the privileged from the suffering of the common person. When the educated *Doctor* or *Baron* in his hilltop castle, his title varying from film to film, disdained the peasants swirling up toward him with their angry torches, his doppelganger monster was inarticulate because, the movies imply, the overly powerful never heed the consequences of their power.

That image has entered the very language of the early 2000s. Genetically modified farm crops are bashed as *Frankenfoods* and contemplated human cloning for spare parts is called a *Frankenstein nightmare*. Shelley has a character say early on, "One man's life or death were but a small price to pay for the acquirement of the knowledge which I sought" (1969, p. 28). That sounds like Victor, but it is Robert, the seeker who learns the limits of seeking. *Frankenstein* is the early twenty-first century's greatest cautionary tale.

ERIC S. RABKIN

SEE ALSO *Autonomous Technology; Brave New World; Playing God; Science Fiction; Science, Technology, and Literature; Shelley, Mary Wollstonecraft.*

BIBLIOGRAPHY

Schoene-Harwood, Berthold, ed. (2000). *Mary Shelley, Frankenstein*. Columbia University Press Critical Guides. New York: Columbia University Press.

Shelley, Mary. (1969 [1818; 1831]). *Frankenstein, or the Modern Prometheus*, ed. M. K. Joseph. The World's Classics series. New York: Oxford University Press.

Shelley, Mary. (2000). *Frankenstein: Complete, Authoritative Text with Biographical, Historical, and Cultural Contexts, Critical History, and Essays from Contemporary Critical Perspectives*, ed. Johanna M. Smith. Boston: Bedford/St. Martin's.

FRAUD AND MISCONDUCT IN SCIENCE

SEE *Misconduct in Science*.

FREE AND INFORMED CONSENT

SEE *Informed Consent*.

FREEDOM

• • •

The concept of freedom or liberty is complex, with political, ethical, and psychological dimensions. In the context of modern science, technology, and ethics, freedom exhibits all of the ambiguity of human experience. The promise of modern science and technology is that the increases in knowledge and the power they afford will expand human freedom in an unqualified sense. But in opposition to this original and continuing justification are questions about the extent to which science and technology may also limit or qualify freedom. Moreover, the professional ethical requirement for the free and informed consent of human participants in scientific research situates the complexities of freedom in the heart of science itself. The issue of "free and informed consent" is a key locus for the discussion of freedom in science and technology.

Human Freedom versus Deterministic Science

The philosophical concept of freedom may be seen in opposition to that of determinism. The determinist holds that there is no freedom. For a hard determinist, all events in nature are strictly determined. As such, the idea of freedom is incompatible with that of the causal determination of all natural events. What is sometimes called soft determinism or compatibilism modifies the hard position by maintaining that freedom is compatible with the determination of natural events. A compatibilist holds that all events in nature are causally determined but that human beings can initiate new series of events and have responsibility for the outcomes of their actions. Thus, moral ideas of praise and blame make sense if people are able to act according to some causality arising from their will or for reasons of their own choosing.

Finally, it should be noted that with the development of quantum mechanics some thinkers allow for indeterminacy at the atomic level. This may allow for a notion of freedom in the sense that an action is not caused, but it may not be able to account for personal responsibility if the action is not determined in some way by the person.

Whether or not human beings are in fact free, most people think and act as if they are. Such acts of freedom have been conceptualized in two basic ways: negative and positive.

Negative Freedom

Negative freedom may be taken as an absence of obstacles to the fulfillment of one's desires or wishes. The view of the English philosopher Thomas Hobbes (1588–1679) is representative of this approach. This form of freedom depends upon the existence of favorable external circumstances for the attainment of a human goal. It can thus be considered a freedom of self-realization. One peculiar implication of this approach is that a person who wishes to be in a prison cell may be said to be free. Nor does it require that there be alternatives from which to choose. If there is only one course of action available, but that is what an individual wishes, then such a person may be said to be free. It also seems to allow for animals to be described as free. A further point is that the obstacles to human desires include physical and social ones. Thus, if persons are physically constrained or constrained by fear, they may not be said to be free to act. If, on the contrary, they are coerced to act in a certain way, they are not considered to be free nor responsible for their actions.

According to this conception, modern science and technology may be construed as eliminating any number of obstacles that have historically restrained human action. Therefore, those taking engineering approaches to science and technology tend to see modern technical methods as enhancing human freedom. With modern methods of communication and travel, for example, time and space seem to shrink in their significance. Many elderly people of the late twentieth and early twenty-first centuries have been able to act without the encumbrances of the maladies of old age that have plagued human beings for millennia. This form of freedom is a freedom from such things as disease, hunger, and fear.

Modern technologies, however, may also be seen as introducing new obstacles to human action. The automobile provides for transportation over great distances, but millions of cars on the roads produce traffic jams that obstruct a person's desire to move. The roadways also block a person's desire to walk if the destination is across a multilane highway. The very complexity of modern technological societies may represent an obstacle to human action. With all of the information that is available through the various media, many persons feel overwhelmed by information overload. Greater knowledge is thought to increase one's freedom to act, but it becomes difficult to act rationally in such an environment. Indeed, the self may become fragmented as it interacts with the technological environment.

This seems to be an outcome that is contrary to the self-realization that is characteristic of the notion of negative freedom. A further problem with the notion of negative freedom is that modern science and technology may be used to manipulate human desires, and so in a sense persons are coerced to act. Thus, propaganda techniques are used to mold consumer desires. Indeed, it has become possible to manipulate human desires pharmacologically. This possibility has been the theme of dystopias such as Aldous Huxley's *Brave New World* (1932). Huxley imagined a society in which a drug called "soma" could be taken that would make a person content in any environment. Anthony Burgess's *A Clockwork Orange* (1962) depicted a cruder reconditioning of human desires. Many thinkers in the humanistic tradition would not consider human beings to be free if there are no obstacles to the fulfillment of a person's desires, but the desires one has result from technical manipulation. It is appropriate from this perspective that B. F. Skinner should have written *Beyond Freedom and Dignity* (1971). The practical application of his behaviorism would make human freedom into an illusion because behaviorism holds that all human behaviors are molded by the environment. The control of nature, as C. S. Lewis (1898–1963) pointed out *The Abolition of Man*, easily leads to the control of some human beings by others.

Positive Freedom

The positive notion of freedom requires that individuals be able not only to act on their desires but also to choose from among the many desires they have to act upon. Such a view of freedom constitutes a theory of self-perfection. According to this conception, some desires may be more worthy than others given a standard of human life that is considered good. Only persons who have acquired virtue or a self-consciousness of their humanity may be said to be free. Contrary to the common view that people have greater freedom to act if they have more choices, in this case ideas of virtue or moral duty may lead individuals to restrict their pursuit of certain desires. Rather than simply doing what one wants, one does what one thinks one ought to do.

Moreover, one may have a desire for freedom itself that requires the subordination of one's physical inclinations. This is an example of second order desire, that is, the desire for certain kinds of desires. Here, freedom is an end to be pursued in itself rather than a means to the pursuit of other ends. A peculiar aspect of the positive notion of freedom is that it seems to require a degree of self-denial, at least the denial of the drives of the lower self for the sake of higher drives or interests. It may be that this is necessary for the fulfillment of the higher self. A certain independence of the self from the social and physical environment is also necessary for the pursuit of this form of freedom. As such, positive freedom does not depend upon external circumstances.

The positive notion of freedom is especially significant in ethical reflections on the impact of science and technology on the quality of human existence. The concern here is whether human existence is degraded by the rationalization of the world associated with modern science and technology. If all of human existence, including human beings themselves, is subject to rational control, then there may be no room for the dignity of persons; in such a scenario, persons will have been reduced to objects of manipulation and control. If technical methods are applied to political action, for example, this tends to transform what has traditionally been considered the "art of the possible" into a matter of technical necessity. Technical rationality is a rationality directed to the efficient determination of means to achieving some end. This form of rationality tends to become the dominant form of rationality in a highly developed technological society in which the only worthy ends that are recognized are those that can be pursued by the technical means available.

Positive freedom, however, seems to require a broader form of rationality that takes into consideration the choice of humanly worthy ends. In the debate concerning human cloning, for example, the President's Council on Bioethics placed special emphasis on "human dignity" by calling one of its reports, *Human Cloning and Human Dignity* (2002). Furthermore, Francis Fukuyama (2003) has described a posthuman future resulting from the genetic manipulation of human beings. If modern science and technology lead to the evolution of a posthuman era, of what value is human freedom?

Dialectical Freedom

Beyond the negative and positive accounts of freedom is a dialectical one, with roots in the work of Georg Wilhelm Friedrich Hegel (1770–1831) and Karl Marx (1818–1883), among others. According to this account, human freedom is to be understood precisely as an opposition to some obstacle. As such, freedom depends on the existence of some resistance against which we struggle. If humanity were to succeed in eliminating all obstacles to the fulfillment of its wishes as per the view of negative freedom, it would also eliminate human freedom. The dialectical approach to freedom recognizes that the obstacles human beings confront take both physical and social forms. As one obstacle is overcome, however, new ones arise, so that human freedom can be seen as developing over time as humans confront new obstacles.

From a dialectical perspective, freedom must be coordinated with the environment in which humans exercise their freedom. The first and historically most fundamental form of freedom in this scenario occurred when human beings struggled against nature. Nature provided both the means of pursuing human desires through the use of tools as well as obstacles to their use. This form of freedom was superseded by a stage in which human beings developed social institutions, which can be seen as "second nature." Social institutions provided protection from the forces of nature but also introduced new human-made obstacles. After the development of this new environment, the desire for freedom had to be directed against social institutions. The dialectical character of this view of freedom can be seen in that the liberty gained with respect to one environment gives rise to new necessities that must be overcome by creating a new form of freedom. In turn, the new form of freedom is also relevant to a new milieu.

The sociologist Karl Mannheim distinguishes a third stage, that of planning. In this stage, the totality of social institutions and other techniques are organized into a systematic whole. For Mannheim, democratic planning is the last stage in the development of human freedom, whereby human beings take conscious control over the social process. Jacques Ellul (1976) depicted this third stage as the stage of technique, which involves a new technological determination of the human person by the systematic application of techniques to human beings. He thus called for a struggle against the technological environment, especially in its ideological aspect.

The Ethics of Freedom in the Scientific and Technological World

In all of its forms—negative, positive, and dialectical—freedom is closely associated with notions of moral autonomy and political democracy. The ideals of moral autonomy and democratic politics depend on persons and citizens not being wholly determined by external forces, able to pursue personal perfection and the public good, in dialectical engagement with others and the world around them. In the contemporary technoscientific milieu, the others and the world exhibit strongly scientific and technological characteristics.

One area in which this is particularly pronounced is in research on human subjects. Especially since World War II both the scientific community and society at large have increasingly stipulated that scientific research on human subjects be limited by requiring free and informed consent of any such subjects. Participants in scientific research must not be constrained to participate either by force (as in Nazi Germany) or by ignorance (as in the Tuskegee Syphilis Study [1932–1972] in the United States); they must be able to see their participation as positive aspects of their own lives; and they will inevitably struggle against the obstacles of disease and perhaps their own lack of understanding in the process. The commitment to such freedom, which respects limitations in science, even when it also limits scientific progress, makes science more human.

In the larger technoscientific world there are further reflections of such efforts to respect freedom in the emergence of individual and collective ethical responses to the artificial environment produced by modern science and technology and the cultural aspirations to use science and technology to transform the world. Thus, Hans Jonas (1984) has called for an ethic of responsibility that posits an ethical imperative to maintain the existence of human beings. This marks a sharp contrast with those who have called for a posthuman age. Further examples include Ellul (1976), who developed an ethic of non-power to counter the technical impulse to augment human power. And, more recently, Bill McKibben has sought limits to the effort to perfect human beings in his 2003 book, *Enough*. All of these observers are concerned with establishing some humanly significant limits to the technological remaking of the world. They recognize that within a dialectical account of freedom, while the reality of human freedom depends upon the overcoming of limits, it also depends upon the recognition of limits. If the technological project has become an attempt to eliminate all limits, it may very well eliminate freedom as well.

DARYL J. WENNEMANN

SEE ALSO *Alienation; Autonomous Technology; Critical Social Theory; Determinism; Dignity; Ellul, Jacques; Free Will; Hegel, Georg Wilhelm Friedrich; Jonas, Hans; Kant, Immanuel; Marx, Karl; Posthumanism; Security; Rand, Ayn; Thoreau, Henry David; Tocqueville, Alexis de; Weil, Simone.*

BIBLIOGRAPHY

Adler, Mortimer J. (1973). *The Idea of Freedom.* 2 vols. Westport, CT: Greenwood.

Berlin, Isaiah. (2002). *Liberty: Incorporating Four Essays on Liberty*, rev. edition, ed. Henry Hardy. Oxford: Oxford University Press. A modern classic of liberalism, Berlin's four essays defended the notion of negative freedom as necessary for maintaining a liberal society.

Borgmann, Albert. (1984). *Technology and the Character of Contemporary Life.* Chicago: University of Chicago Press.

Burgess, Anthony. (1962). *A Clockwork Orange.* New York: Norton.

Campbell, Neil. (2003). *Freedom, Determinism, and Responsibility: Readings in Metaphysics.* Upper Saddle River, NJ: Prentice Hall.

Dennett, Daniel C. (2003). *Freedom Evolves.* New York: Viking.

Ekstrom, Laura Waddell, ed. (2001). *Agency and Responsibility: Essays on the Metaphysics of Freedom.* Boulder, CO: Westview.

Ellul, Jacques. (1976). *The Ethics of Freedom*, trans. and ed. Geoffrey W. Bromiley. Grand Rapids, MI: Eerdmans.

Fukuyama, Francis. (2003). *Our Posthuman Future: Consequences of the Biotechnology Revolution.* New York: Farrar, Straus and Giroux.

Hayek, Friedrich A. (1960). *The Constitution of Liberty.* Chicago: University of Chicago Press. A defense of freedom as an absence of arbitrary coersion.

Hayles, N. Katherine. (1999). *How We Became Posthuman: Virtual Bodies in Cybernetics, Literature, and Informatics.* Chicago: University of Chicago Press.

Hobbes, Thomas. (1994). *Leviathan: With Selected Variants from the Latin Edition of 1668,* ed. Edwin Curley. Indianapolis, IN: Hackett.

Huxley, Aldous. (1932). *Brave New World.* Garden City, NY: Doubleday, Doran.

Jonas, Hans. (1984). *The Imperative of Responsibility: In Search of an Ethics for the Technological Age,* trans. Hans Jonas, with the collaboration of David Herr. Chicago: University of Chicago Press.

Kane, Robert. (1996). *The Significance of Free Will.* New York: Oxford University Press.

Kane, Robert, ed. (2002). *Free Will.* Malden, MA: Blackwell.

Lewis, C. S. *The Abolition of Man.* New York: Macmillan, 1947.

Malinowski, Bronisław. (1976 [1944]). *Freedom and Civilization.* Westport, CT: Greenwood.

Mannheim, Karl. (1940). *Man and Society in an Age of Reconstruction.* New York: Harcourt, Brace & World, Inc.

McKibben, Bill. (2003). *Enough: Staying Human in an Engineered Age.* New York: Times Books.

Mill, John Stuart. (1998). *On Liberty, and Other Essays,* ed. John Gray. Oxford: Oxford University Press. An immensely influential study of the limits of individual liberty and the power of the state.

O'Connor, Timothy, ed. (1995). *Agents, Causes, and Events: Essays on Indeterminism and Free Will.* New York: Oxford University Press.

President's Council on Bioethics. (2002). *Human Cloning and Human Dignity: An Ethical Inquiry.* Washington, DC: Author.

Rutsky, R. L. (1999). *High Technē: Art and Technology from the Machine Aesthetic to the Posthuman.* Minneapolis: University of Minnesota Press.

Skinner, B. F. (1971). *Beyond Freedom and Dignity.* New York: Knopf.

Watson, Gary, ed. (2003). *Free Will,* 2nd edition. Oxford: Oxford University Press.

FREE SOFTWARE

• • •

Proponents of *free* software distinguish *free speech* from *free beer,* and argue that their conception of free software is intended to evoke the former idea. That is, software is a form of speech used by programmers to express technical ideas in very specific language. Free software does not necessarily mean that the price is zero. The *free software* movement is an explicit attempt to encode in technology specific ethical values about how the world should work. The term *free software* was coined by Richard Stallman, and following his lead, free software programmers have written licenses and computer programs that they believe help create liberty.

Freedom to Use, Change, and Expand the Work of Another

In order for programmers' speech to be heard, it must be transmitted to others. Programmers' work is written in *source code,* usually a text file, which is then interpreted or compiled by other programs in order to perform some computation (for example, to calculate a statistical result or to display a web page). A central idea in the free software movement is that programmers' work, their source code, should be made available in its original form to anyone who is interested in it. A related movement refers to this as *open source.* Although in practice open source and free software often refer to the same programs, their emphases are different. Free software focuses on the goal to promote freedom, while open source focuses on the goal to make the source code available to everyone.

In order to guarantee this availability, programmers distribute free software under licenses that prohibit users from denying others the freedoms they have received. Thus free software may be used and shared by anyone who accepts the terms of the license. The most common free software license is the General Public License (GPL). The GPL offers the following:

- The freedom to run the program, for any purpose.

- The freedom to study how the program works, and adapt it to one's needs. Access to the source code is a precondition for this.

- The freedom to redistribute copies so users can help others.

- The freedom to improve the program, and release one's improvements to the public, so the whole community benefits. Access to the source code is a precondition for this.

To use free software licensed under the GPL, one must accept the license terms. If one refuses these freedoms (for example, because one wishes to keep a particular code secret), the right to use free software is forfeited. That is, if a programmer wants to use code from free software in a new application, the new application must carry the same freedoms as the original code. In order to share or distribute free software, one must pass along these same freedoms to the people to whom the software is distributed.

The Origins of the Free Software Movement

Academic computing from the 1950s through the early 1980s had been mostly unconstrained by concerns about copyright. Scientists shared source code with each other, freely commenting and critiquing each others' work. In the early 1980s, markets opened for the commercial development and sale of software, and among the first moves of the new private-sector ventures was to limit the distribution of the original source code. The limitation seemed sensible at the time—why pay programmers to produce something that customers or competitors could take for free?

Some programmers were critical of the new trend to "close the source," or restrict access to source code. The first criticisms were technical: if programmers find a bug in closed-source software, they cannot simply fix the bug, as they had previously been accustomed to doing. Broader critiques soon followed as programmers realized that in this new work environment, they could not easily share code with colleagues in other organizations. It became more difficult to share ideas and experiences.

In January, 1984, Richard Stallman crystallized the discontent with the foundation of the GNU ("Gnu's not Unix," a recursive pun) Project. In his initial announcement, he said that he and his collaborators would write an entire Unix clone from scratch with entirely free software which would be available to anyone who wanted it. The GNU Project succeeded in developing nearly all the parts of an operating system. However, the GNU Project lacked a kernel, the central part of the operating system that manages memory and connections to hardware. Using many of the GNU tools, a Finnish graduate student named Linus Torvalds released *Linux* in 1991, a kernel that provided exactly this component. Over the next five years, the GNU tools and the Linux kernel made free software a practical platform for general purpose computing.

Free software development proceeded rapidly. In the early 1990s, several other free Unixes emerged from a legal battle between free software programmers (mostly in Berkeley, California) and AT&T. The Berkeley programmers replaced nearly all of AT&T's original Unix code. There are a number of descendants of this process, called the "Berkeley Software Distribution" (BSD). In 1995, other programmers re-wrote the original Netscape HTTP web server and named it the Apache HTTP server (the name is a pun: "a patchy server"). At the time of this writing, Apache powers approximately two-thirds of the web servers on the Internet.

By the late 1990s, different positions in the free software community emerged about the relative priority of different goals. Some people felt that the most important aspects of free software was the promotion of a vigorous intellectual community and growth into new areas, especially by convincing businesses to produce free software. In this perspective, the term "free software" was deemed inappropriate because it discouraged potential allies in the corporate world from adopting it. To avoid the perceived anti-business implication of "free," in 1998 this group re-labeled the community "open source." Since then, the term "open source" has grown significantly more quickly than the term "free software." In practice, the terms refer to mostly the same programs, and even to the same licenses, but signify important differences in the license-holder's focus.

Free Software in Practice

Numerous free software programs have been published. There are free operating systems (including GNU/Linux and various versions of BSD), graphical windowing environments (gnome, KDE), Internet browsers (mozilla, konqueror), office software (OpenOffice, gnumeric), a web server (apache), computer languages (C, perl, PHP, python), and scientific software (the R statistical language, grace—a plotting package), to mention only a few of the tens of thousands of free software programs available. It is possible to do almost any computing, on the desktop or on the server, exclusively by using free software, including interoperating with colleagues using proprietary systems (such as those offered by Microsoft or Apple).

Free Software and Scientific Ethics

Many of the technical and ethical values expressed in the free software movement parallel broader values in the scientific and technical community. In particular, free software programmers prize open technical debate in which all the participants have access to the material in question for testing, benchmarking, critique, and for the creation of derivative works.

As described earlier, free software is distributed in a human-readable form called source code, the original form in which programmers write software. By studying the source code, programmers can evaluate the quality of the solution: Does it work? Is it efficient? Is it elegantly written? In this way, free software is transparent and encourages vigorous peer review. Indeed communities of free software programmers usually exist on publicly available Internet mailing lists, newsgroups, and

Internet sites The only requirement for participating in the review process is the skill to be a programmer.

The Ability and Responsibility to Share

Another central idea in free software is that every user is encouraged to share the software with other users. By sharing, programmers build on each other's ideas and accomplishments, and this serves to advance knowledge, another central scientific value. The idea that software *should* be shared is linked to the sense that tinkering with technology is intrinsically valuable, and that the ability to *open the hood* is the first step toward innovation.

However, as implied by the references to freedom, free software programmers explicitly intend their work to advance the cause of human liberty, and so sharing software has several benefits beyond peer review and encouraging exploration. For example, sharing software helps to decentralize control over the access to information technology. With the rise of technology as the mechanism by which most communication is effected, the practice of free speech depends on free access to the means of speech. Decentralizing control of communications software is one way to help to keep virtual space open to everybody.

Sharing free software also helps lower the cash cost of software, which enables more people to be able to use technology to express their ideas. In the world of free software, this means more people can be free.

Free software uses open data standards. Because the internal working of the software is available for any programmer to tinker with, it becomes relatively much easier for other programmers to figure out how to read and write the files used by a particular program. If a free software program's developers decide to change a data format, or if the developers abandon the program (such as when firms go out of business), the users themselves may choose to continue work on the software. Because the source code is available, the users always have the option of becoming developers, if necessary.

Protecting Privacy and Control

Perhaps the most fundamental aspect of free software is that users control their own computers. Users face two challenges to control of their machines. First some governments attempt to monitor or censor their citizens' use of email, the Internet, or other digital communications media. A second challenge is that some media companies (music, movies, electronic books, digital television) would like to monitor who consumes their products, as

well as prevent legal or illegal copying of the content. Accomplishing these goals requires placing monitoring software in the users' computers, and it requires removing the capacity to copy the content from the user. With free software, it is difficult and potentially impossible for users to lose control of their computers in these ways. With a computer running free software, the user can (at least in theory) review all the software on the machine to assure that none of it is spying on him. Similarly if free software can present content to the user, then it can also make copies of that content.

There are a number of differences between free and nonfree software that are debated by software experts. For example, free operating systems have been nearly entirely free of the viruses and worms that plague the world of proprietary software. This may be because free operating systems are more resistant to worms and viruses, or because the virus and worm writers are attracted to more popular consumer computing platforms. Free operating systems have been relatively less frequently cracked by direct attacks, but as with viruses, it is not clear if the free systems are more secure or if attackers are more drawn to proprietary systems. The proponents of the proprietary systems often claim that free systems have no guarantee of functionality or support; proponents of free systems reply that the mere existence of a company charging money for software is no guarantee of support. Finally some charge that free software lacks *user friendliness*.

Programmers write free software because they enjoy pursuing technical challenges, and because they want the respect of their colleagues (Raymond 2001). In short, free software programmers are motivated by the same personal goals that motivate most scientists. The close fit between free software and scientific endeavors is therefore unsurprising. To a scientist or engineer, free software enables a powerful array of tools, of places where one can open the hood and tweak behavior to precise specifications; in a high-performance application, these capacities may outweigh the relatively greater complexity of free software relative to proprietary software. Combined with the richness of the Unix toolset and databases, numerical routines, and statistical software, free software can be the ideal scientific computing environment.

PATRICK BALL

SEE ALSO *Autonomy; Digital Divide; Information Society; Internet; Hardware and Software.*

BIBLIOGRAPHY

Raymond, Eric S. (2001). *The Cathedral and the Bazaar: Musings on Linux and Open Source by an Accidental Revolutionary*. Cambridge, MA: O'Reilly. This book expresses many of the core ideas that motivate many programmers to publish open source/free software.

Torvalds, Linus, and David Diamond. (2001). *Just for Fun: The Story of an Accidental Revolutionary*. New York: HarperCollins. Torvald's memoir of why he wrote the Linux kernel is an engaging example of how programmers write code to satisfy their personal curiosity and then to gain the respect of their peers.

Wayner, Peter. (2000). *Free for All: How Linux and the Free Software Movement Undercut the High-Tech Titans*. New York: Harper Business. Wayner's lively story covers the broad history of the free Unixes—linux, NetBSD, FreeBSD, and later, OpenBSD.

Williams, Sam. (2002). *Free as in Freedom: Richard Stallman's Crusade for Free Software*. Sebastopol, Calif.: O'Reilly. Stallman's biography traces his extraordinary personal life and the growth of the community which has given rise to the ideas and the technology.

INTERNET RESOURCES

Free Software Foundation. Available from http://www.fsf.org. Information on the development of the free software movement. See especially the GNU "General Public License," available from http://www.gnu.org/licenses/licenses.html#GPL.

"Hacker History and Culture." Eric S. Raymond's home page. Available from http://www.catb.org/~esr/faqs.

"Homesteading the Noosphere." Eric S. Raymond's home page. Available from http://www.catb.org/~esr/writings/cathedral-bazaar/homesteading/index.html

SourceForge.net. Available from www.sourceforge.net. Many of the most dynamic free software projects are hosted here.

KernelTrap.org. Available from http:// www.kerneltrap.org. A highly-technical site with interviews with developers from across the community of the free Unixes, and a place to browse Linux kernel mailing list, with commentary by the site's owners.

Open Source. Available from http://www.opensource.org/. A review of the myriad of open source and free software licenses, with critiques.

Python. Available from http://www.python.org/. Access to software, mailing lists, documentation, tutorials, and everything python.

Trusted Computing. Available from the Internet site of British cryptographer Ross Anderson, http://www.cl.cam.ac.uk/~rja14/tcpa-faq.html.

FREE WILL

• • •

To have free will means that in some nontrivial sense persons are able to make choices that are not determined by causes other than themselves, so that each person may be regarded as the unique author of his or her own thoughts and actions. The term *nontrivial* indicates more than the absence of external and future determinants. A snowflake is free to fall until it hits the ground, but this freedom seems trivial. Free will implies the absence of internal or prior determinations.

Notions of free will involve two closely related ideas. Moral freedom is the idea that human being are morally responsible for their actions, and so may legitimately be praised or blamed, rewarded or punished. Metaphysical freedom amounts to the more radical claim that human choosing involves a break in the chain of physical causation. The human being is thus an indeterministic system, producing outcomes that are not wholly caused by previous physical states. Modern controversies over the meaning and possibility of free will tend to pit science against morality. Free will in some sense is thought to be necessary for human dignity, but both versions of free will appear to be at odds with the causality investigated by modern science.

Historical Background

Human free will was not a problem in classical philosophy, for at least two reasons. According to Plato (c. 428–c. 348 B.C.E.), for instance, human freedom is not a given but something to be achieved through education. Most human beings are described as slaves, of their passions if not of other humans. Moreover, for Aristotle (384–322 B.C.E.) nature itself was not seen as a rigid set of causal relations; those things that are by *phusis*, or nature, have their own source of motion and rest, that is, are self-moving. Thus the achievement of human freedom is not opposed to nature but its perfection.

Augustine's *De libero arbitrio* (On free will) is the first extended analysis of the concept. For Augustine (354–430), the early Christian church father, the problem arises not from an opposition between human will and physical causation but between human will and God as the cause of everything. If God is all powerful and all knowing, including predestining humans for salvation and knowing the future, how can humans have free will? The Christian theological solution to this problem is simply to argue that God created humans with free will.

In the modern period, however, it is argued that all human beings are equally free (the democratic proposition) and that nature is a deterministic system of causal relations (the scientific proposition). The ethical implication of these two propositions is that humans should use science to control nature for human benefit (the technological proposition). There nevertheless remains a problem of how to reconcile free will and scientific determinism, in theory if not in practice.

Common Sense and Moral Freedom

Moral freedom is grounded in a commonsense interpretation of choosing (sometimes called folk wisdom). I am persistently conscious of alternatives—rare or medium rare? More importantly, I am subject to temptation—I should not break my promise, but just now I really want to. The impression that I could do either allows for a sense of moral responsibility. If you respect my rights, I ought to respect yours; if I do not, I deserve to be punished. This sense of responsibility in turn becomes the ground of all moral authority. Because I am as capable of it as anyone else, I can be ruled only with my consent. In this way, moral freedom supports the idea of individual dignity that underlies both liberty and democracy.

This folk wisdom view of free will has been vigorously challenged within the modern social sciences. Human beings are subject to any number of influences beyond any individual's control: culturally sanctioned values and taboos, character as formed over a lifetime of interactions, genetic inheritance, and more. When one thinks one is choosing, perhaps one is only expressing these social and biological forces. From this perspective, free will is an illusion. The real authors of one's choices are the various forces of social and natural history.

But it is unclear whether these challenges amount to much. Everyone recognizes powerful outside influences on their will. But our very consciousness of alternatives suggests that these influences never quite add up to a choice. A person is required to complete the action. It may be enough to recognize social forces do not act, people do. Each person stands as a unique pivot point in history, interpreting rather than merely communicating biological and social inputs. This may be an adequate ground for human dignity.

Metaphysical Freedom and Determinism

Unlike moral freedom, which largely abstracts from physical causes, the concept of metaphysical freedom focuses on causation all the way down. A person is metaphysically free only if the sum total of physical forces acting on her, including for example the momentum of every molecule in her brain, is insufficient to determine her choice. This would be to say that human choosing is not in all respects a physically caused event. At first glance modern science would seem to preclude such an account of free will. Much of science presupposes a physically deterministic universe in which the

state of a system at one time rigidly determines its state at any future time. The view that the universe as a whole constitutes such a system is known generally as determinism.

Yet modern science is no longer uniformly deterministic. Quantum physics, in some interpretations, allows that very small events may be physically uncaused. But it is not clear that this does anything to save metaphysical freedom. Quantum events may have no appreciable consequences on the scale of human perception and action, or if they do this would still represent the influence of material constituents on the brain and could not explain how the person as a coherent self makes choices.

Given that metaphysical freedom involves more radical claims than moral freedom, the obvious question is whether the latter depends on the former. Call metaphysical freedom F1 and moral freedom F2. There are then three general positions. Determinists hold that F1 is required for F2, but that F1 does not exist. Thus there can be no free will in either sense. Libertarians accept the dependence of F2 on F1, but argue that F1 is possible. They then try to show how physical indeterminacy can support human choosing.

Finally, compatibilists argue that there can be F2 without F1. In fact, some have argued that F2 requires determinism. It is only because actions are rigidly determined by what a person is that we can praise that person for the actions; otherwise they would be regarded as mere luck. But this is unconvincing. We recognize that a horse's performance on the track results from its breeding and training, but we do not praise the horse for this. We praise an owner because the owner was free to make poorer choices. Compatibilism may save this sort of freedom only as a necessary illusion. We assume we are free precisely because we have no choice in the matter.

Reconciliations

All three positions rest on the assumption that determinism is the primary obstacle to moral freedom because freedom is conceived as whatever wiggle room does or does not exist between the boundaries set by causation. This is probably a mistake for two reasons. First, determinism relies on a concept of rigid causation that is neither required by theory nor possible in practice. While it simplifies our models of many phenomena to assume a perfect determination of events by antecedent states, there is no reason to believe that this perfection is real. And real or not, we can measure anything only to within some degree of precision. Past that point, things can be as messy as they please.

Secondly, the fundamental requirement of moral freedom is that my individual self is the cause of my own thoughts and deeds. To be more precise, I am genuinely free if my conscious choosing is among the causes that determine my choice. Otherwise I am indeed a puppet of forces beyond my control. But determinism is not, in itself, inconsistent with this, because it involves no theory of consciousness. It cannot rule out a role for awareness in the chain of causation. Conversely, libertarians have a hard time explaining how noncaused events can contribute to conscious choosing. If my puppet strings are being pulled by very small particles, it matters little whether those particles themselves are determined or indeterministic. Either way something besides me is in charge.

The real challenge to free will comes not from determinism but from two closely related views of consciousness. Both are examples of reductionism in so far as they attempt to explain an apparently complex thing, in this case the brain, by reference to its simpler material constituents. The epiphenomenalist claims that the conscious mind is an effect of physical events but is in no sense a cause of those events. No conscious state can be responsible for another, so there is no sense trying to think anything through. More radical still, eliminative materialists argue that consciousness does not exist at all. Like a ghost or a mirage, it is a delusion, though who is being deluded is something of a mystery. Moral freedom can scarcely survive any of these claims.

But perhaps it does not have to, because both seem to rest on an untenable dualism. They treat consciousness as something separate from the brain as a whole. A more mature view is possible. Just as sight is not produced by the eyes but is rather the activity of the eyes, nerves, and neurons, so consciousness is precisely an activity of the body and brain working in concert. The mind is a complex whole that functions to gather and store information and translate it into thoughts and actions. Its material constituents, determined or not, participate in this work only by virtue of their integration into the larger whole. It is this larger whole, perhaps, this congress of neurons, that is the seat of government. Consciousness is what happens when congress is in session.

Free will, like vision or flight, may be regarded as a product of mammalian evolution. Evolution can be understood only in the context of real time. The present is the finished product of a now vanished past. The future is, both in theory and practice, open and unpredictable. Trial and error is the engine of evolution, and free will may be understood as a small-scale model of

that engine. Human beings adapt with astounding speed to unforeseen circumstances. Moreover they have constructed moral cultures and political regimes to preserve their successes. Liberal democracy using science for technological benefit is among the most effective of these precisely because it recognizes human beings for what they are. Both determinism and reductionism may have outlived their usefulness as models of the human mind.

Paradoxically, the democratic use of scientific technology may also propose more of a practical than a theoretical threat to free will. Advanced biomedical technologies for the control of human behavior and genetic nature can be interpreted as willful actions that can destroy the will. Recognition of such a possibility might then appeal to the phenomenon of free will as a good to be protected and thus as a moral limit or boundary on technoscientific action.

KENNETH C. BLANCHARD JR.

SEE ALSO *Complexity and Chaos; Determinism; Freedom.*

BIBLIOGRAPHY

Dennett, Daniel C. (2003). *Freedom Evolves.* New York: Viking. Dennett is among the most famous of contemporary compatibilists. He relies on a thoroughly Darwinian account of the human mind in order to build a philosophical account of consciousness and choice. Many critics argue that his version of free will is not genuine and that he is in fact a strict determinist.

Double, Richard. (1991). *The Non-reality of Free Will.* New York: Oxford University Press. Presents a rigorous argument against free will from the standpoint of contemporary analytical philosophy.

Hume, David. 1999 (1748). *An Inquiry concerning Human Understanding,* ed. Charles W. Hendel. Oxford: Oxford University Press. Hume is a famous early modern critic of the notion of free will. Like Dennett, he claims, in effect, to be a compatibilist, but most see his argument as determinist.

Kane, Robert. (1996). *The Significance of Free Will.* Oxford: Oxford University Press. Kane, a libertarian, is a frequent foil for Daniel Dennett. Kane argues that metaphysical freedom is both necessary for moral freedom and that it is possible.

FRENCH PERSPECTIVES

• • •

French intellectual culture, from its Enlightenment heritage, is deeply imbued with a positivist approach to human problems. Modern science and technology are simply assumed to be the proper expressions of human reason. Under such assumptions it would be meaningless to consider the possibility that either science or technology could be intrinsically problematic or that it would be appropriate to try to identify proper limits to their development. Instead, for more than a century the main philosophical debate raised by scientific and technological progress dealt with conflicting political responses to extrinsic problems, such as the uses of technology to exploit the working class.

In France, moreover, academic life is highly centralized and, as a result of their selection and training, professional intellectuals tend to live in a world situated between the Ecole Normale Supérieure and the Sorbonne. Such a context favors the reproduction of existing problems and debates, so that questioning of the intrinsic character of science or technology was at most a minor issue in the history of science. Those few thinkers who took seriously science or technology as issues in themselves remained isolated, their work largely ignored, with students who were interested in such topics systematically discouraged from appropriate programs of study. In consequence, questions of science, technology, and ethics in France during most of the twentieth century were not so much part of a tradition of critical reflection as they were associated with a series of individuals who, in somewhat eccentric manner, undertook to investigate them.

From Henri Bergson to Emmanuel Mounier

The response of Henri Bergson (1859–1941), the leading French philosopher of the first third of the twentieth century, to the disastrous experience of World War I is indicative of the basic attitude during this period. Educated at the Ecole Normale Supérieure, after teaching philosophy at a series of lycées, Bergson became a professor at the College de France, where his lectures attracted not only students and academics but even the general public and tourists. His most original reflections on creativity and time having been completed before the war, afterward Bergson served as a diplomat and worked in support of the League of Nations. His *Les deux sources de la morale et de la religion* (The two sources of morality and religion, 1932) argues a chastened but continuing commitment to the Enlightenment tradition.

Les deux sources acknowledges that there is something frenzied and uncontrolled (*frénétique et emporté*) in the race for material progress. Yet Bergson's perception of the problems raised by the scientific technology that is at the foundations of such progress is surprisingly narrow and shortsighted. He seems mostly sorry about "the search for comfort and luxury which seems to have become humankind's primary concern" (p. 322), although he quickly adds that there is no cause for worry, because humanity has always progressed by oscillating from one extreme attitude to its opposite—from a mysticism oriented toward self-control and self-possession to a materialistic mechanism aspiring to the control and possession of things. This is why "we should engage with no restraint in one direction in order to find out what the result will be: When it will no longer be possible to persist, we shall swing back with all our acquisitions, in the direction we had neglected or abandoned" (p. 321).

The dialectic of progress thus exhibits a kind of fatality that, in due time, can be expected to provide humankind, whose material body has grown dramatically, with a "supplement of soul" (p. 335). Bergson is confident that democracy will enable mechanism to satisfy everyone's true needs. Moreover, he expects that science will liberate the *elan vital* (vital impulse) from its materiality and spiritualize existence: "the material obstacle has almost tumbled down" (p. 337). Material progress fosters spiritual progress and thereby fulfills "the essential function of the universe, which is a machine for making gods" (p. 343). Understandably, a mind that entertains such lofty vistas will not be very sensitive to the concrete problems of everyday life, even those that would lead directly to a new and even more terrible war.

After World War II, French intellectuals were absorbed in the ideological and political debate for or against Marxism and communism. On the margins, such literary and religious thinkers as the Russian émigré Nicholas Berdyaev (1874–1948) and the novelists Georges Bernanos (1888–1948) and Jean Giono (1895–1970) raised pointed criticisms—as exemplified, for example, in the 1947 proceedings from a Geneva conference, *Progrès technique et progrès moral*. Against the threat of such views, Bergson's optimism was reaffirmed and turned into a technological messianism by the French personalist philosopher and founder of the journal *Esprit*, Emmanuel Mounier (1905–1950). His essay *Be Not Afraid* (1948) is a compendium of the irenic technophilia that predominated in French intellectual life until the late 1970s.

In response to the crisis of conscience that Hiroshima caused for some, Mounier dedicated himself to an unconditional justification of technology. For him, the criticisms made of "machinism" are founded on a theoretical error about the relationships between technology and society. The exponents of this view "claim to criticize the essential character of the machine, but in the main they attack the structure of capitalist society which has twisted the first services of the machine to its own ends" (pp. 31–32). Mounier thus summarizes in a nutshell the spirit of the time.

Whether spiritualists or materialists, rationalists or existentialists, most French philosophers were to adopt the Marxist doctrine that states "there is no problem of the machine as such." To the ethically scandalous problems of exploitation, economic inequality, and poor material living standards there are appropriate political responses. Concern for the environment was not yet a serious issue. Thus there was no philosophical problem of technology as such, and the leading French philosophers of the day completely ignored technology or even science as a theme calling for explicit critical assessment. Despite the fact that the work of Martin Heidegger (1889–1976) has been influential in France since the 1930s, there is little to nothing on technology in the work of Jean-Paul Sartre (1905–1980), Maurice Merleau-Ponty (1908–1961), or Albert Camus (1913–1960).

Bernard Charbonneau and Jacques Ellul

Although he does not mention them, Mounier's argument is almost certainly directed in part against the critical position of a small group of "Gascon personalists" led by Bernard Charbonneau (1910–1996) and Jacques Ellul (1912–1994). Born and educated in Bordeaux, under the shadow of World War I, the first truly industrialized war, Charbonneau passed his *agrégation* in both history and geography, but chose not to follow the standard academic career. Instead, he elected employment at a small teachers' college in order to be able to live a rural life in the Pyrenees.

Charbonneau's central intuition is that modern technoscientific development creates what he calls "the great mutation." Early on, Charbonneau became convinced that since the war humankind was experiencing an utterly new phase in its history, one that displays two basic characteristics. First, the Great War (World War I), as a total war, subordinated reality to the logic of industrial and technological imperatives, which require the mobilization of the whole population, resources (industry, agriculture, forests), and space itself. Indeed, the war achieved as well a mobilization of the inner life of the people who, on both sides, were not just affected by the war, but consented to it, thus justifying the anonymous process that would destroy them. The Great War was the first experience of what Charbonneau describes

as "a total social phenomenon," insisting that it does not have to be totalitarian in order to be total.

Second, this great mutation is characterized by auto-acceleration. Human power takes hold of the entire planet at an ever-accelerating pace. This acceleration is a quasi-autonomous process. It is not a collective project, because most of its effects have not been chosen, and there is no pilot, because it simply rushes forward independent of any direct guidance. Technoscientific and industrial development fosters more and more rapid change throughout the world, across all aspects of life, without any respect for cultural meaning or purpose. The result is a radical disruption of society and nature, a state of permanent change.

Charbonneau was convinced that contemporary conflicting ideologies (nationalism, fascism, communism, liberalism) were outdated and provided no purchase on this great mutation, and that the uncontrolled development of industry, technology, and science was the problem and the not the solution. In his major books, written during the 1940s but published much later, Charbonneau insists that the issues of technoscientific development, of totalitarianism, and of ecological disruption are interrelated. In *L'Etat* (1987), he describes how the technological and industrial dynamism of liberal society has created the conditions of a total and technocratic organization of social and individual life. In *Le jardin de Babylone* (1969), he describes how the expansion of human power and of the techno-industrial order into a planetary scale deprives human beings of a harmonious relationship with nature and threatens not only ecological balance but also human freedom. In *Le système et le chaos* (1990), Charbonneau warns that the disorganizing impact of technological, scientific, and industrial development on nature and on society calls for a total organization of social life that will compromise human liberty.

Ellul was likewise born and educated in Bordeaux; together he and Charbonneau developed a version of personalism that promoted small, decentralized, and environmentally focused groups rather than centralized Parisian leadership. Unlike Charbonneau, Ellul elected a more academic career, and following his *agrégation* in Roman Law, became professor of the history of law at the University of Bordeaux.

Ellul is often characterized as a pessimistic Calvinist, urging the rejection of modern technology as an evil runaway power. But although Christian, he is neither Calvinist nor pessimistic; he firmly believes that it is possible to control and direct technological change, and indeed that technological choices are necessary and urgent. This is precisely the great political challenge that humankind must accept, otherwise politics is nothing but vain agitation. But the mastery of technological change is a difficult task, and in order to have any chance of success it is necessary to have a clear vision of the obstacles.

Ellul's analysis of the central role of technology in contemporary society is developed in three books. In *The Technological Society* (1954), he insists that the discussion of the role of machines is no longer relevant, because modern technology is not a mere accumulation of tools and machines; it is a global phenomenon which by means of propaganda, social planning and business management, and the organization of leisure subsumes all areas of individual and social life to the systematic search for efficiency. As a result there is a fundamental ambiguity of technological development, which, on the one hand, emancipates people from natural constraints and, on the other, submits them to a system of abstract and coherent functional constraints that in their own way determine social life. Technological progress fosters a technological society, more and more organized and integrated on the basis of impersonal logics.

In *The Technological System* (1977), Ellul argues that technology is now the environment in which human beings live and to which they must adapt. This technological environment is increasingly exhibiting a systemic cohesion. It is an interconnected network of technological ensembles; it organizes itself and evolves according to a process of "self-augmentation" dictated by its internal needs. This is why it is so resistant to attempts at reorganization from the perspective of non-technological values, whether ethical, political, or aesthetic. This technological system exhibits its own totalizing dynamic and tends to provide the main framework of social life. Nevertheless, Ellul adds that in spite of its capacity for auto-unification, this system is not and cannot be entirely coherent, because irrationalities and dysfunctions occur each time it is in contact with a different environment, natural, human, or social.

In *The Technological Bluff* (1988), Ellul argues that the development of the technological system parallels a cultural inability to address the problems created by technology, and that the suffusion of contemporary *mentalities* by a technicist worldview is one of the major obstacles to the mastery of technology. This is why policies aimed at controlling technological change require, in order to be effective, a change in both collective mentality and individual action. In *Changer la révolution* (1982), Ellul offers some guidelines for this new ethics of political action, which he terms an "ethics of non-power."

Jean Brun's Existentialist Interpretation

Another major contribution to the understanding of technology from an intellectual who lived and worked outside of Parisian institutions is Jean Brun (1919–1994). Like Ellul, Brun was a committed Protestant Christian who taught in the provinces at the University of Dijon. To the analysis of technology he brought an education in Greek and Roman philosophy that enabled him to once again challenge received views.

In *Le rêve et la machine* (1992), which synthesizes his major ideas, Brun maintains that the common understanding of technology as an application of rational and objective knowledge for effectively altering the world in order to satisfy human needs is dramatically one-sided and inadequate for appreciating contemporary problems of science and technology. The formal rationality of technoscientific endeavors is deceiving; it prevents people from recognizing the informal, imaginative, and often unconscious dimensions of technoscientific behavior.

Brun argues that technology is both a force of life and a force of death. On the one hand, without technology of some kind, human life would scarcely be possible. On the other, technology fosters destructive delirium, mechanized hysteria, and the planning of crazy projects. Human use of technology and the way humans develop it is often unreasonable, and its impact on nature and on human beings can be quite violent. For Brun there is a deep connection between technology and irrationality, and the obstacles to its rational uses must be appreciated.

According to Brun, technology manifests two goals: satisfying human *needs* for a better life (motives of pragmatic utility) and responding to *desires* to alter the human condition (existential motives). The study of ancient myths and ancient philosophy convinced Brun that technology is not merely an instrument useful for satisfying human needs, but also a means for empowering human desire for surpassing the ontological foundations of existence, for transmuting and overcoming the human condition. Human beings suffer and have always suffered from their finitude, from the alienation of consciousness, from physical and spiritual limitations, grounded in the necessity of living in space and time.

For Brun, the history of machines has been shaped and fueled by humanity's obstinate attempts to develop technologies of communication and transportation that attempt to break through such limitations. Beneath such obstinacy lies a hidden but fundamental despair within human consciousness regarding its separate and temporal mode of existence. Human technologies are often endowed with the power of discovering doors that open an existential labyrinth. In this respect human techniques are the offspring of human dreams as much as they are the application of positive knowledge. For Brun, "machines are both daughters and mothers of fantasies that we should call metaphysical ... [T]he utilitarian function of the machine is only its diurnal face; we must unveil its nocturnal face" (1992, p. 14).

This unveiling, which he also calls a demystification of technology, is a necessary precondition for any rational control and wise use of science and technology, as it is because humans project onto their technologies their desires for an ontological liberation that they remain fascinated by and addicted to their technologies. For the same reason, people often remain indifferent to technology's negative side effects and tend to transform the means into an end. Along with movies such as *The Fly* (1986) or *eXistenZ* (1999) by the Canadian filmmaker David Cronenberg, Brun argues for examining the ways utilitarian functions of technology are easily contaminated by its symbolic and existential functions.

The Mechanology of Gilbert Simondon

Another and quite different alternative to Enlightenment or positivist approaches to modern technology as applied science is found in the work of Gilbert Simondon (1924–1989), who proposed a general theory of the evolution of technological realities. Simondon was educated as a psychologist and philosopher at the Ecole Normale Supérieure in Paris and worked for the major part of his career in Poitiers and Paris. Because of a long-time interest in the character of machines, he studied what came to be called human factors engineering or ergonomics, which led him to attempt to understand their development somewhat independent of economic or other human interests.

In order to better clarify the human problems raised by machinism, Simondon chose the difficult path of laying the foundations of a kind of natural history of technological evolution. To this end he developed a conceptual framework for understanding the autonomy of technology and its radical alterity or otherness. As with Charbonneau, Ellul, and Brun, for him the category of instrumentality is inadequate for understanding the essential character of the technical order.

In *Du mode d'existence des objets techniques* (1958), Simondon argues that technical objects are not mere embodiments of abstract ideas, that they have their own mode of being or, as he says, of *existing*. Machines and technical objects evolve, and this evolution tends to exhibit a fundamental unity (structure). By analyzing the history of a few artifacts (motors, turbine, lamps, etc.), Simondon demonstrates how engineering practice follows the principle of functional unity, between the parts of the machine and between the machine and the exigencies of the surrounding world. "The technological being evolves by convergence and adaptation to itself. It unifies itself interiorly according to a principle of internal resonance" (p. 20).

Using as an example the evolution of the internal combustion engine, Simondon shows that each element assures the maximum possible of functions rather than attempt to realize a principle in its abstraction. Therefore, it is toward an interdependence of all the parts of the engine that its evolution converges, and it is this that leads to its progressive concretization through an organic-like integration of its diverse technical elements. According to Simondon, "The technological object exists then as a specific type that is found at the end of a convergent series. This series goes from the abstract to the concrete mode. It tends toward a state that would make the technological being a system entirely coherent with itself, entirely unified" (p. 23).

On this analytical basis Simondon develops a general theory of technology which, in the early twenty-first century, provides an intellectual framework for understanding the autonomy of technical objects and of technical systems: They develop according to a relational and reticular logic, obeying inner functional necessities that have little to do with human psychological, economical, social, and political goals. Although human beings produce technology, there is in technology something that is essentially resistant to human projects and values.

Simondon thinks that the solution to problems raised by the technicization of the world cannot be solved by politics, which relies on a poor understanding of the technical order and its dynamism. But for Simondon this is no reason for despair, and most of his subsequent intellectual endeavors aim at bridging the gap between the two cultures: the technoscientific operative one and the humanistic symbolic one. It is worth noting, in this respect, that although the second post-World War II generation of French philosophers such as Michel Foucault (1926–1984) and Jacques Derrida (b. 1930) were as silent about science and technology as their predecessors, some postmodernist authors such as Gilles Deleuze (1925–1995) have been attracted to Simondon. It may also be suggested that even those who do not share Simondon's rather optimistic and technophilic spirit may find in his thought substance for the pursuit of an authentic post-technological culture.

Supplementary Dimensions

The works of these four philosophers and the issues they wished to address were not, during their own time, well received in the French academic world. It is remarkable, for instance, that despite the 1974 French commitment to the development of nuclear power, this led to none of the kinds of public or intellectual debates typical of nuclear power developments in such countries as the United States or Great Britain. Nor has the increased technical powers of the professions of medicine or engineering engendered the kinds of discussions of professional ethics typical, especially, of the United States. Yet in the 1980s things did begin to change. One of these changes was the influence from the English-speaking world of the applied ethics movement, especially the field of bioethics.

In 1983, for instance, French President François Mitterrand created the Comité Consultatif National d'Ethique (CCNE), which consists of forty members, including representatives from different philosophical and religious schools of thought, public figures, and various scientific research institutions. Unlike similar or related commissions in other countries, the CCNE is not designed to be impartial but to elicit different points of view. Also unlike Enquette commissions in Germany or Royal Commissions in Commonwealth countries, the CCNE in not limited to specific topics but is an ongoing body. In 1994, in part as an outgrowth of its opinions, the French National Assembly passed legislation dealing with organ donation, medically assisted reproduction, and prenatal diagnosis.

Another associated activity emphasizing bioethics is the Science Generation Web site, which is cosponsored by the Institute de France, the Aventis Foundation, the Royal Swedish Academy of Engineering Sciences, the Federation of Scientific and Technical Associations, the European Council of Applied Sciences and Engineering, and the European Commission. This Internet site thus serves as a model of interdisciplinary and government-private partnership. But precisely because of their high profiles, neither the CCNE nor Science Generation represents serious critical assessment. Although both manifest an emerging concern for science, technology, and ethics issues, both focus much more on reflecting the opinions of technoscientific experts or the general public.

Another indication of the emerging French interest in science, technology, and ethics has been the stepping out of scholars more consistently devoted to these topics than has previously been the case. One example was an exchange between mathematician and historian of science Michel Serres and science studies ethnographer Bruno Latour (1990), in which the two explore how technoscientific power entails in itself ethical challenges. Still another was the creation in 1992 of the Société pour la Philosophie de la Technique (SPT), which provides an arena where competing philosophic approaches toward technology can be discussed in a constructive way.

Among the contributors to SPT discussions one may take special note of the following: Jean-Jacques Salomon (Conservatoire National des Arts et Métiers), in analyses of relations between science and politics, has raised the issue of democratic control not only of technology but also of scientific research. Dominique Janicaud (Université de Nice), with his theory of *potentialization*, has examined how progress in some types of rationality has created a potential for new forms of dehumanizing irrationalities. Gilbert Hottois (Université Libre de Bruxelles), a Belgian philosopher, has argued the inherently *an-ethicity* and autonomy of technological change, while arguing from the example of bioethics for the possibility an *accompagnement symbolique* for science and technology. And Franck Tinland (Université Paul Valéry, Montpellier) insists from an anthropological point of view on the long term autonomy of technological change and the resulting ethical problems that humankind is now facing.

DANIEL CÉRÉZUELLE

SEE ALSO *Bergson, Henri; Comte; Auguste; Descarte, René Durkheim, Émile; Ellul, Jacques; Enlightenment Social Theory; Existentialism; Foucault, Michel; Levinas, Emmanuel; Lyotard, Jean-François; Paschal, Blaise; Phenomenology; Progress; Rousseau, Jean-Jacques; Verne, Jules.*

BIBLIOGRAPHY

Bergson, Henri. (1932). *Les deux sources de la morale et de la religion* The two sources of morality and religion]. Paris: Félix Alcan.

Brun, Jean. (1992). *Le rêve et la machine* [The dream and the machine]. Paris: Table Ronde.

Charbonneau, Bernard. (1969). *Le jardin de Babylone* [The garden of Babylon]. Paris: Gallimard.

Charbonneau, Bernard. (1987). *L'Etat* [The state]. Paris: Economica.

Charbonneau, Bernard. (1990). *Le systéme et le chaos* [The system and chaos]. Paris: Economica.

Ellul, Jacques. (1964 [1954]). *The Technological Society*, trans. John Wilkinson. New York: Knopf.

Ellul, Jacques. (1967). *The Political Illusion*, trans. Konrad Kellen. New York: Knopf.

Ellul, Jacques. (1980 [1977]). *The Technological System*, trans. Joachim Neugroschel. New York: Continuum.

Ellul, Jacques. (1982). *Changer la révolution: L'inéluctable prolétariat* [To transform the revolution: The unavoidable proletariat]. Paris: Seuil.

Ellul, Jacques. (1990 [1988]). *The Technological Bluff*, trans. Geoffrey W. Bromiley. Grand Rapids, MI: Eerdmans.

Hottois, Gilbert. (1984). *Le signe et la technique* [Symbol and technique]. Paris: Aubier.

Hottois, Gilbert. (1999). *Essais de philosophie bioéthique et biopolitique* [Essays on philosophical bioethics and biopolitics]. Paris: Vrin.

Janicaud, Dominique. (1994). *Powers of the Rational: Science, Technology, and the Future of Thought*, trans. Peg Birmingham and Elizabeth Birmingham. Bloomington: Indiana University Press.

Mounier, Emmanuel. (1951). *Be Not Afraid: Studies in Personalist Sociology*, trans. Cynthia Rowland. London: Rockliff.

Recontres internationales de Geneve. (1948). *Progrès technique et progrès moral* [Technical progress and moral progress]. Neuchatel: La Baconnière.

Salomon, Jean-Jacques. (1973). *Science and Politics*, trans. Noël Lindsay. Cambridge, MA: MIT Press.

Salomon, Jean-Jacques. (1999). *Survivre a la science* [Surviving science]. Paris: Aubier-Montaigne.

Serres, Michel, and Bruno Latour. (1995). *Conversations on Science, Culture, and Time*, trans. Roxanne Lapidus. Ann Arbor: University of Michigan Press.

Simondon, Gilbert. (1958). *Du mode d'existence des objets techniques* [On the mode of the existence of technical objects]. Paris: Aubier.

Tinland, Franck. (1977). *La diférence anthropologique* [The anthropological difference]. Paris: Aubier-Montaigne.

Tinland, Franck. (1997). *L'homme alétoire* [Human otherness]. Paris: Presses Universitaires de France, 1997.

Toulouse, Gérard. (1998). *Regards sur l'ethique des sciences* [Considering the ethics of the sciences]. Paris: Hachette-Littératures. Presents the views of an influential physicist.

Toulouse, Gérard. (2003). *Les scientifiques et les droits de l'Homme* [Scientists and human rights]. Paris: Maison des sciences de l'homme.

FREUD, SIGMUND

• • •

The psychologist Sigmund Freud (1856–1939), who was born in Freiberg (now Príbor in the Czech Republic) on May 6 of Jewish parents and educated as a medical doctor in Vienna, founded the field of depth psychology (which he called psychoanalysis) and became one of the most influential thinkers of the late nineteenth and early twentieth centuries. His studies of the structure of the human psyche, the contents of the unconscious mind, the meaning and interpretation of dreams, repression, anxiety, and the role of the libido in the personality gave rise to many schools of psychological theory and therapy.

Ambivalence toward Science and Technology

Throughout this life Freud maintained a deep-seated belief in the value of scientific inquiry and a deep antipathy toward religion. In *New Introductory Lectures on Psycho-Analysis* (1952 [1932]), Freud stated

> Of the three forces which can dispute the position of science, religion alone is a really serious enemy. Art is almost always harmless and beneficent, it does not seek to be anything else but an illusion. … Philosophy is not opposed to science; it behaves itself as if it were a science, and to a certain extent makes use of the same methods…. Our best hope for the future is that the intellect—the scientific spirit, reason—should in time establish a dictatorship over the human mind…. Whatever, like the ban laid upon thought by religion, opposes such a development is a danger for the future of mankind. (p. 875)

However, Freud seemed ambivalent about the vast achievements of science and technology. On the one hand, he fully endorsed the desirability of human domination of nature. In perhaps his best-known work, *Civilization and Its Discontents* (1961 [1929]), Freud observes: "During the last few generations mankind has made an extraordinary advance in the natural sciences and in their technical application and has established his control over nature in a way never before imagined" (p. 39).

On the other hand, this domination has not brought with it a commensurate increase in human contentment. Human beings, Freud writes in *Civilization and Its Discontents*, "seem to have observed that this newly-won power over space and time, this subjugation of the forces of nature, which is the fulfillment of a longing that goes back thousands of years, has not increased the amount of pleasurable satisfaction which they may expect from life and has not made them feel happier" (p. 39).

Freud's greater worry, however, was the potential for destructive misuse of humankind's new powers. In *The Future of an Illusion* (1961 [1927]) Freud confesses his deep anxiety in a single sentence: "Human creations are easily destroyed, and science and technology, which have built them up, can also be used for their annihilation" (p. 7). This dark theme is taken up again in *Civilization and Its Discontents*, where he states that humans "have gained control over the forces of nature to such an extent that with their help they would have no difficulty in exterminating one another to the last man. They know this, and hence comes a large part of their current unrest, their unhappiness and their mood of anxiety" (p. 112).

Freud's psychoanalytical studies suggested to him that human beings overestimate themselves. In the middle of the calamity of World War I Freud wrote in *Thoughts for the Times on War and Death* (1952 [1915]):

> From the foregoing observations, we may already derive this consolation—that our mortification and our grievous disillusionment regarding the uncivilized behavior of our world-compatriots in this war are shown to be unjustified. They were based on an illusion to which we had abandoned ourselves. In reality our fellow-citizens have not sunk so low as we feared, because they had never risen so high as we believed. (p. 760)

Ethics

Ethics does not constitute an important theme in Freud's major works. In *Civilization and Its Discontents* he suggested that ethics represents an attempt to accommodate the demands of a culture. The pleasure-seeking drive of the id is opposed by social restrictions in the form of the super-ego, and the ego is forced to mediate between these two poles: "Ethics is thus to be regarded as a therapeutic attempt—as an endeavor to achieve, by means of a command of the super-ego, something which has not so far been achieved by means of any other cultural activities" (p. 108)

Freud offers candid, less psychologically-oriented remarks on ethics in letters to a friend, the Swiss pastor Oskar Pfister. Writing in 1918, Freud admits a lack of interest in issues of "good" and "evil" because he has found "little that is 'good' about human beings on the whole. In my experience, most of them are trash, no matter whether they publicly subscribe to this or that ethical doctrine or to none at all. ... If we are to talk of ethics, I subscribe to a high ideal from which most of the human beings I come across depart most lamentably" (pp. 61–62).

In a letter written a decade later Freud characterized ethics as a "kind of highway code for traffic among mankind" (p. 123). His last brief comment on ethics appears in a 1929 letter in which he states that: "ethics are not based on an external world order but on the inescapable exigencies of human cohabitation" (p. 129).

Freud's theories of the mind have been criticized, modified, extended, and even rejected by some schools of thought. Feminist writers, for example, have criticized

Sigmund Freud, 1856–1939. The work of Freud, the Austrian founder of psychoanalysis, marked the beginning of a modern, dynamic psychology by providing the first systematic explanation of the inner mental forces determining human behavior. (*The Library of Congress.*)

Freud's essay on the psychology of women as deeply embedded in the gender stereotypes of his time. Yet even this critical stance must be measured against the strong presence of women in the field of psychoanalysis from its inception; Freud's own daughter Anna extended her father's work into the psychopathology of children.

More than six decades after his death, Freud continues to exert a powerful influence on how people view themselves as individuals and as a culture.

WILLIAM SHIELDS

SEE ALSO *Jung, Carl Gustav; Psychology.*

BIBLIOGRAPHY

Freud, Sigmund. (1952 [1915]). *Thoughts for the Times on War and Death*, trans. E. Colburn Mayne. *Great Books of the Western World*, Volume 54. Chicago: Encyclopedia Britannica. Gloomy essay written in 1915 amidst the onset of World War I.

Freud, Sigmund. (1961 [1927]). *The Future of an Illusion*, trans. James Strachey. New York: Norton and Company. Freud's main assault on religion as the enemy of science.

Freud, Sigmund. (1961 [1929]). *Civilization and Its Discontents,* trans. James Strachey. New York: Norton and Company. Searching inquiry into the causes of human unhappiness viewed from the psychoanalytic viewpoint.

Freud, Sigmund. (1952 [1932]). *New Introductory Lectures on Psycho-Analysis,* trans. W. J. H. Sprott. *Great Books of the Western World,* Volume 54. Chicago: Encyclopedia Britannica. Freud's 1932 revision of his 1915–1917 lectures; essentially the final form of his psychoanalytic theories.

Freud, Sigmund. (1963). *Psychoanalysis and Faith: The Letters of Sigmund Freud and Oskar Pfister,* ed. Heinrich Meng and Ernst L. Freud; trans. Eric Mosbacher. New York: Basic Books. Candid thoughts expressed to a friend, casual but interspersed with sharp observations and criticisms.

Ricoeur, Paul. (1970). *Freud and Philosophy: An Essay on Interpretation.* New Haven, CT: Yale University Press. Penetrating analytic guide to Freud's thought.

Roazen, Paul. (1968). *Freud: Political and Social Thought.* New York: Knopf. Freud's thought applied to politics and social theory.

R. Buckminster Fuller, 1895–1983. The American architect and engineer was in a broad sense a product designer who understood architecture as well as the engineering sciences in relation to mass production and in association with the idea of total environment. (*AP/Wide World Photos.*)

FULLER, R. BUCKMINSTER

• • •

A major contributor to scientific engineering and environmental studies, Richard Buckminster (Bucky) Fuller (1895–1983) was born on July 12 in Milton, Massachusetts, and died July 1 in Los Angeles, California. His epitaph, "TRIMTAB," sums up the worldview of the man who coined the term "spaceship earth." *Trim tab* is an aviator's term that refers to adjusting the wing's surface in order to change direction slightly. "TRIMTAB" refers to Fuller's belief that no one could actually steer the entire spaceship earth, but one could adjust the course slightly and stabilize it in times of turbulence.

Fuller entered Harvard in 1914, only to be expelled twice for "irresponsibility and lack of interest." From this inauspicious educational beginning, Fuller went on to receive forty-four honorary degrees, lecture at more than five hundred universities around the world, author twenty-four books as well as hundreds of articles, travel around the world more than forty times, and hold twenty-six patents.

Fuller was an environmentalist long before the word was popular. In 1927, Fuller designed Dymaxion House, a metal structure hung from a central mast with outer walls of glass. The unique house was heated and cooled by natural means, created its own power, included prefabrication, had rotating closets, was self-vacuuming, and was storm- and earthquake-proof. He built an example of the Dymaxion House in 1946 in Wichita, Kansas.

In naming this contribution, Fuller demonstrated he was also a master of creating neologisms. *Dymaxion* is a combination of "dynamic," "maximum," and "ion." These three properties characterize his design strategy applied to many different problems.

For the 1933 World's Fair in Chicago, Fuller designed and built the Dymaxion Car. It had three wheels, was twenty feet long, carried eleven passengers, got thirty miles to a gallon of gasoline, and obtained a speed of 120 miles per hour. The car could make a u-turn within its own length.

In 1936, Fuller turned his attention to poor sanitation and the high cost of bathrooms. The five-square-foot Dymaxion Bathroom was his solution. The prefabricated bathroom consisted of four sections of either sheet metal or molded plastic. All of the necessary pipes, wires, and appliances were built in so that the entire unit merely required being hooked up. Both the sink and bath/shower allowed easy access by children and seniors.

In 1940, recognizing the need for military housing, Fuller designed and built the Dymaxion Deployment Unit (DDU). The DDU was a circular structure twenty feet in diameter made of corrugated galvanized steel, lined with wallboard on the inside and insulated with fiberglass. The house was naturally air-conditioned. Superheated air rising from the outer steel walls created a vacuum under the house that sucked cool air down the ventilator.

Fuller's Dymaxion Airocean World Map shows the continents on a flat surface without any visible distortion. On this map, the earth appears to be approximately one island surrounded by water. In the March 1, 1943, issue, *Life* magazine published Fuller's world map. That issue sold 3 million copies, the largest circulation of the magazine to that date.

In 1945, the Dymaxion Dwelling Machine house was designed and built. This was a vast improvement on the DDU house. The intention was to create a prefabricated house at low cost whose disassembled parts could be shipped anywhere in the world to meet the housing needs that were emerging at the end of World War II. The house was featured in *Fortune* magazine and generated thousands of unsolicited orders. These orders were never filled because of ethical differences between Fuller and financiers.

In 1948 Fuller created the most well known of his designs, the geodesic dome. A geodesic dome was selected for the United States Pavilion at the 1967 Montreal Exposition, where it still stands.

Buckminster Fuller was an early thinker about the entire earth. His *Operating Manual for Spaceship Earth* (1978) helped to focus world attention on one earth and the growing need to work together for survival. In poetic works such as *No More Secondhand God* (1963), Fuller also imbued technology with religious significance and called on human beings to accept responsibility for their god-like powers. He argued that human beings had to either create utopia or destroy themselves. *Synergetics* and *Synergistics 2* (1975 and 1979) is Fuller's mathematical masterpiece concerning the geometry of nature and the universe.

A truly remarkable man, Fuller's contributions all focused on what he referred to as a "Comprehensive Anticipatory Design Science." In this view, the science is directed to anticipating human problems and solving them by providing more and more support for everyone, with less and less resources. Yet Fuller often expressed himself in a vocabulary that critics sometimes found eccentric if not opaque.

HENRY H. WALBESSER

SEE ALSO *Earth*; *Earth Systems Engineering and Management*; *Efficiency*; *Engineering Ethics*; *Environmentalism*.

BIBLIOGRAPHY

Fuller, Buckminster. (1938). *Nine Chains to the Moon*. New York: J. B. Lippencott. Fuller's solution to the space trajectory to the moon.

Fuller, Buckminster. (1962). *Education Automation*. Carbondale: Southern Illinois Press. Thoughts about educationally effective and environmentally safe applications of automation.

Fuller, Buckminster. (1963). *No More Secondhand God*. Carbondale: Southern Illinois University Press. Fuller's view of how technology and the spiritual are related.

Fuller, Buckminster. (1970 [1928]). *4-D Timelock*. Corrales, NM: The Lama Foundation. Fuller's view of the space-time continuum.

Fuller, Buckminster. (1973). *Operating Manual for Spaceship Earth*. New York: Penguin. A view of earth as a contained spaceship with finite and non-renewable natural resources.

Fuller, Buckminster. (1975). *Synergetics: Explorations in the Geometry of Thinking*. New York: Macmillan. The geometry of nature is presented in this volume and the second volume.

Fuller, Buckminster. (1979). *Synergetics 2: Further Explorations in the Geometry of Thinking*. New York: Macmillan.

Fuller, Buckminster. (1981). *Critical Path*. New York: St. Martin's Press. An economic analysis solving problems within the model of using less to obtain more.

Fuller, Buckminster. (1983). *Grunch of Giants*. New York: St. Martin's Press. An extension of critical path with applications of Fuller's geometry of nature.

Fuller, Buckminster. (1983). *Inventions—The Patented Works of R. Buckminster Fuller*. New York: St. Martin's Press. A compendium of Fuller's inventions.

Fuller, Buckminster. (1992). *Cosmography*. New York: Macmillan. Published posthumously. An extension of Fuller's views of the universe and the earth as one of its elements.

FUTURE GENERATIONS

• • •

Responsibility to future generations appears at first to be an uncomplicated concept, and its widespread appearance in public pronouncements and political rhetoric testifies to its apparently widespread endorsement by public opinion. Moreover, advances in science and technology have directly increased the urgency and relevance of this concept as the present generation becomes ever more aware of its capacity to impact (with industrial chemicals, environmental exploitation, and climate change), destroy (with nuclear and biological weapons), and alter (with genetic engineering) the life conditions of the generations that will follow.

However clear and urgent the concept of responsibility to the future might seem upon casual reflection, as philosophers examine that concept with their typical meticulous analytic scrutiny, numerous puzzles, paradoxes, and quandaries emerge. Questions concerning the ontological, epistemological, and moral status of future persons (stipulated here as having non-concurrent lives with the current generation) are crucial. Most fundamentally, future persons, *qua future*, do not exist *now*, although the burdens of responsibility fall upon the living. Thus the question arises as to the attribution of such moral categories as *rights of* and *duties to* non-actual beings. Moreover, one cannot know future people as *individuals*. Instead, *posterity* is an abstract category containing unnumbered and undifferentiated members. And yet, much moral theory is based upon the principle of "respect for autonomous *individuals*." Additionally, one's relationship with future persons is unidirectional and non-reciprocal. Future persons will be unable to reward or punish the current generation, as the case may be, for the provision for their lives. Finally, because living people are ignorant of the life conditions of future persons, they cannot determine just what might *benefit* future persons—that is, what will or will not be "goods" to them. Clearly, by assigning moral significance to those not yet born, one introduces problems that are unique in moral philosophy.

Four Special Problems

One problem is that of Radical Contingency (or "The Future Persons Paradox"). Attempts in the present to improve the living conditions in the future result in different individuals existing in the future. Accordingly, in the present one cannot improve the lives of any particular future individuals (because any such attempt results in different individuals). Thomas Schwartz, who posed this paradox in 1978, concluded that present generations have no obligations to the future, other than to insure that their lives are, on balance, "worth living" (Schwartz 1978).

A rebuttal position would be to accept the paradox but to conclude that the responsibility to the future is to promote policies that will result in optimum conditions for *alternative populations*. In other words, Policy A is to be preferred to Policy B, if the lives resulting from Policy A are preferable to the lives resulting from Policy B, even though no *particular* future lives are improved thereby (Partridge 1998).

A second problem regards the duties to and rights of future persons. Do future persons have rights to clean air and water, wild areas, a tolerable climate, biodiversity, and energy resources? The question has significant policy implications regarding, for example, the use or conservation of natural resources, the depositing of nuclear wastes, or the reduction in the use of fossil fuels to minimize global warming. Rights claims have stronger moral force than mere *duties of beneficence* that are not correlated with the rights of the beneficiaries.

Accordingly, by acknowledging the rights of future persons, those in the present generations may be morally obliged to accept greater sacrifices.

While many philosophers acknowledge duties to future generations, most who have written on the issue would deny that future persons have rights in the present, for the simple reason that potential persons, because they do not exist and cannot make claims, cannot be said to have rights (deGeorge 1979; Beckerman and Pasek 2001).

A contrary view contends that the denial of the rights of future persons involves an oversimplification of the concept of rights. There are, in fact, several categories of rights. While it is true that future persons, being non-actual, do not now have *active rights* to initiate or forbear activities on their own initiative, they do have *passive rights* not to be deprived of opportunities and not to be harmed. Unlike active rights, passive rights entail no initiative on the part of the rights-holder (future persons) but instead place a burden of responsibility on the correlative duty-bearer (the present generation) (Partridge 1990).

A third problem involves possible people and eventual people. Does a responsibility to future generations entail a duty of procreation—a duty to *create* future people (*possibles*)? Or is it confined to a duty to individuals who will, in any case, exist in the future (*eventuals*)? Clearly, the question has important implications for population policy. If current generations have a duty to bring possibles into existence, then the morally optimal future population will be much larger than if the duty of present persons is confined to eventual people. The issue also entails some deep ontological puzzles. For example, if a person is very pleased to be alive, would that person have been "harmed" if he or she had not been born? If so, then who is the "victim"? By stipulation, there is none. If no victim, then wherein is the harm? And yet, it is generally regarded as irresponsible to conceive a child who is certain to lead a brief and miserable life (e.g., a victim of Tay Sachs disease). Herein lies a paradox that is much discussed by moral philosophers (Warren 1981; Parfit 1984).

Finally, there is the problem of average utility versus total utility. When the utilitarian proposes to "maximize utility" (variously defined), does this mean *average* or *total* utility? With reference to a given (e.g., the current) population, there is no difference: raise the total and the average raises, and *vice versa*. However, the difference arises with the issue of population policy: That is, how many persons should be brought into existence in the future? Full commitment to either average or total utility leads to counter-intuitive conclusions. According to the average utility principle, Adam and Eve alone, before the fall, lived in a better world than a hypothetically later world of thousands or millions of individuals who, though quite happy on average, were slightly less so than the original couple. On the other hand, the total utility principle requires fertile couples to produce children whose lives will be on balance slightly happier than unhappy—an obligation that applies even in an overcrowded world. The average versus total utility dilemma leads to a question that lies at the very foundation of utilitarian philosophy: are those living in the present obliged to create people for happiness (total utility), or should they create happiness for people (average utility)? (Sikora and Barry 1978).

Policy Guidelines

Once one has accepted a moral foundation for a responsibility to future generations, the question remains: How might this moral obligation best be fulfilled?

The question is complicated by current necessary ignorance of the essential needs, of the cultural "goods," and of the technological conditions of future generations. Past generations, out of a sense of responsibility to those of the present, might have uselessly preserved a continuing supply of whale oil (not knowing about petroleum) and taken no heed to preserving semi-conducting substances. How would current generations, similarly, avoid wasting effort and treasure by preserving "goods" that would prove to be of no value to their successors?

One begins by taking inventory of some firm assumptions of the condition and needs of future generations. These assumptions would include:

(a) that they will be humans, with well-known biotic requirements;

(b) therefore, that they will need to be sustained by a functioning ecosystem;

(c) those future persons to whom one has obligations will be *moral agents*, and thus bound by such familiar moral categories as *rights, responsibilities, and the demands of justice;* and

(d) that they will require stable social institutions and a body of knowledge and skills that will allow them to meet and overcome cultural and natural crises that may occur during their lifetimes.

These considerations entail the following three essential policy guidelines:

- *First do no harm.* Because of current generations' ignorance regarding future cultures and technolo-

gies, and considering also the above list of basic needs, it is much easier to identify future harms than future benefits. Accordingly, one should favor policies that mitigate evils over those that promote good. The pains and tribulations of future persons, like those of the currently living, can often be clearly attributed to disruptions in the fundamental biotic, ecosystemic, psychological, and institutional conditions listed above, while their pleasures and satisfactions will come from a future evolution of culture, taste, and technology that one cannot even imagine.

- The *critical Lockean proviso*. John Locke's proviso that one leave "as much and as good" for one's successors, while applicable to the preservation of just institutions and sustainable ecosystems, cannot apply to non-renewable resources. If, for example, current generations were to share fossil fuels equally with all future generations (hypothetically setting aside an ignorance of their numbers), their share might reduce to cup of oil and a lump of coal—in any case, this resource would be useless, and the current industrial civilization would collapse (deGeorge 1979). Instead, the obligation to the future is to supply not fossil fuels but what fossil fuels provide, namely energy. Thus this obligation entails aggressive research and investment in a successor source of energy, presumably renewable. The critical Lockean proviso also entails a utilization of "interest-bearing" renewable resources, such as sustainable forestry, fisheries, and agriculture, and this in turn validates the need to preserve natural ecosystems.

- *Preserve the options.* This rule is clearly entailed by the previous two. While one cannot predict the technological solutions to future resource scarcity, the currently living owe future generations a full range of options and opportunities for research and development of these technologies. This in turn entails a continuing investment in scientific and technical education and research. Happily, such an investment benefits the current generation and that of its immediate successors, as well as the remote future (Partridge 1994).

Historical Background

The issue of the duty to posterity, though recurrent in the history of philosophy, has only recently attracted close scrutiny. Of the approximately one million doctoral dissertations listed in *Dissertation Abstracts* in 2004, the first to contain the either the terms "future generations" or "posterity" in its title was completed in 1976. Of the nearly two hundred entries in *The Philosophers Index* listed under "future generations" and "posterity," all but three have been published since the first Earth Day, April 22, 1970.

An explanation of this sudden appearance of interest in the topic of the responsibility to future generations may be found in an analysis of the concept of *responsibility*. Two criteria that appear to be essential to the concept are *knowledge* of the consequences of an act, and *capacity* to select among alternative anticipated consequences. Accordingly, the issue of responsibility to future generations has arisen with the extraordinary advances in science (knowledge) and technology (capacity).

During the first half of the twentieth century, the very idea that human activities might seriously and permanently affect the global atmosphere and oceans, or the gene pool of the human species and others, seemed preposterous. Now the sciences have disabused humankind of such assurances: technology has produced chemicals and radioactive substances unknown to nature, and evidence proliferates of permanent anthropogenic effects upon the seas, atmosphere, and the global ecosystem. Furthermore, such consequences of industrial civilization as ozone depletion, global warming, the contamination of aquifers, and the deposition of radioactive waste, although the byproducts of benefits to the present generation, exact postponed costs to remote generations. With science providing knowledge of these possible hazards for the future, and technology providing the capacity to deal with them, current generations have the responsibility to act with caution and moral insight, so that they might proceed toward a secure and prosperous future.

ERNEST PARTRIDGE

SEE ALSO *Consequentialism; Deontology; Jonas, Hans; Posthumanism; Responsibility.*

BIBLIOGRAPHY

Beckerman, Wilfred, and Joanna Pasek. (2001). *Justice, Posterity, and the Environment.* Oxford, UK: Oxford University Press. An optimistic view of future prospects by an economist and a philosopher, who deny rights-claims of future generations upon present persons. The authors perceive no great burden of obligation to the future.

deGeorge, Richard. (1979). "The Environment, Rights and Future Generations." *Ethics and Problems of the 21st Century*, ed. Kenneth Goodpaster and Kenneth Sayre. Notre Dame, IN: University of Notre Dame Press. A denial of the rights of future generations combined with an affirma-

tion of a "duty of non-malevolence" toward the future. Argues that current obligations to the future are severely constrained by ignorance of future conditions or the capacity to affect these conditions.

Parfit, Derek. (1984). *Reasons and Persons.* Oxford, UK: Clarendon Press. Through skillful and insightful uses of "thought experiments," Parfit reveals deep paradoxes and puzzles regarding future persons—imaginable, possible, and eventual. These analyses have profound implications for population policy and procreation decisions.

Partridge, Ernest. (1990). "On the Rights of Future Generations." In *Upstream/Downstream: Issues in Environmental Ethics*, ed. Donald Scherer. Philadelphia: Temple University Press. A qualified defense of the present rights-claims of future generations. Concedes that future persons do not currently have "active rights" (to act or to forbear action). However, they have "passive rights" to resources, knowledge, and just institutions that entail moral duties that apply to present persons.

Partridge, Ernest. (1994). "Posterity and the 'Strains of Commitment.'" In *Creating a New History for Future Generations*, ed. Tae-Chang Kim and James Dator. Kyoto, Japan: Institute for the Integrated Study of Future Generations. Explores the psychological capacity of present persons to make moral provision to the future. Lists several policy implications of this capacity.

Partridge, Ernest. (1998). "Should We Seek a Better Future?" *Ethics and the Environment* 3(1): 81–96. A reply to Thomas Schwartz's radical denial of obligation to future generations (see below). Concedes Schwartz's argument that there are no duties to individual persons in the future, but that nonetheless, present persons have a duty to enhance the life prospects of the class of eventual future persons.

Schwartz, Thomas. (1978). "Obligations to Posterity." In *Obligations to Future Generations*, ed. Richard Sikora and Brian Barry. Philadelphia, PA: Temple University Press. An original, influential, and logically compelling presentation of "the future persons paradox": namely, the conclusion that because significant policy decisions cause different individuals to come into existence, no obligations are owed to particular individuals in not-distant future (i.e., approximately six generations).

Sikora, Richard, and Brian Barry, eds. (1978). *Obligations to Future Generations.* Philadelphia, PA: Temple University Press, 1978. An important early collection of essays on the issue of responsibility to future generations. Discusses many issues, most prominently the utilitarian problems of average versus total utility and population policy.

Warren, Mary Anne. (1981). "Do potential people have rights?" In *Responsibilities To Future Generations*, ed. Ernest Partridge. Buffalo, NY: Prometheus Books. Examines "the person-affecting principle"—that the morality of an act is to be evaluated in terms of its benefit or harm to persons. Thus stated, the "principle" seems axiomatic and indisputable. But when applied to "possible" future persons, and to the decision of present persons to procreate, profound ethical and even ontological questions arise. (See also Parfit, above.)

G

GAIA

• • •

First articulated by the British chemist James Lovelock in the 1970s, the Gaia hypothesis (named for the Greek goddess who personified the earth) proposes that the biosphere, atmosphere, oceans, and surface rocks make up a single, self-regulating, homeostatic system (Lovelock 1979). Key observations that Lovelock used in support of Gaia include the long-term stability of chemical disequilibria in the atmosphere and oceans despite both high fluxes of many chemicals within the earth system, and the fact that these persistent (in some cases for billions of years) yet nonequilibrium conditions are particularly well-suited for life as it has evolved. To Lovelock, the implication of these and related observations is that the biosphere must actively modulate the chemical make-up, temperature, pH, and other attributes of the earth system in order to maintain conditions under which life can flourish. In particular, the composition of the atmosphere must be regulated by the biosphere to maintain near-optimal concentrations of chemicals such as hydrogen, oxygen, and nitrogen.

Lovelock and his followers have promoted Gaia as an integrative framework for the study of the earth system. It raises scientific questions and demands experiments that would not be recognized under the traditional disciplinary and reductionist regimens dominant in the earth and environmental sciences. Gaia is thus not only an attempt to specify a unifying framework for the operation of the entire earth system, but also an explicit critique of the existing organization of knowledge inquiry.

Gaia has had little effect on research agendas, however, and the number of working scientists willing to be associated with the hypothesis is small—perhaps less than a dozen. Critique of the hypothesis focuses on three lines of argument. Gaia is said to be tautological because it asserts that life exists under exactly those conditions that are suitable for life. It is said to be teleological because it implies that the earth system must have evolved according to a design concept. And it is said to be trivial because, even so, Gaia adds little to existing knowledge about feedbacks among physical, chemical, and biological processes (Kirchner 2002). In response it is argued that Gaia is an emergent phenomenon that cannot be understood through traditional, disciplinary, and reductionist cause-and-effect reasoning. Lynn Margulis, a forceful advocate of Gaia, suggests: "The Gaian viewpoint is not popular because so many scientists, wishing to continue business as usual, are loath to venture outside of their respective disciplines. At least a generation or so may be required before an understanding of the Gaia hypothesis leads to appropriate research" (Margulis and West 1997, p. 223).

But it remains to be seen if the type of interdisciplinary synthesis that Gaia demands is even possible. Interdisciplinarity founders not just on the administrative boundaries between disciplines, but also on the differences in subject, method, time and spatial scales, types of data, definition of problems, and criteria of proof among various disciplines. These differences cannot easily be transcended or reconciled. This disunity of science is not entirely capricious, but in part reflects the richness and diversity of nature. How actually to move from reductionism and disciplines to Gaian synthesis remains far from clear.

Indeed while the need for interdisciplinarity is accepted by many scientists, strategies in the early twenty-first century—exemplified by the construction of highly complex, mathematical models aimed at simulating the coupled ocean-ice-atmosphere system—are still essentially reductionist in nature, building a story from first principles and supporting bodies of observational data. Gaia's claim is that such approaches can no more yield a comprehensive understanding of the earth system than a mapping of synapses can reveal the workings of an individual's consciousness.

Thus at least at this point in the evolution of science and society, Gaia's greatest impact may be largely metaphorical. On one level this metaphor may continue to challenge science to engage nature more synthetically, just as the Cartesian metaphor of nature as a clockwork helped to advance the cause of reductionist science. But on a broader, societal level Gaia has already been embraced as a cautionary symbol of the earth's complexity, interconnectedness, and inscrutability. Wrote Václav Havel, "Our destiny is not dependent merely on what we do for ourselves but also on what we do for Gaia as a whole. If we endanger her, she will dispense with us in the interests of a higher value—that is, life itself" (Havel 1998, p. 171).

DANIEL SAREWITZ

SEE ALSO Earth; Earth Systems Engineering and Management; Ecology; Environmental Ethics; Environmentalism.

BIBLIOGRAPHY

Havel, Vaclav. (1998). "The Philadelphia Liberty Medal." In The Art of the Impossible. New York: Fromm International.

Kirchner, James W. (2002). "The Gaia Hypothesis: Fact, Theory, and Wishful Thinking." Climatic Change 52: 391–408. A spirited scientific (as opposed to espistemological) critique of the Gaia hypothesis.

Lovelock, James E. (1979). Gaia: A New Look at Life on Earth. Oxford: Oxford University Press. The most complete treatment of the hypothesis, by its creator.

Margulis, Lynn, and Oona West. (1997). "Gaia and the Colonization of Mars." In Slanted Truths: Essays on Gaia, Symbiosis and Evolution, ed. Lynn Margulis and Dorian Sagan. New York: Copernicus.

GALENIC MEDICINE

• • •

Galenic medicine (also called humoralism or Galenism) derives its name from the Greek physician and philosopher Galen (129–c.216 C.E..). Galen's prolific writings were rooted in the Hippocratic corpus as well as the philosophical doctrines of Plato, Aristotle, and the Stoics. Medicine was identified with Galenism for 1,300 years, and was institutionalized in the European universities of the eleventh century after Arabic translations of Galen's writings were retranslated into Latin. Though Galenism was eclipsed in Europe by the rise of modern medicine, it still survives as Unani (Greek) medicine in some parts of India and Pakistan.

The foundation of modern medicine rests on the divorce of medicine from philosophy, two disciplines wedded in Galenism. Both philosophy and medicine were practical arts that sought to answer the Socratic question: How should a person live the good life? (Hadot 2002). The good life demanded a striving toward excellence (arête), in the gymnasium no less than in the symposium. In medicine, health was the excellence expressed by the proper blending of the humors (krasis). In philosophy, virtue required knowing what was moderate or intermediate between excess and deficiency. As such, bodily health was analogous to moral virtue and the physician like the philosopher was a guide to living according to the mean (mesotes)(Tracy 1969).

The Galenic physician could only assist nature (physis) to restore the proper balance in the patient because it was inherently good. Nature thus constituted both the source and the limit of the physician's art. In order to gain insight into the workings of nature, the Galenic physician incorporated the three parts of philosophy (natural philosophy, logic, and ethics) into diagnosis, prognosis, and therapy. How deeply the physician was steeped in the study of philosophy also distinguished true medicine from quackery (Galen 1997).

The study of natural philosophy allowed the physician insight into both human nature and the nature of the universe. The Galenic body was fluid because it was composed of humors—blood, black bile, yellow bile, and phlegm—which were formed by the same elements that constituted the cosmos (fire, water, air, and earth). Disease resulted from the imbalance (dyskrasia) of the humors or the predominance of one or another quality (hot, cold, wet and dry) Humors and their qualities linked humankind to the macrocosm and established a correspondence or proportion between them.

Logic was necessary to make sound judgments about conditions of illness and health. The good physician was urged to train and sharpen the senses, which included not only the five external senses but also the common sense (koine aesthesis) through which the givens of the senses were synthesized and delivered to the intellect (Nutton 1993). Hence both reason and sense experience were essential for the discovery and confirmation of the true nature of things. Logic also led to the search for causes within a teleological cosmology that demanded the inference of the invisible from the visible; the hidden causes from the manifest signs.

Ethics was the domain of human action and conduct. In Galenism, the patient, the physician, and their mutual relation were subjected to an elaborate *askesis* aimed at cultivating certain dispositions and habits (*hexis*) to restore the balance of body and soul (Edelstein 1967). Galenic therapeutics throughout the Middle Ages placed a strong emphasis on *dieta*, the art and craft of moral and somatic virtues. Health required the good ordering of the *naturals* (elements, humors, parts of the body, and faculties) and the regulation of the *non-naturals* (rest, motion, food and drink, evacuation, passions, and errors of the soul). The virtuous character of the physician hastened the healing powers of nature by forging a relationship of trust and friendship (Entralgo 1967).

The union of philosophy and medicine in Galenism was founded on the norms of a teleological nature. The medical art was practiced within the bounds of the natural order that tended toward health and virtue as the right proportion of the humors and the passions. The replacement of Galenism with scientific medicine occurred in the seventeenth century when nature lost its *telos* and was construed as an inert mechanism to be manipulated at will. In Galenism, the foundation in teleological nature made the medical art inherently ethical. By contrast, medical ethics in the twenty-first century is the mere application of established rules to a professional field.

SAMAR FARAGE

SEE ALSO *Acupuncture; Complementary and Alternative Medicine; Health and Disease.*

BIBLIOGRAPHY

Edelstein, Ludwig. (1967). "The Ethics of the Greek Physician." In *Ancient Medicine: Collected Papers of Ludwig Edelstein*, ed. Lilian Temkin and Oswei Temkin. Baltimore, MD: Johns Hopkins University Press.

Entralgo, Lain P. (1967). *Doctor and Patient.* New York: McGraw Hill. Explores the ethics of the doctor-patient relationship from ancient Greece to the present.

Galen. (1997). "The Best Doctor is Also a Philosopher." In *Galen: Selected Works*, trans. Peter N. Singer. New York: Oxford University Press. Galen's text that discusses the nature of the relation between philosophy and medicine.

Hadot, Pierre. (2002). *What is Ancient Philosophy?*, trans. Michael Chase. Cambridge, MA: Harvard University Press. Excellent investigation of ancient philosophy as a way of life and the changes brought about by Christianity and the Middle Ages.

Nutton, Vivian. (1993). "Galen at the Bedside: Methods of a Medical Detective." In *Medicine and the Five Senses*, ed. Roy Porter and William F. Bynum. New York: Cambridge University Press. Nutton shows the intimate relation between experience and reason in Galenic medicine.

Tracy, Theodore. (1969). *Physiological Theory and the Doctrine of the Mean in Plato and Aristotle.* Chicago: Loyola University Press. Exploration of the Greek founding of moral virtues and physical excellences in the notion of the mean.

GALILEI, GALILEO

• • •

Mathematician, astronomer, and natural philosopher, Galileo Galilei (1564–1642), who was born the same year as William Shakespeare in Pisa, Italy, on February 15, contributed fundamentally to the scientific revolution of the sixteenth and seventeenth centuries in which Ptolemaic geocentric cosmology and Aristotelian were successfully challenged by Copernican heliocentric cosmology and a new science of motion. Galileo's participation in the astronomical revolution included the best-known cases of his technological innovation and ethical/religious conflict, namely, his application of the telescope as an astronomical instrument and his engagement with the Roman Catholic Church over matters of biblical interpretation and the Copernican hypothesis. He died on January 8 in Arcetri near Florence where he was living under house arrest that had been imposed following his conflict with the church.

Natural Philosopher and Inventor

Galileo's career as a natural philosopher involved an ongoing study of natural motion, especially inertial motion and that of falling bodies, projectiles, and pendulums. His work helped lay the foundation for the new science of classical mechanics, which found its early modern culmination in the genius of Isaac Newton (1642–1727). In addition to the telescope, which he first turned toward the heavens in 1609, evidence suggests that Galileo contributed to the technological development, improvement, and scientific application of no fewer than eight other scientific instruments. These included the *pulsilogium*, a device that applied a

Galileo Galilei, 1564–1642. The Italian scientist is renowned for his epoch-making contributions to astronomy, physics, and scientific philosophy. *(The Library of Congress.)*

pendulum to measure the human pulse, in 1583; a hydrostatic balance which he developed for his experiments on floating bodies, in 1586; the *thermoscope,* an early thermometer, in 1593; a geometrical and military compass, in 1597; a natural magnet called a loadstone used to further the new science of magnetism, in 1601; the microscope, in 1610; the *giovilabio,* which was an obscure tool developed to compute the distances and periods of revolution of Jupiter's moons, which Galileo had discovered with his telescope in 1612; and finally, a number of *vibration-counters,* some derived from his study of pendulums, which he applied to the mechanisms of clocks by 1637.

That the majority of Galileo's technical instruments were measuring devices is indicative of his philosophical commitment to a quantitative science. Well-know for his aphorism that mathematics is the language in which the book of Nature has been written, Galileo sought mathematical regularities in his scientific description and placed a premium on the collection of quantitatively accurate data. Indeed it was Galileo's unswerving commitment to an ideal of scientific knowledge grounded in rigorous measurement and observation that fostered his commitment to Copernicanism, which led to his famous struggle with authorities of the Roman Catholic Church.

Galileo did not invent the telescope. He learned in 1609, however, that a Dutch lens grinder had secured a patent the previous year for a device that magnified distant objects by combining two lenses. On the strength of this news, Galileo crafted his own telescope and turned the tool, which had originally been conceived for terrestrial purposes, toward the heavens. He reported his observations, which included details of lunar topography, descriptions of previously unobserved stars and constellations, and an account of Jupiter's four principal moons, the following year in his best-selling book, *The Starry Messenger* (1610). By this time Galileo was fully convinced of the truth of Copernican (sun-centered) theory. Hence as his book popularized astronomy, it introduced Copernicanism to the common people. This move did not help his reputation among contemporary natural philosophers.

Argument with the Church

Galileo's Copernicanism placed him in opposition to common sense as well as to reigning scientific and theological opinions. During Galileo's lifetime conclusive scientific evidence sufficient to establish the Copernican system as true was not yet available. Galileo believed his theory of the tides provided the needed empirical proof, but he was mistaken. This error made him overconfident. He was originally attracted to Copernicanism by its mathematical elegance and aesthetic superiority, not because he possessed irrefutable evidence. Galileo's principal opponents, Aristotelian natural philosophers (that is, the scientific community), did not believe that such evidence would ever be found. Moreover many of these opponents disliked Galileo for reasons unrelated to the Copernicanism. Galileo had inherited from his father a feisty spirit and taste for intellectual combat. He had engaged anti-Copernican natural philosophers on other scientific questions related to such matters as floating bodies, sunspots, and a new supernova. In each case he had distanced himself from the established scientific authorities and embittered his opponents, many of whom wished to see Galileo silenced, by the Church if necessary.

The Catholic Church, in the absence of conclusive scientific evidence for Copernicanism, followed the lead of natural philosophers in rejecting it. Moreover this seemed to be in accord with a straightforward reading of relevant Biblical texts (Gen. 1; Eccles. 1:4–5; Josh. 10:12; Ps 19:4–6; Ps 93: 1; Ps 104: 5, 19), the interpretation of which rested with church authorities. As a loyal Catholic, Galileo was interested in persuading church leaders to reject geocentrism and thereby be saved from future embarrassment. To do so, however, would require scientific evidence he could not provide. Just as importantly, it would also require expert theological skill in interpreting those biblical texts that seemed to refute Copernicanism. Although a layman, Galileo attempted to provide such advice on biblical interpretation. His "Letter to the Grand Duchess Christina" (1615) was offered as guidance "concerning the use of Biblical quotations in matters of science." It ranks among the classic statements on the relation of the Bible to science. In it Galileo argued that because God is the author of both the book of nature (i.e. the physical world) and the book of revelation (the Bible), it is not possible for genuine conflict between science and scripture. When there appeared to be such a conflict regarding matters of the physical world, the advice of Cardinal Baronius (1538–1607) should be recalled, namely, "That the intention of the Holy Ghost is to teach us how one goes to heaven, not how the heaven goes" (Drake 1957, p. 186). Because, argued Galileo, the Bible is a religious and moral text, not a scientific text, passages that seemed to treat subjects of scientific inquiry should be interpreted with deference to scientific opinion.

Although Galileo had made a compelling argument, two factors counted against him. First the weight of scientific opinion did not yet favor Copernicanism. Second he was a layman presuming to instruct on principles of biblical interpretation. This was an especially dangerous thing to do in the early seventeenth century in the wake of the Protestant Reformation and the Council of Trent (1545–1563). Predictably and defensibly the Catholic Church acted with prudence by upholding both contemporary scientific judgment and received biblical interpretation. Thus in 1616 the Theological Consultors of the Holy Office (advisors to the Pope) declared the Copernican theory foolish and heretical. This opinion was not uniquely Catholic either: Both Martin Luther and John Calvin disapproved of Copernicanism. After the publication of his *Dialogue Concerning the Two Chief World Systems: Ptolemaic and Copernican* (1632), Galileo was judged by church authorities to be *vehemently suspected of heresy* and sentenced to house arrest. It is interesting and important to note, however, that despite the opposition of so many church leaders to Galileo's Copernicanism, the Church never formally condemned the Copernican theory *ex cathedra*. That is, it never formally made rejection of terrestrial motion a matter of ecclesiastical dogma. Galileo had been punished for transgressing a theological boundary by engaging in biblical interpretation as a layman, even though the question was of scientific relevance. Here a key ethical/philosophical dimension of the affair turned on the question of whether or not Copernicanism was a matter of religious faith. Galileo did not believe that it properly was. Church authorities disagreed. Viewed in such a light, the Galileo affair, ultimately, stands as a religious debate between Roman Catholics about biblical interpretation.

Galileo's achievement as a scientist rests not only on the fact that history has vindicated his Copernicanism. His achievements in the fields of dynamics, technical instrumentation, optics, astronomy, and philosophy of science combine to place him among the greatest of scientific minds. His engagement with the Catholic Church was a complicated affair that testifies to the vigor of his scientific ability and to his Christian faith. Although it has often been presented as a clash between modern science and religion, the Galileo affair is not best understood in this way, because all players were committed churchmen. Rather the affair evidences the complicated religious and ethical dimensions that surface when human beings seek to construct a coherent worldview that also does justice to deeply held convictions about matters both scientific and religious.

MARK A. KALTHOFF

SEE ALSO *Scientific Revolution; Space Exploration.*

BIBLIOGRAPHY

Blackwell, Richard J. (1991). *Galileo, Bellarmine, and the Bible*. Notre Dame, IN: University of Notre Dame Press.

DeSantillana, Giorgio. (1955). *The Crime of Galileo*. Chicago: University of Chicago Press.

Drake, Stillman. (1957). *Discoveries and Opinions of Galileo*. New York: Doubleday Anchor Books. This volume remains the standard English translation of Galileo's *Starry Messenger*, his letter to the Grand Duchess Christina, and other early works.

Galilei, Galileo. (1997). *Galileo on the World Systems: A New Abridged Translation and Guide*, trans. Maurice A. Finocchiaro. Berkeley: University of California Press. An accessible abridged English translation of Galileo's dialogue with a valuable introduction and notes for the modern student.

Howell, Kenneth J. (2002). *God's Two Books: Copernican Cosmology and Biblical Interpretation in Early Modern Science*. Notre Dame, IN: University of Notre Dame Press.

Jaki, Stanley L. (2001). *Galileo Lessons*. Pinckney, MI: Real View Books.

Koyré, Alexandre. (1978). *Galileo Studies*, trans. John Mepham. Atlantic Highlands, NJ: Humanities Press.

Langford, Jerome J. (1992). *Galileo, Science and the Church*, 3rd edition. Ann Arbor: University of Michigan Press. This is the best brief introductory volume on the subject of Galileo's encounter with the catholic church.

Lindberg, David C., and Ronald L. Numbers. (1986). *God and Nature: Historical Essays on the Encounter between Christianity and Science*. Chicago: University of Chicago Press. This classic anthology of scholarly articles includes two important chapters on Copernicanism and Galileo.

Machamer, Peter, ed. (1998). *The Cambridge Companion to Galileo*. New York: Cambridge University Press.

McMullin, Ernan, ed. (1967). *Galileo: Man of Science*. New York: Basic Books.

Shea, William R., and Mariano Artigas. (2003). *Galileo in Rome: The Rise and Fall of a Troublesome Genius*. Oxford, UK: Oxford University Press.

GALTON, FRANCIS

• • •

Francis Galton, 1822-1911. The English scientist, biometrician, and explorer founded the science of eugenics and introduced the theory of the anticyclone in meteorology. (© Corbis-Bettmann.)

Francis Galton (1822–1911), the scientist who created and promoted eugenics, the notion that a fitter human race might be created through selective breeding, was born near Birmingham, England, on February 16, and died in Haslemere, Surrey, England, on January 17. Originally oriented toward a medical career, Galton switched to Cambridge University to study mathematics, graduating with an ordinary degree. But his Cambridge experience was crucial to Galton's future career, during which he attempted to introduce quantitative analysis into whatever problem on which he happened to be working. His quantitative interests led Galton to discover the important statistical concepts of regression and correlation. He applied these in his anthropometric studies whose ultimate goal was to contribute to the improvement of humanity through eugenics, a term coined by Galton, that has profound ethical implications.

Galton's decision to abandon medicine was strongly influenced by his cousin, Charles Darwin (1809–1882), thirteen years his senior. They were grandsons by different marriages of Erasmus Darwin (1731–1802), a physician, scientist, poet, and inventor.

Like Darwin, Galton began his career as an explorer. Several years after graduating from Cambridge, he financed his own expedition and traveled through northern Namibia, a region of Africa not previously visited by Europeans. Galton took careful measurements of latitudes, longitudes, and altitudes, published his results in the *Journal of the Royal Geographical Society* in 1852, and was awarded a gold medal by the Society the same year. He also wrote a nontechnical book about his journey, *Tropical South Africa* (1853), but is best remembered for *The Art of Travel* (1855), an immensely popular guidebook for amateur and professional alike who ventured into the bush. The book went through many editions, grew in size, and Phoenix Press reissued the fifth edition in 2001. Subsequently Galton was active in the Royal Geographical Society for many years commenting frequently at Society meetings. During this part of his career he also became interested in meteorology. This led to his discovery of the anticyclone, a weather feature characteristic of a high-pressure system.

The second part of Galton's career commenced when he read Darwin's *On the Origin of Species* (1859). Galton concluded that it should be possible to improve the human race through selective breeding just as was true for domestic animals and cultivated plants. In 1865 he published a two-part article entitled "Hereditary Talent and Character" in a popular periodical called *MacMillan's Magazine*. The *MacMillan's* article was a precursor for Galton's book *Hereditary Genius* (1869). In both the article and the book Galton attempted to show that what he called *talent and character* were inherited. The book contained sections on judges and statesmen among others. Galton's thesis was that if he picked an eminent judge, for instance, that judge's immediate male relatives (e.g., father and son) were more likely to be eminent than those whose relationship was more distant (e.g., grandfather and grandson). Women were excluded from the analysis. Galton believed that analysis supported his thesis while recognizing, as others argued, that environment (for example, the father might obtain a good position for the son) might also be responsible for the correlation.

Galton was intensely interested in the analysis of quantitative data. By the time he had written *Hereditary Genius* he had become aware of the normal distribution and its application. In the book he used the bell curve to calculate a hypothetical distribution of the estimated 15 million males in the United Kingdom according to their natural abilities. Later Galton described two important new statistical concepts: regression and correlation. In experiments with sweet peas he found that seed diameter was normally distributed, but the diameter of seeds of progeny of large seeded and small seeded plants tended to be closer to the mean of the population as a whole than they did to the parental seed from which they had come. He dubbed this property regression to the mean. Regression to the mean has been documented over and over again since (for instance, in the case of different classes of mutual funds such as ones specializing in growth versus international stocks).

Galton also found he could draw a straight line on a graph comparing the diameters of parental and progeny seeds (Figure 1). This was the first regression line and from it he computed the first regression coefficient. Later he obtained comparable numerical data for humans (e.g., height) in the anthropometric laboratory organized at the International Health Exhibition of 1884 held in South Kensington, London. After the exhibition ended the laboratory reopened in the Science Galleries of the South Kensington Museum. Because Galton collected data on both parents and children, he once more demonstrated regression to the mean (e.g., for height).

FIGURE 1

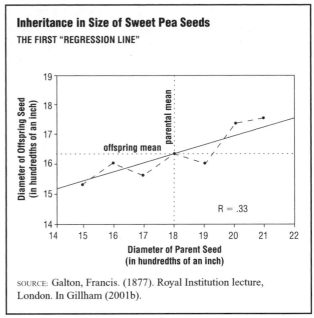

Inheritance in Size of Sweet Pea Seeds
THE FIRST "REGRESSION LINE"

SOURCE: Galton, Francis. (1877). Royal Institution lecture, London. In Gillham (2001b).

While plotting forearm length against height he discovered another important statistical concept, correlation (i.e., tall men have long forearms). He reported the first correlation coefficient, countless numbers of which have been calculated since. Galton also became interested in fingerprints and their classification and used his anthropometric laboratory to collect scores of fingerprints. His work was central to the development of fingerprinting as a forensic technique.

Galton collected many of these important observations together in his book *Natural Inheritance* (1889). He began to acquire disciples. One of these, Karl Pearson (1857–1936), a superb mathematician, was able to develop statistical theory and go far beyond Galton in its formulation.

The Legacy of Eugenics

All the while Galton had been promoting eugenics. The notion that fitter people could be bred through selection began to gain great momentum in the first decade of the twentieth century. Positive eugenics envisioned the selective reproduction of those regarded as fit, while negative eugenics discouraged or prevented the reproduction of those deemed unfit. Sadly negative eugenics prevailed. In the United States eugenic sterilization laws were passed in many states leading to the involuntary sterilization of thousands of people who were thought to be mentally deficient or *feebleminded*. Developments in the United States were followed with interest elsewhere, especially in Germany. When the Nazis came to power they passed an involuntary sterilization law that resulted in the sterilization of hundreds of thousands of individuals. After World War II eugenic sterilization gradually came to an end. Although eugenics is Galton's unfortunate legacy, he also leaves important accomplishments such as statistics and the development of fingerprinting technology.

NICHOLAS WRIGHT GILLHAM

SEE ALSO *Darwin, Charles; Eugenics.*

BIBLIOGRAPHY

Gillham, Nicholas W. (2001a). "Sir Francis Galton and the Birth of Eugenics." *Annual Review of Genetics* 35: 83–101.

Gillham, Nicholas W. (2001b). *A Life of Sir Francis Galton: From African Exploration to the Birth of Eugenics.* New York: Oxford University Press.

Kevles, Daniel. (1995). *In the Name of Eugenics.* Cambridge, MA: Harvard University Press.

Kühl, Stefan. (1994). *The Nazi Connection.* New York: Oxford University Press.

Paul, Diane B. (1995). *Controlling Human Heredity:1865 to the Present.* Atlantic Highlands, NJ: Humanities Press.

Reilly, Philip R. (1991). *The Surgical Solution: A History of Involuntary Sterilization in the United States.* Baltimore, MD: Johns Hopkins University Press.

GAMES

SEE *Video Games.*

GAME THEORY

• • •

Game theory is the analysis of choices made by individuals, institutions, or governments, which are termed players; the results of one player's choice depend on the choices made by the others. Anticipations by players about how others may respond or may anticipate their actions thus influence choices of actions. An important attempt to use game theory involved the formation of nuclear deterrence strategy by the United States during the Cold War (1945–1990). However, game theory has many more general implications that go beyond those involving intentional choice.

Despite the fact that game theory matured only toward the end of the twentieth century, it has become a central tool in some of the behavioral sciences and doubtless will extend its influence into all disciplines that attempt to explain the behavior of living organisms. Indeed, game theory provides a language that transcends and potentially unites the various disciplines that deal with human behavior. Moreover, it provides an experimental methodology that allows for the rigorous construction and testing of strategic interaction because it forces an experimenter to be explicit in defining the actions available to the subjects, the payoffs, and the distribution of information among the subjects.

An Illuminating Example

A fox is chasing a rabbit through a wooded area. Foxes are faster than rabbits, and so if the rabbit runs in a straight line, it will be caught and eaten. The rabbit therefore periodically veers left or right, gaining ground on the fox. If the rabbit changes course too rapidly, its average forward movement will be so slow that it will be caught, but if it changes course too slowly, the fox will be so close that a small misstep by the rabbit will lead to its immediate demise. Therefore, the rabbit must choose the average rate of veering to optimize its probability of escaping.

In game theory it is said that the rabbit has *actions*: R_t = "Veer Right after t seconds" and L_t = "Veer Left after t seconds." The rabbit also wants to randomize its choice of Veer Right and Veer Left, because if the fox discovers a pattern in the rabbit's movement, it may be able to anticipate the rabbit's next move, thereby gaining ground on it. The proper mix of Veer Left and Veer right is doubtless 50 percent Left and 50 percent Right, for the fox potentially could exploit an imbalance in either direction.

However, suppose that there is an open field some distance to the east of the wood and that foxes run much faster than rabbits do in an open field. Then the fox might run constantly a little to the west of the rabbit, forcing the rabbit to turn east more often than it turns west. The rabbit in turn may risk being caught by veering west more frequently than it would otherwise, trying to keep away from the open field. It can be seen that both the rabbit and the fox choose actions to maximize the probability of winning, with each anticipating the effect of its actions on the other. This is the type of situation studied in game theory.

How important is game theory? It is central to understanding life in all its varied forms. This may sound excessive, but one must step back from this interaction between a rabbit and a fox to ask more basic questions. For example, why are rabbits bilaterally symmetrical about the axis along which their movement is most rapid and energy-efficient (left leg and right leg symmetrically placed and equally strong, left eye and right eye symmetrically placed and of equal size and discriminating capacity, and single external body parts such as the nose and tail arrayed along the axis of movement)? The answer is that if rabbits had strength biased to the right, it would be easier for them to jump left than jump right, and that would give an advantage to their natural predators, the foxes. Foxes are bilaterally symmetrical for similar reasons. Game theory thus explains important facts about life that otherwise appear arbitrary and incomprehensible.

This simple game theoretic argument explains a major fact about the organization of life. Animals that run to escape predators or capture prey have body forms that are bilaterally symmetrical about a vertical axis along the direction of their most rapid motion. This applies to most animals and fish but not to plants, which do not run and are radially symmetrical, or to squid, octopuses, and other sea creatures whose primary motion is up and down.

To avoid the conclusion that game theory deals only with conflict, one can consider an example that is called the Cooperation Game. A group of ten hunters in a village spread out in the jungle every day to look for large game. They hunt individually, climbing tall trees and waiting quietly and attentively for long hours until the prey appears. At the end of the day the hunters share the day's kill equally. Of course, each hunter could spend the day sleeping in a tree. Suppose that by working each hunter adds an average of 3,000 calories to the total kill, of which his share is 300, but expends 500 calories of energy hunting as opposed to sleeping. A selfish hunter thus will sleep rather than hunt, saving 200 calories but costing the other group members 2,700 calories. This is a game in which there are n players and each player (i.e., each hunter) has two actions: Work or Shirk. If m hunters Work, the Shirkers' payoff is $3,000m/n$ each, whereas the Workers' payoff is $3,000m/n - 500$.

A *best response* of a player in a game can be defined as a strategy (in this case an action) that maximizes that player's payoff in light of the strategies of the other players. It is easy to see that a self-interested player's best response in this game is to Shirk no matter what the other players do. A *Nash equilibrium* of a game is defined as a choice of strategies made by the players such that each is a best response to the other players' choices. It is clear that in the Cooperation Game there

is only one Nash equilibrium, in which everyone shirks ($m = 0$) and no one eats.

Suppose another rule is added to the game. If a hunter is caught shirking, he is punished by being prohibited from hunting and sharing the kill for two days. Further, suppose the probability of being caught shirking is 0.50. To see that having everyone hunt is now a Nash equilibrium, one must decide whether a single hunter in a group of ten could do better by shirking and risking getting caught. The hunter saves 200 calories by shirking, but half the time he is caught and then loses two days' payoff, which is 5,400 calories. Thus, that hunter loses an average of 2,500 calories a day by shirking, and so his best response to hunt with the others. The conclusion is that with this new punishing mechanism full cooperation by each hunter becomes a Nash equilibrium.

History and Analytics of Game Theory

Game theory presupposes rational choice theory because it assumes that players have rational preferences in regard to the game's outcomes. It also presupposes rational decision theory because choice under conditions of uncertainty is the rule in most game situations. Because rational choice theory and decision theory were codified only in the late twentieth century, it is not surprising that game theory is still an incomplete and rather underdeveloped science. Before about 1950 games were assumed to be zero-sum; that means that what one player loses, the other player wins. The rabbit-fox game described earlier is zero-sum, but the hunter game is not because with the proper strategies all the hunters gain by cooperating.

With the zero-sum assumption cooperation never leas to a gain, and this would undercut some of the major contributions of game theory to the understanding of cooperation in biology and economics. Moreover, the three mathematicians who developed game theory—Ernst Zermelo (1871–1953), Stefan Banach (1892–1945), and John von Neumann (1903–1957)—assumed that each player will choose a strategy that minimizes the maximum gain for an opponent. This so-called *minimax* analysis cannot be extended to more general strategic contexts.

Modern game theory was born in 1950 after the publication of a paper by the young Princeton mathematician John F. Nash, Jr. (b. 1928; winner of a Nobel Prize in economics in 1994), who introduced the novel idea of a game equilibrium as a set of mutual best responses. The central term in modern game theory, the Nash equilibrium, acknowledges his work. Several conceptual problems had to be cleared up before game theory could attain a central position in the behavioral sciences. In 1965 Reinhard Selten (b. 1930; winner of a Nobel Prize in economics in 1994) developed the concept of *equilibrium refinement*, which showed why certain Nash equilibria are likely to be of empirical relevance and others are not. In 1967 and 1968 John Harsanyi (1920–2000; winner of a Nobel Prize in economics in 1994) showed how to apply game theory when the players have incomplete knowledge of the other players and the payoffs.

Until the 1980s it was believed by many people that game theory could be applied only to highly intelligent, so-called rational players because an analysis of the best responses is intellectually demanding. However, in 1972 the biologist John Maynard Smith (1920–2004) applied game theoretic notions to explaining animal conflict, a process that culminated in his publication of *Evolution and the Theory of Games* (1982). The innovation here is the idea that evolution can provide an alternative to high-level mental reasoning. For instance, rabbits veer optimally when chased by foxes not because each rabbit logically compares and empirically tests the alternatives but because running behavior is encoded in a rabbit's genes and those genes which render the rabbit most capable of eluding the fox are favored by natural selection in successive generations of rabbits. Inefficient genes simply become fox food.

The Ultimatum Game and Altruistic Preferences

An example of such research is the *ultimatum game*, in which under conditions of complete anonymity two players separately are shown a sum of money, say, $10. One of the players, called the proposer, is instructed to offer any number of dollars from $1 to $10 to the second player, who is called the responder. The proposer can make only one offer, and the game is never repeated with the same players facing each other. The responder can accept or reject this offer. If the responder accepts the offer, the money is shared accordingly. If the responder rejects the offer, both players receive nothing.

If the responder cares only about her own payoff in the game (it is said that she is *self-regarding* in this case) and the proposer knows or supposes this, the proposer will make the responder the minimum offer of $1, the responder will accept, and the game will be over. However, when the game actually is played, the self-regarding outcome almost never is attained or even approximated. In fact, as many replications of this experiment have documented, under varying conditions and with varying amounts of money, proposers routinely offer responders very substantial amounts (50 percent of the total generally is the modal offer) and responders frequently reject offers below 30 percent (Camerer 2003).

Are these results culturally dependent? Do they have a strong genetic component, or do all "successful" cultures transmit similar values of reciprocity to individuals? Alvin Roth (Roth, Prasnikar, Okuno-Fujiwara, and Zamir 1991) conducted ultimatum games in four different countries (the United States, Yugoslavia, Japan, and Israel) and found that although the level of offers differed slightly in different countries, the probability of an offer being rejected did not. This indicates that both proposers and responders have the same notion of what is considered fair in that society and that proposers adjust their offers to reflect that common notion. The differences in the levels of offers across countries were relatively small.

This ultimatum game result, along with that of many other similar games, suggests that many human subjects are strong reciprocators. Strong reciprocators come to strategic interactions with a propensity to cooperate (*altruistic cooperation*), respond to cooperative behavior by maintaining or increasing their level of cooperation, and responds to noncooperative behavior by punishing the "offenders" even at a cost to themselves and even when they cannot reasonably expect future personal gains to flow from the imposition of such punishment (this is called *altruistic punishment*).

Behavior in the ultimatum game thus conforms to the strong reciprocity model: Fair behavior in the ultimatum game among college students is a fifty-fifty split. Responders reject offers under 40 percent as a form of altruistic punishment of a norm-violating proposer. Proposers offer 50 percent because they are altruistic cooperators or 40 percent because they fear rejection. To support this interpretation it can be noted that if the offers in an ultimatum game are generated by a computer rather than by the proposer and if the respondents know this, low offers very rarely are rejected (Blount 1995). Moreover, in a variant of the game in which a responder's rejection leads to the responder getting nothing but allows the proposer to keep the share she suggested for herself, responders infrequently reject offers and proposers make considerably smaller offers.

The strong reciprocator is not a representative of one of the types of human nature found in traditional political philosophy. A strong reciprocator thus is neither the selfless altruist of utopian theory in the tradition of Jean-Jacques Rousseau (1712–1778) or that of Karl Marx (1818–1883) nor the selfish hedonist found in traditional economics and described by the economist Adam Smith (1723–1790) in *The Wealth of Nations*

GAME THEORY

(1776). Such a person is a conditional cooperator whose penchant for reciprocity can be elicited in circumstances in which pure selfishness would dictate a different action. Indeed, the strong reciprocator is more akin to the empathetic individual found in Adam Smith's other important work, *The Theory of the Moral Sentiments* (1759) except that Smith there emphasizes the sweet side of human nature, playing down the willingness to punish transgressions that is uncovered routinely in behavioral games.

Social Dilemmas

Another important behavioral game that sheds light on human nature and increases people's understanding of human social interaction is the *social dilemma*. A social dilemma is a group interaction in which all the players benefit if they all cooperate but each individual has an incentive to shirk and benefit from the cooperation of others.

An experimental representation of a social dilemma is the so-called *public goods game*. A typical public goods game consists of a number of rounds, say, ten. In each round each subject is grouped with several other subjects, say, three others. Each subject is given a certain amount of money, say, $20. Each subject, unseen by the others, then places a fraction of his or her money in a common account and puts the remainder in his or her private account. The experimenter then tells the subjects how much was contributed to the common account adds to the money in the common account enough so that, when divided among the four players, the private account of each subject can be increased by a fraction, say, 40 percent, of the players' original contribution to in the common account. Thus, if a subject contributes his or her whole $20 to the common account, the experimenter adds an additional $12, so each of the four group members will receive ($20 + $12)/4 = $8 at the end of the round. In effect, by putting the whole endowment into the common account, a player loses $12 and the other three group members gain in total $24 (= $8 × 3).

A self-regarding player will contribute nothing to the common account. However, only a fraction of subjects conform to the self-regarding model. The subjects begin by contributing on average about half of their endowments to the public account. The level of contributions decays over the course of the ten rounds until in the final rounds most players behave in a self-regarding manner. This is exactly what is predicted by the strong reciprocity model. Because they are altruistic contributors, strong reciprocators start out by contributing to the

FIGURES 1–3

Figure 1: The Even-Odd Game

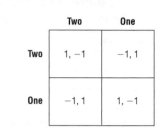

SOURCE: Courtesy of Herbert Gintis.

Figure 2: The Battle of the Sexes

	Gambling	Ballet
Gambling	2, 1	0, 0
Ballet	0, 0	1, 2

SOURCE: Courtesy of Herbert Gintis.

Figure 3: The Centipede Game

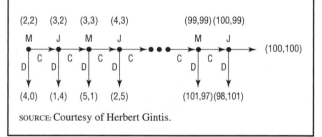

SOURCE: Courtesy of Herbert Gintis.

common pool, but in response to the norm violation on the part of the self-regarding types they begin to refrain from contributing.

How can it be known that the decay of cooperation in the public goods game is due to cooperators punishing free riders by refusing to contribute? Subjects often report this behavior retrospectively. More compelling, however, is the fact that when subjects are given a more constructive way of punishing defectors, they use it in a manner that helps sustain cooperation. For instance, Ernst Fehr and Simon Gächter (2000) set up an experimental situation in which the possibility of punishment for personal gain was removed completely. They used six- and ten-round public goods games with groups of four and with costly punishment allowed at the end of each round, employing three different methods of assigning members to groups.

They found that when costly punishment is permitted, cooperation does not deteriorate; indeed, if the same players stay together for the whole session, despite strict anonymity cooperation increases almost to full cooperation even on the final round. In effect, even though the groups had some selfish players, there was a sufficiently large fraction of strong reciprocators to ensure that it was not in the interest of the selfish to act selfishly.

The Epistemological Foundations of Game Theory

One can characterize the choice situation facing an agent in terms of its level of complexity. The least complex situation occurs when an agent must choose from a set of fixed alternatives. Analytically complete axiomatic models of choice in this situation are well developed and empirically successful. Of intermediate complexity is a situation in which an agent must choose from a set of alternatives, each of which is a *probability distribution* over determinate outcomes. Analytically complete axiomatic models of choice in this situation are also well developed and empirically successful, although some important anomalies in human behavior have been noted in regard to decision theory. The most complex situation is the one described by game theory: An agent's choices affect not only that agent but other agents as well, the other agents also are engaged in making choices that affect themselves and others, and all agents take into account the strategic nature of their interactions. One of the most widely known attempts to illustrate such a game theoretic situation is the Prisoner's Dilemma.

It would be gratifying to have a fully successful analytical model of strategic interaction applicable to the highly complex level, but despite the efforts of theoreticians since the second half of the twentieth century, none exists. Ignoring the Prisoner's Dilemma for now, one can consider three simple games that dramatize the problems in developing such a theory, which then can be used to outline some important contributions to the epistemological underpinnings of game theory.

EVEN-ODD GAME. The first is the simple Even-Odd game. This game has two players, each of whom can show either one finger (One) or two fingers (Two). The two players show their fingers simultaneously, with player 1 winning if his choice matches that of the other

player (i.e., if One-One or Two-Two occurs) and player 2 winning if her choice does not match it (i.e., if One-Two or Two-One occurs). Figure 1 shows the normal form of this game (the normal form specifies the moves that each player can make and the payoffs for each player as a function of the moves of both players).

This game obviously has no Nash equilibria in the "pure" strategies: One and Two. However it does have a unique Nash equilibrium in which each player plays One with probability 1/2 and plays Two with probability 1/2. Doubtless many people remember this solution from schoolyard days, when they learned to "mix up" their choices so that an opponent could not discover their next move. The problem is that this game is played just once (it is a *one-shot* game). Hence, if a player's opponent randomizes as suggested by the Nash equilibrium, it does not matter what the first player does: The expected payoff is zero whether the first player chooses One, Two, or a probability distribution over One and Two. However, the same is true for the opponent. Therefore, there is no reason for either player to randomize, yet that is the solution suggested by game theory.

An important step toward dealing with this problem is to note that each player chooses a best response not to the actual strategy of the other players but to his or her own conjecture about what the other players will do. Robert Aumann and Adam Brandenburger (1995) prove the following theorem for a two-player game. Suppose ϕ_1 is player 1's conjecture concerning player 2's strategy and ϕ_2 is player 2's conjecture concerning player 1's strategy. If both players know each other's conjectures and each knows that the other is rational (i.e., chooses a best response to his or her conjecture), (ϕ_2, ϕ_1) is a Nash equilibrium.

BATTLE OF THE SEXES. This is a fine solution for Odd or Even, which has only one Nash equilibrium. However, one must consider another famous game, the Battle of the Sexes, which is depicted in Figure 2. In this game Rowena and Colin love each other and get one point by being together. However, Rowena loves the ballet and Colin loves gambling. Each gets a point for attending his or her favorite event. Thus, if both go to the opera, Rowena gets 2 and Colin gets 1, whereas if they both go gambling, Colin gets 2 and Rowena gets 1. Moreover, when they are not together, it is assume that they are so unhappy that each gets zero. It is easy to find two Nash equilibria: Both go gambling, and both go to the opera. It turns out that there is also a third Nash equilibrium in which each party goes to his or her favorite place with probability 2/3 and to the other's favorite

FIGURE 4

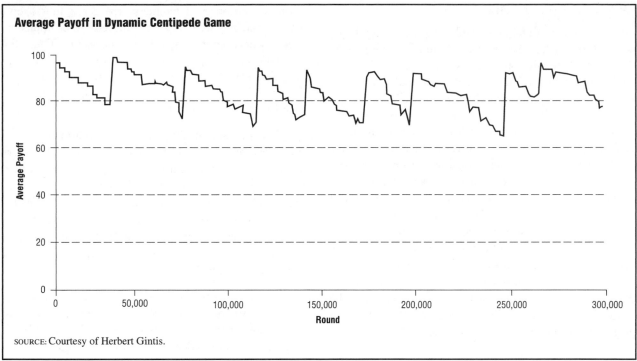

Average Payoff in Dynamic Centipede Game

SOURCE: Courtesy of Herbert Gintis.

place with probability 1/3. This is called a *mixed strategy equilibrium*.

To see that One gambling with probability 2/3 and Two gambling with probability 1/3 is a Nash equilibrium, one should note that the expected payoff to One from gambling equals 2 × 1/3 + 0 × 2/3 = 2/3, whereas the expected payoff to One from ballet equals 0 × 1/3 + 1 × 2/3 = 2/3. Because these probabilities are equal, One can do no better than his probability 2/3 gambling, probability 1/3 ballet strategy, and a similar argument holds for Two.

In the case of Battle of the Sexes it is unreasonable to posit that each player knows the other's conjecture because there is no way of explaining how this mutual knowledge would have come about. Indeed, it is not even plausible to suppose that the players have conjectures concerning what the other will do unless there is more to the social situation than has been explained. Moreover, the players still have no incentive to play according to their partners' conjectures (Binmore 1988).

The problem becomes even more implausible when there are more than two players. In this case Aumann and Brandenburger (1995) show that if all players assign the same probability distribution to player types, it is known mutually that all players are rational (i.e., choose best responses), and the players' conjectures are com-

monly known, these conjectures form a Nash equilibrium. One says that a fact is commonly known if all players know the fact, all know that the others know it, all know that all know that the others know it, and so on (Lewis 1969).

CENTIPEDE GAME. There are simple games in which the very notion of rationality and the adequacy of the concept of the Nash equilibrium are brought into question. Consider, for instance, the Centipede Game. The players, Mutt and Jeff, start out with $2 each, and they alternate rounds. On the first round Mutt can defect (D) by stealing $2 from Jeff, and the game is over. Otherwise Mutt cooperates (C) by not stealing and receives an additional $1. Then Jeff can defect (D) and steal $2 from Mutt, and the game is over, or he can cooperate (C) and receive an additional $1. This continues until one player or the other defects or until each player has $100. The game tree is illustrated in Figure 3.

This game has only one Nash equilibrium outcome, in which Mutt defects on the first round. To see this, let round k be the first round in which either player defects in a Nash equilibrium. If $k > 1$, the other player's best response is to defect on round $k - 1$. Of course, common sense indicates that this is not the way real players would act in this situation, and empirical evidence corroborates this (McKelvey and Palfrey 1992). People in this game will cooperate up to round 90 and beyond before considering defecting.

It would be difficult to fault players for not being rational in this case because they do much better playing the way they do rather than the way dictated by the Nash equilibrium concept. The concept of rationality is problematized for the following reason: If Jeff believes Mutt is rational, Jeff will defect in round 2. This is why Mutt defects in round 1. But suppose Mutt cooperates in round 1. Then Jeff will recognize that his assumption concerning Mutt must be false. Jeff probably will say to himself, "I don't know what strategy Mutt is using, but since he cooperated once, perhaps if I cooperate now, Mutt will cooperate a second time." Thus, Jeff will tend to cooperate in round 2. Now Mutt, who is very smart, can foresee what Jeff will be thinking and hence will cooperate even if he is rational. One can conclude that agents who use best responses will not play the Nash equilibrium in this game. It is easy to see the problem by referring to the analysis of Aumann and Brandenburger (1995): The two players do not know each other's conjectures.

Evolutionary Game Theory

To this point the focus has been on so-called classical game theory, which depicts the strategic interaction among a number of *rational agents*. The interaction is socially disembodied, with the agents having neither history nor substance outside this particular interaction. All socially relevant aspects of the actors must be captured by their beliefs and conjectures, which are totally disembodied and unmotivated. A similar degree of social minimality has given rise to powerful models of decision making when strategic interaction is absent, as in rational choice theory and decision theory. However, this does not extend to game theory, in which a more socially embedded approach is needed to derive plausible results.

The most promising alternative foundation for strategic interaction is known as *evolutionary game theory* (Maynard Smith 1982, Samuelson 1997, Gintis 2000). The central actors in evolutionary game theory are not players but strategies. Suppose a group of agents periodically plays a certain classical game G. One assumes a large population of agents, each of whom adopts a particular strategy in playing G. One does not assume that the strategies represented in the population are in any way optimal, although one does assume that there is enough random variation and mutation across time that all pure strategies are represented.

In each period agents from the population are assigned randomly to play G. Their scores are tallied,

and the change in the population over time is governed by an evolutionary dynamic in the sense that agents whose strategies are very successful tend to be copied by agents whose strategies are less successful. Thus, the population ecology of strategies moves over time in accordance with the notion of survival of the fittest. This is called a *replicator dynamic* (Hofbauer and Sigmund 1998, Gintis 2000).

The fundamental theorem of evolutionary game theory is that every equilibrium point of an evolutionary dynamic is a Nash equilibrium. This provides a justification for the concept of the Nash equilibrium without the need for the epistemological assumptions of classical game theory. Moreover, evolutionary game theory shows that many Nash equilibria of classical game theory are not evolutionarily stable and thus cannot explain observable social behavior.

A case in point is the Centipede Game described earlier in this entry. The author of this entry has created a computer program to simulate the evolution of behavior in the Centipede Game (this is called an *agent-based simulation*). The author created a population of 200 agents, each supplied with a strategy sk of the following form: "cooperate until round k, then defect." Initially, these strategies are assigned randomly to the agents, and they play 300,000 rounds, with a mutation rate of 0.001 (a mutant assumes a random strategy sk where $1 \leq k \leq 101$). The results are shown in Figure 4.

It can be seen that cooperation quickly increases until after only a few rounds the average payoff is more than 95. Then cooperation erodes, as might be expected, until the average payoff dips below 80. At that point a pair of agents who choose strategies near $k = 100$ do very well, and those strategies grow at the expense of the strategies that involve defection on rounds near $k = 80$. Cooperation shoots back up to nearly perfect. The cycle repeats for 300,000 rounds and shows no signs of changing its basic character.

Even though the only Nash equilibrium of the stage game uses strategy s_1, it can be seen that the evolutionary dynamic never remotely approaches this equilibrium. This is the case because the Nash equilibrium involves such poor payoffs that even a small number of mutant players can invade a population of all-defectors, and the system quickly ramps up to almost full cooperation (changing the mutation rate does not alter this result). Thus, evolutionary game theory shows that the behavior observed when people play the Centipede Game is easy to model in a dynamic framework.

Game Theory and Ethics

Game theory has been applied to ethical theory by John Harsanyi (1992). Harsanyi (1920–2000; winner of a Nobel Prize in economics in 1994) develops a theory of justice very close to that of the philosopher John Rawls (1921–2002) and shows that it can be derived from basic game theoretic reasoning. Other important contributions to the game theoretic analysis of ethics include those of Brian Skyrms (1996) and Ken Binmore (1998).

Perhaps the first indication that game theory would be important to ethical theory was the famous tit-for-tat computer competition run by Robert Axelrod (Axelrod and Hamilton 1981). Axelrod asked what a successful strategy in the repeated Prisoner's Dilemma might look like. In that game the dominant strategy is to defect. However, if the game is repeated several times, players may be able to use the threat of defecting in the future to induce their partners to cooperate in the present.

Axelrod recruited fourteen game theorists from economics, mathematics, and the behavioral and computer sciences to submit computerized strategies for playing 200 rounds of the Prisoner's Dilemma. Those strategies were paired with each other in a round robin tournament with the result that the absolutely simplest strategy won. This strategy was *tit-for-tat*, supplied by Anatol Rapoport, a mathematician at the University of Toronto. Tit-for-tat cooperates on the first move and then does whatever its partner did on the previous move. Tit-for-tat is thus a simple reciprocity enforcer, cooperating when its partner cooperates and defecting when its partner defects.

After publishing these results (Axelrod and Hamilton 1981) Axelrod decided to stage a second tournament. More than sixty researchers from six countries submitted new programs, many of which were aimed explicitly at defeating tit-for-tat. Nevertheless, tit-for-tat again won handily.

This result relates to ethical theory because it shows the success of a strategy that is *nice* (never defect first), *punishing* (always retaliate against a defector), and *forgiving* (always revert to cooperating if your partner cooperates). These responses, of course, represent three important ethical principles. A fourth common ethical principle—*always turn the other cheek*—certainly would not fare well in this encounter, as it would be beaten by any program that could detect "wimps" (those who do not punish) and defect consistently in playing against them.

It is clear that the ethical principles behind the strong reciprocity associated with social dilemmas represent a higher development of tit-for-tat. Whereas tit-for-tat applies only to dyadic relationships, strong reciprocity applies to *n*-player social dilemmas.

HERBERT GINTIS

SEE ALSO *Artificial Morality; Choice Behavior; Decision Theory; Rational Choice theory.*

BIBLIOGRAPHY

Aumann, Robert, and Adam Brandenburger. (1995). "Epistemic Conditions for Nash Equilibrium." *Econometrica* 65(5): 1161–1180.

Axelrod, Robert, and William D. Hamilton. (1981). "The Evolution of Cooperation." *Science* 211: 1390–1396.

Binmore, Ken. (1988). "Modelling Rational Players: II." *Economics and Philosophy* 4: 9–55.

Binmore, Ken. (1998). *Game Theory and the Social Contract: Just Playing.* Cambridge, MA: MIT Press.

Blount, Sally. (1995). "When Social Outcomes Aren't Fair: The Effect of Causal Attributions on Preferences." *Organizational Behavior & Human Decision Processes* 63(2): 131–144.

Camerer, Colin. (2003). *Behavioral Game Theory: Experiments in Strategic Interaction.* Princeton, NJ: Princeton University Press.

Fehr, Ernst, and Simon Gächter. (2000). "Cooperation and Punishment." *American Economic Review* 90(4): 980–994.

Gintis, Herbert. (2000). *Game Theory Evolving.* Princeton, NJ: Princeton University Press.

Harsanyi, John C. (1992). "Game and Decision Theoretic Models in Ethics, Vol. I." In *Handbook of Game Theory with Economic Applications,* eds. Robert J. Aumann and Sergiu Hart. Amsterdam and New York: Elsevier Science.

Hofbauer, Josef, and Karl Sigmund. (1998). *Evolutionary Games and Population Dynamics.* Cambridge, UK: Cambridge University Press.

Lewis, David. (1969). *Conventions: A Philosophical Study.* Cambridge, MA: Harvard University Press.

Maynard Smith, John. (1982). *Evolution and the Theory of Games.* Cambridge, UK: Cambridge University Press.

McKelvey, Richard D., and Thomas R. Palfrey. (1992). "An Experimental Study of the Centipede Game." *Econometrica* 60: 803–836.

Roth, Alvin E.; Vesna Prasnikar; Masahiro Okuno-Fujiwara; and Shmuel Zamir. (1991). "Bargaining and Market Behavior in Jerusalem, Ljubljana, Pittsburgh, and Tokyo: An Experimental Study." *American Economic Review* 81(5): 1068–1095.

Samuelson, Larry. (1997). *Evolutionary Games and Equilibrium Selection.* Cambridge, MA: MIT Press.

Skyrms, Brian. (1996). *Evolution of the Social Contract.* Cambridge, UK: Cambridge University Press.

Smith, John Maynard. (1982). *Evolution and the Theory of Games.* Cambridge, UK: Cambridge University Press.

Mohandas Gandhi, 1869–1948. Gandhi was an Indian revolutionary religious leader who used his religious power for political and social reform. Although he held no governmental office, he was the prime mover in the struggle for independence of the world's second-largest nation. (© *Corbis-Bettmann*.)

GANDHI, MOHANDAS

• • •

Mohandas Karamchand Gandhi (1869–1948) was born in Porbandar, Gujarat, India, on October 2, and led India to independence from Great Britain on August 15, 1947, by preaching and practicing nonviolent resistance. After studying jurisprudence at University College, London, Gandhi began practicing law in Durban, South Africa, in 1893. It was here that he started his political career by fighting discrimination against Indians. Following World War I he returned to India and became involved with the Indian National Congress and the movement for national independence. He was repeatedly imprisoned for his use of civil disobedience, fasting, and boycotts as methods of social reform. In addition to his nonviolent opposition to Wes-

tern colonialism and capitalism, Gandhi advocated the reformation of the caste system and the harmonious coexistence of Muslims and Hindus in a unified India. His critiques of modern technoscience also influenced later theoretical developments and social movements. Gandhi was assassinated by a Hindu radical in New Delhi on January 30.

Nonviolence and Westernization

Gandhi initially defined his method of social action as *passive resistance*, but later refined and strengthened his ideals into a principle called Satyagraha. The term is derived from two Sanskrit words highlighting his central beliefs: *satya*, truth, and *agraha*, firmness—but practiced with *ahimsa*, non-injury to living things. As a method of direct social action, Satyagraha is a nonviolent insistence on truth in the political realm. Gandhi employed this principle with its offshoots, noncooperation and civil disobedience, in order to vindicate the truth by inflicting self-suffering rather than forcing his opponents to suffer. His persistence provoked anger in the British, including Winston Churchill, who called Gandhi "a malignant subversive fanatic" (Hardiman 2004, p. 238). The political success of this social reform method demonstrated the efficacy of nonviolence to the world and inspired other peace activists such as Nelson Mandela (b. 1918) and Martin Luther King Jr. (1929–1968).

Gandhi's experiments with Satyagraha made him aware of the economic, social, and political exploitation of people around the world, especially the uneducated and impoverished in South Africa and India. He believed that the root of this oppression and poverty was the culture of violence that resulted from Western materialist values, and he maintained that adopting the culture of nonviolence is the only way to attain truth, peace, and harmony. Thus Gandhi's nonviolent social reform was directly targeted against the globalization of Western values and material culture in the form of capitalism and imperialism.

He described the culture of violence in terms of the *seven social sins* of the world: wealth without work; pleasure without conscience; knowledge without character; commerce without morality; science without humanity; worship without sacrifice; and politics without principles. Gandhi's philosophy of nonviolence requires one to live life as an eternal quest for truth. It is often interpreted dogmatically or rejected as impractical, although it is founded upon the positive and near-universal values of love, respect, understanding, acceptance, and appreciation.

Gandhi believed that the westernization of India would destroy its culture and result in an unequal distribution of wealth and resources. Unlike his political heir, Jawaharlal Nehru (1889–1964), he did not believe that the systems of political organization that develop around Western science and technology could ever promote justice and human dignity. Gandhi maintained that the benefits of westernization would never trickle down to the poor because capitalist technology thrives on exploitation and creates a cycle of greed and consumption that never brings fulfillment.

Gandhi did not espouse communism, and in fact believed that capitalism could work if based on compassion rather than greed. Furthermore he understood that humans have legitimate material needs. The Western model of human development, however, sacrifices morality by overemphasizing materialism. He argued that human relationships ought to be guided by *trusteeship and constructive action*, meaning that human beings do not own their talents but hold them in trust for humanity. This fosters constructive action by helping the disenfranchised achieve greater self-confidence and self-sufficiency.

Gandhi's Reforms

Gandhi's opposition to Western values created an ideological gulf between him and other Indian political leaders. This motivated him to institute several societal reforms (he referred to them as the *constructive program*) even as the country struggled for independence, because he knew that his vision of an agrarian, self-sufficient, and traditional India would not be championed by his successors.

He developed small-scale technologies such as the *charkha*, or spinning wheel that helped liberate poor peasants from England's textile monopoly. Gandhi also helped in the effort to expand and improve basic education. Students learned reading and writing as well as best practices in agriculture. They were exposed to other cultures and religions in order to develop character and foster tolerance.

This education plan was a part of Gandhi's two part social reformation: promoting Hindu-Muslim unity and eradicating the caste system. Acutely aware of the multiethnicity of India and the tensions therein, Gandhi practiced interreligious harmony in his prayers by incorporating hymns from every major religion. His courage in the face of religious and ethnic violence inspired many Muslims to remain in a predominantly Hindu India.

Gandhi worked quietly to eradicate the caste system. He was cautious not to incite bitterness, because he feared the British would capitalize on divisions within India to strengthen their rule. Gandhi wished to change the name of *untouchables* from the derogatory *Bhangi* to the respectful *Harijan* (Children of God). With typical wisdom, he argued that by their suffering the untouchables had earned the right to be called Harijan, but other members of Hindu society will also earn that right when they atone for their sins.

Alternatives to Modern Science and Technology

When asked what he thought of Western civilization, Gandhi famously replied, "I think it would be a great idea." Thus he did not equate increasing scientific and technological sophistication with progress in civilization. In *Hind Swaraj* (1909), one of the earliest critiques of modernity as a development paradigm, Gandhi defined civilization as the ethical performance of one's duty and the attainment of mastery over passion. He also argued that "all research will be useless if it is not allied to internal research, which can link your hearts with those of the millions" (Gupta 2002 Internet site).

Nonetheless admitting there are lessons to be learned from modernity, Gandhi wrote that his "resistance to Western civilisation is really a resistance to its indiscriminate and thoughtless imitation based on the assumption that Asiatics are fit only to copy everything that comes from the West" (Hardiman 2004, p. 71). Gandhi believed that technoscience must be guided toward true human fulfillment and the alleviation of suffering. The fact that it is often used instead in the service of oppression, slavish consumerism, and war fueled Gandhi's conviction that Western values were bankrupt. In 1935, he initiated a movement called Science for People, which sought small-scale technological solutions for the problems faced by the rural poor. This indicated his vision for an alternative Indian future, which influenced especially ideas related to *alternative technology*.

For example, E. F. Schumacher's calls for a more humane economic system built upon small-scale *intermediate technology* were inspired by Gandhi. Likewise the Ghandian economists J. C. Kumarappa and D. R. Gadgil developed the concept of *appropriate technology* to counter the injustices that arise from the application of universal science in mass production processes. Although Gandhi is not explicitly mentioned, parts of Ivan Illich's *Medical Nemesis* (1975) echo Gandhi's *Hind Swaraj*. Gandhi's thought has also informed workers at development nongovernmental organizations (NGOs) such as Oxfam and ecological activists such as Rama-chandra Guha and those participating in the Chipko and Narmada movements.

Faced with the pressures of economic and technos-cientific globalization, Gandhi's vision of a traditional India has largely failed to materialize. It may be that the forces of westernization are too difficult to resist. As Chakravarti Rajagopalachari, a former Governor-General of independent India, wryly assessed, "The glamour of modern technology, money, and power is so seductive that no one . . . can resist it. The handful of Gandhians who still believe in his philosophy of a simple life in a simple society are mostly cranks" (Rushdie Internet site). Yet this does not diminish Gandhi's inspirational legacy or his teaching that life is more than science and technology.

ARUN GANDHI

SEE ALSO *Development Ethics; Indian Perspectives.*

BIBLIOGRAPHY

Andrews, C. F. (1930). *Mahatma Gandhi's Ideas.* Woodstock, VT: Skylight Paths.

Bondurant, Joan. (1965). *Conquest of Violence*, revised edition. Berkeley: University of California Press.

Fischer, Louis, ed. (1962). *The Essential Gandhi: His Life, Work, and Ideas: An Anthology.* New York: Random House.

Gandhi, Mahatma. (1949). *An Autobiography: The Story of My Experiments With Truth*, trans. Mahadev Desai. London: Phoenix Press.

Hardiman, David. (2004). *Gandhi in His Time and Ours: The Global Legacy of His Ideas.* New York: Columbia University Press.

Husain, Abid S. (1959). *The Way of Gandhi and Nehru.* London: Asia House Publishing.

INTERNET RESOURCES

Gupta, Vibha. (2002). "The Economics of Scarcity." Life Positive. Available from http://www.lifepositive.com/Spirit/masters/mahatma-gandhi/gandhian-economics.asp.

Rushdie, Salman. "Mohandas Gandhi." Time. Available from http://www.time.com/time/time100/leaders/profile/gandhi.html.

GATES, BILL

• • •

Born in Seattle, Washington, on October 28, 1955, William Henry Gates III, founder of the Microsoft computer empire, is, in the year 2003, the world's wealthiest

Bill Gates, b. 1955. The co-founder and chief executive officer of Microsoft became the wealthiest man in America and one of the most influential personalities on the ever-evolving information superhighway and computer industry. (© *Jim Lake/Corbis.*)

person, as well as the founder of the world's largest philanthropic foundation. Superlatives and paradoxes stick to Gates. Having scored a perfect 800 on the math portion of the Standard Aptitude Test (SAT), he later dropped out of college. Praising technology for "enhancing our leisure time" (Gates 1996, p. 284), his idea of a slow week (after marrying Melinda French in 1994) was to cut his workday to twelve hours a day during the week and eight hours a day on weekends.

Paradoxes also demarcate his ethical stances, both in business and technology. In 1975 he caused a stir among libertarian computer hackers by arguing in a letter to *Computer Notes* that software programs were "intellectual property" and should be legally protected through copyrights (Lowe 1998, pp. 86–87). As a result of his efforts, copying computer programs became illegal. However over the years a host of other computer companies have complained that he freely borrows their ideas.

A fierce competitor, Gates has said that "business is a good game [with] lots of competition and a minimum of rules" (Lowe 1998, p. 156), yet has been criticized for monopolistic practices. His competitors argue that he "cuts off the oxygen of the competition" (Lowe 1998, p. xiii). In 1998 the U.S. Justice Department sued Microsoft, alleging that the company had forced computer makers to sell its Internet Explorer browser as part of the licensing agreement for its Windows 95 software. Judge Thomas Penfield Jackson declared Microsoft a monopoly a year later. While accused of hardball tactics, it is important to note that Gates votes Democratic (i.e., more rules) more often than Republican and he vows to give away 95 percent of his wealth to charity (Lowe 1998, p. 178)

An unabashed optimist about the future, Gates believes technological "doomsayers vastly underestimate the potential of technology to help us overcome problems" caused by technology (Gates 1996, p. 291). Problems he considers self-correcting with the help of technology include unemployment, overpopulation, environmental dangers, globalization, and virtual reality, as well as privacy and security issues. Quoting H. G. Wells's belief that "human history becomes more and more a race between education and catastrophe" (Gates 1996, p. 293), Gates's foundation invests billions of dollars in the areas of education and global health issues.

Two problems he is less sanguine about include terrorism and artificial intelligence. In the short run terrorism worries him because of the inability of defensive weaponry to keep pace with advances in offensive developments. In the long run he is also concerned that "computers and software could achieve true intelligence" (Gates 1996, p. 290).

But, as a gambler, Gates clearly bets education will trump catastrophe, unlike, for example, Jacques Ellul's dire predictions in his *La technique ou l'enjeu du siecle* [Technology or the Bet of the Century] (Paris: Colin, 1954). The secret to Gates's worldview is poker. Part of Microsoft's startup costs came from Gates's poker winnings at Harvard. As he says, "In poker, a player collects different pieces of information . . . and then crunches all that data together to devise a plan for his own hand. I got pretty good at this kind of information processing" (Gates 1996, p. 43). A extraordinary understatement from a superlative intellect.

JIM GROTE

SEE ALSO *Business Ethics; Computer Ethics.*

BIBLIOGRAPHY

Gates, Bill. (1996). *The Road Ahead.* New York: Penguin.

Gates, Bill. (1999). *Business @ The Speed of Thought: Using a Digital Nervous System.* New York: Warner Books.

Lesinksi, Jeanne. (2000). *Bill Gates.* Minneapolis, MN: Lerner Publications Company.

Lowe, Janet. (1998). *Bill Gates Speaks: Insight from the World's Greatest Entrepreneur.* New York: John Wiley and Sons.

INTERNET RESOURCE

Website of the Microsoft Corporation. Available from http://microsoft.com.

GENDER

SEE *Sex and Gender.*

GENE THERAPY

• • •

Gene therapies (gene transfer technology) involve one or more experimental techniques for correcting or altering genes, including defective genes associated with physiological or psychological disorders. As has historically been the case with many other novel interventions (such as those that depend on drugs or surgery), debates have arisen between those who believe there is a moral obligation to pursue gene transfer research and those who challenge them as illegitimate or unnatural. As yet there is no strong consensus regarding distinctions between what is morally unacceptable, simply permissible, or obligatory.

Technical Aspects

There are a number of approaches to gene alteration including replacing an "abnormal" gene (i.e., DNA sequence) with a "normal" gene through homologous recombination, repairing an "abnormal" gene through selective reverse mutation, and altering the regulation of a particular gene. The term "abnormal" is placed in quotation marks, indicating that there remains room for disagreement about what constitutes a normal gene, is certainly one source of disagreement about these procedures.

Typically, for mostly practical reasons, gene therapy research involves the insertion of a functional gene into a non-specific location in the genome without removal or correction of the disease-causing gene. This can be done in vitro or in vivo. In vitro techniques require cells to be removed from an organism, corrective genetic material added in culture, and the altered cells returned to the organism. The advantages of this approach are twofold. If there is a problem with the genetic manipulation, the altered cells need not be transferred to the organism; also, the risk of unintentionally affecting non-targeted tissues is reduced. Alternatively, the corrective genetic material may be delivered to the targeted cells in vivo using a vector (often a virus that has been altered) to carry the gene into the cells. Retroviruses, adenoviruses, adeno-associated viruses, and herpes simplex viruses are among the viruses altered for vector use.

There are two categories of future gene therapy: somatic cell gene therapy and germ-line gene therapy. Somatic cell therapy involves the genetic alteration of nonfunctioning or malfunctioning somatic (i.e., non-reproductive) cells. These alterations are not passed on to subsequent generations. Germ-line therapy involves the genetic alteration of the germ (i.e., reproductive) cells or the early embryo prior to the development of gonadal tissue. Any resulting genetic changes will be inherited.

Use of gene transfer technology is not limited, however, to therapeutic goals. The technology can also be used for enhancement purposes to improve the functioning of normal genes—for example, the introduction of a growth hormone gene into a person of normal stature.

Ethical Issues

From the beginning, there has been considerable debate about the ethics of future gene therapy (Parens 1995, Walters and Palmer 1997, Stock and Campbell 2000). Some insist that somatic cell gene therapy is a logical extension of available techniques for treating disease; others argue that such genetic interventions are dangerous or inappropriate. Out of this debate emerged a moral demarcation line between somatic and germ-line gene therapy, the latter being widely described as ethically unacceptable because of the risks of physical and social harms (Anderson 1989).

In the late 1980s and early 1990s, with the move to clinical trials involving somatic cell gene transfer (and the possibility of inadvertent germ-line modification), debate about the ethics of germ-line gene transfer resurfaced. Some argued that germ-line gene transfer could be an effective and efficient treatment for diseases that affect many different organs and their cell types (such as cystic fibrosis); for diseases expressed in non-removable or non-dividing cells (such as Lesch-Nyhan syndrome); and for diseases that develop in the very early embryo that could be prevented through germ-line genetic intervention (such as albinism linked to tyrosinase). Indeed, some even argued that in such cases there was a moral obligation to reduce the incidence of disease in subsequent generations using germ-line gene transfer, instead of continuing to treat each successive generation with somatic cell genetic interventions. In opposition, questions have been raised about whether such work is an appropriate use of limited research funds when other efforts might have a more general public benefit, with some also objecting to the pursuit of a kind of biological perfectionism.

Controversial History

The history of gene transfer research in humans is a checkered one, mired in controversy (NRCBL 2002, Johnston and Baylis 2004). Martin Cline of the University of California Los Angeles (UCLA) conducted the first human gene transfer clinical trial in July 1980. The unsuccessful trial involved two patients with thalassemia, one in Israel and the other in Italy. Cline did not inform the UCLA Institutional Review Board (IRB) of his research, and did not fully disclose details of the trial to the Israeli research ethics review committee (at the time, Italy did not have an ethics review system). News of the unauthorized trial became public in a *Los Angeles Times* story published in October 1980. An internal investigation by UCLA and an external investigation by the U.S. National Institutes of Health (NIH) followed, resulting in significant sanctions for Cline. It would be another decade before an officially approved human gene therapy trial would begin in the United States.

The first federally approved gene transfer into humans in the United States came in 1989. The research involved the autologous transfer of gene-marked lymphocytes into five patients with terminal melanoma. The purpose of this research was to demonstrate safety.

A year later, in 1990, the first gene transfer experiment was approved. This research began with four-year-old Ashanthi DeSilva. DeSilva suffered from an adenosine deaminase (ADA) deficiency (a rare immune defect). She was the first of two children to be injected with her own blood cells that had been altered by a retroviral vector to contain functioning ADA genes. DeSilva and the other child research participant also received a new drug, PEG-ADA (a synthetic form of the ADA enzyme).

Around the same time, a similar trial, also involving two patients with ADA deficiency, was conducted by Italian researchers. In both cases the combined interventions proved successful, although the efficacy of the genetic intervention remains unclear since the children continue to receive PEG-ADA therapy.

The pace of gene transfer research picked up after 1990. As reported on the U.S. Center for Disease Control website, gene transfer clinical trials worldwide jumped from one in 1989 and two in 1990 to sixty-six by 1995. Meanwhile, there was growing concern about the hype surrounding gene therapy. In December of 1995, an NIH-appointed ad hoc committee reported that: "[W]hile the expectations and the promise of gene therapy are great, clinical efficacy has not been definitively demonstrated at this time in any gene therapy protocol" (Orkin and Motulsky 1995). In the same report, concerns were raised about the relationship between gene transfer researchers and industry.

A major setback for gene transfer research came at the end of the 1990s with the death of a small number of research participants in different gene therapy trials. The most widely publicized death was that of Jesse Gelsinger, an eighteen-year-old patient with a rare liver condition who was enrolled in a study at the University of Pennsylvania. In September 1999, Gelsinger received an injection of adenovirus vectors designed to carry corrected ornithine transcarbamylase (OTC) genes to his liver. Four days later, he died of multiple organ failure as a result of the experimental intervention (Raper, Chirmule, Lee, et al. 2003).

A few years later, there was yet another major setback. In October 2002, researchers in France with the first apparently unequivocal success in gene transfer research announced that one of the nine boys in their gene transfer trial for X-linked severe combined immunodeficiency disease (X-SCID) had developed a leukemia-like condition. Three months later, in December 2002, a second X-SCID child in the trial was also showing signs of a leukemia-like disease (Johnston and Baylis 2004).

As of early 2005, there had been no unqualified successes in human gene transfer research. Research continues on ways to improve gene control and targeting, effectively integrate DNA into the genome, limit the risk of stimulating an immune response, and avoid the problems with viral vectors that can result in inflammation and toxicity. Only when these problems are resolved will the promise of gene transfer research begin to be realized—assuming the research is deemed morally acceptable.

FRANÇOISE BAYLIS
JASON SCOTT ROBERT

SEE ALSO Genethics; Genetic Counseling; Genetic Research and Technology; Homosexuality Debate; Playing God.

BIBLIOGRAPHY

Anderson, W. French. (1989). "Human Gene Therapy: Why Draw a Line." Journal of Medicine and Philosophy 14: 682–693.

Johnston, Josephine, and Françoise Baylis. (2004). "What Happened to Gene Therapy? A Review of Recent Events." Clinical Researcher 4: 11–15.

Parens, Eric. (1995). "Should We Hold the (Germ) Line?" Journal of Law, Medicine, and Ethics 23: 173–176.

Raper, Steven E.; Chirmule, Naredra; Lee, Frank S.; et al. (2003). "Fatal Systemic Inflammatory Response Syndrome in a Ornithine Transcarbamylase Deficient Patient Following Adenoviral Gene Transfer." Molecular Genetics and Metabolism 80: 148–58.

Stock, Gregory, and John Campbell, eds. (2000). Engineering the Human Germline: An Exploration of the Science and Ethics of Altering the Genes We Pass to Our Children. New York: Oxford University Press.

Walters, LeRoy, and Julie Gage Palmer. (1997). The Ethics of Human Gene Therapy. New York: Oxford University Press.

INTERNET RESOURCES

National Reference Center for Bioethics Literature (NRCBL). (2002). "Scope Note 24: Human Gene Therapy." Available from http://www.georgetown.edu/research/nrcbl/publications/scopenotes/.

Orkin, Stuart H., and Arno G. Motulsky. (1995). "Report and Recommendations of the Panel to Assess the NIH Investment in Research on Gene Therapy." National Institutes of Health, 1995. Available from www.nih.gov/news/panelrep.html.

GENETHICS

• • •

The term genethics first appeared in the literature with the publication of a book of the same title by David Suzuki and Peter Knudtson (1989), a volume that dealt with the moral guidelines for genetic research and engineering. In a second book of the same title, David Heyd (1992) extended the definition to the field that focuses on the mortality of creating people—that is, decisions having to do with people's existence, number, and identity. Since then, the term has spawned several other books (Bayertz 1995, Burley and Harris 2002), a number of periodicals including GenEthics News, and numerous web sites, many of which are no longer active.

Is Genethics Necessary?

There has been some debate over whether the introduction of the term is advisable. While Suzuki and Knudtson and others were arguing for a genethics to deal with the problems raised by the new genetics, John Maddox in a 1993 *Nature* article played down the notion that the sequencing of the genome and related developments in molecular biology created ethical problems that are intrinsically unique. For Maddox, "this new knowledge has not created novel ethical problems, only ethical simplifications" (1993, p. 97). Darryl Macer (1993), in a follow-up letter, agreed that there is no inherent value clash between genetics and human values as Suzuki and Knudtson had proposed. Macer argued that the concept of genethics "should be stopped" and that what is needed instead is "a revival and renewed discussion of ethical values as society interacts with technology, and reassurance that scientists are responsible" (1993, p. 102). Society does not need a new ethics to cope with the impact of genetic technology.

Despite these objections, the term genethics is still in use and its development has received impetus from the Human Genome Project (HGP), the multi-billion dollar public-private, international initiative to map out the entire human genome begun in the 1990s and completed in 2000. Genethics was particularly fostered through the establishment of the Ethical, Legal and Social Implications (ELSI) program, under which the U.S. Department of Energy and the National Institutes of Health devoted 3 to 5 percent of the annual HGP budget toward examining such issues in relation to the availability of genetic information flowing from HGP. Specific areas of funding included the fair use of genetic information, privacy and confidentiality, stigmatization, conceptual and philosophical implications, and clinical and reproductive issues. Through this significant investment, ELSI became the largest bioethics program in the world and spawned similar endeavors elsewhere, often under the genethics moniker.

Although in some quarters the term has become a catchword for ethical issues raised by human interventions only, it is generally used more broadly to encompass the full range of ethical issues raised by advances in the science and technology of genetics and genetic engineering. In this broader sense, genethics cuts across all areas of science and technology related to engineering of genes, from human research and applications, to genetic modification of crops and animals, to other biotechnological applications such as drugs and potential terrorist and warfare uses of this knowledge (Reiss and Straughan 1996; Burley and Harris 2002). It might also be tied to secondary consequences of genetic technology such as eugenics and the link between genes and human behavior alleged to exist in drug or alcohol addiction or violence. Because some specific applications are discussed in other entries, attention here will focus on the issues surrounding human applications, largely the product of the HGP.

Increasing Knowledge

Knowledge of human genetics has undergone an accelerating expansion in the last several decades in large part as a result of the HGP. This increased knowledge and the emerging capacity to apply it for diagnostic and therapeutic purposes promise benefits to individuals and to society as a whole, but they also carry risks. These promises and risks have attracted the interest of bioethicists and social scientists as well as leading researchers. The issues raised in genethics relate directly to the almost daily announcements of new findings in molecular biology and related scientific fields and the development innovative technological applications.

Genetic intervention is especially controversial because of rapid advances in knowledge and the shortened lead time between basic research discoveries and their application. It has been estimated that knowledge in molecular biology is doubling every year, and a cursory survey of journals and Internet sites suggest that, although the shortened lead time might be exaggerated by some observers on either side of the debate, there is a rapid diffusion of applications, giving society less and less time to access their impact. In addition to challenging basic values, human genetics for some persons raises the specter of eugenics and social control (Kevles 1985). References to a "brave new world" scenario, in which human reproduction is a sophisticated manufacturing process and a major instrument for social stability, are commonplace. The notions of designer or made-for-order babies accentuate concern over this apparent quest for the perfect child (McGee 1997). Human genetic engineering is often criticized as playing God or interfering with evolution. Not surprisingly, opposition to genetic and reproductive intervention in this context is frequently intense and pits opponents against the research community and some commercial interests.

Diagnostics and Therapy

A complicating factor is the selective nature of genetic diseases. The success of genetic screening efforts often depends on the ability to isolate high-risk groups. In targeting such groups, however, problems of stigmatization, due process, and invasion of privacy arise. For instance, the early experience with screening for sickle-cell anemia in the early 1970s led to perceived and real threats to the African-American community when they experienced discrimination based on their carrier status by employers, insurance companies, and even the Air Force Academy that denied admission to those identified as having the sickle-cell trait. As DNA tests are developed to identify individuals at heightened risk for alcoholism, personality disorders, aggressive and antisocial behavior, and so forth, the fear of eugenics is bound to reemerge, thus making any attempts to screen most controversial. In this case, however, the "eugenics" is most likely to flow from decisions by individual parents who use the techniques to maximize their children's characteristics, not a social program. Some fear that once the tests become accepted as legitimate by society, it is likely that legislatures and courts will promote professional standards of care that incorporate increasingly intrusive testing.

Following the development of techniques to diagnose genetic disorders are emerging capacities to provide gene therapy. These techniques would act to correct genetic defects by acting directly on the affected DNA and could be directed at either somatic or germ-line cells. This move from diagnostic to therapeutic ends accentuates sensitive issues concerning the role of government in encouraging or discouraging human genome research and applications. The huge financial investment of government in many human genome initiatives clearly demonstrates a commitment to genetic technology and eventually gene therapy. In turn, however, any developments in gene therapy will raise ethical questions concerning safety, parental responsibilities to children, societal perceptions of children, the distribution of social benefits, and definitions of what it means to be human.

Both diagnosis and therapy constitute expansions of genetic knowledge, which can pose ethical challenges both for social and personal use. Socially, there is the problem of discrimination in attitudes not only toward individuals with certain genetic diseases but also toward how individuals might handle such possible knowledge. Personally, some individuals might choose not to know, and it is not clear that this would always be as equally acceptable as knowing.

Immediate genethics issues involved with this expanding genetic knowledge center on problems of discrimination and stigmatization. Genetic information of the type now promised is self-defining and can easily stigmatize individuals, thus enabling others to discriminate against them on the basis of such information. In fact, no information is potentially more invasive of personal privacy than tests that provide precise and inclusive knowledge of a person's genetic makeup. One issue that requires urgent attention concerns access to sensitive information collected through voluntary screening programs. Because such information is potentially embarrassing and humiliating, individuals must be protected from unauthorized disclosure. Even when confidentiality is assured, maintaining the security of genetic records will be difficult, though these are mostly questions of policy not ethics.

This problem is even more difficult, however, because there are circumstances that may warrant disclosure despite risks to patient privacy. Because genetic traits may be present in other family members, one question concerns the possible rights of these family members to any information relevant to their own well-being. Under what circumstances may a genetic counselor or physician disclose genetic information that might affect another family member or even future progeny? These issues of confidentiality and privacy, of course, are heightened significantly if mandatory genetic screening programs are instituted. Given technological developments, genetic tests are soon likely to be routine health indicators, only more precise and accurate than conventional ones. This will lead employers and insurance companies to screen potential employees or those applying for insurance for an array of genetic traits. At the same time, companies might want to include such tests in health promotion or preventive medicine programs with, for instance, persons identified as having a genetic proclivity toward hypertension placed into early diagnosis programs.

When, if ever, is an individual right to genetic privacy to be sacrificed to the interests of an employer? Under what circumstances does the responsibility of a genetic counselor or physician to society outweigh responsibility to the patient? As health care costs continue to escalate, employers will find it attractive to use genetic screening to exclude individuals who might cost them large sums of money in terms of future health bills. This is particularly critical if predictive tests are developed for general health status or for susceptibility to heart disease, cancer, diabetes, or alcoholism. Insurance companies, too, have a stake in data obtained through these methods. Genetic tests could be used either to determine insurability or to establish premium rates on the basis of test results. Life insurance companies traditionally have excluded people who are poor health risks and could easily extend this through tests that place certain individuals at risk for a wide range of conditions or diseases. Likewise, health insurers know that a large proportion of health care costs are attributable to a small proportion of the population, and as tests become available to identify individuals who are genetically predisposed to ill health, this is likely to put pressure on employers to screen prospective employees.

Confidentiality questions become more problematic when DNA or gene data banks are created where thousands of samples of blood, hair, or other tissue are stored for future use. The creation of such banks for criminal investigations elicits intense controversy. The issue is even more complex because unlike traditional fingerprints or other records (medical, credit, criminal) that are currently maintained, the DNA record contains potential as well as actual information. New genetic discoveries permit new information to be decoded from old samples. As science and technology advances, samples collected for a specific use could be used for totally unrelated purposes. Given the uncertainty of just how much and what type of data may be decoded from samples in the future, it is all but impossible to provide fully informed consent. Furthermore, questions remain as to who has proper access to this storehouse of knowledge on potentially millions of individuals.

Commercialization and Allocation of Resources

Although considerable public resources are being invested in human genome initiatives by governments, genetic tests and other applications will largely be influenced by commercial interests. Huge profits are likely to be made, especially as predictive tests for common disease categories are developed. Moreover, it is likely that DNA banking will include a significant entrepreneurial component in both the testing and data development components. Some observers argue that it is critical in light of ethical concerns over record-keeping, confidentiality, and so forth, that the emerging genetic industry be monitored closely and regulated where appropriate to guard sensitive data, control for the possibilities of error, and protect the economic and personal stakes involved.

Other issues inherent in the development of the new genetics involves decisions as to how these resources will be distributed and how high a priority they should be given in funding. Although resource allocation questions have not generally been at the center of genethics, they are becoming more critical because whereas resources are finite, demands and expectations fueled by new technologies have few bounds. While it is premature to speculate about the

relative costs and benefits of yet undeveloped procedures, it is logical to assume that gene therapy will be a complicated, costly procedure. Will access be equitable and coverage universal, and, if so, how will it be funded? Or, will it be yet another reproductive technology available to the affluent but largely denied to persons who lack sufficient resources?

Should these technologies be available to all persons on an equal basis? Maxwell J. Mehlman and Jeffrey R. Botkin (1998) make a persuasive case that access to the benefits of genome technologies is bound to be inequitable. The traditional market-oriented, third-party-payer system leaves out many people. The debate over whether or not the government has a responsibility to facilitate access will intensify as the scope of technological intervention possibilities broadens. What criteria should be used to determine who gets the benefits of the HGP, especially given that considerable research has been financed with public funds?

More broadly, what priority should the search for genetic knowledge and ever-expanding uses of this knowledge have vis-à-vis other strategies and health care areas? What benefits will it hold for the population as a whole, compared to other policy options? In recent decades there has been a proclivity to develop and widely diffuse expensive curative techniques without first critically assessing their overall contribution to health. Similarly, research has been rapidly transferred to the clinical setting, thus blurring the line between experimentation and therapy. In contrast, the availability of effective and inexpensive genetic tests could provide valuable information for disease prevention and health promotion by targeting individuals who are at heightened risk for diseases that could be reduced by early intervention. Therefore, to the extent it furthers preventive efforts, genetic technology could be cost-effective.

The Genethics Controversy

This brief discussion of genethics and the new scientific and technological environment of genetic knowledge and expanding capacities to apply it demonstrate the challenges facing all societies. The revolutionary nature of such developments and the far-reaching implications of how people view themselves and others, requires a reevaluation of how far human genetic intervention should proceed. Additional questions to be addressed more clearly by genethics concern the impact of each potential application of the HGP on society, on individual members, and on the way members of that society relate to each other. Here genethics has been criticized by some observers.

One criticism of genethics and genetic policymaking to date is that they have been largely reactive in scope, pointing out potential problems without assurance they will occur, but offering little in the way of anticipatory solutions in the event they do. Although national commissions or similar bodies have studied these issues and made recommendations in many countries, and the ELSI program has produced innumerable academic studies, most governments have chosen either to take an affirmative stance through funding genome research and encouraging diagnostic and therapeutic applications, or they have attempted to avoid the issues raised.

Another criticism of genethics is that it has been almost exclusively the domain of ethicists and journalists, who in some cases make little effort to communicate with the genetic science and research community and often take a combative stance on the issues (Maddox 1993). Not surprisingly, some in the genetic research community see genethics as an irritant at best and a hostile force against scientific and technological process at worst. In the process, the broader public is often sidelined. Although enlightened public debate over goals and priorities related to the issues raised here seems warranted, it can be argued that genethics has not gone beyond providing a framework for action by clarifying the ethical and moral issues surrounding the science and technology of genetics. While this might be a start, Bartha Maria Knoppers (2000) sees as discouraging the "general failure to develop and include the ethics of public interest, public health, and the notion of civic participation in genetic research for the welfare of the community or for the advancement of science" (p. s38). By focusing on the problems and issues raised by the new genetics, genethics might be overlooking a variety of potential societal benefits. The costs of avoiding admittedly risky technologies out of the fear of potential stigmatization, commodification, or other ethical problems for the individual, then, might be high if it means foreclosing benefits for individuals and society.

In summary, genethics is inextricably related to science and technology and is a product of rapid developments in molecular biology and related fields since the mid-twentieth century. Although one could widen the concept of genethics to include the study of eugenics pre–double helix, the term as applied today represents a direct response to molecular biology and the science and technology surrounding the

genome, and thus it is inextricably tied to and guided by it.

ROBERT H. BLANK

SEE ALSO *Bioethics; Biotech Ethics; Fetal Research; Gene Therapy; Genetic Counseling; Genetic Research and Technology; Health and Disease; Human Genome Organization; In Vitro Fertilization and Genetic Screening; Medical Ethics; Playing God; Privacy.*

BIBLIOGRAPHY

Bayertz, Kurt. (1995). *GenEthics.* Cambridge, UK: Cambridge University Press. Attempts to clarify the ethical dimensions generated by new human reproductive and genetic advancements. Most emphasis is on the reproductive assisting technologies.

Burley, Justine, and John Harris, eds. (2002). *A Companion to Genethics.* Oxford: Blackwell. This 600-page edited volume is a comprehensive look at the philosophical, ethical, social and political dimensions of developments in human genetics.

Heyd, David. (1992). *Genethics: Moral Issues in the Creation of People.* Berkeley: University of California Press. Heyd attempts to resolve many ethical paradoxes in intergenerational justice raised by advances in medicine, genetic engineering, and demographic forecasting.

Kevles, Daniel J. (1985). *In the Name of Eugenics: Genetics and the Use of Human Heredity.* Berkeley: University of California Press. Seminal work on history of eugenics and the eugenic implications of new reproductive technologies.

Knoppers, Bartha Maria. (2000). "From Medical Ethics to 'Genethics.'" *Lancet* 356(suppl. 1): s38. Short article argues for the inclusion of the ethics of public interest, public health and the notion of civic participation in genetic research for the welfare of the community.

Macer, Darryl. (1993). "No to 'Genethics.'" *Nature* 365(6442): 102. Letter in *Nature* that argues that concept of a separate "genethics" should be stopped because the ethical issues raised by the application of genetics are not novel.

Maddox, John. (1993). "New Genetics Means No Ethics." *Nature* 364(6433): 97. Contends that the opinion that genome sequencing will create novel ethical problems is mistaken and that these techniques are unlikely to be any more troublesome than genetic manipulation of bacteria decades ago.

McGee, Glenn. (1997). *The Perfect Baby: A Pragmatic Approach to Genetics.* New York: Rowman and Littlefield. McGee denies the necessity of a "genethics," arguing that the wisdom we need can be found in the everyday experience of parents.

Mehlman, Maxwell J., and Jeffrey R. Botkin. (1998). *Access to the Genome: The Challenge to Equality.* Washington, DC: Georgetown University Press. After summarizing the Human Genome Project, the authors discuss its practical health applications and ethical and policy challenges such as banning them, equal access, genetic handicapping, and genetic lotteries.

Reiss, Michael J., and Roger Straughan. (1996). *Improving Nature? The Science and Ethics of Genetic Engineering.* New York: Cambridge University Press. Covers a broad range of ethical and theological concerns inherent in genetic engineering of microorganisms, plants, animals, and humans.

Suzuki, David, and Peter Knudtson. (1989). *Genethics: The Clash between the New Genetics and Human Values.* Cambridge, MA: Harvard University Press. The authors propose a set of genetic principles that emphasize individual rights and confidentiality with regard to genetic screening, caution in violating boundaries across species, and a ban on biological weapon development and the genetic manipulation of human germ cells.

GENETICALLY MODIFIED FOODS

• • •

The production of genetically modified foods has provoked an ethical debate about whether it is right to use technology to create new forms of plant and animal life that otherwise would not exist. However, throughout human history agricultural crops have been genetically modified. There is nothing "natural" about food crops because most of them would be unable to propagate or survive without human intervention. What have changed over the years are the technologies that have been used to bring about genetic modification.

In general, humans have used three methods to modify plants genetically.

Conventional Breeding

At one time farmers practiced selective breeding and cross-breeding, or what is termed conventional breeding. Conventional breeding is less precise and predictable and therefore arguably less safe than genetic modification or, more correctly, *transgenic* plant breeding. The process has worked well because humans practicing conventional plant breeding have been able to increase yields in agriculture and support a larger population and/or improve human nutrition. The high-yielding dwarf varieties of wheat and rice that produced the Green Revolution were the result of conventional breeding.

Until the twentieth century most plant and animal breeding was largely a matter of selection and cross-breeding. Occasionally crosses between separate species were made as a result of human action or an unexplained "natural" happening. Wheat is a product of two or three different transpecies crosses of plants with different chromosomal structures.

In the 1920s advanced pollination techniques were used to create hybrid maize, a major but accepted genetic modification that far outyielded normal or "natural" maize. However, seed saved from hybrid maize for planting reverts to its original form and yields much less than the hybrid does. This means that a farmer has to buy new seed each year, but the increased yield normally makes that effort worthwhile. Hybrid maize has become the number one food crop in Africa.

Mutagenesis

The next method in this technological continuum involved the use of nuclear radiation or chemical mutagens to bring about mutations. This method is called *mutagenesis* and has the least-predictable outcome of all forms of plant breeding, but the technology is accepted and has escaped the label *genetic modification* presumably because these techniques have been used for more than half a century. The only advantage of the powerful and sometimes lethal genetic mutagens is that they produce a great many more mutations than occur naturally, thus generating the variability that breeders need for finding and introducing new characteristics into their plants. The Food and Agriculture Organization/International Atomic Energy Agency's Mutant Varieties Database Register (December 2000) lists over 2,252 crops in the more than seventy countries in which these mutant varieties are used. Key varieties are grown and/or eaten in virtually every country. Barley used in commercial beers around the world as well as wheats used to make pasta are products of radiation mutation breeding.

Genetic Engineering

With the discovery of the structure of DNA (deoxyribonucleic acid) in the 1950s, followed over the decades by a greater understanding of the process of inheritance, the way became clear for transgenic technology, or genetic engineering. This allowed desirable characteristics expressed by a gene or a small group of genes from any organism to be transferred to another organism. By the early 1980s the first genetically engineered pharmaceuticals were released, and they have been followed by an increasingly sophisticated array of new drugs. By the late 1980s transgenic enzymes and bacteria were involved in the production of cheese, bread, wine, beer, and vitamins that are consumed on a daily basis by numerous people.

Biotechnology is done under precisely controlled conditions in which a gene, together with a marker, is incorporated in plant tissue, which then is grown in tissue culture to produce plants. At this stage the plant is subject to initial evaluation to ensure that the gene has been transferred successfully and stably and produces the desired trait and that there are no unintended effects on plant growth or quality.

The gene transfer process is far more precise than the other accepted procedures and allows desirable plant transformations to be performed that are not possible using conventional breeding.

Benefits

Since their introduction in the mid-1990s transgenic crops engineered for herbicide tolerance, by expressing a protein that is fully digestible by humans and other animals, have brought about a decline in pesticide use, something critics of those crops have long claimed to favor. There have been enormous benefits from plants engineered to resist certain pesticides. Modern conservation tillage (or reduced-, minimum-, or no-tillage) agriculture using pesticides for weed and pest control conserves water, soil, and biodiversity better than does any current or previous form of tillage. In addition, this method saves fuel and therefore releases less carbon into the atmosphere. Conservation tillage is improving soil and soil quality. Planting with a drill, possibly disking the field, preserves soil structure and vegetative cover and the diversity of life therein, such as earthworms and other life forms that often are destroyed by deep plowing and other older forms of conventional agriculture. Conservation tillage has led to a reduction in overall pesticide use as a less toxic broad-spectrum pesticide is substituted for multiple sprayings of an array of targeted pesticides and herbicides.

Popular Fears of the Dangers of Frankenfoods

Genetic modification or engineering of crop plants has generated far more adverse reactions than did the informed guesswork that preceded it. Those products have been called *Frankenfoods*, a pejorative term for genetically modified foods that evokes the film version of Doctor Victor Frankenstein's monster from the novel by Mary Shelley (1797–1851). The fears are based on the extraordinary power of this new technology but concentrate principally on two issues: concern for human health and concern for the environment. Exhaustive tests have been carried out to determine whether genetically modified crops carry an increased risk of allergic reactions or other effects in people who eat them. There is no evidence so far that this or any other adverse reaction or nutritional problem has been caused in consumers of these crops after nearly ten years of production on more than 400 million acres of products consumed by more than 1 billion people.

Damage to the environment has been postulated to be a possible result of growing transgenic crops. Fears include the escape of genes into related wild plants, adverse effects of insect toxins (in the case of crops with the Bt gene) on desirable insects, and transfer of antibiotic resistance. Several factors lessen the likelihood of damage to the environment. Some crop plants and their wild relatives are self-pollinated, and so there is no opportunity for gene transfer to take place. Others have no wild relatives in the local flora, and so the local environment does not have suitable gene recipients. Transfer of antibiotic resistance from transgenic plants into the soil microflora is very unlikely and has not been demonstrated convincingly. Even if there were transfer, these genes already are ubiquitous in the soil microflora.

The most prominent public phobias in developed countries involve *chemicals* (a code word for industrially produced chemicals), which are all assumed to be carcinogenic; and radiation, which is assumed to cause cancer and mutations. One wonders why there has been no outcry about the use of *chemicals* and radiation in plant breeding, particularly in light of the fact that many critics of transgenics also oppose the irradiation of foods to kill microorganisms (a technique that has been used for more than forty years). Starting with a blank slate of public opinion on plant breeding, it would be far easier to frighten people about chemical and radiation breeding than about the insertion of a single gene plus a promoter and a marker. The promoter is simply a DNA sequence that allows the gene to be expressed, whereas current techniques require the use of marker genes.

Conclusion

The process and result of genetic modification have been subject to close scrutiny by some of the world's best scientists. The plants and the foods derived from them are extensively tested to assure consumers that these products are safe for the environment and for humans. In a joint report issued in July 2000 the National Academies of Brazil, China, India, Mexico, the United States, the United Kingdom, and the Third World Academy of Sciences concluded: "It is critical that the potential benefits of GM technology become available to developing countries." They also concluded that "steps must be taken to meet the urgent need for sustainable practices in world agriculture if the demands of an expanding world population are to be met without destroying the environment or natural resource base. In particular, GM technology coupled with important developments in other areas should be used to increase the production of main food staples, improve the efficiency of production, reduce the environmental impact of agriculture and provide access to food for small scale farmers" (Royal Society 2000).

THOMAS R. DEGREGORI

SEE ALSO *Agricultural Ethics; Biotech Ethics; Environmental Ethics; Food Science and Technology; International Relations; Nutrition and Science; Organic Foods*

BIBLIOGRAPHY

Centro Internacional de Mejoramiento de Maiz y Trigo. (2002). *Transgenic Maize in Mexico: Facts and Future Research Needs.* Mexico: International Maize and Wheat Improvement Center.

DeGregori, Thomas R. (2001). *Agriculture and Modern Technology.* Ames: Iowa State University Press.

Harten, A. M. van. (1998). *Mutation Breeding: Theory and Practical Applications.* New York: Cambridge University Press.

McHughen, Alan. (2000). *Pandora's Picnic Basket: The Potential and Hazards of Genetically Modified Foods.* New York: Oxford University Press.

Persley, Gabrielle J., ed. (1999). *Biotechnology for Developing-Country Agriculture: Problems and Opportunities.* Washington, DC: International Food Policy Research Institute.

Persley, Gabrielle J., and M. M. Lantin, eds. (2000). *Agricultural Biotechnology and the Poor.* Washington, DC: Consultitative Group on International Agricultural Research and the U.S. National Academy of Sciences.

Qaim, M., and D. Zilberman. (2003). "Yield Effects of Genetically Modified Crops in Developing Countries." *Science* 299(5608): 900–902.

Royal Society. (2000). *Transgenic Plants and World Agriculture: Report Prepared under the Auspices of the Royal Society of London, the U.S. National Academy of Sciences, the Brazilian Academy of Sciences, the Chinese Academy of Sciences, the Indian National Science Academy, the Mexican Academy of Sciences and the Third World Academy of Sciences.* Washington, DC: National Academy Press.

U.S. Food and Drug Administration. (2000). *Bt Corn: Less Insect Damage, Lower Mycotoxin Levels, Healthier Corn.* Washington, DC: Author, ARS News Service.

INTERNET RESOURCE

International Council for Science. (2003). *New Genetics, Food and Agriculture: Scientific Discoveries—Societal Dilemmas.* Paris: International Council for Science. ICS represents more than 100 science academies, including the U.S. National Academy of Science and the United Kingdom's Royal Society. This study draws together evidence from all leading reviews of genetically modified crops to see where the consensus is. Available from http://www.icsu.org/.

GENETIC COUNSELING
• • •

Genetic counseling is an educational service that aims to help people become informed and responsible consumers of genetic tests and to cope with the results. With nondirectiveness as a basic rule and autonomous decision making its goal, genetic counseling exemplifies a shift of the professional-client relationship from *doctor knows best* to *patient decides best*.

There is a widespread consensus in advanced scientific and technological societies that in order to guarantee a client's informed choice any genetic test, whether prenatal (by amniocentesis or chorion villus sampling) or adult (for example, for hereditary breast cancer), should be prepared for and followed by genetic counseling. Prior to testing, counselors determine a risk profile by examining a client's medical history and family tree for potential genetic risks. The risk profile determines an array of test options with their risks, potential results, and possible actions, all of which are discussed with the client. After genetic testing, a counselor explains the significance of the test result and reviews treatment options. For example, if a prenatal test result shows a fetal chromosomal aberration, the counselor describes the average development of the fetal population in which the unborn child is placed by its cytological anomaly and offers the possibility of terminating the pregnancy. Both before and after testing, the counselor emphasizes that any decision is the client's.

History

The first *hereditary counseling clinics* opened in Germany and Denmark in the 1930s, and in Britain and the United States in the 1940s. Their explicit goal was to improve the population gene pool by avoiding the birth of children probably affected by illnesses or handicaps. For geneticists of the time, all but a few sympathizing with eugenic ideas, giving *marriage advice* was an instrument for breeding a better society. After World War II, when Nazi Germany brought eugenics into public discredit, geneticists shifted their focus from public to individual prevention without losing track of its effects on the population's quality of health.

In order to differentiate individual decision making from state eugenic programs, the geneticist Sheldon Reed coined the term genetic counseling in 1947 (Reed 1947). Ahead of his time, Reed argued that clients should make their own decisions. Most of his colleagues, however, either told their clients what to do or assumed that, after having been enlightened about genetics, they would make the *right choice*. Before amniocentesis was introduced into prenatal care in the 1970s, there were not many options a geneticist could offer anyway: The counselor drew a pedigree and, on the basis of Mendel's laws and empirical data, established the recurrence risk for some disease in question. In cases where the risk was considered high, all the expert could do was advise clients to remain childless. Because people were not yet accustomed to consulting doctors about health problems that might, with some statistical probability, occur in the future, there was no great demand for this kind of expertise.

In 1975 the American Society of Human Genetics adopted a definition of genetic counseling that was purified of all traces of eugenics. The clients' informed decision superseded prevention as the primary goal of the procedure. Genetic counseling was redefined as a *communication process* (Ad-Hoc-Committee on Genetic Counseling 1975) with the aim of informing clients and leading them to a decision that would fit their goals and values. This definition was adopted internationally.

Genetic Counseling in the Early-Twenty-First Century

Demand changed dramatically when chromosomal tests of cells from amniotic fluid and the option of terminating pregnancy allowed geneticists to enter the field of prenatal care. By the end of the 1950s, researchers determined the normal number of human chromosomes and identified deviations such as Trisomie 21 (Down Syndrome). In the 1970s, amniocentesis was introduced into prenatal care and abortion laws liberalized in most Western countries.

Originally intended as special treatment for a defined fraction of pregnancies, namely those diagnosed as being *at risk*, within a few years the chromosomal checkup expanded into a routine procedure. As prenatal monitoring techniques such as ultrasound or maternal serum screening, designed to track down potential risks, became standard, increasing numbers of pregnant women were classified as at risk and in need of professional guidance. Thus the major and still increasing clientele of genetic counselors are pregnant women.

Apart from chromosomal checkups in prenatal care, genetic tests have only limited application in medical practice. Most of them test Mendelian hereditary diseases (such as cystic fibrosis or Huntington's chorea), which are relatively rare. In 1994 a test for *familial breast cancer* opened a new field of counseling activity: the offer to help people cope with test results that cast a shadow over their future. It is estimated that at most 5 to 10 percent of all breast cancer cases can be classified as hereditary. Those who possess a mutation in the BRCA-genes are told that they have a lifetime risk of about 80 percent of actually getting this particular cancer—though further research has provided evidence that these numbers are too high for a general penetrance estimate (Bregg 2002). As a result of human genome research, geneticists expect a growing number of such predictive genetic tests for widespread diseases such as different forms of cancer, coronary heart disease, or Alzheimer's disease.

Risk Communication as Social Technology

Genetic tests go beyond the scope of the traditional doctor-patient relationship because, strictly speaking, there is no medical indication for performing them. Most patients are eligible for testing because they are classified as being at risk. This means, for the most part, they are—and might remain—completely healthy. The test result does not provide a diagnosis in order to determine an appropriate treatment. Instead a positive gene test will leave them with *bad news* about their future without offering any cure. In the case of a prenatal test, the *patient* is not yet born, and the only *therapy* would be an abortion. Predictive tests, such as those for familial breast cancer, result in risk figures for a tomorrow that might never occur.

This heterogeneity between a medical diagnosis and the attribution of a risk profile is generally overlooked. Statistical probabilities express nothing but frequencies in statistical populations. But in the counseling session these numbers jell into risks and chances, indicating to clients a threat to them or to a coming child. Clients expect the counselor to say something relevant about them as individuals, while, by definition, risk measures the frequency with which something happens in the statistical universe from which the sample has been taken.

Because genetic counseling educates clients regarding genetically derived risk figures, it serves as a powerful social technology that individualizes social hazards. Members of various disability communities have criticized such testing as a way to extend prejudices toward those who have only some risk of becoming disabled rather than promoting compassion and social inclusion for those with special needs. Appealing to clients to make autonomous decisions, the counselor invites them to take responsibility for a future that can be statistically assessed but is as yet unknown, so that genetic counseling opens up a completely new possibility for victim blaming: No matter what a client decides, the client becomes responsible.

Professionalization and Ethics

Anticipating the evolving demand for professional guidance provoked by prenatal testing, a two-year masters program was started at Sarah Lawrence College in New York in 1969 to train genetic counselors as collaborators of medical geneticists in hospitals and clinics. Since then, genetic counseling as a profession has grown widely throughout North America and is largely populated by women. In 1979 the National Society of Genetic Counselors (NSGC) was founded. In 1992 NSGC launched its own journal (*Journal of Genetic Counseling*) and adopted a code of ethics (National Society of Genetic Counselors 1992). Genetic counselors have been certified by the American Board of Genetic Counseling since 1993.

In most European countries, genetic counseling is not yet fully professionalized. With the expansion of prenatal testing and, gradually, predictive genetic testing, countries such as the United Kingdom, Norway, and the Netherlands have followed the U.S. model and introduced masters programs for genetic counselors who are not medical doctors. In France and Germany, however, doctors blocked inroads into what they consider their own field of competence. In these countries, medical geneticists usually deal with special cases and predictive testing whereas prenatal diagnostics is left to obstetricians (Godard, Kääriäinen, Kristoffersson et al. 2003).

Genetic counselors insist on nondirectiveness as their basic principle. Originally a psychotherapy precept (Rogers 1951), nondirectiveness has become the cornerstone of a counseling concept that is based on patient autonomy. However there is no consensus about what this actually means in practice. The context and the conception of the encounter between genetic counselor and patient gives rise to different social and ethical conflicts, and so does the nature of the imparted information (Clarke 1994).

In general, an expert's information can cause misunderstandings fraught with consequences for the client. Technical terminology almost inevitably clashes with colloquial language. A term such as *syndrome* can evoke horrifying associations and, as a consequence, clients might expect a child to look monstrous (Chapple, Champion, and May 1997). In order to enable clients to make an informed choice, they are told about test options and their respective risks and benefits. In the case of prenatal testing, women are asked to weigh the probability of detecting a fetal chromosomal or genetic abnormality against the risk of inducing a miscarriage which is about 0.5 percent in case of amniocentesis and at least 1 percent in case of Chorion Villus Sampling. Nevertheless, those interventions are offered as a routine part of prenatal care regardless of women's age and family history, which means that on the long run there are more pregnancies lost than abnormalities detected. Scientific denotation and everyday connotations diverge grossly on the subject of risk figures. Clients inevitably personalize the numbers; they fail to grasp the statistical nature of probabilities and interpret them as personal threats (Rapp 1999, Samerski 2002). This gap between professional information and lay understanding widens with clients from different ethnic backgrounds (Rapp 1999, Browner et al. 2003).

According to their notion of autonomy, genetic counselors are bound to respect both the *right to know* and the *right not to know*. The right to be informed is generally taken for granted because knowing about probabilities, test options, and test results facilitates autonomous decision making. But genetic information may also profoundly change the client's perceptions and lifestyle, and therefore genetic counselors respect confidentiality and the *right not to know*, especially in cases of late-onset diseases (for example, Huntington's chorea) when there is no third party involved. In prenatal diagnostics, test results can only serve to provide grounds for terminating the pregnancy. Even though the moral status of abortion is controversial in most countries, it is generally legalized and socially accepted as pertaining to reproductive autonomy. The decision to abort or not after positive test results is the client's, even though the counselor's judgment might differ considerably. Yet, the options of testing and aborting put new pressures on women: Abnormal children are considered to be avoidable. A new sense of responsibility for the existence of a disabled child after having been offered a choice, fear of stigmatization, and the intimidating effect of professional diagnosis cause most women to terminate their pregnancies in case of abnormal test results. Out of respect for patient autonomy a growing number of genetic counselors would even recognize prenatal sex selection as an acceptable option (Wertz and Fletcher 1998). The call to limit prenatal selection to medically approved conditions is countered by members of the disability community who argue that just like the discrimination of women, "disability" is a social issue. The continuing efforts to track down a "gene for" homosexuality substantiate fears about a new, genetic discrimination of minorities (Schüklenk, Stein, Kerin, and Byne 1997).

There is a growing market of commercial laboratories promising to optimize health and well-being by genetic testing combined with "personalized" guidance on lifestyle, diet, and drugs. But consumer mentality is only one aspect of the seamy side of patient autonomy. The idea of *informed choice* seems to increase autonomy, but could force people to become managerial decision makers on their own behalf and on the behalf of their children. Genetic counseling burdens people with decisions on the basis of statistical probabilities, which makes them responsible for events they cannot control. Wrongful life actions, in which parents argue that the birth of their affected child was an avoidable consequence of misinformation or bad advice, reinforce the idea that misfortune can be avoided by correct information and decision making.

SILJA SAMERSKI

SEE ALSO *Fetal Research; Genethics; Homosexuality Debate; In Vitro Fertilization and Genetic Screening; Playing God.*

BIBLIOGRAPHY

Ad-Hoc-Committee on Genetic Counseling. (1975). "Genetic Counseling." *American Journal of Human Genetics* 27(2): 240–242. First internationally acknowledged definition of genetic counseling.

Bregg, Colin B. (2002). "On the Use of Familial Aggregation in Population-Based Case Probands for Calculating Penetrance." *Journal of the National Cancer Institute* 94: 1221–1226. Critically reviews the early estimates of the lifetime risk of breast cancer associated with BRCA1 and BRCA2 mutations.

Browner, Carole H.; Mabel H. Preloran; Maria Christina Casado; et al. (2003). "Genetic Counseling Gone Awry: Miscommunication between Prenatal Genetic Service Providers and Mexican-origin Clients." *Social Science and Medicine* 56: 1933–1946. Empirical study on misunderstandings by clients due to medical terminology, and non-directiveness, among other factors.

Chapple, Alison; Peter Champion; and Carl May. (1997). "Clinical Terminology: Anxiety and Confusion Amongst Families Undergoing Genetic Counseling." *Patient Education and Counseling* 32(1–2): 81–91. Empirical study that explores clients' understanding of medical terminology, especially the fears and associations evoked by the term syndrome.

Clarke, Angus, ed. (1994). *Genetic Counseling: Practices and Principles.* London, New York: Routledge. Anthology of historical, social, and ethical aspects of genetic counseling, including its eugenic predecessors and the problem of confidentiality.

Godard, Béatrice; Kääriäinen, Helena; Kristoffersson, Ulf; et al. (2003). "Provision of Genetic Services in Europe: Current Practices and Issues." *European Journal of Human Genetics* 11(Suppl. 2): S13–S48. Examines the professional and scientific views on the social, ethical and legal issues that impact on the provision of genetic services in Europe.

National Society of Genetic Counselors. (1992). "National Society of Genetic Counselors Code of Ethics." *Journal of Genetic Counseling* 1(1): 41–43.

Rapp, Rayna. (1999). *Testing Women, Testing the Fetus: The Social Impact of Amniocentesis in America.* New York: Routledge. Study based on a decade of fieldwork on prenatal diagnostics and genetic counseling and interviews with professionals and clients.

Reed, Sheldon. (1974). "A Short History of Genetic Counseling." *Social Biology* 21(4): 332–339. Reed looks back after twenty-seven years and describes the historical background and development of his counseling concept.

Rogers, Carl R. (1951). *Client-centered Therapy.* Boston: Houghton Mifflin. Discusses the concept of nondirectiveness and client-centered psychotherapy and the reconceptualization of genetic counseling as an aid to individual decision making.

Samerski, Silja. (2002). *Die verrechnete Hoffnung: Von der selbstbestimmten Entscheidung durch genetische Beratung* [The mathematization of hope: On autonomous decision-making through genetic counseling]. Münster, Germany: Westfälisches Dampfboot. Study based on participant observation and tape recordings of genetic counseling sessions in Germany. Analyzes how information about risk figures and test options and the demand to choose transforms pregnant women into managerial decision makers.

Schüklenk, Udo; Stein, Edward; Kerin, Jacinta; and Byne William. (1997). "The Ethics of Genetic Research on Sexual Orientation." *Hastings Center Report* 27 (4): 6–13.

Wertz, Dorothy C., and John C. Fletcher. (1998). "Ethical and Social Issues in Prenatal Sex Selection: A Survey of Geneticists in 37 Nations." *Social Science and Medicine* 46(2): 255. Discusses the trend towards prenatal sex selection.

GENETIC ETHICS

SEE *Genethics.*

GENETIC RESEARCH AND TECHNOLOGY

• • •

The early twenty-first century is an era of genetics. Genetic science, genetic technologies, genetically based diseases, animal and human cloning, and genetically modified organisms are regular visitors to the news and entertainment culture. Together with the revolution in information technologies, and sometimes going hand in hand, the biotech revolution promises to transform the world. The well-known successes of molecular biology in the 1950s and 1960s have transformed biology and especially genetics. But because from the very beginning genetics has been intimately involved with human values, the revolutionary changes of this science and technology have challenged moral reflection.

A brief historical review of the development of genetics research will help place such challenges in context. For present purposes this history may conveniently be divided into three periods. The first, and longest, period was one of protogenetics, in which human values played a dominant role. The second period saw the emergence of genetics as a science and its revolutionary research successes. During this period, the science aspired to a complete independence of any specific moral interests that were not directly entailed by the pursuit of scientific knowledge itself. Finally, the third period, although still trying to promote an ideal of value neutrality, may be characterized as making some efforts to bridge science and ethics.

Protogenetics: From Premoderns to the Eighteenth Century

Humans have long interacted with plants and animals, seeking to improve human life through their manipulation. Thus, before there was a formal science of genetics, humans developed tacit or implicit knowledge of how to genetically alter plants and animals for human use. Human needs and values guided these manipulations and search for knowledge. Plants and animals were selectively bred for their usefulness, and microorganisms were used to make food items such as beverages, cheese, and bread.

Early farmers noted that they could improve each succeeding harvest by using seeds from only the best plants of the current crop. They noticed that plants that gave the highest yield, stayed the healthiest during periods of drought or disease, or were easiest to harvest tended to produce future generations with these same characteristics. Through several years of careful seed selection, farmers could maintain and strengthen such desirable traits.

The ancient Greeks also gave careful attention to the heredity of humans. The accounts given were largely speculative, and many aimed at the continuation of noble lineages. Plato (428–347 B.C.E.) in *The Republic*

FIGURE 1

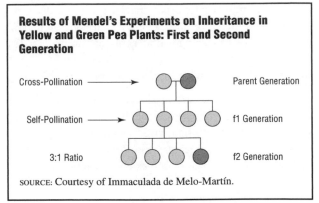

Results of Mendel's Experiments on Inheritance in Yellow and Green Pea Plants: First and Second Generation

Cross-Pollination ——→   Parent Generation

Self-Pollination ——→   f1 Generation

3:1 Ratio   f2 Generation

SOURCE: Courtesy of Immaculada de Melo-Martín.

proposed strict laws governing human reproduction in order to perfect and preserve an ideal state. He presented what is known as the "noble myth," according to which rulers were fashioned from gold, those who would occupy the middle rung in the state were fashioned from silver, and the farmers and artisans were fashioned with bronze. Such an ideology would explain to people that differences between them were in their very nature and needed to be preserved by laws governing procreation.

The fourth century B.C.E. also brought the theory of pangenesis, according to which, the reproductive material included atomic parts that originated in each part of the parental body. This theory was used to explain the transmission of traits from parents to children. Hippocrates (460–377 B.C.E.) also determined that the male contribution to a child's heredity is carried in the semen and argued that because children exhibit traits from both parents, there was a similar fluid in women. Aristotle (384–322 B.C.E.) rejected pangenesis, in part because traits often reappear after generations, which the theory could not explain. He argued that an individual's development was determined by internal nature, and that semen alone determined the baby's form; the mother merely provided the material from which the baby is made.

During Roman and medieval times in Europe, little was added to human understanding of reproduction and heredity. During the seventeenth century, a new conception of natural science began to develop. This new understanding of the scientific enterprise focused on experimental designs and empirical proofs. The belief that the natural sciences were completely value free and, therefore, the best means to understand the natural world began to take root. In this context, the development of the natural sciences brought a renewed attention to human reproduction and heredity. William Har-

FIGURE 2

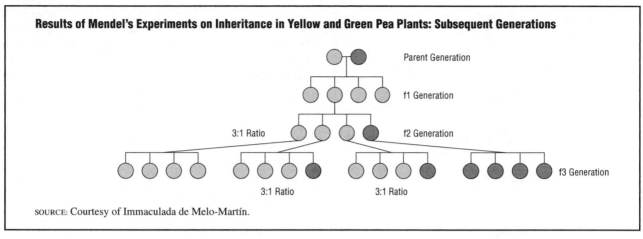

Results of Mendel's Experiments on Inheritance in Yellow and Green Pea Plants: Subsequent Generations

Parent Generation

f1 Generation

3:1 Ratio   f2 Generation

f3 Generation

3:1 Ratio   3:1 Ratio

SOURCE: Courtesy of Immaculada de Melo-Martín.

vey (1578–1657) concluded that plants and animals alike reproduced in a sexual manner and defended the idea of epigenesis, that the organs of the body were assembled and differentiated as produced. Opposing epigenesis, Marcello Malpighi (1628–1694) developed the idea of preformation, according to which new organisms are fully present and preformed within either the egg or the sperm. By the middle of the seventeenth century, however, the idea of preformation was called into question by a variety of scientists. Pierre-Louis Moreau de Maupertuis (1698–1759) rejected preformationism by appealing to observations about the blending of traits. Also, the development of a theory of the cell by Kasper Friedrich Wolff (1734–1794) further supported epigenesis.

The Rise of Modern Genetics: From Mendel to Watson and Crick

The late eighteenth century and the beginning of the nineteenth century in Europe saw the advent of vaccinations, crop rotation involving leguminous crops, and animal-drawn machinery. The growth of modern science and of scientific technologies further contributed to the idea that science should be pursued for its own sake.

MENDELIAN GENETICS. Throughout this period, a number of hypotheses were proposed to explain heredity. The one that would prove most successful was developed by Austrian monk Gregor Johann Mendel (1822–1884). (The part of Austria where Mendel was born and lived is now located in the Czech Republic.) Through a variety of experiments, Mendel realized that certain traits showed up in offspring plants without any blending or mixing of the parent's characteristics. The

traits were not intermediate between those of different parents. This observation was important because it contested the leading theory in biology at the time. Most of the scientists in the nineteenth century, including Charles Robert Darwin (1809–1882), believed that inherited traits blended from generation to generation.

Mendel used common garden pea plants for his research because they could be grown easily in large numbers and their reproduction easily manipulated. Pea plants have both male and female reproductive organs. As a result, they can either self-pollinate or cross-pollinate with another plant. In cross-pollinating plants that produce either yellow or green peas exclusively, Mendel found that the first offspring generation (f1) always had yellow peas. However, the following generation (f2) consistently had a 3:1 ratio of yellow to green (See Figure 1).

This 3:1 ratio occurred in subsequent generations as well. Mendel thus thought that this was the key to understanding the basic mechanisms of inheritance (See Figure 2). He came to four important conclusions from these experimental results:

- that the inheritance of each trait was determined by "units" or "factors" that were passed on to descendents unchanged (now called "genes");

- that an individual inherited one such unit from each parent for each trait (the principle of segregation);

- that a trait might not show up in an individual, but could still be passed on to the next generation;

- that the inheritance of one trait from a particular parent could be independent of inheriting other traits from that same parent (the principle of independent assortment).

FIGURE 3

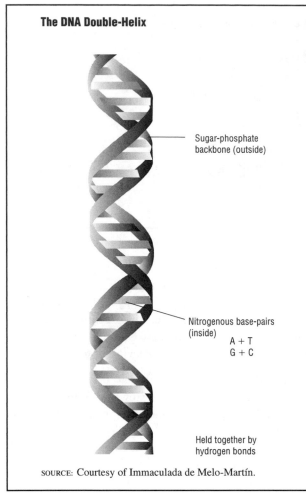

The DNA Double-Helix

Sugar-phosphate
backbone (outside)

Nitrogenous base-pairs
(inside)
A + T
G + C

Held together by
hydrogen bonds

SOURCE: Courtesy of Immaculada de Melo-Martín.

Mendel's ideas were published in 1866. However, they remained unnoticed until 1900, when Hugo Marie de Vries (1838–1945), Erich Von Tschermak-Seysenegg (1871–1962), and Karl Erich Correns (1864–1933) independently published research corroborating Mendel's mechanism of heredity.

POST-MENDEL DEVELOPMENTS. By the late 1800s, the invention of better microscopes allowed biologists to describe specific events of cell division and sexual reproduction. August Friedrich Leopold Weismann (1834–1914), who coined the term "germ-plasm," asserted that the male and female parent each contributed equally to the heredity of the offspring and that sexual reproduction generated new combinations of hereditary factors. He also argued that the chromosomes were the bearers of heredity. Edouard van Beneden (1846–1910) discovered that each species has a fixed number of chromosomes. He later discovered the formation of haploid cells during cell division of sperm and ova.

The publication of Darwin's *The Origin of Species* (1859), together with an incomplete understanding of human heredity, were used as grounds to support the idea of carefully controlling human reproduction to perfect the species. In 1883, Sir Francis Galton (1822–1911) coined the term *eugenics* to refer to the science of improving the human condition through "judicious matings." In the twentieth century, eugenics would be used to justify forced sterilization programs and immigration restrictions in the United States, and human experimentation in Nazi Germany.

After 1900, the pace of advance in genetic science and technology was rapid. During the first decade, William Bateson (1861–1926) coined the terms *genetics*, *allelomorph* (later *allele*), *homozygote*, and *heterozygote*. The cellular and chromosomal basis of heredity (*cytogenetics*) was identified by Theodor Heinrich Boveri (1862–1915) and others. And Sir Archibald Edward Garrod (1837–1936) developed the subspecialty of biochemical genetics by showing that certain human diseases were inborn errors of metabolism, inherited as Mendelian recessive characters.

During his investigations with the fruit fly Drosophila, Thomas Hunt Morgan (1866–1945) proposed that genes located on the same chromosome were linked together and could recombine by exchanging chromosome segments. Alfred Henry Sturtevant (1891–1970) drew the first genetic map, using cross-over frequencies between six sex-linked Drosophila genes to show their relative locations on the X chromosome. And in 1931, Harriet Creighton (1910–2004) and Barbara McClintock (1902–1992), and Curt Stern (1902–1981) working independently, found in cells under the microscope the first direct proof of crossing-over.

THE DISCOVERY OF DNA. In the 1940s, Oswald Theodore Avery (1877–1955), Colin Munro MacLeod (1909–1972), and Maclyn McCarty (1911–2005) offered evidence that DNA was the hereditary material. The challenge then was to determine the structure of this molecule. In 1953, James D. Watson (b. 1928) and Francis Crick (1916–2004) published in *Nature* the three-dimensional molecular structure of DNA, presenting what would be a breakthrough discovery in the biological sciences. They relied on the methods of Linus Pauling (1901–1994) for finding the helical structure in a complex protein and on unpublished x-ray crystallographic data obtained largely by Rosalind Elsie Franklin (1920–1958) and also by Maurice Wilkins (1916–2004). Watson and Crick determined that the DNA molecule was a double helix with phosphate backbones on the outside and the bases on the inside. They also determined that the strands were antiparallel and that there was a specific base pairing, adenine (A) with thymine (T), and guanine (G) with cytosine (C) (see Figure 3).

It is difficult to overstate the importance of the discovery of the structure of the DNA molecule. It has not only revolutionized the field of biology, but has become a cultural icon. The metaphor of the DNA as the "blueprint" of life has become engrained in much talk about human traits, diseases, and development. And with it, the ideology of genetic determinism, the idea that genes alone determine human traits and behaviors, has gained strength, despite the fact that practically every geneticist alive has disavowed it. Indeed, psychologist Susan Oyama has argued that genetic determinism is inherent in the way that what genes do is represented, because they have been given a privileged causal status. To describe and think about DNA in any way other than through this problematic representation of the power of DNA, is ever more difficult.

The Challenge of Genetic Knowledge and Power

The Watson-Crick model of DNA resulted in remarkable theoretical and technological achievements during the next decades. The genetic code was deciphered, the cellular components as well as the biochemical pathways involved in DNA replication, translation, and protein synthesis were carefully described, and the enzymes responsible for catalyzing these processes were isolated.

DNA RESEARCH. A striking result of these theoretical advances was the newly found ability to use a variety of techniques that would allow researchers to control and manipulate DNA. The discovery of restriction enzymes was one of the most important steps in this ability to manipulate DNA material. These enzymes are bacterial proteins that can recognize and cleave specific DNA sequences. They act as a kind of immune system, protecting the cell from the invasion of foreign DNA by acting as chemical knives or scissors. The capacity to cut DNA into distinct fragments was a revolutionary advance. For the first time, scientists could segment the DNA that composed a genome into fragments that were small enough to handle. Human chromosomes range in size from 50 million to 250 million base pairs, and thus are very difficult to work with. Additionally, methods for synthesizing DNA and for using messenger RNA to make DNA copies provided reliable means for obtaining DNA.

Moreover, they now had the opportunity to separate an organism's genes, remove its DNA, rearrange the cut pieces, or add sections from other parts of the DNA or from other organisms. The use of plasmids, extra-chromosomal genetic elements found in a variety of bacterial species, and of bacterial viruses as vectors or vehicles to introduce foreign DNA material into living cells served as a major tool in genetic engineering. Once introduced into the nucleus, the foreign DNA material is inserted, usually at a random site, into the organism's chromosomes by intracellular enzymes. In some rare occasions, however, a foreign DNA molecule carrying a mutated gene is able to replace one of the two copies of the organism's normal gene. These rare events can be used to alter or inactivate genes of interest. This process can be done with stem cells, which will eventually give rise to a new organism with a defective or missing gene, or with somatic cells in order to compensate for a nonfunctioning gene.

No less important for the ability to understand and manipulate genetic material were the development of techniques to sequence DNA, the establishment of the methodology for gene cloning, and the development of the polymerase chain reaction (PCR). With these techniques it was possible to obtain and analyze unlimited amounts of DNA and RNA within a short period of time. Additionally, PCR would prove an invaluable method to identify mutations associated with genetic disease, to detect the presence of unwanted genetic material (for example in cases of bacterial or viral infection), and to use in forensic science.

Researchers working on organisms such as worms developed technologies that allow mapping of their genomes. These mapping techniques permitted the location of the positions of known landmarks throughout the organism's chromosomes. Furthermore, as these molecular techniques improved, their application to cancer studies became more and more common, leading to the discovery of viruses that were able to transform normal cells into cancer cells, the description of oncogenes, cancer suppressor genes, and a variety of other molecules and biochemical pathways involved in the development of cancer.

HUMAN GENOME PROJECT. This new venture traces its origins back to Los Alamos national laboratory and the Manhattan Project. After the atomic bomb was developed and used in Hiroshima and Nagasaki, the U.S. Congress charged the Atomic Energy Commission and the Energy Research and Development Administration, the predecessors of the U.S. Department of Energy (DOE) with studying and evaluating genome damage and repair as well as the consequences of genetic mutations. There was a special interest in focusing the research on genetic damages caused by radiation and chemical by-products of energy production. From this research developed the plan to analyze the entire human genome.

The automation of DNA sequencing in the 1980s brought to the forefront of the scientific community the possibility of not just mapping the human genome, but also sequencing it. Thus, while gene mapping allowed researchers to determine the relative position of genes on a DNA molecule and the distance between them, sequencing let them identify one by one the order of bases along each chromosome.

It was in this context that discussions began about launching a human genome program. During a series of informal meetings, researchers and government officials attempted to assess the feasibility of different aspects of a project to map and sequence the entire three billion bases of the human genome. Although the majority of scientific opinion by the end of the 1980s was that sequencing the entire human genome was feasible, not all researchers were persuaded that such a project was a good idea. Many of them saw it as a massive work in data gathering rather than important research. Many scientists were also worried that the potential huge costs of such a project would diminish the funds dedicated to basic biological research.

In spite of the concerns, in 1990 the Human Genome Project (HGP) was formally launched as a fifteen-year plan coordinated by the U.S. Department of Energy and the National Institutes of Health. James Watson had been asked to head the project and did so until 1992. He resigned then because of his opposition to the patenting of human gene sequences. Francis Collins, who in 2005 is still the director of the National Human Genome Research Institute (NHGRI), replaced him in 1993. The main goals of the project were to identify all the genes in human DNA and to determine the sequences of its three billion chemical base pairs. Other important objectives of this international project were to improve the existent tools for data analysis and store the information obtained about the genome in databases.

The main focus was the human genome. However, important resources were also devoted to sequencing the entire genomes of other organisms, often called "model organisms" and used extensively in biological research, such as mice, fruit flies, and flatworms. The idea was that such efforts would be mutually supportive because most organisms have many similar genes with like functions. Hence, the identification of the sequence or function of a gene in a model organism had the potential to explain a homologous gene in human beings, or in one of the other model organisms.

The International Human Genome Sequencing Consortium published the first draft of the human genome in the journal *Nature* in February 2001, with about 90 percent of the sequence of the entire genome's three billion base pairs completed. Simultaneously the journal *Science* published the human sequence generated by Celera Genomics Corporation headed by Craig Venter.

Although the original expected conclusion date for the project was 2005, in April 2003, coinciding with the fiftieth anniversary of the discovery of the DNA double helix, the full sequence was published in special issues of *Nature* and *Science*. The early conclusion of the program was the result of a strong competition between the public program and the private one directed by Venter. His announcement in 1998 that his company would be able to sequence the entire human genome in just three years, forced the leaders of the public program to increase the pace, so as to not be left behind. The involvement of private capital in a project of this magnitude was a major turning point in science policy because it called into question the common belief since World War II that only the federal government had sufficient resources to fund "big science."

In December 2003, the NHGRI announced the formation of the social and behavioral research branch. This new branch has as its purpose developing approaches to translating the discoveries from the completed human genome into interventions leading to health promotion and disease prevention. The launching of this new branch is evidence of the shift of the NHGRI from genome sequencing to behavioral genetics.

ETHICAL, LEGAL, AND SOCIAL ISSUES. Because of the well-known abuses of eugenics during the beginning decades of the twentieth century in the United States and then in Nazi Germany, there was an unprecedented decision to attend to the possible consequences of the research into the human genome. Thus a significant goal of the HGP was to support research on the ethical, legal, and social issues (ELSI) that might arise from the project. Funds were dedicated to the examination of issues raised by the integration of genetic technologies and information into health care and public health activities and to explore the interaction of genetic knowledge with a variety of philosophical, theological, and ethical perspectives. Similarly, part of the ELSI budget was dedicated to supporting research exploring how racial, ethnic, and socioeconomic factors affect the use, understanding, and interpretation of genetic information; the use of genetic services; and the development of public policy.

Of course, the HGP, and the scientific and technological advances that permitted it, are extremely significant because of the theoretical knowledge it has produced on how, for example, genes work and what their contribution to health and disease is. It is difficult, however, to clearly separate theory and practice in molecular genetics given that this science is very technique intensive. In any case, the research supported by the HGP is also noteworthy because it has grounded the development of a variety of what are now common biotechnologies. Hence, genetic tests and screening for several human diseases such as Tay Sachs, sickle cell anemia, Huntington's disease, and breast cancer are now part of medical practice. Agricultural products such as corn plants genetically modified to produce selective insecticides or tomatoes engineered to prevent expression of a protein involved in the process of repining are common in food markets. Animal cloning does not make the front page anymore. Genetic therapy and pharmacogenetics are more and more often presented as the new medical miracles. And, of course, discussions of genetic enhancement and the hopeful, or frightening, possibility of designer babies are regular features of the news and entertainment media.

Given the increased presence of biotechnologies in people's lives and the significance of the genetic sciences, it is not surprising then that both the so-called theoretical research on human genetics and the practical applications of such knowledge have raised heated debates about ethical, legal, and social implications. Consider, for example, the following issues that have emerged in discussions of medical and agricultural biotechnologies.

GENETIC INFORMATION. The increasing use of genetic knowledge and genetics technologies in medical practice has been a subject of concern, though to different degrees, for both those who support such use and those who are skeptical of its benefits. One of the topics that has attracted the most attention among bioethicists working on ELSI issues is related to the availability and possible abuse of genetic information. Hence, the availability of genetic information has opened discussions about privacy and confidentiality. Questions have arisen about whether medical practitioners have an obligation to inform the family members of a patient with a genetic disease, or whether such information should be available to insurers and employers, for example. The concern for the possibility of genetic discrimination has been such

that many states have proposed and passed legislation prohibiting insurers from discriminating on genetic grounds. Similarly, given past experiences with eugenics, there are good reasons to have some concern about the possible stigmatization of individuals due to their genetic makeup.

GENETIC DIAGNOSIS AND HUMAN RESPONSIBILITY. The use of genetic diagnosis for a variety of medical conditions has received no less attention. Concern about fair access to these technologies, the reliability and usefulness of the tests, the training of health care professionals, the psychological effects they might have on people, and the consequences for family relationships are common. Similarly, many of the tests being developed, and some of the ones already in use, point to genetic susceptibilities or test for complex conditions that are linked to multiple genes and gene-environment interactions. Thus, such tests provide information not of a present or even a future disease, but of an increased risk of suffering such a disease. In many cases, these tests reveal possibilities of disorders, such as Huntington's disease, for which no available treatments exist. Given these issues, concerns about regulation of these tests, whether they should be performed at all, or whether parents have a right or an obligation to test their children for late-onset diseases are certainly justifiable. Moreover, the use of genetic diagnosis techniques in reproductive decision-making can also have serious implications for reproductive rights, our view of human beings, the expectations people might impose on their offspring, and the way we might treat people with disabilities.

The emphasis on people's genetic makeup might also have implications for their ideas of human responsibility, views regarding control of behavior and health status, their notions of health and disease, and their conceptions of treating a disorder or enhancing a trait. Such emphasis also has consequences for the kind of public policies people support regarding education, health promotion, disease prevention, and environmental regulations.

AGRICULTURAL BIODIVERSITY. Discussions about agricultural biotechnologies focus not just on the effects that these technologies might have on human beings, but also the consequences for animals and the natural environment. Genetic recombination techniques are used to create genetically modified organisms (GMOs) and products. These technologies enable the alteration of the genetic makeup of living organisms such as animals, plants, or bacteria, by modifying some of their own genes or by introducing genes from other organisms. GM crops, for example, are now grown commercially or in field trials in more than forty countries and on six continents. Some of these crops, including soybeans, corn, cotton, and canola, are genetically engineered to be herbicide and insecticide-resistant. Other crops grown commercially or field-tested are a sweet potato resistant to a virus that could decimate most of the African harvest, rice with increased iron and vitamins, and a variety of plants able to survive weather extremes. Research is being conducted to create bananas that produce human vaccines against infectious diseases such as hepatitis, fish that mature more quickly, fruit and nut trees that yield years earlier, and plants that produce new plastics with distinctive properties. It is unclear at this point how many of this research lines will be successful.

Questions about whether genetically modified organisms and products are safe for humans, whether they might produce allergens or transfer antibiotic resistance, whether they are safe for the environment, whether there might be an unintended transfer of transgenes through cross-pollination, whether they might have unknown effects on other organisms or result in the loss of floral and faunal biodiversity, for instance, are at the forefront of these debates. But the use of these technologies has also raised concern about possible implications for people's conceptions of other animals and the environment, their views of agricultural production, and their relationships with natural objects. Thus, many have wondered whether the use of these techniques constitutes a violation of natural organisms' intrinsic value, whether humans are unjustifiably tampering with nature by mixing genes among species, or whether the use of animals exclusively for human purposes is immoral. Debates also have been sparked about access to these technologies and the effect that this might have on non-industrialized countries. Some have questioned whether the domination of world food production by a few companies might not be putting food production at risk, and poor farmers in poor countries at an increasing dependence on industrialized nations. Issues about the commercialization of these products through the use of patents, copyrights, and trade secrets are also relevant when analyzing the implications of these technologies. Thus, many have called attention to the accessibility of data and materials.

Assessment

It is important to point out that although the ELSI program of the HGP has certainly had a significant effect on the understanding and evaluation of the conse-quences of new genetic technologies, the prevalent idea that humans must pay attention exclusively to the consequences of scientific or technological advances might be a reason for concern. A focus on consequences reinforces the incorrect view that science and technology are value-neutral. Issues about scientific or technological advances are thus framed as questions related to the implementation of scientific knowledge or technological practices. Hence, under the presumption that such practices are not the problem, but the use that people make of them might be, an evaluation of the scientific practices themselves appears illegitimate. This prevents researchers from trying to analyze the values that might underlie the current focus on genes, or attempting to propose different value assumptions to guide scientific research. Moreover, the emphasis on consequences directs attention to analysis of means and away from an evaluation of ends. Thus, scientists are encouraged to evaluate whether a particular technology is good to solve certain problems, but cannot analyze the goals for which such a technique has been developed. Technical discussions of biotechnology that focus on impacts presuppose that these goals are unquestionable. Thus, attention must be paid to the fact that assessments of new technologies must require not only discussions of risks and benefits—that is, discussions of means—but also reflections about ends. Of course, these issues apply to a variety of bioethical problems and not just to ELSI work.

INMACULADA DE MELO-MARTÍN

SEE ALSO *Bioethics; Biotech Ethics; Fetal Research; Gene Therapy; Genethics; Genetic Counseling; Health and Disease; Human Genome Organization; In Vitro Fertilization and Genetic Screening; Medical Ethics; Playing God; Privacy.*

BIBLIOGRAPHY

Buchanan, Allen; Dan W. Brock; Norman Daniels; and Daniel Wikler. (2001). *From Chance to Choice: Genetics and Justice.* Cambridge, UK: Cambridge University Press.

Davis, Kevin. (2001). *Cracking the Genome: Inside the Race to Unlock Human DNA.* Baltimore: Johns Hopkins University Press.

Dunn, Leslie C. (1965). *A Short History of Genetics: The Development of Some of the Main Lines of Thought: 1864–1939.* New York: McGraw-Hill.

Kass, Leon. (2002). *Life, Liberty, and the Defense of Dignity: The Challenges for Bioethics.* San Francisco: Encounter Books.

Kevles, Daniel J., and Leroy Hood. (1992). *The Code of Codes: Scientific and Social Issues in the Human Genome Project.* Cambridge, MA: Harvard University Press.

Keller, Evelyn Fox. (2000). *The Century of the Gene*. Cambridge, MA: Harvard University Press.

Kristol, William, and Eric Cohen, eds. (2002). *The Future Is Now: America Confronts the New Genetics*. New York: Rowman and Littlefield.

Lee, Keekok. (2003). *Philosophy and Revolutions in Genetics: Deep Science and Deep Technology*. New York: Palgrave MacMillan.

Mahowald, Mary Briody. (2000). *Genes, Women, Equality*. New York: Oxford University Press.

Mayr, Ernest. (1982). *The Growth of Biological Thought: Diversity, Evolution, and Inheritance*. Cambridge, MA: Belknap Press.

Nelkin, Dorothy, and Susan Lindee. (2004). *The DNA Mystique: The Gene as Cultural Icon*, 2nd edition. Ann Arbor: University of Michigan Press.

Oyama, Susan. (1985). *Ontogeny of Information: Developmental Systems and Evolution*. Cambridge, UK: Cambridge University Press.

Sherlock, Richard, and John D. Morrey, eds. (2002). *Ethical Issues in Biotechnology*. Lanham, MD: Rowman and Littlefield.

Stubbe, Hans. (1972). *History of Genetics: From Prehistoric Times to the Rediscovery of Mendel's Laws*, trans. Trevor R. W. Waters. Cambridge, MA: MIT Press.

Tudge, Colin. (2000). *The Impact of the Gene: From Mendel's Peas to Designer Babies*. New York: Hill and Wang.

Watson, James, and Andrew Berry. (2003). *DNA: The Secret of Life*. New York: Knopf.

Watson, James, and Francis Crick. (1953). "A Structure for Deoxyribose Nucleic Acid." *Nature* 171: 737–38.

Watson, James; Michael Gilman; Jan Witkowski; and Mark Zoller. (1992). *Recombinant DNA*, 2nd edition. New York: W.H. Freeman.

Wright, Susan. (1986). "Recombinant DNA Technology and Its Social Transformation 1972–1982." *Osiris* 2: 303–60.

INTERNET RESOURCES

U.S. National Human Genome Research Institute, National Institutes of Health. Available from http://www.genome.gov.

U.S. Department of Energy, Office of Science. "Human Genome Project Information." Available from http://www.ornl.gov/sci/techresources/Human_Genome/home.shtml.

GENETICS AND BEHAVIOR

• • •

Despite longstanding hostility to the biological explanation of human behavior, there are presently three general research programs aimed at the study of genetic influences on behavior: sociobiology and evolutionary psychology, behavioral genetics, and developmental psychobiology. Evolutionary psychology and its forebear, sociobiology, aim to discover species-typical traits that are adaptations (that is, traits that are in most cases the result of natural selection): Why do humans behave aggressively? What is the evolutionary source of altruism? Behavioral genetics aims primarily to uncover and disentangle genetic contributions (as distinct from environmental contributions) to individual differences in behavior: What are the predictors of aggressive versus nonaggressive behavior? Why does one person perform well on an IQ test, and another not? Developmental psychobiology aims to elucidate developmental pathways to particular behaviors: What is the mechanism by which organisms come to behave aggressively? What are the determinants of central nervous system development?

Such sample questions are by no means exhaustive; they are meant simply to illustrate the focal differences between these three approaches to genetics and behavior (see Table 1), the latter two of which will be the focus here. That is, rather than focus on how biological evolution as a whole has affected species-specific behaviors, the emphasis will be on how genetics can account for individual differences within species and on the more detailed pathways by which DNA causally influences human behavior.

Born of Controversy

Both behavior geneticists and developmental psychobiologists aim to move beyond the nature/nurture dichotomy, according to which traits are either genetically influenced (nature) or environmentally influenced (nurture), in favor of some collaborative interaction. What nature-*and*-nurture or nature-*via*-nurture actually means in practice is not always clear, however, because most scientists continue to partition correlational and causal influence in traditional terms (Schaffner 2001, Robert 2003).

The modern roots of the nature–nurture controversy are to be found in the writings of Francis Galton (1822–1911), a cousin of Charles Darwin, in the latter half of the nineteenth century. In 1869 Galton published *Hereditary Genius*, in which he attempted to discern what makes some humans geniuses and others exceptionally stupid. Based in part on anecdotal observations of twins, along with a questionnaire he administered to a small group of twins who were believed to be more similar in their youth than at the time of testing, Galton eventually concluded that "nature prevails enormously over nurture" in explaining variance in cogni-

TABLE 1

Three Approaches to the Study of Genetic Influences on Behavior

Problem domain	Explanatory focus	Content of explanations
Sociobiology and evolutionary psychology	Species-typical social and individual behaviors (adaptations)	Evolutionary vs. cultural, stochastic, or volitional explanations of species functional behaviors
Behavioral genetics	Individual differences, heritabilities	Genetic vs. environmental explanations of variability
Developmental psychobiology	Developmental pathways to phenotypic outcomes	Causal explanations of the role of DNA, other developmental resources, and environments (in evolutionary context)

SOURCE: Courtesy of Jason Scott Robert.

tive outcome (Galton 1875, p. 576). Galton later coined the term *eugenics* as part of a program to increase the number of so-called desirables in a population and to decrease the number of so-called undesirables (Kevles 1985). The "eugenics movement" has, of course, had its own very controversial history—including the rationalization of human rights violations in the United States, Nazi Germany, and other countries.

Since its modern incarnation, then, and however well-intended, behavior genetics has been associated with the justification of class-based and racial prejudice, exemplified more recently with the argument of Arthur R. Jensen (1969) that genetic differences between "races" influence the lower intelligence (or the poorer performance on IQ tests) of blacks as compared with whites. While most behavior geneticists have disowned this and related work, in 1995 the outgoing president of the Behavior Genetics Association (BGA), Glayde Whitney, celebrated Jensen's putatively brilliant and bold 1969 work in his presidential address. Whitney's speech was widely disparaged, and the editor of the BGA journal, *Behavior Genetics*, refused to publish it.

Classical Behavior Genetics

Three key concepts in classical genetics that referred originally not so much to behavioral but to anatomical characteristics that need to be clarified are genotype, phenotypes, and allele. The genotype is simply the genetic make-up of the organism, its complement of DNA. Genes, now known to be sections of chromosomes, manifest themselves as the organism's phenotype, its outward appearance. Any one gene may also come in different or alternative forms called alleles. For example, the founder of genetics, Gregor Mendel (1822–1884), in his research with pea plants, identified that one gene controls seed color, and the two forms of this gene give either green or yellow peas. That is, one allele (for yellow pea color) will be expressed as one phenotype (yellow peas), whereas another allele (for green pea color) will be expressed as another phenotype (green peas). One question for behavior genetics is whether and to what extent there are genotypes with different alleles that control for phenotypical behavior as well as physical characteristics.

The attempt to answer this question through the practice of observing twins continues to this day—though now with considerably more sophistication and computational power. Modern behavior geneticists establish correlations between genes and behavioral outcomes on the basis of two general types of study, involving classical or quantitative genetics (family, twin, and adoption studies) and molecular genetics and genomics (linkage, association, allele sharing, quantitative trait locus mapping, and DNA microarray studies). Although it is not necessary to know the complete meaning of the technical terms here, linkage and association refer to kinds of connections between genes, alleles (as already explained) are different forms of the same gene, trait locus mapping seeks to locate genes at specific points on a chromosome, and DNA microarray studies aim to show which genes are expressed at any given time. Classical studies are used to reveal the relationship between genetic variation and variation in phenotypic outcome.

Twin studies, for instance, are premised on the notion that, on average, identical (monozygotic, or MZ) twins share almost 100 percent of their genes in common, while fraternal (dizygotic, or DZ) twins share approximately 50 percent of their genes in common. A fundamental assumption is that both kinds of twins are affected by their rearing environments in a similar way, and that their "equal environments" cannot make MZ twins any more alike than DZ twins. On the basis of this assumption, behavior geneticists argue that what makes MZ twins more alike than DZ twins is that they are more genetically similar.

In any given population, *heritability* refers to the proportion of phenotypic (or apparent, expressed) variance that can be explained by genotypic (or hidden, genetic) variation, and is quantified as between zero (no variation explained by genetic inheritance) and one (all variation explained by inheritance). In humans, the heritability of having two legs is just about zero: Because almost all humans are born with two legs, there is very little phenotypic variance to be explained. By contrast, the heritability of eye color in a random human population approaches one, inasmuch as the variation in eye color can be explained almost exclusively by genetic variance. T the heritability of height is somewhere in between. Like physical characteristics, behaviors of interest to behavior geneticists have nonzero heritability (often in the range of 0.4 to 0.6), though it is often unclear what inferences are justified on the basis of a heritability estimate (Turkheimer 1998).

Behavior geneticists distinguish between traits that are either present or absent, and those that are continuously distributed. Where presence/absence is appropriate, scientists calculate *concordance rates*. Where the trait is continuous, scientists calculate *correlation coefficients*. So if MZ twins both exhibit some noncontinuous phenotypic outcome (say, depression), they are said to be "concordant" for that trait; and where the concordance rate for MZ twins is greater than that for DZ twins, the greater concordance is attributed to genes. Where MZ and DZ concordance rates are similar, this is attributed to shared environmental influences. And where MZ twins are discordant for a trait, this is attributed to nonshared environmental influences. In many cases, genes, shared environment, and nonshared environment are invoked to partially explain phenotypic differences (Baker 2004, Parens 2004), although nonshared environmental effects remain very difficult to discern (Turkheimer 2000).

Molecular Behavior Genetics

Classical studies can reveal associations between genetic variance and phenotypic variance, but do not identify the particular genes that may generate a trait. In the 1980s, behavior geneticists began to take advantage of emerging molecular techniques to attempt to identify specific genes. *Linkage studies* are employed to detect genes of major effect shared by a disproportionately large number of family members manifesting a condition or trait of interest. Successful linkage studies require three conditions to have been met: that a gene of major effect is implicated; that there is only one such gene segregating in a given family; and that the mode of inheri-

FIGURE 1

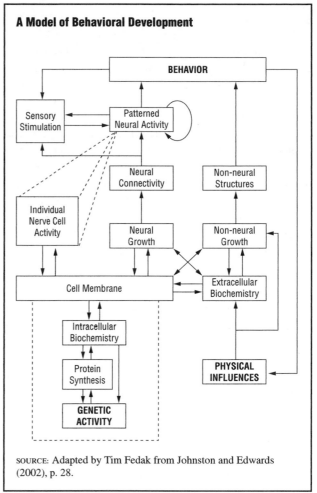

A Model of Behavioral Development

SOURCE: Adapted by Tim Fedak from Johnston and Edwards (2002), p. 28.

tance is known (Robert 2003). For most complex behaviors, at least one of these conditions is violated; for many complex behaviors, all three are violated.

Allelic association studies are employed to discern whether alleles or different forms of particular genes are transmitted preferentially to family members, or whether there are differences in the frequency of alleles between individuals and control populations. These studies avoid the requirement for a single gene of major effect; moreover, in the company of now-possible genome-wide scans, there is no need even to identify candidate genes or regions in order to turn up possible correlations. Further, success with these studies does not depend on knowing a specific mode of inheritance. But correlations are not causes, and allelic association studies risk turning up correlations that are causally spurious. For instance, where an allele is in linkage disequilibrium with another allele, allelic association studies will positively correlate both alleles with the phenotype, even if only one is actually involved in generating the phenotype.

Behavior geneticists are now using still more sophisticated techniques to reveal associations between genes and phenotypic outcomes. These include quantitative trait locus mapping and DNA microarray technology. Most phenotypes, especially of behaviors, are complex combinations of traits and thus governed by more than one gene. Quantitative trait locus mapping attempts to determine in quantitative terms what set of traits define a complex phenotype. DNA microarray technology, using what is variously called a biochip, DNA chip, or gene chip, allows for large-scale gene expression studies in order to identify interacting genes. Progress has nevertheless been slower than initially anticipated (e.g., Hamer 2002), and very few specific genes have been identified.

According to behavior geneticist Michael Rutter (2002), "knowing that a trait is genetically influenced ... is of zero use on its own in understanding causal mechanisms" (p. 4). Some developmental psychobiologists take this as evidence of the sterility of behavior genetics (e.g., Gottlieb 1995). If the focus of behavior genetics is on the establishment of correlations and other associations between inherited genes and particular behavioral outcomes, the focus of developmental psychobiology is on the identification of the developmental pathways that lead to those outcomes. Often, these pathways involve heritable elements, including genes; sometimes, other levels of analysis are more apt to yield developmental insights.

Developmental Psychobiology

Behavior geneticists do not study behavior as such, but rather differences in behavior. Moreover, behavior geneticists study associations between genetic variance, environmental variance, and interactions between the two, not causal relationships between developmental factors. By contrast, developmental psychobiologists seek to unpack genetic and other influences on complex behavioral phenotypes by elucidating causal mechanisms and pathways within the developing organism.

There is a long history of research in animal behavior (ethology) and comparative psychology, including experimental studies of animal behavior. Many historians begin with the work of Konrad Lorenz (1903–1989) on innateness. Lorenz's research was not entirely well-received among "English-speaking ethologists," as he called them, particularly Daniel S. Lehrman (1919–1972). Lehrman's criticisms of Lorenz (1953) continue

to inspire developmental psychobiologists (e.g., Johnston 1987, Lickliter 2000, Oyama 2000), while classical ethology has generally been dislodged by sociobiology and evolutionary psychology. (Developmental criticisms of the concept of innateness preceded the work of Lorenz; see, for instance, Kuo 1921.)

Experimental analyses of animal behavioral development have revealed aspects of development from conception through senescence, including factors, mechanisms, and causal interactions involved in central nervous system development. Coupled with results from brain science, developmental psychobiologists are shedding light on the pathways of neural, cognitive, and motor development in a wide range of animals, including those chosen as models for understanding human development. Nonetheless, developmental psychobiology has yet to yield a fully integrative account of behavioral development, in large part because of the complexity of the task.

Yet a framework for the integrative project is now in place. Timothy D. Johnston and Laura Edwards's series of increasingly specific (or "unpacked") representations of a model of the development of behavior are not intended to specify every molecular or cellular aspect of the complexity of development, but rather to provide "a useful intermediate level of detail that captures that complexity while at the same time rendering it reasonably comprehensible" and open to empirical investigation (Johnston and Edwards 2002, p. 31). Genes, neurons, and experience have indirect and reciprocal effects on the development of behavior, though their activity is mediated through multiple levels of biological, ecological, and social organization. The model is meant to focus investigative attention on developmental interactions and specific mechanisms, as depicted in Figure 1.

Any particular concrete use of this model would represent only a snapshot of a specific developmental moment. The model could also be transformed from two dimensions to three with the addition of information regarding the timing of individual influences on development, though this would obviously make it considerably less easy to represent graphically. This model of behavioral development can be used to organize existing knowledge and to make predictions about behavioral development that can be empirically investigated, yielding support for or requiring alteration of the underlying model.

In using this model of behavioral development in a research context, it is evident that scientists cannot do the kinds of studies with humans that would yield results of interest. There are limits on what is acceptable with human subjects. Accordingly, developmental psycho-biologists (like all developmental researchers) must infer from animal models, a process that is both conceptually and ethically fraught (Gottlieb and Lickliter 2004). Are the behaviors observed (or created) in animal models in fact homologous (or even analogous) to human behaviors? How does a passive–aggressive rat or an alcoholic monkey behave? What can be learned about human neural development from a fruit fly? These challenges beset any attempt to understand human behavioral development on the basis of studies with nonhuman animals.

Ethical and Social Considerations

While both behavior genetics and developmental psychobiology continue to provide important insights into the development of behavior, ethical concerns persist. These range from eugenic fears about the discovery of so-called gay genes and genes predisposing to antisocial behavior, to worries about the possible genetic enhancement of human cognitive function.

Following the mapping and sequencing of the human genome, a project that was sometimes viewed in exaggerated terms, there has been a shift to functional genomics, that is, attempts to determine what genes do and how they interact. Some hope that functional genomics will tell us not just how genes produce certain proteins but also how genes produce phenotypes, including behavior. But according to one policy commentary in *Science* magazine:

> The genetics of behavior offers more opportunity for media sensationalism than any other branch of current science. Frequent news reports claim that researchers have discovered the "gene for" such traits as aggression, intelligence, criminality, homosexuality, feminine intuition, and even bad luck. Rarely is it mentioned that traits involving behavior are likely to have a more complex genetic basis. This is probably because most journalists—in common with most educated laypeople (and some biologists)—tend to have a straightforward, single-gene view of genetics. (McGuffin et al. 2001, p. 1232)

Thus there is clearly a place for the lowering of expectations with regard to behavioral genetics.

More broadly, though, simply to study genetics and behavior by any means is to study what makes humans behaviorally different from one another. For many, any advances in this domain threaten to impinge, at least conceptually, on precisely what it is that distinguishes human from nonhuman nature. While these concerns may be ill-founded, behavioral scientists must take seriously the imperative to assuage these fears by promoting socially responsible public engagement with the science.

JASON SCOTT ROBERT

SEE ALSO *Bioethics; Genethics; Genetic Research and Technology.*

BIBLIOGRAPHY

Baker, Catherine. (2004). *Behavioral Genetics: An Introduction to How Genes and Environments Interact through Development to Shape Differences in Mood, Personality, and Intelligence.* Washington, DC: American Association for the Advancement of Science.

Galton, Francis. (1869). *Hereditary Genius: An Inquiry into Its Laws and Consequences.* London: Macmillan. A classic text that initiated behavioral genetics.

Galton, Francis. (1875). "The History of Twins, as a Criterion of the Relative Powers of Nature and Nurture." *Fraser's Magazine* 12(92): 566–576. Revised version published in *Journal of the Royal Anthropological Institute,* vol. 5, pp. 391–406.

Gottlieb, Gilbert. (1995). "Some Conceptual Deficiencies in 'Developmental' Behavior Genetics." *Human Development* 38(3): 131–141.

Gottlieb, Gilbert, and Robert Lickliter. (2004). "The Various Roles of Animal Models in Understanding Human Development." *Social Development* 13(2): 311–325.

Hamer, Dean. (2002). "Rethinking Behavior Genetics." *Science* 298(5591): 71–72.

Jensen, Arthur R. (1969). "How Much Can We Boost IQ and Scholastic Achievement?" *Harvard Educational Review* 39(1): 1–123.

Johnston, Timothy D. (1987). "The Persistence of Dichotomies in the Study of Behavioral Development." *Developmental Review* 7(2): 149–182.

Johnston, Timothy D., and Laura Edwards. (2002). "Genes, Interactions, and the Development of Behavior." *Psychological Review* 109(1): 26–34.

Kevles, Daniel J. (1985). *In the Name of Eugenics.* New York: Knopf.

Kuo, Zing-Yang. (1921). "Giving Up Instincts in Psychology." *Journal of Philosophy* 18(24): 645–664.

Lehrman, Daniel S. (1953). "A Critique of Konrad Lorenz's Theory of Instinctive Behavior." *Quarterly Review of Biology* 28(4): 337–363.

Lickliter, Robert. (2000). "An Ecological Approach to Behavioral Development: Insights from Comparative Psychology." *Ecological Psychology* 12(4): 319–334.

Lorenz, Konrad. (1957). "The Nature of Instinct." In *Instinctive Behavior: The Development of a Modern Concept,* trans. and ed. Claire H. Schiller. New York: International Universities Press. Originally published, 1937.

McGuffin, Peter; Brien Riley; and Robert Plomin. (2001). "Genomics and Behavior: Toward Behavioral Genomics." *Science* 291(5507): 1232–1233.

Oyama, Susan. (2000). *The Ontogeny of Information: Developmental Systems and Evolution*, 2nd edition. Durham, NC: Duke University Press.

Parens, Erik. (2004). "Genetic Differences and Human Identities: On Why Talking about Behavioral Genetics Is Important and Difficult." *Hastings Center Report*, special supplement, 34(1): S1–S36.

Robert, Jason Scott. (2003). "Developmental Systems and Animal Behaviour." *Biology and Philosophy* 18(3): 477–489.

Rutter, Michael. (2002). "Nature, Nurture, and Development: From Evangelism through Science toward Policy and Practice." *Child Development* 73(1): 1–21.

Schaffner, Kenneth F. (2001). "Nature and Nurture." *Current Opinion in Psychiatry* 14(5): 485–490.

Turkheimer, Eric. (1998). "Heritability and Biological Explanation." *Psychological Review* 105(4): 782–791.

Turkheimer, Eric. (2000). "Three Laws of Behavior Genetics and What They Mean." *Current Directions in Psychological Science* 9(5): 160–164.

Whitney, Glayde. (1995). "Ideology and Censorship in Behavior Genetics." *Mankind Quarterly* 35(4): 327–342.

GENETIC SCREENING

SEE *Genetic Research and Technology; In Vitro Fertilization.*

GENOCIDE

• • •

The word *genocide* is relatively new, even though the act of genocide is not. Yet in part because of its twentieth-century origins, genocide is often associated with the use of modern science and technology. The extent to which this is the case is one of the contentious ethical issues associated with the term.

Origins and Controversies

Polish jurist Raphael Lemkin introduced the term genocide in 1944 to describe the widespread killing of civilians that occurred during the first half of the twentieth century. He created the term as an amalgam of the Greek *genos*, meaning race or kind, and the Latin based suffix *-cide*, indicating killing (Smith 2002, Hinton 2002). At the time genocide was not a distinct crime, but Lemkin lobbied strongly to get it recognized as such.

The result was the 1948 United Nations Convention on the Prevention and Punishment of the Crime of Genocide which 136 countries have ratified. In the convention, genocide is defined as "any of the following acts committed with intent to destroy, in whole or in part, a national, ethnical, racial or religious group, as such: (a) Killing members of the group; (b) Causing serious bodily or mental harm to members of the group; (c) Deliberately inflicting on the group conditions of life calculated to bring about its physical destruction in whole or in part; (d) Imposing measures intended to prevent births within the group; (e) Forcibly transferring children of the group to another group."

As with any legal document, the UN definition of genocide has been scrutinized by scholars and politicians. The current definition, which limits genocide to ethnic, racial, religious, and national identity, describes human characteristics that are inherent to one's person. Race, ethnicity, and to a lesser extent, nationality and religion are determined at birth. Some critics argue that these criteria are too narrow in that they exclude particular social groups, such as political affiliation. Joseph Stalin slaughtered millions in the Soviet Union for largely political reasons, yet his actions do not constitute genocide under the UN definition. Indeed the Soviet Union lobbied the United Nations to remove any reference to political groups that had existed in an earlier draft.

The UN definition also excludes other social groups such as mentally ill or mentally challenged people, of which Nazi Germany exterminated tens of thousands. Homosexuals, bourgeoisie, the educated, and city-dwellers are all social classes that have been victims of genocidal acts although their deaths do not constitute genocide under existing law. Some scholars suggest expanding the definition of genocide to include mass killings in general (Gellately and Kiernan 2003). Others argue that it is beneficial to define mass killings and genocide separately so as to understand the origins of each and learn how to prevent them (Staub 2002). According to Helen Fein, one important component of the UN definition that sets genocide apart from other heinous acts, such as terror, war, oppression or torture, is "the perpetrators' sustained and purposeful attempt to destroy a collectivity" (Hinton 2002, p. 6).

Another phrase related to genocide is *ethnic cleansing*, and sometimes people conflate the two phrases. But as Paul Mojzes explains, "while every genocide is an ethnic cleansing, not every ethnic cleansing is a genocide. If an ethnic cleansing does not genuinely threaten the existence of a group, it would not qualify as genocide" (Mojzes 2002, p. 54). Genocide is also confused with *crimes against humanity*, which describes a "widespread or systematic attack directed against any civilian population" (Rome Statute of the International Criminal Court Internet site), including murder, torture, kidnap, rape, and forced expulsion. Another related phrase is *war crimes* which describes "grave breaches of the Geneva Conventions of 12 August 1949" such as willful killing, torture, unnecessary destruction of property, and denying prisoners of war the right to a fair trial, among others (Rome Statute of the International Criminal Court Internet site). Together genocide, crimes against humanity, and war crimes all fall under the jurisdiction of the International Criminal Court.

How society defines genocide is more than academic; it is a matter of life and death for millions of people. While the international community may respond with force to stop acts of genocide, it may not respond to ethnic cleansing and probably would not respond to mass killings. Thus it is important to understand the moral and ethical consequences of how genocide and related terms are defined, and to clarify the legal basis of controlling them.

Historical Developments

Historical records are rife with accounts of mass killings and genocidal acts perpetrated against tribes, cities, clans, and races in premodern times. The Romans, after defeating Carthage in the Third Punic War, killed the inhabitants, burned the city, and "sowed the ground with salt to symbolize that it should forevermore remain barren" (Alvarez 2001, p. 28). Greeks, Mongols, Christians, Assyrians, and others all committed such acts, yet at the time such killings were an accepted component of war and conquest and not considered a crime against humanity (Rittner et al. 2002). Although acts of genocide have been perpetrated throughout the ages, it was not until the twentieth century that society began to ask whether genocide was wrong. Two reasons explain this process: the rise of science and technology, which enabled acts of genocide on a massive scale, and the growing appreciation of human rights.

The twentieth century began as a century of promise and hope with an expectation that solutions to human problems could be solved through scientific and technological progress. Sadly the century ended as the deadliest in human history. While most persons commonly think of war as the major source of death, and primarily to young men, it was actually genocide that killed more people in the twentieth century than any other human activity, and most of the victims were civilians (Smith 2002). (Others would point out, of course, that more people also survived in the twentieth century.) Some experts place the number of state-sponsored

killings, which includes acts of genocide, at more than 150 million—four times higher than those killed in warfare (Fein 2002).

Science, technology, and the nation-state all contributed to the escalation and scale of genocide. First, the development of more efficient guns, bullets, and bombs enabled perpetrators to kill more people more rapidly. Gun-toting Germans easily slaughtered the Hereros of German Southwest Africa, a primitive culture, in one of the first acts of genocide in the twentieth century (1904–1905). Transportation, improved infrastructure, and bureaucracy enabled Nazi Germany to coordinate and carry out murder more effectively in its attempt to annihilate all Jews (1933–1945). Scientific and technological progress also created new methods of mass killing such as the development and proliferation of weapons of mass destruction (WMDs). Saddam Hussein was the first to use WMDs against his own people (1987–1988), killing thousands of Iraqi Kurds with poison gas. The nation-state, another product of modernity, was very successful at perpetrating genocide on scales that are almost unfathomable. An estimated 20 million civilians died under Stalin's regime in the Soviet Union (1922–1953), and millions more under Mao Zedong (1949–1959) in China and the Pol Pot (1975–1979) regime in Kampuchea. Indeed science, technology, and political institutions of modernity have combined to make genocide possible on a historically unique scale.

Despite efforts by the United Nations and international community to stop genocide, it has not been eliminated. Marginalized groups around the world are increasingly vulnerable, especially with the development of newer and more deadly WMDs. It may be possible in the not-so-distant future to design genetically engineered diseases or poisons that affect only a certain ethnic or racial group that share similar genes. Then again, genocide can also be extremely low-tech, as illustrated in the Rwandan massacres (1994) in which 800,000 Tutsis were slaughtered by machete-wielding Hutus.

ELIZABETH C. MCNIE

SEE ALSO *Holocaust; Human Rights; Race; Weapons of Mass Destruction.*

BIBLIOGRAPHY

Alvarez, Alex. (2001). *Governments, Citizens, and Genocide: A Comparative and Interdisciplinary Approach.* Bloomington: Indiana University Press.

Fein, Helen. (2002). "States of Genocide and Other States." In *Will Genocide Ever End?*, ed. Carol Rittner, John K. Roth, and James M. Smith. St. Paul, MN: Paragon House.

Gellately, Robert, and Ben Kiernan. (2003). "The Study of Mass Murder and Genocide." In *The Specter of Genocide: Mass Murder in Historical Perspective*, ed. Robert Gellately and Ben Kiernan. Cambridge, UK: Cambridge University Press. Provides a good overview on the subject of genocide.

Hinton, Alexander Laban. (2002). "The Dark Side of Modernity: Toward an Anthropology of Genocide." In *Annihilating Difference: The Anthropology of Genocide*, ed. Alexander Laban Hinton. Berkeley: University of California Press.

Mojzes, Paul. (2002). "Ethnic Cleansings." In *Will Genocide Ever End?*, ed. Carol Rittner, John K. Roth, and James M. Smith. St. Paul, MN: Paragon House. A good analysis of genocide vs. ethnic cleansing.

Rittner, Carol; John K. Roth; and James M. Smith, eds. (2002). "Chronology." In *Will Genocide Ever End?*, ed. Carol Rittner, John K. Roth, and James M. Smith. St. Paul, MN: Paragon House. A good place to begin learning about ethics and genocide.

Smith, Roger W. (2002). "As Old as History." In *Will Genocide Ever End?*, ed. Carol Rittner, John K. Roth, and James M. Smith. St. Paul, MN: Paragon House.

Staub, Earl. (2002). "Understanding Genocide." In *Will Genocide Ever End?*, ed. Carol Rittner, John K. Roth, and James M. Smith. St. Paul, MN: Paragon House.

INTERNET RESOURCES

Convention on the Prevention and Punishment of the Crime of Genocide. Office of the High Commissioner for Human Rights-United Nations Office at Geneva. Available from http://www.unhchr.ch/html/menu3/b/p_genoci.htm.

Rome Statute of the International Criminal Court. International Criminal Court. Available from http://www.un.org/law/icc/statute/99_corr/cstatute.htm.

GEOGRAPHIC INFORMATION SYSTEMS

• • •

Geographic information systems (GIS) are computer-based information systems that work with geographic or spatial data. The term GIS is also used to describe the whole discipline dealing with geographic information systems or geographic information science. A GIS can produce maps, but its unique attribute is the ability to integrate and analyze spatial data and related statistical or descriptive data. GIS has been described as "perhaps . . . the most significant event in spatial data handling since the invention of the map" (Pickles 1995, p. 49).

Like maps and any information production system, from writing to scientific research, GIS involves basic ethical questions of truthfulness, equity, and power. Maps are graphical depictions of the nature and spatial relationships of objects—they are generalized, simplified representations of reality. Cartographers strive to produce value-free, objective maps, but maps are also cultural and rhetorical texts imbued with social significance (Harley 2001). The symbols and projection used, the items included and excluded, and the graphic design of maps convey information, but they are also expressions of power that are made all the more effective by being hidden behind a "mask of seemingly objective science" (Harley 2001). The same elements characterize GIS, with the addition of a mask of technology. Cartographic historians and philosophers have developed methods for analyzing the social significance of maps—similar techniques are needed to analyze the statistical and graphic output of GIS.

GIS is founded on developments in computer science, geostatistics, and geography; as well as information from cognitive science, landscape architecture and planning, and many other fields. Roger Tomlinson conceived the architecture of GIS in 1963 and the first system, used to support Canadian national land-use planning, became fully operational in 1971. The U.S. Census Bureau adopted GIS for the 1970 census and was the first to digitize street maps efficiently. As computing power, datasets, and graphical interfaces have improved, GIS has become pervasive in both the public and private sectors.

GIS is not just technology—people are also critical components. Technology may constrain the capacity of a GIS, but the user's choice of data and analytical methods influences the output. Spatial data present complex analytical challenges: very different, but equally valid, results may be obtained by using different analytical methods on the same data. There is an ethical obligation on those using GIS to explain the meaning, limitations, and uncertainty embedded in the output of a GIS. Such explanations may also limit the legal liability of the producer for any subsequent use or interpretation of the information.

Much GIS work deals with the physical infrastructure of the planet and is generally uncontroversial. When GIS is used to examine socioeconomic data, however, its impact can be contentious. Presently there are so many geographic information systems holding large amounts of data, much of it related to individuals, that it is virtually impossible for people to know who holds information about them, the accuracy of that information, and the use to which it is being put. Laws that balance personal privacy with the potential commercial and administrative benefits of comprehensive databases are still being developed.

Approximately 80 percent of government data has a spatial component, so all levels of government are heavy users of GIS. Civilian use of GIS by the U.S. government is strictly regulated. Agencies must have a reason for collecting data. They must protect the privacy of individuals, provide people with access to data pertaining to them and an opportunity to make corrections, and make databases publicly available for the cost of dissemination without copyright restrictions. The U.S. government treats its non-secure databases as a public good; most other countries, and many U.S. state and local governments, regard data as a commodity that may be restricted and sold.

In the United States, GIS use in the private sector is much less regulated than in the federal sector. Marketing companies, realtors, insurance companies, credit-rating agencies, and many other organizations use GIS to assess risk, predict markets, and monitor social changes, among other activities. The private sector holds, and can provide, much of the information on individuals and national security sites that federal agencies go to great lengths to mask. Databases are weakly protected by copyright law in the United States; the European Union provides stronger protection for data compilations.

Military and intelligence use of GIS by governments is difficult to quantify but is known to be extensive. GIS could be described as a non-destructive weapon, a tool that is used to plan and execute actions, to identify targets, to organize infrastructure, and to detect suspicious patterns in individual and group behavior. Information is a global commodity, and many countries and groups monitor and analyze activities both inside and outside their borders. There are concerns that security databases may be used to compromise individual and group liberties.

GIS is not an objective technology, it is a tool used for many ends. Society has not yet found the mechanisms to guarantee the aspirations expressed in Article 1 of the French *Loi No 78-17 du 6 Janvier 1978* which states, "Computer science must be at the service of each citizen; its development has to operate within the framework of international co-operation; it should not damage human identity, human rights, private life or individual and public liberties" (Keane 1991, p. 134).

MAEVE A. BOLAND

SEE ALSO *Computer Ethics; Global Positioning System; Information; Military Ethics; Privacy.*

BIBLIOGRAPHY

Foresman, Timothy W., ed. (1998). *The History of Geographic Information Systems: Perspectives from the Pioneers.* Upper Saddle River, NJ: Prentice Hall PTR.

Harley, J.B., edited By Paul Laxton. (2001). *The New Nature of Maps: Essays in the History of Cartography.* Baltimore, MD: The Johns Hopkins University Press.

Keane, John. (1991). *The Media and Democracy.* Cambridge, UK: Polity Press.

Pickles, John, ed. (1995). *Ground Truth: The Social Implications of Geographic Information Systems.* New York: The Guilford Press.

GEORGIA BASIN FUTURES PROJECT

• • •

The Georgia Basin Futures Project (GBFP) is a five-year regional participatory integrated assessment whose purpose is to combine public values, preferences. and beliefs with expert knowledge in the production of scenarios for the future of the area in western Canada known as the Georgia Basin (see map) over the next forty years. The key goals are to increase public involvement in the discourse about issues of sustainability, explore pathways to sustainability in the region, and create a database of public preferences, values, and acceptable and unacceptable trade-offs that can be analyzed to provide a picture of how participants feel about sustainability issues and evaluate the relationship between the use of computer-based simulation tools and the beliefs, values, and behaviors of the users of those tools (Tansey, Carmichael et al. 2002).

Background

The GBFP is based on a long tradition of futures studies in the environmental field. From the extensive literature associated with this tradition four concerns have been identified that have influenced the project design significantly. The first is a concern with undertaking research that integrates natural and physical science analyses of environmental systems with social science, health science, and humanities research on the human systems that interact with the environment. The second is a focus on the future and on studying the various ways people can work collectively or individually toward bringing about a more sustainable world.

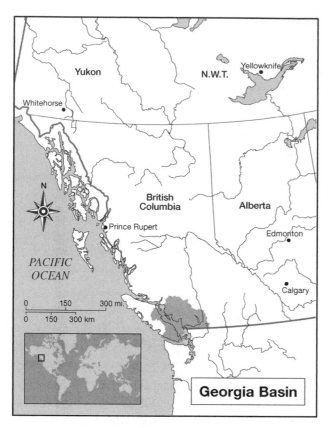

The population of the Georgia Basin is 2.9 million people and the GDP is about C$65 billion.

The third is a growing recognition of the need to involve various interests, or "stakeholders," in the research process. The fourth is a concern with the appropriate temporal and spatial scale of analysis. Although issues such as climate change are inherently global in scope, research that is truly problem-centered, policy-oriented, and connected to users must establish temporal and spatial scales that are relevant for decision makers.

All these strands came together in the development of the conceptual and methodological framework of the Georgia Basin Futures Project, which was funded by the Social Sciences and Humanities Research Council of Canada (SSHRC) in early 1999 and is supported by financial and in-kind contributions from governmental, nongovernmental, and industrial partners in the Georgia Basin region.

Project Design

Research in the project is organized into six major components undertaken by a core team of twenty coinvesti-gators and research collaborators, three research staff members, about thirty graduate students, and several administrative staff members working in conjunction with sixteen nongovernmental organizations, government, and private sector partners in the community.

Using expert analysis of key relationships among the social, ecological, and economic systems in the Georgia Basin and relying on initial consultations with stakeholders, the project has built a number of software tools for engaging stakeholders in sustainability issues. These tools have been used in several interactive processes, including workshops and classroom applications. The effect and effectiveness of this approach to engaging different publics with interactive software tools also are being evaluated.

Model Development: The QUEST Approach

The project's approach to modeling and scenario analysis is based on three key elements:

- A backcasting approach that involves the exploration of the feasibility and consequences of trying to reach desirable futures rather than the prediction of the most likely outcomes (Robinson 2003)

- A design approach to modeling that focuses on the physical flows of matter and energy through the economic system, the economic flows of currency through the economic sectors, and the economic benefits and costs incurred as a result of environmental and socioeconomic decisions (Gault 1987)

- An interactive social science approach to use of the model that requires that the local community be actively involved in both the design and the use of the modeling tool. (Caswill and Shove 2000)

The methodological core of the project is the development and use of the GB-QUEST modeling system (Rothman, Robinson, and Biggs 2002). QUEST is a computer-based system for scenario generation and evaluation that was designed to encourage public participation in thinking about sustainability in a regional context. Through QUEST users explore different scenarios for the future in terms of their social, economic, and environmental characteristics. The goal is to acquaint users with the complex realities of decision making, specifically the uncertainties involved, the necessary trade-offs, and the role of subjective values. For the GB-QUEST modeling system the geographic range encompasses the whole of the Canadian side of the Georgia Basin. The temporal scale is forty years.

Through the adoption of the "feel" and user-friendliness of a computer game, QUEST scenarios actively involve the user in their creation and evaluation. The user-selected scenario choices include choices involving the future patterns of population, economic activity, transportation, the density of urban growth, the style of neighborhoods, agricultural development, forestry practices, and consumption. The consequences of these decisions affect human well-being, environmental quality, economic and social health, and the long-term ability to maintain all these results.

QUEST does not provide a picture of the most likely future and is not intended to reflect a detailed understanding of all the complex systems involved. Instead, it enables users to learn about the linkages between choices and possible consequences and the trade-offs society faces in deciding among available options.

Community Engagement

A critical element of the project relates to the involvement of stakeholders and community partners in the research process. The project builds on the tradition of participatory integrated assessment modeling (Kasemir et al. 2000, van Asselt and Rijkens-Klomp 2002) and has adopted an interactive social science approach that is based on an explicit recognition of the value-laden nature of scientific analysis and modeling and the resultant need to incorporate community-based partners and the interested public directly into the research activities in two ways. First, by working with partner organizations in the community, the project has incorporated public values, preferences, and concerns into the process of model design and implementation. Second, through an elaborate process of community engagement that also involves the partners, the project has included the interested public in the generation of preferred sustainability scenarios using those modeling tools.

The key method for obtaining community engagement is the use of GB-QUEST in various ways, including three regional case studies; expert workshops; classroom use; a large exhibition space at Science World, a local science museum; and Web-based interaction.

The regional case studies involve working with three local municipal or regional governments in the Georgia Basin to use GB-QUEST in workshops with government staff members and stakeholders to explore regional sustainability scenarios, with the goal of contributing to the development of policies for sustainability. These workshops are followed by workshops to explore policy implementation issues, using a conceptual model of policy development that has emerged from the health promotion field. The expert workshops involve working with partner organizations and stakeholder groups to develop desired future scenarios and explore the implementation measures that would be required to realize those scenarios.

A teaching and learning team has tested GB-QUEST in the classroom at the high school level. This group is responsible for creating a set of curriculum guides and resource packages supporting QUEST that focus on sustainability in several classes and at different grade levels.

Since the fall of 2001 a twenty-minute-long video-based version of QUEST has been playing twice per day, five days per week at Science World. Approximately 15,000 people, mostly elementary school students, have played this version of QUEST, using interactive touch pads set into the seat arms of the 200-seat theater at Science World.

Based partly on funding from another project, a Web-based version of GB-QUEST is being developed that will incorporate information visualization and landscape visualization techniques to improve playability and comprehension of the complex contents and results of QUEST scenarios. A prototype was scheduled to be operation in April 2004.

The project also is studying the effect of playing QUEST on the mental models of sustainability, preferences, and behaviors of QUEST users. The GBFP culture and cognition team is holding impact workshops in which QUEST users are interviewed intensively and observed while playing QUEST.

Strategies

An important focus of the Georgia Basin Futures Project (GBFP) is the policy measures required to implement the scenarios that GB-QUEST generates. Both the case study and the expert workshops involve analysis of implementation requirements. In addition, GBFP is creating a database of all the scenarios developed in the project. That database, though limited in quantity, will present an informative picture of the values, preferences, and preferred options of QUEST users with regard to the future of the basin. The project will analyze those scenarios in terms of their policy and implementation requirements.

Other Tools

In addition to GB-QUEST, several interactive software tools have been developed in the GBFP, including the refinement of a personal Climate Change Calculator and the Sustainability Tools and Resources website for helping community groups and individuals establish themselves and interact with other groups. In addition, the GBFP has combined forces with a research group at Natural Resources Canada to develop a prototype of a Georgia Basin Digital Library (GBDL), a Web-based digital library that will be used to integrate natural and social science information (Geographic Information System maps, images, and text) into a comprehensive and interactive information resource to support sustainability research, community-focused decision making, and public consultation activities in the Georgia Basin.

Some Preliminary Results

While the GBFP is still ongoing, some preliminary findings are beginning to emerge. An immense interested has been demonstrated by participants from the general public and local government agencies in exploring desirable futures. Timeframes of forty years are no barrier to participation but the spatial scale of a region the size of the Georgia Basin (about 5.6 million hectares) is a challenge for participants who tend to want to focus on more local issues. In virtually all cases, however, participants are interested in exploring the nature of the choices and consequences of their future scenarios.

The use of interactive tools such as GB-QUEST was found to contribute to community activities to promote sustainability at the municipal scale in several communities. It has been less successful in contributing to the specific needs of regional government policy development. These findings suggest that a preferred audience for such engagement may be individuals and groups, including politicians, who do not have expert knowledge of specific sustainability issues. Classroom pilots of quest-based curriculum indicated a possible significant role for such techniques in school curricula.

Users of GB-QUEST are strongly disposed to make choices about preferred future conditions that reflect a strong environmental ethic. There is a desire to find scenarios that express those values without compromising other goals, such as economic growth or employment. The discussions that ensue explore issues that are not typically part of public and political debates in the region, suggesting a strong latent and unmet demand for such interactive processes.

Science, Technology, Ethics, and Public Policy

The GBFP exists at the interface of science and society. It is intended to combine expert knowledge and public

attitudes, preferences, and values in ways that incorporate the best understanding of complex ecological, social, and economic systems and that will be useful to stakeholder and institutions that are grappling with the practical problems of sustainability.

What distinguishes the GBFP approach is a fundamental commitment to interactivity that recognizes that the role of science in the policy process is inherently value-laden and that stakeholder input into both the development of integrated assessment tools and the development of scenarios is essential for two reasons. First, policy decisions about sustainability are inherently normative. The challenge is to combine those normative considerations with scientific understanding through the use of "boundary objects" such as QUEST and the GBDL. Second, it is clear that a major potential obstacle to achieving sustainability involves public acceptance. Politicians cannot make policy decisions that require significant change without a supportive political constituency. New means of engaging different publics in the complex public policy issues that surround sustainability are essential to build understanding of the policy trade-offs in the public and to learn what trade-offs and choices may be acceptable. In this way a process of community engagement that is appropriately designed may increase the sophistication of discussion about key choices affecting the sustainability of the region and help make explicit the points of conflict between stakeholders in the community that will affected by a decision.

An important question raised by the use of computer-based tools in the GBFP is the degree to which information technology can provide ways to engage large numbers of people in sustainability issues without trivializing the issues or misleading users about the consequences of particular choices. An important danger is the possibility of converting normative questions of deep moral and political significance into technical questions related to the choice of technology or behavior. For this reason the GBFP separates the analysis of the consequences of particular technological and behavioral choices (the realm of the scenario analysis using QUEST) from the discussion of the desirability of those outcomes and the means that may be required to realize them (a discussion that occurs outside the model). In this sense the role of the technology is to provide a basis for stimulating informed discussion of ethical and political questions.

The GBFP is based on the view that science and technology embed normative values that must be made explicit if informed choices are to be made (Jasanoff and

Wynne 1998). The project is testing the idea that complex public policy issues can be illuminated by the development and use of scenario analysis tools that allow citizens to express their views about their preferences and point out the consequences of their choices. The key is that these scenarios are created not by experts but by the users. This makes the process more engaging, creates a higher degree of user buy-in to the process and a greater sense of responsibility for the outcomes, can lead to significant learning, and produces results that embody ethical and moral judgments about the desirability and acceptability of alternative future scenarios.

JOHN B. ROBINSON
JAMES TANSEY

SEE ALSO *Models and Modeling; Participation; Science Policy; Stakeholders; Sustainability and Sustainable Development.*

BIBLIOGRAPHY

Caswill, Chris, and Elizabeth Shove (2000). "Introducing Interactive Social Science." *Science and Public Policy* 27(3): 154–157.

Gault, Fred, et al. (1987). "The Design Approach to Socio-Economic Modelling." *Futures* 19(1): 3–25.

Jasanoff, Shelia, and Brian Wynne (1998). "Science and Decisionmaking." In *Human Choice and Climate Change*, Vol. 1: *The Societal Framework*, ed. Steve Rayner and Elizabeth L. Malone. Columbus, OH: Battelle Press.

Kasemir, Berndt; Daniela Schibli; Susanne Stoll; and Carlo C. Jaeger. (2000). "Involving the Public in Climate and Energy Decisions." *Environmental Monitoring and Assessment* 42(3): 32–42.

Robinson, John (2003). "Future Subjunctive: Backcasting as Social Learning." *Futures* 35: 839–856.

Rothman, Dale; John Robinson; and Dave Biggs. (2002). "Signs of Life: Linking Indicators and Models in the Context of QUEST." In *Implementing Sustainable Development. Integrated Assessment and Participatory Decision-Making Processes*, ed. Hussein Abaza and Andrea Baranzini. Cheltenham, UK: Edward Elgar.

Tansey, James; Jeff Carmichael; Rob Van Wynesberghe; and John Robinson (2002). "The Future Is Not What It Used to Be: Participatory Integrated Assessment in the Georgia Basin." *Global Environmental Change* 12(2): 97–104.

Van Asselt, Marjolein, and Nicole Rijkens-Klomp (2002). "A Look in the Mirror: Reflection on Participation in Integrated Assessment from a Methodological Perspective." *Global Environmental Change* 12: 167–184.

GERMAN PERSPECTIVES

• • •

Contemporary discussions of science, technology, and ethics in Germany take place largely in the context of developments in the philosophy of technology. Although during much of the second half of the twentieth century philosophical discussion of technology was divided up into various schools and approaches, by the beginning of the twenty-first century such divisions were giving way to a new problem-orientated approach that emphasized the social, cultural, human, and ethical dimensions of the production and use of technoscientific knowledge. Reflections on technological development and transfer, for instance, became less ideological and more eclectic, pragmatic, and interdisciplinary than in the past. Nevertheless, discussions of ethics related to the hybridization of science and technology in such fields as information technology and genetic engineering continue to occur against a specifically German philosophical background. Thus the following notes on German approaches to science, technology, and ethics are themselves hybrid introductions to schools and problems, theory and practice.

Background: Gehlen and Heidegger

Arnold Gehlen (1904–1976) and Martin Heidegger (1889–1976) were the two main philosophers to deal with technology during the second half of the twentieth century. Gehlen's anthropological approach was to interpret human beings as deficient beings who use technology to compensate for their organic shortcomings. The characteristic activity of technology involves the creation and use of *Organersatzes,* that is, substitutes for or supersedings of those organs with which humans are endowed by nature. "There are two aspects to this tendency: artificial materials replacing those organically produced; and non-organic energy replacing organic energy" (Gehlen 1980, p. 5). The earliest humans strengthened their hands with wood and stone instruments, then replaced old materials in these instruments with new ones that defined entire ages (the Bronze Age, Iron Age, etc.), a substitution process that has continued into synthetic chemistry. But of even greater significance has been the replacement of human and animal power with coal, oil, electricity, and nuclear power.

Because of this substitution process technology develops a tendency to deny its roots and become independent. The technological world becomes progressively abstract and not tied to any immediate need. This is the starting point for Gehlen's criticism of modern technology as it has developed especially since the Industrial Revolution. According to Gehlen, technology develops an opposition to its previous cultural contexts and tends to become something pursued for its own sake. Coherent social orders decline under a flood of external stimuli, and social institutions lose their stability. Primitivisms such as "sex and drugs and rock-and-roll" become manifest throughout technological civilizations, along with extreme forms of individualism and subjectivism. In response Gehlen becomes a conservative critic of culture. Gehlen's anthropological analysis of the origin of technology leads to a criticism of technological culture.

Heidegger advanced two approaches to technology: first, in *Sein und Zeit* (1927; English trans. *Being and Time*, 1962), that of technology as an implicit or hidden presence in the human lifeworld; second, after the famous *Kehre* (turn), that of technology as a form of truth or revealing. The early Heidegger developed an understanding of (technological) experience in *Being and Time*, paragraphs 14–18. In the analysis of human existence as a *being-in-the-world* he discovered the everyday character of engagement with equipment as prior to any theoretical presence of objects. As is implicit in the Greek naming of objects as *pragmata*, Heidegger argues that technical praxis is the experiential context from which all science is abstracted. It is more accurate to describe science as theoretical technology than technology as applied science. But this *Being and Time* analysis of human interaction with entities or beings is no more than a moment in Heidegger's larger attempt to understand the "meaning of Being."

Turning from the focus on the meaning of Being that predominates in his early work, Heidegger's later thought develops a more explicit philosophy of technology. In "Die Frage nach der Technik" (1954; English trans. *The Question Concerning Technology*, 1977) he argues that technology is not just a practical engagement with the world but a revealing, a disclosure or truth about the world. What modern technology in particular reveals is the world as *Bestand*, that is, stock or resources subject to human manipulation. The coming upon the world as *Bestand* that is operative throughout modern technology as such Heidegger names *Gestell* (enframing), the promotion of which is for contemporary human beings not something that they simply choose to use or not but a *Geschick* (destiny). Like any destiny, however, technology as *Gestell* carries with it both opportunity and danger. The opportunities provided by technology are pervasive in the modern world, but the dangers are more hidden and go deeper than the simple risks so commonly associated with technology, such as the risks of automobile accidents or environmental pollution. The most profound danger is that the disclosure of the world as resource will overwhelm the event of disclosing itself, that the experience of one particular kind of truth will obscure the more primordial truth of Being. The ultimate challenge of modern technology is to be true to the greater human destiny of disclosing in the midst of a technological destiny.

The Frankfurt School and Social Risks

During the 1960s questions of ecological and social risks came to the fore in many discussions of science, technology, and ethics. But in the Frankfurt School it was social risks that held center stage, and a social risk of a particular kind: the risk of failure to use science and technology to realize the Enlightenment ideal of an autonomous humanity for which they were intended.

Criticism of technology in the Frankfurt School is based on the critical theory of Max Horkheimer (1895–1973) and Theodor W. Adorno (1903–1969), especially their post–World War II analysis of what they termed the "dialectic of Enlightenment." Analyzing the social histories of Nazism, Stalinism, and American capitalism they argued that formal rationality—positivism and pragmatism—had been transformed into an instrumental rationality that degraded its users and the things used. In the totalitarianisms of the twentieth century and even in consumer capitalism Enlightenment humanism had been used to justify dehumanization and exploitation. Enlightenment humanism thus runs the risk of becoming its dialectical opposite, a kind of anti-humanism. The science and technology that emerged out of Enlightenment commitments have been used to promote new forms of irrationality and barbarism, which must thus be dialectically criticized in order to save the Enlightenment project.

The critical theory of technology may be summarized in four theses:

(1) Knowledge is power. In the modern world science has become functional and instrumental knowledge, developed in order to achieve the goals of the Enlightenment by establishing human power over nature.

(2) Modern technology leads to technocracy. The Enlightenment values of humanity, emancipation, and social justice are to be realized by means of technical instruments.

(3) But rather than realizing democratic enlightenment, technology develops surrogates for enlight-

enment, especially in the forms of film and advertisement. Entertainment and the culture industry become technological substitutes for the genuine enlightenment to be found in aesthetics and the arts.

(4) Progress thus calls for a dialectic criticism of false enlightenment in the name of true enlightenment. Critical theory points out the ambivalence of progress brought about by technology.

Horkheimer and Adorno thus saw instrumental or calculative rationality (scientific technology) as a paradox: It provided the knowledge and power necessary to liberate human beings from unenlightened subservience to their own superstitions and to nature, enabling them to become autonomous individuals. But instrumental rationality has in fact been deployed by ruling groups to pacify the masses either violently or through material goods and services. The Enlightenment project has failed to prevent itself from being misused. What is needed is a new assertion of the Enlightenment ideal, which Horkheimer and Adorno nevertheless find difficult to derive from their social scientific studies.

It is to this problem that Jürgen Habermas responded with a philosophical deepening of critical theory and an extended reaffirmation of the norms of the Enlightenment ideal in the face of its corruption in contemporary culture. The human lifeworld is characterized by self-reflection, language, labor, and morality. Technological development follows the logic of labor, which is necessary for interacting with nature; technology is not something that can be renounced. At the same time, communicative action through language or a symbolic interaction among human beings engenders social norms. This too is an important aspect of what it means to be human and is not to be renounced. Technological rationality becomes a threat when it overwhelms or obscures symbolic interactions and its cultural traditions from which arise all justifications for using power, whether political or technological. Insofar as Habermas criticizes such a technological colonization of the lifeworld he reiterates Horkheimer and Adorno. But insofar as critical theory only criticizes instrumental rationality, it fails to rehabilitate a sophisticated form of rationality. Only a recovery and articulation of the principles of the communication rationality that is the basis of symbolic interaction can substantiate the critical theory project.

Cybernetics and Systems Theory

Cybernetics and systems theory have developed a scientific conception of technological action in order to control and shape this kind of action. Günter Ropohl's work on "technological systems theory" and "technological enlightenment" is a good extension of this aspect of cybernetics. According to Ropohl, the social dimension of technology is best grasped as an extended action system. It is not technology that formulates aims but certain action systems. These action systems produce technological artifacts, which in turn open up possibilities for new action functions. In this way Ropohl criticizes the ideas of technological determinism or a technological imperative. The physical constraints addressed by technological developments, for instance, are not technical but social in character. According to Ropohl the legitimation crisis of technological progress—that is, public doubts about whether technological change is always for the better—cannot help but promote "enlightenment" about the true character of the technological process (Ropohl 1991).

Klaus Kornwachs has also developed systems theory in ways that can be used to describe technological systems. The principles of any system are as follows: Every system has an author. The term *system* has both descriptive and prescriptive dimensions: Descriptive dimensions involve explaining how a system is to be constructed; prescriptive dimensions involve explicitly identifying the interests a system serves. As people learn to deal with any system it takes on an objective character and can thus become an object of scientific study. The structure of a system is given by the relationships among its elements. Large technological systems can be described at more than one level, and these levels must be integrated in a full description. Paradoxically, expanding systems are often easier to control than systems in equilibrium (Kornwachs 1993).

Contributions from the German Democratic Republic

From 1949 to 1990 the German Democratic Republic (GDR) developed discussions of science, technology, and ethics—and of the philosophy of technology—that were heavily influenced by the thought of Karl Marx (1818–1883), especially as interpreted in the Soviet Union. At the same time, scholars in the GDR attempted to maintain a certain level of independence by analyzing the connection between science and technology against the background of social developments. This in turn was influenced by and influenced the Dresden school of the technological sciences, especially since reunification.

Although its origins are unclear, the term *Technikwissenschaften* (technological sciences) was already in use during the nineteenth century in Germany and the German empire. After what in the Soviet Union was termed the scientific-technological revolution, that is, the unification of science and technology in has also been called "technoscience," the engineering sciences increased in significance for the establishment of socialism. But even though the notion of science implies a (not always realized) degree of stability, the engineering sciences have undergone substantial changes to which engineers must adjust.

The inner structure of any technological science has emerged from a long historical process of analyzing cause-and-effect relations, structures, functions, combinations of materials, and classification principles (Banse and Wendt 1986). In the technological sciences technological rules may be thought of as request systems, which in the process of invention must negotiate oppositions between idea and material possibility. Extending new scientific knowledge into the technological sciences involves the formulation of new technological rules, which are also increasingly required to take into account changing social circumstances. Only in this way can a connection be maintained between technological and social progress. But there is often a tension between technological parameters and those of economic and social effectiveness, not to mention the long-range effects on economy and society.

According to Johannes Müller, who worked for many years with scientists and engineers in the GDR, the technological sciences deal with a class of scientific analyses, operations, procedures, and means for determinate human actions. Their objective is to find solutions for tasks and problems with the help of rules, methodologies, problem-solving operations, procedures, algorithms, and norms. Contemporary construction work has to negotiate the relations among epistemology, technological science, logic, and psychology. Yet the main criterion for technological action and technological design is not truth but fulfillment or, more precisely, the possibility of technological fulfillment or practicality. Scrutiny of the possible realization of technological designs is done on the base of what may be called systematic heuristics (Müller 1990).

Erlangen-Konstanz Constructionism

The universities of Erlangen and Konstanz in Bavaria and Baden-Württemberg, respectively, were in the 1960s sites for the revival of the philosophy of science in Germany. The distinctive approach of philosophers in these two universities was the development of a nonempiricist, constructivist philosophy of science that strongly distinguished itself from logical empiricism. This school of constructivism sought, for instance, to identify a "protophysics" or "prephysics" that could prescribe in advance the measuring instruments necessary to any empirical physics. Peter Janich has added a "protobiology" and "protochemistry" to this prototheory. And from the philosophy of science this type of constructivism, because it focuses on the instrumentization of science, has easily been extended to the interpretation of technology as a way to criticize naturalism, especially in the field of cognitive or information technologies.

Janich has further argued for a constructivism in anthropology that he and Dirk Hartmann (1998) term "methodological culturalism." Along with this "cultural turn" comes the priority of action theory over language philosophy. The claim is that cultural relativism can be rejected on the basis of a preactive and preconscious agreement whenever human beings have achieved a certain level of cultural development. Taking technological development as a model for cultural development, the artificial character of all technological products becomes subject to a means–ends assessment that takes place before subjective or consumer evaluations. That is, the suitability of certain means for certain ends can be judged by their success or failure in achieving or failing to achieve those ends. The success of technological action cannot be reduced to the acceptance or rejection of certain groups but must be demonstrated first by practical reliability at any time in any transcultural context. Rational justification nevertheless remains as a philosophical and ethical issue. The Europäische Akademie zur Erforschung von Folgen wissenschaftlich-technischer Entwicklungen (European Academy for the Study of the Consequences of Scientific and Technological Advances) in Bad Neuenahr-Ahrweiler, under the direction of Carl Friedrich Gethmann, has been inspired by this approach.

Technology Assessment

Extending the social sciences and social philosophy of technology, the basic concern of technology assessment (TA) is systematic research into the preconditions and (potential) consequences for the introduction and use of technologies in order to identify and analyze social conflict areas, especially those that may evolve from the use of technologies. TA thus demonstrates and evaluates action possibilities for the improvement of technologies or their modes of use. The aim of TA is not the obstruction of technological innovations but the reflective design of sociotechnological systems (Petermann 1992).

TA analysis should anticipate conditions of realization and the potential consequences of use of technologies, and thus function as an early warning system. The main theoretical problem of TA is to predict changes caused or influenced by technology. The development of early indicators for effect-chains, which can show the diffusion of technological developments with high reliability, is a major challenge.

A useful assessment of technology should not be satisfied with simply discussing technological innovations but should reflect on the basic human–nature relation as it varies from culture to culture and is practiced in concrete social organizations for action (Bungard and Lenk 1988). The development, production, and initial use of technologies require special knowledge and capital. The elite of the economy, politics, and technological sciences profit from early successful uses of technology, but it is difficult to develop a specific methodological program for the assessment of technologies. There is neither a sophisticated theory of technological consequences nor a well-developed theory of valuation (Ropohl 1996). TA must always contend with unintended, ambivalent, and uncertain consequences. It has to make a functional distinction between scientific identification of possible consequences and their assessment, but must also integrate both steps in a common discourse.

The aforementioned European Academy clearly stresses methodologies related to the technico-philosophical construction of an ethical TA program. Critics from the social sciences reject any such ethical analysis, and thus technological ethics, appealing instead to social pluralism, the differentiation of social subsystems, decentralized technology, and the unpredictability of technological consequences. But surely it is reasonable to pursue ethics as a reflective analysis of right behavior. The responsibility of engineers can at least be based on the way they take concrete actions that result in technological solutions, even if they are subject to a number of influences and basic conditions. The development of technological solutions, equipment, machines, control devices, or consumer goods always includes ideas about users (Grunwald and Saupe 1999) that can be subject to critical assessment.

The Society of German Engineers

The Verein Deutscher Ingenieure (VDI, or Society of German Engineers) has a long history of philosophical ethical reflection on modern technology, as has been surveyed by Alois Huning and Carl Mitcham (1993). In the 1920s the VDI was a locus for extended discussions of the cultural and metaphysical significance of science and technology. In the 1950s it became the primary site for efforts to renew the ethical tradition in German engineering after a period of collaboration with and corruption by the Nazi regime.

As part of this renewal the VDI created a special interdisciplinary "Mensch und Technik" (humanity and technology) study group to examine relations between engineering, the technological sciences, philosophical ethics, and the humanities. Beginning in the 1950s the Mensch und Technik group convened a series of conferences dealing with ethics, industrialization, social impact, education, and philosophy, and issued a wide-ranging series of publications. Out of these discussions—with participation by philosophers such as Huning, Hans Lenk, Friedrich Rapp, and Ropohl—came influential analyses of professional engineering responsibility and technology assessment. Indicative of how Mensch und Technik discussions, even though existing within a professional engineering framework, sought to go beyond what in other national contexts might be considered the appropriate boundaries of engineering interest, were expressed concerns about the way nature was coming to be treated in the same way as artifacts, available simply for human control and manipulation.

During the 1990s a new generation of philosophical contributors to VDI discussions continued their work. Representative of these contributions has been the studies of Christoph Hubig, who argues for an extension of analyses of instrumental action in ways that can lead to a rehabilitation of substantive value ethics (Hubig 1993). For Hubig, the challenge of applied ethics, especially in science and technology, is to build a bridge between principles and specific actions, with an awareness of the complex inner structure of practice. Such a pursuit of ethics in relation to science and technology can be done only by means of interdisciplinary dialogue. Within the technological practice there is always an implicit catalog of values, with conflicts between values being a regular occurrence. The task of discussion-management institutions and organizations is to provide standard approaches for dealing with such conflicts when they occur (Hubig 1997). Taking seriously his own recommendations to work in an interdisciplinary manner, Hubig has worked with the VDI to develop a report on *Ethische Ingenieurverantwortung* (2000), and then led the team that drafted the 2002 VDI code of ethics, *Ethische Grundsätze des Ingenieurberufs*.

Method versus Language, Practice versus Theory

Recent work in the philosophy of technology has tended to emphasize methodology over language, practice over theory. Descriptive propositional knowledge (knowing that) is seen as less important in technology and science than prescriptive skill (knowing how) or productive knowledge. Insofar as this is the case, the explanation–understanding controversy has been replaced by a more pragmatic epistemology (see Zimmerli 1997). From the mid-1980s the expansion of technology has brought with it transformational experiences such as the digitalization of everyday life and associated challenges to tradition and changes in values. Yet it is the lack of practical (not theoretical) orientation in these experiences that gives new life to philosophy. How should we live in the new world we are creating? What should we do with our artifice? During this second modernization the hybridization of technology and science has brought with it a new "dialectic of enlightenment" that is manifested in the philosophy of culture.

In order to address such practical questions philosophers such as Lenk, Walther Zimmerli, and Bernhard Irrgang have been developing a hermeneutic understanding of both technology and ethics. The structures of technological practice, professional activity, and everyday life, together with the background of an implicit technological knowledge, are the basis of collective technological action in a cultural context. The meaning of a technology does not necessarily have to be linguistically articulated in order to be present in a culture. The ways technological practices themselves structure actions include different forms of meaningfulness. This leads to a kind of existential pragmatics of technological action and its models of representation (Corona and Irrgang 1999). Such an approach provides a recursive and reflexive assessment of technological actions. But the impacts of any interpretation of technological actions must also prove successful in psychological, sociological, technical-historical, and cultural-historical terms (Irrgang 2001, 2002). At the same time, reflective modernization depends on the continued existence of such institutions as universities and research centers even as they are altered by globalization.

Reflective modernization must also distinguish the self-understandings of scientific and technical professionals from the external descriptions of their roles. The traditional epistemological foundation for a social role description has been the notion of science as knowledge, but technological science is not another science. Technological science is an action science and thus also contains prescriptive statements as well as descriptive ones. The integration of scientific method into the technological sciences has resulted in new disciplinary formations from more than one perspective: by objects studied, by methods, and by professional fields. A metatheory of the technological sciences is needed to determine the relation of these various disciplinary formations and to search for unity within the technological sciences. A related question concerns the relation between disciplinary, interdisciplinary, and transdisciplinary technoscientific knowledge. Epistemological and professional distinctions ultimately interact with practice-orientated and institutional differentiations in an integrated technology-reflective culture (Irrgang 2003).

Appendix: Ethics in Practice

To this point observations have indicated some of the abstract approaches brought to bear in Germany on issues related to science, technology, and ethics—approaches that serve repeatedly to emphasis the importance of practice. By way of a concluding appendix, it remains to comment on specific practices themselves. In this regard there are at least two practices within technoscience deserving special notice: those having to do with research misconduct and with stem cell research.

RESEARCH MISCONDUCT. In June 2000 the Deutsche Forschungsgemeinschaft (DFG), which is the main Germany research funding agency, after initial allegations of misconduct emerged in 1997, concluded an investigation into the practices of the hematologist and cancer researcher Friedhelm Herrmann of the University of Freiburg Medical Center. According to the DFG report, of Herrman's 347 scientific papers published between 1988 and 1992, at least 52 contained falsifications and another 42 were suspect. A previous investigation of more recent publications had identified 37 papers with falsification and data manipulation. This discovery of such egregious misconduct on the part of a respected member of the scientific community led the DFG in 2002 to require that any institution receiving DFG funds adopt a strong definition of scientific misconduct prohibiting falsification and fabrication of data, unacknowledged data selection, graph and figure manipulation, the inclusion of false information in a curriculum vitae, destruction of primary data, sabotage of others' work, and plagiarism. Previous German policies had been more relaxed; in one step this new policy placed the German scientific research community at the forefront of misconduct policy development.

STEM CELL RESEARCH. As has been explained by Jens G. Reich (2002), among others, the discussion of stem cell research in Germany reflects both philosophical and political history. Philosophically, under the influence of Immanuel Kant (1724–1804), German ethics tends to be strongly deontological, stressing the primacy of treating human beings as ends not as means. Indeed, the first article of the German *Grundgesetz* (Basic Law) of 1949 states that "the dignity of the human being is untouchable." There is also a strong awareness of German failures during the Nazi period to respect human dignity. In a determined stance to respect human dignity in the present, the German Embryo Protection Law of 1990, which was supported by a large majority of the public, explicitly defines human life as beginning at conception. It prohibits manipulation of a human embryo for any purpose other than its implantation into the uterus of the woman from whom the originating ovum was derived. This law thus forbids stem cell creation and applies to privately funded embryo research as well as to publicly funded research.

The law has, however, come under interpretative stress as a result of emerging opportunities for stem cell research. In 2002 the German parliament (Bundestag) reaffirmed the ban on stem cell creation but allowed the importation of stem cells created in other countries provided certain stringent conditions are met. Only stem cell lines created before 2002 are eligible, and then only with the informed consent of the parents of the embryo from which the stem cell line was derived, and on the conditions that the parents have received no payment and that the intention behind the original fertilization was a pregnancy that was abandoned for reasons not related to the embryo—that is, the embryo could not have been rejected as defective. Clearly stem cell research in Germany takes place under more detailed ethical guidelines than in perhaps any other country. It is also worth noting that human cloning, whether for reproductive or therapeutic purposes, is prohibited in Germany, but there are also more liberal positions in bioethics (Irrgang 1997, Irrgang 2005).

BERNHARD IRRGANG
TRANSLATED BY KATRIN FELDHUS

SEE ALSO *Anders, Günther; Central European Perspectives; Dessauer, Friedrich; Existentialism; French Perspectives; Habemas, Jürgen; Hegel, Georg Wilhelm Friedrich; Heidegger, Martin; Husserl, Edmund; Jaspers, Karl; Kant, Immanuel; Leibniz, G. W.; Luhmann, Niklas; Nietzsche, Friedrich W.; Phenomenology; Weber, Max.*

BIBLIOGRAPHY

Banse, Gerhard, and Helge Wendt, eds. (1986). *Erkenntnismethoden in den Technikwissenschaften: Eine methodologishe Analyse und philosophische Diskussion der Erkenntnisprozesse in den Technikwissenschaften* [Epistemological methods in the technological sciences: A methodological analysis and philosophical discussion of epistemological processes in the technological sciences]. Berlin: Verlag Technik.

Bungard, Walter, and Hans Lenk, eds. (1988). *Technikbewertung: Philosophische und psychologische Perspektiven* [Technology assessment: Philosophical and psychological perspectives]. Frankfurt am Main: Suhrkamp.

Corona, Nestor, and Bernhard Irrgang. (1999). *Technik als Geschick? Geschichtsphilosophie der Technik* [Technology as destiny? Historical philosophy of technology]. Dettelbach: Röll.

Gehlen, Arnold. (1980). *Man in the Age of Technology,* trans. Patricia Lipscomb. New York: Columbia University Press. Originally published as *Die Seele im technischen Zeitalter* (Hamburg: Rowohlt, 1957).

Grunwald, Armin, and Stephan Saupe, eds. (1999). *Ethik in der Technikgestaltung: Praktische Relevanz und Legitimation* [Ethics in technological design: Practical relevance and legitimation]. Berlin: Springer.

Habermas, Jürgen. (1984–1987). *The Theory of Communicative Action,* trans. Thomas McCarthy. 2 vols. Vol. 1: *Reason and the Rationalization of Society.* Vol. 2: *Lifeworld and System: A Critique of Functionalist Reason.* Boston: Beacon Press. Originally published as *Theorie des kommunikativen Handelns* (Frankfurt am Main: Suhrkamp, 1981).

Hartmann, Dirk, and Peter Janich, eds. (1998). *Die kulturalistische Wende: Zur Orientierung des philosophischen Selbstverständnisses* [The cultural turn: On the orientation of philosophical self-determination]. Frankfurt am Main: Suhrkamp.

Horkheimer, Max, and Theodor W. Adorno. (2002). *Dialectic of Enlightenment: Philosophical Fragments,* ed. Gunzelin Schmid Noerr; trans. Edmund Jephcott. Stanford, CA: Stanford University Press. Originally published, 1947.

Hubig, Christoph. (1993). *Technik- und Wissenschaftsethik: Ein Leitfaden* [Technological and scientific ethics: A guideline]. Berlin: Springer.

Hubig Christoph. (1997). *Technologische Kultur* [Technological culture]. Leipziger Schriften zur Philosophie 3. Leipzig: Leipziger Universitätsverlag.

Hubig, Christoph, ed. (2000). *Ethische Ingenieurverantwortung: Handlungsspielräume und Perspektiven der Kodifizierung* [Ethical engineering responsibility: Action areas and perspectives on code creation]. Düsseldorf: VDI Verlag.

Huning, Alois, and Carl Mitcham. (1993). "The Historical and Philosophical Development of Engineering Ethics in Germany." *Technology in Society* 15(4): 427–439.

Irrgang, Bernhard. (1997). *Forschungsethik Gentechnik und neue Biotechnologie. Grundlegung unter besonderer Berücksichtigung von gentechnologischen Projekten an Pflanzen, Tieren und Mikroorganismen* [Research ethics in genetic engineering and new biotechnology with a special respect to the projects on the genetic engineerings studies of plants, animals and microorganisms). Stuttgart: S. Hirzel.

Irrgang, Bernhard. (1998). *Praktische Ethik aus hermeneutischer Perspektive* [Practical ethics from the hermeneutical point of view]. Paderborn, Germany: Schöningh.

Irrgang, Bernhard. (2001). *Technische Kultur: Instrumentelles Verstehen und technisches Handeln* [Technological culture: Instrumental understanding and technological action]. Paderborn, Germany: Schöningh.

Irrgang, Bernhard. (2002). *Technische Praxis: Gestaltungsperspektiven technischer Entwicklung* [Technological practice: Design perspectives on technological development]. Paderborn, Germany: Schöningh.

Irrgang, Bernhard. (2003). *Von der Mendelgenetik zur synthetischen Biologie. Epistemologie der Laboratoriumspraxis Biotechnologie* [From Mendel's genetic engineering to synthetic biology. Epistemology of the laboratory practice and biotechnology]. Technikhermeneutik Bd. 3. Dresden: Thelem.

Irrgang Bernhard. (2005). *Einführung in die Bioethik* [Introduction to Bioethics]. UTB 2640. Munich: Fink.

Kornwachs, Klaus. (1993). *Kommunikation und Information: Zur menschengerechten Gestaltung von Technik* [Information and communication: On the human design of technology]. Berlin: Springer.

Lenk, Hans. (1994). *Macht und Machbarkeit der Technik* [Power and realization of technology]. Stuttgart: Reclam.

Lenk, Hans. (1995). *Interpretation und Realität: Vorlesungen über Realismus in der Philosophie der Interpretationskonstrukte* [Interpretation and reality: Lecture on realism in philosophy of construction interpretations]. Frankfurt am Main: Suhrkamp.

Lenk, Hans. (1998). *Konkrete Humanität: Vorlesungen über Verantwortung und Menschlichkeit* [Concrete humanity: Lecture on responsibility and humanity]. Frankfurt am Main: Suhrkamp.

Lenk, Hans, and Matthias Maring. (2001). *Advances and Problems in the Philosophy of Technology.* Münster, Germany: Lit.

Mittelstraß, Jürgen, ed. (1980–1996). *Enzyklopädie Philosophie und Wissenschaftstheorie* [Encyclopedia of philosophy and economy theory]. 4 vols. Mannheim, Germany: Bibliographisches Institut (Vols. 1 and 2); Stuttgart: Metzler (Vols. 3 and 4).

Müller, Johannes. (1990). *Arbeitsmethoden der Technikwissenschaften: Systematik, Heuristik, Kreativität* [Methods of technological science: Systematics, heuristics, creativity]. Berlin: Springer.

Petermann, Thomas, ed. (1992). *Technikfolgen-Abschätzung als Technikforschung und Politikberatung* [Technology assessment as technological research and political advice]. Frankfurt am Main: Campus.

Reich, Jens G. (2002). "Embryonic Stem Cells: The Debate in Germany." *Science* 296(5566): 265.

Ropohl, Günter. (1979). *Eine Systemtheorie der Technik: Zur Grundlegung der Allgemeinen Technologie* [A systems theory of technology: On the foundations of a general technology]. Munich: Hanser.

Ropohl, Günter. (1991). *Technologische Aufklärung: Beiträge zur Technikphilosophie* [Technological enlightenment: Contributions to philosophy of technology]. Frankfurt am Main: Suhrkamp.

Ropohl, Günter. (1996). *Ethik und Technikbewertung* [Ethics and technology assessment]. Frankfurt am Main: Suhrkamp.

Verein Deutscher Ingenieure (VDI). (2002). *Ethische Grundsätze des Ingenieurberufs/Fundamentals of Engineering Ethics.* Düsseldorf: VDI Verlag. This is the basic VDI ethics code.

Zimmerli, Walther. (1997). *Technologie als "Kultur"* [Technology as "culture"]. Hildesheim, Germany: Olms.

GIRARD, RENÉ

• • •

Born in Avignon, France, on Christmas Day, René Girard's (b. 1923) work has been a blend of history, literature, and philosophy with implications for science, technology, and ethics that have only begun to be appreciated. He graduated from the Ecole des Chartes in Paris in 1947 (as a specialist in medieval studies) with a thesis on private life in his hometown of Avignon in the second half of the fifteenth century. A year's trip abroad turned into a Ph.D. in history from Indiana University, after which Girard remained in the United States, where he retired as a professor of French Language, Literature, and Civilization from Stanford University in 1995.

Girard's early historiographic publications soon shifted to an avalanche of literary criticism. His first book, *Deceit, Desire and the Novel* (1966), contrasted the romantic lie of individualism with the novelistic truth of what he called *imitative* or *mimetic desire.* Among five major novelists Girard discovered a triangular structure to desire where the protagonists struggled with the fact that their deepest aspirations were mere imitations of a model or rival. Adultery remains the archetype for this phenomenon as illustrated in Dostoevsky's novella, *The Eternal Husband.* The husband is obsessed by his wife's lovers, who inflame, validate, and aggravate his own desire. Girard's students have likened his discovery of imitation in the social sciences to Newton's discovery of gravity in the physical sciences. The vast secondary literature on mimetic desire now extends these early insights into the diverse fields of economics, sociology, psychology, theology, and anthropology.

His second book, an anthropological study of *Violence and the Sacred* (1977), proposes a rational explanation for sacrificial rituals (as well as religious myths and prohibitions) in what he terms the *victimage mechanism.* Mimetic desire is inevitably conflictual. "Rivalry does not arise because of the fortuitous convergence of two desires on a single object; rather, *the subject desires the object because the rival desires it*" (Girard 1977, p. 145).

Ancient religion developed as an unconscious method of keeping the peace where the mimetic war of all against all is replaced by the more efficient war of all against one—the community's sacrifice of a scapegoat. Sacrifice acts as a kind of vaccination whose small doses of violence inoculate the community against greater violence.

The publication of *Things Hidden Since the Foundation of the World* (1987), a conversation with two French psychiatrists, included discussion of a *founding murder* among mimetically hysterical primates that initiated the long, slow process of hominization as well as sacrificial mechanisms. Girard sheds new light on the often-discarded speculations on primal murders found in Freud's *Totem and Taboo*. He also proposes the controversial thesis that the Judeo-Christian revelation of the victimage mechanism provides the anthropological tools necessary to demythologize pagan religious practices, which for Girard includes much of Western Christianity.

According to Girard, Christ's death was not a sacrifice willed by an angry God to atone for an original sin, but simply a revelation of human brutality and violence by a loving God. The remainder of Girard's major works (aside from a delightful work on Shakespeare) focus on biblical criticism, including *The Scapegoat* (1986), *Job: The Victim of His People* (1987) and *I See Satan Fall Like Lightning* (2001).

For Girard modern science and technology are an inevitable consequence of the demythologization of sacrificial violence and magical thought. Magical thought always seeks a social/moral explanation for pain. For example the Black Plague was often attributed to the Jews poisoning the water supply. As Girard quips, "Those who are suffering are not interested in natural causes" (Girard 1986, p. 53). However, with a loosening of magical thought, the search for natural causes slowly becomes a more reasonable path toward the "relief of man's estate" (Francis Bacon). "The invention of science is not the reason that there are no longer witch hunts, but the fact that there are no longer witch hunts is the reason that science has been invented. The scientific spirit, like the spirit of enterprise in an economy, is a by-product of the profound action of the Gospel text" (Girard 1986, p. 204).

Yet Girard's attitude toward science contains a certain Freudian ambivalence. Science is necessarily part of the Christian concern for victims and is a consequence of this charitable impulse. At the same time, modern technology has an apocalyptic edge to it. With the loosening of ancient sacred restraints and prohibitions, modern technology, like modern economy, unleashes the phenomenon

René Girard, b. 1923. With work encompassing the disciplines of philosophy, literary criticism, theology, and anthropology, Girard is chiefly known for pioneering the mimetic theory of desire and the concept of the "scapegoat mechanism." (© *Bassouls Sophie/Corbis Sygma.*)

of mimetic desire in a wave of consumerism, ethnic rivalry, media frenzy, and politically correct victimology. For Girard it is no accident that names for nuclear weapons are "taken from the direst divinities in Greek mythology, like Titan, Poseidon, and Saturn, the god who devoured his own children" (Girard 1987, p. 256).

JIM GROTE

SEE ALSO *Christian Perspectives: Historical Traditions; Violence.*

BIBLIOGRAPHY

Girard, René. (1965 [1961]). *Mensonge romantique et verite romanesque* [Deceit, desire and the novel], trans. Yvonne Freccero. Baltimore, MD: John Hopkins University Press.

Girard, René. (1977 [1972]). *La Violence et le sacre* [Violence and the sacred], trans. Patrick Gregory. Baltimore, MD: John Hopkins University Press.

Girard, René. (1986 [1982]). *Le Bouc emissaire* [The scapegoat], trans. Yvonne Freccero. Baltimore, MD: John Hopkins University Press.

Girard, René. (1987a [1978]). *Des choses caches depuis la fondation du monde* [Things hidden since the foundation of the world], trans. Michael Metteer (Book I) and Stephen Bann (Books II and III). Stanford, CA: Stanford University Press.

Girard, René. (1987b [1985]). *La route antique des homes pervers* [Job: the victim of his people], trans. Yvonne Freccero. Stanford, CA: Stanford University Press.

Girard, René. (2001 [1999]). *Je vois Satan tomber comme l'éclair* [I see Satan fall like lightning], trans. James G. Williams. Maryknoll, NY: Orbis Books.

Williams, James G., ed. (1996). *The Girard Reader*. New York: Crossroad Publishing Company. The best introduction to Girard's thought.

GLOBAL CLIMATE CHANGE

• • •

Global climate change refers to the ways in which average planetary weather patterns alter over time. The term *global warming*, though common, is a misnomer, for under some scenarios it is possible that part of the earth could cool, even as most of the planet gets warmer. The global climate change debate offers a superb case study of the relations existing in the early twenty-first century among science, technology, politics, and questions of meaning and value.

Defining the Problem

Because of the long timescales involved, climate change is difficult to experience directly; knowledge of meteorological variation generally falls under the classification of "weather." Science and technology—in forms such as the uncovering of the basic physical principles of atmospheric science, geologic evidence such as glacial moraines and plant remains, and determinations of ancient atmospheric concentrations derived from ice cores taken from the Greenland and Antarctic ice sheets—is needed to identify even the possibility of climate change. This fact has encouraged the assumption that both the definition of and the human response to possible climate change should be fundamentally scientific and technological in nature.

Geologists have known since the mid-nineteenth century that local, regional, and global climate undergoes change through time. Indeed, adding the term *change* to climate is nearly a redundancy, because climate varies on all timescales from decades to millions of years. This makes it difficult to clearly distinguish between the concepts of *weather* (transient variations) and *climate* (the long term status of the system).

For instance, the earth experienced an ice age that peaked 18,000 years ago; but considering the larger span of the earth's history, it is still in an ice age. While the norm for humanity, geologic evidence suggests that the earth has had ice on its poles for only a very small fraction of its history.

It was the Swedish chemist Svante Arrhenius (1859–1927) who in 1896 first suggested the possibility of human-induced climate change through the burning of fossil fuels. Climate change came to general notice in the 1970s, when concern was voiced about the possibility of global *cooling* leading to a new ice age. This remains a live possibility: Evidence of ancient climates shows that in the last 800,000 years the planet has seen a series of oscillations between ice ages, of approximately 100,000 years in duration, and interglacials, of around 10,000 years in length. Earth is thus overdue for a cold spell.

The 1980s saw the rise of concern about the "greenhouse effect" caused by increasing levels of human-produced carbon dioxide and other gases that trap heat in the atmosphere. Concern exploded in the summer of 1988, which saw record warmth throughout the United States. This warming trend appears to be continuing: Nine of the ten hottest years since the beginning of record keeping in 1880 have occurred between 1990 and 2003.

Ethical, Political, and Philosophical Issues

What defines climate change as a "problem" at all? This question relates to a long-standing debate within environmental ethics on whether nature has only instrumental value for human beings or has intrinsic value outside of any considerations of its value to humans. The first (anthropocentrist) position claims that concern about the environment should be motivated by an interest in human welfare. The second (ecocentrist) position believes that animals, species, ecosystems, and even rock formations and climate patterns can have qualities that make them the objects of moral concern.

On the first view, climate change is a problem only from the perspective of human wants, needs, and obligations to one another. Rising sea level is a physical event; it is only when it floods New Orleans or the Maldives that it becomes a problem. From this point of view, climate change has become a crisis in two senses in the early 2000s. First, human populations, structures, or the ecosystems societies depend upon may be exposed to climate-induced dangers such as rising sea level, changes in temperature and/or precipitation, changes in the frequency of extreme events such as hurricanes, and changes in vegetation and the growing season. Second, if climate change is partially or wholly human-caused—that is, if it is anthropogenic in nature—then the persons, industries, or societies that have caused these problems may fairly be held accountable.

This latter question has spawned a global debate about the respective responsibilities of developed and developing nations to address climate change. The debate turns on the fact that most of the increase of greenhouse gases to date has been caused by industrial nations, especially the United States, whereas most of the future contribution of greenhouse gases to the atmosphere is likely to come from developing countries such as China. Should developed countries be required to address questions of greenhouse gas emissions first, because they caused the problem, allowing developing nations to pollute more as they develop their industries? Or is such an approach self-negating, in that any real solution to greenhouse gas emissions requires a common global effort?

On another view, however, climate change is a more than a human affair. Climate change is certainly an issue for any species driven to extinction by ecosystem change. It is here that the question of global climate change touches upon core questions within the philosophy of nature. Species come into and go out of existence constantly; does it matter whether a species' extinction is caused by natural climate variability or anthropogenic change? In the mind of some, the difference is crucial: Change (including extinction) that is natural in origin should be tolerated and adapted to, whereas human-caused change or extinction should be addressed and mitigated. Making the question even more vexed are claims that there is no "natural" left in the early twenty-first century. On this view the entire earth, including its atmosphere, has become an artifact through centuries of inhabitation, cultivation, and pollution (Allenby 1999, McKibben 1999). These aspects of the climate change debate point toward religious and metaphysical considerations concerning the status of nature rather than to more and better data and predictions. In ways similar to the current debate concerning genetic engineering, questions are increasingly being asked about whether nature represents a limit that should be acknowledged and in some sense obeyed.

The Scientific Effort

Concerns about global climate change have led to a massive, unprecedented, and worldwide scientific, technological, and political effort to understand the causes and consequences of climate change. The basic assumption underlying all of these efforts is that climate change science is necessary for the devising of climate change policy.

The United States leads the world in climate change research, funding more than half of all the work. Approximately half of the nearly $2 billion annual budget for the U.S. Global Change Research Program (USGCRP, The U.S. Government's Interagency Research Program On Climate Change) is devoted to satellites and other data systems. The rest supports research across a wide range of sciences such as physics, atmospheric chemistry, oceanography, and ecology. A significant part of this research is conducted through computer simulations, the best known of which are global climate models (GCMs) that run on the world's fastest computers. Products of a truly global scientific and technological effort, GCMs have produced sets of predictions concerning the possible state of the atmosphere in 2100. (There is, of course, nothing magical about the year 2100; it was picked for symmetry and because this period was thought to be within the moral horizon of most people. In fact, computer models predict that change will accelerate after this date.)

Research into the social and political aspects of climate change—broadly known as "human contributions and responses to global change"—receives around 2 percent of the USGCRP budget, or $50 million. Even then, the overwhelming majority of this investment goes toward quantitative (often economic) social science research. While questions of ethics and values have often been voiced in public debate, research into such questions has been pursued only at the margins. The overall definition of the problem of climate change thus remains deeply immersed in science: The USGCRP seeks to identify the basic facts of the matter, leaving questions of value and justice to the political realm. More to the point, the assumptions remain quite positivistic: It is assumed that ethical and political solutions will somehow be derived from advances in climate science.

After two decades of concerted research, the community of climate change scientists have reached a high degree of consensus on several basic points: The global climate is warming; this warming is largely anthropogenic in origin; and the consequences of this warming could be quite severe. In the words of the National

Research Council's Committee on the Science of Climate Change, "Greenhouse gases are accumulating in Earth's atmosphere as a result of human activities ... Temperatures are, in fact, rising" (NRC 2001, p. 1).

Science Meets Policy

Climate science research in the United States and other nations (principally the European Union and Japan) feeds into a global political effort to manage the problem of global climate change. The Intergovernmental Panel on Climate Change (IPCC) lies at the center of these efforts. The World Meteorological Organization and the United Nations Environment Programme founded the IPCC in 1988 "to assess scientific, technical and socio-economic information relevant for the understanding of climate change" (IPCC). The IPCC consists of:

- Working Group I, which assesses the scientific aspects of the climate system and climate change

- Working Group II, which focuses on the vulnerability of socioeconomic and natural systems to climate change, the consequences (both negative and positive) of climate change, and possible options for adapting to climate change

- Working Group III, which evaluates options for restricting greenhouse gas emissions and other ways to mitigate climate change

- The Task Force on National Greenhouse Gas Inventories, which runs the IPCC National Greenhouse Gas Inventories Programme

In addition, a series of special reports supports the working groups, the most important being the Special Report on Emissions Scenarios (SRES), which provides baseline sociological, political, and economic parameters for GCMs. Since 1990 the working groups have issued a series of joint assessment reports on a five- to six-year basis. These reports represent a remarkable synthesis of technoscientific research. Each assessment directly involves hundreds of scientists who collectively spend thousands of hours collating and synthesizing the available information on the above topics in a thick set of volumes. After a series of reviews, each volume is then boiled down to a "summary for policymakers" that attempts to extract insights most relevant to decision makers worldwide.

These IPCC reports are created to support the United Nations Framework Convention on Climate Change (UNFCCC), which seeks to devise a global political strategy. In late 1997 the UNFCCC gathered representatives from more than 160 nations in Kyoto, Japan, to negotiate binding limitations on greenhouse gases for developed nations. The resulting Kyoto Protocol called for developed nations to agree to limit their greenhouse gas emissions as compared with the levels emitted in 1990. The bulk of the political efforts to address the challenges of climate change have centered on negotiating the particular provisions of the Kyoto Protocol.

The results, however, have not been encouraging. Even if the Kyoto Protocol were to be ratified—and the Bush Administration announced its rejection of the protocol in 2001—the proposed limitations to greenhouse emissions would not come anywhere near the estimated 50 to 75 percent reduction scientists believe is necessary to stabilize atmospheric levels of carbon dioxide. What is more, the $25 to $30 billion the United States spent on climate change research from the early 1980s to the early 2000s highlights the questionable structure of the existing global climate change debate. Across this twenty-year period, the range of uncertainty for the predicted amount of change in global mean temperatures by 2100 actually *increased*, from 1.4 to 5.4 degrees Celsius in 1980 to 1.4 to 5.8 degrees Celsius in 2001. This increase in the range of possible warming has provided cover for politicians to call for more research instead of devising plans of action.

Future of the Problem

The paradox is that at the same time that a scientific consensus has formed on the reality of climate change, the actual range of future outcomes has increased rather than shrunk. A number of factors contribute to this increase of uncertainty, including a greater appreciation of the complexity and attendant lack of understanding concerning some parts of the climate system (for instance, the behavior of clouds, and the ocean–atmosphere interface), the difficulties in matching differing types of data, and the possibility that a system as complex as world climate is fundamentally unpredictable in nature. But the core difficulty lies elsewhere: The computer simulations used to model the atmosphere for the year 2100 are themselves fundamentally dependent on future sociological and economic indicators that are essentially unknowable. This is the significance of the SRES scenarios, which provide the basic inputs and parameters for the GCMs.

The SRES scenarios consist of six different imagined future patterns of energy use, technological progress, and social, political, and economic development. These six possible development paths explore future choices concerning population, lifestyle, the degree of globalization and economic integration, the development of non-carbon-based energy sources, and the possibility of carbon sequestration—choices that are not predictable in ways analogous to physical systems. Moreover, the point is not just that future social conditions cannot be predicted, but that they are in large part a function of human choices. The future does not simply befall humanity; individually and collectively humans exercise a significant influence over what happens. Rather than treating the future as if it were beyond human control, the challenge of global climate change calls for public debate about desirable futures.

It is thus arguable that while scientific research on climate change has greatly increased the knowledge and appreciation of the problem, the focus of attention should now shift toward two other areas that complement climate science: better understanding the nature of the social, ethical, political, and political dimensions of the problem, and devising ways to increase the resilience of both natural and social systems to a global climate that is already undergoing alteration. This approach would involve a shift in attention away from precisely modeling the climate system and toward devising a "no-regrets" strategy tied to sustainable development, social justice, and the modification of desires. The problem, however, is that such a "soft" approach to global climate change runs up against 300 years of tradition in which humankind has attempted to engineer its way out of problems rather than developing personal and political means for modifying its behavior.

ROBERT FRODEMAN

SEE ALSO *Automobiles; Deforestation and Desertification; Environmental Ethics; International Relations; Oil; Pollution; Rain Forest; United Nations Environmental Program.*

BIBLIOGRAPHY

Allenby, Brad. (1999). "Earth Systems Engineering: The Role of Industrial Ecology in an Engineered World." *Journal of Industrial Ecology* 2(3): 73–93. Argues that climate change should be treated as an engineering problem.

Brunner, Ronald. (2001). "Science and the Climate Change Regime." *Policy Sciences* 34(1): 1–33. Excellent introduction to policy dimensions of the climate debate.

Drake, Frances. (2000). *Global Warming: The Science of Climate Change.* New York: Oxford University Press.

McKibben, Bill. (1999). *The End of Nature,* 2nd edition. New York: Anchor. Well-known and accessible account of cultural dimensions of human impact upon nature.

Miller, Clark A., and Paul N. Edwards, eds. (2001). *Changing the Atmosphere: Expert Knowledge and Environmental Governance.* Cambridge, MA: MIT Press. A rich collection of essays treating different aspects of the climate problem.

National Research Council (NRC). Committee on the Science of Climate Change. (2001). *Climate Change Science: An Analysis of Some Key Questions.* Washington, DC: National Academy Press. An important recent report summarizing the state of scientific knowledge.

Rayner, Steve, and Elizabeth L. Malone, eds. (1998). *Human Choice and Climate Change.* 4 vols. Columbus, OH: Battelle Press. A rich variety of essays on social science perspectives on climate change.

Sarewitz, Daniel, and Roger A. Pielke Jr. (2000). "Breaking the Global-Warming Gridlock." *Atlantic Monthly* 286(1): 54–64. An accessible and innovative treatment of policy questions concerning climate change.

Shackley, Simon; Peter Young; Stuart Parkinson; and Brian Wynne. (1998). "Uncertainty, Complexity, and Concepts of Good Science in Climate Change Modeling: Are GCMs the Best Tools?" *Climatic Change* 38(2): 159–205. A good review of the use of global climate models.

U.S. Climate Change Science Program and the Subcommittee on Global Change Research. (2002). *Our Changing Planet: The Fiscal Year 2003 U.S. Global Change Research Program and Climate Change Research Initiative, A Supplement to the President's Fiscal Year 2003 Budget.* Washington, DC: Author. Federal publication that annually summarizes the state of current research.

U.S. Climate Change Science Program and the Subcommittee on Global Change Research. (2003). *Strategic Plan for the U.S. Climate Change Science Program.* Washington, DC: Author. Another useful governmental report.

Weart, Spencer R. (2003). *The Discovery of Global Warming.* Cambridge, MA: Harvard University Press.

INTERNET RESOURCES

Intergovernmental Panel on Climate Change (IPCC). Available from http://www.ipcc.ch.

National Research Council (NRC). Committee on the Science of Climate Change. (2001). *Climate Change Science: An Analysis of Some Key Questions.* Available from http://books.nap.edu/books/0309075742/html/.

U.S. Climate Change Science Program and the Subcommittee on Global Change Research. (2002). *Our Changing Planet: The Fiscal Year 2003 U.S. Global Change Research Program and Climate Change Research Initiative, A Supplement to the President's Fiscal Year 2003 Budget.* Available from http://www.usgcrp.gov/usgcrp/Library/ocp2003.pdf.

U.S. Climate Change Science Program and the Subcommittee on Global Change Research. (2003). *Strategic Plan for the U.S. Climate Change Science Program.* Available from http://www.climatescience.gov/Library/stratplan2003/final/.

GLOBALISM AND GLOBALIZATION

• • •

Without science neither globalism nor globalization would be conceivable; without technology they would not be practical possibilities. The extent to which the internal ethics of science and the codes of behavior of various engineering professions influence globalism and globalization, or the degree to which independent ethical assessments should be brought to bear on all science, technology, and globalist synergies, remains open to critical discussion. What follows is an analysis that aims to provide a background for such considerations.

Terminology

The terms *globalism* and *globalization* came into use during the last half of the twentieth century. The question of when, and by whom, is contentious. But irrespective of origins the two terms are used in distinct ways. Globalization refers to a multidimensional economic and social process beginning in the late 1970s and early 1980s and that embraces a variety of interlinked economic, communicational, environmental, and political phenomena. Globalism, although it has older roots as a synonym for internationalism, has come to be used as the name of a broad *ideological commitment* in favor of the process of globalization—that is, of a view that sees the process of globalization as entirely or predominantly positive in its implications for humankind (Steger 2002).

Globalists are people who wish the process of globalization to continue, and indeed intensify, although they may also wish to have it politically regulated or controlled in various ways. Globalists are often (though not always) also convinced that globalization, whatever its implications for human welfare, is an *inevitable process* that cannot, and should not, be reversed. They are often contrasted with "localists," who seek to escape or overcome the problems posed by globalization through small-scale forms of economic and cultural development and political organization that minimize involvement in the global economy (Mandle 2003).

In short then, there are theorists and writers on globalization both for and against the process they are analyzing, but those in favor of the process are generally called "globalists" or advocates of "globalism." In the early twenty-first century, enthusiasts for globalization do not call *themselves* "globalists" (this terminology is used only by globalization's opponents), although there is the potential for this to change as the debate unfolds further.

Globalization: Its Characteristics

There are innumerable definitions of the term globalization in the academic literature, but all, in one way or another, refer to essentially the same phenomena. These are:

(1) The increased depth of economic integration or interdependence in the world economy as a whole. Increased depth here usually refers to the integration of different parts of the world and different working populations in the world in the process of economic production itself (Dicken 2003).

(2) The central role played by electronic means of communication and information transmission in facilitating this new deep integration of the world economy.

(3) The much increased importance of global markets in both money and capital in the world economy as a whole (Thurow 1996).

(4) The historically unprecedented scale of international population migration occurring in the world economy in response (primarily) to new work opportunities created by the development of a genuinely global economy.

(5) Sharply increased economic inequalities both within and between different parts of the globe occurring primarily as a result of the very social and spatial "unevenness" of the globalization process.

In addition, there are conceptions of globalization that embrace, but go beyond, these economic aspects of the process to encompass political and cultural phenomena. These include:

(6) The ineluctable spread of a single, materialistic, consumerist culture driven by the Western-dominated global mass media (including both the Internet and television), which in the early twenty-first century forms dominant images of the desirable or good life everywhere on the globe (Castells 1996).

(7) The more or less rapid weakening of the political power of the nation-state in the global economy, a weakening shown by the reduced ability of such states to control crucial economic variables that determine the welfare and standards of living of their populations (Martin and Schumann 1997).

(8) Enhanced cultural and political conflicts in the world caused both by the increasing intermingling of culturally diverse populations in states receiving ever-larger numbers of global labor migrants, and by the so-called clash of cultures or civilizations in

different parts of the world, a clash in part produced by the very information and communications revolution referred to in (2) above. Greatly increased cross-cultural contact also makes different populations aware both of the ever-increasing inequalities among them—see 5 above—and of the different value orientations different cultures may embody. In this conception both global terrorism and the security threats it poses are themselves aspects of globalization (Wade 2001).

Globalization: Its Causes

There is broad unanimity on the origins and causes of globalization. As an economic process globalization dates from the mid- or late 1970s when the postwar "long economic boom" came to an end. The ending of the boom, and the initiation of a much slower growth trajectory for the world economy as a whole, created much more competitive conditions for all firms operating in that economy. The most common firm responses to these heightened competitive conditions were to:

(1) Reduce labor costs by increased automation and "technologization" of production;

(2) Subcontract or "outsource" design, transport, customer service, and even some managerial functions to "independent" consultancy or other firms, thereby reducing "core" labor and payroll costs;

(3) Transfer labor-intensive production activities, that could not be automated to lower wage regions, either in the "home" country or outside the home country altogether.

In addition:

(4) the development and commercial application of computer and information technology from the 1970s onward much facilitated processes (1) to (3) above, and

(5) the ending in roughly the same period (late 1970s and early 1980s) of the postwar Bretton Woods regime of fixed exchange rates facilitated the rapid expansion of global capital and money markets, markets that are themselves deeply dependent on sophisticated information technologies—4 above—for their functioning (Dicken 2003).

In short then, globalization as an economic process dates back no earlier than the mid-1970s, and its political, cultural, and security aspects have also all developed since that time.

Globalization: Its Originality

Although the causality and chronology of contemporary globalization is not disputed, its originality or uniqueness is. Globalization skeptics argue that the nineteenth-century global economy saw flows of investment capital and of international labor migrants that were *proportionately* larger in relation to global economic output or to the then existing world population than contemporary flows are. The nineteenth century also saw very rapid average annual increases in world trade, at periods on occasion larger than contemporary increases. Globalization skeptics even doubt whether modern communications technologies (such as satellite television or the Internet) are any more "revolutionary" in contemporary conditions than was the nineteenth century introduction of the electric telegraph to a world that had previously moved international mail by horse or sail and steamship (Hirst and Thompson 1999).

Although such skeptical arguments have some merit, they understate both the multidimensionality and variety of contemporary communications technologies and the absolute size of current trade, capital, and labor flows. Both the absolute size of the global economy and of the world population are much greater than they were in the nineteenth century. Most importantly of all, such globalization skeptics appear to confuse the "shallow" integration of nineteenth-century economies with the "deep" integration of the contemporary global economy. That is, contemporary international trade is structured (through the massive movement of raw materials and of semifinished goods) so that national economies are tied together *within the production process itself*. The production of everything from cars and other motor vehicles, to electronics, to clothing, footwear, and fashion accessories involves dovetailing inputs from factories located in several different countries through the global trade in goods and services. In this process of deep global economic integration, trade and production become increasingly difficult to distinguish (Dicken 2003). This is a very different situation from that of the nineteenth century, and it makes all countries involved much more vulnerable than ever before to a breakdown, or even to any significant disruption, of the global trade/production system.

Globalization: Its Merits and Demerits

The most discussed and disputed aspect of globalization focuses on the human welfare and economic distributional aspects of the process.

There is broad unanimity that the globalization period in recent history has also been a period of rapidly increasingly income and wealth inequalities both within individual national economies and societies and within the global population as a whole. Agreement ends at this point, however, and there are fierce debates about:

(1) Whether this growing inequality is a product of globalization itself or of the political form globalization has taken—most notably the generally neoliberal political and policy framework—that tends to discourage significant political control or guidance of the process.

(2) Whether this growing inequality matters in any case, if globalization has a tendency to significantly reduce world poverty.

(3) Whether globalization is even achieving poverty reduction, however, is itself a matter of debate, specifically over such matters as how poverty is measured and how increases or reductions in its magnitude are to be assessed (Kitching 2001, Collier and Dollar 2002, Wade 2001).

(4) Whether economic globalization is environmentally sustainable. Here connections are made between economic globalization, especially the spread of industrialization in Asia, Central America, and elsewhere—and such phenomena as global warming.

(5) The strong regional disparities in the spread of globalization and its benefits (and especially the disparity between East and Southeast Asia, on the one hand, and sub-Saharan Africa and the Middle East, on the other).

(6) The very poor labor and environmental conditions existing even in those countries and regions of the world, such as China and East Asia, that are supposedly benefiting from the process. Here it is suggested that regional benefits may not convert into human benefits at all.

(7) Finally, whether there is any connection between the globalization process, and its admitted inequalities, and the upsurge of political terrorism in the world. It is widely admitted, however, that if there is such a connection it is not directly economic. For although contemporary Islamist terrorism is centered in a part of the world (the Middle East) that has fared comparatively poorly in globalization, its militants and activists do not appear to be particularly poor. Moreover there is no terrorist threat emanating from sub-Saharan Africa, the region of the world that is universally admitted to have fared worst in globalization. If there is a connection between globalization and terrorism it is much more likely to be of an indirect cultural and political sort, not of a direct economic sort.

Conclusion: Globalization, Regulation, and Ethics

Conflicting assessments of the merits and demerits of globalization are often tied to different assessments of alternatives to it. The most obvious "total" alternative to globalization is withdrawal of local or regional communities from the world trade/production system into some form of local self-sufficiency or autarky (so-called localism). But this response seems feasible, even in principle, only if populations opting for it are prepared to accept very large reductions in their material standards of living. And whatever may be the situation in the rich parts of the globe, such a policy is unlikely to be attractive to the already poor majority of the world population (Mandle 2003).

In practice therefore, debates and disputes over globalization are most often focused, not on entirely "undoing" its economics, but on the possibility and desirability of politically regulating it so as to reduce its economic volatilities, inequalities, and negative environmental impacts. The central issue at the heart of such debates (aside from whether such regulation is desirable or possible at all) is whether nation-states can continue to be the prime political regulators of the global economy or whether globalization has passed beyond the regulatory capacity of states, so that the task must be turned over to supranational economic and political bodies such as the International Monetary Fund (IMF), World Bank, World Trade Organization (WTO), and International Labour Organization (ILO). But if the latter are to do so, many believe that their responsibilities and powers will have to be enhanced. Advocates of the supranational regulation of globalization are often (though not always) also advocates of a more or less radical restructuring of such bodies in order to make them more genuinely responsive to global public opinion and not simply to the views and preferences of the richest and most powerful states in the world (Stiglitz 2002).

The latter notion recalls the original post-World War II understanding of globalism as a promotion of internationalism in response to the threat of nuclear warfare. Proposals for the international control of nuclear weapons were, for instance, often promoted and stigmatized as one-worldism. To what extent, one may ask, were mid-twentieth century efforts such as the creation of the United Nations and the formulation of the

Universal Declaration of Human Rights the foundations for subsequent economic globalization or institutions and ideals that may help guide it.

From this perspective one may also consider a host of issues related to science, technology, and ethics. Certainly globalization as a phenomenon would not be possible with both science and technology. But does globalization imply or require the universalization of ethics and ethical standards in the same way that it implies and promotes the universalization of technical standards? Can research protocols that are appropriate for HIV/AIDS drugs in Europe and North America be transferred to Africa and Asia? Do professional ethics codes for scientists and engineers function in the same way countries with strong and weak civil society institutions? It is such questions that suggest the importance of both globalism and globalization to the ethical promotion and assessment of science and technology.

GAVIN KITCHING

SEE ALSO *Development Ethics; International Relations; Modernism; Political Risk Assessment; Poverty; Television; Work.*

BIBLIOGRAPHY

Bhagwati, Jagdish. (2004). *In Defense of Globalization.* New York: Oxford University Press. An eloquent defense of globalization from orthodox economic premises.

Castells, Manuel. (1996). *The Information Age: Economy, Society, and Culture,* Vol. 1: *The Rise of the Network Society.* Malden, MA: Blackwell. Probably the best known single work dealing with the information technology dimension of globalization and its possible social and cultural implications.

Collier, Paul, and David Dollar. (2002). *Globalization, Growth, and Poverty: Building an Inclusive World Economy.* Washington, DC: World Bank; New York: Oxford University Press. The standardly optimistic "World Bank" view of globalization.

Dicken, Peter. (2003). *Global Shift: Reshaping the Global Economic Map in the Twenty-First Century,* 4th edition. New York: Guilford Press. An extremely comprehensive and empirically thorough standard textbook on globalization. An excellent non-dogmatic starting point for any beginning student of the subject.

Friedman, Thomas L. (1999). *The Lexus and the Olive Tree.* New York: Farrar, Straus, Giroux. The source of the "Golden Arches" theory of international relations: Countries that are sufficiently capitalistic and consumerist as to have at least one McDonalds franchise do not go to war with each other.

Hirst, Paul, and Grahame Thompson. (1999). *Globalization in Question: The International Economy and the Possibilities of Governance,* 2nd edition. Cambridge, UK: Polity. Perhaps the best singe statement of the skeptical view of globalization as any kind of genuinely new or original phenomenon.

Kitching, Gavin. (2001). *Seeking Social Justice through Globalization: Escaping a Nationalist Perspective.* University Park: Pennsylvania State University Press. Orthodox economic analysis combined with some rather unorthodox political prescriptions and implications.

Mandle, Jay R. (2003). *Globalization and the Poor.* Cambridge, UK: Cambridge University Press. Takes a rather similar view to Kitching but with a much tighter and deeper focus on the issue of poverty and its alleviation.

Martin, Hans-Peter, and Harald Schumann. (1997). *The Global Trap: Globalization and the Assault on Prosperity and Democracy,* trans. Patrick Camiller. London: Zed Books. One of the first and one of the best radical critiques of globalization.

Norberg, Johan. (2003). *In Defense of Global Capitalism.* Washington, DC: Cato Institute. Rather similar to Bhagwati in its analysis, but rather more polemical in tone.

Steger, Manfred B. (2002). *Globalism: The New Market Ideology.* Lanham, MD: Rowman and Littlefield. A text that provides a very useful contrast to Bhagwati and Norberg as an illustration of the levels of ideological polarization among scholars produced by globalization.

Stiglitz, Joseph E. (2002). *Globalization and Its Discontents.* New York: Norton. A very interesting "insiders" view of the financial dimensions of globalization. Deals with some technical economic issues but in a very accessible way.

Thurow, Lester C. (1996). *The Future of Capitalism: How Today's Economic Forces Shape Tomorrow's World.* New York: Morrow. Early text on the globalization phenomenon and still one of the most sophisticated and prescient.

Wade, Robert Hunter. (2001). "The Rising Inequality of World Income Distribution." *Finance and Development* 38(4): 37–39. Very useful statistical compendium on the inequality issue.

GLOBALIZATION

SEE *Globalism and Globalization.*

GLOBAL POSITIONING SYSTEM

• • •

The Global Positioning System (GPS) allows users to pinpoint their location anywhere on Earth to within a few meters. GPS technology was developed for military use, but by the early twenty-first century it had acquired numerous civilian applications including navigation, mapping and surveying, optimizing emergency response systems, and precision agriculture. The major ethical and legal challenges of this technology relate to national control and the potential end-uses of GPS-derived locational data. The U.S. Department of Defense provides the global GPS infrastructure; civilian use is maintained at the discretion of the U.S. government. Personal privacy is a concern because GPS capabilities, embedded in devices such as cell phones, can allow third parties to track the location of individuals. Regulations and laws covering such surveillance are not fully developed.

GPS almost always refers to the NAVSTAR system, the most widely used Global Navigation Satellite System, developed and maintained by the United States government. The U.S. Department of Defense originally developed GPS to locate submarines accurately and thus calculate trajectories for ballistic missile launches. The system depends on twenty-four satellites that continuously broadcast radio signals, positioned in precise orbits approximately eleven nautical miles above Earth. The first satellite was launched in 1978 and the network was completed in 1994. The signals and satellite locations are monitored and corrected as necessary from five ground control stations. A GPS receiver picking up signals from four satellites can compute its location, often to an accuracy of less than ten meters, anywhere on the globe.

GPS depends on the accurate maintenance of the satellites, signals, and related control systems—all of which are entirely under the control of the United States government. The United States deliberately degraded the signal available to civilian users until May 2, 2000. A full-precision civilian signal has since been available to all users, and the United States says that it intends to maintain free worldwide access to the signal. As a result, GPS is increasingly an international utility provided by one nation. The satellites broadcast a separate code for military use, and the U.S. military can jam the civilian signal to selected areas.

GPS itself is an inert provider of locational data. To be used as a tracking device, it must be linked to a communications system. Using GPS in monitoring, surveillance, or intelligence systems raises questions about the invasion of individual privacy, and the legal requirements for warrants and informed consent. GPS-communications devices are often placed on emergency and delivery vehicles to track their locations and optimize their usages. This technology can also be used to track the movements of personal vehicles and to monitor the movements of people including Alzheimer's patients and criminals. The U.S. Federal Communications Commission has directed that cell phones should be locatable in case of an emergency call; placing a GPS link in

cell phones is one way to achieve this. The legal implications of being able to monitor a person's location and movements remotely have not been fully established.

An essential component of modern warfare, GPS is integrated in many advanced weapons and sensors. Combined with communications and geographic information systems, GPS provides comprehensive information on the location and movement of troops and assets, and allows accurate targeting of missiles. Some people have ethical concerns about the military applications of GPS, while others argue that accurate location information lowers collateral damage in warfare.

GPS has evolved from a military system into a widely used global utility, although the basic signal remains available at the discretion of the U.S. National Command Authorities. Individual jurisdictions have yet to decide acceptable parameters for the use of data derived from the GPS signal.

MAEVE A. BOLAND

SEE ALSO *Aviation Regulatory Agencies; Geographic Information Systems.*

BIBLIOGRAPHY

Balough, Richard C. (2001). "Global Positioning System and the Internet: A Combination with Privacy Risks." *CBA Record* (Chicago Bar Association) 15(7): 28–33.

Larijani, L. Casey. (1998). *GPS for Everyone: How the Global Positioning System Can Work for You.* New York: American Interface Corporation.

GLOBAL WARMING

SEE *Global Climate Change.*

GOVERNANCE OF SCIENCE

• • •

Scientific research is a human activity governed by human choice. Governance is exercised at many levels, from the individual scientist deciding how to design an experiment or interpret and report data, to scientific organizations that advocate research funding, to government bureaucrats allocating resources among various projects or programs, to elected representatives establishing budgetary and programmatic priorities, and citizens lobbying to support (or oppose) a particular type of research or technology. Because the consequences of science so powerfully affect the constitution and evolution of society, appropriate governance mechanisms are a key ethical issue for democratic society.

A Republic of Science?

In an influential and powerfully argued paper titled "The Republic of Science, Its Political and Economic Theory" (1962), Michael Polanyi made the case that science was best understood as an autonomous, self-governing activity. Scientists were best positioned not only to understand how to conduct their own research, but also to determine the appropriate directions and levels of effort for new investigations. Likened to the *invisible hand* of the economic marketplace, Polanyi portrayed the governance of science as an emergent consequence of a continual confrontation between an open community of researchers carrying out unconstrained inquiry and nature itself. Interference with this process would lead only to the automatic and inevitable diminution of the ability of science both to advance knowledge and to benefit society.

Polanyi's argument was provoked by attempts in the Soviet Union to subjugate certain scientific disciplines (notably agriculture and genetics) to Marxist dogma, and efforts in England to tie public research agendas more directly to social needs (Polanyi 1964). It also reflected the intellectual conviction that successful scientific endeavor demanded adherence to a clear set of behavioral norms, collectively characterized as "organized skepticism," that were shared by the scientific community as a whole, and which were the only appropriate constraints on the governance of scientific inquiry (Merton 1942).

The practical embodiment of these ideas was articulated by Vannevar Bush, director of the U.S. Office of Scientific Research and Development during World War II. Bush argued, in the seminal policy tract *Science, the Endless Frontier* (1945), that while the public interest would be advanced by a robust, publicly supported science enterprise, the governance of that enterprise was best left entirely in the hands of scientists.

Yet this view, at least in its most extreme form, was explicitly rejected by politicians who believed that no publicly supported enterprise should be fully shielded from democratic accountability (Kevles 1987). Moreover the tremendous expansion of publicly funded research and development enterprises in the United States and other developed nations since the middle of the twentieth century has been accomplished through a variety of political means, in response to a variety of external pressures (notably, the Cold War, but also soci-

etal concerns about health, economic performance, and the environment). The details of this political history utterly vitiate any notion of science advancing according to its own lights, and governed according to its own rules (Greenberg 1967, 2001). Thus, while it is certainly the case that the *conduct* of science is significantly governed by norms and practices that are internal to the research system itself, the more important point is that *directions and velocity* of scientific advance reflect a multitude of factors, many of which are external to science itself (Sarewitz 1996, Kitcher 2001).

Yet the power of Polanyi's position remains strongly in evidence to this day, in the rhetoric used to defend the scientific enterprise from the influence of politics, and in the attitudes of a U.S. public that continues to view science largely as an ungovernable and ungoverned activity whose benefits to society are at once inevitable and unpredictable. For example, National Science Foundation (NSF) survey data consistently show exceptionally strong public support for the statement: "Even if it brings no immediate benefits, scientific research that advances the frontiers of knowledge is necessary and should be supported by the Federal Government" (National Science Foundation, ch. 7).

Documents promoting particular avenues of publicly funded science do so not by invoking the right and obligation of a democratic polity to choose the kind of science it will have, but by repeating what are essentially metaphysical arguments about the autonomous progress of science and its automatic connection to social benefit (Sarewitz 1996). Indeed it is fair to say that a sort of schizophrenia exists between the reality of a science and technology enterprise that is highly governed by decisions made at many levels of society, and the rhetoric of public discourse that perpetuates the illusion of an autonomous, internally governed *Republic of Science* (see, for example, U.S. House Science Committee 1998). This tension is deeply problematic because, concealed by the illusion, is the diverse array of human beings, working in diverse institutions, and ranging from scientists in laboratories to legislators casting votes and corporate executives determining market strategies, that in fact do govern the enterprise by making choices every day about what science to do and how to do it. The persistent notion that science is ungoverned or self-governed, that is, shields from scrutiny those who actually govern.

Political Reality

Nor do different types of research activities—embodied, for example, in the axiomatic taxonomy of unguided basic research, applied research, and development—carry implications about levels or appropriateness of governance. While Polanyi and Bush before him were centrally concerned with an idealized notion of basic research, the politics of science have made no such distinctions. The advance of basic biomedical research has ridden such political campaigns as the *war on cancer* (which was initially much opposed by medical researchers), while such *pure* fields as subatomic physics were justified in practical terms of the Cold War or economic competitiveness. The Republic of Science has, at one time or another, systematically failed to pursue research relevant to vast areas of socially important inquiry, such as diseases characteristic of poor people and regions, and alternative (nonhydrocarbon and nonnuclear) sources of energy. Conversely political action, motivated by interest groups rather than scientists, has been responsible for moving scientific priorities toward areas that had been explicitly avoided by the Republic of Science, for example, research on women's health, and on alternative (non-Western) medicine.

Even the norms and practices of science itself are subject to external governance. Most obviously, the rights of human subjects who participate in scientific experiments are protected by external mechanisms ranging from the Nuremberg Code (a response to Nazi Abuses) and the Helsinki Declaration to decentralized Institutional Review Boards (IRBs) operating in U.S. universities and laboratories (Woodward 1999). These governance mechanisms dictate, for example, that human subjects can participate in experiments only if they have given prior informed consent, a condition that sharply limits the types of science that may be conducted on humans. Additionally, partly in response to political activism that highlighted instances of unnecessary, and unnecessarily cruel, use of animals in research, regulations, norms and practices have progressively evolved in the United States since the 1960s to both reduce the use of, and suffering by, animals in science.

Scientific practice is governed in other arenas as well; for example, national security concerns have dictated where and how certain types of science are conducted, and how scientists can behave in and outside the laboratory. In response to fears of biopiracy, a growing number of nations have passed laws that prohibit foreign scientists from collecting biological samples. The overall point is that, as a societal activity, science is necessarily, appropriately, and unavoidably governed by society. The scientific community, similar to other interest groups, reactively opposes new governance structures, but the scientific enterprise as a whole has demonstrated itself to be remarkably resilient and productive under a wide variety of governance regimes, provided that such regimes do not seek to influence or control the actual results of scientific research (Sarewitz 2003).

Governing the Genome

Some of the most far reaching questions of scientific governance in the early twenty-first century are those associated with human genomics. These questions can only partly be laid at the door of the ongoing debates over abortion and the moral status of embryos. With science already able to intervene in reproductive processes (for example, screening for genetic attributes ranging from sex to particular diseases), and on the verge of a capacity to engineer both individual humans and human germ lines (Stock 2003), profound and complex ethical questions emerge whose resolution may strongly influence future directions of both science and of society (Fukuyama 2002, Wolbring 2003). In most developed countries, these questions are sufficiently conspicuous to command the close attention of government leaders and citizens alike (for example, U.S. presidents Bill Clinton and George W. Bush both convened advisory panels on bioethics), and sufficiently troubling to legitimate the possibility that some lines of scientific endeavor, such as those that could lead to human cloning or manipulation of the human germ-line, should simply not be pursued.

Opposition to a stricter governance of genomics research relies on three lines of argument: first the need to protect freedom of inquiry from societal interference; second the loss of potential social benefits (for instance, enhanced medical treatments); and third the likelihood that even if one country decides to prohibit or restrict a given line of research, others will surely decide to move ahead at full speed.

The first two arguments have little practical validity. Inquiry is never entirely free, and while science surely should be protected from inappropriate societal interference, the definition of what constitutes *appropriate* governance is constantly being renegotiated within society. Similarly choices about what science will be supported by society are continually being made in the public and private sectors, and any such choices entail opportunity costs. There is no reason to believe that the organization of scientific inquiry at any given time will yield optimal results for society.

The third argument is ethically troublesome, but difficult to dismiss in practice. While nations may decide to forego areas of research for moral reasons, the global science enterprise is so institutionally, sectorally, and geographically diverse that uniform compliance with any particular governance decision is likely to be impossible. Despite a fairly broad, global consensus against reproductive cloning, for example, it is inevitable that humans will be cloned at some point simply because the state of the science will allow it to be accomplished. Similarly the vast commercial potential for a wide variety of genetic enhancement and germ-line interventions is likely to be attractive enough to ensure that they will be aggressively pursued somewhere. Of course this likelihood neither justifies participation in such research, nor implies that restraint is without value. For example, the choice not to engage in some lines of research may allow particular nations or cultures to protect cherished values, and could influence choices made by other nations in the more distant future. Moreover, by slowing the advance of science in some areas (just as progress toward reproductive cloning has been slowed), society affords itself more time to develop effective principles and regulations for governance of such unprecedented innovations.

Modulation, not Control

Thus while science is, and will remain, a highly governed activity, this governance should not be confused with control. Rather it is a process by which the momentum and direction of scientific advance are subject to some degree of modulation via human decision making. Particular governance decisions may (or may not) be wise, may (or may not) reflect a commitment to the common good, and so on. The point of this entry is simply to explain that such decisions cannot and *therefore should not* be avoided. As science acquires the capacity to reengineer humanity itself, the choice to slow down, or orient this capacity in particular directions while avoiding others, remains open, but the balance among the attraction of commercial opportunities, the prerogatives claimed on behalf of the Republic of Science, and ethical concerns about the appropriate limits of science remain to be negotiated.

DANIEL SAREWITZ

SEE ALSO *Atlantis, Old and New; Poliical Economy of Science and Technology; Political Risk Assessment.*

BIBLIOGRAPHY

Bush, Vannevar. (1945). *Science the Endless Frontier.* Washington, DC: Government Printing Office. Also available from www.nsf.gov/od/lpa/nsf50/vbush1945.htm.

The seminal science policy document of the twentieth century.

Fukuyama, Francis. (2002). *Our Posthuman Future: Consequences of the Biotechnology Revolution.* New York: Farrar, Straus and Giroux. Conservative political theorist who argues that the implications of biotechnology demand to be regulated.

Greenberg, Daniel S. (1967). *The Politics of Pure Science.* New York: New American Library.

Greenberg, Daniel S. (2001). *Science, Money, and Politics. Political Triumph and Ethical Erosion.* Chicago: University of Chicago Press. Along with his earlier book, provides detailed chronicles of the political realities that underlies the "purity" of science.

Kevles, Daniel J. (1987). *The Physicists: The History of a Scientific Community in Modern America.* Cambridge, MA: Harvard University Press.

Kitcher, Philip (2001). *Science, Truth, and Democracy.* Oxford and New York: Oxford University Press. Elegant treatment of the role of human choice in determining the structure of scientific inquiry.

Merton, Robert K. (1942). "The Normative Structure of Science." *Journal of Legal and Political Sociology* 1: 115–126. Classic discussion of the internal cultural norms of science.

National Science Foundation. (2004). "Science and Technology: Public Attitudes and Understanding." In *Science and Engineering Indicators.* Arlington, VA: National Science Foundation, Division of Science Resources Statistics. Also available from http://www.nsf.gov/sbe/srs/seind04/c7/c7s3.htm#c7s3120. Comprehensive compilation of data about the scientific enterprise.

Polanyi, Michael. (1962). "The Republic of Science: Its Political and Economic Theory." *Minerva* 1(1): 54–73. Classic statement of the argument for scientific autonomy.

Polanyi, Michael. (1964). "Background and Prospect." In *Science, Faith and Society.* Chicago: University of Chicago Press.

Sarewitz, Daniel. (1996). *Frontiers of Illusion: Science, Technology, and the Politics of Progress.* Philadelphia: Temple University Press.

Sarewitz, Daniel. (2003). "Science and Happiness." In *Living with the Genie: Essays on Technology and the Quest for Human Mastery,* eds. Alan Lightman, Daniel Sarewitz, and Christina Desser. Covelo, CA: Island Press.

Stock, Gregory. (2003). *Redesigning Humans: Choosing our Genes, Changing our Future.* Boston: Mariner Books.

U.S. House Science Committee. (1998). *Unlocking our Future: Toward a New National Science Policy.* Washington, DC: Author. Policy report of September 24. Also available from http://www.house.gov/science/science_policy_report.htm.

Wolbring, Gregor. (2003). "Confined to Your Legs." In *Living with the Genie: Essays on Technology and the Quest for Human Mastery,* eds. Alan Lightman, Daniel Sarewitz, and Christina Desser. Covelo, CA: Island Press. Provocative inquiry into the implications of human biotechnology, from the perspective of a disabled person.

Woodward, Beverly. (1999). "Challenges to Human Subject Protection in US Medical Research." *Journal of the American Medical Association* 282(20): 1947–1952.

GRANT, GEORGE

• • •

Philosopher and Canadian nationalist, George Grant (1918–1988), born in Toronto, Ontario on November 13, rose to prominence in the 1960s through his concern that the homogenizing nature of modern technology would lead to the destruction of Canadian independence. He came from a family of prominent Canadian educators. A Rhodes scholar, Grant taught at Dalhousie University in Halifax, Nova Scotia, and McMaster University in Hamilton, Ontario. His meditations on the character of technology led to election to the Royal Society of Canada, several honorary degrees, and an appointment to the Order of Canada.

Grant saw the origins of the Western predicament as follows: Natural law philosophers such as philosopher and religious Thomas Aquinas (c.1225–1274), following the tradition of antiquity, taught that there were moral laws beyond space and time that were absolutely and universally binding on all human beings. In the seventeenth century a British philosopher, Francis Bacon (1561–1626), envisaged a radically new scientific project equally binding: In the future, science was to make human beings the masters of nature. Their moral authority for this dominion was enhanced by the eighteenth century philosopher Immanuel Kant (1724–1804), who maintained that the essential characteristic of human beings was their freedom and that they were bound only by moral rules to which they had freely assented. Aquinas, Bacon, and Kant together had forged the modern world.

Each of their positions seemed, by itself, true and necessary for human well-being. Yet they were, in principle and in practice, incompatible. Grant's philosophical contribution was to reveal the implications of these contradictions and alert his contemporaries to the need for a resolution. Grant was genuinely perplexed. As his early writings show, he understood technology as the dominance over human nature, but the tools it developed—the automobile, the washing machine, penicillin—led to genuine improvements in the human condition and in human freedom. Yet the same technology also brought the holocaust and the atomic bomb. He laid out these contradictions in his first important work, *Philosophy in the Mass Age* (1958), but offered no resolution.

Grant's View of Technology

One quality of modern technology, Grant came to understand, lay in its tendency to impose uniformity. The French philosopher Alexander Kojève (1902–1968) theorized that the whole world was moving relentlessly toward a universal and homogeneous state. For Kojève such an outcome was desirable, since it was a prelude to a universal peace where war between classes or nations no longer existed. In the work that made him famous throughout Canada, *Lament for a Nation* (1965), Grant accepted this understanding of the impact of technology, but for him it was not a cause to rejoice. He maintained that Canada's geographical position next to the dynamic center of technological modernity, the United States, would lead to its eventual disappearance as a independent country, since Canadians and Americans shared the same commitment to technological modernity. "Our culture floundered on the aspirations of the age of progress." (Grant 1965, p. 54)

In *Technology and Empire* (1969), Grant's concerns about the dangers of technology became more intense. Science, he now argued, no longer limited itself to the domination over non-human nature; it now increasingly attempted domination over human nature as well. Some critics of technology believed that it was something *out there* that people could control should they so choose. Grant rejected this view. For him technology was not something outside of people that they could choose to use for good or ill. Human beings lived in a society (and increasingly a world) in which technology determined all existence. "For it is clear that the systematic interference with chance was not simply undertaken for its own sake but for the realisation of freedom ... [but] how do we know what is worth doing with that freedom?" (Grant 1969, p. 138).

The predicament of modernity was that those men and women who were the driving forces behind technological modernity believe that their project promotes "the liberation of mankind" (Grant 1969, p. 27). The older tradition of Plato, Aristotle, and Aquinas held that there were some things that it was absolutely wrong to do and perhaps even wrong to contemplate. By contrast Grant often attributed to J. Robert Oppenheimer (1904–1967) the view that, in modern science, no matter how terrible the possible outcome of an experiment might be, if you see that something is technically balanced, you do it.

When asked whether computers were neutral instruments, Grant observed that their existence required the work of chemists, metallurgists, and mine and factory workers; the use of algebra and other mathematics,

Newtonian and other physics, and electricity; as well as a society in which there are many large corporations. Such a society contains an elite trained to think in a particular way and excludes other forms of society. Technology can never be neutral because of its historical, social, and conceptual preconditions.

Technology for Grant, then, was not just a way of making things or even a way of doing business. It was a way of thinking and it was becoming a way of being. So when the U.S. Supreme Court handed down its historic decision in *Roe* v. *Wade*, 410 U.S. 113 (1973), an abortion case, Grant was profoundly worried. The account of justice given there, influenced as he thought by technological modernity, seemed to put into question what it was to be a person. Consequently modern liberalism seemed unable to answer the question: "What is it about any members of our species that makes the liberal rights of justice their due?" (Grant 1998, p. 78).

Grant never denied that science had delivered the dominance over nature it promised, but it failed in a much more important way. "Brilliant scientists have laid before us an account of how things are, and in that account nothing can be said about justice." (Grant 1986, p. 60) But above all justice mattered. In his last book, *Technology and Justice* (1986), he argued that the technological understanding of the world was fundamentally flawed. Love was a primary fact of human existence; modern human beings "cannot hold in unity the love they experience with what they are being taught in technological science" (Grant 1986, p. 67).

Grant's writings still actively influence Canadian politicians, political scientists, theologians, and scholars interested in technology. Most philosophers are indifferent or hostile.

WILLIAM CHRISTIAN

SEE ALSO *Ellul, Jacques; Heidegger, Martin; Justice.*

BIBLIOGRAPHY

Athanasiadis, Harris. (2001). *George Grant and the Theology of the Cross: The Christian Foundations of His Thought.* Toronto: University Of Toronto Press. Grant's critique of technology was rooted in a sincere and original theological vision. This work examines Grant's philosophy in relation to his understanding of Martin Luther.

Christian, William. (1993). *George Grant: A Biography.* Toronto: Univeristy of Toronto Press. A comprehensive scholarly biography of Grant that made extensive use of archival sources, letters, and personal interviews.

Grant, George. (1965). *Lament for a Nation.* Toronto: McClelland & Stewart. Grant's masterpiece, a complex interweaving of politics, philosophy, and religion that made him a nationally known figure. It explored the idea that technological modernity would lead inexorably to the destruction of Canadian independence.

Grant, George. (1969). *Technology and Empire.* Toronto: House of Anansi. The collection of essays that contained Grant's most concentrated meditations on technological modernity. In these essays Grant came increasingly to see technology in metaphysical terms. He saw it as darkness that prevented human beings from being aware of the absence of divine light in the world.

Grant, George. (1986). *Technology and Justice.* Toronto: House of Anansi. Grant's last collection of essays and his most positive affirmation that Simone Weil's understanding of love shows human beings an alternative to technological modernity.

Grant, George. (1995). *Philosophy in the Mass Age,* ed. William Christian. Toronto: University of Toronto Press. Grant's first book, originally a series of radio lectures, explored the problems of technology as they were arising in the aftermath of the second world war and the new materialism of the 1950s.

Grant, George. (1998). *English-Speaking Justice.* Toronto: House of Anansi. After the *Roe* v. *Wade* decision, Grant worried that technology threatened the very idea of what it was to be human. This work, written in the mid-1970s, attempts an examination of implication of the U.S. Supreme Court's ruling.

GREEN IDEOLOGY

• • •

Green is the color of vegetation, in particular of healthy, growing leaves. At least in the growing season it is the predominant color of undeveloped land in non-polar, non-arid regions. Green as a quality of the landscape was what was destroyed or threatened by the Industrial Revolution in Britain. Thus William Blake, in the poem that has become the hymn *Jerusalem,* contrasted the *green and pleasant land* that England should be with the *dark satanic mills* of his time (early-nineteenth century). And Richard Llewellyn's 1939 novel *How Green Was My Valley* tells the heartbreaking story of the gradual transformation of a rural landscape, where young boys caught trout in the river, to a polluted industrial wasteland where the wastes from coal mining, dumped on the sides of the narrow South Wales valley, threatened to engulf the miners' houses.

Green as undeveloped land, free from industry, is what is evoked by the term *green belt.* Green belt is a planning designation of land around cities or towns intended to prevent urban sprawl, for the benefit of both city and countryside. Green belt land is to be permanently *open,* the presumption being against built development except in special circumstances (UK Office of the Deputy Prime Minister 2001).

Because green is the color of vegetation, and thus plants, it has been linked with agriculture. *Green Europe* was a newsletter on the European common agricultural policy, published by the European Commission. The *green revolution* of the late-1960s and 1970s was about increasing crop yields through the development of new varieties that required high inputs of fertilizers and pesticides. That this form of agriculture was, by the 1990s, considered very un-green is a sign that between the 1970s and 1990s green took on a particular political and philosophical meaning.

Greenpeace was the name taken by a small band of nonviolent, direct activists who, in 1971, tried to take a small boat to Amchitka, an island off the west coast of Alaska where the United States was conducting underground nuclear tests. Greenpeace subsequently became a major environmental nongovernmental organization, campaigning for a *green and peaceful future*. What Greenpeace sees as at stake, threatened by modern technology and economic growth, is not simply a green and pleasant countryside but *the ability of the Earth to nurture life in all its diversity*.

The first political party that took the name *Green* was the West German Green Party, *Die Grünen*. The federal party was formed at the beginning of 1980, but was preceded by numerous local or state-level groups that put up Green or *Rainbow* lists of candidates for elections and, in the case of Bremen Green Slate, won seats in the state parliament. The 5 percent barrier to representation under the West German system of proportional representation meant that there was considerable incentive for a wide variety of different groups to come together as Die Grünen in order to achieve political representation. These groups included those concerned with environmental pollution, protestors against nuclear power, feminists, Marxists, and socialists disillusioned with the Social Democratic Party. They united under the four pillars of ecology, nonviolence, social justice, and grassroots democracy, which have since come to define what it means to be Green.

In the federal elections of 1983 Die Grünen won 5.6 percent of the vote and sent twenty-seven members to the Bundestag. Following this success, parties in other countries with similar philosophies, such as the Ecology Party in the United Kingdom, changed their name to the Green Party. Green parties were also started in other countries, including the United States in 1984. The word *green* evokes rejection of industrialization and protection of life in all its diversity, but also freshness, immaturity, and naivety. The Greens have thus proclaimed themselves to be a fresh force in electoral politics, different from the political elites of the *grey* parties, who the public view as increasingly remote and answerable only to vested interests. Although Greens are often charged with being unrealistic, it is a measure of their success that being green no longer means being naïve.

Newness is also encapsulated in the idea that Green is *neither left nor right but forward*. The influence of anarchism on Green ideology and the resulting rejection of hierarchical structures, results in an emphasis on individual responsibility and initiative akin to that of the right. Greens can also be seen as conservative with respect to technology. They are often skeptical about new technologies that traditional socialism welcomes as enhancing human capacities, defending older technologies and smaller, close-knit communities, though they welcome other innovations, such as solar power and modern wind turbines. However, in their critique of capitalism and the free market, the Greens are firmly on the side of the left. What is new in the green critique is the emphasis on environmental limits: It is the environmental crisis, not the suffering of the proletariat, that makes it imperative to move toward a different economy, technology, and society. This new green society will protect the planet by respecting nature—ecosystems, non-human species, and the rights of animals—and will also be better for the health and well being of humans and their communities.

Green politics and philosophy presents a holistic vision in which monetary reform, participative democracy, meaningful work, social justice, and equality are all of a piece with renewable energy, organic agriculture, protection of wildlife, recycling, and non-polluting technologies. This vision can be sought by the green consumer as well as the voter through boycotting certain goods and buying others (Elkington and Hailes 1988).

Despite this broad holism, green is narrowed in many instances to refer simply to reduced environmental impacts. Thus *green travel plans*, now a condition of many planning permissions in the United Kingdom, are plans introduced by employers to attempt to reduce the use of car transport by their employees. A *green building* is one designed to have reduced impact on the environment during its construction and use.

ANNE CHAPMAN

SEE ALSO *Earth; Ecology; Environmental Ethics; Environmentalism; Green Revolution.*

BIBILOGRAPHY

Dobson, Andrew. (2000). *Green Political Thought*, 3rd edition. New York: Routledge. One of the many publications giving an account of green political thought and philosophy.

Elkington, John, and Julia Hailes. (1988). *The Green Consumer Guide*. London: Victor Gollancz.

Spretnak, Charlene, and Fritjof Capra, in collaboration with Rudiger Lutz.. (1984). *Green Politics: The Global Promise*. London: Paladin. An early account of Green politics in West Germany and the prospects for it in the rest of Europe and North America

INTERNET RESOURCES

Greenpeace. (2005). Available from www.greenpeace.org.

UK Office of the Deputy Prime Minister. "Planning Policy Guidance 2: Green Belts." Available from http://www.odpm.gov.uk/stellent/groups/odpm_planning/documents/page/odpm_plan_606905.hcsp.

GREEN REVOLUTION

• • •

The Green Revolution (not to be confused with "green" as in the environmental movement) was a dramatic increase in grain yields (especially wheat and rice) in the 1960s and 1970s, made possible by the Rockefeller Foundation's development of high-yielding wheat and rice varieties starting in the 1950s. The moral good of producing more food seems unquestionable. Indeed, Norman Borlaug (b. 1914), the scientist who spearheaded the Green Revolution, received the 1970 Nobel Peace Prize for his work. Yet the Green Revolution did spur ethical disputes over the social and environmental changes its technologies produced, especially in the developing world. Proponents argued that increased food supply benefited society generally; opponents pointed to the ways that poorer segments of societies were disproportionately hurt by the Green Revolution. In the early twenty-first century, Green Revolution technologies continue to promote conflict between those who see them as tools in service of society and those who argue that they promote injustice.

Competing Views of Development

The controversy over the social justice of the Green Revolution was apparent from the start of the Rockefeller Foundation work in Mexico. Encouraged by U.S. Vice President Henry Wallace, the Rockefeller Foundation in 1941 offered to send agricultural advisors to Mexico to help improve its wheat crop. The Rockefeller family had both a history of humanitarian work and valuable oil properties in Mexico. Both the family and Wallace were concerned about increasing social unrest in Mexico and sought solutions that would not reawaken interest in the previous Mexican administration's attempts to redistribute land to the poor (Wright 1990). The Rockefeller Foundation officers believed that they could stabilize Mexican society by increasing the supply of cheap, domestically-grown food. The Rockefeller Foundation's survey team of cutting-edge agricultural scientists, including plant breeders and agricultural chemists, unsurprisingly advocated technologies that had proved successful in the United States: the development of new, high-yielding varieties of major crops. North American farmers had profited from this system, despite the increased cost of purchasing new seed stock every year, and Rockefeller expected the same results of *modernization* in Mexico (Fitzgerald 1986).

Critics attacked the plan as inappropriate for small farms, which they believed ought to be the target of any agricultural improvement in Mexico. Carl Sauer, a geographer from the University of California Berkeley, argued that the plan would be disastrous for the peasant economy of Mexico, as peasant farmers would be unable to standardize on expensive new seeds. Other critics argued that by excluding experts on Mexican society from the survey team, Rockefeller risked forcing an inappropriate scientific solution on Mexico. The Rockefeller team fired back that Sauer and other critics simply wanted to keep Mexico backward, and were unwilling to let it modernize (Wright 1990).

Behind this sniping was a fundamental disagreement over how to benefit Mexican society. For Rockefeller's critics, improvement had to target economically-pressed peasants to be beneficial. Rockefeller argued Mexico had to rapidly start producing more food, using the best science and technology available. For Rockefeller, *modern* evoked the moral superiority of doing whatever was necessary, socially or technologically, to produce increases in the food supply. Critics argued that the science and technology should be *appropriate* for the majority of Mexican farmers. Both held moral commitments, but to different visions of the Mexican future.

Expected and Unexpected Consequences

Rockefeller adopted the survey team's recommendations. Borlaug's group employed traditional and novel scientific methods to produce high-yielding semidwarf wheat varieties that exceeded all expectations. Semidwarf varieties are stalky plants that can hold a heavy head of grain. These varieties, used with plentiful water, fertilizers, and pesticides, produced dramatically high crop yields. Interest in semidwarf varieties spread quickly, especially where food security was a concern. The Indian government asked Borlaug to help it develop wheat varieties for India; these were ultimately credited with preventing a major famine (Perkins 1997). Governments lauded the social good of the technology that allowed them to import less food despite growing populations and green revolution science was soon extended to other staple grains, especially rice. Rice-producing countries around the world adopted these new rice varieties as readily as had wheat producers. Those who adopted Green Revolution technologies often experienced increases in their standards of living, although in some places, government-mandated food prices sometimes undercut the economic benefits of higher yields (Leaf 1984).

The fears of critics were also realized, especially in the early years. Medium-sized and large farms could adopt the new technologies easily, and their high yields led to declining food prices. While urban populations benefited, small farmers watched the profits from their own harvests decrease. Some smaller farmers were able to adopt the technologies and improve their standards of living, but others were forced into rural labor or to move to the cities. Because people went hungry despite growing food supplies, critics argued that the Green Revolution could create food, but not relieve hunger (Sen 1981). They pointed to regional inequities, as areas suited to Green Revolution grains and favored by government attention flourished, while poorer regions fell behind. For critics, the Green Revolution failed the test of social justice (Shiva 1991).

Later, unanticipated environmental effects fed ongoing debates about social justice. The issue of mono-cropping highlights the environmental angle. Mono-cropping (producing a single crop in a field) helps produce uniform, high-yielding crops. However, it also produces microenvironments in which crops are more vulnerable to insects. Scientists responded by recommending heavy use of pesticides, with serious systemic consequences: sometimes toxic levels of pesticide exposure for farm laborers (who were often those disenfranchised by the Green Revolution), and rapid adaptation by insects requiring constant innovation and resulting in higher prices. Extensive monocropping sometimes led to less diversity in local food supplies, which critics have argued disproportionately affected the nutrition of the poor. In Green Revolution areas, the poor have come to depend almost exclusively on grains, decreasing the nutritional value of their diet (Shiva 1993). In each critique, the question of justice, whether for the poor or for future generations, is the central concern.

Reconsiderations

The attention that critics have paid to social justice, while sometimes questioned by supporters of the Green Revolution, have not fallen on deaf ears. The agency responsible for the scientific development of Green Revolution crops, the Consultative Group on International Agricultural Research (CGIAR), has responded vigorously. Scientists have decreased the amounts of pesticide needed, reducing risk to farm workers and lowering the cost of inputs. They increased the number of food crops for which they have developed high-yielding varieties, including some crops traditionally cultivated by the poor. Scientists have given attention to developing high yielding crops using less water, an important consideration in arid regions. In the 1990s, scientists began to research ways to introduce Green Revolution technologies to the poor regions of Africa that had been previously bypassed.

Advocates have also argued that making Green Revolution technologies socially just is not only the responsibility of scientists, but also of regional and national governments (Hazell 2003). In places where agricultural credit is accessible, more small farmers have been able to retain or expand their land and benefit from the technologies. Such efforts are not lost on critics, but neither have they quieted the criticism that Green Revolution technologies promote injustice. Supporters are equally steadfast that Green Revolution technologies produce social goods that outweigh shortcomings. A widely agreed-upon ethical judgment of the Green Revolution remains unlikely, because the complex social and environmental consequences of this technology continue to unfold.

SUZANNE M. MOON

SEE ALSO *Agricultural Ethics; Food Science and Technology; Green Ideology; Modernization; Poverty.*

BIBLIOGRAPHY

Fitzgerald, Deborah. (1986). "Exporting American Agriculture: The Rockefeller Foundation in Mexico, 1943–53." *Social Studies of Science* 16: 457–483. Historical look at the work of the Rockefeller Foundation during their earliest interventions in Mexico, exploring the reasons that they based their recommendations to Mexico on the "best practice" of U.S. agricultural science.

Hazell, Peter B. R. (2003). "Green Revolution, Curse or Blessing?" In *The Oxford Encyclopedia of Economic History*, ed. Joel Mokyr. Oxford, UK: Oxford University Press. A critical evaluation of the economics of the Green Revolution that argues that poverty reduction through Green

Revolution technologies is possible when appropriate government policies and credit programs are in place.

Leaf, Murray J. (1984). *Song of Hope: The Green Revolution in a Punjab Village*. New Brunswick, NJ: Rutgers University Press. An anthropological study of a Punjab village that explores the positive social and economic changes of the village since the introduction of Green Revolution technologies.

Perkins, John H. (1997). *Geopolitics and the Green Revolution: Wheat, Genes, and the Cold War*. New York: Oxford University Press. A comprehensive history of the science and politics of the Green Revolution focusing on wheat production.

Sen, Amartya K. (1981). *Poverty and Famines: An Essay on Entitlement and Deprivation*. Oxford, UK: Clarendon. Examines connections between food availability, poverty, and the occurrence of famine. Questions whether increases in the food supply eliminate famine.

Shiva, Vandana. (1991). *The Violence of the Green Revolution: Third World Agriculture, Ecology, and Politics*. Penang, Malaysia: Third World Network. Examines the socioeconomic results of the Green Revolution and argues that it does too much harm to poor communities.

Shiva, Vandana. (1993). *Monocultures of the Mind: Perspectives on Biodiversity and Biotechnology*. London: Zed Press. A critical examination of the issue of monocropping, why it appeals, and its shortcomings as a model for agriculture.

Wright, Angus. (1990). *The Death of Ramón González: The Modern Agricultural Dilemma*. Austin: University of Texas Press. A study of agricultural change in Mexico focusing on sociopolitical background and consequences of the Green Revolution.

H

HABERMAS, JÜRGEN
• • •

Jürgen Habermas (b. 1929) was Germany's foremost social theorist and philosopher in the second half of the twentieth century. Born in Düsseldorf, Germany, on June 18, Habermas is the leading representative of the second generation of the so-called Frankfurt School of critical social theory, taking inspiration from Max Horkheimer, Theodor Adorno, and Herbert Marcuse. At the same time Habermas was strongly influenced by the linguistic turn in analytic philosophy from Ludwig Wittgenstein to John L. Austin and John Searle, as well as by the classics of German thought from Immanuel Kant and Georg W. F. Hegel to Karl Marx and Max Weber. In his magnum opus, *The Theory of Communication Action* (1981), Habermas explained the genesis of modern society in terms of basic categories derived from the philosophical study of language and rationality. This analysis reveals that the processes of rationalization characteristic of modernity have been crucially one-sided, privileging the *instrumental* or *strategic rationality* of selecting the most effective means to ends at the expense of the *communicative rationality* of reaching a shared understanding of ends on the basis of reasons that everyone can accept in free discussion.

Science, Technology, and Politics

A central strand in Habermas's narrative of modernity is thus the intrusion of quasinatural scientific and technological imperatives into the realm of politics. This raises the practical and theoretical issue of the proper relationship between science and politics. Habermas outlines three possible views of this in his early "Technology and Science as 'Ideology'" (1968). On Weber's *decisionistic* model, there is

Jürgen Habermas, b. 1929. The German philosopher and sociologist challenged social science by suggesting that despite appearances to the contrary, human beings are capable of rationality and under some conditions are able to communicate with one another successfully. (© *Darren McCollester/Getty Images.*)

a strict separation between the functions of the politician and the expert: The former makes decisions on the basis of values that are at bottom irrational and the latter carries them out as effectively as possible on the basis of scientific knowledge. *Technocrats*, in contrast, see contemporary politics as bound by objective exigencies of preserving the stability of the system. Experts present policy alternatives as necessary for the achievement of goals like economic growth that are presumed to be grounded in objective needs. Thus whereas decisionists see values as irrational, technocrats consider them irrelevant.

But *techne* cannot be substituted for praxis. Needs must be interpreted in the light of values and cultural meanings before they can guide action. Habermas prefers, therefore, the third, *pragmatist* model of John Dewey. Means and ends are interdependent: On the one hand, the horizon of values in a society guides scientific research, on the other, value convictions persist only insofar as they are connected to potential satisfaction through instrumental action. Consequently technology cannot be value-neutral. Practically relevant scientific achievements must be subjected to free public discussion to make possible a "dialectic of enlightened will and self-conscious potential" (Habermas 1970, p. 73) that both allows new technologies to alter public self-understanding and lets that self-understanding determine the course of future research. Insofar as such discussion is governed by the "unforced force of the better argument," it yields decisions on ends that are rational in a sense decisionists failed to recognize.

Such domestication of technological development is impossible if technology as such amounts to ideology. Marcuse claimed that this is indeed the case since the progress of science and capitalism had undermined the legitimacy once enjoyed by religion and tradition. In partial agreement, Habermas argues in *Knowledge and Human Interests* (1968) that empirical science as such is bound up with an anthropologically deep-seated (and therefore *quasitranscendental*) *technical interest* in potential control and manipulation that is constitutive of its object domain. In contrast to Marcuse, however, he sees this interest as invariant, since it is rooted in the universal conditions of material reproduction of human life. As a result, there is no such thing as *alternative science*.

Normative Issues

Where, then, does one find the normative resources to counteract the insidious form of social domination that legitimizes existing inequalities with an appeal to scientific (such as economic) necessity and placates the public with commercialized mass media and slow but steady growth in material comfort brought about by technological development? Habermas's strategy in his early work is to locate two equally fundamental human cognitive interests pertaining to interaction rather than work. As social beings whose very identity depends on mutual

recognition in linguistic interaction, people have a *practical interest* in solving problems of communication and understanding within and between traditions. This is the task of the hermeneutic or cultural sciences (*Geisteswissenschaften*). The *emancipatory interest* in countering the effects of *systematically distorted communication* through critical reflection is exemplified on the individual level by psychoanalysis and on the social level by critique of ideology that reveals the particular economic, political, and social interests that bias self-understandings embedded in human traditions. The ideological aspect of positivist views of science and technology consists in conflating the practical with the technical and thus obscuring the possibility of rationalization along these other dimensions. The problem is the universalization of instrumental thinking, not instrumental thinking itself.

In later work, Habermas replaces appeals to interests with references to the necessary structures of communication elaborated in *formal pragmatics,* but he remains concerned with the effects of technology on human interaction. *The Future of Human Nature* (2001) addresses the specific problem of *liberal eugenics,* genetic intervention designed not to prevent health problems but to create abilities that parents consider to be useful for the child. Habermas argues that this is ethically unacceptable. First, knowledge that they have been preformed according to someone else's preferences makes it impossible for children to view themselves as the sole ethically responsible authors of their own lives. Second, such engineering introduces a fundamental, irreversible asymmetry among the programmers and the programmed that is contrary to the basic principles of symmetric mutual recognition among free and equal persons that are grounded in the very structure of linguistic interaction.

In sum, Habermas's key contribution to the ethics of science and technology is a plausible theory of intersubjective rationality. Such rationality does not reduce to instrumental efficiency and can therefore be used to set nonarbitrary goals and limits to technical development, if implemented in suitable democratic institutions.

ANTTI KAUPPINEN

SEE ALSO *Critical Social Theory; Discourse Ethics; Marcuse, Herbert.*

BIBLIOGRAPHY

Habermas, Jürgen. (1968). *Technik und Wissenschaft als 'Ideologie'*. Frankfurt am Main, Suhrkamp. Trans. (in part) by

Jeremy J. Shapiro as *Toward a Rational Society: Student Protest, Science and Politics*. Boston: Beacon Press 1970.

Habermas, Jürgen. (1968). *Erkenntnis und Interesse*. Frankfurt am Main: Suhrkamp. Trans. Jeremy J. Shapiro as Knowledge and Human Interests. Boston: Beacon Press, 1971.

Habermas, Jürgen. (1970). "Technology and Science as 'Ideology.'" In *Toward a Rational Society: Student Protest, Science, and Politics*, trans. Jeremy J. Shapiro. Boston: Beacon Press.

Habermas, Jürgen. (1971). *Knowledge and Human Interests*, trans. Jeremy J. Shapiro. Boston: Beacon Press.

Habermas, Jurgen. (1981). *Theorie des kommunikativen Handelns*. Frankfurt am Main: Suhrkamp. 2 vols. Trans. Thomas McCarthy as *The Theory of Communicative Action*. Vol. 1: *Reason and the Rationalization of Society*; Vol. 2: *Lifeworld and System: A Critique of Functionalist Reason*. Boston: Beacon Press, 1984 and 1987.

Habermas, Jurgen. (1981). *Theorie des kommunikativen Handelns*, 2 vols. Frankfurt am Main: Suhrkamp. Translated by Thomas McCarthy as *The Theory of Communicative Action*, Vol. 1: *Reason and the Rationalization of Society*; Vol. 2: *Lifeworld and System: A Critique of Functionalist Reason*. Boston: Beacon Press, 1984 and 1987.

Habermas, Jürgen. (2001). *Die Zukunft der menschlichen Natur. Auf der Weg zu einer liberalen Eugenik?* Frankfurt am Main: Suhrkamp. Trans. Hella Beister and William Rehg as *The Future of Human Nature*. Cambridge, England: Polity Press, 2003.

Habermas, Jurgen. (2003). *The Future of Human Nature*. Cambridge, England: Polity Press.

McCarthy, Thomas. (1978). *The Critical Theory of Jürgen Habermas*. Boston: MIT Press. A classic study and critique of Habermas's early work that begins with a thorough discussion of his views on the scientization of politics and critique of instrumental reason.

HACKER ETHICS

• • •

Originally the term *hacker* was used to refer to someone who is enthusiastic about computing, spends a lot of time figuring out how computers work, and is adept at using computers to accomplish extraordinary feats. *Hacking* referred to the activities of hackers. In the early days of computing hackers were exploring the full potential of computers: They were figuring out what it was possible to achieve with computers, doing things that had never been done before. In this sense hackers were like the imaginative mechanics of the early Industrial Revolution, automotive hot-rodders, barnstorming airplane pilots, and ham radio operators. In those early days there were few laws or policies specifying what individuals were allowed to do or prohibited from doing with computers. Many of the feats that hackers

accomplished subsequently became illegal, for example, breaking into private systems, examining what was in those systems and how the systems worked, copying and distributing information and programs, and telling others how to do the same things.

The meaning of the terms *hacker* and *hacking* changed somewhat over time, and *hacker* began to be used to refer to those who engage in illegal computer activity. Many hackers objected to that usage and insisted that a distinction be made between hackers, who are generally law-abiding, and crackers, who use their computer skills to engage in illegal activity. Currently, the term *hacker* is used in both ways. Occasionally the term "hack" is used more broadly to refer to a playful feat involving scientific or technological expertise, for example, when a group of students break into a campus building undetected and leave visible and fanciful evidence of their success at breaking in (Laszlo 2004).

The Hacker Ethic

Individuals who identify with the original concept of hacking continue to exist and share ideas with one another online. They constitute a subculture that has coalesced around computer technology and the Internet. Members of that subculture share an attitude toward computing and a set of beliefs about how computers and the Internet should be used. This attitude and set of beliefs often is referred to as the hacker ethic.

Although expressions of the hacker ethic have varied over time, at the heart of the subculture is a view of the potential of computing that has two elements: the principle that all information should be free and the belief that access to computers should be unlimited. Surrounding these elements are enthusiasm about computing, a sense that computing is fun and even joyful, and the conviction that computing can be used to bring about positive change in the world by countering mainstream trends toward centralization and privatization. On one Internet site (Raymond 2003) the hacker ethic is defined as follows:

1. The belief that information-sharing is a powerful positive good, and that it is an ethical duty of hackers to share their expertise by writing open-source code and facilitating access to information and to computing resources wherever possible.

2. The belief that system-cracking for fun and exploration is ethically OK as long as the cracker commits no theft, vandalism, or breach of confidentiality.

From an ethical perspective the vision put forward by hackers points to the potential of computing to create a world in which there is no gap, or at least a smaller gap, between the haves (information-rich people) and the have-nots (information-poor people) and in which those who have expertise use it to help others. Moreover, insofar as hackers create open source software and encourage data sharing and access to the Internet, their activities can be seen as furthering the potential of computer technology for social good.

Criticisms and Defenses

The activities of hackers become subject to moral criticism only when hackers engage in illegal activity; using more precise terminology, moral questions arise when hackers become crackers. Once the law is broken, cracking behavior is not just illegal but also seems likely to cause others to be treated unfairly and to harm their interests. For example, when hackers launch viruses that disrupt the use of the Internet, their behavior interferes with the activities of innocent users; when they copy and distribute proprietary software, they are violating the legal rights of individuals to own and license software; and when they break into systems and examine files, they are violating the privacy and property rights of others.

In their defense crackers may argue that (1) they are doing no harm, meaning no physical harm to human beings; (2) they are liberating information that should be free; (3) the laws involving computing are bad and even unjust; or (4) they serve in the role of vigilantes testing and revealing the vulnerabilities of computer systems. All these claims rely on the deeper or prior presumption that sometimes it is permissible to break the law.

In moral philosophy and in democratic theory cases of justifiable law breaking are well recognized. The defense of hacking sometimes is couched in terms of civil disobedience. Acts of civil disobedience are those in which an individual refuses to obey a law either because obeying the law would violate the individual's conscience or because an individual wants to protest the law on the grounds that it is unjust. Although there may be particular acts that fit the definition of civil disobedience, in general cracking does not seem to fit into that category. Indeed, most cracker behavior seems difficult to defend, though there may be particular actions that can be justified.

Cracking behavior is difficult to justify because the laws that have been created around computing, though far from perfect, are aimed at defining the rights and responsibilities of users, and once rights and responsibilities are

allocated, illegal behavior becomes *prima facie* harmful. Viruses disrupt the activities of computer users and force them to invest more resources (time, effort, and money) in securing their systems, resources that could be used in other ways. Pirating software deprives individuals of their legal rights of ownership. Gaining unauthorized access to systems and files violates privacy and property rights.

In recent years scholars have begun to explore new forms of behavior on the Internet that are related to but different from hacking. For example, the term *hacktivism* is used to refer to activists who use their computer skills to make political statements and protest actions by government or industry; in other words, those persons engage in political activism by using computers. Hacktivism may or may not be illegal depending on the actions taken. *Cyberterrorism*, by contrast, refers specifically to political action that involves violence against persons or property.

DEBORAH G. JOHNSON

SEE ALSO *Association for Computer Machinery; Computer Ethics; Computer Viruses/Infections; Digital Divide; Free Software.*

BIBLIOGRAPHY

Himanen, Pekka. (2001). *The Hacker Ethic and the Spirit of the Information Age.* New York: Random House. A somewhat light look at hackers and what they think and do.

Laszlo, Pierre. (2004). "Science As Play." *American Scientist* 92(5): 398–400.

Levy, Steven. (1994). *Hackers: Heroes of the Computer Revolution.* New York: Delta. This book was one of the first to look at hackers and their place in popular culture; while not current, it is important for understanding the early history of hacking.

Thomas, Douglas. (2002). *Hacker Culture.* Minneapolis: University of Minnesota Press. Provides an in-depth look at hackers with a historical perspective and an emphasis on the place of hackers in popular culture.

INTERNET RESOURCES

Denning, Dorothy. (2000). "Cyberterrorism. Testimony before the Special Oversight Panel on Terrorism, Committee on Armed Services, U.S. House of Representatives, May 23, 2000." Available at http://www.cs.georgetown.edu/∽denning/infosec/cyberterror.html. Denning is an expert on cyberspace security and in this testimony speculated on the potential for cyberterrorism.

Raymond, Eric S. (2003). "The Jargon File." Available at www.catb.org/∽esr/jargon/. One of many Internet sites that provide definitions and information about cyberculture.

HALDANE, J. B. S.

● ● ●

John Burdon Sanderson or J. B. S. Haldane (1892–1964) was born in Oxford on November 5 and, as the author of *The Causes of Evolution* (1932), became a founder of what was later called the modern evolutionary synthesis of population genetics. Haldane was also an influential popularizer of science who in essays, fiction, and even verse emphasized the need to develop an ethical framework within which human beings may assimilate emerging technologies. He died on December 1 in Bhubaneswar, India.

With remarkable prescience, Haldane foresaw discoveries in molecular biology and genetic engineering. In *Daedalus or Science and the Future* (1923), he argued that scientific progress in these areas would bring confusion and misery to humankind unless accompanied by progress in ethics. Ideas from *Daedalus* influenced his friend Aldous Huxley's novel *Brave New World* (1932), and Haldane served as the model for the biologist in Huxley's *Antic Hay* (1923). Forty years later, in 1963, Haldane also introduced the concept of clonal reproduction that has since inspired much controversy and discussion in bioethics.

Haldane further maintained that science provides at least one of the key ingredients to moral progress, this being high regard for truth and a refusal to jump to unjustifiable conclusions. Indeed in one statement of this agnostic attitude, Haldane suggested that "the Universe is not only queerer than we suppose, but queerer than we can suppose" (1927, p. 298).

Haldane's views in regard to the ethical influence of science were opposed by Bertrand Russell (1872–1970) in *Icarus or the Future of Science* (1924). Russell argued that technical scientific knowledge does not make people more sensible in their aims or more self-controlled and kind. In his advocacy of a science-based ethical framework, Haldane thought that science would exert an essentially progressive influence on society and politics, and that general agreement could be reached on conceptions of the good, a view that remains highly controversial.

Seeing it in part as a bridge between science and ethics, Haldane was also for years attracted to Marxist Communism, which he embraced during the 1930s. He later abandoned this affiliation when the science of genetics was suppressed in the Soviet Union under the direction of Trofim Lysenko (1898–1976). Ironically that crisis proved one of his own predictions about Soviet science, that "there is ... a very grave danger for

J. B. S. Haldane, 1892–1964. Haldane was an English biologist who utilized mathematical analysis to study genetic phenomena and their relation to evolution. (*The Library of Congress.*)

science in so close an association with the State ... it may lead to dogmatism in science and to the suppression of opinions which run counter to official theories...." (1932, p. 225.)

Another essay by Haldane, "On Being the Right Size" (1927), virtually created analytic morphology. By pointing out, for instance, that exoskeletons can only get so large before the internal organs collapse under their own weight, this essay has influenced fields as diverse as the criticism of mass urbanization, the alternative technology movement, and decentralized economics.

Also important is the fact that Haldane conducted many scientific experiments on himself (Dronamraju 1968, p. 267–275). His ethics precluded making others the subject of experiments when he himself could serve that role, a practice also followed by his father, Oxford physiologist John Scott Haldane (1860–1936).

Throughout his life Haldane emphasized how science and technology create new ethical situations, although different sciences impact ethics in different manners. Physics and biology affect our ethical outlook by altering views about the fundamental nature of the world and the interrelationships between all living beings. For Haldane, Darwinian evolution imposes a new set of ethical values on the relationship between humans and other species. Anthropology shows that any given ethical code is only one of a number practiced with equal conviction and almost equal success. Advanced communication technologies create new duties by pointing out previously unexpected responsibilities for world events.

In 1957 Haldane moved to India, where he was deeply influenced by Hinduism. He saw the Darwinian theory of evolution from a fresh perspective, noting that Christian theologians had drawn a sharp distinction between humans and other species, whereas no such distinction had been made in India. According to Hindu, Buddhist, and Jain ethics, for instance, animals have rights and duties, and the adherents of these religions are duty-bound to adopt a non-violent approach to biological research. He followed this principle in directing the research of his students in India in animal behavior, genetics, human genetics, and the biometry of both animal and plant species.

KRISHNA R. DRONAMRAJU

SEE ALSO *Brave New World*; *Posthumanism*; *Russell, Bertrand*.

BIBLIOGRAPHY

Clark, Ronald William. (1984). *J. B. S.: The Life and Work of J. B. S. Haldane*. New York: Oxford University Press. An extended biography.

Dronamraju, Krishna R. (1985). *Haldane: The Life and Work of J. B. S. Haldane with Special Reference to India*. Aberdeen: Aberdeen University Press. An account of Haldane's life and work in India.

Dronamraju, Krishna R. (1995). *Haldane's Daedalus Revisited*. Oxford: Oxford University Press.

Dronamraju, Krishna R., ed. (1968). *Haldane and Modern Biology*. Baltimore: Johns Hopkins University Press. A memorial tribute by distinguished colleagues.

Haldane, J. B. S. (1923). *Daedalus, or Science and the Future*. London: Kegan Paul.

Haldane, J. B. S. (1927). *Possible Worlds and Other Essays*. London: Chatto & Windus.

Haldane, J. B. S. (1932a). *The Causes of Evolution*. London: Longmans. Reprinted by Princeton University Press (1990). Established the foundation for the synthetic theory of evolution.

Haldane, J. B. S. (1932b). *The Inequality of Man and Other Essays*. London: Chatto and Windus. A collection of some of Haldane's outstanding popular essays.

Haldane, J. B. S. (1959). "An Indian Perspective of Darwin." *Centennial Review* 3: 357–363. Theory of evolution from a Hindu point of view.

Haldane, J. B. S. (1963). "Biological Possibilities for the Human Species in the Next Ten Thousand Years." In *Man and His Future*, ed. Gordon Wolstenholme. Boston: Little Brown.

Haldane, J. B. S. (1964). "The Implications of Genetics for Human Society." In *Genetics Today: Proceedings of the 11th International Congress of Genetics*, vol. 2, ed. S. J. Geerts. Oxford: Pergamon Press.

Haldane, J. B. S. (1985). *On Being the Right Size and Others Essays*, ed. John Maynard Smith. New York: Oxford University Press. Reprints the influential essay "On Being the Right Size" (1927).

Russell, Bertrand. (1924). *Icarus or the Future of Science*. London: Kegan Paul.

HARDIN, GARRETT

• • •

Garrett James Hardin (1915–2003), born in Dallas, Texas, on April 21, was sometimes called the "father of human ecology" for his efforts to popularize a biological understanding of human beings that also draws out ethical implications. He was a strong advocate for controlling population growth and limiting immigration into the United States, because of the ecological implications of these issues. His two best-known essays, "The Tragedy of the Commons" (1968) and "Lifeboat Ethics" (1972), in their description of a problem and presentation of a response, became standard points of reference in bioethics broadly construed. Hardin died in Santa Barbara, California, on September 14.

Hardin earned a B.A. in zoology (University of Chicago, 1936) and a Ph.D. in microbiology (Stanford University, 1941). His most influential mentors were microbiologist Cornelius Bernardus van Niel (1897–1985) and Nobel Prize–winning geneticist George W. Beadle (1903–1989). In 1946 Hardin accepted an appointment in human ecology at the University of California, Santa Barbara, where he spent the next thirty years of his career, retiring in 1976.

In "The Tragedy of the Commons," which was first published in *Science* magazine and then widely reprinted, Hardin employed the historical analogy of the deterioration of common pasturelands in seventeenth-century England to explore the contemporary problems of resource utilization and environmental pollution. When a common resource such as a pasture that will support three cows in good health is available to three families, any one family is tempted to introduce a second cow, because although now all four cows will, like the pasture, be slightly less healthy, the combined value of two modestly healthy cows is greater than one healthy cow. This tendency to exploit a public good for private gain, when the gain belongs to one person but the cost is shared by all, results in the overgrazing and deterioration of the commons.

To solve this problem, personal property ownership must be introduced so that owners have an interest in maintaining the productive capacity of the land because they now share the full costs of any excessive exploitation. The general principle is that individuals will exploit anything that is free to maximize their own gain, with a cost to society. The commons cannot possibly work once the population has become too great. Hardin applied this principle to human reproduction, arguing that people who have many children are imposing a cost on society that they do not fully bear. Hardin argues that coercion is necessary to reduce reproduction of children, just as the freedom to rob a bank is curtailed by criminal law.

In "Lifeboat Ethics," Hardin argues that immigration is a major cause of population increase in the developed world, and he advocates the reduction of immigration to nearly zero. The analogy is that a lifeboat (developed nation) can hold a certain number of people. If more people (developing nation) climb into a boat that is full (to carrying capacity), the lifeboat sinks and everyone drowns. The rational course of action for those already in the boat is to refuse additional passengers.

This is, Hardin admits, a "tough-love ethics" founded on the principle that Earth has a limited carrying capacity for the size of population it can accommodate. Hardin believes the optimum carrying capacity of the United States was reached in the middle of the twentieth century, and that further increases in population will degrade the quality of human life. As the number of people increases, so do pressures on the natural resource base, resulting in suffering and misery.

A further argument in Hardin's work is that multiculturalism provides another reason to reduce immigration. For Hardin, social disorder is promoted by increasing the diversity of the groups encouraged to reside in the United States: "Diversity within a nation destroys unity and leads to civil wars. Immigration, a benefit during the youth of a nation, can act as a disease in its mature state. Too much internal diversity in large nations has led to violence and disintegration" (Hardin 1993, p. 42).

Hardin's prescription for the Third World population explosion is for First World nations to cease food aid, allowing Third World nations to solve their problem of having exceeded their carrying capacity. Food aid leads to more babies being born and surviving,

Garrett Hardin, 1915–2003. Trained as an ecologist and microbiologist, Hardin is best known for his 1968 essay, "The Tragedy of the Commons," now reprinted in over 100 anthologies and widely accepted as a fundamental contribution to ecology, population theory, economics and political science. (© Vic Cox, photographer.)

increasing population size, and requiring more assistance in the future. The only aid First World countries should give to the Third World is information about birth control and contraceptives. If a country is poor and powerless because of too many people, it will become even poorer and more powerless by increasing its population.

Merging biological principles with ethical considerations, Hardin argued for the responsible assessment of the environment to optimize the quality of life for present and future generations. He confronted the human condition and its intricate connection with the natural world in an effort to encourage society to effectively deal with the population-resource equation so that posterity will not be subjected to enforced processes of poverty, starvation, and social disorder.

CRAIG A. STRAUB

SEE ALSO *Environmental Ethics; Population.*

BIBLIOGRAPHY

Hardin, Garrett. (1968). "The Tragedy of the Commons." *Science* 162: 1243–1248.

Hardin, Garrett. (1969). *Science, Conflict and Society.* San Francisco: W.H. Freeman.

Hardin, Garrett. (1993). *Living within Limits.* New York: Oxford University Press.

Hardin, Garrett. (1999). *The Ostrich Factor: Our Population Myopia.* New York: Oxford University Press.

HARDWARE AND SOFTWARE

• • •

The invention of the computer hardware/software distinction is credited to computer scientist John Tukey (1915–2000), who also first used the term *bit* for memory capacity. Many think that the difference between hardware and software is obvious. One rule of thumb defines hardware as the computer stuff one can bump into. But others emphasize the logical equivalence of computer hardware and software: "Any operation performed by software can also be built directly into the hardware … any instruction executed by the hardware can also be simulated in software." (Tanenbaum 1999, p. 8)

Often computer hardware conjures up an image of a central processing unit (CPU) or a memory chip, not the wire that connects the mouse to a keyboard. But all physical entities that are part of a computer should be considered hardware, although some hardware is more directly involved with the symbol manipulation power of a computer than other hardware.

Tangibility and Functionality

Ruminations about the distinctions between hardware and software can lead to interesting contrasts. The hardware is a machine whose state changes as it operates, but whose form is difficult to change. (A light switch alternates between on and off positions, but one rarely changes its constituent parts.) As it executes, a software program remains static (except for self modifying programs, an exception that proves the rule), but the program causes changes in the hardware state (memory) and external devices (such as printers). And that same software that is static during execution is far easier to change between executions than the hardware that constantly changes its state during execution. To better understand the hardware/software distinction, it is useful to consider three distinct aspects of both: tangibility, functionality, and malleability.

Tangibility: If a computing entity is defined by its physical presence, it is hardware. If an entity is independent of any particular physical form, it is software. Notice that a tangible "hardware" can take many physical forms. The double helix spiral of DNA uses proteins as its hardware. The genetic patterns coded therein are software.

Functionality: If a computing entity has as its primary purpose a physical function, it is hardware. If the entity has as its primary purpose a logical function, it is software. Here "logic" is used to mean symbol manipulation, the transformation of bits according to syntactic and semantic rules. A particular set of bits could mean an integer or printable characters, depending on the rules in force. The bits themselves are represented by hardware, but the rules governing their interpretation are software.

Malleability: If a computing entity is relatively easy to change, it is software. If the entity is relatively hard to change, it is hardware. Of the three aspects, this is the one most in flux. The increasing range of options with respect to malleability has led to an intermediate designation, firmware.

Two early examples of computing illustrate the first two distinctions, tangibility and functionality. The first example is the Jacquard loom, the second a Turing machine.

In 1801 Joseph Jacquard invented a weaving loom using stiff pasteboard cards with holes that controlled rods for each step in the weave. These cards led to the punched cards used by Herman Hollerith for computing machines. Jacquard's physical loom, wood and metal, was hardware. The pasteboard of the punched cards was hardware. But the pattern of holes in the cards, and the desired pattern in the cloth, were software. Even in this *ancient* example of computing, there is an interplay between hardware and software. The software of the weaving pattern is realized in and by the loom hardware. In an almost mystical way, the cloth pattern is in, with, and under its hardware implementation. The threads that go through Jacquard's loom are tangible, and fall under the category hardware. But after the loom does its work, the threads become an embodiment of the software pattern represented (indirectly) by the punched holes in the pasteboard cards.

A Turing machine is a theoretical construct in computer science, and is composed of states, a recording tape, and a read/write head (Turing 1936). A computation proceeds by changing states and by reading and writing symbols to the tape. Turing machines are thought experiments, not physical objects, but *could* be manufactured. A recording head and its tape are tangible hardware with a primarily physical function, the recording of symbols. Whatever medium is used to represent different states inside the Turing machine would also be hardware. But the algorithm embodied in the states and the transitions between them is logical, and software. Note that whatever medium is used to embody an algorithm *is* tangible, but that the algorithm itself does not depend on the details of any particular medium. The same algorithm could simultaneously exist in a human brain, on a piece of paper, and in a Turing machine. Each manifestation would simultaneously be physically different and logically identical.

In the same way that a building embodies an architect's plans, a Turing machine embodies an algorithm. The Turing machine analog to an architectural plan is an algorithm; the Turing machine analog to a building is the Turing machine hardware. The architectural analogy is more apt for a Turning machine than for modern, multipurpose computers, because each Turing machine is customized to a single algorithm. Contemporary computers are far more complex than Jacquard's loom or a simple Turing machine. But the relationship between computer hardware and software is consistent with the relationship illustrated in these examples. In all computing machines, hardware implements software, and the software is embodied in the hardware. The software instructs the hardware, and the hardware manifests the actions described in the software.

Despite these examples, controversies remain. There is not much controversy about some hardware/software distinctions in modern computers. CPUs and memory chips are hardware. The algorithms implemented on that hardware are software. But not everyone agrees on other classifications. For example, some people label data as computer software, whereas others explicitly exclude data from being either computer software or hardware. But source code is a widely accepted example of software, and source code is data to the appropriate interpreter or compiler. If data is not software, then the same program is or is not software, depending on how one looks at it.

Some computing scientists and engineers include designs, user manuals, and online help as software, while others explicitly exclude such entities from consideration. It may be misleading to label all documents associated with a program as software. But designs and specification documents are closely related to algorithms. Some designs can be automatically transformed into machine language with minimal human intervention. It may thus be useful to classify as software documents directly related to program development.

Algorithms used when a computer is powered on are typically stored in special memory devices called read-only memory (ROM), that comes in various forms (PROM, EPROM, and EEPROM). These kinds of devices are easier to change than other hardware but harder to change than programs stored on a hard drive. The term *firmware* was coined to designate this intermediate form of malleability. Malleability (or its opposite, resistance to change) is the third criterion for hardware and software: If a computing entity is easy to change, it is software; if an entity is difficult to change, it is hardware; and if it is intermediate in this aspect, it is firmware. A closer argument is required here: The *state* of hardware may be easy to change (much computer hardware is a variation on the on/off switch); but the hardware itself (think of a light switch attached to a wall) is difficult to change. Thus an arithmetic/logic unit is hardware because it has permanently etched silicon algorithms for its calculations, and a C++ program residing on a hard drive is software because it can be more readily modified and recompiled. Although the use of firmware is commonplace, the ethical implications of the hardware/software distinction do not require this middle ground of malleability as a separate category.

Implications of the Hardware/Software Distinction

Some interest in the hardware/software distinction is associated with legal issues. Insofar as a computing entity is a mechanical device (hardware), it is subject to the same body of law that governs ladders and lawnmowers. Insofar as a computer entity functions as an algorithm (software), the laws of intellectual property and professional service are more germane. Hardware designs can be patented, software programs can be copyrighted.

The hardware/software distinction also has ethical implications. On an abstract level, algorithms are pure software. But to have a physical effect, an algorithm is embodied in some physical entity. The nature of the embodiment, the particular hardware chosen, has important ethical implications. For example, if an algorithm is embodied exclusively in a single brain, its ownership as a private thought is uncontroversial; but when the thought is shared as a written document, ethical issues of intellectual property instantly arise. Similarly an algorithmic thought has few consequences for others until it is implemented; when implemented, the algorithm can have important consequences.

When deciding how to embody an algorithm, one must select a location on the malleability continuum. Typically an emphasis on hardware implementation (for example, etched permanently in silicon) will encourage a more reliable implementation and less complex functionality than an implementation in software (such as using a high level programming language). Hardware implementations are more economically feasible when they are mass produced, so widely used algorithms for the many are more likely to be implemented in hardware, whereas customized algorithms for the few are more likely to be implemented in software. The questions of delivery schedules, *how good is good enough*, and a developer's obligations to customers are examples of the ethically charged issues inherent in any implementation decision. As the costs of fabricating hardware fall and the costs of writing software rise, judgments with regard to such issues may have to be reconsidered.

As computing becomes ubiquitous, software tends to replace hardware to deliver certain functionalities. Airplanes provide a dramatic example of this trend. First fighter jets and then commercial airliners have substituted complex computer algorithms for mechanical controls. The computer algorithms enable the newer airplanes to fly more efficiently and economically. But the redundancy of hardware is difficult to replicate in software (software defects tend to reoccur in a way that hardware defects do not), and this has consequences for the reliability of life-critical systems that rely increasingly on software for their safety. These differences result in ethically significant choices between more efficient operations using software and more expensive but safer hardware devices. As these tradeoffs becoming increasingly common, the different traditions of professionals in different engineering fields can become an ethical issue. For example, software engineers are rarely licensed (in the United States, only Texas issues a software engineering license, and corporate software engineers aren't required to obtain that license either), but other engineers in safety critical applications can be licensed. In this case, the hardware/software distinction may help determine the state's interest in certifying professional competence.

A final example of the ethical importance of the hardware/software distinction has less to do with computing professionals and more to do with the non-programming public and how they view problems with computing. Computers can be a handy scapegoat: "We can't help you with that right now; the computer's down." "Impossible. There's no way to type that into the computer." "I don't remember that email. I guess the computer ate it."

Most people know that *software bugs* are the responsibility of programmers. But organizations and individuals can sometimes hide behind the hardware of their machines. Emphasizing the hardware aspects of computers can help create an artificial distance between the general public and human errors in software. This de-emphasis of human accountability is a danger lurking in the hardware/software distinction. The reality is that algorithms are human artifacts for which humans are responsible, no matter how they are implemented.

The ethical challenges lurking within the hardware/software distinction are reflected in legal and political controversies. There is freeware, shareware, and open source software; but there is no parallel movement to declare computer hardware free. The patent system has been, in the main, successful at protecting hardware innovations, but copyright, patent, trademark, and trade secrecy have each proven problematic in a different way when applied to computer software. Controversies over new laws that criminalize what was once considered legitimate reverse engineering of software have highlighted the importance of understanding the differences between hardware and software.

KEITH W. MILLER

SEE ALSO Computer Ethics; Free Software; Information Overload.

BIBLIOGRAPHY

Tanenbaum, Andrew S. (1999). *Structured Computer Organization*, 4th edition. Prentice-Hall, Upper Saddle River, NJ.

Turing, Alan. (1936). "On Computable Numbers, With an Application to the Entscheidungsproblem." *Proceedings of the London Mathematical Society*, Series 2, 42: 230–267.

HASTINGS INSTITUTE
SEE *Bioethics Centers*.

HAZARDOUS WASTE
SEE *Waste*.

HAZARDS
• • •

Hazards are low-probability, high-magnitude phenomena that have the potential to cause large negative impacts on people. While this definition is unavoidably imprecise (what counts as a "phenomenon"? what probabilities qualify as "low"? and what impacts qualify as "large" or even "negative"?), in general hazards can be understood as acting outside of daily human expectations to adversely affect the quality of life of those exposed to them. Hazards refer to a prospect or risk of an occurrence; a particular occurrence of a hazard is more typically termed a "disaster" or sometimes an "extreme event"; when they are technological in origin they may be termed "accidents."

The remains of a trailer park in Miami, Florida, destroyed during Hurricane Andrew. Andrew was one of the most destructive hurricanes ever to hit the U.S., raging from August 16 to August 28, 1992. (© Tony Arruza/Corbis.)

Some types of phenomena—such as hurricanes, earthquakes, landslides, and reactor meltdowns—are unambiguously classified as hazards, whereas others, especially those that are less temporally or spatially discrete, such as droughts, famines, and epidemics, may or may not be included under the term, depending on who does the classifying. Wars and other types of human conflict are generally not categorized as hazards.

A related use of the word "hazard" refers to existing conditions of the environment that may pose a risk to humans, such as a toxic waste site or even the edge of a cliff. Similarly, hazardous materials are those that may create a risk to human or environmental health if exposure to them is not regulated and controlled. This entry, however, focuses on hazards as dynamic phenomena, not as static conditions or material properties.

In the ten-year period 1992 to 2001, hazardous events, or disasters, worldwide were responsible for more than 620,000 deaths. Drought caused almost 45 percent of these deaths; floods, earthquakes, and windstorms caused most of the remainder. An additional 2 billion people required immediate assistance (60% as a result of floods), and the direct costs due to the destruction of infrastructure, crops, homes, and so on was more than $600 billion (with earthquakes, floods, and windstorms making up about 90% of this total). To put these numbers in some perspective, every year hazards seriously disrupt the lives of as many people as the entire population of Brazil or Indonesia, and cost about as much as the entire economic output of Pakistan or Peru (World Health Organization, United Nations Development Programme 2002).

Hazards Are Not Natural

Hazards are commonly divided into two types: natural and technological. Technological hazards are those arising from the failure of technological devices or systems to behave as intended. Natural hazards arise from nonhuman forces and can be subdivided into geophysical hazards, such as volcanoes, earthquakes, and tsunamis, and hydrometeorological hazards, such as hurricanes, floods, and tornadoes. Natural and technological hazards, however, are often related to each other, in that natural disasters may trigger technological failures, for example of power grids or dams. Moreover, natural hazards must be understood not simply as the result of natural phenomena, but as arising from the socioeconomic context within which such phenomena occur.

Human *exposure* to hazards results from humans living in areas where hazards are present; human *vulnerability* to hazards arises from the types of development exposed to hazards. The consequences of hazards are determined as much or more by the extent of exposure and level of vulnerability than by the characteristics of the hazard itself. Thus, for example, when a magnitude 6.9 earthquake struck a densely populated region in Armenia in December 1988, more than 25,000 people died and 1.6 million were directly affected. When, ten months later, a similar magnitude earthquake struck a highly populated region of California (the October 1989 Loma Prieta event near Santa Cruz), sixty-three people died and fewer than 10,000 were affected. This stark difference in impacts was largely a reflection of poor design and construction standards for buildings in Armenia compared to those in California. Moreover, despite Armenia's much lower level of economic development, its economic losses from the 1988 event, estimated at about $14 billion, were greater than the estimated $6-to-$10 billion price tag of Loma Prieta.

The inseparability of hazards from their social context is clearly illustrated by historical trends in disasters, which show a continual and rapid increase in the number of disasters, rising from a worldwide average of about 100 per year in the early 1960s to between 300 and 500 per year by the early 2000s. ("Disasters" here is defined by the World Health Organization's Collaborating Centre for Research on the Epidemiology of Disasters [CRED] as events that kill at least ten people, affect at least 100, result in a call for international assistance, or result in a declaration of emergency.) While some of this increase reflects changes in reporting, most of it arises from increased exposure and vulnerability throughout the world because of growing population, expanding economies, migrations to coasts and other vulnerable regions, increasing urbanization, and related factors. These changes are especially reflected in the costs of major disasters, which according to the German insurance company Munich Re rose more than tenfold in the second half of the twentieth century, from an average—in real (2002) U.S. dollars—of about $4 billion per year in the 1950s to more than $65 billion in the 1990s.

It is important to emphasize that these increases are best explained by changes in social context, not changes in the occurrence or type of hazardous events. For example, it has been well documented that rapidly increasing economic losses from hurricanes striking the U.S. eastern seaboard are caused by growing population and wealth, not by increased frequency or magnitude of storms. The great Miami hurricane of 1926 caused about $76 million in damage (in inflation-adjusted dollars); when Hurricane Andrew, of similar force, struck south Florida in 1992, it caused more than $30 billion in damage (Pielke and Landsea 1998).

Complexity of Hazards

Because hazards are socially embedded, their impacts arise from the complex interaction of many variables. In Armenia, steel that had originally been produced to reinforce buildings was diverted to weapons construction instead, thus revealing cold war geopolitics as one source of vulnerability to the 1988 earthquake (Mileti 1999).

Hurricane Mitch, which in October and November of 1998 killed more than 10,000 people and caused severe economic and social disruption in Central America, was responsible for triggering a mudslide in Nicaragua that killed about 2,000 people (Olson et al. 2001). The mudslide, however, was created not just by the torrential rains brought by Mitch, but also by land-use patterns that led to deforestation of a steep mountain slope, which collapsed when it became saturated with water. Eighteen months later, a debris flow in Manila, Philippines, triggered by normal monsoon rains, killed about 200 people. But in this case the disaster occurred on the flank of a huge landfill where thousands of people scavenged garbage for a living.

In Chicago, a heat wave in the summer of 1995 led to the death of more than 700 people. The temperatures in Chicago were no higher than those regularly experienced in many places; the huge number of casualties was instead caused by a combination of failed social services (for example, insufficient number of emergency vehicles and workers) and the large number of people, mostly poor and elderly, living alone, without resort to social networks (Klinenberg 2002).

Such examples also show that a preliminary event may trigger additional hazards that may themselves be damaging or that may combine with the principal hazard to multiply damages. For example, the Chicago heat wave led to technological failures in the form of power outages and water service interruptions that made it more difficult for people to cope. Major disasters may also trigger disease outbreaks, especially when water supplies are cut off or contaminated. The 1906 San Francisco earthquake is often called the San Francisco Fire because of the disastrous conflagrations it caused throughout the city. These sorts of complexities also underscore the futility of making a clear distinction between natural and technological hazards.

Uneven Distribution of Hazard Impacts

The impacts of hazards are disproportionately borne by poor people living in poor regions and countries; thus, hazards are a manifestation of socioeconomic inequality and an issue of social justice. While the poorest thirty-five countries account for only about 10 percent of the world population, they suffered more than half of the disaster-caused deaths between 1992 and 2001. Of those directly affected by disasters during that decade, almost 90 percent lived in Asia, where dense populations combine with high vulnerability and widespread poverty in nations such as India, China, and Indonesia. As the contrast between the Armenia and Loma Prieta earthquakes starkly shows, the benefits of affluence include a capacity to protect against the most direct and devastating effects of hazards, and a significant component of this capacity is the scientific and technological infrastructure that typically accompanies (and fuels) the growth of affluence.

Not surprisingly, affluent nations suffer the greatest absolute economic losses from hazards. The disproportionately large sizes of their economies create the potential for much greater economic damage from the impacts of hazards. For the decade 1992 to 2001, the forty-five richest countries (making up about 18 percent of global population and accounting for 82 percent of global wealth) experienced about 62 percent of total global economic damage from hazards. As a percentage of gross national product (GNP), however, the economic effects of hazards on poor countries are about 100 times greater than for rich countries. Damages from Hurricane Mitch, for example, were estimated at between $5 billion and $7 billion, which was about the same as the annual combined total GNP of the two most affected nations, Honduras and Nicaragua. The magnitude 6.7 Northridge, California, earthquake of 1994 was the most costly disaster in U.S. history, causing between $20 billion and $40 billion in losses; the total, however, was equivalent to only between about 2 and 4 percent of California's economic activity for that year.

Disparities between rich and poor will compound over time. Global population growth is mostly concentrated in poor countries and leads to rapid urbanization, usually in vulnerable coastal zones, as well as dense rural populations. Unregulated land use translates into widespread environmental degradation, especially deforestation, which in turn exacerbates flooding and related phenomena such as mudflows, debris flows, and landslides. Design and construction standards are typically low, and even when adequate building codes exist, corruption, lack of enforcement, and insufficient resources result in an unsafe built environment. Emergency response capabilities are often inadequate, and hazard insurance is usually unavailable, slowing the recovery process. Technological infrastructure, such as communication and transportation systems, is typically fragile, and capacity to repair damaged systems is limited. Such factors reinforce one another to magnify the vulnerability of poor people and nations to hazards, and they act as a brake on development.

Mitigation

In the affluent world, numerous approaches have been adopted to mitigate the effects of hazards, including building codes that are appropriate to known risks; land-use regulations for floodplains, coastal zones, and seismic zones; and dams, levees, and other engineering interventions for floodplain management. There is little question that such measures, combined with early warning systems for hurricanes, tornadoes, and floods, and coordinated emergency response plans, have limited the human and economic toll of hazards in the developed world. Nevertheless, while the number of people killed and injured has declined for some hazards, and stayed relatively stable for others, the economic costs of hazards appear to be rising at an exponential rate. Absent mitigation efforts, they would be rising more rapidly still.

Despite aggressive mitigation efforts, affluent nations are not exempt from major disasters. The magnitude 7.2 earthquake that struck Kobe, Japan, in January 1995 killed 6,000 people and led to an estimated $100 billion in damages, yet Japan is justifiably considered to have the world's most sophisticated and effective earthquake hazard mitigation practices. In the U.S. Midwest, decades of flood control engineering preceded the 1993 Mississippi River basin floods that caused $18 billion in damages and that arguably constituted, in the aggregate, the worst flood in U.S. history (Changnon 1996).

Such events point to the complexity of mitigating hazards. While mitigation efforts may protect against anticipated or typical hazards, they may also have the effect of attracting more people to live and work in hazardous areas, thus increasing exposure over the long term to even larger events. (This trend is reinforced by the apparent security provided by hazard insurance and disaster relief programs.)

Mitigation of hydrological hazards in particular can alter the function of natural systems in ways that are not sustainable over the long term, both because such altered systems may behave in unanticipated ways and because "unprecedented," and thus unplanned-for, events will inevitably occur at some point, in some areas. Mitigation efforts, it seems—especially those focused on trying to control the behavior of the environment through engineered structures—may have the affect of trading a number of smaller, more manageable events in the short-to-medium term for much greater disasters in the more distant future. This can become a self-perpetuating and self-amplifying process, because after a disaster occurs political pressure inevitably focuses on allowing people to return to their homes and businesses to reopen, which in turn requires increased commitment to environmental control via structural hazard mitigation.

Policy Assessment

While societies have an obligation to limit the negative effects of hazards on people and economies, such action should be informed by the inevitability of hazards, rather than a vain quest to eliminate their impacts or occurrence. Such a perspective focuses on the characteristics of human development, rather than the control of nature, as the cornerstone of effective mitigation. For example, environmental degradation invariably exacerbates hazard damages by altering or destroying natural features that buffer the impacts of hazards—such as forests that stabilize steep slopes, floodplains that allow for dispersion of floodwaters, and coastal lagoons that absorb storm surges. Mitigation policies that keep such features intact, and govern land use in ways that protect them over the long term, are likely to be successful both because they preserve natural function and because they thereby limit human development in particularly hazard-prone areas. In acknowledgement of these realities, after the 1993 floods in the Midwest, the U.S. government increased efforts to *remove* floodplain structures—thus returning some of the natural function of the river—and relocate flood-prone communities to higher ground.

Yet it remains to be seen if it is possible to actually stabilize or reduce the costs of natural hazards in developed countries characterized by continual growth of wealth, infrastructure, urban centers, coastal and wildland development, and overall interconnectedness. Hazards may simply be an unavoidable overhead cost on the growth of affluence.

Outside of the developed world, however, the path to reducing the toll of hazards is clear, if difficult to follow. Poverty and the conditions associated with it—poorly constructed and maintained housing and infrastructure, degraded environmental conditions, rapidly increasing populations, insufficient or ineffective social and emergency services, lack of technical capacity—are the nutrients of hazards. At the global scale, reducing poverty, and the environmental degradation and failures

of governance that accompany it, will continue to be the most effective strategy for hazard mitigation.

DANIEL SAREWITZ

SEE ALSO *Building Destruction and Collapse; Safety Engineering: Practices.*

BIBLIOGRAPHY

Changnon, Stanley A., ed. (1996). *The Great Flood of 1993: Causes, Impacts, and Responses.* Boulder, CO: Westview. Scientific and policy perspectives on this historical event.

Klinenberg, Eric. (2002). *Heat Wave: A Social Autopsy of Disaster in Chicago.* Chicago: University of Chicago Press. An analysis of the social imbeddedness of a hazard and its impacts.

Mileti, Dennis S. (1999). *Disasters by Design: A Reassessment of Natural Hazards in the United States.* Washington, DC: Joseph Henry Press. A comprehensive discussion of the social context of hazards.

Olson, Richard Stuart, et al. (2001). *The Storms of '98: Hurricanes Georges and Mitch; Impacts, Institutional Response, and Disaster Politics in Three Countries.* Boulder: University of Colorado, Natural Hazards Research and Applications Information Center. Basic facts and political interpretation of these catastrophic storms.

Pielke, Roger A., Jr., and Christopher W. Landsea. (1998). "Normalized Hurricane Damages in the United States: 1925–95." *Weather and Forecasting* 13(3): 621–631. Analysis of the causes of increasing coastal hazard losses.

United Nations Development Programme. (2002). *Human Development Report.* New York: Oxford University Press. Source for national economic data.

INTERNET RESOURCES

Munich Re. "Annual Review of Natural Catastrophes, 2002." Available from http://www.munichre.com/2/publications_db_e.asp. Data on costs of hazard losses worldwide.

World Health Organization. Collaborating Centre for Research on the Epidemiology of Disasters (CRED). "EM-DAT: The OFDA/CRED International Disasters Data Base." Available from http://www.em-dat.net/. Unless otherwise noted, all data on hazard losses in the entry are derived from this unique database.

HEALTH AND DISEASE

• • •

Why care about the precise definitions of the words *health*, *disease*, and *illness*? Their meanings seem self-evident: Health is the absence of disease, illness the experience of disease. However the multiple dimensions of these concepts, their moral underpinnings, and the purposes for which they are used are enormously complex, especially in a technological society strongly oriented toward the production of health.

Health and disease are more than just medical terms; they have social, political, moral, and economic dimensions. For example, a pharmaceutical company may advertise its new compound as the cure for a heretofore-unnamed disease such as erectile dysfunction or attention deficit disorder (ADD). Medical or disability coverage is granted or denied based on sociopolitical interpretations of what constitutes a disease or disability. A couple decides to use in vitro fertilization and preimplantation genetic screening to avoid creating a baby with a genetic disease or one who will be a carrier of a diseased gene. Or perhaps a soft drink producer enhances sales by touting the benefits of its new and improved *healthier* beverages. Professional codes of ethics commonly commit engineers to protect public safety, health, and welfare. The concepts of health and disease are invoked in various ways, for purposes weighty and mundane.

Indeed health has been construed not simply as the absence of disease (whatever that is), but much more. In the preamble to its constitution, the World Health Organization (WHO) defines health as "a state of complete physical, mental and social well-being not merely the absence of disease or infirmity." Such statements rely on medico-moral presuppositions of what a disease actually is, how and which diseases ought to be treated, and ultimately on visions of what it means to live the good life.

Recognition of the complexity of the concepts of health and disease has stimulated scholarship in the fields of history and philosophy of medicine, sociology of medicine, and the medicalization of deviance, as well as crucially important policy developments in managed care and resource allocation. Philosophical questions range from clarifying the ontological status of disease (What is a disease?) to understanding particular conditions and the meaning of being diseased. The social sciences, including medical sociology and anthropology, examine the extent to which disease is a value-laden concept shaped and socially constructed. How do power relations influence what is considered to be normal and healthy or abnormal and diseased? On the level of individual experience, still other questions emerge: What is the personal meaning of being healthy or sick? At what point, if any, are the sick blameworthy for their illnesses? What role ought a sick person play in society? How does stigma affect the sick? More broadly framed questions regarding matters of policy ask what responsibilities society has to care for those who are diseased or ill.

Different responses to such questions are associated with diverse historical and philosophical approaches to health and disease. Sociological contributions to the debate and their policy implications also deserve consideration. This simple conceptual breakdown is appropriate for present purposes, but it is important to note that a more holistic picture requires interdisciplinary dialogue.

Historical Sketch

The concepts of health and disease were foundational to the ancient medical arts and bound up with distinct philosophical perspectives. To explain illness or symptoms of disease, as well as to cure the sick, pre-Socratic philosophers and ancient Greek physicians in the Hippocratic tradition (c. 400 B.C.E.) developed a basic explanation of health as balance (*isonomia*). The balance of the four humors—black and yellow bile, phlegm, and blood—in conjunction with environmental and temporal factors was central to the formalized model of health created by the Greek physician Galen (130–199 C.E.). A rudimentary nosology (classification of diseases) was developed around the imbalance of the humors.

Galen's humoral model persisted through the Middle Ages when it was augmented by Christian ideals of salvific suffering. Although the link between disease and sin was not a new development, moral dimensions of health and disease were described in terms of tests from God, punishment for sin, or demonic possession (Gunderman 2000). Toward the end of the Middle Ages, a new model was espoused by the physician-philosopher Paracelsus (Philippus Aureolus Theophrastus Bombastus von Hohenheim 1493–1541) indicating three elemental components (salt, mercury, and sulfur) as critical to healthy physiology. Paracelsus went on to claim that diseases were not simply internal imbalances, but rather resulted from autonomous entities "springing from the body" (Vichow 1981, p. 192). The ontologists—those thinkers who viewed diseases as actual entities—find the roots of their approach in the work of Paracelsus.

Modern concepts of health and disease (and the practice of scientific medicine itself) are grounded in Cartesian dualism. René Descartes (1596–1650) separated the mind and the body, and described the body as a set of parts working together according to mechanical rules. Because disease was the malfunctioning of the bodily machine, treatment consisted of diagnosing the malfunction and repairing the body, bringing it back to normal functioning (von Engelhardt 1995).

Over the next few centuries, the locus of disease shifted from the macroscopic to the microscopic, and eventually to the molecular. Contributions by the anatomists at the University of Padua—in particular Giovanni Morgagni (1682–1771)—opened discussion on pathophysiology and etiology through postmortem dissections of diseased organs. Marie Francis Xavier Bichat (1771–1802) explained the origins of disease in terms of histopathology—disease in tissues. On the cellular level, Rudolf Virchow (1821–1902) synthesized breakthroughs in bacteriology and microbiology and described *mycotic* diseases in terms of *ens morbi* and *causa morbi* (a being with a cause), and as a disruption in interrelated cellular *territories* which, in turn, compound and spread the disease process. With the rediscovery of Mendelian genetics in the late-nineteenth century, *inheritance factors* were singled out as disease entities that caused such disorders as Huntingdon's Disease and sickle cell anemia. Indeed sickle cell anemia was the first modern genetic disease identified as such by Linus Pauling.

In contrast to the ontologists were the nomino-physiologists, such as François-Joseph-Victor Broussais (1772–1838), who opposed the idea that diseases were actual entities. Such entity-based nosologies, he claimed, were not classifications of disease entities but rather were driven by a physician's instrumental and pragmatic need to diagnose or prognosticate. Claude Bernard (1813–1878) emphasized the need for clinical experimentation and observation in describing diseases. Through his diverse research projects, particularly studies of digestion, glycogen function, and vasoconstriction and dilation, Bernard developed physiological models that emphasized homeostasis and feedback loops in the regulation and maintenance of health. So too, the American physiologist Walter Cannon (1871–1945), in *The Wisdom of the Body* (1932), described health and disease in homeostatic terms.

Philosophical Trends

Philosophers of medicine and science began a more formal analysis of the concepts of health and disease during the first half of the twentieth century. The medical history and epistemology of Georges Canguilhem (1904–1995) and his student Michel Foucault (1926–1984) stimulated a renewed discussion of the normal and the pathological. Eventually a cannon of philosophical writings on the concepts of health and disease was formed during the period 1960 to 1981, a development that was driven in part by the birth and development of bioethics and its need for definitional precision for basic medical concepts (Caplan et al. 1981).

During the 1970s, a conceptual dichotomy in philosophy of medicine developed as new accounts of the status of disease took two tracks. First, reminiscent of the earliest philosophical constructions of disease, various versions of naturalism reemerged. Naturalistic accounts explained disease as deviations in natural form and function. As such, a disease was described as an entity or causal factor of that deviation independent of social norms or cultural values. This perspective is sometimes referred to as nonnormativism (Caplan 1988). Christopher Boorse (1975) presented the quintessential nonnormative position by referring to an objective biological framework that guides the identification and diagnosis of disease:

> [B]ehind this conceptual framework of medical practice stands an autonomous framework of medical theory, a body of doctrine that describes the functioning of a healthy body, classifies various deviations from such functioning as diseases, predicts their behavior under various forms of treatment, etc. This theoretical corpus looks in every way continuous with theory in biology and the other natural sciences, and I believe it to be value-free. (Boorse 1975, p. 55.)

In contrast, normativist philosophers point to the value-laden nature of disease constructions, eschewing the possibility that natural is definable and that diseases are value-free. These scholars directly counter the Boorsian model by pointing to research in philosophy and sociology of science that described science and medicine as social endeavors. Because of this social embeddedness, an autonomous and value-free framework of medico-biological theory does not exist independently of values. (Kuhn 1962, Longino 1990, Engelhardt 1981). Arthur Caplan, H. Tristam Engelhardt, and Joseph Margolis are among those who write a defense of moderate normativism. Caplan (1981) points out that, while some objective criteria for defining disease exist, nonnormativism as characterized by Boorse is fraught with conceptual inadequacies. Some conditions generally considered to be normal or natural (e.g., the common cold, dental plaque, acne, and others) are disvalued, while others considered to be abnormal may be valued (for example, dysfunctional gonads in a person who does not wish to reproduce). Margolis (1976) claimed that while certain biological functions may be conceived in universal terms, the actual concept necessarily reflects the state of the technology, social explanations, division of labor, and the environmental conditions of a given population. Engelhardt (1974) describes the pragmatic and value-laden nature of the concept of disease particularly in his historical exposition of the *disease* of masturbation.

The philosophical debate about the nature of somatic disease spills over into the analysis of mental health and disease. With the rise of scientific medicine, the prevailing model of psychiatric illness became biologically based. Mental illness was considered similar to other somatic diseases, rooted in a dysfunction, or even an *ens morbus*. This model was vigorously challenged by physician George Engel (1913–1999) as being overly reductionistic; he offered his own biopsychosocial model to account for the role of relationships and society in health and disease (Engel 1977). Thomas Szasz's *The Myth of Mental Illness* (1961) attacked the notion that mental illness was a disease of the brain or that mental illness existed at all. Szasz claimed that the notion of mental illness was a way to subjugate dissidents of the community's collective ethos or assuage sick individuals of their responsibilities.

In the early-twenty-first century, genetic technology and medicine as well as the results of the Human Genome Project added another level of complexity to analyses of health and disease. A greater understanding of epigenetics and the complexities of gene-environment interactions show it is difficult to identify the genetic causes of diseases that are outside the basic Mendelian framework. Nonetheless health and disease are increasingly described in genetic terms. Reification of genetic anomalies as being diseases has raised the very real possibility that all people are *diseased* in some way (Jüngst 2000).

Sociological Perspectives and the Medicalization of Deviance

Philosophical debates about health and disease as normative concepts grade into descriptive analyses of how society constructs, describes, and reacts to the realities of health, disease, and illness. Talcott Parsons (1902–1979) explained the concepts of health and illness as manifestations of certain role-types.

In framing the *sick role*, Parsons took the first step in describing illness as a form of deviant behavior legitimized by the medical institution (Bosk 1995). The sick role is characterized first by an exemption from social duties, exculpating patients for their illness. Parsons described the physician-patient relationship as analogous to the relationship between a child and parent in which the patient follows doctor's orders in a team effort directed toward the patient's wellness. Often this is a form of social control because a sick person needs to enlist the help of persons who are not sick and their *therapeutic agencies*.

Some social scientists have theorized the construction of disease emerges out of power structures, sanctioned under the guise of medical objectivity. Looking back in history, an early example of this dynamic was the description of drapetomania—a disease of slaves that caused them to try to run away. Physician Samuel Cartwright (1793–1868) presented an account of this disease and potential cures, which included, first, kind treatment, but later various forms of severe bondage and punishment. Since then, health and disease have sometimes been hijacked in the name of ideology or the *betterment of common good*. In hindsight these instances are obvious, for example, eugenics movements during the early-twentieth century in the United States and in Germany in which *diseased* individuals, their families, or their entire race were *treated* (Caplan 1992b). In contrast to these more egregious cases, some social scientists suggest more insidious forms of disease construction have occurred through the medicalization of deviant behavior.

Peter Conrad (1975, 2000) describes how hyperkinesis—now called attention deficit and hyperactivity disorder (ADHD)—became a disease. Conrad explains that, with the invention of the stimulant Ritalin and observation of its paradoxical effect on children, the manufacturer, CIBA, sought to market the compound to parents and teachers of unruly children. The *cure* preceded the *disease*. The administration of a drug that reigned in nonconformist children strengthened the status quo: educational systems not equipped to accommodate certain children and parents released from blame for their children's behavior. Similar examples can be found in feminist accounts of the social construction and medicalization of menopause and premenstrual syndrome (McCrea 1983, Richardson 1995).

Labeling theorists, such as Howard Becker (1928), describe the actions of *moral entrepreneurs* who create and enforce social rules. In medicine, moral entrepreneurs may be physicians who ascribe the label *diseased* to those who break with accepted conventions, thus suppressing or stripping them of opportunity, thereby expanding their own domain of professional influence (Becker 1963, Pfohl 1977, Bosk 1995).

As a result of labeling, stigma is often closely associated with disease. In certain cases, sick people remain closeted because of the stigma of their illness. Norma Ware (1992) offers the example of chronic fatigue syndrome. Delegitimation of the subjective experience of illness leads to further suffering arising from the stigma of the disorder, the alienation resulting from a decision to keep the illness secret, and the shame of being *wrong* in one's own definition of reality.

Broad Policy Implications

The ways in which the concepts of health and disease are framed have significant impact on health policy. In particular, defining what constitutes the appropriate level of medical care provided by a just society has been difficult to determine.

Several nosological frameworks driving insurance schedules and socialized coverage plans have been espoused over the years. Norman Daniels, in *Just Health Care* (1994), proposes a policy framework based not on definitions of disease or health but on *species-typical functioning*. Daniels proposes that normal functioning is an important baseline not because it is *natural*, but because it is a convenient point at which to determine what society should owe to its members. Indeed a consensus of what society ought to give to all its members has been elusive precisely because a common framework of health and disease has been impossible to construct.

Equally important to providing care and treatment of disease is the scientific quest to prevent and cure disease. Operational definitions of health and disease ground biomedical research priorities in government and private funding agencies. The National Institutes of Health (NIH) determine research priorities based on a broad range of criteria related to severity of diseases, epidemiological evidence, cost-benefit analyses, as well as projects that offer promises and opportunities, and interest groups/patient lobbying. Investment in research and development and biotechnology, as well as in allied fields of technology, rest on the social framework and disciplinary matrix within which technicians work. As such core concepts such as health and disease have a profound, albeit overlooked, influence on the trajectory of important advancements in technology.

The concepts of health and disease underlie decisions to fund basic bench research through clinical biomedical research and public health initiatives. Clearly a robust understanding of the complexities of these concepts is crucial for policymakers, clinicians, and patients alike.

DOMINIC A. SISTI

SEE ALSO *Bioengineering Ethics; Cancer; Complementary and Alternative Medicine; Emergent Infectious Diseases; Eugenics; Galenic Medicine; Genethics; HIV/AIDS; Medical Ethics; National Institutes of Health; Therapy versus Enhancement; World Health Organization.*

BIBLIOGRAPHY

Becker, Howard. (1963). *Outsiders: Studies in the Sociology of Deviance*. New York: Free Press.

Bernard, Claude. (1999). *Experimental Medicine* [Introduction à l'étude de la medicine expérimentale], trans. Henry Copley Greene. New Brunswick, NJ: Transaction. First published in 1865.

Boorse, Christopher. (1975). "On the Distinction Between Disease and Illness." *Philosophy and Public Affairs* 5: 49–68.

Bosk, Charles. (1995). "Health and Disease: Sociological Perspectives." In *Encyclopedia of Bioethics*, ed. Warren T. Reich. New York: Simon & Schuster Macmillan.

Broussais, François-Joseph-Victor. (1821). *Examen des doctrines médicales et des systèmes de nosologie* [An examination of the medical doctrines and systems of nosology]. Paris: Mequignon-Marvis.

Cannon, Walter. (1932). *The Wisdom of the Body*. New York: Norton.

Caplan, Arthur L. (1981). "The *Unnaturalness* of Aging—A Sickness Unto Death?" In *Concepts of Health and Disease*, eds. Arthur L. Caplan; H. Tristram Engelhardt; and James J. McCartney. Reading, MA: Addison-Wesley Publishing.

Caplan, Arthur L. (1988). *Am I My Brother's Keeper?* Bloomington: Indiana University Press.

Caplan, Arthur L. (1992a). "If Gene Therapy is the Cure, What is the Disease?" In *Gene Mapping: Using Law and Ethics as Guides*, eds. George Annas, and Sherman Elias. Oxford: Oxford University Press.

Caplan, Arthur L. (1992b). *When Medicine Went Mad: Bioethics and the Holocaust*. Totowa, NJ: Humana Press.

Caplan, Arthur L.; Englehardt, H. Tristram; and James J. McCartney. (1981). *The Concepts of Health and Disease*. Reading, MA: Pearson Addison Wesley.

Cartwright, Samuel. (1981). "Report on the Diseases and Physical Peculiarities of the Negro Race." In *Concepts of Health and Disease*, eds. Arthur L. Caplan, H. Tristram Engelhardt, and James J. McCartney. Reading, MA: Addison-Wesley Publishing. Originally published in the New Orleans Medical and Surgical Journal in 1851, May: 691–715.

Conrad, Peter. (1975). "The Discovery of Hyperkinesis: Notes on the Medicalization of Deviant Behavior." *Social Problems* 23(1): 12–21.

Conrad, Peter, and Deborah Potter. (2000). "From Hyperactive Children to ADHD Adults: Observations on the Expansion of Medical Categories." *Social Problems* 47: 559–582.

Daniels, Norman. (1994). *Just Health Care*. Cambridge, UK: Cambridge University Press.

Engel, George L. (1977). "The Need for a New Medical Model: A Challenge for Biomedicine." *Science* 196: 129–136.

Engelhardt, H. Tristram. (1974). "The Disease of Masturbation: Values and the Concept of Disease." *Bulletin of the History of Medicine* 48: 234–248.

von Engelhardt, Dietrich. (1995). "Health and Disease: History of the Concepts." In *Encyclopedia of Bioethics*, ed. Warren T. Reich. New York: Simon & Schuster Macmillan.

Engelhardt, H. Tristram. (1981). "The Concepts of Health and Disease." In *Concepts of Health and Disease*, eds. Arthur L. Caplan; H. Tristram Engelhardt; and James J. McCartney. Reading, MA: Addison-Wesley Publishing.

Gunderman, Richard. (2000). "Illness as Failure: Blaming Patients." *Hastings Center Report* 30(4): 7–11.

Hofmann, Bjørn. (2001). "Complexity of the Concept of Disease as Shown Through Rival Theoretical Frameworks." *Theoretical Medicine* 22(3): 211–236.

Jüngst, Eric T. (2000). "Concepts of Disease After the Human Genome Project." In *Ethical Issues in Health Care on the Frontiers of the Twenty-First Century*, eds. Stephen Wear, James J. Bono, Gerald Logue, and Adrianne McEvoy. Dordrecht, The Netherlands: Kluwer Academic Publishers.

Kuhn, Thomas. (1962). *The Structure of Scientific Revolutions*. Chicago: University of Chicago Press.

Longino, Helen. (1990). *Science as Social Knowledge: Values and Objectivity in Scientific Inquiry*. Princeton, NJ: Princeton University Press.

Magnus, David. (2004). "The Concept of Genetic Disease." In *Health, Disease, and Illness: Concepts in Medicine*, eds. Arthur L. Caplan, James J. McCartney, and Dominic A. Sisti. Washington, DC: Georgetown University Press.

Margolis, Joseph. (1976). "The Concept of Disease." *Journal of Medicine and Philosophy* 1(3): 238–255.

McCrea, Frances B. (1983). "The Politics of Menopause: The *Discovery* of a Deficiency Disease." *Social Problems* 31(1): 111–123.

Morgagni, Giovanni B. (1981). "The Seats and Causes of Disease: Author's Preface." In *Concepts of Health and Disease*, eds. Arthur L. Caplan, H. Tristram Engelhardt, and James J. McCartney. Reading, MA: Addison-Wesley Publishing. First published in 1827.

Parsons, Talcott. (1951). *The Social System*. New York: Free Press.

Pfohl, Stephen J. (1977). "The *Discovery* of Child Abuse." *Social Problems* 24(3): 310–323.

Richardson, John T. E. (1995). "The Premenstrual Syndrome: A Brief History." *Social Science and Medicine* 41(6): 761–767.

Szasz, Thomas. (1960). "The Myth of Mental Illness." *American Psychologist* 15: 113–118.

Szasz, Thomas. (1961). *The Myth of Mental Illness: Foundations of a Theory of Personal Conduct*. New York: Paul B. Hoeber.

Virchow, Rudolf. (1981). "Excerpt from *One Hundred Years of General Pathology*." In *Concepts of Health and Disease*, eds. Arthur L. Caplan, H. Tristram Engelhardt, and James J. McCartney. Reading, MA: Addison-Wesley Publishing Company. Originally published in 1895.

Ware, Norma C. (1992). "Suffering and the Social Construction of Illness: The Delegitimation of Illness Experience in Chronic Fatigue Syndrome." *Medical Anthropology Quarterly* 6(4): 347–361.

Wolpe, Paul Root. (2002). "Treatment, Enhancement, and the Ethics of Neurotherapeutics." *Brain and Cognition* 50: 387–395.

INTERNET RESOURCE

World Health Organization. (1948). "Preamble to the Constitution." Available from http://whqlibdoc.who.int/hist/official_records/constitution.pdf. WHO Constitution adopted by the International Health Conference, New York, June 19–22, 1946; signed on July 22, 1946 by the representatives of sixty-one states and entered into force on April 7, 1948. Can also be found in the Official Records of the World Health Organization, no. 2, p. 100.

HEAVY METALS

• • •

Heavy metals is a common toxicological term covering a number of metallic substances that acutely damage human beings and ecosystems, and whose atomic weights fall between and 64 and 201. Those responsible for the most injuries and deaths are lead, mercury, and cadmium. Others with toxic properties—for example zinc, beryllium, chromium, aluminum, bismuth, manganese, and copper—are frequently listed as *heavy*, but because their atomic weights fall below 64 are not chemically regarded as such. A term better-suited to all these substances might simply be *toxic*.

Another toxic material, arsenic, is often included among the heavy metals but chemists see arsenic as a semimetal because its chemical and physical properties are only partially metallic. Thus they advocate a separate classification for this substance that since the 1980s has been poisoning well water and damaging the health of hundreds of thousands of villagers in Bangladesh and West Bengal, India.

Origin and Issues

Metals leach into living systems from natural ore deposits. But by far the major sources of toxic entry are emissions and wastes from mining and smelting operations, manufacturing processes, power plant emissions, waste incinerators, and through such consumer items as fuel additives, dental amalgams, toys, paints, light bulbs, plumbing, electronic devices, even vaccines and herbal dietary supplements. Toxic metals are ubiquitous, persistent, and controversial, and because they destroy critical enzymes can be savage in their toxic effects.

Accordingly the regulation of these substances has taken many forms, from public health and consumer protection laws to measures that control air, land, and water contamination. International treaties are probably inevitable, since these metals disperse throughout the ecosphere, cross national boundaries, essentially never degrade, and accumulate to toxic concentrations in fruits, vegetables, farm animals, and seafood. The major practical approach to their control is capture, followed by impounding.

Heavy metals history is replete with stories of environmental injustice and regulatory lethargy. Children and developing fetuses are the most tragic victims, usually suffering from cancer and serious neural disorders such as Parkinsonism and mental retardation. Increasing bodies of evidence indicate that high toxic chemical levels also correlate geographically with high crime rates, raising important legal and ethical questions as to whether polluters should be liable for offenses that promote criminal behavior in persons exposed to metallic emissions.

State, local, and federal regulations over the last three decades of the twentieth century reduced public exposure to these substances. But localized incidents remain frequent in the early-twenty-first century and the legacy of past abuses poses persistent problems through the presence, for example, of industrial waste or Superfund sites that have not yet been cleaned up (or in technical jargon, *remediated*). The history of heavy metals toxicity is a particularly tragic one, marked by bitter conflicts over surreptitious dumping, disposal in areas populated by poor people, exposure to children, lack of equitable compensation of victims, and corporations that are unwilling or unable to pay for control and cleanup.

Mercury

One of the earliest, most heartrending modern instances of heavy metal poisoning was the disaster that occurred in Minamata, Japan, during the 1950s and 1960s when mercury was discharged from a plastics manufacturing plant into the waters of Minamata Bay. The metal, in the form of methyl mercury, accumulated in the bodies of fish that were the food staple for the thousands of persons who lived in that section of southwestern Japan. The pathological result was painful neural disorders that had distressing physiological, social, and psychological effects on the people of Minamata.

Mercury's largest single source is the combustion of coal in power plants, a problem that grows as global industrial economies expand. The challenge is enormous and international health and environmental advisory bodies have urged regulations that call for removal of 90 percent of mercury from such emissions. (Cadmium and lead are also significant emission components.)

Mercury regulation has been a controversial issue in the United States for several years, mainly over government attempts to amend the Clear Air Act in favor of less stringent standards for emissions. Relaxation of standards and regulations has always been under fierce debate in toxic metals regulation, but in the case of mercury, the underlying conflict has been more closely related to the government's market-based approach to regulation as opposed to regulatory procedures specific to conditions near emission sites. The regulatory hope among experts in toxic metals research and regulation is to construct an international treaty similar to the Kyoto protocol that was established to reduce carbon dioxide emissions from industrial operations and thus decrease global warming. In other words, if all industrial operations adhered to low to zero emission standards, environmentalists believe the world would be a much safer place.

In any case, the public and environmental agencies at all levels of government are now acutely aware of the dangers of mercury. Disposal from mining operations remain a problem throughout the world and disputes over health effects and liability generate headlines almost daily. Likewise mercury contamination in ocean fish such as tuna, mackerel, and salmon remains a constant concern. Mercury in dental amalgam was for years a major cause of concern, but due to intense public attention that issue has subsided in recent years.

Lead

Lead contamination is more widely recognized than mercury contamination but vigilance over its dangers has helped to establish broad measures to bring exposure under control. A metal widely used since early times and treasured as a decorative and culinary material in ancient Rome, lead's toxic problems have been known for centuries. Since the mid-twentieth century, thousands of children have suffered the effects of lead poisoning by ingesting or absorbing lead from toys, painted household items, playground soil, and refuse left after the demolition of homes and buildings.

But in the broader sense, it was the overall public health implications of lead in gasoline (in the form of tetraethyl lead) that caused most of the initial furor over the need to control it in the environment. The U.S. petroleum and auto industries successfully fought efforts to end its use. However when the auto industry began installing catalytic converters to comply with U.S. air pollution laws, testing determined that lead rendered the devices inactive. The auto industry had no alternative but to demand development of lead-free gasoline. Leaded gasoline, however, is still in use in many countries.

Lead from mining has always been an environmental and public health problem and remains so in the early-twenty-first century. A typical industrial example is emissions from the smelter at the Bunker Hill lead mine in

Pinehurst, Idaho, during the 1970s. For years fallout from the smelter contaminated the air in the area around Pinehurst. The Center for Disease Control (CDC) tested children in the area for blood-lead levels and found the highest amounts ever recorded in human beings.

On the whole, however, laws, regulations, and a high degree of watchfulness have brought the lead problem relatively under control, though lead poisoning incidents, especially in old housing, continue to be of concern, as do lead emissions from mining and smelting facilities around the world.

Cadmium and Chromium

Cadmium and chromium come from a variety of sources from cigarette smoke to smelting operations to increasingly voluminous waste from electronic devices. They enter living systems from alloys, pigments, batteries, metal coatings, electronic devices, mining operations, and industrial emissions. Cadmium especially affects the kidneys and lungs, but it also causes testicular damage, lung disease, and bone disease.

Chromium, for its part, is an essential nutrient in very small amounts. It is involved in manufacturing chrome-plated materials, tanned leather, dyes and pigments, and wood preservatives. It enters living things mainly through the air and underground water. Extended exposure to chromium can cause asthma, lung cancer, and ulcers.

Controversies

Chemists dislike the term heavy metals because of its inherent imprecision and often urge that it be abandoned. In 2002 the International Union of Pure and Applied Chemistry—the organization that sets the standards for chemistry's precise nomenclature—issued a technical report titled, "Heavy Metals—A Meaningless Term?" that reflected the frustration felt by the chemical community over the term's loose usage by those outside the field of basic chemistry. The term heavy metal, the report pointed out, "has even been applied to semimetals (metalloids) such as arsenic, presumably because of the hidden assumption that 'heaviness' and 'toxicity' are in some ways identical" (p. 796).

The report bemoaned what it called "the persistence of the term and its continuing use in literature, policy, and regulations" (p. 797). It stated,

> There is no similarity in properties between pure tin, which has low toxicity, and tributyltin oxide, which is highly toxic to oysters and dog whelks. Nor is there any similarity in properties between chromium in stainless steel, which is essentially

nontoxic, and the chromate ion which has been associated with causing lung cancer. Thus, the tendency to group certain metals and their compounds together for toxicity assessment under the title 'heavy metals' must lead to fuzzy thinking and is another reason to abandon the term. (p. 799).

Ethical issues surrounding the heavy metals parallel those associated with harm caused by toxic substances in general. Tension always exists between producers of these substances and those exposed to them, often leading to tort damage claims and prolonged litigation. Those who believe industry should be held liable for injuries caused by toxic metal emissions have been turning for support to a relatively new legal theory known as the neurotoxity hypothesis. This hypothesis derives from neurochemical research that suggests that criminal behavior in individuals correlates with high levels of lead, manganese, and cadmium in the bodies of those individuals. Further research reinforcing such new insights could lead to changes in tort law that would impose stricter regulatory standards for these substances and more criminally related penalties for violators.

WIL LEPKOWSKI

SEE ALSO *Environmental Ethics; Environmentalism; Environmental Regulation.*

BIBLIOGRAPHY

Crawford, Colin. (2000). "Criminal Penalties for Creating a Toxic Environment: *Mens rea*, Environmental Criminal Liability Standards, and the Neutotoxicity Hypothesis." *Boston College Law Review* 27(3): 341–290.

Duffus, John H. (2002). "Heavy Metals: A Meaningless Term?" (IUPAC Technical Report). *Pure Applied Chemistry* 74(5): 793–807.

Nriagu, Jerome O. (1966). "A History of Global Metal Pollution." *Science* 272(5259): 223–224.

Van der Voet, Ester; Jeroen B. Guinée; and Helias A. Udo de Haes, eds. (2000). *Heavy Metals: A Problem Solved?: Methods and Models to Evaluate Policy Strategies for Heavy Metals.* Dordrecht, The Netherlands; Boston: Kluwer Academic Publishers.

HEGEL, GEORG WILHELM FRIEDRICH
• • •

German philosopher Georg Wilhelm Friedrich Hegel (1770–1831), born in Stuttgart on August 27 and

Georg Wilhelm Friedrich Hegel, 1770–1831. The German philosopher and educator took all of knowledge as his domain and made original contributions to the understanding of history, law, logic, art, religion, and philosophy. (*The Library of Congress.*)

educated at the University of Tübingen, gained intellectual renown while teaching at the University of Berlin. A thoroughly systematic thinker, Hegel viewed philosophy, natural science, history, ethics, and religion as inherently connected in a whole that included difference while simultaneously transcending it. As a result, he presents the kind of comprehensive interpretation of science, technology, and ethics that is often implicit but seldom articulated in contemporary discussions, which, in light of Hegel, are challenged to move beyond particular case studies. Perhaps most famous for his *Phenomenology of Spirit* (1807), Hegel died suddenly on November 14 during a cholera epidemic.

Science and Technology in Hegel's System

For Hegel, the truths of the empirical (or special) sciences are justified only by the thinking at work in philosophy. Put another way, natural science occupies a middle point between sensation and philosophy. Just as sense experience needs science to grasp its deepest truths, so science requires philosophy.

The relationship between natural science and philosophy is best understood in terms of four modes of consciousness: sense-certainty, perception, understanding, and reason. The empirical sciences build on sense-certainty and perception to establish laws and theories. This move toward universality indicates that understanding is predominant in natural science. What the empirical sciences provide are nevertheless mere facts and concepts that are founded on fixed categories (for example, cause and effect, substance and accidents) that are accepted uncritically. Such a detailed explication of nature has a relative immediacy when viewed from the perspective of self-conscious reason and its characteristic philosophical thinking. It thus becomes the task of philosophy to give final meaning to what the sciences reveal by criticizing their inherent conflicts and contradictions on the way to establishing a unified synthesis in which these differences are preserved while being overcome. Ultimately, the empirical sciences are a necessary and integral phase in the development of consciousness and a crucial first step toward the rational unveiling of what Hegel calls Spirit in nature.

Hegel's view of technology emerges from his defense of the distinctly modern assertion that all knowing involves making. In accordance with this doctrine, Hegel maintains that human beings produce both themselves and their world. Individuals are only insofar as they are productive. In one's relationship with the natural world, such production manifests itself as work, a mediating activity pervaded by the tools one uses. Technology, therefore, emerges as formative for human beings insofar as it allows them to assert themselves over and against their physical environment. Though such is the case with even basic tools, it becomes most evident with the emergence of machines, the effectively self-reliant tools that deceive nature into working toward human ends. Whereas science aids in discovering the Spirit implicit in nature through observation, technology is the human way of actively manifesting Spirit in the natural world, which is continuously transformed through work.

Hegel's Influence

Hegel's initial influence rested with his ability to go beyond the distinction that his predecessor Immanuel Kant (1724–1804) made between phenomenal appearances (which are scientifically knowable) and things-in-themselves (which ground all phenomena, but remain unknowable in all respects other than their actual existence). Against Kant, Hegel argues that systematic philosophical reflection, in grasping the cognitive genesis of scientific knowledge and its contribution to self-consciousness, can indeed know reality in its entirety (that is, both phenomenal appearances and things-in-themselves), because there could not in principle be anything beyond such a synthetic whole.

The first generation of Hegel's followers nevertheless looked more to the practical implications of transcendence, thus proposing a further overcoming of Hegel himself that would make his philosophical synthesis, especially the notion of a self-consciousness that simultaneously makes the world and itself, into a lived reality. It was for this reason that Karl Marx (1818–1883) sought to turn Hegel right-side up and thereby place him on his feet (*The Holy Family*, 1845), not just to understand the world but to change it ("Theses on Feuerbach," 1845). Marx's critique centers around the plight of industrial workers and the alienation they experience in regard to the products of their labor, their work activity, and, above all, their humanity (*Economic and Philosophical Manuscripts*, 1844).

But it was another philosopher, Ernst Kapp (1808–1896), who took the technological implications of Hegel most seriously, and in doing so was the first to speak of a "philosophy of technology." Drawing on Hegel's theory of history, Kapp's materialism took historical evolution to be the result of humanity's various attempts to overcome the constraints of nature (*Vergleichende allgemeine Erdkunde*, 1845). Insofar as such an overcoming necessarily involves technological innovation, Kapp reflected extensively on the nature of tools, construing them as "organ projections" that essentially act as extensions of the human body (*Grundlinien einer Philosophie der Technik*, 1877).

The Master–Slave Dialectic

The historical and ethical import of Hegel's views on technology are best gleaned from his master–slave dialectic, a doctrine interpreted at length by Alexandre Kojève (1902–1968), whose post–World War II lectures, though idiosyncratic, proved influential. For Kojève history begins with the first battle that ends with a victorious master and a vanquished slave. In risking life for genuine human recognition, the master spurns a merely biological existence, thus triumphing over the slaves who, for fear of death, succumb to the master in order to preserve their lives. Through human conquest, the master achieves an independence that, at least for the time being, remains foreign to the slave. The slave works for the master, forced to struggle with an often recalcitrant nature in order to provide for the master's needs.

In spite of the seemingly unenviable position of the slaves, true human progress and genuine freedom would be impossible without them. Masters, freed from dealing with nature, live a life of leisure that consumes the products of nature without any compensatory replenishment. Slaves, by contrast, learn to confront nature, an imposition that obliges them to understand nature in order to control it. It is slaves, then, who develop science and technology and who, unlike masters, are the true creators. Only through such scientific and technological development is progress made and historical change enacted.

Furthermore, the path to true freedom finally becomes apparent as the freedom of the master ultimately reveals itself as false. Though a master achieves a measure of independence from the physical environment, this is an achievement that remains dependent on the activity of the slaves. Slaves, for their part, achieve scientific understanding and create technological innovations that clear the way for a genuine freedom by surmounting nature directly and becoming independent of the services of still other slaves.

Conclusion: The Ethical Dimension

The evolution of science and technology, for Hegel, has direct ethical implications. In marking desire as intrinsic to self-consciousness, Hegel maintains that real human satisfaction can be had only in and through the recognition of another self-conscious subject. Though the master sought such recognition in his relationship with the slave, slavish recognition is necessarily ungratifying insofar as it is given by a slave who is, by definition, less than fully human. Genuine human satisfaction, therefore, will be had only when the master–slave relationship comes to an end and the human beings involved recognize each other as equals.

This ethical ideal of reciprocal recognition is first envisioned by slaves who see how people can free themselves from their merely biological existence and thereby assert their dignity in a way other than the masterly domination of other human beings. Through scientific understanding and the technological mastery of nature, the master–slave relationship can be overcome, reciprocal recognition achieved, and genuine freedom finally won. For Kojève, such an occurrence will mark the end of history because the struggle for recognition, which is the principal cause of historical change, comes to an end.

CRAIG CONDELLA

SEE ALSO *Alienation; Freedom; German Perspectives; Marx, Karl.*

BIBLIOGRAPHY

Harris, Errol E. (1974). "Hegel and the Natural Sciences." In *Beyond Epistemology*, ed. Frederick G. Weiss. The Hauge, Netherlands: Martinus Nijhoff. Situates the natural sciences within Hegel's system, identifying understanding as the mode of consciousness predominantly at work therein.

Hegel, Georg Wilhelm Friedrich. (1970). *Hegel's Philosophy of Nature*, trans. A. V. Miller. Oxford: Clarendon Press. Part Two of Hegel's *Encyclopaedia of the Philosophical Sciences* (1817). This volume deals specifically with modern science, treating figures such as Galileo, Newton, Kepler, Linnaeus, Lamarck, and Lavoisier.

Hegel, Georg Wilhelm Friedrich. (1977 [1807]). *Phenomenology of Spirit*, trans. A. V. Miller. Oxford: Clarendon Press. Intended as an introduction to his philosophical system, Hegel's *Phenomenology* traces the psychological and historical development of human consciousness through various forms of alienation to its final fulfillment in "Absolute Spirit."

Kojève, Alexandre. (1980). *Introduction to the Reading of Hegel*, trans. James H. Nichols Jr. Ithaca, NY: Cornell University Press. This translation of Kojève's *Introduction a la lecture de Hegel* (assembled by Raymond Queneau) analyzes Hegel's *Phenomenology* while simultaneously reinterpreting it, making much of the master–slave dialectic as the driving force of history.

Strauss, Leo. (2000). *On Tyranny*, ed. Victor Gourevitch and Michael S. Roth. Chicago: University of Chicago Press. Taking its cue from Xenophon's *Hiero*, this volume includes Strauss's famous debate with Kojève and deals with the problems that tyranny and the possible end of history entail.

Westphal, Merold. (1998). *History and Truth in Hegel's "Phenomenology."* Bloomington: Indiana University Press. A careful summary of Hegel's *Phenomenlogy* that illuminates what is principally at stake in the text itself while locating Hegel more generally within the Western philosophical tradition.

Martin Heidegger, 1889–1976. The German philosopher has become widely regarded as the most original 20th century philosopher. Recent interpretations of his philosophy closely associate him with existentialism (despite his repudiation of such interpretations) and, controversially, with National Socialist (Nazi) politics. (*AP/Wide World Photos.*)

HEIDEGGER, MARTIN

• • •

Martin Heidegger (1889–1976), who was born in Messkirch, Germany, on September 26 and died there on May 26, was among the most important thinkers of the twentieth century. His significance for science, technology, and ethics may be approached from four directions.

Theoretical Science and Practical Activities

Heidegger's first and still most important book, *Sein und Zeit* (1927; English trans. *Being and Time*, 1962), is a cornerstone of the existentialism that became prominent after World War II. The book's major terms—anxiety, resoluteness, everydayness, authenticity, concern, care, and the like—are concepts Heidegger helps make intellectually cogent. Albert Camus (1913–1960) and Jean-Paul Sartre (1905–1980) work on territory Heidegger opened up philosophically.

Heidegger's own goal, however, was not to outline a theory of human beings as radically insecure or irrationally committed, but to uncover the central openness of human beings to being as such. Humans are the entities for whom how to be is always an issue. This is true for everyone and not merely true generally or abstractly. Heidegger's goal is to clarify the question of being by working out what being is and how it matters for each human being.

Heidegger's analysis in *Being and Time* follows a path that begins with the significance of ordinary human concerns and concludes with the temporal meaning of being. The usual implicit meaning of being is that which is most fully or eternally present. As a result humans conceive all things as essentially static entities with fixed, general characteristics suitable for neutral measuring, spatially and temporally. People objectify even their own selves in this manner. The meaningful present, however, cannot exist apart from the ordinary worlds of significance into which people find themselves thrown. This richer temporality, not static presence, is the heart of being human, and the clue to being as such. There is a historical and temporal motion, indeed, a dizzying abyss beneath all presence.

The relation between theory and practice that Heidegger's analysis suggests has important implications for understanding scientific technology. Purely theoretical enterprises such as natural science or mathematics depend on views of time and space that flatten or narrow the rich meanings of being projected in the ordinary worlds of action and concern. Dealing with things as they are actually used is primary; theoretical and scientific analysis is secondary. The right time and place to use particular tools cannot be determined, for example, from the neutral coordinates of physics, but are inherent in use itself. Instead, physics abstracts from and narrows the richness of tools that do their jobs usefully in the appropriate place and time.

This narrowing does not mean, however, that what science discovers is false in its own realm. The relativism or inordinate human responsibility for meaning that is inseparable from Heidegger's understanding does not imply that everything is magically at human disposal. Rather, what natural science discovers may be correct, but humans must see how it is grounded on the broader truths of being and of human openness to being.

The History of Science

Many of the works of Heidegger and his followers include some notion that use, practice, and everyday concern precede the flattening on which modern science and technology are built. Indeed, this view has served as the basis for Heidegger's influence on academic studies in the history of science. Heidegger's teacher Edmund Husserl (1859–1938) and several of Heidegger's students or those he affected, such as Jacob Klein (1899–1978) and Alexander Koyré (1892–1964), made important contributions to the history of mathematics and science. Klein's *Greek Mathematical Thought and the Origin of Algebra* (1934) and Koyré's *Galileo Studies* (1939) may even be said to have transformed the field, because Heidegger's procedure, which influenced them, involved a relentless search for the experience and understanding at the heart of worn-out philosophical concepts commonly employed by academic history.

To grasp the existential origin of scientific concepts was to uncover their meaning, power, and range.

Heidegger himself explored in various places the original Greek understanding of nature (*phusis*) and the changed understanding of nature and motion that differentiates Aristotelian and Newtonian physics. His 1936 lecture course "Die Frage nach dem Ding" (published in 1962; English trans. *What Is a Thing?* 1967) is especially cogent in this regard.

The Technology Question

Heidegger's most direct discussion of scientific technology is in his "Die Frage nach der Technik," delivered in early versions in the 1940s and published in 1954 (English trans. *The Question concerning Technology*, 1977). His analysis became a basic text for those worried about the power and dominance of contemporary technology. Both directly and indirectly it has influenced thinkers and activists (such as the German Greens) who in the name of the environment opposed growing industrialization and mechanization. Here and in other works, Heidegger's prescient sense of the importance of information science and life chemistry also connects his views to pressing controversies of the day.

Heidegger argues that the essence of technology is nothing technological, that is, that technology is not itself a tool or implement. Rather, the essence of technology involves the manner in which things first present themselves in the contemporary world, namely, as "standing reserve" to be manipulated or rearranged at will. Everything approaches humans as a source of energy, a human "resource," a matter to be organized. Lost in this scenario are the independence of things, their distinctive presence and shape, and the way in which they take place in a meaningful world they help to form. The simple bridge across a river allows the river to meander and stand forth in its own power; the dam that helps to generate electricity transforms this river into an implement interchangeable with other energy resources. Because people see themselves so generally as resources to be manipulated, they become alienated from their roots and traditions, and from the significance of birth and death. Technology sunders human beings from the lifetimes and the times of life that give individuals weight and direction.

Heidegger does not seek to solve the problem of technology directly or to overcome humanity's technological leveling. To do so would make his own effort one more link in the strangling technological chain. Rather, he tries to show that as the predominant presentation of beings today, technology itself must open to and be placed in being as such. The apparent technological annihilation of all other significance becomes a clue to the source of meaning generally. The results of uncovering this source cannot be predicted. But being and human openness to it can be addressed and discussed in the manner of *Being and Time*, or in the more direct yet more elusive way of some of Heidegger's work from the mid-1930s on, in which discussions of poetry and gods come to the fore.

The Nazi Question

Heidegger's work is tainted by his association with the Nazis. He joined the National Socialist Party when he became rector of Freiburg University in May 1933, whereupon he praised Adolf Hitler publicly. The intensity of his support subsequently diminished, and some remarks in his lectures may be read as opposition to the views of Nazi ideologues. Other remarks continued to defend the Nazis, however, and he remained a party member throughout World War II.

The important question for students of Heidegger and of technology is whether his support of the Nazis flows from his philosophical arguments or, rather, stems from personal idiosyncrasy or political naïveté. It would be difficult to take seriously a thinker whose discussions of what it is to be a human being were in no way linked to political actions and judgments; Heidegger's arguments do, in fact, display such a link. Heidegger's thought leads to immoderation and illiberalism because the standpoint from which he confronts issues is too encompassing to allow relevant ethical distinctions to matter or even become clear. Too many issues that to a responsible citizen or political leader involve significant differences between what is just and unjust look, from Heidegger's ontological point of view, to be the same. The substance of his understanding of human openness to being, moreover, with its emphasis on fate, authentic resolve, and the *Volk* (people), allows Heidegger to believe he has found essential links between his thought and the Nazis, and to accommodate his rhetoric to theirs.

It would be incorrect to claim that Heidegger's philosophical immoderation or basic concepts led him inevitably to support the Nazis or to approve all of Hitler's actions. The Nazis, he believed, ultimately failed to live up to what he called in 1935 "the inner truth and greatness of this movement." In the *Introduction to Metaphysics*, the version of 1935 lectures that he published in 1953, he described this "truth" and "greatness" as "the encounter between global technology and modern humanity. This same standpoint, however, led him not only (finally) to question the Nazis but to also treat the substance of Soviet Marxism, American democratic capitalism, and failed Nazism as essentially identical. The ethical and political immoderation to which

Heidegger's view of technology can lead is strikingly captured not only in his political judgment but also in his identification of mechanized agriculture and the Holocaust: "Agriculture is now a motorized food industry, essentially the same as the manufacture of corpses in gas chambers and extermination camps, the same as the blockade and starvation of countries, the same as the manufacture of hydrogen bombs" (Polt 1999, p. 172, translating from Heidegger's "Das Ge-Stell").

Heidegger's thought cannot be reduced to his connection to the Nazis. His understanding of being and being human revitalized the study of philosophy by encouraging an encounter with the phenomena that the great works of Western thought have in view. His central concepts stimulated many to rethink the true sources of human freedom, excellence, and happiness. His view of scientific technology captures its breadth and centrality in a novel and still cogent manner. The paths he helped to open, however, can become closed by dogmatic application of his procedures. Heidegger's politics, moreover, encourage more than ordinary caution in dealing with his insights.

MARK BLITZ

SEE ALSO *Alienation; Arendt, Hannah; Existentialism; German Perspectives; Husserl, Edmund; Phenomenology.*

BIBLIOGRAPHY

Blitz, Mark. (1981). *Heidegger's "Being and Time" and the Possibility of Political Philosophy.* Ithaca, NY: Cornell University Press. A commentary on being and time.

Heidegger, Martin. (1962). *Being and Time,* trans. John Macquarrie and Edward Robinson. New York: Harper and Row. Translation of *Sein und Zeit,* 1927. Heidegger's major work

Heidegger, Martin. (1967). *What Is a Thing?* trans. W. B. Barton Jr. and Vera Deutsch. Chicago: Henry Regnery. Translation of *Die Frage nach dem Ding,* 1962. Includes a comparison of Aristotle and Newton.

Heidegger, Martin. (1977). *The Question concerning Technology,* trans. William Lovitt. New York: Harper and Row. Translation of *Die Frage nach der Technik,* 1954. Heidegger's chief essay on technology.

Heidegger, Martin. (1993). *Basic Writings,* rev. edition, ed. David Farrell Krell. San Francisco: HarperSanFrancisco. Includes "The Question concerning Technology," selections from *Being and Time* and *What Is a Thing?* and other works relevant to Heidegger's understanding of science and technology.

Heidegger, Martin (2000). *Introduction to Metaphysics,* trans. Gregory Fried and Richard Polt. New Haven, CT: Yale University Press. Translation of *Einfuhrung In Die Metaphysik* (1953). Important lectures delivered originally in 1935.

Klein, Jacob. (1968). *Greek Mathematical Thought and the Origin of Algebra*, trans. Eva Brann. Cambridge, MA: MIT Press. Originally published 1934. Early work in the history of science influenced by Heidegger.

Koyré, Alexandre. (1978). *Galileo Studies*, trans. John Mepham. Atlantic Highlands, NJ: Humanities Press. Originally published 1939. Early work in the history of science influenced by Heidegger.

Lovitt, William, and Harriet Brundage Lovitt. (1995). *Modern Technology in the Heideggerian Perspective*. 2 vols. Lewiston, NY: Edwin Mellen Press. Discussions of Heidegger and technology.

Polt, Richard. (1999). *Heidegger: An Introduction*. Ithaca, NY: Cornell University Press. Overall introduction to Heidegger's thought.

HIGHWAYS

SEE *Roads and Highways*.

HINDU PERSPECTIVES

• • •

Hinduism is the oldest of the major world religions, and also apparently one of the most accepting of modern science and technology. It provides a central place to consciousness in its approach to reality, which explains why it has appealed both to scientists looking for a role of observers in physics and biology and also to those who have been critical of standard science for its emphasis on mechanistic explanations.

The origins of Hinduism are not found in a single individual, and its texts go back to antiquity in India. Within the tradition, it is called the Sanātana Dharma or Vedic Dharma (*sanātana* meaning eternal, *veda* meaning knowledge); the term Hindu originally referred to the inhabitants of the Indian subcontinent. The various sects of Hinduism take the Vedas (second millennium B.C.E., or perhaps a bit earlier), which are collections of hymns, to be their canonical texts. But the Vedas are difficult to understand, and for practical reasons, most Hindus rely on later texts such as the Upanishads, the Bhagavad Gītā, and the Epics (first millennium B.C.E.), Sūtras, Āgamas, Shāstras, Purānas (whose time frames range from centuries B.C.E. to texts as late as about 1000 C.E.) for guidance.

Hinduism takes phenomenal reality to be a projection of God (Brahman), who is both transcendent and immanent. In its transcendent form, Brahman is beyond any attributes; in its immanent form it may be visualized

in many different ways, leading to a multiplicity of representations. The evolution of the universe is by laws (*rita*), yet sentient beings have freedom. The law of karma constrains ordinary action, but a realized person is free.

The Vedic texts claim that language cannot describe reality completely, although its mystery may be experienced fully. Knowledge is classified in two ways: the lower or dual; and the higher or unified. The lower knowledge, which describes the objective world, is obtained using logic and it is accessible by language. The higher knowledge concerns the experiencing self and is beyond ordinary language. The seemingly irreconcilable worlds of the material and the conscious are aspects of the same transcendental reality. Hinduism is supportive of all scientific exploration, believing that at its end one becomes aware of its limitations and the need to reach the mystery of the experiencing self. From a personal perspective, Hinduism is concerned with techniques that make self-transformation possible. Hinduism thus endorses both science and technology although not necessarily in their modern forms or for distinctly modern reasons.

Hinduism approaches the world in an ecological sense. Not only humans, but also animals, are conceived as sentient and, therefore, deserving of compassion. The Hindu approach to reality is through *jnāna* (intuitive understanding) that includes subjective and objective knowledge, value and fact, and consciousness and reality. *Jnāna* presupposes *jijnāsā*, a reaching out to understand, that leads to a spark of illumination. *Jnāna* requires the ethics of the individual as an indispensable condition for knowledge, which thus is not value free. Search for truth is a value orientation.

Historical Background

The history of early Hinduism is tied to the history of India. Its chronological time frame is provided by the archaeological record that has been traced, in an unbroken tradition, to about 8000 B.C.E. Prior to this are records of rock paintings believed to be considerably older. The earliest textual source is the Rigveda, which is a compilation of very ancient material. The astronomical references in the Vedic books recall events of the third or the fourth millennium B.C.E. and earlier. The recent discovery that Sarasvati, the preeminent river of the Rigvedic times, went dry around 1900 B.C.E. due to tectonic upheavals suggests that portions of the Rigveda were written prior to this epoch. According to traditional history, the Rigveda was written before 3100 B.C.E.

The other Vedic texts of the Yajurveda, the Sāmaveda, and the Atharvaveda borrow heavily from the Rigveda. The Brahmanas are prose works that describe

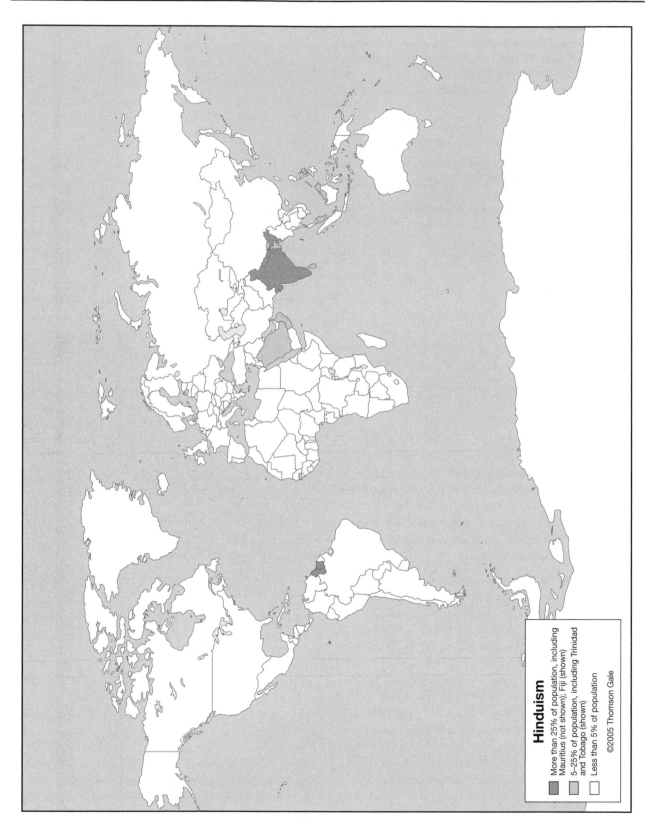

Hinduism

More than 25% of population, including Mauritius (not shown); Fiji (shown)

5–25% of population, including Trinidad and Tobago (shown)

Less than 5% of population

©2005 Thomson Gale

the Vedic ritual, and the Upanishads address philosophical issues. Ethical questions are directly addressed in the Sūtra literature, the Rāmāyana and the Mahābhārata, the Purānas, and the commentaries on these texts that have been written from time to time. Since the medieval times, the Bhagavad Gītā and the Rāmāyana have influenced millions, including Mahatma Gandhi.

Outside India, in the second millennium B.C.E., the ruling Mitannis in West Asia worshiped Vedic gods. The religion of Iran before Zoroastrianism was Vedic. Hindu religion spread to various countries in Southeast Asia in the first millennium B.C.E. and the largest Hindu temple in the world is found in Cambodia. In the twentieth century, Vedanta and Yoga have spread the popularity of Hinduism to Europe and North America.

Academic narratives of Hinduism emphasize issues related to social hierarchy, customs, and sectarian divisions around the worship of Vishnu, Shiva, and the Goddess. In reality, the social classes are not rigid, and most Hindus worship all the deities, although they might personally be more devoted to one or another. To understand why Hindus do not find it troubling to be devoted to more than one deity, it is necessary to examine the common thread of Vedic cosmology running through the tradition.

VEDIC COSMOLOGY. Briefly the Vedic texts present a tripartite and recursive view of the world. The universe is viewed as three regions of earth, space, and sky that in the human being are mirrored in the physical body, breath (*prāna*), and mind. The processes in these regions are connected as the consequence of a binding (*bandhu*) between various inner and outer phenomena. At one level, it means awareness that certain biological cycles, such as menstruation, have the same period as the moon. At another level, equations are postulated, such as the 360 bones of the infant (which fuse into the 206 bones of the adult) that correspond to the number of days in the civil year.

The connection between the outer and inner cosmos is seen most strikingly in the use of the number 108 in Indian religious and artistic expression. Elementary geometrical reasoning establishes that this number is the approximate distance from the earth to the sun and the moon in sun and moon diameters, respectively. The diameter of the sun is also approximately 108 times the diameter of the earth, but that fact is not likely to have been known to the Vedic sages. The number of dance poses given in the Nātya Shāstra is 108, as is the number of beads in a rosary. The *distance* between the body and the inner sun is also 108, which

to span, symbolically, one uses 108 names of the deity in worship. The number of weak points in the body in Āyurveda, the Hindu medicine system, is 107, because in a chain 108 units long, the number of weak points would be one less.

The Vedas are primarily concerned about universal laws related to the inner self (*adhyātma vidyā*) that are true for all times. The Hindu experience is thus not contingent on a particular account of history, or an event that cannot be replicated. Complementing the Veda, which is the *heard* revelation (*shruti*), is the remembered tradition (*smriti*). As custom, *smriti* is considered appropriate for time and location and thus subject to change. This has allowed Hinduism to adapt to change over the millenniums.

VISHNU, SHIVA, AND THE GODDESS. Although the principles of Hinduism may appear very abstract, in practice Hindus relate to a personal deity much like followers of other religions. When viewed as the ethical principle, Brahman is Vishnu; as the inner Self, it is Shiva; and seen as the energy of Nature, it is the Goddess. Although at one level Vishnu and Shiva are the Preserver and the Destroyer; at another level, due to recursion, both Vishnu and Shiva, as well as the Goddess, are each the Creator, the Preserver, and the Destroyer. Furthermore each god has a goddess as consort, emphasizing the complementarity of the two. Shiva and the Goddess are also viewed as a single deity, as half of a whole, called Ardhanārīshvara, and Vishnu and Shiva as a single deity called Harihara.

Hinduism and Science

In Hinduism, the dividing line between objective sciences and *adhyātma vidyā* (spiritual knowledge) is the logical or linguistic paradox. Logical argument and rational proof using Nyāya is the way to obtain correct knowledge. But where paradox (*paroksha*) begins, one must let go of linguistic associations to experience paradox-free, deeper knowledge.

Nyāya's beginnings go back to the Vedic period, but its first systematic elucidation is Akshapāda Gotama's *Nyāya Sūtra*, dated to the third century B.C.E. Its text begins with the nature of doubt and the means of proof, and it considers the nature of self, body, senses, and their objects, cognition and mind.

The Nyāya system supposes that human beings are constructed to seek truth. Their minds are not empty slates; the very constitution of the mind provides some knowledge of the nature of the world. The four *pramānas* through which correct knowledge is acquired are *pratyaksha*, or direct perception; *anumāna*, or inference; *upamāna*, or analogy; and *shabda*, or verbal testimony. Four factors are involved in direct perception: the senses, their objects, the contact of the senses and the objects, and the cognition produced by this contact. The mind mediates between the self and the senses. When the mind is in contact with one sensory organ, it cannot be in contact with another. It is therefore said to be atomic in dimension. It is because of the nature of the mind that one's experiences are essentially linear, although quick succession of impressions may give the appearance of simultaneity.

The Nyāya attacks the Buddhist idea that no knowledge is certain by pointing out that this statement itself contradicts the claim by its certainty. One can check whether cognitions apply to reality by determining if they lead to successful action. Valid knowledge leads to successful action, unlike erroneous knowledge.

The evolution of the universe is ordained by cosmic law. Because it cannot arise out of nothing, the universe must be infinitely old. Because it must evolve, there are cycles of chaos and order or creation and destruction.

According to the atomic doctrine of Kanāda, there are nine classes of substances: ether, space, and time that are continuous; four elementary substances (or particles) called earth, air, water, and fire that are atomic; and two kinds of mind, one omnipresent and another that is the individual. The conscious subject is separate from the material reality but is, nevertheless, able to direct its own evolution.

The Mahābhārata and the Purānas address the question of creation. It is said that humans arose at the end of a chain, at the beginning of which were plants and various kind of animals. In Vedic evolution the urge to evolve into higher forms is taken to be inherent in nature. A system of evolution from inanimate to progressively higher life is a consequence of the different proportions of the three basic attributes of *sattva*, *rajas*, and *tamas*, which represent transparence, activity, and inertia, respectively. In its undeveloped state, cosmic matter has these qualities in equilibrium. As the world evolves, one or another of these becomes preponderant in different objects or beings, giving specific character to each.

Unlike the Abrahamic religions, whose eschatology is centered on the dead rising to the heavens, Hindu visions of the end of the world are naturalistic. For example, the Mahābhārata (Shānti Parva, Chapter 233) speaks of how a dozen suns will begin to burn when the

time comes for universal dissolution a few billion years in the future. First all things mobile and immobile on Earth will disappear merging into the elements, making it, shorn of trees and plants, look as naked as a tortoise shell. Next Earth will melt, and then vaporize and become heat and wind. Then wind will be transformed into space, with its attribute of unheard or unuttered sound. Finally space will withdraw into Mind, ultimately merging into Consciousness, which is the origin of reality.

In Vedic discourse, the cognitive centers of the mind are called *devas*, deities or gods, or luminous loci. The Atharvaveda calls the human body the City of Devas. The number of devas is variously given, the most extravagant estimates are 3.3 million. All devas are taken to embody the same light of consciousness. The mind consists of discrete agents, although it retains a unity. Because each deva reflects primordial consciousness, one can access the mystery of consciousness through any of them.

When the cognitive centers nearer the sense-organs are viewed in anthropomorphic terms, they are called *rishis*, sages. The Yajurveda declares that seven sages reside within the body. The texts also divide the capacities of the mind into various dichotomies, such as high and low, left and right, and masculine and feminine.

MEDICINE. Āyurveda operates in the context that humanity's essential nature is the *ātman*, or Self, which is self-luminous, the source of all power and joy. Actions that aid in the manifestation of the divinity of the soul are beneficial and moral, and those that obstruct it are harmful and immoral. The Āyurvedic physician must help humans and nonhumans in their physical and mental health so that they can fulfill their quest for knowledge.

Āyurveda builds upon the tripartite Vedic approach to the world. Health is maintained through a balance between the three basic humors (*dosha*) of wind (*vāta*), fire (*pitta*), and water (*kapha*). Each of these humors has five varieties. Although literally meaning air, bile, and phlegm, the *doshas* stand for larger principles. The imbalance of these elements leads to illness. The predominance of one or the other leads to different psychological profiles. Charaka and Sushruta are two famous early physicians, and the beginnings of their compendiums have been dated to seventh century B.C.E.. According to Charaka, health and disease are not predetermined and life may be prolonged by human effort. For Sushruta, the purpose of medicine is to cure the diseases of the sick, protect the healthy, and prolong life. Indian surgery was quite advanced, even before 300 B.C.E.. The medical system tells much about the Indian approach to science. There was emphasis on observation and experimentation. The normal length of training appears to have been seven years. Before graduation, the students had to pass a test. Physicians were expected to learn through texts, direct observation, and inference. In addition, they attended meetings where knowledge was exchanged and were enjoined to obtain unusual remedies from herdsmen and forest-dwellers.

SCIENTIFIC IMAGINATION AND MODERN SCIENCE. A remarkable aspect of Indian literature is its scientific speculation. The epic Mahābhārata mentions embryo transplantation, multiple births from the same fetus, battle with extraterrestrials who are wearing airtight suits, and weapons that can destroy the world. The Rāmāyana mentions air travel. The medieval Bhāgavata Purāna has episodes describing how the passage of time can be different for different observers.

Conflict between science and religion has often arisen as a result of creation and end-of-the-world myths. Hindu views on these issues emerged from rational thought and are similar to some scientific views. Erwin Schrödinger, the cocreator of quantum theory, claimed to have been inspired by the Hindu mystical view of the identity of Brahman and the individual Self in his proposal of the quantum universal function that is a superposition of all possibilities. In fact, some philosophers of science see the evolution of quantum theory to be consistent with Vedānta. But because the bases for such beliefs in Hinduism and in modern science are quite different, it could also be argued that such relations are specious.

Hindu Ethics

The Vedas have many passages enjoining ethical behavior. The contemporary Hindu most often consults the Epics, Purānas, and the Bhagavad Gītā for such lessons. The Bhagavad Gītā is about the crisis facing Arjuna, hero of the Pandavas, as he confronts his relatives, the Kauravas, on the battlefield of Kurukshetra. Overcome by despair at the thought of killing his kinsmen in battle, Arjuna lays down his arms. But his charioteer Krishna, who is an incarnation of Vishnu, argues that Arjuna should do his duty and do battle. The human soul is not different from the universal soul and, thus, is immortal. When duties are performed without attachment to success or failure, one is not stained by action. Krishna teaches Arjuna the essence of *karma yoga*

(yoga of works), *jnāna yoga* (yoga of knowledge), and *bhakti yoga* (yoga of devotion). He also teaches that the human being has a free will that permits him to make intelligent choices, which have a bearing on his karma. Using the battlefield of Kurukshetra as a symbol of life's struggles, the lessons of this text can be applied to everyday situations.

Elaboration of the social code is found in the Mahābhārata. The four great aims of human life are *dharma* or righteousness, *artha* or wealth, *kāma* or enjoyment, and *moksha* or spiritual liberation. Life runs through four stages: studentship, householdership, forest dwelling, and wandering ascetic. Society was divided into four classes: the teacher or *brahmin*, the warrior or *kshatriya*, the trader or *vaishya*, and the worker *shūdra*. These four were born from the head, the arms, the thighs, and the feet of *purusha*, the primal man. In reality, the aims of life run somewhat concurrently, and likewise, each individual, having the same *purusha* within, has attributes of each of the four classes.

Patanjali's Yoga Sūtra speaks of a system of eight limbs of which the first two emphasize moral and ethical preparation: moral restraint (*yama*), which includes to do no harm, truthfulness, to refrain from stealing, chastity, and to avoid envy; and discipline (*niyama*), which includes purity, contentment, asceticism, self-study, and devotion to the Lord. The remaining limbs prepare the individual for a mystical union with the Self: posture, breath control, sense withdrawal, concentration, meditation, and absorption. Thus ethical behavior is essential to prepare the individual to receive knowledge. This discipline connects the physical body to the energy sheath, which is the subtle body that envelops it.

Like the Yoga Sūtra, the law book of Gautama lists the following practices for a virtuous person: compassion for all beings, patience, contentedness, purity, earnest endeavor, good thoughts, freedom from greed, and freedom from envy.

Although its diverse texts point to corresponding diversity in practice, a common theme running in the various Hindu traditions is harmony in society and nature, necessitating obligations of different kinds. Humankind is enjoined with the stewardship of nature and a special responsibility towards animals that is symbolically represented in the veneration for the cow, the origins of which veneration rest in the central role of the animal in the economy of the village and because the Sanskrit for "cow" also means "Earth." These attitudes explain why vegetarianism is extolled in many Hindu communities.

PROSPECTS. Because Hinduism makes a distinction between higher and lower knowledge, it has no direct conflict with science, although it would take issue with technologies that do not promote social good. In the Hindu approach, logic and rationality is the means of obtaining outer knowledge that complements the inner science of the self. Hinduism does not contest scientific accounts of creation; in fact, its own accounts of creation and destruction are very similar. The appeal to a cycle of births helps the Hindu find order in events that might otherwise appear chaotic and unjust.

Hinduism recognizes that at one level all creatures are part of a food chain, in which the big fish eats the small. But this physical aspect of life represents the animal self. Hinduism's task is to raise the individual beyond the animal self to a state in which one appreciates the interconnectedness of reality and develops compassion for all beings. Nonviolence is lauded as the highest principle, with the acknowledgement that the real world has violence in it that reflects the level of the development of society.

Regarding the unborn, the Garbha Upanishad claims that the subtle body enters the embryo in the seventh month. Although Hindu law books condemn abortion, the early-twenty-first-century Hindu is likely to defer to the scientist in determining when the fetus is viable. Because the individual is not just the physical body but also the subtle body, cloning the physical body is not problematic. For similar reasons, Hindus are not opposed to stem cell research.

Because Hinduism acknowledges that animals are sentient like humans; it is opposed to the unnecessary medical testing of drugs and procedures on animals. Hindus have opposed genetic modification to crops in advanced countries with the major motivation of greater productivity, because it disrupts farming in the poorer countries and makes it likely that these farmers will become dependent on expensive patented seeds controlled by inaccessible corporations.

In medical practice, the Hindu approach stresses a holistic view to therapy that acknowledges connections between mind and body, which is part of the reason of the increasing popularity of Yoga and Ayurveda. But it is not clear yet to what extent these disciplines will be incorporated in mainstream medicine.

Many Hindus—and this included Mahatma Gandhi—are critical of those technologies that dehumanize the person, treating a human being as a mere cog in a machine, as happens to be the case in certain manufacturing processes. This is why Gandhi praised small-scale industry and urged for self-sufficiency in the village. Hindus believe that science and technology must be harnessed in a manner that furthers humanity's inherent quest for self-knowledge. Because individuals are defined not in isolation, but through their interactions with other persons, this quest cannot ignore the larger good of society, and requires ethical preparation on the part of the individual.

SUBHASH KAK

SEE ALSO *Buddhist Perspectives; Indian Perspectives.*

BIBLIOGRAPHY

Bose, D. M.; S. N. Sen; and B. V. Subbarayappa, eds. (1971). *A Concise History of Science in India.* New Delhi: Indian National Science Academy. Presents a good overview of Indian science.

Crawford, S. Cromwell. (1994). *Dilemmas of Life and Death: Hindu Ethics in North American Context.* Albany: State University of New York Press. Deals with questions of Hinduism and ethics in North America.

Kak, Subhash. (2001). *The Wishing Tree: The Presence and Promise of India.* New Delhi: Munshiram Monoharlal. This collection of essays deals with lesser known aspects of Hindu culture and science.

Kak, Subhash. (2002). *The Gods Within: Mind, Consciousness, and the Vedic Tradition.* New Delhi: Munshiram Manoharlal. Presents the Hindu view of how the Gods constitute the firmament of the inner space.

Klostermaier, Klaus K. (1994). *A Survey of Hinduism.* Albany: State University of New York Press. This is a comprehensive text dealing with the history and practice of Hinduism in its various sects.

Klostermaier, Klaus K. (1998). *A Short Introduction to Hinduism.* Oxford: Oneworld Publications. A very accessible introduction to Hinduism.

Moore, Walter J. (1992). *Schrödinger: Life and Thought.* Cambridge: Cambridge University Press. An excellent resource for the connections between Vedanta Hinduism and modern physics.

Pande, Gobind C., ed. (2001). *Life, Thought, and Culture in India.* New Delhi: Centre for Studies in Civilizations. This text covers the period 600 B.C.E. to 300 C.E. in the more than 80 volumes of the PHISPC series on History of Philosophy, Science, and Culture of India of which Professor D.P. Chattopadhyaya is the general editor.

Subbarayappa, B. V., and S. R. N. Murthy, eds. *Scientific Heritage of India.* Bangalore, India: The Mythic Society. Presents articles on different scientific contributions of India.

HIROSHIMA AND NAGASAKI

• • •

These two cities are etched in the collective consciousness of the world as scenes of utter destruction and inhumanity. The decision of President Harry S Truman to authorize the use of atomic bombs on the Japanese cities of Hiroshima and Nagasaki also remains one of the most contentious issues associated with the conduct of Allied forces in World War II.

Emotional Debates

The deep emotions that people feel toward this decision continue to resonate in American and Japanese life. These emotions were expressed in the reactions to the fiftieth-anniversary exhibitions about the dropping of the bomb at Hiroshima by the Smithsonian Institution's National Museum of American History and an exhibit in Hiroshima in 1995. Professional historians serving as museum curators prepared the Smithsonian exhibit. It was carefully vetted by a wider advisory group of American historians who represented varied views about the rationale and ethics of the American decision to use the atomic bomb. Yet when word leaked to members of Congress about the content of the exhibit, special hearings were held and a firestorm of controversy and publicity resulted in a complete redesign of the exhibit into a much more innocuous display of the *Enola Gay* bomber with a few selected images and commentary about the events of a half-century before.

A widely cited Gallup poll of the American public at the time found 85 percent approving of the use of the bomb on Japanese cities. Various public figures in Japan remonstrated about America's unwillingness to face fully the import of its actions, and public demonstrations occurred in both countries over this contentious exhibit. Yet in a similar manner, considerable controversy occurred in Japan over a new exhibit in Hiroshima on the eve of the fiftieth anniversary that highlighted Japanese aggression in the Pacific and suggested that some Japanese military units had committed war crimes in their prosecution of the war effort. Many Japanese public figures condemned the exhibit and called for its withdrawal.

What Happened

It is impossible from the vantage point of history to fully know what people in the United States and Japan knew, understood, surmised, and most importantly, felt, during the period when these momentous decisions were made. World War II by this point had seen more than 55 million deaths. By 1945 the Japanese military had lost 3 million men, including more than a million in the previous year. U.S. air forces dominated the skies of Japan, and bombers flew sorties in open daylight. More than a million Japanese civilians had been killed in air raids. Yet still the Japanese refused to surrender.

Across the Pacific plans were coming to life as men, materials, ships, communications systems, and so on were all being prepared for a momentous invasion of Japan that would involve in excess of a million troops in the initial assault in the south and another million in a second wave of assaults to the north. Intelligence sources indicated that the Japanese were massing troops all over key points in Japan and preparing to repel an invasion force they were sure was coming. American troops and their leaders who had studied the vicious fighting on Okinawa where U.S. marines suffered 67,000 casualties (about 35 percent of their total fighting force), including 7,700 dead, contemplated what it would be like to now try to take the Japanese homeland where a similarly high casualty rate might be anticipated. Naval personnel recalled the ferocious kamikaze attacks they had already endured and wondered how many thousands of more planes and pilots would be flung at their ships as they entered Japanese home waters and how many more U.S. ships would be sent to the bottom of the sea.

Oral accounts of major actors' thoughts, attitudes, convictions, actions, and beliefs after the fact is colored by those facts as well as the vicissitudes of public opinion such that these recollections may prove unreliable. Historians have amassed considerable written evidence that suggests that all of the following statements hold. Presidents Franklin Delano Roosevelt and Truman always believed the bomb could and should be used. The Soviet Union was already perceived as a major threat to world peace on the conclusion of hostilities against Japan, and containing the Soviet threat was paramount in the minds of America's senior policymakers. Estimations of casualties in the first (ninety-day-long) phase of the invasion of Japan varied widely from a low of 50,000 to a high of 250,000. The United States was willing to let the emperor remain on the throne—even though this was not communicated to the Japanese. Japan had made overtures to surrender through Russian and Swiss contacts as well as directly to General Douglas MacArthur's headquarters in January 1945. General Curtis LeMay, commander of the U.S.

The atomic bomb memorial dome in Hiroshima. The memorial consists of the ruin of the only building to survive the blast. (© *John Hicks/Corbis.*)

strategic bomber forces in the Pacific, was determined to maximize air power effectiveness. The broken Japanese code indicated in July that the emperor was contemplating intervening with the Japanese military to broker a surrender. The United States had advance notice that the Soviets were entering the war against Japan in early August. The atomic bomb possessed a psychological effect well beyond its military effect and was clearly a weapon in a class by itself.

The atomic bomb was dropped on Hiroshima (population 285,000 civilians along with 43,000 soldiers) on August 6, 1945, at 8:15 A.M. local time. The immediate death toll according to estimates from a joint Japanese and American report issued in 1966 was greater than 70,000, including two American prisoners of war, with another 70,000 casualties. Of the city's 76,000 buildings, all but 6,000 were damaged and 48,000 were totally destroyed over an area of about eleven square kilometers. A total of almost 232,000 have died up to the present from disorders and problems linked to this event in Hiroshima, including children from 1945 dying from various cancers caused by the intense radiation.

The bomb dropped on Nagasaki (population 195,000) three days later killed some 36,000 Japanese outright, injured another 40,000, and caused about another 25,000 subsequent deaths due to burns and radiation exposure. By U.S. Army estimates, about 44 percent of the city was destroyed, the remainder being spared by the steep hills and topography of the city. Although Nagasaki was on the list of potential target cities, the selection of Nagasaki was "accidental" that day because clouds obscured the preferred target city of Kokura.

It is important to view the casualty figures in the context of the air war with Japan. The U.S. firebombing of Tokyo on March 9–10, 1945, resulted in more than 100,000 Japanese deaths in a twenty-four-hour period during which ground temperatures reached 1,100 degrees Celsius (the heat at the center of the atomic blasts by contrast briefly equaled that of the interior of the sun). Two subsequent air strikes against Tokyo resulted in more than half the city being completely destroyed by late May. What made the atomic bombs different was the devastation from one single bomb coupled with visual and nonvisual effects that dwarfed nonnuclear devices and long-term effects that could not even be predicted.

Postwar Assessments

The development of the atomic bomb was a major scientific and technical feat that employed at its peak 160,000 people and consumed two-fifths of the entire U.S. war budget while remaining hidden from members of Congress and even most senior military leaders, and prompted considerable angst and second-guessing on its moral appropriateness on the part of many of the scientists intimately connected with its birth.

One scientist, Joseph Rotblat, left the Manhattan Project because of his ethical concerns. Others self-organized and created a series of written documents that expressed their collective ethical and moral concerns about the bomb and its use. Captain Claude Eartherly, a pilot who flew the reconnaissance plane over Hiroshima but did not view the drop itself, later expressed regrets over his involvement and the American decision. This admission was seized on by the German philosopher Gunther Anders in a book called *Burning Conscience* and by advocacy groups to support arguments against both the use of nuclear weapons as well as the American decision to deploy them during the war. Eartherly became somewhat of a hero in communist countries and among "ban the bomb" groups. His wartime colleagues, including his commanding officer colonel Paul Tibbets who flew the B-29 that actually dropped the bomb, viewed Eartherly as a gambler, drunk, and publicity hound. (He spent his later years in a mental health facility.) Brigadier-General Tibbets expressed no regrets over his decision, although his service as deputy director of the U.S. military supply mission to India in the mid-1960s was cut short when the pro-Communist press in India labeled him as the "world's greatest killer."

A small panel of senior military, political, and scientific leaders made the final recommendation to President Truman after an intensive but brief consideration of various options. J. Robert Oppenheimer, lead science director for the project and a participant in these deliberations, later concluded that the military had kept civilians considerably in the dark about the actual state of affairs in the Pacific and the estimated impact of the proposed invasion of Japan.

Admiral William Leahy, Truman's chief of staff, believed throughout the process and after that use of the atomic bomb on two Japanese cities was completely unwarranted. He called for a return to warfare that excluded women, children, and other noncombatants. (The Allies, following the lead of the Japanese in China in the 1930s and the German firebombing of Coventry, England, in November 1940, regularly firebombed Axis cities causing massive civilian casualties on the grounds that this would hasten the end of the war.)

Justifications for the use of the atomic bomb against Japan flowed swiftly after its use, both from the White House and from military press releases. The U.S. public was also reassured that the latent results from this new weapon were modest. The *New York Times* headline of September 13, 1945, amazingly declared, "No Radioactivity in Hiroshima Ruin." Even in the earliest years, however, doubts about the necessity of the bomb as a military option to expedite the surrender of Japan were expressed by senior U.S. military leaders including Supreme Allied Commander Dwight D. Eisenhower, Chief of Staff General George Marshall, and General Henry "Hap" Arnold, commander of the Army Air Forces.

Historians in the ensuing decades have built an extensive, well-documented argument that a complex set of factors determined the decision with a principal facet, as expressed forcefully by Secretary of State James Byrnes, focused on containing the Soviet threat to the postwar world. Demonstrating the bomb against a real target would place the United States and Great Britain in a much more powerful negotiating position with Soviet leader Joseph Stalin at the end of the conflict.

While the necessity of the atomic bomb to end the war with Japan will continue to be debated, as Robert Jay Lifton and Greg Mitchell (1996) noted, "You cannot understand the twentieth century without Hiroshima" (p. xi). The Memorial Cenotaph in Hiroshima Peace Memorial Park declares, "Let all souls here rest in peace; for we shall not repeat the evil." Atomic bombs and the even more powerful thermonuclear weapons that have followed them have spawned a true human capability for omnicide—the wiping out of all life on the planet humans inhabit.

DENNIS W. CHEEK

SEE ALSO *Atomic Bomb; Pugwash Conferences; Weapons of Mass Destruction.*

BIBLIOGRAPHY

Alperovitz, Gar. (1995). *The Decision to Use the Atomic Bomb and the Architecture of an American Myth.* New York: Knopf. One of the better known books that argues that the American decision to use the bomb was motivated by politics not military necessity and how this choice dramatically affected events in the post-war world.

Butow, Robert J. C. (1954). *Japan's Decision to Surrender*. Stanford, CA: Stanford University Press. The best single treatment about the discussions, debates, and maneuvers among military and political leaders in Japan regarding surrender or continued resistance to the allies leading up to and beyond the dropping the atomic bombs.

Goldstein, Donald M.; Katherine V. Dillon; and J. Michael Wenger. (1995). *Rain of Ruin: A Photographic History of Hiroshima and Nagasaki*. Washington, DC: Brassey's. Startling black and white photographs accompanied by text portray the cities before and after the bombings and in the years since, along with photos of survivors and military and political leaders.

Hershey, John. (1989). *Hiroshima*. New York: Vintage. A brief and powerful Pulitzer-Prize winner's account drawing upon first hand survivors recollections.

Lifton, Robert Jay, and Greg Mitchell. (1996). *Hiroshima in America: A Half-Century of Denial*. New York: Quill. An exploration of events surrounding the Hiroshima decision that focuses on its effects in America, including alleged government cover-ups and American insensitivities to violence.

Sekimori, Gaynor, ed. (1986). *Hibakusha: Survivors of Hiroshima and Nagasaki*. Tokyo: Kosei Publishing. "Hibakusha," Japanese for survivors, is a carefully created recollection of activities, emotions, and devastation experienced by the inhabitants of these two cities.

Takaki, Ronald. (1995). *Hiroshima: Why America Dropped the Atomic Bomb*. Boston: Little, Brown. This study by a cultural studies expert, argues that stereotypes of the Japanese influenced racist attitudes on the part of the American public and their leaders and led to the decision to use the atomic bomb to intimidate the Soviet Union.

Thomas, Gordon, and Max Morgan Witts. (1990 [1977]). *Ruin from the Air: The Enola Gay's Atomic Mission to Hiroshima*. Chelsea, MI: Scarborough House. A reconstruction of events leading up to and beyond the moment of the bombing based on interviews, diaries, and documents with a particular focus on the men who flew the Enola Gay.

Walker, J. Samuel. (1997). *Prompt and Utter Destruction: Truman and the Use of Atomic Bombs against Japan*. Chapel Hill: University of North Carolina Press. Concisely argues that the decision to use the bomb was sound and wise militarily, politically, diplomatically, and morally.

HIV/AIDS

• • •

The human immunodeficiency virus/acquired immunodeficiency syndrome (HIV/AIDS) has reached pandemic proportions and has presented a multiple-dimension challenge for science, technology, and ethics. In 2004 approximately 39.4 million people worldwide were infected with HIV/AIDS, among whom about 3.1 million died in that year, including about 510,000 children under age fifteen. The Joint United Nations Programme on HIV/AIDS/World Health Organization (UNAIDS/WHO) estimates that in that year 4.9 million new infections occurred. Impacts have been more severe in southern Africa, where about one-third of the deaths occurred in 2004 and where life expectancies have dropped by more than 20 years in some countries. HIV/AIDS increasingly affects women and children; nearly half of those infected worldwide are female, with even higher infection rates for women in Africa. Infected pregnant and nursing women can pass the disease to their babies.

Between 2001 and 2004 global funding for HIV/AIDS relief tripled to $6.1 billion, with resultant improvements in treatments and services; this figure includes estimates of funding from all sources, ranging from individuals and families to national and international efforts. Like infections, however, services are unevenly distributed, with the poor and stigmatized remaining underserved. Analyzing the ethics and politics of scientific, technological, and other responses is a contentious issue.

Historical Perspectives

It is useful to compare the HIV/AIDS pandemic with the Spanish influenza epidemic of 1917–1918, which also was promoted by the global transportation network at an earlier stage of its evolution. In a little less than two years the Spanish flu is estimated to have killed from 21 to 50 million persons worldwide in a population of approximately 1.8 billion. While HIV/AIDS has not yet killed as large a percentage of the world's population as the Spanish influenza epidemic, HIV/AIDS infections are not self-limiting and infection rates are expected to remain high unless effective prevention programs are developed and implemented.

Mirko Grmek (1990) provides an extensive history of the emergence and identification of HIV/AIDS. In the late 1970s the disease began to appear in the United States and Europe as physicians noticed unusual symptoms in members of homosexual communities in California and New York. Those patients presented with a variety of symptoms, such as pneumonia, mononucleosis, thrush, and Kaposi's sarcoma. Some were relative benign conditions, yet the patients went into a rapid decline, and their immune systems appeared to be compromised. The U.S. Centers for Disease Control announced the disease on June 5, 1981, but the disease was not named acquired immunodeficiency syndrome until the summer of 1982. Most of the early cases involved homosexuals, but other cases developed in intravenous drug users and then in heterosexual males, women, and patients with no history of drug use. The disease eventually was recognized in equatorial Africa, where cases might have appeared as early as 1962.

Scientists eventually identified "Patient Zero," a flight attendant who apparently was responsible for infecting a large number of the early patients in the United States. Spread of the disease thus took advantage of a global transportation network, establishing a pattern that was repeated on a much less dramatic scale with the severe acute respiratory syndrome (SARS) in 2002–2003 and poses an ongoing challenge to world health management. Patient Zero continued to engage in unprotected sexual activity long after being diagnosed, posing questions about responsibility to both patients and the medical community.

Science, Technology, and Responsibility

Since the early 1980s scientific research on HIV/AIDS has been involved in a series of controversies. For instance, immediately after the identification of AIDS researchers began to try to identify the cause. Priority in the 1983 identification of HIV as the infectious agent was claimed by both Robert Gallo at the National Cancer Institute in the United States and Luc Montagnier at the Institut Pasteur in France in what became a widely reported case of questionable scientific conduct. Even after the discovery of HIV a prominent researcher, Peter Duisberg, rejected it as the basic cause of AIDS and was accused of scientific irresponsibility.

HIV/AIDS research has divided scientists and has caused conflicts between scientists and the public about research strategies and priorities. Should the emphasis be on basic immunological science or on clinical treatments? Should treatment research be aimed at preventing human cell infection by HIV or attacking human cells that already are infected? More generally, what are the relative costs and benefits of spending money on HIV/AIDS research instead of on research into another disease, such as malaria or diarrhea? An estimated 300 million people are infected with malaria, among whom 1 to 1.5 million die annually. A fraction of the money spent on HIV/AIDS research and treatment would have a much greater impact on malaria, and the provision of safe supplies of public drinking water would cause a significant reduction in the over 1 million deaths each year from diarrhea.

Research, particularly drug testing, triggers further ethical questions. How much testing should be con-

FIGURE 1

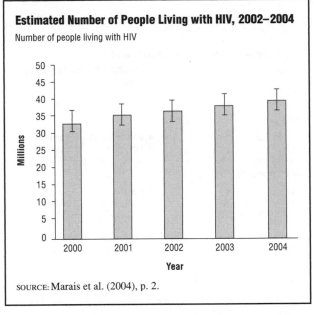

Estimated Number of People Living with HIV, 2002–2004

Number of people living with HIV

SOURCE: Marais et al. (2004), p. 2.

ducted before a potentially lifesaving drug is made available to the public? What rules apply when scientists conduct research in developing countries: the rules of the corporate home nation or the rules in the country where research is conducted? Are some policies, such as informed consent, so basic that they should apply anywhere in the world? Does consent always attach to the individual, or does it extend in some cases to communities with high infection rates? Should subjects and their communities participate in research design? How can information about research be explained effectively to people who are not familiar with scientific research and its implications? Can effective treatment be withheld for the purpose of advancing scientific understanding and the possibility of developing new drugs? How can participants be protected from or compensated for negative unintended consequences of research trials? What obligations do researchers have to provide short- and long-term health care to research subjects and their communities? How should societal needs for research be balanced against the rights of the individual? Vaccine research poses special problems because the subjects subsequently may test positive for HIV/AIDS.

Debate also continues over the relative merits of treatment and prevention. Is it better to ease suffering and prolong the lives of those already infected or to prevent new cases from occurring? Prevention will help only those who are not currently living with HIV/AIDS, whereas treatment is needed for the millions already infected to prolong lives, maintain family incomes, and promote general economic stability. Moreover, infected patients need relief from suffering

TABLE 1

Global Summary of the AIDS Epidemic, December 2004

Number of people living with HIV/AIDS in 2004

Total	**39.4 million (35.9 – 44.3 million)**
Adults	37.2 million (33.8 – 41.7 million)
Women	17.6 million (16.3 – 19.5 million)
Children under 15 years	2.2 million (2.0 – 2.6 million)

People newly infected with HIV in 2004

Total	**4.9 million (4.3 – 6.4 million)**
Adults	4.3 million (3.7 – 5.7 million)
Children under 15 years	640,000 (570,000 – 750,000)

AIDS deaths in 2004

Total	**3.1 million (2.8 – 3.5 million)**
Adults	2.6 million (2.3 – 2.9 million)
Children under 15 years	510,000 (46,000 – 600,000)

The ranges around the estimates in this table define the boundaries within which the actual numbers lie, based on the best available information.

SOURCE: Marais et al. (2004), p. 1.

in addition to treatment to slow the progress of the disease. Should scarce human and financial resources be diverted from prevention and treatment to provide palliative care?

Additionally, some people see HIV/AIDS primarily as a behavioral problem; if the behavior changes, the problem will disappear. Controlling HIV/AIDS is about more than developing drugs and vaccines; social science also plays an important role. New drugs will not reach patients unless medical services and drug delivery systems in poor countries are improved. The public must be educated about both causes and treatment. Researchers should investigate reasons for stigma and develop strategies to reduce discrimination and protect the most vulnerable. Within the prevention camp some advocate abstinence as the only moral alternative, whereas others recognize the reality of sexual activity and believe it is more ethical to promote condom use to reduce infection. In such a complex scientific and technology context what is the proper mix of prevention, treatment, and care?

Social Responsibilities

The infectious nature of HIV/AIDS also raises questions about societal responsibilities to potential victims. Should doctors or health institutions inform others when a patient is diagnosed with HIV/AIDS? How should the need to prevent the spread of a deadly disease be balanced against a patient's right to privacy? Women may be particularly at risk from identification because their subordinate status in many places may subject them to social isolation or deprivation of home or property.

Society often discriminates against people infected with HIV/AIDS. Discrimination may be driven by fear of infection, and education should be provided so that people know that the disease is not spread through casual contact. The general stigma attached to homosexuals, drug users, and the poor also drives discrimination. UNAIDS attributes the lack of political will to deal with the pandemic in part to the high infection rates among "marginalized and stigmatized population groups such as women who sell sex, drug injectors and men who have sex with men."

HIV/AIDS exacerbates gender inequities. Women often lack both information about the disease and the power to refuse sex or demand that their sexual partners use condoms. Identification of affected women puts them at higher risk of stigmatization, expulsion from their families, and deprivation of property and employment. Poor women who lose their spouses to the disease may be unable to support their families. More than 2 million children are infected with HIV/AIDS. Millions of others live with infected family members or have been orphaned by the pandemic.

Earlier in the epidemic the high incidence of HIV/AIDS among American gay males juxtaposed prejudice against homosexuals with the increasing political influence of affluent gay men. The gay community effectively concentrated attention on the emerging disease; that resulted in the allocation of research dollars to develop new treatments. HIV/AIDS is now relatively controllable for those who can afford expensive antiretroviral treatments, but the epidemic continues to spiral out of control because millions of infected poor people cannot afford treatment. The needed antiretrovirals are too expensive for most HIV/AIDS patients, and 90 percent of those who need drugs cannot afford them; most of those patients live in sub-Saharan Africa.

Some countries, such as Brazil, have made antiretroviral drugs available for free or at low cost to poor people who need them. Such programs help current patients but may reduce incentives for future pharmaceutical research. Drug companies engage in research and development to make money; if developing countries can obtain drugs without paying market prices, profits will fall and pharmaceutical companies may be less likely to do research into diseases that occur primarily among the poor. Nevertheless, the World Trade Organization Agreement on Trade-Related Aspects of Intellectual Property Rights (TRIPS Agreement) allows an exception to intellectual property rights in special cases such as emergencies, and that provision has been used to give developing countries access to HIV/AIDS drugs.

Michael Specter maintains that treatment is not enough; only a vaccine can stem the pandemic, yet drug companies lack sufficient incentives to develop a vaccine. This constitutes a case of market failure requiring government intervention.

Pharmaceutical companies in affluent industrialized countries conduct most research on new drugs and vaccines. Do they have a corporate responsibility to spend money on public health problems that may not produce profits? Do their countries have a responsibility to protect the less developed world by providing direct assistance or incentives for drug research? The developed world may have a direct stake in stopping the pandemic to reduce economic and political destabilization in many poor countries.

HIV/AIDS constitutes a global health crisis, but those in greatest need of assistance live in the poorest countries or are among the poorest and most stigmatized members of more affluent societies and lack strong political support. The crisis affects more than individuals; families are disrupted, and societies destabilized: "AIDS is accomplishing a sweeping undoing of past human development advances, especially in southern Africa". The HIV/AIDS pandemic requires strategies to address problems from the individual level to the international level.

MARILYN AVERILL

SEE ALSO *African Perspectives: HIV/AIDS; Drugs; Emergent Infectious Diseases; Health and Disease; Medical Ethics.*

BIBLIOGRAPHY

Attaran, Amir, and Jeffrey Sachs. (2001). "Defining and Refining International Donor Support for Combating the AIDS Pandemic." *Lancet* 357: 57.

Feenberg, Andrew. (1995). *Alternative Modernity: The Technical Turn in Philosophy and Social Theory.* Berkeley: University of California Press.

Grmek, Mirko D. (1990). *History of AIDS: Emergence and Origin of a Modern Pandemic.* Princeton, NJ: Princeton University Press. This book presents a detailed history of the HIV/AIDS pandemic and attempts to identify the emerging disease and considers both biological and social factors contributing to the pandemic.

The Global Aids Policy Coalition. (1992). *AIDS in the World: A Global Report.* Mann, Jonathan M., general editor; Tarantola, Daniel J. M., scientific editor; Netter, Thomas W., managing editor. Cambridge, MA: Harvard University Press, 1992). This comprehensive book provides detailed chapters on "The Impact of the Epidemic," "The Global Response," "Global Vulnerability," and "Critical Issues."

Marais, Hein, et al. (2004). *AIDS Epidemic Update: 2004.* Geneva: UNAIDS. The UNAIDS/WHO web site provides one of the most comprehensive sources for current information about HIV/AIDS and international responses to the pandemic.

Specter, Michael. (2003). "The Vaccine: Has the Race to Save Africa from AIDS Put Western Science at Odds with Western Ethics?" *New Yorker,* February 3.

INTERNET RESOURCE

Wolf, Leslie E., and Bernard Lo. (2001). "Ethical Dimensions of HIV/AIDS." HIV Insite. Available from http://hivinsite.ucsf.edu/InSite?page=kb-08-01-05.

HOBBES, THOMAS

• • •

Thomas Hobbes (1588–1679) was born in Westport, England, on April 5, the son of a clergyman; he was a contemporary of Shakespeare. Hobbes developed a moral and political philosophy that was influenced greatly by geometry and the new sciences of the Enlightenment. After studying at Oxford University Hobbes became a tutor for the Cavendish family and escorted his charges on tours of the European continent. During those travels Hobbes became acquainted with science as it was being developed by Galileo Galilei (1564–1642), René Descartes (1596–1650), and Marin Mersenne (1548–1648), which he found more constructive than the political strife that characterized the English civil war (1639–1651).

Hobbes's political thought first was expounded at length in *The Elements of Law* (1640), which defended the monarchy, although on democratic grounds. He subsequently developed his arguments in *De cive* (1642), *De corpore* (1655), and *De homine* (1658), a trilogy on the state, physics, and anthropology in which Hobbes attempted to build a bridge between the new science and politics. His most widely read book both in his own day and up to the present has been *Leviathan* (1651). He also wrote a scientific dialogue, *Dialogus physicus* (1661), in response to the emerging experimental sciences and Robert Boyle's (1627–1691) work with air pumps. In 1666 Parliament nearly banned *Leviathan* as heretical, and Hobbes continually faced the threat of exile. He spent his later years composing a history of the English civil war and translating the *Odyssey* and *Iliad.* Hobbes died in Hardwick Hall near Chesterfield, England, on December 4.

Thomas Hobbes, 1588–1679. The English philosopher and political theorist was one of the central figures of British empiricism. His major work, *Leviathan*, expressed his principle of materialism and his concept of a social contract forming the basis of society. (*Archive Photos, Inc.*)

Moral and Political Philosophy

The avoidance of civil strife was one of the main intentions of Hobbes's work. His solution made him unpopular with both royalists and parliamentarians. Royalists argued that the king rules on the basis of natural or divine right; parliamentarians advocated democratic rule. Hobbes argued that the king should rule not by nature or divine commandment but because the sovereign is an artificial social construction fashioned by popular human reason motivated by the shared fear of violent death. It was the high probability of that fate in the state of war (or nature) that in earlier times had made life "solitary, poor, nasty, brutish, and short" (*Leviathan*, vol. I, p. 13). For Hobbes civil society is radically conventional because humans are not naturally social. People are compelled to form civil society by the laws of nature, understood as rational instructions on how to cooperate.

Hobbes argued for a subjectivist morality based on psychological egoism (all human action is selfishly motivated), with good and evil as names that signify appetites and aversions, especially those pertaining to self-

preservation and peace. Social peace is possible because all people agree that it is good and are rational enough to cooperate. However, the plurality of tastes and definitions of good and evil means that a state of war will emerge quickly whenever the absolute authority of the sovereign is challenged.

Obedience even to arbitrary government is preferable to the state of war. The commonwealth is formed through social contracts, and the network of those contracts creates the Leviathan (from the Book of Job, meaning "King of the Proud"), or sovereign, which is an artificial "person" responsible for public welfare and social order. The sovereign could be a monarch, as Hobbes preferred, but it also could be a legislature or an assembly of all citizens. Hobbes's notion of the sovereign led to later contractarian philosophies, especially Jean-Jacques Rousseau's (1712–1778) ideal of the general will.

Fear of violent death thus brings humans to reason. In regard to the resulting self-regulating system of passions Hobbes constructed a political philosophy that foreshadowed liberal capitalism and its emphasis on individual rights and the primacy of material self-interest. However, his collectivist image of society comprising the body of the sovereign also has been interpreted as a forerunner of socialist thought. David Gauthier (1969, p. vi) sums up this duality: "Hobbes constructs a political theory which bases unlimited political authority on unlimited individualism." For Leo Strauss (1973) Hobbes marked the beginning of modern political philosophy (foreshadowed by Niccolò Machiavelli [1469–1527]) because he denounced aristocratic distinctions and virtues. He leveled all humans with his theory of natural equality and did not base morality on ideal virtues attainable, if at all, only by the few.

The Role of Science and Technology

A second basic intention in Hobbes's work was to put moral and political philosophy on a scientific basis. His civic science generally is regarded as being based on natural science in both method and material. Human thought and action are explained in mechanistic terms of matter in motion, and thus the laws governing political bodies can be derived from those governing physical bodies. Yet Hobbes held a compatibilist view that causal determination of human conduct is consistent with the freedom required for responsible moral agency.

Even though he worked briefly for the empiricist Francis Bacon (1561–1626), Hobbes was a rationalist who believed that science primarily meant geometry and the methodology of reasoning both from first principles, or causes, to effects and from effects to causes. The purpose of proper philosophy is universal assent attained through absolute certainty, and the first step in arriving at that certainty is an agreement to settle the definitions of words and their precise uses to avoid absurdities and disorder. Science is knowledge of the consequences of words established in that manner. Scholastic and religious reasoning breed controversy because they fail to define terms precisely.

Hobbes's political and natural philosophies are inseparable in the project of establishing consent on what is and how it can be known, thus leading to social order. Human will is the primary force of geometric proofs because humans determine original definitions. Geometry is an instance in which a diverse, subjective, and arbitrary human will has fashioned universal laws and truths by which all people can abide. Just as humans "make" the definitions in geometry (for example, "circle"), so too are the principles of politics (such as authority and justice) fabricated.

Strauss (1973), however, argues that modern natural science distorts Hobbes's moral and civic philosophy. The differences between the modern science of nature and human affairs outweigh the similarities. Indeed, in many places Hobbes stated that physical and political bodies are quite different. Furthermore, he did not take up science until he was forty years old, and he portrayed human nature as mutable and speech, reason, and sociality as products of free will. Vanity (the striving for absolute power) is a peculiarly human trait. Thus, Hobbes has a dualist philosophy (humans can will themselves out of nature) that is hidden by his monist (materialist-deterministic) metaphysics. Hobbes may wish to base his political theory on science because it progresses and produces real power, but a consistent scientific naturalism would ruin his moral philosophy.

The real basis of his philosophy was Hobbes's personal experience of human life. That experience actually has much in common with premodern science in that it proposes to disclose a teleology of human nature, even if a more debased teleology than argued for by the ancients. For that reason, "it can never, in spite of all the temptations of natural science, fall completely into the danger of abstraction from moral life and neglect of moral difference" (Strauss 1973, p. 29). It retains its moral basis precisely because it is not founded on modern science but instead on firsthand experience of humanity. As evidence for his claim Strauss points to the introduction of *Leviathan*, which states that one need not be trained in the physical sciences to formulate the right theory of human nature.

In another account Strauss (1965) argues that Hobbes posited two determinants of human willing—fear of violent death and the pursuit of domination over things—and that this underpins the distinction between the aims of politics and those of natural science. For Hobbes science is the methodical search for causes; in contrast, religion is the unmethodical search for causes. The purpose of science is the conquest of nature to make life more comfortable. It arises from human striving for power and honor, but that inexhaustible urge ensures that what is at stake is not the enjoyment of the object that is desired. Instead, the attainment of objects is only a means to more power: "the end becomes a means, the means becomes an end" (Strauss 1965, p. 89). Even if it is not properly based on science, Hobbes's politics is the foundation of modern technology.

The Politics of Knowledge: Hobbes versus Boyle

Strauss argued that the content of Hobbes's natural science obfuscates his political philosophy. Steven Shapin and Simon Schaffer (1985), however, argue that Hobbes's political theory holds true for the process of science. Both Strauss and Shapin and Schaffer see Hobbes as making constructivism and artifice superior to nature. Strauss uses this to purify Hobbes's politics of natural science; Shapin and Schaffer use it to justify Hobbes's insight that the two are inextricably connected in a single process: "Knowledge as much as the state, is the product of human actions" (Shapin and Schaffer 1985, p. 344).

Contrasting the philosophies of Hobbes and Robert Boyle, Shapin and Schaffer highlight the dynamics of the period when the modern relationship between scientific knowledge and the polity was being formed. The dispute between Hobbes and Boyle can be cast as different notions of what counts as science and legitimate knowledge. Hobbes's science was based on geometry and the deduction of irrefutable (moral and epistemic) truths from distinct first principles. Boyle proposed an experimental science that would be based on empirical observations made by a group with special training. Hobbes attacked this on epistemic grounds, claiming that the "facts" derived from sensory experience are mere "seeming or fancy" because they are too private.

However, this objection to Boyle's science is also moral. Both Boyle and Hobbes offered solutions to the problem of order in terms of ways to produce agreement and consent. Boyle attempted to remove natural philosophy from the "contentious link with civic philosophy" (Shapin and Schaffer 1985, p. 21). Hobbes attempted to erect a philosophy "that allowed no boundaries between the natural, the human, and the social, and which allowed for no dissent within it" (Shapin and Schaffer 1985, p. 21). Boyle's knowledge is produced among a community of experts, and that creates differences in the larger body politic, destroying natural equality, universal assent, and social order. Moreover, Boyle's scientific community allows for dissent about causes within its borders, which Hobbes found to be both a threat to civic order and a sign that it was not a true philosophy. Hobbes saw in Boyle's science the same socially corrosive element that exists in traditional monarchism and religion. The laboratory is a divisive and dangerous form of elitism pretending to a nonartificial hierarchy.

Arguing that "solutions to the problem of knowledge are solutions to the problem of social order," Shapin and Schaffer use the notion of "intellectual space" to distinguish Hobbes from Boyle (Shapin and Schaffer 1985, p. 332). For Hobbes philosophy is not the exclusive domain of professionals. He considered its intellectual space public because its purpose is the establishment of peace and order. In this regard natural science and civic science are the same. In Boyle's experimental science, however, there is a special place for doing natural philosophy—the laboratory—and access to it is quasi-open. In principle anyone could witness the goings-on in that space, but in practice it "was restricted to those who gave their assent to the legitimacy of the game being played within its confines" (Shapin and Schaffer 1985, p. 336). Boyle separates the study of nature, or objects, from the study of human affairs, or subjects. The existence of a separate community producing and legitimating knowledge was anathema to Hobbes, who argued that the philosopher's task was to establish peace and that this separate group threatened civic order. Bruno Latour playfully summed up his interpretation of Hobbes's reaction to Boyle: "we are going to have to put up with this new clique of scholars who are going to start challenging everyone's authority in the name of Nature by invoking wholly fabricated laboratory events!" (Latour 1993, p. 20).

For Hobbes philosophical and political spaces need masters who determine right knowledge and right conduct for all, thus constraining opportunities for interpretation and controversy. A chain is fastened from the lips of the sovereign to the ears of the people. This alleviates the problem of "seeing double" that occurs when loyalties are divided between different professional groups or different personal interpretations of events. Shapin and Schaffer claim that "Hobbes's philosophical truth was to be generated and sustained by absolutism" (p. 339). This

was strictly opposed to Boyle's notion of intellectual space because the foundation of knowledge was considered to be free will. Truth claims are verified by free acts of witnessing. Boyle saw the experimental community neither as tyranny nor as democracy but as a group regulated by conventions of selectively restricted access. The experimental community gained such wide support because it offered solutions to practical problems and because its members presented it as a model of the ideal polity. Nonetheless, this does not deny the fact "that there is a power-structure to truth and a truth-structure to power" (Wolin 1990, p. 12).

In the end Shapin and Schaffer conclude that "Hobbes was right" (p. 344) in the sense that Hobbes's instrumentalism or social constructivism better explains science, society, and their relationship than does Boyle's realism. Knowledge, like society, is conventional and artifactual, and scientists do not produce objective truth claims. Shapin and Schaffer probably exaggerated their instrumentalism to call attention to the increasingly problematic aspects of the "boundary-conventions" that distinguish science from politics. Their main point is that the solution to problems of knowledge is always political in that it requires the establishment of conventions of interaction and rules for determining legitimacy and because the knowledge this community produces becomes an integral part of political action.

Boyle and the experimentalism of the Royal Society "won" not because they reflected nature objectively but because their use of rhetoric garnered the most political power. Even though Hobbes was the first modern mediator between science and society, historians have purified Hobbes of science and Boyle of politics, reinforcing the idea that the two realms are naturally distinct. Shapin and Schaffer work to expose the intellectual and historical roots of that distinction, which increasingly is being questioned on the basis of expanding democracy rather than, as with Strauss, on the basis of a reaffirmation of nature.

ADAM BRIGGLE
CARL MITCHAM

SEE ALSO *Human Nature; Science, Technology, and Society Studies; Scientific Revolution.*

BIBLIOGRAPHY

Gauthier, David P. (1969). *The Logic of Leviathan: The Moral and Political Theory of Thomas Hobbes.* London: Oxford University Press. Covers Hobbes's theories of human nature, morality, sovereignty, authorization, and God.

Latour, Bruno. (1993). *We Have Never Been Modern*. Cambridge, MA: Harvard University Press. Uses the controversy between Hobbes and Boyle over the distribution of scientific and political power to illustrate the beginnings of the "modern constitution," which entails an elaborate separation of nature and society.

Shapin, Steven, and Simon Schaffer. (1985). *Leviathan and the Air-Pump: Hobbes, Boyle, and the Experimental Life*. Princeton, NJ: Princeton University Press. Includes a translation of Hobbes's *Dialogus physicus*.

Strauss, Leo. (1965). *Spinoza's Critique of Religion*. New York: Schocken. Contains a section on Hobbes titled "The Spirit of Physics (Technology) and Religion."

Strauss, Leo. (1973). *The Political Philosophy of Hobbes: Its Basis and Its Genesis*. Chicago: University of Chicago Press. First published in 1936.

Wolin, Sheldon S. (1990). "Hobbes and the Culture of Despotism." In *Thomas Hobbes and Political Theory*, ed. Mary G. Dietz. Lawrence: University Press of Kansas. Argues that Hobbes modernized despotism by founding it on scientific theorization but that he continues a chain of despotism from Plato, through medieval monarchs, to Popper's scientific social engineers.

HOLOCAUST

• • •

The word *holocaust* is derived from the biblical Greek term *holocauston*, meaning a "burnt offering" made in sacrifice to God. The term came to be widely used in the early 1970s to refer to the mass extermination of the Jews in the gas chambers of an organized system of death camps initiated by German dictator Adolf Hitler (1889–1945) and the Nazi Party during World War II. In the 1980s, some scholars argued that the word holocaust imputed more meaning to the event than it deserved and began calling it the *Shoah*, a Hebrew term referring to a time of desolation. The connotations of the latter have come to color even the meaning of the former.

In World War II nearly 30 million people died in combat or as random civilian victims of war. History is filled with wars and massacres, but genocide is something else. While the Turkish attempt to eliminate the Armenians (c. 1915) may be an earlier example of genocide, the Holocaust has come to be described as the archetypal example. Genocide is a systematic, state-sponsored, bureaucratically organized attempt to eliminate an entire people (usually identified in "racial," ethnic, or religious terms) from the face of the earth, not for any strategic military or political advantage but simply because they exist. The Nazi attack on the Jews was not an attempt to eliminate a foreign enemy but its own Jewish citizens first and then all the Jews in Europe. While others were also made victims in the death camps (such as the mentally retarded, homosexuals, and communists), the Jews and Gypsies were the only two peoples targeted for *total* annihilation. Thousands of Gypsies and 6 million Jews were murdered. A third of the world's Jews and two-thirds of Europe's Jews died in the Holocaust.

In the Nazi genocidal project, science was used to provide a biological theory of race that offered ideological justification for genocidal public policies of racial purity, and technology was used to provide the most efficient means to carry out these policies.

Science

The Nazis used English naturalist Charles Darwin's (1809–1882) biological theory of evolution to justify their program of genocide. Darwin posited the evolutionary differentiation of species as the product of "natural selection" in which only those organisms most successful in adapting themselves to a particular ecological niche survive to reproduce themselves and so shape the gene pool. This law of competition came to be known as the "survival of the fittest," meaning the survival of those most successful at adaptation.

In the nineteenth century this scientific theory was transformed into a political ideology known as "social Darwinism" by metaphorically extending Darwin's biological theory into the realm of society. In this way social phenomena such as class conflict or the conflict between nations were imagined to operate by the same laws of "natural selection." It seemed only "natural" to the Nazis to conclude that the ascendancy of the Nazi German nation-state was the outcome of a biological process in which the fittest race had survived to dominate all others, proving the superiority of the Aryan race. The greatest threat to this evolutionary outcome was, in their view, racial pollution—the biological mixing with "inferior races" that would weaken the purity of Aryan blood.

As Robert Jay Lifton (1986) noted, the death camps were viewed as public health projects in which the Jews were considered a cancerous growth on the body of the Aryan race, threatening its organic health (i.e., racial purity), and so had to be cut out in order to restore the body to health. It was no accident that physicians were required to fill the role of those who selected some victims for the work camps while sending others directly to the gas chambers. The doctor, as an elite scientifically trained professional, gave an aura of "scientific" legitimacy to the

Survivors of a concentration camp line up along a wire fence in Dachau, Germany. Many are still wearing the striped uniform of the camp. (AP/Wide World Photos.)

entire genocidal enterprise, and the "scientific theory" of racial purity gave the doctors a rationalization that allowed them to think of killing as a form of healing.

The Nazis had to ideologically twist science to justify their genocidal actions. Biologically all human beings are capable of interbreeding and therefore constitute a single species: There is only one human race. The Nazi "theory" of races was an ideological myth. Moreover, Darwin's theory suggested that it was genetic diversity not genetic uniformity that promoted survival. Finally, "survival of the fittest" had descriptive rather than normative status in Darwin's theory.

Technology

A technological civilization is one shaped by bureaucracies of technical experts who organize society to accomplish all its tasks using the most efficient solutions that

science can discover. Richard Rubenstein (1975) notes that the turning point in the Nazi genocidal program occurred in reaction to Kristallnacht (The Night of Broken Glass, November 9–10, 1938), when Heinrich Himmler (1900–1945), head of the Gestapo, the German secret police, rejected and suppressed the further use of mob violence that been promoted by Joseph Goebbels (1897–1945), German minister of propaganda. Himmler recognized that the only way to efficiently organize mass death was to remove the element of personal emotion and replace it with the cool and efficient operations of the impersonal techno-bureaucratic procedures that typified the death camps.

As the Holocaust well demonstrates, techno-bureaucratic organization is impervious to ethical considerations, because bureaucracy separates ends and means. When persons choose both ends and means they feel the connection in their experience out of which a sense of personal responsibility can emerge. But in a bureaucracy, those higher up are viewed as being in a better position to choose the ends than the technical experts, lower down in the hierarchy, who are charged with developing the means to accomplish them. The Nazi doctors who did the selections in the death camps saw themselves as mere cogs in a complex bureaucratic machine. Even if one refused to participate that would change nothing. Like a replaceable part, one doctor would simply be substituted for by another, more accommodating one. These doctors did not feel responsible because the victims were dead long before they ever got to the camps, declared so by those higher up in the bureaucracy who alone had the authority and responsibility. Indeed, the Nuremberg war-crimes trials that followed World War II (1945–1949) demonstrated the prevalence of this logic in the repeated defense of those accused who plead non-responsibility because they were "just following orders."

Ethics after the Holocaust

The Nuremberg trials, by identifying "crimes against humanity" for prosecution, represent an initial attempt to think globally about ethics. Indeed, the horror of the atrocities of World War II sent a global moral shock through the human race that led to the creation of the first global ethic in history. For the movement for human rights arose in response to the trauma of the Holocaust and the other atrocities of World War II. This movement culminated in the formation of the United Nations (UN) in 1946 and the adoption of the Universal Declaration of Human Rights by the UN in 1948. The preamble to the declaration recalls the "barbarous acts which have outraged the conscience of mankind," preparing the way for the declaration's main body, which strongly affirms the unity of humanity. Consequently this document stands against all ideologies that would divide humanity, racially or otherwise, in order to claim the world and its resources for some superior *volk*—as the Nazis attempted to do.

Unlike the technical and esoteric language of most academic treatises on ethics, human rights language is a language that has spontaneously taken root in cross-cultural public discourse. The language of human rights has become embedded in the language of politics and international relations. Even if, in many cases, the political use of this language is hypocritical, still that is the homage that vice pays to virtue, which means that human rights can be used as a measuring rod for cross-cultural social and political criticism.

Moreover, in the aftermath of World War II, a plethora of both governmental and nongovernmental

organizations committed to preserving and protecting the rights of all human beings across all religions and cultures has emerged, deeply influencing global social policies. Such organizations include the UN itself, especially its Commission on Human Rights and its various subcommissions, as well as the International Court of Justice and regional conventions on human rights in Western Europe, the United States, and Africa. Then there are the governmental offices of individual nations that monitor each other for rights violations and use this information to political advantage. (Motivations of self-interest aside, this political game does keep the pressure on to observe human rights.) Finally, there are nongovernmental voluntary associations committed to human rights such as Amnesty International, the Anti-Slavery Society, and the International Committee of the Red Cross as well as religious organizations, labor organizations, and professional associations.

In the last quarter of the twentieth century and into the twenty-first, the Holocaust or Shoah has become a symbol for the universal call to conscience and responsibility on behalf of the human dignity and human rights of all. In this context the rhetoric of the Holocaust and of human rights, as one might expect, has often become politicized and sometimes trivialized. And yet the moral climate of human history has been unarguably changed by the language of "human rights" and "human dignity" evolving into the global moral language of accountability and by the Holocaust becoming a powerful symbol of everything that would violate such dignity and rights.

DARRELL J. FASCHING

SEE ALSO *Arendt, Hannah; Dignity; Eugenics; Human Rights; Judaism; Levi, Primo; Nazi Medicine; Race; Social Darwinism.*

BIBLIOGRAPHY

Fasching, Darrell J. (1993). *The Ethical Challenge of Auschwitz and Hiroshima: Apocalypse or Utopia?* Albany: State University of New York Press. A study of the emergence of global ethics and its role in shaping public policy after Auschwitz and Hiroshima.

Lifton, Robert Jay. (1986). *The Nazi Doctors: Medical Killing and the Psychology of Genocide.* New York: Basic. A ground-breaking study of the psychological and institutional factors that together shaped the genocidal mentality of the Nazi doctors in the death camps.

Roth, John, ed. (1999). *Ethics after the Holocaust.* St. Paul, MN: Paragon House. An excellent collection of essays by scholars from diverse countries on ethics after the Holocaust.

Encyclopedia of Science, Technology, and Ethics

Rubenstein, Richard. (1975). *The Cunning of History*. New York: Harper and Row. A brilliant and devastating study of the role of scientific and technical bureaucracies in facilitating the Holocaust.

HOMOSEXUALITY DEBATE

• • •

Homosexuality has been a subject of scientific study for many years. Much of the research has focused on whether homosexuality is a product of biology or psychological conditioning. That nature-nurture question often has entered into ethical and political debates about homosexuality. For example, in the early 1990s two studies were released that indicated that homosexuality may be biological. One study identified distinctive neural structures in homosexual men (LeVay 1993). The other correlated a genetic marker with male homosexuality (Hamer and Copeland 1994).

Those studies received significant media attention because they seemed to strike at the heart of the political debate about gay rights. Opponents of gay rights had argued that homosexuality is a choice and that homosexuals seek "special rights" for a deviant and destructive lifestyle. Consequently, gay rights advocates began to argue that the studies mentioned above showed that homosexuality is not a choice but an innate biological characteristic worthy of constitutional protection.

Early Studies of Homosexuality

These debates about homosexuality date back to the mid-nineteenth century, when Karl Heinrich Ulrichs (1825–1895), a German jurist, attempted to theorize homosexuality as a biological condition. Ulrichs believed that the embryo contains female and male "germs" and that as an embryo develops, one of the germs becomes dominant, producing either male or female sexual organs. These sexed germs, he argued, also produce the sex drive, and thus it is possible for the body of one sex to possess the sex drive of the other. Because Ulrichs was a jurist, not a scientist, his primary concern was to secure the civil rights of homosexuals, and he believed a biological theory would facilitate his efforts (Brookey 2002).

Shortly after Ulrichs introduced his theories, they were incorporated into the work of the neurologist Richard von Krafft-Ebing (1840–1902). Krafft-Ebing defined homosexuality as a predetermined sexual attraction brought about by either genetic or situational factors. *Situational homosexuality*, according to Krafft-Ebing,

occurred when men were precluded from sexual intercourse with women or masturbated. He characterized situational homosexuality as an inherited condition that existed as the lingering residue of an animalistic bisexuality that would die out slowly in the process of evolutionary advancement (Brookey 2002). Krafft-Ebing's theories were influential for many years but would be eclipsed when Sigmund Freud introduced his own theories on human sexuality.

Freud argued that children are born into an innate state of bisexuality, but as they develop, this bisexual energy is directed into heterosexuality. However, if a child does not develop proper relationships with his or her parents, sexual development may be arrested and homosexuality can result. Freud did not believe that homosexuality is always the product of psychological pathology. Consequently, he regarded efforts to change homosexuals into heterosexuals with great pessimism (Lewes 1988).

Modern Theories

After World War II American psychoanalysts reinterpreted Freud's theories, particularly those regarding homosexuality. The psychologist Sandor Rado (1890–1972) led that effort when he rejected Freud's theory of innate bisexuality. Rado argued that bisexuality does not exist, rejected the possibility of biological homosexuality, and argued that homosexuality can only be a product of mental pathology. He claimed that homosexuality is a mental pathology and that the possibility for change is much greater than Freud supposed. Edmund Bergler (1889–1962) was a Freudian who advocated psychoanalytic therapy and claimed to have converted homosexuals. Bergler was also an active opponent of the early gay rights movement, and he often testified in government hearings that homosexuals should be precluded from public service.

The psychoanalytic position on homosexuality remained unchallenged until Alfred Kinsey (1894–1956) began publishing his research on human sexuality. Kinsey's work indicated that human sexuality is much more varied and fluid than psychoanalytic theories supposed. Rado's dismissal of bisexuality was challenged by Kinsey's empirical findings, which indicated that a significant number of adults had sexual experiences with persons of both sexes. Consequently, Kinsey's work also challenged psychoanalytic beliefs about homosexual pathology because it recognized that homosexuality was practiced by a variety of individuals and did not treat homosexuals as a distinct or deviant class. The psychiatrist Evelyn Hooker (1907–1996) also challenged many psychoanalytic assumptions about homosexuality. Specifically, Hooker's research concluded that many homosexuals did not suffer from severe mental disturbances and that homosexuals were just as diverse in their behavior and psychological profiles as heterosexuals were.

Kinsey's and Hooker's research established doubt in the psychiatric community about the pathology of homosexuality, and in 1973 the American Psychiatric Association (APA) voted to remove homosexuality from its list of mental diseases. That decision reflected suspicion of psychoanalytic approaches and concern about the use of behavior modification conversion therapy. Many psychoanalysts protested the decision, and *ego-dystonic homosexuality*, a condition experienced by homosexuals who wanted to change their sexual orientation, remained on the list so that therapists could continue to practice conversion therapy. Even that exception, however, was eliminated in 1997 when the APA determined that psychological therapies cannot cure homosexuality.

Judicial Decisions and Ethical Issues

The publication of LeVay's and Hamer's studies has renewed interest in biological explanations of homosexuality. Although gay rights advocates thought that research would yield political advantages, arguments about the biological basis of homosexuality did not acquire legal traction. A biological argument was presented to the Supreme Court in the 1995 hearing on Colorado's Amendment 2, an anti–gay rights initiative. Although the Court ruled against the initiative, the evidence demonstrating a biological basis for homosexuality was not mentioned in its decision (Keen and Goldberg 1998). In addition, the biological argument did not figure in the Court's 2003 decision to strike down state anti-sodomy laws.

Apart from the legal question, there are ethical concerns about the use of biological research to treat homosexuality (Murphy 1997). Would a "homosexual" gene lead to a genetic test for homosexual predisposition? Would couples choose to abort a fetus that tested positive for this genetic predisposition? Could homosexuals seek genetic therapy in order to change their sexual orientation? Could homosexuals be compelled to submit to that therapy? Currently, these ethical questions are moot because additional research has not verified Hamer's and LeVay's research conclusively. Both the legal and the scientific debates about homosexuality have not been resolved.

ROBERT ALAN BROOKEY

SEE ALSO *Feminist Ethics; Gene Therapy; Genetic Counseling; Nature versus Nurture; Sex and Gender.*

BIBLIOGRAPHY

Brookey, Robert Alan. (2002). *Reinventing the Male Homosexual: The Rhetoric and Power of the Gay Gene*. Bloomington: Indiana University Press. Provides an overview of the research on the biological aspect of homosexuality.

Hamer, Dean, and Peter Copeland. (1994). *The Science of Desire: The Search for the Gay Gene and the Biology of Behavior*. New York: Simon & Schuster. Summarizes research on the "gay" gene.

Keen, Lisa, and Suzanne B. Goldberg. (1998). *Strangers to the Law: Gay People on Trial*. Ann Arbor: University of Michigan Press. Covers the Supreme Court decision on *Romer v. Evans.*

LeVay, Simon. (1993). *The Sexual Brain*. Cambridge, MA: MIT Press. Summarizes research on the brain structures of homosexuals.

Lewes, Kennneth. (1988). *The Psychoanalytic Theory of Male Homosexuality*. London: Quartet Books. Discusses the APA decision to declassify homosexuality.

Murphy, Timothy. (1997). *Gay Science: The Ethics of Sexual Orientation Research*. New York: Columbia University Press. Provides an overview of ethical issues related to research on the biological aspect of homosexuality.

HORMESIS

• • •

Hormesis is a dose-response phenomenon in which a low dose of a toxin has the opposite effect on a biological system than a high dose of the same toxin. It is generally characterized as toxic effects that are beneficial at low doses and harmful at high doses. There is some ambiguity in the more precise definition of the term, however, because some speak strictly of low-dose stimulation of biological endpoints (for example, immune system strengthening), whereas others also use it to refer to low-dose inhibition of biological endpoints (such as tumor formation). Hormesis has long been marginalized in medical and environmental fields. A growing body of evidence suggesting hormetic effects across a wide range of biological organisms and systems, however, has brought increased credibility to the topic. The implications of hormesis are potentially huge, especially in terms of risk assessment policies and research paradigms. Skepticism and controversy persist surrounding the future status and impacts of hormesis as new research, aided by advanced technologies, yields uncertainty and more questions than answers.

History

Ideas similar to hormesis have been vaguely formulated for centuries, including Hippocrates' saying that "likes are cured by likes," Paracelsus's notion that "the dose makes the poison," and Friedrich Nietzsche's famous remark that "what does not destroy me makes me stronger." Hugo Schulz, a German pharmacologist who observed that small doses of poisons stimulated the growth of yeast, was the first to systematically describe hormesis in 1888. Rudolph Arndt, a German physician, found similar results in his research on the effects of low doses of drugs on animals. Arndt claimed that toxins in general produced stimulation of biological endpoints such as growth or fertility at low doses, which became known as the Arndt-Schulz law. It lost credibility in the 1920s and 1930s, however, because Arndt was an adherent of homeopathy (Kaiser 2003). Founded by Samuel Hahnemann (1755–1843), homeopathy parallels hormesis in two respects, namely the idea that likes cure likes (symptoms produced by toxic doses can be cured by a remedy prepared from the same substance) and the theory of infinitesimals, which stated that the more dilute a substance is the more potent it can become. The marginalization of Arndt's work meant that hormesis research did not receive federal funding during the formative years of toxicological development.

C. M. Southam and J. Erlich first coined the term hormesis in 1943 in research that showed an antifungal substance had stimulatory effects on fungi when administered in low doses. The term derives from a Greek root meaning to excite, indicating the ability of small amounts of dangerous substances to excite an organism's defense systems, thereby making it healthier than it would be otherwise. Hormetic effects have since been observed in organisms ranging from humans and rats to water fleas and various plants. Yet outside of low-level ionizing radiation studies (the field in which the concept of hormesis is best developed), these observations went largely unexamined and were usually treated as aberrant data.

Edward Calabrese and Linda Baldwin (2001) synthesized these disparate findings in the toxicology literature. They also found that hormetic dose-response curves outnumbered curves showing no effect at the lowest doses by 2.5 to 1 (2003). This coupled with older, extensive literature on the beneficial effects of minute doses of ionizing radiation for animals (Luckey 1980, 1991) sparked increased interest in hormesis. In 1990 a group of scientists representing several federal agencies, the private sector, and academia launched a program of analyses and workshops called Biological Effects of Low Level Exposure (BELLE) and a newsletter devoted to low-dose toxicology.

The U.S. National Research Council (NRC) has sponsored research on radiation hormesis. Researchers in Japan note that victims from the World War II nuclear attacks, if they were sufficiently distant from the blast site, have lower death rates than peers not exposed to the radiation. Researchers at Johns Hopkins University found that tens of thousands of U.S. Navy shipyard workers exposed to radiation in the 1960s and 1970s have fewer cancers than nonexposed workers (Boice 2001). Others have found evidence that lung cancer rates are lowest in areas with the highest levels of radon.

Explanation and Implications

The biological mechanisms underlying the details of hormesis are still poorly understood. In general hormesis is a manifestation of homeostasis, the fundamental property of living organisms to maintain internal conditions that are in a state of (dynamic) equilibrium. Biological systems, even at the molecular level, have adaptive responses to stress that can trigger a variety of effects including increased cellular repair, beneficial apoptosis (programmed cell death), and increased immunological strength (Stebbing 1982). For example, the cellular damage caused by exercise in the short-term stimulates beneficial long-term effects because certain physiological mechanisms overcompensate, thus making the body stronger. Caloric restriction has also been proposed as a hormetic phenomenon. Some researchers have found that low levels of dioxin reduce the occurrence of tumors in rats, low levels of cadmium increase water flea fecundity, and low levels of phosfon (a herbicide) stimulate peppermint plant growth (Kaiser 2003). These results show up as biphasic dose-response curves shaped like a J or an inverted U (see Figure 1). Such dose-response curves are not unique to hormesis, however, because they are found especially in studies of endocrine disruptors that have no beneficial effects at any dose.

When referring to nutrients, hormesis is rather straightforward. Iron, for example, is necessary for transporting oxygen throughout the body, but too much iron is poisonous. The largest scientific and political implications from hormesis come from research that shows beneficial effects from small doses of chemicals long believed to be toxic at any level, such as dioxin and certain pesticides. For example, heavy metals such as mercury spur synthesis of proteins that remove toxic metals from circulation and may prevent some DNA damage caused by free radicals. Because the relationship between dose and effect is the fundamental concept of toxicology, these kinds of results may bring about radical changes in environmental and medical sciences and regulatory practices.

FIGURE 1

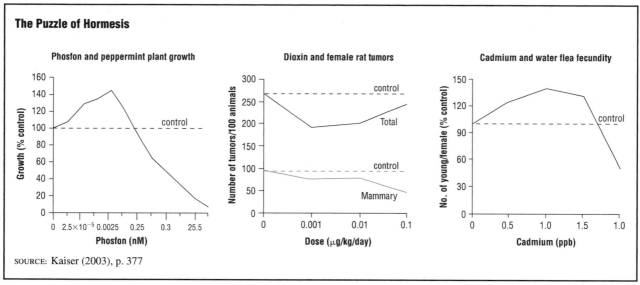

The Puzzle of Hormesis

SOURCE: Kaiser (2003), p. 377

Low doses of phosfon, a herbicide, caused plants to grow better (left); small amounts of dioxin, a carcinogen, reduced tumors in rats (center); and a little cadmium, a toxic metal, caused water fleas to produce more young (right). The effects were reversed at higher doses.

Indeed some suggest that hormesis marks a revolution in toxicology, pharmacology, and risk assessment. The dominant environmental risk assessment model is twofold. For carcinogens, regulatory agencies use a linear, nonthreshold dose-response model that assumes no safe level of exposure. For noncarcinogens, regulatory agencies assume there is a threshold dose, below which there is no risk of harm. Both risk assessment models are riddled with assumptions due to extrapolations from high-dose laboratory experiments to the low doses characteristic of human exposure. Calabrese (2004) argues that the resulting uncertainty has led to a protectionist public health paradigm with stringent environmental standards that often come at high costs.

He claims that these two dose-response models erroneously calculate public health standards, poorly communicate risks to the public, lead to exorbitant cleanup costs, and provide the wrong cues about how to prioritize investments in the environment. Hormesis provides an alternative risk assessment model that harmonizes policies on carcinogens and noncarcinogens, eliminates the need to extrapolate data, and places environmental risk assessment on the same solid empirical grounding as health insurance and other forms of risk estimates. He also claims that hormesis has important implications for clinical medicine. It can improve the selection of dosages and help medical researchers avoid situations in which declining concentrations of drugs in the body (toward the end of treatment, for example) may actually stimulate the microbes or tumors they are intended to eliminate.

Clearly hormesis could radically alter environmental and biomedical practices. For certain carcinogens, for example, the benefits of hormesis may occur at levels higher than the recommended safe doses for humans. It could also change the way scientists perceive and measure risk. But major changes are not likely to occur swiftly. Beneficial hormetic effects differ by individual and are still poorly characterized, and military or industrial interests may compromise the integrity of some hormesis research. Furthermore much of the research done on hormesis has focused too narrowly on single endpoints such as cancer while ignoring others. This may mean that harmful effects at low doses are not registered. Regulators must understand complex interactive effects, which greatly increase the costs of research (Renner 2003). Most importantly Calabrese fails to consider the price paid for eliminating unverifiable extrapolations. Low-dose testing requires long-term experiments with much larger sample sizes than current risk assessment models, because at low doses small signal-to-noise ratios require researchers to collect more data in order to obtain acceptable confidence intervals. The long time periods required for such research are not suited to the needs of decision makers.

As Gary Marchant (2001) argues, the refusal by U.S. Environmental Protection Agency (EPA) regulators to consider the health benefits of ozone in their 1997 revision of air quality standards provides lessons for the regulatory implications of hormesis. First, regulatory agencies are highly resistant to considering hormesis because it is a nonintuitive phenomenon that departs from traditional toxicology assumptions. Second, scientific evidence for hormesis is severely scrutinized, which makes credibility difficult to achieve. Third, judicial review may be an effective mechanism for forcing regulatory agencies to consider hormesis.

The accumulation of scientific data and advances in the techniques of molecular biology have brought the phenomenon of hormesis and the attendant controversies once again to the forefront of science and society. Hormesis carries great economic, environmental, and public health implications, but conclusive data are hard to obtain because of the large sample sizes needed and ethical restrictions on human subjects research. Hormesis supports the argument put forth by Bruno Latour (1998) that science, rather than clearing away societal controversies, actually increases uncertainty. Continued research may resolve conflicts but it may just as well add new uncertainties to those currently generated by extrapolation in risk assessment models.

ADAM BRIGGLE

SEE ALSO *Dose Response Ratios; Radiation; Regulatory Toxicology.*

BIBLIOGRAPHY

Boice Jr., John D. (2001). "Study of Health Effects of Low-Level Radiation in USA Nuclear Shipyard Workers." *Journal of Radiological Protection* 21(4): 400–403.

Calabrese, Edward J. (2004). "Hormesis: A Revolution in Toxicology, Risk Assessment and Medicine." *European Molecular Biology Organization* 5: S37–S40. Proposes a new risk assessment model based on hormetic dose-response curves rather than linear or threshold models.

Calabrese, Edward J., and Baldwin, Linda A. (2001). "The Frequency of U-Shaped Dose Responses in the Toxicological Literature." *Toxicological Sciences* 62(2): 330–338. Surveys the toxicological literature and finds significant evidence that hormetic effects are widespread.

Calabrese, Edward J., and Baldwin, Linda A. (2003). "The Hormetic Dose-Response Model is More Common than the Threshold Model in Toxicology." *Toxicological Sciences* 71(2): 246–250. Tests the validity of dominant dose-response models in toxicology and argues that hormetic models fit the evidence more closely.

Kaiser, Jocelyn. (2003). "Sipping from a Poisoned Chalice." *Science* 302(5644): 376–379.

Latour, Bruno. (1998). "From the World of Science to the World of Research?" *Science* 280(5361): 208–209.

Luckey, Thomas D. (1980). *Hormesis with Ionizing Radiation.* Boca Raton, FL: CRC Press. Reviews the literature pertaining to radiation hormesis.

Luckey, Thomas D. (1991). *Radiation Hormesis.* Boca Raton, FL: CRC Press. The first complete report on radiation hormesis. Shows that many biological functions are stimulated by low doses of ionizing radiation.

Renner, Rebecca. (2003). "Nietzsche's Toxicology." *Scientific American* (September): 28–30.

Southam, C. M., and J. Erlich. (1943). "Effects of Extract of Western Red-Cedar Heartwood on Certain Wood Decaying Fungi in Culture." *Phytopathology* 33: 517–524. First study to coin the term hormesis.

Stebbing, A. (1982). "Hormesis—The Stimulation of Growth by Low Levels of Inhibitors." *Science of the Total Environment* 22(3): 213–234.

INTERNET RESOURCE

Marchant, Gary E. (2001). "A Regulatory Precedent for Hormesis." Biological Effects of Low Level Exposures. Available from http://www.belleonline.com/n8v92.html.

HUMAN CLONING

• • •

Human cloning, which occurs naturally but rarely with the birth of identical twins, became a technological possibility with the development of the technique of somatic cell nuclear transfer (SCNT) to clone the first mammal in 1996. As a result of this scientific advance, the prospect of human cloning quickly became a hotly debated ethical issue. As the debate developed it also became common to distinguish reproductive cloning from therapeutic cloning, each being subject to slightly different ethical assessments.

History and Science

Cloning (from the Greek word *klon*, a twig or slip) is a natural process of asexual reproduction found in many plants and some animals. When strawberry plants send out runners that set roots and turn into new plants, this is an example of a plant naturally cloning itself. Even artificial cloning is not entirely new. For hundreds of years gardeners have taken slips (small shoots or twigs cut from plants) and rooted them to produce new plants in a process that could also be described as cloning. Then in the 1970s scientists began experiments in artificial cloning with frogs and toads, and subsequently with other animal embryos. But it was not until the successful SCNT cloning of the sheep "Dolly," performed in 1996 and formally announced in February 1997 by the Roslin Institute in Scotland, that it became clear something similar might be possible with mammals.

Mammals have two kinds of cells: somatic cells (many of which can reproduce themselves by clonelike division, but only themselves and not a whole organism) and sex cells (which come in two forms, ovum in females and sperm in males). The SCNT process works as follows: The nucleus is removed from a somatic cell of either a female or a male. An unfertilized ovum is taken from a female and has its nucleus removed and then replaced with the somatic cell nucleus. The resulting ovum with a somatic cell nucleus is then stimulated and implanted in a female womb to grow to term. The resulting offspring is genetically identical to the individual that was the source of the original somatic nucleus.

The technology of cloning is thought to be feasible in many mammalian species, including humans. As of 2005, successes in cloning of many species have been achieved. But neither the cloning of primates nor of humans has been successful as yet. Human somatic cell nuclear transfer, if successful in producing offspring, would not be "duplication" because identical genomes do *not* produce identical phenotypes. Nevertheless, Korean scientists have used cloning technology to produce cloned embryos, and subsequent experiments have furthered such technologies, which are aimed at producing embryonic stem cells for research and therapeutic purposes.

The science and technologies of cloning remain in their infancy. Pharmaceutical companies have not expressed great interest in trying to work to clone people because they see much bigger markets in the cloning of animals and cells. Efforts to create a human clone have been limited largely to groups outside the mainstream of science and medicine, and no one knows for sure whether stem cells derived from cloned human embryos really will prove useful as a way to cure diabetes, liver failure, Parkinson's disease, spinal cord injuries, or any other disease or ailment.

Ethical Concerns

It is easy to see why there is so much interest in and concern about human cloning. There is seemingly no end to the parade of people who issue press releases proclaiming that they are close to success in cloning a human baby. And there is certainly a simple fascination with the technical possibility. Proponents of cloning have also suggested it might serve as a new, unusual, but perhaps efficacious treatment for infertility, enabling those unable to pass genes to future generations to do so in a way that is at least analogous to the familial linkage of twins. And, they point out, scientists have created animal clones and at least a small number of human cloned embryos with hardly any oversight or public accountability.

Dolly, the cloned sheep. The result of an experiment by Scottish embryologist Ian Wilmut, Dolly was the first cloned adult mammal. (*Archive Photos, Inc.*)

There are grave risks, however, to any resulting offspring: Mammalian cloning, through the SCNT process, has resulted in the birth of hundreds of organisms. But significantly more nuclear-transfer-generated embryos fail during pregnancy than would fail in sexual reproduction, and a substantial majority of cloned animals who have survived to birth have had some significant birth defect. For these and related reasons President Bill Clinton in 1997 issued a moratorium banning the use of federal funds for human cloning, a position subsequently endorsed by the National Bioethics Advisory Commission.

And for some who believe that any human embryo is a person from the moment of its creation, the fight over human cloning is a fight both about what constitutes membership in the human community and about the morality of abortion. Many opponents of abortion hope that if they can gain legal recognition for cloned human embryos they can then move on to get legal standing for any human embryo or fetus.

One such person is U.S. President George W. Bush. A few months after hearings at the United Nations in

TABLE 1

Chronology of Key Early Events in the Human Cloning Discussion

1932	Aldous Huxley publishes *Brave New World*, including the "Bokanovsky Process" for producing cloned children.
1938	German embryologist Hans Spemann publishes *Embryonic Development and Induction*, in which he speculates about the possibility that the nuclei of fully differentiated cells may be able to initiate normal development in enucleated egg cells.
1952	U.S. embryologists Robert Briggs and Thomas J. King first successfully transfer nuclei from early embryonic cells of leopard frogs to enucleated leopard frog eggs.
1960s and 1970s	British developmental biologist John Gurdon makes further advances in cloning frogs. Debates about the implications of cloning begin.
1966	U.S. biologist Joshua Lederberg publishes an article in *The American Naturalist* titled "Experimental Genetics and Human Evolution," in which he speculates on the implications of cloning humans.
1971	U.S. geneticist James D. Watson testifies before Congress on the subject of human cloning.
July 25, 1978	The birth of Louise Brown, the first baby conceived through in vitro fertilization (IVF), shows that human birth is possible from eggs fertilized outside the body and then implanted in the womb.
1994	The National Institutes of Health Human Embryo Research Panel issues a report that deemed research involving nuclear transplantation, without transfer of the resulting cloned embryo to a uterus, as one type of research acceptable for federal support.
1996	In the U.S. the Dickey Amendment is enacted, which prohibits federal funding to create human embryos for research purposes and research that destroys or discards human embryos.
July 5, 1996	Cloned sheep "Dolly," named after the country singer Dolly Parton, was born using somatic cell nuclear transfer (SCNT).
Feb. 1997	Ian Wilmut et al. (Roslin Institute), "Viable Offspring Derived from Fetal and Adult Mammalian Cells," *Nature*, vol. 385 (27 February), pp. 810–811, announces the birth of Dolly.
March 1997	President Bill Clinton issues moratorium banning the use of federal funds for human cloning, and asks the National Bioethics Advisory Commission (NBAC, also sometimes called the National Bioethics Advisory Board) to analyze the ethical issues involved. It issues its report in June 1997.
August 1997	Clinton Administration proposes legislation banning human cloning for at least five years, in order to give the NBAC sufficient time for reflection.
Sept. 1997	Thousands of U.S. scientists voluntarily commit to a five-year moratorium on human cloning.
Jan. 1998	Nineteen European countries ban human cloning. Dr. Richard Seed, a Chicago physicist, announces plans to clone a human being. U.S. Food and Drug Administration claims authority to regulate human cloning, making it a violation of federal law to attempt cloning without FDA approval.
Nov. 6, 1998	University of Wisconsin biologist James Thomson and Johns Hopkins biologist John Gearhart announce the isolation of human embryonic stem cells, sparking increased interest in therapeutic cloning.
Aug. 2000	President Bill Clinton announces new guidelines for the federal funding of embryo research, but in early 2001 President George W. Bush places them under review before they are implemented.
Nov. 2000	Japan outlaws human reproductive cloning.
July 2001	The U.S. House of Representatives passes the Human Cloning Prohibition Act to outlaw both reproductive and therapeutic cloning, but the bill dies in the Senate.
Nov. 2001	Scientists at Advanced Cell Technology make unverified reports of the first cloned human embryos.
Dec. 2001	Britain outlaws human reproductive cloning.
Feb. 2002	United Nations begins consideration of a world-wide ban on human cloning.
July 2002	The U.S. President's Council on Bioethics issues its report *Human Cloning and Human Dignity: An Ethical Inquiry*.
Sep. 2002	California becomes the first state to pass a law legalizing therapeutic cloning.
Dec. 2002	The Raelians make the unsubstantiated announcement that they successfully cloned a human being.
March 2004	Korean scientists announce they have used SCNT to clone human blastospheres.
June 2004	United Nations Conference on Human Cloning.

SOURCE: Courtesy of Carl Mitcham and Adam Briggle.

February 2002, Bush announced in a speech from the White House's Rose Garden that he favored a ban on all forms of human cloning, including the cloning of human embryos for the purpose of stem cell research (Bush 2002).

Bush warned that in our zeal to find benefits and cures we could also "travel without an ethical compass into a world we could live to regret." Throughout the rest of his speech were salted words and phrases such as "products," "design," "manufacturing," "engineered to custom specifications." Bush was concerned that clon-ing would lead to the literal manufacture of human beings. A few months later, on July 10, the President's Council on Bioethics issued a report concluding that moral concerns about human cloning were sufficient to warrant a complete ban on using cloning to make people and a moratorium of at least four years on using cloning for research purposes.

Bush was hardly acting alone in sounding the tocsin of moral concern about the dangers of cloning. He was simply the most prominent among a long list of conservatives, pro-lifers, and neoconservatives, along with a small number of neo-green thinkers, who saw cloning in general as holding the seeds of the degradation of humanity.

Reproductive Cloning

So is there a strong case against human cloning? Reproductive cloning raises the question: Would it be unethical for anyone to try to clone a human being today or at any point in the future? Those who oppose human cloning point to the repugnance of a style of reproduction with such profound potential for vanity, arguing that the freedom of children and the nature of the family are in danger.

There is little debate concerning the claim of most scientists and ethicists that it would be irresponsible and morally wrong to try to use cloning to make a human being anytime soon. The experience of using cloning to make sheep, cows, pigs, and mice has made it abundantly clear that cloning is dangerous. There is real risk of death for the clone and a high risk of disability, and there are also very real risks for the surrogate mother who carries cloned fetuses to term. Without better safety data from animals, including primates, there is no ethical justification for trying to clone a human being.

But safety, while a very real concern, is not a concern about cloning per se. Presume that cloning were to someday prove safe. Would it still be ethically wrong to use it to make people? Any answer that pins the dangers for early prospective clones on something other than mere physical harms novel to the cloning process can become diffused in two conceptual problems:

- one is attempting to protect future potential persons against harms that might be inflicted by their very existence, and
- societies around the world have indicated that they believe that the early cloning experiments will breach a natural barrier that is moral in character, taking humans into a realm of self-engineering that vastly exceeds any prior experiments with new reproductive technology.

Laws that would regulate the birth of a clone are philosophically difficult in part because they traverse complex jurisprudential ground: protecting an as-yet-nonexistent life against reproductive dangers, in a Western world that, in statutory and case law at least, seems to favor reproductive autonomy.

Many people seem inclined to put those philosophical issues, nonetheless, into a position of primacy in the human cloning debate, including President Bush and

his chief bioethical adviser, Leon R. Kass. But the case against cloning when safety is taken out of the equation is a more difficult case to make than that which pivots upon safety alone. This is true whether one considers merely reproductive cloning or cloning for the sake of embryonic-based stem cell therapies.

One such argument against cloning people is that it is wrong to manufacture people. But cloning human beings is no more manufacturing them than using test-tube baby technology or artificial insemination or even neonatal intensive care. No one feels any less human for having been born in a neonatal unit or delivered by forceps or started up in a petri dish. Clones would be no less people with free will and human dignity than any other person.

Or would they? Some contend that cloning is wrong because everyone is entitled to their own unique genetic endowment. This too is not a strong argument because identical twins and triplets already exist and do quite well despite the existence of another person with the exact same genes. Even if one is worried that parents will try to manipulate or force the clone to behave or develop in certain ways, it has to be said that this is precisely what parents do with their children all the time whether they have a genetic tie to them or not. Should laws against cloning reach into social preferences about how children should be raised that are not enshrined in law?

Therapeutic Cloning

Even if the case against human reproduction by means of cloning is not as strong as it may initially appear, there remains the separate issue of human therapeutic cloning. Therapeutic cloning is not intended to create another human being. It employs SCNT to use the results for other purposes. Is it moral to create cloned human embryos simply to destroy them for the purposes of obtaining stem cells to use in medical research or for other potential uses?

Those who oppose the use of cloned embryos for research or therapeutic purposes do so on the basis of two arguments. First, they may oppose therapeutic human cloning weakly, on the grounds that the cloned embryos are potential human life and as such deserve respect. The opposition here is weak only insofar as it need not entail an opposition without compromise. Second, they may oppose therapeutic human cloning more strongly, on the grounds that embryos have the status of human beings from the moment of conception. Here this opposition is more likely to be one that resists any compromise.

In response to the stronger opposition the fact remains that left in a dish in a lab a cloned human embryo has no potential for personhood unless one assumes the voluntarism of highly trained specialists and of women with empty wombs. Even then such embryos are only dubiously embryonic in that their potential to develop in a human uterus has been anything but established, and their differences from "ordinary" embryos—whether or not one considers such embryos to be persons—have been shown to be abundant and significant. So it is not self-evident that it is immoral to make and destroy cloned embryos on the grounds that this is the same as killing a human being.

Assessment

National debates and those at the United Nations on whether or not to ban human cloning, either outright or merely for reproductive purposes, remain significant venues for science, technology, and ethics interactions. On the one hand, there may be considerable public policy difficulties in implementing any restrictions on reproductive cloning that does not also limit therapeutic cloning, because the initial SCNT technology (or some future technique of a related sort) would be the same for both purposes. On the other hand, it may be that reproductive cloning will remain morally unacceptable simply because it will always be too dangerous or too risky for the future offspring.

At the same time the irony may be that cloned human embryos, which arguably lack true personhood, will remain the best source for stem cells for research and therapeutic uses—uses that may enable humans to respond more effectively to dangers and risks from illness, disease, and injury. Yet because of the potential value of human stem cell research there are also active programs to develop ways to create such cells without involving human embryos. That there might be a technological fix for the moral divide between those in favor and those opposed to stem cell research remains a distinct possibility.

Either way, those who argue about the moral status of human clones and the processes that produce them represent the widest variety of perspectives—in what may almost be called the "kitchen sink" of bioethical debates, involving as they do as many obvious issues about cloning as one could conjecture, as well as a number of subtle issues that depend on careful science and good public policy. On the positive side are proponents such as Michael Fumento (2003) who see human cloning as part of a wave of historically unprecedented benefit and power. On the negative side are critics such as

Francis Fukuyama (2002) who see threats to the very nature of humanity. How well society handles human cloning will demonstrate not only how it handles one of its most extreme and extraordinary cases of conflict in medicine, but also how prepared it is for a world in which different kinds of personhood and parenthood may become as ubiquitous as new kinds of food and transportation.

GLENN MCGEE

SEE ALSO *Embryonic Stem Cells; Eugenics; Playing God; Posthumanism.*

BIBLIOGRAPHY

Bush, George W. (2002). "President Bush's Remarks in Opposition to Human Cloning" (speech in the Rose Garden, Washington, DC, April 10). In *The Future Is Now: America Confronts the New Genetics*, ed. William Kristol and Eric Cohen. A brief articulation of the president's view of human cloning.

Fukuyama, Francis. (2002). *Our Posthuman Future.* New York: Farrar, Straus and Giroux. A defense of the position that cloning is dangerous because it changes human nature.

Fumento, Michael. (2003). *Bioevolution: How Biotechnology Is Changing Our World.* San Francisco: Encounter Books. An exploration of biotechnology and ethics.

Kass, Leon R., and James Q. Wilson. (1998). *The Ethics of Human Cloning.* Washington, DC: AEI Press. Leaders in genetics articulate a strategy for regulation of human cloning.

Klotzko, Arlene Judith, ed. (2001). *The Cloning Sourcebook.* Oxford: Oxford University Press. A compilation of essays concerning the global debate on cloning.

Kristol, William, and Eric Cohen, eds. (2002). *The Future Is Now: America Confronts the New Genetics.* Lanham, MD: Rowman and Littlefield. A conservative position on human cloning.

McGee, Glenn. (2000). *The Perfect Baby*, 2nd edition. Lanham, MD: Rowman and Littlefield. A pragmatic position on human cloning.

McGee, Glenn. (2004). *Beyond Genetics: A User's Guide to DNA.* New York: Harper Perennial. An introduction to cloning and genetic technology.

HUMAN GENOME ORGANIZATION

• • •

The Human Genome Organization (HUGO) is an international society of elected members with an interest in the scientific, commercial, and societal impacts of research on the human genome. HUGO should not be confused with the Human Genome Project (HGP), a U.S. program founded in 1990 and funded by both the U.S. Department of Energy and the National Institutes of Health. HUGO serves as a vehicle for the international coordination of human genome research.

A group of forty-two scientists founded HUGO in September 1988 after a discussion spurred by molecular biologist and Nobel laureate Sydney Brenner (b. 1927) began in April of that year. In the same year the Department of Energy and the National Institutes of Health signed a memorandum of understanding to cooperate in support of human genomic research. An eighteen-member executive council leads the organization, but the complete membership forms a general assembly with ultimate control of the organization. Members of the organization also serve on a number of committees on particular topics, such as ethics and intellectual property rights. New members are elected annually after receiving nominations endorsed by at least five previous or current members.

The purposes of HUGO are to assist the international coordination of research on the human genome, coordinate and facilitate the exchange of data and biomaterials relevant to human genome research, and encourage public debate and provide information and advice on the scientific, ethical, social, legal, and commercial implications of human genome projects (McKusick 1989).

The HUGO Council and its committees have released a number of statements concerning societal impacts, including statements on patenting, cloning, gene therapy, and benefit sharing. In 1996 the HUGO Council approved the first of those statements: "Statement on the Principled Conduct of Genetics Research," which was written by the ethics committee the previous year (Human Genome Organization 1995). The statement includes a general set of recommendations to address concerns about genetic discrimination, information access, and genetic reductionism, among other issues. The recommendations broadly urge the scientific community to meet those concerns through self-oversight and better training.

In statements on patenting in 1995, 1997, and 2000 HUGO argued that expressed sequence tags (ESTs) and single nucleotide polymorphisms (SNPs) do not merit patent protection without detailed knowledge of the biological function of the sequence in question. This position contrasts with patent laws in the United States and Europe, which allow the patenting of those

sequences. HUGO argues that the sequences can be found easily with modern genetics computing but that patent seekers cannot determine the utility of a sequence without doing much more research. HUGO believes that granting patents on ESTs or SNPs prematurely creates disincentives for genetics research.

With regard to cloning HUGO has suggested that no one should attempt reproductive cloning of a human by means of somatic cell nuclear transfer but that basic research using that technique or other cloning techniques and therapeutic cloning should be pursued (Human Genome Organization 1999). However, HUGO also has suggested that embryos should not be created for the purpose of genetic research.

HUGO's statement on gene therapy in 2001 supported the pursuit of somatic gene therapy with strong safeguards, including public oversight and review (HUGO Ethics Committee 2001). The appropriateness of germline therapy that would affect a patient's descendants should be discussed widely. The draft stresses the need for public involvement in setting the limits and ethical principles that should guide gene therapy.

HUGO provides an avenue for the scientific community to communicate its position on the ethical and societal implications of biotechnology research. The organization's international membership includes many preeminent researchers in the field. However, membership is voluntary and the organization has no ability to sanction members or enforce its policies. Its contributions to discussions of the ethical and societal implications of human genome research have been minimal. The organization has not made those issues an important part of its mission.

TIND SHEPPER RYEN

SEE ALSO *Bioethics; Genethics.*

BIBLIOGRAPHY

Human Genome Organization. (1995). "Statement on the Principled Conduct of Genetics Research." *Eubios Journal of Asian and International Bioethics* 6: 59–60.

Human Genome Organization. (1999). "Statement on Cloning." *Eubios Journal of Asian and International Bioethics* 9: 70.

Human Genome Organization Ethics Committee. (2001). "Statement on Gene Therapy Research." *Eubios Journal of Asian and International Bioethics* 11: 98–99.

McKusick, Victor A. (1989). "The Human Genome Organization: History, Purposes, and Membership." *Genomics* 5: 385–387.

INTERNET RESOURCES

Eubios Ethics Institute. Available from http://www.biol.tsu-kuba.ac.jp/~macer/index.html. This institute is run by one of the members of the HUGO Ethics Committee and contains many of the group's ethics statements.

Human Genome Organization. Available from http://www.hugo-international.org/. The official site of the organization.

HUMANISM

• • •

Humanism is a philosophy and way of life (a lifestance) based on empathy, reason, and experience. To humanists, empathy—which is the starting point for compassion and social action—is a product of human nature: the fact that humans are highly developed social animals. Reason is a product of human intelligence that, when combined with experience, leads to the scientific method. And humanists regard the scientific method as the only reliable tool for both acquiring and validating the knowledge necessary to realize the aims of human compassion. To the twentieth-century philosopher Bertrand Russell, the whole concept could be summed up this way: "The good life is one inspired by love and guided by knowledge" (Russell 1957, p. 56).

Given this premise, humanism is an essentially pro-science outlook. And because science becomes socially beneficial primarily through technology, humanists tend to be supportive of technology. Nevertheless, because empathic concerns are basic to humanism, and consequently to humanist ethics, any technology that proves itself more harmful than good in regard to humanity and living nature will be challenged by humanists. This is why humanists have been active in efforts to protect the environment, outlaw certain weapons, ensure product safety, minimize negative social impacts evident in widespread technologies, and so on.

On the other hand, because of the humanist focus on science as the primary means of knowing, there is no place for supernatural belief in humanist thought. Humanism is a completely naturalistic and nontheistic worldview. As such, it leaves humanists with the recognition that humanity alone must take responsibility for making the world better. Along these lines, Humanist Manifesto II (1973) states: "No deity will save us; we must save ourselves." Therefore humanists tend to be relatively fearless in the face of admonitions against scientific hubris and dire warnings that given technologies will allow humans to "play god." In the humanist view, science and technology are tools that allow humans to take charge of their lives, protect themselves from diseases and other dangers, and generally improve the human condition. Therefore, emerging technologies of great promise have tended to be welcomed by humanists rather than feared.

The roots of the humanist worldview are complex, so much so that this background is most clearly understood when pursued as three separate histories: that of the word humanism, the ideas of humanism, and the organized humanist movement.

The Word

The Roman grammarian Aulus Gellius, who flourished circa 160 C.E., noted (in Noctes Atticae [Attic nights] the dual usage of the Latin humanitas (humanity). One usage was comparable to the Greek concept of philanthropia and indicated an attitude of general benevolence or humanitarian sympathies, while the other was comparable to the Greek paideia and indicated the achievement of being humanized (humanissimi) through acquired learning in the liberal arts. Because this latter usage was seen as a capability that separated humans from animals—giving humans the power of independent judgment—it had been favored by the Roman orator and philosopher Cicero (106–43 B.C.E.) and the Roman scholar Varro (116–27 B.C.E.) as a civilizing force.

Such an autonomous, cultured view of life fell largely out of fashion during the Middle Ages, replaced by a notion that human beings were defined players within set hierarchies of the cosmic order, as maintained by the authority of the church, the empire, and the feudal system. But as a few cities and communes gained political independence in the fourteenth century, intellectual independence followed. And with it came a revival of the ancient Greco-Roman spirit. This took the form of a Renaissance literary and philosophic movement of scholars calling themselves humanists. Through a revival of classical letters and a focus on the humanities, Renaissance humanists promoted religious tolerance, worldly ethics, a sense of history, and an interest in nature. In the latter case, what had begun as a revival of humane letters became an impetus for the advancement of science, thus broadening humanism's meaning. Additional broadening occurred as humanist ideas came to be advocated not only by Roman Catholics but also by Protestants, Jews, and nonreligious skeptics.

During the subsequent period of the Enlightenment the term was little used. But in 1853 a democratic organization appeared in England, calling itself the Humanistic Religious Association of London and declaring emancipation "from the ancient compulsory dogmas, myths and ceremonies borrowed of old from Asia and still pervading the ruling churches of our age." Around the same time, in France, the pioneer sociologist Auguste Comte (1798–1857) formulated a "religion of humanity" out of his science-oriented, nontheistic philosophy of positivism.

In 1867 a group of radical Unitarians and freethinkers in the United States formed the Free Religious Association and eventually, by the end of the century, many came to propound what they called *humanistic theism*—essentially a mix of the most liberal Unitarianism, Universalism, and Reformed Judaism of the time together with freethought critiques of more traditional faith. Among the radical Unitarians was Edward Howard Griggs who in 1899 wrote a popular book, *The New Humanism: Studies in Personal and Social Development*, advocating science (particularly Darwinism), "the Greek ideal," Christian spirituality, and social change (including women's rights). These positions were all rolled into an idea for a new religion that would "teach the divinity of common things" and proclaim "the infinite significance of humanity." Another radical Unitarian was the Reverend Frank Carlton Doan, whose 1909 *Religion and the Modern Mind* set forth a more inner-directed, psychological humanism that promoted meditative self-awareness as the starting point for social progress.

Throughout the first three decades of the twentieth century, Irving Babbitt (1865–1933), Paul Elmer More (1864–1937), and Norman Foerster (1887–1972) developed what has been variously termed academic humanism, literary humanism, and the new humanism. This reactionary outlook called for a return to a classics-based education, declared the humanities superior to science, proclaimed human beings superior to nature, and advanced a puritanical morality of decorum. Vestiges of this viewpoint remain in the early twenty-first century among some specialists in the humanities (who sometimes term themselves humanists), often expressed through a distrust of science and technology.

Among philosophers, F. C. S. Schiller in England published *Humanism: Philosophical Essays* in 1903 and *Studies in Humanism* in 1907, advocating a subjectivist form of pragmatism. Later, Jean-Paul Sartre (1905–1980) developed an existential humanism and Jacques Maritain (1882–1973) a theocentric Catholic humanism drawing on the thought of Thomas Aquinas. There have even been both Marxists and Social Darwinists who have taken the humanist label.

While many or all of the above have been regarded as representing different types of humanism, it would be more correct to understand them as different usages of the same word. From this perspective, it is possible to see the current usage of the term humanism as more or less serendipitous and possessing largely superficial rather than substantive connections to the ideas of those who had used the word earlier. The origin of current usage is as follows.

During World War I, the American Unitarian minister John H. Dietrich (1878–1957), having doubts concerning his earlier Christian convictions, adopted a naturalistic, pro-science, ethical worldview linked to a progressive social outlook. But he had no name for this combination of ideas until he read a 1915 article by a positivist, Frederick M. Gould, published in a magazine of the British Ethical Societies. Gould used the term humanism to express a belief and trust in human effort. This was somewhat different from the Renaissance usage already familiar to Dietrich—which suggested that the word could be adapted to his own nontheistic form of Unitarianism. So Dietrich began using it.

Independently, in 1916, another American Unitarian minister, Curtis W. Reese, arrived at similar conclusions. His term of choice, however, was *the religion of democracy*. He argued that democratic religion is human centered in contrast with the authoritarianism of theocratic religion. Edwin H. Wilson, in his 1995 book, *The Genesis of a Humanist Manifesto*, tells how the two men met in 1917 at the annual Western Unitarian Conference: "While Reese was speaking ... on 'The Religion of Democracy,' Dietrich pointed out: 'What you are calling the religion of democracy, I am calling humanism.' It was a momentous convergence of minds—and at that moment, a movement was launched" (pp. 7–8).

The Ideas

In *The Philosophy of Humanism* (1997), Corliss Lamont sees a number of historic ideas, trends, and movements as converging over time to create contemporary humanist thought: these being empirical science, ancient and modern philosophies of materialism and naturalism, free thought, liberal religion, democracy and civil liberties, Renaissance humanism, and literature and the arts—in other words, most of the Western intellectual tradition. There are similar trends in the histories of non-Western cultures, together with cross-pollination with the West, so Lamont also draws attention to relevant intellectual traditions in China, India, and the Middle East. This sort of approach, however, can be accused of creating a pedigree out of ancestors adopted for their compatibility.

Therefore, William F. Schulz, in his 2002 book, *Making the Manifesto,* focuses on more proximate antecedents: nineteenth-century science, the impact of Charles Darwin (1809–1882) and Sigmund Freud (1856–1939), cultural anthropology and the higher criticism of the Bible, free thought and religious modernism, progressivism and the social gospel, and the philosophies of pragmatism and critical realism. Nevertheless, because humanism is not the sum of these things, and because it continues to evolve, it is best described less in terms of its origins and more in terms of what it is: a worldview with the following features.

Humanism's epistemology is derived from the Instrumentalism (the view that the abstract concept of "truth" is best replaced by the more empirical concept of a "warranted assertion") of the American educator and philosopher John Dewey (1859–1952). Metaphysically it is naturalistic (the view that the universe is natural and there is no supernatural). Its worldly ethic is essentially altruistic but because of the humanist commitment to reason, it also involves elements of the Utilitarianism of the English philosopher John Stuart Mill (1806–1873), which holds that acts are good only to the extent that they have practical social benefits that can be rationally decided. Thus humanist ethics are situational (changing with situations) in a context of compassion as well as egoistically consequentialist (taking consequences into account from the standpoint of enlightened self-interest). In the social and political realm this dichotomy reveals itself in a recognition of the inherent conflict between individual liberty and social responsibility, leading to the conclusion that moral dilemmas are real and a necessary part of life and law. Democratic values—including social justice, the enfranchisement of the disenfranchised, and the open society—are central to humanism as an expression of the Golden Rule (do to others as you would have them do to you), which is itself a formula derived from the human capacity for empathy. In matters of personal self-development toward a meaningful life, humanism has been informed by Bertrand Russell's *The Conquest of Happiness* (1930).

The Movement

In 1876 the Society for Ethical Culture was founded by Felix Adler, a Reform Jew who was active in the Free Religious Association. Ethical Culture was a new religion that promoted ethical behavior and social service—deed above creed—with its values derived from neo-Kantian principles. By around 1950, however, the various Ethical Culture societies in the United States and England had evolved Adler's philosophy into humanism or had come to understand it as such. As a result, the American Ethical Union became one of the founding member organizations of the International Humanist and Ethical Union (IHEU), the world coalition of humanists.

In 1916 Reese and Dietrich began preaching humanist ideas from the pulpits of their Unitarian churches. Slowly humanism spread among Unitarians, aided by the creation of the Humanist Fellowship at the University of Chicago in 1927 and the founding of the *New Humanist* magazine one year later. In 1929 the Unitarian minister Charles Francis Potter left the denomination to found the independent First Humanist Society of New York, a church that would eventually count among its members Albert Einstein and Helen Keller.

Meanwhile in India in 1925 Periyar launched Self-Respect, a humanist political and social reform movement devoted to human rights and opposed to the caste system. Openly nontheistic and critical of Hindu and other religious beliefs, it was and remains a proponent of scientific and technological development.

A Humanist Manifesto, published in the *New Humanist* in 1933, was the first major document to lay down the basic principles of humanism. It was signed by prominent academic philosophers (including Dewey), clerics (Ethical Culture, Jewish, Unitarian, and Universalist), educators, journalists, scientists, and social reformers.

In 1941 a number of the manifesto signers founded the American Humanist Association and its magazine, the *Humanist.* Both continue into the twenty-first century, and the organization has counted among its presidents the Nobel Prize–winning geneticist Herman J. Muller and the science popularizer Isaac Asimov.

Following World War II a number of humanist organizations sprung up in Europe, India, and elsewhere. This international growth led to the founding of the IHEU in 1952 at a humanist conclave in Amsterdam chaired by the English biologist Julian Huxley. In the early 2000s the IHEU indirectly represents millions of humanists worldwide in national and local organizations on six continents.

In his 1957 book, *New Bottles for New Wine,* Huxley coined the term *transhumanism* out of a recognition that humanity "*is* in point of fact determining the future direction of evolution on this earth" and therefore a term is needed to signify "man remaining man, but transcending himself, by realizing new possibilities of and for his human nature" (pp. 14, 17). Huxley's word has been taken up by futurist-oriented humanists engaged in exploring the possibilities of radical improvements in the human condition and human capabilities through the likes of cyber-, bio-, and nanotechnology. To foster dialogue and advance this pursuit, the World Transhumanist Association was founded in 1998. Since then a growing number of people have been calling themselves transhumanists.

FRED EDWORDS

SEE ALSO *Humanization and Dehumanization; Science, Technology, and Society Studies.*

BIBLIOGRAPHY

American Humanist Association. (1973). "Humanist Manifesto II." *The Humanist* 33(5): 4–9. Though out of date, this remains the most frequently cited expression of humanism and its social applications.

American Humanist Association. (2003). "Humanism and Its Aspirations: Humanist Manifesto III." *The Humanist* 63(3): 13–14. The most recent basic expression of humanism.

Bragg, Raymond B., ed. (1933). "A Humanist Manifesto." *The New Humanist* 6(3): 1–5. The original, basic expression of humanism, with multiple authors and signatories.

Doan, Frank Carleton. (1909). *Religion and the Modern Mind, and Other Essays in Modernism.* Boston: Sherman, French & Company. An early example of the effort to make personal psychology and emotional well being a starting point for ethics.

Griggs, Edward Howard. (1908). *The New Humanism: Studies in Personal and Social Development*, 6th edition. New York: B.W. Hubesch. Modern religious liberals would find much to agree with in this expression of humanistic theism. First edition published in 1899.

Huxley, Julian. (1957). *New Bottles for New Wine.* New York: Harper & Brothers. Science intersects with moral and social questions in this popular collection of essays.

Kurtz, Paul. (2000). *Embracing the Power of Humanism.* Lanham, MD: Rowman and Littlefield. Offers an exuberant expression of humanism as a positive and personally rewarding lifestyle.

Lamont, Corliss. (1997). *The Philosophy of Humanism*, 8th edition. Amherst, NY: Humanist Press. Recognized as the standard work on humanism. First edition published in 1949.

Olds, Mason. (1996). *American Religious Humanism*, rev. edition. Minneapolis, MN: Fellowship of Religious Humanists. The most complete survey of the early history of religious humanism.

Russell, Bertrand. (1958 [1930]). *The Conquest of Happiness.* New York: Bantam Books, Inc. A practical book on achieving happiness that doesn't insult the intelligence of philosophically-minded readers.

Russell, Bertrand. (1957). *Why I Am Not a Christian and Other Essays on Religion and Related Subjects*, ed. Paul Edwards. New York: Simon & Schuster, Inc. A significant number of humanists and freethinkers name this book as the one most influential in bringing about their break with traditional religion.

Schiller, F. S. C. (1903). *Humanism: Philosophical Essays.* London and New York: Macmillan. The first book to approach humanism academically as a philosophy.

Schiller, F. S. C. (1907). *Studies in Humanism.* London: Macmillan and Co., Limited. The author further develops his philosophy of humanism as a subjective form of pragmatism.

Schulz, William F. (2002). *Making the Manifesto: The Birth of Religious Humanism.* Boston: Skinner House Books. An analytical history of the first Humanist Manifesto by a Unitarian universalist minister who personally interviewed a number of the original signers, all who are now deceased.

Wilson, Edwin H. (1995). *The Genesis of a Humanist Manifesto.* Amherst, NY: Humanist Press. The history of the first Humanist Manifesto told by one of the people most responsible for it.

HUMANITARIAN SCIENCE AND TECHNOLOGY

• • •

Humanitarian was first applied to organizations such as the International Red Cross/Crescent, founded in 1864 by the Swiss philanthropist Jean-Henri Dunant (1828–1910), in response to his experience with wounded soldiers at the Battle of Solferino, Italy, in 1859. From the beginning the term was thus allied with an ethical vision for the use of science and technology (initially in the form of medicine) to benefit human beings who may have previously been harmed by technology (at first in the form of military weapons).

Background

Humanitarianism is an ethical vision closely associated with the creation of the social sciences. During the nineteenth century, modern natural science began to explore social phenomena, in part to deal with the challenges presented by new human powers over the natural world. Industrial technologies created urban centers that needed better management for the benefit of the human beings who lived in them, not as members of some political or religious or ethnic group but simply as human beings, who could also be scientifically studied as such. Public health and public engineering is for the benefit of all, although the "all" was in the first instance understood within a national context.

Humanitarianism thus aims to extend compassion beyond traditional family or village limits, especially through the utilization of science broadly construed. Although this may appear to have been simply a secular version of Christian missionary work—especially since humanitarian organizations often attracted voluntary contributions from believers—the increasing number of middle-class persons involved in providing relief for the victims of warfare and the improvement of urban slums constituted a historically unique social movement (Morehead 1999).

The larger background is that the early-1800s gave science and technology major roles in the construction, organization, and maintenance of both nation-state and colonial empire. First in England and France and later in the United States, centers of raw materials extraction and industrial production also created an exploited working class. Witnessing the living conditions of these people, humanitarian scientists and engineers often responded to alleviate such situations as best they could through technical improvements. After 1830 in Lille, France, humanitarian physicians studied and denounced the deplorable conditions of working class people in order to improve their health and living conditions (Gerard 1999). In 1838 German-born naturalist Robert Schomburgk sought to use his knowledge to reduce the enforced slavery of Indians in British Guiana by establishing a political boundary in harmony with their natural territory (Riviere 1998). Indeed in the 1800s humanitarian science, by emphasizing the unity of all peoples as human beings in the eyes of science, was a significant contributor to abolitionist movements throughout the world.

Across the turn of the nineteenth to twentieth century, international conflicts and natural disasters affecting large populations further spurred efforts to utilize science, technology, and medicine to ameliorate the conditions of wounded and displaced peoples. The Franco-Prussian War (1870–1971), the Ohio and Mississippi River floods (1884), the Spanish-American War (1898), the San Francisco earthquake (1906), and World War I (1914–1918) all provided major tests for the International Red Cross and related humanitarian agencies. The continued involvement of scientists and engineers in humanitarianism was reflected in scientist and inventor Alfred Nobel's creation of the Nobel Prizes at his death in 1896; the first Peace Prize was awarded to Dunant in 1901.

The twentieth century witnessed the further institutionalization of humanitarian activities related to science and technology in labor movements, public health work (including family planning), and immigrant settlement and education (which often emphasized technical education). Finally in response to the horrible uses of science in World War II (1939–1945), especially in the death camps of Nazi Germany, humanitarianism led to adoption of the Universal Declaration of Human Rights (1948), which stipulates "the right freely . . . to share in scientific advancement and its benefits" (Article 27).

Some argue that all science and technology are inherently humanitarian in their basic orientation, which was the view of both early modern scientists and proponents of the Enlightenment. Over the course of the modern period, however, it became increasingly recognized not just by socialists that special efforts are often needed to protect science and technology from dehumanizing distortions caused by economic or political interests. Efforts to liberate the benefits of science and technology from pernicious influences have taken place in national and international regulatory agencies, which may in many instances be styled humanitarian. Especially during the last half of the century humanitarian science and technology were further encouraged by four interrelated phenomena: the consumer movement, the environmental movement, the alternative technology movement, and public interest science.

Engineers especially also have been major contributors to international development work. For instance, the idea for the U.S. Peace Corps originated in 1960 with civil engineer Maurice Albertson, who was also intimately involved in its creation.

However, by the last quarter of the century, disaster and refugee relief had taken on characteristics that exceeded the capacities of many traditional humanitarian organizations. The end of the Cold War (1989) and the subsequent rise of genocide and terrorism as an international threat promoted humanitarian missions by the armed forces, which relied heavily on engineering skills. Increasingly humanitarian action involved scientific and technological developments in psychological counseling, high-tech monitoring (of military movements or weather), and the use of specially designed equipment (mobile power plants, water purification systems, and more). But a further response was the creation of new kinds of not-for-profit and non-governmental organizations oriented toward humanitarian action as part of an emerging international civil society. The failures and inadequacies of post-Cold War ideology of humanitarianism have also been subject to extensive criticism (see, for example, Rieff 2002).

Science and Engineering without Borders

Humanitarian science and technology may be related to what Carl Mitcham (2003) has termed *idealistic activism* among scientists and engineers, as illustrated by organizations such as International Pugwash (founded 1957) and the Union of Concerned Scientists (founded 1969). Among a diverse collection of related organizations seeking to build bridges between humanitarianism and scientific technology are the Responsible Care initiative of the American Chemistry Council and the International Network of Engineers and Scientists for Global Responsibility (INES). Responsible Care, founded in 1988, is a voluntary program to improve environmental health and safety in the chemical and related industries, especially in developing countries. INES, founded at a 1991 international congress in Berlin, is an association of more than ninety organizations in fifty countries promoting the involvement of technical professionals in humanitarian and peace development activities.

In 1971, however, humanitarian science and engineering activism took a new turn with the formation of *Médecins sans Frontières* (MSF or Doctors without Borders). MSF, which has become the largest non-governmental relief agency in the world, grew out of dissatisfaction with the inability of the Red Cross/Crescent to react independently of national government controls, and its tendency to remain within safe boundaries. The idealistic physicians of MSF pioneered new ways to bring medical science and technology to people in crisis and to speak out against human rights abuses. Since its founding, MSF has responded to needs resulting from earthquakes, hurricanes, war, and famine in Central America, Africa, Russia, the Balkans, and the Middle East (Tanguy 1999).

Inspired by MSF, other science and engineering organizations followed suit. Examples include *Avaition sans Frontières* (1980), providing air deployment for humanitarian projects, and ORBIS ophtalmologists (1982), providing preventive and surgical eye care to poor communities throughout the world. In the early-1990s, there also emerged independently a number of groups going under some form of the name Engineers without Borders: *Ingénieurs Sans Frontières—Ingénieurs Assistance Internationale* (Belgium), *Ingeniería sin fronteras* (Spain), *Ingenièrer unden Graenser* (Denmark), *Ingenjörer och Naturvetare utan Gräser-Sverige* (Sweden), *Ingegnería Senza Frontiere* (Italy), and others. In 2003 these groups organized Engineers Without Borders—International as a network to promote "humanitarian engineering . . . for a better world." The process has also led to educational programs in humanitarian engineering, efforts that parallel others in public health and nutrition science, and policy programs that seek comprehensive, interdisciplinary understandings of humanitarian crises.

Undoubtedly one of the personal inspirations for engineering without borders efforts was the life and work of mechanical engineer Fred Cuny (1944–1995). Following relief work in Biafra (1969), Cuny sought to bring his engineering skills to bear in earthquake disasters in Central America (1971 and 1976), Sudan (1985), Iraq (1991), Somalia (1992), Sarajevo (1993–1994), and Chechnya (where he was assassinated). Cuny's book *Disasters and Development* (1983) outlines what became known as the *Cuny approach*, an effort to respond to disasters not just by returning people to their predisaster state, but as opportunities to help them improve their lives beyond what otherwise might have been possible.

Defining the Field

Although subject to continuing debate, the basic dimensions of humanitarian science and technology may be summarized as follows. While advances in science and technology have benefited many persons, they have also often increased rich–poor divides, to which specific organizations have tried to respond. Among these, many emphasize science and engineering expertise. Humanitarian science and technology projects, typically operated on a not-for-profit basis, aim either to provide fundamental needs (such as food, water, shelter, and clothing) when these are missing or inadequate in the developing world, or higher-level needs for underserved communities in the developed world.

In contrast to corporations, which aim for relatively near-term profit, and governments, which fund in light of election cycles and constituent dependencies, humanitarian projects are of longer-term importance for society as a whole. Humanitarian science and engineering ideally engage local communities in direct participation in determining project needs and directions. Additionally they seek strategies, designs, and technologies that promote both the sustainability of natural systems and cultural traditions.

CARL MITCHAM
JUAN LUCENA
SUZANNE MOON

SEE ALSO *Engineering Ethics; Globalism and Globalization; Humanization and Dehumanization; Science, Technology, and Society Studies.*

BIBLIOGRAPHY

Cuny, Frederick C. (1983). *Disasters and Development*, ed. Susan Abrams. New York: Oxford University Press. Highly influential.

Gerard, Alain. (1999). "*Action Humanitaire et Pouvoir Politique: L'Engagement des Medicins Lilliois au XIX Siecle*" [Humanitarian action and political power: the involvement of Lille physicians in the 19th century]. *Revue du Nord* 81(332): 817–835.

Mitcham, Carl. (2003). "Professional Idealism among Scientists and Engineers: A Neglected Tradition in STS Studies." *Technology in Society* 25(2): 249–262.

Morehead, Caroline. (1999). *Dunant's Dream: War, Switzerland, and the History of the Red Cross*. New York: Caroll and Graf. A broad and detailed historical narrative.

Rieff, David. (2002). *A Bed for the Night: Humanitarianism in Crisis*. New York: Simon and Schuster. A critical examination of contradictions in early-twenty-first-century humanitarianism.

Riviere, Peter. (1998). "From Science to Imperialism: Robert Schomburgk's Humanitarianism." *Archives of Natural History* 25(1): 1–8.

Tanguy, Joelle. (1999). "The *Médecins Sans Frontières* Experience." In *A Framework for Survival: Health, Human Rights, and Humanitarian Assistance in Conflicts and Disasters*, revised edition, ed. Kevin M. Cahill. New York: Routledge. The only general history that exists.

HUMANIZATION AND DEHUMANIZATION

• • •

To humanize is to engage with the human. In many instances this involves actions or constructions to accommodate the limits or needs of human beings, as in the "humanization of science and technology." While science and technology have themselves been extolled as humanizing the world, they have also been criticized as in need of humanization—that is, as dehumanizing. Indeed, it is the negative concept that is in more common use and has emerged to play important roles in at least four areas: psychology, theology, art, and social criticism.

Psychology, Theology, and Art

In social psychology dehumanization is defined as the process by which one person or group views others as not worthy of humane treatment. The dehumanization of enemies is common in personal conflict, civil strife, and warfare—and in the case of large-scale warfare perhaps even unavoidable. Extreme dehumanization leads to crimes against humanity and acts of genocide such as the Holocaust, where even technicians and other "innocent" German citizens were culpable in the dehumanization of victims. There are two types of dehumanizing agents here: those who actually commit the crimes and those who passively conform and silently witness them. In both cases, the act of characterizing others as less than human may serve as a coping mechanism to dampen the psychological effects of mass cruelty. The use of dehumanizing names to disparage others is not confined to extreme or fringe situations, however. Such disparaging language can also be found in mainstream elements of society including laws, magazine articles, and scientific journals (Brennan 1995). Research in conflict resolution and peace studies promotes techniques for the rehumanization of enemies (Stein 1996).

Psychological analyses of dehumanization have described it as a process by which individuals or groups project their own faults onto opponents. Dehumanization in this sense is thus a generalization of the scapegoat phenomenon (Girard 1986), which plays an important role in Christian theology. Moreover, in part because of the Enlightenment claims for the humanizing character of science and technology as opposed to the dehumanizing character of religion, religious and theological discussions have developed extended arguments for religion as a humanizing factor in human affairs. For example, Barbara Rumscheidt (1998) argues that the development of socially engaged Christian faith communities can counteract the dehumanizing effects of globalizing capitalism.

Two specific religious contexts in which the question of humanization has taken form are in Marxist-Christian dialogues and liberation theology. In both these cases the problematic of scientific and technological development is also important. For example, the roots of liberation theology stem in part from industrial development in Latin America, which benefited some but marginalized and impoverished others. Subsequent ecclesiastical developments addressed the question of how economic and technological modernization can promote genuine human progress for all.

José Ortega y Gasset (1925) used the concept of dehumanization to characterize art in the early twentieth century, which by abandoning traditions of romanticism and realism, deformed reality and shattered its merely human aspect. In avant garde art, all that is real, natural, and human is purged in favor of purely artistic elements—which, for Ortega, is actually a good thing. Dehumanization in this context is an aristocratic revolt against the industrial massification of culture, an effort to break through to a higher form of civilization, anticipating subsequent notions of post- and transhumanization.

Criticizing Science and Technology

Ortega was also one of the first philosophers to address both the humanizing and dehumanizing aspects of technology. For Ortega technology is an integral part of being human, but by overwhelming human beings with means to transform the world modern technology can undermine the more central human attributes of imagination and intentionality. As if reflecting Ortega's notion, social criticism of science and technology has tended to bemoan both unrealized possibilities and popular acquiescence to inertial trajectories in technoscientific development. Indeed, according to Carl Mitcham (1984), the question of humanization is one of the most broad and synthetic themes in the critical examination of technology. In what ways, and to what extent, do science and technology promote or obstruct human well being? In terms of the individual, this is an ethical question; in terms of social institutions it is a political one.

Three key arguments for science and technology as humanizing forces are as follows. First, science is a natural expansion of human knowledge that promotes material progress as well as intellectual and spiritual fulfillment. This dual humanizing quality of science was famously portrayed by novelist C.P. Snow's "two cultures" argument, in which scientific intellectuals are viewed as more humane than their literary intellectual counterparts.

Second, science has a normative structure that reciprocally reinforces democratic principles and practices, according to sociologist Robert Merton, scientist Michael Polanyi, philosopher Karl Popper, and others. Since the Enlightenment, the structures of the republic of science have often been presented as models for civil society.

Third, technology humanizes by freeing human beings from disease and other burdens of nature. Economist Julian Simon, for instance, has been an outspoken advocate of the view that technology has increased human prosperity and well-being and will continue to do so as long as humans are allowed to freely develop and deploy it. A collateral argument is that computers and artificial intelligence humanize not just nature by placing it under human control but the world of artifice as well by overcoming the limits of machines and making them more human-like.

In opposition there are also three key arguments for science and technology as dehumanizing forces. First, scientific knowledge is said to alienate humans from the natural, organic, or lived experience. Behaviorist psychology and rational actor theories in the social sciences reduce humans to bundles of calculations and reactions. More generally, Edmund Husserl, in analyzing how the sciences interact with the "life world," warned that modern science arose on the basis of a great forgetting of the immediate, which played out in a parallel amnesia in the human sciences (Rajan 1997).

Second, technology creates an artificial world that is even more burdensome than nature. Some versions of this argument lament the spiritual disease and the feelings of anomie and powerlessness engendered by the modern, Western world (for example, Montagu and Matson 1983, Ryan 1972). Technology has increased the tempo of life to a frantic pace and the massification of production processes and media images produce the foreboding by Ralph Waldo Emerson that "Things are in the saddle,/And ride mankind." Indeed social theorist Jacques Ellul argues that technique has shifted from success in the material world toward a broad spectrum of human activities from education to politics, art, and even ethics—each of which it transforms into a technical process aiming at some form of efficiency. For radical educational theorist Paulo Freire (1970), such dehumanization becomes perfected when it is welcomed rather than shunned, and rehumanizing begins by raising consciousness of one's less-than-human existence.

Third, the conquest of nature and the transformation of the social world leads to the conquest of human nature—and thereby its destruction. This argument, as advanced, for instance, by the literary scholar, novelist, and lay theologian C.S. Lewis, has been revised and deepened by, among others, intellectual historian John Hoberman and science policy philosopher Leon Kass. For Hoberman (1992), the use of drugs to enhance performance raises fundamental issues about the structure of human activity and the connection between performance and effort. For Kass, "Human nature itself lies on the operating table, ready for alteration, for eugenic and neuropsychic 'enhancement,' for wholesale redesign" (2002, p. 4). What is most disturbing about this situation, which was foreshadowed in Aldous Huxley's *Brave New World* (1932), is not the lack of freedom or equality, but the dehumanization and degradation of people who choose "nothing humanly richer or higher"—a fate that may emerge in regimes of individualist democratic consumerism more than totalitarian control.

Assessment

As this third critique of technology demonstrates, judgments of both humanization and dehumanization are necessarily based on visions of human nature. They are related to notions of humanism that likewise involve assessments of the character and influence of science and technology. As such, the concepts of humanization and dehumanization fail as primary ethical concepts for the judgment of science and technology, although they often figure in popular discussions as summary presentations of more fundamental views.

Cultural and philosophical visions of human nature can even create fundamentally opposed understandings of humanization and dehumanization. For example, some philosophical anthropologies envision humans as radically circumscribed by the limits of mortality and futility. From this perspective, dehumanization may occur when such limits are drastically altered or surpassed. Other treatments of human nature characterize humans as self-making beings with unbounded potentialities. From this perspective, the imposition or voluntary submission to certain limits could be regarded as dehumanizing acts.

CARL MITCHAM
ADAM BRIGGLE

SEE ALSO Critical Social Theory; Dignity; Human Nature; Human Rights; Weil, Simone.

BIBLIOGRAPHY

Brennan, William. (1995). Dehumanizing the Vulnerable: When Word Games Take Lives. Chicago: Loyola University Press. Examines the use and effect of disparaging language in several contexts.

Freire, Paulo. (1970). Pedagogy of the Oppressed, trans. Myra Bergman Ramos. New York: Seabury Press. Presents a radical educational theory that criticizes the "banking" model of education and argues for a reduction in the teacher-pupil dichotomy.

Girard, René. (1986). The Scapegoat. Baltimore: Johns Hopkins University Press. French original 1982.

Hoberman, John M. (1992). Mortal Engines: The Science of Performance and the Dehumanization of Sport. New York: Free Press. Explores social and philosophical implications of enhancement therapies in sports.

Kass, Leon. (2002). Life, Liberty and the Defense of Dignity. San Francisco: Encounter Books. Presents a "richer bioethics" to deal with issues of humanization and dehumanization raised by biomedical technologies.

Mitcham, Carl. (1984). "Philosophy of Technology." In A Guide to the Culture of Science, Technology, and Medicine, ed. Paul T. Durbin. New York: Free Press, pp. 282–363 and 672–675. See especially the section titled "Toward a Synthesis: The Question of Humanization," pp. 339–344.

Montagu, Ashley, and Floyd Matson. (1983). The Dehumanization of Man. New York: McGraw-Hill. A cultural critique that articulates the psychological and social ills produced by modernity.

Ortega, José y Gasset. (1925). La deshumanización del arte. Ideas sobre la novella. Madrid: Revista de Occidente. English translation: The Dehumanization of Art: And Other Essays on Art, Culture, and Literature. Princeton, NJ: Princeton University Press, 1968.

Rajan, Sundara R. (1997). The Humanization of Transcendental Philosophy: Studies on Husserl, Heidegger, and Merleau-Ponty. New Delhi: Tulika. An analysis of the problematic of humanization and its consequences for science, language, and philosophy.

Rumscheidt, Barbara. (1998). No Room for Grace: Pastoral Theology and Dehumanization in the Global Economy. Grand Rapids, MI: William B. Eerdmans. Offers avenues of action for religious communities to transform dehumanizing economic realities.

Ryan, John Julian. (1972). The Humanization of Man. New York: Newman Press. A cultural critique that diagnoses causes of dehumanization and offers an alternative way of life to humanize modern society.

Stein, Janet Gross. (1996). "Image, Identity and Conflict Resolution." In Managing Global Chaos, ed. Chester Crocker, Fen Hampson, and Pamela Aall. Washington, DC: United States Institute of Peace Press, pp. 93–111.

HUMAN NATURE

• • •

Many ethical judgments make appeals to human nature either as their foundation or as their standard. In the strongest case ethics is argued to be based on human nature; in other instances actions are proscribed if they fail to respect human nature or are recommended because they are said to be in harmony with human nature. Human nature is also an object of scientific investigation, raising questions related to both process and product: whether scientific investigation is undertaken in ways that respect human nature and whether the results of such investigations can contribute to the understanding of human nature in an ethically relevant sense. After a brief review of theories of human nature, the focus in this entry will be on the final question: the extent to which scientific knowledge of human beings can contribute to understanding or assessing these theories, especially in their role as foundations for ethics.

Theories of Human Nature in History

According to Leslie Stevenson (2004), theories of human nature entail theories about the world, human beings, what might be wrong with human beings, and how anything that is wrong might be corrected. Even those who deny any essential human nature in favor of a historical or cultural construction of human nature have views about what kinds of things human beings are and their place in the world. However, with regard to explicit theories of human nature, premodern theories generally viewed humans as properly subordinate to a larger order so that even though people on occasion rebel against that order (by means of what the Greeks calls *hubris* and the Bible calls sin), they are called upon to learn to control such rebellion by means of ethical or religious practices. By contrast, modern theories tend to see human beings as unjustly limited by the larger order and thus encouraged to overcome those limitations, often by means of science or technology.

More specifically, for Plato, in a famous analogy from the *Republic* (p. 437b ff.), the human soul is presented as being composed of three parts: appetite, reason, and spiritedness. Lack of order results whenever appetite or spiritedness predominates and steps outside the guidance provided by reason. In a similar manner, for Aristotle in *Nicomachean Ethics* (vol. I, p. v), human lives can be oriented toward pleasure, politics, or knowledge, but the perfection of human nature resides with rationality in both practice and theory. Thomas Aquinas (1224–1274) further develops this perspective by arguing that the lawful order of nature is manifest in human nature (in a form he terms natural law) in aspirations to life, affective sociability, and the rational pursuit of both politics and science. Although the Jewish, Christian, and Islamic views of human nature seek in some measure to subordinate rationality to faith in revelation, that faith, like reason, ultimately places boundaries on appetitive, political, and even scientific activities. Structurally similar views can be found in the Asian religious and philosophical traditions associated with Hinduism and Buddhism.

More typically modern theories such as those of Thomas Hobbes (1588–1679) and John Locke (1632–1704), even when they offer a materialist and mechanistic analysis of the workings of human nature, argue that humans are improperly constrained by the state of nature. In Hobbes's frequently cited description, the state of nature is one in which human life is "solitary, poor, nasty, brutish, and short" (*Leviathan*, vol. I, p. 13), a condition from which human beings justly seek any means of escape. This notion that people are unjustly constrained by the human condition is repeated and developed in philosophies as diverse as those of Jean-Jacques Rousseau (1712–1778), Immanuel Kant (1724–1804), Georg Hegel (1770–1831), and Karl Marx (1818–1883). According to Rousseau, for instance,

"Man is born free, but everywhere is in chains" (*Social Contract*, vol. I, p. 1). The psychological theory of Sigmund Freud, with its distinction between id, ego, and superego, reverses Plato's theory by suggesting the primacy of id or appetite over both individual self (ego) and social restraint (superego).

Three Basic Approaches to Human Nature

What can science contribute to the assessment or criticism of these diverse theories of human nature? One scientific debate concerns the relative influences of nature and nurture in human affairs. Another focuses on degrees of rationality or nonrationality in human decision and action. Among the most fundamental questions is that concerning whether there is something—a rational or transrational mind or soul—that cannot be accounted for by the same material causes that govern all other things in the natural world.

Materialism (or physicalism) is the position that the physical world is self-contained or closed so that the physical world can be explained only through physical causes and effects. In considering human nature, a materialist would say that human beings must be explained as purely material mechanisms, as physical bodies governed exclusively by physical causes. Consequently, the human mind should be understood as an activity of the physical brain. All the thinking, feeling, and willing of the conscious self must be determined totally by the body, particularly the brain and nervous system.

Against such a materialist view of human nature a dualist would argue that mind is not fully reducible to body, that the mind can act as an immaterial cause on the material brain. An interactionist dualist would agree that the mind depends on the brain as its necessary but not sufficient condition. Thus, if some part of the brain is damaged or ceases to function normally, this can interfere with mental activity. Still, as long as the mind is supported by normal brain activity, the mind can exert its independent power over the brain. When people act through conscious thinking and willing, they use their immaterial minds to control their material brains. A religious believer might go further and claim that the immaterial mind was created by an immaterial God, and thus the mind or soul is supernatural. This supernatural character of the soul could render it immortal so that the human soul could survive the death of the human body.

There are, then, at least three fundamentally distinct views of human nature that are based on three views of the relationship between mind and body. The materialist believes that the mind has no immaterial power to act on the body. The interactionist believes that the immaterial mind interacts with a material body. The supernaturalist believes that the immaterial mind is supernatural and immortal. Each of these views implies more general perspectives on human beings and their place in nature.

Traditional Arguments for Interactionism

These conflicting views run throughout the history of natural science from Socrates to the present. In Plato's *Phaedo* (pp. 96a–100a) Socrates (470–399 B.C.E.) talks with his friends while awaiting execution. He recounts that as a young man he thought that a scientific investigation of nature would explain the causes of everything. He hoped to explain the physical causes of all things coming into being and passing away, including the causes of animal life and the causes of human thinking in the brain. He became frustrated when he found that a complete science of nature as governed by physical causes was beyond his grasp. To explain the world, Socrates insists, it is necessary to understand both physical causes and mental causes. For example, to explain why Socrates is sitting here awaiting his execution, one might describe the physical mechanisms in his body—the bones, muscles, ligaments, and so on—that control his movement. However, although these physical causes are necessary in explaining why he is sitting here, they are not sufficient. It is also necessary to explain how Socrates made up his mind to accept his punishment because this mental decision controls his physical body.

Socrates appeals to a person's ordinary experience of making up his or her mind and then freely choosing to act according to that conscious mental decision; this leads people to think that the mind has a power to act that changes the physical causes of the body. Holding oneself and others morally and legally responsible for their conduct assumes that freedom of thought and choice. People do not hold nonhuman animals or human children morally responsible for their behavior because it is assumed that they lack the moral freedom that is attained only by the development of rational choice in normal human adults through learning and habituation. If human conduct were fully determined by physical causes in the body, it would be impossible to hold people morally or legally responsible for their conduct.

From ancient Greece to the present this kind of Socratic thinking has led many scientists, philosophers, and theologians to conclude that human nature is characterized by a complex interaction of mind and body, mental causes and physical causes. The human mind

acts upon the human body, or the mind exerts an immaterial power that is not reducible to the material causes of the body.

Modern Arguments for Materialism

Socrates was responding to a materialist or physicalist tendency that would become a strong tradition in Western science. That materialist tradition gained great power during the scientific revolution of the sixteenth and seventeenth centuries. Proponents of the new science saw the universe as a mechanism that could be explained by mechanical laws working through purely physical causes. It seemed that much of human nature could be explained similarly without invoking an immaterial soul.

Hobbes saw nature as matter in motion governed by laws of motion such as those discovered by Galileo (1564–1642). Animal life, then, including human life, is "but a motion of limbs." "For what is the heart, but a spring; and the nerves, but so many strings; and the joints, but so many wheels, giving motion to the whole body" (*Leviathan*, Introduction). Animal motion is driven mechanically by selfish passions that goad animals to seek pleasure and avoid pain. Although human beings are moved by some of the same selfish passions, humans are unique in their capacity for reason and speech. However, even this uniquely human intellectual activity can be understood mechanistically as the computational manipulation of informational symbols (*Leviathan*, vol. I, p. 5). The soul or mind cannot be immaterial. It must be the activity of the material body. This must be so if one accepts the claim of natural science that everything in the universe is matter in motion. Because of his materialism Hobbes was denounced by religious and political leaders as a morally corrupting teacher of hedonism, egoism, and atheism.

Hobbes's materialist science of the soul seemed to be confirmed by Thomas Willis's (1621–1675) studies of the brain. Working in England at the same time as Hobbes, Willis compared the anatomy of the human brain with that of other animal brains and combined experiments on brains with medical observations of brain-damaged patients to develop what he called "neurology." He reached five broad conclusions. First, all mental experience arises from the motion of "animal spirits" undergoing chemical changes in the brain. Second, different parts of the brain have different functions. Third, the human brain resembles other animal brains, particularly those of monkeys and apes. Fourth, this science of neurology could be used by medical doctors to cure diseases of the brain through the use of drugs that would alter the chemistry of the brain. Fifth, all this supports the general view of the "mechanical philosophy" of the seventeenth century that the human body and brain are both machines explainable by mechanical laws.

Although Willis was mistaken about many details, his broad conclusions are supported by modern neuroscience. What Willis called animal spirits can be understood as electrical and chemical signaling between neurons. Willis's observation that the brain has specialized functions has been elaborated by studies of the ways neurons are organized into modular networks with distinct functions. Willis's claim that the human brain resembles the brains of other animals can be explained by evolutionary biology. His hope that drugs could cure the diseases of the soul seems to have been fulfilled by modern psychopharmacology in its use of drugs to treat mental disorders and enhance mental function. Finally, Willis's mechanistic account of the mind has been elaborated with computer models of the mind as an information-processing system.

It may appear, then, that the science of the human brain initiated by Willis proves Hobbes's materialist view of the soul. However, Willis was not a strict materialist because he believed that his science showed the existence of two souls. The "sensitive soul" found in all animals was purely material and therefore vulnerable to physical diseases. In contrast, the "rational soul" found only in humans was immaterial and immortal, although it depended on the sensitive soul. Thus, Willis's account of human nature was interactionist in that he thought the material brain and the immaterial soul mutually influence each other. He was also a supernaturalist in that he thought the immaterial soul was created by God to be immortal.

In the early twenty-first century some scientists, such as James Watson (2003), Edward O. Wilson (1998), and Steven Pinker (2002), argue that natural science sustains a purely materialist view of human nature and refutes any belief in the human soul as immaterial or immortal. Those scientists dismiss belief in an immaterial soul as an unscientific superstition. However, other scientists, such as Wilder Penfield (1978) and John Eccles (1994), defend Willis's interactionist view of the mind as an immaterial cause that can act on the brain. Eccles, a Nobel Prize–winning neuroscientist, has argued that modern neuroscience is compatible with belief in the self-conscious mind as an immaterial power for thinking and choosing.

Ethical Implications

What difference do these debates over the science of mind-brain interaction make for an understanding of human nature and morality? Those who argue for an immaterial soul agree with Socrates that the capacity of the mind to act outside the laws of physical nature is necessary for moral freedom. They warn against scientific materialism as a denial of free will that would make it impossible to hold people morally responsible for their conduct. They also warn that a materialistic view of human nature would promote a Hobbesian hedonism in which people would see themselves as animals moved by selfish passions with no spiritual capacity for rising above their material interests. To explain the soul as merely biochemical activity in the brain would seem to deprive human life of any unique moral dignity. Moreover, if scientific materialism teaches that human nature has only limited dignity above the rest of nature and if the ultimate end of modern science is the conquest of nature, people may be tempted to use the technological power of science to alter human nature itself in ways that would be dehumanizing.

The history of eugenics illustrates the potentially corrupting effects of a materialist view of human nature. The Judeo-Christian view of human beings as having been created in God's image with immortal souls has supported the moral principle of the special sanctity of human life. However, by the end of the nineteenth century modern science, particularly Darwinian science, had persuaded many people that human beings are merely highly evolved animals and that they do not have immaterial or supernatural souls that set them above the rest of animal nature.

If human beings are products of an evolutionary process governed by survival of the fittest, it seemed that reproductive fitness would be the only moral value coming from nature. Proponents of eugenics argued that human beings should be bred just as other animals are to improve the genetic quality of the species. As a result many state governments in the United States passed laws that forced individuals regarded as genetically inferior to be sterilized so that they could not reproduce. In Nazi Germany, Adolf Hitler (1889–1945) used policies of eugenics, euthanasia, and genocide to eliminate people whom he identified as belonging to inferior races. Some historians, such as Richard Weikart (2004), have explained the horrors of eugenics and Nazism as having been caused partly by the influence of Darwinian materialism in devaluing human life.

Other philosophers such as Peter Singer (2001) have argued that because religious belief in the sanctity of human life has been refuted by scientific materialism, people may be morally justified in euthanizing infants born with severe deformities. Some posthumanist or transhumanist proponents of biotechnology see no moral limit on the power to use science to redesign human beings, perhaps even to the point of abolishing human nature itself. All this seems to confirm the fears of many people that modern science, insofar as it promotes a materialist view of human nature, subverts traditional morality.

At the same time some scientific reasoning about the human mind may support traditional morality by showing how it is rooted in the brain. In *The Descent of Man* (1871) Charles Darwin (1809–1882) argued that a natural moral sense was implanted in human nature by evolutionary history. As naturally social animals, human beings evolved to have a natural sense of right and wrong that would support social cooperation on the basis of ties of kinship and reciprocity. To reinforce this cooperative behavior they were endowed with emotional propensities to moral emotions such as love, guilt, and indignation and also were endowed with the intellectual capacity to formulate social norms of cooperation rooted in those moral emotions.

Some neuroscientists have found that moral experience depends on the moral emotions sustained by the emotional control centers of the brain and on the moral reasoning carried out in the prefrontal cortex of the brain. If these parts of the brain are not functioning normally, people cannot act as moral beings. For example, psychopathic criminals apparently have an abnormality in their brain circuitry that prevents them from feeling the moral emotions that support the moral conduct of normal human beings. Such scientific research suggests that morality is part of the biological nature of human beings.

LARRY ARNHART

SEE ALSO *Christian Perspectives: Historical Tradition; Dignity; Enlightenment Social Theory; Hobbes, Thomas; Humanization and Dehumanization; Hume, David; Natural Law; Posthumanism.*

BIBLIOGRAPHY

Arnhart, Larry. (1998). *Darwinian Natural Right: The Biological Ethics of Human Nature.* Albany: State University of New York Press.

Barbour, Ian G. (2002). *Nature, Human Nature, and God.* Minneapolis: Fortress Press. A survey of the ethical and religious issues in the scientific study of human nature.

Damasio, Antonio R. (1994). *Descartes' Error: Emotion, Reason, and the Human Brain.* New York: Putnam. A seminal work on the emotional basis of ethics in the brain.

Darwin, Charles. (1871). *The Descent of Man, and Selection in Relation to Sex,* 2 vols. London: J. Murray.

Eccles, John. (1994). *How the Self Controls Its Brain.* Berlin: Springer-Verlag.

Greene, Joshua D.; R. Brian Sommerville; Leigh E. Nystrom, et al. (2001). "An fMRI Investigation of Emotional Engagement in Moral Judgment." *Science* 293: 2105–2108. Images of the brain making moral judgments.

Hobbes, Thomas. (1957). *The Leviathan,* ed. Michael Oakeshott. Oxford: Basil Blackwell.

Penfield, Wilder. (1978). *Mystery of the Mind.* Princeton, NJ: Princeton University Press.

Pinker, Steven. (2002). *The Blank Slate: The Modern Denial of Human Nature.* New York: Viking. A vigorous defense of a scientific theory of human nature against the major criticisms.

Plato. (1961). *The Collected Dialogues of Plato,* ed. Edith Hamilton and Huntington Cairns. Princeton, NJ: Princeton University Press.

Popper, Karl R., and John C. Eccles. (1983). *The Self and Its Brain: An Argument for Interactionism.* London: Routledge & Kegan Paul. A history of the mind-body problem in philosophy and neuroscience and an argument for the immaterial nature of the mind.

Richards, Janet Radcliffe. (2000). *Human Nature after Darwin.* London: Routledge. A study of the philosophic implications of Darwinism for human nature.

Singer, Peter. (2001). *Writings on an Ethical Life.* New York: Ecco.

Stevenson, Leslie Forster, ed. (1999). *The Study of Human Nature: A Reader,* 2nd edition. New York: Oxford University Press. A companion volume to the editor's *Ten Theories of Human Nature,* collecting texts from twenty-four different sources.

Stevenson, Leslie Forster, and David L. Haberman. (2004). *Ten Theories of Human Nature,* 4th edition. New York: Oxford University Press. Includes three ancient religious views (Confucian, Hindu, biblical), five philosophical views (Plato, Kant, Marx, Freud, and Sartre), and two scientific views (behavioral and evolutionary psychology).

Watson, James D. (2003). *DNA: The Secret of Life.* New York: Knopf. A history of how genetics illuminates human nature.

Weikart, Richard. (2004). *From Darwin to Hitler: Evolutionary Ethics, Eugenics, and Racism in Germany.* New York: Palgrave Macmillan.

Wilson, Edward O. (1998). *Consilience: The Unity of Knowledge.* New York: Knopf. A defense of scientific materialism.

Zimmer, Carl. (2004). *Soul Made Flesh: The Discovery of the Brain—and How It Changed the World.* New York: Free Press. A history of Thomas Willis's neurology.

HUMAN RIGHTS

• • •

At first glance human rights might seem to have little relevance for science, but this is not the case. Science is dependent on respect for human rights, particularly freedom of thought and freedom of speech. In many countries, however, the human rights and academic freedoms of scientists are violated by government or by groups that enjoy government support. Science can play an important role in helping to protect and promote human rights. In addition, international human rights law recognizes a substantive right to the freedom necessary for scientific research and a right to have access to the benefits of scientific progress. Yet in some circumstances scientists and health professionals have contributed to human rights violations.

Human Rights

What then are human rights? Rights in moral philosophy and political theory are understood as justified claims. A right is an entitlement of a person or group to some good, service, or liberty. As entitlements, rights differ from ideals, guidelines, or acts of charity. A right creates correlative obligations or duties to secure or not interfere with the enjoyment of that entitlement.

Human rights are a special class of rights, the rights one has by virtue of being a human being. Human rights are predicated on the recognition of the intrinsic value and worth of all human beings. As such, human rights are considered to be universal, vested equally in all persons regardless of their gender, race, nationality, economic status, or social position. Cumulatively human rights represent the minimum conditions for a decent society.

Contemporary twenty-first century conceptions of human rights were formulated at the end of World War II. The Universal Declaration of Human Rights (Universal Declaration), adopted without dissent by the U.N. General Assembly on December 10, 1948, represents an international consensus regarding the core rights and freedoms necessary to realize the inherent dignity and rights of all members of the human family. The Preamble to the Universal Declaration proclaimed "a common standard of achievement for all peoples and nations." As a declaration of the General Assembly, the Universal Declaration does not have direct legal force, but in the past fifty-six years it has become recognized as international common law. Moreover a series of international and regional human rights instruments based on the Universal Declaration are legally binding on countries that ratify them and thus become state parties bound by their provisions.

Many of the rights and standards set out in the Universal Declaration and other human rights instruments are essential to the conduct of science. Science is a worldwide enterprise requiring freedom of thought, communication, and travel, and the freedom to pursue professional activities without interference. The International Covenant on Civil and Political Rights, to which the United States is a state party, recognizes the following rights relevant for scientific inquiry:

- The right of everyone to freedom of thought (article 18);

- The right to hold opinions without interference and the right to freedom of expression, including the right to seek, receive, and impart information and ideas of all kinds, regardless of frontiers, through any media (article 19);

- The right to freedom of association with others (article 22);

- The right to liberty and security of person (article 9);

- The right to liberty of movement and freedom to choose one's residence and to be free to leave any country, including one's own (article 12);

- The right not to be subjected to medical or scientific experimentation without consent (article 6).

Provisions of other international human rights instruments also have important effects on the progress of science. The Universal Declaration includes the following additional rights that have counterpart provisions in the International Covenant on Economic, Social and Cultural Rights to which over 140 countries, but not the United States, are state parties:

- The right to education, including free and compulsory primary education, with technical and professional education generally available and higher education equally accessible to all on the basis of merit (article 26);

- The right to share in scientific advancement and its the benefits (article 27a);

- The right of everyone to the protection of the moral and material interests resulting from any scientific, literary, or artistic production of which he is the author (article 27b).

Protection of Scientists' Rights

Like other members of society, scientists are vested with basic human rights. Those rights are, however, violated in some countries. Scientists are persecuted for their work, for the expression of their opinions or beliefs, and for their peaceful efforts to oppose human rights violations or promote political change. The independent thinking and international connections of members of the scientific community can sometimes seem threatening to repressive or ideologically rigid governments. Scientific reverence for truth, reliance on empirically verifiable facts and measurable data, open dissemination and communication beyond national borders, and universality of discourse and goals by their very nature challenge some regimes.

This potential vulnerability has led scientists in some countries to form networks to protect the international human rights of members of the scientific community and scientific organizations. The Science and Human Rights Program of the American Association for the Advancement of Science (AAAS) is one such organization. Working with AAAS members and affiliated professional societies, the Program conducts casework on behalf of scientists, engineers, and health professionals whose human rights have been violated; prepares statements and reports; convenes meetings on human rights issues of special concern to scientists; organizes humanitarian and fact-finding missions; and assists other scientific organizations with cases and issues of special importance to the scientific community. The Program focuses its individual casework on three main areas: (a) violations of scientific freedom and the professional rights of scientists, engineers, health professionals, students in any of these fields, scientific organizations, and professional groups representing their interests; (b) violations of the human rights of scientists not directly related to the conduct of science, and (c) participation by scientists in practices that infringe on the human rights of others.

Initiated in 1993, the AAAS Human Rights Action Network uses electronic mail to inform AAAS members and other subscribers of cases and developments deserving special attention, and to coordinate the efforts of scientists to appeal to governments on behalf of their colleagues whose human rights are being violated. The network builds on the long-standing tradition of letter writing as an effective means of reminding governments that their transgressions have not gone unnoticed.

Science in the Service of Human Rights

Scientists have unique skills that can help promote and protect the human rights of all people. Scientific applications to human rights involve both utilizing scientific expertise and taking methodologies developed for other purposes and adapting them for human rights uses.

Human rights work requires accurate documentation of violations. Governmental authorities and the general public may be skeptical of reports of human rights violations. In some cases governments may deny that abuses have taken place in their country. Scientific methodologies can help establish the credibility of those who publicize violations and try to bring about change or institute legal action on behalf of victims. Scientifically based methods of data collection, storage, and analysis are particularly necessary when dealing with large volumes of data on human rights violations typical of truth commissions and tribunals. Adaptations of information management technologies for human rights have included specialized research and survey designs, interviewing techniques, database designs, controlled vocabulary structures for database processing, and analytic strategies for quantitative data analysis.

As human rights workers increase their use of electronic media for data storage and electronic communication, they become increasingly vulnerable to a variety of electronic attacks. Cryptographic applications enable human rights groups to secure their information against surveillance, ensure that their communications cannot be faked, and even hide their communications in digital images or sound files.

In the early-twenty-first century, extrajudicial executions and *disappearances* continue to occur in perhaps fifty countries. Independent forensic investigations can be crucial in determining the cause and manner of suspicious death and in proving whether a victim was tortured. Often the judiciary in these countries is reluctant to investigate killings perpetrated by the army or police or other regular security forces, special units outside of the normal chain of military command, paramilitary units, *death squads* sanctioned by the government, or armed groups opposed to the government. To respond to these blatant violations of human rights, forensic pathologists have investigated individual incidents of suspicious deaths by conducting initial and second autopsies, observing official inquests into deaths in detention, and assisting court-ordered investigations of suspicious deaths. In addition, teams of forensic anthropologists exhume mass graves to document murders of groups and communities.

Rights to Scientific Freedom and Access to the Benefits of Science

International human rights law recognizes a substantive right to the freedom necessary for scientific research and a right to have access to the benefits of scientific progress. Building on a parallel provision of the Universal

Declaration, Article 15 of the International Covenant on Economic, Social and Cultural Rights (ICESCR) specifies that state parties "undertake to respect the freedom indispensable for scientific research and creative activity" (Article 15[3]). This article also instructs states parties to "recognize the right of everyone" both "to enjoy the benefits of scientific progress and its applications" (Article 15[1][b]) and "to benefit from the protection of the moral and material interests resulting from any scientific, literary or artistic production of which he is the author" (Article 15[1][c]). To achieve these goals, the text mandates that states parties undertake a series of steps, including "those necessary for the conservation, the development and the diffusion of science and culture" (Article 15[1][c]). More specifically, states parties make the commitment to "recognize the benefits to be derived from the encouragement and development of international contacts and cooperation in the scientific and cultural fields" Article 15[4].

A government can best show respect for the freedom indispensable for scientific research and creative activity by adhering to basic human rights norms recognized in the Universal Declaration and the International Covenant on Civil and Political Rights. In addition, the pursuit of science requires an environment that supports the freedom to pursue scientific research in accordance with ethical and professional standards without undue interference. Conversely the freedom to undertake scientific research and creative activity implies a need for scientific responsibility and self-regulation. Scientific societies in many developed countries have adopted codes of professional ethics in pursuit of these goals. Many of these codes, however, are primarily concerned with the ethics of individual conduct and do not place the scientific enterprise in a broad social context.

Protection against Human Rights Abuses

Much has been written about the challenges posed by science and technology to human rights and human dignity. In the years since the publication of Jacques Ellul's pioneering work *The Technological Society* (1964), for instance, an increasing number of thinkers have called attention to the potential of technology to diminish human dignity and to erode moral values. While the vast majority of health professionals and scientists have sought to be faithful to ethical values, some have been tempted or forced to facilitate harmful practices. Health professionals have been implicated in torture and other cruel and degrading treatment (Amnesty International French Medical Commission and Marange 1989).

Psychiatric institutions have been misused to incarcerate political dissidents. Scientists have developed chemical and biological weapons for regimes that intended to use them on their own populations.

The Universal Declaration on the Human Genome and Human Rights, prepared by UNESCO and then adopted by the U.N. General Assembly in 1999 is an example of an initiative that addresses the potential impact of a new technology on human rights and dignity. It emphasizes that genetic research and applications should fully respect human dignity, freedom, and rights and prohibits all forms of discrimination based on genetic characteristics. The declaration affirms the principle of freedom or research related to the genome (Article 12b), but with the caveat that researchers respect principles of caution, intellectual honesty, and integrity (Article 13). The document assigns responsibility to states to take appropriate measures to foster the intellectual and material conditions that promote freedom in the conduct of research on the human genome and safeguard respect for human rights in the process (Articles 14–16). The declaration further recommends that benefits from advances in biology, genetics and medicine, concerning the genome, should be made available to all (Article 12a).

Human cloning constitutes another issue. In October 2003, the U.N. General Assembly considered a treaty to ban human cloning. Delegates agreed that the treaty should prohibit reproductive cloning, the creation of cloned embryos to produce babies, but they deadlocked on the issue of whether the prohibition should extend to "therapeutic" or "research" cloning. Nor could they agree on going forward with a treaty that only addressed reproductive cloning. Confronted with this disagreement, the General Assembly voted to delay discussion of the treaty until its 2005 session (Aschwanden 2003).

AUDREY R. CHAPMAN

SEE ALSO *American Association for the Advancement of Science; Dignity; Globalism and Globalization; Holocaust; Human Nature; Humanization and Dehumanization; Libertarianism; Natural Law; Rawls, John; Rights Theory; Science, Technology, and Law.*

BIBLIOGRAPHY

Amnesty International French Medical Commission, and Valerie Marange. (1989). *Doctors and Torture: Resistance or Collaboration?* London: Bellew Publishing.

Claude, Richard Pierre. (2002). *Science in the Service of Human Rights*. Philadelphia: University of Pennsylvania Press.

Ellul, Jacques. (1964). *The Technological Society*, trans. Johns Wikinson. New York: Vintage Books.

Lauren, Paul Gordon. (1998). *The Evolution of International Human Rights*. Philadelphia: University of Pennsylvania Press.

Morsink, Hohannes. (2000). *The Universal Declaration of Human Rights: Origins, Drafting, and Intent*. Philadelphia: University of Pennsylvania Press.

United Nations General Assembly. "International Covenant on Civil and Political Rights." G.A. Res. 2200 (XXI), 21 U.N. GAOR, Supp. No. 16, at 52, U.N. Doc. A/6316 (1966). Adopted December 16, 1966, entered into force March 23, 1976.

United Nations General Assembly. "International Covenant on Economic, Social and Cultural Rights." G.A. Res. 2200 (XXI), 21 U.N. GAOR Supp. No. 16, at 49, U.N. Doc. A/6316 (1966). Adopted December 16, 1966, entered into force January 3, 1976.

United Nations General Assembly. "Universal Declaration of Human Rights." G.A. Res. 217A (III), U.N. GAOR, 3rd Sess., U.N. Doc. A/810 (1948).

INTERNET RESOURCES

AAAS Science and Human Rights Program. Human Rights Action Network. Available from http://shr.aaas.org/aaashran. Membership in the network is free; subscriptions are available through the website. Approximately one individual case or issue is circulated each week, with all the information necessary to take action, in a succinct bulletin.

Aschwanden, Christie. (2003). "Nations Fail to Agree on Extent of Human Cloning Ban." *Bulletin Of The World Health Organization*, Vol. 81. no. 11. Available at http://www.who.int/bulletin/volumes/81/11/en/news1103.pdf

UNESCO. "Universal Declaration on the Human Genome and Human Rights." Available from http://portal.unesco.org/en/ev.php@URL_ID=13177&URL_DO=DO_TOPIC&URL_SECTION=201.html.

HUMAN SUBJECTS RESEARCH

• • •

In the field of ethical issues in scientific research, the two most controversial topics concern involve the use of humans as research subjects and the use of non-human animals as research subjects. Each of those debates goes back over a hundred years, to the final decades of the nineteenth century, and thus has a substantial literature that has developed a sophisticated level of discussion. This article will briefly summarize the history of the field first, and then explain some of the regulations that have resulted, and close with identifying some of the most important future issues.

Historical Developments

By 1900 there was ample evidence of an appreciation in the medical and scientific communities of the ethical issues that would have to be resolved before a person was used as a subject in experiments. In Prussia a ministerial directive issued in 1900 restricted research to the use of persons who could benefit from the research, who were told in advance of the risks of participation, and who gave their consent. This was in response to well-known experiments with the leprosy bacillus on unwitting subjects in Prussia around that time.

At around the same time in Cuba, United States General Walter Reed (1851–1902) conducted yellow fever studies but required that both soldiers and civilians volunteer first, be informed of the risks (including the risk of death), and sign a consent form. The form was written in both English and Spanish. This is said to have been the first use of a signed consent form and also could be considered the first example of ethical international research informed by cultural competence. Reed's caution was a response to an experiment in Italy in which five persons were infected with yellow fever without being told and an initial experiment in Cuba by two colleagues of Reed who intentionally infected themselves that led to the death of one of them.

In light of the degree of awareness shown at the beginning of the century, it is surprising that by midcentury some of the most barbaric things ever done in the name of science would come to pass. A combination of factors contributed to that decline in standards, including racism and anti-Semitism, exacerbated by nationalism and xenophobia; those problematic social elements were long established but were pushed to extremes by World War II.

Three examples of well-known and frequently cited unethical research involving human subjects occurred in the middle third of the twentieth century. The Tuskegee experiments, observing the consequences of untreated syphilis in American blacks, began in 1932, when there was no effective treatment, but continued until 1972, long after the discovery of penicillin. The research done by Nazi doctors was by far the most brutal and murderous. Those experiments included testing the limits of human endurance up to and including death from causes such as bullet and knife wounds; decompression at high altitudes, which was tested by putting people in decompression chambers and measuring when their lungs burst; and hypothermia, which was tested by keeping subjects immersed in ice water. Japanese experiments in the notorious unit 731 were just as grievous as the Nazi experiments, though less well known. The thalidomide tragedy revealed the importance of the oversight of drug trials and the recognition of the problems of self-policing by pharmaceutical companies that have a financial investment at stake. That experience propelled the U.S. congressional hearings known as the Kefauver hearings.

Ethically disturbing human experiments were done well after that period. Two examples in the United States were performed on institutionalized populations: testing gamma globulin treatment of hepatitis after infecting children at the Willowbrook State School in Willowbrook, New York, and tracing differences of rejection of live cancer cells in subjects after injecting those cells into people at the Jewish Chronic Disease Hospital in Brooklyn, New York without explaining what was in the injections. These were among twenty-two experiments described by Henry K. Beecher in an influential paper published in the *New England Journal of Medicine* in 1966, "Ethics and Clinical Research."

There are many ironies in this history. For example, the most brutal and murderous research was done in Germany, the country that had promulgated the first modern code for ethical research. Then the country that provided all the judges and all the lawyers at the Nuremberg Trial of Physicians (1946) that led to the Nuremberg Code (1947) acted as if the code did not apply to its citizens in the years after World War II. This history of the field seems to show that some of the lessons need to be learned and relearned periodically and that only revelations of scandals and abuses have the power to restrain research.

Regulations

The last third of the twentieth century saw the codification of many of the lessons that had been learned and left a number of areas of great import that are still very much disputed. Several of those lessons have been accepted widely and codified into U.S. and international law.

In 1964 the original Declaration of Helsinki was passed by the World Medical Association. It reiterated the famous first line of the Nuremberg Code, stating that the voluntary consent of the human subject is absolutely essential, though it still left it up to the researcher to decide what to say, how much to disclose, and how to

document the informed consent process. It has been revised and strengthened a number of times, most recently in 2000. The most important difference from U.S. regulations involves placebo controls, which generally are encouraged in the United States (especially by the Food and Drug Administration) and discouraged (though not forbidden) in the Declaration of Helsinki.

As a result of the public reaction to the Tuskegee experiments in 1974, the U.S. Congress authorized the National Commission for the Protection of Human Subjects of Biomedical and Behavioral Research. The National Commission resulted in the publication of the Belmont Report (1979) and the issuance of federal regulations in 1981 known as 45 CFR 46. Those regulations led to the requirement of prior review of research protocols by independent committees known as Institutional Review Boards (IRBs). This was modeled on prior peer review, which had been required at the National Institutes of Health (NIH) since 1965 and for all NIH-sponsored research since 1966. The basic protections of the regulations (outlined in subpart A) were consolidated into "the Common Rule" in 1991 and adopted by sixteen federal agencies.

IRB oversight, in contrast to peer review, required that there be at least one nonscientist, one community member, and should not be either all men or all women. Although many people still have concerns about the real independence one can expect in light of the fact that most of the members of the committees are usually employees at the same institution where the research is being done, it was an important innovation.

Before approving a proposed research protocol, the IRB must ascertain that the research is scientifically valid (the goals are worthwhile and achievable by the methods proposed) and that the risks to the subjects are kept to a minimum and are justified by the potential benefits to be gained. It also must determine that the selection of subjects has been equitable (no groups are excluded without good reason) and that the subjects have been recruited without any deception or coercion, that the confidentiality of the subjects has been adequately protected, that the subjects have been fully informed about the risks and have given voluntary consent that has been documented, that proper steps have been taken to ensure that the subjects understand all the information they have been given, and that they understand that they can withdraw from the research at any time. The IRB is also responsible for monitoring the research and has the power to stop any study that is dangerous to the participants, a task often assigned to a separate Data and Safety Monitoring Board (DSMB).

An IRB has the responsibility to ensure the voluntary participation of the research subjects as well as their safety. Thus, IRBs often focus on the informed consent form that will be given to potential subjects to ensure that the risks are portrayed realistically and not underplayed and that there is no misleading of the subjects about the likelihood of benefit. Terms such as *the doctor*, *medicine*, and *therapy* can be used by researchers without any intent to deceive yet can be read by subjects as meaning that they are enrolled in an experiment whose purpose is to help them rather than to improve the understanding of a drug or disease process. This is referred to as the therapeutic misconception. The same concern for language has made some IRBs to suggest using the term "participants" instead of "subjects" as a reminder to the researcher that she is seeking the cooperation of well informed volunteers, not passive recruits who don't ask questions. The regulations also require that extra attention be paid before any members of certain groups of persons known as vulnerable populations are used. These groups include children, the mentally handicapped or mentally ill, prisoners, pregnant women, and fetuses.

Ironically, since the 1990s there has been recognition in the United States by the Food and Drug Administration that drugs have been tested disproportionately on white men too frequently and that it would be scientifically helpful to have more studies with women, minorities, and children to test for variations in effectiveness and safety. However, the history of abuse probably has made researchers hesitant to enroll persons in these categories, not to mention the distrust that members of these groups might feel after the historical record at Tuskegee, Willowbrook, and the Jewish Chronic Disease Hospital.

All government funded research with human subjects is required to be reviewed by an IRB. This includes the behavioral and social sciences as well as biomedical sciences. Many of the same ethical issues arise, though the potential harms may be of a psychological nature, such as risk to privacy or to self-image, rather than a physical one. A concern that may occur with greater frequency in psychology is that fully informing a subject of the nature of the research could bias the answers the subject gives. Thus researchers will seek to reveal less of the purpose of the study than would be the case in medical research. This type of purposeful deception will have to be justified to the IRB, and assurances that any risks to the subjects are minimal. Assessing this kind of risk is difficult, as seen in the fact that the highly innovative and influential milgram experiments conducted in the

1960s are deemed controversial by some commentators to this day. The primary harm to the subjects was a loss of self-esteem as they reflected on their own willingness to submit to the orders of an authority figure and inflict pain on strangers. But it would not have been possible to do the experiment had the consent process told them in advance that the strangers in apparent pain were only actors. An honest debriefing, with counseling if necessary, may help to alleviate possible harms in cases where some initial deception cannot be avoided.

This also brings up the question of non-government funded research. Much pharmaceutical research and research on medical devices is funded by the FDA, and so falls under the common rule. But beyond government funding sources, there is currently no review needed in the U.S. for privately funded research. Should private enterprise, from marketing research to genetics and biotechnology, be unencumbered by regulations whose intent is to ensure the safety of citizens? Should civil rights and human rights be allowed to set restrictions on private companies in cases where there is, as yet, little risk identified? When one pictures marketing questionnaires, it is easy to be swayed towards a libertarian distrust of unnecessary and intrusive government regulations. But when one considers the potential profits from genetics and biotechnology research, there may be more reason to consider preemptive regulation, such as already exists with state commissions in many European Union countries concerning IVF.

Future Issues

Soon after the Belmont Report the Council for the International Organization of Medical Sciences (CIOMS) produced a report on the special issues that occur in international research. The beginning years of the twenty-first century have seen growth in funding for international research. Although some of this increase in funding could be due to economic globalization and the lessening of national identity for multinational corporations, there may be more ominous motivations. For example, funding sources for pharmaceutical research are often in first world countries such as the United States, the United Kingdom, France, Germany, Belgium, and Switzerland. However, when an even larger proportion of research in is done developing nations, it could be because of lax regulations (including ethical regulations) in the developing world.

A second topic that inevitably will grow in importance is the range of new research resulting from the Human Genome Project. That project was completed in less time than originally planned and has provided an enormous amount of raw data with which biologists hope to map a deeper understanding of normal development and pathogenesis. However, all genetic information has ethically complex properties, such as providing information about the relatives of research subjects as well as about the persons who volunteered to be involved in the research.

Another challenging ethical issue unique to genetics is the possibility of curing a disease by means of germline gene therapy, removing the disease from human history but at the risk of altering the human genome. Similarly, genetic interventions have the potential to blur the intuitive distinction between medical treatment for an illness or dysfunction and enhancement of traits which a person may find unsatisfactory yet fall within the normal range of human beings. Either way we are on the cusp of gaining the knowledge of the human genome that would allow genetic engineering with the purpose of improving the race (using Nazi terminology, creating a new master race). Might we soon enter a phase of deliberate evolution, or worse, develop into two sub-species, the feral and the enhanced?

The third topic of concern is stem cell research and the related issue of human cloning. Advances in invitro fertilization (IVF) and other assisted reproduction technologies (ARTs) have made the possibility of human cloning real. Many species of mammals already have been cloned, and it may be only a matter of time before a human is cloned. Although some people have argued that this should be considered an alternative technique for infertile couples to have a child, it has been outlawed in many countries as threatening the dignity inherent in the uniqueness of each life.

Stem cell research, which would find its best source of human embryonic stem cells in the excess embryos created by IVF programs, also has been opposed by critics who believe it violates the respect owed to human embryos or treats them as means rather than ends. However, attempts at broad bans have been less successful than with cloning for a number of reasons: The therapeutic potential could benefit many more people, and the majority of scholars and researchers in both ethics and developmental biology believe that there is a fundamental moral difference between a preimplantation embryo and an embryo or fetus that has been implanted successfully in a human womb.

Beyond issues related to transnational experimentation, genetics, and stem cells research, one might suggest that as the scientific and technological enterprise advances, all people become the subjects of scientific research. Mike Martin and Roland Schinzinger (1996) have argued for understanding engineering as a form of social experimentation. But even more broadly, the increasing use of medicines that often create therapeutic dependencies, unregulated uses of IVF and frozen embryos, and the popularization of plastic surgeries and advanced prosthetics all point toward people treating themselves (not just scientists treating people) as human subjects in scientifically based actions the full outcomes of which remain uncertain.

JEFFREY P. SPIKE

SEE ALSO Fetal Research; Nazi Medicine; Responsible Conduct of Research; Scientific Integrity; Tuskegee Experiment.

BIBLIOGRAPHY

Altmann, Lawrence K. (1998). Who Goes First? Berkeley: University of California Press. A book that's fun to read, on an approach to choosing research subjects that was once quite honorable but is now strongly discouraged: try it on yourself first.

Dunn, Cynthia McGuire, and Gary L. Chadwick. (2004). Protecting Study Volunteers in Research: A Manual for Investigative Sites, 3rd edition. Boston: Thompson-Centerwatch. A very complete users manual for IRB coordinators and members.

Emanuel, Ezekiel E.; Robert A. Crouch; John D. Arras; Jonathan D. Moreno, and Christine Grady. (2003). Ethical and Regulatory Aspects of Clinical Research: Readings and Commentary. Baltimore: Johns Hopkins University Press. An excellent collection of primary sources; includes the Belmont report, Helsinki, CIOMS, and the Henry K. Beecher article mentioned in the text.

Martin, Mike W., and Roland Schinzinger. (1996). Ethics in Engineering, 3rd edition. New York: McGraw-Hill.

Milgram, Stanley. (1974). Obediance to Authority: An Experimental View. New York: Harper and Row.

Shamoo, Adil E., and David B. Resnick. (2003). Responsible Conduct of Research. New York: Oxford University Press. A useful textbook for a graduate level course in research ethics; includes many topics besides human subjects.

Zoloth, Laurie, Jane Maienschein, and Ronald M. Green, "Ethics of Stem Cell Research: A Target Article and Open Peer Commentaries." American Journal of Bioethics 2(1): 1–59. Three introductory target articles are followed by 19 diverse short commentaries.

HUME, DAVID

• • •

David Hume (1711–1776) is one of the most influential philosophers of the modern period. He was born in Edinburgh, Scotland, on April 26. His first and most

David Hume, 1711–1776. The Scottish philosopher developed a philosophy of "mitigated skepticism," which remains a viable alternative to the systems of rationalism, empiricism, and idealism. (© Corbis.)

important work, *A Treatise of Human Nature* (published in two installments in 1739 and 1740, before Hume turned thirty years old), was supplemented in later life by *Essays, Moral and Political* (two volumes, 1741–1742), *An Enquiry Concerning Human Understanding* (1748), and *An Enquiry Concerning the Principles of Morals* (1751). The latter two books restate in more accessible form the arguments of the *Treatise*. He also wrote a six-volume history of England (1754–1762) and *Dialogues Concerning Natural Religion*, published posthumously in 1779. Hume, who died in Edinburgh on August 26, applied what he considered the experimental method of science to an examination of ideas and morals, thereby developing an ethics that bases moral judgments on feelings. Because emotivism is so frequently assumed in the contemporary West, to read Hume can be an exercise in cultural self-understanding.

Empiricism

Hume begin his *Treatise* arguing that human knowledge is limited to sense-experience. The contents of sense-experience can be distinguished into impressions and ideas. Impressions, which include all sensations and passions, are more forceful and lively than ideas, which are "the faint images of these in thinking and reasoning" (Hume 1888 [1739–1740], p. 1). Ideas are thus epistemologically inferior to impressions, and the secondary status that Hume gives them remains characteristic of popular denigrations of their relative impotence. This distinction also suggests that the logical analysis of conceptual relations is less important than the knowledge of matters of fact.

Hume further distinguishes between the simple and complex. Simple impressions and ideas, such as the seeing or imagining of a particular shade of red, admit of neither distinction nor separation. Complex impressions and ideas, such as the seeing or imagining of an apple, can be analyzed into their component parts. Whereas all simple ideas are derived from and exactly represent simple impressions, many complex ideas are not, and so their veracity must be called into question. In *Enquiry Concerning Human Understanding*, Hume remarks, "When we entertain, therefore, any suspicion that a philosophical term is employed without any meaning or idea (as is but too frequent) we need but enquire, from what impression is that supposed idea derived? And if it be impossible to assign any, this will serve to confirm our suspicion" (Hume 1894 [1748], p. 22). Something like this view is often employed when people appeal to science in rejecting ideas of God or the soul.

But the most famous subject of Hume's criticism is the relation of cause and effect. Philosophers and scientists traditionally believed that to know something fully requires knowledge of the cause on which it depends. For Hume, such knowledge is impossible. Although the causal relationship provides the basis for all reasonings concerning matters of fact, all such reasoning is quite contingent. This is because one can always imagine, without contradiction, the contrary of every matter of fact (e.g., "the sun will not rise tomorrow" neither is nor implies a contradiction). For Hume, the causal relationship between any two objects is based strictly on experience, and all that experience establishes concerning causal relationships is that the cause is prior in time and contiguous to its effect. Experience cannot establish a necessary connection between cause and effect, because one can imagine without contradiction a case in which the cause does not produce its usual effect (e.g., one can imagine that a cue ball violently strikes another billiard ball and then, instead of causing the billiard ball to move, the cue ball bounces off it in some random direction). The reason why a person might mistakenly infer that there is something in the cause that necessarily produces its effect is because past experiences have habituated the person to think in this way (see *Treatise*, Book

I, Part III; first *Enquiry*, secs. IV–V). In thus arguing that humans have no direct impression of anything more than spatial and temporal contiguity, Hume sees himself extending empirical science. At the same time, he reduces science's epistemic power by depriving it of any deep knowledge about what lies beyond experience.

Theory of Morals

Hume's argument with regard to morals is similar. For Hume, moral distinctions are derived from feelings of pleasure and pain of a special sort, and not—as held by many Western philosophers since Socrates—from reason. Working from the empiricist principle that the mind is essentially passive, Hume argues that reason by itself can never prevent or produce any action or affection. Because morals concerns actions and affections, it cannot be based on reason.

Reason can influence human conduct in only two ways. First, reason can inform a person of the existence of something that is the proper object of a passion, and thereby excite it. Second, reason can deliberate about means to an end that a person already desires. But should reason be in error in either of these areas (for instance., by mistaking an unpleasant object for one that is pleasant, or by mistakenly selecting the wrong means to a desired end), it is not a moral but an intellectual failing. As a final point, Hume argues for a distinction between facts and values. According to Hume, one cannot infer conclusions about what *ought to* or *ought not* be the case based on premises of what *is* or *is not* (see *Treatise*, Book III, Part I, sec. 1).

Because moral distinctions are not based on reason, Hume infers that they are based on sentiments that are felt by what he calls a "moral sense." When a person describes an action, sentiment, or character as virtuous or vicious, it is because its view causes a pleasure or pain of a particular kind. Hume is well aware that not all pleasures and pains lead to moral judgments (for example, the pleasure of drinking good wine). Rather, it is "only when a character is considered in general, without reference to our particular interest, that it causes such a feeling or sentiment, as denominates it morally good or evil" (Hume 1888 [1739–1740], p. 472). Finally, Hume argues that even though moral distinctions are based on feelings, this does not lead to moral relativism. This is because the general moral principles and the moral sense faculty that recognizes them are common to all human beings.

Influence

As indicated, Hume's view that the source of moral approval and disapproval is not reason but the senti-

ments that are felt has been widely influential. In the twentieth century this view was restated as the emotive theory of ethics. According to A. J. Ayer's *Language, Truth, and Logic* (1936), once statements of the form "X is wrong" are distilled of their factual components, they merely evince the speaker's moral disapproval, for example, "Boo X!"

In contemporary times, such a view is often deployed against anyone who attempts to make ethical criticisms of science or technology, with the claim that critics are simply stating their own personal preferences. Abandoning Hume's belief in a moral sense faculty common to all humans as itself unjustified by empirical science, it is argued that in a pluralistic society, with many different sentiments and preferences, scientists and engineers should be at liberty to research or invent as they see fit—with perhaps the sole proviso that they do not materially harm other persons. Whether or to what extent this is an adequate ethics for science and technology is a question that Hume's philosophy obliges us to ponder.

SHANE DREFCINSKI

SEE ALSO *Enlightentment Social Theory; Human Nature; Locke, John; Risk and Emotion; Scientific Ethics.*

BIBLIOGRAPHY

Ayer, A. J. (1936). *Language, Truth, and Logic.* London: Gollancz.

Ayer, A. J. (1980). *Hume.* Oxford: Oxford University Press.

Hendel, Charles William. (1925). *Studies in the Philosophy of David Hume.* Princeton, NJ: Princeton University Press.

Hume, David. (1741–1742). *Essays, Moral and Political,* 2 vols. Edinburgh.

Hume, David. (1754–1762). *History of England,* 6 vols. London.

Hume, David. (1777). *The Life of David Hume, Esq., Written by Himself,* ed. Adam Smith. London. Included in the 1955 Macmillan edition of the *Enquiry Concerning Human Understanding.*

Hume, David. (1888 [1739–1740]). *A Treatise of Human Nature,* 3 vols., ed. L. A. Selby-Bigge. Oxford: Oxford University Press.

Hume, David. (1894 [1748 and 1751]). *Enquiries Concerning Human Understanding and the Principles of Morals,* ed. L. A. Selby-Bigge. Oxford: Oxford University Press.

Hume, David. (1935 [1779]). *Dialogues Concerning Natural Religion,* ed. Norman Kemp Smith. Oxford: Clarendon Press.

Smith, Norman Kemp. (1941). *The Philosophy of David Hume.* London: Macmillan.

HUSSERL, EDMUND

• • •

Born in Prossnitz, Moravia (now Prostêjov, Czech Republic) on April 8, Edmund Husserl (1859–1938) inaugurated the phenomenological movement in philosophy. Trained as a mathematician at Vienna, where he received his Ph.D. in 1883, Husserl began studying philosophy in 1884 under Franz Brentano (1838–1917) and went on to teach in the philosophy faculties at Halle an der Saale, Göttingen, and Freiburg. His most notable works—*Logical Investigations, Ideas* (Volumes I, II, and III), *Cartesian Meditations,* and *The Crisis of European Sciences and Transcendental Phenomenology*—seek a philosophical grounding for mathematics, logic, and science by analyzing the intentional or essential structures of consciousness in its relation to objects in the world relations between subjectivity and objectivity. After his death on April 26 in Freiburg, a substantial body of posthumously published work extended his account of subjectivity and its correlative world into the domain of intersubjective experience, and the development of an ethical system that exhorts a fully rational human existence in which all persons repeatedly justify their beliefs and actions.

The fundamental method of phenomenology is the "reduction," which entails suspending the philosopher's own participation in our natural beliefs about the world. Not a denial of the external world, the reduction simply neutralizes dogmatic assumptions about experience in order to examine more closely experience and its objects just as they are given; hence, phenomenology calls itself a "presuppositionless" enterprise.

Husserl's most overtly relevant work for science, technology, and ethics, *The Crisis* (1936), argues that science and technology constitute a nonneutral transformation of life rather than a simple neutral extension of ahistorical human concerns. Neither pro– nor anti–science and technology, Husserl's *Crisis* suspends the typically modern commitment to science in order to disclose and examine the repercussions of those unreflectively accepted scientific presuppositions and practices that transform the prescientific life-world of human experience. Husserl values the way science tests and retests experience, thereby contributing to a fuller sense of objectivity than everyday judging. In their great success, however, science and technology create "fact-minded" citizens blinded by promises of objectivity and control. In their narrow view of reason as mere calculation, science and technology consider themselves value neutral and thus exempt from responsibly advising about how to make difficult decisions arising from the means

Edmund Husserl, 1859–1938. The German philosopher is considered the father of phenomenology, one of the most important trends in 20th-century philosophy. (*The Granger Collection, New York.*)

they produce. Moreover, one could argue, science and technology evolve in rarefied discourses unavailable to most citizens and beyond democratic control. Followers of Husserl thus are able to argue that humankind's historical circumstance marks a crisis in which science and technology develop independently of value questions and democratic voice, yet are unreflectively and passively received and deployed.

To philosophers of technology, however, Husserl's corrective measure in the form of a relentless search by the subject for a fuller sense of evidence to justify beliefs and actions often appears to be a formal, abstract quest for ideal essences. Ethical discussions of science and technology thus often disregard Husserl's phenomenology. Husserl's protégé, Martin Heidegger (1889–1976), for example, believes Husserl's emphasis on cognition lands him squarely in the path of human technological domination of the world. The phenomenological reduction, Heidegger argues, "reduces" the world to human "intentional" activities and sacrifices world independence to consciousness's drive to explain and predict experience with absolute certainty. American pragmatist philosopher Larry A. Hickman (2001) argues that privileging conscious reflection and increased objectivity over lived experience renders phenomenological inquiry a private enterprise tied to "ideal essences." Unable to reconfigure its ideals, Hickman finds phenomenology incapable of a providing a viable program for the reform of technology. And the American post-phenomenologist Don Ihde (1990) reiterates Heideggerian and pragmatist criticisms. Because Husserl neglected the inseparability of sense-extending technologies from scientific discovery, Ihde argues he never reached beyond an intimation of a philosophy of technology.

Yet Husserl's contribution to the philosophy of technology can be found in these criticized notions intentionality and objectivity, which form the basis of his ethics of a self-conscious community founded on intuitionally fulfilled beliefs and actions, and provide the basis for a critical assessment of technology. For Husserl, consciousness, in its very nature as activity, is intentional. In its care for and interest in the world, consciousness transcends itself. Always outside of itself, a subject experiences the world in a public and inter-subjective rather than private and solipsistic way. Intuitional fulfillment denotes the correlation of a subject's intentional anticipations with the evidence found in experience. When experience does not confirm a subject's anticipation, the intention goes "unfulfilled" and demands that the subject revise prior beliefs, thus achieving a degree of objectivity. When experience confirms a subject's anticipation, the intention gets "fulfilled," again achieving a degree of objectivity. Because Husserl advocates self-critique and reflection as a lifelong task, even fulfilled intentions require further experiential confirmation over time and across subjects. Rather than a fixed ideal, objectivity remains open to reconfiguration according to experiential evidence given in the fluxing relation between subject and world.

An interesting instance of the kind of self-critical agency that Husserl advocates can be found in the life of the Polish scientist Joseph Rotblat (b. 1908), who worked on the atomic bomb. Rotblat initially justified his participation by reasoning that only Allied bomb development would counter potential German development. After the German defeat, Rotblat reflected on the standard attitude of the scientists working on the project—many of whom believed it was not their job to advise about how the atomic bomb should be used—leading him to leave the project before the first testing and use of the bomb. Rotblat resolved to henceforth carefully choose each of his future projects, accepting only assignments he judged of definite bene-

fit for humanity. Rotblat's revised outlook on his career as a scientist follows in the spirit of Husserl's ethics based on a subject's vow to live a life guided by a repeated and critical evaluation of beliefs. Rotblat's decision to withstand the heedless activity that Husserl believes characterizes the contemporary relation to science and technology exemplifies the self-reflection and self-responsibility for which Husserl argues when he exhorts subjects to continuously assess their experiences.

MICHAEL R. KELLY

SEE ALSO Axiology; Existentialism; German Perspectives; Heidegger, Martin; Leibniz, G. W.; Phenomenology.

BIBLIOGRAPHY

Hickman, Larry A. (2001). *Philosophical Tools for Technological Culture*. Bloomington: Indiana University Press. Develops John Dewey's pragmatist philosophy of technology, and criticizes phenomenology along the way.

Husserl, Edmund. (1973). *The Crisis of European Sciences and Transcendental Phenomenology: An Introduction to Phenomenological Philosophy*, trans. David Carr. Evanston, IL: Northwestern University Press. Discusses the value of reason and the life-task of seeking intuitional fulfillment and increased objectivity beyond its mere facts and calculation.

Husserl, Edmund. (1982). *Ideas Pertaining to a Pure Phenomenology and to a Phenomenological Philosophy*, Book 1: General Introduction to a Pure Phenomenology, trans. Fred Kersten. The Hague, Netherlands: Martinus Nijhoff. Extended, detailed discussion of the phenomenological "reduction."

Ihde, Don. (1990). *Technology and the Lifeworld: From Garden to Earth*. Bloomington: Indiana University Press. A clearly written overview and analysis of relevant phenomenological literature.

Melle, Ullrich. (1991). "The Development of Husserl's Ethics." *Études Phénoménologiques* 13–14: 115–135. Overview of the chronological development of Husserl's ethical thought.

Melle, Ullrich. (1998). "Responsibility and the Crisis of Technological Civilization: A Husserlian Meditation on Hans Jonas." *Human Studies* 21(4): 329–345. In place of Jonas's heuristic of fear, Melle discusses Husserl's notions of self-critique, objectivity, evidence, and intuitional fulfillment as methods for ensuring responsible engagement with technology.

Rotblat, Joseph. (1986). "Leaving the Bomb Project." In *Assessing the Nuclear Age: Selections from the Bulletin of Atomic Scientists*, ed. Len Ackland and Steven McGuire. Chicago: University of Chicago Press.

Aldous Huxley, 1894–1963. The novels, short stories, and essays of the English author Huxley explore crucial questions of science, religion, and philosophy.

HUXLEY, ALDOUS

• • •

Aldous Leonard Huxley (1894–1963) was a British writer best known for his 1932 novel *Brave New World*, which portrays the dehumanizing aspects of scientific and technological progress. Born in Godalming, Surrey, England on July 26, Huxley's poor eyesight kept him from an early goal of becoming a scientist. After attending Oxford University and achieving fame as the author of several novels, in 1937 Huxley moved to California, where he became a screenwriter. Later he experimented with psychedelic drugs and incorporated mysticism into his work. Huxley died from throat cancer in Hollywood on November 22.

A moralist, social satirist, and interdisciplinary intellectual, in *The Perennial Philosophy* (1942) Huxley sought to identify the origin of being, prior to the fragmentation of experience into diverse languages, religions, and systems of knowledge. In the present age, however, he realized that reconnecting with such a foundation would involve reconciling humanity to the social and spiritual consequences brought about by science and technology (Murray 2002). To this end, Huxley often used literature

to advance the causes of social sanity and personal enlightenment. Three themes are central to this life-long project: relations between literature and science; science, technology, and the abuse of power in emerging mass societies; and the potential for science and technology to enrich or corrode human nature.

Science and Literature

Huxley believed it was crucial to connect science and literature. Indeed, his novel *Point Counter Point* (1928) has been described as an "application of the theory of relativity to the art of fiction" (Deery 1996, p. 31). But he also felt it mistaken to define literary theory as a progressive, systematic, and verifiable body of knowledge employing the scientific method.

Huxley sought to reclaim a unified human experience by achieving the proper balance between different forms of knowledge. In this respect, he was the intellectual descendant of the debates about science and humanism held between his grandfather, Thomas Henry Huxley (1825–1895), and his great uncle, Matthew Arnold (1822–1888). The issue was particularly impelling because the secularization of society placed a great burden on literature to uphold the humanist tradition just when scientific discoveries were undermining traditional understandings of the world and the human place within it.

Huxley also saw literature as a vehicle for critiquing the social and moral consequences of scientific progress. The seriousness with which Huxley took up this task distinguished him from contemporaries, most of whom distanced themselves from social criticism. He held that "one of the prime duties of the twentieth-century artist is to draw attention to the evil ends for which a morally neutral science is being used" (Deery 1996, p. 25).

Science, Technology, and Power

Huxley believed that the "most profoundly important sociological factor of modern times [is] the growth of technology and what may be called the technicization of every aspect of human life" (1978, p. 18). Indeed, it was Huxley who caused Jacques Ellul's *The Technological Society* to be translated into English in 1964. Although he portrayed the relationship between science, technology, and social control in *Brave New World*, Huxley also examined the issue in essays such as "Science, Liberty and Peace" (1946), where he argued that science and technology tend to perpetuate and intensify inequalities and threaten peace and freedom. Mass production and mass media are used by the few to manipulate and control the many, as the rationalization of society reduces citizens to mere cogs in the machine. Huxley also recommended that scientists boycott harmful work and take action to foster positive scientific research. In *Brave New World Revisited* (1958), he argued that individual liberties must be protected from abuses of power.

This issue also dominates Huxley's last novel, *Island* (1962), which portrays a utopian society. It is a small, self-sustaining community removed from the pernicious effects of industrialization and the materialistic mindset of a scientific culture. Education, tranquility, and spirituality take the place of indoctrination, consumerism, and carnality portrayed in *Brave New World*. Multi-parent families and disciplined sexual practice replace machines and artificial stimulation. It is a society characterized by the pursuit of personal fulfillment and selfless care for the community. Although technology is not dominant, *Island* is not a pre-modern utopia. Its technologies serve community and spiritual flourishing rather than social power and personal distraction.

Science, Technology, and Human Nature

The difference between the drug "soma," in *Brave New World* and "moksha," in *Island* raises a basic question in Huxley's work and suggests the connections between his work and later developments in biomedical technology. Whereas soma flattens and attenuates human experience, moksha enhances and enlightens it, posing the question of what it means to be truly human. In fact, many of the central themes of Huxley's work (love, family, mortality, happiness, authenticity, consciousness, and the human spirit) highlight this basic issue. Huxley was aware that technoscience, especially biomedical science, could fundamentally alter these aspects of life.

There is disagreement about whether the *Brave New World* scenario of a dehumanized, or post-human, future is a likely consequence of biomedical technologies such as psychotropic drugs and germline engineering. Some argue that as long as individuals freely choose these technologies, there is no threat to human dignity (Blackford 2004). Others claim that human nature itself is under threat, even if these technologies are adopted within a liberal democratic system (Kass 2004). Hard, top-down attempts to control human behavior are not the only threats; there are also soft, bottom-up threats that appeal to the lowest common denominators in human desire.

Huxley also considers the dehumanizing potentials of scientific and technological change in other works. In *Point Counter Point*, he argues that liberal democracies and autocracies share a common faith in the powers of science and technology to deliver human happiness. Realizing that this happiness is oftentimes shallow and inauthentic, the protagonist in the novel proclaims that the only difference is "whether we shall go to hell by communist express train or capitalist racing motor car" (p. 414). In *Antic Hay* (1923), Huxley lampooned a decadent society of lost souls searching for meaning and true happiness, but only on the surface of the latest fads. In *Ape and Essence* (1948), he warned of the catastrophes that can result from humanity's hubristic search for knowledge and control of nature. He also satirized the scientific quest for immortality in *After Many a Summer Dies the Swan* (1939).

Huxley's most telling interpretation of the proper use of technology to enhance rather than corrupt human nature comes from his two books about psychedelic drugs, *The Doors of Perception* (1954) and *Heaven and Hell* (1956). These works present a philosophy of the prudent use of technology as an aid in the search for truth, goodness, and beauty, which Huxley believed to be the purpose of human life. Drugs can assist someone in this search, but he warned that they must be used cautiously. They are not an excuse to forgo the responsibilities that come with freedom. Rather, "Ethical and cognitive effort is needed if the experience is to go forward from this one-shot experience to permanent enlightenment" (Deery 1996, p. 109). As John the Savage remarked in *Brave New World*, experience has to "cost" us—it has to be difficult—if it is going to be truly meaningful.

Huxley was not opposed to new developments in science and technology. His message is that these developments must be guided by moral inquiry and held to the standards of individual dignity and enlightenment as well as social sanity and peace. They must further be directed and adopted by a free and well-educated populous, and not forced by the hand of technocrats or the mantras of mass society. As his utopian *Island* illustrates, this will mean science and technology play much smaller roles in human life, not because they are inherently nefarious, but rather because they can only go so far in assisting the good life. It is a thin line, after all, between enhancing the human soul and erasing it.

ADAM BRIGGLE

SEE ALSO *Brave New World; Haldane, J. B. S.; Science Fiction; Science, Technology, and Literature; Utopia and Dystopia.*

BIBLIOGRAPHY

Bowering, Peter. (1969). *Aldous Huxley: A Study of the Major Novels*. New York: Oxford University Press. Contains critical essays on nine of Huxley's novels as well as introductory material on the "novel of ideas" and morality and a conclusion on morality and art.

Deery, June. (1996). *Aldous Huxley and the Mysticism of Science*. New York: St. Martin's Press. Focuses on Huxley's view of the connection between literature, science, and mysticism.

Huxley, Aldous. (1923). *Antic Hay*. London: Chatto and Windus.

Huxley, Aldous. (1928). *Point Counter Point*. London: Chatto and Windus

Huxley, Aldous. (1932). *Brave New World*. London: Chatto and Windus.

Huxley, Aldous. (1939). *After Many a Summer Dies the Swan*. London: Chatto and Windus.

Huxley, Aldous. (1942). *The Perennial Philosophy*. New York: Harper.

Huxley, Aldous. (1946). *Science, Liberty and Peace*. New York: Harper & Brothers.

Huxley, Aldous. (1948). *Ape and Essence*. New York: Harper.

Huxley, Aldous. (1954). *The Doors of Perception*. New York: Harper.

Huxley, Aldous. (1956). *Heaven and Hell*. New York: Harper.

Huxley, Aldous. (1958). *Brave New World Revisited*. London: Chatto and Windus.

Huxley, Aldous. (1962). *Island*. London: Chatto and Windus.

Huxley, Aldous. (1963). *Literature and Science*. New York: Harper and Row.

Huxley, Aldous. (1978). *The Human Situation: Lectures at Santa Barbara*, ed. Piero Ferrucci. New York: Harper and Row.

James, Clive. (2003). "Out of Sight: The Curious Career of Aldous Huxley," *The New Yorker* March 17. A lucid insight into Huxley's world, with particular attention paid to his political beliefs and their consequences for social organization and the regulation of technology.

Kass, Leon R. (2004). *Life, Liberty, and the Defense of Dignity: The Challenge for Bioethics*, 2nd ed. San Francisco: Encounter Books. Advances the argument that certain biomedical technologies threaten human dignity. Also discusses the meaning of liberty, as opposed to libertarianism, in an attempt to formulate a way for liberal democracies to answer the challenges posed by technologies.

Murray, Nicholas. (2002). *Aldous Huxley: An English Intellectual*. London: Little, Brown. First new biography of Huxley for thirty years, draws upon unpublished sources and interviews with family and friends.

INTERNET RESOURCES

Blackford, Russell. (2004). "Debunking the Brave New World." Betterhumans. Available from: http://www.better-humans.com. Argues that biomedical technologies do not endanger human dignity.

Pearce, David. (1998). "Brave New World? A Defence of Paradise Engineering." Available from: http://www.huxley.net. Claims that *Brave New World* has become a pernicious slogan of fear that prohibits humans from unlocking their full potentials through biomedical technology.

HYPERTEXT

• • •

Hypertext is a way to organize information in a digital format that makes use of traditional text structures (words, sentences, pages, articles or chapters, books, and libraries) as enhanced by the multiple linkages (words to words, words to sentences, sentences to sentences, sentences to pages, pages to pages, pages to chapters, and so on) possible in cyberspace. When hypertexts further employ graphics, images, audio, and video, they become hypermedia. By both enhancing and subverting traditional assumptions about the linear reading of a text (i.e., word after word, sentence after sentence, page after page) hypertexts also both expand ethical reflection and create ethical issues related to access, the implications of linking choices, and more.

Structures and Opportunities

The architecture of information in the digital context consists of three basic elements: nodes, links, and anchors. In hypertext, information is distributed in building units called nodes. Nodes store a large amount of information, anything from a printed page to an entire book. Nodes can include text, graphics, images, and sounds (hypermedia). They are connected by links; a link between two nodes allows the reader to switch from one to another.

Anchors allow readers or users to determine whether a link exists and if so, to access it.. The reader can switch from one informational unit to another by clicking an anchor zone. Anchor zones are identifiable by some kind of emphasis; they may be a different color than other text, cause changes in the shape of the cursor, appear as icons, and so on, and usually give an indication of the link destination.

With these three basic construction elements, among others, designers can build simple and reduced hypertextual organizations, as well as large, complex ones. Well-designed hypertextual organizations are of great help in translating information to the computer screen. If providers of digital support were limited to using traditional methods of information dissemination, such as print matter, the efficiency and value of that support would be severely reduced.

Providing information through hypertext creates different ways of reading. Readers have more paths open to them because nodes offer a variety of links. Sequential or traditional reading does not allow such multiplicity. Thus hypertextual reading is termed *navigation*.

Historical Development

Vannevar Bush presented a precursor to modern hypertext technology in "As We May Think" (1945). Using the technology available at that time, Bush proposed the Memex, a device that could present independent documents in much the same way that memory works, jumping from one to another. In 1965 Theodor (Ted) H. Nelson coined the term *hypertext* and discussed the *Docuverse*, a universe composed of a range of documents, including international literary works. He argued that one should be able to navigate through all the documents and their interrelated fragments and parts. The very same year J. C. R. Licklider published *Libraries of the Future*.

These ideas and concepts could not be realized until devices to implement them were created. Douglas Engelbart, for example, not only proposed theoretical concepts but played a key role in inventing devices which are now integral parts of the modern computer, including the mouse, computer windows, and other graphic interfaces.

Hypertext is the result of technological achievements in hardware and software as well as the creativity of authors who experimented with different structures. Hypertext requires communication networks, computers, authoring tools, and browsers that allow readers to see the hypertext on the computer screen and interact with it. Hypertext also requires continued exploration of the possibilities in this new information framework.

Developers, designers, and inventors have achieved major technological advances in this nascent field. Among them are Tim Berners-Lee's invention of the World Wide Web (www, the largest and best-known hypertextual construct) and HyperText Markup Language (HTML); Peter Brown's development of the first software guide for hypertext production in personal computers, accessible to computer users of all levels of expertise; and Bill Atkinson's design of the HyperCard for Macintosh, which uses the programming language HyperTalk.

Achievements and Ethics

Information on the www is like an unbound book. Any author can add to the work by using a link. Readers navigate through this information and each binds the material into an individual *book* composed of different authors' pages. Boundaries that define the notions of intellectual property are difficult to maintain and traditional methods of protecting copyrights are becoming obsolete. New legal and cultural tools are needed to deal with such changes.

The Wiki Wiki Web is an example of a hypertext construct based on the unrestricted access of users. Each user contributes to the collective work and decides where to create links. There are no webmasters or any central control. Each reader is an author and has the power to eliminate or change the contributions of others. Individual responsibility and self-control and a sense of collaboration on a collective work are guiding forces in these activities, one of which is the continuous creation of the online Wikipedia. Robert McHenry (2004), however, has challenged the quality of this "faith-based encyclopedia."

Hypertext technology allows virtually unrestricted linking of information nodes. Links to information that is clearly related to the subject matter of a particular text are certainly acceptable. But when the destination of a link is not visible, or when readers are diverted to a destination despite their intent to go elsewhere, ethical issues arise.

Likewise decisions to link to certain materials or web sites and not to others, while understandable and arguably defensible, could result in the marginalization of groups with less scientific or social prestige and power. The need to discriminate among the vast amount of information available on the Internet could lead to *cherry-picking* sources of information and experts in fields, thus virtually excluding access to other sources and experts. This situation raises the potential for what has been called *a balkanization of the global village* by Marshall Van Alstyne and Erik Brynjolfsson.

Transcending the barriers of traditional text is certainly an achievement with positive implications that are still being explored. However the potential misuse of hypertext technology or the unforeseen negative results of its use are causes for concern and thoughtful examination.

ANTONIO RODRÍGUEZ DE LAS HERAS
CARL MITCHAM

SEE ALSO *Computer Ethics; Cyberculture; Free Software; Internet; Networks; Science, Technology, and Literature.*

BIBLIOGRAPHY

Landow, George P. (1997). *Hypertext 2.0.* Baltimore, MD: Johns Hopkins University Press. A revised version of *Hypertext: The Convergence of Contemporary Critical Theory and Technology* (1992).

Licklider, J. C. R. (1965). *Libraries of the Future.* Cambridge, MA: MIT Press.

Van Alstyne, Marshall, and Erik Brynjolfsson. (1996). "Could the Internet Balkanize Science?" *Science* 274(5272): 1479–1480.

INTERNET RESOURCES

Bush, Vannevar. (1945). "As We May Think." Originally published in the *Atlantic Monthly* (July 1945), the text is available on a number of Internet sites, including http://ccat.sas.upenn.edu/∽jod/texts/vannevar.bush.html.

McHenry, Robert. (2004). "The Faith-Based Encyclopedia." Available at Tech Central Station, http://www.techcentralstation.com/111504A.html.

Nelson, Theodor Holm. The best single source for information on Ted Nelson and his ideas, with links to other pages, is an article on the Wikipedia at http://en.wikipedia.org/wiki/Ted_Nelson.

I

IBERO-AMERICAN PERSPECTIVES

• • •

To introduce a Spanish- and Portuguese-speaking perspective on science, technology, and ethics is difficult and somewhat artificial. From the beginning it must be acknowledged that Spain and Portugal on the Iberian Peninsula of Europe together with the more than twenty Spanish- and Portuguese-speaking countries that can be identified in the Americas compose a heterogeneous group. In many respects differences outweigh similarities. Nevertheless, the differences are perhaps no greater than those present in other large-scale linguistic or cultural perspectives such as are represented by Africa, China, or India. Provided that this introduction is not taken as a substitute for more particular assessments of the situations in Argentina, Brazil, Chile, Colombia, Cuba, Ecuador, Guatemala, Mexico, Peru, Portugal, Spain, and Venezuela (to mention only a dozen of the most populous countries), it may serve to highlight some modest commonalities that do in fact exist.

Context

Understanding relations between science, technology, and ethics in the Ibero-American countries requires some appreciation of the historical relations between Spain and Portugal on the Iberian peninsula and those countries in the Americas that emerged from Iberian colonization. The sixteenth and seventeenth century Iberian colonizations of the Americas brought with them ideals of the Counter Reformation rather than the ideals of liberalism the practice of exclusion that were more characteristic of English colonialism. From the very beginning, there was thus little enthusiasm for science and technology in themselves, and even considerable skepticism regarding their benefits. The local cultures that emerged in the eighteenth century and then sought independence in the nineteenth century adopted a sense of being on the periphery that was reinforced, especially in Spain, by its sense of separation from Europe and then the loss of its last major possessions to the United States in Spanish-American War of 1898.

Subsequent early twentieth century attempts by Latin American countries to modernize and become players in international affairs had to struggle with the increasing influence of the United States and continuing marginalization in the mother countries of Spain and Portugal. Virtually all Ibero-American countries were also afflicted until the 1970s with civil wars and economic difficulties. The last decades of the twentieth century were then characterized by attempts to recover cultural roots and establish regional identities, often through ambitious political projects of international cooperation and development such as the Alliance for Progress (which was proposed by American president John Fitzgerald Kennedy [1917–1963] in 1961 but petered out by the late 1960s), or through more modest academic projects, including the formation of regional networks promoting scientific education and research. The failures of major development programs to achieve their stated goals, and the difficulties that emerging cadres of scientists and engineers had in securing adequate employment in their home countries, nevertheless sponsored an ongoing sense of skepticism and dissatisfaction with scientific and technological initiatives.

Against such a background it is thus appropriate to review in slightly more detail various indicators concerning the role of science and technology in various Ibero-American situations. This will be followed by an assessment of social and academic attitudes toward science and technology, including those manifested in Latin American social thought. Finally it will be appropriate to comment on the reception and development of science, technology, and society (STS) studies in the region, and to review recent initiatives to promote a proper regional reflection on the social meaning of science and technology.

Science and Technology in Ibero-American Countries

Until the latter decades of the twentieth century, the role of scientific research and technological development (R&D) in Ibero-American countries can be encapsulated in a well-known phrase from Miguel de Unamuno y Jugo (1864–1936): "¡Qué inventen ellos!" (Let others invent!). Unamuno was one of the leading members of the "Generation of 1898," the year that Spain lost the last of its major colonies, and a philosopher who struggled to come to a new self-understanding of what it meant to be Spanish. Unamuno's manifesto was to make a virtue of history: Spain should not compete with others in science and technology, but seek a non-technical identity in its cultural traditions. Although Unamuno himself was adamantly opposed to the traditionalists who made up the base for Francisco Franco (1892–1975) during the Spanish Civil War (1936–1939), the fascist triumph can be interpreted as an initial victory for such an ideology. Only in the 1950s did this victory evolve into a kind of technocratic development that, after Franco's death in 1975, could serve as a foundation for major scientific and technological change. (Changes of a comparable character took place in Portugal after the death of António de Oliveira Salazar [1889–1970].)

The increasing social and political belief in a link between economic development and technoscience that characterized the last half of the Franco regime was also reflected in Latin America in emerging public policies for the promotion of R&D. It was especially at the exhaustion of the development model known as "industrialization by import substitution" during the 1980s, when a large number of national science and technology organizations were created, that many governments began to recognize a need to support their science and technology systems. The loss of a dream of self-sufficiency in the midst of globalization was coincident with the diffusion of new discussions of innovation. The new discourse has nevertheless brought its own worries, especially a tendency to subsume science policies under economic policies—a view that at the beginning of the twenty-first century serves as the guiding principle for the reorganization of R&D in many Latin American countries. In such a context, economics trumps science—as well as ethics.

Yet good intentions have seldom equaled actions. In the decade of the 2000s, it has remained the case that only around 0.5 percent of the gross domestic product (GDP) is allocated to R&D in most Latin American countries. Because there is little privately supported higher education, universities absorb the major portion of public R&D funding; there is no significant demand for R&D in the private sector. In spite of public declarations and formal documents, Unamuno's spirit remains strong. Indeed, with regard to science and technology, the inequality of Latin America in relation to other regions is even more pronounced than the much better known economic inequalities. This is well documented by a wide variety of indicators: funding, active researchers, science students, scientific publications, patents, and more.

Relevant data is available on the Ibero-American and Inter-American Network on Science and Technology Indicators Website (www.ricyt.org), as well at regular United Nations Educational, Scientific, and Cultural Organization (UNESCO) and Organisation for Economic Co-operation and Development (OECD) publications on the state of science and technology throughout the world. For example, while in developed countries about fifty percent of the student-age population pursues some level of higher education, in Latin American this number is below twenty percent. This is after a doubling of university graduates during the 1990s. From this scarce percentage, in 1997 only eleven percent were graduates in mathematics, science, or engineering. It is against this scarcity and imbalance that the efforts of research groups in Brazil, Colombia, Cuba, Mexico, and Venezuela must be appraised.

A certain imbalance among these countries must also be recognized. Over seventy percent of Latin American scientific researchers are concentrated in three countries: Argentina, Brazil, and Mexico. While some countries, such as Brazil and Cuba, are making a strong economic and political effort to promote R&D, others, such as Peru and El Salvador, were not even investing 0.1 percent of their GDP in science and technology as of the late 1990s. The situation in Spain and Portugal is also distinctive. Particularly since joining the European Union, Spain and Portugal have worked to reach European standards with regard to science, technology, and innovation indicators. Although they have yet to match the general European standards, especially in relation to their weak public funding and poor investment from the private sector, their indicators are significantly better than those of most Latin American countries. For example, publications from Spain included in the Science Citation Index (around 25,000 in year 2000) were almost as great as those from all of Latin America (around 28,000 in the same year)—but still far below a single North American country such as Canada (with 38,000 in 2000).

Clearly much work remains to raise the profile of science and technology in all the Ibero-American countries. Mere awareness of the problem is not enough. Instead, decisive steps are required from many social actors in order to promote science and technology and to develop their economic potential. At some level, such work will rest on an ethical assessment of the value of science and technology that does not ignore their potential dangers. Indeed, the issues concerning relationships between science, technology, and development have been themes of critical social reflection in Latin America—a tradition of reflection that is in the process of being modified by the regional emergence of STS studies.

Latin American Social Thought

Relevant in the present context is the evolution of a distinctive school of Latin American social thought on science and technology, especially as reflected in a number of thinkers concerned with both the foundations of science and regional political change. Although the most significant of these were born and based in Argentina, they had a much wider influence during the 1970s and since. What follows is a brief review of the work of three representatives of this school.

OSCAR VARSAVSKY. Oscar Varsavsky (1920–1976) was an Argentinean mathematician and physicist who was also one of the most politically engaged scientists of his generation. He developed a criticism of what he called "scientism," particularly in Latin America: that is, the ideological attitude often assumed in science in which scientists focus their professional interest on their own careers, adopting them to the patterns operative in leading foreign scientific centers, thus developing an external dependence while ignoring immediate social needs and the political meanings of their work. According to Varsavsky, it is a prevailing obsession for quantitative methods and an illusion of freedom in research that obscures the scientists' dependency on capitalist economic forces and market laws.

Adopting a relativist viewpoint, Varsavsky held that there is more than one way to do science and technology. There are different styles in science and technology, linked to different national projects and eventually to different social values. Varsavsky thus developed a normative criticism of contemporary science, rejecting the linear model of innovation as dependent on basic science—a science policy ideology that became very influential in Latin America during the 1970s. In a more positive vein, Varsavsky argued for a new style in science and technology in Latin America: a science for the people or, better, a science from the people, providing the region with a certain scientific and technological autonomy, and linked to a style of society that he called *socialismo nacional creativo* (creative national socialism).

JORGE SÁBATO. Jorge Alberto Sábato (1924–1983) was an Argentinean metallurgist and self-educated physicist who had an important role as the promoter of research in the Argentine Atomic Energy Commission. He was also influential in the creation of the Physics Institute in Bariloche, Argentina. A sharp and lucid thinker, Sábato had a strong influence from the 1960s concerning the way to conceptualize scientific and technological development in Latin America. His most widely cited contribution is his 1968 paper "La ciencia y la tecnología en el desarrollo de América Latina" (Science and technology in the development of Latin America), coauthored with Natalio Botana. In this paper he introduces the metaphor of the triangle of scientific and technological development, whose three vertices are government, the production sector, and the knowledge-generation sector. This has come to be known as the "Sábato triangle," which he used as a heuristic tool for analyzing problems posed by the lack of innovation in the periphery.

According to Sábato, it is the weak connections between those three vertices, in contrast to the situation in developed countries where they constitute a system, that explains the weakness of innovation capacities in Latin America and its technological dependency. These ideas were contrary to the then-prevalent linear model of innovation, and clearly anticipated forthcoming theories on systems of innovation.

AMÍLCAR HERRERA. Amílcar Herrera (1920–1995) was an Argentinean geologist and eldest, but also longest-lived, of the three the thinkers under review. His main book, *América Latina: Ciencia y tecnología en el desarrollo de la sociedad* (1970), an edited volume that included Sábato's 1968 paper, outlines his primary intellectual orientation: developing a Latin American view about the problems of underdevelopment and their relation to science and technology. Immediately afterward, his *Ciencia, tecnología y desarrollo social: ciencia y política en América Latina* (Science, technology, and social development: science and politics in Latin America, 1971), critically analyzed the social and historical context of science and science policy in the Latin American region. It is in this second volume that Herrera, adopting a structural and contextual approach, introduced a now widely used distinction between explicit and implicit science policies. An explicit science policy is the one that can be found in standard formal documents, a modernizing and progressive policy in accordance to universal ideals. The implicit science policy is the one really at work, characteristically at the service of the ruling social classes.

Herrera also criticized the use of conventional socioeconomic indicators for development in Latin America, and argued in favor of an orientation of the scientific and technological capacities toward proper regional problems such as those of undernourishment, misery, and ignorance. Finally, shortly before his death, his *Las nuevas tecnologías y el futuro de América Latina: Riesgo y oportunidad* (New technologies and the future of Latin America: risk and chance, 1994) proposed a strategy for scientific and technological development appropriate to the Latin American countries and sensitive to the type of society to be pursued.

Of course, none of these authors considered himself a STS scholar. They were simply critical scientists interested in the social realities of Latin America, as connected to science, technology, and innovation, and as such they anticipated some of the ideas that could subsequently find a home in STS scholarship. Indeed, they created a social thought tradition that has molded the later reception of STS authors and ideas.

Moreover, they are not alone in the movement of Latin American social thought on science and technology. Others who deserve mention are the Chilean Fernando Fajnzylber (who focused his work in the study of the relationship between economic development and inequity) and the Venezuelan Marcel Roche (founder of the journal *Interciencia* and promoter of science studies in his country). Still others have also made lively contributions to STS research, in countries such as Brazil, Colombia, Cuba, and Uruguay.

The tradition of Latin American social thought was not, however, particularly influential in promoting an ethical assessment of science and technology in relation to developments in either Spain or Portugal. In Spain one primary influence was the work of José Ortega y Gasset (1883–1955) who, as a member of the "Generation of 1927" (the generation associated with the second Republic), criticized the views of Unamuno. Ortega's *Meditación de la técnica* (1939) provided a positive but critical analysis of technology as central to human life. Another influential philosopher of Spanish origin, Juan David García Bacca (1901–1992), who spent most of his adult life in Ecuador, Mexico, and Venezuela, adopted an even more positivist perspective that virtually ignored any negative political implications of scientific and technological progress. More recently the critical phenomenological analyses of the Venezuelan Ernesto Mayz Vallenilla on the tendencies of technology to be transformed into what he terms a meta-technology have also had some limited influence.

STS in Ibero-American Countries

It is within the previously noted contexts that STS studies—as the basic framework within which discussions of science, technology, and ethics have been manifest—have emerged in Spain, Portugal, and Latin America. Before turning to this emergence, however, it is necessary to provide some commentary on the underlying interpretation of science studies in these countries.

In Spanish- and Portuguese-speaking countries there is a certain ambiguity concerning how to interpret and translate the English acronym "STS." Some translate it as "science and technology studies" (*estudios sobre ciencia y tecnología*); others take it to stand for "science, technology, and society" (*ciencia, tecnología, y sociedad*). The well-known distinction between the two STS subcultures—an academic subculture focused on the study of technoscientific change as a social process, and the social factors that might be rendered responsible for shaping such a change, versus an activist subculture focused on the social and environmental effects of technoscientific products, upon their educational, ethical, or political aspects—is repeated in the Ibero-American perspective. But this repetition is a weak one, and the fact is that in Latin America especially the two approaches have tended to merge even when the interpretation of STS as "ciencia, tecnología, and sociedad" predominates.

The STS subcultures, whether disciplinary or activist, originated in the late 1960s and the early 1970s in the United States and the United Kingdom, and from there were transferred to other industrialized countries mostly in continental Europe. It was during the 1980s and 1990s that STS penetrated the academic and educational institutions of more peripheral European countries, such as Spain or Portugal, and other peripheral regions, such as Latin America. In Spain, Colombia, and Cuba, it was only in the late 1980s that such things as social constructivism, technology assessment, gender issues in scientific research, along with new trends in science education, began to be pursued. The academic and institutional consolidation of STS, however, did not reach the region until the 1990s, and even then in a slow and hesitant way that has continued into the twenty-first century.

There are nevertheless some exceptions worth mentioning, both in research and education. With regard to research, a number of groups linked to universities have had some important results. Examples include

- the STS postgraduate program and research group organized by José Sanmartín in the University of Valencia, which started the first formal STS education program in Spain in the 1980s;

- the group led by Mario Albornoz at the Ibero-American Science and Technology Indicators Network (RICYT) in Buenos Aires, which gathers scholars from many Latin American countries;

- the Hebe Vessuri group at the Venezuelan Institute of Scientific Research (IVIC) in Caracas, with its tradition of collaboration with UNESCO;

- the team arranged by Javier Echeverría and Emilio Muñoz at the STS Department, Spanish Research Council, Madrid;

- the scholars gathering around Renato Dagnino in the University of Campinas, near Sao Paulo in Brazil;

- the research group linked to Maria Eduarda Gonçalves and José Luís Garcia at the Institute of Social Sciences at the University of Lisbon;

- the network of Jorge Núñez, Director of Postgraduate Studies, Havana University, Cuba;

- the research group led by León Olivé and Rosaura Ruiz in the National Autonomous University of Mexico (UNAM), hosting many editing and teaching STS activities.

Not included in this list are other important researchers who have made no less significant contributions in countries such as Colombia (Mauricio Nieto, Carlos Osorio) and Uruguay (Judith Sutz, Rodrigo Arocena).

With regard to education, STS has been making a strong impact on the Spanish secondary school system and on higher education in Cuba since the mid-1990s. More modest impacts are to be found in Mexican secondary education, where a certain implementation of STS content has taken place in natural sciences subjects and is underway in technological education. There are also a number of particular universities where STS research groups have flourished when linked to diverse graduate or postgraduate programs (see above). However, these are rather exceptions to the general rule of slow consolidation of a regional STS scholarship.

The case of Cuba is worth special note. After the end of the Cold War, reforms in Cuba began also to affect education. Under the title of "Social problems of science and technology," the content of STS experienced an impressive expansion in the Cuban system of higher education. STS largely replaced the previously obligatory study of Marxism, and so is now taught as part of practically all university degrees. It constitutes a compulsory examination for Ph.D. candidates and for any scholar applying for promotion within the faculty system.

Discourse Transfer Issues

STS and related discussions of science, technology, and ethics can be understood as cultural constructs. Such discourses arose initially in more economically and technologically developed countries in response to certain social demands. These demands included calls for alterations in the academic image of science, desires to increase scientific literacy among non-scientist citizens, needs for reforms in science education, political efforts to extend public control over technological change, concerns for social accountability related to science and technology policies, and more. Discussions of the professional ethical responsibilities of scientists and engineers and efforts to enhance the responsible conduct of scientific research were especially associated with the intensified interactions between science, technology, economics, and politics. The transfer of such discourses to the more peripheral Ibero-American countries, despite the differences that exist among them, has confronted a number of common problems.

First, an obvious but important fact is that many of the social demands out of which STS originally emerged in the Anglo-American center of scientific and technological advance in the 1960s did not exist until much more recently in Spain, Portugal, or Latin America. With no significant interest in the classic sociology of science, one should not expect there to be much interest in the sociology of scientific knowledge and related analyses of the academic status of science. Where large sections of the population are illiterate, it is unlikely that there will be desires to increase scientific literacy. Without democracy, it would be nonsensical to argue for an extension of democracy into the regulation of science and technology.

Second, the constitution of a critical mass of STS scholars in every country requires an established research infrastructure in the natural and social sciences. It depends on reasonable input and output indicators in those fields, as well as institutional structures for facilitating interdisciplinary research. Unfortunately, in Ibero-American countries there has traditionally been an important lack in both respects. At the same time, the creation of small national groups of STS scholars, who could be put together and form a critical mass in the region as a whole, has faced serious difficulties because of severe restrictions on academic support and communication.

Finally, third, there has been an excessive peripheral focusing on the English-speaking center. Spanish STS scholars, for example, tend to read STS literature in English, produced by American, British, Dutch, or French authors. They thus largely ignore what their cultural neighbors are doing in Colombia, Venezuela, or even Portugal.

Fortunately, the situation in Ibero-American countries is changing. The effort of a number of international governmental organizations, such as UNESCO, the Programa Iberoamericano de Ciencia y Tecnología para el Desarrollo (CYTED), and the Organización de Estados Iberoamericanos (OEI), as well as some national science teachers associations (such as those in Chile or the Brazil), are promoting research and breaking down communication barriers. Traditionally isolated local universities and research centers are increasingly cooperating to promote STS throughout the region. This will undoubtedly stimulate discussions of science, technology, and ethics, as well.

Recent Initiatives

Two significant recent examples of recent initiatives are the creation of an Ibero-American STS Thematic Network and the promotion of a number of university Science, Technology, Society, and Innovation (STS&I) chairs, in both cases as initiatives of the OEI—an intergovernmental organization that depends on the ministries of education of the Spanish- and Portuguese-speaking Latin American countries, plus Spain and Portugal.

The Ibero-American STS Thematic Network gathers STS scholars from some fifteen countries in the region, focusing their work around typical STS subjects such as science and gender, social impact indicators for R&D activities, or ethical aspects of new technologies. The central goal of this network is to stimulate an endogenous STS scholarship in the Ibero-American region, while promoting a constructive dialogue with the international forefront in the field. Among the tools that are already in use are the support of STS publications (in Spanish and Portuguese), electronic diffusion and distance learning courses, and the sponsoring of STS conferences and meetings in the region (see http://www.oei.es/cts.htm).

The network draws applications to the fields of science education, communication, and management. For example, in the field of science and technology management, the OEI has made use of network resources and included a STS orientation in Science Administration courses that have been organized since 1998. These courses are addressed to young high officials of the Latin American ministries of science (or whatever ministry holds science policy competencies) as well as national organizations responsible for science policy in the region. The inclusion of STS content in these courses, comprising fifteen to thirty percent of the lecture time, has been well received.

As to the research guidelines of this initiative, its critical focus has been the urgent need to promote economic development in the region and a central place for science and technology in such a process. According to this view and reflecting the critical tradition of Latin American thought, a social critique of science should be compatible with the encouragement of science and science policies. In more practical terms, policy, ethics, and history-based applied analyses, often assuming the form of interdisciplinary studies, have taken precedence over theory-oriented and disciplinary stances. "Science, technology, and society" has dominated over "science and technology studies."

The creation of STS+I Chairs in Ibero-America exhibits similar tendencies. STS+I chairs are an OEI initiative in collaboration with national science and technology agencies, and in some cases ministries of education. Basically, the idea underlying STS+I Chairs is to constitute networks of universities (both public and private) that, duly supported by other public organizations, will be able to strengthen particular lines or research and education (linked to STS and innovation issues), thus making better use of the potentialities of participant institutions. To date, chairs have been established in El Salvador (September 2000), Argentina-Uruguay (April 2001), Colombia (September 2001),

Cuba (November 2001), Costa Rica (July 2002), Panama (April 2003), Mexico (May 2003), and Peru (June 2003).

The organizational model is different in each case, respecting each country's characteristics (with a strong or weak national differentiation, with one or another higher education system, etc.). But basically a STS+I chair is constituted by a named professorship with supplementary funds to support education and research activities. What unifies the various STS+I chairs is the attempt to promote a dialogue between the scientific and humanistic cultures, as well as the social projection of scientific knowledge generated in the university by means of teaching seminars and other initiatives of knowledge diffusion, as well as the support of research. The general idea is the creation of a common working ground for higher education and research institutions, a common space conceived for sharing and rationalizing human and material resources. Not only banks and corporations, but also education and research institutions, need to establish alliances and common projects in order to be competitive and make an optimal use of their potentialities.

STS+I Studies

The STS+I acronym emphasizes the particular perspective in which STS studies are being developed in the Ibero-American region, receiving international STS scholarship and adapting it to the tradition of critical thought on science and public policy represented by Varsavsky, Sábato, and Herrera. The STS+I perspective also tries to cope with the two major challenges of the so-called knowledge society, as seen from a regional perspective: the appropriation of such knowledge by the production sector, and its appropriation by the civil society.

A pragmatic approach to the region's sensibilities is perhaps the best way to characterize these fields and their interrelation in the present geographic and cultural context. On the one hand, in the STS field, through the study of academic themes such as science and gender, science education, or engineering ethics, the goal is to achieve an understanding of the relationships between science and technology in their social context in order to promote social interests for scientific issues, scientific and technological literacy, and public participation in public policies related to science and technology. On the other hand, in the innovation field, through the study of themes such as university-corporation relationships, national systems of innovation, and technological management, the goal is to understand institutional and socioeconomic conditions underlying the phenomenon of innovation in order to support innovation and the creation of innovation systems in countries of the region. The great challenge and novelty of the STS+I approach has been the combination of these lines of work in a common framework of interdisciplinary reflection, with a strong practical or policy orientation.

"Society" and "innovation" are the key terms of the so-called "Declaration of Santo Domingo," from a regional summit on science and technology held in March 1999 as a preparatory meeting of the World Congress on Science, arranged by UNESCO and the International Council of Scientific Unions (ICSU) and held in Budapest in June-July 1999. It is not by chance that these two points were also emphasized in the final declaration of the Third Ibero-American Course for Science and Technology Administrators, held also in March 1999 in Bogota, Colombia, which gathered participants from twelve Latin American countries.

In fact, in the contemporary world, and especially in Latin America, these two goals—to open science and technology systems to social sensitivities and public participation, and to reorient these systems toward economic development—are not only compatible but mutually interdependent. Technological innovation, the process that begins with the organized creation of an idea and concludes with the social diffusion of its material realization, requires social support and participation for its feasibility and consolidation. Just as a country with half its population in poverty cannot pretend to be internationally competitive or to enjoy sustainable economic development, the consolidation of such growth and competitiveness requires public interest, democratic support, and confidence in institutions among all the citizens. Moreover, from the perspective of the periphery, technological innovation is necessary for national economic competitiveness and also for the creation of the material conditions that make possible, among other things, the modernization of political and administrative structures and the generation of a participatory culture.

Acknowledgment

Although only the author is responsible of any omissions or mistakes, this article draws on a number of experiences and personal communications involving the STS Network of the OEI. Due credit must be given to the Network scholars who provided many clues and much information. Particularly helpful have been Elsa

Beatriz Acevedo (Colombia), Rodrigo Arocena (Uruguay), Renato Dagnino (Brazil), José Luis Luján (Spain), Jorge Núñez (Cuba), and Sara Rietti (Argentina).

JOSÉ A. LÓPEZ CEREZO

BIBLIOGRAPHY

Arocena, Rodrigo, and Judith Sutz. (2003). *Subdesarrollo e innovación: navegando contra el viento* [Underdevelopment and innovation: sailing against the wind]. Madrid: Cambridge University Press/OEI.

Bazzo, Walter Antonio. (1998). *Ciência, tecnologia e sociedade, e o contexto da educação tecnológica* [Science, technology, and society in the context of technological education]. Florianópolis, Brazil: UFSC.

Buch, Tomás. (1999). *Sistemas tecnológicos: contribuciones a una teoría general de la artificialidad* [Technological systems: contributions to a general theory of artificiality]. Buenos Aires: Aique.

"Ciencia, tecnología y sociedad ante la educación" [Science, technology, and society before education]. (1998). Monographical issue of *Revista Iberoamericana de Educación* 18(Sept.-Dec.).

"Ciencia, tecnología y sociedad" [Science, technology, and society]. (1999). Special issue, *Revista Pensamiento Educativo* 24(July). Pontificia Universidad Católica de Chile.

Dagnino, Renato, ed. (2000). *Amílcar Herrera: un intelectual latino-americano* [Amílcar Herrera: a Latin American intellectual]. Campinas, Brazil: UNICAMP.

Echeverría Ezponda, Javier. (2003). *La revolución tecnocientífica* [The technoscience revolution]. Madrid: FCE.

Funtowicz, Silvio, and Jerome Ravetz. (1993). *La ciencia posnormal: ciencia con la gente* [Post-normal science: science with the people]. Buenos Aires: Centro Editor de América Latina.

González García, Marta; José Antonio López Cerezo; José Luis Luján López; et al. (1996). *Ciencia, Tecnología y Sociedad: Una introducción al estudio social de la ciencia y la tecnología* [Science, technology, and society: An introduction to the social study of science and technology]. Madrid: Tecnos.

Herrera, Amílcar, ed. (1970). *América Latina: Ciencia y tecnología en el desarrollo de la sociedad* [Latin America: Science and technology in the development of society]. Santiago de Chile: Editorial Universidad.

Herrera, Amílcar. (1971). *Ciencia, tecnología y desarrollo social: Ciencia y política en América Latina* [Science, technology, and social development: Science and politics in Latin America]. Mexico City: Siglo XXI.

Herrera, Amílcar. (1994). *Las nuevas tecnologías y el futuro de América Latina: Riesgo y oportunidad* [New technologies and the future of Latin America: Risks and opportunities]. Mexico City: Editorial de la Universidad de las Naciones Unidas and Siglo XXI.

Hoyos Vásquez, Guillermo. (2000). *Ciencia, tecnología y ética* [Science, technology, and ethics]. Medellín, Colombia: Instituto Tecnológico Metropolitano.

"La educación ciencia-tecnología-sociedad" [Science-technology-society education]. (1995). Special issue, *Alambique. Didáctica de las Ciencias Experimentales* 3(January).

López Cerezo, José Antonio, and Andoni Ibarra, eds. (2003). "Studies in Science, Technology and Society (STS): North and South." Special theme issue, *Technology in Society* 25(2).

Martínez, Eduardo, ed. (1994). *Ciencia, tecnología y desarrollo: interrelaciones teóricas y metodológicas* [Science, technology, and development: theoretical and methodological interrelations]. Caracas: Nueva Sociedad.

Martins, Herminio, and José Luís Garcia. (2003). *Dilemas da civilizaçao tecnológica* [Dilemmas of technological civilization]. Lisbon: ICS/Universidade de Lisboa.

Muñoz, Emilio. (2002). *Biotecnología y sociedad* [Biotechnology and society]. Madrid: Cambridge University Press/OEI.

Nieto Olarte, Mauricio. (2000). *Remedios para el Imperio: historia natural y la apropiación del nuevo mundo* [Remedies for the empire: natural history and the appropriation of the New World]. Bogota: Instituto Colombiano de Antropología e Historia.

Núñez Jover, Jorge. (1999). *La ciencia y la tecnología como procesos sociales: lo que la educación científica no debería olvidar* [Science and technology as social processes: what scientific education should not forget]. Havana: Ed. Félix Varela.

Olivé, León. (2000). *El bien, el mal y la razón: facetas de la ciencia y de la tecnología* [Good, bad, and reason: aspects of science and technology]. Mexico City: Paidós.

Queraltó Moreno, Ramón. (2003). *Ética, tecnología y valores en la sociedad global* [Ethics, technology, and values in a global society]. Madrid: Tecnos.

Pérez Sedeño, Eulalia, ed. (2001). *Las mujeres en el sistema de ciencia y tecnología: estudios de casos* [Women in the science and technology system: case studies]. Madrid: OEI.

Sábato, Jorge, and Natalio Botana. (1970). "La ciencia y la tecnología en el desarrollo de América Latina" [Science and technology in Latin American development]. In *Ciencia, tecnología y desarrollo social: Ciencia y política en América Latina* [Science, technology, and social development: science and politics in Latin America], ed. Amílcar Herrera. Mexico City: Siglo XXI.

Sábato, Jorge, ed. (1975). *El pensamiento latinoamericano en la problemática ciencia-tecnología-desarrollo-dependencia* [Latin American thought about the issues on science-technology-development-dependency]. Buenos Aires: Paidós.

Sábato, Jorge, and Michael Mackenzie. (1982). *La producción de tecnología. Autónoma o transnacional* [The production of technology. Autonomous or transnational]. Mexico City: Nueva Imagen,

Sanmartín, José. (1990). *Tecnología y futuro humano* [Technology and human future]. Barcelona: Anthropos.

Santos, Lucy; Elisa Ichikawa; Paulo V. Sendin; and Doralice Cargano, eds. (2002). *Ciência, tecnologia e sociedade: O desafio da interação* [Science, technology, and society: The challenge of interaction]. Londrina, Brazil: IAPAR.

Vaccarezza, Leonardo, and Juan Zabala. (2002). *La construcción de la utilidad social de la ciencia* [The construction of the social utility of science]. Buenos Aires: Universidad Nacional de Quilmes.

Varsavsky, Oscar. (1969). *Ciencia, política y cientifismo* [Science, politics and scientism]. Buenos Aires: Centro Editor.

Varsavsky, Oscar. (1972). *Hacia una política científica nacional* [Toward a national science policy]. Buenos Aires: Ediciones Periferia.

Varsavsky, Oscar. (1974). *Estilos tecnológicos* [Technological styles]. Buenos Aires: Ediciones Periferia.

ILLICH, IVAN

• • •

Most well known as a 1970s social critic of the technologies of schooling, development, and health, Ivan Illich (1926–2002) was born in Vienna, Austria, on September 4, and died in Bremen, Germany, on December 2. In the 1980s Illich shifted from social criticism to cultural archeology, that is, an effort to expose *modern certainties* or assumptions, in order to provide an ethical perspective on the ways technology has transformed human experience in the late twentieth and early twenty-first centuries.

Early Years and the Centro Intercultural de Documentación

Illich was born in Vienna, of French and Serbo-Croatian descent. During World War II he was in some danger because of the Jewish heritage on his mother's side. After the war he undertook studies in science, philosophy, theology, and history; was ordained a Catholic priest; and in the 1950s was posted to the United States, where he became a protégé of the conservative Cardinal Spellman, head of the New York archdiocese. There he acted as a pastor to Puerto Rican immigrants, and as a result of sympathies with their plight, was appointed Vice-Rector of the recently established Catholic University of Puerto Rico. His work in Puerto Rico galvanized an emerging criticism of policies promoting economic and technological development, and led him in the 1960s to establish the Centro Intercultural de Documentación (CIDOC) in Cuernavaca, Mexico, as an institutional base for the exploration of alternatives. CIDOC became a locus for visits by many dissatisfied with technosocial trends and the inspiration for a generation of social critics, from Paul Goodman (1911–1972) to Paolo Frerire (1921–1997). Accused by the

Ivan Illich, 1926-2002. Theologian, educator, and social critic Illich sought bridges between cultures and explored the bases of people's views of history and reality. (*The Library of Congress.*)

Vatican of thereby becoming a scandal to the Church, Illich resigned his institutional ministry, although he was never laicized or married.

It was from CIDOC that Illich published his most widely read books: *Deschooling Society* (1971), *Tools for Conviviality* (1973), and *Medical Nemesis* (1976). In each case Illich identified what he termed the phenomenon of *counterproductivity*: that is, the pursuit of a technical process to the point where it undermines its original goals. The system of public schooling, originally conceived to advance learning, had become an impediment to real education. Advanced technological tools were at odds with autonomous human development and the culture of friendship, in the name of which they were often invented. High-tech health care was making people sick. Iatrogenic illnesses, that is, illnesses caused by physicians—as when patients have negative reactions to drugs, are harmed by diagnostic x-ray treatments, or are otherwise misdiagnosed and mistreated—had, he argued, become a counterproductive epidemic.

The correct response, for Illich, was to learn to practice a more disciplined and limited use of science and technology, and to invent alternative, especially low-scale, technologies. In many instances, however, the practice of such an ethical imperative was made difficult by what Illich termed *radical monopolies*: Although no car manufacturer has a monopoly on the automobile market, cars themselves have a overwhelming monopoly on roads so as to crowd out pedestrians and bicycles.

Living His Theory: After CIDOC

Practicing what he preached, and fearing that CIDOC itself might become counterproductive, Illich closed the center in 1976. He divided up its accumulated assets equally among all those who worked there, from the teachers to the gardeners, and became for the remainder of his life an itinerant scholar. During this period he held posts as visiting professor at a number of universities, from the University of California at Berkeley and the United Nations University in Tokyo to Pennsylvania State University and the University of Bremen, Germany. Two early collections from these years—*Toward a History of Needs* (1978) and *Shadow Work* (1981)—stress counterproductivity in the economics of scarcity, or the presumption that economies function to remedy scarcities rather than to promote community sharing of available goods. Technoeconomic progress was, Illich argued, actually undermining society and culture, the possibilities for friendship and solidarity, and specifically increasing the gap between the rich and the poor both within developed countries and between developed and developing countries.

Toward a History of Needs also hints at a new project in historical archeology that takes its first full-bodied shape in *Gender* (1982), an attempt to recover those social experiences of female/male complementary obscured by the modern economic regime of sex. *H_2O and the Waters of Forgetfulness* (1985) explores the possibility of a history of *stuff*. *ABC: The Alphabetization of the Popular Mind* (1988) carries historical archeology forward into the area of literacy, as does *In the Vineyard of the Text* (1993). Both examine how the techniques of reading transform human beings' experience of themselves and each other, thus inviting contemporary consumers of automobiles and computers to consider that they might not be wholly unaffected users of neutral technologies. Modern technology, for Illich, tends to emerge from and then reinforce a distinctive ethos, the recognition of which is best appreciated by investigations into the moral environments of previous techniques.

In the 1980s Illich became afflicted by a mascular tumor for which, again in accord with his beliefs, he refused high-tech medical treatment. Although he was in increasing pain during the last two decades of his life, he sought to practice what he understood as the traditional arts of suffering, and continued to develop his ideas. He was in his last years especially critical of the notions of "environmental responsibility" and what he saw as the new ideology of "life." Calls for environmental responsibility were, he argued, often just another excuse for advancing technological management of the world, and even the Christian pro-life movement gave too much ground to science insofar as it defined human life in terms of a molecular-biological genesis that cannot be directly experienced. What was at work in history was a counterproductivity writ large that he often described with a Latin phrase, *corruptio optimi que est pessima*, the corruption of the best is the worst. Just as the sweetest flowers, when they rot, smell worse than weeds, scientific and technological attempts to better the human condition, not to mention Christian efforts to institutionalize the friendship of charity, ultimately undermine their own ends.

Illich's criticism itself often was criticized as being overstated and polemical—too much a radical, anarchistic prophesy to be taken seriously. Many of his specific historical claims seemed exaggerated to more sober historians, and he was sometimes unfair to those who questioned his ideas. Yet popular recognition of counterproductivities in government regulation were an ironic echo of Illich's more sweeping analyses. And precisely because of his efforts to live friendship as a fundamental human good, he remained until his death at a friend's home in Bremen, Germany, a charismatic figure who continued to influence cultural criticism and to inspire students seeking alternatives to the standard paths of worldly success.

CARL MITCHAM

SEE ALSO *Bioethics; Development Ethics; Science, Technology, and Society Studies.*

BIBLIOGRAPHY

Cayley, David. (1992). *Ivan Illich in Conversation*. Concord, Ontario, Canada: Anansi. Personal and autobiographical in character.

Hoinacki, Lee, and Carl Mitcham, eds. (2002). *The Challenges of Ivan Illich: A Collective Reflection*. Albany: State University of New York Press. Includes nineteen original reflections by close Illich associates, with an epilog by Illich, and an annotated bibliography.

Illich, Ivan. (1971). *Deschooling Society*. New York: Harper and Row.

Illich, Ivan. (1973). *Tools for Conviviality*. New York: Harper and Row.

Illich, Ivan. (1976). *Medical Nemisis*. New York: Pantheon.

Illich, Ivan. (1978). *Toward a History of Needs*. New York: Pantheon. Includes the essay "Energy and Equity," which argues the counterproductivity of increased energy consumption.

Illich, Ivan. (1981). *Shadow Work*. London: Marion Boyars.

Illich, Ivan. (1982). *Gender*. New York: Pantheon.

Illich, Ivan. (1985). *H₂O and the Waters of Forgetfulness: Reflections on the Historicity of "Stuff."* Dallas, TX: Dallas Institute of Humanities and Culture.

Illich, Ivan. (1993). *In the Vineyard of the Text: A Commentary to Hugh's "Didascalicon."* Chicago: University of Chicago Press.

Illich, Ivan, and Barry Sanders. (1988). *ABC: The Alphabetization of the Popular Mind*. San Francisco: North Point Press.

IMMATERIALIZATION

SEE *Dematerialization and Immatrialization.*

INCREMENTALISM

• • •

In its most generic form, incrementalism is a normative theory of problem solving and decision making. Incrementalist strategies favor small-scale changes, monitoring, flexible positions, and decentralized organization. Incrementalists have been inspired by the epistemology of Karl Popper (1902–1994) and the economic views of Friedrich Hayek (1899–1992). These connections tie the theory of incremental development to controversies over democratic versus totalitarian forms of government and over socialist versus capitalist economic systems. Incrementalist principles thus have wide application but are explored particularly in the search for solutions to social problems and more specifically in the effort to intelligently control technology.

Basic Arguments

As a means of addressing social problems, Robert Dahl and Charles Lindblom give incrementalism a clear standing relative to other approaches.

> Incrementalism is a method of social action that takes existing reality as one alternative and compares the probable gains and losses of closely related alternatives by making relatively small

adjustments in existing reality, or making larger adjustments about whose consequences approximately as much is known as about the consequences of existing reality, or both. Where small increments will clearly not achieve desired goals, the consequences of large increments are not fully known, and existing reality is clearly undesirable, incrementalism may have to give way to a calculated risk. Thus scientific methods, incrementalism, and calculated risks are on a continuum of policy methods. (Dahl and Lindblom 1953, p. 82)

Incrementalism is here conceived as one of several processes that facilitate rational calculation by reducing sources of complexity. By emphasizing alternatives that differ from existing reality by only small degrees, prediction of consequences is improved, identification of causes is made possible, reversibility of decisions is maintained, and the cost of altering organizational hierarchies is avoided.

David Braybrooke and Charles Lindblom (1963) develop these views by criticizing deductive systems, welfare functions, and other synoptic models (including Bayesian methods) that achieve quantitative rigor by assuming that options and states of the world can be exhaustively and exclusively specified. Such models are too formal, centralized, and idealistic to apply to practical decision making. By laying out two intersecting axes, one for degree of knowledge (low to high) and one for size of proposed change (small to large), four quadrants of decision making are established (Figure 1). Incrementalist strategies flourish in the quadrant defined by small changes in the context of low degrees of knowledge. The strategy of *disjointed incrementalism* proceeds through partisan mutual adjustment by which agreements are negotiated among individuals with no need of an overarching design. Lindblom's classic statements (1959, 1979) characterize incrementalism as *muddling through* (a term delivered to incrementalists via Popper's critique of its denigration by Karl Mannheim [1940]). These statements often present a more extreme form of incrementalism, no longer merely one of several alternatives (including calculated risk) but rather as the only viable approach to resolving social problems.

For David Collingridge, incremental advances are prudent where consequences of choice are unclear. Such is the case in the emergence of new technologies, and here incrementalist concepts provide a framework for controlling technological development. Collingridge's early work (1980, 1982) attacks the Bayesian account of decision making and articulates the role of flexibility and corrigibility in decision making. In particular the

FIGURE 1

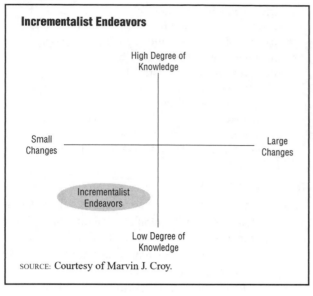

Incrementalist Endeavors

High Degree of Knowledge

Small Changes — Large Changes

Incrementalist Endeavors

Low Degree of Knowledge

SOURCE: Courtesy of Marvin J. Croy.

Bayesian model is blind to the fact that earlier choices, prescribed by the model, may serve to prevent the adoption of new, better options at a later date. Yet one should choose such that one's future flexibility is not precluded by current choices. The art of choosing options that maintain flexibility thus becomes the cornerstone of Collingridge's theory of decision making and technology control. Controlling technology depends upon two factors, the ability to anticipate undesirable consequences and the ability to avoid such consequences once predicted, but this generates the *dilemma of control.*

> Attempting to control a technology is difficult, and not rarely impossible, because during its early stages, when it can be controlled, not enough can be known about its harmful social consequences to warrant controlling its development; but by the time these consequences are apparent, control has become costly and slow. (Collingridge 1980, p. 19)

This dilemma, which has sometimes been termed the *Collingridge dilemma,* can be resolved either by improving predictability or by increasing controllability. Efforts to improve predictive reliability in the context of infant technologies are absolutely hopeless on Collingridge's view, so resolution occurs only by focusing on the control issue. The manner in which technologies become resistant to change must be understood. Entrenchment occurs because technologies become intertwined so that changing one requires changing many, thus making change costly and slow, if even possible. The solution is to develop technologies in ways that avoid rigidity and maintain flexibility. This may be achieved by implementing corrigible technologies whose flaws can be detected quickly and corrected easily. Continuous monitoring with the aim of finding error is thus imperative. Decisions that keep future options open should always be favored. Collingridge's later work (1992) provides close analyses of several contemporary technologies. These analyses show that the cost of inevitable mistakes can be further reduced through decentralized decision making and non-hierarchical organizational structures. Significantly concern about the unpredictability of technology is no longer limited to the emergence of *new* technologies. The inevitability of error and of predictive unreliability are seen as general conditions of human existence. "We are indeed poor naked creatures ... People have to make choices under great adversity, where the levels of uncertainty seem bottomless" (Collingridge 1992, p. 3). Like Lindblom, Collingridge's later work takes a more extreme view of the nature and promise of incrementalism.

Additional support for incrementalist principles can be found in the work of Joseph Morone and Edward Woodhouse (1986) where the aim is to explain the *infrequency* of technological disasters. Several strategies facilitate intelligent control of technology, one of which is to "Be actively prepared to learn from error, rather than naively expecting to analyze risks in advance or passively waiting for feedback to emerge" (Morone and Woodhouse 1986, p. 160). In the realm of business operations, James Quinn (1980) concludes that, while most companies have formal planning structures, formal planning has little to do with effective operations. Major strategic decision making occurs outside of the formal planning process. Managers maintain flexibility and avoid premature decisions, delay action as long as feasible to increase feedback and communication, and promote interactive learning. So incremental development may account for successful business management as well as technology development per se.

Theory and Criticism

As indicated, incrementalist thinking emphasizes flexibility and responsiveness, values with roots in the philosophy of Popper and the economic theory of Hayek. Popper's epistemology (1972) emphasizes the inevitability of error and the necessity of devising effective means for learning from mistakes. This is achieved by maximizing opportunities for feedback via experience, especially by subjecting proposals to critical tests and by continuously reshaping ideas in the face of failed predictions. Hayek (1960) opposes the concentration of power in the hands of the few and top-down management via pre-planned, large-scale solutions. Rather he argues that market activity conveys information so as to exploit knowledge distributed throughout society, which results in a bottom-up problem solving. Even effective social institutions can arise by means of decentralized action. These concepts of decentralized decision making, continuous expectation of error, and iterative improvement through active exploitation of mistakes resonate throughout incrementalist thinking.

In respect to criticisms of incrementalism, Collingridge (1992) identifies two categories: critiques that point to successful, non-incremental development and critiques that point to unsuccessful incremental change. Ian Lustick (1980), for instance, argues that non-incremental approaches are superior in achieving safety in nuclear power. Other examples of this type include Paul Schulman (1975, 1980) and Jennifer Hochschild (1984). By contrast, incrementalism is also criticized as too plodding to resolve certain social problems. Sometimes radical innovations are called for in response to radical socioeconomic contingencies. Problems whose severity has quantitative measures, such as air pollution, prove that small-scale change allows desirable goals to gradually slip further away. Or threshold and *sleeper* effects may occur in which large unpredictable changes result from small steps (Dryzek 1987; Mushkat 1987).

A related worry concerns the extent to which incrementalism is too extreme in its distrust of prediction and knowledge. This distrust may derive from concerns expressed by Adam Ferguson and Baron de Montesquieu, and later echoed by Hayek, over the unanticipated consequences that attend all technological innovations. In its extreme forms, incrementalism justifies limiting technological or social programs on the basis of general unpredictability and inadequate knowledge. Nevertheless more than mere inevitability of error is required to justify restricted development. The fact that unexpected consequences, even undesired consequences, will occur fails to make the case. What is needed is some assurance that the magnitude of the unexpected, undesirable consequences will outweigh the magnitude of the expected, desirable consequences. For this, substantial and reliable predictability is required.

Collingridge admits, for example, that not all resources can be committed to maintaining flexibility and that corrigibility and monitoring have costs. These costs must be weighed against that of making an error. Yet if predictability is forsaken, such *error costs* cannot be accurately estimated (Croy 1996). By granting accurate estimation of error costs, Collingridge's case assumes predictive reliability and lapses into inconsistency.

Incrementalism has also been criticized for insensitivity to the political process, both in the attempt to develop technology and to solve social problems. Collingridge has been admonished for not recognizing that determining what counts as an error or mistake is essentially a politically driven judgment (Johnston 1984), and more recent work on incrementalist theory takes pains to deal with the deleterious effect of special interest groups on incremental development (Hayes 2001). In each case, political process complicates the speedy responsiveness to error so crucial for flexibility.

Critiques such as these reveal the connection between incrementalism and controversies surrounding attempts at social progress, particularly those that pit utopian reform against incremental development. Popper's distinction between piecemeal social engineering and utopian engineering paves the way for this connection and for the wide reach of incrementalist concepts. When taken in its less extreme form as one problem solving strategy among many, one that is warranted or not by the nature of the problem confronted, incrementalism withstands critical scrutiny, provides a helpful methodological tool in the quest for improving society, and stimulates questions about the nature of social reform.

MARVIN J. CROY

SEE ALSO *Popper, Karl; Social Engineering.*

BIBLIOGRAPHY

Braybrooke, David, and Charles Lindblom. (1963). *A Strategy of Decision; Policy Evaluation as a Social Process.* New York: Free Press of Glencoe. Collaboration between the philosopher, Braybrooke, and the social scientist, Lindblom, brings early incrementalist thinking into focus. Lindblom's chapters foreshadow future development of incrementalist concepts.

Collingridge, David. (1980). *The Social Control of Technology.* New York: St. Martin's Press.

Collingridge, David. (1982). *Critical Decision Making.* New York: St. Martin's Press.

Collingridge, David. (1992). *The Management of Scale: Big Organizations, Big Decisions, Big Mistakes*. London: Routledge.

Croy, Marvin. (1996). "Collingridge and the Control of Educational Computer Technology." *Techné: Journal of the Society for Philosophy and Technology* 1(3–4): 1–14.

Dahl, Robert, and Charles Lindblom. (1953). *Politics, Economics, and Welfare: Planning and Politico-Economic Systems Resolved into Basic Social Processes*. New York: Harper and Row. Seminal, lucid characterization of incrementalism and its relation to other strategies.

Dryzek, John. (1987). "Complexity and Rationality in Public Life." *Political Studies* 35: 424–442.

Hayek, Friedrich. (1960). *The Constitution of Liberty*. London: Routledge and Kegan Paul.

Hayes, Michael. (2001). *The Limits of Policy Change: Incrementalism, Worldview, and the Rule of Law*. Washington, DC: Georgetown University Press. This work provides a contemporary elaboration of incrementalist principles critically compared against recent case studies. The contributions of Popper and Hayek are clearly delineated.

Hochschild, Jennifer. (1984). *The New American Dilemma: Liberal Democracy and School Desegregation*. New Haven, CT: Yale University Press.

Johnston, Ron. (1984). "Controlling Technology; An Issue for the Social Studies of Science." *Social Studies of Science* 14: 97–113.

Lindblom, Charles. (1959). "The Science of Muddling Through." *Public Administration Review* 19: 79–88.

Lindblom, Charles. (1979). "Still Muddling, Not Yet Through." *Public Administration Review* 39: 517–526.

Lustick, Ian. (1980). "Explaining the Variable Utility of Disjointed Incrementalism: Four Propositions." *The American Political Science Review* 74: 342–353.

Mannheim, Karl. (1967 [1940]). *Man and Society in an Age of Reconstruction; Studies in Modern Social Structure*. New York: Harcourt, Brace & World.

Morone, Joseph, and Edward Woodhouse. (1986). *Averting Catastrophe: Strategies for Regulating Risky Technologies*. Berkeley: University of California Press.

Mushkat, Miron. (1987). "Towards Non-Incremental Strategies in Developing Public Products and Services." *European Journal of Marketing* 21: 66–73.

Popper, Karl. (1972). *Objective Knowledge: An Evolutionary Approach*. London: Oxford University Press.

Quinn, James. (1984). *Strategies for Change: Logical Incrementalism*. New Haven, CT: Yale University Press.

Schulman, Paul. (1975). "Non-Incremental Policy Making." *The American Political Science Review* 69: 1354–1370.

Schulman, Paul. (1980). *Large-Scale Policy Making*. New York: Elsevier.

INDIAN PERSPECTIVES

• • •

India (along with China and Egypt) is home to one of the oldest and perhaps the most continuous cultural tradition on the earth. Although it occupies only 2.4 percent of the global land area, it is home to fifteen percent of the population, and by 2050 is projected to be the most populous country in the world. India spends approximately six billion dollars every year on science and technology; science and technology have been central to the country's development since its independence in 1947, while themselves being subject to distinctive assessments and adaptations.

Historical and Cultural Context

Knowledge enjoys sacred stature in Indian culture and civilization. Saraswati, the goddess of knowledge, occupies a place of pride in the Hindu pantheon, while India's much-reviled caste system accorded the highest social status to Brahmins, whose profession was to create and disseminate knowledge. Ancient India's sometimes contested scientific contributions—including theories of gravity, the age of the universe, modern numerals, trigonometry, and the conception of zero—were often first described in religious scriptures. Utilitarian and empirical observations about agriculture and medicine that survived generations were often couched in idioms and expressions with religious connotations. Even during Mughal rule (1526–1707), respect for Indian mathematics was instrumental in its spread to places as far as Central Asia, Spain, and North Africa (Teresi 2002).

Respect for knowledge workers—scientists and doctors—turned to awe during British rule, when science as practiced in Europe took hold through the Geological, Botanical, and Trigonometric Surveys established through the efforts of the Asiatic Society (founded 1784). Although the first voices of dissent—notably of the philosopher Ananda Kentish Coomaraswamy (1877–1947), poet and literature Nobel laureate Sir Rabindranath Tagore (1861–1941), and the father of India's freedom struggle, Mohandas Karamchand Gandhi (1869–1948)—against this surrender to Western science were voiced as early as 1905, they had to wait till the 1970s and 1980s to gain traction through democratic people's movements.

When India became independent after two hundred years of British rule, Gandhi anointed Jawaharlal Nehru (1889–1964) the country's first prime minister. While

Nehru was popular in his own right and received three overwhelming electoral mandates following his appointment in 1947, his elevation to the highest office—although a foregone conclusion—was a curious one.

It was curious because Nehru's and Gandhi's visions of independent India could not have been more different. Nehru's vision of India was that of a highly industrialized and progressive economy where dams, laboratories, industrial facilities, and mechanization would be revered as "temples of modern India" (Nehru 1958, p. 3). Gandhi's vision, conscious of India's predominantly rural base, focused on the rural village as the central element of development and opposed all products of science and technology that displaced human labor. Few, however, shared Gandhi's economic views in the Congress party, and a desire to undertake rapid (state-sponsored) industrialization was articulated as early as 1931 (Chandra, Mukherjee, and Mukherjee 1999).

To Nehru's credit, then, goes the rapid growth of India's industrial infrastructure, the creation of the Council of Scientific and Industrial Research (CSIR), the world's largest chain of publicly-funded research laboratories, the founding of the Indian Institutes of Technology, the country's nuclear and space research programs, and most importantly, the faith that "science alone ... could solve ... problems of hunger and poverty, of insanitation and illiteracy, of superstition and deadening custom and tradition, of vast resources running to waste, of a rich country inhabited by starving people" (Gopal 1972, p. 807).

Nehru's investment in science and technology produced the Green Revolution, which is arguably the first significant achievement of mainstream Indian science. India's Green Revolution refers to the enormous improvement in agricultural productivity the nation achieved starting mid-1960s. Thanks to Green Revolution's introduction of a high-yield variety of seeds, fertilizers, and scientific agricultural practices, India's production of food grains increased by thirty-five percent between 1967 and 1970. India, which imported 10.3 million tons of food grains in 1966, had food grain reserves of 128.8 million tons in 1984 and exported 4.8 million tons of food grains in 2001 (Chandra et al. 1999).

While its ecological legacy is sometimes disparaged, the Green Revolution vindicated Nehru's "temples of modern India" and established their legitimacy as effective instruments of development. These institutions have since notched several accomplishments, including super-computers in response to technology denial from the United States; the production, launch, and utilization of satellite technology; processes to produce raw materials for fuels and textile fibers; and a cheap but effective telecommunication network (Parthasarathi 2003).

Yet academics complain that scientific work accomplished entirely in India is yet to win a Nobel Prize, and the number of peer-reviewed papers decreased by almost twenty percent between 1980 and 2000, even as the number of universities and research institutions almost doubled and funding grew seventeen times (Balaram 2002). Further, corporate innovation and science-based entrepreneurship, notwithstanding several promising efforts, has been limited in scope and success (Turaga 2000). At the same time, China, South Korea, Brazil, and Israel have registered impressive growth, leaving critics to suggest that mainstream Indian science is of a mediocre quality.

Technocracy versus People's Science

Even so, the rapid advent of globalization has dulled dissent and absolute devotion to the technocrat was witnessed as late as 2002, when the renowned missile scientist A. P. J. Abdul Kalam (b. 1931) became independent India's eleventh president. Although elected indirectly, Kalam's nomination received a landslide vote, nationwide support, and near fanatical endorsement from India's educated middle class.

Kalam's disheveled long hair, soft-spoken demeanor, and spartan lifestyle (a bachelor, he lived in a one-bedroom government apartment until he became president) reinforced stereotypes of scholarship and suggested integrity uncommon to recent Indian public life. Kalam became a national icon and household name following India's May 1998 nuclear tests, of which he was the widely recognized scientific architect. What fired the imagination of the nation's educated, however, were Kalam's dreams of a *developed* India constructed through the apolitical pursuit of science and technology as an entirely objective and value-free activity (Kalam 2002).

In sharp contrast to Kalam is the articulate activist Sunita Narain, chairperson of the New Delhi-based radical environmental advocacy group Centre for Science and Environment. Narain has marshaled scientific research, data, and opinion to create immensely popular media campaigns for clean air, water, and food that have eventually influenced public policy. Narain commands enough influence for *India Today*, a leading Indian newsmagazine, to list her as one of India's fifty most powerful and influential citizens in 2004. However, "development is not a road" for Narain, who is severely critical of India's scientific, political, and social establishment (Narain 2003).

The lopsided battle being fought at the crossroads of these conflicting definitions of *development* constitutes a central theme in the emerging interdisciplinary field of Science, Technology, and Society (STS) studies in India. The stronger side in this battle is the statist version of science promoted by the likes of Kalam, whose vision of development is sanitized, crystalline, and sees power plants, dams, roads, factories, and software firms as both instruments and milestones in the quest for India's development. The rapidly growing ranks of the country's educated middle class see in Kalam an unprecedented opportunity to achieve this vision.

Cast against this powerful technocracy is a motley crowd of academics, environmentalists, and social critics with diverse but strong intellectual views. These are people who agitate against dams because of their inhumane consequences on marginalized tribal groups, picket government offices to protest power plants in protected forests, and advocate indigenous, small-scale technologies to harvest water and energy. Not half as focused or strong resource-wise as the statist agenda, these constituents of civil society have covered ground using imaginative ideas, rich rhetoric, moral leadership, articulate spokespersons, and successful grassroots political action.

Critics also question why a developing country like India should invest in supercomputers, satellites, and atomic energy, especially when more Indians sleep hungry than elsewhere in the world, one in three is illiterate and subsists on less than a dollar a day, infant mortality is at sixty-eight per one thousand births, nearly half of all Indian children are malnourished, access to affordable drugs is heterogeneous and available to every other Indian in the best of communities, and only twenty-eight percent of India's population has access to improved sanitation (United Nations 2003, pp. 237–339). That the Indian discourse of ethics in science and technology should raise these questions indicates that Nehru's "temples of modern India" have not been successful enough.

According to critics, the Nehruvian model was never suited to address these problems in the first place and instead has aggravated them. The Booker Prize-winning author Arundhati Roy, for example, estimates that the 3,600 dams India has built have "displaced maybe up to 56 million people" from their farms and livelihoods to the growing ranks of the urban poor (Roy 2001, p. 10). Things would have been different in Gandhi's village-based economy, they argue. Gandhi, however, was not alone critiquing the application of science and technology in the Indian context.

The Swadeshi Movement

The role and effects of modern science on Indian traditions, people, and society was intensively debated as early as 1905, during the *Swadeshi* (local, native, indigenous) Movement, when "the boycott of foreign goods ... met with the greatest visible success at the practical and popular level" (Chandra, Mukherjee, Mukherjee, et al. 1989, p. 129). Although Swadeshi was a political movement belonging to the larger freedom struggle, "it was accompanied by an efflorescence of cultural debates ... around the civilizational question of science and state" (Visvanathan 1987, p. 15).

Coomaraswamy was a leading figure in this debate; he was concerned that, lacking concerted effort, India's great craft traditions and art cultures would be lost to modern science. Intermediate technologists such as the British civil servant and founder of the Indian National Congress, Allan Octavian Hume (1829–1912), appreciated, if reluctantly, the rationality of traditional technologies but questioned their viability against the "onslaught of modernity, capitalism, and imperialism" (Visvanathan 1987, p. 17). If intermediate technologists exhorted blending both medieval and modern technological traditions to facilitate meaningful industrialization, Tagore was convinced that the two cultures could converse only after the differences between them were first recognized. It was to facilitate such studies that Tagore created Visva-Bharati University at Santiniketan in eastern India in 1925.

Most of these Swadeshi arguments, however, have gone unaddressed and modern India would disappoint Coomaraswamy, Hume, and Tagore. India's current and future economic growth rests on exporting software services, rendered by engineers educated at institutions (for example, the Indian Institutes of Technology) built with the support of Western universities. Curricula at such universities rarely include STS studies or the traditional technologies that Coomaraswamy wanted to preserve. Globalization and liberalization have relentlessly destroyed Indian communities practicing traditional agriculture and medicine, art, and handicrafts. Although governmental and voluntary initiatives seek to preserve the few remaining bastions of India's cultural traditions, they are a far cry from the "gene pools of an alternative imagination which had to be sustained and eventually made available to the West" (Visvanathan 1987, p. 16).

Future Prospects

Not all is lost, however, and there is cause for optimism in contemporary India. One heartening illustration is the pioneering work of Sulabh International, which has worked with local governments, communities, and vendors to develop a low-cost, environmentally sustainable, and socially acceptable sanitation system for both rural and urban communities. In the past thirty-five years, Sulabh has created fifty thousand jobs through its one million latrine units that have served over ten million people. Similar efforts by several voluntary outfits, people's science movements, and public-spirited initiatives have helped achieve social equity, improved literacy, and better and affordable public health care (United Nations 2003, p. 105).

Even the mainstream scientific establishment is better engaging traditional and indigenous knowledge systems. In the mid-1990s, CSIR successfully contested and overturned a U.S. patent on the use of turmeric powder to heal wounds. The U.S. Patent and Trademark Office upheld CSIR's claims that turmeric has been used in India for centuries and its medical properties are well ingrained in Indian folklore.

Indian scientific agencies have since aggressively espoused the intellectual property rights of India's indigenous communities and encourage research, development, and commercialization of traditional knowledge. Fundamentally, however, India has made a decisive shift towards the Western scientific and technological traditions to derive the same economic and human development benefits realized by developed nations. Thus, when CSIR succeeded in overturning the turmeric patent, it chose to project the victory as the best possible evidence of the integrity, transparency, and objectivity of the international patenting regime, which India began conforming to completely starting 2005.

India is, however, yet to embrace STS concepts such as risk assessment, informed consent, engineering ethics, right to information, and transparency to the extent they are ingrained in the practice of science in the developed West. This will change with economic and technological development, which is occurring rapidly, as well as grassroots people's movements. A greater impetus, however, might come from Western collaborators, who are increasingly using India's modern infrastructure, engineering talent, and large population to cheaply develop products, design cars and factories, and conduct clinical research (Turaga 2003). Illustrating this growing trend is a 2001 controversy involving Johns Hopkins University and a cancer hospital in southern India, where some patients participating in a clinical trial were not informed of the new drug's consequences (Bidwai 2001).

The foremost of Indian polity and society's concerns for the future relate to advancing the quality of its people's economic status, health care, and education. Science and technology are now widely accepted as important to such development. This unquestioning acceptance has been tempered to some extent with grassroots activism and people's movements, which have their origin in India's successful practice of and absolute commitment to democracy. Further, India's globalization will enable the quick assimilation in its public policy of Western principles shaping scientific and technological progress. Thus, the relationship between India—one of the world's most profound civilizations—and science, technology, and ethics will be shaped by two important trends that differ in size and methods, but have goals that share some philosophical similarity.

UDAY T. TURAGA

SEE ALSO *Buddhist Perspectives; Hindu Perspectives.*

BIBLIOGRAPHY

Balaram, P. (2002). "Science in India: Signs of Stagnation." *Current Science* 83(3): 193–194.

Bidwai, P. (2001). "The Misuse of Science." *Frontline* 18(17): 53–55.

Chandra, Bipan, M. Mukherjee, A. Mukherjee, et al. (1989). *India's Struggle for Independence, 1857–1947.* New Delhi: Penguin.

Chandra, Bipan, Aditra Mukherjee, and Mridula Mukherjee. (1999). *India after Independence.* New Delhi: Penguin.

Gopal, Sarvepalli. (1972). *Selected Works of Jawaharlal Nehru,* 15 vol. New Delhi: Oxford University Press.

Kalam, A. P. J. Abdul, and Y. S. Rajan. (1998). *India 2020: A Vision for the New Millennium.* New Delhi: Viking.

Kalam, A. P. J. Abdul. (2002). *Ignited Minds: Unleashing the Power Within.* New Delhi: Viking.

Narain, Sunita. (2003). "Development is Not a Road." *Down To Earth* December 15: 1.

Nehru, Jawaharlal. (1958). *Jawaharlal Nehru's Speeches, March 1953—August 1957,* Volume III. Calcutta: Publications Division, Ministry of Information and Broadcasting, Government of India.

Parthasarathi, Ashok. (2003). "India: A Champion of New Technologies." *Nature* 422: 17–18.

Roy, Arundhati. (2001). "Scimitars in the Sun: N. Ram Interviews Arundhati Roy on a Writer's Place in Politics." *Frontline* January 19: 4–18.

Teresi, Dick. (2002). *Lost Discoveries: The Ancient Roots of Modern Science—from the Babylonians to the Maya.* New York: Simon & Schuster.

Turaga, U. (2000). "India's Techno-Economic Revolution." *Chemical Innovation* 30(8): 43–49.

Turaga, U. T. (2003). "Outsourcing R&D." *Chemical Engineering Progress* 99(9): 5.

United Nations Development Programme. (2003). *Human Development Report 2003*. New York: Oxford University Press.

Visvanathan, Shiv. (1987). "On Ancestors and Epigones." *Seminar* 330: 14–24.

INDICATORS

SEE *Science and Engineering Indicators; Social Indicators*.

INDIGENOUS PEOPLES' PERSPECTIVES

• • •

The term *indigenous* is used to refer to the original inhabitants in a region. With regard to human populations, this term can be politically ambiguous, but the concept is still helpful in referring to small-scale societies with distinct languages, mythic narratives, sacred places, and kinship systems. Located on all the major continents (except Antarctica) as well as the Pacific Ocean areas, more than 500 million peoples are considered indigenous. In many contemporary settings these native societies are so marginalized within their nation-state settings and so subject to the extractive exploitation of multinational corporations that their existence is threatened. In these traditional societies the distinctive activities of understanding nature, the technology of subsistence, and an ethics of balance are not separate from one another. Rather, in diverse ways in these different native settings, the interactive relationships of knowing, producing, and thinking about behavior constitute coherent social wholes that can be called worldviews. Indigenous worldviews change over time, yet they also manifest symbols shared with the larger human community in rituals and myths that bind the quest for personal identity, the spirit of community, and ways of knowing the cosmos.

The term *lifeway* is used here to indicate this cultural integration of thought, production, and distribution among indigenous societies. These diverse and integrated perspectives of native peoples have often been dismissed as animism, or *failed epistemologies*, that posited a vitality or life force within the world that entered into all technological activities and ethical considerations. From a social science perspective, no such life force could be measured or consistently observed, and, thus, the world-views, ethics, and technologies of native peoples were seen as too limited for attention by modern urban societies. However in the early-twenty-first century, the philosophical subtlety and social creativity evident in such native technologies as astronomical and ethnobotanical knowledge, healing therapies, cosmological narratives, and aesthetics of performance evident in ritual performances and rock art petroglyphs (rock incisions) and pictographs (applied paint) are being reassessed.

Approaches to Indigenous Peoples

Early encounters by Western Europeans with indigenous peoples were generally interpreted in the context of the Bible. When indigenous peoples manifested empirical knowledge, productive technology, or disciplined behavior, observers judged the achievements to be God-given and their genesis related to Western scriptures. Thus a naïve view of native peoples as prelapsarian, or living in the original innocence of the edenic paradise, gave rise to a romantic view of indigenous peoples as *noble savages*. From a similar but negative biblical perspective, indigenous peoples, their arts, and their activities were seen as spawned by the devil and deprived of the divine grace of the Western civilized arts. Thus any striking architecture, such as the mounds of the river valleys of Ohio or the Mesoamerican pyramids, was attributed to *lost* biblical tribes, or prehistoric Caucasian influences from Viking navigators or Irish monks. Lacking a coherent social science, early encounter-period European views dismissed as childlike the petroglyphs and pictographs of indigenous peoples. Thus the lyrical hunting scenes in the cave art of Zimbabwe, Botswana, and South Africa, or the numinous presences manifest in the cave art of Australian indigenous peoples was largely interpreted as psychological projection, sympathetic hunting magic, or primitive aesthetic. For indigenous peoples, however, these varied forms of symbolic expression symbolically made present their commitments to place, the numinous forces in local regions, and often their knowledge base regarding animals, plants, land, and weather.

Beginning with the sixteenth-century early modern period, new intellectual perspectives in Western Europe associated with critical, skeptical thought allowed for innovative views of indigenous lifeways. Influenced by the *Jesuit Relations* (1632–1673) as well as limited exchanges with Brazilian native peoples, Michel Montaigne (1533–1592) rejected the idea of native peoples as morally depraved and favorably compared the reported cannibalism of indigenous peoples with the savage brutality of the religious wars of Europe of his

Petroglyphs. These images had deep cultural and religious significance for the societies that created them. *(Field Mark Publications.)*

day. Baron de La Brède et de Montesquieu (1689–1755) in *The Spirit of the Laws* (1748) proposed that the spirit of native societies also resulted in laws, political structures, and social decorum.

By the early-twentieth century, the philosopher Lucien Levy-Bruhl (1857–1939) proposed that indigenous worldviews emerged from a prelogical mentality, intellectually different than the rational, logical Western mind, characterized by *mystical participation* in a pervasive life force (Levy-Bruhl 1923, 1985). His thesis is sharply questioned for projecting a universal mindset on very different peoples, but his emphasis on a cultural logic brought to descriptions of the world is now widely accepted. For native peoples, their perception, knowledge, and explanation of the world relates to their immediate technological-environmental circumstances as well as their linguistic and ideological heritage.

Franz Boas (1858–1942) emphasized *cultural relativity* and oriented a new generation of anthropologists to investigate the knowledge, technologies, and ethics of indigenous peoples as whole systems, or cultures. The anthropologist Claude Levi-Strauss (b. 1908), in *The Savage Mind* (1962), observed that the science and technology of native peoples follows from a mental structure evident in mythologies in which perception and attention to the natural world gradually lead to a cultural

world. From a religious perspective, Mircea Eliade (1907–1986) proposed in the 1950s that indigenous peoples embodied technologies and ways of living that were based on seasonal and cosmological cycles rather than linear, historical understandings of reality.

Faced with the description of native North American peoples as the *first ecologists*, scientists in the 1980s questioned the roles of indigenous peoples in the extinction of large mammals, which occurred when native peoples were believed to have migrated to the American hemisphere (Martin and Klein 1984). The scientific understanding of indigenous knowledge continues into the present often including the voices of indigenous elders, artists, and intellectuals who seriously challenge the extinction theory. Acknowledging the roles of native hunters in mammoth and mastodon die-off evident in Clovis and Folsom spear-point technologies, they propose broader considerations of both anthropogenic and natural causes such as climatic change, disease pathogens, and fire (Deloria 1995, Wong 2001).

Indigenous Perspectives

Indigenous perspectives suggest that the art of knowing, or science, and the forces of production, or technology, as well as the sense of appropriate behavior, or ethics, weave together social and cosmological values. That is, knowledge of the world, tools for work, and reflection on one's behavior are properties of persons who are actively engaged with a living environment. Human persons interact with a world alive with dynamic forces that are powerful persons watchful of human behavior. Science, technology, and ethics are not transmitted in traditional thought as ways of controlling nature but primarily as modes of interaction with these other-than-human persons. Indigenous science results from maturing attention to nature as beings-in-the-world having capacities to interact with humans in person-to-person exchanges.

Technology is a way of creating the world, in relation to a task, that a person comes to gradually and internally as much as productively and externally. Ethics among indigenous peoples embodies a cultural relationship with specific places and forms of life in a local region that matures as the person ages. Through ritual and performance arts, such as rock art, basketry, canoe making, beading, and habitat construction, indigenous people express personal and social identity. These coherent, integrated activities place the human person in relation to powerful other-than-human spirit beings that inhabit the cosmos. Thus the personal subjectivity of humans, in indigenous perspectives, is brought to fruition through intersubjectivity with the world of animate forces. Paraphrasing the observations of Thomas Berry (1988), the weave of indigenous science, technology, and ethics is evident in their recognition that the universe is not a collection of objects, but a communion of subjects.

The social and cosmological basis of science, technology, and ethics within indigenous thought stands in sharp contrast to nonnative, European, Western, Marxist, capitalist, or other current globalization views. Broadly speaking, in modern standpoints technology has been identified as technical or mechanical manipulation of inert matter related to work as production. Ethics, following this paradigm, comes before action as intentional thought brought to fruition in activity. In all three acts, namely, science as knowing, technology as work, and ethics as intention, the human is central. The contemporary global ethos associated with urban, industrial societies is wholly anthropocentric. In the indigenous perspective the roles of science, technology, and ethics are integrated into the formation of persons and communities (Ingold 2000). Science, technology, and ethics are not simply anthropocentric acts that psychologically orient individuals and communities inward as the source of ultimate value. Rather indigenous perspectives foster an anthropocosmic orientation in which the living world is central, and the human seeks to balance inner identity and meaning in relation to a holistic outer world.

Indigenous intellectual knowledge exemplified in such inventions as the canoe, the bow and arrow, ritual ceremonies of seasonal renewal, and shamanistic therapies all involve complex interactions of place, spirit persons, and symbolic language. Coupled with the striking traditional environmental knowledge evident, for example, in the extraction and blending of plants to produce the ritual hallucinogen, ayahuasca, they affirm the prowess of science and technology among indigenous peoples. Rarely, however, have observers determined that material, human, and spiritual worlds are separated by the indigenous ethics implicate in those inventions. Becoming an authentic human in indigenous views involves relationship with and treatment of the natural world-as-person. Knowing and using the world implicates one's own body, social setting, and larger cosmological forces.

One Example from the Yekuana Peoples of South America

Among the Yekuana peoples of Venezuela traditional environmental knowledge gives rise to technical skills that foster an ethics, constructed in relation to mythological stories, for progressing gradually into mature personhood. Technical developments, such as the press for extracting yucca, the large circular community houses, as well as forms of social life are considered to have come from the culture hero, Wanadi; whereas all the troublesome, corruptible, dangerous aspects of nature and human life come from Odosha. The complex stories of the birth of Odosha from Wanadi's afterbirth, which was improperly buried, and the consequent yearnings and desires embedded within the natural world serve to teach Yekuana traditional environmental ethics. Each individual Yekuana participates in both the cosmic struggle of Wanadi and Odosha, as well as in the creative presence of Wanadi, for example, in the knowledge, skill, and intention of making yucca presses and especially baskets.

The Yekuana have developed a complex set of ethical teachings connecting the emergence of designs for baskets, the materials for making baskets, and limits on collecting those materials. Set within mythological stories of Wanadi and Odosha, the tense and ambiguous weave of the actual human condition is likened to those cosmological webs of relationships. Among the Yekuana the pragmatic use and location of grasses and roots for basket making are hedged with ethical warnings of the allure of those spirit beings who inhabit the grasses as well as the danger of inappropriate and unlimited use.

The knowledge of these grasses, the technical skills used in weaving them into baskets, and the complex of stories associated with their presence in the region are also directly related to personal maturing and social status (Guss 1989).

These complex cosmological stories braid cognitive-intellectual and affective-emotional realms of human experience into a learned and embodied practice of restraint. In effect, the weaving of baskets among the Yekuana is considered an aesthetic and contemplative skill in which individuals mature in their self-realization of society and bioregion. Thus Yekuana ethics springs from an inherent knowledge of limits with regard to natural consumption.

Conclusion

Indigenous knowledge is traditional in that it informs technical means not as a separate ethical mode but as the cosmological weave of storied knowledge, natural materials, and a respect for beings-in-the-world that limits consumption. No doubt ethical teachings emerged among indigenous peoples because there were those who overstepped cultural boundaries. The examples given here are not descriptive of all individuals within any one particular native community, nor of the diverse ways of knowing, embodying technical skills, and implementing ethical teachings among indigenous peoples. Yet there are shared indigenous perspectives, or *family resemblances*, embodied in science, technology, and ethics as ways of living that arise from the mutual dialogues of body, society, and place in the larger cosmological whole.

JOHN A. GRIM

SEE ALSO *Development Ethics; Globalism and Globalization; Modernization.*

BIBLIOGRAPHY

Basso, Keith. (1996). *Wisdom Sits in Places: Landscape and Language among the Western Apache.* Albuquerque: University of New Mexico Press.

Berry, Thomas. (1988). *The Dream of the Earth.* San Francisco: Sierra Club Books.

Bettinger, Robert. (1991). *Hunter-Gatherers: Archaeological and Evolutionary Theory.* New York: Plenum Press.

Boas, Franz. (1938). *The Mind of Primitive Man.* New York: Macmillan.

Boas, Franz. (2004). *Anthropology and Modern Life.* New Brunswick, NJ: Transaction Publishers.

Deloria, Vine, Jr. (1995). *Red Earth, White Lies: Native Americans and the Myth of Scientific Fact.* New York: Scribner.

Eliade, Mircea. (1985). *Cosmos and History: The Myth of the Eternal Return.* New York: Garland. Originally published in 1959.

French, Michael. (1988). *Invention and Evolution: Design in Nature and Engineering.* London: Longman.

Grim, John, ed. (2001). *Indigenous Traditions and Ecology: The Interbeing of Cosmology and Community.* Cambridge, MA: Harvard Divinity School Center for the Study of World Religions.

Guss, David. (1989). *To Weave and Sing: Art, Symbol, and Narrative in the South American Rain Forest.* Berkeley: University of California Press.

Harvey, Graham, ed. (2000). *Indigenous Religions: A Companion.* London, New York: Cassell.

Hutchins, Edwin. (1995). *Cognition in the Wild.* Cambridge, MA: MIT Press.

Ingold, Tim. (2000). *Perception of the Environment: Essays in Livelihood, Dwelling & Skill.* New York: Routledge.

Levi-Strauss, Claude. (1966). *The Savage Mind.* Chicago: The University of Chicago Press.

Levy-Bruhl, Lucien. (1923). *Primitive Mentality,* trans. Lilian A. Clare. New York: Macmillan.

Levy-Bruhl, Lucien. (1985). *How Natives Think,* trans. Lilian A. Clare. Princeton, N.J.: Princeton University Press. Originally published in 1910.

Martin, Paul, and Richard Klein, eds. (1984). *Quaternary Extinctions: A Prehistoric Revolution.* Tucson: University of Arizona Press.

Montesquieu, Baron de La Brède et de. (1994). *The Spirit of the Laws,* trans. G. D. H. Cole. Chicago: Encyclopedia Britannica.

Sahlins, Marshall. (1972). *Stone Age Economics.* Chicago: Aldine-Atherton.

Wong, Kate. (2001) "Mammoth Kill." *Scientific American* 284(2): 22.

INDUSTRIAL REVOLUTION

• • •

The concept of an industrial revolution denotes an economic transition in which the means of production become increasingly specialized, mechanized, and organized. This process uses technology, in some association with science, to create large increases in the productive capacity of an economy, which in turn eventually transforms society as a whole. Industrial revolution is less violent or dramatic than political revolution and has roots that extend into the preindustrial agrarian past as well as consequences that continue to influence distant places and times. Great Britain inaugurated the Industrial Revolution in the late eighteenth century, and other nations have undergone similar revolutions in

A puddling furnace. Iron production was the first pillar of the Industrial Revolution. *(Hulton Archive/Getty Images.)*

subsequent years, continuing to the present. This process may be described as a single ongoing Industrial Revolution or as a series of separate revolutions that influence one another. Either way, the Industrial Revolution is without question one of the most important transformations in human history, and it is best understood through an appreciation of its complex origins, its evolution and spread, and its ethical and political influences.

Historical Origins

Most human societies have passed through several broadly defined stages marked by major turning points or revolutions. The transition from nomadic hunting and gathering to settled agriculture (farming and herding) that first occurred in the Near East is often called the Neolithic revolution. By enabling humans to live in one area, grow more numerous, and produce sufficient food surpluses to support nonfarming vocations such as artisanship and soldiery, the Neolithic revolution laid the groundwork for the next stage in societal evolution, the urban revolution. Human history is largely the history of cities and nations, and the gathering of populations into concentrated areas is responsible for many political, cultural, technological, scientific, and other developments. The Industrial Revolution is a third major societal transition point that follows and was made possible by the first two revolutions.

An industrial revolution requires a confluence of favorable labor, capital, technological, and ideological conditions. One vital component of industrialization is a populous labor supply that receives support from an agricultural sector capable of feeding it, and that possesses the necessary skills and discipline for manufacturing work. Capital is vital for covering the start-up and operating expenses that accompany new industrial endeavors, such as the purchase of land, facilities, and machinery; the preparation of stock on hand; the establishment of accounts receivable; and salary payments. Industrialization also depends on technological developments in manufacturing, power generation and transmission, transportation, and raw materials processing. Finally, an industrial revolution is facilitated by the development of political and philosophical ideologies that justify or mandate human organization and control over the natural environment. After many centuries of heterogeneous worldwide population growth, economic development, and technological advancement, all of these conditions converged for the first time in eighteenth-century Great Britain.

The Original Industrial Revolution

A variety of conditions caused Britain to experience moderate economic and manufacturing growth in the early eighteenth century, but these factors produced the greatest effects after 1760. By the 1780s, the British Empire's population, mechanization, and productive output were dramatically expanding. The term "Industrial Revolution" was first formulated by British historian Arnold Toynbee (1884), who considered this period of industrial and technological change more historically significant than political events such as the French Revolution.

Some of the preconditions of the British Industrial Revolution span or even predate the eighteenth century. New agricultural practices, such as the enclosures policies that brought more land under development, Jethro Tull's mechanical drill for sowing seed (c. 1701), Lord Townshend's four-year crop rotation system, advances in animal breeding, and the cultivation of the potato in Ireland, made possible a period of steady population growth. This population included a large supply of available laborers who started to concentrate in towns or cities.

Prior to the existence of large manufacturing establishments, Great Britain fostered a rich craft tradition that provided technological infrastructure and a substantial pool of skilled labor. Farmers comprised more than 90 percent of the preindustrial population, but artisans played a vital economic role. Indeed, while specialized artisans often congregated in cities, many farmers themselves practiced diverse craft trades or produced domestic manufactures in the evening or during winter months, serving as a vast pool of potential labor. This labor was increasingly tapped by enterprising merchants through the putting-out system, which involved the coordination of decentralized part-time laborers and led to regional specialization and the promotion of markets and towns. Early manufacturing networks introduced organizational, managerial, and business strategies that fostered the division of labor, specialization, and greater cooperation between workers or firms.

Great Britain also benefited from a convergence of advantageous economic, environmental, and technical factors. It possessed ample supplies of natural resources such as waterpower and coal, and its efficient transportation networks, including turnpike roads and water transport, further aided development. The commanding British navy and merchant network facilitated the shipment of raw materials to the mother country and carried British products to distant colonies or foreign markets. Described as a "nation of shopkeepers," Britain was founded on commerce, and its many merchants and middlemen fostered the spread of the market and funded manufacturing endeavors. Investment capital could also be raised and distributed through an advanced banking system and institutions such as the London Stock Exchange, and favorable regulatory policies (especially in comparison with European practices) enabled British manufacturers to practice their trades with a minimum of government interference. Two hundred years of British economic growth produced a relatively high level of prosperity, a widespread market economy, and a large potential demand for manufactured goods. And because the Industrial Revolution first took place within a capitalist economy, the pursuit of private profit drove the technological and industrial transformation.

What made it possible to take advantage of this confluence in material factors was the contemporary development of new ideals about how human beings could best realize their humanity. A sense of human beings as having the right to dominate the nonhuman world through technology, which had been emerging within a Christian theological framework in Europe, was given new secular articulation by, for instance, Francis Bacon (1561–1626) and his followers. Bacon's ethical vision of "the conquest of nature" for the "relief of man's estate" both justified and encouraged those activities that merged historical changes into a revolution in human industrial activity.

The takeoff of the British Industrial Revolution arose when several key productive sectors used new technologies to increase quantities of low-priced manufactured goods, change employment patterns, and expand technological networks that aided technical innovation and adoption. As the first nation to industrialize, Britain could not receive capital or technological aid from others. Fortunately, the technological challenges of the early Industrial Revolution were relatively simple and were certainly addressable via decentralized and informal experimentation and tinkering.

Iron production was the first technology to influence the British Industrial Revolution, in conjunction with developments in coal processing. Prior to the eighteenth century, British iron production had been increasingly limited by scarce supplies of wood, which was used to make charcoal. Coal was unusable in blast furnaces for various reasons, but in 1709 Abraham Darby discovered that coke, a burnable substance produced from coal, could be used. Technical barriers and quality control issues proved very limiting until 1760, at which point the British iron industry rapidly expanded.

Steam engines served as a second pillar of the Industrial Revolution and had close ties to coal mining and iron production. The Newcomen steam engine, invented by Thomas Newcomen in 1712, was a bulky and inefficient apparatus requiring enormous quantities of coal fuel. These limitations did not deter coal mine operators, who used it to pump water from deep mine shafts. Steam engines also became increasingly important for the iron industry, where they pumped water for water-powered bellows beginning in 1742, drove air

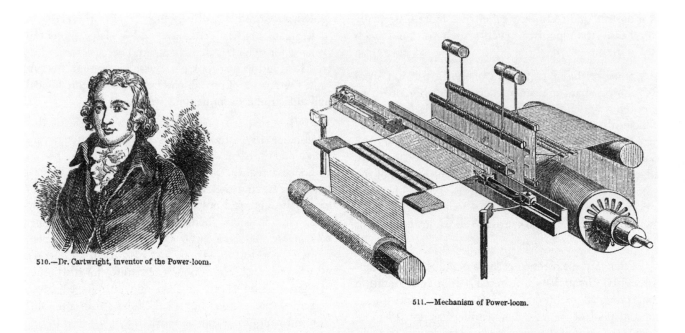

510.—Dr. Cartwright, inventor of the Power-loom.

511.—Mechanism of Power-loom.

Dr. Edmund Cartwright (left), the inventor of the power loom (right). The textile industry was the third pillar of the Industrial Revolution. (© *Hulton Getty/Liaison Agency.*)

bellows a few years later, and then directly pumped air into furnaces after 1776 via the far more efficient Boulton-Watt steam engine (produced by James Watt and Matthew Boulton). Steam engines freed blast furnaces from the restrictions of water power and were used in different types of factories by the early 1780s.

The third and most visible British technology was the textile industry, which became increasingly mechanized throughout the eighteenth century as self-acting machinery replaced hand manufactures. The weaving process underwent steady productivity increases from early inventions such as the 1733 hand loom and flying shuttle, which caused weaving to outpace yarn production and create yarn shortages. The situation was corrected by subsequent inventions that automated the spinning process, such as James Hargreaves's spinning jenny (c. 1764) and Richard Arkwright's 1769 water frame. Samuel Crompton's 1779 spinning mule combined aspects of earlier spinning technologies and enabled yarn production to outpace weaving technology. This in turn inspired Edmund Cartwright to make a powered weaving loom in 1785. In addition to this technological escalation, the imposition of new organizational schemes in increasingly large textile factories greatly facilitated productivity increases as well as more exacting standards for the production of uniform thread and woven products.

As a result of these industrial developments, relatively high-quality and inexpensive British goods seized control of the home market and led to enormous increases in the demand for manufactured goods and in the standard of living. Mass production (a term first introduced to describe early-twentieth-century industrialization in the United States) helped inspire mass consumption. In addition to the large and steady domestic market, British goods also dominated many overseas markets, aided by Great Britain's colonization efforts, powerful navy, and aggressive merchant network. Great Britain also spurred industry through wartime purchases.

Britain appreciated the benefits it incurred from its sizable technological lead and attempted to guard and maintain this advantage through mercantile policies and the strict prohibition of technology transfer. Of course, other nations attempted to compete with Britain, which led to industrial espionage, the emigration of British technicians, and industrialization in other nations.

Waves of Industrialization

Although Britain led the world in industrial growth through the 1830s, the Industrial Revolution soon spread to other countries. A second wave of industrialization took place from the 1810s to the 1870s in Belgium, France, Germany, and the United States; and a third wave swept through Russia, Japan, Sweden, Italy, and other nations in the decades surrounding 1900. Latecomer nations have several advantages over industrial

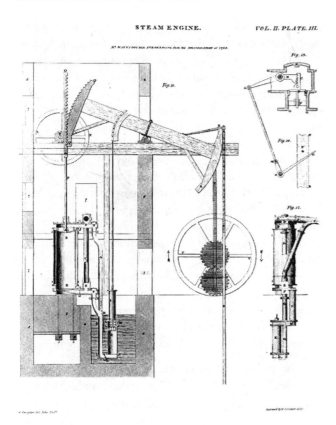

Diagram of the Watt steam engine. The steam engine is seen as the second pillar of the Industrial Revolution. (*The Library of Congress.*)

the creation of institutions to collect and distribute investment capital.

Also in the mid-nineteenth century, the Prussian government took an active role in the sponsorship and funding of large-scale industry, and a close family of German banks offered capital and advice to support new industrial ventures. German industrialization truly began after the 1871 unification of the German states, but powerful agricultural interests successfully protected agrarian subsidies at the expense of the industrialists well into the 1890s. German industry also pioneered the inclusion of research laboratories as a well-funded and influential component of manufacturing endeavors, strengthening the link between science and technology.

Finally, industrialization in the United States was hampered by its small, sparse, and rural population; the lack of a strong economy or banking system; and competition from British goods. Many of these inhibiting factors had been reduced or removed by the mid-nineteenth century, and industrialization was aided in the United States by booming population growth, plentiful natural resources, increased access to investment capital, and the import and modification of technologies from Britain.

The end of the nineteenth century introduced an array of new technological products such as chemicals, bicycles, automobiles, and electrical networks; new methods of mass production and factory mechanization; dramatic increases in the quantity of capital required to launch new manufacturing endeavors; and the corresponding development of new capital-raising strategies such as large-scale stock subscriptions and direct government subsidies. Russia and Japan were the two largest economies to industrialize during this third wave of the Industrial Revolution, following Russia's abolition of serfdom and Japan's increasing degree of interaction with foreign nations. Both governments directly and unhesitatingly supported industrialization by running pilot companies, raising taxes or requesting foreign loans to produce investment capital, and establishing pro-industry policies. During the twentieth century the Industrial Revolution continued to evolve and spread to new regions such as China and India.

Indeed, as a result of post–World War II developments in automation, cybernation, and computerization, people began to speak of a second industrial revolution originating in the United States and spreading to other parts of the world. The phenomenon of globalization, which depends on advances in transportation and communication, could also be described as an extension of

pioneers: Governments recognize the advantages of industrialization and develop supportive policies; investment capital is often available from individuals or institutions in more advanced economies; and technological expertise can often be borrowed or appropriated from the industrial powers. In addition to the iron, coal, and textile industries, railroads emerged as a fundamental technology of later industrialization. The Industrial Revolution continued to catalyze changes in technological development, managerial and labor organization strategies, economic policy, and consumer behavior.

As with the British example, the nations in the second wave of industrialization experienced long periods of gradual population growth fostered by agricultural improvements, economic and commercial expansion, and technological development that promoted a rapid industrial takeoff. Despite an overall manufacturing output that, as late as the 1780s, was not that far behind Britain, French industrialization was hindered by strong conservative craft and agrarian traditions and setbacks from the French Revolution and Napoleonic Era. France's mid-nineteenth-century growth was driven by widespread rural industry and thriving local markets, and was greatly aided by new government policies and the industrialization that began in eighteenth-century Great Britain.

Ethics and Politics

The Industrial Revolution affected everyone and everything on the globe, starting with irrevocable alterations to societal development. Individuals and families increasingly left behind their rural agrarian life to gather in urban centers that offered increased access to a staggering variety of jobs, services, and goods, at the cost of health risks and a very different way of life. While the increased productivity of industrialization generally led to rising standards of living and increased consumption, societies became highly stratified and the newly created wealth and luxury items were not shared equally.

Industrial laborers often endured horrible working conditions, such as bad air quality, deafening noise, poor lighting, cramped conditions, lack of sanitation and resultant disease, repetitive work, and dangerous equipment that could cause mutilation and death. Industrialization also imposed a new system of managerial regulation, increased discipline, and the removal of skilled laborers' privileges. When laborers resented or resisted new workplace policies, employers considered them lazy and responded by structuring wages in a manner that forced employees to work long hours at a rapid pace in order to earn a living. This often resulted in the employment of entire families, especially in the textile industry. Unskilled workers frequently lived under the constant threat of unemployment, and even when they were employed their living conditions were often squalid.

The Industrial Revolution may have most affected the lives of women and children. Although advocates of industrialization asserted that contemporary children worked long hours on the farm, children working in factories routinely endured truly nightmarish work environments. Labor laws and other responses to unpleasant child labor conditions gradually shifted the focus of childhood from productivity to education. And although industrialization often forced women to work under horrible conditions for less pay than their male counterparts, this was sometimes mitigated by new opportunities for employed women, such as freedom from the toil or drudgery of farm labor, increased personal and economic freedom, and exposure to urban influences. The Industrial Revolution steadily pushed work out of the family setting and redefined gender and child roles.

These changes inspired extensive commentary from contemporary participants, particularly when the impacts were experienced for the first time in Great Britain. Romantic poets such as William Blake (1757–1827),

Victorian novelists such as Charles Dickens (1812–1870), and socialist philosophers such as Friedrich Engels (1820–1895) approached this problem from different perspectives but were united in their association of industrialization with corruption, exploitation, poverty, and other social evils that primarily affected members of the laboring classes. Responses to industrialization included the Luddites' destruction of textile machinery as a means of protesting technological displacement of workers; the promotion of socialist ideals by philosophers such as Engels and Karl Marx (1818–1883); and efforts by Edwin Chadwick (1800–1890) to use the public health movement to establish scientific and technological principles for the improvement of housing and sanitation systems. But on balance, especially under the influence of such ameliorative initiatives, industrialization also clearly improved the material qualities of human life. Versions of these initiatives have been manifested and criticized in other industrializing nations, and debates over the positive and negative impacts continue into the present.

The Industrial Revolution also permanently altered the global power balance. The earliest industrializing nations exerted a substantial and lasting economic and military influence on the nonindustrial world. The growth of industrial economies and trade networks often promoted deindustrialization in less advanced countries that had previously benefited from the sale of handicrafts or other goods. Most nineteenth-century industrial powers practiced imperialism and colonialism, which yielded new supplies of raw materials and new markets and propagated capitalist and Western values throughout the world. In addition, the Industrial Revolution inspired many governments to shift their political philosophy from laissez-faire policies that favored traditional landed interests to proactive social and economic reforms.

Finally, the Industrial Revolution produced previously unimaginable effects on the human–environment relationship. The Industrial Revolution removed many barriers to population growth and accelerated the ability of farmers to produce food more efficiently, leading to an ever-increasing world population. And by increasing fuel use, the supply and demand of manufactured goods, and the scope of extractive tools and machinery, industrialization led to astronomical levels of raw material harvesting and ensuing environmental consequences such as deforestation and air and water pollution. At the same time, the Industrial Revolution firmly connected the scientific tradition to technological development, leading to increased industrial research and development, new standards of education, superior scientific equipment, government funding of science, and renewed support for the increase of human knowledge.

ROBERT MARTELLO

SEE ALSO *Affluence; Christian Perspectives; Colonialism and Postcolonialism; Modernization; Science, Technology, and Society Studies; Work; Urbanization.*

BIBLIOGRAPHY

Ashton, T. S. (1997). *The Industrial Revolution, 1760–1830*, rev. edition, with a new preface and bibliography by Pat Hudson. Oxford: Oxford University Press. This concise study focuses on the beneficial and progress-oriented aspects of the Industrial Revolution, with emphases upon innovation and the revolution's historical context.

Breunig, Charles, and Matthew Levinger. (2002). *The Revolutionary Era, 1789–1850*, 3rd edition. New York: Norton. Offers a general overview of the industrial revolution as well as a contextual study of relevant trends and events in contemporary Europe.

Diamond, Jared. (1999). *Guns, Germs, and Steel: The Fates of Human Societies*. New York: Norton. This expansive work explores the geographical, environmental, and demographic factors that caused human civilizations to develop differently. While it does not explicitly discuss the Industrial Revolution, its explanation of the Neolithic and urban revolutions provide interesting background and a global perspective.

Hobsbawm, Eric. (1962). *The Age of Revolution: Europe, 1789–1848*. London: Weidenfeld and Nicolson. This survey work adopts an economic approach to European history that is sympathetic to workers and the labor movement. Hobsbawm explores the causes and impacts of the industrial revolution at length and also discusses other European events in the same era.

Hobsbawm, Eric. (1999). *Industry and Empire: From 1750 to the Present Day*, revised and updated with Chris Wrigley. New York: New Press. This detailed study explores the political, social, and economic history of the industrial revolution (extending into the late twentieth century), primarily from a Western perspective.

Hughes, Thomas P. (1989). *The American Genesis: A Century of Invention and Technological Enthusiasm, 1870–1970*. New York: Viking. Praises the achievements of technological creativity in the United States during the second wave of industrialization as equivalent in significance to those of the Renaissance.

Landes, David S. (2003). *The Unbound Prometheus: Technical Change and Industrial Development in Western Europe from 1750 to the Present*, 2nd edition. Cambridge, UK: Cambridge University Press. Classic and detailed analysis of the economic and societal causes and impacts of the first industrial revolution, emphasizing the importance of technological enhancements to productive potential and the associated enrollment of natural, human, and financial resources.

Mokyr, Joel. (1990). *The Lever of Riches*. New York: Oxford University Press. A historical approach to technological creativity, exploring the many factors that give some nations innovative advantages over others, as well as the economic and social impacts of creativity.

Pacey, Arnold. (1990). *Technology in World Civilization: A Thousand-Year History*. Cambridge, MA: MIT Press. A global approach to the history of technology, concisely highlighting the impact of technology transfers and industrial development upon different civilizations' economies and societies.

Simon, Julian L., ed. (1995). *The State Of Humanity*. Oxford: Blackwell. Collects fifty-eight original articles arguing that industrialization has created across-the-board material benefits for all segments of society.

Toynbee, Arnold. (1884). *Lectures on The Industrial Revolution in England*. London: Rivingtons. First use of the phrase "industrial revolution" to describe the great industrial and economic changes faced by great Britain at the end of the eighteenth century.

INFORMATION
• • •

Science, technology, and ethics are all forms of information that depend *on* information to work. Furthermore there exist sciences, technologies, and ethics *of* information. To disentangle some of the main relations among these aspects of information, it is helpful to start with a simple example.

Monday morning. John turns the ignition key of his car, but nothing happens: The engine does not even cough. Not surprisingly the low-battery indicator is flashing. After a few more unsuccessful attempts, John calls the garage and explains that, last night, his wife had forgotten to turn off the car's lights—this is a lie, John did but is too ashamed to admit it—and now the battery is dead. John is told that the car's operation manual explains how to use jumper cables to start the engine. Luckily his neighbor has everything John needs. He follows the instructions, starts the car, and drives to the office.

This everyday example illustrates the many ways in which people understand one of their most important resources: information. The information galaxy is vast, and this entry will explore only two main areas: information as content and information as communication. The reader interested in knowing more about the philosophical analysis of the concept should consult the work of Jaakko Hintikka and Patrick Suppes (1970), Philip P. Hanson (1990), and Fred I. Dretske (1999).

Information as Content

It is common to think of information as consisting of *data* (Floridi 2005). An intuitive way of grasping the notion of data is to imagine an answer without a question. Ultimately data may be described as relational differences: a 0 instead of a 1; a red light flashing; a high or low charge in a battery.

To become information, data need to be *well-formed* and *meaningful*. Well-formed means that data are clustered together correctly, according to the rules (syntax) of the chosen language or code. For example, the operation manual from the example above shows the batteries of two cars placed one next to, not one on top of, the other. Meaningful indicates that the data must also comply with the meanings (semantics) of the chosen language or code. So the operation manual contains illustrations that are immediately recognizable.

When meaningful and well-formed data are used to talk about the world and describe it, the result is *semantic content* (Bar-Hillel and Carnap 1953, Bar-Hillel 1964). Semantic content has a twofold function. Like a pair of pincers, it picks up from or about a situation, a fact, or a state of affairs *f*, and models or describes *f*. *The battery is dead* carves and extracts this piece of information—that the battery of the car is dead—and uses it to model reality into a semantic world in which the battery is dead. Whether the work done by the specific pair of pincers is satisfactory depends on the resource *f* (realism) and on the purpose for which the pincers are being used (teleologism). Realistically *the battery is dead* is true. Teleologically it is successful given the goal of communicating to the garage the nature of the problem. *The battery is dead* would be realistically false and teleologically unsatisfactory if it were used, for instance, to provide an example of something being deceased.

INFORMATION AS TRUE SEMANTIC CONTENT. *True semantic content* is perhaps the most common sense in which information can be understood (Floridi 2005). It is also one of the most important ways, since information as true semantic content is a necessary condition for knowledge. Some elaboration of this concept is in order. First the data that constitute information allow or invite certain constructs and resist or impede others. Data in this respect work as *constraining affordances*. Second the data are never accessed and elaborated independently of a *level of abstraction* (LoA). An LoA is like an interface that establishes the scope and type of data that will be available as a resource for the generation of information (Floridi and Sanders 2004). *The battery is what provides electricity to the car* is a typical example of information elaborated at a driver's LoA. An engineer's LoA may output something like *a 12-volt lead-acid battery is made up of six cells, each cell producing approximately 2.1 volts*, and an economist's LoA may suggest that *a good quality car battery will cost between $50 and $100 and, if properly maintained, it should last five years or more*. Data as constraining affordances—answers waiting for the relevant questions—are transformed into information by being processed semantically at a given LoA (alternatively the right question is associated to the right data at a given LoA).

Once information is available, knowledge can be built in terms of *justified* or *explained information*, thus providing the basis of any further scientific investigation. One knows that the battery is dead not by merely guessing correctly, but because one sees the red light of the low-battery indicator flashing and perceives that the engine does not start. The fact that data count as *resources* for information, and hence for knowledge, rather than *sources*, provides a constructionist argument against any representationalist theory that interprets knowledge as a sort of picture of the world.

An instance of *misinformation* arises when some semantic content is false (untrue) (Fox 1983). If the source of the misinformation is aware that the semantic content is false, one may speak of *disinformation*, for example *my wife left the lights on*. Disinformation and misinformation are ethically censurable but may be successful teleologically: If one tells the mechanic that one's wife left the lights on last night, the mechanic will still be able to provide the right advice. Likewise information may fail to be teleologically successful; just imagine telling the mechanic that one's car is out of order.

INSTRUCTIONAL INFORMATION. True semantic content is not the only type of information. The operation manual, for example, also provides *instructional information*, either imperatively—in the form of a recipe: First do this, then do that—or conditionally—in the form of some inferential procedure: If such and such is the case do this, otherwise do that. Instructional information is not about *f* and does not model *f*: It constitutes or instantiates f, that is, it is supposed to make *f* happen. The printed score of a musical composition or the digital files of a program are typical cases of instructional information. The latter clearly has a semantic side. And semantic and instructional information may be joined in performative contexts, such as christening a vessel—for example, "this ship is now called HMS *The Informer*"—or programming—for example, when declaring the type of a variable. Finally the two types of information may come together in magic spells, where

FIGURE 1

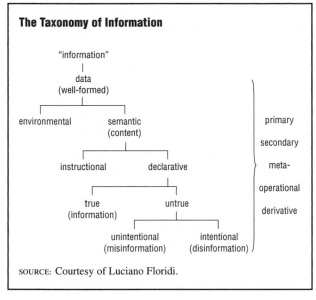

The Taxonomy of Information

"information"
|
data
(well-formed)

environmental   semantic
(content)

instructional   declarative

true   untrue
(information)

unintentional   intentional
(misinformation)   (disinformation)

primary

secondary

meta-

operational

derivative

SOURCE: Courtesy of Luciano Floridi.

The figure illustrates how some of the main types of information may be related into a tree-like structure.

semantic modeling is confused with instructional power and control. Yet, as a test, one should recall that instructional information does not qualify alethically (from *aletheia,* the Greek word for truth). In the example, it would be silly to ask whether *only use batteries with the same-rated voltage* is true or false.

ENVIRONMENTAL INFORMATION. When John turned the ignition key, the low-battery indicator flashed. He translated the flashing into (a) semantic information: The battery is dead; and (b) instructional information: The battery needs to be charged or replaced. However the flashing of the indicator is actually an example of *environmental information.*

Environmental information may be described as *natural data:* It requires two systems *a* and *b* to be coupled in such a way that *a* being (of type, or in state) *F* is correlated to *b* being (of type, or in state) *G,* thus carrying to the observer the information that *b* is *G* (Jon Barwise and Jerry Seligman provide a similar analysis based on Dretske 1999). The correlation is usually *nomical* (it follows some law). It may be engineered—as in the case of the low-battery indicator (*a*) whose flashing (*F*) is triggered by, and hence is informative about, the battery (*b*) being dead (*G*). Or it may be natural, as when litmus—a coloring matter from lichens—is used as an acid-alkali indicator (litmus turns red in acid solutions and blue in alkaline solutions). Other typical examples include the correlation between fingerprints and personal identification, or between the age of a plant and its growth rings.

One may be so used to equating the low-battery indicator flashing with the information (that is, meaning) that the battery is dead as to find it hard to distinguish sufficiently between environmental and semantic information. However it is important to remember that environmental information may require or involve no semantics at all. It may consist of correlated data understood as mere differences or constraining affordances. Plants (e.g., a sunflower), animals (e.g., an amoeba) and mechanisms (e.g., a photocell) are certainly capable of making practical use of environmental information even in the absence of any (semantic processing of) meaningful data. Figure 1 summarizes the main distinctions introduced so far.

FIVE TYPES OF INFORMATION. More detail may now be added. First it should be emphasized that the actual *format, medium,* and *language* in which information is encoded is often irrelevant. The same semantic, instructional, and environmental information may be analog or digital, printed on paper or viewed on a screen, or in English or some other language. Second thus far it has been implicitly assumed that *primary* information is the central issue: things like the low-battery indicator flashing, or the words *the battery is dead* spoken over the phone. But remember how John discovered that the battery was dead. The engine failed to make any of the usual noises. Likewise in Sir Arthur Conan Doyle's *Silver Blaze* (1892), Sherlock Holmes solves the case by noting something that has escaped everybody else's attention, the unusual silence of the dog. Clearly silence may be very informative. This is a peculiarity of information: Its absence may also be informative. When it is, the difference may be explained by speaking of *secondary information.*

Apart from secondary information, three other typologies are worth some explanation since they are quite common (the terminology is still far from being standard or fixed, but see Floridi 1999b). *Metainformation* is information about the nature of information. *"The battery is dead is encoded in English"* is a simple example. *Operational information* is information about the dynamics of information. Suppose the car has a yellow light that, when flashing, indicates the entire system that checks that the electronic components of the car is malfunctioning. The fact that the light is off indicates that the low-battery indicator is working properly, thus confirming that the battery is indeed dead. Finally *derivative information* is information that can be extracted from any form of information whenever the latter is used as a source in search of patterns, clues, or inferential evidence, namely for comparative and quantitative analyses. From a credit card bill concerning the purchase of gasoline, one may derive information about the cardholder's whereabouts at a given time.

Information as Communication

Also important is the concept of information as communication, as in the sense of a transmitted message (Cherry 1978). Some features of information are intuitively quantitative. Information can be *encoded*, *stored*, and *transmitted*. One also expects it to be *additive* (information a + information b = information $a + b$) and *nonnegative*. Similar properties of information are investigated by the *mathematical theory of communication* (MTC, also known as *information theory*; for an accessible introduction, see Jones 1979).

MTC was developed by Claude E. Shannon (Shannon and Weaver 1998 [1949]) with the primary aim of devising efficient ways of encoding and transferring data. Its two fundamental problems are the ultimate level of data compression (how small can a message be, given the same amount of information to be encoded?) and the ultimate rate of data transmission (how fast can data be transmitted over a channel?). To understand this approach, consider the telephone call to the garage.

The telephone communication with the mechanic is a specific case of a general communication model. The model is described in Figure 2.

John is the *informer*, the mechanic is the *informee*, *the battery is dead* is the message (*the informant*), there is a coding and decoding procedure through a language (English), a channel of communication (the telephone system), and some possible noise. Informer and informee share the same background knowledge about the collection of usable symbols (the *alphabet*).

MTC treats information as only a selection of symbols from a set of possible symbols, so a simple way of grasping how MTC quantifies *raw information* is by considering the number of yes/no questions required to guess what the informer is communicating. When a fair coin is tossed, one question is sufficient to guess whether the outcome is heads (h) or tails (t). Therefore a binary source, like a coin, is said to produce one bit of information. A two-fair-coins system produces four ordered outputs: <h, h, h, t, t, h, t, t> and therefore requires two questions, each output containing two bits of information, and so on. In the example, the low-battery indicator is also a binary device: If it works properly, it either flashes or it does not, exactly like a tossed coin. And since it is more unlikely that it flashes, when it does, the red light is very informative. More generally the lower the probability of p the more informative the occurrence of p is (unfortunately

this leads to the paradoxical view that a contradiction—which has probability 0—is the most informative of all contents, unless one maintains that, to qualify as information, p needs to be true [Floridi 2004]).

Before the coin is tossed, the informee does not *know* which symbol the device will actually produce, so it is in a state of *data deficit* equal to 1 (Shannon's *uncertainty*). Once the coin has been tossed, the system produces an amount of raw information that is a function of the possible outputs, in this case two equiprobable symbols, and equal to the data deficit that it removes. The reasoning applies equally well to the letters used in your telephone conversation with the mechanic.

The analysis can be generalized. Call the number of possible symbols N. For $N = 1$, the amount of information produced by a unary device is 0. For $N = 2$, by producing an equiprobable symbol, the device delivers one unit of information. And for $N = 4$, by producing an equiprobable symbol, the device delivers the sum of the amount of information provided by coin A plus the amount of information provided by coin B, that is two units of information. Given an alphabet of N equiprobable symbols, it is possible to rephrase some examples more precisely by using the following equation: $\log 2$ (N) = bits of information per symbol.

Things are made more complicated by the fact that real coins are always *biased*, and so are low-battery indicators. Likewise in John's conversation with the mechanic a word like *batter* will make y as the next letter almost certain. To calculate how much information a biased device produces, one must rely on the frequency of the occurrences of symbols in a finite series of occurrences, or on their probabilities, if the occurrences are supposed to go on indefinitely. Once probabilities are taken into account, the previous equation becomes Shannon's formula (where H = uncertainty, what has been called above data deficit):

$$H = -\sum_{i=1}^{N} P_i \log P_i (\text{bits per symbol})$$

The quantitative approach just outlined plays a fundamental role in coding theory, hence in cryptography, and in data storage and transmission techniques, which are based on the same principles and concepts. Two of them are so important as to deserve a brief explanation: *redundancy* and noise.

Redundancy refers to the difference between the physical representation of a message and the mathematical representation of the same message that uses no more bits than necessary. It is basically what can be taken away from a message without loss in

FIGURE 2

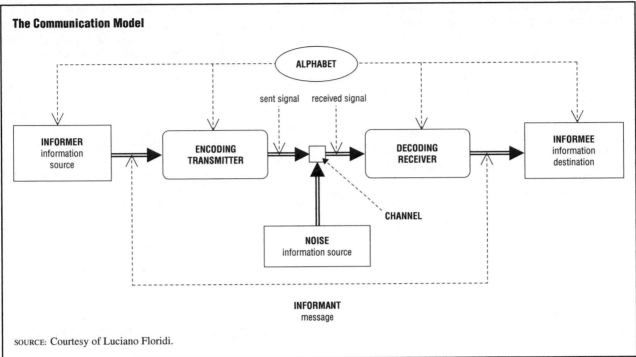

The Communication Model

sent signal   received signal

INFORMER
information
source

ENCODING
TRANSMITTER

ALPHABET

DECODING
RECEIVER

INFORMEE
information
destination

CHANNEL

NOISE
information source

INFORMANT
message

SOURCE: Courtesy of Luciano Floridi.

The figure illustrates the main elements and their relations in the communication model according to Shannon's approach.

communication. John's statement that his wife was responsible for the dead battery was redundant.

Compression procedures work by reducing data redundancy, but redundancy is not always a bad thing, for it can help to counteract *equivocation* (data sent but never received) and noise (received but unwanted data, like some interference). A message + noise contains more data than the original message by itself, but the aim of a communication process is *fidelity*, the accurate transfer of the original message from sender to receiver, not data increase. The informee is more likely to reconstruct a message correctly at the end of the transmission if some degree of redundancy counterbalances the inevitable noise and equivocation introduced by the physical process of communication and the environment. This is why, over the phone, John said that *the battery is dead* and that *the lights were left on last night*. It was the *by whom* that was uselessly redundant.

MTC is not a theory of information in the ordinary sense of the word. The term raw information has been used to stress the fact that in MTC information has an entirely technical meaning. Two equiprobable *yeses* contain the same quantity of raw information, regardless of whether their corresponding questions are

Is the battery dead? or *Is your wife missing?* Likewise if one knows that a device could send with equal probabilities either this whole encyclopedia or just a quote for its price, by receiving one or the other message one would receive very different quantities of data bytes but only one bit of raw information. Since MTC is a theory of information without meaning, and since information – meaning = data, *mathematical theory of data communication* is a far more appropriate description than *information theory*.

MTC deals not with semantic information itself but with messages constituted by uninterpreted symbols encoded in well-formed strings of signals, so it is commonly described as a study of information at the *syntactic* level. This generates some confusion because one may think the syntactic versus semantic dichotomy to be exhaustive. Clearly MTC can be applied in information and communication technologies (ICT) successfully because computers are syntactical devices. It is often through MTC that information becomes a central concept and topic of research in disciplines like chemistry, biology, physics, cognitive science, neuroscience, the philosophy of information (Floridi 2002, Floridi 2004a), and computer ethics (Floridi 1999a).

LUCIANO FLORIDI

SEE ALSO *Computer Ethics; Cybernetics; Digital Libraries; Geographic Information Systems; Information Overload; Information Society; Internet; Wiener, Norbert.*

BIBLIOGRAPHY

Bar-Hillel, Yehoshua. (1964). *Language and Information: Selected Essays on Their Theory and Application.* Reading, MA; London: Addison-Wesley. Important collection of relevant essays by one of the philosophers who first tried to apply information theory to semantic information.

Bar-Hillel, Yehoshua, and Rudolf Carnap. (1953). "An Outline of a Theory of Semantic Information." In *Language and Information: Selected Essays on Their Theory and Application* Reading, MA; London: Addison-Wesley. Influential attempt to quantify semantic information.

Barwise, Jon, and Jerry Seligman. (1997). *Information Flow: The Logic of Distributed Systems.* Cambridge, UK: Cambridge University Press. Applies situation logic to the study of the dynamics of information.

Cherry, Colin. (1978). *On Human Communication: A Review, a Survey, and a Criticism,* 3rd edition. Cambridge, MA: MIT Press.

Dretske, Fred I. (1999). *Knowledge and the Flow of Information.* Stanford, CA: CSLI Publications. Originally published in 1981, Cambridge, MA: MIT Press. A simple and informative introduction to the mathematical theory of communication, even if no longer up-to-date.

Floridi, Luciano. (1999a). "Information Ethics: On the Theoretical Foundations of Computer Ethics." *Ethics and Information Technology* 1(1): 37–56. Provides a first foundation for an approach in computer ethics known as information ethics.

Floridi, Luciano. (1999b). *Philosophy and Computing: An Introduction.* London, New York: Routledge. Textbook introduction for philosophy students to computer science and its conceptual challenges.

Floridi, Luciano. (2002). "What Is the Philosophy of Information?" *Metaphilosophy* 33(1–2): 123–145. Provides a definition of the philosophy of information as a area of research.

Floridi, Luciano. (2004). "Outline of a Theory of Strongly Semantic Information." *Minds and Machines* 14(2): 197–222. Defends a theory of semantic information based on a veridical interpretation of information, in order to solve the Bar-Hillel-Carnap paradox.

Floridi, Luciano, and J. W. Sanders. (2004). "The Method of Abstraction" In *Yearbook of the Artificial: Nature, Culture and Technology: Models in Contemporary Sciences,* ed. Massimo Negrotti. Bern: Peter Lang. Develops the method of abstraction for philosophical analysis.

Floridi, Luciano. (2005). "Is Information Meaningful Data?" *Philosophy and Phenomenological Research* 70(2): 351–370. Defends the thesis that information encapsulates truth, hence that "false" in "false information" is to be understood as meaning "not authentic."

Fox, Christopher J. (1983). *Information and Misinformation: An Investigation of the Notions of Information, Misinformation, Informing, and Misinforming.* Westport, CT: Greenwood Press. Overview of the concepts mentioned in the title from an information science perspective.

Hanson, Philip P., ed. (1990). *Information, Language, and Cognition.* Vancouver: University of British Columbia Press. Proceedings of an influential conference on semantic information.

Hintikka, Jaakko, and Patrick Suppes, eds. (1970). *Information and Inference.* Dordrecht, The Netherlands: Reidel. Influential collection of essays on the philosohy of semantic information.

Jones, Douglas Samuel. (1979). *Elementary Information Theory.* Oxford: Clarendon Press. Excellent introduction to the mathematical theory of communication for the mathematically uninitiated.

Shannon, Claude E., and Warren Weaver. (1998 [1949]). *The Mathematical Theory of Communication.* Urbana: University of Illinois Press. Classic presentation of the fundamental results in the mathematical theory of communication.

INFORMATION ETHICS

• • •

Information ethics is a field of applied ethics that addresses the uses and abuses of information, information technology, and information systems for personal, professional, and public decision making. For example, is it okay to download someone else's intellectual property like pictures or music? Should librarians ever remove controversial books from the shelves or monitor users' Internet searching? Should a scientist post the genome for the Ebola virus on the Internet?

Information ethics provides a framework for critical reflection on the creation, control, and use of information. It raises questions about information ownership and access to intellectual property, the rights of people to read and to explore the World Wide Web as they choose. Information ethicists explore and evaluate the development of moral values, the creation of new power structures, information myths, and the resolution of ethical conflicts in the information society (Capurro 2001). If bioethics addresses living systems, then information ethics similarly covers information systems. Where bioethics evolved from medical ethics after World War II to engage the broader implications of societal changes such as informed consent and reproductive rights, information ethics grew out of the professional ethics traditions of librarians and early information professionals in order to describe and evaluate the competing interests that sought to control the information assets of a high-tech society (Smith 1997). Like other areas of applied ethics in science and technology, information ethics focuses on social responsibility and the meaning of humanity in relation to machines.

Built from the codes and commitments of professional librarians to protect the right to read, fight censorship, protect patron privacy, assure confidentiality of library records, and provide service for everyone, information ethics has extended these traditions into cyberspace. The term information ethics first appeared in the literature of library and information science in the late 1980's (Hauptman) alongside other terms such as information technology ethics, cataloging ethics, and archival ethics. In the next few years, information ethics grew to encompass dilemmas facing librarians and information professionals (Mason, Mason, and Culnan 1995) as they introduced new information and communications technologies (ICTs) to public, academic, and special libraries and also into publishing, healthcare, and the new information industry.

Today information ethics encompasses a wide range of issues involving the creation, acquisition, organization, management, translation, duplication, storage, retrieval, and any other processes involving printed or digital texts, graphics, voice, and video. Information ethics can address any issue relating to the *Information Society* or the *Knowledge Economy*. As a field of applied ethics, it draws upon historical and philosophical insights (Floridi 1999) in order to describe current problems such as bridging the digital divide and to craft normative solutions for personal and professional conduct and for public policy (Tavani 2003).

The Historical Context

In the mid-fifteenth century, Johannes Gutenberg's invention of the movable type printing press altered the parameters of information access and control and began to change the world. Widespread dissemination of printed information helped to change the balance of power in Europe, notably contributing to the sixteenth-century Protestant Reformation, disruptions to the political power of the Roman Catholic Church, and the rise of the nation-state.

In the mid-twentieth century, Claude Shannon (1948) and others developed elegant mathematical theories that made modern information technologies possible while other advances, such as the development of the atomic bomb, made the risks and rewards of widespread scientific and technological knowledge more significant and more visible in everyday life. Since then the increasing volume of digitized information and the exponential improvements in digital processing, storage, and communication have again altered the landscape of information access and control.

FIGURE 1

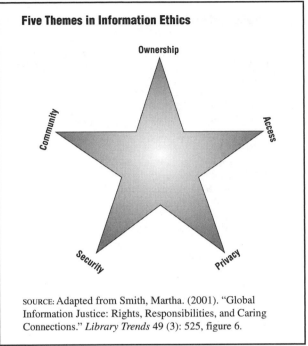

SOURCE: Adapted from Smith, Martha. (2001). "Global Information Justice: Rights, Responsibilities, and Caring Connections." *Library Trends* 49 (3): 525, figure 6.

The COAPS star suggests the potential both for overlap and conflict amongst five infoethical themes.

Alongside the technological advances that have occurred since the mid-twentieth century, formal consideration of the uses and abuses of information began even before it was designated information ethics, or *infoethics*. The UN General Assembly raised many infoethical themes in the Universal Declaration of Human Rights (1948) including information access (Article 19), intellectual property (Article 27), privacy (Article 12), security (Articles 17 and 27), community (Article 27), and education (Article 26). Since then, the role of information in government, healthcare, and business, and concerns about the uses of that information, have continued to fuel public policy debates. The UN Educational, Scientific and Cultural Organization (UNESCO) uses the term information ethics to focus attention on global problems ranging from literacy, including cell phone access in the developing world, the need to protect local cultures and languages from the dominance of English on the Internet, and the ramifications of expanding databases of genetic information.

In the last fifteen years, information ethics has also evolved within and beyond its early professional and academic communities. Its academic vitality is evident in the formation of scholarly associations such as the International Society for Ethics and Information Technology (INSEIT), scholarly websites such as the International

TABLE 1

Most Challenged Books, 2002

Title and Author	Sexual Content	Offensive Language	Unsuitable to Age	Wizardry, Occult	Racism	Insensitivity	Violence	Disobedience
Harry Potter series, J.K. Rowling A young wizard studies magic and battles evil.				✓				
Alice series, Phyllis Reynolds Naylor Alice searches for a female role model.	✓	✓	✓					
The Chocolate War, Robert Cormier Jerry challenges the high school power structure.		✓	✓					
I Know Why the Caged Bird Sings, Maya Angelou Autobiography of an African American poet.	✓	✓	✓		✓		✓	
Taming the Star Runner, S. E. Hinton A talented, urban punk exiled to a farm.		✓						
Captain Underpants, Dav Pilkey Comic battles with Dr. Diaper and talking toilets.			✓			✓		✓
The Adventures of Huckleberry Finn, Mark Twain Classic novel of a boy's journey down the Mississippi.		✓			✓	✓		
Bridge to Terabithia, Katherine Paterson Friends reign in a fantasy kingdom in the woods.	✓	✓		✓				
Roll of Thunder, Hear My Cry, Mildred D. Taylor African-American family struggles to stay together in the 1930s South.		✓			✓	✓		
Julie of the Wolves, Jean Craighead George Can Julie/Miyax survive with wolves in the Alaskan wilderness?	✓	✓	✓				✓	

SOURCE: Adapted from ALA Office of Information Freedom. (2003). "Ten Most Frequently Challenged Books for 2002." Available from http://www.ala.org/ala/oif.

Book challenges illustrate the ethical tension between freedom of information and other values.

Center for Info Ethics (ICIE), and journals such as *The Information Society* (1981), *Journal of Information Ethics* (1992), *Science and Engineering Ethics* (1995), *Ethics and Information Technology* (1999), and *International Review of Information Ethics* (2004). The growing number of books and journal articles that address ethics in academic and professional literature indicates the expanding recognition of and participation in the field.

Key Ethical Themes

From the perspective of information ethics, there are five important themes to be considered: *community, ownership, access, privacy,* and *security* (COAPS; see Figure 1). As a framework, the COAPS themes help to guide ethical analysis and aid the discovery of underlying conflicts, as illustrated by ethical questions that have emerged since the mid-twentieth century.

> Does the anonymity of the web encourage or detract from *community* formation online?

> Who *owns* e-mail messages on a corporate e-mail server, and who can read them?

> Do patients have a right of *access* to information about a terminal illness?

> Do libraries and librarians have an obligation to protect the *privacy* of patron records?

> Does personal *security* warrant the widespread use of surveillance cameras in public places?

Community

In an 1813 letter, Thomas Jefferson distinguished goods that are lessened and ideas that are multiplied when shared:

> He who receives an idea from me, receives instruction himself without lessening mine; as he who lights his taper at mine receives light without darkening me.

The distinction has become increasingly salient over time. Future creative work builds on past creative work. All branches of science have flourished since the Royal Society of London first published the *Philosophical Transactions* in 1765, establishing a creative commons of scientific work for scrutiny, criticism, and derivation.

TABLE 2

Post 9/11 U.S. Government Legislation and Programs

Legislation or Program Name	Summary
Terrorism Information Awareness Program (TIA)	.. "search[ing] for indications of terrorist activities in vast quantities of data."
Uniting and Strengthening America by Providing Appropriate Tools Required to Intercept and Obstruct Terrorism Act of 2001 (USA PATRIOT)	Grants law enforcement broad rights of search and surveillance with limited judicial oversight.
Computer Assisted Passenger Pre-screening System II (CAPPS II)	Focused on identifying and computing risk score for airline passengers.

SOURCE: Defense Advanced Research Program Agency (DARPA), http://www.darpa.mil; Electronic Privacy Information Center (EPIC), http://www.epic.org; American Civil Liberties Union (ACLU), http://www.aclu.org.

Legislation and government programs illustrate the ethical tensions that arise between the search for security and the desire for privacy.

While ideas on paper may be expensive to reproduce and awkward to distribute, they have demonstrated great power. Creativity requires a balance of access, to make future creative work possible, and control to make creative work worthwhile. The U.S. Constitution, Article 1, Section 8, establishes such a balance by granting inventors limited-term, exclusive rights to exploit their inventions, in exchange for full disclosure for the benefit of future inventors. Lawrence Lessig (2001) has written and spoken extensively about the intellectual and creative commons. In 2002, Lessig and others founded Creative Commons (http://www.creativecommons.org), "devoted to expanding the range of creative work available for others to build upon and share."

For software, the *open source* movement, described by Eric Raymond (1999), encourages community and collaboration by requiring programmers to share software source code and to allow the creation of derivative works. The widely deployed Linux operating system and Apache web server demonstrate the multiplicative benefits of a creative software commons.

Ownership

Modern technology, practice, and law allow tight control over the communication of and access to ideas, threatening the creative commons and future creative works. For example, while Charles Dickens's *Oliver Twist* (1837) exists in the public domain, *digital rights management technology* allows a publisher to prevent a buyer from sharing, copying, or printing the e-book version, a level of control that becomes more significant when fewer printed copies of a work exist. In practice, librarians balance owning paper journals against licensing electronic journals. Web-based, electronic journals offer economy and powerful access capabilities but also carry the risk of complete loss when the license expires. In law, the United States has extended the period of

copyright protection, once fourteen years after publication, to seventy years after the author's death, seriously restricting the creation of derivative works.

The Internet hosts a dynamic evolution of morals, ethics, and laws related to information ownership and use. Freed from the limitations of identity, distance, and substance, Internet users have not always transplanted their behavioral norms directly from the *real* to the *virtual world*. Individuals and legislators face novel situations when the concept of theft is separated from both physical location and physical loss. Peer-to-peer file-sharing networks allow complete strangers to share perfect copies of digitized songs across vast distances while a presumed anonymity frees them from social constraints they might feel off-line.

Access

The First Amendment to the U.S. Constitution prohibits Congress from making laws "abridging the freedom of speech or of the press." The Universal Declaration of Human Rights, Article 19, begins "Everyone has the right to freedom of opinion and expression." These declarations codify ethical principles that recognize the value of expressing multiple points of view.

But freedom of speech, while widely recognized as a fundamental right, remains controversial in detail and execution. Because members of a pluralistic society may hold different values, there are frequent conflicts about what information should be publicly available and what information should not be. The American Library Association (ALA) Code of Ethics states, "We uphold the principles of intellectual freedom and resist all efforts to censor library resources." That commitment conflicts with the values of those who challenge the availability of some books in school and public libraries. The ALA Office for Intellectual Freedom reports over 6,000 book challenges (that is, " an attempt to remove or restrict materials, based upon the objections of a person or group") between 1990

TABLE 3

Information Ethics in Popular Culture

Film, Story, or Book	Dilemma
Frankenstein, Mary Shelley (fiction, 1818)	Ownership
1984, George Orwell (fiction, 1949)	Privacy
"The Enormous Radio," John Cheever (fiction, 1953)	Privacy
Fahrenheit 451, Ray Bradbury (fiction, 1954)	Access
The Gods Must Be Crazy (film, 1980)	Ownership
Blade Runner (film, 1982)	Security
The Electric Grandmother (film, 1982)	Community
"Melancholy Elephants," Spider Robinson (fiction, 1984)	Ownership
Neuromancer, William Gibson (fiction, 1984)	Access
The Handmaid's Tale, Margaret Atwood (fiction, 1986)	Community
Gattaca (film, 1997)	Privacy
AI: Artificial Intelligence (film, 2001)	Community
Minority Report (film, 2002)	Security

SOURCE: Courtesy of Ed Elrod and Martha Smith.

Popular books and films frequently draw on infoethical dilemmas for dramatic conflict.

and 2000. Table 1 lists the most frequently challenged books of 2002 and the reasons for the challenge.

Privacy and Security

Competing values and interests in public policy and government activities also lead to ethical tensions. Terrorist attacks, whether in Madrid, London, Tel Aviv, Kashmir, Tokyo, or New York, place governments in unfamiliar ethical territories as they develop responses in the form of new laws, policies, and programs that are in turn subject to the critical appraisal of civil liberties and human rights groups. James Moor (1998) describes such circumstances in terms of *conceptual muddles* and *policy vacuums* that arise when new situations (such as terrorism) and emerging capabilities (data mining) lead to new behaviors (widespread surveillance) with concomitant ethical questions of whether familiar concepts (privacy) apply and whether the new behaviors are acceptable. Table 2 presents a selection of U.S. government actions that have raised serious ethical dilemmas of privacy versus security and that illustrate an ongoing struggle between secrecy and accountability.

To the extent that such programs occur in secrecy, they leave their scope, policies, methods, activities, and even underlying data insulated from review and criticism. They leave the participants unaccountable outside their bailiwicks. As Joseph Pulitzer observed,

> There is not a crime, there is not a dodge, there is not a trick, there is not a swindle, there is not a vice which does not live by secrecy. (Brin 1998)

While secrecy does not presuppose malicious intent, it reduces the opportunity for accountability and opens the door for individual and institutional misuse of information.

Information professionals face dilemmas when balancing their ethical and legal obligations. For example, the USA PATRIOT Act grants law enforcement agencies broad rights to examine the records of library patrons. The ALA Privacy Toolkit describes privacy as "essential to the exercise of free speech, free thought, and free association" and urges libraries to adopt routine patron privacy and record retention policies in support of the library mission. At the same time, library policies may conflict with fulfilling the surveillance mission of law enforcement agencies.

Government responses to terrorism provide the opportunity for both practical and philosophical consideration. Practically it is reasonable to consider how much these actions enhance security, how much they impinge upon privacy, and what are the relative weights to be applied on either side of the equation. Philosophically it is valuable to ponder how government efforts to ensure security conflict with guaranteed civil rights.

Information Ethics in Popular Culture

Fiction and films frequently illustrate information ethical dilemmas, illuminating significant points that may not be apparent in everyday life. The entertainment value of emphasizing particular dilemmas and their consequences in fictional settings does not reduce the value of ethical exploration by way of popular culture.

Machines have long mimicked and extended human physical capabilities. But a physical aid such as a snow shovel presents few consequential dilemmas and appears only infrequently as the dramatic centerpiece of a film or book. At the other extreme, information technologies mimic and extend the human mind—popularly regarded as the essence of being human. The role of self-aware creations in fiction and film has increased as information and information technology permeate everyday life. Consider the Terminator (1984, 1991, 2003) and Matrix (1999, 2003, 2003) trilogies which project the ethical dilemmas that arise when the roles of information processing machines conflict with the needs, even the survival, of human society. Table 3 lists examples of films and fiction that highlight infoethical dilemmas drawn from the COAPS framework.

Professional Ethics

Ethical dilemmas also arise in the course of professional activities. When individuals adopt professional roles, they assume obligations beyond and sometimes in conflict with their personal beliefs. Librarians who order

TABLE 4

Professional Codes of Ethics—A Sample

Professional Organization	Of Particular Note
American Association of University Professors (CSEP)	Resolution on covert intelligence.
American Library Association (http://www.ala.org)	Explicit commitment to intellectual freedom, privacy, and service.
American Medical Association (CSEP)	Patient right to receive information.
American Society for Information Science and Technology (http://www.asist.org)	Multiple responsibilities to employers, clients, users, profession, and society.
American Society for Public Administration (CSEP)	*Whistle blower* policy statement.
Association for Computing Machinery (http://www.acm.org)	Identifies 24 imperatives as the elements of a personal commitment to ethical professional conduct. Supported by detailed guidelines.
Chartered Institute of Libraries and Information Professionals— UK (http://www.cilip.org.uk)	Statement of principles and multi-dimensioned responsibilities.
Dutch Association of Information Scientists (CSEP)	Multiple responsibilities to self, profession, employer, and society.
Institute of Electrical and Electronics Engineers (http://www.ieee.org)	Commitment "to seek, accept, and offer honest criticism of technical work, to acknowledge and correct errors, …"
International Federation of Journalists (CSEP)	Primacy of respect for *truth*.

SOURCE: Courtesy of Ed Elrod and Martha Smith.

only books and materials supporting their political views about capital punishment are not exercising their professional obligations to build balanced collections and to provide services for a diverse, multicultural public. *Professional neutrality* refers to the commitment to separate professional obligations and personal beliefs.

Many professional groups have developed formal statements to guide decision making and behavior in situations common to their professions. These are often called *codes of ethics* to reflect their deliberate and conscious origins. Table 4 presents a sample of professional organizations with published ethical codes in fields related to the use of information.

Ethical decision making is neither straightforward nor predictable. Codes provide public statements of ideals and intentions. However they are only the starting point for decision making in professional activities. Codes cannot foresee every situation, yet professionalism often calls for decision making and action in unclear situations. Such ambiguity can require a delicate balancing act among stakeholder beliefs and priorities, the demands of professional obligations, and short-term, long-term, and unintended consequences.

Future Prospects for Information Ethics

The published literature of information ethics intertwines with other areas of applied ethics such as computer ethics, cyberethics, journalism, communications, and media ethics, image ethics, Internet ethics, engineering ethics, and business ethics, reflecting its broad philosophical underpinnings and practical applications

far beyond academia. Information ethics contributes to society when it addresses problems that affect the quality of life. Looming ethical questions may seem to arise more from science fiction than science and technology, but science fiction quickly becomes everyday fact. For example, witness the confluence of technology, biology, and national security in the increasing use of biometric identification methods. Looking forward to future technologies and ethical debates:

Will single-issue, virtual *communities* focused on abortion or animal rights, for example, reduce the tolerance for other points of view?

What new business models will arise if intellectual property *ownership* withers in the face of unstoppable copying?

Who will have *access* to the research information about cloning a human?

Will the *privacy* rights of consumers be renegotiable with every credit card transaction?

After the poliovirus has been successfully synthesized from its constituent chemical building blocks, does publishing the gene sequences for deadly viruses on the Internet pose a threat to worldwide *security*?

The future is arriving quickly in the emerging field of *bioinfoethics*. It signals a fresh arena for exploration using the combined insights of bioethics and information ethics. It encompasses recent discussions of reproductive ethics, genetics ethics, healthcare ethics, and computer ethics. Bioinfoethics promises to shape

FIGURE 2

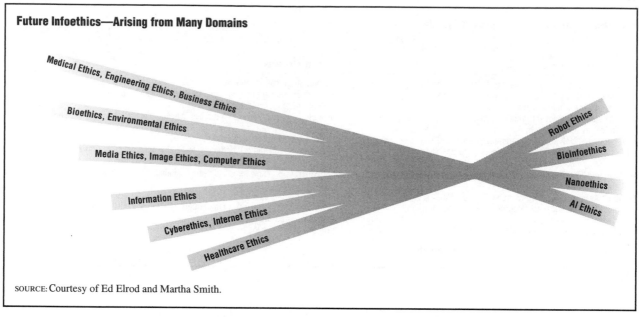

Future Infoethics—Arising from Many Domains

Medical Ethics, Engineering Ethics, Business Ethics

Bioethics, Environmental Ethics

Media Ethics, Image Ethics, Computer Ethics

Information Ethics

Cyberethics, Internet Ethics

Healthcare Ethics

Robot Ethics

Bioinfoethics

Nanoethics

AI Ethics

SOURCE: Courtesy of Ed Elrod and Martha Smith.

Infoethics embraces many ethical domains. It provides a rich foundation for evaluating the ethics of emerging technologies and practices.

personal decisions, professional practice, and public policy. Beyond that, new infoethical domains will continue to emerge wherever new technologies and practices raise new dilemmas that might include applications of robots, nanotechnology, and artificial intelligence. Figure 2 illustrates the contributions of many, diverse domains of ethical analysis to bioinfoethics and to other emerging ethical domains in the future.

An Icelandic genetic mapping project illustrates such a bioinfoethical dilemma. With parliamentary approval, a private company has begun collecting and analyzing genealogical, medical, and genetic data about the people of Iceland in the hope of uncovering diseases with genetic bases and then developing profitable new drugs to treat those diseases. Such research holds the potential for immense medical benefit and immense privacy intrusion. Genetic mapping is likely to become more widespread, thereby expanding the relevance of the bioinfoethical debate.

The COAPS framework (Figure 1) suggests bioinfoethical questions about such a database. How should *communities* organize and negotiate to assure that the use and benefits of genetic databases best reflect the community interests? Should *ownership* of the genetic and medical data lie with the individuals or the company? What financial benefits accrue to the individuals if they do *own* the data? Should there be widespread *access* to the data to maximize the scientific benefit? Does one-way identity coding sufficiently protect individual *privacy* when the records carry other medically relevant but

potentially traceable information? What *security* procedures are demanded for the centralized accumulation of immense amounts of personal and medical data? The Association of Icelanders for Ethics in Science and Medicine (Mannvernd) maintains a broad collection of information about genetic practices and the corresponding ethical considerations.

The Icelandic genetic database represents the leading edge of converging medical, social, government, and information technology practices. The associated bioinfoethical dilemmas explore frontiers of emerging ethical debates and demonstrate the relevance of information ethics to everyone.

EDWIN M. ELROD
MARTHA M. SMITH

SEE ALSO *Association for Computing Machinery; Communications Ethics; Computer Ethics; Cyberspace; Digital Divide; Geographic Information Systems; Hypertext; Information; Institute of Electrical and Electronics Engineers; Intellectual Property; Internet; Monitoring and Surveillance; Movies; Museums of Science and Technology; Popular Culture; Privacy; Science, Technology, and Law; Science, Technology, and Literature; Security; Terrorism; Virtual Reality.*

BIBLIOGRAPHY

Brin, David. (1998). *The Transparent Society.* Reading, MA: Perseus Books.

Floridi, Luciano. (1999). "Information Ethics: On the Theoretical Foundations of Computer Ethics." *Ethics and Information Technology* 1(1): 37–56.

Hauptman, Robert. (1988). *Ethical Challenges in Librarianship*. Phoenix, AZ: Oryx Press.

Hauptman, Robert, ed. (1992). *Journal of Information Ethics*. Jefferson, NC: MacFarland and Company.

Lessig, Lawrence. (2001). The Future of Ideas: The Fate of the Commons in a Connected World. New York: Random House.

Mason, Richard; Florence Mason; and Mary Culnan. (1995). *Ethics Of Information Management*. Thousand Oaks, CA: Sage Publications.

Moor, James. (1998). "Reason, Relativity, and Responsibility in Computer Ethics." *Computers and Society* 28(1): 14–21.

Raymond, Eric S. (1999). *The Cathedral and the Bazaar: Musings on Linux and Open Source by an Accidental Revolutionary.* Beijing; Sebastopol, CA: O'Reilly.

Shannon, Claude E. (1948). "A Mathematical Theory of Communication." *Bell System Technical Journal* 27: 379–423, 623–656.

Smith, Martha M. (1997). "Information Ethics." *Annual Review Of Information Science and Technology* 32: 339–366.

Smith, Martha M. (2001). "Information Ethics." In *Advances in Librarianship*, Vol. 25, ed. Frederick Lynden. New York: Academic Press.

Tavani, Herman T. (2003). *Ethics and Technology: Ethical Issues in an Age of Information and Communication Technology*. New York: John Wiley and Sons.

INTERNET RESOURCES

Capurro, Rafael. (2001). "The Field." Center for Information Ethics. Available from http://icie.zkm.de.

Center for the Study of Ethics in the Professions (CSEP). Illinois Institute of Technology (IIT). "Codes of Ethics Online." Available from http://ethics.iit.edu/codes.

Jefferson, Thomas. "No Patent on Ideas," Letter To Isaac McPherson, August 13, 1813. Available from HTTP://etext.lib.Virginia.edu.

Mannvernd, Association of Icelanders for Ethics in Science and Medicine. Available from http://www.mannvernd.is/english/index.html.

INFORMATION OVERLOAD

• • •

First comprehensively treated by the futurologist Alvin Toffler (1970), *information overload* refers to excessive flows and amounts of data or information that can lead to detrimental computational, physical, psychological, and social effects. For the vast majority of human history, information was scarce and its production, dissemination, and retrieval were nearly unqualified goods that could improve culture, develop commerce, and promote personal autonomy. The advance of information and communication technologies especially since World War II has transformed this scarcity into an abundance. For example, Peter Lyman and Hal Varian (2003) estimated that print, film, magnetic, and optical storage media produced roughly five exabytes of new information in 2002, equivalent to the information that could be stored in 37,000 libraries the size of the Library of Congress. This doubled the amount of new information that had been stored just three years earlier. The glut of information takes several forms and raises many concerns. Indeed it is ironic that information technologies, envisioned by many of their progenitors as devices for organizing information, improving understanding, and boosting productivity often also contribute to disorders, inefficiencies, and confusion.

Causes and Types

Technology, the free-market, and democracy have nearly erased the limits that once caused only the most important information to be published and distributed. Computers, cell phones, the Internet, optical cables, and wireless and satellite transmissions are just a few key technologies fueling the information age. People have become increasingly dependent on such technologies in both their professional and private lives, making information overload nearly unavoidable. The ease and low cost of online publishing and electronic mailing swells the amount of available information, including irrelevant and low quality information.

Information overload occurs in several forms. The term is frequently used in computer theory when so much information has entered an information-processing system that the system cannot easily, if at all, process it. This is usually due to hardware or software limitations, and the idea parallels findings by psychologists that cognitive constraints limit human capacities to process information. Information overload has also been utilized by cognitive scientists in their explanations of intelligent activity. One example is Herbert Simon's concept of *near decomposability* (where short-run behavior of components is independent of other components in the same system). An organism's visual subsystem, for example, can suffer from information overload, while the overall organism does not. In turn, the overall organism can suffer information overload, because it may lack the architectural structure to manage the information gathered and transmitted by each of its subsystems. Another more general concept useful in describing information overload is the decline in the *signal-to-noise* ratio, which denotes the proportion of useful information to all information present in some particular context.

Information overload is commonly experienced in the workplace, especially by managers and government officials who must synthesize growing streams of data. Academics and others who perform research are also negatively impacted by excessive flows of information that make it hard to discern high from low quality knowledge. Finally, information overload is a general experience shared by citizens in developed nations, where streams of information from a variety of media are unavoidable in daily life. Human beings have limited cognitive capacities to store and render information meaningful, and the blitz of information made available by modern technology can easily overwhelm these capacities. Spam, unsolicited commercial bulk E-mail, and its attendant aggravations and lawsuits highlight one specific instance of the personal and social ramifications of information overload.

Effects and Responses

Although information overload in computers can cause technical and social problems, its most detrimental effects usually occur when individual humans must cope with excess information. Indeed Toffler summarized one of the most pernicious effects with his term *information anxiety*. Richard Saul Wurman (1989) explains that, "Information anxiety is produced by the ever-widening gap between what we understand and what we think we should understand. Information anxiety is the black hole between data and knowledge. It happens when information doesn't tell us what we want or need to know" (p. 222). Showing the close connection between information overload and the overwhelming speed of modern social change, Wurman warns that information anxiety limits people to being only seekers of knowledge, because there is no time to reflect on the meaning of that knowledge for one's life. Many people become so obsessive in this quest that they experience what some have called an *information addiction* (Reuters 1997).

The printing press and its many unintended social consequences are often cited as precursors to such ethical implications of increased information. The sociologist Georg Simmel pointed to information overload in several of his studies. For example, he noted that some city dwellers developed the habit of hardly noticing individuals when moving through a crowd. In the 1960s, James G. Miller researched the psycho-pathological effects of information overload, and Karl Deutsch described information overload as a disease of cities that limits freedom as well as efficient communication and

transport. In his 1986 *Overload and Boredom*, the sociologist Orrin E. Klapp argued that the second law of thermodynamics applies to information and culture as well as energy: The greater a social system's information and culture output the greater the system's disorganization in the form of information overload. This yields noise, banality, alienation, despair, anxiety, disenchantment, anomie, feelings of illegitimacy, and absurdity. Boredom results not from the absence of stimulation, but by its excess and repetition. Irrational or poor decisions can also result from information overload. Indeed, some researchers in choice behavior argue that too much choice can be a bad thing (Schwartz 2004).

Walter Kerr (1962) argued that modern societies erode pleasure because the information made available nearly anywhere (now via cell phones and portable computers) enables work to impinge on leisure time. On the positive side, this can improve work and enhance communication with loved ones. In a similar vein, some research suggests that children exposed to computer-enriched environments develop higher-order thinking abilities to a significantly greater degree than those not so exposed (Hopson 2001). Finally, recent philosophers (such as Braybrooke 1998) have conceptualized social information overload as a central element in the logic and processes of social change more generally.

A 1996 survey conducted by Reuters is just one of many reports investigating the effect of information overload. Surveying more than 1,000 managers, this report found that increasing numbers of people suffer ill health due to the stress of information overload and important decisions are delayed by excessive information. David Lewis proposed the term *Information Fatigue Syndrome* for the symptoms uncovered in this report, including poor decision making, difficulties in remembering, reduced attention span, and stress. Nearly half of those surveyed predicted that the Internet would play a primary role in aggravating the problem further. Yet in a follow-up report two years later, researchers at Reuters found that only 19 percent of respondents felt the Internet was making the situation worse, while nearly half felt it was improving the situation. More broadly, this report concluded that the age of information overload is waning, because although some economies were still struggling with it (for example, Southeast Asia), others (such as the United States, Japan, and Western Europe) were beginning to overcome it.

Timely, relevant, and accurate information is crucial for much of the government and many sectors of the economy (although opinions on the degree to which information is important for different tasks vary across the globe). So, if the solution is not to simply tune out, how are the problems posed by information overload resolved? Solutions can be categorized under the broad heading of information management. Entities implementing information management policies (according to the 1998 Reuters report) experienced marked increases in productivity and decision-making capability. Information management in this context connotes methods for evaluating, prioritizing, and processing information (for example, the ranking operations performed by many search engines). Technology (especially e-mail and the Internet) is increasingly regarded as enabling information management rather than exacerbating information overload. Work practices are being adapted to use information technologies more effectively and businesses increasingly rely on a single, trusted source of comprehensive information in order to improve efficiency.

Perhaps a computer-neural interface will be developed to improve cognitive abilities to process and store information, but will this necessarily enhance the capacity to understand and control nature and society? One conundrum raised by the issue of information overload is that infinitely increasing both information and the capacity to use that information does not guarantee better decisions leading to desired outcomes. After all, information is often irrelevant, either because people are simply set in their ways, or natural and social systems are too unpredictable, or people's ability to act is somehow restrained. What is required, then, is not just skill in prioritizing information, but an understanding of when information is not needed. These cases do not point to insoluble problems. Instead they raise a more appropriate question most eloquently stated by T. S. Eliot (1952) in *The Rock*: "Where is the wisdom lost in knowledge and where is the knowledge lost in information?"

A. PABLO IANNONE
ADAM BRIGGLE

SEE ALSO *Communication Systems; Cyberculture; Hardware and Software; Information; Information Society.*

BIBLIOGRAPHY

Braybrooke, David. (1998). *Moral Objectives, Rules, and the Forms of Social Change.* Toronto: University of Toronto Press.

Eliot, T. S. (1952). *Complete Poems and Plays.* New York: Harcourt and Brace.

Hopson, Michael H. (2001). "Using a Technology-Enriched Environment to Improve Higher-Order Thinking Skills." *Journal of Research on Technology and Education* 34(2): 109–119.

Klapp, Orrin Edgar. (1986). *Overload and Boredom: Essays on the Quality of Life in the Information Society.* New York: Greenwood Press.

Reuters. (1996). "Dying for Information? An Investigation into the Effects of Information Overload in the U.K. and Worldwide," news release.

Reuters. (1997). "Glued to the Screen: An Investigation into Information Addiction Worldwide," news release.

Reuters. (1998). "Out of the Abyss: Surviving the Information Age, news release.

Schwartz, Barry. (2004). *The Paradox of Choice: Why More is Less.* New York: Ecco. Challenges the view of humans as rational utility maximizers and argues that unlimited choice can lead to suffering.

Toffler, Alvin. (1970). *Future Shock.* New York: Random House. Argues that there can be too much change in too short of a time, and therefore people need to improve their ability to wisely regulate, moderate, and apply technology to serve human ends. Information overload is discussed on pages 311–315.

Wurman, Richard Saul. (1989). *Information Anxiety.* New York: Doubleday.

INFORMATION SOCIETY
• • •

The term *information society* refers to a set of developments in the global human environment that began during the last quarter of the twentieth century. These developments entailed increasingly intensive use of technologies of information and communication, from desktop computers and a plethora of sensing and "smart" devices to mobile telephony and portable handheld electronics, all progressively interconnected. In a cursory sense, an information society is simply one heavily dependent on these technologies for human interactions and transactions, though no clear threshold exists for classifying a society as informational at any particular stage of technological development. In a more important and complex sense, an information society is one in which use of various technologies has produced or is producing substantial change in the ways people live, learn, work, socialize, and govern themselves.

A New Context for Ethics

As societies around the globe integrate various technologies into economic, social, educational, personal, and governmental practices, the resulting changes create new contexts for ethics in the mundane sense of the customary guidelines for their engagement. The altered context involves new linkages among individuals and

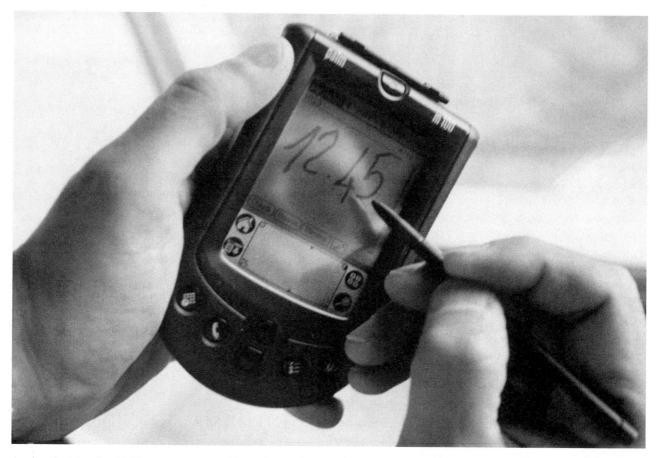

A palm pilot. These hand-held computers are one of the products of the Information Society. (© SIE Prouductions/Corbis.)

organizations, along with transformations in the identity and structure of human collectivities. One hallmark is decentralization in some forms of control and decision making. In economic transactions, this involves reduced reliance on hierarchical structures in favor of more distributed, flexible, horizontal links among organizations able to communicate and coordinate with less centralization. At the same time that certain organizational structures are becoming less centralized, possibilities for the collection of highly centralized data about citizens and their activities are expanding. This means that increasingly centralized bodies of information about citizens are available both to government and to the new forms of decentralized organizations in economic and other realms.

In the social arena, such changes involve more flexible, complex patterns of association with decreased dependence on physical proximity or identification with well-bounded groups or communities. One extreme gives rise to self-organizing groups and associations that may be temporary in nature. Another is the direct exchange among individuals of digitized artifacts such as

moving or still photography, music, books, or other narratives in the presence of little or no coordinating authorities and brokers.

The trend toward information societies is sometimes described in terms of the replacement of industrial-age human structures with *networks*. These networks involve adaptable, flexible, complex communication and interactions with many points of intersection. One of the classic descriptions of the information society is the trilogy of books by Manuel Castells collectively entitled *The Information Age: Economy, Society, and Culture* (1996, 1998, 1999). An earlier and widely influential statement that anticipated some of the developments of the information society is Daniel Bell's *The Coming of Post-Industrial Society* (1973).

Origins of the Information Society

Several factors contributed to the trend toward the information society. The intellectual threads of the information society can be traced back as far as Gottfried Wilhelm Leibnitz (1646–1716), who in the seventeenth century postulated a machine that might manipulate mathematical representations of human thought. More concretely, it was the rapid technological innovation by twentieth century corporations and universities that resulted in the production of a stream of fast, inexpensive, portable computing and communication technologies. These innovations began with the development of digital computers from the 1940s through the 1960s. Following establishment of digital computing as a field, progressive miniaturization from the 1970s through the 1990s promoted the design of smaller and more powerful information and communication devices.

Another interpretation traces the technological origins of the information society back to the industrial revolution. James Beniger (1986) argues that the demands for control and management of information associated with industrial activity and socioeconomic structure paved the way for the eventual rise of the information society.

A second economic factor contributing to information society trends was a wave of privatization of previously state-dominated media in many countries during the last two decades of the twentieth century. Both the technological developments and the restructuring of media-state relations fed economic globalization. Intensified economic and financial linkages among nations through corporate multinationalism and national trade policies created tighter interdependencies among societies and provided for means of cultural and social change. These developments were concentrated in the Americas, Europe, Australia, and non-authoritarian Asian nations. In many societies with authoritarian governments, such as Saudi Arabia and China, various government policies limited the deployment of information technologies.

Ethical Issues in the Information Society

Most technological innovations, beginning with the first use of tools made of stone or bone, have permitted humans to engage in new kinds of actions and to restructure relationships with one another. These altered actions and relationships raise important questions of ethics, social order, and governance. Which new possibilities for human agency and for restructuring of societies are desirable? How should such questions be decided—by individuals, by markets, by states, by religious institutions? Technologies of the information society precipitated many such questions. For instance, access to the technological infrastructure of the information society is highly unequal across nations and across groups and among individuals within nations, raising the possibility that information societies will be

exclusionary. Additionally, the digitization and centralization of information and the density of interconnections among people allow for far greater possibilities of privacy violations by other individuals, corporations, and governments in information societies than in industrial ones. One of the most important ethical issues in information societies involves challenges to traditional conceptions of information ownership, which in earlier periods was defined partly by practical constraints on its replication and exchange.

Because cultures are sustained and altered by communication and the preservation of certain artifacts and information, information society changes create cultural mixing and shifts. Especially controversial is the transfer of Euro-American cultural norms and practices to non-Euro-American societies. Even within Euro-American societies, debates about the regulation of speech that were once largely settled matters have been revisited, due to the vastly increased capacity for people to communicate material or ideas that were previously limited by such simple constraints as cost.

It is difficult to predict how far trends toward information societies will extend. It is important to observe that the rise of industrial societies in the nineteenth and twentieth centuries did not bring about the cessation of agricultural activities or the end of agrarian ways of life. Instead they produced a shift in the locus of human activity for many people in many societies. Similarly, the rise of the information society will not entail the end of industrial activity or the termination of industrial-age social structures, cultural practices, and economics, but rather a transition to altered human arrangements across this spectrum.

BRUCE BIMBER

SEE ALSO Bell, Daniel; Communication Ethics; Digital Divide; Information; Information Overload; Leibniz, G. W.; Networks; Privacy.

BIBLIOGRAPHY

Bell, Daniel. (1973). The Coming of Post-Industrial Society: A Venture in Social Forecasting. New York: Basic Books.

Beniger, James R. (1986). The Control Revolution: Technological and Economic Origins of the Information Society. Cambridge, MA: Harvard University Press.

Castells, Manuel. (1996, rev. 2000). The Rise of the Network Society—The Information Age: Economy, Society, and Culture, Vol. 1. Malden, MA: Blackwell.

Castells, Manuel. (1997). The Power of Identity—The Information Age: Economy, Society, and Culture, Vol. 2. Malden, MA: Blackwell.

Castells, Manuel. (1998, rev. 2000). *End of Millennium—The Information Age: Economy, Society, and Culture, Vol. 3.* Malden, MA: Blackwell.

INFORMED CONSENT

• • •

Informed consent is an individual's voluntary agreement, based on adequate understanding of relevant facts, to permit some type of intervention by a second party. This term is most commonly used in medical contexts to refer to individuals' agreements to undergo medical treatment or to participate in research. In most cases, informed consent is required both ethically and legally prior to the commencement of treatment or enrollment in research.

Recent History

The ethical and legal mandate for informed consent as understood in the early twenty-first century was not established until the latter half of the twentieth century. Before that time, a paternalistic paradigm governed the relationship between patient and health care provider. However, driven by landmark cases, revelations of abuse, and a changing professional ethic, there has been a shift toward patient autonomy and away from physician paternalism. The establishment of a requirement for informed consent occurred independently but concurrently in the two contexts of medical treatment and research with human subjects.

MEDICAL TREATMENT. U.S. courts first recognized the need for patients to give consent for medical treatment in *Schloendorff* v. *Society of New York Hospital* in 1914. It was not until *Salgo* v. *Leland Stanford, Jr. University Board of Trustees* in 1957, however, that the additional provision requiring physicians to give patients information relevant to their treatment decisions was established. This requirement for physician disclosure was expanded, developed, and solidified by *Natanson* v. *Kline* (1960), *Mitchell* v. *Robinson* (1960) and *Canterbury* v. *Spence* (1972). These precedents were then incorporated into statements by the Judicial Council of the American Medical Association (AMA) in 1981 and the President's Commission for the Study of Ethical Problems in Medicine and Biomedical and Behavioral Research in 1982. The need to obtain patients' informed consent has since been incorporated into the medical practice guidelines of numerous national and international organizations of medical professionals.

RESEARCH SUBJECTS. The evolution of informed consent in research with human subjects was spurred not by legal decisions but by public and professional reaction to several cases in which people were used as research subjects without their knowledge or permission. The Nuremberg Code of 1947 established general guidelines for human subjects research in response to the revelation of the Nazi medical experiments, stating that the informed and voluntary consent of subjects was "absolutely essential." In an effort to create more specific ethical guidelines for research, the World Medical Association (WMA) adopted the first version of the Declaration of Helsinki in 1964, which also held the subjects' informed consent to be a necessary element of ethical research.

In 1966, Henry K. Beecher published an article in the *New England Journal of Medicine* identifying twenty-two ethically problematic studies involving human subjects, including studies at the Jewish Chronic Disease Hospital and Willowbrook State Hospital. Beecher concluded that patients must give informed and voluntary consent before participating in research. The uncovering of the Tuskegee syphilis study that took place between 1932 and 1972 brought widespread attention to violations of the rights of human subjects. In the Tuskegee case, poor and uneducated African American men were enrolled in a study of the progression of untreated syphilis without their knowledge or consent. At least partially in response to these abuses, The Belmont Report, published in 1978, and finally the federal Common Rule (45 CFR 46) in 1991 incorporated the requirement for informed consent into United States regulation.

PHILOSOPHICAL FOUNDATIONS. The moral requirement for informed consent can be grounded in both deontological and consequentialist ethical theory. Immanuel Kant (1724–1804) held that moral worth is based upon the ability to reason and that the ability to reason must be respected by others. Rational choices are expressions of the ability to reason and so have intrinsic value. As a result, people have obligations to make rational choices and others are obligated to respect those choices. Giving (or refusing to give) informed consent is a form of rational choice and so therefore has intrinsic value within a Kantian deontological framework.

The value of informed consent can also be derived from consequentialist ethical theory. Consequentialists hold that something is good if it produces good outcomes. In most cases, people know their own goals and values better than anyone else, and therefore are in the best position to decide how to promote their own good. Even though people may, on occasion, be mistaken about what is good for them, they benefit overall from exercising self-determination. As a result, the best outcomes are brought about when people make decisions for themselves. The requirement for informed consent is one way to protect and encourage self-determination and therefore to bring about good consequences.

Five Elements of Informed Consent

There are at least five necessary elements of informed consent: disclosure, understanding, capacity or competency, voluntariness, and assent. These elements can take different forms in research and treatment contexts and can entail various ethical and legal standards.

DISCLOSURE. Informed consent can only be given if the person consenting is adequately informed. The first part of this process involves the disclosure of information. The physician, researcher, or in some cases another individual, must make available to the patient or potential subject sufficient information to make a decision about treatment or participation in research.

What constitutes sufficient disclosure is ambiguous, but there are three plausible ways this can be interpreted. The *professional practice standard* of disclosure requires physicians to give patients as much information as is generally disclosed by other medical professionals about a particular procedure or research protocol. The *reasonable person standard* sets the disclosure requirement at whatever a reasonable person would want to know in a given situation. A final disclosure standard is the *subjective standard*, which states that a physician should tell a patient whatever that subject would want to know. Each of these views on disclosure has advantages and disadvantages.

There is no consensus on which standard best describes the ethical obligation of disclosure. Generally, however, disclosure must include at least a description of the treatment or procedure, the material risks and benefits, and the available alternatives. In research contexts, additional information must be provided to the individual considering participation. Examples of such additional information include: a statement about the experimental nature of the procedures, information about confidentiality of the subject's records, information about what to do in case of injury from the study, and a statement that participation in the research is voluntary.

Legally, state jurisdictions are approximately evenly split between using the professional practice standard and the reasonable person standard in treatment contexts. Only a few jurisdictions hold physicians to the subjective disclosure standard. The U.S. Common Rule provides an itemized list of the information that must be conveyed to potential subjects within research contexts.

UNDERSTANDING. In order for an individual to be informed in the ethically relevant sense, that individual must respond to the disclosure in an appropriate way. That is, the individual must internalize the information that has been made available though the disclosure process. If a patient or potential subject is unable to understand the provided information, informed consent is not possible. It is the responsibility of a physician or researcher to make an effort to maximize the understanding of the patient or potential subject. For example, a researcher should convey the relevant information in language that the potential subject can comprehend and should answer clearly any questions that the subject asks about the protocol. In practice, formal assessment of an individual's level of understanding is rare. Instead, patients or potential subjects may simply be asked if they understand the information they have been given or if they have any questions.

CAPACITY AND COMPETENCY. Capacity refers to an individual's ability to appropriately manipulate the information that has been understood. There are a number of different ways that decision-making capacity could be defined and by which the presence or absence of capacity could be assessed. The ability to appreciate the consequences of one's life options, to weigh the various considerations and come to a decision, to reason logically about one's situation, and to evaluate the situation in light of one's own values could all be used as indicators of capacity. There is no ethical or legal consensus on which of these definitions should be used.

Capacity is a task-specific concept, meaning that the level of decision-making capacity needed to make a given decision varies depending on the nature of the decision itself. As a result, at any given time one may have the capacity to make some decisions but not others. Generally, the higher the risk posed by a procedure, the more capacity one must have to make a decision to undergo that procedure. For example, a patient may have capacity to consent to having an IV inserted but not to having invasive surgery.

Individuals without the capacity to make a given decision about treatment or research cannot give informed consent to undergo that treatment or research. When a person lacks decision-making capacity, informed consent is solicited from a surrogate decision maker, that is, a family member or other individual appointed to made decisions on behalf of that person.

Competency is the legal analogue to capacity. Adults are presumed to have competency unless it has been demonstrated to a court that they are unable to make autonomous decisions, in which case the court declares the adult to be incompetent. At that time, a legally authorized representative is appointed for that individual. In contrast, children and adolescents under the age of eighteen do not have competency to make their own decisions unless a court decides otherwise.

VOLUNTARINESS. An individual's decision to undergo treatment or to participate in research must be voluntary. That is, the individual must not be coerced or unduly influenced by either external or internal factors. Threats of unwanted consequences such as physical harm or withdrawal of medical care are obvious examples of coercion. More subtle challenges to voluntariness include the provision of substantial incentives and the manipulation of an individual's decision-making process through the biased presentation of information. Because of the importance of voluntariness, informed consent is often denominated "free and informed consent" or just "free informed consent."

A physician or researcher may not coerce or unduly influence a patient or potential subject to make a desired decision. The conditions under which and the manner in which the physician or researcher solicits consent should be designed to minimize the possibility that voluntariness will be compromised.

ASSENT. The final element of informed consent is the decision made about undergoing treatment or participating in research. Inherent in the idea of informed consent is a positive decision—one gives informed consent to undergo a particular treatment or procedure. A negative decision—that is, a decision not to undergo the treatment or procedure—constitutes an informed refusal.

Generally, verbal agreement is sufficient for low-risk treatment decisions. When treatment methods involve higher levels of risk, however, the patient may be required to sign a consent form. The form summarizes the relevant information and states that the individual is voluntarily agreeing to the treatment or procedure. In research contexts, an individual's consent to participate in the research protocol must almost always be documented by the individual's signature on a consent form.

Exceptions to Informed Consent

There are a few exceptions to the requirement for informed consent for medical treatment. In emergency situations, treatment can be administered without the patient's consent because it is presumed the patient would consent if given the opportunity. Other exceptions include cases in which an individual poses a threat

to public health. In such cases, treatment may be forced on that individual without consent. For example, a person with tuberculosis may be compelled to undergo treatment. Individuals may also waive their right to informed consent, stating that they do not wish to be informed of a diagnosis or to make decisions about their own treatment. Finally, children and incompetent adults do not give informed consent for treatment, although consent must be obtained from parents or guardians.

Informed consent is almost always required prior to enrollment in research. However, federal regulation allows individuals to be enrolled without their consent in research protocols in some emergency situations if obtaining consent would be impossible. It further allows emergent use of an investigational drug or procedure on a case-by-case basis if it is believed that doing so will have therapeutic value for the patient. A second exception to the requirement for informed consent in research contexts enables parents or guardians to give consent for the participation of children and incompetent adults. It has, however, been recommended that physicians and researchers seek the assent of these individuals when possible.

Informed Consent and Science and Technology

Although the concept of informed consent is most thoroughly developed within the contexts of medical treatment and biomedical research, it has ethical implications for the development and use of the products of science and technology more broadly defined. Research into and implementation of innovations in fields such as civil engineering, nuclear energy, genetic engineering, and nanotechnology have inherent risks. In many cases, the members of the community in which these innovations are being developed and put into use are exposed to these risks. The ethical requirement for informed consent, however, suggests that these individuals should not have to bear this burden without their knowledge and voluntary consent.

In most cases, the process of obtaining consent for medical treatment or for enrollment in biomedical research is dyadic, consisting of a dialogue between a physician or investigator and a subject. In non-medical contexts, however, this model of obtaining consent is often not feasible. Practically, it would be impossible to obtain individual consent from each member of the community that could be exposed to risk. Further, many of those who may be affected by these innovations could not even theoretically be asked for consent, such as members of future generations.

Despite these difficulties, the requirement for informed consent generates ethical obligations for those who develop and implement the products of science and technology. These obligations may be discharged through various community consent mechanisms, such as allowing public participation in the creation of policies that govern innovations, consultation with community leaders, and assessment of public opinion. The use of these and other community consent methods may help to ensure that science and technology move forward in an ethically appropriate way, and therefore that the goods that they produce are not achieved at too great a cost.

JANET MALEK

SEE ALSO *Bioethics; Human Subjects Research; Medical Ethics; Sociological Ethics*.

BIBLIOGRAPHY

Beauchamp, Tom L., and Ruth R. Faden. (2004). "Informed Consent: Meaning and Elements." In *Encyclopedia of Bioethics*, 3rd edition, ed. Stephen G. Post. New York: Macmillan. A concise and updated review of the elements of informed consent.

Beecher, Henry K. (1966). "Ethics and Clinical Research." *The New England Journal of Medicine* 274(24): 1354–1360. A landmark article identifying numerous ethically problematic research protocols.

Berg, Jessica W., et al. (2001). *Informed Consent: Legal Theory and Clinical Practice*, 2nd edition. Oxford, UK: Oxford University Press. Comprehensive review of the legal history and current practical applications of informed consent.

Committee on Bioethics, American Academy of Pediatrics. (1995). "Informed Consent, Parental Permission, and Assent in Pediatric Practice." *Pediatrics* 95(2): 314–317. Statement establishing the ethical requirement for children's assent for participation in research.

Faden, Ruth R.; Tom L. Beauchamp; and Nancy P. King. (1986). *A History and Theory of Informed Consent*. New York: Oxford University Press. Reviews the development of informed consent and provides a detailed analysis of its elements.

Grisso, Thomas, and Paul S. Appelbaum. (1995). "Comparison of Standards for Assessing Patients' Capacities to Make Treatment Decisions." *American Journal of Psychiatry* 152(7): 1033–1037. Discusses the possible standards and their implications for evaluating capacity.

Jones, James H. (1981). *Bad Blood: The Tuskegee Syphilis Experiment*. New York: Free Press. A historical report and analysis of the Tuskegee syphilis experiment.

Levine, Robert J. (1986). *Ethics and Regulation of Clinical Research*, 2nd edition. Baltimore, MD: Urban and Schwartenberg. Detailed review of the ethical requirements for clinical research.

Levine, Robert J. (2004). "Informed Consent: Consent Issues in Human Research." In *Encyclopedia of Bioethics*, 3rd edition, ed. Stephen G. Post. New York: Macmillan. Concise review of the concept of informed consent as it applies to research with human subjects.

Shrader-Frechette, Kristin. (1991). *Risk and Rationality*. Berkeley: University of California Press. Exploration of the conditions under which risks may be imposed on societies.

Shrader-Frechette, Kristin. (1994). *Ethics of Scientific Research*. Lanham, MD: Rowman & Littlefield. Discussion of scientific research emphasizing the importance of public welfare and public decision making.

United States Department of Health and Human Services. (1991). "Protection of Human Subjects." *Code of Federal Regulations*, Title 45, Part 46. United States federal regulation governing research involving human subjects; also known as the "common rule."

INNOVATION

SEE *Technological Innovation*.

INSTITUTE OF ELECTRICAL AND ELECTRONICS ENGINEERS

• • •

The Institute of Electrical and Electronics Engineers (IEEE) is the largest technical society in the world with more than 375,000 members in 150 nations; it publishes 30 percent of the global technical literature in electrical and computer engineering. The organization was formed in 1963 through a merger of the American Institute of Electrical Engineers (AIEE, founded in 1884) and the Institute of Radio Engineers (IRE, formed in 1912 when two local organizations founded in Boston and New York were merged).

In its early years the AIEE struggled to espouse professionalism in engineering despite strong pressure to the contrary from the businesses (mostly electric utilities) that employed the great majority of its members (Layton, 1986). Indeed, the famous engineer and socialist Charles Steinmetz served as president of the AIEE in 1901–1902. By the late twenties, however, business interests dominated the AIEE, evidenced by a lower membership standards that admitted business executives in the utility industry, restriction of the activities of local sections to purely technical matters, censorship of publications critical of business practices, stifling of dissent and public discussion of the profession through restrictions in the code of ethics, and abandonment of open elections in favor of a nominating committee. Many observers would argue that the AIEE's predisposition toward business interests was carried over to the IEEE and prevails to the present day (Herkert 2003).

The IRE was founded in part out of dissatisfaction with the growing dominance of business interests in the IEEE's affairs and in part due to the strong scientific basis and rapid growth of the field of radio engineering, which resulted in a higher sense of professionalism (McMahon 1984, Layton 1986). The IRE also aspired to become an international organization. Ironically, however, the IRE shied away from speaking for its members on professional and policy matters (Layton 1986). By the time of the merger the IRE had surpassed the AIEE in membership, buoyed by the explosive growth in electronics following World War II. The merger, an inevitable result of this development, resulted in a blending of the two institutional cultures that incorporated the IRE's decentralized management structure and professional groups, now known as technical societies (IEEE History Center 1984).

In 1973 the IEEE amended its constitution changing it from a strictly "learned" society to one that also represented the professional interests of its members. As a result, the United States Activities Board (USAB) was formed to represent the interests of U.S. members. (IEEE History Center 1984, McMahon 1984). The USAB and its successor organizations have played an important role in ethics activities of the IEEE and in promoting policy favorable to the U.S. engineering and business community. The affect of the USAB's presence on efforts to globalize the IEEE has been more controversial.

Codes of Ethics

The AIEE promulgated one of the earliest codes of engineering ethics in 1912. The code provided that the "first professional obligation" was to protect the interests of the engineer's clients or employers (Layton 1986). In 1950 the AIEE code was revised to incorporate the cannons of the code of ethics of the engineers' council for professional development, including a provision that the engineer "will have due regard for the safety of life and health of the public and employees who may be affected by the work for which he is responsible" (CSEP 2004). The first IEEE code of ethics was adopted in the 1970s (Unger 1994) following revisions in 1979 and 1987 (CSEP 2004). The current IEEE code of ethics (adopted 1990), in parallel with other contemporary engineering codes, pledges its members to protect the "safety, health and welfare of the public." Unlike others,

however, the IEEE code also includes specific language regarding ethics support, committing its members "to assist colleagues and co-workers in their professional development and to support them in following this code of ethics."

Ethics Activities

The IEEE has long enjoyed a reputation as one of the more proactive professional engineering societies in the ethics area. This positive image derives primarily from ethics activity in the1970s, including preparation of a friend of the court brief supporting the three whistle-blowing engineers in the Bay Area Rapid Transit (BART) case. Much of this activity was encouraged by the formation of a Committee on Social Implications of Technology by Stephen Unger and other organizational activists, which evolved into the IEEE Society on Social Implications of Technology (SSIT). The SSIT, though only 2,000 members strong, has remained an important voice in the IEEE for ethical responsibility and concern for societal implications of technology. The SSIT publishes a quarterly journal, *IEEE Technology and Society Magazine*, hosts an annual conference, and periodically bestows its Carl Barus Award for Outstanding Service in the Public Interest on engineers who uphold the highest ethical standards of the profession. As noted earlier, the IEEE sub-unit that represents the interests of U.S. members has also been active in ethics issues. At the level of the parent organization, however, ethics activity was generally dormant between the late-1970s and mid-1990s (Unger 1994).

The IEEE reputation for promoting engineering ethics was, in the opinion of many observers, seriously tarnished by events that began to unfold in the late 1990s when a staff and volunteer leader backlash crushed gains in ethics support (Unger 1999, Herkert 2003). Prior to 1995, the only committee of the IEEE Board of Directors (BoD) charged with dealing with ethics was the Member Conduct Committee (MCC), founded in 1978. The MCC's purpose was twofold: to recommend disciplinary action for violation of the Code of Ethics and to recommend support for members who when following the Code encountered difficulties such as employer sanctions.

A BoD-level Ethics Committee, formed in 1995 as a result of efforts by members to elevate the prominence of ethics in the organization, was intended to provide information to members and advise the BoD on ethics-related policies and concerns. As one of its first actions, the Ethics Committee, whose membership included Stephen Unger, in 1996 established an Ethics Hotline designed to provide information and advice on ethical matters to professionals in IEEE fields of interest. Cases brought to the attention of the Ethics Hotline included falsification of quality tests, violations of intellectual property rights, and design and testing flaws that could result in threats to public safety. In some instances, such cases were referred to and acted on by the MCC (Unger 1999).

The Executive Committee of the BoD suspended the Ethics Hotline in 1997 after less than a year of operation. In 1998 the Executive Committee rejected and suppressed its own task force report, which recommended reactivation of the hotline. In the same year, the IEEE implemented bylaw changes that reduced the terms in office of members of the MCC and Ethics Committee, and, in apparent disregard of the IEEE Code of Ethics, prohibited the Ethics Committee from offering advice to any individuals including IEEE members. The cycle was complete in 2001 when the Ethics Committee and the MCC were merged. Like the old MCC, the combined committee has a dual-charge of member discipline and ethics support, but its activities are limited by IEEE Bylaw I-306.6: "Neither the Ethics and Member Conduct Committee nor any of its members shall solicit or otherwise invite complaints, nor shall they provide advice to individuals."

In another example of what one IEEE member describes as *ethical timidity*, in 2002 the IEEE denied membership benefits to its members in Iran and several other nations on the grounds that such action was required by U.S. trade restrictions, a position that was not shared by most other U.S.-based scientific and technical societies. Compounding the blow to the IEEE ethics profile, the IEEE leadership initially sought to conceal this action on a *need to know basis* (Gaffney 2003). Though the IEEE later claimed to be vindicated by a government exemption permitting editing and publication of papers submitted by Iranians, the ruling imposed restrictions on collaboration with Iranian scientists and left unchanged the IEEE's suspension of the membership benefits of residents of the sanctioned countries (Foster 2004)

JOSEPH R. HERKERT

SEE ALSO *Engineering Ethics; Professional Engineering Organizations.*

BIBLIOGRAPHY

Foster, Kenneth. (2004). "Call for Action to Protect Free Exchange of Ideas." *Nature* 429: 343. Letter to the editor calling for response of scientific and engineering community to U.S. government restrictions on publishing and other activities of scientific societies.

Gaffney, Owen. (2003). "IEEE Under Fire for Withdrawing Iranian Members' Benefits." *Science* 301: 1646. News article regarding the IEEE's controversial actions with respect to the U.S. Treasury's Office of Foreign Assets Control.

Herkert, Joseph. (2003). "Biting the Apple (But Not Inhaling): Lessons from Engineering Ethics for Alternative Dispute Resolution Ethics." *Penn State Law Review* 108: 119–136. Review of recent developments in the field of engineering ethics including discussion of the IEEE's changing stance on ethics support.

Layton, Edwin. (1986). *The Revolt of the Engineers*, 2nd edition. Baltimore: Johns Hopkins. Seminal history of the development of the engineering profession in the first half of the nineteenth century that focuses on the continuing conflict between business and professional values.

Mcmahon, A. Michael. (1984). *The Making of a Profession: A Century of Electrical Engineering in America.* New York: IEEE Press. History of the electrical engineering profession commissioned by the IEEE for its centennial.

Unger, Stephen. (1994). *Controlling Technology: Ethics and the Responsible Engineer*, 2nd edition. New York: Wiley. Well-known text on engineering ethics that gives extensive coverage to the IEEE's activities in the ethics arena from the perspective of a long-time participant in many of them.

Unger, Stephen. (1999). "The Assault on IEEE Ethics Support." *IEEE Technology and Society Magazine* 18(1): 36–40. Review of the IEEE backlash to ethics committee activities in mid-1990s by a prominent member of the ethics committee.

INTERNET RESOURCES

Center for the Study of Ethics in the Professions (CSEP). (2004). "Codes Of Ethics Online." Available from http://www.iit.edu/departments/csep/PublicWWW/codes/. Archive of more than 850 codes of professional ethics and other material on codes.

IEEE History Center. (1984). "Origins of the IEEE." Available from http://www.ieee.org/organizations/history_center/history_of_ieee.html. Brief history of the AIEE and the IRE and their merger to form the IEEE.

IEEE Society on Social Implications of Technology. (2004). Available from http://ieee.org/ssit. Internet site of the IEEE "technical" society concerned with social, ethical, and policy implications of technology; publisher of the *IEEE Technology and Society Magazine* and sponsor of the international symposium on technology and society.

Institute of Electrical and Electronics Engineers. (1990). "IEEE Code of Ethics." Available from http://www.ieee.org/about/whatis/code.html. The IEEE's current Code of Ethics; this code has a somewhat different format from other contemporary codes of engineering ethics.

Institute of Electrical and Electronics Engineers. (2001). "Bylaw I-306." Available from http://www.ieee.org/about/whatis/bylaws/i-306.html. Bylaw describing functions and membership of IEEE standing committees and boards; Section 7 applies to the Ethics and Member Conduct Committee.

INSTITUTE OF PROFESSIONAL ENGINEERS NEW ZEALAND

• • •

The first engineering society in New Zealand was the Institute of Local Government Engineers, which was founded in 1912. In 1914 it merged with the New Zealand Society of Civil Engineers, founded in 1913, which in 1982 became the Institute of Professional Engineers New Zealand (IPENZ). IPENZ is open to all engineering professionals.

The terms *engineering profession*, *professional engineer*, and *professional engineering* are used by IPENZ in the most general possible manner, to include all those who use a systematic process of analysis, design/synthesis, and implementation; strive to operate in a responsible way; are governed by a code of ethics set by their peers; and engage in continuing professional development to maintain the currency of their competence. IPENZ publishes the peer-reviewed print journals *e.nz* and *Engineering treNz* as well as the member newsletter, *engineering dimension*. Membership is currently about 9,000.

Because of New Zealand's unique geology—it is prone to floods, earthquakes, and volcanic eruptions—IPENZ has a strong focus on natural hazard and risk management. It also supports a heritage project, whose goal is described on its web site as "To inspire and teach present and future generations by preserving the legacy of the past through the identification, maintenance and promotion of New Zealand's engineering heritage."

The IPENZ code of ethics, perhaps under the influence of Engineers for Social Responsibility, gives high priority to social and environmental responsibility; along with the Australian Institution of Professional Engineers, IPENZ was one of the first engineering societies to do so. While it would be incorrect to describe it as an activist organization, IPENZ has on occasion taken strong public stands on issues such as dam safety.

ALISTAIR GUNN

SEE ALSO *Australian and New Zealand Perspectives; Professional Engineering Organizations.*

BIBLIOGRAPHY

INTERNET RESOURCE

Institute for Professional Engineers New Zealand. Available from www.ipenz.org.nz.

INSTITUTIONAL BIOSAFETY COMMITTEES

• • •

Institutional Biosafety Committees (IBCs) are review boards appointed by an institution to evaluate and approve potentially biohazardous lines of research. IBCs were established in 1976 by the National Institutes of Health (NIH) Guidelines for Research Involving Recombinant DNA Molecules (Guidelines). Their function is to provide local institutional oversight and approval of nearly all forms of NIH-sponsored research utilizing recombinant DNA (rDNA) in order to ensure that such research is in compliance with the Guidelines. IBCs were developed in response to fears about the risks posed by genetic engineering and guided by principles considered at the Asilomar Conference on recombinant DNA molecules.

Although IBCs still serve as the cornerstone for oversight of this research, their role has also been expanded to include review and supervision of a variety of experiments involving biological materials and other potentially hazardous agents. The potential threats posed by "dual use research" has prompted the National Science Advisory Board for Biosecurity (NSABB) to consider further expanding the role of IBCs to monitor research that may have implications for bioterrorism. There are doubts, however, about the ability of some IBCs to perform this expanded function. Furthermore, controversy exists not only about the performance of certain IBCs in their main role of ensuring safety, but also about how transparent their work should be. Judging the validity of these concerns is hampered by a general paucity of evaluations and assessments of individual IBCs and the system as a whole.

Background, Development, and Institutionalization

The risks presented by emerging techniques in rDNA research during the early 1970s led scientists to implement a brief self-imposed moratorium on this work. Research with rDNA eventually continued under the principles and guidelines established at the 1975 Asilomar Conference. A mechanism for institutionalizing review and approval of proposed research was considered but not formally adopted at Asilomar. However, in 1976 such a mechanism was created by the NIH Guidelines in the form of IBCs. The model of local, decentralized review committees created and guided by a mandate at the federal level already existed in the form of Institutional Review Boards (IRBs), formalized by the 1974 National Research Act, to monitor human subjects research. Like IRBs, IBCs serve as a mechanism for delegating oversight and approval responsibilities to the local institutions performing research supported by federal grants.

The 1976 Guidelines created Institutional Biohazards Committees, but in 1978 there was a formal shift in focus from "biohazards" to "biosafety." The same year also brought other changes, including more emphasis on ensuring appropriate review, the appointment of a Biological Safety Officer (BSO) at each institution, and improved training protocols and implementation procedures. In 1984 IBCs became responsible for oversight and approval of human gene transfer research. Two years later, the IBC system formally incorporated the "points to consider" developed by the Recombinant DNA Advisory Committee (RAC) Working Group on Human Gene Therapy, and in 1990 the emphasis was shifted to gene transfer rather than gene therapy. The NIH Guidelines were further expanded with additional appendices during the 1980s and 1990s as emerging techniques presented novel regulatory requirements. In 2000 the gene transfer protocols were amended to require RAC review prior to IBC approval (Grilley and Gee 2003).

Although the core responsibilities (review and oversight of rDNA research) of IBCs have remained stable throughout these changes, they have also been expanded to include oversight of other potentially hazardous research, including work on such materials as infectious agents and carcinogens. In March 2004, the U.S. government announced plans to create a National Science Advisory Board for Biosecurity (NSABB), which would identify possible "dual use research" (legitimate scientific work that could be misused to threaten public health or national security) and develop guidelines and recommendations for oversight. The task of implementing the board's recommendations will fall mostly on the roughly 400 IBCs (Couzin 2004).

Each IBC must be composed of no fewer than five members, and at least two members must be unaffiliated with the institution and represent the environmental and public health interests of the surrounding community. Members must be selected in a manner that ensures adequate expertise in rDNA technology and competence in assessing potential risks of proposed research. The functions and responsibilities of IBCs include: assessing containment levels required by the Guidelines for the proposed research; implementing contingency plans; maintaining proper facilities; ensuring adequate training of personnel; ensuring compliance with all surveillance, data reporting, and adverse event reporting; and additional responsibilities for human gene transfer experiments in accordance with Appendix M of the Guidelines. The NIH Office of Biotechnology Activities (OBA) manages and evaluates the conduct of the IBCs as part of its broader mandate to implement oversight mechanisms and information resources to promote the science, safety, and ethics of rDNA research (Shipp and Patterson 2003).

Criticisms and Assessments

Transparency, or openness to public review, is the most contentious issue surrounding both the conduct of individual IBCs and the system as a whole. Proprietary rights and privacy issues often conflict with demands for information about research that could threaten public health. IBCs have been targeted by several watchdog organizations, including the Sunshine Project, which investigates activities that could undermine the 1972 Biological and Toxin Weapons Convention and the 1993 Chemical Weapons Convention. The question of transparency is especially contentious when it involves biodefense research. Such activities require secrecy, yet in the absence of public oversight they could cross over into offensive research or generate new risks (Enserink 2004). The increasing awareness that terrorists could misuse some of the research regulated by IBCs only intensifies the conflict as some call for tighter controls on information and others demand increased public involvement.

There are additional concerns that several IBCs are not only reluctant to publicize information but may be lax in their oversight responsibilities. The Sunshine Project (Enserink 2004) and Diana Dutton and John Hochheimer (1982) accused some IBCs of meeting too infrequently or informally to adequately fulfill their duties. This raises further doubts about the ability of certain IBCs to take on the additional responsibilities of monitoring dual use research. The charge could also be made that a more neutral body should be responsible for oversight and evaluation of IBCs, because the OBA is housed within the NIH, which may raise conflict-of-interest issues.

Dutton and Hochheimer (1982) and Philip Bereano (1984) carried out detailed evaluations of IBCs. Both sets of researchers agreed that IBCs represent novel and promising experiments in the joint regulation of technology by lay and technical communities. However, both also argued that several shortcomings in the IBC system have severely limited its potential to forge consensual judgments about the acceptable risks of scientific research. Bereano argued that the Guidelines were developed primarily by the group being regulated, which narrowed their scope and unduly constrained the purpose of the IBC system. In a related critique, he claimed that IBCs are often dominated by rDNA scientists, which leads to a narrow perception of risk and a general hostility toward regulation. Dutton and Hochheimer argued that IBCs rarely realize their potential for genuine public participation, both for the reasons Bereano outlined and because IBCs often lack adequate resources.

Follow-up on these evaluations has been relatively sparse, which may reflect the difficulty in assessing a decentralized system designed to tailor oversight responsibilities to specific project proposals. The paucity of neutral, comprehensive evaluations, however, also means that many criticisms of IBCs are difficult to substantiate and operationalize into reforms that could improve the regulatory system.

ADAM BRIGGLE

SEE ALSO *Bioethics Commissions and Committees; Biological Weapons; Biosecurity; Institutional Review Boards.*

BIBLIOGRAPHY

Bereano, Philip. (1984). "Institutional Biosafety Committees and the Inadequacy of Risk Regulation." *Science, Technology, and Human Values* 9(4): 16–34. Provides a thorough critique of IBCs and argues that they are still primarily a self-serving mechanism for science rather than a genuine opportunity to involve the public in the process of broadly judging the risks of science from political, social, and environmental angles.

Couzin, Jennifer. (2004). "U.S. Agencies Unveil Plan for Biosecurity Peer Review." *Science* 303(5664): 1595. Discusses the formation of the NSABB as a mechanism for regulating dual use research and the impacts it may have on IBCs.

Dutton, Diana, and John Hochheimer. (1982). "Institutional Biosafety Committees and Public Participation: Assessing an Experiment." *Nature* 297: 11–15. Study of twenty IBCs in California as experiments in local participation in the regulation of science. Revealed great diversity in performance, but concluded that IBCs do little to support public participation.

Enserink, Martin. (2004). "Activist Throws a Bright Light on Institutes' Biosafety Panels." *Science* 305(5685): 768–769. Examines the efforts of Edward Hammond and the Sunshine Project to ensure that IBCs are transparent.

Grilley, B.J., and A.P. Gee. (2003). "Gene Transfer: Regulatory Issues and their Impact on the Clinical Investigator and the Good Manufacturing Production Facility." *Cytotherapy* 5(3): 197–207. Discusses regulatory oversight of human gene-transfer studies and comments on good manufacturing practice issues.

Shipp, Allan, and Amy Patterson. (2003). "The National Institutes of Health System for Enhancing the Science, Safety, and Ethics of Recombinant DNA Research." *Comparative Medicine* 53(2): 159–164. An overview of the NIH Guidelines, IBCs, and RACs, with a commentary on their relationship to animal research.

INTERNET RESOURCE

National Institutes of Health. "Guidelines for Research Involving Recombinant DNA Molecules." Available from: http://www.nih.gov.

INSTITUTIONAL REVIEW BOARDS

• • •

Established by Congress in the 1974 National Research Act, institutional review boards (IRBs) are decentralized committees that review and monitor nearly all federally funded research projects involving human subjects in the United States. In most other nations these groups are called research ethics committees (RECs). The purpose of IRBs is to ensure that research conforms to ethical standards and protects the rights and welfare of the people who participate as research subjects. This is accomplished through the IRB Review of Research process, which involves the review of protocols, informed consent documents, and related materials for proposed research. Although flawed and contentious, the IRB regulatory framework is improving in its ability to assure the upholding of ethical standards in a rapidly evolving research context.

Background

The unethical practices of Nazi doctors at concentration camps spurred several attempts to formulate ethical principles for the conduct of research involving human subjects and institutionalize political mechanisms capable of upholding those principles. The most notable international efforts include the 1948 Nuremberg Code and the 1964 Declaration of Helsinki made by the World Medical Association. In 1975 the Helsinki Declaration was revised to include a statement recommending that independent committees review research proposals. The declaration has been revised five more times (1983, 1989, 1996, 2000, and 2002), but the role of ethical review committees has remained central.

In the United States the first federal document requiring committee review was issued in 1953, but it applied only to research conducted at one National Institutes of Health (NIH) facility. In 1966 the U.S. Public Health Service required recipients of its grants to establish committees to review the ethical merits of proposed research involving human subjects. In the early 1970s the U.S. Food and Drug Administration (FDA) and the U.S. Department of Health, Education, and Welfare (DHEW) (forerunner to the Department of Health and Human Services [DHHS]) both promulgated regulations that required committee review of research conducted in institutions.

In 1974, one year after the unethical Tuskegee syphilis study was discontinued, the National Research Act established a statutory requirement for review of FDA- and DHEW-funded research by a committee to which it called an institutional review board (IRB National Research Act 1976). That act also created the National Commission for the Protection of Human Subjects of Biomedical and Behavioral Research (hereafter the Commission) to identify the basic ethical principles that should underlie the conduct of research involving human subjects and to develop guidelines to assure that that research is conducted in accordance with those principles. In 1978 the Commission added a requirement to ensure the equitable selection of research subjects.

In the next year the Commission issued its basic ethical principles and guidelines in the Belmont Report, which resulted from a four-day period of discussions held in February 1976 at the Smithsonian Institution's Belmont Conference Center. The three basic ethical principles identified in the report are justice, beneficence, and respect for persons. The Belmont Report did not make specific recommendations for administrative action by the Secretary of DHEW. Instead, the Commission recommended that the report be adopted in its entirety as a statement of the department's policy. The subsequent adoption of the Belmont Report represents a rare instance of the federal government formally accepting a moral theory as the foundation for legislation (Callahan 2003).

Each of the three principles outlined in the Belmont Report has engendered specific regulations for the practice of research involving human subjects. The principle of justice focuses on the question of who should receive the benefits and bear the burdens of research. It has given rise to both federal and NIH regulations that ensure that the selection of research subjects is equitable (that is, no discrimination against such groups as women, children, and minorities) and that research subjects not be coerced or manipulated

in any way. The principle of beneficence entails producing the greatest good while minimizing harm. This principle is reflected in federal regulations that require risk-benefit assessments. The principle of respect for persons highlights researchers' responsibility to treat autonomous persons as such and to protect those with diminished autonomy. The first aspect of that principle is reflected in the regulation requiring informed consent from potential participants. The second is embodied in special regulations designed to protect vulnerable populations, including children, fetuses, and prisoners.

Institutionalization and Criticism

In 1981 the FDA and the U.S. Department of Health and Human Services issued regulations in reaction to the Belmont Report. In that year the FDA created non-institutional review boards (NRBs) to accommodate the increased scope of the review process. In 1991 more than a dozen federal departments and agencies adopted the IRB process as the official Federal Policy for the Protection of Human Subjects, or "Common Rule." The Common Rule includes requirements for (1) assessing compliance; (2) informed consent; (3) IRB membership, function, operations, review of research, and record keeping; and (4) protection for vulnerable research subjects.

The IRB system has improved research practices by making researchers aware of ethical norms and exercising the power to withhold approval for substandard proposals. It is essential for the protection of human subjects and "is an important structural innovation in the social control of science" (Robertson 1979, p. 29). Nonetheless, the IRB system is "under strain" and "in need of reform," and "significant doubt exists regarding [its] capacity to meet its core objectives" (Federman et al. 2003, p. 5).

Central to this debate is whether the regulations unduly inhibit scientific output and progress. Before this question can be answered, however, more data about the impacts of IRBs must be collected. Frustrating this task is the absence of a national registry of all subjects participating in biomedical or social science research. Also, many people claim that the system is too strict in regard to less invasive social science projects and too lenient in regard to more risky research. A failure to balance risk-benefit ratios often hurts the credibility of the IRB system, and this weakens its capacities to achieve its goals (Levine 1986).

A third contentious issue is the decentralized structure of the system and the difficulty of applying general guidelines to specific research projects. Although the decentralized system allows IRBs to remain close to ongoing research, there may be too much local discretion and inadequate oversight of both researchers and individual IRBs. Without adequate assurance of compliance, research institutions may utilize IRBs to protect themselves and researchers rather than to protect subjects.

A fourth area of debate concerns the proper scope of IRB authority. An example from this set of issues is the question of whether IRBs should have the authority to approve or disapprove the scientific design of research protocols.

Assessment

IRBs have been the subject of intense scrutiny, and in 1979 the Hastings Center established a journal, *IRB: A Review of Human Subject Research*, devoted exclusively to issues raised by and within the system. As with any regulatory framework, the IRB system has had a host of administrative and structural challenges, yet it has proved to be resilient and adaptable. One example is the membership structure of IRBs. Early review committees were limited to immediate peer groups within the research community, but subsequent reforms have led to the requirements of gender diversity, the presence of at least one nonscientist, and the inclusion of at least one member not affiliated with the institution. Further reform efforts are improving the ethics education and certification requirements for IRB members.

The charge that the IRB system may impede scientific output is dubious in light of the rapid development of new drugs and other products and the fact that very few research proposals are rejected by IRBs. Daniel Callahan states that current scientific practice is motivated more and more by the imperative to do research and less and less by the quest for meaningful, life-enhancing knowledge and products (Callahan 2003).

RECs in other countries may offer lessons for reform of the U.S. IRB system. One example is the use of regional, national, and even international committees in other parts of the world. For example, in contrast to the U.S. commitment to local IRBs, many European RECs are regional (McNeill 1989). One issue that will always plague RECs and IRBs, however, is the difficulty of establishing objective criteria by which to evaluate their effectiveness. Most likely, assessment will remain a contested topic that is as much philosophical as it is empirical in nature.

ADAM BRIGGLE

SEE ALSO *Bioethics; Bioethics Commissions and Committees; Institutional Biosafety Committees.*

BIBLIOGRAPHY

Callahan, Daniel. (2003). *What Price Better Health? Hazards of the Research Imperative.* Berkeley, CA: University of California Press. Examines the "research imperative" in the United States and the ways in which it can be distorted into harmful practices.

Curran, William J. (1970). "Governmental Regulation of the Use of Human Subjects in Medical Research: The Approach of Two Federal Agencies." In *Experimentation with Human Subjects*, ed. Paul Freund. New York: George Braziller. Compares the FDA and the NIH review protocols and argues that their differences stem from the fact that they are different types of agencies: One is regulatory, and the other funds exogenous research.

Federman, Daniel D.; Kathi E. Hanna; and Laura L. Rodriguez, eds. (2003). *Responsible Research: A Systems Approach to Protecting Research Participants.* Washington DC: National Academies Press. An Institute of Medicine study commissioned by the DHHS to assess the national system for providing protection for research subjects. An excellent overview with insightful recommendations.

IRB National Research Act. (1976). *Code of Federal Regulations*, Title 45, Pt. 46.

Levine, Robert J. (1986). *Ethics and Regulation of Clinical Research*, 2nd edition. New Haven, CT: Yale University Press.

MacKay, Charles R. (1995). "The Evolution of the Institutional Review Board: A Brief Overview." *Clinical Research and Regulatory Affairs* 12(2): 65–94.

McNeill, Paul M. (1989). "Research Ethics Committees in Australia, Europe, and North America." *IRB: A Review of Human Subjects Research* 11(3): 4–7. A look at RECs worldwide.

Robertson, John A. (1979). "Ten Ways to Improve IRBs." *Hastings Center Report* 9(1): 29–33. A brief summary of the performance of IRBs followed by ten recommendations for improvement. Categorizes defects into two types: administrative and structural.

Sherman, Max, and John D. Van Vleet. (1991). "The History of Institutional Review Boards." *Regulatory Affairs Journal* 3: 615–27.

U.S. National Commission for the Protection of Human Subjects of Biomedical and Behavioral Research. (1978). *Report and Recommendations: Institutional Review Boards.* Washington, DC: U.S. Government Printing Office.

Williams, Peter C. (1984). "Success in Spite of Failure: Why IRBs Falter in Reviewing Risks and Benefits." *IRB: A Review of Human Subjects Research* 6(3): 1–4. An assessment of inherent defects in the IRB system.

INSTRUMENTATION

• • •

Instrumentation refers to the use or application of instruments or specialized technologies for observation, measurement, control, or production. In the last sense one even speaks of the instrumentation of a piece of music, meaning its adaptation to being produced or played by a particular set of musical instruments. Technologies in the form of instrumentation have also played a crucial role in the production of human knowledge science prehistory. In all these senses, instrumentation calls for general philosophical reflection, including ethical reflection.

Instrumentation, Ancient and Modern

The usual story about the origins of science cite ancient Greek philosophical speculations such as the prescient hypothesis of Democritus (460–370 B.C.E.) that there must be ultimate small bits of matter, which he termed "atoms," that constitute the most basic things of the world. Plato (428–347 B.C.E.), in opposition, developed an alternative hypothesis of a finite set of ideal geometrical forms into which the universe fits, a finite number of polyhedron shapes or Platonic solids at the base of things. Yet neither Democritus nor Plato produced any concrete, verifiable knowledge about the physical universe through their speculations. It was not until the later Hellenic period of Greek antiquity that heirs to the intellectual tradition initiated by Democritus and Plato began to produce lasting scientific knowledge of physical phenomena by developing measuring instrumentation.

When Robert Crease (2003) asked physicists to identify what ten experiments in the history of science were the most "beautiful," number seven turned out to be the measurement of the circumference of the Earth by Eratosthenes of Cyrene (c. 276–c. 194 B.C.E.). Combining a shrewd set of assumptions with a simple instrument, a gnomon or variation on a sundial, and mathematical measurements, Eratosthenes made a reasonable estimate of planetary size. Assuming a spherical Earth and a Sun at great distance, when the shadow of a gnomon was vertical at Syene he instrumentally measured its angle some 800 kilometers away at Alexandria; then using angular geometry he calculated the curvature necessary to account for such a difference, and extended this to reach an estimate of the circumference that remains respectable to the present day.

The vote that confirmed this as a "beautiful experiment" should come as no surprise, because it simulta-neously validates popular belief in the genius of the Greeks, confirms human nostalgia for this particular history, and emphasizes the geometrical thinking that characterizes what later became modern science. But this experiment neither stands alone, nor is it even close to being the most ancient example of knowledge production embodied in technological instrumentation. A multicultural survey of almost any pre-historic set of peoples would show that instrumentation played a role in the knowledge of natural processes that in the early twenty-first century would be called *astronomy* or even *cosmology*. Virtually all larger cultures of the past were sky-watchers and developed often deep knowledge of celestial motions, solstices, seasons, moon cycles, eclipses, sometimes parallax, and other complex astronomical phenomena. These were recorded upon calendars, some of which were superior to calendars within Western traditions until very recent times. Moreover, although sometimes simple, most such observations were made through instrumental mediations. Indeed, *archeoastronomy*, the study of ancient astronomies, has led to the recognition that many of the stone circles of antiquity and prehistory had instrumental uses for establishing solstices, moon and sun cycles, and the like. Examples include Stonehenge, Mesoamerican equivalents, and even ancient North American sites in the Mississippian cultures. Calendar signs of moon cycles can be recognized on antler markings that go back to the last Ice Ages of more than ten millennia ago. Thus technologies have been incorporated into the production of what would now be called "scientific" knowledge from pre-history and within multiple cultures, all using instruments.

Within what some would term the Western master narrative, much is made of a seventeenth century "scientific revolution" as the turning point of early modern science. Yet it is possible to reframe this episode in the accelerated production of scientific knowledge as a second high point in the crucial development of instrumentation as well. Its predecessors were the Renaissance, itself a revival of ancient knowledge, much of it developed and conveyed by Islamic cultures that had perfected instruments and preserved ancient texts, thus creating an instrument-saturated and instrument-fascinated epoch.

Instrumental Perception

Instruments embody *measuring perceptions*. Those previously mentioned entail *visual* sightings, using some stable feature (the instrument) to make repeated observations. Of course the motivation and human contexts for performing such practices differed across cultures. Edmund Husserl (1970) recognized that a simple geometry arose out of the *lifeworld* practices of re-measuring agricultural fields after the annual floods of the Nile in Egypt. In other contexts, the annual renewal of kingships (as in ancient Sumeria) called for accurate dates and times. Islamic cultures needed accurate instruments to identify directions to Mecca, instruments such as the astrolabe and world maps with mathematical grids allowed such measurements, but later were also applied to navigation in the age of exploration. In the early 2000s, with space exploration such as that of the Cassini spacecraft orbiting Saturn, much more precise instrumentation is called for.

Measuring perceptions are not restricted to visual perceptions. Auditory perception has also been mediated through a variety of instruments. Listening tubes, later stethoscopes, amplify the capacity of auditory perception to determine interiors, including voids and shapes. More complex and later acoustic devices, including early sonar, remained auditory but gave way to a preference for visualization in scientific culture. Contemporary radar and sonar produces visual imagery on screens.

Further, various animal-analogues became technologically produced, one example being the development of thermal imaging. Thermal perception is common with reptiles, particularly snakes, which can even sense the shapes of prey through thermal awareness. Thermal awareness in the human case does have a moment in Western science. William Herschel (1792–1871), experimenting with a prism, detected warmth beyond the edge of the red end of the spectrum and correctly inferred the radiation that became known as *infrared*. Thermal imaging in military instruments has become highly sophisticated. But again, the tendency is to translate the thermal image into a visual one, such as obtains with certain types of night-vision instruments (other night vision instrumentation amplifies ambient light).

Tactile instrumentation plays especially important roles in medical practices. The setting of broken bones traditionally employed direct physical, bodily manipulations, and even with early instrumentation, the trained surgeon could "feel" what he was doing through the instrument. In dentistry, for example, the tools used to examine teeth reveal the cracks, soft areas, and cavities that are of dental interest. These perceptual experiences are *mediated* through the instruments, or the instruments are embodied by the practitioner. Contemporary instruments, however, often change previous practice. For example, laparoscopy, or even more extremely, distance

surgery, entails practices that more resemble video games than earlier forms of surgery. Here miniscule tubes outfitted with imaging devices and connected to microsurgical tools are operated by the surgeon through skilled eye-hand coordination to perform the operation (sometimes called "Nintendo surgery").

Instrumental Hermeneutics

To this point, instrumentation has been described in relation to the way in which bodily-perceptual capacities are amplified or magnified. A different set of instrumentations, again going back to antiquity, relates more obviously to the human capacities for making and reading *inscriptions*, that is, instrumentation that engages interpretive or *hermeneutic* practices. Inscriptions found on reindeer antlers, dated as much as 30,000 years ago, have been found with twenty-eight cycle patterns, thus likely signifying a lunar cycle. Abstract hatch marks and other inscriptions have been found alongside the highly isomorphic or "realistic" depictions of animals in the cave regions of France and Spain have also been found (18,000 to 15,000 years ago). With early modernity, calculating machinery began to be employed, usually with counters inscribed with numbers or letters and driven by complex gearing. Dials, gauges, readable panels, all are forms of instrumentation engaging "reading" or hermeneutic skills.

The recognition of perceptual patterns, particularly as *images*, and the recognition of inscriptions in number-like (counting) or letter-like (reading) form, are both instrumental. The philosopher of science Peter Galison (1997) calls these the *image and logic* traditions that dominate late modern physics. But the data-to-image-to-data inversions are also a newly dominant form of instrumentation in contemporary science.

Technoscientific Instrumentation

Contemporary science is *technoscience*—that is, a science thoroughly embodied in its technologies and instruments. Only since the middle of the twentieth century has astronomy broken the bounds of both ancient "eyeball" and then early modern optical instrumentation. First, with the breakthrough provided by radio astronomy (associated with the development of radar), then through forms of spectroscopy that range from very short gamma-waves to very long radio-waves, has the limitation of optical wave frequencies been exceeded. In the early twenty-first century, slices of the microwave spectrum, such as x-ray imaging, can show pulsars in action, or map the dark emissions of the radio spectrum. In a parallel fashion, medical imaging, ranging from photography through the x-ray devices from 1895, to MRI and PET scans since the 1970s, perform the same function with respect to imaging the human body.

These processes are made possible through: (a) the data-image conversions possible through computer tomography and computer-aided technology (CAT) processes, (b) modeling and simulation techniques again employing computations, and (c) the algorithmic projection of imagery such as may be instanced in fractal patterning. Thus contemporary imaging may be either processed as data (numbers, counting, calculational) or as imaging (picture-like), and each form is transposable to the other. More important, however, is that the range of phenomena detectable through contemporary forms of sensing may not only be remote, but it exceeds all ordinary bodily perceivability, as has been analyzed at length by Ernesto Mayz Vallenilla (1990).

Yet, indirectly or in the form of new mediations, such instrumentation *translates* its results into countable/readable ones. Contemporary Mars and Saturn missions image the surface of Mars or the rings of Saturn, close up, and return these to the earth-bound observer for perceivable, close-up results. Or in the case of the Chandra X-ray source, images of the explosive nebulae through "false color" depictions can be perceptually grasped in human visual form. Instrumentation provides science with its own sensorium.

Popular Instrumentation

While the above overview of instrumentation has been focused upon various science practices, the same or similar types of instrumentation have more common manifestations. Some have said that the twenty-first century will be the century of one big and one little technology. The big technology is the home entertainment and work center, containing a high-definition screen for television, computing, and communicating, and a multimedia, multi-tasking station that incorporates the Internet, word processing, communications, and entertainment. The small, mobile technology that incorporates digital photography, mini-screen, for everything from cell phoning to email to reading barcodes for purchases is the other extreme of the big/little technofantasy. This is a not unrealistic extrapolation from extant technologies that are also social-cultural-economic instruments.

In one sense, these technologies are the same as those noted in science. Each transforms the texture of human experience. If contemporary astronomy produces near-distance with its images of multi-billion-year-old galaxies, so does electronic communication make near-

distance of every electronically accessible spot on the globe. If the technologies are state-of-the-art audio-video ones, then any online place can produce conference interchanges. Or if lapsed time is used, as in videos, cinemas, or Internet technologies, then the result is even more like the galaxy example, and lapsed-time phenomena are made into present-time phenomena. Academic experience is illustrative: Many first time contacts are electronic, by email, or telephone. Arrangements for conferences, lectures, travels, are almost always arranged electronically—including air tickets. First person contact may or may not follow, and when it does, the follow-up reverts to the electronic instrumentation. Academic globalization is already electronically embodied and actual.

Ethics of Instrumentation

This communication-entertainment-information instrumentation also entails complex ethical-political, cultural-economic dimensions. Especially in the area of medical instrumentation, a primary question concerns safety. In the area of communications instrumentation more generally, a primary issue is privacy.

But more generally still one can examine the social justice of who has access to the whole communication-entertainment-information instrumentation complex. Is the globally interconnected world merely another elite? Is the trajectory centralist or decentralist? Many have noticed the extreme irony of the Internet—originally designed to be a fail-safe mode of communication for a military-university elite, it has become a diffuse, world-connecting instrument for everything from spam, electronic scams, and virtual romances to instant political dissemination of news and politics—and a new mode of campaign financing. No one knows if the outcome will be more democratic or more totalitarian. Yet by virtue of both the unpredictability and the indeterminacy (or, better, underdetermined) qualities produced by these new instruments, new opportunities have clearly come into being.

While prognosis is ambiguous, in part because all technologies display multiple possible developments and uses, the human-instrument relationship exhibits its own multiple dimensions. Many contemporary instruments are complex and characterized as "high tech" machines, implying the need for a highly skilled, technically informed set of users—technocrats and technically trained individuals. But although some subset of technically proficient persons is needed for the infrastructure of such technologies, a different set of skills is required for instrumental uses. For example, generational differences are often remarked upon in that young children quickly become computer literate whereas older people often display reluctance or "technophobias" regarding these technologies. Yet the child is not so much a technician as a skilled user. One need not know computer programming to play a video game, any more than one needs to know physics to ride a bicycle. Yet it is also interesting that the emergence of both many software developments and the location of much worldwide hacking and virus development is associated with countries once thought to be underdeveloped or under-technologized.

Instrumentation, whether in knowledge production, communication, commerce, entertainment, and much of the full range of human activity, is a means by which human perceptual and interpretive activity is embodied. As the above examples show, instrumentation may be very simple (a gnomon) or very complex (Internet), but the diffusion, adaptability, and spread of instrumentation technologies is more dependent upon the easy adaptability into human use practices—which then change—than the degree or type of complexity built into the technology. Historically, photography, radio, cinema, and television all were rapidly diffused, whereas modern agricultural and transportation technologies were not, or took much longer to be adapted. One possible reason for this may be the ease with which bodily-perceptual actions are quickly and without much technical training brought into play. To hear a radio and recognize a voice, to see a movie, to recognize a photograph is an almost immediate phenomenon. Contrarily, to transfer a set of agricultural practices or ship building skills is much more complex. Instrumentation, in the very contemporary sense, entails both kinds of complexity. The evening news, or the Cassini image of Saturn's rings, both involve large, complex infrastructures and global or even interplanetary connections—but both yield perceivable results as focal outcomes of instrumentation.

DON IHDE

SEE ALSO Body; Experimentation; Scientific Review.

BIBLIOGRAPHY

Crease, Robert P. (2003). *The Prism and the Pendulum: The Ten Most Beautiful Experiments in Science.* New York: Random House. Robert Crease, philosopher of science and historian for the Brookhaven National Laboratory, discusses the history and science of ten historical experiments in Western science.

Galison, Peter. (1997). *Image and Logic: A Material Culture of Microphysics.* Chicago: University of Chicago Press. Mallinckrodt Professor of the History of Science at Harvard University, Peter Galison is particularly noted for his philosophical style which more deeply incorporates the role of technologies into science practice. This book shows that two traditions, imagers and counters, pervade late modern physics.

Husserl, Edmund. (1970). *The Crisis of European Sciences and Transcendental Phenomenology,* trans. David Carr. Evanston, IL: Northwestern University Press. This "classic" of phenomenology, the last major book published by Husserl, introduces the notions of *Lifeworld* and science arising from practices now more common in science studies.

Ihde, Don. (1990). *Technology and the Lifeworld: From Garden to Earth.* Bloomington: Indiana University Press. This systematic philosophy of technology introduces the roles of different human-technology relations, cultural hermeneutics and contemporary trajectories to philosophy of technology.

Ihde, Don. (1991). *Instrumental Realism: The Interface Between Philosophy of Science and Philosophy of Technology.* Bloomington: Indiana University Press. The first English language book to relate philosophy of technology to the "newer" technologically sensitive philosophers of science.

Mayz Vallenilla, Ernesto. (1990). *Fundamentos de la meta-técnica.* Caracas: Monte Avila. A Spanish language philosopher of technology is one of the few to deeply appreciate and work with contemporary instrumentation in science. English translation: *The Foundations of Meta-Technics,* trans. C. Mitcham, Lanham, MD: University Press of America, 2004.

INTEGRITY

SEE *Research Integrity*.

INTELLECTUAL PROPERTY

• • •

The concept of *property* is as old as civilization. As people acquired possessions or inhabited land or shelters, they sought to secure these items for personal or collective use. Customs and rules evolved to define ownership and specify the rights and responsibilities that attached to ownership. In conjunction especially with developments in science and technology, property has taken on intellectual forms that embody ethical stances and have policy implications.

From Property to Intellectual Property

The definition of property evolved as society invented or identified new things that can be owned. *Property rights* began with the physical or concrete, such as land, and eventually expanded to include more intangible or abstract phenomena (Horwitz 1992). Interference with such rights shifted from a physical invasion to interference with a proprietary right or a decrease in market value.

Property rights are a series of formal and informal rules governing what owners are allowed to do with their property and the degree to which they can exclude others from its use. Such rights describe relations "not between an owner and a thing, but between the owner and other individuals in reference to things. Property rights reflect societal values of how wealth should be distributed and protected.

Intellectual property is abstract and refers to the products of human intellect such as inventions, literary works, music, and art. Many societies historically have not recognized ownership in intellectual property. Others have associated names with achievements but have not provided serious protection. As societies industrialized, they found a need to protect intellectual property, especially the valuable products of science and technology. Intellectual property rights (IPRs) describe a bundle of rights or privileges that, like other property rights, allow the owner to use, derive income from, and transfer the ownership of the property.

Intellectual Property Rights

IPRs define the rights and privileges attached to ownership of intellectual property. Such rights allow owners to exercise a temporary monopoly over the use of their creations; they have exclusive rights, for a limited time, to decide who may use a product or work and under what conditions. Such rights define ownership and specify the degree to which inventors and creators may profit from their work, the access others may have to the works themselves or to information about them, and how others may use or improve upon existing works.

IPRs involve issues of wealth distribution, incentives for innovation and creativity, access to information, and basic human rights. Ethical issues attach to questions of what should be publicly or privately owned, how ownership is established, how much and how long the owner can control the property, and whether public policy should create exceptions to intellectual property rules to serve social interests.

IPRs encourage innovation by protecting new work from appropriation by others and allows people and institutions to profit from their work. Such rights promote the communication of information; as long as the right is in place, information can be published without fear of loss. IPRs also define public rights by indicating when private protections expire.

Rationales for IPRs fall into two categories, "instrumental rationales, which view intellectual property in terms of its benefits to society as a whole, and natural rights which stresses the inherent authority of innovators to control works they have created" (Schecter and Thomas 2003, p. 7). *Instrumental* rationales focus on the need for protection to promote societal goals, such as economic growth or technological innovation. *Natural rights* arguments, grounded in the philosophy of John Locke, assert that people are entitled to protection for the products of their minds, regardless of whether the protections serve other societal goals. The two rationales may lead to different policy decisions about the appropriate type and level of intellectual property protection.

Intellectual property protection is regarded as a basic human right. According to Article 27 of the Universal Declaration of Human Rights (1948), "Everyone has the right to the protection of the moral and material interests resulting from any scientific, literary or artistic production of which he is the author." The reach of the right differs across nations, with industrial nations generally providing higher levels of protection than developing nations.

Constructing an IPRs system requires the balancing of often conflicting societal values and needs, such as the need to promote innovation; concerns for equitable distributions of wealth, information, and other benefits; and the desire to allow authors or inventors to profit from the fruits of their labor and imagination. Increasing protection in one area often detracts from another.

Types of Intellectual Property Law

The international system recognizes two types of intellectual property: industrial property, including but not limited to inventions, trade secrets, and trademarks; and copyright. Systems of IPRs laws differ across nations but often include the following.

Copyrights protect works of creativity and authorship such as literary, musical, and visual art, as well as audio recordings, choreography, and computer software. Laws specifying such protection must consider issues such as fair balance between private and public use, the need for public access to information, when protection should begin and how it can be triggered, and how to enforce protections.

Patents protect innovative products and processes by allowing the patent holder to control use of the invention for a limited period of time. In exchange for patent protection, inventors generally must agree to disclose information about their inventions to the public. Patents are generally restricted to inventions, including both products and processes, although the restrictions on patents have narrowed in recent years. Products of nature have generally not been patentable but improvements in biotechnology have challenged definitions of what is natural.

Trademarks identify the origin of products or services and are used to promote them. Trademark protection prevents others from using a trademark to promote a product or service of a different origin. Such protection prevents the appropriation of the competitive advantage trademarks are intended to provide.

Trade secret law protects proprietary business information from misappropriation. Some information, such as an industrial process, might be eligible for protection either under patent law or trade secret law but not both. Trade secrets must be protected from disclosure, while the patent process generally requires making information public.

Science, Technology, and Intellectual Property

Science and technology provide many societal benefits, such as the enhancement of economic growth or quality of life. They also can produce negative, unintended consequences. Most societies promote science and technology, but this can be costly. Establishment of IPRs that protect new works and give innovators the right to profit from their creations provides incentives for expensive innovation without the need for direct government subsidies (Posner 2004). At the same time, IPRs may maintain or aggravate wealth inequities.

Rights have little meaning unless they can be enforced and modern technology has made IPRs enforcement increasingly difficult. Photocopiers make it possible for anyone with access to a machine to reproduce works entitled to copyright protection and the Internet allows anyone to make literary or musical works available to the world.

Science and technology challenge intellectual property systems, particularly patent laws. New fields such as information technology and genetic engineering force courts to decide how to apply laws made before such technologies were contemplated. As knowledge itself becomes more valuable, people and institutions seek additional protection for control of the knowledge and its profits. At the same time, society has an increasing need for access to some kinds of knowledge and protection from the use of others.

Abstract ideas cannot be patented but their applications can qualify for patent protection. For example, "Einstein could not patent his celebrated law that $E = MC^2$; nor could Newton have patented the law of gravity. Such discoveries are 'manifestations of Nature, free to all men and reserved exclusively to none.'" (Diamond v. Chakrabarty, p. 309, quoting Funk Brothers Seed Co. v. Kalo Inoculant Co., 333 U.S. 127, 130, 1948). General ideas remain in the public domain but their applications may be privatized through the patenting process.

Biotechnology, perhaps more than any other field, has challenged courts and lawmakers to reconsider intellectual property laws. In 1972 Ananda Chakrabarty, a microbiologist, sought a U.S. patent for a genetically engineered bacterium. The U.S. Patent Office denied the application because bacteria are products of nature, and living things cannot be patented under U.S. law. The case was appealed and eventually reached the U.S. Supreme Court. The Court restated the principle that natural phenomena cannot be patented, but found that Chakrabarty's bacterium was "a product of human ingenuity," and therefore was patentable under U.S. law.

So many biotechnology patents have been issued for such small innovations that some fear the creation of a *tragedy of the anti-commons* in which new innovations involve so many existing patents that innovation is discouraged. At least one study has found the anti-commons is not yet a significant deterrent to innovation, but that the situation should be monitored.

IPRs can be attached to writings or products regarded as dangerous or immoral, and IPRs tend to legitimize such works by implying social approval. Societies must decide whether to provide protection for harmful or otherwise objectionable work. New technologies, particularly those that create or replicate life, often trigger debate over whether the work should be done at all, much less be protected by law. IPRs also establish ownership of particular innovations, which may help to determine liability if a product causes harm. This raises questions of whether innovators should be held responsible for their products, particularly when the products are used in unintended ways.

Public funding for science and technology further complicate intellectual property issues. Who should benefit from works developed under public funding, the creator or the public? What balance of public/private benefits best serves societal goals?

Academics build their reputations by producing intellectual works. They seek recognition for their accomplishments, control over any economic benefits, and protection against plagiarism. IPRs promote release of information to the public by assuring the author of protection for the work, even after it is made public. IPRs protect authors from possible appropriation of ideas by others, including peer reviewers, before the work has actually been published.

Ownership can be a major IPRs issue. Who owns the product of collaborative work? At what point does a contribution by a supervisor, graduate student, or coworker deserve coauthorship? When the creator works for a corporation or a university, does ownership lie with the creator or the institution? What about funding agencies? In many cases, ownership or authorship is established by disciplinary customs or by agreements among the parties (Kennedy 1997).

Plagiarism is professionally unacceptable and sometimes illegal, but timing is critical to determining whether plagiarism has occurred. According to Donald Kennedy, "To take someone else's idea and use it before it has been placed in the public domain is a form of theft ... [t]o make further use of someone else's idea after it has been published is scholarship" (1997, p. 212). Of course attribution is critical even, or especially, in scholarship, whether or not a work is protected.

International Intellectual Property Rights

The absence of an international sovereign makes a global IPRs system problematic. Every nation has different intellectual property laws, making cooperation difficult, although many international IPRs agreements have been developed. Which nation's standards should apply? Most international agreements take a national approach in which a country agrees to provide foreign innovators with the same protection provided to its domestic citizens. Creators of intellectual property generally must seek protection separately in each jurisdiction, a cumbersome process.

The United Nations World Intellectual Property Organisation (WIPO) provides support for the international intellectual property system. Its mission is "to promote through international cooperation the creation, dissemination, use and protection of works of the human mind for the economic, cultural and social progress of all mankind."

Globalization has increased the need for more international IPRs coordination. Multinational organizations seek consistent laws across borders and inventors want universal protection for their inventions. The World Trade Organization Agreement on Trade-Related Aspects of Intellectual Property Rights (TRIPS Agreement) attempts to provide a more standard IPRs

system and sets minimum protection that must be provided by all member states.

Ethical issues become particularly important at the international level (Rischard 2002). Some fear that increasing IPRs protection will increase inequities between the developed and the developing world. Others are concerned that IPRs deny access to products desperately needed by the poor or powerless. Still others believe adequate IPRs standards are critical to promoting technology transfer and foreign investment.

IPRs can deny access to essential products and information to those who need them most, particularly in developing countries. Drug research and development is extremely expensive, and pharmaceutical countries price drugs to recoup expenses and make a profit. No one else is allowed to manufacture drugs protected under patents. Those who need the drugs often have little money. Is it fair to allow people to die because they cannot afford drugs that could prolong their lives?

The TRIPS Agreement allows for compulsory licenses, an exception to IPRs in special cases such as emergencies, that give developing countries access to essential drugs for major health problems such as HIV/AIDS or malaria. Such policies may have a boomerang effect; pharmaceutical companies may be less likely to invest in research to develop drugs for conditions found primarily in poor countries if there is no profit to be made. The answer to the drug access problem may be better addressed by turning to solutions unrelated to intellectual property rights, such as foreign aid. Some pharmaceutical companies have made drugs available at drastically reduced rates to those who cannot afford them.

Inspiration for new products often comes from local or traditional knowledge. Who should benefit when a drug company develops a new drug based on knowledge about the properties of a plant gained from an indigenous tribe in a remote region? Is the company that developed a commercial drug entitled to all the profits or should it share revenues with the people who supplied the information or with the country from which the plants are harvested?

Conclusion

Consensus exists over the need for IPRs but not over the content of such rights. Countries that produce more science and technology and other intellectual property support more protection than other nations. Globalization requires more consistency in IPRs than has traditionally been available. IPRs help to promote innovation and the communication of information, but questions remain about the appropriate balance between public and private rights, the nature of ownership, and the equitable provision of access to products and information. Debates continue over the types of intellectual property that should be protected by law. New technologies intensify such debates, particularly technologies that create new or duplicate old life forms.

MARILYN AVERILL

SEE ALSO *Human Rights; International Relations; Property; Science, Technology, and Law.*

BIBLIOGRAPHY

Horwitz, Morton J. (1992). *The Transformation of American Law: 1870–1960.* Oxford: Oxford University Press. General property rights are addressed throughout this treatment of the development of American law over a ninety-year period. IPRs are not specifically treated.

Kennedy, Donald. (1997). *Academic Duty.* Cambridge, MA: Harvard University Press.

North, Douglass C. (1990). *Institution, Institutional Change and Economic Performance.* Cambridge, England: Cambridge University Press. This book discusses how formal rules develop, including property rights; it does not specifically discuss IPRs.

Posner, Richard A. (2004). *Catastrophe: Risk and Response.* Oxford: Oxford University Press.

Rischard, Jean-François. (2002). *High Noon: 20 Global Problems, 20 Years to Solve Them.* New York: Basic Books.

Schechter, Roger E., and John R Thomas. (2003). *Intellectual Property: The Law of Copyrights, Patents and Trademarks.* St. Paul: West. This comprehensive reference covers American intellectual property law in great detail, including the history of, rationales for, and elements of each right.

Walsh, John P.; Ashish Arora; and Wesley M. Cohen. (2003). "Effects of Research Tool Patents and Licensing on Biomedical Innovation." In *Patents in the Knowledge-based Economy,* ed. Wesley M. Cohen and Stephen A. Merrill. Washington, DC: National Academies Press.

INTERNET RESOURCES

World Intellectual Property Organization. (2005). *WIPO Intellectual Property Handbook: Policy, Law and Use.* Available from http://www.wipo.int/about-ip/en/iprm/. General organization site can be found at http://www.wipo.int/. This site provides a wealth of materials on intellectual property rights.

World Trade Organization. TRIPS Material on the WTO Website. Available from http://www.wto.org/english/tratop_e/trips_e/trips_e.htm.

INTERDISCIPLINARITY

• • •

Any attempt to consider relations among science, technology, and ethics is by definition interdisciplinary. This entry distinguishes among basic approaches to terminology, reflects on the intersection of interdisciplinarity and ethics, and assesses future prospects. Because it provides an existing model, with both strengths and weaknesses, for examining science, technology, and ethics interactions, references will often be made to the existing interdisciplinary field of science, technology, and society (STS) studies.

Forms of Interdisciplinarity

Interdisciplinarity has both broad and more restricted meanings. In the broad sense it includes a number of different forms, one of which is interdisciplinarity in a more narrow sense. There are three forms of interdisciplinarity in the broad sense that are important to distinguish, and that provide a framework for discussions of types of interdisciplinary interactions.

Multidisciplinarity juxtaposes separate disciplinary approaches around a common interest, adding breadth of knowledge and approaches. But the disciplines continue to speak as separate voices in encyclopedic alignment. Underlying assumptions are not examined, and the status quo remains intact.

The major disciplines in STS multidisciplinarity have traditionally been history, philosophy, and sociology of science (Cozzens 2001). Studies of science in literature and scientific literature also received attention, and anthropology became prominent in the 1990s. Although disciplinary identities remain strong, there are specialized interdisciplinary bridges, such as alliances of economists of scientific research and technological development with historians and sociologists of technology interested in technological innovations. Gary Bowden (1995) distinguishes three methods of explanation in STS: topic-, issue-, and combined-focus. Topic-focus is the most common, using methods and techniques of one discipline to study an aspect of science or technology. The result is an amalgamation of contextualist approaches. Both Bowden and Susan Cozzens characterize STS as a multidisciplinary array of activities.

Interdisciplinarity integrates separate data, methods, tools, concepts, theories, and perspectives in order to answer a question, solve a problem, or address a topic that is too broad or complex to be dealt with by one discipline (Klein and Newell 1997). In education, content is revised and the curriculum is restructured around a theme, problem, or issue. In research, the task at hand becomes the primary focus, and in interdisciplinary fields a new body of knowledge emerges.

Several added distinctions appear. *Instrumental, strategic, pragmatic,* or *opportunistic* forms tend to focus on economic and technological problems. During the 1980s, interdisciplinarity gained heightened visibility in science-based areas of intense economic competition such as computers, biotechnology, and manufacturing. In this instance, interdisciplinarity served the political economy of the market and national needs.

Critical and *reflexive* interdisciplinarities differ. They interrogate the existing structure of knowledge and education, raising questions of value and purpose that are silent in instrumentalist forms (Weingart 2001, Klein 2001). Bowden also aligns interdisciplinarity with *combined focus methods,* marked by a common culture of investigation and a coherently integrated package of analytic resources and, often, new concepts. Insofar as STS becomes interdisciplinary, Cozzens adds, it ceases to be anchored in constituent disciplines.

Transdisciplinarity was initially defined as an overarching synthesis that transcends the narrow scope of disciplinary worldviews. General systems theory, structuralism, Marxism, the policy sciences, feminism, and complexity theory are leading examples. Likewise, sustainability and science, technology, and society reorganize and further develop knowledge and education around new synthetic frameworks. The term has also been a descriptor for broad fields (for example area studies, cultural studies), and synoptic disciplines (philosophy, geography). Recently in humanities it has been aligned with new critical paradigms, and it is a label on web sites in areas as varied as education, health care, and engineering sciences. In the 1980s and 1990s, three new connotations appeared: a new structure of unity informed by the worldview of complexity in science (Nicolescu 1996); a new mode of knowledge production that fosters synthetic reconfiguration and recontextualization by drawing on expertise from a wide range of organizations; and collaborative partnerships for sustainability that cross the boundaries of social sectors as well (Klein et al. 2001). Bowden also associates transdisciplinarity in STS with analytic *issue-focused methods* that emphasize a particular theoretical issue. They are not limited to the particulars of a specific substantive topic. The problem of reflexivity, for instance, is not unique to social studies of science and technology. Postmodernism also appears across a wide range of subjects.

Relation to Science, Technology, and Ethics

These basic distinctions are apparent in science, technology, and ethics (STE). In multidisciplinary STE, science, engineering, and ethics all retain their distinctive superdisciplinary features. Science is a superdiscipline that includes physics, chemistry, biology, geology, and other kindred disciplines. Comparably, engineering encompasses mechanical, chemical, electrical, and other fields of engineering. Ethics, in turn, encompasses distinctive forms that range from consequentialism and deontologism to virtue ethics. Applied ethics in its first incarnation took ethics as is and put it to work in and for science and engineering. In interdisciplinary STE, new fields emerged that combined a science and ethics. Applied ethics in its second incarnation appeared as fields of computer ethics, engineering ethics, environmental ethics, bioethics, and so on. In an instance of transdisciplinary STE, some philosophers also attempted to create a general ethics of technology that transcends any one type of science or engineering while subsuming other forms of ethics. Examples include Hans Jonas's argument in *The Imperative of Responsibility* for an overarching ethical obligation to protect the future of human and all life. Other examples include proposals for sustainability as a moral obligation and the precautionary principle as a general guideline for scientific research and technological innovation.

Interdisciplinarity intersects with ethics in science and technology in many ways. During the 1960s and 1970s, a renewal of ethics occurred in philosophy, driven by new problems of justice, fairness, and values in professional practice. In the ensuing decades, new categories of moral thought and action emerged, the moral and ethical dimensions of every field began to be explored and, in general education, related issues were incorporated into disciplinary and interdisciplinary core courses. As David Edge (1995) observes, it is not accidental that new critical approaches evolved hand in hand with new developments in training technical experts. Such developments were part of a broad shift from positivist models and programmed research on applied problems toward critical scrutiny of their implications. The distinction is not absolute though. Research on problems of the environment and health, for instance, often combines programmed problem solving with critique of current practices and institutional structures.

The interdisciplinary character of STS also fostered greater attention to implications and consequences. Before the 1980s, Bowden recounts, social science and humanities research in the field was primarily historical, philosophical, and, to a lesser degree, sociological. Science and technology were treated as autonomous entities separate from social context. Philosophers examined the logic of the scientific method, historians documented the evolution of ideas and technological artifacts, and sociologists looked at the institutional structure and internal patterns of science. In the mid 1960s, especially among historians of technology and some in engineering education, notions of autonomous technology and the neutrality of technology were challenged by new understandings of technology as a complex enterprise in specific contexts that are shaped by, and in turn shape, human values.

This development generated a sizable literature on ethics and values in relation to technology. The new discourse of problem understanding and political choice placed greater emphasis on social impacts and policy as well. In the mid-1970s, developments in philosophy and history of science opened up the content of scientific knowledge to sociological scrutiny, fostering empirical examination of social bases of scientific knowledge and challenging the authority and epistemological privileging of science. In the late 1980s, a turn toward technology occurred. The first two developments involved conceptual reformulations that contextualized science and technology and the manner in which context affects creation of scientific knowledge and the impact of science and technology on society.

Analogously multidisciplinarity, science, engineering, and ethics retain their distinctive features. *Applied* ethics takes ethics as it is and puts it to work in and for science and engineering. In science-technology-ethics (STE) interdisciplinarity, new fields that combine a science and ethics emerge, producing areas such as computer ethics, engineering ethics, environmental ethics, and bioethics. In STE transdisciplinarity, some philosophers and ethicists have created a general ethics of technology that transcends any one type of science or engineering and subsumes other forms of ethics.

Assessment

Interdisciplinarity and STS are both conflicted discourses, marked by unresolved questions and differing positions. Disagreements center on key issues and problems, the role of disciplines, and priorities of integration versus critique. Moreover, the full range of options exists simultaneously, from multidisciplinary juxtapositions to interdisciplinary integrations to transdisciplinary frameworks. Both interdisciplinarity and STS are also maturing movements. Knowledge is

widely considered to be increasingly interdisciplinary and, Bowden observes, the scholarly endeavor of STS has come of age. Nonetheless the widely touted interdisciplinary transformation of the university has not occurred. Multidisciplinary approaches are more common, institutional impediments retain their force, and Cozzens concludes, the integrated whole of STS thought is more of an ideal than a pervasive reality. The practice of STS often remains discipline-bound and removed from the world of practice. Edge concurs, asking whether "the heady sense of interdisciplinary adventure" and "seductive combination of academic priority and practice urgency" has disappeared (Edge 1995, p. 3).

There is also a constant tension between the particular and the general. Interdisciplinary STE is often criticized for trying to be too general: for instance in comments that "there is no such thing as 'technology' but only 'technologies'" and "all general principles are vacuous." At the same time, applied ethics fields such as computer ethics and biomedical ethics are criticized for reinventing the wheel: for instance in talk about risk analysis or informed consent and in their failure to synergize achievements from different applications.

In existence only since the early-1970s, STS has attained an expanding presence and established a platform for greater interdisciplinarity. An identifiable group of scholars and teachers has formal affiliations with the field, and an infrastructure for communication is in place. Cozzens highlights, in particular, the generation emerging from interdisciplinary STS programs in the early-twenty-first century. They are less bound to disciplinary identities than their professors and more prepared to move in the direction of *postdisciplinary* research that goes beyond narrowly circumscribed conceptual categories and analytical practices while often critiquing underlying premises of disciplinarity as well. Yet much work remains. The gains that have been made must be secured and the field must continue to develop on its own terms, not as the cumulative sum of its disciplinary parts. Doing so will require diligence to insure sufficient economic and symbolic capital; inclusion in funding categories of research agencies; an adequate number and scale of programs; full-time appointments in STS programs and departments; secure locations in organizational hierarchies; and autonomy in decisions about curriculum, budget, and staffing.

In both STS and STE, there is constant tension between the particular and the general. Both, moreover, raise the same question that all interdisciplinary fields raise. Where do they *fit?* The problem of fit, Lynton Caldwell (1983) advises from the experience of environmental studies, prejudges the epistemological problem at stake. Interdisciplinary categories arose because of a perceived misfit among need, experience, information, and the prevailing structure of knowledge. If the structure must be changed to accommodate the new field, perhaps the structure itself is part of the problem.

JULIE THOMPSON KLEIN

SEE ALSO *Two Cultures; Science, Technology, and Society Studies.*

BIBLIOGRAPHY

Bowden, Gary. (1995). "Coming of Age in STS." In *Handbook of Science and Technology Studies*, eds. Sheila Jasanoff, Gerald E. Markle, James Petersen and Trevor Pinch. A diagnosis of three distinct methods of explanation in STS (topic-, issue-, and combined focus), the visions of the field they imply, formative experiences that defined STS's identity, and assessment of the appropriateness of the methods.

Caldwell, Lynton K. (1983). "Environmental Studies: Discipline or Metadiscipline?" *Environmental Professional* 5: 247–259. A genealogy of the field that accounts for its rise and historical shifts, structural and curricular forms, intellectual and practical affiliations, and multidisciplinary versus interdisciplinary dimensions.

Cozzens, Susan E. (2001). "Making Disciplines Disappear in STS." In *Visions of STS: Counterpoints in Science, Technology, and Society Studies*, eds. Stephen H. Cutcliffe and Carl Mitcham. Albany: State University of New York Press. A mapping of STS as a movement and multidisciplinary network of discipline-bound theory and practice that also calls for a more integrated and interconnected body of knowledge responsive to the interdisciplinarity of STS problems.

Edge, David. (1995). "Reinventing The Wheel." In *Handbook of Science and Technology Studies*, eds. Sheila Jasanoff, Gerald E. Markle, James Petersen, et al. Newbury Park, CA; London: Sage. An overview recalling the field's origin in the mid-1960s, early research and educational considerations, the democratic impulse of the field, subsequent growth and differentiation, and current possibilities for a creative reconciliation of tensions.

Jasonoff, Sheila; Gerald E. Markle; James C. Petersen; and Trevor Pinch, eds. (1995). *Handbook of Science and Technology Studies*. Thousand Oaks, CA: Sage. A comprehensive collection of essays on the field by leading scholars that defines, summarizes, and synthesizes the literature and provides theoretical, historical, and policy essays plus case studies on the cutting edge of STS.

Klein, Julie Thompson. (2001). "A Conceptual Vocabulary of Interdisciplinary Science." In *Practicing Interdisciplinarity*, eds. Peter Weingart and Nico Stehr. Toronto: University of Toronto Press. An overview of definitions and related forms of discourse, patterns of activity, the role of complexity and hybridity, practices of borrowing and problem solving, and the conflicted status of disciplines and integration.

Klein, Julie Thompson, ed. (2002). *Interdisciplinary Education in K-12 and College: A Foundation for K-16 Dialogue*. New York: College Board. A collection of essays on the history and forms of interdisciplinary approaches, use of technology in teaching, integrative processes, political stakes in curriculum reform, and strategies of course design, team teaching, administration, and evaluation.

Klein, Julie Thompson, and William H. Newell. (1997). "Advancing Interdisciplinary Studies." In *Handbook of the Undergraduate Curriculum: A Comprehensive Guide to Purposes, Structures, Practices, and Change*, eds. Jerry Gaff and James Ratcliff. San Francisco, CA: Jossey-Bass. An overview of the history, motivations, structures, and forms of interdisciplinary study with analysis of the core concept of integration and its implications for course design, approaches to teaching and learning, and assessment.

Klein, Julie Thompson, et al., eds. (2001). *Transdisciplinarity: Joint Problem Solving among Science, Technology, and Society*. Basel, Switzerland: Birkhauser.

Newell, William H., ed. (1998). *Interdisciplinarity: Essays from the Literature*. New York: College Board. An anthology of essays on the nature of interdisciplinary study, its administration, disciplinary contexts, lessons from interdisciplinary fields, and theory and practice in social sciences, humanities and fine arts, and natural sciences.

Nicolescu, Basarab. (1996). *La transdisciplinarité: manifeste* [Manifesto of interdisciplinarity]. Paris: Editions du Rocher. English edition, Albany: State University of New York Press, 2001; trans. K. C. Voss. Also available from the web site of the Centre International de Recherches et Etudes Transdisciplinaires at http://perso.club-internet.fr/nicol/ciret/. A defining proposal for a new scientific and cultural approach informed by the new worldview of complexity in science and based on the three pillars of: complexity, multiple levels of reality, and the logic of the included middle.

Weingart, Peter. (2001). "Interdisciplinarity: The Paradoxical Discourse." In *Practicing Interdisciplinarity*, eds. Peter Weingart and Nico Stehr. Toronto: University of Toronto Press. An analysis of the apparent paradox of the intensified discourse of interdisciplinarity in the face of increasing specialization, links with innovation, and the realities of funding agencies and research institutions in Germany.

Weingart, Peter, and Nico Stehr, eds. (2001). *Practicing Interdisciplinarity*. Toronto: University of Toronto Press. A collection of essays on changes in the traditional order of knowledge in Europe, North America, and Australia that includes the role of funding bodies, current social and economic contexts, and case examples of practices.

INTERNATIONAL AFFAIRS

SEE *International Relations*.

INTERNATIONAL COMMISSION ON RADIOLOGICAL PROTECTION

• • •

The International Commission on Radiological Protection (ICRP) is a non-governmental organization that issues recommendations for radiation protection from ionizing radiation. With Wilhelm Roentgen's 1895 discovery of x-rays that, unlike the rays of visible light or of radio transmissions, tend to break down or ionize atomic structures, a new phenomenon was added to human experience. As this phenomenon became increasingly utilized especially in medical work, its dangers were likewise progressively recognized. The recommendations issued by the ICRP are used by many national and international radiation protection agencies to deal with such dangers and have a profound influence on radiation protection all over the world.

History and Activities

The ICRP was established in 1928 by the Second International Congress of Radiology, in order to address health and safety issues concerning radiation used for medical purposes. Until 1950 it was called the International X-ray and Radium Protection Committee. The new name reflected a widened scope to include all aspects of protection against ionizing radiation.

The ICRP functions as an advisory body to national and international agencies in the field of radiation protection. According to its constitution, the ICRP shall provide recommendations and guidance on all aspects of radiation protection and consider the fundamental principles and quantitative bases for radiation protection, while leaving to national bodies the responsibility of formulating specific advice, codes of practice, or regulations best suited for each country. No country or international organization is obliged to follow the recommendations of the ICRP. International organizations that use the ICRP recommendations include the International Atomic Energy Agency (IAEA), the World Health Organization (WHO), the International Labor Organization (ILO), and the Nuclear Energy Agency of the Organization for Economic Cooperation and Development (OECD).

The ICRP is registered as an independent charity in the United Kingdom and is mainly financed by voluntary contributions from international and national bodies with an interest in radiation protection. The organization consists of the Main Commission and five standing committees. The Main Commission has twelve members and a chair. The Main Commission elects itself, and three to five members of the Main Commission are replaced after each four-year period. According to the constitution of the ICRP, members shall be chosen on the basis of their recognized activity within professional fields of relevance to radiation protection. The standing committees are chaired by members of the Main Commission and consist of fifteen to twenty experts (mostly biologists, physicians, and physicists) appointed by the Main Commission. The committees are Committee 1 (radiation effects), Committee 2 (doses from radiation exposure), Committee 3 (protection in medicine), Committee 4 (application of ICRP recommendations), and Committee 5 (protection of non-human organisms). In addition to these committees, the ICRP also appoints task groups comprised of radiation protection experts outside the ICRP. At any given time, about 100 scientists are involved in ICRP work.

The ICRP publishes reports containing guidelines on a variety of topics related to radiation protection. Examples of such reports include: "Radiological Protection in Biomedical Research" ICRP Publication 62, 1993), "Radiological Protection Policy for the Disposal of Radioactive Waste" (ICRP Publication 77, 1998) and "Principles for Intervention for the Protection of the Public in a Radiological Emergency" (ICRP Publication 63, 1993). The *ICRP Recommendations* are special reports containing fundamental principles for radiation protection advocated by the ICRP. The main objective of these recommendations is "to provide an appropriate standard of protection for man without unduly limiting the beneficial practices giving rise to radiation exposure" (ICRP 1991, p. 3). The ICRP recognizes that this objective cannot be achieved solely on the basis of scientific data, but must also include value judgments and ethical considerations.

ICRP Recommendations

The basic principles of the ICRP recommendations for radiation protection have evolved considerably over time. In 1928 the first ICRP report on health effects concerned primarily damage to the skin and the destruction of blood forming tissues, that is, injuries caused by massive cell death following exposure to high levels of ionizing radiation. There is a *threshold dose* for these effects, which means that they occur only when sufficient numbers of cells are destroyed. The first ICRP report aimed to prevent these kinds of effects by providing recommendations on working practices and guidelines for use, but due to problems of defining a relevant dose measure, a dose limit was not included. In a subsequent 1934 report, however, the ICRP did recommend a dose limit, called a *tolerable dose*, which added a margin of safety to the threshold dose.

The system of tolerable doses was retained into the 1950s when a new appreciation of the risks from ionizing radiation altered the foundation for radiation protection. Previously it had been assumed that in the absence of no immediate negative health effects below a threshold level, there were also no long-term effects. But evidence had accumulated that ionizing radiation could also cause cancer and hereditary defects. Such longer-term results are called *stochastic effects* and are caused by modification, rather than the destruction, of cells, and occur with a certain probability, which was taken to be proportional to the dose. It was argued likely that no threshold existed for these kinds of effects. This meant that every dose implied a risk—that there was no completely safe level for ionizing radiation. Ever since, radiation protection has had to deal with the implications.

In 1950 the ICRP recognized the potential for cancer and hereditary effects from ionizing radiation, and recommended new, lower dose limits, called maximum *permissible doses*. But if there is no wholly safe dose, the concept of permissible dose becomes problematic. What is permissible or not? The ICRP based its judgments on a comparison with other hazards in life. The ICRP also recommended that exposure to ionizing radiation should be reduced to the lowest possible level, meaning that doses should be kept as low as practicable and that any unnecessary exposure should be avoided. Eventually this evolved into the principle that doses should be kept *as low as reasonably achievable* (ALARA), which became known as the *ALARA-principle*.

The next major step was taken in 1977 when ICRP introduced a protection system consisting of three BASIC principles. No practice involving exposure to radiation should be adopted unless it produces a positive net benefit (the *justification principle*). All exposures should be kept as low as reasonably achievable, economic and social factors being taken into account (the *optimization* or ALARA-principle). Doses to individuals should not exceed specified dose-limits (the *dose-limitation principle*). The emphasis was no longer on permissi-

ble doses, but on the requirement that doses should be kept as low as reasonably achievable (optimization). Mere compliance with dose limits was not sufficient— exposure must also be justified and optimized. The ICRP recommended that the optimization procedure should operate on the *collective dose*, defined as the product of the number of exposed individuals and their average dose.

Subsequent recommendations were adopted in 1990 (ICRP 1991) retaining the overall structure from the recommendations of 1977. The emphasis was still on the optimization principle, but in order to limit inequities that could follow from application of its three principles, the ICRP introduced a restriction on the optimization process. The reason for this was to prevent situations where the optimization principle would advocate a protection alternative (that is, the lowest collective dose) where, although all individuals would be below the dose limits, a few individuals would also be exposed to much higher doses than the rest of the exposed population. This is obviously a problem if there is no threshold for the risk from exposure to ionizing radiation. To avoid this the ICRP recommended additional individual limits, usually much lower than the old dose limits, called *dose constraints*. The concept of dose limits was retained but the definition was changed in order to define a boundary above which individual risk was considered unacceptable. Another difference was that the dose constraints were *source-related*, while the dose limits included exposure from all relevant sources.

The recommendations from the ICRP have been updated at intervals of ten to fifteen years, and the ICRP plans to deliver the next general recommendations in 2005. The proposed recommendations (ICRP 2003) involve further emphasis on the concept of dose constraints. The new system is based on the idea that constraints should be applied for each individual. The starting point for selecting the level of these constraints should, according to the proposal, be the *concern that can reasonably be felt* about the annual dose from natural sources. After applying the dose constraints there will still be a requirement to reduce doses even further. The proposal also suggests less emphasis than previously on the application of the collective dose and that individual doses below a fraction of the average annual dose from natural sources should be excluded from the system of protection.

The proposal for the new recommendations has been publicly discussed by the ICRP since 1999. Critics claim that the 1990 recommendations work well and that no substantial change to the basic system is needed. It has also been argued that the previous application of the collective dose ought to be retained, and that the introduction of a general exclusion level for very small doses has not been satisfactory justified.

PER WIKMAN

SEE ALSO Radiation; Regulatory Toxicology.

BIBLIOGRAPHY

International Commission on Radiological Protection. (1991). 1990 Recommendations of the International Commission on Radiological Protection–ICRP Publication 60. Oxford: Pergamon Press. Contains the basic principles of the ICRP 1990 recommendations, which have had a profound influence on radiation protection internationally.

International Commission on Radiological Protection. (1998). International Commission on Radiological Protection: History, Policies, Procedures. Oxford: Elsevier Science. A comprehensive work on the history and operations of the ICRP.

International Commission on Radiological Protection. (2003). "The Evolution of the System of Radiological Protection: The Justification for the New ICRP Recommendations." Journal of Radiological Protection 23: 129–142. Describes the proposed ICRP 2005 recommendations.

Mould, Richard F. (1993). A Century of X-rays and Radioactivity in Medicine: with Emphasis on Photographic Records of the Early Years. Bristol, PA: Institute of Physics Publishing. A thorough description of the history and development of radiation protection.

Sowby, David, and Jack Valentin. (2003). "Forty Years On: How Radiological Protection Has Evolved Internationally." Journal of Radiological Protection 23: 151–157. Gives an overview of international organizations in radiation protection and describes the development of the ICRP recommendations since 1950.

Taylor, Lauriston S. (1979). Organization for Radiation Protection: The Operations of the ICRP and NCRP 1928–1974. Springfield, VA: National Technical Information Service. Provides a detailed account of the early relations between the ICRP and the U.S. organization NCRP (National Council on Radiation Protection and Measurements).

Wikman, Per. (2004). "Trivial Risks and the New Radiation Protection System." Journal of Radiological Protection 24: 3–11. Contains a review of the proposed ICRP 2005 recommendations from an ethical and philosophical perspective.

INTERNET RESOURCE

International Commission on Radiological Protection. Available from www.icrp.org.

INTERNATIONAL COUNCIL FOR SCIENCE

• • •

The International Council for Science, still known by the initials of its former name, International Council of Scientific Unions or ICSU, is a nongovernmental organization (NGO) that includes more than one hundred national scientific bodies and close to thirty international scientific unions. The ICSU mission is to:

• Identify and address major issues of importance to science and society.

• Facilitate interaction among scientists across all disciplines and from all countries.

• Promote the participation of all scientists—regardless of race, citizenship, language, political stance, or gender—in the international scientific endeavor.

• Provide independent, authoritative advice to stimulate constructive dialogue between the scientific community and governments, civil society, and the private sector.

The main philosophy of the organization is perhaps best reflected in section 5 of its statutes, where the principle of the universality of science is expressed:

This principle entails freedom of association and expression, access to data and information, and freedom of communication and movement in connection with international scientific activities, without any discrimination on the basis of such factors as citizenship, religion, creed, political stance, ethnic origin, race, colour, language, age or sex. ICSU shall recognize and respect the independence of the internal science policies of its National Scientific Members. ICSU shall not permit any of its activities to be disturbed by statements or actions of a political nature.

History

ICSU was founded in Brussels in 1931, originally under the name International Council of Scientific Unions. It emerged as an extension of two earlier bodies, the International Association of Academies (1899–1914) and the International Research Council (1919–1931). The main change brought about through the founding of ICSU was the dual membership: Both national scientific bodies (initially forty) and international scientific unions (initially eight) make up the membership, and the unions received a more prominent and independent role.

World War II marked an interruption in ICSU activities. But after the war ICSU was the first NGO with which the newly founded United Nations Educational, Scientific and Cultural Organization (UNESCO) signed an agreement.

In light of wartime experiences and the new political prominence of science and technology, Joseph Needham (1900–1995), then Head of the Natural Sciences Division of the Preparatory Commission of UNESCO, addressed the ICSU Committee on Science and Its Social Relations, outlining the prospects of postwar scientific cooperation. This was discussed during ICSU's London General Assembly of 1946, and the first agreement between UNESCO and a non-governmental organization, i.e. ICSU, was signed shortly thereafter. Topics discussed included a plea for the elimination of military secrecy, a hope for increased international collaboration in applied science especially with regard to atomic power, a request for scientific "frankness, openness and integrity" so as to promote the common good, and advancement of the public understanding of science.

During the ensuing cold war period a new challenge emerged within the ICSU structure, namely the free circulation of scientists across national borders. Prewar ICSU statements already expressed the universality of science. For instance, in 1934, ICSU president George Ellery Hale proclaimed: "We welcome to our meetings the man of science in all countries and we appreciate the opportunity to join with them in the pursuit of our common object" (Greenaway 1996, p. 93). With the creation of the North Atlantic Treaty Organization (NATO) in 1949, however, realities became very different. For instance, East German scientists were refused visas for entry into NATO countries thus effectively blocking their participation in scientific meetings in these countries.

In 1963 ICSU formed the Standing Committee on the Free Circulation of Scientists (SCFCS), which in 1993 was renamed the Standing Committee on Freedom in the Conduct of Science and given an expanded mandate. The work of this committee became increasingly important as political tensions increased. The SCFCS worked primarily by correspondence contact with key persons in countries that either prevented entry or exit of individual scientists. The balance between safeguarding free scientific communication and keeping a politically neutral position was always a delicate one, and necessitated low-key action. By and large, the SCFCS managed to fill its watchdog role. In 1976 the SCFCS published its first edition of the "blue book," which is currently entitled "Universality of Science" and contains the principles pertaining to the rights of scientists and their freedom of movement.

Structure

The main decision-making body within ICSU is the General Assembly, which convenes every three years at various locations around the world upon invitation from a host country. Currently the General Assembly is assisted by an Executive Board, which consists of six executive officers and eight ordinary members, four from the unions and four from national members. The Executive Board is assisted by a permanent Secretariat, headed by an executive director.

Since 1972 the ICSU Secretariat has been based in Paris with French government support. A small structure was built up under the leadership of Julia Marton-Lefèvre (1978–1997) and has become a cornerstone in ICSU activities. Since 2002 ICSU has been headed by Thomas Rosswall as executive director. Compared to other international bodies or to its national members, the ICSU Secretariat of twelve people is strikingly small in size.

Activities

ICSU activities are varied and have changed character over the years. Some of its activities serve as examples of international scientific cooperation, despite political situations that at times seem to render them impossible. One such example was the International Geophysical Year (IGY), 1957–1958, which involved sixty-seven nations. The IGY established the principle that "expeditions and explorations in the remoter parts of the earth are now geophysical in intention" (Greenaway 1996, p. 156). An International Polar Year is planned for 2007–2008.

ICSU also engaged in other areas of common concern for international science. ICSU in 1966 set up its interdisciplinary Committee on Data for Science and Technology (CODATA) aimed at making scientific data of various kinds accessible to scientists beyond their origin. In 1969 the Scientific Committee on Problems of the Environment (SCOPE) was established to plan and facilitate, among other things, a global monitoring network and a training program for future environmental managers. SCOPE contributed to the Untied Nations (UN) Conference on the Human Environment in Stockholm, Sweden, in 1972 and the International Geosphere-Biosphere Programme (IGBP), which was initiated in 1986.

Such activities strengthened the ICSU role in the area of global environment and development, and led to close collaboration with various UN bodies. The International Conference on an Agenda of Science for Environment and Development into the 21st Century (ASCEND 21), held in Vienna in 1991, contributed to "Agenda 21: Science for Sustainable Development," the major document to emerge from the United Nations Conference on Environment and Development in Rio de Janeiro in 1992 (commonly called the Earth Summit). When the follow-up World Summit on Sustainable Development was held in Johannesburg in 2002, ICSU was again among the key NGOs addressing scientific issues.

ICSU now sponsors three global observing systems (GOS)—the Global Ocean Observing System, the Global Climate Observing System, and the Global Terrestrial Observing System—in collaboration with partner organizations such as UNESCO, the World Meteorological Organization, the Food and Agriculture Organization of the United Nations, and the United Nations Environment Programme. The goal of the GOS is improved monitoring of the global Earth system.

ICSU links with the social sciences and engineering remain relatively weak. Of the member unions in ICSU, four can be counted as belonging to the social sciences, among them the International Union of the History and Philosophy of Science (IUHPS). Already during the 1980s and early 1990s it was recognized that the global problems facing humankind required cooperative efforts from scientists, social scientists, and engineers. Efforts were made to bring these various fields together through closer cooperation between ICSU and the International Social Science Council (ISCC). In 1996, then, ICSU, ISCC, and other organizations became cosponsors of the International Human Dimensions Programme on Global Environmental Change (IHDP), originally established in 1990. In the early 2000s the IHDP, IGBP, and related programs were brought together under the banner of the Earth System Science Partnership (ESSP) to promote international and interdisciplinary research within four focal areas: carbon, food, water, and human health. It remains to be seen how the challenge of multi- and interdisciplinarity across the various fields will be met in practice.

Standing Committee on Responsibility and Ethics of Science (SCRES)

At the end of the 1980s and the beginning of the 1990s under the presidency of M. G. K. Menon the ICSU Executive Board took up issues of the ethics of science.

Two observations spurred this discussion. First, previous views that simply identified progress in science with social progress were more and more difficult to uphold. In the light of environmental and developmental issues, science was seen as not only part of the solution but to some extent as part of the problem. Second, scientific activities need to be guided by a sense of social responsibility. While ICSU already had established a mechanism to deal with the rights (freedom) of scientists, it lacked a platform to deal effectively with scientific responsibilities.

Following these discussions IUHPS was contacted for further suggestions on how to deal with this challenge. L. Jonathan Cohen (Oxford University), then secretary-general of ICSU and member of IUHPS, and Jens Erik Fenstad (University of Oslo), member of the Executive Board and former president of IUHPS, were among the driving forces in this effort. In collaboration with ICSU a workshop in London on ethical issues in science was arranged by Philip Kitcher (Columbia University) and Nancy Cartwright (London School of Economics and Political Science) in 1994 on behalf of the Philosophy of Science section of IUHPS (with contributions eventually published in *Perspectives on Science*, 1996). IUHPS then focused its activities on ethics of science, leading to a special session on this topic during the 1995 International Congress on Logic, Methodology, and Philosophy of Science in Florence, Italy (see Dalla Chiara et al. 1997). As a general outcome of these activities ICSU set up an informal working group that proposed a Standing Committee on Responsibility and Ethics of Science (SCRES). This proposal was endorsed by the General Assembly in Washington, DC, in 1996.

The remit of SCRES included:

- to act as a focus within ICSU and with outside partners for questions pertaining to scientific responsibility and ethics;

- to clarify issues of moral principle which affect the choice of policies for scientific research . . .;

- to raise awareness of important ethical issues among scientists, policy makers and the general public . . . (ICSU documents GA 1996)

An offer from the Norwegian Academy of Science and Letters led to SCRES being based in Oslo and sharing offices with the National Committees for Research Ethics.

SCRES was a small committee, compared with the more established Standing Committee on Freedom in the Conduct of Science, and it struggled to define its agenda. This took a new turn in the planning of the

World Conference on Science (WCS) that was jointly hosted by ICSU and UNESCO in Budapest, Hungary, in 1999. Cooperation with the UNESCO World Commission on the Ethics of Scientific Knowledge and Technology (COMEST) led to a special WCS session on "Science, Ethics and Responsibility." Indeed, SCRES prepared a WCS background document that was one of only two such documents distributed to all speakers, chairs, and rapporteurs (ICSU-SCRES 2000).

The WCS also placed a new topic on the SCRES agenda. The WCS keynote speech of Joseph Rotblat (b. 1908), the Polish-born physicist and international activist, called for a universal oath or pledge to be taken by scientists when receiving a degree in science. Such a "Hippocratic oath" would make explicit the commitment to social responsibility in science. This proposal spurred intense discussions, and while it proved impossible to include Rotblat's suggestion in the final endorsed documents of the WCS, section 3.2 of the "Science Agenda—Framework for Action" calls for COMEST and SCRES to follow up with a view to encourage young scientists to "respect and adhere to the basic ethical principles and responsibilities of science."

In response, SCRES produced a study of 115 ethical guidelines and codes of ethics that was presented to the ICSU General Assembly at its Rio de Janeiro meeting in 2002. At the same time SCRES presented an evaluation of its own activities and suggested that ICSU reconsider how best to place the ethics of science within its structure. SCRES pointed out that a body of its kind and structure could not meet the expectations expressed in its remit, especially regarding public awareness of science and society issues. Its impact remained peripheral, perhaps with the exception of China where SCRES activities spurred a major influence at the national level.

SCRES furthermore suggested that a better balance be found for ad hoc activities directed at special areas of wide ethical interest and addressed through cooperation with other partners, while retaining the continuity and identity that a standing committee can provide. ICSU was asked to consider whether a revised and renewed SCFCS with an explicit mandate for ethics might not be a better framework. As a result SCRES was dissolved in 2002, and ICSU established a strategic review committee to work out suggestions for the future of ethics within ICSU. While the importance of ethics of science is widely recognized by many of the ICSU members and by the Executive Board, ICSU still needs to find its own profile in this area that would not duplicate activities of other bodies, but at the same time provide a voice for global and international concerns.

MATTHIAS KAISER

SEE ALSO *American Association for the Advancement of Science; Nongovernmental Organizations; Pugwash Conferences; Royal Society.*

BIBLIOGRAPHY

Cetto, Ana María, ed. (2000). *World Conference on Science: Science for the Twenty-First Century, A New Commitment.* Paris: United Nations Educational, Scientific and Cultural Organization.

Dalla Chiara, Maria Luisa; Kees Doets; Daniele Mundici; and Johan van Benthem, eds. (1997). *The Tenth International Congress of Logic, Methodology, and Philosophy of Science, Florence, August 1995,* Vol. 2: *Structures and Norms in Science.* Dordrecht, Netherlands: Kluwer Academic.

Greenaway, Frank. (1996). *Science International: A History of the International Council of Scientific Unions.* Cambridge, UK: Cambridge University Press.

International Council for Science. (1996). ICSU documents GA 1996: ICSU Report 25th General Assembly of ICSU (26GA/99/3.1, point 9.c), and Resolutions of the 25th General Assembly of ICSU, 25GA/99/3.1.i, point 11, and Recommendation to establish a standing Committee on Responsibility and ethics in science, 25/GA/96/9.3.

International Council for Science (ICSU). Standing Committee on Responsibility and Ethics in Science (SCRES). (2000). "Ethics and the Responsibility of Science: Background Paper for the World Science Conference." *Science and Engineering Ethics* 6(1): 131–142.

Kitcher, Philip, and Nancy Cartwright. "Science and Ethics: Reclaiming Some Neglected Questions." *Perspectives on Science* 4(2).

INTERNET RESOURCES

International Council for Science. Available from http://www.icsu.org/.

"Science Agenda—Framework for Action." United Nations Educational, Scientific and Cultural Organization. Available from http://www.unesco.org/science/wcs/eng/framework.htm.

INTERNATIONAL RELATIONS

• • •

The term "international relations"—subsuming "international affairs" and "foreign affairs"—refers to interactions among nation states, and includes such diverse topics as international law, international trade, and the international monetary system. Although international corporations and non-governmental organizations influence these interactions, and international bodies such as the United Nations help manage them, the primary actors remain nation states. Insofar as nations carry and articulate values, and find their powers conditioned by changes in science and technology (from military effectiveness and productivity to means of communication and bureaucratic organization), international relations also function as an important site for science, technology, and ethics interactions.

Historical Transformations

Following the Peace of Wespthalia (1648) and acceptance of the nation state as the sovereign arbiter of values and power within its borders, questions arose about how to manage interstate relations. The assumption, shared more by theorists than political leaders, from the seventeenth through the nineteenth centuries was that all nations desired peace, which was to be achieved through international law, which laid out the rules of the game for managing the balance of power through international treaties. The failure of this system in World War I, in which technological destructiveness exceeded civilized control, and the subsequent rise of state actors empowered by new techniques of organization, driven by aggressive ideologies in Russia and especially in Germany committed to the marshaling of science and technology for violent conquest, challenged the classic consensus. As Hans J. Morgenthau (1948) observed, peace and security is the ideology of satisfied powers.

The study of international relations grew after World War II into a major focus of social science to encompass these new realities, new states, and new issues, and developed in two directions. In the first case, social scientific studies endeavor to understand why state actors behave as they do, including how technology helps to determine their capabilities. In the second case, advances in science and technology became integral aspects of the relations between and among states including, among others, their role in war and peace, in the management of conflict, in the promotion of economic development, and in the analysis of decision marking.

Other less spectacular but equally far-reaching changes have been the ability to reach any telephone instantaneously and inexpensively worldwide, increased dependence of weapons systems on competitive techno-

logical innovation, the relevance of scientific competence to national economies, and the immediacy and global reach of television. Advances in the technologies of transportation, communication, and information thus contributed measurably to such phenomena as the fall of Soviet Communism and the end of the Cold War (1990), public demands for international humanitarian action, the increased unification of Europe, and economic globalization. Still others underline causal connections between local actions and global consequences, such as destruction of stratospheric ozone as a result of the widespread use of chlorofluorocarbons (CFCs), the far-reaching consequences of a disruption in energy supplies or a failure of information systems, and the climatic effects of the accumulation in the atmosphere of waste gases. (International response to the CFC problem in the form of the Montreal Protocol for their elimination has become one of the success stories of multi-state cooperation in response to issues both engendered and identified by science and technology.) The transnational impacts of space exploration and environmental issues, plus the post-Cold War rise of non-state actors adapting technologies for terrorism are further examples of new science and technology-related issues altering international affairs.

Yet the international significance of science and technology goes beyond physical power. The intellectual currents of the Enlightenment, which was largely a product of the experimentation and rationality of the scientific revolution, have stimulated massive forces for change in the West—and have been interpreted as forces involved in a post-Cold War "clash of civilizations" (Huntington 1996).

Moreover, science and technology are not static. By 2003 worldwide investment in research and development (R&D) had risen to $750-800 billion per year, leading to rates of innovation that defy accurate forecasting, let alone estimation of their social effects. There is now in place a formidable and growing system for dedicating human ingenuity to the rapid expansion of knowledge and the production of new technologies to serve perceived or speculative needs. Not only do the results of this system have significant international implications, its very operation favors the creation of global markets. Science and technology may not *cause* changes in international affairs, but their interaction with a mosaic of social, economic, and political factors clearly does so.

The present and future implications for the international system may be summed up in a cliché: Advances in science and technology and their application have led to an unprecedented degree of interaction and mutual dependence among nations in their economies, social structures, and security relationships. The result has moved nations to a new level of interdependence. Nevertheless, the fundamental principles and organization of the international system have *not* been altered substantially. Although multilateral and transnational organizations have increasingly important roles to play across the spectrum of issues from security to economies, this does not imply the end of nation-states. The world is still organized as a system that retains the basic structure of states, each jealous of its independence, seeing itself in competition with others, attempting to maintain maximum freedom of action, and committed to enhancing national welfare and influence. At the same time the state capacity to act as an independent unit increasingly depends on the breadth and depth of its links to other states. Indeed, degrees to which states are intertwined with others may affect internal matters as well. And the frequency with which domestic and foreign policies related to science and technology are confronted with ethical issues concerning the effect of policies on other states and peoples is a product of such intensive linkages.

Ethical and Political Issues

Changes in international relations resulting from interactions with science and technology have raised ethical and political issues that range in scale and consequence from minor inconveniences in travel or communication to decisions that may dictate the immediate violent deaths of thousands of people or choices that have long-term, potentially large, but uncertain effects.

WEAPONS SYSTEMS. Perhaps the most obvious instance arises from the development of weapons systems that directly or inadvertently target civilian populations as well as military forces. The most dramatic are the nuclear weapons used by the United States against Japan to end World War II but not used since. In 1945 there was some debate in government circles and the scientific community about using a weapon with such destructive power and unleashing a means of warfare that would have a profound effect on international relations.

That decision remains controversial, but at the time the imperative to end the war and avoid large losses of American lives in an invasion of Japan was irresistible to the U.S. president. Moreover, a different technological weapon—incendiary bombs—had already been used against both Germany and Japan with equivalent loss of

life; the atomic bomb did not appear radically different in terms of the number of lives at risk. There are other arguments about the moral use of this weapon, but these were decisive at the time (Alperovitz 1996).

The decision to proceed with development of the hydrogen (fusion) bomb in 1950 was likewise fraught with moral and political consequences because of the extent of the destruction it could unleash (Bundy 1988).

The policy of nuclear deterrence that is based on the destructive power of nuclear weapons—that is, the paradoxical threat of use in order to avoid use—has been highly controversial. Conventional weapons systems that cause considerable "collateral" damage—the death and destruction of noncombatants—raise moral issues as well, though on a smaller scale.

Other weapons-related programs and policies that have been proposed and questioned on ethical grounds include nuclear test ban treaties and ballistic missile defense. The Limited Nuclear Test Ban Treaty (1963) and subsequent proposals to limit testing in space and underground have necessarily involved politicians working closely with scientists and engineers on programs that had wide moral support. In the case of President Reagan's Strategic Defense Initiative or "Star Wars" program to create a shield against nuclear armed ballistics missiles, a program revived by President George W. Bush as the National Missile Defense, there have been important questions about feasibility and functionality in which science, technology, and ethics are intimately intertwined.

GLOBAL CLIMATE CHANGE. A major environment-related ethical issue of international significance is the threat of global climate change or warming, which (like CFC emissions) became an issue only as a result of theoretical calculations made by scientists, not evidence of actual damage. Based on computer models and solid evidence of the accumulation of carbon dioxide and other atmospheric greenhouse gases, scientists have warned that more solar radiation will lead to a growing heat burden for the planet. Depending on the timing and magnitude of the effects, the impact could be very large, with a major effect on low-lying nations (because of sea-level rise) and on agricultural production, especially in developing countries. The calculations of the scientists are controversial, but the relevant scientific community has accepted the validity of the threat. The Intergovernmental Panel on Climate Change (IPCC), an international panel of scientists from many countries charged by governments to assess the danger, increasingly accepts the existence of the phenomenon and in its last assessment predicted a temperature rise between 1.4 and 5.8 degrees Centigrade by the end of the twenty-first century (Intergovernmental Panel on Climate Change 2001).

International negotiations have been proceeding since the "Earth Summit" in Rio in 1992, itself a major science and technology related international event, with a Framework Convention on Climate Change that was negotiated that same year and entered into force in 1994; in 1997 the Kyoto Protocol was accepted for ratification (Skolnikoff 1999, O'Riordan and Jager 1996). The United States under President Clinton signed the protocol, but President George W. Bush withdrew the signature and has refused to consider ratification. The U.S. Administration argument is that the science is not proven, the developing countries that eventually will be major producers of carbon dioxide have no obligations under the protocol, and the costs to the American economy would be too great. Modest alternative policies, largely voluntary, have been pursued instead by the Administration. Regardless of the merits of the general arguments, the ethical issue is stark: Does the United States, which is by far the major producer of greenhouse gases (25 percent or more of global emissions), have the right to ignore an issue that could have a catastrophic effect on other countries and peoples? The United States will suffer from global warming, but its wealth will make it relatively easy to adapt to the effects of changes in climate. That is not true for other countries, especially the poorer ones.

GENETICALLY MODIFIED ORGANISMS. Genetically modified organisms (GMOs) raise significant ethical questions. These organisms, which so far have been used largely in the agriculture domain, are familiar crop strains (corn, soybeans, wheat, cotton) modified by biotechnological techniques to have valuable new characteristics, such as reduced sensitivity to herbicides and better cold-weather stamina (Thompson 2002). The new strains have been introduced widely in the United States but have been resisted in some other countries, particularly in Europe.

Companies that market GMO products in the United States assert that the resultant food is indistinguishable at the consumer level from unmodified food; Europeans respond that the evidence is inconclusive. Moreover, consumers in Europe insist that food should be labeled so that they have a choice about whether to buy modified food. The United States takes the position that labeling would destroy the market for the American-produced food, that there is no scientific evidence

of danger, and that the European position is a ploy to protect European agriculture from less expensive imports. Some African countries, desperately in need of food aid such as Zambia, Zimbabwe, Malawi, Mozambique, and Angola have refused U.S. food on the grounds that their crops would become "contaminated" and thus unable to be exported to Europe (Bohannon 2002). The United States is taking the issue to the World Trade Organization (WTO) on the grounds that the E.U. policy is a form of protectionism. Yet Europeans argue that the United States is attempting to impose its values in an area that will be irreversible once the modified crop strains are in widespread use. Does one nation have the right to make such a decision regardless of the validity of the political and economic arguments?

FOREIGN WORKERS. An issue that is a perennial focus of criticism of multinational corporations is variance in the standards of treatment of workers in different countries. Is it ethically appropriate for corporations to follow identical standards regardless of local wages or living and employment conditions, or should there be differences that take account of variations in income or environment? U.S. corporations often have been the focus of protest, especially when they pay workers in developing countries wages much below American scales or do not provide equivalent working conditions.

The subcommission for protecting and promoting human rights of the United Nations Commission on Human Rights has been drafting a code on norms of responsibility for multinational corporations, the draft of which was approved in August 2003 (Draft Norms on the Responsibilities of Transnational Corporations and Other Business Enterprises with Regard to Human Rights 2003). If the draft ultimately is approved by the full commission and accepted by the member states, it will for the first time create a standard for the ethical behavior of multinational corporations. Final approval will not create an enforcement mechanism but should have considerable influence, particularly on larger corporations that are vulnerable to public pressure and protest.

INTELLECTUAL PROPERTY RIGHTS. An issue with similar characteristics is the general subject of intellectual property rights (IPR). Patents, including those in the pharmaceutical industry, copyrights, and trademarks are issued to provide a protected monetary return for an inventor or artist and thus to encourage innovation and performance. Ethical issues arise when intellectual property is pirated or when royalties or fees are too high for developing countries.

Often new technologies are not available in developing countries because of the cost. When copying of intellectual property is easy and low-cost, as in the case of copying videos or music records without paying the copyright fee, the result has been wholesale reproduction and sale at a fraction of the original price. This would seem to be clearly unethical. Many argue, however, that it is the IPR regime that is unethical and that intellectual property should be considered a public good, freely available or available at a low cost, to anyone. That position is not likely to be accepted in countries that produce most of the intellectual property, which argue that without a chance to recoup costs, innovation and artistry would dry up. It is particularly important for the United States, which is increasingly dependent on high-technology and innovation-intensive goods. The Trade Related Intellectual Property Rights Agreement (TRIPS) of the WTO represents an attempt to reach international agreement on this issue, so far with limited success.

Another IPR debate focuses on the patenting of genetically engineered organisms and of products found in the wild for use in pharmaceutical research and development. In the case of genetically engineered organisms, the European Union is much more restrictive on this practice than the United States, thus raising an IPR issue that requires international harmonization. The patenting of biological discoveries in what are sometimes called "gene-rich" poor countries by corporations based in so-called "gene-poor" rich countries has been criticized as a form of "biopiracy" that fails adequately to compensate the country from which these new resources are derived.

TERRORISM. A more recent issue has arisen from the fear of terrorists' use of scientific data. Since September 11, 2001, the U.S. government has sought to limit the publication of the results of research that might benefit terrorists. This has revived issues of the proper boundaries of government imposed scientific secrecy that were prominent during the Cold War but had abated since, and is particularly relevant in the case of fast-moving biological research but also affects other areas with weapons potential, particularly in the nuclear and chemical fields. Scientists are resisting such regulations on the grounds that they would degrade the scientific enterprise and make it difficult to counter possible weapons development or acquisition by terrorists. Should it be possible to publish in a journal or on the Internet any information, such as the methodology for producing biological agents or the design of a nuclear weapon, that could be misused

even though the information is otherwise available and is not classified? What is the ethical (and political) judgment? The issue has not been settled (Skolnikoff 2002), although a number of biology journals have agreed to institute a review process to flag potentially dangerous articles and consider how the suspect material might be reduced or eliminated. No recent cases of "prior" censorship outside classified areas have reached the courts.

Other Issues

Weapons systems, global climate change, genetically modified organisms, foreign workers, intellectual property rights, and terrorism constitute six representative international relations issues intimately engaged with science and technology. Many others might be mentioned, from population growth, economic development (the rich/poor divide and the proper parameters of foreign aid), and world health, to biodiversity loss, the allocation of resources in international waters (as provided for in the Law of the Sea Treaty, 1982) and space (including communication satellite orbits), and remote sensing of countries and individuals from space without their permission.

Issues of these kinds arise ubiquitously and are a natural product of advances in science and technology and the use of those advances in national and international policies. In recognition of this fact, the U.S. National Research Council (1999) argued strongly for major innovations in the department of state to more effectively deal with these issues. Improved education and personnel policies for regular foreign service officers, creation of a new post of science adviser to the Secretary, and recruitment of more scientists and engineers to the department's ranks were advocated and most of the recommendations approved by the then Secretary. Additionally, all sciences and technologies are "dual use" in the sense that they can be used for benign or malevolent purposes. Inevitably, they will often pose choices that raise ethical as well as social, political, and economic considerations. Some of those choices will be minor and insignificant, but others will require careful thought and almost surely will be controversial.

EUGENE B. SKOLNIKOFF

SEE ALSO *Atomic Bomb; Baruch Plan; Global Climate Change; Genetically Modified Food; Globalism and Globalization; Intellectual Property; Limited Nuclear Test Ban Treaty; Montreal Protocol; Terrorism.*

BIBLIOGRAPHY

Alperovitz, Gar. (1996). *The Decision to Use the Atomic Bomb*. New York: Vintage. A controversial analysis of the decision to drop the a-bomb that argues the primary purpose was to influence future negotiations with the Soviet Union.

Bohannon, John. (2002). "Zambia Rejects GM Corn on Scientists' Advice." *Science* 298(8): 1153–1154.

Bundy, McGeorge. (1988). *Danger and Survival: Choices about the Bomb in the First Fifty Years*. New York: Random House. Superb account of the policy process and decisions surrounding nuclear weapons by the special assistant for national security affairs to Presidents Kennedy and Johnson.

Intergovernmental Panel on Climate Change. (2001). *Climate Change 2001: The Scientific Basis*. Cambridge, UK: Author. The third assessment report to the international community from scientists appointed by some fifty governments. The most authoritative single voice, though controversial among a minority of scientists.

Keatley, Anne G., ed. (1985). *Technological Frontiers and Foreign Relations*. Washington, DC: National Academy Press. Of special interest are Simon Ramo's "The Foreign Dimensions of National Technology Policy," Richard N. Cooper and Ann L. Hollick's "International Relations in a Technologically Advanced Future," and C. W. Robinson's "Technological Advances—Their Impact on U.S. Foreign Policy Relative to the Developing Nations."

Lee, Thomas H., and Proctor P. Reid. (1991). *National Interests in an Age of Global Technology*. Washington, DC: National Academy Press. The emerging global technical enterprise is altering relations among nations.

Morgenthau, Hans J. (1948). *Politics among Nations: The Struggle for Power and Peace*. New York: Knopf.

O'Riordan, Tim, and Jill Jager. (1996). *Politics of Climate Change: A European Perspective*. London: Routledge. An analysis of European views by two knowledgeable environmentalists.

Skolnikoff, Eugene B. (1993). *The Elusive Transformation: Science, Technology and the Evolution of International Politics*. A Council of Foreign Relations Book. Princeton, NJ: Princeton University Press. On how international politics has been changed by science and technology advance since World War II. For an earlier, more restricted overview: *Science, Technology, and American Foreign Policy* (Cambridge, MA: MIT Press, 1967).

Skolnikoff, Eugene B. (1999). "The Role of Science in Policy: The Climate Change Debate in the United States." *Environment* 41(5): 42–44. A political analysis of the climate change debate in the U.S. that shows how and why the scientific evidence plays a much lesser role than would be expected in this quintessential science-related subject.

Skolnikoff, Eugene B. (2002). "Research Universities and National Security: Can Traditional Values Survive?" In *Science and Technology in a Vulnerable World*, ed. Albert H. Teich, Stephen D. Nelson, and Stephen J. Lita. Washington, DC: Association for the Advancement of Science. On threats to American research universities posed by the

attempts to keep potentially dangerous scientific information out of the hands of terrorists.

Thomson, Jennifer A. (2002). *Genes for Africa: Genetically Modified Crops in the Developing World*. Landsdowne, South Africa: University of Capetown Press.

United Nations Commission on Human Rights. (2003). *Draft Norms on the Responsibilities of Transnational Corporations and Other Business Enterprises with Regard to Human Rights*. E/CN.4/Sub.2/2003/12 Minneapolis: Human Rights Library, University of Minnesota.

U.S. National Research Council. (1999). *The Pervasive Role of Science, Technology, and Health in Foreign Policy: Imperatives for the Department of State*. Washington, DC: National Academy Press.

Wines, Michael. (2004). "Angola's Plan To Turn Away Altered Food Imperils Aid." *New York Times*: March 30, p. A3.

INTERNET

• • •

Emerging from the integration of computer and communications technologies, the Internet is a text- and graphics-based communications system that supports people and organizations in the performance of multiple activities. As such it has the potential to transform the worlds of work in industry, government, education, and entertainment as well as everyday life. A variety of ethical issues arise with this technology, involving not only individual users, but also corporations and governments.

There are two basic meanings associated with the word Internet. In a narrow sense, the Internet is a global network *inter*-connecting computer *networks*, from which the word derives. Hence it is a complex network connecting large numbers of devices such as computers, file servers, and video cameras, by means of telephone lines, satellites, and wireless networks. In a broad sense, the Internet also includes that which such technological infrastructure makes possible, which some refer to as *cyberspace*.

For present purposes the Internet will be characterized as constituting a *digital habitat* where people increasingly live. Habitat denotes here an environment in which people carry out activities, possibly in interaction with other people, involving specific actions and things. Because the kinds of things people interact with in the Internet are not *material* in the usual sense of the term, but rather electronic and digital, the Internet may be termed a digital habitat (Stefik 1996).

Emergence and Development of the Internet

Initial development of Internet technology was supported in the 1960s by the Advanced Research Projects Agency (ARPA) of the U.S. Department of Defense in the context of the Cold War between the United States and the former Soviet Union. ARPA's task was to establish the technological and military superiority of the United States. But the agency gave considerable freedom of action to researchers and the development of Internet technology was carried out mainly at university research laboratories by academics whose primary agenda was to develop technologies to allow computers to communicate with each other (Castells 2001).

In 1969 the first nodes of the ARPANET, a packet-switching network, became operational. Subsequently, to deal with the proliferation of computer networks that had appeared in the United States and other countries, additional technology was developed during the 1970s and 1980s to interconnect any kind of network, as long as certain preestablished rules of communication were followed. It is in this context that the Internet, as a network of computer networks, was born (Abbate 1999).

Initially the Internet was used primarily at universities, for the purposes of exchanging electronic mail and for transferring files. It was not until the 1990s, with the development of the World Wide Web—a particular kind of Internet application (Berners-Lee and Fischetti 1999)—that a massive use of the Internet became possible. By mutually reinforcing each other, factors such as an increasing number of users, a growing number of services provided through the Internet, and increasing investment in technologies led to an explosive growth of the Internet. What had started as the ARPANET with four nodes in 1969, had become the Internet with millions of users by the end of the twentieth century.

Ethical Issues

Some have suggested that the ethical issues of the Internet are the same ones that arise in preexisting practices. Another position maintains that although these issues have a correspondence with well known, preexisting dilemmas they nonetheless constitute novel and significant variations (Johnson 2000). Their novelty arises from the very special properties of the entities that populate the Internet.

Because the Internet is composed of digital representations of text, data, music, and software, it can be characterized as a digital habitat. Because of the powerful capabilities of computers and networks, these entities can be *reproduced* and *transferred* with minimal effort and delay. One consequence of these properties is a notable characteristic of the Internet that can be called *virtual nearness*. Every public entity embedded in the Internet, within certain limits, is immediately available to the user—is *near* in a virtual way. This characteristic makes the emergence of virtual communities possible.

People perform activities in the Internet by means of *digital actions* carried out by digital programs. The specific steps programs perform can be easily recorded to leave a trace of the actions. In addition, because actions are carried out by programs, there is a question as to who is ultimately behind them, leading to certain forms of anonymity.

Privacy Issues

People carry out an increasing number of activities on the Internet, including exchanging email messages, visiting Internet sites, and buying goods. Transactions with government are increasingly done through the Internet. Medical records are created and made available online. In all of these activities sensitive information about people is gathered and stored. Because of its interconnectivity, the Internet makes it possible to transfer, combine, and cross-reference personal information at a much higher level than was previously achievable. The existence of multiple databases containing information on individuals about health, education, tax, and police matters, as well as on shopping patterns, enables the development of detailed profiles of individuals. Such profiles can be used for making decisions about them, for example, to grant or deny loans, to grant or deny medical insurance, to hire or not to hire, possibly leading to certain forms of discrimination.

Personal information is routinely used for purposes other than those originally intended, in most cases without the knowledge of the people involved. This situation constitutes a significant erosion of privacy.

Although there is a wide consensus that privacy—in particular, medical privacy—has been negatively affected by Internet technology (for example, Etzioni 1999; Johnson 2000; Parenti 2003), there is less agreement on how to confront the situation. Corporations claim they need personal information on their customers in order to be more efficient and profitable. Government agencies claim they need access to personal information for law enforcement purposes. For some theorists, then, the issue is to find a balance between the desires of individuals to keep information about themselves private and the desires of corporations and government to freely access that information. For others this perspective is too narrow because it transforms privacy issues into the balancing of competing claims. In a broader sense, privacy refers to a fundamental aspect of the human condition. Etymologically it is related to the Latin word *privus*, meaning single, alone. While human beings cannot be understood apart from the communities they belong to, they cannot be understood, either, unless it is recognized that they are unique individuals and have the potential to become increasingly autonomous.

By autonomy is meant the capacity to understand the sources, meanings, and consequences of actions and to exercise that understanding in deciding what actions to take. When information is collected and processed by others, autonomy is endangered in the sense that others can openly or surreptitiously attempt to influence actions on the basis of that information. In this respect, an important consequence of the availability of large amounts of personal information to corporations and government is that it increases their relative power with respect to that of individuals, possibly upsetting a delicate societal balance. For this reason, privacy is not only relevant to individuals but it should also be considered a social good, relevant to society as a whole.

Further discrepancies exist on how to deal with the erosion of privacy. Those who assign an intrinsic value to privacy tend to favor an approach in which individuals must provide explicit consent for the exchange of personal information among corporations, coupled with legislation enforcing such procedures. They claim that existing legislation—such as the Fair Credit Reporting Act, the Driver's Privacy Protection Act, and the Electronic Communications Privacy Act—has been developed piecemeal, and propose stronger forms of regulation similar to those existing in some European countries. Others favor a mixture of self-regulation by companies, use of technology to control access to information, and institutional changes leading to practices where information is less exposed to misuse.

An approach increasingly followed by companies is to develop privacy policies that are made available to their customers, indicating how information about them is used and with what other organizations it will be shared, and offering certain privacy options to customers. But without appropriate legislation many are skeptical that corporations can truly police themselves.

Two factors will exacerbate the erosion of privacy in the future. First, given the pace of technological development it is likely that increasing amounts of personal information will be available online. Second, the fight against terrorism triggered by the attacks that destroyed the World Trade Center on September 11, 2001, will put significant pressure on government agen-

cies to acquire and make use of that information, by wiretapping or other means, to detect terrorism-related activities. The Patriot Act enacted by Congress in October 2001 points strongly in this direction (Hentoff 2003). To conclude, a significant, multi-pronged effort will be required to deal with the erosion of privacy underway at the beginning of the twenty first century and to a large extent catalyzed by Internet technology.

Intellectual Property Issues

Intellectual property differs from tangible forms of property, such as cars and other goods, in that it is easily reproducible. Given that in the context of the Internet intellectual property, such as software and music, is stored in electronic files, people can reproduce and transfer it with minimal effort. It is precisely this notable characteristic of the Internet that is at the source of contentious issues regarding intellectual property. The case of Napster—a company that facilitated the global sharing of music files over the Internet and was shut down in 2001 as a consequence of a lawsuit brought against it by the recording industry—is important because it brought to light subtle issues, both at the core of the notion of intellectual property and on why and how the law protects it.

Ideas, literary works, and music are forms of speech. Freedom of speech, in one sense, implies the freedom to formulate and propagate ideas, as well as to have unfettered access to ideas and forms of speech produced by others. In the latter case, the authors of these works regard them as property and would like to be fairly compensated for their use. In addition, the free flow of ideas, for example of those that emerge in the context of science and technology, is regarded as beneficial to society as a whole. How can the tension between freedom of speech and *progress*, on one hand, and ownership of intellectual works, on the other, be resolved?

The Constitution itself lays out a basic framework for dealing with these issues, and gives Congress the power to enact legislation. Copyright law emerged in this context. An important distinction is established between ideas and expression of ideas, such that only the latter can be owned, and for a limited time.

Copyright law grants exclusive rights of copy to owners of intellectual property or to those whom owners grant permission, but through the notion of *fair use* it also establishes limits on this exclusivity. If a person buys a compact disk containing music, it is considered fair use to make extra copies of the disk for use in a car and for backup purposes. This is also true with regard to software. The law imposes additional restrictions on what can be copyrighted, including that the expression of ideas be novel and developed independently by its author.

Given these subtle distinctions, limits and restrictions imposed on intellectual property, the determination of whether copyrights have been infringed, and the enforcement of these rights are very difficult matters. The advent of the Internet has complicated the issues. The Napster case illustrated how the Internet made the copy and dissemination of music possible on a grand scale. While the recording industry considered it a form of electronic thievery, for some the exchange of files may have been an extreme case of fair use.

Supplemental Ethical Issues

Because the Internet makes it easy for people and groups to publish electronically, and given the potentially large audience that can be reached, the issue of what can be expressed on and accessed through the Internet arises. Again conflicting demands come into play. For example, freedom of access to public information conflicts with the desire to limit the availability of material that many regard as unacceptable.

Specifically impeding access to pornography by children in public libraries through the Internet could interfere with access to those same materials by adults. The Communications Decency Act passed by Congress in 1996 addressed that issue. A year later, the Supreme Court declared the act unconstitutional, siding with freedom of access and against censorship.

As already discussed, virtual nearness makes the emergence of virtual communities in the Internet possible, giving rise to virtual community (Turkle 1995). Some communities, in which people are represented by icons and fictitious names, provide opportunities for socializing in novel ways. In particular anonymity allows for the possibility of altering important elements of one's identity including gender, age, and race. What range of behavior is permissible in these situations? What would count as *violence*, as being too *close* to another person, as an attempted *rape* (Johnson 2000)?

Global Issues

A more global view raises two sociopolitical questions. First, given that the Internet facilitates the association of people with shared views, in particular, political views, and that it allows for the communication of those views to large numbers of people, does the Internet promote democracy as some have suggested? Second, considering that geographical barriers have little or no effect on the Internet, could the Internet contribute to undermine nation-states?

To a large extent, the answer to these questions depends on what the Internet becomes in the future. The Internet could remain as it is in the early-twenty-first century, except that almost everybody, everywhere, would have access to it and more activities would be carried out with its support. Or the Internet could become primarily a *global entertainment machine* by the convergence of radio, television, and the film, recording, and computer game industries. Or finally the reach of the Internet could be extended by ubiquity, wirelessness, and wearable computers.

In the context of these scenarios, the question of promotion of democracy answers itself: Although the possibility of performing political actions through the Internet would continue to exist, in the last two scenarios—the most likely—given the amount of noise that a global entertainment machine and the various extensions to the Internet would put into circulation, anything else would become barely audible and visible, including political action. In addition, the erosion of privacy mentioned earlier could contribute to undermining autonomy with, possibly, negative consequences for democracy.

With respect to the second question, about nation-states, the pressure to have common rules and laws, for electronic commerce, intellectual property, and privacy, that would facilitate the migration of activities to the Internet could undermine the sovereignty of less powerful countries. Although nation-states could try to control what regions of the Internet are accessible to its citizens (Hamelink 2000), given the connectivity of the Internet the effort would fail.

Finally the third scenario posed above leads to a fundamental philosophical question that can only be set out in this entry. Is it possible that the pervasive and substantial intermediation of human activities by the Internet—which would amount to a massive migration from material habitats to a global digital habitat—could invite essential transformations of the way human beings *are*? And what kinds of transformations would they be? But importantly, do people still have the ability to actually ask this question, or will the increasing noise make such questioning impossible?

AGUSTIN A. ARAYA

SEE ALSO *Computer Ethics; Computer Virusus; Cyberspace; Digital Divide; Hypertext; Networks.*

BIBLIOGRAPHY

Abbate, Janet. (1999). *Inventing the Internet*. Cambridge, MA: The MIT Press.

Berners-Lee, Tim, and Mark Fischetti. (1999). *Weaving the Web: The Original Design and Ultimate Destiny of the World Wide Web*. San Francisco: Harper.

Castells, Manuel. (2001). *The Internet Galaxy: Reflections on the Internet, Business, and Society*. Oxford: Oxford University Press.

Dreyfus, Hubert. (2001). *On the Internet*. New York: Routledge. The author presents a critique of the Internet reminiscent of his previous critique of Artificial Intelligence. Focusing on possibilities opened up by the Internet such as navigation with hyper-links, distance learning, disembodied telepresence, and anonymity the author shows them to give rise to weaker forms of experience when compared with the way in which people deal with the world without the intermediation of the Internet.

Etzioni, Amitai. (1999). *The Limits of Privacy*. New York, NY: Basic Books.

Halbert, Terry, and Elaine Ingulli, eds. (2002). *CyberEthics*. Cincinnati, OH: West-Thomson Learning. See especially in this collection John Perry Barlow's "The Economy of Ideas: A Framework for Patents and Copyrights in the Digital Age."

Hamelink, Cees J. (2000). *The Ethics of Cyberspace*. London: SAGE Publications.

Hentoff, Nat. (2003). *The War on the Bill of Rights and the Gathering Resistance*. New York: Seven Stories Press.

Johnson, Deborah G. (2000). *Computer Ethics*, 3rd edition. Upper Saddle River, NJ: Prentice Hall.

Parenti, Christian. (2003). *The Soft Cage: Surveillance in America from Slavery to the War on Terror*. Cambridge, MA: Basic Books.

Poster, Mark. (2001). *What's the Matter with the Internet?* Minneapolis: University of Minnesota Press. In this ambitious work, the Internet and its transformative power is examined from a variety of perspectives ranging from Heideggerian meditations on technology to Foucauldian considerations of subjectivity to post-Marxian analyses of capitalism and the nation-state. By juxtaposing these perspectives a rich understanding of the Internet emerges.

Stefik, Mark. (1999). *The Internet Edge: Social, Legal, and Technological Challenges for a Networked World*. Cambridge, MA: The MIT Press.

Stefik, Mark, ed. (1996). *Internet Dreams: Archetypes, Myths, and Metaphors*. Cambridge, MA: The MIT Press.

Turkle, Sherry. (1995). *Life on the Screen: Identity in the Age of the Internet*. New York: Simon & Schuster.

INVENTION

• • •

Invention (from the Latin *invenire*, to find or to discover) in a broad sense refers to any novel idea or the process of its creation. In the technological sense it means the identification of a science or technology potential matching a specific human need or the result of this process: a novel technical product.

Because any invention implies a use, it is intrinsically value-laden and thus of ethical interest. This applies to the intended purpose as well as to the unintended side effects of production or use, the possibilities of misuse, and of so-called dual use (when the function of a product may be employed for either good or bad use). The social promotion or regulation of the inventive process also has ethical dimensions.

Basic Distinctions

Originally there was no distinction between invention and discovery. Invention could refer to theoretical cognition as well as to technical designing. However, beginning with the twentieth century, these concepts usually are distinguished. To discover is to recognize an existing but previously unknown phenomenon. To invent is to conceive of a novel and previously not existing phenomenon.

Invention is the starting point for a new technical development. An innovative cognition in science or technology may precede invention but not necessarily. What is decisive is the notion that some natural or technical effect might function as an artifact that could replace or enhance some human activity or operation. Any invention creates a new means for some human end.

According to the German engineer and writer Max Eyth (1905), there are four types of invention. One is a new means for a new end; an example would be television. Second is a new means for a preexisting end; an example is the transistor as replacement for the electronic vacuum tube. Third is an existing means put to a new end; an example is the telephone to transmit written materials, as in the documents in a telefax. Finally fourth is an existing means for an existing end; an example is when music CDs are used to store data. Completely new inventions are rare; frequently, an invention is a mere combination of elements already in existence. The distinction between these different types of invention naturally raises questions about whether some types might present more serious ethical challenges than others, and whether ethics might be differentially related to different types of invention.

Inventors may apply for patents, which will protect the idea against illegal imitation. Usually, however, the invention by itself is not immediately ready for everyday use. Lengthy designing, testing, and improving are required before a properly functioning form is achieved.

This part of development is called the *innovation process*, and results, if technically and economically successful, in an *innovation* (in the narrow sense). The period between invention and innovation may take years or even decades. The question concerning whether in modern technology this period tends to progressively shorten, is highly debated.

Where Inventions Come From

At one time, the ability to invent was ascribed to the ingenious talent of gifted engineers who were regarded more as artists than as skilled experts. The art of inventing was explained by so-called creativity, an ability limited to only a few exceptional persons. Traditional histories of technology glorified the uniqueness of the inventor by drawing up long lists, in which important inventions were assigned to specific dates and famous names. The phenomenon of multiple inventions, however, disturbed this individualistic view. When both a technological potential and human need are in existence, the idea of bringing them together in an invention readily occurs to several persons at the same time. Although the aura of the individual ingenious creator may be shaken by this phenomenon, the process of inventing itself acquires a more solid explication.

According to John Guilford (1950), cognitive psychology explains creativity as a specific mixture of individual mental activities, partly conscious and partly subconscious. In the conscious stage, a person collects all knowledge available regarding certain problems and possible solutions (*preparation*). This knowledge sinks down to the subconscious, where it is stored, processed, and accidentally combined with additional tacit knowledge, without any explicit awareness on the person's part (*incubation*). Suddenly a new combination of knowledge and imagination emerges from the subconscious, and is identified as the perfect solution to a problem (*illumination*). In a final stage this new idea has to be tested and elaborated explicitly by rational thinking (*verification*).

In design theory, a modern branch of engineering research, the art of inventing is methodologically reconstructed. Instead of accumulating technical knowledge in an accidental and unsystematic way, design theory suggests systematic patterns arranging all the elements of possible solutions according to basic functional and structural features. This procedure, design theory claims, results in the totality of possible solutions to a given problem, and the only remaining difficulty is to choose the optimal solution among hundreds or even thousands of feasible combinations. Thus the associating and combining process, originally hidden in the subconscious, is objectified and rationalized, and is even accessible to computer programming.

Whether this rational strategy of inventing is actually feasible is debatable. Some observers hold that on principle the role of intuition and tacit knowledge in inventing is indispensable. For others the rationalistic approach seems promising for social interaction in teamwork, because individual intuitions from the subconscious are hard to communicate. Also invention cannot be reduced to personal performance alone, but obviously has social implications. Often it depends on the sociocultural context, which technical potentials an individual inventor takes into account, and which human needs and purposes are being realized. Furthermore the inventing activity depends on an innovative social climate and on economic incentives to motivate persons and corporations. Some hold that in the early twenty-first century the majority of inventions are made by large corporations, but there remain many individuals who also perfect basic inventions.

Ethical Issues

Recognizing that numerous inventions are ambiguous or even harmful to environment and society, several critics have considered whether an effective assessment and approval of the innovation process might be instituted. Some of them refer to historic examples, when certain inventions, in ancient Greece or medieval eastern Asia, had been suppressed systematically on ethical grounds, either by the very inventor or by political forces. The German economist Werner Sombart (1934) made the radical suggestion that every invention ought to be submitted to a National Council of Culture, which would release only such inventions as prove beneficial without question. Less radical approaches to improve the ecological and social quality of inventions are discussed at present in *engineering ethics*, which focuses on the professional responsibility of individual inventors, and in *technology assessment*, which concentrates on industrial strategies and political regulations.

Individual refusal—like that of the father of cybernetics, Norbert Wiener, who in 1947 rebelled against doing any further work for the military—usually is not very effective, because nearly always there will be found others to continue a questionable project. Therefore, moral sensitivity of the individuals has to be supported by corporate and political institutions such as those of technology assessment, which proves to be the social organization of teleological ethics.

Some commentators question the ever growing rate of inventions and innovations, mostly driven by economic forces, which possibly threaten natural environment, the stability of cultural traditions, and personal self-fulfillment. Such views are, of course, at odds with the dominant innovation tendencies in modern industrial and information society.

GÜNTER ROPOHL

SEE ALSO *Political Economy of Science and Technology; Technological Innovation.*

BIBLIOGRAPHY

Eyth, Max. (1905). "Zur Philosophie des Erfindens." In *Lebendige Kraefte*. Berlin: Springer, 249–284.

Gilfillan, S. Colum. (1935). *The Sociology of Invention.* Cambridge, MA: MIT Press. Analyzes social conditions and consequences of inventions.

Guilford, John P. (1950). "Creativity." *American Psychologist* 5: 444–454.

Hubka, Vladimir, and W. Ernst Eder. (1988). *Theory of Technical Systems: A Total Concept Theory for Engineering Design.* Berlin: Springer.

Porter, Alan L.; Frederick A. Rossini; Stanley R. Carpenter, et al. (1980). *A Guidebook for Technology Assessment and Impact Analysis.* New York and Oxford: North Holland.

Ropohl, Guenter. (1999). *Allgemeine Technologie.* Munich and Vienna: Hanser. Provides a comprehensive understanding of technology and analyzes thoroughly the process of technological development.

Schumpeter, Joseph A. (1939). *Business Cycles.* New York and London: McGraw Hill. Introduces the fundamental notions of invention and innovation from an economic point of view.

Sombart, Werner. (1934). *Deutscher Sozialismus.* Berlin: Buchholz & Weisswange. A peculiar attempt to suggest leftist ideas to German "national socialism", but important because of the perspicacious criticism of technological development.

Wiener, Norbert. (1993). *Invention: The Care and Feeding of Ideas.* Cambridge, MA: MIT Press. A book found in manuscript and published forty years after it was written, which provides a broad analysis of invention in relation to most of the themes touched on in this article.

IN VITRO FERTILIZATION AND GENETIC SCREENING

• • •

The first birth following in vitro fertilization (IVF) took place in the United Kingdom in 1978, and the number of IVF births per year has increased steadily since then. More than 35,000 infants were born with the help of IVF in 2000 in the United States alone, and more than 1 million infants have been born worldwide following IVF. Although IVF has become an integral part of fertility medicine, ethical and policy issues continue to be debated as technologies change and IVF becomes more common. Among the topics debated are those relating to the moral status of embryos, disposition of frozen embryos, use of genetic testing of embryos to detect the presence of moderate rather than serious genetic disorders, and the adequacy of regulation.

Technologies

For an IVF cycle, physicians stimulate a female patient with hormones to induce the release of more than one egg. When tests show the eggs are ready to be released, physicians remove the eggs in an office procedure, fertilize them in vitro (in glass) with spermatozoa from the male partner or a donor, culture the fertilized eggs for two to three days to at least the stage of a four-cell embryo, and transfer the embryos to the woman's uterus for possible pregnancy.

Although IVF was primarily designed for women with blocked fallopian tubes who could benefit from the way IVF bypasses these tubes, advances over the years have extended the versatility of IVF as a method for circumventing infertility. For example women who do not ovulate can use donated eggs, and men with extremely low sperm counts can be aided by the manual injection of a single spermatozoan into an egg in a technique known as intracytoplasmic sperm injection (ICSI).

Another technique used in conjunction with IVF is pre-implantation genetic diagnosis (PGD), which is available for couples at risk for passing serious genetic diseases, such as Tay-Sachs disease and cystic fibrosis, to their offspring. In one form of PGD, the embryo biopsy, technicians remove a single cell from an embryo created through IVF and amplify the DNA to detect the presence of the disease-linked gene in question. Physicians then selectively transfer only those embryos without the anomaly to the woman's uterus. PGD is also used to detect chromosomal abnormalities that cause serious disorders in offspring or that interfere with conception. The first birth following IVF/PGD occurred in 1990. More than 1,000 infants had been born worldwide by 2002, with a pregnancy rate of about 24 percent (Robertson 2003).

Moral Status of Embryos

Perspectives about the moral status of embryos differ significantly among individuals. Some believe that early

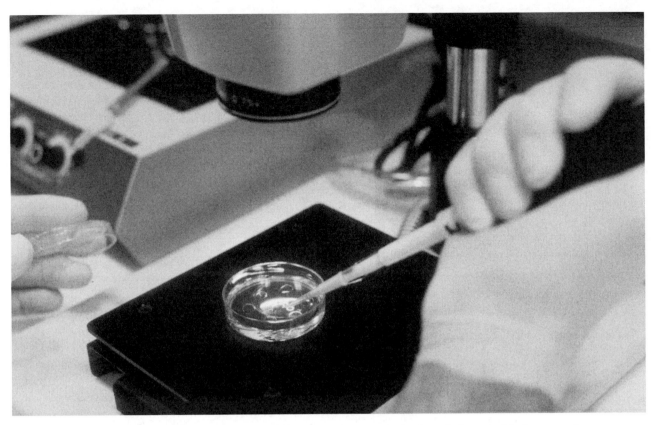

In vitro fertilization. Many ethical questions surround the process, in which egg cells are fertilized outside the mother's body in cases where conception is impossible through normal intercourse. (© Owen Franken/Corbis.)

stage embryos are human beings with the moral status of persons that must be protected from injury or destruction. Others believe embryos are potential human beings warranting special respect but not the moral status of persons. Policy advisory groups in the United States have generally adopted the latter perspective (Ethics Advisory Board 1979). Due to a lack of consensus about the status of embryos, however, as of 2004 federal funds cannot under law be used to finance research in which human embryos are injured or destroyed. To the extent that investigators study human embryos, they do so with private research funds.

Disposition of Frozen Embryos

When couples undergo IVF or IVF/PGD, extra embryos are often created and frozen for later thaw and transfer. More than 100,000 embryos were frozen in the United States alone in 2002. Couples who no longer want or need their spare embryos can direct that the embryos be discarded, donated for research and eventual destruction, or donated to other couples. Difficulties can arise, however, if a couple divorces and has no prior written agreement about what should be done with the embryos

or if one party seeks to nullify the agreement. The first appellate court to rule on this matter held that the person who wants to avoid parenthood (by not transferring the embryos) generally ought to prevail over the person who wants to achieve parenthood (by transferring the embryos) (*Davis* v. *Davis* 1992). Judges rely on case-by-case rule making in frozen embryo cases. In general they accept the principle established in *Davis* v. *Davis*, but differ on whether they will enforce prior agreements (Elster 2002).

Extending PGD

PGD is generally regarded as an ethically acceptable way of preventing human suffering when the disease in question is serious or fatal. Some have voiced reservations, however, about the potential for tests that can be used to detect less serious diseases such as deafness or predispositions to diseases such as Alzheimer's disease or breast cancer. The concern is that this will discourage tolerance for imperfections and devalue the inherent worth of individuals. Another concern is that negative selection (discarding affected embryos) will, when technologies allow it, set the stage for positive selection (seeking embryos with socially desirable traits), which would magnify differences between the rich and the less well-off; have eugenic overtones; and contribute to the mindset that people can be made to order, like commodities.

Those who do not share these concerns argue that IVF/PGD is so costly and intrusive that only a small number of people will use it. They point out that PGD is an alternative to prenatal testing for at-risk couples who know they will not terminate the pregnancy of a fetus with serious disorders and who welcome the opportunity to transfer only unaffected embryos for a potential pregnancy. In addition supporters of PGD question the wisdom of interfering with a technique that could prevent the birth of babies with serious disorders now on the basis of speculative concerns about possible future uses of PGD.

Policy Issues

IVF and other reproductive technologies are governed in a decentralized manner in the United States. Debates continue about whether more oversight is needed and, if so, what forms it should take. One point of view is that the system of oversight, which is based on state laws, medical licensing requirements, tort law, self-regulation by professional associations, administrative rule making, and the power of the marketplace, is thorough and effective (Adamson 2002). One federal law directs the government, in conjunction with professional associations, to collect and publicize data from fertility clinics to educate patients and the public about clinic performance. Those who believe that the oversight system is sufficient point to statistics on healthy children and improved birth rates for IVF as indicators of effective regulation and professional caution. They argue that concerns, such as those questioning the sizeable number of twin, triplet, and higher order births following IVF, can be addressed by professional self-management and improved technologies.

From another point of view, the government should take a more active role in monitoring IVF/PGD practice by developing a centralized oversight system and taking other steps to protect the health of patients and offspring. According to this view, the government should develop a centralized data gathering system or, at least, a national level forum for debating issues relating to infertility treatment. It should also enact laws to address specific concerns; for example, to limit the number of multiple births, and regulate by law the number of embryos that can be transferred per IVF cycle (International Society for Law and Technology [ISLAT] Working Group 1998).

The ability of the federal government to regulate IVF is limited by constitutional protections of reproductive liberty. In addition political controversies over the status of embryos make legislation difficult to enact. The likelihood of enacting in the United States a central oversight board for assisted reproductive technologies, as exists in the United Kingdom, is slim. In the meantime researchers are engaged in data gathering to assess the long-term safety of IVF, state legislatures are considering various forms of regulation, and practitioners are continuing to produce practice guidelines as part of self-regulatory policies.

Conclusion

IVF has led to the birth of more than 1 million children who may not otherwise have been born to couples experiencing various infertility problems. Issues about the status of embryos, disposition of frozen embryos, proper reach of PGD, and optimal forms of oversight have recurred in the years since 1978. New ethical issues arise as the technologies and applications change. For example what payment is appropriate for egg donors? Should practitioners accept single persons as patients? What should be done with embryos abandoned by couples? What issues are raised when egg or sperm donors are related to the recipients? Should PGD be used to determine predisposition to disease? Should children conceived with donor eggs, sperm, or embryos be told how they were conceived?

Although no central forum exists for debating these issues in the United States, the public fascination with IVF ensures that the issues are aired and discussed. While it is tempting to call for governmental controls, the issues raised by rapidly changing technologies are not easily amenable to preemptive legislation, which can be rigid and easily dated. Moreover government policy precludes funding research in which embryos are injured or destroyed, which removes the *power of the purse* as a vehicle for oversight. Consequently, robust discussion, public education, regulations of medical facilities in general, and self-regulation by professional associations all contribute to oversight. Though complex and decentralized, this system allows monitoring while also respecting the reproductive liberty of couples seeking the services of fertility clinics.

ANDREA L. BONNICKSEN

SEE ALSO *Assisted Reproduction Technology; Embryonic Stem Cells; Eugenics; Fetal Research; Genethics; Genetic Counseling; Playing God.*

BIBLIOGRAPHY

Adamson, David. (2002). "Regulation of Assisted Reproductive Technologies in the United States." *Fertility and Sterility* 78: 932–940. An overview of numerous professional guidelines and government laws and policies that make up the regulatory framework for assisted reproductive technologies in the United States.

Davis v. Davis, 842 S.W.2d 588 (Tenn. 1992). First court case involving a dispute between male and female partners over the disposition of their frozen and stored embryos.

Elster, Nanette R. (2002). "ARTistic Licence: Should Assisted Reproductive Technologies Be Regulated?" In *Assisted Reproductive Technology: Accomplishments and New Horizons*, eds. Susan M. Avery and Peter R. Brinsden. New York: Cambridge University Press. Review of court decisions, state and federal laws, and self-regulation of assisted reproductive technologies, including what Elster considers to be regulatory gaps.

Ethics Advisory Board. U.S. Department of Health, Education, and Welfare. (1979). *Report and Conclusions: HEW Support of Research Involving Human In Vitro Fertilization and Embryo Transfer.* Washington, DC: Government Printing Office. First report by a governmental advisory body on the ethics of in vitro fertilization; report concluded that research into IVF was ethically acceptable, with limits.

International Society for Law and Technology (ISLAT) Working Group. (1998). "ART into Science: Regulation of Fertility Techniques." *Science* 281: 65–66. Brief review of regulation of assisted reproductive technologies with recommendations for increased governmetnal oversight.

Robertson, John A. (2003). "Extending Preimplantation Genetic Diagnosis: The Ethical Debate." *Human Reproduction* 18: 1–7. Review of situations in which pre-implantation genetic diagnosis can be used; discussion of ethical and legal issues relating to these uses.

Society for Assisted Reproductive Technology and the American Society for Reproductive Medicine. (2002). "Assisted Reproductive Technology in the United States: 1999 Results Generated from the American Society for Reproductive Medicine/Society for Assisted Reproductive Technology Registry." *Fertility and Sterility* 78: 918–931.

IQ DEBATE

• • •

In 1905 two Frenchmen, Alfred Binet (1857–1911) and Theophil Simon (1873–1961), invented the IQ (Intelligence Quotient) test to distinguish between mentally retarded and normal school children. They set tasks that normal children could do; for example, five-year-olds were asked to compare two weights, copy a square, repeat a sentence of ten syllables, count four pennies, and unite the halves of a divided rectangle.

By 2005 there were thousands of tests but two have special significance. The first, Raven's Progressive Matrices, measures on-the-spot problem solving where no previously learned method is applicable. It presents a pattern of shapes from which one piece is missing, offers six alternative missing pieces, and then asks the examinee to choose the correct one (Raven 2000). The second, the Wechsler Intelligence Scale for Children (WISC), supplements Raven's by using ten to twelve subtests to measure a variety of cognitive skills. These tests constitute technologies that raise significant ethical issues.

What IQ Tests Measure

Various cognitive skills go into problem solving. One such skill is mental acuity, which involves both solving problems without a previously learned method and the active creation of alternative solutions. The WISC subtest called Similarities measures mental acuity: The subject must decide what certain things, such as dawn and dusk, have in common. Similar subtests include Block Design, Picture Concepts, and of course Matrices. Another set of subtests are quite different. Clearly, a wide range of basic knowledge and a large vocabulary enhance problem-solving ability. These are measured by the Vocabulary and Verbal Comprehension subtests and, until recently, by the Information and Arithmetic subtests that were dropped in the fourth edition of the WISC. Although there is learned content in these subtests, it is the kind of learning that intelligent people will master more easily and more thoroughly. A third kind of relevant skill is speed of information processing—which is measured by the Coding and Symbol Search subtests. Finally, that ability called memory, which allows individuals to access accumulated knowledge, is tested by the Digit Span (the number of digits a person can repeat after they are read out —and the ability to repeat them in reverse order) and the Letter-Number Sequences subtests.

Given that the WISC tests cover the cognitive skills that go into problem solving, it may seem surprising that there is so much debate about whether IQ tests measure intelligence. There are several reasons why the controversy endures.

Attitudes affect cognitive skills because people invest mental energy into problems only if they feel they are significant. Attitude shifts over time have enhanced performance on some subtests more than on others (Flynn 2003). Members of a street gang may see little point in problems that appear to lack practical significance. Lots of noncognitive skills contribute to problem solving such as empathy, tact, setting people at ease, and being a good listener. In addition, IQ tests do not measure a host of attributes regarded as important, such as artistic and musical ability, honesty, and generosity.

Most debate about what IQ tests measure consists in endless repetition of these points and inventing a host of intelligences, such as emotional intelligence, social intelligence, surviving-in-a-wilderness intelligence, and musical intelligence, among others (Jensen 1998). This sterile debate can perhaps be circumvented by a modest claim: IQ tests measure *cognitive* skills relevant to problems encountered in the *mainstream of industrial societies*; and test the *basic knowledge* needed to function in those societies. However, there is a caveat: IQ tests cannot determine when a person scores better than others because of *attitudes* friendlier toward the kind of problems that are to be solved.

Uses of IQ Tests

IQ tests perform three main roles: comparing individuals for cognitive skills; comparing groups; and measuring cognitive skill trends over time, this last being a special case of comparing groups because it entails comparing one generation with another.

IQ scores give each person a percentile rank using Standard Deviations (SDs) as the link. An IQ of 100 is average for any particular age and is at the 50th percentile. An IQ of 130 is two SDs above the mean (an SD = 15) and is at the 98th percentile (only 2.3% of the subject's peers have a higher score); an IQ of 110 is 0.67 SDs above the mean and is at the 75th percentile; an IQ of 70 is two SDs below the mean and equals the 2nd percentile (only 2.3% of the subject's peers have a lower score). Certain IQ scores set the threshold for performing certain social roles. Few people with IQs below 130 will receive a Ph.D. from an academically superior university; few with IQs below 110 will enter the elite professions, that is, medicine, law, accounting, natural science, and engineering; and few with IQs below 100 will hold a professional, managerial, or technical post of any kind. Those with IQs below 70 are often regarded as being unable to cope with normal life and are labeled mentally retarded.

Race Differences

The existence of IQ thresholds for occupations generates group comparisons unfavorable to blacks. The mean IQ of white Americans is 100, while black Americans have a mean IQ of 85 or one SD below whites. The pool of potential professionals, managers, and technicians has a threshold of 100. Therefore, 50 per cent of whites

would qualify but only the highest scoring 16 per cent of blacks (a score of 100 is at their 84th percentile). The Berkeley psychologist Arthur Jensen suggests that even if environments were equalized, blacks would still have a mean IQ of only 90 (Jensen 1973, p. 363). If he is correct, even then, only 25 percent of blacks would qualify.

Some believe scholars should not debate whether ethnic groups show genetic differences for intelligence. This moral advice will fail and should fail. Those who read Jensen will quickly find that he has an argument that must be answered, high professional standards, and no trace of racial bias. Thus the only reason not to test his hypothesis is that it would be unpleasant if it were true. In addition, if those who have offered evidence in favor of genetic equality were to opt out of the debate, Jensen's hypothesis would remain undisputed, a sort of unilateral disarmament. The debate should proceed and be conducted purely along evidential lines. The strongest evidence supporting a genetic hypothesis is the under performance, both on IQ tests and academically, of children of the black middle and upper classes—who do fall at least 10 IQ points short of their white counterparts (Herrnstein and Murray 1994, p. 288). The strongest evidence in favor of an environmental hypothesis was obtained as the result of an historical event: the U.S. military occupation of Germany after World War II, which removed thousands of black males from the American environment. The U.S. army left behind many illegitimate children. The mean IQs of those with black fathers and those with white fathers were the same (Flynn 1999).

Whatever the causes of the IQ gap between black and white Americans, it exists. When standardized tests are used as screening devices, the lesser representation of blacks leaves the realm of theory and becomes fact. The debate as to whether affirmative action should be used to redress the balance is complex. Opponents point to cases of underprivileged whites who are rejected in favor of the child of a black professional, lower performance in key areas such as police protection, and the fact that blacks may actually suffer harm, for example, by being admitted to universities where they are doomed to fail (Herrnstein and Murray 1994).

Proponents argue that black Americans suffer from their group membership in many ways, ranging from police behavior toward them, higher consumer prices in the ghetto, discrimination in housing and employment, and an unfavorable marriage market. White men very rarely marry black women. Therefore, black women are restricted to marrying black men and many are unlikely to find permanent partners—because too many black men die young, are imprisoned, or are not regularly employed. Therefore, more than one-half of black children are raised in solo-mother homes, often below the poverty line (Flynn 2000, pp. 148–149). Supporters of affirmative action also contend that most efficiency gains would accrue if standardized tests were only used to disqualify those without essential skills and if job-related criteria were substituted to rank applicants above that level. They cite data showing that when blacks admitted to elite universities (for which they would not normally qualify) are matched with blacks who went to other universities, the graduation rates are similar—and that the former profit by earning higher incomes (Kane 1998).

Genes and Environment

Studies of identical twins separated at birth and raised apart show that, at adulthood, twin and co-twin are far more alike in IQ than randomly selected individuals. This appears to be because of their identical genes—and does that not mean that genes are far more potent than environment? Jensen calculated that if environment were in fact this weak, no plausible environmental difference within a society such as America could account for a one SD IQ gap—which is the gap between the IQs of blacks and whites (Jensen 1973, pp. 166–169).

In 1987 James R. Flynn, a moral philosopher at the University of Otago, challenged this reasoning with evidence showing the existence of massive IQ gains over time. For example, the Dutch gained fully 20 IQ points on Raven's Matrices from one generation to the next, that is, from 1952 to 1982, a result replicated in several nations. Since there can be little genetic upgrading in a single generation, Flynn contended that these huge gains must have been due to environment (Flynn 2003). Thus, a paradox arose that baffled the discipline for many years: How can twin studies show environment to be so weak, while IQ gains over time show environment to be so enormously potent?

In 2001 William T. Dickens, an economist at the Brookings Institution, and Flynn offered *reciprocal causation* as a possible solution. Imagine identical twins who were separated at birth and raised apart in a basketball-mad state such as Indiana. Their identical genes dictate that they are born both a bit taller and quicker than average. Thus, although raised in different cities, both tend to be picked for informal basketball games at school. The extra play upgrades their skill advantage and they both get picked for the school team. They then play a rigorous schedule and get professional coaching, which upgrades their skill advantage further. At adulthood, they end up with basketball skills that are remarkably similar and well above average—and their identical genes get all the credit. But that assumption is a mistake. It overlooks the fact that these identical twins also had atypically similar basketball environments—their genes are getting credit for shared factors like more practice, playing on a team, and professional coaching. The kinship studies mask the potency of environment.

Skill gains over time show the true strength of environment. In 1950 TV brought basketball into American homes and basketball put baseball into the shadows—those close-ups look so good even on the small screen. Suddenly everyone was playing basketball and skills escalated. At first, to be better than average, a player needed merely to pass and shoot well. However, the rising quality of the average performance became a powerful factor in its own right. To excel, a few people learned to shoot with both hands. Then everyone who wanted to compete had to try to do the same, which pushed the mean up further. Soon a few people learned to pass with both hands and then, everyone had to try to do that. Every rise in the average performance encouraged a further rise.

So now this has resolved the gene-environment paradox: The key is reciprocal causation as a potent multiplier of skill differences. Within a generation, genes drive the feedback process and get credit for the environmental input—which gives the illusion of environmental weakness. Between generations, a persistent environmental factor (the rising popularity of basketball) drives the feedback process—and shows how environment can produce huge skill differences between groups separated by only a few years of time.

New Spectacles

The concept of reciprocal causation provides spectacles that improve our perception of what may cause group IQ differences. Do blacks start with what may be a modest but significant genetic disadvantage, one that gets multiplied into a 15-point IQ deficit? Or are there persistent environmental factors that divide black and white, analogous to belonging to the pre-TV and post-TV generations? Some have attempted to identify the kind of factors that might inhibit black academic achievement and IQ test performance: that they feel threatened by intellectual competition with whites; that black males are ambivalent about intellectual success and may even strive to fall below the class mean (so blacks would have negative multipliers!); and, as has been seen, that the problems of black males affect black children, so that a majority of them are raised by solo-mothers struggling to avoid poverty.

The brute fact that average IQ scores increase over time adds a new dimension to another debate: whether IQ tests should be used to classify people as mentally retarded. IQ gains mean that subjects will get higher IQs on an out-of-date test. If someone was average when compared to the test performance of their peers today (and therefore gets an IQ of 100), they would automatically be better than average compared to their peers of 20 years ago (and therefore get an IQ well above 100). After all, the fact that the average performance was worse in the past is what constitutes IQ gains over time. There is no doubt that people have been denied special education or have been executed on death row because taking obsolete tests inflated their IQs above 70, the usual cut-off point for mental retardation (Kanaya et al. 2003). These facts strengthen the argument of those who believe in purely behavioral criteria for mental retardation: School children should be classified as such if they cannot understand the rules of games they play frequently; prisoners should be executed only if their life histories show they can cope with the usual activities of everyday life, for example, by qualifying for a driver's license.

Are IQ Gains Real?

The United States and other nations have been making massive IQ gains for at least as far back as the 1930s. Are these really intelligence gains? The answer is that they are piecemeal cognitive skill gains that affect the real world—but they are not gains in terms of the kind of general intelligence IQ tests are designed to measure.

When an IQ test measures individuals competing with one another, certain people tend to do better than average on all or most of the WISC subtests—which is to say part of what is being measured is a better functioning brain that gives someone an advantage for most cognitive skills. Society does not upgrade average brain quality from one generation to another because it does not run radical experiments in selective breeding. What it does do is manipulate environmental factors that have a differential effect on various cognitive skills. If Americans fill more leisure time with cognitively demanding games, and fill more professional positions in which they must make decisions rather than simply following rules, scores on the Similarities subtest should rise—and they have enormously. If efforts to improve reading in the United States have not made people love books, and if visual entertainment of a largely escapist sort tempts people away from books, one would not expect better ability to read serious literature, or bigger non-specialized vocabularies, or the command of more general

information—and the relevant WISC subtests show that this is indeed the case (Flynn 2003).

In sum, IQ tests are good tools for comparing the cognitive skills of individuals and alerting researchers to group differences. However, finding causes and solutions for those differences involves the totality of social science. The general intelligence factor that IQ tests are designed to measure may indicate which mind competes best with other minds at a certain time and place. But it is a crude measure of what society is doing to a wide variety of cognitive skills over time. We must free our minds of it and look at trends on the various WISC subtests. They reveal the intellectual history of these times.

JAMES R. FLYNN

SEE ALSO Emotional Intelligence; Eugenics; Race.

BIBLIOGRAPHY

Deary, Ian J. (2001). Intelligence: A Very Short Introduction. New York: Oxford University Press. A good introduction to IQ tests and their significance.

Dickens, William T., and James R. Flynn. (2001). "Great Leap Forward." New Scientist 170(2287): 44–47. Spells out the concept of multipliers and how they clarify the roles of genes and environment.

Flynn, James R. (1999). "Searching for Justice: The Discovery of IQ Gains Over Time." American Psychologist 54: 5–20. Discusses the social issues that IQ tests raise.

Flynn, James R. (2000). How to Defend Humane Ideals: Substitutes for Objectivity. Lincoln: University of Nebraska Press. Makes a case for affirmative action as necessary even in the absence of racism.

Flynn, James R. (2003). "Movies about Intelligence: The Limitations of g." Current Directions in Psychological Science 12: 95–99. Argues that IQ gains represent real gains in certain cognitive skills even if they are not general intelligence gains.

Herrnstein, Richard J., and Charles Murray. (1994). The Bell Curve: Intelligence and Class in American Life. New York: Free Press. A case that social progress is producing an underclass without the genetic potential to contribute to society; for a rebuttal, see Flynn (2000).

Jensen, Arthur R. (1973). Educability and Group Differences. New York: Harper and Row. A case that blacks on average have a lower potential for intelligence than whites; for a rebuttal, see Flynn (1999).

Jensen, Arthur R. (1998). The g Factor: The Science of Mental Ability. Westport, CT: Praeger. A case that the concept of general intelligence, g, is central to the study of human cognition—for reservations, see Flynn (2003).

Kanaya, Tomoe; Matthew H. Scullen; and Stephen J. Ceci. (2003). "The Flynn Effect and U.S. Policies: The Impact of Rising IQ Scores on American Society via Mental Retardation Diagnoses." *American Psychologist* 58: 778–790. Spells out how IQ gains over time have made a lottery out of classifying people as mentally retarded.

Kane, Thomas J. (1998). "Racial and Ethnic Preferences in College Admissions." In *The Black-White Test Score Gap*, eds. Christopher Jencks and Meredith Phillips. Washington, DC: Brookings Institution Press. Presents data that show that blacks profit from affirmative action at the university level.

Raven, John. (2000). "The Raven's Progressive Matrices: Change and Stability over Culture and Time." *Cognitive Psychology* 41: 1–48. The "architect" of Raven's Progressive Matrices describes its role in measuring cognitive trends throughout the world.

ISLAMIC PERSPECTIVES

• • •

Islam is at once a religion, a community, and a civilization. In all three senses, Islam is a source of unique perspectives on relations between science, technology, and ethics. As a religion, Islam upholds knowledge as the key to both individual and societal salvation. With the idea of unity of reality and knowledge as a guiding principle it refuses to entertain any distinction between the religious and the secular in the realm of knowledge. Science and technology are as relevant as the so-called religious sciences to the human pursuit of the divine. As a community, Islam stresses on the divine law as the most important source of ethics to guide human actions in all sectors of personal and public life and as the most visible expression of Muslim cultural identity. This law is generally viewed as not only all-embracing in the scope of its applications but also as dynamic enough to be adaptable to the changing needs of space and time. Science and technology are to be regulated by ethics embodied in this law. As a civilization, Islam seeks to promote the interests of all humanity by standing up for the perspectives of universalism, the common good and inter-faith understanding. As so many of Islam's thinkers have asserted over the centuries science and technology are the most powerful and the most enduring universal elements in human civilization and should be pursued for the sake of the common good and inter-faith peace, Islam places strict limits on technology and subordinates scientific rationality to revelation. As a community, Islam is more concerned to adapt science and technology for practical benefit.

Historical Background

Islam was born in Mecca, Arabia, in 610 C.E. when Muhammad, an illiterate but highly respected member of Arabia's most powerful tribe, the Quraysh, claimed he had received revelations from God. During one of his regular spiritual retreats in a cave on the outskirts of Mecca, the archangel Gabriel appeared before him instructing him to recite a few verses in Arabic and proclaiming him God's new messenger to humankind. That initial revelation was essentially about the true spirit of human learning: Seeking knowledge is to be done in the name of God who is humanity's best teacher, and the best human instrument of knowledge is the intellect as symbolized by the pen. This tenet supported the new religion's claim to be essentially *a way of knowledge*.

The Prophet, as every generation of Muhammad's followers call him, received further revelations intermittently over a period of twenty-three years until just before his death in 632 C.E. These revelations were systematically compiled into a book known as the *Qur'ān* (literally meaning *The Recitation*). The precise arrangement of the *Qur'ān* itself is traditionally thought to be divinely inspired. This book, believed sacred both in text and meaning, is the most authentic and the most important source of Islamic teachings. The names Islam for the religion and Muslims for its followers are set out in the *Qur'ān*. Islam means both *submission to God's will* and *peace*, while Muslim means *one who submits to the divine will*. More than anything else the *Qur'ān* is a source of guidance in the domain of knowledge. Muslims believe that the *Qur'ān* contains the principles of all sciences. Islam claims to revive the pure monotheism of Abraham while presenting itself as the synthesis of all previously revealed religions, which has helped foster a positive attitude among Muslims toward the intellectual and cultural legacies of other civilizations.

As a full-fledged religious community (*ummah*) with distinctive characteristics as envisioned in the *Qur'ān*, Islam was founded in Medina, formerly known as Yathrib, in 622 C.E. (although the nucleus of the community had formed earlier in Mecca). The Prophet and his followers migrated to Medina to escape persecution following his uncompromising stand on idol worship. This flight, known as the *hijrah*, marked a major turning point in the history of Islam. The original group grew to become a worldwide community that is estimated at 1.2 billion followers in the early-twenty-first century. As an extension of his community, the Prophet established a city-state that he named *Madinat al-Nabiy* (City of the Prophet) or simply *al-Madinah* (The City). This pluralistic city-state, multiethnic and multireligious, reflected

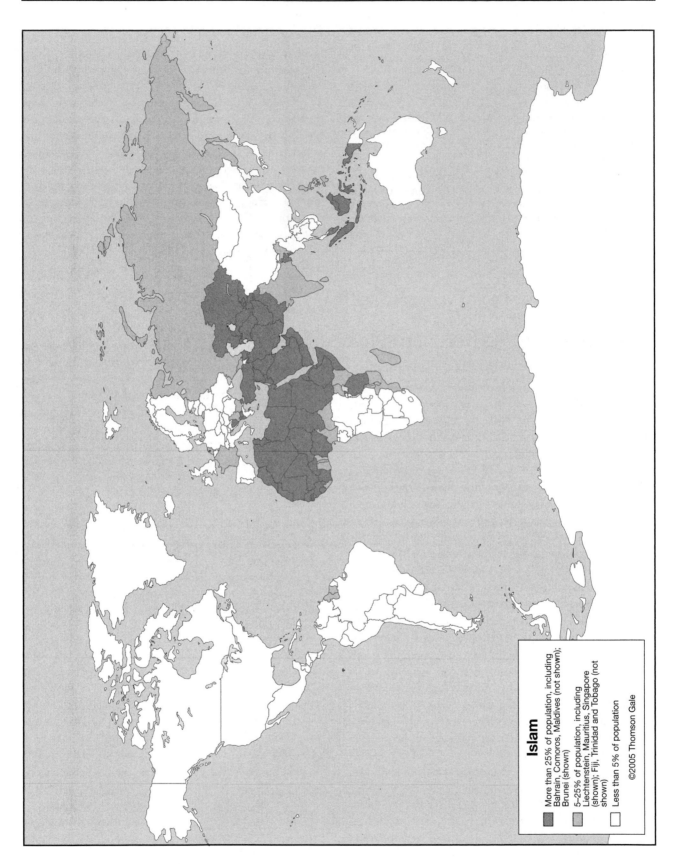

Islam

More than 25% of population, including Bahrain, Comoros, Maldives (not shown); Brunei (shown)

5-25% of population, including Liechtenstein, Mauritius, Singapore (shown); Fiji, Trinidad and Tobago (not shown)

Less than 5% of population

©2005 Thomson Gale

the moral and ethical basis of the sociopolitical teachings of Islam. In postprophetic Muslim history, Medina is an enduring model of Islamic polity.

As a civilization (*tamaddun*), Islam manifested itself when the community organized all aspects of daily life in accordance with the spiritual and ethical values set out in the *Qur'ān* and as interpreted by the Prophet. The cultural identity of Muslims became easily visible in the way they cultivated *a knowledge culture* that did not separate the religious and the secular, envisioned and practiced moderation in religious life, merged temporal life with the spiritual, championed social justice, permeated ethical concern in all individual and societal activities, engaged pluralism, and approached relations with other faiths. But the happenings in Medina merely lay the foundation of Islam. Fuller development of the civilization occurred after the religion spread throughout the world, encountering other civilizations, and the *ummah* grew into a more ethnically and culturally diverse circle of believers.

Islam and the world did not have to wait long to see the realization of a civilization that was innovative, unique, and unrivaled in brilliance for its times. The spread of Islam to distant places was astoundingly swift. Within a century from the death of the Prophet, Islam swept through North Africa reaching Spain in the west and central Asia in the east, and even became a minority religion in China. With a generally positive attitude toward the cultural and scientific legacies of past and contemporary civilizations, Islam tried to create a new civilization by merging the best of these traditions with its own resources. The hallmark of Islam, the civilization, is *the grand synthesis*. Islam, the religion, inspires the Muslim mind to create a human civilization that is basically *synthetic* in nature.

Islam, Science, and Technology

This historical background provides a context to understand science, technology, and ethics in Muslim culture and civilization. Muslims believe the *Qur'ān* affirms the supreme role of knowledge in ordering human life and thought and delivering success. Knowledge is regarded as the key to human salvation and to human happiness in this world and in the afterlife. But knowledge that saves must be sacred in nature. Sacredness is not defined in terms of primacy of revelation over reason. Among Muslim philosophers and scientists the distinction between revelation and reason is rather blurred. This is because reason is regarded as a minor revelation given to every human individual and as such is itself sacred in nature even if many humans are not aware of it. By

Man holding an ancient copy of the Koran. Muslims believe that the Koran is the literal word of God and culmination of God's revelation to mankind, revealed to the Prophet Muhammad over a period of 22 years by the archangel Gabriel. (© Nik Wheeler/Corbis.)

sacred knowledge the *Qur'ān* means knowledge that is related in some way to God, pursued in the name of God, and used and applied in the name of God. As Muslims see it, human knowledge, including science, possesses a sacred character because God is the ultimate source of all knowledge regardless of whether humans acquire it empirically or otherwise. The *Qur'ān* speaks of God as the All-Knower and the giver of knowledge to humans through various avenues ranging from physical senses to intellectual reflection, dream interpretation to divine revelation. The Muslim idea of sacred knowledge is contained in the very first revelation Muhammad received.

The *Qur'ān* also maintains that the ultimate purpose of human knowledge is to know God. This objective is attainable because human knowledge of creation will lead to knowledge of the divine reality, which is considered to be the highest form of knowledge possible. The *Qur'ān* is emphatic in acknowledging that God is knowable. Muslims approach the study of different branches of knowledge, including science and technology, with this spiritual objective in mind. Scientists view their study of the natural world as a form of religious worship, but the lesser objectives of knowledge are duly recognized. Knowledge helps humans to fulfill their rational and mental needs, such as clarity of mind, certitude of thought, and rational explanations of both natural and social phenomena, as well as those material needs that can be met by technology. In the traditional Muslim pursuit of knowledge, the deepest theoretical understanding of things goes hand in hand with an earnest appreciation of their practical utility.

It was the Prophet who inspired Muslims to pursue knowledge of things for both their theoretical and practical considerations. He encouraged his followers to reflect and contemplate natural phenomena pursuant to the *Qur'ān* with a view toward deepening understanding of divine power and wisdom in creation. But the Prophet also compared knowledge that had no practical benefits to a tree without fruit. He often prayed to God seeking protection from *useless knowledge*. On the basis of this tradition, Muslim scholars progressively sought to articulate ideas, concepts, and theories on the broader issue of the ethics of knowledge as activities of knowledge production and applications in the new civilization expanded and became more complex. Major issues included clarifying the meanings of beneficial and harmful knowledge in the perspective of Islamic law and determining the general criteria for each type of knowledge. Muslim preoccupation with the knowledge culture took many different forms. One was classification of knowledge, which proved to be a good way of keeping track of the state of knowledge at any given time. Classification of knowledge divided the sciences into thematic groups of well-defined disciplines, and preserved their hierarchy.

The Arab philosopher al-Kindi (c. 801–873) authored the first Muslim classification of the sciences in the ninth century. Since then many scholars have devoted considerable effort to expositions of this theme. The last significant work on the subject is the classification written by the Indian theologian Shah Waliallah of Delhi (1703–1762) in the eighteenth century. The importance and popularity of classification of the sciences was evident not only from the number of books written on the subject but also from the diverse nature of the scholarly community that produced them. Theologians, philosophers, scientists, historians, and jurists, among both Sunnis and Shiites, were represented in this unique enterprise. Classifications had been particularly

useful to the organization of educational curricula. Interestingly there appeared to be a correlation between the rate of production of classifications of knowledge and the intensity of knowledge expansion. The interest in classifications was at its height during the era when Muslims were the most productive in terms of adding new scientific disciplines to the existing body of human knowledge. After the sixteenth century when intellectual and scientific innovations began to decline in most parts of the Muslim world, work on classifications dropped sharply. The fact that hardly any work has appeared on the subject since the eighteenth century testifies to the reduced importance of the role of knowledge among Muslims in the early-twenty-first century world.

It is clear from past classifications that Muslims were concerned with the need for a balanced approach to both theoretical and practical knowledge. In addition, Muslims accord relative importance to each science in the context of human knowledge as a whole. Generally scholars use three criteria to determine the epistemic position of each science in what is traditionally called the hierarchy of knowledge. The criteria are defined in terms of the relative excellence of the objects of study, methods of study, and benefits of study. Some sciences may be viewed as more laudable than others on the basis of one or more of these criteria. The greatest science in light of the three criteria is the science of God or theology in the true sense of the word.

Islamic Culture, Science, and Technology

As clearly reflected in classifications over the centuries, Muslims do not consider science and technology to be the most important branch of knowledge, as do many people in Europe and North America who view science as the sole basis for reliable knowledge and technology as the best means to solve human problems. From the Muslim perspective, science could never take the place of metaphysics and theology in either temporal or moral importance because the latter have God and the divine realities as their *object of study* whereas science and technology focus on natural objects created by God. Additionally technology could never replace divine law (*shari'ah*) as the best provider of efficacious solutions to human individual and societal problems. Despite these beliefs, at the apex of their cultural influence, Muslims demonstrated a degree of appreciation of science and technology unseen in earlier times. Such appreciation was contextual, as dictated by the *shari'ah* itself.

Muslims distinguish between two types of obligatory knowledge. The first type is *fard 'ayn*, meaning obligatory for everyone to have as, for example, in the case of knowledge of canonical prayer. The second type is *fard kifayah*, meaning obligatory for society to possess, though the task of acquiring it may be left to certain individuals or groups. Implicit in the meaning of this category of knowledge is that without it a society would lack something that is important to its well being. *Shari'ah* confers the status of *fard kifayah* knowledge to science and technology on the basis of their immense benefits to human society. A society without a level of science and technology proportionate to its problems is considered unhealthy. Political philosophers like al-Farabi (870–950) went so far as to claim that science and technology are necessary ingredients in the pursuit of human happiness. But to Muslims, science and technology serve society best when pursued and employed in the light of ethical-legal principles of *shari'ah*.

Muslims believe both *shari'ah* and science and technology are necessary to societal salvation, and that the two should be joined within the ethical and legal framework of *shari'ah*. *Shari'ah*, which is primarily based on the teachings of the *Qur'ān* and the prophetic *hadiths*, is considered by Muslims to be the most important source of ethical values and principles to guide human actions and conduct. In the case of the Shiites, the *hadiths* extend to embrace the teachings of their supreme spiritual leaders known as *Imams*. *Shari'ah* refuses to separate between ethical and legal thought. What is legal has to be ethical, and vice versa. The religious significance of scientific and technological activities resides in the fact that the *shari'ah* divides all human actions into five categories. These categories are the obligatory (*wajib*), the meritorious or the recommended, the indifferent (*mubah*), the forbidden (*haram*), and the reprehensible (*makruh*). The main significance of these ethical categorizations for science and technology in Muslim culture is that society and the state are in broad agreement on what ought to be the priorities in scientific and technological pursuits. Obviously scientific and technological products and activities in the obligatory and meritorious categories are given the greatest priority. At the same time *shari'ah* is ever present to remind society and the state of the need to refrain from indulging in scientific and technological activities belonging to the forbidden category because *haram* would be harmful to society. *Shari'ah's* general objectives, namely to protect religion, reason, life, progeny and property, and its specific exhortations pertaining to both worship and social duties determine the types and scopes of scientific and technological activities to be encouraged or shunned. Muslim science and technology over the centuries had more or less developed along the ethical track that *shari'ah* pro-

vided. Muslims emphasized sciences like mathematics, astronomy, geography, medicine, botany, and agriculture because of their practical relevance to *shari'ah*. For the same reason, Muslims developed civil engineering and medical, agricultural, and navigational technology to new heights in the medieval period. But on the whole, harmony between science, technology and ethics was rarely shattered.

Contemporary Issues

In many early-twenty-first century Muslim societies worldwide, the traditional bond between divine law and technology has been severed. For various reasons, *shari'ah* is no longer seen as relevant to the shaping of technological pursuits. Muslims face the ethical challenge of dealing with science and technology issues that are largely not of their own making, and that pose numerous challenges to traditional Islamic ethics.

Perhaps the most serious challenge derives from military technology and biotechnology including medical technology that enables humans to, literally, determine life and death. Modern military technology in the form of weapons of mass destruction, such as nuclear and biological weapons, clearly transgresses the limits of traditional Islamic war ethics. Some Muslim states are defending the right to acquire such weapons on what they claim to be *Islamic grounds*, although it seems clear that their motive is primarily political. Many scholars in Sunni Pakistan defend that country's *Islamic bomb* on the basis of geopolitical considerations. In Shiite Iran clerics are divided on the issue of possessing nuclear weapons with President Seyed Mohamed Khatami (elected 1997) taking the stand that such weapons are contrary to Islamic ethical teachings. Muslims throughout the world are divided on the issue not along theological or jurisprudential grounds but by political, ideological perspectives. However one thing is clear: Pronuclear weapons advocates have been able to sustain their views largely by appealing to political considerations rather than to the more fundamental Islamic ethics on the conduct of war. Proponents of the supremacy of Islamic political power are likely to endorse such weapons.

Biomedical technology has impacted the social fabric of Europe and North America in an unprecedented way and has sent shock waves into the Muslim world. The range of biomedical technology currently employed in Muslim countries is still limited. But that limited use is apparently dictated far more by economics than by perceptions of ethical incompatibility with Islam. But the few richer ones as well as Muslim minorities in the west have helped Muslims to keep abreast with ethical issues arising from modern biomedical practice. In countries such as Nalaysia, Indonesia, Turkey, and Kuwait issues in biomedical ethics such as debated in the west are likewise discussed in the medical profession and the academia. The Islamic Organization for Medical Sciences based in Kuwait is exceptionally active in organizing international meetings of Muslim medical doctors to discuss implications of contemporary biomedical technology for Islamic societal values. Quite often experts in Islamic law are invited to these meetings for religious consultation. This meeting of Muslim scientific and religious minds has been successful in coming up with well-defined criteria for Muslim acceptance of biomedical technology. There is a particular concern for the impact of biomedical technology on traditional family values and institutions. The general Muslim view is that while that technology is not the cause of the breakdown in traditional family and marriage institutions, it nonetheless has created new possibilities that allow the viability of alternative lifestyles. Life-support machines that call into question the traditional definition of death, technology that uncovers information about babies still in the womb, sperm banks, and artificial insemination are major examples of modern-day scientific and technological innovations that have attracted the attention of Muslim ethicists. Debates on those issues had hardly settled when the more serious ethical issue of cloning emerged.

On some issues such as the technology associated with prenatal information and artificial insemination the Muslim debate has been fairly brief as religious experts and political authorities quickly find satisfactory answers to initial Muslim grievances on the possible misuse of the technology. On other issues such as the life-supporting machines the debate rages on. The majority view is that as traditionally held the community of believers should help to facilitate "easy and peaceful" death of the dying and not to prolong agony and suffering such as through the use of the life-supporting machine. The traditional belief is that death, a passage to afterlife, is itself a suffering. To be in a state of neither life nor death is viewed as being in a state of suffering. The traditional way of facilitating peaceful death is recitation of verses from the Qur'ān. The minority view is that use of the machine is permissible because religion also teaches the saving of every human life through every possible means. While debates on such issues rage on the more serious ethical issue of cloning emerged. Muslims are unanimous in rejecting human cloning. But they are deeply divided on the use of stem cells for research. The overwhelming majority oppose using human embryonic stem cells for research. But many Muslim groups consider use of adult stem cells as religiously permissible.

The following patterns emerge in the still-fluid Muslim response to bioethical issues. First Muslims are increasingly turning to Islam's inner resources as found in the Qur'ān, prophetic traditions, and traditional ethics in looking for answers to dilemmas posed by new technologies. Second Muslims are evaluating the potential value of new technologies while remaining committed to defending shari'ah-sanctioned social institutions. They are likely to adopt new technologies within the constraints of shari'ah as they have already done in many cases. For example, Muslim jurists have permitted artificial insemination as long as the couple is legally married according to Islamic law and the semen is that of the husband. Third Muslims are questioning whether humanity needs to have better and more encompassing ethical ideas than just those that appeal to research interests or search for medical cures in order to justify controversial, new scientific research and biomedical technology. As Muslims become more immersed in technological matters they more often find the need to consult the ethics of shari'ah.

A deep interest in ethical issues in science and technology presupposes a certain level of scientific and technological progress. As things are, most Muslim countries have hardly attained that level of progress. Many factors ranging from the religious and the political have contributed to the present Muslim lack of progress in science and technology. One of these is the neglect in Muslim education of that dimension of Islamic teachings favorable to scientific and technological progress. The current lack of interest in the ethics of science and technology in Muslim societies is thus understandable. But this lack of interest does not at all reflect the intellectual richness that characterizes the traditional treasury of Islamic ethical wisdom. Students of the shari'ah and the ethical dimension of Islamic science and technology when it was at its best are quite aware that Islamic ethical thought remains largely relevant to many of the contemporary ethical issues. There is nothing more glaring than the example of environmental ethics to illustrate the wide discrepancy between Islam's actual teachings and the current index of Muslim environmental awareness. The Qur'ān is replete with verses of environmental significance. Traditional Islamic architecture and urban planning has been one of the best Muslim attempts to embody the ideals of Islamic environmentalism as taught by the Qur'ān. Yet in the early twenty-first century Muslim countries are plagued with environmental pollution and urban degradation.

A promising Muslim country is Malaysia. It is one of the most advanced Muslim countries in science and technology. While seeking to reap the benefits of modern western science and technology Malaysia has also shown much interest in Islamic values as a contributing factor to scientific and technological progress in the twenty-first century. There is a visible attempt in the country to create a new synthesis of tradition and modernity not only in science and technology but also in other fields of civilization. The Malaysian government has created several institutions with that goal in mind. The most well known is perhaps the Malaysian Institute of Islamic Understanding, which has organized many programs on ethical issues in science and technology. Malaysia is quite advanced in genetic engineering. For a country noted for its Islamic fervor it is rather interesting that Islam does not appear to be a hindrance to the progress of genetic engineering. The new Badawi administration (succeeding that of Mahathir in 2003) has unveiled an agricultural policy that places great emphasis on genetic engineering and biotechnology. Interestingly, Badawi views this agricultural policy as an integral part of his Islam policy now known as civilizational Islam.

The case of Malaysia is important. It is not Arab but predominantly Malay like its neighbor Indonesia, which is the largest Muslim nation on earth. And yet in the early 2000s Malaysia appears to be more vocal than all the Arab states in championing modern Islamic issues. And many Muslims do make a careful distinction between *Islamic* and *Arabic'* while acknowledging the Arabic coloring of Islam by virtue of the Muslim belief that God has revealed the *Qur'ān* in Arabic. Islamic issues as distinct from Arabic are those that concern all Muslims transcending ethnic barriers. The Islamic organization in Kuwait may be led by Arabs but the ethical issues they discuss are Islamic issues of importance to all Muslims. Similarly the Malaysian institute of Islamic understanding is led by Malays who are non-Arabs but its programs on ethics in science and technology have the participation of Muslims from various parts of the world including Arabs.

Muslim attitudes toward modern science and technology are far more positive in the early twenty-first century than in the colonial period when they generally equated modernization with Westernization. From Morocco in the western wing of the Muslim world to Indonesia in its eastern most part colonial attempts at modernization such as in education, agriculture, and business often found stiff resistance from the Muslim populace. Such attitudes became the legacy of post-independence leaders in the Muslim world. But in the last several decades Islam has also emerged as an important source of positive influence on the Muslim thinking on science and technology. Many Muslims now see the possibility of merging the best of modern scientific and technological culture with the best of Islamic intellectual and cultural tradition.

OSMAN BAKAR

SEE ALSO *Christian Perspectives: Historical Traditions; Jewish Perspectives; Scientific Revolution.*

BIBLIOGRAPHY

Anees, M. A. (1989). *Islam and Biological Futures: Ethics, Gender, and Technology.* London: Mansell. One of the few Muslim discussions of biomedical ethical issues in an Islamic context. It gives a good treatment of the implications of advances in biomedical technology for Muslim beliefs and practices including approaches to issues of gender.

Anees, M. A. (1994). "Human Clones and God's Trust: An Islamic View." *New Perspectives Quarterly* 2: 1. A pioneering treatment on the subject of human clones by a Muslim writer. It has also been published as part of a chapter entitled "God and the Clone" in the book *The Human Cloning Debate,* ed. Glenn McGee and Arthur Caplan, Berkeley, CA: Berkeley Hills Books (2004).

Bakar, Osman. (1986). "Islam and Bioethics." *Greek Orthodox Theological Review* 31(2): 157–179. A detailed discussion of the relevance of Islamic law (*shari'ah*) to Muslim understanding of bioethical issues.

Bakar, Osman. (1997). *History and Philosophy of Islamic Science.* Cambridge, UK: Islamic Texts Society. A good introduction to the relations between science and religion according to Islamic perspectives.

Bakar, Osman. (1998). *Classification of Knowledge in Islam.* Cambridge, England: Islamic Texts Society. Perhaps the most comprehensive study of Muslim classifications of the sciences ever taken. It is a useful source of information on the place of science and technology in Muslim intellectual culture in different periods of Islamic history.

Hill, Donald, R. (1996). *Islamic Science and Engineering.* Chicago: Kazi Publications. Provides an insightful account of various branches of medieval Islamic engineering and its contribution to the development of Western technology.

Hill, Donald R, and Ahmed Yusef al-Hasan. (1992). *Islamic Technology: An Illustrated Study.* Lanham, MD: Cambridge University Press. Impressive illustrated study of medieval Islamic technology providing readers with invaluable manuscripts and artifacts illustrations.

King, David. (1995). *Islamic Astronomical Instruments.* London: Variorum Reprints. Collection of papers the author has published in various journals on astronomical instruments originating or widely used in medieval Islam such as the astrolabe, quadrant and sundial.

McClellan, James E., and Harold Dorn. (1999). *Science and Technology in World History: An Introduction*. Baltimore, MD: Johns Hopkins University Press. Good introduction for lay readers and undergraduate students to the history of science and technology in different civilizations in the East and in the West.

Nasr, Seyyed Hossein. (1976). *Islam Science: An Illustrated Study*. London: World of Islam Festival Publishing. The first illustrated study of the whole of Islamic science ever undertaken and thus provides a good companion to *Science and Civilization in Islam* (see below). The book combines an account of the morphology and brief history of the various sciences with illustrations drawn from sources spread throughout the Islamic world.

Nasr, Seyyed Hossein. (1987). *Science and Civilization in Islam*. Cambridge, England: Islamic Texts Society. Excellent overview of the theory and practice of Islamic science. It provides a useful context for understanding Muslim appreciation of science when Islam was at its best.

Sayili, Aydin. (1981). *The Observatory in Islam*. New York: Arno Press. One of the best historical studies of an important scientific institution in Islam, namely the observatory, which initiated a clear inter-dependence between scientific research and technological innovations in the field of astronomy.

ITALIAN PERSPECTIVES

• • •

The Italian cultural tradition has historically belittled the cultural, ethical, and social roles of science and technology. This is surprising given that an Italian, Galileo Galilei (1564–1642), was one of the founders of modern science, and that his *Dialogues Concerning Two New Sciences* (1638) praised the cultural role of technology and the philosophical importance of science. In the last half of the twentieth century, Italian appreciation of Galileo's theories increased, especially in relation to ethical discussions of science and technology, along with recognition of the philosophical importance due to technics and scientific thought.

Historical Background

Italian tradition was biased by the circumstances of Galileo's 1633 trial by the Holy Office of the Catholic Church. Despite his defense of science and technology, the trial ended with the Pisan scientist recanting his beliefs and being sentenced to house arrest for life. This condemnation long hindered the free development of scientific research and, together with the Counter-Reformation climate and Italy's difficult economic and political evolution, effectively sidelined the develop-

ment of science and technology. Even though a few thinkers continued to maintain the importance of scientific knowledge and technological innovation, as a whole Italian intellectual culture became centered around literary, artistic, historical, and political activities.

This attitude was reinforced, in the first half of the twentieth century, by the hegemony of the neo-Hegelian idealism of Benedetto Croce (1866–1952) and Giovanni Gentile (1875–1944), who saw science as possessing no philosophical significance. Croce contended that science produces only *pseudo-concepts* of practical utility. Such concepts were subordinate to truth, which was, in his opinion, the exclusive province of the *sciences of the Spirit* (namely art, literature, philosophy, and history), of which philosophy was the crown jewel. *True* knowledge rises above science, which is irremediably tied to a practical horizon. Giovanni Gentile similarly devalued science, which he saw as oscillating between art and religion, unable to unify the two in a higher synthesis such as that achieved by philosophy. For Gentile, science combined the defects of art, objectivity and universality, with those of religion, subjectivity and rationality, and was thus the fruit of multiple errors and devoid of any autonomous historical development.

This negation of science by Croce and Gentile proved widely influential, both because it was set in a traditionally antiscientific culture and because these two neo-idealists played leading roles in the opposing political movements of liberalism and fascism. Their thinking exerted an almost dictatorial authority and aggravated the general cultural devaluation of science and technology.

Post World War II

Following World War II, the social and economic crisis in Italy contributed to the decline of the theories of Croce and Gentile. A new generation of intellectuals rejected neo-idealism, attacking its ambiguous cultural categories and sterile antinaturalistic, antiscientific polemics. In this climate, a dialog emerged among proponents of various ideologies including neopositivist philosophy, developed in Vienna by Moritz Schlich and Rudolf Carnap, the early ideas of Ludwig Wittgenstein, and the mathematical logic of Bertrand Russell. This led to the formation of a neo-enlightenment movement (Dal Pra and Minazzi 1992), with the participation most notably of Ludovico Geymonat (1908–1991) and Giulio Preti (1911–1972). Geymonat and Preti—through numerous studies, books, translations, and reviews—

critically introduced neopositive issues into Italian thinking, arguing both the cultural value of science and the importance of technology.

Geymonat, beginning with his *Studies for a New Rationalism* (1945), delineated a neo-enlightenment philosophy centered in the philosophy of science, logic, and the history of science and technology, arguing for replacement of *static* with *dynamic* studies of scientific theories. Geymonat became, in 1956, the first Italian to hold a chair in philosophy of science (at the University of Milan) and, in 1974, to win the *Médaille Koyré* for history of science, awarded by the Académie Internationale d'Histoire des Sciences in Paris. He also was mentor to a group of young scholars working in these fields. Geymonat's own work culminated in the publication of the highly regarded, seven-volume *History of Philosophical and Scientific Thought* (1970–1976) and *Science and Realism* (1977). In these works, he developed a materialistic-dialectic perspective and placed the fundamental role of the *scientific-technical legacy* at the heart of critical comprehension of knowledge and of the historical development of society. Preti, in a series of books including *Idealism and Realism* (1943), *The History of Scientific Thought* (1957), and especially *Praxis and Empirism* (1957), related neopositivist themes to both the pragmatism of John Dewey and the philosophy of the young Karl Marx.

Parallel with work conducted by the neo-enlightenment thinkers was that of Valerio Tonini (1901–1992), a Catholic engineer and philosophy of science scholar. After working in the field of engineering for many years, Tonini turned to information theory, epistemology, the sociology of work, and bioethics. A member of the *Académie Internationale de Philosophie des Sciences* (International Academy of Philosophy of Science), in 1950 Tonini founded the *Società Italiana di Logica e di Filosofia della Scienza* (Italian Society of Logic and Philosophy of Science) and, in 1955, started a review of human sciences and philosophy of science called *La Nuova Critica* (The New Critic), which he edited until his death. Tonini also raised important issues regarding the philosophy of technology, to which he devoted a book titled *Structures of Technology* (1968). In the ambit of what was described as his *long march to scientific realism*, Tonini defined technology as the *science of praxis*. He argued that technology implemented processes that modify the environment and, as a *new science*, was capable of achieving semantic precision, synthetic rigor, and verification of its theories. It created a direct link to communication theory, information theory, cybernetics, control theory, process theory, and systems theory.

Contemporary Contributions

In the last quarter of the twentieth century, Italian scholars became particularly interested in science, technology, and ethics. Discussion of biomedical ethics, not only from a Catholic perspective, broadened, with reflections on nuclear weapons and environmental ethics. In 2001, the Council of Genetic Rights was founded in Rome by Mario Capanna.

In the early-twenty-first century, two of Italy's most influential thinkers in the area of science, technology, and ethics are Evandro Agazzi and Luciano Floridi. Agazzi especially has made important contributions to the critical study of these issues. Born in Bergamo in 1934, Agazzi studied philosophy at the Catholic University in Milan, and continued his education, in physics and philosophy, at Marburg, Oxford, and Münster. Agazzi was part of the logical-mathematical team founded by Geymonat in the 1960s. He thereafter became a professor at universities in both Genoa and Fribourg (Switzerland), and published a number of studies on mathematical logic, including *Introduction to Axiomatic Problems* (1961), *Symbolic Logic* (1964), and *Themes and Problems of the Philosophy of Physics* (1969), in which he outlined an original objectivist and realistic epistemological perspective.

Agazzi's positive philosophical revaluation of technology is rooted in the antitheoreticism with which he reacted to the epistemology of the neopositivists and Karl Popper (cf., his philosophical dialogue with Geymonat in *Philosophy, Science, and Truth* [1989]). He developed his own interpretation of the hermeneutic dimension of science, embodied in *Wisdom the Technique* (1986) and most influentially in *Right, Wrong and Science: The Ethical Dimensions of the Techno-Scientific Enterprise* (1992).

The merits of Agazzi's analysis rest with his arguments regarding the ethical dimensions of the scientific-technological undertaking. Agazzi proposed to distinguish between technics (know-how that works without an awareness of its purpose), technology (which he used to denote, by contrast, effective action that has an awareness of its purpose), and science (knowledge capable of explaining empirical facts by adducing reasons that explain *why* reality is configured in a given way). Technology represents the result of the development of science, and Agazzi stresses the subtlety of the interconnections between science, technics, and technology, analyzing the scientific ideology, technological system, and complex encounter between ethics, norms, and values within human action. By defending a dynamic model of knowledge, Agazzi opts for a systemic approach in which the regulation of research is configured as a projection of responsibility. From this perspective, his science and technic studies are closely entwined with those devoted to bioethics, fostering a debate between Catholic and secular thinking that has contributed to the development of a freer and more responsible society.

Luciano Floridi, a professor of philosophy (at the University of Bari in 2004), has done influential work on the relationship between philosophy and computing from an ethical perspective. For Floridi Information Ethics represents the philosophical foundational counterpart of Computer Ethics which is thought as a nonstandard, object-oriented and ontocentric theory.

FABIO MINAZZI

SEE ALSO *Axiology; Galilei, Galileo; Pareto, Vilfredo.*

BIBLIOGRAPHY

Agamben, Giorgio. (1998). *Homo Sacer: Sovereign Power and Bare Life*, tr. Daniel Heller-Roazen. Stanford, CA: Stanford University Press. Original Italian publication, 1995.

Agazzi, Evandro. (1961). *Introduzione ai problemi dell'assiomatica* [Introduction to the problem of axiomatics]. Milan: Vita e Pensiero.

Agazzi, Evandro. (1964). *La logica simbolica* [The symbolic logic]. Brescia: La Scuola.

Agazzi, Evandro. (1969). *Temi e problemi di filosofia della fisica* [Themes and problems of the philosophy of physics]. Milan: Manfredi.

Agazzi, Evandro. (1986). *Weisheit im Technischen* [Wisdom in the technique]. Luszern: Verlag Hans Erni-Stiftung.

Agazzi, Evandro. (1992). *Il bene, il male e la scienza* Milan: Rusconi. English Translation: *Right, Wrong and Science: The Ethical Dimension of the Techno-Scientific Enterprise*, ed. Craig Dilworth. Amsterdam and New York: Rodopi (2004).

Agazzi, Evandro, Fabio Minazzi, and Ludovico Geymonat. (1989). *Filosofia, scienza e verità* [Philosophy, science, and truth]. Milan: Rusconi.

Dal Pra, Mario, and Fabio Minazzi. (1992). *Ragione e storia: Mezzo secolo di filosofia italiana* [Reason and history: a half-century of Italian philosophy]. Milan: Rusconi.

Floridi, Luciano. (1999). *Philosophy and Computing. An Introduction*. London, New York: Routledge.

Geymonat, Ludovico. (1970–1976). *Storia del pensiero filosofico e scientifico* [History of philosophical and scientific thought], 7 vols. Milan: Garzanti.

Geymonat, Ludovico. (1977). *Scienza e realismo* [Science and realism]. Milan: Feltrinelli.

Preti, Giulio. (1943). *Idealismo e positivismo* [Idealism and positivism]. Milan: V. Bompiani

Preti, Giulio. (1957). *Storia del pensiero scientifico* [History of scientific thought]. Milan: Mondadori.

Preti, Giulio. (1975). *Praxis ed empirismo* [Praxis and empiricism]. Turin: Einaudi.

Tonini, Valerio. (1968). *Strutture della tecnologia* [Structures of technology]. Rome: Armando Editore.

Tonini, Valerio, and Minazzi, Fabio. (1989). *La realtà della natura e la storia dell'uomo* [The reality of nature and the history of man]. Milan: Franco Angeli.

J

JAPANESE PERSPECTIVES

• • •

In the early years of the twenty-first century ethical concerns related to scientific and technological developments are receiving a great deal of attention in Japan. A focus on globalization has resulted in a renewed concern with the impact of traditional values on technology, as well as in the adaptation of some western perspectives on ethical issues. Currently evolving discussions, in areas ranging from bioethics to nuclear power, make an excellent case study of how a society's ethical considerations both arise out of a given historical context and interact with a wider global context.

Japan is an ancient nation of 127 million people (2003) living mostly on four mountainous islands in the Northern Pacific off the coast of Asia. Records of inhabitance date back to the early centuries of the Common Era. After a long history of isolation followed by tentative openings, during the period of the Tokugawa Shogunate (1603–1868), Japan almost totally closed itself off from the outside world and consequently also from the influences of Western scientific and technological developments. It even successfully abolished the production and use of firearms, thus becoming one of the few examples where a more advanced technology, after having been widely utilized, was suppressed for an extended period of time. Toward the end of the Shogunate, however, it became clear that Japan would have to adopt Western technology in order to survive as an independent state, as was made evident in 1853 by the arrival of Commodore Matthew C. Perry in his black ships with their superior firepower, demanding an opening of trade. The subsequent Meiji Restoration of the emperor in 1868 accelerated a period of change in Japan, during which Western science and technology were rapidly integrated into an agrarian social system in flux. The slogan for the process of adoption was *wakon yōsai* or *Japanese soul with Western technology*, indicating an unwillingness to identify modernization with a transformation of the national cultural characteristics.

Historical Evolution of Ethical Issues

An initial movement to bring in experts from throughout the world and send students abroad, while adapting foreign learning to the Japanese cultural context and improving on it, set the pattern for much of the twentieth century. Japan became known as a society that emphasized incremental improvements on revolutionary innovations developed elsewhere. This reflected a societal objective of catching up to European and North American powers in economic and military strength, where the national government assumed the primary leadership role in building up the infrastructure necessary for scientific and technological growth. In this process, Japan became the first country to establish a college of engineering within a university system. As early as the 1870s, the Imperial University (later the University of Tokyo) established a Faculty of Engineering with its own service departments in the sciences. Ever since, the university system has produced many more engineering graduates than ones in the sciences. Under this system relatively less attention was given to basic or pure scientific research; the dominant focus was on applied science and technology for industrial development. As a result, the demarcation between science and technology has not been as evident in Japan as in the Western tradition. Neither has been the close cooperation between corporations and universities typical in the United States.

To understand how the historical evolution of science and engineering is connected to ethics in Japan, it is necessary to gain some insight into Japanese social values, which are influenced both by the general Asian traditions of Buddhism, Daoism, and Confucianism, and by the native Shintō religious perspective. As a unified value system, these social values have resulted in an emphasis on the group over the individual; a focus on family and clan, with priority being given to loyalty and hierarchy; sacredness associated with the elements of nature, and an integrated perspective on body and spirit. In addition, there are still religious connotations associated with the emperor and the land of Japan itself as having divine origins. All of these values, in turn, have influenced Japanese conceptions of ethics, which in general are dominated by relativistic and situational group norms.

During the late-nineteenth and early-twentieth centuries, a large percentage of engineers and scientists came from the samurai class. Thus, as Nitobe Inazo pointed out in his influential *Bushido: The Soul of Japan* (1900), the heritage of science and engineering ethics in Meiji Japan was associated with Japanese ideals of *chivalry*. However, perhaps as a result of Japan's ethnic and linguistic homogeneity, a written code of ethics for engineers and scientists was not introduced until 1938, when the Japan Society of Civil Engineers adopted the first one. This code, based upon its U.S. counterpart, was a pioneering work largely authored by Aoyama Akira (1878–1963), a leading engineer of the time, who had worked on the construction of the Panama Canal and had a well-developed international perspective, reflected in his humanitarian philosophy and his Christian beliefs. As Imperial Japan was hastening toward World War II, however, his work must be considered to have been well ahead of its time.

Ethical Concerns in the Postwar Recovery Period

In the postwar recovery period, during which first priority was given to materialistic goals, Japan experienced a tremendous turmoil in thought as Western idealism and democracy rapidly replaced prewar ultranationalistic values and the associated ethical framework. These were denied in large part because they were identified in the minds of the people with the political stand of Imperial Japan. In particular, the memories of Hiroshima and Nagasaki had a critical impact on how Japanese scientists, especially physicists, viewed their role in society. The Science Council of Japan (SCJ) declared at its first assembly after its establishment in 1949 that the aim of scientific research should be to contribute to the welfare of humankind and to world peace. When

the government officially made budgetary arrangements for utilizing nuclear power in 1954, the SCJ demanded that research on and the use of nuclear energy be conducted on the principles of "openness, democracy, and independence." The first Japanese Nobel Laureate, the theoretical physicist Yukawa Hideki (1907–1981), was one of the signatories of the Russell-Einstein Manifesto (1955). After recommending to the government in 1962 and 1976 that it establish the Basic Act on Scientific Research, the SCJ proposed a Charter for Scientific Researchers, reemphasizing basic values such as human welfare, world peace, freedom of scientific research, safety, and internationalism. SCJ efforts to emphasize the social responsibility of scientists were important historically; however, because most members of the organization are senior scholars and researchers, its statements appear to have had limited influence on young scientists and engineers with career ambitions.

Changes in engineering ethics had a similarly limited impact. Modeled after the American system of consulting and professional engineers, and the British system of chartered engineers, the Japanese version of engineering licensing was legally institutionalized in 1957, and the Institution of Professional Engineers, Japan (IPEJ), formed in 1951. IPEJ adopted a code of ethics in 1961. However because of the limited number of licensed engineers (approximately 40,000 since 1958) and the general lack of interest in engineering ethics, this code was not widely promoted. In addition, the concept of *engineering as a profession* is unequivocally absent in Japan, most likely because the development of engineering was dominated by the state and industry, rather than by public forces. The Japanese employment system has also encouraged engineers to develop identities with their company rather than as part of a professional association.

Aside from such attempts to formalize ethical concerns, the postwar period could well be characterized as an *ethical vacuum*, in which traditional values dominated, but without an underlying ethical framework. The situation in bioethics perhaps best illustrates the difference between traditional and Western perspectives. The medical establishment is quite paternalistic in its approach. Informed consent has been recognized, but is not well institutionalized, with physicians sometimes using patients in experimental procedures without their knowledge. Truth-telling and patient autonomy are only slowly being recognized as significant values. Traditionally concealing the truth from patients, and more rarely from their families, has been seen as protecting the health of even dying patients. The assumption has been that physicians are authority figures, so that explanations to patients are not necessary. Only recently, for example, have physicians been held to account for practicing involuntary euthanasia.

On the societal level the impact of Japanese values has also been influential in the medical field. Despite legalization, religious and social norms have prevented any significant use of organ transplants. Conceptions of human nature have resulted in a hesitancy to adopt Western standards of brain death, further inhibiting both transplants and a widespread *death with dignity* movement. At the same time, abortion is commonly practiced in Japan without social stigma, both because the woman and fetus are considered to be one entity and because contraceptive pills are not generally available.

However any assessment of the state of scientific and engineering ethics in Japan must recognize that the society is entering a period of structural change, which has already begun to influence discussion about a variety of ethical issues. During the entire postwar period developments in technology were considered issues of national security and survival. National interests took priority over popular consumer desires. In order to spur economic development, the government took a central role in technological planning activities and in guiding research. Major corporations adopted systems of lifetime employment and seniority-based pay to foster workforce loyalty. Japan quickly became an economic juggernaut based on the total commitment of its workers and on the innovative use of management and production strategies such as quality circles and just-in-time supply procurement.

A New Emphasis on Ethics for the Twenty-First Century

Then came the decade-long recession of the 1990s, resulting in fundamental changes in corporate life and public attitudes. Japanese increasingly accepted the need for more global approaches, a move away from governmental direction, and more attention being given to the public. The impacts of these changes are evident in a variety of new discussions of ethical issues. In the area of bioethics, for example, there is a burgeoning patient rights movement and an increased emphasis on physician accountability.

Many of the cutting edge technological innovations in Japan have come from corporations rather than out of the university system. Consequently any changes in the corporate environment tend to influence discussions of research ethics. For example, notions of intellectual property are undergoing testing. Traditionally researchers received little monetary reward. However as Japan is moving toward more mobility in its professional class,

with the weakening of lifetime employment and seniority-based pay, researchers are increasingly seeking a greater ownership stake in their work. University researchers are likewise being granted greater independence with a shift away from government direction of the university system as a whole. University science departments, operating on the chair system, in the past have been awarded a set amount of research funding rather than operating on a competitive grant basis. With change to a more merit based system, it can be expected that research priorities will be different and that increased coordination between university and corporate researchers will be established, in turn resulting in new discussions about ethical issues.

Another area that is undergoing change is concern about the natural environment. Although respect for nature is a dominant factor in the Japanese value system, during the period of economic expansion environmental preservation was considered secondary to economic growth. Since the late-twentieth century, especially after the signing of the Kyoto protocol in Japan, a renewed concern with the environment has been in evidence. Japanese are moving away from an ethics that emphasizes disposal to a recycling culture. There is also increased recognition of the global nature of environmental issues such as the heavy use of wood products in Japan and the lack of suitable disposal opportunities for refuse.

The 1990s was also a decade of awakening for engineering ethics. Various incidents and accidents having to do with engineering practice occurred, including a major sodium leak at the Monju fast-breeder reactor in 1995, the sarin gas attack in the Tokyo subway system that same year (by members of a religious cult who were educated as engineers and scientists), and the disastrous nuclear criticality accident in Tokaimura in 1999. These prompted increased interest in engineering ethics and major engineering societies established codes of ethics one after the other, starting with the Information Processing Society of Japan in 1996. The Japan Society of Civil Engineers revised its code honoring the spirit of Aoyama's contribution in 1999. By 2003 most of the major engineering societies had adopted codes, which in general include fundamental values such as giving first priority to the safety of the public, in common with their North American counterparts.

The process of globalization has had great impact on engineering ethics. In 1999 the Japan Accreditation Board for Engineering Education (JABEE) was established to harmonize engineering education with international standards, to enable participation in mutual recognition of engineering qualifications. This required ethics education as one of its components and set in motion a flurry of activity, ranging from short courses on the subject, to conferences, to modification of engineering curricula to include required courses on engineering ethics. All of this activity is financially well supported by the government, so that large numbers of people are involved in what is essentially a new area of inquiry in Japan. In this work there is a twofold emphasis on application to specific ethical problems and on theoretical philosophical analysis. Given the scientific-technological heritage of Japan, the emphasis in the discussions tends to be broader than it has been in the United States, leaning more toward a science, technology, and society (STS) perspective than one that emphasizes strictly professional responsibilities. This is in part because Japan has an existing tradition of STS studies and lacks a tradition of professional identification. The JABEE accreditation criteria therefore require the study of engineering ethics conceptualized as "understanding of the effects and impact of technology on society and nature, and of engineers' social responsibilities," as opposed to the U.S. standards that emphasize "professional and ethical responsibility" and put these in a separate category from the need to "understand the impact of engineering solutions in a global and societal context."

Given the attention to engineering ethics present in Japan, it can be expected that the discussion will increasingly impact the overall consideration of ethical concerns in Japanese society and its scientific community. The population as a whole appears to be seeking new standards of accountability in many areas of life, including business, government, and universities, and in relation to the environment. These discussions will be influenced by both local traditions and a more global outlook.

HEINZ C. LUEGENBIEHL
JUN FUDANO

SEE ALSO Buddhist Perspectives; Engineering Ethics.

BIBLIOGRAPHY

Fudano, Jun; Heinz C. Luegenbiehl; et al. (2004). *Gijyutusha Rinri* [Introduction to engineering ethics]. Tokyo: Society for the Promotion of the University of the Air. Textbook written for the first Japanese television course on engineering ethics, which is offered by the National Broadcasting University. Codes of ethics are given special attention.

Low, Morris Fraser; Shigeru Nakayama; and Hitoshi Yoshioka. (1999). *Science, Technology and Society in Contemporary Japan.* Cambridge and New York: Cambridge University Press. Discusses a variety of technological issues from the perspective of various stakeholders.

Morishima Michio. (1982). *Why Has Japan 'Succeeded'?: Western Technology and the Japanese Ethos.* Cambridge, UK, and New York: Cambridge University Press. Relates the Japanese value system to contemporary emphases in technological development.

Morris-Suzuki, Tessa. (1994). *The Technological Transformation of Japan: From the Seventeenth to the Twenty-First Century.* Cambridge, UK, and New York: Cambridge University Press, 1994. Provides an introduction to the evolution of technology in the Japanese societal context.

Nitobe, Inazo. (1900). *Bushido, the Soul of Japan.* Philadelphia: Leeds and Biddle Company. The major exposition of the code of ethics of the Samurai class. It was the foundation of much of the ethical perspective of scientists and engineers, most of whom came from that class.

JASPERS, KARL

• • •

Psychiatrist and philosopher Karl Jaspers (1883–1969), who was born in Oldenburg, Germany on February 23, became one of the most important representatives of existential philosophy. He died in Basel, Switzerland on February 26.

Jaspers developed an existential analysis of technology in two distinct phases. His early conception of technology, which he put forth in *Man in the Modern Age* (1931), revolved around the transformation of human society into a mass, mechanized culture. His initial assessment of this transformation was negative. He wrote of the *demonism of technology*, describing technology as an independent power that had been summoned into existence by human beings but that now has turned against them. According to Jaspers, technology transforms human society into a mass culture, alienating human beings from themselves and from the world around them.

Jaspers considered mass-rule a byproduct of the close interaction between technological development and population growth, which results in a vast number of human beings whose existence becomes utterly dependent on technology. This dependency requires a quite specific social and cultural formation. Besides a mechanization of labor, society needs a smoothly operating bureaucratic organization in order to keep functioning. Society becomes a machine itself, described by Jaspers as *The Apparatus*.

This apparatus of workers, machines, and bureaucracy increasingly determines how human beings carry out their daily lives. It has two different but related effects. First its system of mass production fosters a homogenization of the material environment in which human beings live. No attachment is possible to mass produced objects, which only exist as exemplars of a general form and are primarily present in terms of their functionality. Second the apparatus approaches human beings not as unique individuals, but as fulfillers of functions who are in principle interchangeable. Both effects of the technological transformation of society impede human beings from being present as authentic existences, and from living their lives authentically and in existential proximity to the world around them. From an existential point of view, therefore, technology deprives human beings of their highest possibilities.

After World War II, Jaspers's analysis of technology changed course. Rather than viewing technology as a threat to authentic human existence, in *The Origin and Goal of History* (1949) and *The Atom Bomb and the Future of Man* (1958), Jaspers saw technology as what was *at stake* in it. He concluded that technology is ultimately neutral or no more than a means for human goals, because it is incapable of generating its own goals. This neutrality makes human beings responsible for what they make of technology: Technology requires human guidance.

Jaspers no longer considered demonism to be an intrinsic property of technology, but a result of the fact that humans have handled it as an end in itself, rather than a means for human ends. To overcome this demonism, therefore, humanity needs to ask itself the question of what it wants to do with technology. The task for human beings is to reassert sovereignty over technology.

This sovereignty, according to Jaspers, requires a reversal in thinking in which technological thought, or *intellect* (*Verstand*), is transformed into an existential way of thinking that he calls *reason* (*Vernunft*), and in which individuals are present authentically as themselves. Only this way of thinking will allow humans to experience the situation in which they find themselves as *their* situation, for which they are *responsible*. Reason can turn the contemporary situation into a task, and allow humanity to seek new goals for applying technology.

Jaspers's later perspective allowed him to discern not only a threatening side of technology but also ways in which it opened up new existential possibilities. These include new proximity to reality, by understanding the laws of nature lying behind the functioning of technology; recognition of the beauty of technological constructs; and making use of the possibilities opened up by media and transportation technologies, which allow humans to experience the Earth as one whole for which they can feel responsible.

Karl Jaspers, 1883–1969. The German philosopher wrote important works on psychopathology, systematic philosophy, and historical interpretation. (*David E. Scherman/Getty Images.*)

Jaspers's analysis is important as an existential philosophy of technology. Yet in light of later understandings, his separation of technology and society—with autonomous technology dominating society or a sovereign society guiding technology—has become problematic. An existential analysis of technology should take as a starting point the interrelationship of human existence and technology, and investigate how technologies mediate the ways in which human beings realize their existence, by impeding specific aspects of human existence and creating space for new ones.

PETER-PAUL VERBEEK

SEE ALSO *Existentialism; German Perspectives.*

BIBLIOGRAPHY

Jaspers, Karl. (1931). *Die geistige Situation der Zeit.* Berlin: Göschen (Band 1000). Published in English as *Man in the Modern Age,* trans. Eden and Cedar Paul. (1973). New York: AMS Press.

Jaspers, Karl. (1932). *Philosophie.* Berlin: Springer (3 volumes). Published in English as *Philosophy,* trans. E. B. Ashton (1969–1971). Chicago: University of Chicago Press (3 volumes).

Thomas Jefferson, 1743–1826. The American philosopher and statesman was the third president of the United States. A man of broad interests and activity, he exerted an immense influence on the political and intellectual life of the new nation. (*The Library of Congress.*)

Jaspers, Karl. (1949). *Vom Ursprung und Ziel der Geschichte*. Zurich: Artemis. Published in English as *The Origin and Goal of History*, trans. Michael Bullock (1953). London: Routledge; New Haven, Conn.: Yale University Press.

Jaspers, Karl. (1958). *Die Atombombe und die Zukunft des Menschen*. Munich: Piper. Published in English as *The Future of Mankind*; also as *The Atom Bomb and the Future of Man*, trans. E. B. Ashton. (1961). Chicago: University of Chicago Press.

Jaspers, Karl. (1994). *Basic Philosophical Writings*, 2nd edition. Atlantic Highlands, NJ: Humanities Press.

JEFFERSON, THOMAS

• • •

The early American political philosopher and politician Thomas Jefferson (1743–1826), was born in Albemarle Country, Virginia on April 13. By the time of his death at his home of Monticello just outside Charlottesville, Virginia on July 4, Jefferson considered his three greatest achievements to be writing the Declaration of Independence, writing the Statute of Virginia for Religious Freedom, and founding the University of Virginia. It is nevertheless also the case that Jefferson's views on science, politics, and ethics present a uniquely American perspective on technological progress as flowing from individual liberty, economic freedom, and personal Christian morality.

This "American System" of viewing advances in scientific knowledge as part of political freedom and moral development, remains a distinctive approach to the social issues of economic development, education, crime, religious freedom, and personal happiness. Its confidence in technological and scientific progress tempered by religious and ethical considerations is the basis for American concerns with problems of medical/genetic ethics, environmentalism versus economic development, and private rights versus social responsibility. Its enthusiasm for the free individual and for relatively unrestrained international expansion of these American values has, at times, caused it to be accused of imperialism, hegemony, and disregard for traditional nontechnological and more hierarchical societies (including Islamic, African, and Asian societies) and for socialist economics. Much of contemporary world conflict, such as terrorism, is to some extent an extension of the debate over this "Jeffersonian" worldview of progress, knowledge, religious liberty, democracy, and individual freedom.

Jefferson as Scientist and Inventor

Jefferson's scientific and technological interests were wide ranging. He investigated every branch of science, from botany to biology, meteorology, archaeology, astronomy, chemistry, geology, mathematics, paleontology, and ethnology. He designed the curriculum at the new University of Virginia (1819) to revolve around a core of natural philosophy (science), including physics, engineering, and mineralogy, when most American colleges still focused exclusively on the liberal arts and divinity. He wished to develop as a discipline "the science of the mind" (contemporary psychology), calling it "moral zoology." Throughout his life, Jefferson conducted scientific studies and collected data. He studied new methods for determining the heights of mountains (using mathematical calculations with barometer measurements), tested atmospheric moisture with a hygrometer, and used double-refraction optical instruments to measure small angles, eclipses, lunar movement, and Earth's longitude. Jefferson was a close observer of nature, recording the appearance of many plants, animals, and birds on his Monticello estate and wherever his travels took him. He kept weather data all his life and shared it with other meteorological observers around the country.

Not confining his scientific interests to observation alone, Jefferson invented several useful products. His most famous invention was a new design for a moldboard plow, the simple and efficient design of which drew attention throughout the Association of Agricultural Societies in America and within England's Board of Agriculture. He also invented a swivel chair, a writing desk that could be placed on one's lap, a walking cane that converted to a chair, and a copying machine that duplicated letters as they were being written. He enthusiastically supported other inventions, including the hot-air balloon, dry docks for ships, the submarine, fireproofing for houses, telescopes, the camera obscura, carriage odometers, and personal pedometers. He was an advocate of the decimal system of American currency.

While U.S. minister to France (1785–1789), Jefferson consulted with European scientists on new inventions and the natural environment of the Old World. When he moved to Philadelphia as vice president in 1797, Jefferson brought a box of prehistoric bones for the American Philosophical Society museum. As U.S. President (1801–1809), Jefferson conducted botanical expeditions around the Washington, DC, area and distributed European seeds to the local vegetable markets. In the White House, he displayed scientific instruments, globes, charts, a dry-dock model, a mockingbird, and a grizzly bear (in the garden) brought back by the Lewis and Clark expedition (1803–1806), which he had commissioned. He led discussions on the serious cowpox disease and presented an evening slide show on "The Natural History of French Parrots."

Jefferson's Science Policy

Jefferson's main interest in science was as technology, or for its *usefulness*. The practical benefits to humanity, economic development, and individual happiness were always foremost in his mind. This explains his special devotion to agriculture, because food production was, for him, the basis of all other social wealth. For the same reason, he believed in the free sharing of scientific knowledge: that it would enhance the prosperity of all people in the world. He gave every new discovery to his neighbors without charge, showing that such shared knowledge "is the great parent of science and of virtue; . . . a nation will be great in both, always in proportion as it is free" (Letter to Joseph Willard, March 14, 1789). Therefore, the advance of science and technology, for Jefferson, necessitated economic freedom (capitalism, free markets) and intellectual freedom (freedom of speech, press, and academic inquiry), including religious freedom. Thus, political democracy is integral to technological advances.

Jefferson's intellectual attitudes and scientific interests sometimes earned him ridicule, especially from his political opponents (who caricatured them as "philosophical fogs"). But his own international reputation for scientific inquiry raised the prestige of American science throughout the world. Jefferson was elected to the Institut de France, the Dutch Royal Institute of Sciences, the Board of Agriculture in England, the Agronomic Society of Bavaria, and the Linnaean Society of Paris. His comparative study of European and North American animals refuted the French naturalist Buffon's claim of New World degeneracy (proving, for example, that North American otters weigh more than their European counterparts).

The cosmological foundations of Jefferson's scientific ethics may be described as "deistic science." That is, he believed (after Aristotle, Thomas Aquinas, Isaac Newton, and John Locke) that a divinity created the universe, rather than that the world emerged out of itself randomly. "[I]t is impossible for the human mind," Jefferson wrote, "not to perceive and feel a conviction of design, consummate skill, and infinite power in every atom . . . up to an ultimate cause, a Fabricator of all things from matter and motion, their Preserver and Regulator . . . an eternal pre-existence of a Creator" (Letter to John Adams, April 11, 1823). Such Creationist ethics for Jefferson implied that all of nature, including humankind, exists within God's laws. This commends, for him, a humble, reverent appreciation of the universe and shows the limits of human knowledge. Such divine, moral limitations serve as checks on scientific presumption and hubris, or human pride. Ethical concerns regarding genetic engineering, embryonic research, euthanasia, and nuclear power in the early twenty-first century reflect such Jeffersonian ethical sensibilities.

Jefferson's ethical philosophy reflected his scientific empiricism by placing values in a human "moral sense" (akin to other physical senses such as sight and hearing). Though of divine origin, this moral sense provides for Jefferson a biological basis for ethics, or knowledge of good and evil, justice and injustice. As with Aristotle's teleological ethics, however, this human capacity is innate but undeveloped. Society must educate and refine this ethical faculty, especially through religion, politics, and law. "I consider ethics, as well as religion, as supplements to law in the government of man," Jefferson wrote (Letter to Judge Augustus B. Woodward, March 24, 1824). The highest ethics for him was "the ethics of Jesus," or what he called "the most sublime and benevolent code of morals which has ever been offered to man" (Letter to John Adams, October 12, 1813). This consisted of a simple Christian ethics, such as that presented in Jesus' Sermon on the Mount. But the best means of learning these ethics, for Jefferson, was freedom of religion—the liberty of every individual to investigate, proclaim, and believe religious truth, and the freedom to change religious faiths on the basis of personal conscience. Jefferson believed that such religious freedom, like freedom of intellectual inquiry, economic activity, and scientific advancement, would produce the most prosperous, happy people.

GARRETT WARD SHELDON

SEE ALSO *Agrarianism; Democracy.*

BIBLIOGRAPHY

Bedini, Silvio A. (1990). *Thomas Jefferson: Statesman of Science.* New York: Macmillan.

Bedini, Silvio A. (2002). *Jefferson and Science.* Charlottesville, VA: Thomas Jefferson Foundation.

Sheldon, Garrett Ward. (1993). *The Political Philosophy of Thomas Jefferson.* Baltimore: Johns Hopkins University Press.

JEWISH PERSPECTIVES

• • •

Judaism is the most ancient of three Abrahamic religions (the other two being Christianity and Islam) that are distinct from other world religions in at least three respects: they are all strongly monotheistic; they claim divine or supernatural intervention (revelation) into the world through their historical founders in ways that are in tension with natural reason; and they place special authority on one or more written texts. Judaism (like Christianity) also has a close historical relation with modern science and technology; historians of science have argued that in its origins science was dependent on a view of the world as well ordered and subject to human investigation and control precisely in the ways presented by the Jewish revelation, and certainly Jewish scientists especially are disproportionately represented in the technical community. At the same time, science and technology have presented specific challenges to Jewish tradition and identity, the responses to which offer special contributions to more general discussions of science, technology, and ethics.

Approaches to Judaism

Individuals explain their adoption of a Jewish designation by their adherence in various degrees to one or more facets of the "Jewish way of life." Among the most important aspects are the beliefs that there is only one God; that the first five books of the Hebrew Scriptures (known as the Torah, containing 613 commandments and canonized between 700 and 200 B.C.E.) were handed down from God to Moses around 1500 B.C.E.; and that Jews should follow both the oral and written laws that have been handed down through the generations. These laws, which number in the thousands and whose varied selection or adoption accounts for the varieties of Judaism, are found in a number of tracts:

- The Mishnah (the oral law handed down from Moses and put into writing in six volumes about 1800 years ago).

- The Gemara, comprising commentaries on the Mishnah and other aspects of Jewish life and stories, found as part of the Talmud.

- The Talmud (of which there are at least two versions: Babylonian, with about 2.5 million words, and Jerusalem, about one-eighth the size), which is a commentary on the Mishnah and Gemara; it was compiled and redacted (canonized) between 300 and 500 C.E.

- The Midrash (also considered a part of the Talmud), a commentary on the first five books of the Old Testament.

- The Kabala, a book that emphasizes the mystical relationships between humans, God, heaven and its inhabitants, and hell with its entourage.

- The remaining thirty-four books of the English Old Testament, referred to as the Prophets and the Writings.

- The Apocrypha, which contains the books that were left out of the Bible when the latter was canonized to include additional sections on the prophets and writings (a process that began with Ezra in 530 B.C.E. and continued until the fall of the second temple in 70 C.E.).

- The Shulchan Auruch, a summary of the laws drawn up in the sixteenth century.

- The Haggadah, a story of the Exodus of the Jews from Egypt in about 1450 B.C.E., whose formulation began in pre-temple times (1000 B.C.E.); put into its conventional form in the thirteenth century but still provides the basis of numerous modern variants.

In addition to belief in the holiness of the above writings and the requirement to follow all or a selection of the laws, Jews may also define themselves in relation to:

- their descent from other Jews (in particular a Jewish mother, although in biblical times it is clear that patrilineal descent also pertained);

- their conversion;

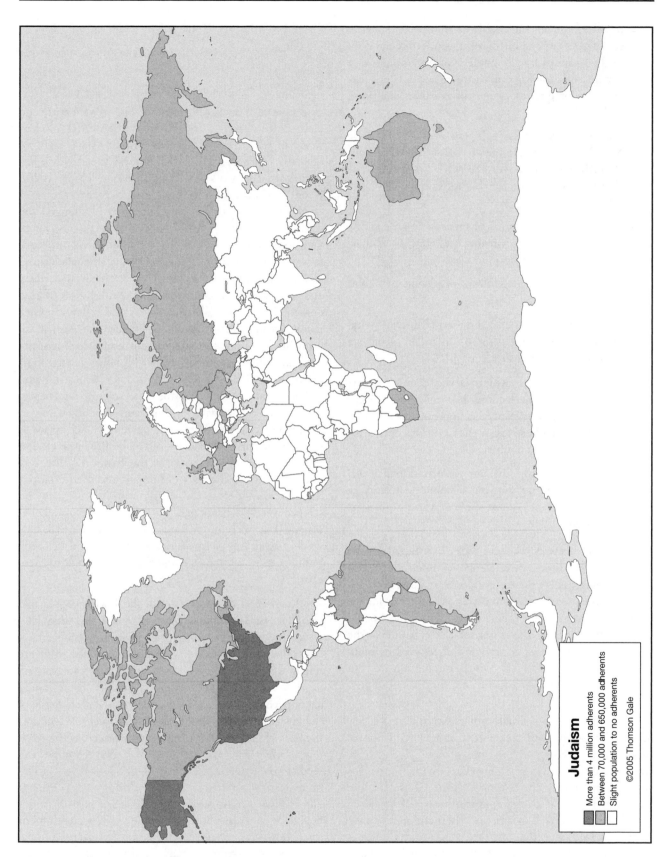

Judaism

More than 4 million adherents

Between 70,000 and 650,000 adherents

Slight population to no adherents

©2005 Thomson Gale

- Jewish traditions such as those that pertain at rites of passage such as birth (plus penile circumcision in the event of a male child), confirmation as in a bar mitzvah for boys at thirteen and recently bat mitzvah for girls at twelve or thirteen, marriage, and death;

- the annual calendar of religious events such as the New Year (Rosh Hashanah), Day of Atonement (Yom Kippur), Passover (Pesach), Festival of Lots (Purim), Festival of the Lights (Chanukah), and others;

- culture defined in terms of types of food, cooking methods, respect for learning and education, charity, style of clothing, and modesty;

- the acceptance of the rulings of a court of Rabbis referred to as a Beth Din;

- the need to have at least ten men (a minyan—and recently, may count women) in order to have a fully competent prayer meeting;

- the State of Israel, which is the country in the world where a persecuted Jew may seek succor without further fear of the pogroms or selective legislation that has been a characteristic of the history of most other countries;

- or a membership in an internationally dispersed community that has a common history or treatment in the hands of a variety of host communities.

In structural terms, a Jew who seeks to follow the laws may refer to the literature cited above or consult a rabbi. There is an extensive correspondence extant that consists of individuals or communities asking for guidance from the most eminent rabbis of the day. The *responsa* that result constitute the norm for the behavior of the respondent. This worked well for ghettoized communities living in relatively static circumstances, but history since the mid-eighteenth century has been anything but static.

On the basis of which tenets an individual Jew adopts, he or she will associate (or not) with one or more of the recognized religious groups. These range from the ultra-orthodox (themselves divided into sects such as the Lubavich, Satmars, Aish, Chasidim, and Chaderim) who reject the opportunities of the modern world and generally do not permit their children to view television (although they may make use of the Internet for Midrashic discussions, with Web sites such as http://www.vbm-torah.org), to the Liberal Progressives with whom virtually anything goes. In between there are the

Orthodox, Masorti, Conservative, Reform, and Liberal groupings.

Science, Technology, and Judaism

The European Enlightenment of the seventeenth and eighteenth centuries sowed the seeds of the modern world in which science and technology have changed both the way people think and the way they live. Beginning with the works of Francis Bacon (1561–1626), René Descartes (1596–1650), Isaac Newton (1642–1727), and others, the Enlightenment challenged the Jewish community as it did other religious groups. Those who were in occupations that brought them into contact with prominent business people, politicians, or royalty rapidly learned the language of the host country and became both educated and secular to differing degrees. In the late 1700s and early 1800s, Jews in Germany, Poland, Russia, Holland, France, and Austria set up schools where the medium of education was the national language and where Yiddish (or Jewish German) was in some cases outlawed for education and business transactions. At this time science was beginning to make a showing in these curricula, especially at the secondary level. As time advanced, science began to provide secular explanations of the biblical miracles, of the creation of the universe, of the creation of life, and of the creation and nature of humans and their relationships to the rest of the living world. Not only did science provide challenges to the intellect and belief system, but technology and engineering offered new ways of working, of traveling, of writing, and of doing business. How did Judaism and the Jews respond to these changes?

In the contemporary world, the Jewish people live either in Israel or outside Israel in the so-called Diaspora. In the early twenty-first century in Israel about one third of the Israelis are secular, another third are religious and follow the dictates of the laws with varying degrees of observance, while a middle third would acknowledge a belief in God and do not follow many of the laws in their day-to-day lives, but observe them during rites of passage or special occasions such as the reading of the Haggadah at Passover. Nevertheless, the secular government of Israel does not generally legislate on matters of a religious nature. While the government allows Jews a right of return to Israel, it has not so far made a legal definition of who is a Jew. The government does, however, require a religious marriage for official dealings; nevertheless, foreign civil marriages are recognized. Local authorities, however, may choose to operate transport systems on the Sabbath or may ban them as being contrary to religious laws that forbid travel on the Sabbath. Similarly, erotically suggestive advertisements may be banned by some localities while accepted in others. Work on the Sabbath is generally banned nationally, although particular industries may obtain special dispensations from the government. Those industries that are essential to the economy such as defense, food, and health care find it easy to obtain licenses to operate, as do industries that rely on continuous processes, an interruption to which will disrupt production with considerable economic loss.

The introduction of new technology has presented religiously disposed Israelis and Diaspora Jews with many concerns. This is because the laws as defined by that body of literature that is accepted as the Halakhah expressly forbid many of the applications that are made possible by contemporary machines and devices. There are four main areas where such concerns are expressed. The first relates to the observance of the laws pertaining to work on the Sabbath. A second concerns determinations as to whether certain food preparations are in compliance with the religious laws of kashruth—that is whether they are, or are not, *kosher*. This latter term derives from the biblical laws of what foods are allowable (Lev. 11:2–47); for example, it is allowable to eat meat from cloven hoofed animals that chew the cud but not shellfish, a calf may not be cooked in its mother's milk, and creatures that crawl on their bellies are forbidden. A third set of issues relate to health care and medicine. Finally, a fourth area of concern focuses on changes occurring in agriculture.

The fourth commandment requires Jews to keep the Sabbath holy and to do no work on that day. But what is work? This is often held to be activities of a constructive nature such as preparing food, making a tool or object, giving professional advice, teaching (but learning is acceptable), and doing anything that creates fire, such as making a spark whenever an electrical contact is made. Similar laws apply on holy days.

These prohibitions are managed in a number of ways. First, one may appeal to an overriding statement by God in the Torah (*Deut.* 30:19): *Therefore choose life ...*''; if work is effected in an effort to save life, it would be acceptable. Secondly, it is possible to employ a non-Jew to do the constructive work on the Sabbath, such as to make the fire, heat food, or run a factory. A third option is to use an automatic device such as an electrical timer switch. A battery of these switches may be programmed and used to effect the daily routine jobs that require electrical equipment (heating, lighting, cooking, communicating, elevators, and alarms). It is

moot as to whether a modern computer can be used as part of this automation process or whether its use is proscribed because it is an instrument of writing.

To engage in more detail with those issues where a technological fix can obviate a religious prohibition, the Institute for Science and Halakhah was founded in Jerusalem. This body seeks to use sensor systems, robotics, computers, and information devices to loop around the traditional laws and accomplish ends that would otherwise have been forbidden. Its work is proving so successful that this independently-funded body has been adopted as an element of the national government.

Whether or not food is kosher is defined by the rabbis of the local jurisdiction or on appeal to a more respected rabbi with international stature. Clearly, because food is now purchased as pre-prepared items or is made as a composition in tins, it is difficult to know whether or not such material is kosher. While many food producers act under the supervision of the rabbinate, it is possible to produce kosher foods outside this restriction. A food ingredients list is helpful, but it does not specify the way the ingredients are produced in sufficient detail to satisfy a rabbi that non-kosher material was not been prepared with the same equipment and the washing process was effected with sufficient (and often excessive) thoroughness that it could be used for kosher manufacture.

In addition to pig insulin, pig heart valves are generally deemed acceptable for transplantation into observant Jews. As and when pigs are reared that are immunologically compatible with human immune systems, the transplantation of pig hearts, livers, lungs, kidneys, and other organs might also be deemed acceptable by the orthodox Jew.

However, there are medical issues in the area of abortion and *in vitro* fertilization that exercise the minds of those seeking ethical acceptance from a Judaic standpoint. Facing infertility, an orthodox Jewish couple could receive a dispensation from a rabbi for *in vitro* fertilization, even if this means creating extra embryos that are eventually killed. Abortion, however is generally forbidden unless the health of the mother is threatened. There are other issues that raise concern, such as blood transfusion: many religious people believe that a person's life is in the blood, and to accept another person's life (albeit in part) is not allowable (*Lev.* 17:13–15).

The relevant agricultural restriction is that it is forbidden to plant two different kinds of seed in the same field. From this standpoint, genetically manipulated seeds do not present a problem nor do trees that are grafted because the stock and the plant are of the same type. However, the production of hybrid plants that derive from clearly different stocks does cause difficulty and some religious kibbutzim (Israeli agricultural settlements) do not permit themselves the advantages that hybrid vigor provides.

Where science challenges religion most is in those areas that have to do with origins and miracles. Judaism seems to be able to ride the resulting intellectual issues with aplomb. It takes evolution in its stride by asserting that Darwin's ideas are but hypotheses; they have not been, nor can they be, proven. The account of creation in the Torah, however, is a truth as it was given to Moses by God and this constitutes the "gold standard" of knowledge. A mere hypothesis cannot seriously challenge such a truth. The miracles may be treated similarly. There may well be scientific explanations for some of the miracles. For example, the turning of the river Nile into blood by Moses may be explained by the emergence of a bloom of a euglenoid alga that has lost its chlorophyll and appears red by virtue of its red carotenoids. It yet remains possible that God performed the event to provide Pharaoh with evidence of his powers to effect miracles.

When it comes to metaphysical considerations such as the nature and origin of matter, Judaism relies on a belief in an all-powerful God who created all things. Theories of the big bang still leave dangling the origin of the matter that made the "bang" possible, or the process whereby all the matter in the universe was made in an unimaginably short time. The possibility of God creating other universes is not considered, although there is no reason to uphold the claim that humans (and maybe others) inhabit the only universe. Since the beginning of the twentieth century, humans have come close to understanding how an abiotic (lifeless) world some four billion years ago gave rise to a molecule that evidenced the properties of life. The story of the evolution of this notional entity to humans, is also well thought out. Nevertheless, those who profess a strict adherence to the literature and the codes of Judaism will not brook such thinkings because they adhere to the letters and words of Genesis.

R.E. SPIER

SEE ALSO *Arendt, Hannah; Holocaust; Virtue Ethics.*

BIBLIOGRAPHY

Ausubel, Nathan. (1964). *The Book of Jewish Knowledge.* New York: Crown Publishers.

Rohl, David M. (2002). *The Lost Testament: From Eden to Exile: The Five-Thousand Year History of the People of the Bible.* London: Century; New York: Random House.

Wigoder, Geoffrey, et al, ed. (1997). *Encyclopedia Judaica,* CD-ROM edition. Jerusalem: Keter Publishing House

INTERNET RESOURCES

Bible Gateway. Available at http://www.biblegateway.com. Contains fourteen searchable versions of the Bible.

Internet Sacred Text Archive. Available at http://www.sacred-texts.com/jud/. Contains an English translation of the Babylonian Talmud and other texts as referred to in the article.

The Jewish Chronicle. Available at http://www.thejc.com. Weekly UK newspaper available internationally and on the Internet.

JONAS, HANS

• • •

The intellectual heritage of Hans Jonas (1903–1993) spans and reflects the twentieth century. Born in Mönchengladbach, Germany, on October 4, he died in New Rochelle, New York, on February 5, having become one of the most important contributors to philosophical reflection on science, technology, and ethics. For more than half a century, Jonas worked consistently to develop a persuasive alternative to modern nihilism in its diverse existentialist, positivist, scientific, and technological manifestations.

Life and Works

In Germany Jonas studied with the major figures of philosophy such as Edmund Husserl (1859–1938), Martin Heidegger (1889–1976), and the Protestant theologian Rudolf Bultmann (1884–1976). His doctoral dissertation adapted Heidegger's Dasein analysis from *Time and Being* (1927) to demythologize Gnostic texts from the early centuries of the Common Era, revealing the extreme dualism and world estrangement of this ancient religious literature. Increasingly aware of the social estrangement of Jews in Europe (his mother would be murdered in Auschwitz), Jonas joined the Zionist movement and, as the Nazi's came to power, left Germany for Palestine. During World War II he joined the Jewish Brigade of the British forces in Italy as an artillery soldier; in 1948 he fought in the Israeli War of Independence.

During this period Jonas also began to reflect on the philosophical problems of modern science, especially biology, distancing himself from Heidegger and Gnosticism by noting the parallels between the inimical cosmos of Gnostic belief and the conception of an indifferent nature found in science. In 1949 he left Israel for Canada, and after a few years moved to New York, where he taught at the New School for Social Research until his retirement. From the 1960s on, Jonas made a number of visits to Germany and as a result published frequently in the German language. He became influential in the land of his birth, especially in the Green movement. He received European recognition for his work, beginning with the *Friedenspreis*, the peace prize awarded by the German Book Trade, in 1987.

Jonas first major book was *The Phenomenon of Life* (1966), his initial foray into a phenomenological interpretation of biology that might disclose the metaphysical significance of organic phenomena. *Philosophical Essays: From Ancient Creed to Technological Man* (1974) contains his first essays on the ethics of technology. His single most important book, *The Imperative of Responsibility: In Search of an Ethics for the Technological Age* (1979), brings together these two lines in an attempt to ground his ethics of technology in a philosophy of nature. Over the last two decades of his life, Jonas sought to extend the practical applications of his thinking while deepening its cosmological and theological foundations in such works as *Technik, Medizin und Ethik* (Technology, medicine, and ethic) (1985) and *Mortality and Morality* (1996). *Errinnerungen* (Recollections) (2003) is a collection of autobiographical interviews.

Responsibility for Integrity and Sustainability

In his central work, *The Imperative of Responsibility*, Jonas spells out the need for an early formulation of the precautionary principle that he calls the *heuristics of fear*, which gives "prevalence to the bad over the good prognosis" in case of unforeseeable and irreversible technological risks to the future of humankind (Jonas 1984, p. 31). For Jonas, such a procedure is justified by the ontological idea of humanity as that being which is able to bear responsibility. Because of this capacity, Jonas argues that humans have an unlimited responsibility to preserve life on Earth, in which they, as those who bear responsibility, may be primary, but which encompasses all of nature. This responsibility is total, continuous, and future oriented. Parental responsibility for children is archetypical, although in this case there is a terminus: Children grow up and become adult bearers of responsibility themselves. But with regard to nature, responsibility does not cease. The imperative of that responsibility associated with technology is to pass on responsibility, or more generally, to

Hans Jonas, 1903–1993. The German-born philosopher is best known for his influential work, *The Imperative of Responsibility*. His work centers on social and ethical problems created by technology. (© *Bettmann/Corbis.*)

safeguard conditions for the continual existence of responsibility on Earth. Indeed, for Jonas, "The presence of [human beings] in the world is demanded to ensure the very premise of responsibility—the existence of mere candidates for a moral order" (Jonas 1984, p. 10).

Until the modern period, responsibility for the integrity of life on Earth was not a human imperative, because nature took adequate care of itself. But the human relation to nature has decisively changed. Human responsibility is disclosed in its new intensity by the vulnerability of nature to human destruction and to the potential mutilation of the human genetic heritage by long-range effects of modern science and technology. The world now needs human care to a degree previously unexperienced in the history of humankind. This theory holds insofar as one accepts Jonas's argument that a striving, teleological nature, revealed in the attempts at self-preservation among even the most primitive forms of organic life, constitutes an objective affirmation of good that is infinitely superior to a cold and indifferent universe.

Criticisms

Four main criticisms have been leveled against Jonas's ethics of technological responsibility. A first is that his responsibility is too general or formal. Who is responsible for what? Jonas maintains that humanity ought to exist. But Richard Bernstein (1995) replies that even if one accepts the general goodness of organic nature as a whole, no obligation to exist follows for humanity, nor does obligatory human existence imply any specific moral guidance for medical or environmental practices.

But the imperative of responsibility is not meant to be part of a deductive system. Instead the heuristics of fear and criticism of utopianism offer more practical counsel. According to Jonas, utopianism is a form of idolatry that the heuristics of fear counters by pointing out how in the technological pursuit of utopian goals the integrity of natural species or even of human existence may itself be at stake. Categorical responsibility functions as the overriding argument for preservation.

A second criticism asks whether Jonas's ethics is compatible with democracy and personal autonomy. Jonas has little faith that democratic politics works beyond short-term interests. Eventually a noble tyrant might have to avert the apocalypse. There is a parallel in medical ethics in which Jonas treats the requirement of informed consent as a problem instead of a solution. In both cases, Jonas seems to hold the view that fallible autonomous subjects need to be protected from themselves.

This reflects the asymmetry of the concept of responsibility. For Jonas, morality is not based on a social contract made up by self-reliant individuals, but originates with the call for protection from vulnerable beings. Human beings have to work to ensure the welfare of future generations because those generations cannot do it for themselves. In medical experiments, the sick should be the last to be recruited as subjects because they are the most vulnerable and dependent. Nevertheless the implied paternalism, though restricted to negative injunctions (*Do not* or *Refrain from doing*) represents an unpopular and therefore important perspective.

Third is whether the restoration of a metaphysical ethics is necessary to answer questions posed by modern technology. Karl Otto Apel (1994) strongly rejects Jonas's metaphysical principle of responsibility as incompatible with justice. The survival of humanity might entail the starvation of many people in developing countries, which Apel refers to as a social Darwinist solution.

But as in the case of democracy and autonomy, Jonas is well aware of the dilemma. Moreover he does not dismiss the demands of justice but relates their obligating force to the still higher duty of sustainability. Whereas Apel argues that there is no meaning in survival without justice, Jonas replies that there is no

meaning in justice without survival. According to Jonas, sustainability is finally a metaphysical issue. Prevailing attempts in the ethics of technology based on nonmetaphysical, symmetrical rationality seem unable to enter substantive discussion on topics involving individual liberties. Therefore it becomes impossible to put a hold on the insatiable demands of the modern individual for justice, safety, health, and welfare. Jonas meets this vacuum with his *first rule*: that no future condition should be accepted that would affect the integrity of humanity.

Finally the boldest aspect of Jonas's ethical theory, involving the move from *is* to *ought*, is his claim that living nature objectively appeals to human responsibility to heed its integrity. Lawrence Vogel, however, criticizes such *cosmic deontology* as unnecessary in his 1996 introduction to Jonas's *Mortality and Morality*. Jonas clearly aims to replace Kantian deontology with an equally categorical imperative in which nature serves as a good in itself. If this were not the case, human obligation might be illusory. But perhaps it is not cosmology that teaches people to be responsible for living nature and the future. Maybe the reverse is the case: A basically self-evident responsibility teaches respect for a cosmos that brought forth life in its manifold of species and in its depth of subjective intensity. Yet while others argue for an ethics rooted solely in the social world, Jonas deliberately invokes an argument that overarches both the social and the natural domains. When considerations of the limits of progress lead to discussions of the limits of the human condition, people have to proceed from ethics to metaphysics in a new attempt to answer eternal questions regarding poverty, illness, and evil, in both natural and human forms.

TON VAN DER VALK

SEE ALSO *Deontology; Freedom; Future Generations; German Persepctives.*

BIBLIOGRAPHY

Apel, Karl Otto. (1994). "Die ökologische Krise als Herausforderung für die Diskursethik" [The ecological crisis as a challenge to discourse ethics]. In *Ethik für die Zukunft: Im Diskurs mit Hans Jonas,* ed. Dietrich Böhler. Munich: Beck.

Bernstein, Richard. (1995). "Rethinking Responsibility." In *The Legacy of Hans Jonas,* ed. Strachan Donnelley. *Hastings Center Report* 25(7). A special issue of the journal.

Jonas, Hans. (1963). *The Gnostic Religion: The Message of the Alien God and the Beginnings of Christianity.* Boston: Beacon Press. A translation of Jonas's doctoral dissertation from 1934.

Jonas, Hans. (1966). *The Phenomenon of Life: Toward a Philosophical Biology*. New York: Dell.

Jonas, Hans. (1974). *Philosophical Essays: From Ancient Creed to Technological Man*. Englewood Cliffs, NJ: Prentice Hall.

Jonas, Hans. (1984). *The Imperative of Responsibility: In Search of an Ethics for the Technological Age*. Chicago: University of Chicago Press. Unites two volumes first published in German: *Das Prinzip Verantwortung* [The imperative of responsibility] (1979) and *Macht oder Ohnmacht der Subjektivität* [Power or importance of subjectivity] (1981).

Jonas, Hans. (1985). *Technik, Medizin und Ethik: Zur Praxis des Prinzips Verantwortung* [Technology, medicine, and ethic: On the practice of the principle of responsibility]. Frankfurt: Suhrkamp.

Jonas, Hans. (1992). *Philosophische Untersuchungen und metaphysische Vermutungen* [Philosophical investigations and metaphysical suppositions]. Frankfurt: Insel.

Jonas, Hans. (1996). *Mortality and Morality: A Search for the Good after Auschwitz*, ed. Lawrence Vogel. Evanston, IL: Northwestern University Press.

Jonas, Hans. (2003). *Erinnerungen* [Recollections], ed. Christian Wiese. Frankfurt: Insel.

Levy, David J. (2002). *Hans Jonas: The Integrity of Thinking*. Columbia: University of Missouri Press.

JOURNALISM ETHICS

• • •

Journalism is the profession of writing, editing, and publishing high-frequency periodicals that aim to report and comment on events of public interest, commonly called *news*, with its frontline practitioners those who gather the data—reporters, photographers, videographers—and those who approve the data and prepare the collection of text and visuals for presentation—editors and producers. The unique role-related responsibility of journalists, which includes all of these practitioners, in democracy is to communicate to citizens information needed for self-governance. Self-governance includes the most mundane of decisions, such as what weather to prepare for when driving to work, and the most complex of choices, such as voting on referendums or candidates for public office.

As a profession journalism is dependent on certain ethical standards to maintain the credibility needed to perform its role-related responsibilities. The professional acts of discovering, reporting, and disseminating the news is dependent on various technologies. Thus insofar as both changes in science and technology alter the practice of journalism and journalists report on scientific and technological news, journalism ethics is of relevance to science, technology, and ethics, and vice versa.

Origins and Ethics

Journalism has emerged parallel with the development of technologies for the rapid, mass dissemination of written texts and broadcast messages. Although anticipations can be found in serial official announcements such as the *Acta diurnal* (Daily proceedings) of the Roman Empire or the *Tching-pao* (Palace news) of the Chinese T'ang dynasty, the first modern news sheets appeared in Germany in the 1450s, where Johann Guttenberg invented the printing press. The first true newspaper was probably the *Gazette de France*, which began publication in Paris in 1631. Since then both Germany and France have maintained strong journalistic traditions, which after World War II exhibited special expertise in reporting on science and technology in relation to, for instance, nuclear weapons and environmental issues. Indeed one can argue that the strength of the environmental movement in Europe rests in part on such reporting.

The early 1700s is sometimes described as the golden age of English journalism, with what are now classified as more literary journalist-publishers such as Joseph Addison (1672–1719) and Richard Steele (1672–1729), among others, developing the occasional general interest essay in the *Spectator* and the *Tatler*. Such essays are no doubt ancestors of the personal columns and op-ed perspective pieces of the present. In another development, when the London *Times*, founded initially in 1785 as the *Daily Universal Register*, published dispatches from correspondents at the front during the Napoleonic Wars (1793–1815), it was the first time the public was able to read about the results of military battles from other than government sources.

In the United States the rise of the journalism profession is strongly associated with the writing and publishing of early patriots such as James Franklin's *New England Courant* and his younger brother Benjamin Franklin's *Pennsylvania Gazette*. In part because of the contributions of the press to successful revolutionary politics, the first amendment of the U.S. Constitution (1791) guaranteed freedom of the press to a historically unprecedented degree. The development of this freedom during the mid-1800s drew on new technologies to create a pluralistic, mass circulation *penny press*, which in the late-1800s began to be consolidated into a set of newspaper chains that themselves drew on new means of communication such as the telegraph. These major newspapers subsequently separated themselves into the high-standards press (*New York Times*, *Washington Post*, among others) and more popular publications that practiced what was criticized as *yellow journalism*.

Reaction to the distortion and sensationalism of yellow journalism, with its power to influence events through muckraking exposes and jingoistic politics, led to efforts to professionalize the field. In 1892 Joseph Pulitzer proposed the creation of a school of journalism at Columbia University (which did not happen, however, until twenty years later). At virtually the same time, in 1909, reporters themselves established their first professional association (the Sigma Delta Chi fraternity, which in 1988 changed its name to the Society of Professional Journalists). It was in these two contexts that ethics began a process of explicit development, with the first code of ethics for professional journalism written by members of Sigma Delta Chi in 1926. Two organizations focused on science writing emerged at about the same time. The American Medical Writers Association traces its origins to 1924, with the development of its own code of ethics in 1976. The National Association of Science Writers was formed in 1934 to promote the dissemination of accurate information regarding science.

Following from the interdependence of technical and professional growth, wire services contributed to the development of common journalistic standards. Wire services, which sent a single story or photograph to multiple outlets via telegraph, then telephone lines, then satellite, both reflected and influenced subscriber news organization standards. The service had to meet the professional demands of its subscribers, but it also served as a model for local news organizations.

Journalism ethics at the macro level describes and criticizes the practices of news organizations and the role journalism plays in society. Drawing on the disciplines of history, sociology, philosophy, and political theory, scholars work to distinguish those practices that are ethically obligatory, desirable, and proscribed. At the micro level journalism ethics both describes and argues for normative behaviors of individual practitioners and the profession.

In a democracy, journalists play a central role in providing citizens with the information they need to practice self-governance. In a highly scientific and technological democracy this responsibility extends to accurate reporting on science, technology, and engineering. This role-related responsibility in journalism to present informative accounts of issues and events, including of science and technology, serves as the basis for a cornucopia of ethical issues.

At the macro level such issues include critical assessment of (a) domination of media attention and story spin by the most powerful; (b) the presence of less powerful individuals and groups often not considered immediately newsworthy; and (c) the determination of events, issues, and people as newsworthy based on audience interest, government promotion, or corporate influence. These macro issues are apparent in which scientific actions get reported and which get ignored. The science that finds its way into the public press is that most easily distilled, most eagerly promoted by articulate spokespersons, and which attracts funding or policy discussions.

At the micro level issues include (a) conflicts between media exposure and individual desires to limit such exposure; and (b) conflicts between professional journalist responsibilities and recognized or unrecognized bias by reporters.

Scientific and Technological Change

While the Internet has made it possible for all people with computer access to broadcast their messages, recognized news outlets remain in the hands of a few corporate owners. "In Britain now, 85% of the national daily press is in the hands of four groups ... In the United States ... six companies control most of the media" (Bertrand 2003, p. 5). Technology has offered the tools for true participatory democracy, but technology has also limited the countries and corporations that can reach the world through satellites in geostationary orbit.

Since the early-1800s, technology has influenced how journalism is practiced, produced, and presented. Technological advances that have affected journalism include methods of recording events as well as methods of data transmission from the field to the news organization and from the news organization to its audiences. The challenge for the profession is to use evolving technology to meet the institution's unique role-related responsibilities. Technology also makes some unethical acts, such as fabricating photos or recorded quotes, easier to perform and more difficult to detect.

The standard of *objective reporting*, for example, finds its origins in the development of the wire service in the early-twentieth century. For the first time, it was possible for reporters, and then photographers, to be present at a distant scene and disseminate coverage of the event to large numbers of news organizations at the same time. What sold best to audiences in a variety of markets was journalism that appealed to the broadest possible interests. Journalists covering the story could not make assumptions about the political, religious, or cultural beliefs of readers and viewers as they might have when reporting for a specific hometown audience. Thus the reporting that worked best for the most general audience became the standard. Generations of students in journalism schools learned to report the five W's and an H—who, what, when, where, why and how—with the importance of each obvious in its order of appearance in the news product. The technology of precomputer pagination dictated an inverted pyramid style of reporting that put the most important facts at the top of the story so that the layout staff could lop off from the bottom of an account material that did not fit into limited space.

While these technologically influenced norms served as standards for the field of journalism, they did not necessarily assist in meeting role-related responsibilities. For example, one general interpretation of objective balancing of facts is the myth that each story has two sides that must be accurately presented. Complicated stories involving policy decisions have many sides. When a story is reported as a two-sided issue, the reporting itself creates a polarized debate rather than a nuanced public discussion. The attempts to establish a national healthcare system in the United States in 1994, for example, was reported as a political debate between the Clinton White House and the Republican-controlled Congress. The story of the need for uninsured citizens to access needed healthcare was overpowered by the win-lose style of its presentation. It took another decade before the public issue of developing a new healthcare policy could be discussed without the goal being lost in the reporting. Technological advances during the 1990s added to technology-accommodating norms, such as photo-transmitting cell phones. Digital cameras and satellite transmission made the delivery of information from the field to the news organization instantaneous. In homes the introduction of cable and satellite television and the World Wide Web (WWW) allowed for multichannel broadcast, 24-hour news channels, and instantaneous transmission of material from the news organization to the audience. Indeed, in an era of live coverage, the news organization itself is bypassed by journalists and nonjournalists who are on the scene, broadcasting and making their own decisions about what to reveal and what is and is not news.

The resulting norms, as questionable as the striving for two-sided objective news coverage, include the following:

1. an assumption that on-the-scene coverage is the best;

2. accessible information is synonymous with news;

3. news is a never-ending evolution of first impressions or viewable dramatic events—while interpretation and context building may get viewers and broadcasters through quiet periods, it is access to new and dramatic pictures that creates *breaking news*;

4. mediated reality is reality.

The first news team on the scene is more likely to report speculation than fact. Turning a camera to a scene and flooding viewers' homes with dramatic images creates mediated events, not news.

News stories developed for print dissemination or electronic news packages are more than recordings of slices of reality. If information is to be useful to citizens for self-governance, they need to understand the context and meaning behind events. Citizens are dependent on journalism to know what is happening in the world, but it is easy to confuse mediated reality with reality.

Experiencing the events of September 11, 2001 in New York City, or at the Pentagon, was far different from watching the scenes played out on television. Yet most viewers felt they *experienced* the terrorist attacks through the media. *Watching* the second plane hit the South Tower, *watching* the towers tumble, *watching* those on the scene scramble for safety was possible for everyone with access to a television screen, what one author calls *mass interpersonal communication*. (Newton 2001, p. 153). But making sense of a mediated event is limited by what the videographer, story producer, and news organization has chosen to show the audience.

American journalism has cultural domination of broadcast media in that it serves as primary source material for historians and others who create records of contemporary events (Winch 1997, p. 4). The importance of these accounts create the ethical necessity for journalists to use technology to enhance their ability to meet role responsibilities rather than allowing technology to create standards that interfere with meeting those responsibilities. The technological worldwide domination of American journalism also creates the ethical necessity for journalists to perceive of themselves as representing global, not national, interests. Reality, if left unrecorded, is not available for public consideration or discussion.

According to communication scholar Paul Ansah, a problem with the domination of technology and news is "the paucity of the horizontal flow of news among developing countries in the South, thus compelling people in those countries to see one another from the perspective of foreign correspondents whose value systems, ideological options and even prejudices are reflected in the reports" (Ansah 1986, p. 66).

The Internet gives every person with access to that technology the opportunity for free expression and access to a world of ideas. In twenty-first century university life in the United States, where professors expect students and colleagues to exist in a wired world, it is easy to forget that such access actually exists only for the privileged few. According to a 2003 UN report, 91 percent of Internet users represent 19 percent of the world's population.

Yet in a world in which anyone with access can find an audience—what might be called information anarchy—credible journalism is more necessary than ever to sustain democracy. Citizens "need a guarantee of authenticity.... There is an ever greater need for competent, honest journalists to filter, check, and comment upon the information available" (Bertrand 2003, p. 4).

Specific Ethical Concerns of Reporting on Science, Technology, and Engineering

Science coverage rose steadily from the mid-twentieth century into the early-twenty-first century. The explosive growth in technology and in medical knowledge fueled a steady stream of science news. The need for average citizens to achieve a higher degree of science literacy so that they could understand and operate new technological equipment and so that they could understand and access advanced medical technology created a greater and sustained need for mediation between experts and general public.

Increasing awareness and concern for environmental impact on the part of scientists, policy makers, and the public created the same need for the development of environmental journalism as a specialization. Journalism education responded with the development of science writing courses and curriculum.

A 1978 directory (Friedman, Goodell and Verbit) found fifty-nine colleges and universities teaching 104 science communication courses including those in general science, technical writing, environmental journalism, and agricultural journalism. A mid-1990s update of Sharon Dunwoody's directory found an increase in the number of programs, courses, and specializations. For example, specialized communication courses were offered in risk, engineering, cyberspace, marine science, and earth sciences, in addition to general science, and technical, environmental, and medical writing.

Journalists and scientists continue to recognize the need for collaboration between the professions and to understand the different professional conventions that make such collaboration difficult. Professional societies, web resources, and workshops for scientists and journalists are necessary to create a communication bridge between science and the public it affects.

DENI ELLIOTT

SEE ALSO *Communications Ethics.*

BIBLIOGRAPHY

Ansah, Paul A. V. (1986). "The Struggle for Rights and Values in Communication." In *The Myth of the Information Revolution*, ed. Michael Traber. Beverly Hills, CA: SAGE Publications.

Bertrand, Claude-Jean. (2003). *An Arsenal for Democracy.* Cresskill, NJ: Hampton Press.

Bucchi, Massimiano. (1998). *Science and the Media: Alternative Routes in Science Communication.* New York: Routledge.

Dunwoody, Sharon. (1993). *Reconstructing Science for Public Consumption: Journalism as Science Education.* New York: Hyperion Books.

Dunwoody, Sharon, and Ellen Wartella. (1979). "A Survey of the Structure of Science and Environmental Writing Courses." *Journal of Environmental Education* 10(3): 29–39.

Friedman, Sharon M.; Rae Goodell; and Lawrence Verbit. (1978). *Directory of Science Communication Courses and Programs.* Binghamton: Science Communication Directory, Department of Chemistry, State University of New York at Binghamton.

Friedman, Sharon M.; Sharon Dunwoody; and Carol. L. Rogers, eds. (1999). *Communicating Uncertainty: Media Coverage of New and Controversial Science.* Mahwah, NJ: Erlbaum.

Friedman, Sharon M.; Sharon Dunwoody; and Carol L. Rogers, eds. (1986). *Scientists and Journalists: Reporting Science as News.* New York: Free Press.

Levi, Ragnar. (2001). *Medical Journalism: Exposing Fact, Fiction, Fraud.* Ames: Iowa State University Press.

Nelkin, Dorothy. (1995). *Selling Science: How the Press Covers Science and Technology.* New York: W. H. Freedman.

Newton, Julianne H. (2001). *The Burden of Visual Truth.* Mahwah, NJ: Lawrence Erlbaum Associates.

Paradis, James G., and Muriel L. Zimmerman. (2002). *The MIT Guide to Science and Engineering Communication.* Cambridge, MA: MIT Press.

Street, John. (1992). *Politics and Technology.* New York: Guilford Press.

Winch, Samuel P. (1997). *Mapping the Cultural Space of Journalism.* Westport, CT: Praeger Press.

JUANA INÉS DE LA CRUZ

• • •

Born in Nepantla, near Mexico City, Sor Juana Inés de la Cruz (1648 or 1651–1695) is best known as one of the greatest Baroque poets and as the iconic forerunner of Hispanic feminism. However, the significance of her work and life in studies of the relationships among gender, science, and society in New Spain (Mexico) and colonial Spanish America has been gaining greater recognition.

In 1662 Sor Juana, then known by her birth name, Juana de Asbaje y Ramírez, was admitted into the service of the viceroy's wife, the marquise of Mancera, who became her protector, a role later filled by the wife of the succeeding viceroy, the countess of Paredes. Believing that a religious life was most compatible with her intellectual pursuits, Sor Juana entered a Carmelite convent in 1667 but left after three months, eventually joining the more lenient order of San Jerónimo in 1669. In the convent Sor Juana pursued her scientific studies—of which little is known—and wrote the bulk of her literary works despite the opposition of her confessor and the archbishop of Mexico.

The first volume of her collected works was published in Madrid in 1689, with the publication of the second volume occurring in 1692. In 1694, under ecclesiastical pressure, Sor Juana renounced all literary activity, sold her library and scientific instruments, and signed in blood a profession of faith in which she described herself as "the worst of all." She died on April 17, 1695, during an epidemic. An unfinished poem and some money were found in her cell. The third volume of her collected works was published in 1700.

Her poetry, especially "Dream" (1692), which is less a description of a dream than an allegory of the acquisition of knowledge, has been read as a feminist interpretation of Cartesian thought and, alternatively, as the most complex instance of the confluence of hermetic science—as exemplified by the works of the German Jesuit Athanasius Kircher (1601–1680)—and literature in the Baroque period. However, it is in her autobiographical works, such as the "Letter of Monterrey" (1681), addressed to her confessor, and the public "Response to Sor Filotea de la Cruz" (1691), a true *apologia pro vita sua*, that her most explicit critique of the limitations placed on the intellectual and scientific endeavors of women by the colonial patriarchal religious and political hierarchies can be found. In defense of her right to engage in intellectual activity, Sor Juana identifies in the "Response" a genealogy of women intellectuals—including such diverse examples as Hypatia of Alexandria (370–415), Saint Gertrude the Great (1256–1311), and Queen Christina of Sweden—and argues that humanistic and scientific pursuits are compatible with theology and necessary for its comprehension. Sor Juana also defended the importance of what in her time were spaces and activities for scientific knowledge, claiming that "Aristotle would

Sor Juana Inés de la Cruz, 1651–1695. Sor Juana Inés de la Cruz was a Mexican nun renowned for her phenomenal knowledge of the arts and sciences of her day, her devotion to scientific inquiry, and her lyric poetry. (*Philadelphia Museum of Art/Corbis-Bettmann. Reproduced by permission.*).

have written more if he had cooked" (Sor Juana 1951–1957, p. 460).

Although Sor Juana's tragic fate demonstrates that her words were ignored by the misogynist and antirational establishment of seventeenth-century colonial Mexico, her criticisms of the ethical limitations of patriarchal science and knowledge have begun to be acknowledged as prefiguring feminist approaches to the study and history of science.

JUAN E. DE CASTRO

SEE ALSO *Colonialism and Postcolonialism; Feminist Ethics.*

BIBLIOGRAPHY

Hill, Ruth Ann. (2000). *Sceptres and Sciences in the Spains: Four Humanists and the New Philosophy (Ca. 1680–1740)*. Liverpool, England: Liverpool University Press. In this study, the author analyzes the influence of the scientific philosophies developed by Francis Bacon and Pierre Gassendi, among others, on Sor Juana and three other Spanish and Spanish-American humanists (Gabriel Álvarez de Toledo, Pedro de Peralta Barnuevo, and Francis Botello de Moraes).

Juana Inés de la Cruz, Sor. (1951–1957). *Obras Completas de Sor Juana Inés de la Cruz*, ed. Alfonso Méndez Plancarte. Mexico City, Mexico: Fondo de Cultura Económica. This is the standard Spanish-language edition of Sor Juana's complete works.

Juana Inés de la Cruz, Sor. (1997). *Poems, Protest, and a Dream: Selected Writings*, trans. Margaret Sayers Peden. New York: Penguin. A good translation of some of Sor Juana's key texts.

Merrim, Stephanie, ed. (1991). *Feminist Perspectives on Sor Juana Inés de la Cruz*. Detroit: Wayne State University Press. An important collection of articles that study Sor Juana from feminist perspectives.

Paz, Octavio. (1988). *Sor Juana, or, the Traps of Faith*, trans. Margaret Sayers Peden. Cambridge, MA: Harvard University Press. Written by Mexico's best-known poet and essayist and a Nobel Prize winner himself, this book is a convincing reconstruction of Sor Juana's intellectual and social environment.

Trabulse, Elías. (1995). "El universo científico de Sor Juana Inés de la Cruz." *Colonial Latin American Review* 4(2): 40–50. Trabulse is one of the foremost historians of science during the Mexican colonial period and this is probably the most readily available in the United States of his articles on Sor Juana.

INTERNET RESOURCES

"The Sor Juana Inés de la Cruz Project." Available from http://www.dartmouth.edu/~sorjuana. On this Website one can find Sor Juana's complete works, as well as some of the most important critical articles written on her.

JUNG, CARL GUSTAV

• • •

Psychologist Carl Gustav Jung (1875–1961), who was born in the village of Kessweil, Switzerland on July 26, and died on June 6 in Zurich was, along with Sigmund Freud (1856–1939), a creator of depth psychology. His controversial research in this area has ethical implications for both makers and users of modern technology. Jung received an undergraduate degree in psychiatry at the University of Basel and completed his doctoral studies at Burghölzli mental hospital in 1902. In 1907 he achieved international recognition with his seminal study of dementia praecox (schizophrenia), leading to a five-year collaboration with Freud, the originator of psychoanalysis. By 1912, however, Jung found his ideas diverging from those of Freud, and from that point until the end of his life, Jung's intellectual journey was both creative and independent.

Like his former mentor, Jung was determined to penetrate and comprehend the human psyche at the deepest possible level. Unlike Freud, who emphasized the

Carl Jung, 1875–1961. A Swiss psychologist and psychiatrist, Jung was a founder of modern depth psychology. (© *Bettmann/Corbis.*)

central importance of childhood experience in the understanding of neuroses, Jung focused on adult psychology, treating patients whose neuroses did not seem rooted in infantile experiences and fantasies. Among Jung's now-familiar concepts are the personality traits of introversion and extroversion; psychological types (which lead to the standardized Myers-Briggs typology test); stage of life distinctions, including description of the mid-life crisis; primitive mental frameworks called archetypes embedded in a collective unconscious; and the notion of the *Shadow*, a part of the psyche all but inaccessible to the conscious mind but often revealed in dreams.

Jung's body of work, together with that of Freud and Alfred Adler (1870–1937), formed the basis of modern psychoanalytic techniques. These methods of treating mental disorders are today used alongside behavioral and cognitive therapy and (increasingly) psychoactive drugs. Criticism of Jung has tended to focus on the teleological (i.e., that psychic events have a purpose towards future development) and mystical elements of his thought, a significant source of the latter being his explorations of his own complex psyche. His belief in *synchronicity*, a non-causal linkage of mental and physical phenomena, has also been criticized as speculative and without scientific foundation.

Can the concepts of the collective unconscious and the Shadow help people to better understand their connection to the natural world and to their own technological creations? Prominent Jungian psychologists James Hillman (b. 1926), Stephen Aizenstat, Marie-Louise von Franz (1915–1998), and Robert Sardello have postulated that psychological health in the modern world may demand less focus on the narrow confines of the human mind and more on the connection of the human mind, both conscious and unconscious, with the rest of the natural and technological world. Historian Theodore Roszak (b. 1933) has suggested that an *ecological unconscious* links the human psyche with the natural world just as Jung's collective unconscious links human beings with each other, while biologist Edward O. Wilson (b. 1929) has argued that evolution has built into human beings an innate connection with and affinity for the natural world that should be explored by psychologists.

Jung himself was much concerned with the impacts of modern life on the psyche. Four years before his death, he published *The Undiscovered Self*, in which he argues that European civilization's obsession with the externalities of life had left largely untouched the mysteries of the human mind.

The psyche, which is primarily responsible for all the historical changes wrought by the human hand on the face of this planet, remains an insoluble puzzle and an incomprehensible wonder, an object of abiding perplexity—a feature it shares with all of Nature's secrets. In regard to the latter, says Jung, human beings still have hope of making more discoveries and finding answers to the most difficult questions. But in regard to the psyche and psychology there seems to be a curious hesitancy to explore.

Jung's fear was that humankind's collective Shadow, empowered by modern technology, could be released destructively in all its irrational fury. "The more power man had over nature, the more his knowledge and skill went to his head, and the deeper became his contempt for the merely natural and accidental, for all irrational data—including the objective psyche, which is everything that consciousness is not" (Jung 1957, p. 47).

Failure to advance self-understanding thus becomes, in Jung's view, a dangerous moral problem:

It is not that present-day man is capable of greater evil than the man of antiquity or the primitive. He merely has incomparably more effective means with which to realize his propensity to evil. As his consciousness has broadened and

differentiated, so his moral nature has lagged behind. That is the great problem before us today. *Reason alone no longer suffices*. (Jung 1957, p. 54, Jung's emphasis)

In *Memories, Dreams, Reflections* (1961), Jung's personal memoir completed just weeks before his death, he stresses that the solution to the problem of evil lies in self-knowledge, to be arrived at through psychological inquiry:

Today we need psychology for reasons that involve our very existence ... [W]e stand face to face with the terrible question of evil and do not know what is before us, let alone what to pit against it. And even if we did know, we still could not understand "how it could happen here." (Jung 1961, p. 331)

His argument for the necessity of such psychological knowledge remains a basic challenge for the future development of scientific technology. For Jung, solutions to the problems of evil do not lie in simply extending power over nature, but in better understanding humankind and its place in the universe.

WILLIAM M. SHIELDS

SEE ALSO *Freud, Sigmund; Psychology.*

BIBLIOGRAPHY

Bair, Dierdre. (2003). *Carl Jung: A Biography.* Boston: Little, Brown.

Jung, Carl Gustav. (1969). *Collected Works,* trans. Richard Francis Carrington Hull. Princeton, NJ: Princeton University Press. Definitive English edition of Jung's complete works and letters.

Jung, Carl Gustav. (1989 [1961]). *Memories, Dreams, Reflections,* trans. Richard Winston, and Clara Winston. New York: Random House. Jung's autobiography, oriented towards personal transformations and inner discoveries rather than narrative.

Jung, Carl Gustav. (1990 [1957]). *The Undiscovered Self,* trans. Richard Francis Carrington Hull. Princeton, NJ: Princeton University Press. Jung stresses the importance of exploring the human psyche.

Main, Roderick. (1997). *Jung on Synchronicity and the Paranormal.* Princeton, NJ: Princeton University Press.

Roszak, Theodore, Mary E. Gomes, and Allen D. Kanner, eds. (1995). *Ecopsychology: Restoring the Earth and Healing the Mind.* San Francisco: Sierra Club Books. Collection of essays, with introductions by Lester Brown and James Hillman, exploring how the health of the earth affects the minds of human beings.

Roszak, Theodore. (2002). *The Voice of the Earth: An Exploration of Ecopsychology,* 2nd edition. Kimball, MI: Phanes Press. The historian's latest work on the connection between the human psyche and the planet Earth.

Wilson, Edward O. (1986). *Biophilia*. Cambridge, MA: Harvard University Press. Prominent biologist argues that love of all life can be expressed by the "conservation ethic."

JÜNGER, ERNST

• • •

Ernst Jünger (1895–1998) was a German soldier and a controversial author who was best known for his militarism and prophetic descriptions of a new world being created by the interplay of nationalism, industrialization, and advances in technology. Born in Heidelberg on March 29, Jünger served on the western front in World War I. During the interwar years he studied entomology, contributed to several right-wing journals, and criticized both the Weimar Republic and the National Socialists. Although politically opposed to many aspects of Adolf Hitler's regime, Jünger served as an officer in the German army in World War II. After the war he continued to write novels, including prescient depictions of dystopias, and pioneered the prose style now called magic realism. His work was independent and dispassionate, indifferently observing and commenting on historical and social developments. A longtime friend of and influence on the philosopher Martin Heidegger (1889–1976), Jünger died in Wilflingen, Germany, on February 17.

Ernst Jünger, 1895–1998. German author Jünger was one of the most original and influential German writers and intellectuals of the 20th century. (© *Sophie Bassouls/Corbis Sygma*.)

World War and Mobilization

World War I left a lasting impression on Jünger. Three characteristics of that conflict shaped his view of the world: the destructive power of the new armaments, their lethality, and the consequential subordination of individual courage to the power of machines. In the end whoever made the best use of the war industry would be victorious. The new weapons changed the character of killing and dying because violence was inflicted at a distance and on a massive scale. The person who falls is not seen, his last breath is not heard, and his blood does not splatter the aggressor. At a distance death is wrapped in indifference and anonymity. The slaughter becomes more sudden, massive, and above all reciprocal.

Jünger's first book, *In Stahlgewittern* (1920), is a memoir of his four years on the western front. In this work he showed his ideological embrace of technology even as he struggled with the tension between human will and the power of mechanized warfare. His interpretation of the larger meaning of the war is presented in

Die Totale Mobilmachung (1931). The title refers to the fact that the mobilization of all forces, including industrial and productive capacity, becomes decisive in the definition of conflicts. Jünger read these phenomena as signs of a historical transition. A new reality was emerging, dominated by the "figure of the worker."

In his single most influential work, *Der Arbeiter: Herrschaft und Gestalt* (1932), Jünger developed his vision of a radically antibourgeois future based on total mobilization. This work often is interpreted as a totalitarian or authoritarian rebuttal of the bourgeois conception of freedom, the market economy, and the liberal nation-state. In it Jünger envisioned the "worker" as the destiny of the coming age, to be characterized by technocratic control in place of the anarchy of liberal individualism. The bourgeois individual will be replaced by the worker "type" in an "organically constructed" political order. Freedom will become identical to obedience. Individuals will be folded into the unity of the whole. Both this metaphysical substructure, or gestalt, of the worker and Jünger's political philosophy of detachment deeply influenced Heidegger.

Der Arbeiter also is predicated on Jünger's concept of "heroic realism," which seeks out the danger that bourgeois reason domesticates by making all risk calculable. In opposition to bourgeois concerns for comfort and convenience, modern technology has an inner destructive character as "the way in which the gestalt of the worker mobilizes the world" (Junger 1932, p. 156). The conversion of all activity into some kind of work is a manifestation of the predominance of this work character. Indeed, the term *worker* does not so much designate a class or social affiliation as it defines a *Lebenstand*, or "state of life," to which Jünger attributed the formative power emerging in history. Jünger thus disassociates his conception from the proletariat of Marxism. It is indicative of Jünger's political complexity that *Der Arbeiter* was regarded by the right as communistic and by the left as fascist.

Total mobilization and the predominance of the worker express a new reality in which the efficacy of an action has priority over its legitimacy. In this sense Jünger's philosophy is aligned with Friedrich Nietzsche's (1844–1900) "active nihilism" and Heidegger's "empire of technics." In fact, Jünger's greatest influence on Heidegger stems from this metaphysical analysis of technology as an essential way of being in the world.

Outside National Socialism

Jünger's *Auf den Marmorklippen* (1939) is a covert criticism of National Socialist tyranny. A poetic and obscure book that seems to aestheticize violence, it presents types more than concrete characters and in that way achieves a general critique of totalitarianism. Indeed, by the time of the 1938 *Krystall Nacht* (the Nazi attack on Jewish businesses in Germany) it was evident to Jünger that the National Socialist regime was essentially the same crude form of proletariat totalitarianism as the Bolshevik regime in Russia.

Gläserne Bienen (1957) raised the moral dilemma of the use of technology in society and foreshadowed modern developments in robotics and nanotechnology, presenting a world where "even the molecules were controlled." The novel questioned how people might retain a sense of place and identity in light of the accelerating pace at which the old is replaced by the new. It also expressed a growing contempt for both an impersonalized, bureaucratized society and the scientific, materialistic worldview that discredits meaning and purpose and cosiders humans to be lowly cosmic accidents.

Jünger did not produce a systematic philosophy, but his complex, inconsistent, and fierce independence often captured an emerging technoscientific world in an indifferent but therefore critical gaze. Jünger disdained any nostalgic form of antitechnology but refused to hail a world of sustained technological progress culminating in rationality and moral decency. His heroic realism is a qualified yes that comes out of an encounter with the emerging: It is as useless to attempt to avoid the power of modern technology as it is naive to ignore its enormous potential for destruction.

MARCOS GARCÍA DE LA HUERTA
TRANSLATED BY JAMES A. LYNCH

SEE ALSO *German Perspectives; Heidegger, Martin.*

BIBLIOGRAPHY

Jünger, Ernst. (1920). *In Stahlgewittern.* [Storm of steel]. Translated into English by Basil Creighton. (1929). *Storm of Steel: From the Diary of a German Storm-Troop Officer on the Western Front.* Garden City, NY: Doubleday.

Jünger, Ernst. (1931). *Die Totale Mobilmachung.* Translated into English by Joel Golb and Richard Wolin. (1993). "Total Mobilization." In *The Heidegger Controversy: A Critical Reader,* ed. Richard Wolin. Cambridge, MA: MIT Press.

Jünger, Ernst. (1932). *Der Arbeiter: Herrschaft und Gestalt.* Hamburg: Hanseatische Verlagsanstalt. Sections 44 through 57 are translated as "Technology as the Mobilization of the World through the Gestalt of the Worker" in Carl Mitcham and Robert Mackey, eds. (1983). *Philosophy and Technology: Readings in the Philosophical Problems of Technology.* New York: Free Press.

Jünger, Ernst. (1939). *Auf den Marmorklippen.* Translated into English by Stuart Hood. (1947). *On the Marble Cliffs.* London: John Lehmann.

Jünger, Ernst. (1957). *Gläserne Bienen.* Translated into English by Louise Bogan and Elizabeth Mayer. (1960). *The Glass Bees.* New York: Noonday.

Jünger, Ernst. (1977). *Eumeswil.* Translated into English by Joachim Neugroschil. (1993). *Eumeswil.* New York: Marsilio.

Nevin, Thomas. (1996). *Ernst Jünger and Germany: Into the Abyss 1914–1945.* London: Constable. A biography that covers the first half of Jünger's life.

JUSTICE

• • •

Justice has to do with the distribution of benefits and burdens, rewards and punishments. Among the most important benefits and burdens of contemporary society are science and technology, their products and their costs. Although science and technology are involved with the administration of legal justice in many ways—from their uses in forensics to identify and prosecute criminals to the testimony of scientific and engineering experts in civil cases—the primary focus in this entry will be on the nature of justice in its own right, pointing out some implications for science and technology.

Versions of Justice

As an instrument for the distribution of benefits and burdens, the general concept is clear, but the various interpretations of the concept, and its applications are more contentious. Is justice a transcendent reality, as Plato held? A formal property having to do with proportional distribution, as Aristotle contended? Simply what contracting parties invent in mutually self-interested agreements, as Thomas Hobbes argued? An artificial construct as David Hume maintained? Or does justice have to do with ownership, a rendering to each according to one's due, as Polemarchus reports in Plato's *Republic* (331e) was the definition of the poet Simonides—a view also advanced by the Roman legal philosophers Cicero and Ulpian, as well as Thomas Aquinas? Is it possible that scientific and technological progress promote justice, especially the just power of human beings over the unjust forces of nature, as Francis Bacon argued? Or is a kind of natural justice thereby diminished, as Socrates in the *Republic* (372e) and Jean-Jacques Rousseau, in quite different ways, both proposed?

The traditional symbol of justice is a woman wearing a blindfold, holding a pair of equally balanced scales in one hand, and a sword in the other. The metaphor points to the symmetry between the quality of human judgment on one side and the rewards or punishments on the other. Justice is blind to all irrelevant considerations such as birth or social status or race or gender, and is concerned only with giving one what is deserved.

The earliest definition of justice in the West is the Simonides quote from Plato's *Republic:* "Justice is to render each person his due," giving to each person what each deserves, based on the person's character traits, including ability, virtues, and vices. If one is excellent, a suitable reward is appropriate. If one is vicious, punishment is warranted. A mediocre individual earns a mediocre benefit. Indeed Plato's *Republic* describes a meritocracy, made up of people in three classes, categorized according to their abilities.

The classic conception applies both to distributive and retributive versions of justice. Distributive justice concerns the distribution of benefits and burdens. Retributive justice deals with punishments and rewards.

Immanuel Kant argued that not only should people who are good be rewarded with happiness in proportion to their goodness, but people who willfully do bad things should be unhappy in proportion to their bad intentions. Following this thought, he argued that crimes such as murder justified imposition of the death penalty. Kant used this thinking as a premise for the existence of God and life after death, arguing that justice required a god and a future existence for persons to receive their just rewards and punishments.

This classic view has been held by many philosophers throughout history. It is found in the Hindu and Buddhist idea of karma, which holds that each person will be reincarnated according to individual moral character, and in the Bible, which states, "whatsoever a man soweth that shall he also reap" (*Gal.* 6:7). Somewhat unexpectedly, even Karl Marx in his labor theory of value (a worker should be rewarded for the full value of his work) seems to share the classical theory of just desert. The utilitarian philosopher John Stuart Mill also advocated a version of this doctrine, deeming it the central meaning of justice, which in turn signifies simply the most stringent requirements of utilitarian morality. Is justice simply the secular analogue to the religious doctrine of rewards and punishment according to merit? Contemporary political philosophers, such as John Rawls and Derek Parfit rejected or qualified the salience of this classic conception of justice as desert by arguing more egalitarian or need-based conceptions.

Contemporary Conceptions

In current discussions Rawls's *A Theory of Justice* (1971) and Robert Nozick's *Anarchy, State, and Utopia* (1974) remain common reference points. Rawls argues a view of *justice as fairness* defined by that impartial, hypothetical contract that people would adopt from behind a veil of ignorance regarding with what benefits or burdens they might begin their lives in a social order. Extending a perspective developed in John Hospers's *Libertarianism* (1971), Nozick defends justice as grounded in rights to liberty and ownership. Other contemporary analyses of justice include arguments by Parfit (1984), that justice requires some consideration of need; and by Michael Walzer (1983) and Nicholas Rescher (2002) that justice is not a single concept, but a plurality of concepts relative to different social contexts.

According to Hume, questions of justice typically arise when, in situations of scarcity, human beings seek to adjudicate between competing claims for limited goods. Such goods might be material benefits, social prestige, or power—any of which could be closely associated with science or technology. Suppose 100 competitors apply for a highly desirable position such as candidate at a leading graduate program in science or director of a major engineering project. What are the correct moral and legal criteria by which to decide who should be granted the position? Should selection be based on technical knowledge, need, utility, previous effort, likely contribution to be made? Should market forces be a factor? Race, ethnicity, or gender? If in the past blacks or women or the disabled were systematically discriminated against, should affirmative action come into play?

Or consider the use of kidney dialysis machines in a county hospital that can afford only five machines, but has a waiting list of twenty or thirty people. How should doctors decide which five people should be treated? By lottery? By a process of first come first served? By greatest need? By merit? By desert? By utility, for example, if one of the candidates is the mayor of a town that is part of the county and who has served the community well for many years? Or should a complex set of factors (including age, contribution, responsibilities, merit, and need) be used?

The most significant controversial issue in the debate over distributive justice is that of economic justice. How should wealth be divided up in society? Should the free enterprise system determine how much money and wealth people end up with or should an effort be made to redistribute wealth through some sort of income tax policy? Should there be a vigorous welfare program, ensuring that no one falls below a certain economic threshold?

Types of Justice: Formal and Material

Theories of justice may be divided into *formal* and *material* types. A formal theory of justice provides the formula or definition of justice without directly filling in the content or criteria of application. Material theories of justice specify the relevant content to be inserted into the formulas. They dictate what the relevant criterion is. The classical principle of formal justice, based on Book V of Aristotle's *Nicomachean Ethics* is that "equals should be treated equally and unequals unequally." The formula is one of proportionality:

A has X of P = A should have X of Q

B has Y of P = B should have Y of Q

That is, if person A has X units of a relevant property P, and B has Y units (where Y is more or less than X), then A should pay proportionally more or less of the relevant burden Q than B. For example, if A has worked eight

hours at a job and B only four hours, and time worked is the relevant criterion for reward, A should be paid twice as much as B.

The formal principle is used in law in the guise of *stare decisis*, the rule of precedent—like cases should be decided in like manner. The principle applies not only to the case of distributive justice, but also of retributive justice or punishment and commutative justice, in which obligation is based on a promise or contract that requires fulfillment.

The formal principle of justice seems reducible to the principle of universalizability: Treat like cases similarly unless there is a relevant difference, which itself is simply the principle of consistency. Be consistent in decisions. If there is no relevant difference between agents, treat them similarly. Insofar as there is no relevant moral difference between the sexes, this applies to the morality of sexual relations. If it is all right for Jack to engage in premarital sex, then it is also all right for Jill to engage in premarital sex; but if it is immoral for Jill to engage in premarital sex, it is also immoral for Jack. The formal principle of justice does not indicate whether some act is right or wrong, but simply calls for consistency. If people were content to live only with the formal principle, they might treat others very badly and still be considered just. As player Henry Jordan once said of Vince Lombardi, the legendary coach of football's Green Bay Packers, "He treated us all the same—like dogs."

Some philosophers, such as Stanley Benn, believe that the formal principle of equal treatment for equals implies a kind of *presumption* of equal treatment of people. But there are problems with this viewpoint. As Joel Feinberg (1970) points out, sometimes the presumption is for unequal treatment of people. Suppose that a father suddenly decides to share his fortune and divides it in two, giving half to his oldest son and half to his neighbor's oldest son, but nothing to his other children. This kind of impartiality is arguably misguided and, in reality, unjust. Society must determine in which respect people are equal and so deserve the same kind of treatment; this seems to be a material problem, not a purely formal one. In other words, Benn confuses an *exceptive principle* (Treat all people alike except when there are relevant differences among them) that is formal with a *presumptive principle* (Treat all people alike *until it can be shown* that there are relevant differences among them).

The formal principle does not tell which qualities determine which kinds of distribution of goods or treatment. Thus material principles are needed to supplement the formal definition. Aristotle's own material

Encyclopedia of Science, Technology, and Ethics

principle involved merit: People are to be given what they deserve. A coach could justifiably treat his players like dogs only if they were doglike; otherwise, he should treat them more humanely.

Types of Justice: Patterned and Nonpatterned

Material theories of justice may be divided into *patterned* and *nonpatterned* types of justice. A patterned principle chooses some trait(s) that indicates how the proper distribution is to be accomplished. It has the form:

To each according to —————.

Robert Nozick (1974) rejects patterned types of principles, such as those of Aquinas, Rawls, and Rescher, because this type of attempt to regulate distribution constitutes a violation of liberty. The point can be illustrated by considering how a great inventor can justly upset the patterned balance. Suppose the existence of a patterned situation of justice based on equality. Imagine also that there is a great demand for some inventor's product and that people are willing to pay the inventor well for it. If millions of people pay for the product, the inventor takes home a great deal more than the patterned formula allows, but seems to have a right to it. Nozick's point is that, in order to maintain a pattern, one must either interfere to prevent people from allocating resources as they wish, or intervene to take from people resources that others have transferred to them.

Nozick argues for a libertarian view of nonpatterned justice, which he calls the *theory of entitlement*. A distribution is just if all people have those things to which they are entitled. In determining what people are entitled to, the original position of holdings or possessions is an important factor, as is what constitutes a just transfer of holdings. Borrowing from John Locke's theory of property rights, Nozick argues that people have a right to any possession so long as ownership does not worsen the position of anyone else.

Continuing Debates

As in the past, justice in the early twenty-first century remains a widely contested concept. The main current rival positions are the classic theory of just desert, egalitarian theory of distribution according to need, and rights theories. The challenge for political philosophy is to sort out the competing claims of such theories and make sense of people's deepest but conflicting intuitions—especially with regard to the uses and influences of science and technology.

With regard to retributive or criminal justice, the scientific study of human behavior has, for instance, raised important questions about levels of human accountability. To what extent should psychology and neuroscience inform the legal justice system? Forensics and studies of evidence that, for instance, question the reliability of eyewitness accounts, along with increased reliance on scientific experts, likewise have implications for court procedures. Some philosophers such as Brian Barry (1989) argue the importance of the sciences of game theory and decision theory to analyses of justice.

With regard to distributive justice, science and technology, by their discoveries and inventions especially in the areas of new drugs and lifesaving medical devices, create new challenges for justice. How shall society use these drugs and therapies? Should drugs for AIDS be distributed gratis to African countries that cannot afford to pay the market price? Is it just for pharmaceutical companies, which produced the drugs, to charge the same price to all buyers, or should allowances be made for depth of need and relative ability to pay?

With regard to science and technology in general, what constitutes a just distribution of the benefits of scientific discoveries and engineering inventions? Do owners of patents have an obligation to make some sacrifice in foregoing potential profits from their work to enhance distribution? Or does justice allow them to sell their work to the highest bidder, independent of the social result? Does the state promote justice through the regulation of science and technology, or is regulation properly constrained by respect for liberty and property? In advanced technological societies where, according to Langdon Winner (1986), technological design can be a hidden form of politics, and for Ulrich Beck (1986), the avoidance of risk is now a scarce commodity, do different theories of justice imply different responsibilities for scientists, engineers, citizens, politicians, or corporations? Indeed in a social system in which corporations are granted the status of legal persons, and serve as major vehicles for scientific and technological research, development, and innovation, what concept of justice best enlightens responsibilities in the public realm?

Finally because of technological transformations of the public realm, questions of justice have been extended both spatially and temporally. Increased telecommunications promotes questions of international justice. Increased ability to impact future generations raises questions of intergenerational justice.

LOUIS P. POJMAN

SEE ALSO *Death Penalty; Environmental Justice; Equality; Grant, George; Rawls, John; Science, Technology, and Law.*

BIBLIOGRAPHY

Barry, Brian. (1989). *Theories of Justice*. Berkeley: University of California Press.

Beck, Ulrich. (1986). *Riskogesellschaft: Auf dem Weg in eine andere Moderne*. Frankfurt: Suhrkamp. English translation: *Risk Society: Towards a New Modernity*, London: Sage, 1992.

Benn, Stanley. (1967). "Justice." In *Encyclopedia of Philosophy*, ed. Paul Edwards. New York: Macmillan.

Feinberg, Joel. (1970). *Doing and Deserving: Essays in the Theory of Responsibility*. Princeton, NJ: Princeton University Press.

Hospers, John. (1971). *Libertarianism: A Political Philosophy for Tomorrow*. Los Angeles: Nash Publishers.

Nozick, Robert. (1974). *Anarchy, State, and Utopia*. New York: Basic Books.

Parfit, Derek. (1984). *Reasons and Persons*. New York: Oxford University Press.

Rawls, John. (1999). *A Theory of Justice*, revised edition. Cambridge, MA: Harvard University Press. Originally published in 1971.

Rescher, Nicholas. (2002). *Fairness: Theory and Practice of Distributive Justice*. Somerset, NJ: Transaction Publishers.

Walzer, Michael. (1983). *Spheres of Justice: A Defense of Pluralism and Equality*. New York: Basic Books.

Winner, Langdon. (1986). *The Whale and the Reactor: A Search for Limits in an Age of High Technology*. Chicago: University of Chicago Press.

JUST WAR

* * *

The term *just war* refers to the major moral tradition of Western culture that deals with the justification and limitation of the use of force by public authority. Just war tradition has particular relevance for moral reflection about many scientific and technological developments related to military affairs.

Historical Background

Just war tradition can be traced back to Saint Augustine (354–430) in the fourth and fifth centuries and through him to the Old Testament and the ideas and practices of classical Greece and Rome. Augustine, however, did not write systematically or at length about the idea of just war; his treatment of these issues is found in passages about the use of force in works on various topics. A coherent, systematic body of thought and practice on just war did not emerge until the Middle Ages. The thought of Augustine and other earlier Christian writers was drawn together by the canonist Johannes Gratian, whose *Decretum* dates to the middle of the twelfth century. Two generations of canonists who built on Gratian's work, the Decretists and the Decretalists, took the development of the just war idea into the thirteenth century. In the second half of that century theologians, including most notably Thomas Aquinas (1224–1274), placed the canonical materials in an overarching theological framework that showed both a strong dependence on Augustine's thinking and a new effort to give ideas about just war a footing in natural law.

During the thirteenth century but more during the fourteenth and fifteenth centuries, secular factors began to reshape this canonical and theological concept into a broad cultural consensus. These factors were the growing study of Roman law, especially the idea of *jus gentium* (law of peoples or nations); the maturation of the chivalric code as a guide to the conduct in arms of the international brotherhood of knights; and increased reflection on the experience of governing found in works dealing with the characteristics of a good ruler.

By the end of the Hundred Years War in the mid-fifteenth century the resulting synthesis (seen particularly in writers such as the theologian and scholar Honoré Bonet [1340–1410] and the poet and historian Christine de Pisan [1363–1430]) had defined a cultural consensus in western Europe on the justified use of armed force and the restraints to be observed in using that force. This consensus included the major factors that continue to define the idea of a just war. From canon law and theology came the requirements that for a resort to armed force to be just it must be undertaken on the authority of a sovereign and for the public good; be for a just cause, defined as defending the common good, retaking that which had been taken wrongly, and punishing evil; and right intention, defined negatively as the avoidance of self-aggrandizement, bullying, implacable hatred, and so on, and positively as aiming to restore the peace that had been violated.

The chivalric code joined canon law to provide two kinds of restraint on the employment of force: noncombatant immunity, defined by lists of persons not normally involved in war and thus not to be subjected to direct harm in war, and limits on means, defined by efforts to ban certain weapons (specifically arrows and siege machines) as *mala in se*. The *jus gentium* and the growing consolidation of political authority reinforced these developments in useful ways: the former by placing them in a broader theoretical framework to define relationships among autonomous political communities and the latter by sovereigns' adoption of these rules both in the use of force to maintain public order and in warfare against external threats.

In this manner the just war tradition was passed to the modern era. Theological and secular theorists of the law of nations, including the theologian Francisco de Vitoria (1492–1596) in the sixteenth century and the jurist Hugo Grotius (1583–1645) in the seventeenth, placed the inherited just war tradition in the context of a general theory of international law based on natural law and the *jus gentium*. After Grotius and as a result of the international order created by the Peace of Westphalia (1648), emphasis on the former part of the tradition, by then called the *jus ad bellum*, began to be reduced as sovereigns' rights to use force were redefined as *compétence de guerre* at the same time that a new emphasis was placed on the restraints to be observed in the use of force, the *jus in bello*.

This has been the pattern of the development of the just war tradition during the modern period. Beginning in the 1860s with the work of Francis Lieber and the U.S. Army's General Orders No. 100 of 1863 and, at almost the same time, the international adoption of the First Geneva Convention, positive international law has played a major role in defining the just war *jus in bello*. Through much of the nineteenth century and continuing into the nuclear age, moral thought on war has focused on efforts to rule out recourse to armed force by states, in effect denying that a *jus ad bellum*, a justification of the resort to armed force, exists any longer, or severely restricting the terms of such justification. During this period, because of its concentration on eliminating war, moral thought effectively lost sight of the just war *jus in bello*. At the same time, however, the increasing codification of international law reframed the tradition's *jus in bello* as positive-law rules for the conduct of nations in war.

The law of armed conflict in international law remains one of the important arenas for the efforts to restrain war first defined in the just war tradition. In moral thought, largely as the result of work by the theologian Paul Ramsey (1913–1988) and the political philosopher Michael Walzer (b. 1935) and public debate occasioned by the U.S. Catholic bishops' 1983 pastoral *The Challenge of Peace*, just war thinking has reemerged in American and some European debates over the use of armed force, informing not only the religious and philosophical spheres but also public policy discussions and professional military education. Just war is studied in all the service academies and the war colleges and by military lawyers, and it is a common topic in academic and policy-oriented conferences and workshops on military issues.

Science and Technology

Both historically and in recent debates just war tradition has responded to developments in the science and technology of the use of force. In the Middle Ages this involved efforts to eliminate the use of weapons that were deemed too harmful or destructive. Specifically, there was an effort to ban crossbows and bows and arrows, which could penetrate armor and kill, whereas the normal weapons of knights—swords, maces, and lances—were likely to injure but not kill armored opponents. Siege weapons capable of causing heavy and indiscriminate damage when used against fortified places were also the target of a ban.

These themes were carried forward into efforts to restrict or eliminate certain weapons or uses of weapons in positive international law. The first Hague Conference (1899) sought to ban exploding bullets for being too lethal and tending to inflict especially cruel wounds. That conference sought to ban asphyxiating gases, though this did not become positive law until the 1925 Geneva Protocol on gas warfare. Various efforts, beginning from the first Hague Conference, have been made to prohibit bombardment of unfortified population centers from the land, sea, and air. Since World War II international conventions have been adopted prohibiting the use of chemical and biological weapons as "weapons of mass destruction," and the nuclear proliferation treaty has sought to restrict possession of nuclear weapons as a way to limit the likelihood of their use. A 1980 United Nations Convention prohibits or restricts the use of certain conventional weapons "deemed to be excessively injurious or to have indiscriminate effects." The 1997 Ottawa Convention, responding to technologies that have made antipersonnel mines cheap, difficult to detect, and ubiquitous, formally prohibits their production, stockpiling, transfer, and use.

These are all examples from positive international law, a major modern carrier of the just war tradition. In the moral debate some have argued that the entire technology of contemporary warfare—not only weapons of mass destruction, including nuclear weapons, but also conventional weapons because of their ability to produce widespread death and destruction—is disproportionately and often indiscriminately harmful. This position, often called "modern-war pacifism" (including nuclear pacifism as one of its forms) holds that the technology of modern warfare is so destructive that the moral requirements of the *jus in bello*, avoidance of direct harm to noncombatants and of disproportionate destruction, cannot be met, and so there can be no just resort to force.

Opponents of this position, including Ramsey, Walzer, and James Turner Johnson (b. 1938), distinguish between the availability of highly destructive weaponry and the decision about how to fight: The latter is a moral decision, and it implies moral control over whatever means are available. In the debates over nuclear weapons during the early 1980s this difference of judgment about the technology of warfare led to two sharply different policy conclusions. Nuclear pacifists argued against nuclear weapons as inherently immoral and against the development of targeting technologies intended to make them more accurate and thus more discriminating. Others argued that development of such capabilities was a moral imperative both because it could reduce direct harm to noncombatants and because it opened the door to the development of lower-yield warheads, including conventional explosives, that could perform the same strategic and tactical functions as high-yield nuclear and thermonuclear warheads.

Questions of Technological Superiority

The policy decision at that time was to continue developing more accurate targeting technologies and delivery systems. Since then this line of development has matured progressively to produce a "revolution in military affairs" characterized by laser- and satellite-guided bombs and missiles, stealth technology that allows airplanes to get close enough to their targets to enable direct guidance of weaponry onto a target, drone airplanes and satellite imaging to identify and target enemy armed forces without collateral damage to noncombatants, and increasingly sophisticated means of gathering enemy intelligence to lower the levels of force needed for combat.

These developments first became general knowledge with publicity over the "smart bombs" of the 1991 Persian Gulf War. The use of such technology also marked the bombing of Serbia in the conflict over Kosovo (1999), and it was both ubiquitous and decisive in the conflicts in Afghanistan (2001) and Iraq (2003), where in the latter the technological superiority of the U.S. and British forces made possible a campaign that used far lower numbers of troops than previously would have been necessary, destroyed the Iraqi army while coalition forces suffered only a small number of casualties, and allowed bombs and missiles to destroy major Iraqi government targets with unprecedentedly low levels of collateral damage.

All this is morally significant from the standpoint of the just war tradition, for even in an age of weapons of massive destructive power such technology allows armed force to be used in a way that honors the just war requirements of noncombatant immunity and as low a level of destruction as possible. At the same time, from the perspective of the technologically inferior, the use of superior technology may appear to represent a refusal to accept an equal playing field in which courage and loyalty to opposing causes have a fair chance to compete with each other. What is to be made of this objection?

The latter argument cannot be used to justify means of fighting that disregard moral and legal restraints. In the moral terms of the just war tradition as well as the legal terms of the law of armed conflict, technologically superior and inferior adversaries are equally bound by the same rules. Technological inferiority is no excuse, for example, for terrorist actions against civilians or the Fedayeen Saddam's use of noncombatants as human shields in the 2003 Iraq war, both of which were clear violations of the moral concept of noncombatant immunity and the legal restrictions laid down in international law. In a conflict involving technologically asymmetrical adversaries each force is restricted, both morally and legally, to means that do not violate noncombatant immunity and do not involve prohibited weapons, such as weapons of mass destruction.

Technological asymmetry is not a new problem ushered in by precision-guided munitions. In earlier ages technological superiority was conferred by the use of Greek fire, firearms, rifled handguns and artillery, repeating rifles, the use of railroads for military transport, semaphore signaling systems and later the telegraph and radio, and the development of armored fighting vehicles. A technologically inferior armed force faces an enormous practical problem: how to match or overcome an enemy that is technologically superior. However, this is a practical problem, not a moral one. The idea of a "level playing field" means that both adversaries must play by the same rules; it does not mean that within the framework of those rules neither side may use means that it alone possesses.

The possession of superior technology, it may be argued, imposes a special moral responsibility to use that technology in ways that honor the *jus in bello* restraints. The moral rule of double effect has long been used to determine when collateral harm to noncombatants is morally allowed; by this rule such harm is allowed only when it is the indirect, formally unintended result of an attack on a legitimate military target that cannot be attacked except with such collateral harm. Thus, when an enemy places artillery next to a school or deploys troops with rifles to fire from the windows of a hospital, the artillery and the troops can be attacked despite the harm to the school and hospital and the noncombatant persons who may be inside.

However, Michael Walzer (1977) has argued that the rule of double effect also should be understood to impose a proportionality criterion; therefore, a projected attack against an otherwise legitimate target should not go forward if the collateral harm to noncombatants is judged to be disproportionate to the ends to be gained from the attack. In such cases, an alternative weapon or another means of neutralizing the target should be used or the target should be bypassed. This reasoning seems to have been employed in the targeting decisions made by U.S. forces in the 2003 Iraq conflict, in which the choice of weapons systems, the angle of attack, the time of day, fuse timing, and other factors were employed to avoid or reduce collateral damage. The possession of superior technology thus imposes an added moral burden: to use that technology to avoid harm that would be allowed in its absence.

This means that from a moral standpoint based on the just war tradition the question of the technology of warfare does not stand alone. It is also necessary to consider whether overall planning and policy, strategy, rules of engagement, means of command and control, tactics, and military training allow the use of the available technology in ways consonant with the aims of discrimination and proportionality. Not only does the U.S. military in the early twenty-first century have a virtual monopoly on the technology of the "revolution in military affairs," it is the only national military that has made operational all these elements in the channel of decision that leads toward conducting military actions within the framework required by the *jus in bello*. Arguably, the ability to conduct war more closely in accordance with just war requirements implies the moral obligation to do so. For example, carpet bombing of a mixed combatant-noncombatant area to destroy a legitimate target cannot be the moral option if precision guidance technology allows that target to be destroyed without harming noncombatants.

The question is what this implies for societies that lack such technology: Do they have the obligation to develop it, or may they not fight wars anymore? On just war reasoning, they have the moral obligation to use whatever means they have in the most moral way possible; they do not, for example, have the moral right to target civilians directly or use weapons of mass destruction, which are both indiscriminate and disproportionate. Beyond this they are obliged to try to develop more discriminate and proportionate means of fighting within the capabilities available to them and taking into

account their other responsibilities. If they cannot fight according to the minimum standards of noncombatant immunity and avoidance of weapons *mala in se*, by just war reasoning they should not fight. However, the question whether to engage in armed conflict with a technologically superior adversary is not one of morality but one of political prudence.

The moral obligation to develop more discriminating and proportionate means of fighting extends also to technologically advanced militaries. During the Vietnam War Paul Ramsey (1968) argued for the use of incapacitating gases as morally preferable to the use of weapons such as napalm and even bullets because those gases could incapacitate soldiers without killing them or producing lasting harm. The United States Defense Advanced Research Products Administration has been encouraging research and development in nonlethal weapons technologies. Just war reasoning tends to support the development and use of such weapons in principle, though any particular weapon, even if nonlethal, still would have to be judged by the standards of the *jus in bello*.

In summary, just war tradition places the use of armed force in a moral framework in which some technologies are good and others are bad. The criterion is whether a specific technology makes it possible to use military force, when justified and used on public authority for the common good, in ways that honor the principles of noncombatant immunity and minimal overall destructiveness.

JAMES TURNER JOHNSON

SEE ALSO *Aggression; Atomic Bomb; Augustine; Biological Weapons; Chemical Weapons; Military Ethics; Science, Technology, and Law; Thomas Aquinas; Weapons of Mass Destruction.*

BIBLIOGRAPHY

Best, Geoffrey. (1980). *Humanity in Warfare.* New York: Columbia University Press. A historical study of the development of international laws on war, peace, and neutrality from the eighteenth through the twentieth centuries.

Elshtain, Jean Bethke. (2003). *Just War against Terror.* New York: Basic Books. An argument for the justification of the war on terror from a just war standpoint.

Johnson, James Turner. (1981). *Just War Tradition and the Restraint of War.* Princeton, NJ, and Guildford, Surrey, UK: Princeton University Press. A historical and thematic study of the just war tradition and its relation to the conduct of war from the Middle Ages through the twentieth century.

Johnson, James Turner. (1999). *Morality and Contemporary Warfare*. New Haven, CT, and London: Yale University Press. A just war analysis of contemporary warfare.

National Conference of Catholic Bishops. (1983). *The Challenge of Peace: God's Promise and Our Response*. Washington, DC: United States Catholic Conference. A landmark pastoral letter examining the Catholic tradition on war and peace in the context of the Reagan-era debate over nuclear weapons.

Ramsey, Paul. (1961). *War and the Christian Conscience*. Durham, NC: Duke University Press. A landmark study drawing Christian just war theory from the idea of love of neighbor and applying this theory to nuclear war.

Ramsey, Paul. (1968). *The Just War: Force and Political Responsibility*. New York: Charles Scribner's Sons. A collection of essays on political ethics, the idea of just war, and moral conduct applied to nuclear war and insurgency.

Russell, Frederick H. (1975). *The Just War in the Middle Ages*. Cambridge, UK, and New York: Cambridge University Press. A thorough and detailed historical study of the development of the just war idea from the twelfth-century canonists through Thomas Aquinas and his circle in the late thirteenth century.

Walzer, Michael. (1977). *Just and Unjust Wars*. New York: Basic Books. A reconstruction of the just war idea on the basis of philosophical analysis with historical illustrations, aimed at recapturing this idea for political and moral theory.

K

KANT, IMMANUEL

• • •

Immanuel Kant (1724–1804) was born in Köningsberg, East Prussia (now Kaliningrad, Russia), on April 22 and died there on February 12, having lived such an uneventful life that one early commentator questioned whether he had one. Yet his critical philosophy constituted a watershed in Western intellectual history. For science, technology, and ethics the significance of the Kantian watershed lies in the analysis of human experience as constructive and the argument that reason has insight only into that which it produces according to its own plan. With this argument Kant developed a new critical interpretation of scientific knowledge and of ethical reason that presents both as exhibiting constructive, not to say technological, dimensions.

Prior to Kant, modern philosophy was characterized by a contest between rationalism and empiricism. Rationalists such as René Descartes (1596–1650) and Gottfried Wilhelm Leibniz (1646–1716) considered reason to be the origin of all true knowledge, sensation merely a degraded form of thought or source of illusion. By contrast, empiricists such as Francis Bacon (1561–1626) and John Locke (1632–1704) argued that all knowledge derived from the senses, with thought being no more than an extension of sense perception. Kant's precritical writings included works in natural philosophy, aesthetics, and ethics reflective of the rationalist tradition. But reading the British empiricist David Hume (1711–1776) awakened Kant from what he described as his "dogmatic slumbers." This awakening led, in turn, to a synthesis of these two approaches in his major work, *The Critique of Pure Reason*, which argued that the form of human experience is constructed a

Immanuel Kant, 1724–1804. The major works of this German philosopher offer an analysis of speculative and moral reason and the faculty of human judgment. He exerted an immense influence on the intellectual movements of the 19th and 20th centuries. (© *Corbis-Bettmann. Reproduced by permission.*)

priori by reason while its material content arises a posteriori from sensation. This is the core of Kant's transcendental or critical idealism, which he subsequently extended into ethics and aesthetics in order to respond to what he considered the three main questions of philosophy: What can I know? What ought I to do? What can I hope for? He later added a fourth question that synthesized the first three: What is the human being?

The Critique of Pure Reason (1781)

Kant's major work undertakes what he terms a "Copernican revolution" in philosophy. Whereas traditionally philosophy had begun with particular objects of experience, Kant's transcendental method begins with experience in general and tries to uncover the "transcendental preconditions" that make such experience possible. For Kant, objects are seen as fitting into human representational structures rather than representational structures simply arising from objects. As Hume had shown, the necessity that these representational structures possess, the fact that all objects must appear in space and time, simply cannot be derived from sensory experience. According to Kant, then, space and time are the *a priori* forms of sensibility, the ideal or transcendental forms that make it possible for human beings to experience any object.

Only as a manifold of content within space and time is sense intuition or experience possible. But objects that first appear to the senses within the necessary structures of space and time are further known using concepts such as substance and causality. For Kant, the expectation that events necessarily have causes is not so much derived from experience as brought to experience, although of course the particular causes are determined by experience. Experience would not be what it is, would not be intelligible or knowable, without these *a prior* pure concepts of the understanding. The justification of these categories rests with their constitutive role in human experience and the fact that they work to make experience possible.

What is it that is known when sensation and understanding cooperate in this way to make experience scientifically intelligible? The answer is phenomena. Perhaps the single most important distinction in Kant's thought is that between phenomena and noumena, things as they appear to people and things in themselves, respectively. The former are open to positive knowledge, whereas the latter can be thought but never known in a positive or scientific sense.

The human mind nevertheless has a tendency to try to extend itself beyond phenomena to things-in-themselves. This includes claiming to have positive knowledge of supersensible realities such as God, the soul, and freedom, the topics of traditional metaphysics. These ideals of pure reason can never be scientifically verified. Thus Kant argued that traditional metaphysics, which focuses on objects that transcend experience rather than the transcendental preconditions of experience, is not an authentic form of knowledge. Yet, although the ideals of pure reason cannot be experienced they can be thought, and in their thinking serve what Kant calls a regulative function.

Critique of Practical Reason (1788)

The second critique turns from science to ethics and deals with practical or moral reasoning. This book was preceded by an introductory *Foundations of the Metaphysics of Morals* (1785), which developed a deontological theory of ethics, that is, one based on the primacy of duty. For Kant, the only unconditional good is a good will, one that wills to do what is a duty merely because it is a duty, or to choose duty for its own sake. The philosophical challenge for ethics is to explicate what this means, and to identify the transcendental preconditions of its possibility.

Kant thus approaches ethics not in terms of the consequences of actions or whether decisions make a person happy, but in terms of moral obligation. The idea of duty leads Kant to the idea of freedom as its basis. Although with regard to many actions the will may be influenced by factors outside itself, that is, be heteronomous, at least in some instances the will is able to decide for itself, that is, act autonomously. In exercising its own decision-making capacity, the will may also reason according to hypothetical imperatives (If one wants X then do Y) or categorical imperatives (Do Y, no matter what). Practical reason at the highest level displays a spontaneity that makes its own law for itself, simply because this is the right way to act, independent of any particular consequences.

Hypothetical reasoning may be described as the basis of technological thinking. Indeed, Kant calls one form of a hypothetical imperative a technical imperative, which focuses on discovering the means to achieve some end. Categorical reasoning, by contrast, focuses on the identification of worthy ends. According to Kant the most worthy end, and thus categorical imperative, is to act according to a maxim that is universal, that is, applies to all, or to treat all persons as ends in themselves. Human beings have an inherent worth or dignity, unlike objects that have exchange value. To recognize this and act accordingly is to begin to construct something more than a traditional society or state, which presumes people acting out of self-interest and treats others as means to their own ends, and to begin to construct instead a new kind of social order that Kant calls a "kingdom of ends." This moral ideal has been applied widely to a range of ethical issues related to science and technology, from the treatment of human subjects in medical research, to privacy in the use of computers and debates about the permissibility of human reproductive cloning.

The second critique postulates freedom, the existence of God, and immortality of the soul as necessary presuppositions of moral experience. Freedom is necessary to make sense of the human experience of moral responsibility, God to guarantee the ultimate triumph of moral order, and immortality to allow for the final realization of the good will. In this regard, practical reason provides access to a supersensible reality closed to science, though in a manner that can only be an issue of rational faith.

The Third Critique and Kant's Influence

Kant's *Critique of Judgment* (1790) attempted to show how theoretical and practical reason—science and ethics—are unified in the sense of beauty. For Kant, the judgments of beauty and purpose provide a sensible symbol of the supersensible realm. They suggest that the natural and ethical realms make up a unified whole. The purposeful structures humans observe especially in organic bodies and their beauty provide clues to the further understanding of nature. The idea that nature is purposeful lies behind the human belief that a system of laws of nature is possible. It can also lead to the extension of humanity's empirical investigations of nature. Judgments of beauty are based on a subjective feeling of delight in an object, but this feeling has a universal validity deriving from the harmony of the faculties of imagination and understanding. The feeling of the sublime depends upon the moral feeling Kant supposed common to all of humanity.

Taken together, Kant's three critiques thus answer what he takes to be the basic questions of philosophy. The fourth question was to find its answer in a study of anthropology. What can be known are intelligent constructions of science that constitute the basic form of knowledge. What ought to be done is to treat human beings as ends in themselves in order to establish a kingdom of ends. For the individual human being there is hope for personal immortality in order to be able to make infinite moral progress. For the human race there is the hope that human progress will be instantiated in the moralization of the human race, so that the advance of human capacities, including humankind's scientific

understanding of the world, may contribute to the construction of a harmonious moral social order.

Kant's influence is inestimable. Although in the next generation Georg Wilhelm Friedrich Hegel (1770–1831) challenged Kant's distinction between phenomena and noumena, Hegel's alternative system never became as influential as Kant's. Future efforts to explicate the unique power and limitations of science and the independent validity of ethics have repeatedly returned to formulations of what have become known as various forms of neo-Kantianism. Ernst Cassirer (1874–1945), for instance, widened Kant's appreciation of human construction in science to include the entire range of cultural symbolic production, including the realms of language, myth, and religion. Bernard Gert's *Morality: Its Nature and Justification* (2004) develops a Kantian-like set of moral rules, often explicitly considering issues related especially to biomedical technologies.

More generally, Friedrich Dessauer (1881–1963) developed a broadly Kantian interpretation of technology, going so far as to propose a fourth Kantian critique of the transcendental preconditions of technological invention (Mitcham 1994). More recently, Ernesto Mayz Vallenilla (1989) has provided an analysis of the transformations of technical rationality brought about by new instrumentations that reflects a Kantian and phenomenological heritage. Finally, detached from its transcendental moorings, Kant's approach may also be seen as supporting contemporary social constructivist interpretations of science and technology (Bijker, Hughes, and Pinch 1987).

In his first critique, Kant sought to limit positive scientific knowledge to phenomenal reality so that noumena may be posited without interference by rational faith and that ethics may be able to rest on its own foundations. In this way, ethics could be freed from the dogmatic assumptions and skepticism associated with traditional metaphysics.

> [E]ven the *assumption*—as made on behalf of the necessary practical employment of my reason—of *God, freedom,* and *immortality* is not permissible unless at the same time speculative reason be deprived of its pretensions to transcendent insight ... thus rendering all *practical extension* of pure reason impossible. I have therefore found it necessary to deny *knowledge,* in order to make room for *faith.* The dogmatism of metaphysics, that is, the preconception that it is possible to make headway in metaphysics without a previous criticism of pure reason, is the source of all that unbelief, always very dogmatic, which wars against morality. (Kant 1965 [1781], pp. Bxxix–Bxxx)

Kant's philosophy sought the harmonious development of human faculties but ended in separating scientific intellection and ethical reflection. Both exhibit the free and spontaneous constructive activity of the human mind. Yet Kant did not foresee how scientific (and technical) development could outpace the application of ethical reflection. As a result, ethical thought often appears to lag behind technoscientific achievements. To what extent should ethical concerns establish limits on scientific inquiry? This question manifests itself repeatedly in contemporary discussions of advancing science, new technologies, and ethics.

DARYL J. WENNEMANN

SEE ALSO Axiology; Deontology; Discourse Ethics; Freedom; Leibniz, G. W.; Risk and Emotion; Scientific Ethics.

BIBLIOGRAPHY

Bijker, Wiebe E.; Thomas P. Hughes; and Trevor J. Pinch, eds. (1987). *The Social Construction of Technological Systems: New Directions in the Sociology and History of Technology.* Cambridge, MA: MIT Press. A collection of essays exploring the social construction of technology.

Cassirer, Ernst. (1953, 1955, 1957, and 1996). *Philosophy of Symbolic Forms,* trans. Ralph Manheim and John Michael Krois. 4 vols. New Haven, CT: Yale University Press.

Gert, Bernard. (2004). *Morality: Its Nature and Justification,* rev. edition. Oxford and New York: Oxford University Press. The second extensive revision of Gert's *The Moral Rules: A New Rational Foundation for Morality* (1970).

Kant, Immanuel. (1902–1997). *Gesammelte Schriften* [Collected works], ed. the Royal Prussian Academy of Sciences (and its successors). 29 vols. Berlin: Georg Reimer (subsequently Walter de Gruyter). The standard German edition of Kant's works.

Kant, Immanuel. (1965 [1781/1787]). *Critique of Pure Reason,* trans. Norman Kemp Smith, unabridged edition. New York: St. Martin's Press.

Kant, Immanuel. (1990 [1785]). *Foundations of the Metaphysics of Morals,* trans. Lewis White Beck, 2nd edition. New York: Macmillan.

Kant, Immanuel. (1993 [1788]). *Critique of Practical Reason,* trans. Lewis White Beck, 3rd edition. New York: Macmillan.

Kant, Immanuel. (2000 [1790]). *Critique of the Power of Judgment,* edited by Paul Guyer, translated by Paul Guyer and Eric Matthews. Cambridge, UK: Cambridge University Press.

Körner, Stephan. (1955). *Kant.* Harmondsworth, UK: Penguin. A basic introduction to Kant's philosophy.

Mayz Vallenilla, Ernesto. (1989). *Fundamentos de la meta-técnica.* Caracas, Venezuela: Monte Avila Editores. English translation: *The Foundations of Meta-Technics,* tr. Carl Mitcham. Laham, MD: University Press of America, 2004.

A phenomenological analysis of the transition from the technical to the meta-technical forms of space and time.

Mitcham, Carl. (1994). *Thinking through Technology: The Path between Engineering and Philosophy.* Chicago: University of Chicago Press. A critical introduction to the philosophy of technology.

KENNEDY INSTITUTE

SEE *Bioethics Centers.*

KIERKEGAARD, SØREN

• • •

Søren Aabye Kierkegaard (1813–1855) was born in Copenhagen, Denmark, on May 5. A prolific author, he produced an impressive series of books devoted to philosophical and religious themes, including a parallel series published under various pseudonyms. He is perhaps best known for his critical engagement with the guiding values of Protestant Christendom in the mid-nineteenth century. Fearing that Christianity had become dangerously enmeshed in the bourgeois malaise sweeping Europe at the time, he urged his readers to aspire to lives of greater passion, intensity, inwardness, and faith. In a sustained provocation that won him few contemporary admirers, he vowed to reintroduce the practice of Christianity into Christendom.

Kierkegaard's most influential pseudonymous work, *Fear and Trembling* (1843), challenges the primacy assigned to the universality of ethical life. With specific reference to the biblical story of Abraham on Mount Moriah, Kierkegaard raises the possibility that some religious obligations may actually trump the recognized ethical obligations of contemporary Christian practice. As indicated, supposedly, by the trial of Abraham, the pursuit of faith may eventually oblige individuals to seek the truth of their existence *beyond* the ethical universal, in the religious sphere. Through his pseudonym, Johannes de silentio, Kierkegaard alleges that the "greatness" of Abraham remains an anomaly within contemporary Christian belief and practice. Abraham can be considered "great" only by virtue of his faith, and the most compelling expression of his faith was his decision to obey his God's command to sacrifice his only son Isaac. If Johannes is correct in his analysis, then the "greatness" of Abraham is inextricably linked to his willingness to perform what Johannes calls a "teleological suspension of the ethical," that is, an abrogation of

Søren Kierkegaard, 1813–1855. The Danish philosopher and religious thinker was the progenitor of 20th-century existential philosophy.

his moral obligations in the service of a higher, religious obligation.

Some readers insist at this point that Kierkegaard simply misidentifies or exaggerates the "greatness" of Abraham. Still others allow that Christians continue to honor Abraham only as a symbol of their Judaic prehistory. Yet, the point Kierkegaard raises bears further consideration: Do people not, at least occasionally, admire individuals who exempt themselves from acknowledged moral conventions? If so, how can people persist in their avowed allegiance to ethical universality as the highest expression of human flourishing? Do people not in fact reserve an even higher status for those "knights of faith" who, like Kierkegaard's Abraham, sacrifice morality for a supposedly higher purpose?

As these questions indicate, Kierkegaard's critical engagement with conventional morality was motivated in large part by the overriding value he attaches to the life of authentic individuality. Although conventional morality serves most people, most of the time, as a perfectly adequate expression of their humanity, it proves to be inadequate, and even inhospitable, to those who seek an authentic, singular existence. The individuals whom Kierkegaard most admired find the truth of their existence not outside themselves (for example, in public expressions of the ethical universal), but *within* themselves, in the passion and spirit that constitute their essential inwardness. The greatest expression of inwardness, he further believed, is faith, wherein the individual is raised above the ethical universal and placed in an absolute relationship to God. Kierkegaard thus concluded that conventional morality may actually pose a formidable obstacle to the pursuit of a life of faith.

Kierkegaard rarely commented directly on the rise of modern technology, but his writings are peppered with insights into the subtle ways in which emerging technologies contribute to the overall leveling of social life. The busyness that defines life in the modern epoch is both supported and exacerbated by the introduction of technological wonders, which enable modern people to distract themselves ever more effectively from their spiritual emptiness. While not the cause of the spiritual poverty that Kierkegaard detects around him, technology encourages people to postpone indefinitely the difficult regimen of self-examination and introspection that he prescribed.

Toward the end of his life, Kierkegaard engaged in an increasingly vituperative attack on the Danish state church, which, he believed, had fallen captive to the dispassionate values of bourgeois modernity. Owing in part to the fallout from this attack, he died in disrepute on November 11. Since the time of his death, however, his philosophical reputation has grown steadily. In the early twenty-first century he is widely read for his pioneering contributions to depth psychology; his prescient criticisms of the spread of bourgeois values; his fresh interpretations of Christian faith and practice; his astute observations on contemporary political life; his challenge to ethical universality; and, perhaps most prominently, his spirited defense of authentic individuality.

DANIEL CONWAY

SEE ALSO *Alienation; Existentialism.*

BIBLIOGRAPHY

Arbaugh, George E., and George B. Arbaugh. (1967). *Kierkegaard's Authorship: A Guide to the Writings of Kierkegaard.* Rock Island, IL: Augustana College Library.

Hannay, Alastair. (2001). *Kierkegaard: A Biography.* New York: Cambridge University Press.

Kierkegaard, Søren. (1985). *Fear and Trembling,* trans. and introduction by Alastair Hannay. New York: Viking Penguin.

Kierkegaard, Søren. (1989). *Sickness unto Death*, trans. and introduction by Alastair Hannay. New York: Viking Penguin.

Kierkegaard, Søren. (1997). *Christian Discourses: The Crisis and a Crisis in the Life of an Actress*, ed. and trans. with introduction and notes by Howard V. Hong and Edna H. Hong. Princeton, NJ: Princeton University Press.

KUHN, THOMAS

• • •

Historian and philosopher of science, Thomas Kuhn (1922–1996), who was born in Cincinnati, Ohio, on July 18, was perhaps the most influential theorist of science in the second half of the twentieth century. Kuhn received all his degrees (in physics) and his first job at Harvard University, though he failed to be awarded tenure there in 1956, shortly after the departure of his mentor, Harvard President James Bryant Conant. Kuhn was finally tenured at Princeton University in 1964, on the basis of what remains his best known book, *The Structure of Scientific Revolutions* (1962). In 1979 Kuhn moved to the Massachusetts Institute of Technology (MIT), where he eventually retired as Laurence Rockefeller Professor of Philosophy and Linguistics. Essays from Kuhn's Harvard and Princeton years appear in *The Essential Tension* (1977). Essays from his MIT years are collected in *The Road Since Structure* (2000). At the time of his death, in Cambridge, Massachusetts, on June 17, Kuhn had been long working on an update of the perspective first developed in *Structure*.

Kuhn's influence rests mainly on *Structure*, his second book, which departs from the then prominent logical empiricist efforts to understand science through its rational reconstruction in favor of a more historically based appreciation of its internal dynamics. Kuhn presents a theory of scientific change as a cycle of relatively clearly defined phases, centered on the creation, development, and destruction of a *paradigm*, a word that has entered the general vocabulary in the early twenty-first century. For Kuhn, the distinctiveness of science lies in the ability of its practitioners to take hold of the means of knowledge production by agreeing on a theoretical framework, methods, and suitable problems to pursue. Kuhn's protean use of paradigm to cover every aspect of this process has led to much confusion. Nevertheless the overall thrust of his account is clear. *Normal science*, the rather routine pursuit of paradigmatic puzzles, is the heart of the scientific enterprise, and the source of whatever progress science displays. Kuhn's picture was very much at odds with the more heroic Galilean image of scientists as bold destroyers of tradition. On the contrary, for Kuhn, scientists themselves worked within strict traditions of practice that were typically passed down through apprenticeship with master practitioners.

Kuhn's image of science is profoundly conservative, a point overlooked by most of his supporters. To be sure, *revolution* figures in the title of Kuhn's first two books—the first being *The Copernican Revolution* (1957)—and is the basis on which many readers have imagined him to be a *radical* thinker. Nevertheless Kuhn draws on a conception of revolution received from the conservative political tradition, whereby a revolution eventuates in a restoration of natural order. Thus, for Kuhn, revolutions in science happen only as a last resort, when the paradigm can no longer solve the problems it has set for itself. In that case a *crisis* ensues, the result of which is a new paradigm that then provides the basis for a new kind of normal science. Philosophically inspired criticism of fundamental assumptions in science is licensed only once a paradigm is in crisis. Under normal circumstances, scientists take a more *heads-down* approach to their work.

The widespread misunderstanding of Kuhn's theory has been an ironic source of its influence. Although Kuhn himself was careful to restrict the evidence base of his theory to roughly three centuries of the history of the physical sciences (1620–1920), he was quickly read as referring to a pattern of change that could be found in all sciences—even the humanities—across all periods. This misreading is partly due to the fact that Kuhn does not distinguish science by reference to its technological applications or material impact on the world. On the contrary, for Kuhn, a field becomes scientific by becoming autonomous from such external concerns. Thus physics is a science not because it produces real-world effects but because physicists are in full control of the physics research agenda. Many of Kuhn's hopeful readers outside of physics drew the conclusion that their own fields could similarly acquire the status of science by generating their own paradigms. Thus in the early 2000s virtually every discipline outside physics has at least one theorist or methodologist whose reputation is based on the claim of having founded a paradigm of some sort.

In first two decades after it was written, *The Structure of Scientific Revolutions* was subject to much philosophical criticism, especially from Karl Popper and his followers. They questioned the normative backdrop to Kuhn's history of science: Was Kuhn effectively valorizing the most conformist elements of scientific practice? The answer appeared to be yes, but that did not prevent the book from entering the philosophical canon after 1980. Eventually most philosophers took for granted Kuhn's overall account of scientific change, especially his methodological assumption that science needs to be understood from the *inside*, so to speak. A mark of Kuhn's influence on contemporary discussions in the history, philosophy, and sociology of science is the preoccupation with demonstrating one's mastery of the inner workings of a science. In his later years, Kuhn grew closer to the standard philosophical understanding of these matters, while openly dissociating himself from *relativist* and *constructivist* sociologists who claimed to have been inspired by his work.

In taking the measure of Kuhn's legacy, it is puzzling how a physicist with an amateur understanding of the history, philosophy, and sociology of science could have had such a profound impact on these fields, which already enjoyed a relatively high degree of sophistication. In effect, Kuhn's *Structure* offered a historian's sense of philosophy, a philosopher's sense of sociology, and a sociologist's sense of history. That this particular book should have such an enduring impact cannot be explained simply by its content, because many of its supposedly distinctive theses could also be found in the work of contemporaries such as Norwood Russell Hanson, Paul Feyerabend, and Stephen Toulmin. However, unlike them, Kuhn singularly benefited from the patronage of Conant, to whom *Structure* is dedicated. *Structure* was written while Kuhn taught in a general education program that Conant had created to instill faith in science as an autonomous enterprise in a time—the Cold War—when it would be increasingly subject to public scrutiny. This helps explain Kuhn's peculiar inclusions and omissions. As conceptual horizons become detached from Kuhn's Cold War moorings, his work will probably lose its hold on the meta-scientific imagination.

STEVE FULLER

SEE ALSO *Progress; Scientific Revolution.*

BIBLIOGRAPHY

Fuller, Steve. (2000). *Thomas Kuhn: A Philosophical History for Our Times*. Chicago: University of Chicago Press.

Hoyningen-Huene, Paul. (1993). *Reconstructing Scientific Revolutions*. Chicago: University of Chicago Press.

Kuhn, Thomas. (1957). *The Copernican Revolution*. Chicago: University of Chicago Press.

Kuhn, Thomas. (1970). *The Structure of Scientific Revolutions*, 2nd edition. Chicago: University of Chicago Press. First edition published in 1962.

Kuhn, Thomas. (1977). *The Essential Tension*. Chicago: University of Chicago Press.

Kuhn, Thomas. (2000). *The Road Since Structure*. Chicago: University of Chicago Press.

Lakatos, Imre, and Alan Musgrave, eds. (1970). *Criticism and the Growth of Knowledge*. Cambridge, UK: Cambridge University Press.